Jp Biz-Jet & 98

By Brian Gates

31st Edition
ISBN 1-898779-15-5
48 Colour photographs

Published and Distributed by:

BUCHair (U.K.) Ltd

P.O. Box 89, Reigate, Surrey, RH2 7FG
Tel: +44 (0)1737 224747
Fax: +44 (0)1737 226777
Email: buchair_uk@compuserve.com
www: http://www.buchair.rotor.com

Printed in Great Britain
Copyright © BUCHair (U.K.) Ltd 1998

Where to get the products of BUCHair (U.K.) Ltd and Bucher & Co. Publikationen.?

If you cannot get our products at your local bookstore or if you need information about publication dates or prices, please contact the following organisations in, or nearest to, your country.

Australia

Mr James L Bell
BUCHER PUBLICATIONS (Australia)
PO Box 232
Riddell, Victoria 3431

Phone: (3) 54 28 63 12
Fax: (3) 93 79 07 67

Canada

Aviation World
195 Carlingview Drive
Rexdale
Ontario, M9W 5E8

Phone: (416) 674-5959
Fax: (416) 674-5915

http://www.interlog.com/~avworld
avworld@interlog.com

Denmark

NYBODER BOGHANDEL, ApS
114 Store Kongensgade
DK-1264 Copenhagen K

Phone: 33 32 33 20
Fax: 33 32 33 62

Deutschland

BUCHair (Deutschland)
PO Box 750210
D-60532 Frankfurt a.M.

Phone: (06105) 951 700
Fax: (06105) 951 701

http://www.photopart.com
buchair.d@photopart.com

Finland

AVIATION SHOP
Kajanuksenkatu 12
SF-00250 Helsinki

Phone: (09) 449 801
Fax: (09) 149 6163

France

LA MAISON DU LIVRE
76 Boulevard Malesherbes
F-75008 Paris

Phone: (1) 45 22 74 16
Fax: (1) 42 93 81 23

Great Britain

BUCHair (U.K.) Ltd
PO Box 89
Reigate
Surrey
RH2 7FG

Phone: (0) 1737 224 747
Fax: (0) 1737 226 777

http://www.buchair.rotor.com
buchair_uk@compuserve.com

Ireland

Irish Air Letter
20 Kempton Way
Navan Road
Dublin 7

Phone: (01) 838 06 29

Italy

LA BANCARELLA
AERONAUTICA S.A.S.
Corso Duca Degli Abruzzi 12
I-10128 Torino

Phone: (011) 53 13 41
Fax: (011) 562 93 59

Japan

NISHIYAMA YOSHO Co. Ltd
Takin Building 3F
12 4-7-11, Ginza, Chuo-ku
Tokyo 104

Phone: (03) 3562-0820
Fax: (03) 3562-0828

Netherlands

BOEKHANDEL VENSTRA B.V.
Binnenhof 50
Postbus 77
NL-1180 AB Amstelveen

Phone: (020) 641 98 80
Fax: (020) 640 02 52

Norway

Scanbook Norway
Automobilla, A/S
Automobilla Bookshop
Uranienborgveien 25
Postboks 7035, Homansbyen
N-0306 Oslo
Phone: 22 60 86 00
Fax: 22 60 86 02

Osterreich

BUCHER & CO., PUBLIKATIONEN
Postfach 44
CH-8058 Zurich-Flughafen
Schweiz

Phone: (01) 874 1 747
Fax: (01) 874 1 757

jp@buchair.ch

Schweiz
(Bestellungen)

BUCHER & CO., PUBLIKATIONEN
Postfach 44
CH-8058 Zurich-Flughafen

Phone: (01) 874 1 747
Fax: (01) 874 1 757

jp@buchair.ch

Schweiz
(Laden - Glattbrugg)

BUCHairSHOP
Bucher & Co., Publikationen
Schaffhauserstrasse 76
CH-8152 Glattbrugg

Phone: (01) 874 1 747
Fax: (01) 874 1 757

jp@buchair.ch

Schweiz
(Laden - Flughafen)

BUCHairSHOP
Bucher & Co., Publikationen
Terrasse, Terminal B
CH-8058 Zurich-Flughafen

Phone: (01) 874 1 747
Fax: (01) 874 1 757

jp@buchair.ch

South Africa

Mr Karel Zayman
SKY NEEDS
PO Box 1385
Kempton Park 1620

Phone: (011) 472 32 55
Fax: (011) 792 83 18

Spain

LA AERTOTECA
LIBRERIA MIGUEL CREUS
C/Congost, 11
E-08024 Barcelona

Phone: (93) 210 54 07
Fax: (93) 210 59 92

USA

BUCHair (USA), Inc.
PO Box 750515
Forest Hills
NY 11375-0515

Phone: (718) 263 8748
Fax: (718) 275 6190

Buchair@mail.idt.net

INTRODUCTION

A growing demand for flexible transportation as an alternative to airline travel, and the ever present concern over safety measures, coupled with strong worldwide economies with the exception of Asia has produced sustained sales and utilisation of business aircraft.

The fractional ownership market continues to grow apace. All major manufacturers are now represented. NetJets Europe is now established, and a similar type operation for the Middle East is pending.

Gulfstream Aerospace achieved its 1000th Gulfstream delivery, as well as establishing numerous world records with its flagship Gulfstream V demonstrator, and also commenced initial deliveries. Bombardier expect to close the gap on its rival by certifying the Global Express soon followed by first deliveries mid-1998. Competition for this market sector is expected to strengthen with the advent of the Boeing Business Jet and the Airbus A319CJ. Bombardier delivered its 400th Challenger and maintained a strong demand for its 604 model. Sales of the Learjet 31A and 60 remain steady, whilst the Learjet 45 certified in August has now commenced first customer deliveries. Dassault, now producing 5 models, the Falcon 50 and 50EX, the Falcon 900B and 900EX and Falcon 2000, have sold over 70 aircraft this year, and anticipate similar demand for 1998. Raytheon's Premier 1 first flight is now expected mid-1998 with reported orders of 50. Hawker 800XP sales remain strong, whilst the Beechjet has slowed a little now that the 180 USAF aircraft contract is complete. Cessna continue to be as prolific as ever with strong unit sales of its CitationJet and Ultra, and deliveries of the Bravo and Citation X now in full swing, and the Excel with its 200 order backlog coming on stream in 1998. Israel Aircraft Industries new mid-size Galaxy first flew on Christmas Day, certification and first deliveries are expected in 1999. Their established Astra SPX has had a good year with strong demand. Swearingen's SJ30-2 first flew in September with its new FJ44-2 engines, certification again expected in 1999. A further candidate for the 1999 certification date is Visionaire's single-engined Vantage jet.

This 31st edition maintains the format of previous years. Any comments with regard to errors, omissions and content please contact the editor on:

Fax	+44 0118 973 6400
Email	briangates@compuserve.com

January 1998

ACKNOWLEDGEMENTS

My sincere thanks to the following for their invaluable help: Air Britain News, Aviation Letter, Corporate Monthly, Mach III Internet, Anton Heumann for Swiss allocations, Pierre Parvaud for French information, Christian Gundelfinger for German update. A.Dann, J W Parkin, J Rose, P Suter, D Metherell, and to many other enthusiasts who provided a little but essential data.

Bombardier Inc of Montreal, Canada, provided the front cover illustration.

CONTENTS

JETS

TURBOPROPS

REFERENCE

BUSINESS JETS
Country Index

BUSINESS JETS
By Country

AP = Pakistan

Civil

Reg	Yr	Type	c/n	Owner/Operator	Prev Regn
□ AP-BEH	92	B 737-33A	25504	Government of Pakistan, Karachi.	
□ AP-BEK	92	Learjet 31A	31A-062	Government of Baluchistan, Quetta.	N25997
□ AP-BEX	93	Beechjet 400A	RK-80	Government of Pakistan, Karachi.	N8180Q

Military

Reg	Yr	Type	c/n	Owner/Operator	Prev Regn
□ 0233	93	Citation V	560-0233	Pakistan Army, Qasimaab, Rawalpindi.	N1288A
□ 68-19635	68	B 707-351C	19635	Pakistan Air Force, Islamabad.	AP-BAA
□ 68-19866	68	B 707-340C	19866	Pakistan Air Force, Islamabad.	AP-AWY
□ J 468	82	Falcon 20F	468	Pakistan Air Force, Sargodha.	F-WMKG
□ J 469	82	Falcon 20F	469	Pakistan Air Force, Sargodha.	F-WMKI
□ J 753	72	Falcon 20E	277	Pakistan Air Force, 12 Squadron Chakala, Islamabad.	F-WPXD

A2 = Botswana *(Republic of Botswana)*

Civil

Reg	Yr	Type	c/n	Owner/Operator	Prev Regn
□ A2-AGM	82	Citation 1/SP	501-0245	Executive Charter P/L. Gaborone.	ZS-LDO

Military

Reg	Yr	Type	c/n	Owner/Operator	Prev Regn
□ 0K1	91	Gulfstream 4	1173	Government/Botswana Defence Force, Gaborone.	N17587

A4O = Oman *(Sultanate of Oman)*

Civil

Reg	Yr	Type	c/n	Owner/Operator	Prev Regn
□ A4O-AB	91	Gulfstream 4	1168	Government of Oman, Seeb. (r/c Oman 3).	N462GA
□ A4O-AC	92	Gulfstream 4	1196	Government of Oman, Seeb. (r/c Oman 4).	N420GA
□ A4O-SC	84	Citation II	550-0486	Oman Aviation Services Co. Seeb.	(N1253Y)
□ A4O-SO	79	B 747SP-27	21785	Government of Oman, Seeb. (r/c Oman 2).	N351AS
□ A4O-SP	80	B 747SP-27	21992	Government of Oman, Seeb. (r/c Oman 1).	N150UA

A6 = United Arab Emirates

Civil

Reg	Yr	Type	c/n	Owner/Operator	Prev Regn
□ A6-AUH	90	Falcon 900	84	Government of Abu Dhabi.	F-WWFD
□ A6-ESH	81	B 737-2W8	22628	Government of Sharjah.	N180RN
□ A6-HEH	82	Gulfstream 3	356	Dubai Air Wing, Dubai.	
□ A6-HHH	87	Gulfstream 4	1011	Dubai Air Wing, Dubai.	N17581
□ A6-PFD	85	Airbus A300C4-620	374	Government of Abu Dhabi.	F-WWAJ
□ A6-SHK	87	BAe 146 Statesman	E1091	Government of Abu Dhabi.	G-BOMA
□ A6-SHZ	84	Airbus A300B4-620	354	Government of Abu Dhabi.	F-ODRM
□ A6-SMM	80	B 747SP-31	21963	Dubai Air Wing, Dubai.	N602AA
□ A6-SMR	79	B 747SP-31	21961	Government of Dubai.	N58201
□ A6-SMS	97	Learjet 60	60-094	Fujairah National Group, Fujairah.	N60LR
□ A6-UAE	90	Falcon 900	86	Government of Abu Dhabi.	F-WWFE
□ A6-ZKM	88	Falcon 900	47	Government of Abu Dhabi.	F-WWFA
□ A6-ZSN	87	B 747SP-31	23610	Government of United Arab Emirates.	N60697

A7 = Qatar *(State of Qatar)*

Civil

Reg	Yr	Type	c/n	Owner/Operator	Prev Regn
□ A7-...	88	Airbus A310-304	473	Government of Qatar, Doha.	F-ODSV
□ A7-AAA	77	B 707-3PIC	21334	Qatari Government, Doha.	
□ A7-AAD	91	Falcon 900	91	Qatari Government, Doha.	F-WWFH
□ A7-AAE	91	Falcon 900	94	Qatari Government, Doha.	F-WWFC
□ A7-AHM	79	B 747SP-27	21786	Qatari Government, Doha.	A7-ABM
□ A7-HHK	93	Airbus A340-211	026	Qatari Government, Doha.	F-WWJQ

A9C = Bahrain *(State of Bahrain)*

Civil

Reg	Yr	Type	c/n	Owner/Operator	Prev Regn
□ A9C-BA	80	B 727-2M7	21824	Government of Bahrain. (r/c Bahraini One).'Al Bahrain'	N740RW
□ A9C-BB	83	Gulfstream 3	393	Government of Bahrain. (r/c Bahraini Three).	N17587
□ A9C-BG	77	Gulfstream 2TT	202	Government of Bahrain. (r/c Bahraini Two).	N17586

B = China *(People's Republic of China)*

Civil

Reg	Yr	Type	c/n	Owner/Operator	Prev Regn
□ B-...	96	Challenger SE	7149	Poly Technologies Inc. Beijing.	C-FZIS
□ B-...	96	Challenger SE	7138	Poly Technologies Inc. Beijing.	C-FZAT
□ B-...	97	Challenger SE	7180	Poly Technologies Inc. Beijing.	C-GATM
□ B-...	97	Challenger SE	7189	Poly Technologies Inc. Beijing.	C-GATY
□ B-...	97	Challenger SE	7193	Poly Technologies Inc. Beijing.	C-GBFR
□ B-...	97	CitationJet	525-0204	Broad Air Conditioning/China Southern Airways, Baiyun.	(N323LM)
□ B-3980	82	Learjet 55	55-027	Hainan Airlines, Haikou.	N227A
□ B-3981	95	Learjet 60	60-053	Hainan Airlines, Haikou.	N5053Y
Reg	*Yr*	*Type*	*c/n*	*Owner/Operator*	*Prev Regn*

Reg	Yr	Type	c/n	Owner/Operator	Prev Regn
☐ B-4101	85	Citation S/II	S550-0049	Airborne Remote Sensing Centre, Beijing.	N1270K
☐ B-4102	85	Citation S/II	S550-0050	Airborne Remote Sensing Centre, Beijing.	N1270S
☐ B-4103	81	Citation II	550-0301	Zhongfei Airlines, Xian.	091
☐ B-4104	82	Citation II	550-0297	Zhongfei Airlines, Xian.	092
☐ B-4105	81	Citation II	550-0305	Zhongfei Airlines, Xian.	090
☐ B-4106	92	Citation VI	650-0220	Airborne Remote Sensing Services, Beijing.	N6830T
☐ B-4107	92	Citation VI	650-0221	Airborne Remote Sensing Services, Beijing.	N1301A
☐ HY-984	85	Learjet 36A	36A-053	Geological Survey, Chinese Government, Beijing.	N39418
☐ HY-985	78	Learjet 36A	36A-034	Geological Survey, Chinese Government, Beijing.	N763R
☐ HY-986	85	Learjet 35A	35A-601	Poly Technologies Inc. Beijing.	
☐ HY-987	85	Learjet 35A	35A-602	Poly Technologies Inc. Beijing.	
☐ HY-988	85	Learjet 35A	35A-603	Poly Technologies Inc. Beijing.	

B-H = Hong Kong (Chinese Colony of Hong Kong)

Civil

Reg	Yr	Type	c/n	Owner/Operator	Prev Regn
☐ B-HSS	82	HS 125/700B	7169	Helicopters (Hong Kong) Ltd.	VR-HSS

B = Taiwan (Republic of China)

Civil

Reg	Yr	Type	c/n	Owner/Operator	Prev Regn
☐ B-98183	89	Learjet 35A	35A-654	Golden Eagle Air Transport, Taipei.	(ZS-NSB)

Military

Reg	Yr	Type	c/n	Owner/Operator	Prev Regn
☐ 2721	67	B 727-109	19399	Taiwan Air Force, Taipei.	B-1818
☐ 2722	67	B 727-109	19520	Taiwan Air Force, Taipei.	B-1820
☐ 2723	69	B 727-109C	20111	Taiwan Air Force, Taipei.	B-1822
☐ 2724	67	B 727-121C	19818	Taiwan Air Force, Taipei.	B-188

C = Canada

Civil

Reg	Yr	Type	c/n	Owner/Operator	Prev Regn
☐ C-....	97	Global Express	9006	Bombardier Inc. Montreal, PQ.	
☐ C-FALI	87	Citation S/II	S550-0142	Irving Oil Transport Ltd. St John, NB.	N542CC
☐ C-FANS	97	Citation Bravo	550-0807	Ainsworth Lumber Co. Kamloops, BC.	N52144
☐ C-FBCL	72	Citation	500-0042	Sky Service FBO Inc. Dorval, PQ. (status ?).	N542CC
☐ C-FBEL	84	Challenger 601	3028	Bell Canada Enterprises Inc/Execaire, Montreal, PQ.	C-GLXB
☐ C-FBFP	75	Learjet 35	35-038	Appeal Enterprises/Canada Jet Charters Ltd. Vancouver, BC.	VH-ELJ
☐ C-FBGX	96	Global Express	9001	Bombardier Inc. Montreal, PQ. (Ff 13 Oct 96).	
☐ C-FBLJ	96	Learjet 60	60-092	Bombardier Inc. Montreal, PQ.	N5092R
☐ C-FBMG	79	HS 125/700A	NA0248	Irving Oil Transport Ltd. Saint John, NB.	C-GKCI
☐ C-FBOM	93	Challenger 601-3A	5124	Execaire/Bank of Montreal, Toronto, ON.	C-G...
☐ C-FCFP	96	Astra-1125SPX	079	Canfor Corp/Canadian Forest Products Ltd. Vancouver, BC.	4X-CUX
☐ C-FCLJ	85	Learjet 55	55-118	Canada Jet Charters Ltd. Vancouver, BC.	
☐ C-FDDD	86	Citation S/II	S550-0115	Chartright Air Inc. Toronto, ON.	N520RP
☐ C-FDYL	82	Citation II	550-0341	Sobeys Inc. Halifax, NS.	N182U
☐ C-FEMA	85	Citation S/II	S550-0040	Manitoba Emergency Aero-Medical Services, Winnipeg, MT.	(N1269D)
☐ C-FETB	61	B 720-023B	18024	Pratt & Whitney Canada Inc. Montreal. PQ. (PW530 testbed).	OD-AFQ
☐ C-FFCL	79	Citation II	550-0069	Flexi-Coil Ltd/West Wind Aviation Inc. Saskatoon, SK.	N550AB
☐ C-FFTM	89	BAe 125/800A	NA0437	Execaire/Canadian Pacific Forest Products, Montreal, PQ.	G-BPXW
☐ C-FGAT	81	Citation II	550-0229	102662 Canada Inc/Expressair, Gatineau, PQ.	OY-GRC
☐ C-FGGH	87	Westwind-1124	431	Chartright Air Inc. Toronto, ON.	N431AM
☐ C-FHFP	69	F 28-1000	11016	Peregrine Air Charter/Canadian Forest Products, Vancouver.	VR-BNC
☐ C-FHGX	97	Global Express	9002	Bombardier Inc. Montreal, PQ.	
☐ C-FHJB	85	Learjet 55	55-122	Execaire Inc/Interconinvest Canada Inc. Toronto, ON.	N99KV
☐ C-FHPM	71	Gulfstream 2	104	Barrick Gold Corp. Toronto, ON.	N858W
☐ C-FICU	79	Learjet 35A	35A-249	Keewatin Air Ltd. Winnipeg, MT.	N300DA
☐ C-FIGD	67	Falcon 20C	109	Government/National Research Council, Ottawa, ON.	117506
☐ C-FIMO	84	Citation III	650-0065	Imperial Oil Ltd. Toronto, ON.	N500E
☐ C-FIPE	97	Hawker 800XP	8319	IPL Energy Inc. Calgary, AB.	N2291X
☐ C-FJBO	97	Citation Bravo	550-0812	Execaire Inc. Montreal, PQ.	N5223P
☐ C-FJCZ	92	Citation II	550-0700	Department of Transport, Ottawa, ON.	
☐ C-FJGX	97	Global Express	9003	Bombardier Inc. Montreal, PQ.	
☐ C-FJJC	86	Citation III	650-0116	Sky Service FBO Inc. Dorval, PQ.	N788BA
☐ C-FJWZ	91	Citation II	550-0685	Department of Transport, Moncton.	(N6778T)
☐ C-FJXN	91	Citation II	550-0684	Department of Transport, Montreal, PQ.	(N6778T)
☐ C-FKCE	91	Citation II	550-0686	Department of Transport, Edmonton, AB.	(N6778V)
☐ C-FKCI	81	Falcon 50	63	Irving Oil Transport Ltd. Saint John, NB.	N48GP
☐ C-FKDX	91	Citation II	550-0687	Department of Transport, Toronto, ON.	(N6778Y)
☐ C-FKEB	91	Citation II	550-0688	Department of Transport, Ottawa, ON.	(N6779D)
☐ C-FKGX	97	Global Express	9004	Bombardier Inc. Montreal, PQ.	
☐ C-FKLB	92	Citation II	550-0699	Department of Transport, Toronto, ON.	
☐ C-FKMC	73	Citation	500-0073	Klemke & Son Construction Ltd. Edmonton, AB.	(N881M)
☐ C-FLPC	87	Challenger 601-3A	5006	Canadian Pacific Ltd. Calgary, AB.	C-G...
☐ C-FLPD	80	Citation II	550-0234	AeroPro/2553-4330 Quebec Inc. Ste Foy, PQ.	N511WC
☐ C-FLTL	83	Citation III	650-0007	Odessey Aviation Ltd. Mississauga, ON.	N929DS
☐ C-FLZA	92	Citation II	550-0701	Department of Transport, Ottawa, ON.	
☐ C-FMCI	97	Citation Bravo	550-0816	Mountain Cabelevision Ltd. Hamilton, ON.	N5225K
☐ C-FMFL	82	Falcon 50	96	McCain Foods Ltd. Florenceville, NB.	N4AC
Reg	*Yr*	*Type*	*c/n*	*Owner/Operator*	*Prev Regn*

Reg	Yr	Type	c/n	Owner/Operator	Prev Regn
☐ C-FMFM	92	Citation II	550-0702	Department of Transport, Montreal, PQ.	
☐ C-FMPP	82	Citation II	550-0411	RCMP - GRC Air Services, Ottawa, ON.	N200YM
☐ C-FMTC	66	HS 125/1A-522	25104	Cypress Jetprop Charter Ltd. Vancouver, BC. (status ?).	N140AK
☐ C-FNCG		Gulfstream 2	208	Sugra Ltd/Norcan Energy, Toronto, ON.	N62CB
☐ C-FNNC	87	Falcon 50	176	Mylight Aircraft Inc/Westair Aviation, Carp, ON.	VR-CFI
☐ C-FNNT	90	Challenger 601-3A	5068	Sky Service FBO Inc. Dorval, PQ.	N113WA
☐ C-FNRG	84	Learjet 55	55-101	Noranda Inc. Toronto, ON.	N251NG
☐ C-FONX	70	Falcon 20D	225	Knighthawk Air Express, Ottawa, ON.	N102AD
☐ C-FPCP	87	BAe 125/800A	8095	Pancanadian Petroleum Ltd. Calgary, AB.	(N500LL)
☐ C-FPEL	78	Citation II	550-0042	Air Charters Inc/Air Montreal, Montreal, PQ.	N66AT
☐ C-FPXD	68	B 727-171C	19859	Corp Air Inc. Edmonton, AB.	N1727T
☐ C-FRJZ	97	Astra-1125SPX	087	611897 Alberta Ltd. Calgary, AB.	4X-CUU
☐ C-FROY	87	Westwind-1124	429	Royal Group Technologies Ltd. Toronto, ON.	4X-CUK
☐ C-FRPP	88	BAe 125/800A	NA0414	Execaire/Repap Enterprises Corp. Montreal, PQ.	G-BOTX
☐ C-FSCI	87	BAe 125/800A	NA0406	Shaw Communications Inc. Calgary, AB.	C-GKLB
☐ C-FSKC	72	Citation	500-0018	Skyward Aviation Ltd. Thompson, MT.	N70841
☐ C-FTAM	82	Citation II	550-0406	Exco Technologies Ltd. Mississauga, ON.	N551CE
☐ C-FTBZ	94	Challenger 604	5991	Bombardier Inc. Montreal, PQ. (prototype Ff 18 Sep 94).	C-GLYA
☐ C-FTEN	75	Falcon 10	45	Noranda Inc. Toronto, ON.	N110CG
☐ C-FTOM	81	Citation II	550-0292	Kal Air Ltd. Vernon, BC.	(N63FS)
☐ C-FURG	86	Challenger 601	3063	Government of Quebec, Sainte-Foy, PQ. (Air Ambulance).	C-GLYH
☐ C-FWCE	89	BAe 125/800A	NA0443	West Coast Energy Inc. Vancouver, BC.	C-GMOL
☐ C-FYMM	95	Citation V Ultra	560-0314	Syncrude Canada Ltd. Fort McMurray, AB.	N314CV
☐ C-FYYH	96	Challenger 604	5318	Stolinich Bank, Moscow, Russia..	(TC-DHE)
☐ C-FZEI	88	Westwind-Two	441	CSI Aviation Inc. Toronto, ON.	HK-3893X
☐ C-FZHT	70	Learjet 24B	24B-217	Samaritan Air Service Ltd. Toronto, ON.	N217AT
☐ C-FZHU	71	Learjet 25C	25C-070	Samaritan Air Service Ltd. Toronto, ON.	N32SM
☐ C-FZQP	78	Learjet 35A	35A-168	Sky Service FBO Inc. Dorval, PQ.	C-GPDO
☐ C-GAAA	82	HS 125/700A	NA0327	Chartright Air Inc. Mississauga, ON.	N810SC
☐ C-GABX	78	HS 125/700A	NA0233	Execaire/BCE Inc-Nova an Alberta Corp. Calgary, AB.	G-BFZI
☐ C-GAGU	90	BAe 125/800A	NA0450	Agrium Inc. Calgary, AB.	N355WG
☐ C-GAJS	81	Learjet 35A	35A-380	Perimeter Airlines (Inland) Ltd. Winnipeg, MT.	N903WJ
☐ C-GAPC	89	Citation V	560-0033	AMOCO Canada Petroleum Co. Calgary, AB.	N4333W
☐ C-GAPD	91	Citation II	550-0691	AMOCO Canada Petroleum Co. Calgary, AB.	N910H
☐ C-GAPT	90	Citation III	650-0195	Centaero Aviation Ltd. Windsor, ON.	N411BP
☐ C-GAWH	87	BAe 125/800A	8087	Execaire/IMASCO Ltd. Montreal, PQ.	N800TR
☐ C-GAZU	92	Falcon 50	228	Cathton Holdings Ltd. Edmonton, AB.	VR-CAE
☐ C-GBDK	97	Challenger 604	5347	Bombardier Inc. Montreal, PQ.	C-GLXW
☐ C-GBFP	74	Learjet 25B	25B-167	Adlair Aviation (1983) Ltd. Cambridge Bay, NT.	
☐ C-GBJA	97	Challenger 604	5341	Bombardier Inc. Montreal, PQ.	C-G...
☐ C-GBKB	82	Challenger 600S	1045	Maxwell Ward/Morningstar Air Express Inc. Edmonton, AB.	N900FC
☐ C-GBKE	97	Challenger 604	5352	Bombardier Inc. Montreal, PQ.	C-G...
☐ C-GBNE	78	Citation	500-0378	Province of Manitoba, Winnipeg, MT.	N3156M
☐ C-GBOT	72	Learjet 25B	25B-090	Samaritan Air Service Ltd. Mississauga, ON.	N112ME
☐ C-GBRQ	97	Challenger 604	5358	Bombardier Inc. Montreal, PQ.	C-GLXS
☐ C-GCFG	84	Challenger 601	3022	Govt of Canada, DoT Aircraft Services Directorate, Ottawa.	C-G...
☐ C-GCFI	84	Challenger 601	3020	Govt of Canada, DoT Aircraft Services Directorate, Ottawa.	C-G...
☐ C-GCGS	88	BAe 125/800A	NA0416	Perimeter Airlines (Inland) Ltd. Winnipeg, MT.	N353WG
☐ C-GCGT	79	Challenger 601	3991	Bombardier Inc. Montreal, PQ. (-X suffix for test flights).	
☐ C-GCIB	85	BAe 125/800A	8048	Canadian Imperial Bank of Commerce, Toronto, ON.	(N125BA)
☐ C-GCLQ	77	Citation	500-0348	Air 500 Ltd. Toronto, ON.	N301HC
☐ C-GCRP	88	BAe 125/800A	NA0417	Chartright Air Inc. Mississauga, ON.	N801NW
☐ C-GCUL	80	Citation II	550-0122	Canadian Utilities Ltd. Edmonton, AB.	N221GA
☐ C-GCUW	90	Citation V	560-0053	Canadian Utilities Ltd. Edmonton, AB.	N111CF
☐ C-GDDR	82	Challenger 600S	1048	Pal Air Ltd. Winnipeg, MT.	C-FSIP
☐ C-GDJH	80	Learjet 35A	35A-353	Canada Jet Charters Ltd. Vancouver, BC.	N3819G
☐ C-GDLR	79	Citation II	550-0062	Toma Jetprop Ltd. Calgary, AB.	(N77SF)
☐ C-GDMF	79	Citation II	550-0113	Air 500 Ltd. Toronto, ON.	N227PC
☐ C-GDPF	94	Challenger 601-3R	5145	Execaire/P G Desmarais-Power Corp of Canada, Montreal, PQ.	C-GPGD
☐ C-GEEN	65	Learjet 24	24-087	Air Alma Inc. Alma, PQ.	D-IKAB
☐ C-GEGX	97	Global Express	9005	Bombardier Inc. Montreal, PQ.	
☐ C-GESR	83	Challenger 601	3003	E S Rogers Communications Corp/Execaire Inc. Toronto, ON.	N500TD
☐ C-GFCI	94	Citation II	550-0733	Sunwest International Aviation Services Ltd. Calgary, AB.	N1213Z
☐ C-GFCL	96	Citation V Ultra	560-0346	Flexi-Coil Ltd. Saskatoon, SK.	N346CC
☐ C-GFEE	80	Citation 1/SP	501-0169	ANS Alliance Nord Sud Inc. Dorval, PQ.	D-IBWG
☐ C-GGFW	81	Citation II	550-0276	British Columbia Telephone Co., Richmond, BC.	N68649
☐ C-GGPM	80	Gulfstream 3	307	Barrick Gold Corp. Toronto, ON.	C-GSBR
☐ C-GGSP	80	Citation II	550-0131	Sunwest International/Pasley Aviation Corp. Calgary, AB.	N177CJ
☐ C-GHEC	90	Citation V	560-0084	British Columbia Telephone Co. Vancouver, BC.	N16NM
☐ C-GHKY	97	Challenger 604	5343	Husky Injection Moulding Systems Ltd. Bolton, ON.	C-GAUK
☐ C-GHKY	94	Learjet 60	60-043	Husky Injection Moulding Systems Ltd. Bolton, ON.	N5043D
☐ C-GHOM	80	Citation II	550-0187	Toma Jetprop Ltd. Calgary, AB.	N57CK
☐ C-GINT	94	CitationJet	525-0062	Interforest Ltd. Hanover, ON.	C-FVRE
☐ C-GIOH	89	Challenger 601-3A	5034	Imperial Oil Ltd. Mississauga, ON.	C-G...
☐ C-GIRE	74	Learjet 35	35-004	Sky Service FBO Inc. Montreal, PQ.	N74MJ
☐ C-GIWD	81	Learjet 35A	35A-407	Delta Mike Inc. Toronto, ON.	N407MR
Reg	*Yr*	*Type*	*c/n*	*Owner/Operator*	*Prev Regn*

Reg	Yr	Type	c/n	Owner/Operator	Prev Regn
☐ C-GJBJ	83	HS 125/700A	NA0335	Sears Canada Inc. Toronto, ON.	N702E
☐ C-GJCD	76	Learjet 35	35-055	Canada Jet Charters Ltd. Vancouver, BC. (status ?).	N255JH
☐ C-GJPG	83	Challenger 601	3017	Jim Pattison Industries Ltd. Vancouver, BC.	VR-CAR
☐ C-GKHA	65	Falcon 20C	19	Knighthawk Air Express, Ottawa, ON.	(N41PD)
☐ C-GKPM	78	HS 125/700A	NA0239	2847-7404 Quebec Inc/American Barrick Resources, Toronto, ON	N33BK
☐ C-GKTM	83	Learjet 55	55-076	Samuel Son & Co Ltd. Toronto, ON.	N30GL
☐ C-GLIG	83	Diamond 1A	A090SA	Lignum Ltd. Vancouver, BC.	N300LG
☐ C-GLJQ	90	Learjet 35A	35A-660	Sky Service FBO Inc. Montreal, PQ.	N660L
☐ C-GLMK	79	Citation II/SP	551-0143	Northwest International Airways, Yellowknife. (was 550-0100)	N140DA
☐ C-GLRJ	92	Learjet 31A	31A-043	Bombardier Inc. Montreal, PQ.	N5009V
☐ C-GLRS	96	Learjet 60	60-077	Bombardier Inc. Montreal, PQ.	N677LJ
☐ C-GLTG	82	Citation II	550-0364	Air 500 Ltd. Mississauga, ON.	N550AV
☐ C-GLYC	97	Challenger 604	5354	Bombardier Inc. Montreal.	
☐ C-GMAJ	75	Citation	500-0247	Myrand Aviation Inc. Saint Foy, PQ.	XA-JUA
☐ C-GMBA	83	HS 125/700A	NA0344	McMillan Bloedel Ltd. Vancouver, BC.	C-FWCE
☐ C-GMEA	68	HS 125/3A-RA	NA702	Metro Jet Inc. Collingwood, ON.	N813PR
☐ C-GMGB	96	Citation V Ultra	560-0390	Magna International Inc. Markham, ON.	N390CV
☐ C-GMII	92	Falcon 50	229	Magna International/543236 Alberta Inc. Buttonville, ON.	N114FJ
☐ C-GMLR	84	BAe 125/800B	8025	Millar Western Industries Ltd. Edmonton, AB.	N7C
☐ C-GMMY	88	Learjet 35A	35A-644	Wal-Mart Canada Inc. Mississauga, ON.	N54SB
☐ C-GMPF	83	Westwind-1124	391	RCMP - GRC Air Services, Ottawa, ON.	N24VH
☐ C-GMPQ	83	Citation II	550-0456	GRC Air Services - RCMP, Ottawa, ON.	(N20RF)
☐ C-GMSM	89	Citation II	550-0603	Canfor Corp/Canadian Forest Products Ltd. Richmond, BC.	N603CJ
☐ C-GMTR	89	BAe 125/800A	NA0435	Alberta Energy Co Ltd. Calgary, AB.	N800BA
☐ C-GMTV	85	Citation S/II	S550-0015	Alberta Energy Co Ltd. Calgary, AB.	(N1258U)
☐ C-GNAZ	78	HS 125/700A	NA0222	North American Airlines Ltd. Calgary, AB.	HB-VLJ
☐ C-GNCA	90	Falcon 50	203	Nova Corp of Alberta, Calgary, AB.	VR-CCQ
☐ C-GNPT	87	Learjet 35A	35A-626	Northwood Pulp & Timber Ltd. Prince George, BC.	N711NF
☐ C-GNSA	74	Citation	500-0160	Air Spray 1967 Ltd. Edmonton, AB.	N59TS
☐ C-GNWM	82	Citation II	550-0410	Northwest International Airways Ltd. Yellowknife, NT.	N46MF
☐ C-GOCM	81	HS 125/700A	NA0325	Carl Mullard/Morningstar Air Express Inc. Edmonton, AB.	N26H
☐ C-GPAW	94	Citation V	560-0260	Pratt & Whitney Aircraft of Canada Ltd. St Hubert, PQ.	N260CV
☐ C-GPCS	82	Learjet 55	55-035	Potash Corp. Saskatoon, SK.	N97AF
☐ C-GPDO	92	Learjet 35A	35A-673	Premdor Inc/Skyservice FBO Inc. Dorval, PQ.	C-FBDH
☐ C-GPFC	81	Westwind-Two	328	PFC Aviation Inc. Edmonton, AB.	N819JA
☐ C-GPGD	96	Challenger 604	5310	Execaire/P G Desmarais-Power Corp of Canada, Montreal, PQ.	C-FCCP
☐ C-GPOP	84	Citation III	650-0042	Expressair, Ottawa-Gatineau, PQ.	N342AS
☐ C-GPOT	91	Challenger 601-3A	5088	Potash Corp of Saskatchewan Inc. Saskatoon, SK.	N601HC
☐ C-GQBR	84	Learjet 55	55-105	Execaire Inc. Dorval, PQ.	N274
☐ C-GRDP	76	Westwind-1124	188	McCain Foods Ltd. Florenceville, NB.	N1124G
☐ C-GRFO	77	Learjet 35A	35A-100	Sky Service FBO Inc. Dorval, PQ.	N558E
☐ C-GRGE	81	Westwind-Two	353	Chartright Air Inc. Mississauga, ON.	N89UH
☐ C-GRHC	79	Citation II	550-0046	Chevron Canada Resources Ltd. Calgary, AB.	(N3292M)
☐ C-GRIO	97	Challenger 604	5353	Bombardier Inc. Montreal, PQ.	C-GLXH
☐ C-GRIS	73	Falcon 10	2	Skycharter Ltd. Toronto, ON.	N103JM
☐ C-GRMJ	90	Learjet 35A	35A-664	Canada Jet Charters Ltd. Vancouver, BC.	N117RY
☐ C-GRPF	94	Challenger 601-3R	5168	Royal Group Technolgies Ltd. Woodbridge, ON.	C-GLYC
☐ C-GRSD	68	Falcon 20C	157	Knighthawk Air Express, Ottawa, ON.	C-GRSD-X
☐ C-GRVJ	95	Learjet 31A	31A-111	611897 Alberta Ltd. Calgary, AB.	N50114
☐ C-GSAS	73	Learjet 25B	25B-109	Skycharter Ltd. Toronto, ON.	N333HP
☐ C-GSCL	83	Falcon 200	495	Shell Canada Ltd. Calgary, AB.	N800HM
☐ C-GSCX	82	Citation II	550-0348	Shell Canada Ltd. Calgary, AB.	N550CA
☐ C-GSKL	74	Learjet 25B	25B-179	Skycharter Ltd. Toronto, ON.	C-GBQC
☐ C-GSKN	66	Falcon 20C	65	Skycharter Ltd. Toronto, ON.	N5052U
☐ C-GSKQ	66	Falcon 20C	40	Skycharter Ltd. Toronto, ON.	N65LE
☐ C-GSKV	67	HS 125/3A-RA	NA704	Royal Aviation Inc. Vancouver, BC.	N888WK
☐ C-GSQI	91	Challenger 601-3A	5096	Execaire Inc. Montreal, PQ.	HB-IKW
☐ C-GSSS	89	Falcon 900	78	Chartright Air Inc. Toronto, ON.	N332MC
☐ C-GSUM	79	Citation 1/SP	501-0100	Suncor Energy Inc. Calgary, AB.	C-GSUN
☐ C-GSUN	97	Citation V Ultra	560-0448	Suncor Energy Inc. Calgary, AB.	
☐ C-GTAK	69	Falcon 20D	197	Knighthawk Air Express. Ottawa, ON.	N399SW
☐ C-GTCP	87	Falcon 900	29	Trans Canada Pipelines Ltd. Calgary, AB.	N421FJ
☐ C-GTDE	76	Learjet 35	35-057	Sky Service FBO Inc. Montreal, PQ.	N35MR
☐ C-GTJL	77	Learjet 35A	35A-124	Toma Jetprop Ltd. Calgary, AB.	N8LA
☐ C-GTLG	82	HS 125/700A	NA0321	Loewen Group/Corporate Jet Charters Inc. Vancouver, BC.	N469JR
☐ C-GTPL	83	Falcon 50	137	Trans Canada Pipelines Ltd. Calgary, AB.	(N119FJ)
☐ C-GTTS	66	HS 125/731	25108	PartnerJet Inc. Barrie, ON.	N25LA
☐ C-GTVO	78	Falcon 10	137	Fraser Paper Inc. Edmundston, NB.	N837F
☐ C-GUUU	82	Citation II	550-0423	LID Brokerage & Realty Co (1977) Ltd. Saskatoon, SK.	N45MC
☐ C-GVCA	76	Learjet 35	35-043	414660 Alberta Ltd/Business Flights Ltd. Calgary, AB.	
☐ C-GVER	78	Citation	500-0369	West Wind Aviation Inc. Saskatoon, SK.	N3132M
☐ C-GVVA	74	Learjet 35	35-002	Sunwest International Aviation Services Ltd. Calgary, AB.	N35SC
☐ C-GWCR	96	Citation V Ultra	560-0379	Weldwood of Canada Ltd. Vancouver, BC.	N5101J
☐ C-GWFG	72	Learjet 24D	24D-256	West Fraser Air Ltd. Vancouver, BC.	N703J
☐ C-GWFM	84	BAe 125/800A	8015	Sky Service FBO Inc. Dorval, PQ.	C-FTLA
☐ C-GWKF	79	Westwind-1124	271	Kelowna Flightcraft Air Charter Ltd. Kelowna, BC.	N218SC

| *Reg* | *Yr Type* | | *c/n* | *Owner/Operator* | *Prev Regn* |

Reg	Yr Type	c/n	Owner/Operator	Prev Regn
☐ C-GYPH	83 BAe 125/800B	8007	Power Financial Corp. Montreal, PQ.	C-GKRL
Military				
☐ 144601	82 Challenger 600S	1040	CC144, DND, 412 (Transport) Squadron, Uplands, ON.	C-GLYM
☐ 144602	82 Challenger 600S	1065	CC144, DND, 434 Squadron, 14 Wing, Greenwood, NS.	C-GBVE
☐ 144603	80 Challenger 600S	1006	CE144A, DND, 434 Squadron, 14 Wing, Greenwood, NS.	C-GCSN
☐ 144604	80 Challenger 600S	1007	CC144, DND, 434 Squadron, 14 Wing, Greenwood, NS.	C-GBKC
☐ 144605	80 Challenger 600S	1008	CC144, DND, 434 Squadron, 14 Wing, Greenwood, NS.	C-GBEY
☐ 144606	80 Challenger 600S	1009	CC144, DND, 434 Squadron, 14 Wing, Greenwood, NS.	C-GCVQ
☐ 144607	80 Challenger 600S	1014	CE144A, DND, 434 Squadron, 14 Wing, Greenwood, NS.	C-GBLL
☐ 144608	80 Challenger 600S	1015	CC144, DND, 434 Squadron, 14 Wing, Greenwood, NS.	C-GBLN
☐ 144609	81 Challenger 600S	1017	CC144, DND, 434 Squadron, 14 Wing, Greenwood, NS.	C-GBPX
☐ 144610	81 Challenger 600S	1022	CC144, DND, 434 Squadron, 14 Wing, Greenwood, NS.	C-GOGO
☐ 144611	81 Challenger 600S	1030	CE144A, DND, 434 Squadron, 14 Wing, Greenwood, NS.	C-GCZU
☐ 144612	79 Challenger 600S	1002	CX144, DND, AETE, Cold Lake, AB. (status ?).	C-GCGS-X
☐ 144614	85 Challenger 601	3036	CC144, DND, 412 (Transport) Squadron, Uplands, ON.	C-GCUP
☐ 144615	85 Challenger 601	3037	CC144, DND, 412 (Transport) Squadron, Uplands, ON.	C-GCUR
☐ 144616	85 Challenger 601	3038	CC144, DND, 412 (Transport) Squadron, Uplands, ON.	C-GCUT
☐ 15001	87 Airbus A310-304	446	Canadian Armed Forces, 437 Squadron, 8 Wing, Trenton.	C-GBWD
☐ 15002	88 Airbus A310-304	482	Canadian Armed Forces, 437 Squadron, 8 Wing, Trenton.	C-GLWD
☐ 15003	86 Airbus A310-304	425	Canadian Armed Forces, 437 Squadron, 8 Wing, Trenton.	C-FWDX
☐ 15004	87 Airbus A310-304	444	Canadian Armed Forces, 437 Squadron, 8 Wing, Trenton.	F-WQCQ
☐ 15005	87 Airbus A310-304	441	Canadian Armed Forces, 437 Squadron, 8 Wing, Trenton.	HS-TIF

CC = Chile (Republic of Chile)

Civil

Reg	Yr Type	c/n	Owner/Operator	Prev Regn
☐ CC-CTC	76 Sabre-60	306-112	Aguos Clara Inc. Santiago.	N740RC
☐ CC-CWW	84 Citation S/II	S550-0002	Cardal AG/Aerocardal, Santiago.	N211VP
☐ CC-DAC	93 Citation VI	650-0233	Directorate of Civil Aviation, Santiago.	N1303H
☐ CC-DGA	90 Citation II	550-0657	Directorate of Civil Aviation, Los Cerillos.	N3986G
☐ CC-LLM	83 Citation II/SP	551-0481	Club Aereo de Carabineros de Chile, Santiago.	N531A
☐ CC-PGL	94 Citation VII	650-7045	Rimac SA. Santiago.	N95CM
☐ CC-PZM	74 Citation	500-0203	ENAER, Santiago-Tobalaba.	N101HB
Military				
☐ 301	80 Citation II	550-0104	Chilean Navy, Santiago.	E-301
☐ 301	79 Falcon 200	401	VP-1, Chilean Navy. Santiago.	F-GEXF
☐ 303	87 Citation III	650-0131	Chilean Navy, Santiago.	E-303
☐ 351	76 Learjet 35	35-050	Fuerza Aerea Chilena, Santiago.	CC-ECO
☐ 352	76 Learjet 35	35-066	Fuerza Aerea Chilena, Santiago.	CC-ECP
☐ 902	67 B 707-351C	19443	Fuerza Aerea Chilena, Santiago.	CC-CCK
☐ 903	65 B 707-330B	18926	Fuerza Aerea Chilena, Santiago.	CC-CEA
☐ 904	67 B 707-321B	19374	Fuerza Aerea Chilena, Santiago.	CC-CYO
☐ 911	85 Gulfstream 3	465	Fuerza Aerea Chilena, Santiago.	N465GA
☐ 921	70 B 737-58N	28866	Fuerza Aerea Chilena, Santiago.	

CN = Morocco (Kingdom of Morocco)

Civil

Reg	Yr Type	c/n	Owner/Operator	Prev Regn
☐ CN-TDE	74 Corvette	5	Monsieur Sari/CASA Air Services, Casablanca.	F-BVPA
☐ CN-TFU	91 Falcon 900B	105	Daewoo Cars Morocco, Casablanca.	F-WQBJ
☐ CN-TNA	88 Falcon 100	212	SA Majeste Hassan II, Rabat.	F-WZGU
Military				
☐ CNA-NL	76 Gulfstream 2TT	182	Government of Morocco, Rabat.	N17589
☐ CNA-NM	69 Falcon 20ECM	165	Ministry of Defence, Kenitra.	CN-MBH
☐ CNA-NN	68 Falcon 20ECM	152	Ministry of Defence, Kenitra.	CN-MBG
☐ CNA-NO	80 Falcon 50	12	Government of Morocco, Rabat.	F-WZHC
☐ CNA-NR	79 B 707-3W6C	21956	Government of Morocco, Rabat.	N707QT
☐ CNA-NS	61 B 707-138B	18334	Government of Morocco, Rabat. 'Africa Crown'	N58937
☐ CNA-NU	82 Gulfstream 3	365	Government of Morocco, Rabat.	HZ-AFO
☐ CNA-NV	89 Citation V	560-0025	Government of Morocco. Rabat.	(N12285)
☐ CNA-NW	89 Citation V	560-0039	Government of Morocco. Rabat.	

CP = Bolivia (Republic of Bolivia)

Civil

Reg	Yr Type	c/n	Owner/Operator	Prev Regn
☐ CP-2259	67 Jet Commander	95	Aerojet SA. Cochabamba.	N95JK
☐ CP-2263	66 Jet Commander	58	Aerojet SA. Cochabamba.	N957RC
☐ CP-2317	74 Sabre-40A	282-136	Aerojet SA. Cochabamba.	N112ML
Military				
☐ FAB 001	76 Sabre-60	306-115	President of Bolivia, La Paz.	(XA-LEI)
☐ FAB 008	75 Learjet 25B	25B-192	Fuerza Aerea Boliviana, La Paz. (photographic survey)	
☐ FAB 009	78 Learjet 35A	35A-152	Fuerza Aerea Boliviana, La Paz.	N964CL
☐ FAB 010	76 Learjet 25D	25D-211	Fuerza Aerea Boliviana, La Paz. (status ?).	

CS = Portugal (Portuguese Republic)

Civil

Reg	Yr Type	c/n	Owner/Operator	Prev Regn
☐ CS-ATG	72 Falcon 20F	264	Heliavia Transporte Aereo Ltda. Lisbon. 'Bezas'	F-GJJS

Reg	*Yr Type*	*c/n*	*Owner/Operator*	*Prev Regn*

Reg	Yr	Type	c/n	Owner/Operator	Prev Regn
☐ CS-AYY	80	Citation 1/SP	501-0183	Air Luxor Lda. Lisbon. (sold USA 12/97 ?).	ZS-KPA
☐ CS-DBM	74	Citation	500-0200	Air Luxor Lda. Lisbon.	N96EA
☐ CS-DCA	74	Citation	500-0157	Initiative SA. Luxembourg.	N190AB
☐ CS-DNA	85	Citation S/II	S550-0032	Air Luxor Lda/European NetJets, Lisbon.	N232QS
☐ CS-DNB	85	Citation S/II	S550-0051	Air Luxor Lda/European NetJets, Lisbon.	N251QS
☐ CS-DNC	85	Citation S/II	S550-0077	Air Luxor Lda/European NetJets, Lisbon.	N277QS
☐ CS-DNF	97	Citation VII	650-7080	Air Luxor Lda/European NetJets, Lisbon.	N780QS
☐ CS-DNG	97	Citation VII	650-7081	Air Luxor Lda/European NetJets, Lisbon.	N781QS
☐ CS-MAC	95	Challenger 601-3R	5178	STDM/Jet Asia Ltd. Macau.	C-FWGE
☐ CS-TMJ	89	Falcon 50	190	Heliavia Transporte Aereo Ltda. Lisbon.	I-CAFE
☐ CS-TMK	90	Falcon 900	66	Vinair Aeroservicos SA. Lisbon.	F-GJPM
Military					
☐ 17103	69	Falcon 20D	217	Portuguese Air Force, Esc 504, Montijo, Lisbon.	8103
☐ 17401	89	Falcon 50	195	Portuguese Air Force, Esc 504, Montijo, Lisbon.	7401
☐ 17402	89	Falcon 50	198	Portuguese Air Force, Esc 504, Montijo, Lisbon.	7402
☐ 17403	91	Falcon 50	221	Portuguese Air Force, Esc 504, Montijo, Lisbon.	7403

CX = Uruguay (Republic of Uruguay)

Civil

Reg	Yr	Type	c/n	Owner/Operator	Prev Regn
☐ CX-ECO	77	Learjet 25D	25D-229		LV-MBP

C6 = Bahamas (Commonwealth of the Bahamas)

Civil

Reg	Yr	Type	c/n	Owner/Operator	Prev Regn
☐ C6-BEV	62	MS 760 Paris-2	111	Petrolclor Services Inc. Nice, France.	I-FINR

D = Germany (Federal Republic of Germany)

Civil

Reg	Yr	Type	c/n	Owner/Operator	Prev Regn
☐ D-ACTU	91	Challenger 601-3A	5085	Transalpina Flugzeughalter GmbH. Munich.	N618CC
☐ D-ADAM	78	VFW 614	G17	DLR Flugbetrieb, Oberpfaffenhofen. (status ?)	D-BABP
☐ D-ALME	91	Falcon 900	101	Metras-Medien Spedition/Heinrich Bauer Verlag, Hamburg.	N466FJ
☐ D-AMIM	96	Challenger 604	5317	Flow Tex Technologie/Flugbereitschaft GmbH. Soellingen.	C-FYXC
☐ D-BERT	91	Falcon 50	218	Bertelsmann AG. Paderborn.	N218WA
☐ D-BEST	97	Falcon 2000	50	Bauhaus GmbH. Mannheim.	F-WWME
☐ D-BFAR	00	Falcon 50	16	Dornier Luftfahrt GmbH. Oberpfaffenhofen.	D-BIRD
☐ D-BOOK	91	Falcon 50	215	Bertelsmann AG. Paderborn.	XA-SIM
☐ D-BSNA	82	Challenger 600S	1066	Fondsprojekt Josef Esch GmbH/Challenge Air GmbH. Cologne.	N51TJ
☐ D-BUSY	82	Challenger 600S	1070	MTM Aviation GmbH. Munich.	N670CL
☐ D-CACM	94	Citation VII	650-7039	ACM Air Charter Minninger GmbH. Baden-Baden.	N12643
☐ D-CAPO	78	Learjet 35A	35A-159	Phoenix Air GmbH. Munich.	(N135CK)
☐ D-CARL	81	Learjet 35A	35A-387	GFD GmbH. Hohn AFB.	
☐ D-CATL	83	Learjet 55	55-051	MTM Aviation GmbH. Munich.	N55KD
☐ D-CAVE	81	Learjet 35A	35A-423	Avia Luftreederei GmbH/Deutsche Rettungsflugwacht, Stuttgart	(N335GA)
☐ D-CAWA	88	Citation II	550-0596	Windrose Air Flugcharter GmbH. Berlin.	N96TD
☐ D-CAWU	90	Citation V	560-0042	Adolf Wurth GmbH. Schwaebisch Hall-Hessental.	N42CV
☐ D-CBEN	94	Citation V Ultra	560-0282	Wuerth Leasing GmbH. Schwaebisch Hall-Hessental.	
☐ D-CBMW	89	BAe 125/800A	8155	BMW Flight Service GmbH. Munich.	G-5-628
☐ D-CBNA	67	Falcon 20C	63	Naske-Air GmbH. Magdeburg.	PH-LPS
☐ D-CBWW	92	BAe 1000B	9028	Hanseatische Leasing GmbH. Hamburg.	G-5-749
☐ D-CCAA	80	Learjet 35A	35A-315	Deutsche Rettungsflugwacht, Cologne.	N662AA
☐ D-CCAB	97	Citation Bravo	550-0827	Airtrans Flugzeugmietungs GmbH, Schwaebisch Hall.	
☐ D-CCAT	92	Astra-1125SP	059	Preussag AG/Wiking Flight Service GmbH. Hannover.	N4341S
☐ D-CCAY	77	Learjet 35A	35A-112	Georg Paetzold/Taunus Air GmbH. Frankfurt.	3810G
☐ D-CCCA	78	Learjet 35A	35A-160	Georg Paetzold/Taunus Air GmbH. Frankfurt.	
☐ D-CCCB	90	Learjet 35A	35A-663	Maschinenfabrik E Mollers GmbH/Aero Dienst GmbH. Nuremberg.	N91480
☐ D-CCCF	80	Citation II	550-0189	CCF Manager Airline GmbH. Cologne.	HB-VGP
☐ D-CCGN	82	Learjet 55	55-017	Werner Robert Schweigert/Quick Air Service GmbH. Cologne.	N760AQ
☐ D-CCHB	76	Learjet 35A	35A-089	Bauhaus GmbH. Mannheim.	N3547F
☐ D-CCON	84	Learjet 55	55-098	Primair Executive Charter GmbH. Munich.	N726L
☐ D-CDEN	91	Learjet 31A	31A-049	HDE Air-Hermann Dieter Eschmann/Air Evex GmbH. Dusseldorf.	
☐ D-CDWN	78	Learjet 35A	35A-175	Abens Flugzeugvermietung GmbH/Aero Dienst GmbH. Nuremberg.	
☐ D-CEIS	90	Beechjet 400A	RK-10	Schoeller Lebensmittel GmbH. Nuremberg.	(D-CLSG)
☐ D-CFAN	87	BAe 125/800B	8094	TS-Transport Services, Dusseldorf.	G-5-576
☐ D-CFCF	81	Learjet 35A	35A-413	Senator Aviation Charter GmbH. Hamburg.	N27KG
☐ D-CFFB	97	Learjet 60	60-107	F & F Burda GmbH. Baden Baden.	N107LJ
☐ D-CFGA	78	Learjet 35A	35A-179	Aero-Dienst GmbH. Nuremberg.	N801PF
☐ D-CFGB	79	Learjet 35A	35A-268	Aero Dienst GmbH. Nuremberg.	N2U
☐ D-CFGC	80	Learjet 35A	35A-331	Aero-Dienst GmbH. Nuremberg.	N435JW
☐ D-CFOX	94	Citation V Ultra	560-0277	VEI GmbH. Moenchen-Gladbach.	
☐ D-CFTG	78	Learjet 35A	35A-204	Juliane Griesemann/Quick Air Service GmbH. Cologne.	(N277AM)
☐ D-CFUX	82	Learjet 55	55-061	Fuchs Gewurze GmbH. Dissen. 'Spirit of Spice'	N132EL
☐ D-CGAS	82	Citation II	550-0443	Interjet Charter GmbH. Stadtlohn-Wenningfeld.	OY-CYT
☐ D-CGFD	77	Learjet 35A	35A-139	GFD GmbH. Hohn.	N15SC
☐ D-CGFE	89	Learjet 36A	36A-062	GFD GmbH. Hohn.	N4291N
☐ D-CGFF	89	Learjet 36A	36A-063	GFD GmbH. Hohn.	N1048X
☐ D-CGGG	91	Learjet 31A	31A-042	Franz Gausepohl, Munster-Osnabruck.	
Reg	Yr	Type	c/n	Owner/Operator	Prev Regn

Reg	Yr	Type	c/n	Owner/Operator	Prev Regn
□ D-CGIN	84	Learjet 55	55-090	Neue Presse Verlags GmbH. Minich.	N723H
□ D-CGPD	78	Learjet 35A	35A-202	Air Traffic GmbH. Dusseldorf.	N499G
□ D-CGRC	79	Learjet 35A	35A-223	Georg Paetzold/Taunus Air GmbH. Frankfurt.	N215JW
□ D-CHAL	77	Westwind-1124	207	HAL Holstenair Luebeck, Luebeck.	N666K
□ D-CHBL	77	Westwind-1124	226	HAL Holstenair Luebeck, Luebeck.	N120S
□ D-CHCL	80	Westwind-1124	277	HAL Holstenair Luebeck, Luebeck.	N504JC
□ D-CHDE	89	Citation V	560-0031	BBP-Kunstoffwerk Marbach Baier GmbH. Marbach.	N1229F
□ D-CHDL	76	Westwind-1124	199	HAL Holstenair Luebeck, Luebeck.	N999MS
□ D-CHEF	85	BAe 125/800A	8038	PJC Private Jet Charter GmbH. Dusseldorf.	N550RH
□ D-CHPD	80	Learjet 35A	35A-309	Air Traffic GmbH. Dusseldorf.	N100MN
□ D-CHSW	94	Beechjet 400A	RK-84	Hilde Schmitt-Woelfe, Augsburg.	N8138M
□ D-CIEL	80	Falcon 10	155	Hertie Stiftung, Frankfurt.	(N220FJ)
□ D-CIFA	79	Citation II/SP	550-0074	FAI Airservice AG. Nuremberg. (was 551-0109).	(N81AJ)
□ D-CILL	90	Citation II	550-0660	Illbruck GmbH. Koln-Bonn.	N550JF
□ D-CION	81	Learjet 55	55-015	Primair Executive Charter GmbH. Munich.	N550LJ
□ D-CITY	78	Learjet 35A	35A-177	Helmut Idzkowiak/Senator Aviation Charter GmbH.Hamburg.	N174CP
□ D-CJPG	77	Learjet 35A	35A-108	Juliane Griesemann/Quick Air Service GmbH. Cologne.	(N86PQ)
□ D-CLBA	91	Beechjet 400A	RK-25	LBA-Luftfahrt Bundesamt, Braunschweig.	(VR-CDA)
□ D-CLIP	81	Learjet 55	55-004	MTM Aviation GmbH. Munich.	N24CK
□ D-CLUB	84	Learjet 55	55-092	Ratioflug GmbH. Frankfurt.	N400JT
□ D-CLUE	89	Citation III	650-0174	GEG Grundstuecksentwicklungs H H Goettsch KG. Cologne.	N674CC
□ D-CMAD	90	Learjet 55C	55C-143	Quelle Flug GmbH. Nuremberg.	N10871
□ D-CMAX	68	Falcon 20D	158	Canos Grundstuecksverwaltung GmbH. Koblenz.	F-WMKJ
□ D-CMEI	91	Citation V	560-0117	Siegfried Mansur, Ausberg/Lech-Air, Lech AFB.	(N6804F)
□ D-CMET	75	Falcon 20E	329	DLR Flugbetrieb, Oberpfaffenhofen.	F-WRQV
□ D-CMMM	76	Learjet 24D	24D-328	Franz Gausepohl, Munster-Osnabruck.	D-IMMM
□ D-CNCI	90	Citation V	560-0061	Karl Hertel GmbH/Aerostar Aircraft Charter GmbH. Nuremberg.	N2701J
□ D-COKE	81	Learjet 35A	35A-447	Private Wings Flugcharter GmbH. Berlin-Tempelhof.	N300FN
□ D-COOL	83	Learjet 55	55-052	MTM Aviation GmbH. Munich.	N551DB
□ D-COSA	71	HFB 320	1056	M-B-B GmbH. Munich.	
□ D-COSY	81	Learjet 35A	35A-415	Phoenix Air GmbH. Munich.	N415DJ
□ D-CSAP	92	Learjet 31A	31A-057	SAP AG Systeme, Mannheim.	N9147Q
□ D-CSFD	88	Citation S/II	S550-0148	Eheim/SFD Stuttgarter Flugdienst GmbH. Stuttgart.	N170RD
□ D-CTAN	91	Citation V	560-0150	H D E Air-Hermann Dieter Eschmann/Air Evex GmbH. Dusseldorf.	(6879L)
□ D-CURE	89	Learjet 35A	35A-652	Ronald Ernst/RE GmbH. Mannheim.	N99FN
□ D-CVHA	94	Citation V Ultra	560-0275	Viessmann Werke GmbH. Allendorf.	
□ D-CVIP	84	Learjet 55	55-109	MTM Aviation GmbH. Munich.	N348HM
□ D-CWAY	84	Learjet 55	55-107	MFG Milan-Flug GmbH. Hannover.	N304AT
□ D-CZAR	91	Citation V	560-0114	Viktor Bondarenko/Interdean International Spedition, Munich.	OE-GPS
□ D-I...	97	CitationJet	525-0193		N193CJ
□ D-IALL	96	CitationJet	525-0143	Allkauf SB-Warenhaus GmbH. Moenchengladbach.	D-IOMP
□ D-IAMM	93	CitationJet	525-0041	MSK Verpackungssyteme GmbH. Kiev, Ukraine.	HB-VJQ
□ D-IANE	79	Citation 1/SP	501-0106	Kuri-Flugdienst KG. Friedrichshafen.	OE-FYF
□ D-IATC	73	Citation	500-0116	EFS Flug Service GmbH. Dusseldorf.	EC-CJH
□ D-IAVB	97	CitationJet	525-0172	AVB GmbH. Westerland/Sylt Flug Charter Gmbh.	N172CJ
□ D-IAWA	82	Citation II	551-0421	Gerd Brandecker, Schwalbach. (was s/n 550-0422).	N550RD
□ D-ICCC	75	Citation	500-0269	EFS Flug Service GmbH. Dusseldorf.	(D-IKUC)
□ D-ICEE	95	CitationJet	525-0096	Krause Bautraeger Holding GmbH. Bayreuth.	(EC-...)
□ D-ICGT	96	CitationJet	525-0164	AGIB fuer Grundbesitz u Industriebeteiligung mbH. Bielefeld.	
□ D-ICTA	81	Citation II/SP	551-0051	Flugbereitschaft GmbH. Karlsruhe.	(D-IHAT)
□ D-ICUR	80	Citation II/SP	551-0029	Kuri-Flugdienst KG. Friedrichshafen.	N500ER
□ D-IDAG	96	CitationJet	525-0144	DAS-Duscholux Flugbetrieb GmbH. Mannheim.	
□ D-IEAR	79	Citation II	551-0018	Rieker Air Service GmbH. Stuttgart. (was 550-0373).	N387MA
□ D-IEIR	83	Citation 1/SP	501-0259	Alfred Eisele/Rieker Air Service GmbH. Stuttgart.	N501MS
□ D-IEWS	97	CitationJet	525-0217	Mietkauf GmbH/Losch AG. Muelheim.	N5202D
□ D-IFAN	97	CitationJet	525-0214	Michael Nixdorf, Hoevelhof.	
□ D-IFUP	95	CitationJet	525-0126		N14TV
□ D-IGAS	97	CitationJet	525-0223		
□ D-IHEB	94	CitationJet	525-0064	Heberger Bau GmbH. Schifferstadt.	
□ D-IHHS	94	CitationJet	525-0082	Adolf Reuter, Soest.	N5090Y
□ D-IHOL	98	CitationJet	525-0229		
□ D-IJYP	96	CitationJet	525-0165	Air Evex GmbH. Dusseldorf.	
□ D-IKOP	93	CitationJet	525-0016	Bau und Umwelttechnologie Kahlen & Partner GmbH. Maastricht.	N216CJ
□ D-ILAN	89	Citation II/SP	551-0614	SFD Stuttgarter Flugdienst GmbH. Stuttgart.	
□ D-ILAT	97	CitationJet	525-0209	Liebherr-Aerospace Lindenberg GmbH. Lindenberg.	
□ D-ILCC	81	Citation II	551-0335	Wiederschein & Rettig GbR. Mannheim. (was 550-0298).	N431DS
□ D-ILTC	89	Citation II/SP	551-0617	LTO-Lufttransport Osnabruck-Munster GmbH.	N617CM
□ D-IMMD	97	CitationJet	525-0211	Makro-Medien-Dienst GmbH. Ostfildern.	
□ D-IMME	82	Citation II/SP	551-0400	AGV Mobilienvermietungs GmbH. Wiesbaden. (was 550-0359).	N280JS
□ D-IMRX	85	Citation 1/SP	501-0688	Rena Informationstechnik GmbH/Heron Flugdienst, Munich.	
□ D-IMTM	79	Citation II/SP	551-0009	Then Air KG. Coburg.	N1959E
□ D-IOBO	93	CitationJet	525-0025	OBO Bettermann OHG. Menden.	N9LR
□ D-IRKE	95	CitationJet	525-0123	H D E Air-Hermann Dieter Eschmann/Air Evex GmbH. Dusseldorf.	N52462
□ D-IRON	96	CitationJet	525-0168	Geisers Stahlbau GmbH. Dusseldorf.	
□ D-IRWR	95	CitationJet	525-0118	SLOVGOLD GmbH. Vienna, Austria.	N118AZ
□ D-ISGW	94	CitationJet	525-0070	Siemag Verwaltungs GmbH. Hilchenbach.	

Reg   Yr   Type   c/n   Owner/Operator   Prev Regn

Reg	Yr	Type	c/n	Owner/Operator	Prev Regn
□ D-ISIS	81	Citation 1/SP	501-0182	Hanseatische Jet Charter GmbH. Bremen.	(N125CA)
□ D-ITSV	94	CitationJet	525-0084	Brunner & Co/Comfort Air GmbH. Munich.	
□ D-IURH	97	CitationJet	525-0196	Hekuma Herbst Maschinenbau GmbH. Eching.	
□ D-IVHA	95	CitationJet	525-0103	Viessmann Werke GmbH. Allendorf.	N5204D
□ D-IVIN	97	CitationJet	525-0188	Westflug Aachen Luftfahrt GmbH. Aachen-Marzbrueck.	
□ D-IVOB	80	Citation II/SP	551-0191	Westflug Aachen Luftfahrt GmbH. Merzbrueck.	N107SB
□ D-IWHL	93	CitationJet	525-0029	ABC Nordflug GmbH. Luebeck-Blankensee.	N13308
□ D-IWIL	97	CitationJet	525-0221	Rademacher Charter GmbH. Stadtholm.	
Military					
□ 10+01	68	B 707-307C	19997	Luftwaffe, FBS, Koln-Wahn. 'Otto Lilienthal'	(68-11071)
□ 10+02	68	B 707-307C	19998	Luftwaffe, FBS, Koln-Wahn. 'Hans Grade'	(68-11072)
□ 10+03	68	B 707-307C	19999	Luftwaffe, FBS, Koln-Wahn. 'August Euler'	(68-11073)
□ 10+04	68	B 707-307C	20000	Luftwaffe, FBS, Koln-Wahn. 'Hermann Koehl'	(68-11074)
□ 10+21	89	Airbus A310-304	498	Luftwaffe, FBS, Koln-Wahn. 'Konrad Adenauer'	D-AOAA
□ 10+22	89	Airbus A310-304	499	Luftwaffe, FBS, Koln-Wahn. 'Theodor Heuss'	D-AOAB
□ 10+23	89	Airbus A310-304	503	Luftwaffe, FBS, Koln-Wahn. 'Kurt Schumacher'	D-AOAC
□ 10+24	87	Airbus A310-304	434	Luftwaffe, FBS, Koln-Wahn.	D-AIDA
□ 10+25	88	Airbus A310-304	484	Luftwaffe, FBS, Koln-Wahn.	D-AIDB
□ 12+01	84	Challenger 601	3031	Luftwaffe, FBS, Koln-Wahn.	C-GTCB
□ 12+02	85	Challenger 601	3040	Luftwaffe, FBS, Koln-Wahn.	N608CL
□ 12+03	85	Challenger 601	3043	Luftwaffe, FBS, Koln-Wahn.	N609CL
□ 12+04	85	Challenger 601	3049	Luftwaffe, FBS, Koln-Wahn.	C-FQYT
□ 12+05	86	Challenger 601	3053	Luftwaffe, FBS, Koln-Wahn.	N604CL
□ 12+06	86	Challenger 601	3056	Luftwaffe, FBS, Koln-Wahn.	N612CL
□ 12+07	86	Challenger 601	3059	Luftwaffe, FBS, Koln-Wahn.	N614CL
□ 16+08	66	HFB 320	1025	Luftwaffe, Erprobungstelle 61, Manching.	D-9537
□ 17+01	77	VFW 614	G14	Luftwaffe, FBS, Koln-Wahn.	
□ 17+02	77	VFW 614	G18	Luftwaffe, FBS, Koln-Wahn.	
□ 17+03	77	VFW 614	G19	Luftwaffe, FBS, Koln-Wahn.	

D2 = Angola (Republic of Angola)

Civil

Reg	Yr	Type	c/n	Owner/Operator	Prev Regn
□ D2-ECB	85	Gulfstream 3	474	LAA/Government of Angola, Luanda. 'Cunene'	N311GA
□ D2-ESV	72	Falcon 20F-5	262	SONANGOL Aeronautica - Helipetrol, Luanda.	F-GHVR
□ D2-EXR	70	HS 125/403B	25215	Madlba Air, Lanseria, RSA.	ZS-NPV
□ D2-TPR	74	B 707-3J6B	20715	Government of Angola, Luanda.	B-2404

EC = Spain (Kingdom of Spain)

Civil

Reg	Yr	Type	c/n	Owner/Operator	Prev Regn
□ EC-DQE	76	Corvette	26	Alfajet International SA. Madrid.	F-GDAY
□ EC-DQG	76	Corvette	27	Dominguez Toledo SA. Malaga.	F-BVPH
□ EC-EDC	65	Falcon 20C	6	Audeli Air Express, Madrid.	N750SS
□ EC-EDN	77	Citation 1/SP	501-0010	Instituto Carto Grafico de Cataluna, Barcelona.	VH-POZ
□ EC-EGY	85	Learjet 25D	25D-373	MAC Aviation SL. Zaragoza.	N29EW
□ EC-EHC	66	Falcon 20DC	46	Audeli Air Express, Madrid.	N46WG
□ EC-EKK	67	Falcon 20C	106	Audeli Air Express, Madrid. (status ?).	N31V
□ EC-ELK	85	BAe 125/800B	8022	Teire SA. Madrid.	G-5-569
□ EC-EQP	68	Falcon 20C	149	Delta Aviation SA. Madrid. (status ?).	EC-263
□ EC-FES	91	Citation II	550-0678	Gestair/Industrias Titan SA. Barcelona.	EC-777
□ EC-FGX	65	JetStar-731	5062	Transaviation SA. Madrid.	EC-697
□ EC-FPI	92	Falcon 900B	115	Gestair Executive Jet SA. Santiago de Compestella.	EC-235
□ EC-FQX	76	JetStar 2	5202	Transaviation SA. Madrid.	EC-232
□ EC-FRV	79	Gulfstream 2B	237	Gestair Executive Jet SA. Madrid.	EC-363
□ EC-FZP	94	CitationJet	525-0065	Gestair Executive Jet SA. Madrid.	EC-704
□ EC-GIB	81	Westwind-Two	335	Gestair Executive Jet SA. Madrid.	EC-354
□ EC-GIE	96	CitationJet	525-0133	Gestair Executive Jet SA. Madrid.	EC-261
□ EC-GJF	79	Citation 1/SP	501-0107	Jose Maria Caballe Morta, Alicante.	(N75471)
□ EC-GLM	90	Citation V	560-0062	Gestair/Banco Zaragozano SA. Zaragosa.	EC-411
□ EC-GMO	96	Falcon 900EX	6	Gestair/Banco de Santander, Madrid.	F-OIBL
□ EC-GNK	96	Falcon 2000	37	Grupo March SA/Gestair Executive Jet SA. Madrid.	F-WWMH
□ EC-GOV	97	Citation V Ultra	560-0419	Gestair Executive Jet SA. Madrid.	F-WWMH
□ EC-GPN	90	Falcon 50	204	Gestair Executive Jet SA. Madrid.	VP-CGP
Military					
□ T 11-1	71	Falcon 20E	253	EC-ZCJ, 45-02, Ed1A/45 Grupo, Torrejon. (prev code 401-02).	EC-BZV
□ T 11-5	83	Falcon 20F	475	EC-ZCN, 45-05, Ed1A/45 Grupo, Torrejon.	F-WJML
□ T 16-1	82	Falcon 50	84	EC-ZCP, 45-20, Ed1A/45 Grupo, Torrejon. (prev code 401-09).	(N2711B)
□ T 17-1	68	B 707-331B	20060	EC-ZC., 45-10, Ed1A/45 Grupo, Torrejon.	N275B
□ T 17-2	64	B 707-331C	18757	EC-ZC., 45-11, Ed1A/45 Grupo, Torrejon.	N792TW
□ T 17-3	77	B 707-368C	21367	EC-ZC., 45-12, Ed1A/45 Grupo, Torrejon.	N7667B
□ T 18-1	88	Falcon 900	38	EC-ZC., 45-40, Ed1A/45 Grupo, Torrejon.	F-WWFE
□ T 18-2	91	Falcon 900	90	EC-ZC., 45-41, Ed1A/45 Grupo, Torrejon.	F-WWFG
□ TM 11-2	71	Falcon 20D	222	EC-ZCK, 45-03, Ed1A/45 Grupo, Torrejon. (prev code 401-03).	EC-BXV
□ TM 11-3	70	Falcon 20D	219	EC-ZCL, 45-04, Ed1A/45 Grupo, Torrejon. (prev code 401-04).	EC-BVV
□ TM 11-4	75	Falcon 20E	332	EC-ZCM. 408-12, Ed1A/45 Grupo, Torrejon.(prev 45-01)	EC-CTV
□ TR 20-02	92	Citation V	560-0193	403-12,	N1282K

Reg	Yr	Type	c/n	Owner/Operator	Prev Regn
□ U 20-1	82	Citation II	550-0425	01-405, Armada, 4a Escuadrilla, Rota.	(LN-FOX)
□ U 20-2	83	Citation II	550-0446	01-406, Armada, 4a Escuadrilla, Rota.	N1248N
□ U 20-3	88	Citation II	550-0592	01-407, Armada, 4a Escuadrilla, Rota.	N1302N

EI = Eire (Republic of Eire)
Civil

Reg	Yr	Type	c/n	Owner/Operator	Prev Regn
□ EI-CIR	80	Citation II/SP	551-0174	Air Group Finance Ltd. Dublin. (was 550-0128).	N60AR
□ EI-LJR	96	Falcon 2000	18	Executive Air Transport AG/I C Kosmos, Moscow, Russia.	F-GMPR
□ EI-MAS	96	Challenger 601-3R	5194	Executive Air Charter & Services, Stansted, UK. 'Monette'	C-FXIP
□ EI-SXT	94	Challenger 601-3R	5159	Sextant Avionique, Villacoublay-Paris, France.	C-FTNN
□ EI-WDC	67	HS 125/3B	25132	Westair Aviation Ltd. Shannon.	G-OCBA
□ EI-WGV	96	Gulfstream V	505	Westair Aviation Ltd. Shannon.	N505GV
Military					
□ 251	91	Gulfstream 4	1160	Irish Air Corps. Casement-Dublin.	N17584

EL = Liberia (Republic of Liberia)
Civil

Reg	Yr	Type	c/n	Owner/Operator	Prev Regn
□ EL-LIB	67	BAC 1-11/412EB	111	Eagle Aviation, Paris, France.	F-WQFM
□ EL-VDY	71	Falcon 20E	245	Government of Liberia, Monrovia.	HB-VDY

EP = Iran (Islamic Republic of Iran)
Civil

Reg	Yr	Type	c/n	Owner/Operator	Prev Regn
□ EP-AGA	77	B 737-286	21317	Government of Iran, Teheran.	
□ EP-AGY	73	Falcon 20E	286	Iran Asseman Airlines, Teheran.	F-WRQU
□ EP-AKC	74	Falcon 20E	301	Iran Asseman Airlines, Teheran.	F-WNGL
□ EP-FIC	75	Falcon 20E	334	CAO/Iran Asseman Airlines, Teheran.	F-WRQU
□ EP-FID	75	Falcon 20E	338	CAO/Iran Asseman Airlines, Teheran.	F-WMKG
□ EP-FIE	71	Falcon 20E	251	Iran Asseman Airlines, Teheran.	EP-VAP
□ EP-FIF	75	Falcon 20E	320	Iran Asseman Airlines, Teheran.	YI-AHG
□ EP-NHY	78	B 707-386C	21396	Government of Iran, Teheran.	1001
□ EP-PAZ	76	F 28-1000	11104	Iran Asseman Airlines, Teheran.	F-GIAK
□ EP-PLN	64	B 727-30	18363	Government of Iran, Teheran. "Palestine"	EP-SHP
□ EP-SEA	77	Falcon 20E	367	Iran Asseman Airlines, Teheran.	F-WRQR
Military					
□ 1003	76	JetStar 2	5203	Government of Iran, Teheran. (status ?).	EP-VLP
□ 1004	69	JetStar-8	5137	Government of Iran, Teheran.	EP-VRP
□ 15-2235	75	Falcon 20E	333	Ministry of SEPHA L.D. D.A.L. Teheran.	5-2801
□ 15-2533	75	Falcon 20E	318	Ministry of SEPHA L.D. D.A.L. Teheran.	EP-FIG
□ 5-2802	75	Falcon 20E	336	Iranian Navy, Mehrabad.	F-WRQP
□ 5-2803	76	Falcon 20E	340	Iranian Navy, Mehrabad.	F-WRQX
□ 5-2804	76	Falcon 20E	346	Iranian Navy, Mehrabad.	F-WRQP
□ 5-3021	76	Falcon 20E	350	Iranian Air Army, Mehrabad.	5-4040
□ 5-9001	76	Falcon 20F	351	Iranian Air Force, 1st Transport Base, Mehrabad.	F-WMKJ
□ 5-9002	76	Falcon 20F	353	Iranian Air Force, 1st Transport Base, Mehrabad.	F-WRQP
□ 5-9003	76	Falcon 20F	354	Iranian Air Force, 1st Transport Base, Mehrabad.	F-WRQR
□ 5-90..	82	Falcon 50	101	Government of Iran, Teheran. (5-9011/9012 or 9013).	YI-ALC
□ 5-90..	83	Falcon 50	120	Government of Iran, Teheran. (5-9011/9012 or 9013).	YI-ALD
□ 5-90..	83	Falcon 50	122	Government of Iran, Teheran. (5-9011/9012 or 9013).	YI-ALE

EZ = Turkmenia (Türkmenistan)
Civil

Reg	Yr	Type	c/n	Owner/Operator	Prev Regn
□ EZ-A001	93	B 737-341	26855	Government of Turkmenistan, Ashkhabad.	EK-A001
□ EZ-A010	94	B 757-23A	25345	Government of Turkmenistan, Ashkhabad.	
□ EZ-B021	92	BAe 1000B	9029	National State Aviacompany, Ashkhabad.	G-5-751

F = France (French Republic)
Civil

Reg	Yr	Type	c/n	Owner/Operator	Prev Regn
□ F-BJET	59	MS 760 Paris-1A	39	Euralair/Ste Groupe M.S.C. Paris-Le Bourget.	F-WJAA
□ F-BJLH	73	Falcon 10	1	Leadair Unijet SA. Paris-Le Bourget.	F-WJLH
□ F-BLKL	64	MS 760 Paris-3	01	Euralair International, Paris-Le Bourget.	F-WLKL
□ F-BMSS	65	Falcon 20F	2	IGN France, Creil.	F-WMSS
□ F-BSIM	67	HS 125/3B	25130	Ste. Joelle/Groupe Ariel, Paris-Le Bourget.	TR-LXO
□ F-BTEL	80	Citation II	550-0190	S N C Air Delta 79/Air Gama, Niort-Souche.	N98715
□ F-BUQN	73	Corvette	3	S.N.I.A.S., Toulouse-Blagnac.	F-WUQN
□ F-BUQP	74	Corvette	4	Aero Vision SARL. Toulouse-Blagnac.	F-WUQP
□ F-BVPB	74	Corvette	6	S.F.A.C.T., Saint Cyr.	F-OGJL
□ F-BVPG	76	Corvette	25	Aero Stock, Paris-Le Bourget.	F-OBZV
□ F-BVPK	74	Corvette	7	SFACT Operations Aeriennes, Saint Cyr.	N611AC
□ F-BVPN	75	Falcon 20E-5	311	Manufacture Francaise PNEUS Michelin, Clermont-Aulnat.	F-WRQS
□ F-BVPR	75	Falcon 10	5	Dassault Falcon Service, Paris-Le Bourget.	F-WVPR
□ F-BVPT	75	Corvette	16	SFACT Operations Aeriennes, Saint Cyr.	
□ F-BXAS	75	AC 690A-TU	11240	Ste. Turbomeca, Pau.	F-WXAS
□ F-BXPT	65	Learjet 23	23-014	BNP Bail/Darta/SARL Aerojet Prive, Paris-Le Bourget.	(HB-VEL)
□ F-BXQL	61	MS 760 Paris-2B	105	Euralair International, Paris-Le Bourget.	N760Q
Reg	Yr	Type	c/n	Owner/Operator	Prev Regn

Reg	Yr	Type	c/n	Owner/Operator	Prev Regn
☐ F-BYFB	82	HS 125/700B	7166	Ste. Bouyghes/Groupement International de Commerce, Paris.	G-5-18
☐ F-GBRF	75	Falcon 10	38	Ste. Roquette Freres, Merville-Callonne.	N20ET
☐ F-GBTM	78	Falcon 20GF	397	Dassault Falcon Service, Paris-Le Bourget.	F-WBTM
☐ F-GDLR	78	Falcon 10	121	Leadair Unijet SA. Paris-Le Bourget.	HB-VFT
☐ F-GDLU	74	Falcon 20E	314	Leadair Unijet SA. Paris-Le Bourget.	D-COTT
☐ F-GECR	67	HS 125/3B	25128	Aero Stock, Paris-le Bourget.	G-AVOI
☐ F-GELT	86	Falcon 100	211	CATEX SA. St Etienne.	F-WZGT
☐ F-GEPL	74	Citation	500-0164	Segi Aero Services/Euralair, Paris-Le Bourget.	N4209K
☐ F-GEPQ	75	Corvette	50	Laboratoire de Recherche, Avignon-Caumont.	F-SEBH
☐ F-GERO	81	Falcon 10	179	Sogedim/Biscuiterie de Pont Aven, Quimper.	(F-GGRA)
☐ F-GFDH	75	Corvette	13	CALIF/SNIAS Eurocopter, Marseilles.	N601AN
☐ F-GFGB	81	Falcon 10	177	Air BG, Paris-Le Bourget.	N533CS
☐ F-GFMD	81	Falcon 10	136	CIP Transports/Aero Services Executive, Paris-Le Bourget.	F-WZGS
☐ F-GFMP	67	HS 125/3B	25125	Air Trema, Paris-Le Bourget.	G-AVAI
☐ F-GFPF	75	Falcon 10	68	CIP Transports/Aero Services Executive, Paris-Le Bourget.	N80MP
☐ F-GGAL	86	Citation III	650-0117	S.F.E.C. Aero + Ste Carat Publicite, Paris-Le Bourget,	N1321N
☐ F-GGCP	80	Falcon 50	9	Ste Sporto-CCME/Aero Services Executive, Paris-Le Bourget.	(N100WJ)
☐ F-GGGA	88	Citation II	550-0586	Air Gama, Niort-Souche.	N1301N
☐ F-GGGT	89	Citation II	550-0611	Timac SA. Dinard.	(N1242K)
☐ F-GGRH	86	Falcon 900	5	Korreda Ltd/Air Enterprise International, Paris-Le Bourget.	VH-BGF
☐ F-GGVB	79	Falcon 50	11	AXA Reassurance SA/Leadair Unijet SA. Paris-Le Bourget.	N5739
☐ F-GHDT	69	Falcon 20-5	176	DFS/Soc. Fiat Aviazione SpA. Turin.	F-WGTM
☐ F-GHDX	79	Falcon 10	140	Agropar-Ste West Reefer Line/Doux, Quimper.	N88WL
☐ F-GHER	76	Falcon 10	88	ATP SA/Soltralentz, Nancy.	F-GKCD
☐ F-GHPA	69	Falcon 20C	170	DFS/Soc. Aviazione Fiat SpA. Turin.	I-EKET
☐ F-GHPB	88	Falcon 100	215	THS Helicopteres, Lyon-Bron.	F-WZGY
☐ F-GHSK	89	Falcon 100	218	Watwill'Air SNC. Paris-Le Bourget.	F-WZGB
☐ F-GICN	90	Falcon 50	210	CGE=Compagnie Generale d'Electricite, Paris-Le Bourget.	F-WWHL
☐ F-GIDE	84	Falcon 900B	1	Dassault Falcon Service, Paris-Le Bourget.	F-WIDE
☐ F-GILM	76	Corvette	32	Mariette Gillon/Aero Services Executive, Paris-Le Bourget.	EC-DUF
☐ F-GIPH	82	Falcon 100	194	Regourd Aviation, Paris-Le Bourget.	N61FC
☐ F-GIRH	75	Corvette	14	Aero Stock, Paris-Le Bourget.	SP-FOA
☐ F-GIVT	66	Falcon 20C	32	Aero Stock, Paris-Le Bourget.	TL-AJK
☐ F-GJAP	76	Corvette	31	Soc Aerospatiale, Paris-Le Bourget.	EC-DYE
☐ F-GJAS	74	Corvette	8	CECICO Enterprises/S.N.C. Evasion, Paris-Le Bourget.	6V-AAH
☐ F-GJCC	67	Falcon 20C	72	Air Entreprise International, Paris-Le Bourget. (status ?).	N725P
☐ F-GJDB	67	Falcon 20C	76	Philip Morris Inc/Meridian Air, Moscow, Russia.	F-GGFO
☐ F-GJDE	68	HS 125/3B	25131	Uni Air SA/Ste Conquest, Cannes.	3A-MDB
☐ F-GJDG	76	Citation	500-0312	SNC GIN Finances/Air Normandie, Le Havre.	N82AT
☐ F-GJLB	78	Corvette	39	Ste Algair/Ste FISA, Savigny sur Orge.	TL-RCA
☐ F-GJLL	74	Falcon 10	22	Leadair Unijet SA/Air Inter Ivoire, Abidjan, Ivory Coast.	N48JC
☐ F-GJMA	78	Falcon 10	116	Chaillotine Air Service, Auxerre-Branches. (status ?).	N525RC
☐ F-GJPR	65	Falcon 20C	5	Air Provence International, Marseilles.	N747W
☐ F-GJXX	89	Citation V	560-0070	Ste J C Decaux, Toussus.	
☐ F-GJYD	82	Citation II	550-0415	DARTA Aero Charter, Paris-Le Bourget.	N1949M
☐ F-GKAE	87	Falcon 100	213	Air Entreprise International, Paris-Le Bourget.	ZK-MAZ
☐ F-GKBC	77	Falcon 10	99	Michelet International/DARTA Aero Charter, Paris-Le Bourget.	N67JW
☐ F-GKBZ	88	Falcon 50	185	G I E Air BG, Paris-Le Bourget.	N238Y
☐ F-GKDB	73	Falcon 20E	271	Air Entreprise International, Paris-Le Bourget.	F-GHPO
☐ F-GKGA	75	Corvette	11	Gallic Aviation, Paris-Le Bourget.	F-WFPD
☐ F-GKGD	76	Corvette	34	Gallic Aviation, Paris-Le Bourget.	SE-DEE
☐ F-GKGL	90	Citation V	560-0058	Ste Jean Lion/Canal/GTM Entrepose-Euralair, Paris-Le Bourget	N2686Y
☐ F-GKHL	90	Citation V	560-0059	Loca Cio + SARI + Groupe MSC + Upsa/Euralair, Paris.	N2687L
☐ F-GKID	76	Citation	500-0319	Sinair, Grenoble.	N94MA
☐ F-GKIR	77	Citation	500-0361	Soc de Location pour l'Industrie Aerienne-Solid Air, Paris.	N90EB
☐ F-GKIS	74	Falcon 20E	307	SEJP-Ste de Location de Jets Prives/Darta, Paris-le Bourget.	OE-GLL
☐ F-GKJL	90	Citation V	560-0093	Soc Perigourdine de Participations, Paris-Le Bourget.	(N6785C)
☐ F-GLEC	76	Corvette	30	CALIF/SNIAS Eurocopter, Paris-Le Bourget.	(F-GKGB)
☐ F-GLGY	87	Falcon 900	11	E F Education/Dassault Aviation, Paris-Le Bourget.	UN-09002
☐ F-GLHI	97	Falcon 900B	166	Dassault Aviation, Paris-Le Bourget.	N900FJ
☐ F-GLIM	91	Citation V	560-0119	Ste SICA LG Services-Limagrain, Clermont Ferrand.	(N6804N)
☐ F-GLMD	67	Falcon 20-5	117	THS Helicopteres, Lyon-Bron.	EC-FJP
☐ F-GLMM	67	Falcon 20-5	116	Meridian, Moscow, Russia.	F-WLMM
☐ F-GLMT	71	Falcon 20F	246	Regourd Aviation, Paris-Le Bourget.	F-BSTR
☐ F-GLTK	89	Citation II	550-0609	Knauf Trade SNC. Colmar.	N344A
☐ F-GLYC	92	Citation V	560-0205	Ste Gaspard Fournitures de Bureau, Valenciennes-Denain.	
☐ F-GMCI	79	Citation II	550-0050	Air Vendee Investissements, Rouen.	F-ODUT
☐ F-GMCK	97	Falcon 2000	46	Air Entreprise International, Paris-Le Bourget.	(N220JM)
☐ F-GMGA	81	Falcon 50	51	Korreda Ltd/Air Entreprise International, Paris-Le Bourget.	N52DQ
☐ F-GNAF	75	Corvette	15	Global Systeme International/Aero Stock, Paris-le Bourget.	D6-ECB
☐ F-GNCJ	93	CitationJet	525-0024	Jet Investissement SNC (Jet Services), Lyon.	(N13291)
☐ F-GNCP	78	Citation II	550-0004	Ste Sporto/Aero Services Executive, Paris-Le Bourget.	N312GA
☐ F-GNDA	90	Falcon 900	88	LVMH-Louis Vuitton Moet Hennessy Services, Paris-le Bourget.	VR-BLT
☐ F-GNDZ	74	Falcon 10	17	Leadair Unijet SA. Paris-Le Bourget.	EC-949
☐ F-GOAL	82	Falcon 50	131	Exair/Continent Hypermarches SNC. Paris-Le Bourget.	F-WGTF
☐ F-GOBZ	73	Falcon 20E-5	293	Bizjet SA. Geneva/Dassault Falcon Service, Paris-le Bourget.	F-WQBN
Reg	Yr	Type	c/n	Owner/Operator	Prev Regn

Reg	Yr	Type	c/n	Owner/Operator	Prev Regn
☐ F-GOJT	85	Falcon 200	501	Dassault Falcon Service, Paris-Le Bourget.	F-WWGP
☐ F-GOPM	74	Falcon 20E	302	Manufacture Francaise PNEUS Michelin, Clermont-Ferrand.	F-WQBM
☐ F-GOYA	96	Falcon 900EX	11	Pinaud SA/SNC Artemis Conseil, Paris-Le Bourget.	F-WWFI
☐ F-GPAA	67	Falcon 20EW	103	AVDEF-Aviation Defense Service, Nimes-Garons.	G-FRAV
☐ F-GPAB	71	Falcon 20EW	254	AVDEF-Aviation Defense Service, Nimes-Garons.	G-FRAC
☐ F-GPAK	88	Gulfstream 4	1061	H R H Prince Karim Aga Khan, Paris-Le Bourget.	N457GA
☐ F-GPFC	95	CitationJet	525-0101	Promod SA. Marcq en Baroeul.	N5157E
☐ F-GPFD	90	Falcon 100	221	GIE Avion ECCO/THS Helicopteres, Lyon-Bron.	OE-GHA
☐ F-GPIM	66	Falcon 20C	30	Chaillotine Air Service, Chailley.	CS-ATD
☐ F-GPLA	76	Corvette	28	Aero Vision SARL. Toulouse-Blagnac.	(OO-TTL)
☐ F-GPPF	81	Falcon 50	65	Laboratoires Pierre Fabre SA/Air Toulouse, Blagnac.	N1EV
☐ F-GPSA	83	Falcon 50	123	Ste Gefco & Cie/Air Gefco, Paris-Le Bourget.	N211EF
☐ F-GROC	73	Falcon 20E	279	Regourd Aviation, Paris-Le Bourget.	N279AL
☐ F-GSAA	96	Falcon 2000	36	Dassault Aviation, Paris-Le Bourget.	F-WWMA
☐ F-GSER	78	Falcon 50	2	Regourd Aviation/ELF Gabon, Port Gentil, Gabon.	F-BINR
☐ F-GSMC	76	Citation	500-0308	Ste Air Stream, Tours/Airlec Aviation, Bordeaux..	F-GMLH
☐ F-GUEQ	97	Falcon 900B	167	DFS/Government of Equatorial Guinea, Malabo.	F-WWFO
☐ F-GVDN	97	Falcon 50EX	264	V DeNora/Petrolclor, Nice.	F-WWHO
☐ F-GYSL	75	Falcon 20F-5	341	ASE/Berlys Aero SA-Yves St Laurent, Luxembourg.	F-OHCJ
☐ F-ODSR	76	Corvette	35	Monique Arimanana, Antananarivo, Malagasy.	5R-MVD
☐ F-OGVA	82	Citation II/SP	551-0378	Air Claims France, Raizet, Guadaloupe. (status ?).	(F-OGUO)
☐ F-OHRU	96	Citation V Ultra	560-0407	Jetperle SNC. Tahiti.	N1218Y
☐ F-OKSY	97	Falcon 50EX	257	BASF AG. Speyer, Germany.	F-WWHE
☐ F-WDFJ	77	Falcon 20G	362	AMD-BA., Bordeaux. (20G prototype).	F-WATF
☐ F-WNAV	93	Falcon 2000	1	Dassault Aviation, Paris-Le Bourget.	F-WNAV
☐ F-WWFP	97	Falcon 900B	169	Dassault Aviation, Paris-Le Bourget.	
☐ F-WWFR	97	Falcon 900B	170	Dassault Aviation, Bordeaux.	
☐ F-WWFS	97	Falcon 900EX	23	Dassault Aviation, Bordeaux.	
☐ F-WWFZ	98	Falcon 900EX	28	Dassault Aviation, Bordeaux.	
☐ F-WWHQ	97	Falcon 50EX	266	Dassault Aviation, Paris-Le Bourget.	
☐ F-WWHR	97	Falcon 50EX	267	Dassault Aviation, Bordeaux.	
☐ F-WWHS	97	Falcon 50EX	268	Dassault Aviation, Bordeaux.	
☐ F-WWHT	97	Falcon 50EX	269	Air BG, Paris-le Bourget.	
☐ F-WWML	97	Falcon 2000	54	Sony Europe GmbH. Cologne, Germany.	
☐ F-WWMM	97	Falcon 2000	55	IBM Euroflight, Paris-Le Bourget.	
Military					
☐ 32	75	Falcon 10MER	32	Marine Nationale, 57S, Landivisiau.	F-W...
☐ 48	82	Gardian	448	Marine Nationale, 12S, Papeete-Tahiti.	F-ZWVF
☐ 65	83	Gardian	465	Marine Nationale, 9S, Noumea-New Caledonia.	F-ZJTS
☐ 72	83	Gardian	472	Marine Nationale, 12S, Papeete-Tahiti.	F-Z...
☐ 77	83	Gardian	477	Marine Nationale, 9S, Noumea-New Caledonia.	F-Z...
☐ 80	84	Gardian	480	Marine Nationale, 12S, Papeete-Tahiti.	F-ZJSA
☐ 101	77	Falcon 10MER	101	Marine Nationale, 57S, Landivisiau.	F-WPXJ
☐ 129	79	Falcon 10MER	129	Marine Nationale, 57S, Landivisiau.	F-WZGA
☐ 133	79	Falcon 10MER	133	Marine Nationale, 57S, Landivisiau.	F-ZGTI
☐ 143	80	Falcon 10MER	143	Marine Nationale, 3S, Hyeres.	F-WZGO
☐ 185	81	Falcon 10MER	185	Marine Nationale, 3S, Hyeres.	F-WZGQ
☐ F-RADA	87	Airbus A310-304	421	A de l'Air, COTAM, Esterel Squadron, Paris.	F-ODVD
☐ F-RADB	87	Airbus A310-304	422	A de l'Air, COTAM, Esterel Squadron, Paris.	F-ODVE
☐ F-RAEA	72	Falcon 20E	260	A de l'Air, ET 1/65, Villacoublay.	F-WMKJ
☐ F-RAEB	68	Falcon 20C	167	A de l'Air, ET 1/65, Villacoublay.	F-RAFL
☐ F-RAEC	76	Falcon 20F	342	A de l'Air, ET 1/65, Villacoublay.	F-RAEG
☐ F-RAED	67	Falcon 20C	93	A de l'Air, ET 1/65, Villacoublay.	F-RAEC
☐ F-RAEE	71	Falcon 20C	238	A de l'Air, ET 1/65, Villacoublay.	F-RAFM
☐ F-RAEF	72	Falcon 20E	268	A de l'Air, ET 1/65, Villacoublay.	F-RAFK
☐ F-RAEG	74	Falcon 20E	291	A de l'Air, ET 3/65, Villacoublay.	F-RAEC
☐ F-RAEH	80	Falcon 20F	422	A de l'Air, ET 3/65, Villacoublay.	F-RCAL
☐ F-RAFC	65	DC 8F-55F	45819	A de l'Air, ET 3/60, Roissy.	F-BNLD
☐ F-RAFD	69	DC 8-72CF	46043	A de l'Air, SARIGUE ELINT, EE/51, Evreaux-Fauville.	OH-LFV
☐ F-RAFE		DC 8-53	45570	A de l'Air, SARIGUE ELINT, EE/51, Evreaux-Fauville.	F-ZARK
☐ F-RAFF	69	DC 8-72CF	46130	A de l'Air, ET 3/60, Roissy.	OH-LFY
☐ F-RAFG	69	DC 8-72CF	46013	A de l'Air, ET 3/60, Roissy.	OH-LFT
☐ F-RAFI	79	Falcon 50	5	A de l'Air, ET 1/60, Villacoublay.	F-WZHB
☐ F-RAFJ	82	Falcon 50	78	A de l'Air, ET 1/60, Villacoublay.	F-GEOY
☐ F-RAFK	81	Falcon 50	27	A de l'Air, ET 1/60, Villacoublay.	F-WGTG
☐ F-RAFL	80	Falcon 50	34	A de l'Air, ET 1/60, Villacoublay.	F-WEFS
☐ F-RAFP	85	Falcon 900	2	A de l'Air, ET 1/60, Villacoublay.	F-GFJC
☐ F-RAFQ	86	Falcon 900	4	A de l'Air, ET 1/60, Villacoublay.	F-WWFA
☐ F-RHFA	66	Falcon 20C	49	A de l'Air, SIET 98/120, Cazaux.	F-TEOA
☐ F-SEBI	74	Falcon 20E-5	315	CNET Operations Aeriennes, Lannion-Servel.	F-GDLO
☐ F-UKJA	70	Falcon 20SP	182	A de l'Air, CIFAS-328, Bordeaux-Merignac.	F-BVFV
☐ F-UKJC	81	Falcon 20SNA	451	A de l'Air, 339-JC, CITAC-339, Luxeuil. 'Fil d'Ariane'	F-UGWN
☐ F-UKJE/463	70	Falcon 20SNA	186	A de l'Air, 339-JE, CITAC-339, Luxeuil.	F-UGWM
☐ F-UKJG	67	Falcon 20SNA	115	A de l'Air, 339-JG, CITAC-339, Luxeuil.	F-UGWL
☐ F-UKJI	83	Falcon 20F	483	A de l'Air, 339-JI, CITAC-339, Luxeuil. 'L'Oeil des Fees'	F-UGWO
☐ F-WQFZ	80	Falcon 50	30	Marine Nationale,	I-SNAC
Reg	*Yr Type*		*c/n*	*Owner/Operator*	*Prev Regn*

Reg	Yr	Type	c/n	Owner/Operator	Prev Regn
☐ F-WWHZ	80	Falcon 50	36	Marine Nationale,	N59GS
☐ F-ZACA	71	Falcon 20E	252	CEV,	I-GIAZ
☐ F-ZACB	71	Falcon 10	02	CEV, Bretigny. (wfu ?).	F-ZJTA
☐ F-ZACB	67	Falcon 20C	96	CEV,	F-GERT
☐ F-ZACC	68	Falcon 20C	124	CEV, Bretigny.	F-WJMJ
☐ F-ZACD	68	Falcon 20C	131	CEV, Bretigny.	F-WJMK
☐ F-ZACG	67	Falcon 20C	86	CEV, Cazaux.	F-WRGQ
☐ F-ZACR	68	Falcon 20C	138	CEV, Cazaux.	F-BUIC
☐ F-ZACS	66	Falcon 20C	22	CEV, Istres.	F-BMKK
☐ F-ZACT	67	Falcon 20C	79	CEV, Bretigny.	F-BNRH
☐ F-ZACU	68	Falcon 20C	145	CEV, Bretigny.	F-GCGY
☐ F-ZACV	73	Falcon 20E	288	CEV,	F-BUYE
☐ F-ZACW	67	Falcon 20C	104	CEV, Bretigny.	F-BOXV
☐ F-ZACX	69	Falcon 20C	188	CEV, Melun-Villaroche.	F-BRPK
☐ F-ZACY	72	Falcon 20E	263	CEV,	F-BSBU
☐ F-ZACZ	78	Falcon 20F	375	CEV,	F-GBMD
☐ F-ZVMV	72	Corvette	1	CEV, Bretigny.	F-BUAS
☐ F-ZVMW	73	Corvette	2	CEV, Bretigny.	F-BNRZ
☐ F-ZVMX	74	Corvette	10	CEV,	F-GFEJ

G = United Kingdom (U.K. of Great Britain & Northern Ireland)

Civil

Reg	Yr	Type	c/n	Owner/Operator	Prev Regn
☐ G-....	97	CitationJet	525-0177	Kestrel Aviation Ltd/Eagle Airways Ltd. Guernsey, C.I.	(RP-C717)
☐ G-....	89	Falcon 50	196	Shell Aircraft Ltd. Rotterdam, Holland.	D-BNTH
☐ G-ATPD	66	HS 125/1B-522	25085	Wessex Air (Holdings) Ltd/Westminster Aviation, Hurn.	5N-AGU
☐ G-AXDM	69	HS 125/403B	25194	GEC-Marconi Avionics (Holdings) Ltd. Edinburgh.	
☐ G-BFRM	78	Citation II	550-0027	Marshalls of Cambridge Aerospace Ltd/Hawkair, Cambridge	N527CC
☐ G-BGOP	79	Falcon 20F	406	Falcon Jet Centre/Nissan (UK) Ltd. Heathrow.	F-WMKF
☐ G-BGYR	75	HS 125/F600B	6045	British Aerospace (Operations) Ltd. Warton.	G-5-11
☐ G-BJDJ	81	HS 125/700B	7142	Falcon Jet Centre Ltd. Heathrow.	G-RCDI
☐ G-BJIR	81	Citation II	550-0296	Gator Aviation Ltd/Aviation Beauport Ltd. Jersey, C.I.	N6888C
☐ G-BKBH	75	HS 125/600B	6052	Beamalong Ltd. Tortola, BVI.	5N-DNL
☐ G-BKRL	88	Leopard	001	Chichester-Miles Consultants Ltd. Cranfield.	
☐ G-BLSM	83	HS 125/700B	7208	Dravidian Air Services Ltd. Heathrow.	G-5-19
☐ G-BLTP	83	HS 125/700B	7210	Dravidian Air Services Ltd. Heathrow.	G-5-18
☐ G-BMIH	80	HS 125/700B	7115	Surewings Ltd. London.	G-5-502
☐ G-BNFW	80	HS 125/700B	7100	Lynton Aviation Ltd. Kidlington-Oxford.	G-5-549
☐ G-BRNM	97	Leopard	002	Chichester-Miles Consultants Ltd. Cranfield.	
☐ G-BTAB	87	BAe 125/800B	8088	ARAVCO/BAT-Dean Finance Co Ltd. Farnborough.	G-BOOA
☐ G-BTSI	91	BAe 1000B	9007	Shell Aircraft Ltd. London.	
☐ G-BVCM	93	CitationJet	525-0022	Kwik Fit plc. Edinburgh.	(N1329N)
☐ G-BWKL	80	HS 125/700B	7118	Kemekod Exports Ltd. London.	5N-AVJ
☐ G-BWOM	91	Citation II	550-0671	Ferron Trading Ltd/Aviation Beauport Ltd. Jersey, C.I.	N671EA
☐ G-BWSY	91	BAe 125/800B	8201	British Aerospace Airbus Ltd. Filton.	G-OCCI
☐ G-CZAR	90	Citation V	560-0046	Chauffair (CI) Ltd. Farnborough.	(N26656)
☐ G-DEZC	79	HS 125/700B	7070	Frewton Ltd/Aviation Beauport Ltd. Jersey, C.I.	(G-JETG)
☐ G-DJLW	68	HS 125/3B-RA	25140	Osprey Aviation Ltd. Southampton.	G-AVVB
☐ G-DNVT	89	Gulfstream 4	1078	Shell Aircraft Ltd. London.	N17589
☐ G-ETOM	88	BAe 125/800B	8130	TWR Group Ltd. Kidlington-Oxford.	G-BVFC
☐ G-EVES	97	Falcon 900B	165	Northern Executive Aviation Ltd. Manchester.	F-WWFD
☐ G-FFRA	68	Falcon 20EW	132	F R Aviation Ltd. Hurn.	N902FR
☐ G-FJET	82	Citation II	550-0419	Worldstage Ltd/London Executive Aviation, Stapleford.	G-DCFR
☐ G-FRAD	75	Falcon 20EW	304	F R Aviation Ltd. Hurn.	G-BCYF
☐ G-FRAE	73	Falcon 20EW	280	F R Aviation Ltd. Hurn.	N910FR
☐ G-FRAF	74	Falcon 20EW	295	F R Aviation Ltd. Hurn.	N911FR
☐ G-FRAH	69	Falcon 20EW	223	F R Aviation Ltd. Hurn.	G-60-01
☐ G-FRAI	72	Falcon 20EW	270	F R Aviation Ltd. Hurn.	N901FR
☐ G-FRAJ	66	Falcon 20EW	20	F R Aviation Ltd. Hurn.	N903FR
☐ G-FRAK	69	Falcon 20EW	213	F R Aviation Ltd. Hurn.	N905FR
☐ G-FRAL	68	Falcon 20EW	151	F R Aviation Ltd. Hurn.	N904FR
☐ G-FRAM	70	Falcon 20EW	224	F R Aviation Ltd. Hurn.	N907FR
☐ G-FRAO	69	Falcon 20EW	214	F R Aviation Ltd. Hurn.	N906FR
☐ G-FRAP	69	Falcon 20EW	207	F R Aviation Ltd. Hurn.	N908FR
☐ G-FRAR	69	Falcon 20EW	209	F R Aviation Ltd. Hurn.	N909FR
☐ G-FRAS	67	Falcon 20EW	82	F R Aviation Ltd. Hurn.	117501
☐ G-FRAT	67	Falcon 20EW	87	F R Aviation Ltd. Hurn.	117502
☐ G-FRAU	67	Falcon 20EW	97	F R Aviation Ltd. Hurn.	117504
☐ G-FRAW	67	Falcon 20EW	114	F R Aviation Ltd. Hurn.	117507
☐ G-FRBA	70	Falcon 20EW	178	F R Aviation Ltd. Hurn.	OH-FFA
☐ G-GAUL	80	Citation II	550-0127	Chauffair (CI) Ltd. Farnborough.	N550TJ
☐ G-GDEZ	92	BAe 1000B	9026	Frewton Ltd/Aviation Beauport Ltd. Jersey, C.I.	G-5-743
☐ G-GJET	81	Learjet 35A	35A-365	GAMA Aviation Ltd. Fairoaks.	G-CJET
☐ G-HARF	90	Gulfstream 4	1117	Fayair (Jersey) Ltd. Jersey, Channel Islands.	N1761J
☐ G-HNRY	92	Citation VI	650-0219	Quantel Ltd. Farnborough.	N219CC
☐ G-HUGG	95	Learjet 35A	35A-432	1427 Ltd/Airtours International Ltd. Manchester.	VR-CAD
☐ G-ICFR	86	BAe 125/800B	8050	Chauffair (CI) Ltd. Farnborough.	G-5-503
Reg	Yr	Type	c/n	Owner/Operator	Prev Regn

Reg	Yr	Type	c/n	Owner/Operator	Prev Regn
☐ G-IFTC	69	HS 125/F3B-RA	25171	Interflight Ltd. Gatwick.	G-OPOL
☐ G-IFTE	78	HS 125/700B	7037	Interflight Ltd. Gatwick.	G-BFVI
☐ G-JCBI	96	Falcon 2000	27	J C Bamford (Excavators) Ltd. East Midlands.	F-WWMM
☐ G-JCFR	81	Citation II	550-0282	Chauffair (CI) Ltd. Farnborough.	G-JETC
☐ G-JEAN	76	Citation	500-0339	Foster Associates Ltd/Oxaero, Oxford.	N300EC
☐ G-JETA	79	Citation II	550-0094	IDS Aircraft Ltd. Bournemouth-Hurn.	(N26630)
☐ G-JETI	86	BAe 125/800B	8056	Ford Motor Co. Stansted.	G-5-509
☐ G-JETJ	80	Citation II	550-0154	GAMA Aviation Ltd. Heathrow.	G-EJET
☐ G-JETN	80	Learjet 35A	35A-324	GAMA Aviation Ltd. Heathrow.	G-JJSG
☐ G-LEAR	79	Learjet 35A	35A-265	Northern Executive Aviation Ltd. Manchester.	(G-ZEST)
☐ G-LJET	88	Learjet 35A	35A-643	GAMA Aviation Ltd. Heathrow.	N39418
☐ G-LOFT	76	Citation	500-0331	Atlantic Air Transport Ltd/Oxaero, Coventry.	LN-NAT
☐ G-MLTI	96	Falcon 900B	164	Multi Flight Ltd. Leeds-Bradford.	F-WWFC
☐ G-NCFR	79	HS 125/700B	7054	Chauffair (C I) Ltd. Farnborough.	G-BVJY
☐ G-OCAA	80	HS 125/700B	7091	CAA/Magec Aviation Ltd. Luton.	G-OMGD
☐ G-OCDB	88	Citation II	550-0601	Chris de Burgh/Paycourt Ltd. Belfast.	G-ELOT
☐ G-OCFR	86	Learjet 35A	35A-614	Chauffair (CI) Ltd. Farnborough.	G-VIPS
☐ G-OEJA	75	Citation	500-0264	Eurojet Aviation Ltd. Birmingham.	G-BWFL
☐ G-OFOA	83	BAe 146-100	E-1006	Formula One Administration, Biggin Hill.	EI-COF
☐ G-OICE	93	CitationJet	525-0028	Iceland Frozen Foods PLC. Hawarden-Chester.	N1330S
☐ G-OJPB	71	HS 125/F600B	25258	Hawkair, Cambridge.	VP-CJP
☐ G-OMGD	82	HS 125/700B	7184	Magec Aviation Ltd. Luton.	9K-AGA
☐ G-OMGE	91	BAe 125/800B	8197	Magec Aviation Ltd. Luton.	G-BTMG
☐ G-OMGG	86	BAe 125/800B	8058	Magec Aviation Ltd. Luton.	N125JW
☐ G-OPWH	95	Falcon 900B	151	P B W Hamlyn/Lynton Corporate Jet Ltd. Kidlington-Oxford.	F-WWFK
☐ G-ORHE	74	Citation	500-0220	R H Everett/Earthline Aircraft Ltd. Thruxton.	(N619EA)
☐ G-ORJB	77	Citation	500-0364	Martin Brundle Racing-L'Equipe Air Ltd/Oxaero, Gamston.	G-OKSP
☐ G-OSCA	75	Citation	500-0270	Oscar Aviation Ltd/Oxaero, Oxford.	G-SWET
☐ G-PLGI	78	HS 125/700B	7034	Interflight Ltd/Polygram Record Operations Ltd. Gatwick.	N510HS
☐ G-RAAR	91	BAe 125/800B	8210	Millenium Executive Charter Ltd. Southampton.	G-5-705
☐ G-RAFF	83	Learjet 35A	35A-504	Graff Aviation/Aerocharter Midlands Ltd. Coventry.	N8568B
☐ G-RAHL	93	Beechjet 400A	RK-61	Air Hanson Aircraft Sales Ltd. Blackbushe.	N82378
☐ G-RCEJ	85	BAe 125/800B	8021	ARAVCO Ltd. Farnborough.	VR-CEJ
☐ G-SHEC	93	BAe 1000B	9037	Shell Aircraft Ltd. London.	G-SCCC
☐ G-SUFC	74	HS 125/600B	6035	ICN Pharmaceuticals Ltd. Glasgow.	G-BETV
☐ G-SVLB	80	HS 125/700B	7112	ARAVCO/Solvalub Trading Ltd. Jersey, Channel Islands.	(D-CLUB)
☐ G-TCAP	88	BAe 125/800B	8115	British Aerospace (Operations) Ltd. Warton.	G-5-599
☐ G-TCDI	71	HS 125/F400B	25248	ARAVCO Ltd. Farnborough.	N792A
☐ G-THCL	87	Citation II	550-0563	Tower House Consultants Ltd. St Lawrence, Jersey, C.I.	N1298P
☐ G-TJHI	77	Citation	500-0354	Trustair Ltd. Blackpool.	G-CCCL
☐ G-TSAM	91	BAe 125/800B	8028	British Aerospace (Operations) Ltd. Warton.	G-5-12
☐ G-VIPI	92	BAe 125/800B	8222	Yeates of Leicester Ltd. Souhampton. 'VIP One'	G-5-745
☐ G-WBPR	87	BAe 125/800B	8085	Granada Group plc. Luton.	G-5-551
☐ G-XRMC	90	BAe 125/800B	8180	RMC Group Services Ltd. Farnborough.	G-5-675
☐ G-ZAPI	80	Citation	500-0404	Titan Airways Ltd. Stansted.	G-BHTT
☐ G-ZENO	81	Learjet 35A	35A-429	Northern Executive Aviation Ltd. Manchester.	G-GAYL
Military					
☐ XS709/M	65	Dominie T1	25011	RAF, 3 FTS, Cranwell.	
☐ XS712/A	65	Dominie T2	25040	RAF, 3 FTS, Cranwell.	
☐ XS713/C	65	Dominie T1	25041	RAF, 3 FTS, Cranwell.	
☐ XS727/D	65	Dominie T1	25045	RAF, 3 FTS, Cranwell.	
☐ XS728/E	65	Dominie T2	25048	RAF, 3 FTS, Cranwell.	
☐ XS730/H	65	Dominie T2	25050	RAF, 3 FTS, Cranwell.	
☐ XS731/J	66	Dominie T1	25055	RAF, 3 FTS, Cranwell.	
☐ XS736/S	66	Dominie T2	25072	RAF, 3 FTS, Cranwell.	
☐ XS737/K	66	Dominie T2	25076	RAF, 3 FTS, Cranwell.	
☐ XS739/F	66	Dominie T2	25081	RAF, 3 FTS, Cranwell.	
☐ XX507	73	HS 125/CC2A	6006	RAF, 32 (The Royal) Squadron, Northolt.	
☐ XX508	73	HS 125/CC2A	6008	RAF, 32 (The Royal) Squadron, Northolt.	
☐ ZD620	82	BAe 125/CC3	7181	RAF, 32 (The Royal) Squadron, Northolt.	G-5-16
☐ ZD621	82	BAe 125/CC3	7190	RAF, 32 (The Royal) Squadron, Northolt.	
☐ ZD703	82	BAe 125/CC3	7183	RAF, 32 (The Royal) Squadron, Northolt.	G-5-20
☐ ZD704	82	BAe 125/CC3	7194	RAF, 32 (The Royal) Squadron, Northolt.	G-5-870
☐ ZE395	83	BAe 125/CC3	7205	RAF, 32 (The Royal) Squadron, Northolt.	G-5-19
☐ ZE396	84	BAe 125/CC3	7211	RAF, 32 (The Royal) Squadron, Northolt.	G-5-12
☐ ZE700	86	BAe 146/CC2	E1021	RAF, 32 (The Royal) Squadron, Northolt.	G-5-507
☐ ZE701	86	BAe 146/CC2	E1029	RAF, 32 (The Royal) Squadron, Northolt.	G-5-03
☐ ZE702	89	BAe 146/CC2	E1124	RAF, 32 (The Royal) Squadron, Northolt.	G-6-124
☐ ZF130	76	HS 125/600B	6059	MODPE/BAe, Dunsfold. (has Royal Navy titles).	(9M-DMF)
☐ ZH763	80	BAC 1-11/539GL	263	Ministry of Defence, Boscome Down.	G-BGKE

HB = Switzerland (Swiss Confederation)

Civil

Reg	Yr	Type	c/n	Owner/Operator	Prev Regn
☐ HB-AAS	76	Fokker F 28-2000	11110	AVISTO Ltd. Oberglatt/Sirte Oil Co. N Africa.	F-GDUX
☐ HB-I..	97	Gulfstream V	517	Sky Unlimited AG. Zug.	N517GA
☐ HB-IAB	87	Falcon 900	9	GFTA Trendanalysen, Dusseldorf, Germany.	(PH-ILC)
Reg	*Yr*	*Type*	*c/n*	*Owner/Operator*	*Prev Regn*

Reg	Yr	Type	c/n	Owner/Operator	Prev Regn
☐ HB-IAD	88	Falcon 900	35	Aeroleasing SA. Geneva.	F-WWFC
☐ HB-IAI	89	Falcon 900	75	JABJ/Socavia AG. Zurich.	C-FWSC
☐ HB-IAW	95	Falcon 2000	16	Starjet Establishment for Aviation, Vaduz.	F-WWMB
☐ HB-IAX	96	Falcon 2000	33	Rabbit Air AG. Zurich.	F-WWME
☐ HB-IAY	96	Falcon 2000	34	Rabbit Air AG. Zurich.	F-WWMF
☐ HB-IAZ	96	Falcon 2000	30	ALAG, Zurich/Aeroleasing SA. Geneva.	F-WWMG
☐ HB-IBH	96	Falcon 2000	42	Aeroleasing SA. Geneva.	F-WWMG
☐ HB-IBX	93	Gulfstream 4	1183	JABJ/Jet Club Ltd. Farnborough, UK.	VR-BDC
☐ HB-IBY	88	Falcon 900	44	Need-Air AG. Chur/Jet Aviation, Zurich.	N914JL
☐ HB-IDJ	97	Challenger SE	7136	ALG Aeroleasing SA. Geneva.	VP-CRJ
☐ HB-IEE	89	B 757-23A	24527	Petrolair SA. Geneva.	HB-IHU
☐ HB-IEJ	90	Gulfstream 4	1148	JABJ/S R Transportation SA. Geneva.	(JA6380)
☐ HB-IES	81	Falcon 50	61	Logarcheo Anstalt Vaduz, Zurich.	F-WZHI
☐ HB-IKJ	96	Challenger 604	5327	Jet Aviation Business Jets AG. Zurich.	N609CC
☐ HB-IKS	89	Challenger 601-3A	5042	Air Charter AG/Kraus & Naimer, Basle, Switzerland.	C-FEUV
☐ HB-IKT	87	Challenger 601-3A	5003	Swiss Air Ambulance, Zurich.	N778XX
☐ HB-ILH	81	Challenger 600S	1025	Jet Aviation Business Jets, Basle.	N111J
☐ HB-ILK	84	Challenger 601	3033	Impala Air-Trimjet Ltd/JABJ. Basle.	N601TJ
☐ HB-IMX	81	Gulfstream 3	335	Bright Star Estab. Singapore.	
☐ HB-IMY	89	Gulfstream 4	1084	Sit Set AG. Geneva.	(N448GA)
☐ HB-ITX	89	Gulfstream 4	1093	PrivatAir SA. Geneva.	VR-BLC
☐ HB-IVL	97	Gulfstream V	513	Interjet AG/JABJ, Basle.	N513GA
☐ HB-V..	79	Citation II/SP	551-0133	Speedwings SA. Geneva.	I-JESA
☐ HB-V..	80	Sabre-65	465-32	Speedwings SA. Geneva.	N303A
☐ HB-VEV	74	Falcon 20E	317	E Schwarz BeteiligungenAG/Jetcom SA. Geneva.	N88FE
☐ HB-VFZ	79	Learjet 35A	35A-222	Aeroleasing AG. Geneva.	N90AL
☐ HB-VGS	80	Citation II	550-0183	Jet Aviation Business Jets, Zurich.	(XC-DUF)
☐ HB-VHH	85	Citation S/II	S550-0028	JABJ/Smelt Intag, Zurich.	(N1260L)
☐ HB-VHV	89	BAe 125/800A	8153	CAT Aviation AG. Zurich.	G-5-627
☐ HB-VIA	84	Diamond 1A	A087SA	Hezemans Air AG. Zurich.	N487DM
☐ HB-VIC	78	Citation 1/SP	501-0098	Gruezi Air Services Ltd. & Lions Air AG. Zurich.	(N144AB)
☐ HB-VIF	86	Learjet 36A	36A-057	Air Glaciers SA. Sion.	N39994
☐ HB-VIK	87	BAe 125/800B	8091	Swiss Air Ambulance, Zurich.	G-5-560
☐ HB-VIL	87	BAe 125/800B	8097	Swiss Air Ambulance, Zurich.	G-5-567
☐ HB-VIO	80	Citation II/SP	551-0205	Sky Work AG. Berne. (was 550-0161)	N342DA
☐ HB-VIP	83	Citation II	550-0469	Jet Holdings Ltd-IOM/Aeroleasing SA. Geneva.	VR-BIZ
☐ HB-VIS	82	Citation II	550-0447	CAT Aviation AG. Zurich.	(N447CJ)
☐ HB-VIT	80	Citation II	550-0197	Aeroleasing SA. Geneva.	N44FC
☐ HB-VIZ	80	Citation II	550-0188	Jet Aviation Business Jets, Zurich.	VH-SWL
☐ HB-VJB	78	Citation 1/SP	501-0067	AFLAG/Sius AG. Zurich.	VR-BLW
☐ HB-VJI	89	Learjet 31	31-011	Birdy SA. Jouxtens-Mezery/Aeroleasing SA. Geneva.	N3803G
☐ HB-VJJ	89	Learjet 35A	35A-649	ALG Leasing SA/Executive Jet Aviation SA. Geneva.	N10870
☐ HB-VJK	89	Learjet 35A	35A-651	Aeroleasing SA. Geneva.	
☐ HB-VJL	89	Learjet 35A	35A-653	Aeroleasing SA. Geneva.	
☐ HB-VJV	70	Falcon 20D-5	237	Aeroleasing SA. Singapore.	VR-BKH
☐ HB-VJY	89	BAe 125/800A	8176	JABJ/Tricom Inc. Lugano.	G-5-652
☐ HB-VKB	93	CitationJet	525-0037	Executive Air Transport AG. Zurich.	N1820E
☐ HB-VKE	74	Falcon 10	7	ALG/Starways SA. Lugano.	D-CASH
☐ HB-VKI	93	Learjet 60	60-019	Jet-Airservice AG. Baar.	N40366
☐ HB-VKJ	79	HS 125/700A	7067	RAR SA. Lisbon, Portugal.	N144DJ
☐ HB-VKK	74	Citation	500-0178	Leopair SA. Geneva.	(HB-VJP)
☐ HB-VKO	71	Falcon 20F-5	257	Aeroleasing SA. Geneva.	F-GKDD
☐ HB-VKP	89	Citation II	550-0622	Pleuss-Staufer AG. Oftringen.	N826EW
☐ HB-VKR	84	Falcon 100	209	Cx HB- 1 Dec 97 to ?	(N312AR)
☐ HB-VKS	82	Citation II	550-0441	Giangiorgio Spiess, Lugano/JABJ, Basel.	VR-CCE
☐ HB-VKT	81	Citation II	550-0274	AVCON AG/Jetag AG. Zurich.	N14GA
☐ HB-VKV	92	BAe 125/800B	8228	Jacques Lejeune-Cologny/ALG Aeroleasing SA. Geneva.	G-5-758
☐ HB-VKW	93	BAe 125/800B	8246	Sky Jet AG. Zurich.	N387H
☐ HB-VLE	76	Citation Eagle	500-0313	Nicolaus Springer, Gstaad/ Sky Work AG. Berne.	N313BA
☐ HB-VLF	94	Hawker 800	8264	Robert Bosch GmbH. Stuttgart, Germany.	G-5-806
☐ HB-VLG	94	Hawker 800	8265	Robert Bosch GmbH. Stuttgart, Germany.	G-5-809
☐ HB-VLL	80	HS 125/700A	NA0274	Jet Club AG/MWM AG. Zug.	N700DE
☐ HB-VLM	94	Beechjet 400A	RK-66	Corpavia Jets SA. Geneva.	N400Y
☐ HB-VLN	94	Beechjet 400A	RK-94	Corpavia Jets SA. Geneva.	N3051S
☐ HB-VLP	95	Citation VII	650-7064	Sirius AG. Zurich.	N52626
☐ HB-VLQ	81	Citation II	550-0324	Speedwings SA. Geneva.	N23W
☐ HB-VLR	96	Learjet 31A	31A-127	Liebherr-Aerospace Toulouse SA/Aeroleasing SA. Geneva.	N80727
☐ HB-VLS	80	Citation II	550-0196	Sky Work AG. Berne.	N400DK
☐ HB-VLT	93	BAe 125/800B	8240	Privat Air SA. Geneva.	G-SHEA
☐ HB-VLV	90	Citation V	560-0077	Eagle Air Service AG. Berne.	N42NA
☐ HB-VLW	95	Beechjet 400A	RK-103	Sirius AG. Zurich.	D-CIGM
☐ HB-VLY	82	Citation II	550-0430	Servair Private Charter AG. Zurich.	N1278
☐ HB-VLZ	97	Citation V Ultra	560-0446	Sarena Jet AG. Gstaad/Sky Work AG. Belp.	
Military					
☐ T-781	76	Learjet 35A	35A-068	Swiss Air Force/STAC, Berne.	HB-VEM
☐ T-783	81	Falcon 50	67	Swiss Air Force/STAC, Berne.	HB-IEP
Reg	*Yr*	*Type*	*c/n*	*Owner/Operator*	*Prev Regn*

HC = Ecuador (Republic of Ecuador)

Civil

Reg	Yr	Type	c/n	Owner/Operator	Prev Regn
HC-BSS	68	Falcon 20C	150	TRANS AM Cia Aero Express DHL. Guayaquil.	TG-RBW
HC-BSZ	80	Learjet 35A	35A-311	Banco del Pacifico SA. Guayaquil.	N311BP
HC-BTJ	78	Citation II	550-0016	Ecuavia Ltda. Guayaquil.	(N216VP)
HC-BTN	81	Learjet 35A	35A-417	Jaboneria Nacional SA. Guayaquil.	N90RK
HC-BTT	70	HS 125/400A	NA751	Servicios Aereos Colaereos SA.	N79B
HC-BUP	69	Falcon 20D	200	TRANS AM Cia Aereo Express DHL. Guayaquil.	YV-200C
HC-BVH	83	Falcon 200	490	INDEGA-Industrial de Gaseosas Cia Ltda.	(N208RT)
HC-BVP	79	Citation	500-0389		ANE-201
HC-BVX	84	Westwind-1124	411		N47LR

Military

Reg	Yr	Type	c/n	Owner/Operator	Prev Regn
FAE-001A	76	Sabre-60	306-117	Aviacion del Ejercito,	N22MY
FAE-043	65	Sabre-40R	282-43	Ministry of National Defence, Quito.	N4469F
FAE-045	76	Sabre-75A	380-45	Ministry of National Defence, Quito. (status ?).	N753TW
FAE-047	72	Sabre-40A	282-109	Ministry of National Defence, Quito.	N77AT
FAE-049	73	Sabre-60	306-68	Ministry of National Defence, Quito. (status ?).	N265DP
IGM-628	90	Citation II	550-0628	Aviacion del Ejercito,	N183AB

HI = Dominican Republic

Civil

Reg	Yr	Type	c/n	Owner/Operator	Prev Regn
HI-581SP	81	Citation 1/SP	501-0172	Santo Domingo Motors Co C.A/Ambair SA. Santo Domingo.	N907KH
HI-646SP	83	Diamond 1A	A064SA	Urbanizadora los Reyes C.A. Santiago.	N800LE

HK = Colombia (Republic of Colombia)

Civil

Reg	Yr	Type	c/n	Owner/Operator	Prev Regn
HK-2485W	79	Westwind-Two	239	Ingenio del Cauca SA. Bogota.	HK-2485
HK-2624P	81	Learjet 25D	25D-339	Carlos Eduardo Restrepo Gaviria, Bogota.	HK-2624X
HK-3400	82	Citation II	550-0363	Helitaxi Ltda. Bogota.	HK-3400X
HK-3653	70	HS 125/400A	NA743	SAS-Servicio Aereo de Santander Ltda.	HK-3653X
HK-3971X	80	Westwind-Two	306	Central Charter de Colombia SA/Aeroejecutivos Ltda.	(N722W)
HK-3983X	79	Learjet 35A	35A-259	APSA-Aeroexpresso Bogota SA. Bogota.	N259JC
HK-4016X	82	Learjet 55	55-041	Central Charter de Colombia SA. Bogota.	N550RH
HK-4128W	87	Citation II	550-0550	Central Charter de Colombia SA. Bogota.	N550FM

Military

Reg	Yr	Type	c/n	Owner/Operator	Prev Regn
FAC-0001		F 28-1000	11992	Fuerza Aerea Colombiana, ETE, Bogota-El Dorado.	FAC-001
FAC-1201	67	B 707-373C	19716	Fuerza Aerea Colombiana, Esc 711, Bogota-El Dorado.	HL7425
FAC-1211	88	Citation II	550-0582	Fuerza Aerea Colombiana, Esc 713, Bogota-El Dorado.	(N1301A)

HL = South Korea (Republic of Korea)

Civil

Reg	Yr	Type	c/n	Owner/Operator	Prev Regn
HL7222	92	Gulfstream 4	1188	Korean Air Lines, Seoul.	N482GA
HL7277	77	Citation	500-0327	Korean Ministry of Transport, Seoul.	N5327J
HL7301	95	Falcon 900B	156	Samsung Aerospace Industries Ltd. Seoul.	N202FJ
HL7501	95	Citation V Ultra	560-0292	Korean Air Lines, Cheju.	N1295N
HL7502	95	Citation V Ultra	560-0294	Korean Air Lines, Cheju.	N1295Y
HL7503	95	Citation V Ultra	560-0297	Korean Air Lines, Cheju.	N1296N
HL7504	95	Citation V Ultra	560-0300	Korean Air Lines, Cheju.	N1297V
HL7522	96	Challenger 604	5303	SsangYong Oil Refining Co. Seoul.	C-FXKE
HL7577	95	Challenger 601-3R	5182	Ministry of Construction & Transport, Seoul.	C-FVZC

Military

Reg	Yr	Type	c/n	Owner/Operator	Prev Regn
85101	85	B 737-3Z8	23152	Government of South Korea, Seoul.	

HP = Panama (Republic of Panama)

Civil

Reg	Yr	Type	c/n	Owner/Operator	Prev Regn
HP-1A	69	Gulfstream 2	78	Government of Panama, Panama City.	N90HH

Military

Reg	Yr	Type	c/n	Owner/Operator	Prev Regn
HP-500A	65	B 727-44C	18894	Fuerzas de Panama, Panama City.	FAP 400

HR = Honduras (Republic of Honduras)

Civil

Reg	Yr	Type	c/n	Owner/Operator	Prev Regn
HR-AMO	66	BAC 1-11/401AK	086	Coastal Corp. Houston, Tx. USA.	N950CC
HR-CEF	81	Westwind-1124	333	Government of Honduras, Tegucigalpa.	HR-002

HS = Thailand (Kingdom of Thailand)

Civil

Reg	Yr	Type	c/n	Owner/Operator	Prev Regn
HS-DCG	96	Citation VII	650-7071	Directorate of Civil Aviation, Bangkok.	N1130N
HS-JJA	95	Challenger 601-3R	5188	Skyserv Co. Bangkok.	N614CC
HS-UCM	94	Beechjet 400A	RK-95	Thai Flying Service Co. Bangkok.	N3114X

Military

Reg	Yr	Type	c/n	Owner/Operator	Prev Regn
22-222	83	B 737-2Z6	23059	Royal Thai Air Force, Bangkok.	N45733
44-444	91	Airbus A310-324	591	HS-TYQ, Royal Thai Air Force, Bangkok.	F-WWCH
Reg	Yr	Type	c/n	Owner/Operator	Prev Regn

Reg	Yr	Type	c/n	Owner/Operator	Prev Regn
□ 55-555	95	B 737-4Z6	27906	HS-RTA, Royal Thai Air Force, Bangkok.	
□ 60504	87	Learjet 35A	35A-623	Royal Thai Air Force, 605 Sqn 6 Wing, Bangkok-Don Muang.	N7260Q
□ 60505	88	Learjet 35A	35A-635	Royal Thai Air Force, 605 Sqn 6 Wing, Bangkok-Don Muang.	N1471B

HZ = Saudi Arabia (Kingdom of Saudi Arabia)

Civil

Reg	Yr	Type	c/n	Owner/Operator	Prev Regn
□ HZ-AB3	80	B 727-2U5	22362	Al-Anwa Establishment,	V8-BG1
□ HZ-ABM2	66	BAC 1-11/401AK	060	AMC-Aviation Management Consortium, Jeddah.	HZ-MAA
□ HZ-ADC	88	Gulfstream 4	1037	Raytheon Middle East Systems, Jeddah.	VR-BKE
□ HZ-AFA2	96	Challenger 604	5320	Saudia Special Flight Services, Jeddah.	N605CC
□ HZ-AFH	75	Gulfstream 2	171	Saudia Special Flight Services, Jeddah.	N17586
□ HZ-AFI	77	Gulfstream 2TT	201	Saudia Special Flight Services, Jeddah.	N17585
□ HZ-AFJ	77	Gulfstream 2TT	203	Saudia Special Flight Services, Jeddah.	N17587
□ HZ-AFK	79	Gulfstream 2TT	239	Saudia Special Flight Services, Jeddah.	N17582
□ HZ-AFN	82	Gulfstream 3	364	Saudia Special Flight Services, Jeddah.	N1761D
□ HZ-AFP	83	Citation II	550-0472	Saudia Special Flight Services, Jeddah.	N12511
□ HZ-AFQ	83	Citation II	550-0473	Saudia Special Flight Services, Jeddah.	N12513
□ HZ-AFR	83	Gulfstream 3	410	Saudia Special Flight Services, Jeddah.	N350GA
□ HZ-AFT	87	Falcon 900	21	Saudia Special Flight Services, Jeddah.	(HZ-R4A)
□ HZ-AFU	87	Gulfstream 4	1031	Saudia Special Flight Services, Jeddah.	N434GA
□ HZ-AFV	87	Gulfstream 4	1035	Saudia Special Flight Services, Jeddah.	N435GA
□ HZ-AFW	88	Gulfstream 4	1038	Saudia Special Flight Services, Jeddah.	N438GA
□ HZ-AFX	90	Gulfstream 4	1143	Saudia Special Flight Services, Jeddah.	N410GA
□ HZ-AFY	91	Gulfstream 4	1166	Saudia Special Flight Services, Jeddah.	HZ-SAR
□ HZ-AFZ	89	Falcon 900	61	Saudia Special Flight Services, Jeddah.	HZ-AB2
□ HZ-AIJ	82	B 747SP-68	22750	Saudi Royal Family, Riyadh.	
□ HZ-BL1	66	BAC 1-11/401AK	080	Sheikh bin Laden/AMC Aviation, Jeddah.	HZ-MFA
□ HZ-BL2	88	BAe 125/800A	NA0419	Sheikh Bin Laden, Jeddah.	N564BA
□ HZ-DA4	81	HS 125/700B	7124	Dallah AVCO, Jeddah.	
□ HZ-DG1	66	B 727-51	19124	Dallah Al Baraka, Jeddah.	N604NA
□ HZ-DG2	80	Gulfstream 3	317	Dallah Al Baraka, Jeddah.	N83D
□ HZ-FMA	66	HS 125/1B	25105	Saudi Arabian Carpets, Jeddah.	G-AYRY
□ HZ-HA1	78	Gulfstream 2TT	216	Harth Trading Establishment, Jeddah.	HZ-ND1
□ HZ HE4	68	B 727-29C	19987	Al Anwae Aviation/Sheikh Hassan Enany, Riyadh.	N444SA
□ HZ-HM11	69	DC 8-72	46084	Saudi Armed Forces Medical Services, Riyadh.	HZ-MS11
□ HZ-HM1A	83	B 747-3G1	23070	Saudi Royal Family, Riyadh.	N1784B
□ HZ-HM1B	78	B 747SP-68	21652	Saudi Royal Family, Riyadh.	HZ-HM1
□ HZ-HM2	75	B 707-368C	21081	Saudi Royal Family, Riyadh.	HZ-HM1
□ HZ-HM3	77	B 707-368C	21368	Saudi Royal Family, Riyadh.	HZ-ACK
□ HZ-HM5	83	TriStar 500	1250	Saudi Royal Family, Riyadh.	7T-VRA
□ HZ-HM6	83	TriStar 500	1249	Saudi Royal Family, Riyadh.	VR-CZZ
□ HZ-HMED	94	B 757-23A	25495	Saudi Armed Forces Medical Services, Riyadh.	
□ HZ-HR1	80	B 727-2K5	21853	R B Hariri/Saudi Oger Ltd. Riyadh.	LX-MMM
□ HZ-HR2	82	Gulfstream 3	346	R B Hariri/Saudi Oger Ltd. Riyadh.	HZ-RH2
□ HZ-HR3	82	B 727-2Y4	22968	R B Hariri/Saudi Oger Ltd. Riyadh.	HZ-RH3
□ HZ-KA4	62	B 720-047B	18453	Sheikh Kamal Adham, Jeddah.	N93147
□ HZ-KA5	75	HS 125/600B	6049	Sheikh Kamal Adham, Jeddah.	G-BCXL
□ HZ-KA7	80	BAC 1-11/492GM	260	Sheikh Kamal Adham, Jeddah.	G-BLHD
□ HZ-KAA	96	Gulfstream 4SP	1294	Mawarid Ltd. Farnborough, UK.	(HZ-MAL)
□ HZ-KB1	71	BAC 1-11/414EG	158	Al Tameer Co Ltd.	HZ-AB1
□ HZ-MA1	67	JetStar-8	5105	Sheikh Ashmawi Aviation, Malaga, Spain.	N17005
□ HZ-MAJ	66	BAC 1-11/401AK	088	Jarallah Corp. (stored GVA since 8/97).	HZ-NIR
□ HZ-MFL	90	Gulfstream 4	1128	Saudia Special Flight Services, Jeddah.	N429GA
□ HZ-MIS	81	B 737-2K5	22600	Sheikh Mustafa Idrees,	D-AHLH
□ HZ-MS1	82	Learjet 35A	35A-467	Saudi Armed Forces Medical Services, Riyadh.	N3796Q
□ HZ-MS3	83	Gulfstream 3	385	Saudi Armed Forces Medical Services, Riyadh.	N1761K
□ HZ-MS4	71	Gulfstream 2	103	Saudi Armed Forces Medical Services, Riyadh.	N833GA
□ HZ-MSD	79	Gulfstream 2	256	Saudi Armed Forces Medical Services, Riyadh.	N17581
□ HZ-NES	69	Falcon 20C	174	Nesma Co Ltd. Jeddah.	HZ-KA3
□ HZ-NR1	79	Sabre-75A	380-71	Rashid Engineering, Riyadh.	
□ HZ-NR2	83	Gulfstream 3	415	R B Hariri/Saudi Oger Ltd. Riyadh.	HZ-HR4
□ HZ-OCV	66	B 727-21	19006	Salem bin Zaid Ahmed Al Hassan, Jeddah.	HZ-TFA
□ HZ-OFC3	94	Falcon 900B	133	Olayan Finance Co. Dhahran.	F-GODE
□ HZ-PCA	76	Gulfstream 2	179	ARABASCO-Arabian Aircraft Services Co Ltd. Jeddah.	HZ-CAD
□ HZ-RC3	81	Gulfstream 3	331	Saudi Special Flight Services, Jeddah.	N17LB
□ HZ-SAB2	92	Falcon 900B	113	SABIC-Saudi Arabian Basic Industries Corp. Riyadh.	F-WWFB
□ HZ-SAK1	63	B 707-351C	18586	Al Wizar Trading, Riyadh.	VR-BOR
□ HZ-SFA	91	Citation V	560-0132	Saudi FAL Group, Riyadh.	N6811Z
□ HZ-SJP2	92	BAe 1000A	9012	Jouanou & Parskevaides Saudi Arabia Ltd. Farnborough, UK.	G-5-726
□ HZ-TBA	86	B 737-205	23468	H R H Talal bin Abdul Aziz,	N891FS
□ HZ-TNA	68	JetStar-731	5120	Prince Turki bin Nasser, Jeddah.	N40DC
□ HZ-WBT	66	B 727-95	19252	Khalid bin Al Waleed Foundation, Riyadh. 'KR-2'	N740EV
□ HZ-WT2	82	Challenger 600S	1074	Kingdom Establishment, Riyadh.	HZ-SAA

Military

Reg	Yr	Type	c/n	Owner/Operator	Prev Regn
□ 101	69	JetStar-8	5129	Royal Saudi Air Force, 1 Squadron, Riyadh.	N7974S
□ 102	69	JetStar-8	5130	Royal Saudi Air Force, 1 Squadron, Riyadh.	103
Reg	Yr	Type	c/n	Owner/Operator	Prev Regn

Reg	Yr	Type	c/n	Owner/Operator	Prev Regn
□ 105	88	BAe 125/800B	8118	Royal Saudi Air Force, Riyadh.	G-5-605
□ 130	89	BAe 125/800B	8164	Royal Saudi Air Force, Riyadh.	G-5-654
□ HZ-103	85	Gulfstream 3	453	RESA=Royal Embassy of Saudi Arabia, Riyadh.	103
□ HZ-106	81	Learjet 35A	35A-374	Royal Saudi Air Force, Riyadh.	
□ HZ-107	81	Learjet 35A	35A-375	Royal Saudi Air Force, Riyadh.	
□ HZ-108	82	Gulfstream 3	353	RESA=Royal Embassy of Saudi Arabia, Riyadh.	HZ-BSA
□ HZ-109	89	BAe 125/800A	8146	Royal Saudi Air Force, Riyadh.	G-5-825
□ HZ-110	89	BAe 125/800B	8148	Royal Saudi Air Force, Riyadh.	(G-BPYE)
□ HZ-123	59	B 707-138B	17696	RESA=Royal Embassy of Saudi Arabia, Riyadh.	N138MJ
□ HZ-124	92	Airbus 340-211	004	RESA=Royal Embassy of Saudi Arabia, Riyadh.	F-WWBA

I = Italy (Italian Republic)

Civil

Reg	Yr	Type	c/n	Owner/Operator	Prev Regn
□ I-ACTL	82	Falcon 20F	427	Far Airlines SRL. Milan.	5N-AYN
□ I-AEAL	72	Citation	500-0053	Air Columbia SRL. Pescara.	HB-VGO
□ I-AGEC	70	Falcon 20F	239	Eurojet Italia SRL. Milan.	N134CJ
□ I-AGEN	82	Learjet 35A	35A-491	Eurojet Italia SRL. Milan.	N485
□ I-AIRW	90	Learjet 31	31-025	Soc. Air Vallee SpA. Aosta.	N39399
□ I-ALKA	82	Citation II	550-0351	Air One SpA. Rome.	(N167WE)
□ I-ALPG	81	Citation II/SP	551-0355	Action Air SRL. Milan.	N551AS
□ I-ALPR	83	Learjet 55	55-078	Soc. ALPI Eagles SpA. Thiene (Vicenza).	N56TG
□ I-ALSI	87	Beechjet 400	RJ-31	Soc. Aliserio SpA. Milan.	N5450M
□ I-AMAW	73	Citation	500-0095	Noman SRL. Roma-Ciampino.	N500KP
□ I-AMCU	73	Citation	500-0109	Panair Compagnia Aerea Mediterranea SRL. Milan.	G-RAVY
□ I-AMCY	74	Citation	500-0192	Aerotop SRL/Soc. Aviomar SRL. Rome-Urbe.	(N70WA)
□ I-ASAZ	82	Citation II	550-0432	Soc. Benetton/Benair SpA. Treviso.	N432CC
□ I-AUNY	81	Citation 1/SP	501-0213	Panair Compagnia Aereo Mediterranea SRL. Palermo.	N6785D
□ I-AVGM	84	Citation II	550-0492	A A A V Aerea Radiomisure, Rome.	(N1254G)
□ I-AVJE	78	Learjet 25D	25D-254	Soc. Avioriprese Jet Executive SpA. Capodichino-Napoli.	N973
□ I-AVJG	78	Learjet 35A	35A-189	Soc. Avioriprese Jet Executive SpA. Naples.	N727JP
□ I-AVRM	84	Citation II	550-0491	A A A V Aerea Radiomisure, Rome.	(N1254D)
□ I-AVVM	85	Citation S/II	S550-0062	A A A V Aerea Radiomisure, Rome.	N12715
□ I-BAEL	80	Falcon 20F	426	Soc. ALBA Servizi Trasport SpA. Milan.	N416RM
□ I-BEAU	87	Falcon 900	23	Soc. SARAS, Milan.	F-WWFK
□ I-BETV	86	Citation III	650-0104	Soc. Benetton/Benair SpA. Treviso.	N13195
□ I-BEWW	88	Challenger 601-3A	5020	Soc. Benetton/Benair SpA. Treviso.	C-FBKR
□ I-BLUB	92	Citation VI	650-0216	Soc. Barilla Servizi Finanziari SpA. Parma.	N68269
□ I-BMFE	73	Learjet 25C	25C-146	Gruppo Compagnie Aeronautiche SRL. Parma.	N6KJ
□ I-CAFD	88	Falcon 50	183	CAF-Cosorzio Aeromobili Fiat SRL. Turin.	F-WWHF
□ I-CIGB	80	Citation 1/SP	501-0163	Jonathan SARL/Datamat.	(I-AGIK)
□ I-CIGH	83	HS 125/700A	NA0341	Del Air SRL. Sant'Elpidio a Mare.	N2KW
□ I-CIST	85	Citation III	650-0085	Soc. CISET-Cia Italiana Servizi Tecnici SpA. Brescia.	N650DA
□ I-CMUT	79	Falcon 20F	389	Soc. Aer Marche SpA. Ancona.	F-WRQV
□ I-CREM	80	Falcon 10	161	Soc. Interjet SRL. Bologna.	N50SL
□ I-DENR	82	Falcon 50	125	Soc. Flynor Jet Spa/De Nora, Ciampino.	N711KT
□ I-DIES	88	Falcon 900	30	Soc. CAI, Rome.	F-WGTH
□ I-DLON	80	Learjet 35A	35A-346	Soc. Nauta SRL. Treviso.	N35AJ
□ I-DRIB	69	Falcon 20D	201	Soc. Aliadria SRL. Roma-Ciampino.	D-CELL
□ I-EDEM	96	CitationJet	525-0155	Ritmo SRL.	N155CJ
□ I-EDIK	83	Falcon 50	132	Soc. Isefi Internazionale SpA. Bologna.	F-WPXF
□ I-FARN	80	Citation	500-0401	Soc. Icaro SRL/Soc. Villa Rossa SRL. Morrovalle.	N2651
□ I-FFRI	83	Learjet 35A	35A-493	Soc. Fri Fly SpA. Trieste.	N482SG
□ I-FICV	88	Falcon 900	54	Soc. CAI, Rome.	F-WWFC
□ I-FJTO	91	Citation II	550-0679	Fly Jet SRL. Turin.	(N6776P)
□ I-FLYA	79	Citation 1/SP	501-0099	Soc. Alma Fly SRL. Como.	N3170NA
□ I-FLYB	79	Citation	500-0392	Soc. Eurofly Service SpA. Turin.	N26461
□ I-FLYC	80	Learjet 35A	35A-298	Soc. Eurofly Service SpA. Turin.	
□ I-FLYD	81	Citation II	550-0393	Panair Compagnie Aerea Mediterranea SRL. Palermo.	N12GK
□ I-FLYH	83	Learjet 35A	35A-498	Soc. Eurofly Service SpA. Turin.	N8564P
□ I-FLYJ	83	Learjet 55	55-084	Soc. Eurofly Service SpA. Turin.	N740AC
□ I-FLYK	71	Falcon 20E	241	Soc. Eurofly Service SpA. Turin.	HZ-PL7
□ I-GASD	84	Citation III	650-0037	Soc. Gretair SRL. Vergiate.	N37VP
□ I-GCFA	88	Beechjet 400A	RJ-44	Soc. Orion SpA. Milan.	N3144A
□ I-GIRL	81	Diamond 1A	A012SA	Soc. Orion SpA. Milan.	N107T
□ I-IDAG	95	CitationJet	525-0093	Eurofly Service SpA. Turin.	N5151S
□ I-IPIZ	91	Beechjet 400A	RK-29	Aliparma SRL. Parma.	N15693
□ I-ITPR	78	Falcon 10	115	Orion Viaggi SRL.	F-GGAR
□ I-JESO	81	Citation II	550-0255	Soc. Genavia SRL. Genoa.	N28GA
□ I-KELM	82	Learjet 35A	35A-406	Soc. Kelemata SpA. Turin.	N35Q
□ I-KWYJ	77	Citation 1/SP	501-0040	Fly Jet SRL. Turin.	N85FS
□ I-LIAB	68	Falcon 20C	172	Soc. ALI Capitol SRL. Rome.	F-BRHB
□ I-LIAC	70	Falcon 20D	234	Soc. ALI Capitol SRL. Rome.	D-COLL
□ I-LIAD	77	Learjet 35A	35A-111	Soc. ALI Capitol SRL. Rome.	OE-GMA
□ I-LOOK	82	Learjet 55	55-021	Soc. Eurofly Service SpA. Turin.	EI-BSA
□ I-LUBI	90	Gulfstream 4	1123	Soc. ALBA Servizi Trasporti SpA. Milan.	N457GA
□ I-LXGR	94	Gulfstream 4SP	1234	Soc. Luxottica SpA. Tesseria-Venezia.	N924ML
Reg	Yr	Type	c/n	Owner/Operator	Prev Regn

Reg	Yr	Type	c/n	Owner/Operator	Prev Regn
☐ I-MADU	84	Gulfstream 3	448	Soc. ALBA Servizi Trasporti SpA. Milan.	N255SB
☐ I-MESK	78	Citation II/SP	551-0003	Giorgio Forno, Nairobi, Kenya.	N72RC
☐ I-MILK	96	Challenger 604	5304	Soc Parmalat SpA/Soc Eli-Air SRL. Milan. Fontana-Parma.	C-FXHE
☐ I-MOCO	81	Learjet 35A	35A-445	E A S-Executive Aviation Services, Vicenza.	HB-VHG
☐ I-NIAR	81	Citation II/SP	551-0056	Euraviation SRL. Milan.	N214AM
☐ I-NUMI	91	Falcon 900	89	Soc. CAI, Rome.	F-WWFB
☐ I-OTEL	77	Citation 1/SP	501-0048	Soc. Italfy SRL. Caproni.	(I-DAEP)
☐ I-PAPE	82	Citation 1/SP	501-0256	Soc. Ariete 21 SRL. Milan.	D-ILLL
☐ I-PEGA	73	Citation	500-0081	Soc. Icaro SRL. Forli.	HB-VDA
☐ I-PIAL	69	PD 808	504	Soc. Rinaldo Piaggio, Genoa.	
☐ I-PNCA	81	Citation II	550-0235	Panair Compagnia Aereo Mediterranea SRL. Ciampino-Roma.	N67SG
☐ I-REAL	72	Falcon 20E	267	Soc. Rusconi Editore SpA. Milan.	F-WRQZ
☐ I-RELT	73	Sabre-40A	282-133	Soc. Elettronica SpA. Rome.	N41NR
☐ I-ROST	78	Citation	500-0381	Euraviation SRL/Orio Aircargo SRL. Orio Al Serio.	N445CC
☐ I-SAME	80	Falcon 50	37	Soc. CAI, Rome.	(I-CAIK)
☐ I-SDFG	88	BAe 125/800A	8136	Trevi SpA. Treviglio. (was NA0424).	N452SM
☐ I-SNAB	86	Falcon 50	169	Soc. SNAM SpA. Rome.	F-WPXD
☐ I-SNAW	95	Falcon 2000	12	Soc. SNAM SpA. Milan.	F-WWMM
☐ I-SNAX	89	Falcon 900	69	Soc. SNAM SpA. Milan.	F-WWFD
☐ I-TOIO	82	Citation 1/SP	501-0252	Soc. Jolly Hotels, Valdagno (Vicenza).	N574CC
☐ I-TYKE	96	Learjet 31A	31A-120	Soc. Eurofly Service SpA. Turin.	N5020Y
☐ I-UUNY	77	Citation	500-0358	Soc. Sicil Fly SARL. Fontanarossa, Catania.	SE-DEP
☐ I-VIGI	81	Diamond 1	A013SA	Air Kontiki SRL. Tassignano.	N81HH
☐ I-VULC	81	Learjet 35A	35A-421		N413JP
☐ I-ZOOM	77	Learjet 35A	35A-135	Aliven SRL. Verona.	N11AK

Military

Reg	Yr	Type	c/n	Owner/Operator	Prev Regn
☐ MM577		PD 808-TA	501	AMI, RS-38, RSV=Reparto Sperimentale Volo, Pratica di Mare.	
☐ MM578		PD 808-TA	502	AMI, RS-5, RSV=Reparto Sperimentale Volo, Pratica di Mare.	
☐ MM61948	69	PD 808-VIP	506	AMI, 14 Stormo, 8 Gruppo, Pratica di Mare.	
☐ MM61949	69	PD 808-VIP	507	AMI, 14 Stormo, 8 Gruppo, Pratica di Mare.	
☐ MM61950	69	PD 808-VIP	508	AMI, 14 Stormo, 8 Gruppo, Pratica di Mare.	
☐ MM61951	69	PD 808-VIP	509	AMI, 14 Stormo, 8 Gruppo, Pratica di Mare.	
☐ MM61952		PD 808-GE2	510	AMI, 14 Stormo, 71 Gruppo, Pratica di Mare.	
☐ MM61954		PD 808-TP	512	AMI, 14 Stormo, 8 Gruppo, Pratica di Mare.	
☐ MM61955		PD 808-GE2	513	AMI, 14 Stormo, 71 Gruppo, Pratica di Mare.	
☐ MM61956		PD 808-TP	514	AMI, 14 Stormo, 8 Gruppo, Pratica di Mare.	
☐ MM61957		PD 808-TP	515	AMI, 14 Stormo, 8 Gruppo, Pratica di Mare.	
☐ MM61958		PD 808-GE1	505	AMI, 14 Stormo, 71 Gruppo, Pratica di Mare.	
☐ MM61959		PD 808-GE1	516	AMI, 14 Stormo, 71 Gruppo, Pratica di Mare.	
☐ MM61960		PD 808-GE1	517	AMI, 14 Stormo, 71 Gruppo, Pratica di Mare.	
☐ MM61961		PD 808-GE1	518	AMI, 14 Stormo, 71 Gruppo, Pratica di Mare.	
☐ MM61962		PD 808-GE1	519	AMI, 14 Stormo, 71 Gruppo, Pratica di Mare.	
☐ MM61963		PD 808-GE1	520	AMI, 14 Stormo, 71 Gruppo, Pratica di Mare. (status ?).	
☐ MM62012	74	DC 9-32	47595	AMI, 31 Stormo, Roma-Ciampino. (ex 31-12).	N54635
☐ MM62013	74	DC 9-32	47600	AMI, 31 Stormo, Roma-Ciampino. (ex 31-13).	
☐ MM62014		PD 808-RM	521	AMI, 14 Stormo, 8 Gruppo, Pratica di Mare.	
☐ MM62015		PD 808-RM	522	AMI, 14 Stormo, 8 Gruppo, Pratica di Mare.	I-PIAY
☐ MM62016		PD 808-RM	523	AMI, 14 Stormo, 8 Gruppo, Pratica di Mare.	
☐ MM62017		PD 808-RM	524	AMI, 14 Stormo, 8 Gruppo, Pratica di Mare.	
☐ MM62020	85	Falcon 50	151	AMI, 31 Stormo, Roma-Ciampino.	F-WPXD
☐ MM62021	85	Falcon 50	155	AMI, 31 Stormo, Roma-Ciampino.	F-WPXH
☐ MM62022	85	Gulfstream 3	451	AMI, 31 Stormo, Roma-Ciampino.	N330GA
☐ MM62025	86	Gulfstream 3	479	AMI, 31 Stormo, Roma-Ciampino.	N319GA
☐ MM62026	89	Falcon 50	193	AMI, 31 Stormo, Roma-Ciampino.	F-WWHH
☐ MM62029	91	Falcon 50	211	AMI, 31 Stormo, Roma-Ciampino.	F-WWHR

JA = Japan

Civil

Reg	Yr	Type	c/n	Owner/Operator	Prev Regn
☐ JA001A	96	Citation V Ultra	560-0349	Asahi Shimbun Publishing Co. Tokyo.	N1127P
☐ JA001G	92	Gulfstream 4	1190	JCAB=Japanese Civil Aviation Bureau, Tokyo.	N403GA
☐ JA002G	94	Gulfstream 4SP	1244	JCAB=Japanese Civil Aviation Bureau, Tokyo.	N404GA
☐ JA01TM	97	Citation V Ultra	560-0403	Nozaki & Co Ltd. Tokyo.	N1202D
☐ JA119N	90	Citation V	560-0067	Naha Nikon Koku, Tokyo.	N45BA
☐ JA30DA	83	Diamond 1	A053SA		D-CDRB
☐ JA50TH	96	Falcon 900EX	3	Sony Corp. Tokyo.	N903FJ
☐ JA8248	79	Diamond 1	A002SA	Mitsubishi Heavy Industries, Nagoya.	JQ8003
☐ JA8380	76	Citation 1/SP	501-0027	Konno Kiseki & Partner, Miyagi.	N54DS
☐ JA8420	94	CitationJet	525-0056	Itogumi Construction Co. Sapporo Hokkaido.	N56NZ
☐ JA8431	74	Gulfstream 2	141	Diamond Air Service/Mitsubishi Heavy Industries, Nagoya.	N17584
☐ JA8438	76	Citation	500-0321	Asahi Shimbun Publishing Co. Tokyo. (status ?).	N5321J
☐ JA8474	81	Citation	500-0415	Asahi Shimbun Publishing Co. Tokyo.	N2072A
☐ JA8493	83	Citation 1/SP	501-0324	San-Kei Press Ltd. Haneda.	N2651J
☐ JA8570	88	Falcon 900	53	Japanese Maritime Safety Agency, Haneda Air Station.	N438FJ
☐ JA8571	88	Falcon 900	56	Japanese Maritime Safety Agency, Haneda Air Station.	N440FJ
☐ JA8576	90	Citation V	560-0080	Nozaki & Co Ltd. Tokyo.	N2746E

Reg	Yr	Type	c/n	Owner/Operator	Prev Regn

Military

Reg	Yr	Type	c/n	Owner/Operator	Prev Regn
☐ 20-1101	90	B 747-47C	24730	Government of Japan, Tokyo.	JA8091
☐ 20-1102	91	B 747-47C	24731	Government of Japan, Tokyo.	JA8092
☐ 29-3041	92	BAe U-125A	8215	041. JASDF-Japanese Air Self Defence Force, Iruma.	G-JFCX
☐ 39-3042	92	BAe U-125A	8227	042. JASDF-Japanese Air Self Defence Force, Iruma.	G-BUUW
☐ 41-5051	93	Beechjet 400T	TX-1	051, JASDF-Japanese Air Self Defence Force, Miho.	N82884
☐ 41-5052	93	Beechjet 400T	TX-2	052, JASDF-Japanese Air Self Defence Force, Miho.	N82885
☐ 41-5053	93	Beechjet 400T	TX-3	053, JASDF-Japanese Air Self Defence Force, Miho.	N82886
☐ 41-5054	94	Beechjet 400T	TX-4	054, JASDF-Japanese Air Self Defence Force, Miho.	N3195K
☐ 41-5055	94	Beechjet 400T	TX-5	055, JASDF-Japanese Air Self Defence Force, Miho.	N3195Q
☐ 49-3043	94	BAe U 125A	8242	043. JASDF-Japanese Air Self Defence Force, Iruma.	G-BVFE
☐ 51-5056	95	Beechjet 400T	TX-6	056, JASDF-Japanese Air Self Defence Force, Miho.	N3195X
☐ 51-5057	95	Beechjet 400T	TX-7	057, JASDF-Japanese Air Self Defence Force, Miho.	N3228M
☐ 51-5058	95	Beechjet 400T	TX-8	058, JASDF-Japanese Air Self Defence Force, Miho.	N3228V
☐ 52-3001	93	BAe U-125A	8245	001. JASDF-Japanese Air Self Defence Force, Komaki.	G-JHSX
☐ 52-3002	93	BAe U-125A	8247	002. JASDF-Japanese Air Self Defence Force, Komaki.	G-5-813
☐ 52-3003	94	BAe U-125A	8250	003. JASDF-Japanese Air Self Defence Force, Komaki.	G-BVRG
☐ 52-3004	95	BAe U-125A	8268	004. JASDF-Japanese Air Self Defence Force, Komaki.	N809H
☐ 71-5059	95	Beechjet 400T	TX-9	059, JASDF-Japanese Air Self Defence Force, Miho.	N1069L
☐ 72-3005	96	BAe U-125A	8288	005. JASDF-Japanese Air Self Defence Force, Komaki.	N816H
☐ 72-3006	96	BAe U-125A	8305	006. JASDF-Japanese Air Self Defence Force, Komaki.	N305XP
☐ 75-3251	95	Gulfstream U-4	1270	JASDF, 2nd Air Transport Group, Iruma	N442GA
☐ 75-3252	95	Gulfstream U-4	1271	JASDF, 2nd Air Transport Group, Iruma.	N452GA
☐ 75-3253	95	Gulfstream U-4	1303	JASDF, 2nd Air Transport Group, Iruma.	N435GA
☐ 9201	86	Learjet U36A	36A-054	JMSDF, 81 Kokutai, Iwakuni.	N1087Z
☐ 9202	86	Learjet U36A	36A-056	JMSDF, 81 Kokutai, Iwakuni.	N3802G
☐ 9204	90	Learjet U36A	36A-059	JMSDF, 81 Kokutai, Iwakuni.	N1087Z
☐ 9205	90	Learjet U36A	36A-060	JMSDF, 81 Kokutai, Iwakuni.	N1088A
☐ 9206	93	Learjet U36A	36A-061	JMSDF, 81 Kokutai, Iwakuni.	N2601B
☐ N1103U	97	BAe U-125A	8306	JASDF-Japanese Air Self Defence Force, Komaki.	
☐ N1112N	97	BAe U-125A	8325	JASDF-Japanese Air Self Defence Force, Komaki.	

JY = Jordan (Hashemite Kingdom of Jordan)

Civil

Reg	Yr	Type	c/n	Owner/Operator	Prev Regn
☐ JY-ADP	71	B 707-3D3C	20495	Government of Jordan, Amman.	
☐ JY-AFH	77	Sabre-75A	380-57	Arab Wings Ltd. Amman.	HZ-RBH
☐ JY-AFP	78	Sabre-75A	380-62	Arab Wings Ltd. Amman.	
☐ JY-HAH	85	Gulfstream 3	467	Government of Jordan, Amman.	N341GA
☐ JY-HKJ	83	TriStar 500	1247	Governemnt of Jordan, Amman.	N64854
☐ JY-HS1	69	B 727-76	20228	HS Aviation, Paris, France.	VR-CHS
☐ JY-HS2	75	B 727-2L4	21010	HS Aviation, Paris, France.	VR-CCA

J2 = Djibouti (Republic of Djibouti)

Civil

Reg	Yr	Type	c/n	Owner/Operator	Prev Regn
☐ J2-KBA	81	Falcon 50	71	Government of Djibouti, Djibouti.	YI-ALB

LN = Norway (Kingdom of Norway)

Civil

Reg	Yr	Type	c/n	Owner/Operator	Prev Regn
☐ LN-AAD	81	Citation II	550-0236	European Helicopter Center AS. Sandefjord.	N711VR
☐ LN-NLA	80	Citation II	550-0123	Valdesfly A/S-Norsk Luftambulanse A/S. Oslo.	SE-DEV
☐ LN-NLC	84	Citation III	650-0028	Norsk Luftambulanse A/S. Oslo.	N328QS
☐ LN-NLD	84	Citation III	650-0070	Norsk Luftambulanse A/S. Oslo.	N370TG

Military

Reg	Yr	Type	c/n	Owner/Operator	Prev Regn
☐ 0125	67	Falcon 20-5B	125	RNAF, 335 Squadron, Gardemoen-Oslo.	LN-FOE
☐ 041	66	Falcon 20-5ECM	41	RNAF, 335 Squadron, Gardemoen-Oslo.	LN-FOI
☐ 053	66	Falcon 20-5ECM	53	RNAF, 335 Squadron, Gardemoen-Oslo.	LN-FOD

LQ/LV = Argentina (Republic of Argentina)

Civil

Reg	Yr	Type	c/n	Owner/Operator	Prev Regn
☐ LV-...	67	Jet Commander	119	Aerocat SA. Catamarca.	N119AC
☐ LV-...	73	Sabre-75A	380-4	Air VIP SA. Buenos Aires.	N11887
☐ LV-...	75	Westwind-1123	182	Southern Flights Inc. Buenos Aires.	(N123SE)
☐ LV-AIT	81	Learjet 35A	35A-408	Direccion de Aeronautica, Tierra del Fuego, Ushuaia.	LV-POG
☐ LV-APL	81	Citation II/SP	551-0361	TIA S.A/Citi Bank NA. Buenos Aires. (was 550-0330).	LV-PNB
☐ LV-AXZ	71	HS 125/400B	25251	SASA-Sudamericana de Aviacion SA. Buenos Aires.	5-T-30-0653
☐ LV-JTZ	70	Learjet 24D	24D-234	National Bank of Lavoro, Buenos Aires.	LV-PRA
☐ LV-JXA	71	Learjet 24D	24D-240	Argenflight SA. Buenos Aires.	LV-PRB
☐ LV-LRC	75	Learjet 24D	24D-316	Presidencia de la Nacion, Buenos Aires.	T-03
☐ LV-LZR	76	Citation	500-0332	Fernando Jorge Vallejo, Buenos Aires.	LV-PUY
☐ LQ-MRM	78	Citation	500-0386	Argentine Federal Police, Buenos Aires.	LV-PAX
☐ LV-OEL	80	Learjet 25D	25D-307	Benjamin Colunga Lopez, Buenos Aires.	LV-PEU
☐ LV-RED	91	Citation V	560-0126	Sudacia SA. Buenos Aires.	LV-PFN
☐ LV-VFY	92	Citation V	560-0190	Viajaire SA. Buenos Aires.	LV-PGC

Reg	Yr Type	c/n	Owner/Operator	Prev Regn

Reg	Yr	Type	c/n	Owner/Operator	Prev Regn
☐ LV-WBP	81	Learjet 25G	25G-337	American Jet SA. Buenos Aires.	N337GL
☐ LV-WDR	93	Citation V	560-0227	Lanolec SA.	LV-PGR
☐ LV-WEJ	93	Citation II	550-0724	Gobierno Provincia de Buenos Aires.	LV-PGU
☐ LV-WFM	94	Learjet 60	60-024	Macair Jet SA.	LV-PGX
☐ LV-WGO	94	Citation V	560-0251	Cia Interamericana de Automov. Buenos Aires.	LV-PGZ
☐ LV-WGY	94	Citation V	560-0246	TIA SA. Buenos Aires.	LV-PHD
☐ LV-WHY	93	Citation VI	650-0231	Taxval SA. Buenos Aires.	LV-P..
☐ LV-WHZ	67	Jet Commander	108	Empresa Aerea Falcon SRL. Buenos Aires, Argentina.	N77ST
☐ LV-WIJ	94	Citation V Ultra	560-0265	LAPA, Buenos Aires.	LV-P..
☐ LV-WJN	87	Citation II	550-0558	Banco del Buen Ayre SA. Buenos Aires.	N558AG
☐ LV-WJO	94	Citation II	550-0728	Gobierno Provincia de Corrientes.	LV-PHN
☐ LV-WJU	75	Westwind-1123	179	Flymell SA. Buenos Aires.	N1123Y
☐ LV-WLG	81	Learjet 25D	25D-345	Medical Jet SA. Buenos Aires.	LV-PHU
☐ LV-WLR	72	Westwind-1123	183	Charter Jet SRL. Buenos Aires.	N51990
☐ LV-WLS	95	Citation V Ultra	560-0289	Gobierno Provincia de Santa Cruz.	LV-PHY
☐ LV-WMF	65	Falcon 20C	9	Air Service SA. Buenos Aires.	LV-PLC
☐ LV-WMM	66	Falcon 20C	29	Air Service SA. Buenos Aires.	LV-PLD
☐ LV-WMT	95	Citation V Ultra	560-0305	Lanolec SA. Buenos Aires.	LV-PLE
☐ LV-WND	73	Sabre-40A	282-131	Executive Jet SA. Buenos Aires.	N82R
☐ LV-WOC	79	Learjet 25D	25D-269	Heli Air SA. Buenos Aires.	LV-PLL
☐ LV-WOE	95	Citation V Ultra	560-0319	Flymell SA. Buenos Aires.	LV-P..
☐ LV-WOF	68	Sabre-60	306-25	Alas del Sur SA. Neuquen.	(OB-1550)
☐ LV-WOM	95	Gulfstream 4SP	1274	Sevel Argentina SA/MAC Air Jet, Buenos Aires.	LV-P..
☐ LV-WOZ	89	Citation II	550-0626		LV-PLR
☐ LV-WPE	95	Beechjet 400A	RK-104	Gobierno Provincia de Chabut.	LV-PLT
☐ LV-WPO	87	Sabre-60	306-3	TranSky, Buenos Aires.	N160CF
☐ LV-WPZ	92	Learjet 35A	35A-671	Gobierno Provincia de Salta.	LV-PLV
☐ LV-WRE	81	Learjet 25D	25D-355	Helijet SA. Buenos Aires.	N355AM
☐ LV-WSS	96	Gulfstream 4SP	1297	Amilcar Cabral Internacional CA. 'Santa Marguerita III'	N420GA
☐ LV-WTN	95	Citation VII	650-7054	Ledesma S A A I, Buenos Aires.	N7243U
☐ LV-WTP	96	Beechjet 400A	RK-118		LV-PMH
☐ LV-WXD	82	Citation II	551-0396		N45GA
☐ LV-WXJ	74	Citation	500-0148		LV-PMP
☐ LV-WXN	97	Learjet 60	60-102	Servicios Aereos Sudamericanos SA. Buenos Aires.	N8082B
☐ LV-WXV	88	Falcon 50	188	Loma Negra Cia Industrial, Buenos Aires.	PT-WAN
☐ LV-WXX	74	Sabre-60	306-91	Patria Cargas Aereas SA. Buenos Aires.	(N45MM)
☐ LV-WXY	82	Learjet 25D	25D-357	Helijet SA. Buenos Aires, Argentina.	N27KG
☐ LV-WYH	97	Citation Bravo	550-0818	I M S A=Industrial Metalurgica Sud Americana, Buenos Aires.	LV-PMV

Military

Reg	Yr	Type	c/n	Owner/Operator	Prev Regn
☐ 5-T-10-0740		F 28-3000M	11147	Armada-Argentina, Buenos Aires. 'Stella Maris'	PH-EXW
☐ 5-T-20-0741		F 28-3000C	11145	Armada-Argentina, Buenos Aires. 'Canal de Beagle'	LV-RRA
☐ 5-T-21-0742		F 28-3000M	11150	Armada-Argentina, Buenos Aires. 'Islas Malvinas'	PH-EXX
☐ AE-175	74	Sabre-75A	380-13	Ejercito/Comision Especial do Adquisiciones, Buenos Aires.	N65761
☐ AE-185	77	Citation	500-0356	Ejercito/Instituto Geografico Militar, Buenos Aires.	N36848
☐ T-02	72	F 28-1000	11048	FAA=Fuerza Aerea Argentina, Buenos Aires.	T-03
☐ T-03	70	F 28-1000	11028	FAA=Fuerza Aerea Argentina, Buenos Aires.	T-04
☐ T-10	74	Sabre-75A	380-3	FAA=Fuerza Aerea Argentina, Buenos Aires.	N8467N
☐ T-21	77	Learjet 35A	35A-115	Grupo 1 de Aerofotografico, II Brigada Aerea, Parana.	
☐ T-22	77	Learjet 35A	35A-136	Grupo 1 de Aerofotografico, II Brigada Aerea, Parana.	
☐ T-23	80	Learjet 35A	35A-319	Grupo 1 de Aerofotografico, II Brigada Aerea, Parana.	
☐ TP-01	92	B 757-23A	25487	FAA=Fuerza Aerea Argentina, Buenos Aires.	T-01
☐ VR-17	81	Learjet 35A	35A-369	Escuadron Verificacion Radio Ayudos, Moron.	
☐ VR-18	82	Learjet 35A	35A-484	Escuadron Verificacion Radio Ayudos, Moron.	

LX = Luxembourg (Grand Duchy of Luxembourg)

Civil

Reg	Yr	Type	c/n	Owner/Operator	Prev Regn
☐ LX-DPA	78	Falcon 10	113	Partner Aviation, Luxembourg.	F-GFHH
☐ LX-EPA	75	Falcon 10	48	Partner Aviation, Luxembourg.	F-GHRV
☐ LX-FMR	86	Falcon 50	165	Partner Aviation, Luxembourg.	HZ-HM3
☐ LX-GDL	78	Citation II	550-0033	Luxaviation SA. Luxembourg.	F-WPLT
☐ LX-GED	81	Falcon 50	54	Gedair SA/Gedeam Investment Group SA. Monaco.	(OY-GDA)
☐ LX-JCG	80	Falcon 10	160		F-GFFP
☐ LX-MMB	94	Challenger 601-3R	5146	Bouygues Transport Air Service, Paris-Le Bourget, France.	C-FRQA
☐ LX-PCT	95	Learjet 31A	31A-112	PCT-Powder Coating Technologies International SA. Luxembourg	N5082S
☐ LX-TRG	74	Falcon 10	19	Robert Giori, Cannes, France.	3A-MGT
☐ LX-YKH	73	Citation	500-0086	Husky Luxembourg/Cirrus Luftfahrt GmbH. Ensheim.	C-FMAN

LY = Lithuania (Republic of Lithuania)

Civil

Reg	Yr	Type	c/n	Owner/Operator	Prev Regn
☐ LY-AMB	73	JetStar-731	5161	Lithuanian Airlines, Vilnius.	N99VR

N = U S A (United States of America)

Civil

Reg	Yr	Type	c/n	Owner/Operator	Prev Regn
☐ N.....	97	Challenger 604	5349	Bombardier Business Jet Solutions LLC. Dallas, Tx.	(N331TP)
☐ N.....	67	Falcon 20C	94	Cx F- 30 Dec 97 to ?	F-GLNL
Reg	Yr	Type	c/n	Owner/Operator	Prev Regn

Reg	Yr	Type	c/n	Owner/Operator	Prev Regn
□ N.....	83	Learjet 55	55-060	Atlantic Corp. Miami, Fl.	PT-OUG
□ N1	89	Gulfstream 4	1071	FAA, Washington, DC. 'Spirit of America'	N410GA
□ N1AB	93	BAe 1000A	9036	Northern Tankers (Cyprus) Ltd.	N161BA
□ N1AP	95	Citation X	750-0003	Arnold Palmer, Youngstown, Pa.	N5223D
□ N1AT	88	Citation II	551-0591	Betaco Inc. Indianapolis, In. (was 550-0591).	N672CA
□ N1AT	87	Citation II/SP	551-0559	Betaco Inc. Indianapolis, In.	D-ICHE
□ N1BN	96	Gulfstream 4SP	1300	Odyssey Aviation Inc. West Palm Beach, Fl.	N432GA
□ N1BX	81	Falcon 50	47	Baxter Healthcare Corp/Allegiance Healthcare, Waukegan, Il.	N23AQ
□ N1C	96	Falcon 2000	40	Sears Roebuck & Co. Chicago, Il.	F-W...
□ N1CA	95	Learjet 60	60-055	C A Leasing Corp. St Louis, Mo.	N5055F
□ N1CC	79	Sabre-65	465-6	Carlson Companies Inc. Minneapolis, Mn.	N65SR
□ N1CF	79	Sabre-65	465-3	Intermet Corp. Troy, Mi.	N170CC
□ N1CH	94	Citation V Ultra	560-0283	CMH Homes Inc/Clayton Homes, Knoxville, Tn.	N560JC
□ N1DA	75	Citation	500-0288	Donald Anderson, Roswell, NM.	(N502BA)
□ N1DC	79	Learjet 35A	35A-246	Dallas Cowboys/Ark Air Flight Inc. Fort Smith, Ar.	N555GB
□ N1DE	90	Learjet 31	31-016	Champion Air Inc. Mooresville, NC.	N92LJ
□ N1ED	81	Learjet 35A	35A-392	Edward J DeBartolo Corp. Youngstown, Oh.	N931GL
□ N1GC	96	Challenger 604	5329	McCaw Communications Inc. Seattle, Wa.	N8MC
□ N1GH	80	Citation II	550-0201	Contrarian Group Inc & OrionAir Inc. Corvallis, Or.	N566TX
□ N1GN	95	Gulfstream 4SP	1284	Greg Norman/Great White Shark Enterprises, W Palm Beach.	N475GA
□ N1GT	96	Gulfstream 4SP	1292	GTE Service Corp. Beverly, Ma.	N413GA
□ N1HA	78	Citation 1/SP	501-0072	Aker Plastics Co. Plymouth, In.	N3110M
□ N1HF	82	Falcon 20F-5	474	Harbison-Fischer Inc. Fort Worth, Tx.	N211HF
□ N1HP	75	Learjet 35	35-039	Helmerich & Payne Inc. Tulsa, Ok.	
□ N1HS	95	Beechjet 400A	RK-106	Erickson Petroleum Corp. Minneapolis, Mn.	N3246H
□ N1JN	94	Gulfstream 4SP	1239	Jack Nicklaus/Air Bear Inc. West Palm Beach, Fl.	N484GA
□ N1JX	70	Sabre-60	306-61	Technographics International Inc. Nashua, NH.	N1JN
□ N1KE	88	Gulfstream 4	1033	Nike Inc/Philip Knight, Beaverton, Or.	N173LP
□ N1KT	78	Westwind-1124	230	Med Jet/McDonald Group Inc. Birmingham, Al.	N102U
□ N1MC	91	Astra-1125SP	055	Marathon Corp. Birmingham, Al.	N1125Z
□ N1NA	80	Gulfstream 3	309	NASA, Washington, DC.	N18LB
□ N1PB	82	Falcon 100	198	Baldwin & Baldwin Inc. Marshall, Tx.	N100RB
□ N1PC	78	B 737-2P6	21613	Airtran Airways Inc. Orlando, Fl.	A6-AAA
□ N1PG	94	Gulfstream 4SP	1259	Procter & Gamble Co. Cincinnati, Oh.	N495GA
□ N1PR	81	Gulfstream 3	341	Paragon Ranch Inc. Broomfield, Co.	N263C
□ N1S	83	Falcon 50	139	U S Fidelity & Guarantee Co. Baltimore, Md.	N96CE
□ N1SC	75	Learjet 35	35-016	South Carolina Aeronautics Commission, Columbia, SC.	N9CN
□ N1SF	88	Gulfstream 4	1060	Samuel Fly/Gulf States Toyota Inc. Houston, Tx.	N427GA
□ N1SL	91	Gulfstream 4	1167	Sara Lee Corp. Ypsilanti, Mi.	N17586
□ N1SN	91	Citation V	560-0153	Sky Night LLC. Greeneville, Tn.	N502F
□ N1SV	80	Citation II	550-0150	Viersen Air Services LLC. Okmulgee, Ok.	N2668A
□ N1TJ	74	Falcon 10	18	Tyler Jet LLC. Tyler, Tx.	N80CC
□ N1TK	88	Challenger 601-3A	5025	Consolidated Charter Service Inc. Fort Wayne, In.	N605HJ
□ N1TM	89	Gulfstream 4	1087	Airflite Inc/Toyota Motor Credit Corp. Long Beach, Ca.	N310SL
□ N1TS	83	Learjet 35A	35A-499	Top Sales Co. Charlotte, NC.	N499WJ
□ N1U	85	Learjet 25G	25G-371	Nugenco Nevada Inc. Rollins, Mt.	N1WT
□ N1UA	80	Citation II	550-0162	University of Alabama, Tuscaloosa, Al.	N550KP
□ N1UL	85	Citation S/II	S550-0057	Carroll's Foods Inc. Warsaw, NC.	N1271D
□ N1UP	92	Citation VI	650-0224	The Upjohn Co. Kalamazoo, Mi.	N224CD
□ N1VA	87	Citation S/II	S550-0143	Commonwealth of Virginia, Sandston, Va.	N143QS
□ N1WP	87	Gulfstream 4	1030	AMSI/Wm Pennington/WNP Aviation Inc. Reno, Nv.	(N811JK)
□ N1WS	79	Westwind-1124	252	Spirit Aviation & Tamburro & Purwin GP. Van Nuys, Ca.	(N9WW)
□ N1XT	96	CitationJet	525-0162	J B Aero Services Inc. Dover, De.	
□ N1ZC	84	Citation III	650-0031	H B Zachary Co. San Antonio, Tx.	N631CC
□ N2	92	Learjet 31A	31A-063	FAA, Washington, DC.	N27
□ N2AT	97	Falcon 2000	51	ALLTEL Corp. Little Rock, Ar.	N82AT
□ N2AV	81	Westwind-Two	322	ChipWind Aviation LLC, Carson City, Nv.	4X-CRP
□ N2BA	76	Learjet 35	35-051	San Juan Jet Charter Inc. Hato Rey, PR.	(N123MJ)
□ N2BG	82	Westwind-Two	365	California Hotel & Casino, Las Vegas, Nv.	N185MB
□ N2BT	77	Citation 1/SP	501-0054	S T Air Services, Oakhurst, Ca.	N98563
□ N2FE	91	Challenger 601-3A	5095	Federal Express Corp. Memphis, Tn.	N95FE
□ N2FU	90	Learjet 31	31-027	Motor Racing Development Corp. Biggin Hill, UK.	N30LJ
□ N2G	97	Hawker 800XP	8354	GenCorp. Canton, Oh.	
□ N2GG	81	Citation II	550-0286	Pro Aviation Inc. Wadsworth, Il.	N78BA
□ N2HF	78	Gulfstream 2SP	221	JABJ-Sun Life/Hospitality Franchise Systems, Parsippany, NJ.	N575SE
□ N2HP	82	HS 125/700A	NA0328	Helmerich & Payne Inc. Tulsa, Ok.	N710BG
□ N2HZ	81	Westwind-1124	314	Herzog Contracting Corp. St Joseph, Mo.	N84PH
□ N2JR	73	Gulfstream 2B	131	Budget Jet Inc. New Castle, De.	N759A
□ N2JW	92	Citation V	560-0191	J W Charter Inc. Houston, Tx.	PT-ORT
□ N2N	81	Sabre-65	465-63	Occidental Petroleum Inc. Bakersfield, Ca.	N605Y
□ N2NT	94	Citation II	550-0730	Northcote Inc. Bend, Or.	N730BR
□ N2RC	81	Citation II	550-0311	Rico Marketing Corp. Flint, Mi.	N121CP
□ N2S	75	Gulfstream 2	158	N2S Corp/Giant Group Ltd. Van Nuys, Ca.	N401M
□ N2SG	92	BAe 1000A	NA1007	HM Anglo-American Ltd & HM Industries Inc. Morristown, NJ.	N792H
□ N2T	86	Falcon 50	167	X Inc/Oakley Inc. Hillsboro, Or.	N186HG
□ N2TF	88	Citation III	650-0176	Gulf States Toyota Inc. Houston, Tx.	N1526L

Reg	Yr	Type	c/n	Owner/Operator	Prev Regn
☐ N2UP	93	Citation VI	650-0227	The Upjohn Co. Kalamazoo, Mi.	N1302C
☐ N2WL	89	Gulfstream 4	1118	Warner Lambert Co. Morristown, NJ.	N1526M
☐ N3AS	78	Learjet	28-001	Shamrock Foods Co. Phoenix, Az.	N128MA
☐ N3AV	81	Westwind-Two	361	Avjet Corp. Burbank, Ca.	N610HC
☐ N3BL	64	Learjet 23	23-003	Freedom Leasing Inc. Warren, Oh.	N3BL
☐ N3BM	96	Falcon 2000	38	Morris Communications Corp. Augusta, Ga.	F-WWMI
☐ N3BY	82	Falcon 100	193	TCBY Enterprises Inc. Little Rock, Ar.	N259FJ
☐ N3D	85	Citation S/II	S550-0038	Aromatique Inc. Heber Springs, Ar.	
☐ N3FA	79	Citation II	550-0112	Truck Body Aviation Inc. Lynchburg, Va.	N550PS
☐ N3FE	89	Challenger 601-3A	5054	Federal Express Corp. Memphis, Tn.	N619FE
☐ N3GL	83	HS 125/700A	7197	Giddings & Lewis Inc. Fond du Lac, Wi.	N207RC
☐ N3GN	78	Citation 1/SP	501-0090	Esper Petersen/Wisconsin Aviation Inc. Watertown, Wi.	N41JP
☐ N3GT	90	Citation V	560-0091	GTE North Inc. San Angelo, Tx.	N56GT
☐ N3H	88	MD-87	49670	Otter Corp. Kirkland, Wa.	D-ALLG
☐ N3HB	82	Challenger 600S	1059	Hamilton Oil Corp. Denver, Co.	N227GL
☐ N3HX	89	Gulfstream 4	1092	Iridium Inc. Washington, DC.	N3H
☐ N3JL	83	Challenger 600S	1080	Lupton Co LLC. Chattanooga, Tn.	N800CC
☐ N3M	87	Gulfstream 4	1021	3M Co. Minneapolis, Mn. (r/c Mining One).	N412GA
☐ N3MB	82	Citation II	550-0367	Trebuchet Transport LLC. Waukegan, Il.	N17LV
☐ N3PC	90	Challenger 601-3A	5066	Trinity Broadcasting of Florida Inc. Van Nuys, Ca.	N100KT
☐ N3PG	95	Gulfstream 4SP	1260	Procter & Gamble Co. Cincinnati, Oh.	N461GA
☐ N3PY	81	Gulfstream 3	336	Pillowtex Corp. Dallas, Tx.	(N523TX)
☐ N3QE	74	Citation	500-0121	QE Air Inc. Chesapeake, Va.	OY-SUJ
☐ N3QS	65	JetStar-731	5064	Kuse Enterprises Inc. Atlanta, Ga.	N184GP
☐ N3RC	78	Westwind-1124	222	RCR Inc. Welcome, NC.	N700R
☐ N3VF	81	Westwind-Two	324	Vanity Fair Corp/V F Corp. Wyomissing, Pa.	4X-CRR
☐ N3VJ	91	Learjet 31A	31A-035	Bergen Jet LLC. Mahwah, NJ.	N618RF
☐ N3VL	75	Westwind-1123	180	Van Lewis Inc. Henderson, Tx.	N192LH
☐ N4	91	Citation V	560-0113	Federal Aviation Administration, Washington, DC.	N26
☐ N4AZ		McDonnell 220	1	Greco Air Inc. El Paso, Tx.	N220N
☐ N4CP	95	Gulfstream 4SP	1257	Pfizer Inc. Mercer County Airport, NJ.	N448GA
☐ N4CR	66	HS 125/1A-522	25109	Fant Air LLC. Birmingham, Al.	N201H
☐ N4CS	81	Citation II	550-0220	M S I Inc. Austell, Ga.	N80513
☐ N4EF	96	CitationJet	525-0167	Capital Technologies Inc. Newton Centre, Ma.	
☐ N4EG	95	Citation VII	650-7059	Magnet Realty Corp. NYC.	N95CC
☐ N4EM	90	Astra-1125SP	050	North Transport Inc. Caracas, Venezuela.	
☐ N4J	77	Learjet 35A	35A-110	RR Investments Inc/Million Air, Dallas, Tx.	(N12EP)
☐ N4JS	92	Citation V	560-0196	John Scarpa Inc. Merrimack, NH.	N560JG
☐ N4LG	67	Sabre-60	306-9	Southern Building Products Inc. West Palm Beach, Fl.	N5071L
☐ N4LK	82	Diamond 1	A021SA	C C Medflight Inc. Lawrenceville, Ga.	N222Q
☐ N4MB	85	Falcon 50	156	Mellon Bank NA. Pittsburgh, Pa.	N5733
☐ N4MH	78	Westwind-1124	232	Amscray Air LLC. Newark, De.	N773AW
☐ N4MM	82	Citation II/SP	551-0431	Morgan McClure Motorsports Inc. Abingdon, Va.	N59CC
☐ N4NR	79	Gulfstream 2B	255	Rockwell International & Boeing N A Inc. Long Beach, Ca.	N442A
☐ N4NT	70	Sabre-60	306-48	W A Thomas Co. Taylor, Mi.	N4228A
☐ N4PC	82	Gulfstream 3	340	Paramount Communications, Van Nuys, Ca.	N340GA
☐ N4PG	89	Challenger 601-3A	5052	Procter & Gamble Co. Cincinnati, Oh.	C-G...
☐ N4QB	71	HS 125/F400A	25255	N102AB Inc. Van Nuys, Ca.	(N255TS)
☐ N4SG	92	Challenger 601-3A	5111	HM Industries Inc North America, Palm Springs, Ca.	N46SG
☐ N4TL	95	Citation V Ultra	560-0334	Amory Garment Co. Amory, Ms.	N5109W
☐ N4TS	72	HS 125/600A	6004	R & G Aviation Inc. Staten Island, NY.	VR-BRS
☐ N4VF	85	Citation III	650-0082	Blue Bell Inc. Greensboro, NC.	N82VP
☐ N4WC	72	HS 125/F400A	NA771	W C Aviation LLC. Vail, Co.	N298NM
☐ N4WG	76	Westwind-1124	200	Owners Jet Services Ltd. Las Vegas, Nv.	N1124X
☐ N5DL	78	Gulfstream 2	226	RELCO/Regal Quad Inc. Cincinnati, Oh.	N1902L
☐ N5EJ	83	Diamond 1A	A033SA	Line Power Manufacturing Corp. Bristol, Va.	N223S
☐ N5ES	92	BAe 1000B	9024	Raytheon E-Systems Inc. Greenville, Tx.	G-BUIX
☐ N5FW	74	Citation	500-0187	Far West Capital Inc. REno, Nv.	(N95TJ)
☐ N5G	85	BAe 125/800A	8053	GenCorp Inc. Akron, Oh.	N361BA
☐ N5GF	95	Gulfstream 4SP	1277	685TA Corp/American Home Products, Teterboro, NJ.	N426GA
☐ N5JY	76	Falcon 10	74	Harlekin USA/Augsburg Air GmbH. Augsburg, Germany.	N108MR
☐ N5MC	93	Gulfstream 4SP	1218	McCaw Communications Inc. Seattle, Wa.	N418SP
☐ N5NC	85	Learjet 25D	25D-372	Noland Co. Newport News, Va.	N72606
☐ N5PG	89	Challenger 601-3A	5053	Procter & Gamble Co. Cincinnati, Oh.	C-G...
☐ N5RD	74	Gulfstream 2	142	RDC Marine Inc. Houston, Tx.	N60CC
☐ N5T	80	Citation II/SP	551-0031	Walter Tintron, Austin, Tx.	N75TG
☐ N5TM	90	Challenger 601-3A	5076	Airflite/Toyota Motor Credit Corp. Long Beach, Ca.	C-G...
☐ N5TR	81	Citation II	551-0351	N5TR LLC. Agness, Or. (was 550-0322).	N322CS
☐ N5UU	87	Falcon 200	513	Southern Aircraft Leasing Inc. Rockville, Md.	N881JT
☐ N5VF	86	Falcon 50	166	VF Corp. Wyomissing, Pa.	N316PA
☐ N5VG	90	Learjet 31	31-014	Wal-Mart Stores Inc. Rogers, Ar.	PT-OFJ
☐ N5VH	86	Falcon 50	163	Commerce Leasing, Norcross, Ga.	N5VF
☐ N5VP	77	Citation 1/SP	501-0046	Sunday Unlimited, El Monte, Ca.	N405CC
☐ N5WF	78	Citation 1/SP	501-0082	Mociva Inc. Del Mar, Ca.	XB-ERX
☐ N5WM	79	B 737-268	22050	Interlease Aviation Investors II, Northfield, Il.	HZ-HM4
☐ N5XP	97	Citation V Ultra	560-0422	Xpress Air Inc. Chattanooga, Tn.	N20SB
Reg	*Yr*	*Type*	*c/n*	*Owner/Operator*	*Prev Regn*

Reg	Yr	Type	c/n	Owner/Operator	Prev Regn
☐ N5XR	87	Citation II	550-0553	Kenneth Anderson, Memphis, Tn.	(N555SL)
☐ N6BX	89	Falcon 900B '	79	Baxter Healthcare Corp. Deerfield, Il.	N901FJ
☐ N6FE	89	Citation V	560-0028	Federal Express Corp. Memphis, Tn.	(N1229A)
☐ N6FR	97	Citation Bravo	550-0828	Cessna Aircraft Co. Wichita, Ks.	
☐ N6GV	79	Sabre-65	465-9	AG Atlantic Investment Inc. Ocean City, Md.	(N769EG)
☐ N6HF	80	Citation II	550-0260	HCF Realty Inc. St Clair Shores, Mi.	N8CF
☐ N6HT	77	Citation 1/SP	501-0008	Ottesen Prop & Accessories, Phoenix, Az.	(N501DB)
☐ N6JB	93	Challenger 601-3A	5131	Fuqua National Corp. Atlanta, Ga.	N602JB
☐ N6JL	81	Citation II	550-0252	Pine Ridge Aviation Inc. Sulphur, La.	N507GP
☐ N6JW	73	Gulfstream 2	138	Walter Industries Inc. Tampa, Fl.	
☐ N6I.L	86	Citation S/II	S550-0094	H & S Aircraft LLC. Troy, Al.	N560AJ
☐ N6NB	96	Gulfstream 4SP	1290	NationsBanc Services Inc. Charlotte, NC.	N408GA
☐ N6NE	61	JetStar-731	5006	Wfu. Located Southampton, UK.	(VR-CCC)
☐ N6NR	80	Sabre-65	465-29	Rockwell International & Boeing N A Inc. Long Beach, Ca.	
☐ N6PC	69	Gulfstream 2B	775	Paramount Communications Inc. Van Nuys, Ca.	N723J
☐ N6SG	89	Challenger 601-3A	5046	Executive Flight Management Inc. Las Vegas, Nv.	C-G...
☐ N6SS	66	HS 125/1A-522	25100	Robert Krillich/Rainbow Air Corp. Naperville, Il.	N44TQ
☐ N6TM	85	BAe 125/800A	8026	Torchmark Corp. Birmingham, Al.	N800HS
☐ N6VF	84	Falcon 20F	486	V F Corp. Wyomissing, Pa.	F-GEFS
☐ N6VG	75	Falcon 10	62	P & G Investments LC. Burbank, Ca.	N12LB
☐ N7AB	90	Citation II	550-0633	American Buildings Co. Eufala, AL.	N388FA
☐ N7AU	85	Learjet 55	55-112	AUS Aviation Inc. Kristiansand, Norway.	LN-VIP
☐ N7CC	93	CitationJet	525-0004	INTRUST Financial Corp. Wichita, Ks.	N24CJ
☐ N7DJ	79	Westwind-1124	265	Del Webb Corp. Phoenix, Az.	N167J
☐ N7EN	96	CitationJet	525-0151	Cirrus Development Corp. Las Vegas, Nv.	(N151TT)
☐ N7FE	90	Citation V	560-0063	Federal Express Corp. Memphis, Tn.	
☐ N7GF	66	Learjet 23	23-093	Great Basin Aircraft Leasing Inc. Las Vegas, Nv.	N80775
☐ N7GP	65	Learjet 23	23-082	RBS Aviation Group Inc. Dover, De.	(N700NP)
☐ N7HM	79	Westwind-1124	266	Wild Turkey LLC. Midland, Tx.	N50DR
☐ N7KG	73	Sabre-40A	282-111	Kalamazoo Group Inc. Mattawan, MI.	(N246GS)
☐ N7NR	80	Sabre-65	465-44	Allen Bradley Co. Milwaukee, Wi.	
☐ N7PW	82	Diamond 1	A027SA	By-Car Inc. Austin, Tx.	N27TJ
☐ N7RP	88	Gulfstream 4	1064	Petersen Aviation Inc. Van Nuys, Ca.	HB-ITT
☐ N7SJ	71	HS 125/731	25250	American Garment Finishers Corp. El Paso, Tx.	(N125SJ)
☐ N7SN	81	Citation II	550-0280	Sevenson Enviromental Services Inc. Niagara Falls, NY.	N300TC
☐ N7TJ	68	Gulfstream 2	23	Tyler Jet Aircraft Sales Inc. Tyler, Tx.	USCG01
☐ N7TK	79	Citation 1/SP	501-0116	Timothy Mellon, Lyme, Ct.	(N90MT)
☐ N7US	93	Learjet 60	60-014	JABJ/United States Aviation Underwriters, NYC.	
☐ N7WC	78	HS 125/700A	NA0226	Wolf Camera Inc. Atlanta, Ga.	N42SR
☐ N7WG	81	HS 125/700A	NA0293	Jones & Granger, Houston, Tx.	N7CT
☐ N7WY	89	Citation II	550-0598	Randy Robinson, Portland, Or.	XC-HOO
☐ N7ZU	82	Citation II	550-0433	Jagenberg Inc. Enfield, Ct.	N131GA
☐ N8AE	76	Learjet 24E	24E-335	Time Compression Inc. New Castle, De.	N2DD
☐ N8AF	64	Sabre-40	282-24	American Horizons Ltd. Fort Wayne, In.	N40DW
☐ N8BX	91	Falcon 900B	111	Baxter Healthcare Corp/Allegiance Healthcare, Waukegan, Il.	N472FJ
☐ N8DE	74	Citation	500-0166	Air Ambulance by Air Trek Inc. Punta Gorda, Fl.	(N29MW)
☐ N8DX	76	Citation Eagle	500-0303	Abiquiu Development Inc. San Diego, Ca.	C-GDWS
☐ N8GA	80	Falcon 10	127	Reliant Airlines Inc. Ypsilanti, Mi.	I-CALC
☐ N8JC	97	Citation X	750-0020	Jepson Associates Inc. Savannah, Ga.	N95CM
☐ N8KG	82	HS 125/700A	7189	Apache Hawker LC & & Gaines-Hawker Inc. Palm Beach, Fl.	N81HH
☐ N8LE	83	Diamond 1A	A042SA	TPI Restaurants Inc. Palm Beach Gardens, Fl.	N420TJ
☐ N8LT	80	Falcon 10	173	Aviation Enterprises Inc. West Columbia, SC.	(N34LT)
☐ N8MQ	72	Learjet 25B	25B-085	AirNet Systems Inc. Columbus, Oh.	N8MA
☐ N8NR	78	Sabre-60	306-141	T-5 LLC. St Louis, Mo.	(N8NF)
☐ N8TG	75	Citation	500-0253	Miyasaka & Soilfume Inc. Watsonville, Ca.	N722US
☐ N8UP	87	BAe 125/800A	8083	Union Pacific Resources, Fort Worth, Tx.	N5C
☐ N9AX	72	Citation Eagle	500-0016	Charles White, Horseshoe Bay, Tx.	N7110K
☐ N9AZ	76	HS 125/600A	6063	Wilder Aviation Sales & Leasing Inc. Clearwater, Fl.	5N-...
☐ N9BX	81	Falcon 50	45	Baxter Healthcare Corp. Waukegan, Il.	N731F
☐ N9CH	92	Learjet 31A	31A-041	C W Hurd Jr. Timberon, NM.	N131TA
☐ N9CU	96	Learjet 60	60-075	PMG Acquisition Corp. Fort Lauderdale, Fl.	N675LJ
☐ N9EE	78	Learjet 35A	35A-193	Executive Flight Inc. East Wenatchee, Wa.	N2743T
☐ N9GT	88	Citation S/II	S550-0159	General Telephone of NW/GTE North Inc. Westfield, In.	N289CC
☐ N9KL	85	Citation III	650-0068	L & L Manufcaturing Co. Van NUys, Ca.	N985M
☐ N9LD	76	Learjet 24F	24F-336	Rocket Aviation LLC. Fircrest, Wa.	N162J
☐ N9NB	97	Hawker 800XP	8317	NationsBank NA. Charlotte, NC.	N2173X
☐ N9RA	66	Learjet 23	23-095	Royal Air Freight Inc. Pontiac, Mi.	N5D
☐ N9RD	77	Westwind-1124	220	Dowdy Plane Aviation Sales Inc. Orlando, Fl.	N106BC
☐ N9SS	80	Citation II/SP	551-0214	Thomas Grange Somermeier Jr. Van Nys, Ca. (was 550-0163).	N178HH
☐ N9UP	89	BAe 125/800A	NA0433	Union Pacific Resources Co. Fort Worth, Tx.	N919P
☐ N9VF	90	Citation II	550-0646	Vanity Fair Mills Inc. Monroeville, Al.	
☐ N9VL	88	Astra-1125	026	Laredo National Bank, Laredo, Tx.	N120WS
☐ N10AH	90	Learjet 35A	35A-657	Wal-Mart Stores Inc. Rogers, Ar.	N1CA
☐ N10AZ	76	Learjet 35A	35A-080	Anschutz Corp. Denver, Co.	N23HB
☐ N10BD	83	Learjet 35A	35A-506	Daniels Holdings Inc. Denver, Co.	N317BG
☐ N10C	78	HS 125/700A	NA0235	Dennis O'Connor, Victoria, Tx.	N700GB

Reg	*Yr*	*Type*	*c/n*	*Owner/Operator*	*Prev Regn*

Reg	Yr	Type	c/n	Owner/Operator	Prev Regn
☐ N10CN	81	HS 125/700A	NA0308	Coleman Co. Wichita, Ks.	N1843S
☐ N10D	65	HS 125/1A-522	25029	Eagle Foundation Inc. Dover, De.	N391DA
☐ N10EG	79	Citation II	550-0055	EG&G/United States Department of Energy, Las Vegas, Nv.	(N1466K)
☐ N10EH	85	Gulfstream 3	436	Sinclair Oil Corp. Salt Lake City, Ut.	N436GA
☐ N10F	74	Falcon 10	12	DB Aviation Inc. Houston, Tx.	N3100X
☐ N10FG	76	Citation Eagle	500-0295	Hanseatic Air Inc. Coldwater, Mi.	N44HC
☐ N10FL	90	Beechjet 400A	RK-27	Food Lion Inc. Salisbury, NC.	
☐ N10FN	75	Learjet 36	36-015	Flight Intl/Maritime Sales & Leasing Inc. Hockessin, De.	N14CF
☐ N10HE	77	Falcon 10	111	Augie Aviation Inc. Wilmington, De.	N820CE
☐ N10J	91	Beechjet 400A	RK-31	Carpenter Technology Corp. Reading, Pa.	
☐ N10JK	80	Citation II	550-0175	Sub-Sonic Jetprop Ltd. Butler, Md.	
☐ N10JM	96	Citation X	750-0022	Interstate Equipment Leasing Inc/Swift Air, Phoenix, Az.	(N5116)
☐ N10JP	79	Citation II	550-0066	Pattco Inc. Louisville, Ky.	N783H
☐ N10LR	79	Citation II/SP	551-0122	DJL Properties Inc. Baker, Or. (was 550-0059).	(N2662F)
☐ N10LX	70	Sabre-60	306-59	Loral Defence Systems, Goodyear, Az.	N20GX
☐ N10LY	83	Citation II	550-0466	Aircraft Leasing Group LLC. Greenville, SC.	N412MA
☐ N10M	71	Sabre-75	370-2	Ten Mike Inc. Dover, De.	N80K
☐ N10MV	80	Westwind-Two	300	Bloomberg Services LLC. NYC.	(N20NW)
☐ N10MZ	87	Falcon 900	32	ZWA Inc. Teterboro, NJ.	N500BL
☐ N10NB	97	Hawker 800XP	8331	NationsBank Corp. Charlotte, NC.	
☐ N10NC	80	Falcon 10	172	Hayward Enterprises Inc. Fort Lauderdale, Fl.	N172CP
☐ N10NM	82	Diamond 1A	A016SA	Advance Grocery Systems Inc. Plattsmouth, Ne.	N530RD
☐ N10PN	83	Citation III	650-0020	Fountainhead Sales & Leasing Inc. Wilmington, De.	N488JT
☐ N10PP	92	Falcon 50	231	A T Aviation Inc. Atlanta, Ga.	N10PQ
☐ N10RU	83	Citation II	550-0470	Omni Services Inc. Brandy Station, Va.	F-GFJL
☐ N10RZ	68	Falcon 20C	161	Pioneer Claim Management Inc. Rock Hill, NY.	N10PP
☐ N10S	80	Westwind-1124	278	J S C Enterprises Inc. Belleville, Il.	C-GJLK
☐ N10SL	64	Sabre-40	282-11	Commercial Aviation Enterprises, Fort Lauderdale. (wfu ?).	XA-...
☐ N10ST	71	Learjet 31A	31A-039	SouthTrust Corp. Birmingham, Al.	N90LJ
☐ N10TC	87	BAe 125/800A	8096	Southern Pacific Rail Corp. Dallas, Tx.	N800UP
☐ N10TD	91	Citation V	560-0096	Teledyne Industries Inc. Van Nuys, Ca.	(N96JJ)
☐ N10TN	80	HS 125/700A	7085	Asheville Jet Charter/Hawker 700 Co. Asheville, NC.	RP-C1714
☐ N10UC	80	HS 125/700A	NA0284	Urohealth Systems Inc. Newport Beach, Ca.	N125AP
☐ N10UH	76	Citation	500-0304	University Alabama Critical Care Transport, Birmingham, Al.	N70U
☐ N10UJ	77	Westwind-1124	204	WW191 Inc. Santa Monica, Ca.	N26TJ
☐ N10VG	73	Learjet 25B	25B-125	Vince Granatelli Racing Inc. Phoenix, Az.	(N11MC)
☐ N10WJ	75	Learjet 24D	24D-311	P W Air Inc/Perry Williams, Amarillo, Tx.	N748GM
☐ N10YJ	87	BAe 125/800A	8099	International Aviation Ltd. Englewood, Co.	OY-MCL
☐ N11AM	80	Learjet 35A	35A-340	International Association of Machinists, Washington, DC.	
☐ N11AQ	68	Learjet 24	24-129	VEC Corp of Delaware, Teaneck, NJ.	N723JW
☐ N11FH	78	Citation II	550-0012	Express Airlines I Inc/Northwest Airlink, Atlanta, Ga.	C-GHOL
☐ N11HJ	72	Citation	500-0034	Downtown Airpark Inc. Oklahoma City, Ok.	(N111FS)
☐ N11LC	97	Citation V Ultra	560-0427	Lowes Companies Inc. North Wilkesboro, NC.	
☐ N11LX	74	Sabre-60	306-75	Centurion Investments Inc. St Louis, Mo.	N509AB
☐ N11MN	76	Citation	500-0266	Minnesota Department of Transportation, St Paul, Mn.	N40RF
☐ N11NZ	78	Gulfstream 2	214	JABJ/Lion Advisors, Teterboro, NJ.	OMAF 601
☐ N11SS	82	Citation II/SP	551-0436	D R Johnson Lumber Co/CO-GEN Co. Riddle, Or. (was 550-0437)	N437CF
☐ N11TK	75	Learjet 25B	25B-201	Sales Operating Control Services Inc. Littleton, Co.	N111AD
☐ N11TS	86	Citation S/II	S550-0119	NTS Development Co. Louisville, Ky.	N700SV
☐ N11UF	76	Falcon 20F	356	GAT II Inc. Wilmington, De.	N111F
☐ N11WF	84	Diamond 1A	A075SA	Flowers Industries Inc. Thomasville, Ga.	(N13WF)
☐ N12AM	74	Citation	500-0235	Gutmann Leather Co. Chicago, Il.	N235CC
☐ N12BW	72	Sabre-40A	282-99	Barry Wehmiller Group Inc. St Louis, Mo.	N100FG
☐ N12CQ	93	Citation V	560-0231	Emerson Electric Co. St Louis, Mo.	N501E
☐ N12CV	78	Citation 1/SP	501-0081	Worldwide Charters LLC. Stevensville, Md.	(N12CQ)
☐ N12EP	80	Falcon 10	175	Kenneth Padgett, Vero Beach, Fl.	XA-LOK
☐ N12F	90	BAe 125/800B	8182	Wolfe Enterprises Inc. Columbus, Oh.	N128RS
☐ N12FN	75	Learjet 36	36-016	Flight Intl/Maritime Sales & Leasing Inc. Hockessin, De.	N616DJ
☐ N12FU	94	Learjet 60	60-027	Motor Racing Development Corp. Biggin Hill, UK.	N4230S
☐ N12L	88	Citation II	550-0583	Twin City/Burnet Inc. Minneapolis, Mn.	(N583VP)
☐ N12MB	77	Falcon 10	112	Wing Corp. Hobby-Houston, Tx.	N12XX
☐ N12NM	82	Citation 1/SP	501-0257	Pektron Aviation Inc. Gamston, UK.	OE-FLY
☐ N12PA	93	CitationJet	525-0012	Poly-Flex Inc. Grand Prairie, Tx.	(N1327Z)
☐ N12PB	68	Sabre-60	306-18	Airstream Aviation Inc. Del Ray Beach, Fl.	N11AQ
☐ N12RN	95	Citation V Ultra	560-0316	Island Aircraft Associates Inc. Dallas, Tx.	
☐ N12RP	79	Learjet 35A	35A-278	Stevens Aviation Inc. Greer, SC.	N17GL
☐ N12ST	89	Citation V	560-0014	SouthTrust Corp. Birmingham, Al.	N88TJ
☐ N12TV	96	Citation V Ultra	560-0358	Photo Electronics Corp. West Palm Beach, Fl.	N358CV
☐ N12TX	76	Falcon 10	90	TXI Aviation Inc & others, Dallas, Tx.	N14U
☐ N12U	89	Gulfstream 4	1112	United Technologies Corp. Hartford, Ct.	N12UT
☐ N12WF	84	Diamond 1A	A083SA	Flowers Industries Inc. Thomasville, Ga.	N383DM
☐ N12WH	78	Citation 1/SP	501-0064	Nuevo Aviation Inc. Wilmington, De.	N96DS
☐ N12WW	70	Learjet 25	25-051	Corporate Aircraft Inc. Wilmington, De.	PT-LPT
☐ N12YS	69	HS 125/731	NA721	Nemco Aviation LLC. Palm Beach, Fl.	HR-AMD
☐ N13BK	76	Falcon 10	94	Steen Aviation Inc. Shreveport, La.	N54RS
☐ N13BT	78	Citation 1/SP	501-0078	5K Aircraft Sales Inc. Fort Myers, Fl.	ZS-MZO
Reg	*Yr*	*Type*	*c/n*	*Owner/Operator*	*Prev Regn*

Reg	Yr	Type	c/n	Owner/Operator	Prev Regn
N13FE	66	DC 9-14	45706	Personal Growth Production Inc. Van Nuys, Ca.	N5NE
N13GW	73	Westwind-1123	162	Jet Set Aircraft, Bogota, Colombia.	N13GW
N13NH	74	Sabre-75A	380-25	Executive Aircraft Corp. Newton, Ks.	N16LF
N13ST	78	Citation 1/SP	501-0285	Shuttleworth Inc. Huntington, In. (was 500-0366).	N100BX
N13VG	81	Learjet 35A	35A-386	Corporate Aviation Services Inc. Tulsa, Ok.	
N14AZ	67	B 707-336C	19498	Grecoair Inc. El Paso, Tx.	9G-ACX
N14CG	82	Falcon 50	100	Beta Aircraft Corp. Teterboro, NJ.	N102FJ
N14CJ	84	Falcon 200	499	Vartec Properties Inc. Dallas, Tx.	N565A
N14DM	76	Learjet 24E	24E-341	Future Care Consultants Inc. Fort Lauderdale, Fl.	(N103JW)
N14FN	73	Learjet 25C	25C-126	Flight International/Maritime Sales & Leasing, Hockessin, De	(N162AC)
N14GD	87	Challenger 601-3A	5005	GG Aircraft/Gordon Gund, West Trenton, NJ.	HB-IKU
N14HH	66	HS 125/731	25118	Textile Rubber & Chemical Co. Dixville Notch, NH.	N227HF
N14PT	69	Learjet 24B	24B-208	SVW Air Inc. San Antonio, Tx.	N444AQ
N14R	96	Challenger 604	5319	Rainin Instrument Co. Oakland, Ca.	N604KR
N14RM	80	Citation II	551-0169	Guardian American Security, Southfield, Mi. (was 550-0126).	N700YM
N14SR	86	Astra-1125	015	Newair Inc. Miami, Fl.	N46UP
N14SY	83	Gulfstream 3	391	Sybron International Corp. Milwaukee, Wi.	N94BN
N14T	85	Learjet 35A	35A-608	Alumax Inc. Norcross, Ga.	N111SF
N14TN	75	Sabre-75A	380-40	Thunder Air Inc/JTF Corp. Chattanooga, Tn.	N820DY
N14UH	94	Gulfstream 4SP	1247	United Healthcare Corp. St Paul, Mn.	(N990UH)
N14VF	91	Citation V	560-0130	Blue Bell Inc/Modern Globe Inc. N Wilkesboro, NC.	N130CV
N15AS	94	Falcon 2000	3	Contran Corp. Dallas, Tx.	N2000A
N15AW	73	Citation	500-0139	Russell Schneider, Windemere, Fl.	N3771U
N15CQ	73	Citation	500-0101	New York Central Mutual Fire Insurance Co. Edmeston, NY.	N15CC
N15CV	80	Citation 1/SP	501-0152	Cache Valley Electric Co. Logan, Ut.	N48FJ
N15DJ	75	Sabre-60	306-97	Dominion Jet Partnership LLC. Glen Allen, Va.	N97NL
N15EA	83	Citation II	550-0450	Mount Hood Transport Inc. Salem, Or.	N1249K
N15EH	77	Learjet 35A	35A-126	Sinclair Marketing Inc. Sinclair, Wy.	N744GL
N15ER	79	Learjet 25D	25D-267	E W Richardson, Albuquerque, NM.	
N15HF	70	Sabre-60	306-60	National Flight Services Inc. Swanton, Oh.	N15H
N15NA	81	Citation II	550-0266	Dennis Quaid,	OE-GEC
N15RH	83	Learjet 35A	35A-497	Delaware Machinery & Tool Co. Muncie, In.	N21DA
N15RL	84	Citation II	550-0489	Levi, Ray & Shoup Inc. Springfield, Il.	N63CC
N15SP	88	Citation II	550-0566	SPX Corp. Muskegon, Mi.	(N1299B)
N15TT	90	Citation III	650-0192	Cleo J Thompson, Ozona, Tx.	N2622C
N15TV	83	Citation II	550-0459	711 Air Corp. Teterboro, NJ.	N15TW
N15TW	82	Falcon 50	73	World Publishing Co. Tulsa, Ok.	N48TW
N15UC	76	Gulfstream 2	176	The United Co. Blountville, Tn.	N176SB
N15VF	83	Citation III	650-0012	H D Lee Co. Shawnee Mission, Ks.	N1305V
N15WH	76	Learjet 35A	35A-085	AirNet Systems Inc. Columbus, Oh.	
N15XM	81	Citation II	550-0308	John Lawson Rock & Oil Inc. Fresno, Ca.	N30SA
N16AJ	84	Citation III	650-0075	Concord Jet Service Inc. Troutdale, Or.	N1315T
N16AS	84	Citation III	650-0055	A & S Air Service Inc. Dover, De.	N515VC
N16AZ	72	JetStar-8	5156	David Topokh, El Paso, Tx.	XB-DBT
N16DD	77	Falcon 10	105	Regency Aircraft Sales, Arlington, Tx.	N711MT
N16HC	66	Learjet 24	24-126	RBS Aviation Group Inc. Fort Myers, Fl.	(N345SF)
N16LG	74	Citation Eagle	500-0174	JH690 Inc. Lake St Louis, Mo.	N19AJ
N16MK	66	Jet Commander	84	Westar Aviation Inc. Miami, Fl.	N600ER
N16NK	82	Gulfstream 3	354	NT Air Inc/Castor Trading, Miami, Fl.	3D-AAI
N16NL	77	Citation 1/SP	501-0043	Saturn Machine & Welding Co. Sturgis, Ky.	N10NL
N16RP	85	Citation S/II	S550-0047	Briarthorne LLC. South Plainfield, NJ.	I-CEFI
N16SU	84	Citation III	650-0025	Speciality Travel Services Inc. Oak Creek, Wi.	(N522GS)
N16TS	81	Challenger 600S	1016	Randco Investments Ltd. Fort Lauderdale, Fl.	N757MC
N16VG	80	Citation 1/SP	501-0157	Gordon Rosenburg, San Ardo, Ca.	(N88BR)
N17A	80	Learjet 36A	36A-046	Air Jet Inc/Aeromet Inc. Tulsa, Ok.	N146MJ
N17AH	80	Learjet 25D	25D-316	Aero Charter Nashville Inc. Nashville, Tn.	(N782JR)
N17AN	84	Citation III	650-0054	AON Aviation Inc. Palwaukee, Il.	(N26RG)
N17CN	84	Challenger 601	3027	PACPAL LLC. Seattle/National Charter Network, Wa.	N5402X
N17DM	82	Citation II	550-0417	Capital Aircraft Inc. Springfield, Il.	ZS-LHW
N17FL	90	Citation II	550-0627	Towne Management Co. Youngstown, Oh.	(N1256N)
N17GV	97	Gulfstream V	520	Gulfstream Aerospace Corp. Savannah, Ga.	N596GA
N17JT	68	Falcon 20D	179	Basic Capital Management Inc. Dallas, Tx.	N12MF
N17LK	90	Citation V	560-0037	Washington Aircraft Corp. Wilmington, De.	
N17MK	65	BAC 1-11/410AQ	054	Yucaipa Management Co. Van Nuys, Ca.	N17VK
N17ND	69	Gulfstream 2	63	James A Morse/Jetaway Air Service, Muskegon, Mi.	N149JW
N17TE	87	Challenger 601-3A	5007	Flying T LLC. Gallatin Gateway, Mt.	(N607CZ)
N17TJ	75	Falcon 10	43	Deerfleet LLC. Miami, Fl.	F-GIQP
N17TN	83	Citation III	650-0011	Stanley Smith Drywall Co. Chelsea, Al.	N17TE
N17UC	94	Citation VI	650-0239	The United Co. Blountville, Tn.	N1304G
N17VB	97	Citationjet	525-0206	Star Light Air LLC/INCO Express Inc. Seattle, Wa.	
N18AN	93	Gulfstream 4SP	1228	AON Aviation Inc. Palwaukee, Il.	N464GA
N18BA	82	HS 125/700A	NA0316	Laminated Papers Inc. Holyoke, Ma.	(N501F)
N18BH	67	JetStar-731	5099	Bob L Hope, Burbank, Ca.	N62KK
N18CG	84	Learjet 55	55-104	Corning Inc. Corning, NY.	
N18FM	72	Citation	500-0014	N18FM Inc. Tulsa, Ok.	N800W
N18FN	77	Learjet 35A	35A-105	Proflight Inc. Englewod, Co.	(D-CHRC)

Reg	Yr	Type	c/n	Owner/Operator	Prev Regn

Reg	Yr	Type	c/n	Owner/Operator	Prev Regn
N18GA	97	CitationJet	525-0215	Griffin Industries Inc. Cold Springs, Ky.	
N18GB	94	Citation VII	650-7048	Green Bay Packaging Inc. Green Bay, Wi.	N51176
N18HC	81	Citation 1/SP	501-0223	Dillon Companies Inc. Hutchinson, Ks.	(N26HA)
N18HH	66	B 727-30	18936	Helmsley Spear Inc/Geni Aircraft Corp. Georgetown, De.	N5073L
N18HJ	88	Citation II	550-0587	Kinnarps AB. Falkoping, Sweden.	N1301S
N18LH	81	Learjet 35A	35A-379	American Jet International Corp. Houston, Tx.	N23VG
N18MX	78	Falcon 10	117	Boy-Cen Air Co LLC. Boyne Falls, Mi.	N923DS
N18NM	76	Learjet 25D	25D-209	Chariot Air LLC. Dallas, Tx.	N770AQ
N18PV	86	Citation III	650-0127	Bell Helicopter Textron Inc. Fort Worth, Tx.	N723BH
N18RF	97	Falcon 900EX	18	Ivanhoe Capital Aviation LLC. Vancouver, BC. Canada.	N918EX
N18SH	81	HS 125/700A	NA0310	Quantum Chemical Co. Cincinnati, Oh.	N2640
N18ST	80	Learjet 35A	35A-316	East Coast Jet Center, Fort Lauderdale, Fl.	N35AH
N18T	83	Diamond 1A	A061SA	Addison Products Co. Addison, Mi.	N348DM
N18TM	93	Gulfstream 4SP	1224	SDA Enterprises Inc. West Palm Beach. Fl.	N454GA
N18WE	81	Learjet 35A	35A-377	Bandag Inc. Muscatine, Ia.	N10WF
N19AJ	80	Citation II	550-0171	Elias Rodriguez, Fairfax, Va.	EI-BYN
N19ER	79	Citation II	550-0048	Nie Citation LLC. Hailey, Id.	N10BF
N19H	71	HS 125/731	NA763	Hubbard Broadcasting Inc. St Paul, Mn.	N62TC
N19HF	83	Challenger 600S	1081	Hershey Foods Corp. Middletown, Pa.	C-GLWZ
N19LT	90	Learjet 31	31-019	Wal-Mart Stores Inc. Rogers, Ar.	PT-OFL
N19MK	96	Citation V Ultra	560-0395	Moki Corp. Vero Beach, Fl.	N5093Y
N19NW	75	Learjet 35	35-019	Northwestern Aircraft Capital Corp. Mercer Island, Wa.	PT-LGF
N19PV	97	Citation V Ultra	560-0416	Provident Companies Inc. Chattanooga, Tn.	
N19R	83	Diamond 1A	A043SA	Delaware Aviation, Muncie, In.	N322DM
N19RP	81	Learjet 35A	35A-363	Rail Management & Consulting, Wilmington, De.	N183JC
N19SV	92	Citation VII	650-7002	SuperValu Inc. Minneapolis, Mn.	N95CC
N19TJ	90	Learjet 31	31-018	Inductoheat Inc. Madison Heights, Mi.	(N20LL)
N19UC	94	Citation VI	650-0238	The United Co. Blountville, Tn.	(N9UC)
N20AE	76	Falcon 20F	258	Double Wharf Co. East Haven, Ct.	N544X
N20CF	77	Falcon 10	106	Contract Freighters/Orcas Aircraft Leasing, Eastsound, Wa.	N103MM
N20CL	84	Falcon 200	497	Werner Aire Inc. Omaha, Ne.	N720HC
N20CR	83	Learjet 55	55-097	A T & T Global Information, Dayton, Oh.	N40CR
N20DA	61	MS 760 Paris-2B	102	Four Square Ltd. Venice, Fl.	N99HB
N20DK	78	Learjet 35A	35A-143	PNC Leasing Corp. Pittsburgh, Pa.	OE-GER
N20ES	73	Sabre-40A	282-124	National Wholesale Co Inc. Lexington, NC.	N40JE
N20FJ	67	Falcon 20C	119	Owners Jet Services Ltd. St Charles, Il.	F-GHFP
N20FL	94	CitationJet	525-0069	Food Lion Inc. Salisbury, NC.	N169CJ
N20G	83	Challenger 601-3R	5136	Goodyear Tire & Rubber Co. North Canton, Oh.	N51GY
N20H	69	Gulfstream 2	51	Hubbard Broadcasting Inc. St Paul, Mn.	N7C
N20HJ	69	Learjet 25	25-024	Winfair Aviation Ltd/Hop A Jet Inc. Fort Lauderdale, Fl.	N137BC
N20KH	78	Westwind-1124	223	Kitty Hawk Aircargo Inc. DFW Airport, Tx.	N303PC
N20NW	72	Learjet 25B	25B-096	Airstar Aviation Inc. San Francisco, Ca.	N235JW
N20NY	66	Falcon 20C	61	Air Force Systems Command, Bedford, Ma.	N299NW
N20PA	85	Diamond 1A	A089SA	Double L Manufacturing Inc. Colorado Springs, Co.	N88CR
N20PL	67	Falcon 20C	83	C A L Aviation Inc. Kansas City, Mo.	N68JK
N20RD	87	Citation III	650-0142	Roger Dean Enterprises Inc. West Palm Beach, Fl.	N142CC
N20RM	77	Citation 1/SP	501-0025	C I A Ltd. Las Vegas, Nv.	N21BS
N20T	78	Falcon 20F-5	381	JABJ/Tuthill Corp. Hinsdale, Il.	N20TZ
N20TA	65	Learjet 23	23-062	American International Airways, Morristown, Tn.	N670MF
N20TV	80	Citation II	550-0184	Next Century Aviation Inc. San Francisco, Ca.	N200NC
N20TX	73	Falcon 20F-5	296	Healthcare Airplane Group LLC. Nashville, Tn.	N19TX
N20WP	74	Falcon 10	23	J B Tool & Machine Inc. New Knoxville, Oh.	N91MH
N20XY	88	Gulfstream 4	1080	Occidental Petroleum Corp. Bakersfield, Ca.	N447GA
N21AC	95	Learjet 60	60-070	Alex Campbell Jr. Delray Beach, Fl.	N5035R
N21AM	72	Gulfstream 2SP	110	Hughes Aircraft Co. Van Nuys, Ca.	N200PB
N21CC	73	Citation	500-0099	TSI Turbine Source International, Dallas, Tx.	N599CC
N21CV	95	Citation V Ultra	560-0340	Dale Jensen, Lincoln, Ne.	
N21CZ	90	Gulfstream 4	1137	Cognizant Transportation Services, Westport, Ct.	N299DB
N21EG	86	Citation S/II	S550-0087	Kola Air Inc. Hazelton, Pa.	N1274Z
N21JJ	90	Citation V	560-0055	Jackson Air Services LLC. Jackson, WV.	N282RH
N21KR	90	Gulfstream 4	1139	Emerald Aviation Inc/Capital Air, Seattle, Wa.	N99WJ
N21NG	80	Learjet 35A	35A-343	Northrop Grumman Aviation Inc. Hawthorne, Ca.	N21NA
N21NY	93	Challenger 601-3A	5126	21 Aviation Corp/Foamex Aviation. Windsor Locks, Ct.	N21CL
N21VC	95	CitationJet	525-0106	International Veneer Co. South Hill, Va.	N5211A
N22	88	Gulfstream 4	1042	Motorola Inc. Schaumburg, Il.	N400GA
N22AF	91	Citation V	560-0129	American Fiber & Finishing Inc. Burlington, Ma.	
N22BN	82	Diamond 1	A023SA	Insurance Services Inc. Reading, Pa.	N17TJ
N22CS	96	Falcon 900EX	10	CSX Transportation Inc. Jacksonville, Fl.	N910FJ
N22FM	83	Citation II	550-0461	Federal Mogul Corp. Southfield, Mi.	N12507
N22G	93	Learjet 60	60-022	Goodyear Tire & Rubber Co. Akron, Oh.	N2602Z
N22GA	78	Citation II	550-0031	Air Storage Inc. Boise, Id.	RP-C296
N22HP	81	Citation II	550-0289	Hamilton Ranches Inc. Itasca, Il.	N820SA
N22KH	78	HS 125/700A	NA0232	Charter Communications Inc. St Louis, Mo.	N22EH
N22KW	94	Citation V	560-0256	Citation 22KW Inc/Kent B Williams, Columbus, Mt.	
N22LP	90	Citation V	560-0083	Peterson Industries Inc. Decatur, Ar.	(N2747R)
N22LZ	78	Westwind-1124	236	Zilberto Zanchet, Sao Paulo, Brazil.	(N236TS)
Reg	Yr	Type	c/n	Owner/Operator	Prev Regn

Reg	Yr	Type	c/n	Owner/Operator	Prev Regn
☐ N22MB	76	Citation	500-0337	Soc Aeronautique Auboise, Troyes-Barberey, France.	(F-GNAB)
☐ N22MS	78	Learjet 35A	35A-209	Evergreen International Aviation, McMinnville, Or.	N711DS
☐ N22NB	77	Sabre-75A	380-56	Nelson Aviation Services LLC. Birmingham, Al.	N14JD
☐ N22NJ	73	Learjet 25C	25C-097	National Jets Inc/WER Aviation Corp. Tamarac, Fl.	I-SFER
☐ N22RD	67	Jet Commander	99	N22RD Inc/David L Perry & Assoc. San Antonio, Tx.	N22RT
☐ N22RG	97	Citation X	750-0031	Greenhill Aviation LLC. NYC.	N5061W
☐ N22SF	96	Learjet 60	31A-126	State Farm Mutual Auto Insurance Inc. Bloomington, Il.	N8066P
☐ N22SN	93	Learjet 35A	35A-674	Southaire Inc. Memphis, Tn.	N22SF
☐ N22T	92	Falcon 900B	119	Aplomado Inc. Portland, Or.	N477FJ
☐ N22UL	85	Citation S/II	S550-0039	Universal Leaf Tobacco Inc. Sandston, Va.	
☐ N23A	77	Falcon 20F-5B	368	BCJ Aviation LLC. Wilmington, De.	N107LW
☐ N23BJ	83	HS 125/700A	NA0340	Air Laurel Inc. Laurel, Ms.	XA-SSY
☐ N23BX	81	Sabre-65	465-61	Corporate Property Investors, Windsor Locks, Ct.	
☐ N23CJ	67	HS 125/3A-RA	25152	Vartec Properties Inc. Dallas, Tx.	N50MJ
☐ N23ET	84	Gulfstream 3	434	Ed Tillson Management/Bright Flight Inc. Albuquerque, NM.	XB-FXD
☐ N23M	87	Gulfstream 4	1022	3M Co. Minneapolis, Mn.	(N63M)
☐ N23NM	86	Citation S/II	S550-0121	Ness, Motley & Partners, Barnwell, SC.	D-CLOU
☐ N23SB	90	Challenger 601-3A	5074	United States Tobacco Co. White Plains, NY.	C-G...
☐ N23SK	77	HS 125/700A	NA0216	PNC Leasing Corp. Louisville, Ky.	(N23SN)
☐ N23TJ	88	Astra-1125	023	Alsois Aviation Corp. Miami, Fl.	N125GB
☐ N23UD	97	Citation V Ultra	560-0442	United Dominion Industries Inc. Charlotte, NC.	N5226B
☐ N23YZ	79	Citation 1/SP	501-0088	YZ Corp. Germantown, Tn.	N86MT
☐ N24AJ	74	Citation	500-0221	JMB Air Services Inc. Mobile, Al.	XC-GUH
☐ N24CK	72	Learjet 24D	24D-258	RFS Aviation Corp. Sellersburg, In.	(N24DZ)
☐ N24E	90	Citation II	550-0651	Blue Cross & Blue Shield, Richmond, Va.	
☐ N24G	93	Learjet 60	60-018	Goodyear Tire & Rubber Co. Akron, Oh.	N4016G
☐ N24JG	82	Learjet 35A	35A-477	Jeff Gordon Inc. Harrisburg, NC.	(N477MS)
☐ N24JK	92	Challenger 601-3A	5118	Star Bank NA/Central Investment Corp. Cincinnati, Oh.	N824JK
☐ N24KL	79	Westwind-1124	237	Orlando Financial Corp. Wilmington, De.	N28TJ
☐ N24KT	87	Citation III	650-0132	Jostens Inc. Minneapolis, Mn.	N1323V
☐ N24LG	65	Learjet 24A	24A-011	Younkin Boreing Inc. Hot Springs, Ar.	N225LJ
☐ N24MN	84	Westwind-Two	414	Morris Newspaper Corp. Savannah, Ga.	N980AW
☐ N24NB	95	Citation VII	650-7052	La Vencedora SA. Asuncion, Paraguay.	N752CM
☐ N24PH	84	Citation S/II	S550-0026	Pamplemousse Corp. Dover, De.	(N126LP)
☐ N24RF	80	Sabre-65	465-28	Sunrise Aviation/Ray Floyd Enterprises Inc. Miami, Fl.	N742R
☐ N24S	74	Learjet 24D	24D-297	Metropolitan Air Inc. Baltimore, Md.	N8094U
☐ N24TE	72	Learjet 24D	24D-246	Tyler Jet LLC. Tyler, Tx.	(N444HE)
☐ N24TW	67	Falcon 20C	80	Tailwind Aviation Corp. Bridgeview, Il.	N76MB
☐ N24UD	97	Citation V Ultra	560-0443	United Dominion Industries Inc. Charlotte, NC.	N5228Z
☐ N24VB	86	Citation III	650-0110	Von Hoffman Press Inc. Jefferson City, Mo.	N121AG
☐ N24WF	67	Learjet 24	24-143	Ozark Management Inc. Jefferson City, Mo.	N724LG
☐ N24WX	66	Learjet 24	24-101	Robert & Carol Mace Aviation, Prineville, Or.	XA-SGU
☐ N25AM	80	Learjet 25D	25D-321	CPN Technologies Inc. Clearwater, Fl.	
☐ N25AZ	65	B 727-30	18370	David Topokh/Aviation Consultants, El Paso, Tx.	Z-WYY
☐ N25BX	76	Sabre-75A	380-47	Air N25BX Inc. Southfield, Mi.	N25BH
☐ N25CZ	80	Learjet 25D	25D-301	Airplanes of Boca Inc. Boca Raton, Fl.	(N888JA)
☐ N25EC	69	Learjet 25	25-026	M & M Executive Aircraft Holdings Ltd. New York, NY.	N281R
☐ N25FM	70	Learjet 25	25-063	Jets Inc. Wilmington, De.	N24LT
☐ N25FN	81	Learjet 25G	25G-352	Nevada Air Charter Inc. Incline Village, Nv.	XA-MMO
☐ N25FS	97	Citation Bravo	550-0823	Filter Specialists Inc. Michigan City, In.	(N823CB)
☐ N25GJ	68	Learjet 25	25-015	Commonwealth Aviation Corp. Daytona Beach, Fl.	N25FN
☐ N25GT	81	Citation 1/SP	501-0173	Ramsey Air Meds LLC. Mesa, Az.	OE-FPH
☐ N25HA	73	Learjet 25XR	25XR-141	Proflight Inc. Englewood, Co.	N94RS
☐ N25HS	81	Citation 1/SP	501-0222	Space City Aviation, Austin, Tx.	N6781R
☐ N25HV	97	Citation Bravo	550-0825	A Bar V Cattle & Commerce Corp. Prescott, Az.	
☐ N25KL	81	Sabre-65	465-69	Execujet, McMurray, Pa.	N400KV
☐ N25MB	87	Falcon 50	184	Mellon Bank NA. Pittsburgh, Pa.	N89FC
☐ N25MD	70	Learjet 25	25-054	S I Aircraft Sales Inc. Arlington, Tx.	N509G
☐ N25ME	70	Learjet 25	25-062	Walter Conlogue/World Aviation Co. Wilmington, NC.	N27FN
☐ N25MJ	68	HS 125/731	NA705	Executive Aircraft Services Inc. Scottsdale, Az.	(N60AM)
☐ N25MK	83	Citation II	550-0458	Taft Sales & Leasing Corp. Dallas, Tx.	XA-SET
☐ N25MR	73	Learjet 25C	25C-129	Rasmark Jet Charter Inc. El Paso, Tx.	(N193DR)
☐ N25NB	80	Learjet 25D	25D-326	North Coast Aviation Inc. Wilmington, De.	N771CB
☐ N25NH	79	Citation II/SP	551-0026	Nanticoke Homes Inc. Greenwood, De. (was 550-0377).	N12TV
☐ N25NY	80	Learjet 25D	25D-304	Savage Aviation Inc. Lebanon, Tn.	
☐ N25QT	88	Citation II/SP	551-0584	QuikTrip Corp. Tulsa, Ok.	N550WV
☐ N25RE	77	Learjet 25D	25D-227	RAR Investments Inc. Alexandria, La.	N227EW
☐ N25SB	92	Challenger 601-3A	5115	United States Tobacco Co. White Plains, NY.	C-GLYC
☐ N25TE	72	Learjet 25C	25C-087	Tensorex Co. Indian Harbour Beach, Fl.	N99XZ
☐ N25TK	72	Learjet 25B	25B-100	Rasmark Jet Charter Inc. El Paso, Tx.	(N59AC)
☐ N25UB	95	Falcon 50	248	United Dominion Industries Inc. Charlotte, NC.	N25UD
☐ N25W	92	BAe 125/800A	NA0473	Watkins Motor Lines Inc. Lakeland, Fl.	N675BA
☐ N25WX	94	Hawker 800XP	8359	Raytheon Aircraft Co. Wichita, Ks.	
☐ N26AT	73	Learjet 25B	25B-130	Air Transport Inc/Co-Motion Inc/NKL Inc. El Paso, Tx.	N25PL
☐ N26CB	95	CitationJet	525-0117	Brown Transport Inc. Holland, Mi.	N217CJ
☐ N26GP	78	Learjet 35A	35A-157	Pappas Telecasting Companies, Visalia, Ca	ZS-MWW

| Reg | Yr Type | | c/n | Owner/Operator | Prev Regn |

32

Reg	Yr	Type	c/n	Owner/Operator	Prev Regn
□ N26JP	93	Beechjet 400A	RK-74	Jefferson-Pilot Corp. Greensboro, NC.	N8146J
□ N26KL	84	Westwind-Two	409	Noram Energy Corp. Houston, Tx.	N217BM
□ N26LB	88	Falcon 900	51	American Money Management Corp. Cincinnati, Oh.	N59LB
□ N26LC	89	Learjet 31	31-006	Cappelli Development Corp. Valhalla, NY.	
□ N26ME	82	HS 125/700A	NA0315	Marathon Electric Manufacturing Corp. Wausau, Wi.	N869KM
□ N26SC	80	HS 125/700A	NA0283	Swiss Colony Inc. Monroe, Wi.	N93GR
□ N26SD	85	Citation III	650-0099	Square D Co. Waukagan, Il. 'We Respond'	(N555EW)
□ N26SQ	78	Sabre-60	306-140	Flight Specialists Inc. North Canton, Oh.	N26SC
□ N26T	80	Westwind-1124	293	B E & K Inc. Birmingham, Al.	N26TV
□ N26VF	84	Westwind-Two	410	VF Corp. Wyomissing, Pa.	N22BG
□ N26WJ	79	Falcon 10	126		F-GFHG
□ N27AJ	74	Falcon 10	31	International Union of Bricklayers & Allied Craftworkers, DC	(N29AA)
□ N27AT	74	Westwind-1123	176	Faith Landmarks Ministries, Richmond, Va.	N661MP
□ N27BH	86	Astra-1125	012	Ben Hill Griffin Inc. Frostproof, Fl.	N312W
□ N27BJ	71	Learjet 24B	24B-227	Key Lime Air. Englewood, Co.	N28AT
□ N27BL	78	Learjet 35A	35A-163	AirNet Systems Inc. Columbus, Oh.	YV-173CP
□ N27CD	90	Citation 4	1136	Schering Plough Inc. Morristown, NJ.	N401GA
□ N27FP	85	Citation S/II	S550-0027	Flying Partners CV. Antwerp, Belgium.	N27EA
□ N27KL	81	HS 125/700A	NA0289	Kalin Ltd. Midland, Mi.	N369G
□ N27L	72	Citation	500-0038	Athena Air & AnLaCo Air Inc. Dallas, Tx.	(N207L)
□ N27LT	77	Sabre-75A	380-59	J & M Air Inc. Westwood, Ks.	N911CR
□ N27MJ	77	Learjet 36A	36A-027	Medical Jets International Inc. Rochester, Mn.	N484HB
□ N27R	94	Falcon 2000	5	R J Reynolds Tobacco Co. Winston Salem, NC.	F-WWMB
□ N27RC	66	JetStar-731	5086	Seagull Aircraft Corp. Fort Lauderdale, Fl.	(N65JW)
□ N27SF	73	Citation	500-0064	Seneca Flight Operations/S S Pierce Co. Dundee, NY.	N564CC
□ N27SL	70	Gulfstream 2	84	Wiley Sanders Truck Lines Inc. Troy, Al.	N5101T
□ N27TS	80	Citation Eagle	501-0147	Eagle SP 147 Inc. Stansted, UK.	N392DA
□ N27TT	77	Learjet 35A	35A-122	AirNet Systems Inc. Columbus, Oh.	OE-GMP
□ N27WP	96	Falcon 2000	35	Weyerhaeuser Co & Paccar Inc. Tacoma, Wa.	F-WWMG
□ N27WW	77	Citation	500-0353	Wildwood Industries Inc. Bloomington, Il.	N353WB
□ N28C	79	Falcon 20F	404	Edwin L Cox, Athens, Tx.	N404F
□ N28CK	69	Learjet 25	25-045	Kalitta Flying Service Inc. Morristown, Tn.	N24FN
□ N28GA	97	CitationJet	525-0216	Griffin Industries Inc. Cold Springs, Ky.	
□ N28GC	78	Citation 1/SP	501-0074	S & K Aviation, Elk Grove, Il.	N717RB
□ N28GP	81	HS 125/700A	NA0296	Genuine Parts Co. Atlanta, Ga.	N31AS
□ N28JG	81	Citation 1/SP	501-0194	The W W Williams Co. Columbus, Oh.	N6781L
□ N28LR	79	Learjet	28-003	Echelon Services Inc. Mobile, Al.	N14QG
□ N28MJ	79	Learjet 35A	35A-224	Medjet International Inc. Birmingham, Al.	N40RW
□ N28NP	93	Astra-1125SP	067	Heitman Holdings Ltd. Palwaukee, Il.	N20FE
□ N28R	95	Falcon 2000	7	R J Reynolds Tobacco Co. Winston Salem, NC.	F-WWME
□ N28RC	81	Citation II	550-0269	R D C Inc. Mukilteo, Wa.	N760
□ N28TS	73	HS 125/600A	6009	John Netto, Salt Lake City, Ut.	(N183RM)
□ N28TX	92	Citation VII	650-7007	Bell Helicopter Textron Inc. Fort Worth, Tx.	N944L
□ N29AF	67	DC 9-15F	45826	U S Department of Energy, Albuquerque, NM.	CF-TON
□ N29AU	87	Citation III	650-0145	Associated Aviation Underwriters Inc. Short Hills, NJ.	N650AF
□ N29B	80	Citation II/SP	551-0181	American Air Travelers Inc. Stuart, Fl.	N137CF
□ N29CL	84	Westwind-Two	404	MAC Aircraft Inc. Las Vegas, Nv.	N404W
□ N29GD	79	HS 125/700A	NA0250	Genuine Parts Co. Atlanta, Ga.	N29GP
□ N29GP	97	Hawker 800XP	8344	Genuine Parts Co. Atlanta, Ga.	
□ N29PC	79	Westwind-1124	263	Pittway Corp/PJF Corp & Joel Hirsch, Chicago, Il.	4X-CNI
□ N29RE	95	Learjet 31A	31A-106	Allen Investments Inc. Pinehurst, NC.	
□ N29TS	75	Learjet 25B	25B-198	Thomas Pride Inc. Scottsdale, Az.	N20DK
□ N29UF	76	Westwind-1124	201	USF Holland Inc. Holland, Mi.	N300TE
□ N29WE	92	Citation V	560-0185	Whelen Engineering Co. Chester, Ct.	N1281K
□ N29XA	86	Citation S/II	S550-0096	Interflight Inc. West Palm Beach, Fl. (rebuild ?).	N29X
□ N30AB	78	Westwind-1124	235	Albert Biedenharn Jr/Osborn Heirs Co. San Antonio, Tx.	N65A
□ N30AD	69	Jet Commander-C	143	Washington State Department of Transport, Seattle, Wa.	N41
□ N30AV	78	Citation II	550-0026	Hersey Mountain Air Inc. Concord, NH.	(N2231B)
□ N30CX	84	Citation S/II	S550-0007	G P Aeronautics Inc. Encino, Ca.	TC-SAM
□ N30DK	80	Learjet 35A	35A-345	Corporate Jets Inc. West Mifflin, Pa.	N345LJ
□ N30EJ	79	Citation II/SP	551-0022	Gateway Chevrolet Inc. Fargo, ND.	N30FJ
□ N30FT	80	Falcon 20F-5	377	Intercon Inc. NYC.	D-CCMB
□ N30FW	77	Gulfstream 2	210	Rochester Aviation Inc. Rochester, NY.	8P-LAD
□ N30GR	90	Citation II	550-0656	Oil & Gas Rental Services Inc. Morgan City, La.	
□ N30HD	80	Diamond 1A	A005SA	Executive Beechcraft STL Inc. Chesterfield, Mo.	N700LP
□ N30HJ	79	Learjet 35A	35A-226	Winfair Aviation Ltd. Fort Lauderdale, Fl.	N1127M
□ N30JD	80	Citation II	550-0205	WWB Inc. Minneapolis, Mn.	(N88727)
□ N30JN	75	Citation	500-0272	Automotive Consultants Inc. Wilmington, De.	N30SB
□ N30LB	96	Falcon 900EX	8	Great American Insurance Co. Cincinnati, Oh.	F-WWFB
□ N30MR	78	Westwind-1124	225	Mike Rutherford Oil Co. Houston, Tx.	N1124U
□ N30PA	79	Learjet 35A	35A-245	Big Sur Waterbeds Inc. Denver, Co.	N1526L
□ N30PC	90	Citation V	560-0090	Southern Company Services Inc. Atlanta, Ga.	
□ N30PR	68	Gulfstream 2	35	Rutherford Oil Co. Houston, Tx.	N830TL
□ N30RL	90	Citation II	550-0653	Roseburg Lumber Co. Roseburg, Or.	N36854
□ N30SJ	91	SJ 30-2	001	Sino Swearingen Aircraft Co. San Antonio, tx.	
□ N30TH	90	Falcon 50	212	Sony Aviation Inc. Teterboro, NJ.	N50FJ
Reg	*Yr*	*Type*	*c/n*	*Owner/Operator*	*Prev Regn*

Reg	Yr	Type	c/n	Owner/Operator	Prev Regn
☐ N30TK	94	Learjet 31A	31A-096	Beach Air Travel Inc. Floyd Knobs, In.	N30LX
☐ N30W	95	Learjet 60	60-065	Worrell Enterprises Inc. Boca Raton, Fl.	N5006V
☐ N30WR	83	Gulfstream 3	380	Rollins Inc/LOR Inc. Atlanta, Ga.	N159B
☐ N31AA	69	Learjet 25	25-041	Nevada Desert Holdings Inc. Minden, Nv.	(N25RE)
☐ N31BG	75	Learjet 24D	24D-301	American Aircraft Sales Co/Robert H Coutches, Hayward, Ca.	(N87MJ)
☐ N31BP	68	JetStar-731	5125	Executive Aircraft Corp. Newton, Ks.	N48UC
☐ N31CK	65	Learjet 23	23-079	Kalitta Flying Service Inc. Detroit, Mi.	N240AQ
☐ N31DA	79	Citation II/SP	551-0010	Duncan Aircraft Sales, Venice, Fl.	YV-137CP
☐ N31DP	76	Learjet 35	35-062	Innovation Data Processing Inc. Little Falls, NJ.	N310BA
☐ N31HD	97	CitationJet	525-0261	Cessna Aircraft Co. Wichita, Ks.	
☐ N31HY	95	Learjet 31A	31A-107	Husky Injection Molding Systems Inc. Buffalo, NY.	
☐ N31LT	66	Falcon 20C	69	Imperial Oil Co. Tampa, Fl.	N176BN
☐ N31NR	97	Learjet 31A	31A-150	Learjet Inc. Wichita, Ks.	
☐ N31PV	97	Learjet 31A	31A-130	Feral Investment Ltda. Miami, Fl.	N5013N
☐ N31RC	89	Citation V	560-0023	Warehouse Management Inc. Cincinnati, Oh.	N560JM
☐ N31SG	97	CitationJet	525-0207	Savage Fuels Inc. Billings, Mt.	
☐ N31TJ	69	HS 125/400A	NA731	Alpine Air Inc. Charleston, WV.	(N700PG)
☐ N31TK	92	Learjet 31A	31A-059	Consolidated Charter Service Inc. Fort Wayne, In.	(N67MP)
☐ N31TR	79	B 727-212RE	21948	Triangle Aircraft Services Co. Stewart, NY.	VR-COJ
☐ N31V	97	Learjet 45	45-015	Learjet Inc. Wichita, Ks.	
☐ N31WR	80	Learjet 35A	35A-313	AirNet Systems Inc. Columbus, Oh.	TR-LZI
☐ N31WS	75	Learjet 35	35-027	Denveraire Inc/Windstar Aviation Corp. Englewood, Co.	
☐ N32B	88	Falcon 900	59	Black & Decker Corp. Baltimore, Md.	N442FJ
☐ N32BC	96	Hawker 800XP	8321	Brunswick Corp. Waukegan, Il.	N691H
☐ N32CA	66	Learjet 24	24-132	Condor Aviation Co. Midland, Tx.	N238R
☐ N32DD	76	Learjet 24E	24E-331	GRV Aviation Inc. New Orleans, La.	XA-REA
☐ N32FM	81	Citation 1/SP	501-0210	Imperial Transport Inc. NYC.	N67848
☐ N32HJ	82	Learjet 35A	35A-463	Hop-A-Jet Inc. Fort Lauderdale, Fl.	N68LL
☐ N32HM	78	Learjet 35A	35A-187	Original Honey Baked Ham Co. Atlanta, Ga.	N755GL
☐ N32HP	83	Diamond 1A	A074SA	ST Air LLC/Stoughton Trailers, Stoughton, Wi.	N19GA
☐ N32JA	76	Learjet 36	36-017	Manning Leasing Inc. Dover, De.	N17LJ
☐ N32JJ	89	Citation III	650-0170	Archer Daniels Midland/ADM Milling Co. Decatur, Il.	N170CC
☐ N32KB	72	HS 125/731	NA773	S S K Hawker Group LLC. Shelton, Ct.	C-FPPN
☐ N32KJ	84	Learjet 55	55-093	Air Path Inc. Memphis, Tn.	N725K
☐ N32KR	77	JetStar 2	5220	Aspen Trading Corp. Wilmington, De.	N5546L
☐ N32MG	83	Citation III	650-0016	Magellan Health Services Inc. Atlanta, Ga.	(N45US)
☐ N32PA	77	Learjet 36A	36A-025	Phoenix Air Group Inc. Cartersville, Ga.	N800BL
☐ N32PE	84	Learjet 35A	35A-327	Blue Canyon Inc. Smyrna, Tn.	N32PF
☐ N32TC	91	Falcon 50	225	Georgetown Management Inc. Kirkland, Wa.	N50FJ
☐ N33BC	95	Hawker 800XP	8292	Brunswick Corp. Waukegan, Il.	N673H
☐ N33BQ	84	Citation III	650-0047	Pacific Connection Inc. Phoenix, Az.	N33BC
☐ N33CX	78	Citation 1/SP	501-0079	Executive Aircraft Services Inc. Scottsdale, Az.	N555EW
☐ N33D	68	Falcon 20C-5	166	Dow Chemical Co. Freeland, Mi.	N4368F
☐ N33DT	94	CitationJet	525-0080	Taylor Industries Inc. Des Moines, Ia.	N80CJ
☐ N33EK	81	Citation II	550-0281	King Radio Corp/Peabody Coal Co. Flagstaff, Az.	N31RK
☐ N33FW	97	CitationJet	525-0203	Air Trebuchet LLC. Cincinnati, Oh.	N525GP
☐ N33GG	90	Falcon 900	87	George Gund III Co. San Francisco, Ca.	N462FJ
☐ N33GK	95	Citation VII	650-7050	Gold Kist Inc. Atlanta, Ga.	N6150B
☐ N33GQ	82	Falcon 50	97	Dassault Falcon Jet Corp. Teterbor, NJ.	N33GG
☐ N33HC	79	Citation 1/SP	501-0104	Scientific Flight Corp. Oakbrook Terrace, Il.	N29CA
☐ N33JW	74	Sabre-60	306-92	Starflight LLC. Belle Chasse, La.	N74AB
☐ N33L	69	Falcon 20D-5	202	Dow Chemical Co. Freeland, Mi.	N814PA
☐ N33LX	97	Citation V Ultra	560-0433	Lennox Industries Inc. Richardson, Tx.	
☐ N33M	88	Gulfstream 4	1056	3M Co. Minneapolis, Mn.	N436GA
☐ N33MK	82	Westwind-Two	374	Titan Indemnity Co. San Antonio, Tx.	N248H
☐ N33MM	82	Diamond 1A	A017SA	MCC Air Inc. Virginia Beach, Va.	N75BL
☐ N33NJ	80	Learjet 35A	35A-305	National Jets Inc. Fort Lauderdale, Fl.	N3VG
☐ N33PF	69	Learjet 25	25-028	Anthony Aiello, Palwaukee, Il.	N277L
☐ N33RH	77	HS 125/700A	NA0207	BeautiControl Cosmetics Inc. Wilmington, De.	N500FC
☐ N33SJ	66	JetStar-731	5087	Craig Aviation Inc. Del Mar, Ca.	N75MG
☐ N33TP	75	Learjet 24D	24D-321	Mundet Inc. Colonial Heights, Va.	C-FRNR
☐ N33TR	70	Sabre-60	306-54	Trinity Industries Inc. Dallas, Tx.	(N100EU)
☐ N33TW	73	Learjet 25B	25B-124	Gayle Schroder, Houston, Tx.	(N400DB)
☐ N33TY	93	Falcon 50	240	Tyco International Ltd. Portsmouth, NH.	N780F
☐ N33UL	88	Citation III	650-0160	Universal Leaf Tobacco Co. Sandston, Va.	(N830GB)
☐ N33UT		Sabreliner CT-39A	276-34	University of Tennessee, Tullahoma, Tn.	62-4481
☐ N33WW	78	Citation 1/SP	501-0065	W R J Inc. Bloomfield Hills, Mi.	N2888A
☐ N34AM	65	Sabre-40	282-31	Sabre Investments Ltd. Chesterfield, Mo.	N577VM
☐ N34CD	93	Challenger 601-3R	5139	Schering-Plough Corp. Morristown, NJ.	N139CD
☐ N34CW	74	Falcon 20F	305	Winn Exploration Co. Eagle Pass, Tx.	N282U
☐ N34GB	84	Learjet 55	55-114	Konfara Co. Scottsdale, Az.	N72608
☐ N34SW	67	Jet Commander	97	OK Turbines Inc. Hollister, Ca.	N3082B
☐ N34TJ	79	Learjet 35A	35A-225	Keystone Air Inc. Melbourne, Fl.	TG-JAY
☐ N34TN	78	Learjet 25D	25D-249	Avastar Jet Charter & Management Services Inc. Waterford, Mi	XA-FMU
☐ N34TR	65	Learjet 23	23-069	C G Aviation, Lantana, Fl.	N37BL
☐ N34WR	77	JetStar 2	5207	Orkin Extermination Co. Atlanta, Ga.	N176BN

| *Reg* | *Yr* | *Type* | *c/n* | *Owner/Operator* | *·Prev Regn* |

Reg	Yr	Type	c/n	Owner/Operator	Prev Regn
N35AK	80	Learjet 35A	35A-314	The Lowell Dunn, Hialeah, Fl.	(N118GM)
N35AW	79	Learjet 35A	35A-233	Anheuser Busch Companies Inc. St Louis, Mo.	N35SL
N35AZ	78	Learjet 35A	35A-201	Comm Aviation LLC. Nashville, Tn.	XA-PIN
N35BG	81	Learjet 35A	35A-402	Tashi Corp. Carson City, Nv.	N7AB
N35BN	75	Learjet 35	35-013	Avcon Industries Inc. Newton, Ks. (ventral delta fin demo).	N35JN
N35CC	66	Sabre-40	282-79	Crown Controls Corp. New Knoxville, Oh.	N701NC
N35CZ	80	Learjet 35	35A-352	S P Aviation Inc. Hayward, Ca.	(N999JA)
N35D	72	Westwind-1123	156	G W Taylor/Taylor Aircraft, Halfway, Mo.	N566MP
N35DL	80	Learjet 35A	35A-348	Panoramic Flight Svc/Cameron-Henkind Corp. Westchester, NY.	N35TL
N35ED	79	Learjet 35A	35A-215	E H Darby & Co. Sheffield, Al.	N80GD
N35GC	79	Learjet 35A	35A-266	DLS Enterprises Inc. Houston, Tx.	N922GL
N35HC	80	Citation II	550-0151	Skyflight Services Inc. Portland, Or.	(N98528)
N35HS	96	Citation VII	650-7072	Rex Realty Co. St Louis, Mo.	N8494C
N35HW	83	Learjet 35A	35A-501	Rockit Aviation Corp. Philadelphia, Pa.	N711PR
N35LH	85	Westwind-Two	413	Liberty Homes Inc. Goshen, In.	N413WW
N35LW	81	Learjet 35A	35A-439	Palmer Aviation Leasing LLC. Germantown, Tn.	N55RZ
N35NW	76	Learjet 35A	35A-069	Northwestern Aircraft Capital Corp. Mercer Island, Wa.	N10AQ
N35NX	80	Learjet 35A	35A-328	Amerada Hess Corp. West Trenton, NJ.	N35NY
N35RG	62	Sabreliner CT-39A	276-25	Ronald Green,	62-4472
N35RZ	82	Falcon 50	113	Ritz Camera Centers Inc. Beltsville, Md.	N394U
N35SM	81	Learjet 35A	35A-419	State of Mississippi, Jackson, Ms.	N53JM
N35TJ	77	Learjet 35A	35A-137	Western New Mexico Tlephone Co. Silver City, NM.	N41FN
N35TN	82	Learjet 35A	35A-472	Avstar Jet Charter & Management Services Inc. Waterford, Mi.	(N472AS)
N35UK	90	Learjet 35A	35A-662	Wal-Mart Stores Inc. Rogers, Ar.	G-BUSX
N35WB	80	Learjet 35A	35A-350	Wagner & Brown, Midland, Tx.	(N88NE)
N35WR	79	Learjet 35A	35A-234	SP Aviation Inc. Hayward, Ca.	
N36GS	80	HS 125/700A	NA0267	R Gene Smith Inc. Louisville, Ky.	N745TH
N36H	96	Hawker 800XP	8332	Halstead Industries Inc. Greensboro, NC.	N332XP
N36LX	97	Citation V Ultra	560-0436	Lennox Industries Inc. Richardson, Tx.	
N36MJ	78	Learjet 36A	36A-036	Apple Computers/ACM Aviation Inc. San Jose, Ca.	N610GE
N36PN	68	Gulfstream 2B	42	Peninsular Marine Service Inc. New Bern, SC.	N1164A
N36RR	67	Gulfstream 2B	4	RGR Technologies Inc. Mansfield, Oh.	N8490P
N36SJ	76	Citation	500-0306	R P Air Inc. Mendota Heights, Mn.	N36CJ
N36TJ	80	Learjet 35A	35A-289	TAVAERO Inc. Houston, Tx.	N289LJ
N37BE	83	Westwind-1124	396	Baldor Electric Co. Fort Smith, Ar.	8P-BAR
N37DG	95	CitationJet	525-0109	Delaware Leasing Inc. South Hackensack, NJ.	N52113
N37FA	76	Learjet 35A	35A-091	Aspen Base Operation Inc. Aspen, Co.	D-CIRS
N37FN	79	Learjet 35A	35A-263	Career Aviation Academy Inc. Oakdale, Ca.	N4577Q
N37HE	85	Gulfstream 3	466	Hercules Inc. New Castle, De.	N325GA
N37LA	78	Citation 1/SP	501-0080	Jet Manager Inc. Wilmington, De.	N347DA
N37MH	80	Citation II	550-0153	Charlie Brown Air Corp. State College, Pa.	N27MH
N37NY	88	B 737-4YO	23976	ITT Flight Operations Inc/New York Knicks, Allentown, Pa.	N773RA
N37SG	73	HS 125/600A	6021	Segrac Inc. Los Angeles, Ca.	XA-SNH
N37SJ	65	Jet Commander	38	B J Johnson LLC. Wooster, Oh. (status ?).	N106CJ
N37SV	82	Learjet 35A	35A-492	SV Leasing Company of Florida Inc. Coconut Grove, Fl.	N35NP
N37TA	75	Learjet 35	35-034	RLO Aviation Inc. Peoria, Il.	
N37WH	76	Gulfstream 2	180	Huizenga Holdings/WACO Services Inc. Fort Lauderdale, Fl.	N359K
N37WP	94	Citation V	560-0259	Paccar Inc/Weyerhauser Corp. Seattle, Wa.	N1293Z
N38AE	81	Westwind-Two	318	Time Compression Inc. New Castle, De.	N10FG
N38AM	78	Learjet 35A	35A-174	AmVestors Financial Corp. Topeka, Ks.	N82883
N38BG	77	JetStar 2	5208	Konfara Co. Scottsdale, Az.	(N29TC)
N38D	82	Learjet 55	55-068	Peters Corp & E V Thaw of New Mexico Inc. Santa Fe, NM.	
N38DA	75	Falcon 10	27	Hotel Aviation LLC. Atlanta, Ga.	OK-EEH
N38DD	82	Citation II	550-0340	JRW Aviation Inc. Dallas, Tx.	ZP-TWN
N38EC	86	Citation S/II	S550-0109	Empress River Casino Corp. Joliet, Il.	N7QC
N38GL	68	Gulfstream 2B	16	Sonic Financial Corp. Charlotte, NC.	N711MT
N38LB	71	HS 125/731	NA770	Fredrick Barron, Dallas, Tx.	N7170J
N38MH	75	Citation	500-0265	Sawyer Brown Inc. Nashville, Tn.	XA-VYF
N38PA	77	HS 125/700A	NA0208	Fan 4 Inc. Fort Lauderdale, Fl.	N37PL
N38SA	76	Citation	500-0297	Tropic Winds Hotel/Stanley L Allen, Myrtle Beach, SC.	N48DA
N38SW	83	Challenger 601	3008	SLW Aviation Inc. Houston, Tx.	(N999SW)
N38TS	69	HS 125/400A	NA722	Camel Aviation Inc. Eutawville, SC.	N209NC
N38TT	81	Citation II/SP	551-0311	Litton Systems Inc. Van Nuys, Ca. (was 550-0268)	N500FX
N38WP	84	Citation III	650-0032	Weyerhaeuser Co. Seattle, Wa.	N54WC
N39CB	76	Sabre-60	306-116	Aero Del Inc. Laredo, Tx.	N44WD
N39CD	84	Challenger 601	3030	Global Aviation P/L-China Southern Airlines, Guangzhou.	N34CD
N39CJ	93	CitationJet	525-0039	Harold Bagwell, Raleigh, NC.	(N1959E)
N39CK	68	Learjet 25	25-005	Kalitta Flying Service Inc. Morristown, Tn.	XA-SDQ
N39DK	82	Learjet 35A	35A-480	AirNet Systems Inc. Columbus, Oh.	(N484)
N39EL	72	Learjet 24D	24D-251	Executive Aviation Logistics Inc. Chino, Ca.	N69XW
N39FN	74	Learjet 35	35-006	Flight International Inc. Newport News, Va.	N39DM
N39FS	62	Sabreliner CT-39A	276-33	Tracor Inc/AMLECO Inc. Tucson, Az.	N24480
N39H	91	Citation	650-0206	East Tennessee Equipment Finance Inc. Memphis, Tn.	PT-OKV
N39HD	82	Citation II	550-0448	David Wilson, Orange, Ca.	N93DW
N39J	74	Citation	500-0207	Trees Unlimited Inc. Amelia Island, Fl.	N929A
N39JN	79	Westwind-1124	261	DBN Investments Inc. Dover, De.	N87GS

| *Reg* | *Yr Type* | | *c/n* | *Owner/Operator* | *Prev Regn* |

35

Reg	Yr	Type	c/n	Owner/Operator	Prev Regn
☐ N39KM	69	Learjet 24B	24B-198	Cirrus Air International Inc. Fort Lauderdale, Fl.	N21XB
☐ N39LH	73	Citation	500-0089	Indiana Trading Co. Elkhart, In.	EC-EBR
☐ N39ML	78	Citation II/SP	551-0002	Shanair Inc. Memphis, Tn.	XA-SLD
☐ N39N	93	Citation V	560-0243	Union Carbide Corp. White Plains, NY.	
☐ N39RE	81	Westwind-Two	342	AiRush Inc. Uvalde, Tx.	N342AJ
☐ N39TF	87	Citation S/II	S550-0139	Teleflex Inc. Plymouth Meeting, Pa.	N706SB
☐ N39TW	91	Learjet 31A	31A-047	N Squared Aviation LLC. Griffith, In.	N31UK
☐ N39WH	94	Gulfstream 4SP	1243	Southern Aircraft Services Inc. Fort Lauderdale, Fl.	VP-CBL
☐ N39WP	84	Citation III	650-0039	Weyerhaeuser Inc. Seattle, Wa.	N81TC
☐ N40	68	B 727-25QC	19854	FAA R&D Flight Program, Atlantic City, NJ.	N8171G
☐ N40AJ	89	Astra-1125	031	Gilbert Amelio, Los Altos Hills, Ca.	4X-CU.
☐ N40AN	79	Learjet 35A	35A-271	AirNet Systems Inc. Columbus, Oh.	LV-OAS
☐ N40AS	86	Falcon 50	171	AlliedSignal Inc. Morristown, NJ.	N171FJ
☐ N40BD	77	Learjet 35A	35A-140	Newcastle Corp. Wichita Falls, Tx.	N72TP
☐ N40CH	83	Gulfstream 3	377	Granite Properties Inc. Granite Bay, Ca.	N342GA
☐ N40CN	82	Falcon 50	92	Champion International Corp. Stamford, Ct.	N85A
☐ N40CR	90	Learjet 55C	55C-144	A T & T Global Information, Dayton, Oh.	PT-OJH
☐ N40DK	78	Learjet 35A	35A-171	Corporate Jets Inc. Pittsburgh, Pa.	(N48DK)
☐ N40FC	87	Citation III	650-0143	Frank's Casing Crew, & Rental Tools, Lafayette, La.	N312CF
☐ N40GA	83	Diamond 1A	A040SA	ANCO Management Services Inc + co-owners, Florence, Al.	N188ST
☐ N40GG	78	Westwind-1124	229	Golden Greens Inc. Wilmington, De.	N162E
☐ N40GT	73	Sabre-40A	282-126	Big Sky Aviation Inc. Overland Park, Ks.	XA-SNI
☐ N40HP	73	Citation	500-0104	Rocky Mountain Sunshine Inc. Denver, Co.	C-GPJW
☐ N40KM	84	Citation S/II	S550-0008	Kreuter Engineering Inc. New Paris, In.	N204A
☐ N40LB	65	Sabre-40	282-36	Commercial Aviation Enterprises Inc. Delray Beach, Fl.	N200MP
☐ N40MT	81	Citation II	550-0215	McNeilus Truck & Manufacturing Co. Dodge Center, Mn.	(N240MC)
☐ N40N	93	Citation VII	650-7031	Union Carbide Corp. White Plains, NY.	N12636
☐ N40PD	73	Citation	500-0059	East Alabama Feed Supplements Co. Smyrna, Ga.	N40RD
☐ N40PH	90	Citation VI	650-0201	Nordic Gaming Corp. Las Vegas, Nv.	(N26264)
☐ N40PK	79	Learjet 35A	35A-260	Porta Kamp Manufacturing Co. Houston, Tx.	
☐ N40SH	66	DC 9-15	45775	KEB Aircraft Sales Inc. Danville, Ca.	N89SM
☐ N40TA	76	Westwind-1124	194	Air Methods Corp/Colorado Air Group LLC. Englewood, Co.	N124FM
☐ N40TL	75	Sabre-60	306-103	ASK Inc. Adrian, Mi.	N11UL
☐ N40WJ	75	Falcon 10	21	Architectural Air II LLC. Washington, DC.	3D-ACB
☐ N40WP	91	Citation V	560-0155	Weyerhaeuser Co/Paccar Inc. Tacoma, Wa.	
☐ N40YC	87	Citation II	550-0554	Ribeiro Corp. Las Vegas, Nv.	N40FC
☐ N40ZA	73	Sabre-40A	282-112	Z Air Inc. Laredo, Tx.	N164DN
☐ N41BH	87	Citation II	550-0567	Jitney-Jungle Stores Inc. Jackson, Ms. (was 551-0567).	N321F
☐ N41BP	68	Falcon 20D	177	IBP=Iowa Beef Processors Inc. Sioux City, Ia.	N14FG
☐ N41C	81	Citation II	550-0291	Chevron USA Inc. Houston, Tx.	N550CD
☐ N41CP	91	Gulfstream 4	1179	Pfizer Inc. Mercer County Airport, NJ.	N471GA
☐ N41DP	96	Learjet 45	45-010	Dean Phillips Inc. Las Vegas, Nv.	(D-CWER)
☐ N41EB	95	CitationJet	525-0116	Musikantow Consulting & Management Co. Chicago, Il.	N52144
☐ N41GT	79	Citation 1/SP	501-0297	Tobin Aviation Enterprises, Abilene, Tx. (was 500-0394).	N35LD
☐ N41H	77	Learjet 25D	25D-217	L J Associates, Blairesville, Pa.	
☐ N41ME	91	Beechjet 400A	RK-19	MidAmerican Energy & Export Packaging, Des Moines, Ia.	(N8ME)
☐ N41MH	65	Falcon 20C	14	Elite Air Inc. Cartersville, Ga.	N91JF
☐ N41MP	67	Learjet 24	24-148	Marpar Air NA Inc. Wilmington, De.	(N47NR)
☐ N41NK	97	CitationJet	525-0190	Premier Ink Systems Inc. Harrison, Oh.	
☐ N41PG	96	CitationJet	525-0175	Davco Consulting Inc. Greely Wold County, Co.	N175CP
☐ N41PJ	75	Learjet 35	35-041	Northwestern Aircraft Capital Corp. Van Nuys, Ca.	N711BH
☐ N41PR	93	Gulfstream 4SP	1226	Prime Resources LLC. Carson City, Nv.	N460GA
☐ N41SH	75	Citation	500-0267	Windrise Corp/Joseph Bianco, NYC.	(N4090P)
☐ N41SM	81	Citation II	550-0231	McKenzie Development Corp. Cleveland, Tn.	N148DR
☐ N41TC	81	Learjet 25D	25D-346	B & C Aviation, Nashville, Tn.	N300WG
☐ N41ZP	79	Learjet 25D	25D-279	Bernstein/Rein Advertising Inc. Kansas City, Mo.	
☐ N42		Convair 880	55	FAA, Oklahoma City, Ok.	N112
☐ N42B	79	Learjet 35A	35A-277	Robert Wright Aircraft Inc. Tampa, Fl.	N27TJ
☐ N42CK	65	HS 125/731	25038	Pegasus North Inc. Dover, De.	N28M
☐ N42CM	76	Westwind-1124	189	AGG Aircraft Sales & Leasing Inc. Beverly, Ma.	N200DL
☐ N42FD	65	HS 125/731	25042	Glieberman Aviation LLC. Novi, Mi.	N79TS
☐ N42G	74	Falcon 10	20	Philip R Palm, Great Falls, Mt.	N113FJ
☐ N42HP	83	Learjet 35A	35A-507	Heritage Network Inc. Southgate, Mi.	N35HP
☐ N42LL	81	Learjet 35A	35A-427	Hop-A-Jet Inc. Fort Lauderdale, Fl.	VH-FOX
☐ N42PG	72	Learjet 24D	24D-247	Jeffrey Gieger, Laurel, Ms.	N23AM
☐ N42PH	81	Citation II	550-0304	Harron Communications Corp. Frazer, Pa.	(N70PH)
☐ N42SK	96	Beechjet 400A	RK-111	Kellett Investment Corp. Atlanta, Ga.	
☐ N42US	80	Falcon 10	171	U S Aircraft Sales Inc. McLean, Va.	PT-OIC
☐ N43AR	72	JetStar-8	5154	Tyler Jet LLC. Tyler, Tx.	XC-SRH
☐ N43BE	81	Falcon 50	49	Edward J DeBartolo Corp. Youngstown, Oh.	N43ES
☐ N43EC	80	Falcon 10	168	Sequoia Properties Inc. Chicago, Il.	N175BL
☐ N43JG	74	Sabre-60	306-79	Wheel-Air Inc. Anoka County, Mn.	(N7682V)
☐ N43KS	87	Gulfstream 4	1018	SME Aircraft Leasing Co & KALCO Corp. Raleigh, NC.	N300L
☐ N43M	88	Gulfstream 4	1057	3M Co. Minneapolis, Mn.	N437GA
☐ N43PR	87	Challenger 601-3A	5002	Town & Country Food Markets Inc. Wivhita, Ks.	N585UC
☐ N43R	97	Challenger 604	5334	Rockwell International Corp. Seal Beach, Ca.	N604RC
Reg	Yr	Type	c/n	Owner/Operator	Prev Regn

Reg	Yr	Type	c/n	Owner/Operator	Prev Regn
☐ N43RC	87	Citation S/II	S550-0149	Rohrer Corp. Wadsworth, Oh.	N810V
☐ N43RJ	69	Gulfstream 2B	64	Trump Hotels & Casino Resorts Holdings LP. NYC.	N82CK
☐ N43RK	93	Challenger 601-3A	5134	Home Depot Inc. Atlanta, Ga.	N43R
☐ N43SF	91	Learjet 31A	31A-048	State Farm Mutual Auto Insurance Inc. Bloomington, Il.	
☐ N43SP	82	Citation 1/SP	501-0243	Spidela Inc. Kenner, La.	N2624Z
☐ N43TJ	77	Learjet 35A	35A-121	Eric Barnum & Richard Landrum, Maumee, Oh.	(D-CFVG)
☐ N43TR	88	Learjet 35A	35A-645	Aeromanagement Inc. Herndon, Va.	
☐ N43VS	85	Citation S/II	S550-0069	Kurt Manufacturing Co. Minneapolis, Mn.	(N12720)
☐ N43W	64	Sabre-40	282-15	H L Brown Jr. Midland, Tx.	N21PF
☐ N44AB	82	Learjet 35A	35A-473	Aero Boise Co. Boise, Id.	N3UJ
☐ N44AS	79	Citation II	550-0047	Julrich Aviation Inc. Loveland, Co.	N66VM
☐ N44CP	69	Learjet 24B	24B-185	Rasmark Jet Charter Inc. El Paso, Tx.	N754M
☐ N44EL	94	Learjet 60	60-036	U S Epperson Underwriting Co. Boca Raton, Fl.	N60LR
☐ N44EV	76	Learjet 36A	36A-022	Hawk Flight Inc. Wilmington, De.	N36PD
☐ N44FG	97	Citation V Ultra	560-0470	AFG Industries Inc. Kingsport, Tn.	
☐ N44FJ	93	CitationJet	525-0003	Cessna Aircraft Co. Wichita, Ks.	(N1326B)
☐ N44FM	79	Citation 1/SP	501-0156	Procedures Inc. Santa Monica, Ca.	N123FG
☐ N44GT	93	Citation V	560-0252	Groendyke Transport Inc. Enid, Ok.	N252CV
☐ N44HG	78	Learjet 35A	35A-180	H S Geneen/HSG Aviation Inc. NYC.	N35CX
☐ N44HH	92	BAe 125/800A	NA0474	Integrated Health Services of Skyview, Owings Mills, Md.	N622AD
☐ N44JC	82	Falcon 20F-5	471	John L Cox, Midland, Tx.	N478F
☐ N44KB	71	Learjet 24D	24D-245	General Transport Services Inc. Las Vegas, Nv.	N275E
☐ N44LC	90	Citation II	550-0649	Lowe's Companies Inc. North Wilkesboro, NC.	N4320P
☐ N44LF	81	Citation II	550-0277	Scott Madden & Associates Inc. Raleigh, NC.	(N550MT)
☐ N44M	84	Citation III	650-0050	Corporate Jets/Seward Prosser Mellon, Pittsburgh, Pa.	(N1311K)
☐ N44MD	67	B 727-44	19318	Davis Oil Co. Los Angeles, Ca.	N727EC
☐ N44PA	73	Learjet 25B	25B-144	Flight Management & Chaver Aviation Inc. Tulsa. (status ?).	N10NT
☐ N44PR	73	Westwind-1123	169	N44PR Inc. Corpus Christi, Tx.	N1100D
☐ N44QG	92	Learjet 31A	31A-053	Quad/Graphics Inc. Pewaukee, Wi.	N9173Q
☐ N44SF	92	Learjet 31A	31A-065	State Farm Mutual Auto Insurance Inc. Bloomington, Il.	N26005
☐ N44SK	81	Learjet 35A	35A-444	Stephen Kesler Enterprises, Warsaw, In.	N1U
☐ N44SW	80	Citation 1/SP	501-0162	Steel Warehouse Co. South Bend, In.	N55H
☐ N44TT	78	Learjet 35A	35A-211	Apple Jet Inc. Athena, Ga.	N998JP
☐ N45AE	81	Learjet 35A	35A-422	Chesrown Aviation Inc. Denver, Co.	N86BL
☐ N45AF	81	HS 125/700A	NA0291	Willis, Stein Advisors LLC. Chicago, Il.	XA-MSH
☐ N45AQ	77	Citation 1/SP	501-0029	Alpha Quest LLC. Tucson, Az.	N31JM
☐ N45BK	69	Learjet 25	25-036	Barken International/Avior Technologies Inc. Nairobi, Kenya.	N15M
☐ N45BR	97	Citation X	750-0045	Cessna Aircraft Co. Wichita, Ks.	
☐ N45CP	72	Learjet 25XR	25XR-073	Air Geronimo Inc/Concrete Pipe & Products Co. Richmond, Va.	N888DB
☐ N45ED	66	Learjet 24	24-104	E H Darby & Co. Sheffield, Al.	N924ED
☐ N45FG	75	Learjet 36	36-010	Fremont Group/Bechtel Corp. Oakland, Ca.	N50SF
☐ N45GP	86	Citation S/II	S550-0110	Gate Asphalt Co. Jacksonville, Fl.	N1291V
☐ N45H	79	Sabre-65	465-2	45 Hotel Corp. Columbus, Oh.	N251JE
☐ N45HG	90	Learjet 31	31-026	Asphalt Materials Inc. Indianapolis, In.	N39TJ
☐ N45KK	92	Learjet 35A	35A-672	Blanco Oil Co. San Antonio, Tx.	N672DK
☐ N45ME	79	Citation II	550-0080	Suncrest Farms Inc. Phoenix, Az.	N22511
☐ N45MK	80	Citation 1/SP	501-0193	Business Resources International, Winnetka, Il.	N39300
☐ N45MR	67	Falcon 20C	123	Rasmark Jet Charter Inc. El Paso, Tx.	N513T
☐ N45MS	90	Astra-1125	041	Silver Spur Trading Inc. Long Beach, Ca.	VR-BME
☐ N45NC	93	Falcon 50	232	National City Corp/NCC Services Inc. Richmond Heights, Oh.	N244FJ
☐ N45NP	80	Sabre-65	465-42	Executive Aircraft Corp. Newton, Ks.	N64SL
☐ N45NS	83	Citation II	550-0479	Seaman Corp. Wooster, Oh.	PT-OOM
☐ N45PH	83	Challenger 601	3004	Pizza Hut Inc. Dallas, Tx.	N501PC
☐ N45PK	91	Learjet 31	31-034	Floyd S Pike Electrical Contractor, Mount Airy, NC.	N604LJ
☐ N45RC	90	Citation V	560-0071A	Thorn Americas Inc. Wichita, Ks.	N2728N
☐ N45RK	92	Beechjet 400A	RK-43	Boder Corp/Explorer Van, Warsaw, In.	N56400
☐ N45SC	97	Falcon 2000	45	Siecor Corp. Hickory, NC.	F-WWMM
☐ N45SJ	96	Falcon 900EX	7	Sid Richardson Carbon & Gasoline Co. Fort Worth, Tx.	N907FJ
☐ N45TL	77	Citation 1/SP	501-0016	TLR Inc. Costa Mesa, Ca.	N17TJ
☐ N45UF	93	Learjet 31A	31A-072	Unifi Inc. Greensboro, NC.	N31LJ
☐ N45XL	95	Learjet 45	45-001	Learjet Inc. Wichita, Ks. (Ff 7 Oct 95)	
☐ N45Y	88	BAe 125/800A	NA0427	Schuller International Inc. Denver, Co.	N581BA
☐ N46BK	77	Westwind-1124	214	Barken International of NC Inc. Salt Lake City, Ut.	(N248H)
☐ N46F	89	Challenger 601-3A	5055	Hunt Oil Co. Dallas, Tx.	N601HC
☐ N46FE	91	Beechjet 400A	RK-16	Excel Industries Inc. Elkhart, In.	N71FE
☐ N46KB	78	Learjet 35A	35A-206	General Transport Services Inc. Las Vegas, Nv.	N38PS
☐ N46MK	83	Falcon 100	206	Merillat Industries Inc. Adrian, Mi.	N367F
☐ N46MT	94	Citation V	560-0253	Manchester Tank & Equipment Co. Dallas, Tx.	N253CV
☐ N46PL	79	HS 125/700B	7055	B L Yachts Inc. Eaton Park, Fl.	F-GGHG
☐ N46SC	79	Citation 1/SP	501-0137	Southeastern Mechanical Contractors, Greensboro, NC.	N26503
☐ N46SR	82	Challenger 600S	1046	J Larry Fugate Enterprises/EAC, Wichita, Ks.	C-GTXV
☐ N46UF	93	Learjet 31A	31A-073	Unifi Inc. Greensboro, NC.	
☐ N46WC	83	HS 125/700A	NA0334	Weldbend Corp. Chicago, Il.	N93GC
☐ N47CM	88	Citation III	650-0153	Carolina Mills Inc. Maiden, NC.	N653CC
☐ N47EC	78	Gulfstream 2	231	Eastman Chemical Co. Blountville, Tn.	N205K
☐ N47FH	93	CitationJet	525-0047	Wheeler Development Corp. Parker, Az.	N47TH
Reg	*Yr*	*Type*	*c/n*	*Owner/Operator*	*Prev Regn*

Reg	Yr	Type	c/n	Owner/Operator	Prev Regn
☐ N47HV	73	HS 125/600A	6014	Monument Aircraft Inc. Granite Bay, Ca.	N47HW
☐ N47JR	75	Learjet 35	35-007	BYR737/Beck & Yeaggy, Cincinnati, Oh.	(N65FN)
☐ N47LP	81	Falcon 20F	457	High Valley Air Service Inc. Colorado Springs, Co.	N4362M
☐ N47MR	72	Learjet 25B	25B-101	Rasmark Jet Charter Inc. El Paso, Tx.	N821AW
☐ N47PB	83	Diamond 1	A047SA	MKJ Aviation Inc. Steamboat Springs, Co.	N45NP
☐ N47RK	80	Falcon 10	162	M & A Properties Inc. Bandera, Tx.	N796MA
☐ N47SE	80	Sabre-65	465-34	AIG Aviation Inc. Atlanta, Ga.	N65TS
☐ N47SW	82	Citation II	550-0444	Villages Travel Co. Lady Lake, Fl.	N67MP
☐ N47TH	95	CitationJet	525-0119	Effingham Air Inc/Air Lex-Sprite Flight Jets, Lexington, Ky.	N5264E
☐ N47TL	81	Citation 1/SP	501-0200	Corporate Aircraft Services, Meadville, Pa.	7Q-YTL
☐ N47TW	83	Citation II	550-0477	Toyota of Greenville Inc. Greenville, SC.	N649WW
☐ N47UF	80	Falcon 50	28	Unifli Inc. Greensboro, NC.	PH-LEM
☐ N47VC	95	Citation V Ultra	560-0304	Vencor Inc. Louisville, Ky.	N5226B
☐ N47WT	74	Learjet 24XR	24XR-283	Rasmark Jet Charter Inc. El Paso, Tx.	(N711SC)
☐ N48AH	80	Westwind-1124	288	ARJA LLC=JAE Endorsements Inc & ARH Ltd. Englewood, Co.	N94AT
☐ N48CC	76	Gulfstream 2	181	Centex Service Co. Dallas, Tx.	N924DS
☐ N48CG	96	Falcon 2000	41	Corning Inc. Horseheads, NY.	N2073
☐ N48CK	87	Beechjet 400	RJ-22	Causey Aviation Service Inc. Liberty, NC.	N724AA
☐ N48CT	73	Learjet 24XR	24XR-274	FSB-Ut/Aero Prodin S.A. Guatamala City.	(TG-...)
☐ N48EC	67	Gulfstream 2B	9	Eastman Chemical Co. Blountville, Tn.	N343K
☐ N48FN	71	Learjet 24D	24D-238	Flight International/Maritime Sales & Leasing, Hockessin, De	N49DM
☐ N48FU	88	Challenger 601-3A	5021	Ventura Air Services/Fantasy Unlimited, Port Washington, NY.	N122WF
☐ N48HB	92	Falcon 50	233	Huntington Bancshares Inc. Columbus, Oh.	N232FJ
☐ N48HC	81	Learjet 55	55-012	Stanwich Aviation Inc. West Mifflin, Pa.	N104BS
☐ N48L	66	Learjet 24A	24A-107	Royal Air Freight Inc. Waterford, Mi.	
☐ N48MC	76	Sabre-60	306-119	JMD Corp. Wilmington, De.	XA-JMD
☐ N48PL	97	Beechjet 400A	RK-138	F P Lathrop Properties Inc/Shand Air, Portland, Or..	N40PL
☐ N48SD	83	Westwind-Two	399	Valy Aviation Corp. Wilmington, De.	N78WW
☐ N48SE	92	Beechjet 400A	RK-48	Investment Capital Group LLC. Sacramento, Ca.	N94HT
☐ N48SR	88	Beechjet 400	RJ-39	Susan S Root (Trustee), Daytona Beach, Fl.	N3239K
☐ N48TT	86	Citation III	650-0105	Texas Utilities Services Inc. Dallas, Tx.	N15TT
☐ N48WA	73	Learjet 25B	25B-136	Fleet Unlimited Inc/Lear Group Inc. Van Nuys, Ca.	N753CA
☐ N48WS	77	Sabre-60A	306-124	Whiteco Industries Inc. Merrillville, In.	N60RS
☐ N48Y	84	BAe 125/800A	8009	Davison Transport Inc. Ruston, La.	N45Y
☐ N49BE	78	Learjet 35A	35A-192	Mayo Aviation Inc/CWW Inc. Englewood, Co.	N49PE
☐ N49CJ	94	CitationJet	525-0049	Smithfield Foods Inc. Smithfield, Va.	(N37201)
☐ N49CK	65	Learjet 23	23-009	Kalitta Flying Service Inc. Morristown, Tn.	N13SN
☐ N49HS	80	Citation II	550-0158	Huntco Steel/HSI Aviation Inc. Springfield, Mo.	N2JW
☐ N49LD	92	Citation V	560-0175	McKee Foods Corp. Collegedale, Tn.	N1279Z
☐ N49MJ	95	Citation V Ultra	560-0306	Melvin L Joseph Construction Co. Georgetown, De.	N5231S
☐ N49MW	88	Astra-1125	019	Dassault Falcon Jet Corp. Teterboro, NJ.	N30AJ
☐ N49RF	94	Gulfstream 4SP	1246	U S Department of Commerce, NO&AA, MacDill AFB. Fl.	N407GA
☐ N49RJ	66	Sabre-40	282-69	Airstream Aviation Inc. Delray Beach, Fl.	N777VZ
☐ N49U	79	Citation II	550-0082	Helicopters Inc. Calokia, Il.	N21DA
☐ N49UR	87	Challenger 601-3A	5016	Home Depot Inc. Atlanta, Ga.	N604CC
☐ N49US	84	Falcon 200	494	Falcon Air LLC. Newport Beach, Ca.	LV-BAI
☐ N49WA	73	Learjet 25B	25B-142	Fleet Unlimited Inc/Wolfe Air, Van Nuys, Ca.	N70CE
☐ N49WL	81	Learjet 35A	35A-457	O & S LLC. Traverse City, Mi.	N113LB
☐ N50AF	82	Learjet 55	55-038	Polygon Air Corp. White Plains, NY.	N551HB
☐ N50AH	82	Falcon 50	111	Monument Aircraft Services Inc. Granite Bay, Ca.	VR-CDF
☐ N50AJ	90	Astra-1125SP	044	Alpine Cascade Corp. Goleta, Ca.	
☐ N50AK	78	Learjet 35A	35A-172	Anker Inc. Daytona Beach, Fl.	N32JA
☐ N50AM	72	Citation	500-0041	G B Boots Smith Corp. Laurel, Ms.	N50AS
☐ N50AZ	81	Citation II	550-0261	H P Z Investments Inc. Wilmington, De.	N261SS
☐ N50BH	77	Falcon 20F	365	Crystal Jet Aviation Inc. Scotia, NY.	N777TX
☐ N50BK	85	Citation S/II	S550-0031	Woodland Aircraft Leasing Inc. St Paul, Mn.	N54WJ
☐ N50CK	74	Learjet 25B	25B-157	American International Airways, Morristown, Tn.	N57CK
☐ N50CR	69	Sabre-50	287-1	Rockwell Collins Inc. Cedar Rapids, Ia.	N287NA
☐ N50CS	90	Falcon 50	207	Charles Schwab & Co Inc. San Francisco, Ca.	N55BP
☐ N50DG	68	Sabre-60	306-19	Sabre Investments Ltd. Chesterfield, Mo.	N8000U
☐ N50DS	85	Citation III	650-0078	First Southeast Aviation Corp. Milwaukee, Wi.	N650WJ
☐ N50EF	84	Diamond 1A	A081SA	Eastern Foods Inc/Robert Brooks, Atlanta, Ga.	N750TJ
☐ N50ET	97	CitationJet	525-0251	Cessna Aircraft Co. Wichita, Ks.	
☐ N50FD	82	Westwind-1124	381	Knight Financial Corp. Naples, Fl.	N381W
☐ N50FH	81	Falcon 50	62	First Hawaiian Bank, Honolulu, Hi.	N292BC
☐ N50FJ	97	Falcon 50EX	254	Gulf Island Leasing Inc. Redmond, Wa.	F-WWHB
☐ N50HC	90	Falcon 50	208	Group Holdings OR Inc. Salem, Or.	N50AE
☐ N50HE	80	Falcon 50	7	Timberland Aviation Inc. Stratham, NH.	N5DL
☐ N50HM	85	Falcon 50	153	Health Management Associates Inc. Naples, Fl.	N16CP
☐ N50HS	84	Westwind-Two	412	HealthSouth Aviation Inc. Birmingham, Al.	N50XX
☐ N50KD	85	Falcon 50	145	Oakmont Corp. TCW Group & Vincente Management Co. Bend, Or.	(TC-...)
☐ N50KH	93	Beechjet 400A	RK-59	Kayhall Corp. Los Angeles, Ca.	N80544
☐ N50M	81	Westwind-1124	327	First Bank System, Minneapolis, Mn.	4X-CRU
☐ N50MG	85	Falcon 200	507	OMG Jett Inc. Cleveland, Oh.	N79MB
☐ N50MJ	78	Learjet 35A	35A-164	K & K Jets LLC. Teterboro, NJ.	N248HM
Reg	Yr	Type	c/n	Owner/Operator	Prev Regn

Reg	Yr	Type	c/n	Owner/Operator	Prev Regn
☐ N50MM	73	Citation	500-0118	Jet Services of Iowa Inc. Peosta, Ia.	N972GW
☐ N50MW	85	Falcon 200	503	Millard Refrigerated Services Inc. Omaha, Ne.	N300HA
☐ N50NF	90	Citation II	550-0636	Ohio National Life Insurance Co. Cincinnati, Oh.	N4EW
☐ N50PH	87	Citation III	650-0148	Pacesetter Corp. Omaha, Ne.	N55SQ
☐ N50PL	81	Westwind-Two	338	Lahaye Lab Inc/Panda Leasing Co. Redmond, Wa.	N114WL
☐ N50RD	75	Citation	500-0260	Dodson Aviation Inc. Ottawa, Ks.	OK-FKA
☐ N50RW	74	Learjet 25B	25B-135	Randolph Williams/LRW Aircraft Sales Inc. McLean, Va.	(N1RW)
☐ N50SF	88	Falcon 50	180	Fremont Group/Bechtel Corp. Oakland, Ca.	N2254S
☐ N50SJ	81	Falcon 50	80	Ocean Air Charters, Carlsbad, Ca.	N80WE
☐ N50SL	79	Westwind-1124	269	O'Sullivan Corp. Winchester, Va.	N3031
☐ N50TC	82	Falcon 50	115	Dune Jet Services LP. NYC.	N777MJ
☐ N50TG	93	Astra-1125SP	065	Lake Creek Research LLC. Chicago, Il.	N75TT
☐ N50TY	76	Falcon 10	72	Falcon 10 LLC. Grand Haven, Mi.	N31SJ
☐ N50US	78	Citation 1/SP	501-0044	Bergeron Marine Service Inc. Metairie, La.	N131SY
☐ N51B	83	Diamond 1A	A063SA	NACCO Industries Inc. Mayfield Heights, Oh.	N54BE
☐ N51BP	75	Falcon 10	51	IBP=Iowa Beef Processors Inc. Sioux City, Ia.	N137FJ
☐ N51CD	96	CitationJet	525-0163	Carefree Capital Inc. Minneapolis, Mn.	
☐ N51EB	87	Beechjet 400	RJ-28	News Press & Gazette Co. St Joseph, Mo.	N700LP
☐ N51ET	78	Citation 1/SP	501-0075	Hammill Manufacturing Co. Toledo, Oh.	(N325PM)
☐ N51FT	80	Citation II	550-0121	Sierra Industries Inc. Uvalde, Tx.	N850BA
☐ N51LC	80	Learjet 35A	35A-302	LAICO Inc/Corporate Wings, Cleveland, Oh.	N631CW
☐ N51MN	76	Westwind-1124	198	Media News Group Inc. Denver, Co.	N98TS
☐ N51TV	83	Westwind-Two	402	Texas Television Inc. Corpus Christi, Tx.	N999LC
☐ N51V	85	Learjet 55	55-116	Tennessee Gas Pipeline Co. Houston, Tx.	N116DA
☐ N51WP	80	Citation 1/SP	501-0133	Weber Plywood & Lumber Co. Tustin, Ca.	(N955WP)
☐ N52AJ	73	Citation	500-0061	Sawyer Aviation Inc. Phoenix, Az.	N490EA
☐ N52AN	72	Citation	500-0030	Aviation Resources Inc. Portsmouth, NH. (status ?).	N530CC
☐ N52CK	86	Citation S/II	S550-0124	Koury Aviation Inc. Liberty, NC.	N1867W
☐ N52CT	87	Learjet 55B	55B-131	Coyne Textiles/Textron Financial Corp. Fort Worth, Tx.	N7260K
☐ N52DA	81	Learjet 25D	25D-327	Flyaway Inc. Odessa, Tx.	(N327BC)
☐ N52DC	82	Falcon 50	126	Dow Chemical Co. Freeland, Mi.	N931G
☐ N52DD	76	Learjet 24E	24E-339	Colvin Air Charter Inc. Athens, Ga.	N1TJ
☐ N52GA	88	Beechjet 400	RJ-36	Larned Investments Inc. Detroit, Mi.	VR-BNV
☐ N52JA	68	Learjet 25	25-007	Nevada Desert Holdings Inc. Gardnesville, Nv,.	(N58JA)
☐ N52JJ	90	Falcon 50	205	ADM Milling Co. Decatur, Il.	N59EL
☐ N52N	85	Falcon 100	197	Long Island Airlines LLC. E Hampton, NY.	N888G
☐ N52PK	93	CitationJet	525-0052	North Park Transportation Co. Billings, Mt.	N252CJ
☐ N52PM	74	Citation	500-0222	Multiple Systems Inc. Amarillo, Tx.	N636SC
☐ N52RF	78	Citation II	550-0021	U S Dept of Commerce, MacDill AFB. Fl.	N900LJ
☐ N52SD	73	Learjet 25B	25B-110	Briedan Aviation Inc. Chicago, Il.	N110HA
☐ N52SM	92	BAe 1000A	NA1009	Sierra Pacific Industries Inc. Redding, Ca.	N125CJ
☐ N52SN	93	Citation V	560-0207	SONAT Services Inc. Birmingham, Al.	N207CV
☐ N52TC	76	Citation	500-0324	Thomas F Harter, Naperville, Il.	N324C
☐ N52TJ	73	Falcon 10	3	Universal Pacific Investment, Bend, Or.	(N10PN)
☐ N52TL	77	Citation 1/SP	501-0053	Citizens Telephone Co. Brevard, NC.	N14EA
☐ N53CG	97	CitationJet	525-0233	Cessna Aircraft Co. Wichita, Ks.	N233CJ
☐ N53DF	90	Challenger 601-3A	5078	JABJ/Sierra Land Group Inc. Burbank, Ca.	N601MD
☐ N53EZ	79	Citation 1/SP	501-0119	Gerald Bruce, Roseburg, Or.	N77GJ
☐ N53FP	82	Citation II	550-0434	Fawn Enterprises Inc. Middlesex, NC.	(D-CVAU)
☐ N53GH	82	HS 125/700A	NA0314	G Howard Associates Inc. NYC.	(N106AE)
☐ N53HJ	82	Learjet 55	55-037	Hop A Jet Inc. Fort Lauderdale, Fl.	PT-OBR
☐ N53M	89	Gulfstream 4	1089	3M Co. Minneapolis, Mn.	N465GA
☐ N53MS	93	Beechjet 400 + co-owners, Warsaw, In.	RK-64	Image Air LLC	N8164M
☐ N53PJ	60	MS 760 Paris-1R	53	Airborne Turbine Inc. Ca.	No 53
☐ N53RG	77	Citation	500-0368	GWC LLC. Olathe, Ks.	N104AB
☐ N53WA	75	Falcon 10	53	Exact Science Inc. Woodside, Ca.	I-LCJG
☐ N53WW	83	Westwind-1124	393	Rocky Mountain Aviation Inc. Irvine, Ca.	4X-CUK
☐ N54	93	Learjet 60	60-009	Federal Aviation Administration, Oklahoma City, Ok.	N26029
☐ N54CC	80	Citation II	550-0083	V T Inc. Decatur, Il.	N98718
☐ N54CG	84	Citation 1/SP	501-0677	Craig Goess Inc. Greenville, NC.	C-GAAA
☐ N54CJ	94	CitationJet	525-0054	Kennedy Rice Dryers Inc. Angelfire, NM.	
☐ N54DA	89	Falcon 50	201	Adam Acquisition Corp. Bryan, Tx.	N41TH
☐ N54DC	87	Falcon 900	22	Dow Chemical Co. Freeland, Mi.	N416FJ
☐ N54FN	72	Learjet 25C	25C-083	Flight International Inc. Tampa, Fl.	N200MH
☐ N54HC	85	Citation III	650-0098	AMI Aviation II LLC. Arlington, Tx.	N389W
☐ N54HP	92	Beechjet 400A	RK-49	AMPCO Inc. Wilmington, De.	N8060Y
☐ N54J	76	Gulfstream 2	193	W W Grainger Inc. Wheeling, Il.	N26LT
☐ N54JA	78	Learjet 36A	36A-044	Quicksilver Aviation Inc. Philadelphia, Pa.	(N77JW)
☐ N54NW	82	Learjet 55	55-054	Northwestern Aircraft Capital Corp. Van Nuys, Ca.	HB-VHL
☐ N54RM	83	Diamond 1	A065SA	Carmar Group Inc. Carthage, Mo.	(N65EB)
☐ N54SB	94	Hawker 800	8258		G-BVJI
☐ N54TS	75	Citation 1/SP	501-0643	Peter L Sturdivant Inc. Las Vegas, Nv. (was 500-0293).	N54CM
☐ N54YR	85	Falcon 50	158	Phifer Wire Products Inc. Tuscaloosa, Al.	N142FJ
☐ N55	93	Learjet 60	60-013	FAA, Oklahoma City, Ok.	N26011
☐ N55AR	82	Challenger 600S	1044	AAR Corp. Elk Grove Village, Il.	N205MM
☐ N55AS	90	Falcon 50	214	Boise Cascade Corp/Albertsons Inc. Boise, Id.	N296FJ

| *Reg* | *Yr* | *Type* | *c/n* | *Owner/Operator* | *Prev Regn* |

Reg	Yr	Type	c/n	Owner/Operator	Prev Regn
☐ N55BH	84	Citation III	650-0041	International Industries Inc. Gilbert, WV.	
☐ N55BM	81	Citation 1/SP	501-0201	Carnegie Construction Co. Santa Ana, Ca.	N123SF
☐ N55CJ	74	Learjet 36	36-003	Premier Jets Inc. Portland, Or.	N36TA
☐ N55DG	93	CitationJet	525-0044	Griffin Ag Inc. Durham, NC.	XA-SKW
☐ N55F	78	Learjet 35A	35A-147	AirNet Systems Inc. Columbus, Oh.	N717W
☐ N55G	68	HS 125/731	NA709	Mayo Aviation Inc. Englewood, Co.	(N2G)
☐ N55GM	79	Learjet 55C	55C-139A	Scott Brothers Aviation Corp. Deerfield Beach. (was 55-002)	(N55ZT)
☐ N55GR	74	Citation	500-0217	K3C Inc. Reno, Nv.	N217S
☐ N55HF	95	Challenger 601-3R	5183	Hudson Foods Inc. Rogers, Ar.	N601HF
☐ N55HL	82	Learjet 55	55-046	San Francisco Aviation Co. Salinas, Ca.	N3HB
☐ N55KS	94	Learjet 60	60-029	Goodyear Tire & Rubber Co. Akron, Oh.	N4029P
☐ N55LC	95	Citation V Ultra	560-0324	Lowe's Companies Inc. North Wilkesboro, NC.	N5100J
☐ N55LJ	82	Learjet 55	55-030	Drei T's LLC. O'Fallon, Mo.	(N155CD)
☐ N55LK	85	Learjet 55	55-120	Aviation Enterprises Inc. Bedford, Tx.	(N486)
☐ N55LS	79	Citation II/SP	551-0021	Exec Three Inc. Cincinnati, Oh.	N551CF
☐ N55MT	78	HS 125/700B	7046	Metrotrans Corp. Griffin, Ga.	N746TS
☐ N55NM	83	Learjet 55	55-085	HM International Inc. Tulsa, Ok.	N58FM
☐ N55NT	82	Falcon 50	87	Northern Telecom Inc. Nashville, Tn.	N283K
☐ N55NY	82	Learjet 55	55-020	Air New York LLC. Fresh Meadows, NY.	N35PF
☐ N55PD	73	Learjet 25B	25B-105	Vailjet Inc. Vail, Co.	XA-SXD
☐ N55RF	65	HS 125/731	25020	National Aircraft Leasing LLC. Anchorage, Ak.	N125TJ
☐ N55RG	66	Gulfstream 2SP	1	R W Galvin/Motorola Inc. Wheeling, Il.	N801GA
☐ N55RT	84	Learjet 55	55-095	Townsend Engineering Co. Des Moines, Ia.	N8565Z
☐ N55SC	95	Citation VII	650-7060	Scharbauer Cattle Co. Midland, Tx.	N5218T
☐ N55SK	94	CitationJet	525-0063	Skyline Corp. Elkhart, In.	
☐ N55SL	77	Learjet 25XR	25XR-219	Sutherland Lumber Southwest Inc. Kansas City, Mo.	
☐ N55SN	88	Falcon 50	189	SONAT Services Inc. Birmingham, Al.	N50WG
☐ N55TH	65	Falcon 20C	17	Tollman Hotels/SD Travel Inc. Mahwah, NJ.	N5CE
☐ N55VC	86	Learjet 55B	55B-130	R T Vanderbilt Co. White Plains, NY.	
☐ N55WL	80	Citation II	550-0140	Trebuchet Transport LLC. Waukegan, Il.	N2646Z
☐ N55ZM	74	Sabre-60	306-84	Zimair Corp. Miami, Fl.	N383TS
☐ N56	94	Learjet 60	60-033	Federal Aviation Administration, Washington, DC.	N4031A
☐ N56AG	90	Astra-1125SP	043	Atlanta Gas Light Co. Atlanta, Ga.	
☐ N56BE	91	Beechjet 400A	RK-13	B-200 Inc. East Farmingdale, NY.	
☐ N56BP	79	Westwind-1124	268	Bradley Petroleum Inc. Denver, Co.	N41WH
☐ N56D	79	Gulfstream 2B	257	Port of Call Aviation Ltd. Minneapolis, Mn.	N911WW
☐ N56EM	77	Learjet 35A	35A-144	McMillin Aircraft Inc. Winter Park, Fl.	N56HF
☐ N56JA	80	Learjet 35A	35A-342	LaDane Aircraft Sales & Leases Inc. Rougemont, NC.	YV-15CP
☐ N56K	93	CitationJet	525-0005	Kempthorn Inc. Canton, Oh.	(N1326G)
☐ N56L	93	Gulfstream 4SP	1213	Newsflight Inc. El Segundo, Ca.	N416GA
☐ N56LF	92	Learjet 31A	31A-056	Ooltewah Aviation Inc. Chattanooga, Tn.	N303WB
☐ N56LP	80	Falcon 10	165	Pulaski Bank & Trust, Little Rock, Ar.	N111WW
☐ N56LW	81	Citation 1/SP	501-0314	Larry Phillips, Fruitland Park, Fl.	N56MC
☐ N56MK	77	Citation 1/SP	501-0023	M L Kuhn Enterprises Inc. Whitesboro, Tx.	(N501FB)
☐ N56MM	76	Learjet 24F	24F-332	Northeastern Aviation Corp. Wilmington, De.	N13KL
☐ N56PA	77	Learjet 36A	36A-023	Phoenix Air Group Inc. Cartersville, Ga.	N6YY
☐ N56RD	74	Learjet 24XR	24XR-286	RDC Marine Inc. Houston, Tx.	N86GC
☐ N56RN	76	Sabre-60A	306-122	Reserve National Insurance Co. Oklahoma City, Ok.	N168H
☐ N56SN	91	Falcon 50	216	SONAT Services Inc. Birmingham, Al.	N180AR
☐ N56WD	97	Learjet 45	45-019	Global Industries Inc. Lafayette, La.	
☐ N56WE	78	Citation 1/SP	501-0056	Wings Aviation International Inc. Franklin Lakes, NJ.	CC-CTE
☐ N56WJ	75	Falcon 10	56	Astec Industries Inc. Chattanooga, Tn.	N16DD
☐ N57	94	Learjet 60	60-039	Federal Aviation Administration, Washington, DC.	N5003X
☐ N57B	92	Beechjet 400A	RK-36	Bruno's Inc. Birmingham, Al.	N56327
☐ N57BJ	85	Citation S/II	S550-0052	Charter Flight Inc. Charlotte, NC.	N27GD
☐ N57CE	90	Citation V	560-0048	Chapman Exploration Inc. Tulsa, Ok.	N74TL
☐ N57DC	82	Falcon 50	119	Dow Chemical Corp. Freeland, Mi.	N83FC
☐ N57EL	95	Falcon 900B	153	Enterprise Rent-A-Car Co. St Louis, Mo.	N153FJ
☐ N57FL	72	Learjet 24XR	24XR-243	Flight Operations Leasing Inc. Westlake, Oh.	N56WS
☐ N57HC	96	CitationJet	525-0157	Bridgeport Associates Inc. Brookline, Ma.	N1115V
☐ N57HJ	76	Gulfstream 2	194	Flying Squirrel/GFS Manassas Air Services Inc. Arlington, Va	N194WA
☐ N57LL	72	Citation	500-0025	McMullen Aircraft Inc. Erie, Pa.	N220W
☐ N57MB	95	Citation V Ultra	560-0286	Mathis Brothers Furniture Co. Oklahoma City, Ok.	N286CV
☐ N57MC	81	Citation 1/SP	501-0229	Massman Construction Co. Kansas City, Mo.	N636CC
☐ N57MH	85	Learjet 55	55-113	Marriott International Inc. Washington, DC.	N236HR
☐ N57MK	91	Citation V	560-0145	Klein Tools Inc. Chicago, Il.	(N6877L)
☐ N57MQ	79	Sabre-65	465-11	Coastal Corp. Houston, Tx.	N5739
☐ N57NP	68	JetStar-731	5123	Nicholas Price, Jupiter, Fl.	N123GN
☐ N57PM	92	BAe 125/800A	NA0471	Philip Morris Inc. Milwaukee, Wi.	N673BA
☐ N57ST	96	Citation V Ultra	560-0383	Beehawk Aviation Inc. Smyrna, Ga.	N51038
☐ N57TT	84	Citation III	650-0046	Thompson Tractor Co. Birmingham, Al.	N658CC
☐ N58	95	Learjet 60	60-057	FAA, Washington, DC.	N50050
☐ N58AU	88	Beechjet 400	RJ-45	Upchurch Associates/Highland Aviation LLC. Birmingham, Al.	N3145F
☐ N58BT	73	Citation	500-0100	Robert Lee Turley, Duluth, Ga.	N80AJ
☐ N58CG	86	Learjet 55	55-124	Corning Inc. Corning, NY.	
☐ N58EM	76	Learjet 35	35-046	B L Bennett Aviation Inc. Nashville, Tn.	VH-LJL

Reg	*Yr* *Type*		*c/n*	*Owner/Operator*	*Prev Regn*

Reg	Yr	Type	c/n	Owner/Operator	Prev Regn
☐ N58FN	69	Learjet 24B	24B-184	Flight International/Maritime Sales & Leasing, Hockessin, De	(N58FN)
☐ N58GL	85	Learjet 35A	35A-599	Global Aviation Inc. Portland, Or.	N40144
☐ N58HC	81	Learjet 25D	25D-341	Bankair Inc. West Columbia, SC.	XA-SAE
☐ N58HT	81	Sabre-65	465-70	Air Melinda LLC. Concord, NH.	N58CM
☐ N58JF	69	Gulfstream 2	65	Sterling Jet Inc & Richmor Aviation, Hudson, NY.	(N300FN)
☐ N58LC	92	Citation II	550-0711	Lusardi Construction Co. San Marcos, Ca.	N711CN
☐ N58MM	79	Learjet 35A	35A-261	Aspen Base Operation Inc. Aspen, Co.	N63DH
☐ N58PM	92	BAe 125/800A	NA0472	Philip Morris Inc. Milwaukee, Wi.	N674BA
☐ N58T	77	Citation 1/SP	501-0262	Aircraft Operators Inc. Lafayette, La.	N794EZ
☐ N58TC	75	Citation	500-0261	Hawk Equipment LLC. Bloomington, Mn.	N711SF
☐ N58TS	66	JetStar-731	5079	Camel Aviation Inc. Wilmington, De.	XA-MAZ
☐ N59AP	83	Westwind-1124	398	All-Phase Electric Supply Co. Benton Harbor, Mi.	N98WW
☐ N59CC	73	Falcon 10	6	Cavenaugh Charters Inc. Dallas, Tx.	N32VC
☐ N59CF	90	Falcon 900	98	JABJ/Compass Foods Inc. Montvale, NJ.	(N903FJ)
☐ N59DF	91	Citation V	560-0098	Sierra Land Group Inc. Burbank, Ca.	(N18SK)
☐ N59EC	85	Citation S/II	S550-0034	Imperial Eagle Corp. Worthington, Oh.	N34CJ
☐ N59FD	96	Learjet 60	60-084	Flour Daniel Inc. Greenville, SC.	N684LJ
☐ N59FL	74	Learjet 25B	25B-169	Flight Operations Leasing Inc. Westlake, Oh.	N983WA
☐ N59GB	82	Citation II/SP	551-0060	Watts Brothers Farms, Patterson, Wa.	(N60HW)
☐ N59GU	81	Citation II	550-0315	Bison Air Corp. Hope Hull, Al.	N618DB
☐ N59JR	86	Gulfstream 4	1007	Fightertown Inc. Miami, Fl.	N100GJ
☐ N59K	74	Sabre-60	306-82	Marotta Scientific Controls & co-owners, Montville,, NJ	N60SL
☐ N59KG	96	Citation V Ultra	560-0363	Krause Gentle Corp. West Des Moines, Ia.	N5180K
☐ N59MA	77	Citation 1/SP	501-0050	Edco Disposal Corp. Lemon Grove, Ca.	N750LA
☐ N59PM	87	Falcon 50	178	Forethought Life Insurance Co. Batesville, In.	N239R
☐ N59TJ	74	Falcon 10	14	Tyler Jet LLC. Tyler, Tx.	(N50B)
☐ N60AE	95	Citation V Ultra	560-0343	Atwood Enterprises Inc. Charleston, SC.	N343CV
☐ N60AJ	94	Astra-1125SP	071	Durandal Inc. Chicago, Il.	4X-CUW
☐ N60AV	79	Westwind-1124	254	Avjet Corp. Burbank, Ca.	N72HB
☐ N60CC	78	Citation II	550-0034	Tafco Refrigeration Inc. Birmingham, Al.	N697A
☐ N60CD	65	Jet Commander	44	Command Air/Chet Duncan, Lawndale, Ca.	N69GT
☐ N60CE	95	Learjet 60	60-069	Columbia Gas System Service Corp. Washington, DC.	N50234
☐ N60CN	81	Falcon 50	79	Champion International Corp. Stamford, Ct.	N86FJ
☐ N60CT	96	Challenger 604	5325	USI American Holdings Inc. Iselin, NJ.	C-GLXW
☐ N60DK	78	Learjet 25D	25D-245	U S Motorsports Inc. Statesville, NC.	N245DK
☐ N60EF	84	Diamond 1A	A070SA	Hooters of America Inc. Atlanta, Ga.	N84GA
☐ N60ES	94	CitationJet	525-0053	Starjet Air Inc. Park City, Ut.	N66ES
☐ N60EW	82	Citation 1/SP	501-0319	Jet Star Inc/IPM E F Weisert GmbH. Nuremberg, Germany.	N124KC
☐ N60FM	67	B 727-27	19535	Airfreight Services Inc. Dover, De.	N7294
☐ N60GL	92	Citation V	560-0187	Mazda Distributor/Transnational Motors, Grand Rapids, Mi.	N12812
☐ N60GT	58	MS 760 Paris-1A	8	David Bennett, Colorado Springs, Co.	G-APRU
☐ N60HJ	82	Challenger 600S	1058	Hop-A-Jet Inc. Fort Lauderdale, Fl.	N4000X
☐ N60JC	70	Sabre-60	306-51	Commercial Aviation Enterprises Inc. Delray Beach, Fl.	N141JA
☐ N60JM	77	JetStar 2	5213	Jet Management Group Inc. Syracuse, NY.	N501J
☐ N60LJ	97	Learjet 60	60-110	Learjet Inc. Wichita, Ks.	
☐ N60LR	97	Learjet 60	60-125	Learjet Inc. Wichita, Ks.	
☐ N60PL	95	Citation VII	650-7056	Omicron Transportation Inc. Reading, Pa.	N6781C
☐ N60PM	90	BAe 125/800A	NA0453	Philip Morris Inc. Richmond, Va.	N615BA
☐ N60PT	94	Gulfstream 4SP	1251	Penske Jet Inc. Detroit, Mi.	(N321PT)
☐ N60RD	93	Citation V	560-0244	JABJ/R R Donnelley & Sons Co. Chicago, Il.	N244CV
☐ N60RV	79	Westwind-1124	250	Winnebago RV Inc. Forest City, Ia.	N29995
☐ N60S	90	Citation V	560-0066	Richard Schaden PC. Birmingham, Mi.	
☐ N60SB	93	Learjet 60	60-023	TXI-60SB Training LLC/Richards Group Inc. Dallas, Tx.	N40323
☐ N60T	93	Learjet 60	60-011	Jefferson Smurfit Corp. St Louis, Mo.	
☐ N60TC	89	BAe 125/800A	NA0448	Tekloc Enterprises/Tristram C Colket Jr. Paoli, Pa.	N125BA
☐ N60UK	92	Learjet 60	60-004	LightAir Ltd. Twinsburg, Oh.	
☐ N60VE	92	Learjet 60	60-006	Valero Management Co. San Antonio, Tx.	
☐ N60WL	81	Learjet 35A	35A-382	Executive Services Inc. Winfield, Ks.	OE-GAF
☐ N60WM	94	Learjet 60	60-045	WMX Technologies Inc. Chicago, Il.	N5045S
☐ N61BP	77	Falcon 10	102	IBP=Iowa Beef Processors Inc. Sioux City, Ia.	N178FJ
☐ N61CF	81	Citation II	550-0312	Arrowcrest Group P/L. Adelaide, SA. Australia.	(VH-ARZ)
☐ N61CK	81	Citation II	550-0195	Causey Aviation Service & Koury Aviation Inc. Liberty, NC.	N60JD
☐ N61DF	81	Sabre-65	465-59	Cintas Corp. Cincinnati, Oh.	N59SR
☐ N61DT	80	Citation 1/SP	501-0188	Richard Thurman, Louisville, Ky.	(N351DT)
☐ N61EW	74	Learjet 25B	25B-161	Corporate Aviation Services Inc. Tulsa, Ok.	N4VC
☐ N61FB	74	Sabre-60	306-80	Fine Airlines Inc. Miami, Fl.	PT-KOT
☐ N61HA	95	Citation V Ultra	560-0295	First Union Corp. Charlotte, NC.	N295CV
☐ N61LH	69	Gulfstream 2	61	Limerick Holdings LLC. Denver, Co.	N57BG
☐ N61MA	80	Citation II	550-0176	Magnolia Aviation LLC. Mobile, Al.	(N24TR)
☐ N61RH	65	Sabre-40	282-27	Apex Aviation Co. Wilmington, De.	N111EA
☐ N61RS	83	Westwind-Two	384	Stark Aviation Ltd. Fort Dodge, Ia.	N48WW
☐ N61SH	95	CitationJet	525-0095	Sunseeker Air Inc. Wilmington, De.	(N5153K)
☐ N61SM	92	Beechjet 400A	RK-60	Superior Metal Products Inc. Lima, Oh.	N8260L
☐ N61TJ	75	Falcon 10	41	Tyler Jet LLC. Tyler, Tx.	F-GKLV
☐ N61TW	89	Citation V	560-0019	Tom Wood Inc. Indianapolis, In.	OE-GRW
☐ N61WH	69	Gulfstream 2B	48	Huizenga Holdings Inc. Fort Lauderdale, Fl.	N711MC

Reg	Yr	Type	c/n	Owner/Operator	Prev Regn
☐ N62BR	73	Citation	500-0093	Avjet Corp. Burbank, Ca.	G-OCPI
☐ N62CH	84	Diamond 1A	A082SA	C & H Travel LLC. Little Rock, Ar.	N555FA
☐ N62DM	69	Learjet 24B	24B-194	Dorr Aviation Inc. Marlboro, Ma.	(N62FN)
☐ N62FJ	88	Falcon 900	62	Case Corp. Racine, Wi.	F-GIVR
☐ N62HA	92	Citation V	560-0189	First Union Corp. Charlotte, NC.	N189CV
☐ N62MB	80	Learjet 35A	35A-282	West Scenic Management Co. West Bend, Wi.	N444CM
☐ N62MS	94	Gulfstream 4SP	1248	Melvin Simon & Associates Inc. Indianapolis, In.	N422GA
☐ N62ND	81	Westwind-1124	379	Goodtimes Aircraft Management Services Inc. NYC.	N52FC
☐ N62PG	77	Learjet 36A	36A-031	Phoenix Air Group Inc. Cartersville, Ga.	N20UG
☐ N62RG	81	Citation 1/SP	501-0199	R & G Aviation Inc. Mobile, Al.	N441JT
☐ N62TC	93	BAe 125/800A	8239	Thiokol Corp. Ogden, Ut.	N904H
☐ N62WA	96	Citation V Ultra	560-0360	Muscatine Corp. Muscatine, Ia.	N6780A
☐ N62WH	66	BAC 1-11/401AK	078	Palm Beach Aerospace Group Inc. West Palm Beach, Fl.	(N167X)
☐ N63BL	95	Learjet 60	60-051	Electronic Data Systems Corp. Dallas, Tx.	N5051X
☐ N63CG	79	Citation 1/SP	501-0135	Robert Rounsavall III, Prospect, NY.	N49MP
☐ N63CK	67	Learjet 24	24-119	American International Airways, Morristown, Tn. (status ?).	N61CK
☐ N63DR	83	Diamond 1A	A067SA	Renshaw & Renshaw Management Group, Madisonville, Ky.	N65SA
☐ N63GA	71	Learjet 24D	24D-201	Starfire Express Inc/General Automation, Skokie, Il.	N363BC
☐ N63HA	92	Citation V	560-0199	First Union Corp. Charlotte, NC.	N1283K
☐ N63HJ	81	Challenger 600S	1021	Hop-A-Jet Inc. Fort Lauderdale, Fl.	N914XA
☐ N63HS	94	Gulfstream 4SP	1249	Melvin Simon & Associates Inc. Indianapolis, In.	N423GA
☐ N63JG	85	Citation S/II	S550-0036	Jay Gee Holdings Inc. Hazelton, Pa.	N27B
☐ N63LB	96	CitationJet	525-0127	Lee Beverage Co. Oshkosh, Wi.	N127CJ
☐ N63LF	79	Learjet 35A	35A-250	Guild Investments LLC. Chicago, Il.	N63LE
☐ N63M	90	Gulfstream 4	1152	3M Co. St Paul, Mn.	N446GA
☐ N63PM	90	BAe 125/800A	NA0451	Professional Funeral Associate, Lufkin, Tx.	N90PM
☐ N63TM	83	Citation II	550-0457	F T Stent, Stone Mountain, Ga.	N457CF
☐ N63TS	75	Falcon 10	66	First Security Bank NA. Salt Lake City, Ut.	YV-70CP
☐ N64AH	67	Jet Commander	94	EM Travels & Sales, Miami, Fl.	N94WA
☐ N64BH	83	Citation 1/SP	501-0325	Bob Howard Automall, Oklahoma City, Ok.	N501LM
☐ N64CE	69	Learjet 24B	24B-205	Royal Air Freight Inc. Waterford, Mi.	(N721J)
☐ N64CF	82	Learjet 35A	35A-461	CF Industries Inc. Chicago, Il.	
☐ N64CP	79	Learjet 35A	35A-264	AirNet Systems Inc. Columbus, Oh.	VR-CDI
☐ N64DH	66	Sabre-40R	282-52	Frank Haws, Huntsville, Al.	N282MC
☐ N64FG	78	Westwind-1124	227	AFG Industries Inc. Blountville, Tn.	N250PM
☐ N64HA	91	Citation V	560-0127	First Union Corp. Charlotte, NC.	N6809T
☐ N64HB	67	Learjet 24	24-149	Parmley Aviation Services Inc. Council Bluffs, Ia.	N300LB
☐ N64LF	97	CitationJet	525-0218	Lamplight Farms Inc. Menomonee Falls, Wi.	
☐ N64MA	65	Sabre-40	282-44	Las Vegas Asset Management Trust, Fort Lauderdale, Fl.	N600JS
☐ N64MP	82	Learjet 35A	35A-490	L J Ventures, Chicago, Il.	
☐ N64RT	81	Citation 1/SP	501-0191	Rubaiyat Trading Co. Birmingham, Al.	N98ME
☐ N64TF	79	Citation II	550-0064	KS LLC. Indianapolis, In.	(N336CP)
☐ N64VM	85	Beechjet 400	RJ-1	Verco Manufacturing/Nektor Industries Inc. Phoenix, Az.	
☐ N64YP	85	Falcon 200	502	TEC Air Inc. Princeton, NJ.	N232F
☐ N65A	73	Learjet 25B	25B-134	World Changers Church International, College Park, Ga.	N26FN
☐ N65AF	81	Sabre-65	465-62	AFLAC Inc. Columbus, Oh.	N56NW
☐ N65AK	80	Sabre-65	465-35	Universal Forest Products Inc. Grand Rapids, Mi.	N2590E
☐ N65B	79	Falcon 50	10	Borden Inc. Columbus, Oh.	N50FG
☐ N65BL	95	Learjet 60	60-054	Electronic Data Systems Corp. Farnborough, UK.	
☐ N65DL	72	HS 125/731	NA780	Chestnut Ridge Air Ltd. Wilmington, De.	N72HC
☐ N65FF	80	Sabre-65	465-46	Frederick P Furth, Healdsburg, Ca.	N79CD
☐ N65HA	91	Citation V	560-0143	First Union Corp. Charlotte, NC.	N6877C
☐ N65HJ	82	Challenger 600S	1038	Hop-A-Jet Inc. Fort Lauderdale, Fl.	N1045X
☐ N65KB	90	Citation III	650-0199	B & D Aviation Inc. San Jose, Ca.	(N2626X)
☐ N65L	81	Sabre-65	465-76	Acopian Technical Co. Easton, Pa.	
☐ N65R	75	Sabre-65	306-114	Winn Dixie Stores Inc. Jacksonville, Fl.	N60TF
☐ N65RA	85	Beechjet 400	RJ-9	Polycom Huntsman Inc. Washington, Pa.	N800FT
☐ N65SA	73	Citation	500-0114	Adkison Interests, Henderson, Tx.	I-AMCT
☐ N65SD	96	Falcon 2000	32	Square D Co. Waukegan, Il.	F-WWMD
☐ N65SR	81	Sabre-65	465-54	Arrow Molded Plastics Inc. Circleville, Oh.	N1909R
☐ N65ST	67	Gulfstream 2	5	Badger Air Inc/Charles Schwab & Co. Van Nuys, Ca.	N100PJ
☐ N65TC	80	Sabre-65	465-30	Thornton Corp. Van Nuys, Ca.	N89MM
☐ N65TD	97	Astra-1125SPX	093	Helios Ltd/Vitol BV. Luton, UK.	
☐ N65TJ	80	Sabre-65	465-45	R J Aircraft II Inc. Minneapolis, Mn.	N448WT
☐ N65TS	65	HS 125/1A-522	25043	Nevada Desert Holdings Inc. Minden, Nv.	N522HA
☐ N65WH	72	Learjet 25B	25B-086	Dolphin Aviation Inc. Sarasota, Fl.	N23DB
☐ N65WL	86	Citation III	650-0122	Lane Industries Inc. Northbrook, Il.	(N650MT)
☐ N66AF	67	DC 9-15RC	47152	U S Department of Energy, Albuquerque, NM.	N65AF
☐ N66AG	79	Citation 1/SP	501-0136	T W Inc. Ashland, Ky.	N66AT
☐ N66AM	96	CitationJet	525-0160	GECC, Charlotte, NC.	
☐ N66CC	73	Citation	500-0066	Aduddell Roofing & Sheet Metal Inc. Moore, Ok.	N566CC
☐ N66CF	75	Falcon 10	65	Nashville Air Associates Inc. Galletin, Tn.	(F-GJMA)
☐ N66DD	86	Gulfstream 3	483	Duchossois Enterprises/D-Aire Inc. Advance, NC.	N309GA
☐ N66EH	88	Citation S/II	S550-0158	SIT-JET Inc. Wilmington, De.	N158QS
☐ N66ES	97	CitationJet	525-0244	Starjet Air Inc. Park City, Ut.	
☐ N66GE	75	Sabre-60	306-99	Grey Eagle Distributors Inc. Chesterfield, Mo.	N16PN
Reg	*Yr*	*Type*	*c/n*	*Owner/Operator*	*Prev Regn*

Reg	Yr	Type	c/n	Owner/Operator	Prev Regn
□ N66KG	97	Astra-1125SPX	096	Galaxy Aerospace Corp. Fort Worth, Tx.	
□ N66LE	74	Citation Eagle	500-0170	NU-REX Inc. Hilton Head Island, SC.	N818R
□ N66LX	83	Westwind-Two	375	NJE Aircraft Corp & WWI Inc. Bethesda, Md.	N79AP
□ N66MC	81	Citation II	550-0239	TCB Inc. Eden Prairie, Mn.	N4720T
□ N66MS	81	Citation II/SP	551-0053	Farmers Equipment Rental Inc. Fargo, ND.	(N6890C)
□ N66NJ	80	Learjet 35A	35A-296	National Jets Inc. Fort Lauderdale, Fl.	N51JA
□ N66WM	90	Learjet 55C	55C-145	WMX Technologies Inc. Chicago, Il.	
□ N67BF	74	Citation	500-0184	McDonald Oil Co. Lagrange, Ga.	N67BE
□ N67CK	67	Learjet 24	24-147	Kalitta Flying Service Inc. Morristown, Tn.	N825AA
□ N67GH	96	CitationJet	525-0149	Hughes & Hughes Investment Corp. West Bountiful, Ut.	
□ N67GW	95	Citation V Ultra	560-0355	Gary Williams Energy Corp. Denver, Co.	N355CV
□ N67HB	75	Learjet 25B	25B-189	Water Soft Inc. Saxonburg, Pa.	N888DF
□ N67JR	81	Gulfstream 3	324	Coast Video Duplicating Inc/Mediacopy, San Leandro, Ca.	N44200
□ N67PA	78	Learjet 35A	35A-208	Dickerson Associates, West Columbia, SC.	(N39DJ)
□ N67PC	92	Citation II	550-0696	Prent Corp. Janesville, Wi.	N29PF
□ N67PR	69	Gulfstream 2	67	Intersouth Inc. Daytona Beach, Fl.	N400JD
□ N67PW	81	HS 125/700A	NA0303	Bosques Corp. NYC.	N70PN
□ N67SB	91	Learjet 31A	31A-045	Avionics International Inc/Bin Laden Aviation, Jeddah.	
□ N67SF	84	Citation III	650-0045	Canandaigua Wine Co & Seneca Foods Corp. Penn Yan, NY.	N84G
□ N67TS	66	HS 125/1A-522	25097	Air Diversified Inc. Miami, Fl.	N89HB
□ N67VW	97	CitationJet	525-0212	West Quality Food Service Inc. Laurel, Ms.	
□ N67WB	97	Falcon 900EX	16	BDA/US Services Ltd & WRBC Transportation Inc. Greenwich, Ct	N916EX
□ N68BC	68	Falcon 20C	155	Beef Products Inc & BPI Technology Inc. Dakota Dunes, SD.	N404R
□ N68BK	81	Citation II/SP	551-0369	Beckley Flying Service Inc. Mount Hope, NY.	N998GP
□ N68CB	71	HS 125/731	NA765	Cracker Barrel Old Country Store Inc. Lebanon, Tn.	N61MS
□ N68CJ	96	CitationJet	525-0169	BLC Corp. Foster City, Ca.	
□ N68DM	68	Gulfstream 2	28	Morningstar Group Inc. Dallas, Tx.	N85EQ
□ N68EA	78	Citation 1/SP	501-0068	Snap-Tite Inc. Erie, Pa.	N50GT
□ N68GA	75	HS 125/600A	6047	BrenAir Aviation Inc & Shasta Minerals Inc. Dover, De.	XA-RYK
□ N68GT	88	Falcon 100	217	Walnut Hill Cellular Telephone Co. Wilmington, De.	F-GIFL
□ N68HC	94	Citation V Ultra	560-0270	HCA-Hospital Corp of America, Nashville, Tn.	
□ N68HR	86	BAe 125/800A	8068	Sky King H & Sky King R Inc. NYC.	N68GP
□ N68LP	97	Citation X	750-0040	Cessna Aircraft Co. Wichita, Ka.	
□ N68LU	68	Learjet 24	24-163	Lewis University, Romeoville, Il. (wfu ?).	N65WM
□ N68MA	91	Citation V	560-0159	Pacelli Detoiligungs GmbH. Munich, Germany.	
□ N68MJ	86	Learjet 35A	35A-607	Jervis B Webb Co. Waterford, Mi.	PT-LIJ
□ N68SD	82	Challenger 600	1062	Consolidated Aircraft Leasing Inc. Caracas, Venezuela.	N62BL
□ N68SK	92	Citation VII	650-7016	Safety-Kleen Aviation Inc. Elgin, Il.	N18SK
□ N68TS	67	Falcon 20C	129	New Valley Corp. Miami, Fl.	N666DA
□ N68WW	83	Westwind-1124	386	Remuda Resources Inc/Dan J Harrison III, Catarina, Tx.	(VH-JPL)
□ N69BH	79	Learjet 35A	35A-276	FlightSolutions Inc. Gatlin, Tn.	N613RR
□ N69EC	85	Falcon 200	498	Societe Generale Financial/Ensearch Corp. Dallas, Tx.	N422L
□ N69GF	78	Learjet 25D	25D-265	Chaparral Leasing Inc. Lubbock, Tx.	N31WT
□ N69LD	85	Citation III	650-0080	McKee Baking Co. Collegedale, Tn.	N1316E
□ N69MT	67	JetStar-8	5107	OK Aircraft Parts Inc. Hollister, Ca. (status ?).	N7788
□ N69WJ	75	Falcon 10	60	Producers Pipeline Corp/PPC Flight Inc. Fort Worth, Tx.	SE-DKD
□ N69X	60	MS 760 Paris	90	RPJ Energy Fund Management Inc. Burnsville, Mn.	(N5TA)
□ N70AG	83	Gulfstream 3	376	A G Spanos Co/San Diego Chargers Football Co. Stockton, Ca.	A6-HHS
□ N70AP	72	HS 125/400A	25271	Global Aviation Inc. Wilmington, De.	N400DP
□ N70BJ	92	Beechjet 400A	RK-39	Stroehmann Bakeries Inc. Coopersburg, Pa.	N34VP
□ N70CA	75	Citation	500-0234	ABI LLC. Laguna Hills, Ca.	PH-CTG
□ N70CK	68	Falcon 20C	128	Kalitta Flying Service, Morristown, Tn.	N228CK
□ N70EW	87	Falcon 900	25	East West Air Inc. Teterboro, NJ.	N418FJ
□ N70HF	80	HS 125/700B	7082	BRBF Inc/Taughannock Aviation Corp. Ithaca, NY.	XA-TCB
□ N70HJ	70	Learjet 25	25-049	Winfair Aviation Ltd/Hop A Jet Inc. Fort Lauderdale, Fl.	N900Q
□ N70HL	76	Sabre-60	306-102	LaValle Developers Inc. Youngstown, Oh.	N265TJ
□ N70HS	95	Falcon 900B	140	First Security Bank, Salt Lake City, Ut.	VR-CES
□ N70JF	79	Learjet 25D	25D-278	Career Aviation Academy Inc. Oakdale, Ca.	
□ N70KS	95	Falcon 2000	14	Eaglestone Transportation Services, Billings, Mt.	(PT-...)
□ N70KW	94	CitationJet	525-0050	Wilair Inc. Raleigh, NC.	(N4614N)
□ N70LF	96	Falcon 900EX	9	G C I/Gary Comer Inc. Waukesha, Wi.	N909FJ
□ N70MG	73	Citation	500-0063	AGG Aircraft Sales & Leasing Inc. Delray Beach, Fl.	OO-RST
□ N70NE	87	BAe 125/800A	NA0407	Banc One Corp. Columbus, Oh.	N552BA
□ N70PC	90	Citation II	550-0664	Southern Company Services, Birmingham, Al.	XA-RYR
□ N70PL	70	Falcon 20F	247	Pelican Leasing Inc. Little Rock, Ar.	N95JR
□ N70PM	93	BAe 125/800A	8238	Philip Morris Management Corp. White Plains, NY.	N163BA
□ N70TH	92	Falcon 900B	117	Sony Aviation Inc. Teterboro, NJ.	N476FJ
□ N70TJ	69	Learjet 24B	24B-199	Yelvington Transport Inc. Daytona Beach, Fl.	N444HC
□ N70TT	84	Citation III	650-0029	JMK Holdings Inc/CJE Inc. Fort Worth, Tx.	N600GH
□ N70U	78	Falcon 20-5B	399	Anheuser-Busch Companies Inc & Earthgrains Co. St Louis, Mo.	N70NF
□ N70VT	84	Diamond 1A	A085SA	Bec-Faye LLC. Greenville, NC.	(N911JJ)
□ N70WA	76	Citation	500-0320	Wrangler Aviation Inc. Norman, Ok.	I-NORT
□ N70X	78	Citation II	550-0008	Spectrum Air Services Inc. Atlanta, Ga.	N550JF
□ N70XX	83	Diamond 1	A052SA	Lovair Inc. Dover, De.	I-FRAB
□ N71CC	73	Sabre-60A	306-71	Sampey Scientific Ltd. Uniontown, Pa.	N1028Y
□ N71CK	77	Learjet 36A	36A-035	American International Airways Inc. Morristown, Tn.	VH-BIB

Reg *Yr* *Type* *c/n* *Owner/Operator* *Prev Regn*

Reg	Yr	Type	c/n	Owner/Operator	Prev Regn
☐ N71E	80	Learjet 35A	35A-306	Apex Aviation Inc. Haslet, Tx.	N1110S
☐ N71GH	83	Diamond 1A	A071SA	Excel Air Inc. Jacksonville, Fl.	N70GA
☐ N71GW	94	CitationJet	525-0059	Whitworth Air Inc. Medford, Or.	N923GW
☐ N71HB	75	Citation	500-0275	Rioja Holding Co. Boca Raton, Fl.	N102HF
☐ N71HS	80	Learjet 35A	35A-287	Hi-Stat Manufacturing Co. Mansfield, Oh.	N17EM
☐ N71L	82	Citation 1/SP	501-0242	Lindair Inc. Sarasota, Fl.	N500BK
☐ N71LP	81	Citation II/SP	551-0039	Landis Plastics Inc. Chicago Ridge, Il.	EC-DOH
☐ N71M	86	Falcon 100	208	BAC Inc. Jackson, Wy.	I-OANN
☐ N71MA	80	HS 125/700B	7107	HWP Aviation LLC. Dallas, Tx.	VR-CVD
☐ N71MT	92	BAe 125/800A	8230	Michelin North America Inc. Greenville, SC.	N678BA
☐ N71NK	89	Citation V	560-0040	Flint Ink Corp. Detroit, Mi.	
☐ N71NP	85	BAe 125/800A	8041	Nationwide Mutual Insurance Co. Columbus, Oh.	N819AA
☐ N71PG	75	Learjet 36	36-013	Phoenix Air Group, Cartersville, Ga.	D-CBRD
☐ N71RP	93	Gulfstream 4SP	1222	Time Warner Inc. Burbank, Ca.	N452GA
☐ N71WF	87	Westwind-Two	442	Mulberry Street Investment Co/LTMC Inc. Macon, Ga.	N830
☐ N72AM	84	Citation S/II	S550-0004	Avondale Mills Inc. Monroe, Ga.	N554CA
☐ N72CK	78	Learjet 35A	35A-165	Kalitta Flying Service Inc. Morristown, Tn.	N16BJ
☐ N72CT	90	Citation V	560-0072	Business Investment Properties, Medford, Or.	N572CV
☐ N72DA	77	Learjet 35A	35A-098	Duncan Aviation Inc. Lincoln, Ne.	(N998DJ)
☐ N72FC	95	Citation V Ultra	560-0347	Best Aviation Inc. Baldwin, Ga.	
☐ N72FL	81	Citation II	550-0193	Aquila LLC. Bloomington, In.	N492ST
☐ N72JF	77	Learjet 35A	35A-088	AirNet Systems Inc. Columbus, Oh.	OE-GBR
☐ N72LE	84	Citation III	650-0063	Leeson Electric Corp. Grafton, Wi.	N14ST
☐ N72PS	87	Falcon 900	18	American International Aviation Corp. Teterboro, NJ.	N413FJ
☐ N72RK	91	Gulfstream 4	1171	Reebok Aviation Inc. Bedford, Ma.	N465GA
☐ N72SL	84	Citation II	550-0504	Stock Lumber Inc. Green Bay, Wi.	N979G
☐ N72ST	85	Citation III	650-0072	Pioneer Private Aviation Inc. Minneapolis, Mn.	N651CN
☐ N72TQ	64	Jet Commander	4	ADF Aircraft Inc. Miami, Fl.	N72TC
☐ N72WS	97	Falcon 900EX	14	American International Aviation Corp. Teterboro, NJ.	F-WWFN
☐ N73AW	72	Gulfstream 2	109	Dealaero Inc. Wilmington, De.	N862CE
☐ N73B	76	Falcon 10	79	The Kroger Co. Cincinnati, Oh.	N160FJ
☐ N73BE	86	Beechjet 400	RJ-15	Schneider National & Monroe Aviation Inc. Green Bay, Wi.	N73BL
☐ N73BL	93	Beechjet 400A	RK-71	Badger Liquor Co. Fond du Lac, Wi.	N777ND
☐ N73CE	65	Learjet 23	23-068	Argentum Air Corp. Coronado, Ca.	XB-GRR
☐ N73DJ	79	Learjet 25D	25D-273	Delaware Flights Inc. Summerfield, Fl.	N321AS
☐ N73DR	73	Sabre-40A	282-120	Dialysis Laboratories Management, Deland, Fl.	N73HP
☐ N73FW	80	Citation 1/SP	501-0150	Stuart & Margaret Conrad, Banner Elk, NC.	N95MJ
☐ N73GP	86	Learjet 55B	55B-127	Gerber Products Co. Fremont, Mi.	HZ-AM2
☐ N73KH	93	Citation V	560-0220	Killam & Hurd, San Antonio, Tx.	N23UB
☐ N73LJ	73	Learjet 25B	25B-138	Richardson Industries Inc. Sheboygan Falls, Wi.	(N2HE)
☐ N73LP	85	Learjet 35A	35A-604	Louisiana Pacific Corp. Hillsboro, Or.	N604BL
☐ N73TJ	75	Learjet 35	35-042	Sound Shop/Central South Music Sales Inc. Nashville, Tn.	N221UE
☐ N73TW	74	Learjet 25C	25C-181	Wesiskopf Enterprises Inc/73TW Inc. Scottsdale, Az.	N94PK
☐ N73WF	88	BAe 125/800A	NA0429	Westran Services Corp. San Diego, Ca.	N106GC
☐ N74A	68	Gulfstream 2B	36	Aviation Technologies Inc/Aeromet Inc. Tulsa, Ok.	N901KB
☐ N74AG	66	JetStar-731	5072	Pal-Waukee Aviation Inc. Wheeling, Il.	N500Z
☐ N74B	77	HS 125/700A	NA0209	Borden Inc. Columbus, Oh.	N586JR
☐ N74BJ	85	Citation S/II	S550-0041	Arkansas Aircraft Inc. Jonesboro, Ar.	N692M
☐ N74BS	73	Sabre-60	306-64	Sabre Sixty-Four Inc. North Miami, Fl.	N96CP
☐ N74FH	79	Citation Eagle	501-0138	West Coast Charters/All-Way Leasing, Redding, Ca.	N501CE
☐ N74FS	90	Falcon 900	85	F S Air Inc. Houston, Tx.	N461FJ
☐ N74G	78	Citation II	550-0035	NorAm Energy Corp. Shreveport, La.	(N50GG)
☐ N74GR	77	Westwind-1124	218	Accrued Investments Inc. Houston, Tx.	N218DJ
☐ N74HH	69	Gulfstream 2	74	MAC Aircraft Inc. Las Vegas, Nv.	N74TJ
☐ N74JA	82	Challenger 600S	1060	J A Interests Inc. Versailles, Ky.	N22AZ
☐ N74ND	86	BAe 125/800A	8063	Nordam Group Inc. Tulsa, Ok.	N684C
☐ N74NP	89	BAe 125/800A	NA0440	Nationwide Mutual Insurance Co. Columbus, Oh.	N593BA
☐ N74PC	89	BAe 125/800A	NA0439	PNC Bank Corp. Pittsburgh, Pa.	N592BA
☐ N74PM	79	Citation 1/SP	501-0112	Huktra UK Ltd. Hawarden-Chester, UK.	N14TV
☐ N74RD	78	Learjet 25D	25D-260	Private Jets LLC. Oklahoma City, Ok.	N43783
☐ N74RP	88	Gulfstream 4	1040	Time Warner Inc. Burbank, Ca.	N423GA
☐ N74RT	70	HS 125/731	NA742	Cable Holdings Aviation Corp. Westchester, NY.	(N87DC)
☐ N74RY	82	Learjet 55	55-063	A-OK Jets/Ryder System Inc. Miami, Fl.	N1744P
☐ N74TC	79	Citation II	550-0067	Executive Aircraft Corp. Newton, Ks.	N81TC
☐ N74TS	74	Westwind-1124	174	Air Travel Services Inc. Nashville, Tn.	N124VF
☐ N74VC	80	Sabre-65	465-17	Computer Sciences Corp. El Segundo, Ca.	N32290
☐ N74VF	88	Citation III	650-0156	Vernon Faulconer Inc. Lafayette, La.	N209A
☐ N75AK	75	Sabre-75A	380-38	Western Jet Sales Inc. Fort Worth, Tx.	N95TJ
☐ N75B	91	Citation V	560-0156	Siegel-Robert Inc. St Louis, Mo.	N560L
☐ N75BC	86	Westwind-1124	426	Ball Corp. Muncie, In.	N426WW
☐ N75BS	74	Sabre-75A	380-12	Aero Trading International Inc. Boca Raton, Fl.	(N4WJ)
☐ N75CC	72	Gulfstream 2	117	Crown Controls Corp. New Bremen, Oh.	N888SW
☐ N75CK	78	Learjet 25D	25D-256	American International Airways, Morristown, Tn.	N6LL
☐ N75CN	75	Sabre-75A	380-31	Choctaw Nation of Oklahoma, Durant, Ok.	N62
☐ N75CS	86	BAe 125/800A	8066	CSX Transportation Inc. Richmond, Va.	N369BA
☐ N75CV	90	Citation V	560-0075	McClaskey Aviation & Columbia Forest Products, Portland, Or.	(N2745L)

| *Reg* | *Yr Type* | | *c/n* | *Owner/Operator* | *Prev Regn* |

Reg	Yr	Type	c/n	Owner/Operator	Prev Regn
N75F	91	Citation V	560-0139	Consumers Power Co. Jackson, Mi.	N561A
N75G	91	Citation V	560-0140	Consumers Power Co. Jackson, Mi.	N562E
N75GP	86	Learjet 55B	55B-129	Metlife Capital/Gerber Products Co. Fremont, Mi.	
N75GW	75	Citation	500-0257	Senate Press Inc. Grand Haven, Mi.	N75FN
N75HS	97	Citation X	750-0037	Cessna Aircraft Co. Wichita, Ks.	
N75LM	78	Learjet 25D	25D-233	Manderson & Assocs Inc. Atlanta, Ga.	N55LJ
N75MC	92	Learjet 31A	31A-052	Stoneridge Inc. Warren, Oh.	N301AS
N75MG	79	Gulfstream 2	247	EG Air LLC. Detroit, Mi/MG-75 Inc. Troy, Mi.	N73MG
N75MT	92	BAe 125/800A	8231	Michelin North America Inc. Greenville, SC.	N685BA
N75NP	89	BAe 125/800A	NA0441	Nationwide Mutual Insurance Co. Columbus, Oh.	N594BA
N75RJ	91	Citation II	550-0692	K T & C Ltd. West Columbia, SC.	N692TT
N75RP	88	Gulfstream 4	1073	Time Warner Inc. Ronkonkoma, NY.	
N75RS	78	Sabre-75A	380-63	Randall Stores Inc. Mitchell, SD.	
N75TJ	76	Learjet 25D	25D-210	WRH Aircraft Inc. Germantown, Tn.	N97FT
N75TP	81	Citation II	550-0256	Crimson Aviation LLC. Tuscaloosa, Al.	N75HS
N75V	87	Falcon 900	13	CMS Energy Corp. Dearborn, Mi.	N328K
N75VC	81	Sabre-65	465-71	B-S Investment Inc. Dover, De.	N75GL
N75WE	84	Falcon 50	152	Aetna Life Insurance Co. Windsor Locks, Ct.	N75W
N75Z	96	Citation V Ultra	560-0345	Florida Pipe & Supply Co. Orlando, Fl.	N345CV
N75ZA	81	Citation II	550-0214	Laboratory Corp of America Holdings, Burlington, NC.	N75Z
N76AE	95	CitationJet	525-0139	EDCO Products Inc. Hopkins, Mn.	
N76CK	96	Citation V Ultra	560-0372	Kirkland Aviation Inc. Jackson, Tn.	N372CV
N76CS	92	Challenger 601-3A	5103	CSX Transportation Inc. Jacksonville, Fl.	C-GLWV
N76D	96	Citation X	750-0006	Dayton Hudson Corp. Minneapolis, Mn.	
N76ER	82	Westwind-Two	369	KHSS Inc/Philadelphia 76'ers, Philadelphia, Pa.	N85WC
N76FD	81	Falcon 50	41	Ronaele Aviation Inc/Scott Paper, Philadelphia, Pa.	(N760DL)
N76HG	66	JetStar-731	5076	Pro Flights Inc. Santurce, PR.	N69ME
N76MB	76	Falcon 10	83	Impact International Inc. Carson City, Nv.	N67TJ
N77A	95	Falcon 2000	17	AMP Inc. Harrisburg, Pa.	N2035
N77BT	67	HS 125/3A-RA	25155	Nevada Desert Holdings Inc. Gardnerville, Nv.	N333CJ
N77C	79	JetStar 2	5232	Trenton Foods/Parn Aviation Corp. Dover, De.	N90QP
N77CE	87	Falcon 900	12	CalEnergy Co. Omaha, Ne.	N991AS
N77CP	92	Gulfstream 4	1194	Pfizer Inc. Mercer County Airport, NJ.	N415GA
N77CS	86	BAe 125/800A	8065	CSX Transportation Inc. Jacksonville, Fl.	N368BA
N77D	80	HS 125/700A	NA0272	Bitz Aviation Inc. Baden, Pa.	N14SY
N77FD	82	Citation 1/SP	501-0250	Ingles Markets Inc. Asheville, NC.	
N77FK	82	Gulfstream 3	363	K Services Inc. Teterboro, NJ.	N83AL
N77GA	85	Beechjet 400	RJ-5	TIC Aviation Inc. Dallas, Tx. (was A1005SA)	
N77HF	94	Citation VII	650-7036	Hudson Foods Inc. Rogers, Ar.	N95HF
N77HH	67	Jet Commander	103	H & R Ltd. Ardmore, Ok.	N13AD
N77HW	66	JetStar-6	5080	Delta Omni Corp. Corona, Ca.	N914P
N77LA	85	BAe 125/800A	8029	L2 Aviation Group LLC. Indianapolis, In.	N600HS
N77MR	77	Learjet 24E	24E-351	Goldenwings Inc. Bloomington, Il.	(N94BD)
N77ND	78	Citation II	550-0005	University of N Dakota, Grand Forks, ND.	OE-GKP
N77NJ	69	Learjet 25	25-033	National Jets Inc. Fort Lauderdale, Fl.	YV-88CP
N77NR	83	Learjet 35A	35A-503	Nedra Roney, Mapleton, Ut.	N8567A
N77PH	94	Learjet 31A	31A-089	Parker Hannifin Corp/Travel 17325 Inc. Cleveland, Oh.	N5009L
N77PR	81	Citation II	550-0211	David Albin, Santa Fe, NM.	N77PH
N77RC	80	Citation II	550-0130	Mason C Rudd Enterprises, Louisville, Ky.	(N88845)
N77SF	79	Falcon 10	141	Seneca Flight/Canandaigua Wine Co. Penn Yan, NY.	N900D
N77SW	93	Gulfstream 4SP	1207	Joseph E Seagram & Sons Inc. NYC.	C-FJES
N77TC	95	Hawker 800	8275	The Timken Co. Canton, Oh.	N905H
N77TE	82	Falcon 50	110	Temple Inland Forest Products, Diboll, Tx.	VR-BJA
N77UB	82	Citation 1/SP	501-0240	United Builders Service Inc. Broomfield, Co.	N62864
N77VJ	65	Learjet 23	23-041	RBS Aviation Group Inc. Fort Myers, Fl.	C-GDDB
N77W	89	BAe 125/800A	NA0434	AMP Inc. Harrisburg, Pa.	N588BA
N77WD	68	Learjet 24	24-174	Jeffreys Aviation Inc. Spokane, Wa.	N321GL
N78AG	66	HS 125/731	25101	VBI, Van Nuys, Ca.	N251LA
N78AM	90	Citation V	560-0056	Alice Manufacturing Co. Greenville, SC.	N2682F
N78CK	81	Citation II	550-0319	Central Pennsylvania Cellular Telephone Co. Wilmington, De.	N76CK
N78FK	84	Citation II	550-0498	Hancor Leasing Corp. Findlay, Oh.	N1823B
N78GA	79	Citation II/SP	551-0132	Kinnarps AB. Falkoping, Sweden. (was 550-0087).	C-GTBR
N78GJ	80	Westwind-1124	310	Mid Oaks Investments LLC. Buffalo Grove, Il.	D-CBBD
N78MC	77	Learjet 35A	35A-117	Midlantic Jet Charters Inc. Cardiff, NJ.	N3155B
N78NP	77	Citation V	560-0107	Wausau Insurance Companies, Mosinee, Wi.	(N6801L)
N78PH	94	Learjet 31A	31A-090	Parker Hannifin Corp/Travel 17325 Inc. Cleveland, Oh.	N9173X
N78PT	90	Citation III	650-0187	Cessna Aircraft Co. Wichita, Ks.	N70PT
N79AD	93	Challenger 601-3R	5140	Arthur S DeMoss Foundation Inc. West Palm Beach, Fl.	N1016D
N79HC	79	HS 125/700A	NA0246	Harsco Corp. Camp Hill, Pa.	N130BB
N79PF	88	Falcon 50	174	Associates Information Services Inc. Irving, Tx.	HB-IAG
N79PM	97	Citation V Ultra	560-0459	Cessna Aircraft Co. Wichita, Ks.	
N79RP	93	Gulfstream 4SP	1220	Time Warner Inc. Ronkonkoma, NY.	N449GA
N79RS	73	Citation	500-0107	FBN Holding Corp. Incline Village, Nv.	N40RW
N79SE	88	Citation II	550-0585	Southeast Air Transportation Inc. Montgomery, Al.	N94AF
N79SF	78	Learjet 36A	36A-041	Phoenix Air Group Inc. Cartersville, Ga.	N60AF
N79SL	67	DC 9-15F	47011	U S Department of Energy, Albuquerque, NM.	
Reg	*Yr*	*Type*	*c/n*	*Owner/Operator*	*Prev Regn*

Reg	Yr	Type	c/n	Owner/Operator	Prev Regn
□ N79TJ	79	Falcon 10	148	PSC Inc. Wichita, Ks.	N103PJ
□ N80AB	92	Citation V	560-0169	Southern Company Services Inc. Atlanta, Ga.	N6888C
□ N80AG	75	Gulfstream 2SP	164	AGI Holding Corp. Van Nuys, Ca.	N93LA
□ N80AP	75	Learjet 24D	24D-312	Air Ambulance Professionals Inc. Fort Lauderdale, Fl.	N312NA
□ N80AR	81	Learjet 35A	35A-454	HME Enterprises inc. Boca Raton, Fl.	(N80KR)
□ N80AT	90	Gulfstream 4	1151	Taubman Air Inc. Waterford, Mi.	N375GA
□ N80AW	80	Citation II	550-0186	Erie Airways Inc & EZ Holdings Inc. Erie, Pa.	YV-187CP
□ N80BF	93	Challenger 601-3A	5117	NWT Aircraft Co. Lexington, Ky.	N606CC
□ N80BL	82	Falcon 100	200	BNB LLC. Dover, De.	(N682D)
□ N80CK	74	Learjet 24D	24D-309	Kalitta Flying Service Inc. Morristown, Tn.	N789AA
□ N80CL	82	HS 125/700A	NA0318	JABJ/Compass Foods Inc. Montvale, NJ.	N819M
□ N80CN	82	Falcon 50	105	Champion International Corp. Stamford, Ct.	N106FJ
□ N80F	86	Falcon 900	6	Anheuser-Busch Companies Inc. St Louis, Mo.	N405FJ
□ N80FD	79	Westwind-1124	260	Fisher Development Co. Lancaster, Pa.	(N503RH)
□ N80GM	80	Citation II	550-0147	Avcorp Inc. Portland, Or.	(N155JK)
□ N80HG	74	Sabre-75A	380-19	Herring Group Inc. San Antonio, Tx.	N54HH
□ N80J	84	Gulfstream 3	441	USX Corp. Pittsburgh, Pa.	N306GA
□ N80K	81	HS 125/700A	NA0298	USX Corp. Pittsburgh, Pa.	N125G
□ N80KM	92	Beechjet 400A	RK-50	K Mart Corp. Troy, Mi.	
□ N80L	83	Gulfstream 3	406	USX Corp. Pittsburgh, Pa,	N356GA
□ N80PG	76	Learjet 35	35-063	Phoenix Air Group Inc. Cartersville, Ga.	N663CA
□ N80PM	93	BAe 125/800A	8236	Philip Morris Inc. Milwaukee, Wi.	N162BA
□ N80R	81	Sabre-65	465-53	Winn-Dixie Stores Inc. Jacksonville, Fl.	N76NX
□ N80SF	81	Citation 1/SP	501-0189	Flightworks Inc & Prodelin Corp. Hickory, NC.	(N500SS)
□ N80SL	79	Citation 1/SP	501-0294	Howe/Tolladay Aviation LLC. Fresno, Ca. (was 500-0391).	N8EH
□ N80TF	94	CitationJet	525-0076	Friedkin International Aviation Services Co. Sugar Land, Tx.	
□ N80TR	80	Falcon 50	32	TRT Aeronautical Inc. Tucson, Az.	F-WZHJ
□ N80WD	70	Gulfstream 2B	88	Weeks Davies Aviation Inc. West Palm Beach, Fl.	N901AS
□ N80WJ	83	Falcon 100	202	Air Del Inc. Jackson, Ms.	3D-ADR
□ N81CC	79	Citation II	550-0089	J F Wilbur Jr Inc. Tucson, Az.	N43RW
□ N81EB	77	Citation 1/SP	501-0003	T C Mueller Oil & Gas/TCM Air Inc. Fort Smith, Ar.	N781L
□ N81GD	81	Citation II/SP	551-0046	Sierra Air Inc. Smiths Creek, Mi.	N34YL
□ N81HH	93	BAe 1000B	9034	JABJ/Horsehead Industries Inc. NYC.	N290H
□ N81KA	78	HS 125/700A	NA0227	Crusader Aviation Inc & DeGeorge Aviation Inc. Oxford, Ct.	N10CZ
□ N81MR	86	Learjet 35A	35A-622	Birndorf & Preston LLC. San Diego, Ca.	N610R
□ N81MW	79	Learjet 25D	25D-277	Rocketown Tours Inc/I'll Lead You Home LLC. Nashville, Tn.	N321GL
□ N81P	79	Falcon 10	153	Pilot Corp. Knoxville, Tn.	N344A
□ N81PJ	61	MS 760 Paris-1R	81	Airborne Turbine Inc. Ca.	No 81
□ N81QV	79	HS 125/700A	NA0245	Amrash Aviation Ltd. Richardson, Tx.	N810V
□ N81RA	97	CitationJet	525-0194	Regency Aviation Co. Jacksonville, Fl.	
□ N81RR	79	Gulfstream 2TT	246	Geneva International Ltd. Arlington, Va.	N14LT
□ N81SH	96	Citation V Ultra	560-0357	Conair Group Inc. Pittsburgh, Pa.	N5148N
□ N81TJ	90	Beechjet 400A	RK-14	Tyler Jet LLC. Tyler, Tx.	(N414RK)
□ N81TT	93	Beechjet 400A	RK-65	Tom-Tom Aviation LLC. Charlotte, NC.	N39HF
□ N82A	89	Gulfstream 4	1068	Prudential Insurance Co. Newark, NJ.	N90AE
□ N82AE	81	Gulfstream 3	342	Wings Aviation International Inc. Franklin Lakes, NJ.	N82A
□ N82AF	74	Sabre-80A	380-21	Promotional Researchers Inc. Charlottesville, Va.	N647JP
□ N82AJ	78	Citation	501-0282	European Air Service Inc. Dover, De. (was 500-0376).	XA-SQX
□ N82CR	80	Sabre-65	465-49	Rockwell Collins Inc. Cedar Rapids, Ia.	N500RR
□ N82CW	82	Challenger 600S	1050	Costco Wholesale/N Pacific Enterprises Inc. Seattle, Wa.	N600MK
□ N82KK	92	Learjet 31A	31A-054	Krispy Kreme Doughnut Corp. Winston-Salem, NC.	N92FD
□ N82LS	84	Citation 1/SP	501-0681	Les Schwab Warehouse Center Inc. Prineville, Or.	N2616G
□ N82P	81	Citation 1/SP	501-0208	Arno Bohn, Toussus Le Noble, France.	(N25M)
□ N82RP	92	Falcon 900B	116	Rich Aviation Inc. Lake Worth, Fl.	N5VN
□ N82TS	74	Learjet 25B	25B-154	Crayson Corp. Wilmington, De.	N47DK
□ N83AG	84	Citation II	550-0483	Allan Gardiner/Erich Aviation Inc. Concord, Ca.	N483AS
□ N83CE	65	Learjet 23	23-074	Aircraft Exchange, Carlsbad, Ca.	XB-GRQ
□ N83CK	74	Learjet 25B	25B-183	American International Airways, Lakeview, Or.	N5LL
□ N83DM	81	Citation 1/SP	501-0227	DDM Holdings Inc. Wilmington, De.	N47CF
□ N83FJ	81	Falcon 50	74	Anheuser Busch Companies Inc. St. Louis, Mo.	F-WZHA
□ N83FN	75	Learjet 36	36-007	Flight International Inc. Tampa, Fl.	N83DM
□ N83HC	69	Learjet 24B	24B-193	Vailjet Inc. Dallas, Tx.	N83H
□ N83LC	88	Challenger 601-3A	5029	K L W Kansas Aircraft Services Inc. Leawood, Ks.	N67MR
□ N83MD	76	Falcon 10	78	G M N Inc. Seattle, Wa.	(N83MF)
□ N83ND	81	Citation 1/SP	501-0178	Jet Manager Inc. Wilmington, De.	N4246A
□ N83RR	92	Citation V	560-0183	Universal Enterprises Inc. Mansfield, Oh.	N12807
□ N83SA	82	Diamond 1A	A030SA	Southern Aircraft Leasing Inc. Rockville, Md.	N301P
□ N83SD	82	Learjet 55	55-032	Wesdix Corp. Northbrook, Il.	N11TS
□ N83TE	81	Learjet 25D	25D-329	Blixseth-Hedinger Aviation LLC. Portland, Or.	N613GL
□ N83TF	80	Sabre-65	465-43	Building & Construction Trades Department, Washington, DC.	N228LS
□ N83TJ	76	HS 125/F600A	6070	Beartooth Communications Co. Helena, Mt.	N411TP
□ N83TR	97	CitationJet	525-0185	Tom Ryan Distributing Co. Flint, Mi.	(RP-C8288)
□ N83WM	97	Learjet 60	60-104	Ward Equipment Co. Cockeysville, Md.	N104LJ
□ N83WN	93	Learjet 31A	31A-081	Bombardier Business Jet Solutions LLC. Dallas, Tx.	N83WM
□ N84BA	97	Hawker 800XP	8313	Bell Atlantic Network Services Inc. Arlington, Va.	N313XP
□ N84BJ	86	Beechjet 400	RJ-17	McRae's Inc. Jackson, Ms.	N94BJ
Reg	*Yr*	*Type*	*c/n*	*Owner/Operator*	*Prev Regn*

Reg	Yr	Type	c/n	Owner/Operator	Prev Regn
☐ N84CF	82	Citation 1/SP	501-0254	Carlton Forge Works, Paramount, Ca.	N2629Z
☐ N84EA	83	Citation II	550-0484	Executive Air Services Ltd. N Canton, Oh.	N1253N
☐ N84EB	83	Citation II	550-0488	Elkhart Brass Manufacturing Co. Elkhart, In.	N12536
☐ N84FA	85	BAe 125/800A	8047	Bell Atlantic Corp. Philadelphia, Pa.	N84BA
☐ N84FG	97	CitationJet	525-0192	AFG Industries Inc. Kingsport, Tn.	
☐ N84FN	74	Learjet 36	36-002	Flight Intl/Maritime Sales & Leasing Inc. Hockessin, De.	N84DM
☐ N84GC	84	Citation II	550-0493	J R Miller Enterprises LLC. Park City, Ut.	N84AW
☐ N84HP	81	Falcon 50	56	Hewlett-Packard Co. San Jose, Ca.	N112FJ
☐ N84NG	77	Sabre-75A	380-52	PAJMM Inc. Heathrow, Fl.	N70KM
☐ N84TJ	81	Falcon 10	188	Lakeview Transportation Corp. Lakeview, Or.	I-TFLY
☐ N84WU	82	Westwind-1124	383	Provost & Umphrey Law Firm, Beaumont, Tx.	N20DH
☐ N84WW	83	Westwind-1124	401	MAC Aviation Enterprises Inc. Tampa, Fl.	4X-CUQ
☐ N85	93	Challenger 601-3R	5138	FAA, Washington, DC.	N138CC
☐ N85BN	85	Beechjet 400	RJ-7	Purcell Tires Ventures LLC. Potosi, Mo. (was A1007SA).	N25BN
☐ N85CC	72	Sabre-40A	282-108	Crown Controls Corp. New Bremen, Oh.	N306CW
☐ N85CR	91	Beechjet 400A	RK-22	Rockwell International Corp. Cedar Rapids, Ia.	N51ML
☐ N85D	87	Falcon 900	28	Doyle Foods & Castle & Cooke Aviation Inc., Van Nuys, Ca.	N420FJ
☐ N85DA	85	Citation II	650-0073	Duncan Aviation Inc. Lincoln, Ne.	N673JS
☐ N85DW	85	BAe 125/800B	8034	DW Aviation Co. Phoenix, Az.	G-5-595
☐ N85GW	79	Learjet 35A	35A-227	Great Gibraltar II Inc. Cleveland, Oh.	N88NE
☐ N85HR	71	Learjet 25XR	25XR-079	Insurance Investors Inc. Austin, Tx.	N36CC
☐ N85HS	69	Sabre-60	306-23	Fine Air Services Inc. Miami, Fl.	N68MA
☐ N85KC	91	Citation V	560-0128	Schwitzer U S Inc. Asheville, NC.	N19ME
☐ N85M	87	Gulfstream 4	1023	MBNA Corp. Highland Heights, Oh.	N778W
☐ N85MP	85	Citation S/II	S550-0016	Steiner Corp. Salt Lake City, Ut.	(N99VC)
☐ N85NC	83	Learjet 55	55-077	Novacare Services Inc. King of Prussia, Pa.	N58M
☐ N85SV	80	Learjet 35A	35A-347	Randall Aviation Inc. Fort Lauderdale, Fl.	OE-GNP
☐ N85TW	78	Learjet 25D	25D-251	Terry Air Corp. New Orleans, La.	TG-VOC
☐ N85V	85	Gulfstream 3	449	Aviation Enterprises Inc. Bedford, Tx.	N7C
☐ N85WN	84	Falcon 100	210	Wabash National Corp. Lafayette, In.	N812AM
☐ N85WP	77	Citation Eagle	501-0032	WP Enterprises Ltd. Stamford, Ct.	N550L
☐ N86AJ	97	Citation Bravo	550-0842	Cessna Aircraft Co. Wichita, Ks.	
☐ N86AK	81	Falcon 50	52	Air Ketchum Inc. Hailey, Id.	N18G
☐ N06BA	84	Citation S/II	S550-0001	Unlimited Aircraft Service Inc. San Juan, PR.	N151DD
☐ N86BE	78	Learjet 35A	35A-194	Bullock Charter Inc. Princeton, Ma.	N86BL
☐ N86CP	74	Sabre-60	306-76	AvBase Aviation Ltd. Cleveland, Oh.	(N760SA)
☐ N86MC	83	Falcon 50	141	McCaw Communicartions Inc. Seattle, Wa.	(N222MC)
☐ N86PC	85	Citation S/II	S550-0017	C L Swanson Corp. Madison, Wi.	N88GD
☐ N86QS	86	Citation S/II	S550-0086	EJI Inc/NetJets-Telxon Corp. Columbus, Oh.	N900RB
☐ N86RX	82	Learjet 35A	35A-458	Rolex/Woodhill Aviation Corp. Palwaukee, Il.	YV-997CP
☐ N86SG	82	Citation II	550-0350	Seymour Grubman, Beverly Hills, Ca.	
☐ N86WP	85	Citation III	650-0089	Washington Water Power Co. Spokane, Wa.	N650JC
☐ N87BA	97	Citation S/II	S550-0131	Strange Bird Inc/Sails in Concert Inc. Tallahassee, Fl.	D-CHJH
☐ N87BL	67	BAC 1-11/419EP	120	The Jet Place Inc. Tulsa, Ok.	N524AC
☐ N87EC	85	BAe 125/800A	8052	Albemarle Corp. Richmond, Va.	N360BA
☐ N87FL	85	Citation S/II	S550-0055	Flying Lion Inc. Nashville, Tn.	N374GC
☐ N87GS	85	Westwind-Two	422	Shinn Enterprises Inc. Charlotte, NC.	N251SP
☐ N87PT	80	Citation II	550-0174	Hargray Holdings Corp. Hilton Head, SC.	N666WW
☐ N87SF	79	Citation II	550-0096	Seneca Foods Corp. Penn Yan, NY.	N30UC
☐ N87TD	68	Gulfstream 2	39	Leaman Air Services Inc. Exton, Pa.	N87HB
☐ N87TH	81	Falcon 10	178	CMI Aviation Inc. Champaign, Il.	N79BP
☐ N88B	65	Learjet 24	24-015	Tucson Air Museum Foundation of Pima County, Tucson, Az.	
☐ N88BF	81	Sabre-65	465-60	Bassett Furniture Industries, Bassett, Va.	XA-...
☐ N88BG	76	Learjet 35A	35A-490	AirNet Systems Inc. Columbus, Oh.	I-FIMI
☐ N88BY	74	Learjet 25B	25B-168	Bankair Inc. West Columbia, SC.	N88BT
☐ N88DJ	78	Learjet 25D	25D-234	Dale Jarrett Inc. Hickory, NC.	N39BL
☐ N88G	93	Citation V	560-0208	Burlington Industries Inc. Greensboro, NC.	N892SB
☐ N88GJ	74	Citation	500-0155	Tillson Aircraft Management/Bright World Inc. Santa Fe, NM.	N5ZZ
☐ N88HF	91	Citation V	560-0133	Hudson Foods Inc. Rogers, Ar.	N77HF
☐ N88JA	77	Learjet 35A	35A-118	U S Navy, Naples, Italy.	N50MT
☐ N88JJ	88	Citation III	650-0169	Jim Clark & Associates Inc. Oklahoma City, Ok.	(N749DC)
☐ N88LD	97	CitationJet	525-0181	Econo Inns Inc. Marion, Il.	N181CJ
☐ N88ME	83	Diamond 1A	A066SA	Aerodynamics Inc. Pontiac, Mi.	N88MF
☐ N88MF	90	Astra-1125SP	048	Murphy Farms Inc. Rose Hill, NC.	N1125V
☐ N88MM	85	Citation S/II	501-0689	McMahans Furniture Co of Las Vegas Inc. Carlsbad, Ca.	N75TJ
☐ N88MX	80	HS 125/700A	NA0264	Maxim Financial Corp. Boulder, Co.	N299CT
☐ N88NW	76	Citation	500-0309	Peninsula Group Inc. Turnwater, Wa.	N791MA
☐ N88TB	77	Citation 1/SP	501-0002	Gateway Aviation Inc. Carlsbad, Ca.	(N501WK)
☐ N88ZL	65	B 707-330B	18928	Zestmo Ltd-Oman/Lowa Ltd. Boston, Ma.	N5381X
☐ N89AM	83	Westwind-Two	389	Atherton & Murphy Investment Co. Tulsa, Ok.	N812G
☐ N89D	79	Citation II	550-0056	Valley Air Services Inc. Altoona, Pa.	N444FJ
☐ N89EC	77	Falcon 10	109	Citadel Communications Corp. Fresno, Ca.	N69EC
☐ N89KM	92	Beechjet 400A	RK-56	K-Mart Corp. Troy, Mi.	
☐ N89LS	89	Citation II	550-0623	Les Schwab Warehouse Center Inc. Prineville, Or.	(N1255L)
☐ N89MD	81	HS 125/700A	NA0295	McDermott Inc. New Orleans, La.	N888SW
☐ N89NC	87	BAe 125/800A	NA0403	National City Corp. Highland Heights, Oh.	(N89KT)

Reg	Yr	Type	c/n	Owner/Operator	Prev Regn

47

Reg	Yr	Type	c/n	Owner/Operator	Prev Regn
☐ N89SR	72	HS 125/731	NA778	Fant Air LLC. Birmingham, NY.	N67EC
☐ N89TC	75	Learjet 35	35-026	N A Degerstrom Inc. Spokane, Wa.	N54754
☐ N89TD	85	Citation S/II	S550-0076	Teledyne Industries Inc. Huntsville, Al.	C-GQMH
☐ N90AG	94	Learjet 60	60-042	AGCO Corp. Duluth, Ga.	N4010K
☐ N90AH	75	Learjet 35	35-036	AWH Corp. Winston Salem, NC.	N76GP
☐ N90AJ	90	Astra-1125SP	052	Industrian Metalurgicas Pescarmona SAIC YF, Pcia de Mendoza.	
☐ N90AM	81	Falcon 50	53	TMP Worldwide Inc. NYC.	N22TZ
☐ N90BJ	92	Citation II	550-0710	James Ridings, Fort Mill, SC.	N510VP
☐ N90BL	91	Citation II	550-0682	Bourland & Leverich Aviation Inc. Pampa, Tx.	N682CJ
☐ N90CF	81	Citation 1/SP	501-0184	West Coast Charters/Reuland Electric Co. Industry, Ca.	N67786
☐ N90CP	78	Gulfstream 2IT	224	CONOPCO Inc/Unilever USA Inc. White Plains, NY.	N631SC
☐ N90EW	87	Falcon 900	27	East West Air Inc. San Jose, Ca.	N419FJ
☐ N90GW	75	Sabre-75A	380-27	Grove Manufacturing/Kidde Industries Inc. Shady Grove, Pa.	N90GM
☐ N90J	65	Learjet 24	24-060	Aeroflight 1 Inc. Orland Park, Il.	N899WF
☐ N90JJ	88	Citation II	550-0571	Southwestern Public Service Co. Amarillo, Tx.	(N1299T)
☐ N90KC	81	Westwind-Two	339	Teterboro Transportation Enterprises, Teterboro, NJ.	N74AG
☐ N90LC	82	Gulfstream 3	360	Lockheed Martin Corp. Baltimore, Md.	N341GA
☐ N90MA	80	Citation II	550-0103	Martinaire of Oklahoma Inc. Irving, Tx.	XA-JEZ
☐ N90ME	87	BAe 125/800A	8082	Sulzermedica USA Inc. Angleton, Tx.	N499SC
☐ N90MF	90	Citation V	560-0060	Mclean-Fogg/Caledonia Leasing Partnership, Mundelein, Il.	(N2689B)
☐ N90NE	83	Learjet 55	55-075	FLC Co LP. Cleveland, Oh.	N55GH
☐ N90PG	89	Citation V	560-0002	Hunt Building Corp. El Paso, Tx.	N562CV
☐ N90UC	81	Challenger 600S	1023	Union Camp Corp. Morristown, NJ.	N680M
☐ N90UG	89	Astra-1125	030	Universal Underwriters Group, Kansas City, Mo.	N90U
☐ N90WA	90	Learjet 31	31-028	Wal-Mart Stores Inc. Bentonville, Ar.	
☐ N90WR	75	Learjet 35	35-022	F S Air Service Inc. Anchorage, Ak.	OY-BLG
☐ N90Z	82	Citation II	550-0336	Northern Jet, Iron Mountain, Mi.	N6830Z
☐ N91AP	79	Citation 1/SP	501-0117	Euroflite Inc. Dover, De.	LV-MZG
☐ N91B	80	Citation II	550-0194	Beckett Enterprises/Scott Fetzer Co. Lakewood, Oh.	N88723
☐ N91BZ	80	Sabre-65	465-19	General Foam Plastics Corp. Mount Gilead, NC.	N65RC
☐ N91CH	85	BAe 125/800A	8030	CNC Aircraft Inc. Louisville, Ky.	N10WF
☐ N91CM	80	HS 125/700A	NA0265	MyPac LLC. Rancho Murietta, Ca.	N783M
☐ N91CV	66	Falcon 20C-5	48	Cyprus Amax Minerals Co. Englewood, Co.	N910Y
☐ N91DP	93	Learjet 31A	31A-079	Dean Phillips Inc. Baker, Mt.	N41DP
☐ N91DV	92	BAe 125/800B	8211	Sequoia Two LLC. Menlo Park, Ca.	PT-OSB
☐ N91DZ	76	Citation	500-0343	Janet Payne, Wiesbaden, Germany.	HB-VJR
☐ N91KH	72	HS 125/600A	6003	Rhema Bible Church/Kenneth Hagin Ministries, Tulsa, Ok.	N42TS
☐ N91LA	77	Gulfstream 2B	198	Leucadia Inc. Salt Lake City, Ut.	N3652
☐ N91ML	87	Citation S/II	S550-0132	Koro Aviation Inc. Hazleton, Pa.	N91ME
☐ N91MT	87	Beechjet 400	RJ-26	Maleco, Salem, Or.	N426MD
☐ N91PN	72	Learjet 25B	25B-091	Aero Nash Inc. Tucson, Az.	VR-CCH
☐ N91SA	84	Westwind-Two	420	Westwind LLC. City of Industry, Ca.	N420W
☐ N91TH	88	Falcon 900	60	Sony Aviation Inc. Teterboro, NJ.	N900FJ
☐ N91UC	82	Challenger 600S	1051	Forcenergy Inc/Trinity Aviation Services, Fort Lauderdale.	N601SR
☐ N91WZ	79	Citation 1/SP	501-0132	Prince Georg Waldburg-Zeil, Leutkirch-Unterzel, Germany.	N50US
☐ N91YC	91	Citation V	560-0115	York International Corp. York, Pa.	I-NEWY
☐ N92AE	96	Gulfstream 4SP	1301	American Express Co. Stewart, NY.	N433GA
☐ N92B	74	Citation Eagle	500-0212	Lease Air Inc. Cleveland, Oh.	N223MC
☐ N92BD	88	Citation II	550-0588	Dillard Department Stores Inc. Little Rock, Ar.	N255CC
☐ N92BL	95	Learjet 60	60-058	Electronic Data Systems Corp. Dallas, Tx.	N5016V
☐ N92CC	77	Citation 1/SP	501-0026	Skylane Farms Inc. Woodburn, Or.	N36891
☐ N92DF	66	Learjet 24XR	24XR-117	Boomerang Air Inc. Piedmont, Ok.	N90DH
☐ N92EC	90	Learjet 31	31-024	Virginia Air Corp. Glenn Allen, Pa.	N90PB
☐ N92FE	80	Westwind-1124	286	Air Lake Lines Inc. Brooklyn Center, Mn.	N4447T
☐ N92GS	62	B 720-047B	18452	Blue Metals Inc. Miami, Fl. (status ?).	N93146
☐ N92LA	73	Gulfstream 2B	125	Leucadia Aviation Inc. Salt Lake City, Ut.	N3643
☐ N92ME	85	Citation S/II	S550-0044	Old Dominion Freight Line Inc. High Point, NC.	
☐ N92MG	82	Learjet 55	55-025	RDM Commerce Inc & Wilsonart International Inc. Temple, Tx.	N57FM
☐ N92ND	77	CitationJet	525-0186	Nordam Group Inc. Tulsa, Ok.	N186CJ
☐ N92NE	77	Learjet 35A	35A-092	F & A Marketing Corp. Dallas, Tx.	N46931
☐ N92QS	86	Citation S/II	S550-0092	EJI Inc/NetJets-BCP Service Corp. Columbus, Oh.	N1275A
☐ N92RP	78	HS 125/700A	7022	Petersen Aviation/Raleigh Jet Enterprises, Van Nuys, Ca.	N34RE
☐ N92RW	85	Beechjet 400	RJ-4	A S K Leasing Inc. Atlanta, Ga. (was A1004SA).	(N401TJ)
☐ N92SM	73	Citation	500-0124	Sharon Shelton Shay/Sierra Mike Air Charter Inc. Dallas, Tx	N8FC
☐ N92SS	96	Citation V Ultra	560-0388	Southern States Co-operative Inc. Richmond, Va.	
☐ N92UG	92	Learjet 31A	31A-050	Zurich Holding Co of America, Overland Park, Ks.	
☐ N92WW	80	Westwind-1124	296	C P Horizons Corp. Clearwater, Fl.	N770JJ
☐ N93AE	96	Gulfstream 4SP	1302	American Express Co. Stewart, NY.	(N98AE)
☐ N93AT	70	Gulfstream 2SP	85	Aircraft Services Group, Teterboro, NJ.	N931CW
☐ N93BD	82	Citation II	550-0454	Dillard Department Stores Inc. Little Rock, Ar.	N12490
☐ N93BP	68	Learjet 24	24-169	Kalitta Flying Service, Morristown, Tn.	N927AA
☐ N93CT	85	BAe 125/800A	8049	Century Service Group Inc. Monroe, La.	N24SP
☐ N93CV	93	Citation V	560-0239	Carolina Power & Light Co. Raleigh, NC.	N1GC
☐ N93CX	80	Gulfstream 3	314	Cyprus Amax Minerals Co. Englewood, Co.	N1640
☐ N93DK	86	Citation III	650-0112	93 Delta Kilo Corp. Oak Brook, Il.	N60BE
☐ N93FR	83	HS 125/700A	NA0342	Fisher Rosemount Inc. Minneapolis, Mn.	N8400E

Reg   *Yr Type*   *c/n*   *Owner/Operator*   *Prev Regn*

Reg	Yr	Type	c/n	Owner/Operator	Prev Regn
☐ N93GH	95	Falcon 2000	6	Allstate Insurance Co. Wheeling, Il.	F-WQBL
☐ N93GR	87	Falcon 900	24	BDA/US Services Ltd & WRBC Transportation Inc. Greenwich, Ct	N67WB
☐ N93KE	81	Westwind-1124	316	Kennecott Energy & Coal Co. Gillette, Wy.	VH-ASR
☐ N93LE	84	Learjet 35A	35A-592	Land's End Inc. Madison, Wi.	N45KK
☐ N93QS	86	Citation S/II	S550-0093	EJI Inc/NetJets-KCI Air Inc. Columbus, Oh.	N33DS
☐ N93RM	66	Jet Commander	74	Holli Aero Inc. Hollister, Ca.	N300DH
☐ N93SC	67	Jet Commander	90	93SC LLC. Pleasant View, Tn.	N1121E
☐ N93TJ	77	Citation 1/SP	501-0006	Martin Aero Corp. Tucson, Az.	N5016P
☐ N93TX	92	Citation VII	650-7009	Textron Inc. Providence, RI.	N12596
☐ N93WW	81	Westwind-1124	321	Worldwide Aircraft Sales Inc. Wilmington, De.	N83CT
☐ N94	89	BAe 125/C-29A	8129	Federal Aviation Administration, Oklahoma City, Ok.	88-0269
☐ N94AE	96	Gulfstream 4SP	1307	American Express Co. Stewart, NY.	N443GA
☐ N94AF	77	Learjet 35A	35A-094	Ameriflight Inc. Burbank, Ca.	(N35PF)
☐ N94AJ	72	Citation	500-0024	Air Jet Inc. Cary, NC.	VH-ICN
☐ N94BA	94	Challenger 601-3R	5160	Bell Atlantic Network Services Inc. Philadelphia, Pa.	N710HM
☐ N94BJ	93	Falcon 50	237	McRaes Inc. Jackson, Ms.	(N5425)
☐ N94FL	84	Gulfstream 3	424	Runicom Ltd. Gibraltar.	N228G
☐ N94HE	94	Beechjet 400A	RK-89	Hughes-Ergon Co. Jackson, Ms.	N1560G
☐ N94LH	83	Diamond 1A	A073SA	CCC Holding-Cameron Communications Corp. Sulphur, La.	N1715G
☐ N94LT	96	Gulfstream 4SP	1313	Lucent Technologies Inc. Morristown, NJ.	N455GA
☐ N94MZ	95	CitationJet	525-0094	JSM Inc/Cannondale Corp. Georgetown, Ct.	(N51522)
☐ N94NB	93	BAe 125/800A	8241	NationsBank NA. Charlotte, NC.	N125CJ
☐ N94RL	77	Learjet 35A	35A-096	LeTourneau Inc. Longview, Tx.	N96FA
☐ N94RT	85	Citation S/II	S550-0023	ROCK-TENN Converting Co. Norcross, Ga.	(F-OHAI)
☐ N94SL	79	Citation 1/SP	501-0111	Rocky Mountain Bancorporation Inc. Billings, Mt.	(N59WP)
☐ N94TJ	69	Gulfstream 2B	75	Treasure Solutions Inc. Iselin, NJ.	N760U
☐ N94TX	93	Citation V	560-0247	Provident Companies Inc. Chattanooga, Tn.	N800BS
☐ N94WN	84	BAe 125/800A	8014	Western National Life Insurance Co. Houston, Tx.	88-0271
☐ N95	89	BAe 125/C-29A	8131	Federal Aviation Administration, Oklahoma City, Ok.	N103TC
☐ N95AB	70	Learjet 24B	24B-213	Martin DeHaan, Norfolk, Va.	N599BA
☐ N95AE	89	BAe 125/800A	NA0447	Transamerica Finance Corp. San Francisco, Ca.	N319LJ
☐ N95AG	96	Learjet 60	60-079	AGCO Corp. Duluth, Ga.	N42825
☐ N95BP	80	Learjet 25D	25D-314	Russell Enterprises LLC & Ravenair LLC. Hamilton, Mt.	(N703VP)
☐ N95CC	93	Citation VII	650-7030	Cessna Aircraft Co. Wichita, Ks.	N900WA
☐ N95CK	78	Learjet 25D	25D-248	Kalitta Flying Service Inc. Morristown, Tn.	
☐ N95CM	97	Citation X	750-0042	Cessna Aircraft Co. Wichita, Ks.	N532CJ
☐ N95DJ	93	CitationJet	525-0032	Dick Stebbins, Longview, Ix.	N50FJ
☐ N95HC	94	Falcon 50	244	Harsco Corp. New Cumberland, Pa.	N293PC
☐ N95HE	92	Citation II	550-0713	Van Ness Plastic Molding Co. Clifton, NJ.	N17UC
☐ N95JK	80	Westwind-1124	283	Capital Air Services Inc. Hallandale, Fl.	N800CJ
☐ N95NB	93	BAe 125/800A	8244	NationsBank NA. Charlotte, NC.	N3199Q
☐ N95PA	95	Beechjet 400A	RK-99	First Alabama Bank, Birmingham, Al.	N10LT
☐ N95PH	89	Falcon 50	194	Pioneer Hi-Bred International Inc. Johnston, Ia.	N11KA
☐ N95Q	73	Citation	500-0119	Captain Bly Inc. Rochester, Mn.	N749UP
☐ N95RC	77	Sabre-60	306-129	Engineered Plastic Components Inc. Benton Harbor, Mi.	PT-WLC
☐ N95RX	93	Citation VII	650-7035	CVS Corp. Woonsocket, RI.	N737CC
☐ N95TX	94	Citation VII	650-7037	Textron Inc. Warwick, RI.	N10NL
☐ N95WJ	82	Falcon 100	195	Western Jets International Inc. Arlington, Tx.	N17GL
☐ N95WK	84	Learjet 55	55-099	Cole/TDI Aviation LLC. White Plains, NY.	
☐ N95ZC	95	Learjet 60	60-068	Ziegler Inc. Minneapolis, Mn.	
☐ N96	89	BAe 125/C-29A	8134	Federal Aviation Administration, Oklahoma City, Ok.	88-0270
☐ N96AA	67	Learjet 24	24-139	Kalitta Flying Service, Morristown, Tn.	N481EZ
☐ N96AE	87	Gulfstream 4	1024	American Express Co. NYC.	N412GA
☐ N96AT	95	Citation V Ultra	560-0325	HealthTrust Inc. Nashville, Tn.	N5101J
☐ N96BB	64	JetStar-6	5049	Glenn Cunningham, East Freetown, Ma.	N96B
☐ N96CK	65	Learjet 23	23-016	Kalitta Flying Service, Morristown, Tn.	N7GF
☐ N96CP	87	Citation III	650-0139	Central Parking System Inc. Raleigh, NC.	N4EG
☐ N96DS	90	Falcon 50	209	SES Holdings Inc. Denver, Co.	N1902W
☐ N96FB	73	Citation	500-0094	MTN Wings Inc. Wilmington, De.	N80GB
☐ N96FG	96	Falcon 2000	25	U S Fidelity & Guarantee Co. Baltimore, Md.	N2042
☐ N96FN	78	Learjet 35A	35A-186	Flight International Inc/Skyjet, Volkel, Holland.	N96DM
☐ N96G	93	CitationJet	525-0034	Reading Pretzel Machinery Corp. Robesonia, Pa.	(N1772E)
☐ N96GD	93	CitationJet	525-0042	Wachovia Leasing Corp. Winston-Salem, NC.	(N3230M)
☐ N96GM	95	CitationJet	525-0114	Greenwood Mills Inc. Greenwood, SC.	N5214L
☐ N96GS	85	Learjet 35A	35A-606	Wings Service LP. Wilmington, De.	N3WP
☐ N96JA	77	Gulfstream 2SP	213	Jordan Industries-SLC/JI Aviation Inc. Deerfield, Il.	(N96BK)
☐ N96LT	89	Falcon 50	192	Lucent Technologies Inc. Morristown, NJ.	N212T
☐ N96MT	89	Citation V	560-0032	James French & co-owners, Sheboygan, Wi.	G-DBII
☐ N96NB	94	Citation V Ultra	560-0267	NationsBank NA. Charlotte, NC.	N1295J
☐ N96PR	81	HS 125/700A	NA0304	Corporate Air Management Inc. West Mifflin, Pa.	N99JD
☐ N96RE	81	Sabre-65	465-52	Reliance Electric Co. Cleveland, Oh.	N500E
☐ N96RT	68	Falcon 20C	159	WCC/Capital Pacific Holdings Inc. Newport Beach, Ca.	N96WC
☐ N96SG	65	HS 125/1A-522	25060	Gulf Air Group Corp. Houston, Tx.	XB-CXZ
☐ N96SK	79	Citation 1/SP	501-0127	SK Foods LP. Lemoore, Ca.	N88BM
☐ N96TS	73	Westwind-1123	159	JARAMAR Ltd. Houston, Tx.	N344CK
☐ N96UH	81	Falcon 50	55	UIM Aircraft Inc. Denver, Co.	N300CR
Reg	*Yr* *Type*		*c/n*	*Owner/Operator*	*Prev Regn*

Reg	Yr	Type	c/n	Owner/Operator	Prev Regn
☐ N96VR	82	Falcon 100	199	K F S Management Inc. Watertown, SD.	N486MJ
☐ N96WW	87	Beechjet 400	RJ-34	Duro-Last Inc. Saginaw, Mi.	I-ALSO
☐ N97	89	BAe 125/C-29A	8154	Federal Aviation Administration, Oklahoma City, Ok.	88-0272
☐ N97AG	83	Gulfstream 3	401	Affinity Group Holding, Ventura, Ca.	F-400
☐ N97AM	71	Learjet 25C	25C-071	Fairway Aircraft Inc. Dover, De.	YV-132CP
☐ N97AN	81	Learjet 35A	35A-373	I J Knight Inc. New Tripoli, Pa.	XA-RUY
☐ N97BG	82	Citation II	550-0428	Brasfield & Game General Contractor, Birmingham, Al.	N107WV
☐ N97BH	95	Citation V Ultra	560-0290	Air Finance Corp. Van Nuys, Ca.	N5145P
☐ N97BP	86	Citation S/II	S550-0090	R A Beeler Leasing Co. Carrizzo Springs, Tx.	N76FC
☐ N97BZ	82	Falcon 50	89	BDA/US Services Ltd & WRBC Transportation Inc. Greenwich, Ct	N120TJ
☐ N97CT	86	Citation S/II	S550-0125	Century Service Group Inc. Monroe, La.	N122CG
☐ N97DD	74	Citation	500-0159	MRF Ltd. Odessa, Tx.	N881JT
☐ N97DK	97	Citation X	750-0035	Cessna Aircraft Co. Wichita, Ks.	
☐ N97FB	97	Beechjet 400A	RK-152	Action Transport Inc. Tupelo, Ms.	N2252Q
☐ N97FD	80	Citation 1/SP	501-0149	Frank Branson PC. Dallas, Tx.	N104CF
☐ N97FF	96	Beechjet 400A	RK-117	Raytheon Aircraft Co. Wichita, Ks.	(N97FB)
☐ N97FN	67	Learjet 25	25-003	Flight International/Maritime Sales & Leasing, Hockessin, De	N97DM
☐ N97HW	80	Westwind-1124	312	Accrued Investments Inc. Houston, Tx.	N300LH
☐ N97LB	85	Citation S/II	S550-0079	Africair Inc. Miami, Fl.	N97AJ
☐ N97LE	89	Learjet 35A	35A-648	Land's End Inc. Madison, Wi.	N648J
☐ N97LT	90	Falcon 50	202	Lucent Technologies Inc. Morristown, NJ.	N212N
☐ N97NB	97	Citation V Ultra	560-0399	NationsBanc Services Inc. Charlotte, NC.	
☐ N97PJ	62	MS 760 Paris-1R	97	Airborne Turbine Inc. Ca.	No 97
☐ N97QA	80	Learjet 35A	35A-304	Business Jets LLC. St James, NY.	N53GL
☐ N97S	81	Citation II	550-0238	Corporate Flight Inc. Detroit, Mi.	N67999
☐ N97SC	75	Sabre-75A	380-34	Sabreliner Corp. St Louis, Mo.	FAE-034
☐ N97SH	95	Hawker 800XP	8277	Sierra Health Services Inc. Las Vegas, Nv.	G-BVYW
☐ N97SK	76	Citation	500-0316	Jim Clark & Associates Inc. Oklahoma City, Ok.	N398RP
☐ N97SM	80	Westwind-Two	307	JB Associates - Indiana LLC. Carmel, In.	N925Z
☐ N97VF	96	CitationJet	525-0171	AMSCO Transportation inc. Ponca City, Ok.	
☐ N97WJ	67	Falcon 20C	101	J P Aviation Inc. Charlotte, NC.	N342F
☐ N98	89	BAe 125/C-29A	8156	Federal Aviation Administration, Oklahoma City, Ok.	88-0273
☐ N98AC	80	Learjet 35A	35A-301	Aero Care-Methodist Hospital/TLC Investments, Lubbock, Tx.	N190VE
☐ N98AD	97	Astra-1125SPX	095	Galaxy Aerospace Corp. Fort Worth, Tx.	
☐ N98BL	95	Learjet 60	60-061	Devonwood LLC. Durham, NC.	(N63BL)
☐ N98E	91	Citation V	560-0103	Southern Company Services, Atlanta, Ga.	(N67989)
☐ N98FT	75	Gulfstream 2TT	173	Rocky Mountain Aviation, Chino, Ca.	XA-SQU
☐ N98LC	76	Learjet 35A	35A-077	AirNet Systems Inc. Columbus, Oh.	ZS-NRZ
☐ N98LT	95	Gulfstream 4SP	1278	Lucent Technologies Inc. Morristown, NJ.	VR-CTA
☐ N98MB	72	Citation	500-0054	Ozark Management Inc. Jefferson City, Mo.	(N28U)
☐ N98Q	72	Citation	500-0040	Northwest Partners LLC. Troutdale, Or.	N600WM
☐ N98QS	86	Citation S/II	S550-0098	F & A Marketing Corp. Oklahoma City, Ok.	(N1290E)
☐ N98RG	81	Citation 1/SP	501-0209	Russelectric Inc. Hingham, Ma.	N36FD
☐ N98RP	81	Gulfstream 3	328	Petersen Aviation, Van Nuys, Ca.	N78RP
☐ N98RS	74	Learjet 25XR	25XR-148	Stern Holdings Inc. Dallas, Tx.	(N98JA)
☐ N98SC	65	Jet Commander	32	Richard Sollon, Canonsburg, Pa.	N101BU
☐ N98TX	97	Citation X	750-0041	Cessna Aircraft Co. Wichita, Ks.	(N22NG)
☐ N98WJ	73	Learjet 24D	24D-268	Northlake Foods Inc. Atlanta, Ga.	N58BL
☐ N99	89	BAe 125/C-29A	8158	Federal Aviation Administration, Oklahoma City, Ok.	88-0274
☐ N99BB	96	Citation X	750-0005	Harris Air Inc. Logan, Ut.	N5223Q
☐ N99BC	79	Falcon 10	128	Interstate Equipment Inc. Bozeman, Mt.	N99MC
☐ N99BL	76	Falcon 10	87	Aero Toy Store Inc. Fort Lauderdale, Fl.	N80TS
☐ N99DE	81	Citation II/SP	551-0065	Metropolitan Leasing Co. Charlotte, NC.	N8AD
☐ N99GA	92	Gulfstream 4	1198	Viad Corp. Phoenix, Az.	N425GA
☐ N99JD	82	Falcon 50	129	Central Financial Services Inc. Golden Valley, Mn.	N4903W
☐ N99KW	96	Learjet 60	60-085	Florida Wings Inc. Boca Raton, Fl.	N685LJ
☐ N99NJ	77	Learjet 25XR	25XR-220	National Jets Inc. Fort Lauderdale, Fl.	N220NJ
☐ N99RS	78	Learjet 36A	36A-039	Tudor Corp. Delray Beach, Fl.	N25PK
☐ N99ST	69	HS 125/731	NA734	Speciality Retailers Inc. Houston, Tx.	(N38TS)
☐ N99TK	89	Citation II	550-0621	Thermo King Sales & Service Inc. St Paul, Mn.	N102PA
☐ N99W	65	Jet Commander	46	ADF Airways, Miami, Fl. (status ?).	N202ST
☐ N99WA	79	Falcon 10	150	Wheels Aviation Inc/Wings of Africa, Harare, Zimbabwe.	N212NC
☐ N100A	94	Gulfstream 4SP	1235	Exxon Corp. Dallas, Tx.	N477GA
☐ N100AC	77	Falcon 20F	366	Energy Aviation Services LLC. Charlottesville, Va.	N300CT
☐ N100AG	97	Beechjet 400A	RK-150	R S Allen Aviation Inc. San Diego, Ca.	N1135U
☐ N100AK	85	Westwind-Two	436	Pacific Diversified Investments Inc. Anchorage, Ak.	N1904G
☐ N100AR	89	Gulfstream 4	1100	ARCO-Atlantic Richfield Co. Burbank, Ca.	
☐ N100AY	82	Citation II	550-0365	Raytheon Aircraft Co. Wichita, Ks.	N100AG
☐ N100BC	87	Westwind-1124	438	Ball Corp. Muncie, In.	N438AM
☐ N100BP	76	Sabre-75A	380-48	PSI Aviation Inc. Livonia, Mi.	N132DB
☐ N100CH	84	Westwind-1124	406	Coachmen Industries Inc. Elkhart, In.	N651E
☐ N100CJ	80	Citation II	550-0167	South Street Aviation LC. Baltimore, Md.	N88737
☐ N100CK	69	Falcon 100	222	Chesapeake Corp. Sandston, Va.	N125FJ
☐ N100DL	69	Learjet 24B	24B-201	AJM Airplane Co. Naples, Fl.	C-GTFA
☐ N100DS	90	Citation II	550-0639	DBS Transit Inc. Harrisburg, Pa.	N62RG
☐ N100EJ	73	Sabre-75A	380-1	Centurion Investments Inc. (status ?).	N30GB

| Reg | Yr | Type | c/n | Owner/Operator | Prev Regn |

Reg	Yr	Type	c/n	Owner/Operator	Prev Regn
N100ES	90	Gulfstream 4	1135	Earth Star Inc/Walt Disney Co. Burbank, Ca.	N500MM
N100FF	97	Citation X	750-0028	Frederick Furth, Healdsburg, Ca.	N728CX
N100FJ	77	Falcon 10	100	Smith Flooring Inc. Mountain View, Mo.	(N217CP)
N100GN	94	Gulfstream 4SP	1236	Gannett Co. Washington, DC.	N478GA
N100GP	76	Learjet 35	35-064	E & J Partnership Inc. Wheeling, Il.	N291BC
N100HB	79	Citation II	550-0058	Ciralsky & Associates, Toledo, Oh.	N71CJ
N100KK	81	Learjet 35A	35A-420	Kohler Co. Kohler, Wi.	N35PT
N100KU	97	Citation Bravo	550-0813	University of Kansas, Lawrence, Ks.	N813CB
N100LR	82	Challenger 600S	1064	Lars Aviation Inc/L A Rosenthal, Cleveland, Oh.	N75B
N100LX	81	Citation 1/SP	501-0220	Lexicon Inc. Little Rock, Ar.	N100QH
N100MS	77	Learjet 35A	35A-138	Douglas Albrecht Revocable Trust, St Louis, Mo.	N35WH
N100QR	82	Challenger 600S	1043	Quick & Reilly Group Inc. Farmingdale, NY.	N43NW
N100RH	69	HS 125/731	NA727	Syndacon Inc. Bethesda, Md.	N117RH
N100RR	91	Falcon 50	220	Avocet Aviation LLC. Washington, DC.	N75RD
N100RS	82	Diamond 1A	A029SA	The Stoll Companies, Toledo, Oh.	N89TJ
N100SC	90	Citation V	560-0054	560 Inc. Chattanooga, Tn.	N531F
N100SQ	66	Learjet 24	24-113	Brandis Aircraft, Springfield, Il. (status ?).	N204Y
N100SR	89	Astra-1125	037	Steven Rayman, Big Rock, Il.	N589TB
N100T	77	Falcon 10	107	Semitool Inc. Kalispell, Mt.	N160TJ
N100U	91	BAe 1000A	NA1001	United Technologies Corp. Hartford, Ct.	G-BTTG
N100UF	68	Falcon 20C	160	JTJ Inc. Wilmington, De.	N48BT
N100VA	82	Learjet 55	55-029	Northwestern Aircraft Capital Corp. Mercer Island, Wa.	N82JA
N100VQ	67	Learjet 24	24-140	National Aviation Academy of Mississippi Inc. Clearwater, Fl	N100VC
N100WN	79	Learjet 25D	25D-288	AirNet Systems Inc. Columbus, Oh.	(N40BC)
N100WP	90	Citation V	560-0073	Arawak Air Inc. Newport Beach, Ca.	
N101AF	94	Citation II	550-0732	American Freightways Inc. Harrison, Ar.	N1213S
N101AJ	75	Learjet 36	36-008	Enviromental Research Institute, Ann Arbor, Mi.	(N43A)
N101AR	73	Learjet 24D	24D-279	Andalex Resources Inc. Louisville, Ky.	N75CJ
N101BX	80	Citation II	550-0157	Rose Aviation LLC. Secaucus, NJ.	N550K
N101CC	87	Beechjet 400	RJ-19	Crown Crafts Inc. Fulton County, Ga.	N24BA
N101FC	80	HS 125/700A	NA0263	Field Container Aviation LLC. Elk Grove Village, Il.	(N130AE)
N101HF	77	HS 125/700A	7013	Hardees Food Systems Inc. Rocky Mount, NC.	N75ST
N101HG	74	Citation	500-0213	Argentum Air Corp. Coronado, Ca.	N100UH
N101HS	69	HS 125/731	NA730	H C Investments LLC. Grand Rapids, Mi.	N730TS
N101HW	81	Learjet 35A	35A-403	RMSC of West Palm Beach Inc. Delray Beach, Fl.	N100NR
N101KK	81	Citation II	550-0209	Kohler Co. Kohler, Wi.	(N877GB)
N101LD	78	Citation 1/SP	501-0083	Luxury Wheels, Grand Junction, Co.	N420RC
N101PG	86	Citation III	650-0126	The Pape' Group Inc. Coburg, Or.	N311MA
N101QS	86	Citation S/II	S550-0101	EJI Inc/NetJets-General Signal Corp. Columbus, Oh.	(N1290Y)
N101RR	82	Citation 1/SP	501-0241	Coral Aircraft LLC. North Platte, Ne.	C-GSTR
N101SK	90	Challenger 601-3A	5058	Avn Methods/Charles E Smith Management Inc. Leesburg, Va.	N404SK
N101SV	78	Westwind-1124	246	Medic Air Corp. Reno, NV.	4X-CMR
N101US	83	Learjet 35A	35A-500	First USA Bank, Wilmington, De.	N66LN
N101VS	70	Learjet 24B	24B-218	Advanced Technology Center, Buffalo, NY.	N682LJ
N101YC	84	Citation III	650-0057	Yankee Candle Co. South Deerfield, Ma.	N368G
N102AB	69	Gulfstream 2	53	World Changers Church International, College Park, Ga.	N167A
N102AD	75	Citation	500-0280	EPPS Air Service Inc. Atlanta, Ga.	N100HP
N102AF	95	CitationJet	525-0122	American Freightways, Harrison, Ar.	N5264U
N102AR	68	Learjet 25	25-012	Air Response Inc. Fort Plain, NY.	N846YC
N102BW	73	Westwind-1123	165	Marlin Air Inc. Detroit, Mi.	C-GWSH
N102C	76	Learjet 24E	24E-343	Drake, Goodwin & Graham Inc. NYC.	N102B
N102CX	71	Gulfstream 2B	102	Clorox Co/Kaiserair, Oakland, Ca.	N400CC
N102HB	82	Citation II	550-0409	Buzz-Air of Indiana Inc. East Chicago, In.	VR-CIT
N102MC	88	Beechjet 400	RJ-50	Medusa Corp. Cleveland, Oh.	N56GA
N102MU	90	Gulfstream 4	1145	CHK Management Ltd. Teterboro, NJ.	N416GA
N102ST	83	Learjet 55	55-069	Bontona Aviation Inc. Fort Lauderdale, Fl.	N551UT
N102VS	74	Learjet 25B	25B-180	Arvin Calspan Corp. Buffalo, NY.	N266BS
N103CF	80	Learjet 35A	35A-318	Actus Corp. Napa, Ca.	N444WB
N103CL	79	Learjet 35A	35A-273	ACX Technologies Inc. Broomfield, Co.	N103C
N103M	81	Citation II/SP	551-0038	A T Massey Coal Co Inc. Richmond, Va.	N550DA
N103QS	86	Citation S/II	S550-0103	EJI Inc/NetJets-National Car Rental System, Columbus, Oh.	N12900
N103RR	70	HS 125/731	NA746	R R Aviation Inc. New Orleans, La.	N74WF
N103VF	85	Citation II	S550-0046	Geo M Martin Co. Emeryville, Ca.	N760NB
N104AR	85	Gulfstream 3	461	Andalex Resources/Tower Resources Inc. Louisville, Ky.	N323GA
N104BK	77	JetStar 2	5219	N14BK Inc. Bloomfield Hills, Mi.	N219MF
N104CE	67	JetStar-8	5108	Southwest Jet Inc. Belton, Mo.	XA-SWD
N104DD	77	Falcon 10	110	U S Aircraft Sales Inc. McLean, Va.	N712US
N104GA	66	Learjet 24	24-112	CVS Systems Inc. Marion, In.	N112DJ
N104JG	79	HS 125/700A-2	NA0243	GCA Aviation Inc. Allison Park, Pa.	XA-JRF
N104RS	79	Westwind-1124	273	C 23 Ltd. Lewes, De.	(N566PG)
N104SL	72	Sabre-40A	282-104	LRC Credit Corp. Livingston, Mt.	XA-SEU
N105BA	74	Learjet 25XR	25XR-152	Bonaero Inc. St Charles, Il.	XA-JSC
N105BG	86	Citation S/II	S550-0105	Barnes Group Inc. Bristol, Ct.	(N12907)
N105BN	91	Challenger 601-3A	5101	Barnes & Noble/B & N Aircraft Co. Teterboro, NJ.	N604CC
N105DM	69	Sabre-60	306-27	Alas Air Leasing Ltd. Wilmington, De.	(N55ME)
N105FX	93	Learjet 31A	31A-086	FlexJets, Dallas, Tx.	N867JS
Reg	*Yr*	*Type*	*c/n*	*Owner/Operator*	*Prev Regn*

Reg	Yr	Type	c/n	Owner/Operator	Prev Regn
☐ N105GA	66	Learjet 24A	24A-116	Walter & Marguerite Kostich/W & M Co. Jenks, Co.	N51B
☐ N105HS	65	HS 125/1A	25031	Dr Paul Madison, Michigan City, In.	N79AE
☐ N105TB	68	Gulfstream 2	31	MIT Lincoln Labs/Air Force Material Command, Bedford, Ma.	N200CC
☐ N105TF	73	HFB 320	1055	Kalitta Flying Service Inc. Morristown, Tn.	N7865T
☐ N105TW	73	Falcon 20F	289	Titan Wheel International Inc. Quincy, Il.	N211HF
☐ N105UP	86	Challenger 601	3066	ARAMARK Services Inc. Philadelphia, Pa.	VP-CLE
☐ N105Y	83	Gulfstream 3	412	Occidental International Exploration, Bakersfield, Ca.	N610CC
☐ N106CG	86	Beechjet 400	RJ-12	Southwest Bank, St Louis, Mo.	N3112K
☐ N106CJ	93	CitationJet	525 0006	Northwest Stamping Inc. Eugene, Or.	(N1326P)
☐ N106EA	79	Citation 1/SP	501-0101	Navaho Air Inc. Wilmington, De.	N100TW
☐ N106FX	93	Learjet 31A	31A-087	FlexJets, Dallas, Tx.	N868JS
☐ N106JL	84	BAe 125/800A	8012	Air Management Systems LLC/Ham Marine Inc. Pascagoula, Ms.	N80BR
☐ N106KC	96	Beechjet 400A	RK-132	Cambata Aviation inc. Millboro, Va.	N1087Z
☐ N106QS	86	Citation S/II	S550-0106	EJI Inc/NetJets-Galey & Lord Industries Inc. Columbus, Oh.	(N12909)
☐ N106SP	82	Citation II	550-0346	Stan Partee, Big Springs, Tx.	N550CF
☐ N106TF	69	HFB 320	1042	Kalitta Flying Service Inc. Morristown, Tn.	N7648X
☐ N106TJ	76	Learjet 24E	24E-340	Aero Charter Nashville Inc. Nashville, Tn.	(N95CP)
☐ N106TW	76	Falcon 10	84	Titan Wheel International Inc. Quincy, Il.	N192MC
☐ N107A	88	Gulfstream 4	1070	ARAMCO Associated Co. Dhahran, Saudi Arabia.	N407GA
☐ N107CG	91	Citation VI	650-0207	Vesta Fire Insurance Corp. Birmingham, Al.	N334WC
☐ N107FX	95	Learjet 31A	31A-102	FlexJets/CII Holdings Inc. Dallas, Tx.	N5002D
☐ N107HF	69	Learjet 25XR	25XR-029	Chipola Aviation Inc. Marianna, Fl.	N28LA
☐ N107LT	81	HS 125/700A	NA0301	Lifetouch Inc/LT Flight Services Inc. Minneapolis, Mn.	N421SZ
☐ N107RC	87	Citation S/II	S550-0150	American Yard Products Inc. Augusta, Ga.	N150CJ
☐ N107RM	83	Learjet 25D	25D-362	Walbridge, Aldinger Co & M Air Inc. Detroit, Mi.	(N107MS)
☐ N107TB	76	Falcon 10	77	Tom Brown Inc. Midland, Tx.	N53TS
☐ N108CJ	95	CitationJet	525-0108	Vulcan Northwest Inc. Bellevue, Wa.	N5211Q
☐ N108CR	97	CitationJet	525-0258	Cessna Aircraft Co. Wichita, Ks.	
☐ N108DB	72	Gulfstream 2	112	Richard Blum & Associates LP. San Francisco, Ca.	N909L
☐ N108FX	95	Learjet 31A	31A-104	FlexJets/Herman Miller Inc. Dallas, Tx.	N4010N
☐ N108GD	77	Falcon 10	114	Global Link/Geruda Midwest Inc. Fairfield, Ia.	N555DH
☐ N108KC	74	Falcon 10	8	Keller Companies Inc. Manchester, NH.	N88ME
☐ N108LJ	95	Citation V Ultra	560-0337	Jacques Admiralty Law Firm PC. Detroit, Mi.	
☐ N108MC	76	Citation	500-0322	Mitchell Co. Mobile, Al.	N1AP
☐ N108QS	86	Citation S/II	S550-0108	Shilo Management Corp. Portland, Or.	(N1291K)
☐ N108R	67	Falcon 20DC	108	Reliant Airlines Inc. Ypsilanti, Mi.	N101ZE
☐ N108RA	70	B 707-347C	20315	Cx USA 9/96 to ?	HR-AMN
☐ N108RB	77	Learjet 35A	35A-097	Miami Valley Aviation Inc. Middleton, Oh.	N135J
☐ N108RF	97	Citation Bravo	550-0805	Reading's Fun Ltd. Fairfield, Ia.	
☐ N108WV	91	Citation	650-0204	Willamette Valley Co. Eugene, Or.	N811JT
☐ N109AL	72	Citation	500-0037	Southern Cross Aviation Inc. Fort Lauderdale, Fl.	(SE-DPL)
☐ N109FX	95	Learjet 31A	31A-105	FlexJets/KVOA Communications Inc. Dallas, Tx.	N5005K
☐ N109G	82	HS 125/700A	NA0322	Chevron Corp. San Francisco, Ca.	G-5-20
☐ N109GA	80	Citation II	550-0124	Wheel-Air, Mentor, Oh.	N124CR
☐ N109JC	79	Citation II	550-0099	Stroud Aviation Inc. Stroud, Ok.	N2664L
☐ N109JR	77	Learjet 35A	35A-101	Ruan Cab Co. Des Moines, Ia.	N40149
☐ N109KM	75	B 727-251	21155	U S Marshals Service, Arlington, Va.	N276US
☐ N109NC	92	Challenger 601-3A	5112	JABJ/Nalco Chemical Co. Chicago, Il.	N112NC
☐ N109PW	83	Diamond 1A	A046SA	Air Charlotte Inc. Oak Park, Mi.	N900BT
☐ N109TD	96	Hawker 800XP	8307	USA Waste Services Inc. Houston, Tx.	N307XP
☐ N109VP	91	Citation V	560-0109	U S Sugar Corp. Clewiston, Fl.	N27
☐ N110AB	75	Citation	500-0262	Air Foley Inc. Tavares, Fl.	N110AF
☐ N110AE	78	Learjet 35A	35A-155	International Capital Associates LLC. Grand Junction, Co.	(N1001L)
☐ N110AJ	78	Sabre-75A	380-70	SI Aircraft Sales Inc. Arlington, Tx.	N1NR
☐ N110BR	80	Gulfstream 3	301	CSC Transport Inc. White Plains, NY.	(N100P)
☐ N110EJ	80	HS 125/700A	NA0273	J & E Air Inc. Waterville, Me.	N46TJ
☐ N110ET	82	Learjet 55	55-023	Morse Operations Inc. Fort Lauderdale, Fl.	N7784
☐ N110FS	66	Sabre-40	282-58	Executive Aircraft Corp. Newton, Ks.	N1101G
☐ N110FT	82	Learjet 35A	35A-471	Milam International Inc. Englewood, Co.	N95AP
☐ N110FX	95	Learjet 31A	31A-108	FlexJets/Callaway Golf Co. Dallas, Tx.	
☐ N110J	79	Falcon 10	139	Foster Poultry Farms, Livingston, Ca.	(N610J)
☐ N110LE	96	Gulfstream V	506	Arepo Corp. Teterboro, NJ.	N158AF
☐ N110MH	88	BAe 125/800A	NA0425	Humana Inc. Louisville, Ky.	N733K
☐ N110MT	84	Gulfstream 3	444	Morton International Inc. Waukegan, Il.	N328GA
☐ N110ST	69	Jet Commander-A	129	Chapperel Trucking Inc. Tumwater, Wa.	N525AW
☐ N110TD	82	Challenger 600S	1052	Thermadyne Holdings Corp. Chesterfield, Mo.	N110M
☐ N110TG	96	Beechjet 400A	RK-123	Raytheon Aircraft Co. Wichita, Ks.	N1123Z
☐ N110WA	82	Citation II	550-0408	Musco Corp. Oskaloosa, Ia.	N400TX
☐ N111AD	80	Sabre-65	465-27	Arapahoe Development Inc & Eaton Metal Products, Englewood.	N351AF
☐ N111AM	95	CitationJet	525-0113	Ambanc Group LP. Wilmington, De.	N5214K
☐ N111BA	86	Beechjet 400	RJ-11	Willamette Industries, Owensboro, Ky. (was A1011SA).	N114DM
☐ N111BB	75	Citation	500-0248	B & B Sales, Dallas, Tx.	(N70PB)
☐ N111BP	67	Falcon 20C	111	IBP=Iowa Beef Processors Inc. Sioux City, Ia.	N111AM
☐ N111F	77	Sabre-60	306-126	Joseph G Fabick/JGF Farms, Koeltztown, Mo.	HC-BUN
☐ N111G	85	Gulfstream 3	454	GTE Service Corp. Beverly, Ma.	N273G
☐ N111GD	75	Gulfstream 2	170	GDL Aviation Inc. Teterboro, NJ.	(N318GD)

| *Reg* | *Yr* | *Type* | *c/n* | *Owner/Operator* | *Prev Regn* |

Reg	Yr	Type	c/n	Owner/Operator	Prev Regn
N111HZ	96	Citation VII	650-7068	Hertz Corp. Park Ridge, NJ.	N51160
N111JL	66	B 727-21	18998	Continental Aviation Services Inc. Naples, Fl.	N7271P
N111KK	81	Learjet 35A	35A-425	Kohler Co. Kohler, Wi.	
N111LM	69	Learjet 25	25-025	Sagittarius Leasing Inc. Pensacola, Fl.	N225DS
N111LR	97	CitationJet	525-0222	Alpha Wolf Enterprises LLC. Twin Bridges, Mt.	
N111ME	74	Citation	500-0146	Ed Unicome, Spokane, Wa.	N194AT
N111MP	73	Learjet 25XR	25XR-139	Ryan Air Inc/Medical Air Services Association, Ft Lauderdale	N605NE
N111NF	73	Westwind-1123	168	Martinaire/Bruce Martin, Dallas, Tx.	N66SM
N111NL	92	Gulfstream 4	1184	Bugle Boy Industries Inc. Van Nuys, Ca.	N477GA
N111QS	86	Citation S/II	S550-0111	EJI Inc/NetJets-Texaco Air Services Inc. Columbus, Oh.	(N1291Y)
N111US	90	Learjet 55C	55C-147	First USA Management Inc. Dallas, Tx.	N499SC
N111VP	89	Citation V	560-0044	Joseph Canion, Houston, Tx.	N2665S
N111VV	90	Learjet 31	31-023	Inductotherm Industries Inc. Rancocas, NJ.	
N111VW	74	Gulfstream 2	153	Volkswagen of America Inc. Waterford, Mi.	(N602CM)
N111WB	74	Learjet 35	35-003	Warm Air Inc. Cincinnati, Oh.	N703MA
N111Y	92	Citation VI	650-0223	Ingram Industries Inc. Nashville, Tn.	N1301D
N111ZN	96	Hawker 800XP	8327	Zenith National Insurance Corp. Woodland Hills, Ca.	N327XP
N111ZS	79	HS 125/700A	7076	Wings Aviation International Inc. Franklin Lakes, NJ.	N111ZN
N111ZT	89	Gulfstream 4	1111	Arrow Aviation Corp. Dover, De.	N111JL
N112CM	93	Learjet 31A	31A-078	Cast Masters Inc. Salina, Ks.	XA-SNM
N112CP	74	Citation	500-0183	Dr Joe Allegra, Atlanta, Ga.	N1880S
N112CT	69	Falcon 20-5B	168	Peter Kiewit Sons Inc. Omaha, Ne.	N731RG
N112FX	95	Learjet 31A	31A-116	FlexJets/ATM Corp of America, Dallas, Tx.	N113FX
N112JS	78	Citation II	550-0032	Flamingo 500 Inc/Air Citation Ltd, Luton, UK.	N905EM
N112K	85	BAe 125/800A	8042	Wells Fargo Bank/Kaiserair Inc. Oakland, Ca.	N20S
N112M	69	HS 125/731	NA736	WRH Aircraft Sales Inc. Germantown, Pa.	N30PP
N112NW	88	BAe 125/800B	8112	Northwestern Aircraft Capital Corp. Mercer Island, Wa.	PT-OBT
N112PV	79	Sabre-65	465-12	Vulture One Corp. Portland, Or.	N112PR
N112QS	86	Citation S/II	S550-0112	EJI Inc/NetJets-H&D Leasing Inc. Columbus, Oh.	(N12910)
N112TJ	61	JetStar-731	5029	Omnimex Aviation Inc. Fort Worth, Tx.	N166AC
N113	66	B 727-30	18935	DoJ-U S Marshals Service, Oklahoma City, Ok.	N18G
N113ES	72	Learjet 25B	25B-092	Suzan Charters Inc. Stuart, Fl.	N18AK
N113SH	75	Citation	500-0285	Canary Aircraft Sales Inc. Ft Smith, Ar.	N86SS
N113T	76	Sabre-60	306-113	Casco Motor Cars Inc. Sarasota, Fl.	N2626M
N114CP	90	Citation V	560-0018	CP Air Inc/Crown Pacific Island Lumber LLP. Portland, Or.	N164DW
N114DS	81	Citation II	550-0397	Delta Sierra Inc. Harbor Springs, Mi.	N68888
N114FX	96	Learjet 31A	31A-119	FlexJets/N1LJI I LLC & 6D Air LLC. Dallas, Tx.	
N114HC	73	Learjet 25B	25B-114	JWS-SLH Inc. Tulsa, Ok.	N25JD
N114LG	81	Sabre-65	465-51	Global Air Inc. Hackensack, NJ.	N3QM
N114MX	66	BAC 1-11/422EQ	119	W A Moncrief Jr/Montex Drilling, Fort Worth, Tx.	N114M
N115BP	84	Westwind-Two	417	Blandin Paper Co. Grand Rapids, Mn.	N700WE
N115CR	59	Sabre-60	306-43	A S Aviation Inc. Wilmington, De.	N10UM
N115FX	96	Learjet 31A	31A-129	FlexJets/John T Walter Jr. Dallas, Tx.	N8079Q
N115MC	73	Gulfstream 2SP	137	J S Aviation Inc. Las Vegas, Nv.	N23AH
N116AM	77	Learjet 35A	35A-116	Aeromanagement Inc. Herndon, Va.	I-MMAE
N116AS	94	Learjet 60	60-035	Ali-Air Inc. Gainesville, Fl.	VR-BST
N116DD	74	HS 125/600A	6044	S E Distributing Inc. Wilmington, De.	N9282Y
N116FX	97	Learjet 31A	31A-132	FlexJets/Monumant Aircraft Services-Vestar, Dallas, Tx.	
N116K	80	Citation II	550-0149	Kaiserair Inc. Oakland, Ca.	
N116KX	66	Jet Commander	87	Brittney Inc. Tenopah, Az.	N430DC
N116LJ	97	Learjet 60	60-116	Learjet Inc. Wichita, Ks.	
N116MA	77	Learjet 36A	36A-029	Cal-Air Charter Inc/Dutch Navy, Valkenburg, Holland.	N16MA
N116RM	69	Learjet 24B	24B-206	San Antonio Redi-Mix Inc. Baytown, Tx.	F-BTYV
N116SC	62	Sabre-40R	282-1	Sabreliner Corp. St Louis, Mo.	(N351JM)
N117AE	76	Learjet 24E	24E-346	Averitt Inc. Cookeville, Tn.	N117AJ
N117AH	81	Westwind-Two	352	James D Wolfensohn Inc. Teterboro, NJ.	N117JW
N117CC	95	Citation V Ultra	560-0287	Circus Circus Enterprises Inc. Las Vegas, Nv.	N287CV
N117FJ	78	Gulfstream 2	229	J T Aviation Corp. NYC.	N702H
N117FX	97	Learjet 31A	31A-133	FlexJets/WNB Asset Management Inc. Dallas, Tx.	N8073Y
N117GL	78	Gulfstream 2	220	G&L Aviation Inc. Los Angeles, Ca.	N315TS
N117JJ	75	Gulfstream 2SP	163	Gavilan Corp. Fort Lauderdale, Fl. 'El Condor'	N117JA
N117JL	77	Sabre-60	306-128	Silver Lining Aviation LLC. Westlake, Oh.	N100CE
N117K	73	Learjet 24D	24D-272	Koch Industries Inc. Wichita, Ks.	N51GL
N117MB	77	Sabre-65	465-1	Centurion Investments Inc. St Louis, Mo. (was 306-136).	N65KJ
N117RB	77	Learjet 35A	35A-154	Gillman/Universal-Ramsbar Inc. Houston, Tx.	N650NL
N117TA	83	Citation II	550-0464	Williamson Oil Co. Fort Payne, Al.	XA-SQR
N117TF	88	Falcon 900	42	Tudor Investment Corp. Ronkonkoma, NY.	N42FJ
N118AD	78	Falcon 10	118	Cooper/T Smith Stevedoring Co. Mobile, Al.	(N97RJ)
N118AF	74	Westwind-1123	177	Seawinds Leasing Corp. Singer Island, Fl.	(N114ED)
N118B	77	JetStar 2	5211	Four Star International Inc. Laredo, Tx.	N821MD
N118DA	76	Learjet 35A	35A-081	AirNet Systems Inc. Columbus, Oh.	N3523F
N118EA	80	Citation II	550-0118	Solomon Associates Inc. Dallas, Tx.	EC-FIL
N118FX	97	Learjet 31A	31A-134	FlexJets/General Atlantic Service Corp. Dallas, Tx.	N5014E
N118MP	81	Westwind-Two	340	MedPartners Aviation Inc. Birmingham, Al.	N340PM
N118NP	67	Gulfstream 2SP	7	Jet Aviation California Inc. Burbank, Ca.	N93QQ
N118R	88	Gulfstream 4	1066	R J Reynolds Tobacco Co. Winston Salem, NC.	N443GA
Reg	Yr	Type	c/n	Owner/Operator	Prev Regn

Reg	Yr	Type	c/n	Owner/Operator	Prev Regn
N119AM	92	Falcon 50	226	Allianz AG. Munich, Germany.	F-WWHC
N119BA	65	Learjet 23	23-084	Barr-Clay Auto Sales Inc. Glenmoore, Pa.	N101JR
N119FX	97	Learjet 31A	31A-136	FlexJets/Paging Network Inc. Dallas, Tx.	
N119GA	66	BAC 1-11/401AK	072	Ackerley Comms/SS Aviation Inc. Seattle, Wa.	N310EL
N119LJ	97	Learjet 60	60-119	Learjet Inc. Wichita, Ks.	
N119MA	69	Learjet 24B	24B-200	Tango Juliet LLC. Dover, De.	N246CM
N119MH	83	Diamond 1A	A057SA	Wye Air Corp. Easton, Md.	N342DM
N119PH	79	Falcon 50	8	Hasbro Inc. Pawtucket, Rl.	(N119HT)
N119R	86	Gulfstream 4	1008	R J Reynolds Tobacco Co. Winston Salem, NC.	N412GA
N119RM	92	Citation VII	650-7018	B J McCombs, San Antonio, Tx.	N718VP
N120AF	65	Falcon 20DC	16	Career Aviation Academy, Stockton, Ca.	N122CA
N120DP	77	Citation 1/SP	501-0042	Turbo Aviation, Stuart, Fl.	I-TOSC
N120FX	97	Learjet 31A	31A-137	FlexJets-Ariel Corp. Dallas, Tx.	
N120GS	75	Gulfstream 2SP	167	Guess? Inc. Van Nuys, Ca.	N683FM
N120HC	88	Citation II	550-0577	Inland Paperboard & Packaging Inc. Indianapolis, In.	N100CX
N120JC	79	HS 125/700A	NA0247	Southwestern Jet Charter Inc. St Louis, Mo.	(N530TE)
N120JP	83	Citation II	550-0468	Kruse Inc. Auburn, In.	N123FH
N120LJ	97	Learjet 60	60-120	Learjet Inc. Wichita, Ks.	
N120MB	80	Learjet 35A	35A-307	AND Inc. Stamford, Ct.	(N119HB)
N120MH	78	HS 125/700A	NA0238	M A Hanna Co/Hanna Mining Co. Cleveland, Oh.	N130MH
N120MP	84	Challenger 601	3034	MedPartners Aviation Inc. Birmingham, Al.	N372PG
N120NE	66	DC 9-15	45731	Genesis Aeronautics Inc. Van Nuys, Ca.	HB-IFA
N120Q	82	Citation II	550-0332	Therm-O-Disc Inc. Mansfield, Oh.	N12CQ
N120RA	67	Learjet 24	24-153	Royal Air Freight Inc. Waterford, Mi.	N153BR
N120SL	73	Learjet 25B	25B-120	Silver Lining Aviation LLC. Westlake, Oh.	N101FU
N120TF	78	Falcon 20F	384	Inland Container Corp. Indianapolis, In.	N120CG
N120WH	78	Falcon 20	385	Inland Paperboard & Packaging Inc. Indianapolis, In.	G-FRAA
N120YB	80	HS 125/700A	NA0282	Bemis Co Inc. Minneapolis, Mn.	N1982G
N121AT	89	Falcon 100	226	Anthony Timberlands Inc. Bearden, Ar.	XA-RLX
N121CG	86	Citation S/II	S550-0123	Phantom Sales Inc. Plantation, Fl.	N1293A
N121CK	65	Learjet 23	23-039	Fieldtech Avionics & Instruments Inc. Fort Worth, Tx.	(N9JJ)
N121CP	94	CitationJet	525-0083	Nonami Aircraft Associates, Atlanta, Ga.	N34TC
N121DF	93	Challenger 601-3A	5133	Cintas Corp. Cincinnati, Oh.	N53DF
N121EL	68	Learjet 25	25-010	AirNet Systems Inc. Columbus, Oh.	(N82UH)
N121FX	97	Learjet 31A	31A-141	FlexJets, Dallas, Tx.	
N121GV	96	Astra-1125SPX	082	Starship Enterprise Leasing/Greenspun Inc. Las Vegas, Nv.	
N121JE	67	Sabre-60	306-4	Falcon General Inc. Plant City, Fl.	N1210
N121JJ	88	Gulfstream 4	1075	Liamaj Aviation Inc. Houston, Tx.	N901K
N121JM	81	Gulfstream 3	332	Aviation Methods/RIM Air, Menlo Park, Ca.	N65BE
N121LJ	96	Learjet 31A	31A-121	Renfro Corp. Mount Airy, NC.	
N121TW	81	Challenger 600S	1012	Transworld Corporate Aviation Inc.	N121VA
N121US	85	Learjet 55	55-123	First USA Management Inc. Dallas, Tx.	N44EL
N122AP	73	Citation Eagle	500-0122	G B L Corp. Las Vegas, Nv.	N122LM
N122DU	67	Gulfstream 2	6	Pincervale Ltd-UK/Jet Services Corp. Houston, Tx.	N122DJ
N122FX	07	Learjet 31A	31A-143	FlexJets, Dallas, Tx.	
N122G	79	Citation II	550-0110	Dakota Investment Corp. Wilmington, De.	N222SG
N122HM	80	Citation II	550-0129	Heilig Meyers Furniture Co. Richmond, Va.	N122MM
N122JW	79	Learjet 35A	35A-217	AirNet Systems Inc. Columbus, Oh.	N111RF
N122M	65	Learjet 23	23-065A	Aero Flight Service Inc. Henderson, Nv.	(N156AG)
N122MP	83	Westwind-1124	390	MedPartners Aviation Inc. Birmingham, Al.	N59SM
N122SP	82	Citation II/SP	551-0393	Sklar & Phillips Oil Co. Shreveport, La.	(N18CC)
N122ST	57	Jet Commander-B	122	BTU Energy Inc. Bellevue, Wa.	XA-SCV
N122SU	87	Learjet 55B	55B-132	ESA Management Inc. Fort Lauderdale, Fl.	(N133SU)
N122WS	86	Citation S/II	S550-0122	Weather Shield Manufacturing Inc. Medford, Wi.	I-TALG
N123AV	97	CitationJet	525-0180	Marlis Aviation Inc. Sparta, Mi.	
N123CC	85	Gulfstream 3	455	Circus Circus Enterprises Inc. Las Vegas, Nv.	N103GC
N123CD	64	Sabre-40	282-23	Falcon General Inc. Plant City, Fl.	(N55ME)
N123CG	79	Learjet 25D	25D-270	General Transport Services Inc. Las Vegas, Nv.	N842GL
N123CV	74	Westwind-1123	178	International Computer Collections, Bowman, ND.	N999U
N123DG	76	Learjet 24F	24F-342	Glynn Air Inc. Bartlesville, Ok.	N824GA
N123EB	77	Citation 1/SP	501-0020	CPM Research West Inc.Orange, Ca.	N32JJ
N123FG	74	Sabre-60	306-90	Centurion Investments, St Louis, Mo.	N148JP
N123H	70	BAC 1-11/414EG	163	Earthwinds/BAC 1-11 Corp. Las Vegas, Nv.	D-AILY
N123JN	93	CitationJet	525-0046	Union Air LLC. Camp Douglas, Wi.	(N36881)
N123KD	81	Citation 1/SP	501-0195	Kentucky Derby Hosiery Co. Hickory, NC.	N109DC
N123LC	82	Learjet 55	55-034	Summit Aviation Corp. Republic, NY.	N334JW
N123SL	86	Citation III	650-0134	Silverline Windows & Doors/SL Systems Inc. N Brunswick, NJ.	N27SD
N123VP	79	Citation II	550-0111	Goodwyn Sales LLC. Wilmington, De.	(N3184Z)
N124DC	84	Sabre-60	306-95	Drummond Company Inc. Birmingham, Al.	N999DC
N124HM	83	Falcon 50	117	BLC Corp/Heilig-Myers Furniture Co. Richmond, Va.	N50TG
N124JL	66	Learjet 24	24-127	Dolphin Aviation Inc. Sarasota, Fl.	(N6462)
N124MA	73	Learjet 25B	25B-118	B & C Aviation Co Inc. Nashville, Tn.	(N800JA)
N124TS	71	Learjet 24XR	24XR-233	Kittyhawk Enterprises LLC. Batavia, Il.	(N56GH)
N124UF	79	Westwind-1124	257	USF Dugan Inc. Wichita, Ks.	N942FA
N124VS	66	Jet Commander	64	International Fruit & Produce Co Ltd. Hong Kong.	N124JB
N125AC	82	Challenger 600S	1072	Avia Carriers Ltd. Luton, UK.	N331FP

| Reg | Yr | Type | c/n | Owner/Operator | Prev Regn |

Reg	Yr	Type	c/n	Owner/Operator	Prev Regn
N125AD	65	HS 125/1A-522	25046	Walter Conlogue/Air Wilmington Inc. Wilmington, NC.	N812TT
N125AN	82	Challenger 600S	1073	Avia Carriers Ltd. Luton, UK.	(N512AC)
N125AS	82	HS 125/700A	NA0333	American Saw & Manufacturing Co. East Longmeadow, Ma.	(N301AS)
N125BH	65	HS 125/1A	25027	Schultz Family Trust, Newport Beach, Ca.	N227DH
N125CA	82	Falcon 100	196	Air Waukegan LLC. Waukegan, Il.	N573J
N125CK	71	HS 125/F400A	25266	Kalitta Flying Service Inc. Morristown, Tn.	N135CK
N125CS	77	HS 125/700A	NA0212	Clintondale Aviation, Tashkent, Uzbekistan.	N662JB
N125DC	69	Gulfstream 2	55	Drummond Co Inc. Birmingham, Al.	N225SE
N125DS	75	Citation	500-0258	Data Sales Financial Inc. Burnsville, Mn.	N886CA
N125DT	74	TriStar 100	1079	Ultimate Air Corp/Donald Trump, NYC.	C-GIFE
N125EA	80	Citation 1/SP	501-0125	Eastern Alloys Inc. Maybrook, NY.	N69EP
N125F	67	HS 125/3A-RA	25151	World Jet Aircraft International Sales, Fort Lauderdale, Fl.	G-AVTY
N125FD	66	HS 125/3A	25123	T M Aviation Inc. Charlotte, NC.	N44PW
N125GK	67	HS 125/F3B	25127	First Hawker Holdings Inc. Wilmington, De.	G-KASS
N125GM	93	BAe 1000A	9038	Western Aviation Inc. Key Biscayne, Fl.	N167BA
N125GS	76	HS 125/600A	6055	Great Southern Wood Preserving Inc. Abbeville, Al.	N100QP
N125HF	77	HS 125/600A	6064	Henig Furs Inc. Montgomery, Al.	N600SN
N125JJ	68	Gulfstream 2	15	JABJ/Jack Kent Cooke Inc. Middleburg, Va.	N416SH
N125JL	72	Learjet 25B	25B-088	Universal Jet Aviation Inc. Boca Raton, Fl.	N42FE
N125JR	65	HS 125/1A	25052	Aero Charter Services Inc. Wilmington, De.	N388WM
N125N	85	Challenger 601	3044	Natural Gas Pipeline Co/Midcon Corp. Chicago, Il.	N921K
N125NT	66	HS 125/3A	25078	Wings for Christ Inc. Beach Grove, In.	N448DC
N125NX	71	Sabre-75	370-3	Select Leasing Inc. Brookfield, Wi.	N125N
N125PS	86	Challenger 601	3058	Omni Restaurant Consulting Co. Newport Beach, Ca.	C-G...
N125Q	86	Citation III	650-0128	Milliken & Co. Greenville, SC.	N628CC
N125RH	91	Citation V	560-0147	International Jet Traders Inc. Fort Lauderdale, Fl.	XA-RKX
N125RJ	87	Beechjet 400	RJ-25	Raytheon Aircraft Holdings Inc. Wichita, Ks.	I-MPIZ
N125RM	75	Learjet 25B	25B-193	APC Delaware Inc. San Antonio, Tx.	N350DH
N125RT	69	HS 125/731	NA733	Promotions Unlimited Corp. Bannockburn, Il.	N31GT
N125SB	85	BAe 125/800A	8046	McClatchy Newspapers Inc. Sacramento, Ca.	N800BA
N125SJ	80	HS 125/700A	NA0275	Starstruck Jet Inc. Gallatin, Tn.	N404CF
N125TJ	66	HS 125/731	25121	Global Aircraft Holdings Inc. Alajuela, Cost Rica.(status ?)	XB-GHC
N125WJ	75	HS 125/600B	6053	World Jet Inc. Fort Lauderdale, Fl.	HC-BUR
N125XX	80	HS 125/700A	NA0254	Surewings Inc. Luton, UK.	N125AR
N126GA	83	Diamond 1A	A059SA	Tennessee Aluminum Processors Inc. Columbia, Tn.	I-DOCA
N126KC	95	Hawker 800	8276	Key Corp Aviation Co. Cleveland, Oh.	N667H
N126KD	84	Learjet 55	55-096	Kardan Inc. Northbrook, Il.	N8010X
N126QS	87	Citation S/II	S550-0126	EJI Inc/NetJets-LaFarge Corp. Columbus, Oh.	N1293K
N126R	68	Falcon 20C	126	Reliant Airlines Inc. Ypsilanti, Mi.	N102ZE
N126TF	97	Citation Bravo	550-0815	Fox Lumber Sales Inc. Hamilton, Mt.	
N127BU	80	Citation II/SP	551-0179	Dolphin Express Corp. Dover, De. (was 550-0134).	N203BE
N127EM	88	Falcon 900	63	Kvaerner Inc/Executive Jet Charter. Farnborough, UK.	N90TH
N127GT	83	Learjet 55	55-067	127 GT Corp. Dover, De.	N120EL
N127KC	94	Hawker 800	8255	Key Corp. Cleveland, Oh.	N946H
N127SA	86	Westwind-Two	440	WEKEL SA. Bogota, Colombia.	N440WW
N127V	91	Learjet 31A	31A-036	El Paso Natural Gas Co. El Paso, Tx.	N316LJ
N128CA	79	Learjet 35A	35A-248	Ameriflight Inc. Burbank, Ca.	C-GBFA
N128JS	68	Learjet 25	25-017	Air Stewart Inc. Sarasota, Fl.	N128DM
N128TS	73	Gulfstream 2	128	Services Group of America Inc. Seattle, Wa.	N367EG
N129BA	81	Challenger 600S	1013	Flexible Airline Inc. Harrietta, Mi.	N601SA
N129DM	69	Learjet 24B	24B-187	B B Barr Enterprises, Dallas, Tx. (status ?).	N5WJ
N129JE	67	Falcon 20-5B	113	Sackett Corp. Rye, NY.	N731F
N129MC	96	Beechjet 400A	RK-129	Mississippi Chemical Corp. Yazoo City, Ms.	N1129X
N129ME	79	Learjet 24F	24F-357	Kingswood Aviation Inc. Los Gatos, Ca.	N288J
N129PJ	93	Citation V	560-0235	Flying Squall Inc. Scottsdale, Az.	N52RG
N129RH	93	Challenger 601-3A	5129	Alesco Inc. Tulsa, Ok.	C-G...
N129SP	82	Learjet 55	55-058	J V E Corp. Kilgore, Tx.	N200PC
N129TS	81	Citation II	550-0316	Services Group of America Inc. Seattle, Wa.	N42KC
N130F	75	Learjet 35	35-044	AirNet Systems Inc. Columbus, Oh.	(N44VW)
N130MW	68	HFB 320	1033	Air National Aircraft Sales & Service, Carmel, Ca. (wfu ?).	N132MW
N130RS	67	Learjet 24	24-138	Stern Holdings Inc. Dallas, Tx.	N94JJ
N130YB	80	HS 125/700A	NA0285	Curwood Inc. Oshkosh, Wi.	(N14WJ)
N131AP	86	Beechjet 400	RJ-10	Merlin's Odyssey Inc. West Palm Beach, Fl. (was A1010SA).	I-ALSE
N131AR	97	Learjet 31A	31A-139	Alan Ritchey Inc. Valley View, Tx.	N139LJ
N131BH	64	Sabre-40	282-18	Little Falls Surgical Clinic Inc. Gallatin Gateway, Mt.	N15TS
N131JR	77	Sabre-75A	306-131	Executive Aircraft Corp. Newton, Ks.	N61DF
N131LA	70	HS 125/400A	NA750	Southwest Jet Inc. Kansas City, Mo.	XA-RWN
N131RG	96	CitationJet	525-0159	Administrative Concepts, Long Beach, Ca.	
N131SB	75	Citation	500-0256	Sunbird Air Corp. Harrison, NY.	PT-OZT
N132AH	96	CitationJet	525-0132	Arbor Health Care Co. Lima, Oh.	
N132AP	74	Falcon 20F	312	132AP Aviation Inc. Bloomfield Hills, Mi.	N1971R
N132EP	82	Falcon 20F	463	Prince Transportation Inc. Holland, Mi.	N134JA
N132LA	69	Jet Commander-B	133	Alberto Herreros, Miami, Fl.	XB-GBZ
N132MW	68	HFB 320	1032	Air National Aircraft Sales & Service, Carmel, Ca. (wfu ?).	N130MW
N133BL	66	Learjet 24	24-133	Vail Jet Inc. Vail, Co.	N133DF
N133EJ	77	Learjet 35A	35A-133	NAVCOM Aviation Inc. Manassas, Va.	N133GJ

Reg	Yr	Type	c/n	Owner/Operator	Prev Regn
☐ N133W	65	Learjet 23	23-021	National Jets Inc. Tulsa, Ok. (status ?).	N427NJ
☐ N134BJ	97	Beechjet 400A	RK-134	Raytheon Aircraft Co. Wichita, Ks.	N1094D
☐ N134CM	97	Beechjet 400A	RK-144	Croft Metals Inc. Lumber Bridge, NC.	
☐ N134MJ	86	Citation III	650-0109	Tyler Jet LLC. Tyler, Tx.	N134M
☐ N134QS	87	Citation S/II	S550-0134	EJI Inc/NetJets-Reams Asset Management LLC. Columbus, Oh.	N1294M
☐ N134RG	83	Diamond 1A	A037SA	CheckFree Corp. Norcross, Ga.	N109TW
☐ N135AG	77	Learjet 35A	35A-132	Bankair Inc. West Columbia, SC.	N37TJ
☐ N135BC	90	Challenger 601-3A	5080	Burrell Professional Laboratories Inc. Crown Point, In.	N601DB
☐ N135BJ	90	Beechjet 400A	RK-135	DJMD Corp/Daniel Lewis, Coconut Grove, Fl.	N1135A
☐ N135DE	91	Learjet 35A	35A-667	U S Department of Energy, Albuquerque, NM.	N91566
☐ N135FA	76	Learjet 35A	35A-067	Crow Executive Air Inc. Millbury, Oh.	(N52FL)
☐ N135HC	88	Citation III	650-0158	Southern Bag Corp Ltd & others, Madison, Ms.	N121AT
☐ N135ST	78	Learjet 35A	35A-169	Southeast Toyota Distributors, Pompano Beach, Ca.	
☐ N136JP	80	Learjet 35A	35A-359	Jet Charter Inc.Van Nuys, Ca.	HB-VHB
☐ N136MW	69	HFB 320	1036	Kalitta Flying Service Inc. Morristown, Tn.	(N92047)
☐ N136SA	74	Citation	500-0136	C & H Aviation LLC. North Charleston, SC.	F-BUUL
☐ N137AL	92	CitationJet	525-0002	Adalid Corp. Alto, NM.	N25CJ
☐ N137CL	93	Challenger 601-3R	5137	Bombardier Capital Inc. Colchester, Vt.	C-G...
☐ N137MR	88	Citation III	650-0163	Motorola Inc. Schaumberg, Il.	N137M
☐ N137TA	83	Falcon 200	487	A2 Jet Leasing LC. Miami, Fl.	I-SOBE
☐ N137WC	76	Citation	500-0305	Collins Investment Corp. St Paul, Mn.	C-GMLC
☐ N138DM	81	Falcon 10	181	Jabil Circuit Inc. St Petersburg, Fl.	F-GJHG
☐ N138F	81	Falcon 50	57	First International Aviation/Dr Hassan, Caracas, Venezuela.	N57B
☐ N138FJ	77	Falcon 20F	369	Ethox Chemicals Inc. High Point, NC.	N420J
☐ N138MR	88	Citation III	650-0164	Motorola Inc. Schaumberg, Il.	N138M
☐ N138SA	73	Citation	500-0138	Miller Management Group Inc. Erie, Pa.	N3056R
☐ N138SR	59	B 707-138B	17697	Comtran Inc. San Antonio, Tx.	(N138MJ)
☐ N139M	97	Hawker 800XP	8330	Brunswick Corp. Waukegan, Il.	N330XP
☐ N139N	87	Citation III	650-0149	EJI Inc. Columbus, Oh.	N139M
☐ N140AK	82	Diamond 1	A026SA	Corporate Aircraft Partners Inc. Cleveland, Oh.	N900DW
☐ N140CA	73	Learjet 25B	25B-140	James Donaldson, Farmersville, Tx.	N403AC
☐ N140CH	89	Challenger 601-3A	5047	Wayfarer Ketch Corp. White Plains, NY.	C-G...
☐ N140DR	78	Westwind-1124	242	RAL Capital LLC. Scottsdale, Az.	N340DR
☐ N140GC	77	Learjet 25D	25D-225	Global Air Rescue Inc. Clearwater, Fl.	N808DS
☐ N140JC	97	Learjet 60	60-106	North American Plastics Inc. Madison, Ms.	N106LJ
☐ N140JS	68	HS 125/3A-RA	NA703	MAK Air Transport Inc. Winchester, Va.	N22GE
☐ N140LJ	97	Learjet 31A	31A-140	Learjet Inc. Wichita, Ks.	
☐ N140RC	65	Learjet 23	23-048	RBS Aviation Group Inc. Fort Myers, Fl.	N48MW
☐ N140RF	66	Sabre-40A	282-67	C L Frates & Co. Oklahoma City, Ok.	N711T
☐ N140RT	97	Falcon 50EX	261	Fort Howard Corp. Green Bay, Wi.	F-WWHL
☐ N140SC	74	TriStar 500	1067	Orbital Sciences Pegasus ALB. Bakersfield, Ca.	C-FTNJ
☐ N141AQ	91	Citation V	560-0141	Aeroquip Corp. Maumee, Oh.	N6876S
☐ N141H	83	Diamond 1A	A054SA	Harmon Industries Inc. Blue Springs, Mo.	N850TJ
☐ N141SM	82	Learjet 55	55-019	Barnett Investments Inc. Laverne, Ca.	YV-41CP
☐ N142B	90	Challenger 601-3A	5062	Blount Inc. Montgomery, Al.	N548W
☐ N142BJ	97	Beechjet 400A	RK-142	Raytheon Aircraft Credit Corp. Wichita, Ks.	
☐ N142DA	77	Citation 1/SP	501-0004	Amerijet International Inc. Fort Lauderdale, Fl.	N86JJ
☐ N142LG	78	Learjet 35A	35A-216	Life Guard Alaska/Era Aviation Inc. Anchorage, Ak.	N39MB
☐ N143BP	86	Citation S/II	S550-0085	Cornerstone Aviation LLC. Aspen, Co.	N8BG
☐ N143CK	73	Learjet 25B	25B-143	Pacific Flights Inc. Medford, Or.	N113RF
☐ N144AD	82	Falcon 50	112	Archer Daniels Midland Co. Decatur, Il.	N107FJ
☐ N144GA	79	Citation II	550-0065	Kingsland Aviation LLC. South Burlington, Vt.	ZS-RCS
☐ N144HE	79	Falcon 10	144	Super Shops Inc. San Bernardino, Ca.	N101TF
☐ N144JP	75	Citation	500-0281	GEM Aircraft Leasing Inc. Scottsdale, Az.	N72WC
☐ N144LJ	97	Learjet 31A	31A-144	Learjet Inc. Wichita, Ks.	
☐ N144PK	85	Gulfstream 3	447	Petersen Aviation/P K Aire Inc. Van Nuys, Ca.	N186DC
☐ N145AM	76	Learjet 35A	35A-078	A/M Transport Inc. Cincinnati, Oh.	N45AW
☐ N145CJ	96	CitationJet	525-0145	United Supermarkets of Oklahoma, Altus, Ok.	N.....
☐ N145CM	74	Citation	500-0224	Southern Pine Plantations of Georgia Inc. Macon, Ga.	N5FG
☐ N145DF	78	Citation 1/SP	501-0055	Star Diamond Co Ltd/Ambrion Aviation, Luton, UK.	N400PG
☐ N145G	66	Sabre-40	282-65	D E C Air Inc. Clinton Township, Mi.	N2232B
☐ N145LJ	90	Learjet 31A	31A-144	Learjet Inc. Wichita, Ks.	(N124FX)
☐ N145SH	73	Learjet 25B	25B-145	Valdosta Mall Inc. Duluth, Ga.	N2127E
☐ N145ST	98	Learjet 45	45-024	JM Family Enterprises Inc. Deerfield Beach, Fl.	
☐ N145W	87	Falcon 900	40	Jetflight Aviation Inc. Lugano, Switzerland.	N904M
☐ N146CF	84	Falcon 200	488	ChemFirst Inc. Jackson, Ms.	C-GTNT
☐ N146TA	97	Beechjet 400A	RK-146	Raytheon Travel Air Co/James Johnson, Wichita, Ks.	
☐ N147A	80	Westwind-1124	294	Salter Labs, Fresno, Ca.	HK-3884X
☐ N147BJ	97	Beechjet 400A	RK-147	Raytheon Aircraft Credit Corp. Wichita, Ks.	
☐ N147BP	73	Learjet 25B	25B-147	Echo Aviation Sales & Leasing, Delray Beach, Fl.	N911JG
☐ N147G	87	Falcon 100	214	W W Grainger Inc. Wheeling, Il.	N275FJ
☐ N147TA	85	Falcon 200	506	First Security Bank NA. Salt Lake City, Ut.	I-CNEF
☐ N147TW	14	Learjet 25	25-023	Sierra American Corp. Dallas, Tx.	N767SC
☐ N147VC	95	Citation V Ultra	560-0285	Vencor Inc. Louisville, Ky.	N285CV
☐ N147WS	72	Citation	500-0009	K3C Inc. Reno, Nv.	(N147DA)
☐ N147X	69	Falcon 20-5B	185	DX Service Co/Morian Aviation Inc. Houston, Tx.	N3WN

| Reg | Yr | Type | c/n | Owner/Operator | Prev Regn |

56

Reg	Yr	Type	c/n	Owner/Operator	Prev Regn
☐ N148C	94	Learjet 31A	31A-098	CTI of North Carolina Inc. Savannah, Ga.	N50378
☐ N148CJ	96	CitationJet	525-0148	FJL Leasing Corp. Cincinnati, Oh.	
☐ N148ED	80	Citation 1/SP	501-0148	Edwin Davis Inc. Daytona Beach, Fl.	N148EA
☐ N148H	77	Westwind-1124	206	Holly Corp. Dallas, Tx.	N100ME
☐ N148LJ	97	Learjet 31A	31A-148	Learjet Inc. Wichita, Ks.	
☐ N148MC	80	Falcon 20-5B	428	United of Omaha Life Insurance Co. Omaha, Ne.	N98R
☐ N148WC	68	Falcon 20C	148	West Coast Charter-Mega Air Inc. El Paso, Tx.	N888WS
☐ N149HP	79	Falcon 10	154	White Industries Inc. Bates City, Mo.	N777FJ
☐ N149MC	80	Falcon 20-5B	429	United of Ohama Life Insurance Co. Omaha, Ne.	N4286A
☐ N149TJ	73	Falcon 10	9	Tyler Jet LLC. Tyler, Tx.	N10TX
☐ N150CA	80	HS 125/700A	NA0286	Crown American Air Charters Inc. Johnstown, Pa.	(N156K)
☐ N150F	87	Citation III	650-0150	Dana Flight Operations Inc. Swanton, Oh.	(N1326D)
☐ N150JT	81	Falcon 50	40	ASC/Ventura Air Services Inc. Port Washington, NY.	N695ST
☐ N150K	82	Falcon 50	108	Koch Industries Inc. Wichita, Ks.	N350X
☐ N150MS	82	Learjet 55	55-049	Martin Sprocket & Gear Inc. Arlington, Tx.	D-CCHS
☐ N150RD	80	Citation II	550-0143	Transtar Industries Inc. Cleveland, Oh.	PT-LQW
☐ N150RM	78	Citation 1/SP	501-0076	RMM Transportation Services Inc. Lafayette, La.	N315MP
☐ N150RS	74	Learjet 25XR	25XR-162	Stern Holdings Inc. Dallas, Tx.	N97JJ
☐ N150SA	71	HS 125/731	NA766	Ray Thurston, Scottsdale, Az.	VH-PAB
☐ N150TT	74	Citation	500-0176	J-Air Inc. Mobile, Al.	G-TEFH
☐ N150TX	80	Falcon 50	13	TXI Aviation Inc. Dallas, Tx.	(N150NW)
☐ N150UC	82	Falcon 50	86	Trans UCU Inc. Kansas City, Mo.	HB-IAT
☐ N151AE	96	Falcon 2000	39	Aetna Life & Casualty, Windsor Locks, Ct.	N2061
☐ N151AG	66	Learjet 24	24-137	High Country Helicopters Inc. Montrose, Co.	N72FP
☐ N151CC	94	Challenger 601-3R	5167	FAA, Greenville, Tx. (for outfitting).	C-GLYA
☐ N151PJ	83	Learjet 55	55-074	Hampton Airways Inc. Louisville, Ky.	N74GL
☐ N151Q	87	Citation S/II	S550-0151	Southaire Inc. Memphis, Tn.	N151QS
☐ N151SG	65	HS 125/1A	25035	Aircraft Exchange, Carlsbad, Ca.	(N57TS)
☐ N151SP	77	Citation 1/SP	501-0021	Summa Peto LLC. Portola Valley, Ca.	ZS-MGL
☐ N151TB	74	Sabre-80A	380-11	Baldwin Aircraft Corp. Raleigh, NC.	N265SR
☐ N151WW	68	Learjet 24	24-170	Addison Aviation Services Inc. Dallas, Tx.	N200DH
☐ N152A	87	Gulfstream 4	1036	ITT Flight Operations Inc. Allentown, Pa.	
☐ N152KB	91	Gulfstream 4	1149	Commercial Financial Services Inc. Tulsa, Ok.	N149GU
☐ N152KC	96	CitationJet	525-0152	Crews Inc. Memphis, Tn.	
☐ N152NS	90	BAe 125/800A	NA0456	Norfolk Southern Railway Co. Norfolk, Va.	N618BA
☐ N153AG	65	Learjet 23	23-058	Great Oaks Institute of Technology, Cincinnati, Oh.	N7FJ
☐ N153BJ	97	Beechjet 400A	RK-153	Raythcon Aircraft Co. Wichita, Ks.	
☐ N153JP	74	Citation	500-0153	Coastal Atlantic Aviation Investments, Virginia Beach, Va.	N153WB
☐ N153NS	89	Challenger 601-3A	5056	Norfolk Southern Railway Co. Norfolk, Va.	N614CC
☐ N153RA	87	Gulfstream 4	1050	ITT Flight Operations Inc. Allentown, Pa.	
☐ N153TW	70	Learjet 25	25-053	Sierra American Corp. Dallas, Tx.	N37GB
☐ N154AG	65	Learjet 23	23-034	Aero Flight Service Inc. Ypsilanti, Mi.	N24FF
☐ N154C	79	Gulfstream 2	253	CONSOL Inc. Pittsburgh, Pa.	N15TG
☐ N154CC	85	Citation 1/SP	501-0687	On Time Aviation Inc. Englewood Cliffs, NJ.	N321FM
☐ N154DD	89	Astra-1125	029	Danka Business Systems Inc. Sparks, Nv.	N94TW
☐ N154JD	78	HS 125/700A	NA0223	JHD Aircraft Sales Co. Wilmington, De.	N154JS
☐ N154NS	94	Challenger 601-3R	5169	Norfolk Southern Railway Co. Norfolk, Va.	N773A
☐ N154PA	85	Falcon 50	154	Anschutz Corp. Denver, Co.	N404E
☐ N154SC	80	Citation 1/SP	501-0154	Edgar Thomas & Flying Tiger Films Inc. Santa Monica, Ca.	CC-CWW
☐ N154VP	91	Citation V	560-0154	Moran Foods Inc/Save-A-Lot Ltd. St Louis, Mo.	N503T
☐ N155AC	88	Citation II	550-0573	S E Aircraft Sales Inc. Jacksonville, Fl.	PT-OKM
☐ N155AG	69	Learjet 25	25-037	Aero Flight Service Inc. Ypsilanti, Mi.	N28AA
☐ N155AM	77	Learjet 35A	35A-131	MGA Transport Co. Amelia, Oh.	N26GD
☐ N155AV	67	JetStar-731	5104	ALG Transportation Inc. Kansas City. 'Christopher Columbus'.	N902KB
☐ N155BC	85	Learjet 55	55-115	M M Coal Co & B & S Resources Inc. Columbus, Oh.	N633AC
☐ N155DB	90	Learjet 55C	55C-141	Dal Briar, Dallas, Tx.	
☐ N155GB	88	Citation S/II	S550-0155	Comcar Services Inc. Auburndale, Fl.	N155QS
☐ N155HM	82	Learjet 55	55-036	Hicks-Muse/Berg Electronics Inc. St Louis, Mo.	N236JW
☐ N155J	69	Learjet 24B	24B-182	MIG Magic Inc. Tualatin, Or.	N500ZH
☐ N155JC	82	Learjet 55	55-071	Cole National Group Inc. Cleveland, Oh.	(N155UT)
☐ N155MP	81	Learjet 55	55-014	SMC Corp. Harrisburg, Or.	N550RH
☐ N155PT	94	Citation V	560-0257	Rig Corp/P J Taggares Co. Othello, Wa.	N1293L
☐ N155RB	85	Learjet 55	55-117	Rooney Brothers Co. Tulsa, Ok.	N255MB
☐ N155SP	89	Learjet 55C	55C-137	Mirage Enterprises Inc. Van Nuys, Ca.	N95SC
☐ N155TJ	80	Citation II	550-0155	Pixie Ice Cream P/L. Toowoomba, QLD. Australia.	N31RC
☐ N156ML	96	CitationJet	525-0156	Mark & Diana Levy, Paradise Valley, Az.	
☐ N156SC	92	Learjet 31A	31A-060	Society Aviation Co. Wilmington, De.	N2600Z
☐ N157CM	96	Citation VII	650-7057	Interstate Equipment leasing Inc. Troutdale, Or.	N5216A
☐ N157H	93	Gulfstream 4SP	1209	H J Heinz Co. Pittsburgh, Pa. 'Collegiality'	N445GA
☐ N157LH	78	Gulfstream 2TT	228	Hawk Aviation Inc. Wilmington, De.	(N30B)
☐ N157QS	88	Citation S/II	S550-0157	EJI Inc/NetJets-Swedish Match N A Inc. Columbus, Oh.	N2639N
☐ N158JA	97	Falcon 900EX	20	Tavistock Aviation Inc. Nassau, Bahamas.	N920EX
☐ N158RA	97	Gulfstream V	522	ITT Flight Operations Inc. Allentown, Pa.	N595GA
☐ N159LC	79	Citation 1/SP	501-0094	Five-O-One LLC. Misooka, Il.	N59CC
☐ N159MR	86	Citation III	650-0130	Motorola Inc. Schaumburg, Il.	N159M
☐ N159NB	74	Gulfstream 2B	140	NB Aviation Inc. Anchorage, Ak.	N730TK

Reg	Yr	Type	c/n	Owner/Operator	Prev Regn
☐ N160AG	68	HS 125/3A-RA	NA707	Aero Flight Service Inc. Fort Lauderdale, Fl.	SE-DHH
☐ N160GC	77	Learjet 36A	36A-030	Global Air Rescue Inc. Tampa, Fl.	(N36AX)
☐ N160H	84	Diamond 1A	A084SA	Forum Air Inc. Cincinnati, Oh.	N160S
☐ N160LC	82	Challenger 600S	1068	Royal Pacific Aviation LP/Executive Flight, Wenatchee, Wa.	N938WH
☐ N160W	72	Sabre-40A	282-101	Northrop Grumman Corp. Los Angeles, Ca.	N101RR
☐ N161AC	77	Learjet 25D	25D-230	USA Aviation Sales Inc. Melbourne, Fl.	N16LJ
☐ N161CC	74	Citation	500-0161	Interplanetary Aviation & Park City, Ut.	C-GHEC
☐ N161CM	67	Sabre-60	306-5	CMI Corp. Oklahoma City, Ok.	(N477JM)
☐ N161MA	86	Learjet 35A	35A-610	Richards Aviation Inc & RB Aviation LC. Memphis, Tn.	N101AR
☐ N161MM	86	BAe 125/800A	8061	JABJ/Berkshire Inn Inc-Goodman Co. Palm Beach, Fl.	N365BA
☐ N161RB	79	Learjet 25D	25D-294	Superior Transport Service Inc. Palm Springs, Ca.	N125TJ
☐ N161WC	73	Citation	500-0117	Washington Corporations, Missoula, Mt.	N90BA
☐ N161X	79	Westwind-1124	234	Westwind Co. Pottstown, Pa.	N1124Z
☐ N162JB	70	Sabre-60	306-62	Executive Aircraft Corp. Newton, Ks.	N62CF
☐ N162JC	83	Gulfstream 3	373	Eugene Air Inc. Los Angeles, Ca.	VP-BAB
☐ N162W	66	BAC 1-11/401AK	087	Northrop Grumman Corp. Los Angeles, Ca.	N173FE
☐ N163A	74	Learjet 35A	35A-073	Corporate Aviation Services Inc. Tulsa, Ok.	(N64FN)
☐ N163AG	69	HS 125/3A-RA	25169	Aero Flight Service Inc. Fort Lauderdale, Fl.	N122AW
☐ N163AV	80	Falcon 10	163	Aerovertigo Inc/Aero Air Inc. Hillsboro, Or.	N163CH
☐ N163DC	67	Jet Commander	89	Covenant Sales & Leasing Inc. New Orleans, La.	N10BK
☐ N163DL	73	Westwind-1124	163	East Coast Jet Center, Fort Lauderdale, Fl. (status ?).	N47DC
☐ N163MR	92	Challenger 601-3A	5113	Motorola Inc. Schaumburg, Il.	N163M
☐ N163WC	77	Westwind-1124	217	Washington Corporations, Missoula, Mt.	N217WC
☐ N164MA	86	Falcon 50	164	Cie IBM France, Paris-Le Bourget, France.	HB-IAM
☐ N164RJ	86	Gulfstream 3	482	Bausch & Lomb. Rochester. NY.	N600BL
☐ N164W	66	BAC 1-11/401AK	090	Northrop Grumman Corp. Los Angeles, Ca.	G-AXCK
☐ N164WC	81	HS 125/700A	NA0326	Washington Corporations, Missoula, Mt.	(N702TJ)
☐ N165CM	77	Learjet 24E	24E-355	ECA Inc/La Tour Air Inc. New Baltimore, Mi.	N500NH
☐ N165G	83	Gulfstream 3	414	Key Air Inc/Jet Flight Corp. Oxford, Ct.	N165ST
☐ N165HB	94	Beechjet 400A	RK-90	M & M Air Inc. Raleigh, NC.	N1570L
☐ N165MC	80	Citation II	550-0182	Groupe Valfond, Paris-Le Bourget, France.	N30XX
☐ N165ST	88	Gulfstream 4	1053	JM Family Enterprises Inc. Jacksonville, Fl.	N91AE
☐ N165W	68	B 737-247	19605	Northrop Grumman Corp. Los Angeles, Ca.	N4508W
☐ N166HL	79	Learjet 35A	35A-235	Schooner Inc. East Aurora, NY.	N256MA
☐ N166KB	96	Citation V Ultra	560-0374	Cessna Aircraft Co. Wichita, Ks.	N7728T
☐ N166MA	80	Citation II/SP	551-0180	Miller Aviation Inc. Johnson City, NY. (was 550-0136).	D-ICAB
☐ N166RM	90	Astra-1125SP	047	Veloz Corp. Santa Barbara, Ca.	N30AJ
☐ N166WC	83	Gulfstream 3	413	Jet Charter Services P/L. Melbourne, VIC. Australia	249
☐ N167G	77	JetStar 2	5212	Flying Franci's, Houston, Tx.	(N95SR)
☐ N168CV	92	Citation V	560-0168	Lee Enterprises Inc. Davenport, Ia.	
☐ N168W	65	Sabre-40	282-33	Northrop Grumman Corp. Los Angeles, Ca.	N903KB
☐ N168WC	86	Gulfstream 4	1002	Washington Corporations, Missoula, Mt.	N440GA
☐ N169CA	94	Gulfstream 4SP	1241	Computer Associates International Inc. Islandia, Ny.	N487GA
☐ N169P	75	Gulfstream 2	169	G L Nemirow Inc/Terry Hines & Assocs. Burbank, Ca.	N7155P
☐ N169TA	80	HS 125/700A	NA0281	Truman Arnold Companies, Texarkana, Tx.	C-GXYN
☐ N169US	74	Learjet 24D	24D-298	MAL Aircraft Sales Inc. Memphis, Tn.	N470TR
☐ N170BG	97	CitationJet	525-0170	C M Gatton, Linville, NC.	
☐ N171GA	69	HFB 320	1039	Grand Aire Express Inc. Monroe, Mi.	N208MM
☐ N171SG	78	JetStar 2	5227	Gilrichco Inc/Style Air, Van Nuys, Ca.	N110AN
☐ N171TS	76	HS 125/600A	6071	Hardees Food Systems Inc. Rocky Mount, NC.	N121SG
☐ N172CV	92	Citation V	560-0172	Cessna Aircraft Co. Wichita, Ks.	
☐ N172MA	73	Citation	500-0110	BMUS Corp. Tempe, Az.	N368K
☐ N173A	80	Sabre-65	465-20	Versa Technologies Inc. Racine, Wi.	N2544E
☐ N173GA	71	HFB 320	1052	Grand Aire Express Inc. Monroe, Mi.	PT-IDW
☐ N173HH	73	Citation	500-0082	Eagle Consulting Inc. Miami, Fl.	N178HH
☐ N173PS	89	Learjet 31	31-009	Ossian Airways III Inc & others, Rochester, NY.	OO-JBA
☐ N173RD	75	Learjet 24XR	24XR-319	Dodson International Parts Inc. Ottawa, Ks.	XC-SUP
☐ N173VP	89	Citation III	650-0173	Mercury Communications Inc. Coudersport, Pa.	N843G
☐ N173W	94	Astra-1125SPX	073	Sherwin-Williams Co. Cleveland, Oh.	4X-WIX
☐ N174A	89	BAe 125/800A	NA0445	BankAmerica Special Assets Corp. Oakland, Ca.	N597BA
☐ N174B	79	Falcon 10	142	Kroger Co. Cincinnati, Oh.	N5LP
☐ N174FJ	81	Falcon 10	96	Tyler Jet LLC. Tyler, Tx.	I-LCJT
☐ N174GA	66	Falcon 20C-5	27	Grand Aire Express Inc. Monroe, Mi.	N33TP
☐ N174SA	84	Diamond 1A	A072SA	Spitfire Aviation, Goldsboro, NC.	PT-LGD
☐ N174SJ	91	Gulfstream 4	1174	Sky Jet Aviation Estab/Vad-Air Service SA. Lugano.	HB-IEQ
☐ N175BG	83	Gulfstream 3	396	Bentley Aviation LLC. Van Nuys, Ca.	N800MK
☐ N175BJ	97	Beechjet 400A	RK-175	Raytheon Aircraft Co. Wichita, Ks.	
☐ N175FJ	77	Falcon 10	97	Marmac Corp. Parkersburg, WV.	F-WPXF
☐ N175FS	65	Learjet 24A	24A-031	Furmanski Imaging Inc. S Las Vegas, Nv.	N202BA
☐ N175GA	66	Falcon 20C	45	Grand Aire Express Inc. Monroe, Mi.	N202KH
☐ N175J	89	Citation III	650-0168	Dana Flight Operations Inc. Swanton, Oh.	N1314H
☐ N175PS	94	CitationJet	525-0087	C I O S, Memphis, Tn.	
☐ N175ST	88	Challenger 601-3A	5023	JM Family Enterprises Inc. Fort Lauderdale, Fl.	N601CJ
☐ N175VB	81	Citation II	550-0202	Bancstar, Phoenix, Az.	(N5GA)
☐ N176CF	96	Falcon 900B	160	CIGNA Corp. Windsor Locks, Ct.	F-WWFA
☐ N176FB	82	Citation 1/SP	501-0244	Fountain Powerboats Inc. Washington, NC.	N244SL

Reg	*Yr*	*Type*	*c/n*	*Owner/Operator*	*Prev Regn*

Reg	Yr	Type	c/n	Owner/Operator	Prev Regn
N176GA	71	HFB 320	1053	Grand Aire Express Inc. Monroe, Mi.	PT-IOB
N177BC	74	Falcon 10	25	Lewis Computer Services Inc. Baton Rouge, La.	N719AL
N177JB	96	CitationJet	525-0182	Classic Auto Campus LLC. Wilmington, De. ·	(N740JB)
N177JC	67	Jet Commander	77	Centennial Machine/Harrison Haynes, Gainesville, Ga.	N121JC
N177RE	93	CitationJet	525-0030	UTE Management Corp. Oklahoma City, Ok.	N1331X
N177SB	81	Learjet 35A	35A-401	Seven Bar Flying Service Inc. Albuquerque, NM.	N66LJ
N178CP	74	Learjet 35	35-005	Epps Air Service Inc. Atlanta, Ga.	N175J
N178GA	68	Falcon 20D	163	Grand Aire Express Inc. Monroe, Mi.	N500LD
N178HH	81	Westwind-Two	347	World Marketing Alliance Inc. Norcross, Ga.	N666CP
N179GA	67	Falcon 20C	100	Grand Aire Express Inc. Monroe, Mi.	I-VEPA
N179T	70	Gulfstream 2B	86	Texas Eastern Transmission Corp. Houston, Tx.	N880GA
N180AR	74	Gulfstream 2B	148	JABJ/ASARCO Inc. Teterboro, NJ.	N2815
N180CP	96	Learjet 60	60-081	Pohlad Companies Inc. Minneapolis, Mn.	N60LJ
N180GC	74	Learjet 36	36-004	Global Air Rescue Inc. Clearwater, Fl.	N50DT
N180MC	78	Learjet 35A	35A-212	Integrated Control Systems Inc. Punta Gorda, Fl.	N3803G
N180NA	77	Sabre-75A	380-51	Corporate Flight Inc. Detroit, Mi.	N80LX
N181G	84	Citation S/II	S550-0006	Boston Post Leasing Corp. Manchester, NH.	N65DT
N181JC	95	Challenger 601-3R	5173	Johnson Controls Inc. Milwaukee, Wi.	C-GLXY
N181JT	94	CitationJet	525-0081	CJ Aviation LLC. Niwot, Co.	N5090V
N181RB	66	Falcon 20C	66	International Group Inc. Horseheads, NY.	N109RK
N181SG	92	Citation V	560-0181	Garaventa Co. Concord, Ca.	N1280R
N181WT	84	Falcon 20F-5	478	Flagstar Corp. Spartanburg, SC.	N161WT
N182GA	68	Falcon 20C	146	Grand Aire Express Inc. Monroe, Mi.	C-FCDS
N182K	80	Learjet 35A	35A-293	Koch Industries Inc. Wichita, Ks.	
N183FD	78	Learjet 35A	35A-183	Carlsbad Air Services Inc. Encinitas, Ca.	(N137TS)
N183GA	68	Falcon 20C	147	Grand Aire Express Inc. Monroe, MI.	N41154
N183SC	70	Gulfstream 2SP	91	Sequent Computer Systems Inc. Hillsboro, Or.	N99ST
N183TS	74	Falcon 20-5B	313	B & C Co. Columbus, Ga.	N212PB
N184GA	72	Falcon 20E	266	Grand Aire Express Inc. Monroe, Mi.	N4115B
N184TB	80	HS 125/700A	NA0260	Thomas & Betts Holdings Inc. Memphis, Tn.	N965JC
N185BA	75	Learjet 35	35-025	N185BA Lear Inc. Van Nuys, Ca.	(N188JA)
N186DS	90	Gulfstream 4	1154	DSC Communications Corp. Addison, Tx.	N151GX
N186G	84	BAe 125/800A	8011	EAF Aviation Corp. Wilmington, De.	N1BG
N186SC	74	Citation	500-0186	Cumberland Aircraft Sales & Leasing Corp. Portland, Me.	N186MW
N186VP	90	Citation III	650-0186	Dana Flight Operations Inc. Swanton, Oh.	PT-OAK
N187PH	78	Gulfstream 2	218	Medallion Enterprises LLC & Rifkin & Assocs Inc. Denver, Co.	N218GA
N187TA	81	Citation II	550-0257	Westair Corp. Norcross, Ga.	XA-SQQ
N188BC	72	Learjet 25B	25B-078	Connerty Group Inc. Atlanta, Ga.	N276LE
N188PS	73	Sabre-40A	282-122	Southland Enviromental Services Inc. Jacksonville, Fl.	N40JW
N188R	80	Learjet 25D	25D-305	Ruby Tuesday inc. Mobile, Al.	N53TC
N188TA	79	Learjet 25D	25D-276	CRL Inc. Englewood, Co.	N188TQ
N188TC	96	Learjet 60	60-078	Ackerley Communications/T C Aviation Inc. Seattle, Wa.	N5068F
N188TJ	83	Gulfstream 3	399	Tyler Jet LLC. Tyler, Tx.	7T-VRD
N189CM	97	CitationJet	525-0189	Shoffner Industries Inc. Burlington, NC.	
N189H	89	Citation V	560-0004	Honeywell Inc. Phoenix, Az.	
N189K	90	Challenger 601-3A	5083	Crawford Fitting Co. Cleveland, Oh.	C-G...
N189MM	81	Falcon 20F	453	Garrett Aviation Services, Phoenix, Az.	N25S
N190BD	65	Falcon 20C	8	Grand Aire Express Inc. Monroe, Mi.	N1500
N190BP	69	Learjet 24B	24B-190	Zephyr Aviation LLC. Houston, Tx.	(N190DB)
N190DA	78	Learjet 35A	35A-156	DavisAir Inc. West Mifflin, Pa.	N190EB
N190EK	95	Challenger 601-3R	5190	Bombardier Aerospace Corp. Hartford, Ct. (for FAA).	C-GLWT
N190GC	75	Learjet 35	35-014	Global Air Rescue Inc. Clearwater, Fl.	N77LJ
N190H	77	Falcon 10	71	Owen Holdings Inc. Carter Lake, Ia.	(N203PV)
N190JJ	90	Citation III	650-0190	J J Gumberg Co. Pittsburgh, Pa.	N142B
N190K	81	Citation 1/SP	501-0192	Robert Klabzuba, Fort Worth, Tx.	(N6781G)
N190KL	96	Citation V Ultra	560-0380	Kwik Lok Corp/KLC Transportation Ltd. Yakima, Wa.	N380CV
N190L	81	Falcon 10	190	Stephen Haight, Sioux Falls, SD.	N36BG
N190MC	80	Falcon 50	26	MASCO Corp. Taylor, Mi.	N52FJ
N190WC	82	HS 125/700A	NA0329	Walker Aviation Inc. Jackson, Ms.	N512GP
N190WW	76	Westwind-1124	190	Carlisle Air Corp. Portsmouth, NH.	N890WW
N191BE	96	Challenger 601-3R	5191	Bombardier Capital Inc. Colchester, Vt.	C-G...
N191NC	97	Beechjet 400A	RK-143	National Century Financial Enterprises Inc. Dublin, Oh.	
N192G	74	Citation	500-0163	Southeastern Aviation LLC. Germantown, Tn.	
N192R	69	Falcon 20D	192	Reliant Airlines Inc. Ypsilanti, Mi.	N910W
N193DQ	82	Challenger 600S	1041	Duquesne Aviation Inc. Pittsburgh, Pa.	(N95DQ)
N193G	81	Citation V	560-0137	Childish Creations International Inc. Norcross, Ga.	N733H
N193RC	79	HS 125/700A	NA0258	Raymond Corp. Greene, NY.	N812M
N193SS	88	Citation II	550-0572	P & B Inc. Caroloina, PR.	(N12992)
N193TR	85	BAe 125/800A	8039	Redmond Products Inc. Chanhassen, Mn.	N200KF
N194BJ	98	Beechjet 400A	RK-194	Raytheon Aircraft Co. Wichita, Ks.	
N194DC	85	Citation III	650-0074	Yellow Pages/GTE Directories Corp. Dallas, Tx.	N234YP
N194JS	93	BAe 125/800A	8251	J R Simplot Co. Boise, Id.	N937H
N194MC	68	Falcon 20-5	135	MASCO Corp. Taylor, Mi.	N9999E
N194RC	80	Citation 1/SP	501-0146	Houston Street Properties LLC. Marshall, Tx.	N1VU
N194SA	93	Citation V	560-0238	I M P Inc. Paoli, Pa.	N46WB
N194WM	97	Challenger 604	5340	Challenger Administration LLC. Kirkland, Wa.	N606CC

Reg	Yr	Type	c/n	Owner/Operator	Prev Regn

Reg	Yr	Type	c/n	Owner/Operator	Prev Regn
☐ N195FC	89	Astra-1125	036	Frontier Corp. Rochester, NY.	N82RT
☐ N195JH	89	Beechjet 400	RJ-64	Jack Henry & Associates Inc. Monett, Mo.	N1564B
☐ N195KC	90	BAe 125/800A	NA0455	Kansas City Life Insurance Co. Kansas City, Mo.	N6UB
☐ N195L	92	BAe 1000A	9008	WHX Corp. NYC.	HZ-OFC2
☐ N195ME	95	CitationJet	525-0110	Methode Electronics Inc. Chicago, Il.	N5213S
☐ N195SV	93	Falcon 50	236	Silver Ventures Inc. San Antonio, Tx.	XA-HGF
☐ N196CF	69	Learjet 24B	24B-186	Air Ambulance Care Flight International Inc. Clearwater, Fl.	N73PS
☐ N196HA	97	CitationJet	525-0256	Cessna Aircraft Co. Wichita, Ks.	
☐ N196JH	88	Beechjet 400	RJ-52	Jack Henry & Associates Inc. Monett, Mo.	(N52EB)
☐ N196MC	87	BAe 125/800A	8081	Landex Inc. Romulus, Mi.	(N196MG)
☐ N196RJ	81	Citation II	550-0207	First Security Bank NA. Salt Lake City, Ut.	N207BA
☐ N196SA	96	Citation V Ultra	560-0384	IMP Inc. Paoli, Pa.	
☐ N196SD	81	Learjet 35A	35A-414	S D Aviation Investment Inc. Villanova, Pa.	N414TJ
☐ N196TB	69	Learjet 24B	24B-196	Brundage Management Co. San Antonio, Tx.	N196AF
☐ N197CF	76	Learjet 25B	25B-197	Air Ambulance Care Flight International Inc. Seminole, Fl.	N197WC
☐ N197HF	92	Citation II	550-0704	Hormel Foods Corp. Austin, Mn.	(N197GH)
☐ N197JS	65	JetStar-731	5069	Kalitta Flying Service Inc. Ypsilanti, Mi. (status ?).	XA-PGO
☐ N197JT	85	Learjet 35A	35A-597	CB Enterprises Inc. Duncan, SC.	N597BL
☐ N197LS	97	Learjet 25D	25D-363	La Stella Inc. Pueblo, Co.	XA-RSU
☐ N197PF	92	Beechjet 400A	RK-33	P I E Mutual Insurance Co. Cleveland, Oh.	N60B
☐ N197SD	96	Beechjet 400A	RK-126	BUR-CON LLC/Burklund Distributors Inc. East Peoria, Il.	N3226B
☐ N197SL	83	Diamond 1A	A069SA	ARCH Air Medical Services, St Louis, Mo.	N250GP
☐ N198GJ	78	Learjet 35A	35A-198	Colvin Aviation/Apple Jet Inc. Athens, Ga.	I-ALPT
☐ N198HF	91	Astra-1125SP	054	Hormel Foods Corp. Austin, Mn.	N70AJ
☐ N198JH	97	CitationJet	525-0265	Cessna Aircraft Co. Wichita, Ks.	
☐ N198T	76	Learjet 35A	35A-074	Premier Jets Inc. Hillsboro, Or.	N100T
☐ N199BT	80	Learjet 25D	25D-311	The Mortgage Banc Construction Co. Portland, Or.	ZS-NJH
☐ N199CJ	76	Learjet 35A	35A-071	U S Navy/Corporate Jets Inc. Pittsburgh, Pa.	N82GA
☐ N199CK	74	Citation	500-0216	Albatross Aviation Inc. Marshfield, Ma.	N99CK
☐ N199HF	88	Astra-1125	027	George A Hormel & Co/Hormel Foods Corp. Austin, Mn.	N199GH
☐ N199QS	89	Gulfstream 4	1099	EJI/Gulfstream Shares. Montvale, NJ.	N299FB
☐ N199SC	97	Learjet 60	60-114	Learjet Inc. Wichita, Ks.	N3014R
☐ N199SP	74	Citation	500-0199	Sverdrup Corp. St Peters, Mo.	
☐ N200A	90	Gulfstream 4	1138	Exxon Center Services, Dallas, Tx.	N403GA
☐ N200AB	69	Gulfstream 2	71	7-N Aircorp. Dallas, Tx.	N47A
☐ N200BA	81	Learjet 55	55-011	Austin Jet Ltd. Horseshoe Bay, Tx.	D-CREW
☐ N200BL	91	Beechjet 400A	RK-23	Bausch & Lomb Inc. Rochester, NY.	N107BJ
☐ N200CK	95	Citation V Ultra	560-0298	Wisconsin Tissue Mills Inc. Appleton, Wi.	N5166U
☐ N200CV	85	Citation S/II	S550-0064	Summit Seafood Supply Inc. Wilmington, De.	N200CX
☐ N200DE	87	Challenger 601-3A	5015	Dunavant Enterprises Inc. Memphis, Tn.	N601KR
☐ N200DW	65	JetStar-731	5058	Whiteco Industries Inc. Bartlett, Il.	N131EL
☐ N200GF	87	Citation II/SP	551-0556	Golden Flake Snack Foods Inc. Birmingham, Al.	(N12979)
☐ N200GH	71	Gulfstream 2SP	108	Meridian Services Inc. Teterboro, NJ.	N801GA
☐ N200GM	74	Citation	500-0142	King Leasing Corp. Carson City, Nv.	N650TF
☐ N200GP	92	Beechjet 400A	RK-53	General Parts Inc. Raleigh, NC.	N62KM
☐ N200GT	68	Falcon 20C	137	Allied Signal Inc. Phoenix, Az.	N777PV
☐ N200GY	78	HS 125/700A	NA0224	W Temple Webber Jr. Houston, Tx.	N200GX
☐ N200J	79	Falcon 20-5B	410	Anheuser Busch Companies Inc. St Louis, Mo.	N200CP
☐ N200JE	68	Falcon 20C	133	Jet Ex Inc. San Ramon, Ca.	N133FJ
☐ N200L	96	Falcon 900EX	2	Anheuser-Busch Companies Inc. Chesterfield, Mo.	F-WWFA
☐ N200LC	88	Gulfstream 4	1067	Loral Travel Services Inc. Teterboro, NJ.	N145ST
☐ N200LH	92	Citation VII	650-7005	De Luxe Check Printers Inc. Minneapolis, Mn.	N1259S
☐ N200LP	81	Diamond 1	A006SA	KPA LLC. Springfield, Mo.	N750TJ
☐ N200LS	92	Gulfstream 4	1182	Limited Stores Inc. Columbus, Oh.	N475GA
☐ N200LX	85	Citation S/II	S550-0061	Lexicon Inc. Little Rock, Ar.	N53JM
☐ N200MT	78	Corvette	40	Meyer Tool Inc. Cincinnati, Oh.	N601CV
☐ N200NE	95	Falcon 2000	22	Banc One Corp. Columbus, Oh.	F-WWMF
☐ N200NK	86	Citation S/II	S550-0095	Enkay Corp. Shreveport, La.	N1275H
☐ N200PM	90	Gulfstream 4	1147	Philip Morris Management Corp. White Plains, NY.	N419GA
☐ N200RT	87	Falcon 50	175	Semitool Inc. Kalispell, Mt.	N330MC
☐ N200SG	93	Falcon 50	239	Hill Air Corp. Dallas, Tx.	N239FJ
☐ N200SK	80	Gulfstream 3	319	SK Travel LLC. Raleigh, NC.	N319Z
☐ N200TW	81	Learjet 35A	35A-397	Top of the World LLC. Wilmington, De.	D-CLAN
☐ N200WK	72	Falcon 20-5B	261	Lawrence Rockefeller/Wayfarer Ketch, NYC.	N4368F
☐ N200WY	86	Falcon 200	509	TXI Aviation Inc/C & S Aviation Ltd. Dallas, Tx.	N70TH
☐ N200XR	76	HS 125/F600A	6058	M J Enterprises LLC. Wilmington, De.	(N658TS)
☐ N201PM	82	HS 125/700A	NA0337	Lukens Steel Co. Coatesville, Pa.	N2015M
☐ N202CE	79	Citation II	550-0097	Condor Express Corp. Mount Kiscoe, NY.	N404E
☐ N202CF	78	Citation 1/SP	501-0061	Corporate Flight/G R M Holdings LLC. St Clair Shores, Mi.	N1401L
☐ N202ES	76	JetStar 2	5204	618BR LLC. Pleasant View, Tx.	N25WZ
☐ N202JK	85	Citation III	650-0100	Royal Oak Enterprises Inc. Atlanta, Ga.	N200LL
☐ N202JS	73	Learjet 24D	24D-278	J R S Aviation Inc. St Petersburg, Fl.	N5695H
☐ N202LC	97	Learjet 31A	31A-147	Learjet Inc. Wichita, Ks.	N198KF
☐ N202TS	81	Citation II	550-0253	John Fabick Tractor Co. Fenton, Mo. (was 551-0308).	N75EC
☐ N202WR	78	Learjet 35A	35A-190	Dumont Associates Inc. Morris PLains, NJ.	N202VS
☐ N203A	70	Gulfstream 2	89	The Jet Place Inc. Dallas, Tx.	N100A

Reg	Yr	Type	c/n	Owner/Operator	Prev Regn
☐ N203AL	82	Learjet 35A	35A-483	Air Liquide America Corp. Houston, Tx.	N202BT
☐ N203BA	85	Beechjet 400	RJ-3	P & I Inc. Leawood, Ks.	N508DM
☐ N203JK	80	Sabre-65	465-39	Royal Oak Enterprises Inc. Atlanta, Ga.	N551FA
☐ N203JL	69	Learjet 24B	24B-203	Bruce Leven, Mercer Island, Wa.	N203CK
☐ N203RW	78	Learjet 35A	35A-203	American Jet International Corp. Houston, Tx.	VR-CUC
☐ N204AB	82	Citation II	550-0335	Snake River Air Services LLC. Jackson, Wy.	N667CG
☐ N204CA	78	Citation 1/SP	501-0283	James M Krueger, Newport Beach, Ca. (was 500-0377).	C-GPTC
☐ N204JC	89	BAe 125/800A	8175	St Thomas Energy Exports/Johnson Controls Inc. Milwaukee.	VR-BLQ
☐ N204JP	66	Falcon 20-5	24	Hidden Bridge Aviation Inc. Trappe, Md.	N25TX
☐ N204R	92	BAe 1000B	9017	Raytheon Co. Lexington, Ma.	N963H
☐ N204SM	88	BAe 125/800A	NA0423	ServiceMaster Aviation Corp. Lake in the Hills, Il.	(N241SM)
☐ N204TM	81	Westwind-1124	320	Globe Life & Accident Insurance Co. Dallas, Tx.	N60JP
☐ N204TW	69	Falcon 20DC	204	Sierra American Corp. Dallas, Tx.	EC-EGM
☐ N205AJ	77	Westwind-1124	205	Eagle Leasing Inc. Alpharetta, Ga.	SE-DLL
☐ N205BE	90	Citation II	550-0647	Basin Electric Power Cooperative, Bismark, ND.	N647CC
☐ N205BS	78	HS 125/700A	NA0234	Henderson Jet Sales Inc. Henderson, Nv.	N711YP
☐ N205CM	94	Citation V	560-0250	Charlotte-Macklenberg Hospital Authority, Charlotte, NC.	N1291Y
☐ N205FX	93	Learjet 60	60-005	FlexJets, Dallas, Tx.	N869JS
☐ N205K	74	Falcon 20F	319	Koch Gateway Pipeline Co. Wichita, Ks.	N77LA
☐ N205PC	89	Citation V	560-0010	PacifiCorp Trans Inc. Portland, Or.	
☐ N205R	91	Beechjet 400A	RK-30	Raytheon Co. Lexington, Ma.	
☐ N205SC	80	Citation II	550-0156	BellSouth Telecommunications Inc. Atlanta, Ga.	N205SG
☐ N205TS	66	Falcon 20-5	54	Aero Toy Store Inc. Fort Lauderdale, Fl.	N100HG
☐ N206FX	94	Learjet 60	60-028	FlexJets, Dallas, Tx.	N870JS
☐ N206MD	68	Gulfstream 2	22	JABJ/G B H Aviation, Jackson, Wy.	(N655JH)
☐ N206PC	97	Citation X	750-0016	PacifiCorp Trans Inc. Portland, Or.	
☐ N207BS	97	CitationJet	525-0241	Cessna Aircraft Co. Wichita, Ks.	(N241WS)
☐ N207CA	68	Falcon 20D	153	Jim Donaldson, Farmersville, Tx.	N70MD
☐ N207FX	95	Learjet 60	60-050	FlexJets/Sammons Corp. Dallas, Tx.	N50450
☐ N207JC	76	Learjet 25D	25D-207	B L Bennett Aviation LLC. Nashville, Tn.	I-LEAR
☐ N207MJ	58	MS 760 Paris-2	2	Wayne Cozad, Xenia, Oh.	N1EP
☐ N208FX	95	Learjet 60	60-059	FlexJets/Herman Miller Inc. Dallas, Tx.	N5059J
☐ N208JV	97	CitationJet	525-0208	Juliet Victor Airway LLC. Hot Springs, Ar.	
☐ N208PC	90	Citation V	560-0050	PacifiCorp Trans Inc. Portland, Or.	(N26771)
☐ N208R	91	BAe 1000A	NA1003	Raytheon Co. Lexington, Ma.	N14GD
☐ N209CV	93	Citation V	560-0209	209 Aircraft Co LLC. Monticello, Mo.	
☐ N209FX	95	Learjet 60	60-060	FlexJets/Calloway Golf Co. Dallas, Tx.	N50602
☐ N210F	84	Citation III	650-0067	Flint Aviation Corp. Tulsa, Ok.	(N1314X)
☐ N210FX	95	Learjet 60	60-064	FlexJets/Newport News Shipbuilding & Dry Dock Co. Dallas.	
☐ N210WL	78	Learjet 35A	35A-210	D & D Aviation LC. Salt lake City, Ut.	XB-FNF
☐ N211CC	91	Citation VI	650-0211	Crain Communications Inc. Detroit, Mi.	N333WC
☐ N211DH	79	Gulfstream 2TT	236	Hagadone Corp. Coeur d'Alene, Id.	N54BM
☐ N211EC	80	Falcon 10	166	Jet Flight II LLC. Easton, Md.	(N166SS)
☐ N211FX	96	Learjet 60	60-076	FlexJets/Aspenwood Aviation Inc. Dallas, Tx.	
☐ N211GA	81	Diamond 1A	A011SA	Corporate Leasing Co. Springfield, Mo.	(N77GA)
☐ N211JC	80	Learjet 25D	25D-310	Smail Aviation, Greensburg, Pa.	N211PD
☐ N211MA	89	Citation V	560-0022	ADT Aviations Inc. Wilmington, De.	(N1228Y)
☐ N211MT	69	Gulfstream 2	52	Global Trading Ltd. Bristol, UK.	N711MT
☐ N211QS	84	Citation S/II	S550-0011	RSM Leasing Co. Jackson, Mi.	N68SK
☐ N211ST	80	Westwind-Two	303	Stewart Title Co. Houston, Tx.	N50QJ
☐ N211TJ	74	Falcon 10	11	Refreshment Services Inc. Springfield, Il.	(N190DB)
☐ N211TS	65	Learjet 23	23-066	A J Leasing Inc. Wilmington, De.	XA-SDP
☐ N211WH	79	Learjet 35A	35A-269	Bontona Aviation Inc. Fort Lauderdale, Fl.	N225F
☐ N211X	78	Citation 1/SP	501-0058	BCR LLC. Galesburg, Il.	N211EF
☐ N212AP	72	JetStar-8	5147	KayDee Aviation Inc. Clayton, Mo. (status ?).	N718R
☐ N212AT	93	Gulfstream 4SP	1204	A T & T Corp. Morristown, NJ.	N435GA
☐ N212BD	95	Citation V Ultra	560-0309	Bill Heard Enterprises Inc. Columbus, Ga.	N52352
☐ N212FJ	79	Falcon 10	147	Aviation Charter Inc. Sioux City, Ia.	N125GA
☐ N212FX	97	Learjet 60	60-099	FlexJets/Dell USA LP. Dallas, Tx.	
☐ N212GA	80	Learjet 35A	35A-354	Apple Jet Inc. Athens, Ga.	D-CART
☐ N212H	79	Citation II/SP	550-0488	Sierra-Napa Citation Partners, Sparks, Nv. (was 551-0140).	N17S
☐ N212JP	82	Falcon 50	94	A T & T Corp. Morristown, NJ.	N82NC
☐ N212K	93	Gulfstream 4	1192	A T & T Corp. Morristown, NJ.	N407GA
☐ N212PA	83	Diamond 1	A038SA	Pioneer Aviation LLC. Eugene, Or.	N42SR
☐ N212R	69	Falcon 20DC	212	Reliant Airlines Inc. Ypsilanti, Mi.	N31FE
☐ N212T	97	Falcon 2000	52	Dassault Falcon Jet Corp. Teterboro, NJ.	F-WWMI
☐ N212TJ	68	Gulfstream 2SP	12	Tyler Jet LLC. Tyler, Tx.	N160WC
☐ N213C	84	HS 125/700A	7213	Vulcan Materials Co. Birmingham, Al.	G-BLEK
☐ N213CA	78	Learjet 25D	25D-241	Delaware DI Properties LLC. Dallas, Tx.	N713LJ
☐ N213JS	89	Citation II	550-0597	Oriole 550 Inc-JS Aviation Ltd. Luton, UK.	N400EX
☐ N213LS	67	Falcon 20C	107	LESEA-Lester Sumrall Evangelistic Assoc. South Bend, In.	(N91LS)
☐ N213MC	95	Challenger 601-3R	5171	Cape Clear Capital Corp. Manchester, NH.	C-GLXW
☐ N214CA	74	Citation	500-0214	Fox Air Inc. Wilmington, De.	(N371W)
☐ N214FX	94	Learjet 60	60-046	FlexJets/Alcan Management Services USA, Dallas, Tx.	(PT-WGB)
☐ N214QS	84	Citation S/II	S550-0014	EJI Inc/NetJets-First Security Bank, Salt Lake City, Ut.	N32TJ
☐ N214RW	83	Citation II	550-0478	Midwest Security Services Inc. Onalaska, Wi.	N57BC

Reg	Yr Type	c/n	Owner/Operator	Prev Regn

Reg	Yr Type	c/n	Owner/Operator	Prev Regn
□ N215DL	77 JetStar 2	5215	Tyler Jet LLC. Tyler, Tx.	XA-FHS
□ N215FX	97 Learjet 60	60-101	FlexJets/Entrepreneur Inc. Wichita, Ks.	
□ N215RS	78 HS 125/700A	NA0215	Stephens Institute/Academy of Art College, San Francisco, Ca	N195XP
□ N216CA	65 Falcon 20C	11	Delaware DI Properties LLC. Dallas, Tx.	N983AJ
□ N216FX	97 Learjet 60	60-103	FlexJets/Monsanto Co. Dallas, Tx.	
□ N217A	74 HS 125/600B	6030	Arnoni Aviation Inc. Houston, Tx. (for wfu ?).	G-TOMI
□ N217AJ	68 Falcon 20D	171	Real World Tours Inc. Nashville, Tn.	F-GICB
□ N217FS	81 Citation II	550-0273	Peter Crowell, West Springfield, NH.	N68637
□ N217FX	97 Learjet 60	60-105	FlexJets/Amway Corp. Dallas, Tx.	
□ N217JS	72 Gulfstream 2	113	Elite Charter Services Inc. West Palm Beach, Fl.	N211BL
□ N217RM	84 BAe 125/800A	8017	Current Aviation Group Inc.	N810P
□ N217RN	76 Sabre-60	306-94	Platina Investment Corp. Oklahoma City, Ok.	N217RM
□ N217RR	85 Citation III	650-0079	Raymond J Rutter, Irvine, Ca.	N59CD
□ N217SA	81 Citation II	550-0217	C A S Leasing LLC. Tappahannock, Va.	(D-IMME)
□ N218CA	69 Falcon 20DC	218	Delaware DI Properties LLC. Dallas, Tx.	EC-EEU
□ N218FX	97 Learjet 60	60-098	FlexJets/Jet Set Inc. Dallas, Tx.	N50578
□ N218JG	81 Citation 1/SP	501-0218	Curtis Transportation Inc. Bloomington, In.	N218AM
□ N218NB	81 Learjet 31A	31A-146	Beckair Co. Elkhart, In.	N30046
□ N218NR	83 Learjet 25D	25D-361	Beckair Co. Elkhart, In.	N218NB
□ N218TJ	65 HS 125/731	25018	Tyler Jet LLC. Tyler, Tx.	N118TS
□ N219CA	69 Falcon 20D	193	Delaware DI Properties LLC. Dallas, Tx.	9Q-CTT
□ N219CJ	97 CitationJet	525-0219	Dr F David Prentice, Kerrville, Tx.	N5197A
□ N219EC	70 HS 125/F400A	25219	Dodson Aviation Inc. Ottawa, Ks.	RA-02805
□ N219FX	92 Learjet 60	60-007	FlexJets/Business Jet Solutions LLC. Dallas, Tx.	N448HM
□ N220AB	90 Citation II	550-0630	Mid-America Dairymen Inc. Springfield, Mo.	PH-CSA
□ N220AU	71 DC 10-10ER	46501	Project Orbis International Inc. NYC.	G-GCAL
□ N220CA	69 Falcon 20DC	220	Addison Aviation Services Inc. Dallas, Tx.	EC-EDL
□ N220CJ	97 CitationJet	525-0220	Cessna Aircraft Co. Wichita, Ks.	
□ N220FX	97 Learjet 60	60-108	FlexJets-John McHale, Dallas, Tx.	N108LJ
□ N220JM	97 Falcon 2000	47	Interstate Equipment Leasing/Allied Signal Inc. Phoenix, Az.	F-WWMB
□ N220JR	69 Gulfstream 2	50	Air Tiger Inc/Key Air Inc. Oxford, Ct.	N220FL
□ N220JT	94 Citation V Ultra	560-0272	Champlain Air Inc. Plattsburgh, NY.	
□ N220LC	82 Challenger 600S	1071	HCE Leasing LLC. Cleveland, Tn.	N711DB
□ N220M	74 Falcon 10	34	Olan Mills Inc. Chattanooga, Tn.	N110M
□ N221CM	81 Gulfstream 3	343	Admerex Inc/Mardenaire Inc. NYC.	N400AL
□ N221EB	82 Citation 1/SP	501-0221	MFA Inc. Columbia, Mo.	N555HR
□ N221FJ	86 Falcon 200	505	Dodson's International Parts Inc. Ottawa, Ks, (status ?).	XA-SKO
□ N221FX	97 Learjet 60	60-111	FlexJets, Dallas, Tx.	
□ N221PA	87 Astra-1125	016	Rational Software Corp. Santa Clara, Ca.	N221DT
□ N221RE	88 BAe 125/800A	NA0413	Raytheon Engineers Constructors International, Philadelphia.	N203R
□ N221SG	78 Learjet 35A	35A-182	Path Corp. Rehoboth Beach, De.	N3HA
□ N221TR	79 Learjet 35A	35A-221	Career Aviation Academy, Stockton, Ca.	VH-FSY
□ N221TW	69 Falcon 20DC	221	Kitty Hawk Air Cargo Inc. Dallas, Tx.	EC-EIV
□ N222	90 Gulfstream 4	1142	Jetstream Inc. Seattle, Wa.	N142NW
□ N222B	69 Learjet 25	25-047	Kalitta Flying Service, Morristown, Tn.	
□ N222BE	82 Learjet 35A	35A-489	Mayo Aviation Inc/Cyprus Amax Minerals Co. Englewood, Co.	
□ N222BG	81 Learjet 35A	35A-448	Goody's Family Clothing Inc. Knoxville, Tn.	N48MJ
□ N222FA	93 Citation II	550-0725	Information Technology Inc. Lincoln, Ne.	N1207Z
□ N222FX	93 Learjet 60	60-008	FlexJets-First Security Bank, Dallas, Tx.	N608LJ
□ N222GT	85 Citation III	650-0093	Air Luxor Lda/European NetJets, Lisbon.	N93VP
□ N222HL	80 HS 125/700B	7088	Atsinger Aviation LLC. Van Nuys, Ca.	N29RP
□ N222LH	77 Westwind-1124	209	Lewis Hyman Inc. Van Nuys, Ca.	N988WH
□ N222MC	87 Falcon 50	179	A T & T Wireless Flight Operations Inc. Seattle, Wa.	HL7386
□ N222MF	78 JetStar 2	5229	Trey Aviation LLC. Minneapolis, Mn.	VR-CNM
□ N222MS	88 BAe 125/800A	NA0422	Stavola Aviation Inc. Anthony, Fl.	N125TR
□ N222MU	80 Falcon 10	164	Mustang Fuel Corp. Oklahoma City, Ok.	N228FJ
□ N222MW	79 Westwind-1124	255	McWane Inc. Birmingham, Al.	4X-CNA
□ N222NB	79 Gulfstream 2B	245	222 Aviation LLC. Burbank, Ca.	N99WJ
□ N222NG	65 HS 125/1A	25016	Rhino of Wilmington Inc. Wrightsville Beach, NC.	N4997E
□ N222TW	68 Learjet 24	24-161	Sierra American Corp. Dallas, Tx.	N24KF
□ N222WA	77 Citation 1/SP	501-0007	Unitco Air/Ralph Kiewit Jr. Van Nuys, Ca.	N5360J
□ N223DD	82 Falcon 50	128	Novus Credit Services Inc. Riverwoods, Il.	N733E
□ N223LB	80 Sabre-65	465-23	Smithco Leasing Inc. Camp Hill, Pa.	(N904KB)
□ N223WA	86 Westwind-1124	423	Texas International Gas & Oil Co. El Paso, Tx.	4X-CUC
□ N224CJ	97 CitationJet	525-0224	Air Management Inc. Wilmington, De.	
□ N224F	86 Challenger 601	3064	Freeman Air Charter Service Inc. Stowe, Vt.	N224N
□ N224KC	86 Citation S/II	S550-0104	Tri-State Executive Air Inc. Raleigh, NC.	N12903
□ N224MC	97 Beechjet 400A	RK-165	Midmark Corp. Versailles, Oh.	N2225Y
□ N225BC	75 Citation	500-0274	Hughes Unlimited Inc. Beaumont, Tx.	XB-ETE
□ N225BJ	78 HS 125/700A	NA0231	Jerbo Holdings VIII Inc. Chicago, Il.	N125G
□ N225JL	74 Learjet 25B	25B-182	Midlantic Jet Charters, Atlantic City Airport, NJ.	F-GFMZ
□ N225LS	65 Sabre-40	282-51	Sabre Investments, St Louis, Mo.	(N51MN)
□ N225LY	96 Challenger 604	5311	JABJ/Dacion Corp. Teterboro, NJ.	C-GLWR
□ N225N	81 Westwind-Two	319	JBW Aircraft Leasing Co. Seattle, Wa.	N200KC
□ N225SF	84 Gulfstream 3	423	Chevron Corp. Oakland, Ca.	N7134E
□ N225TR	78 Gulfstream 2	225	Rollins Properties Inc. New Castle, De.	N289K

Reg	Yr Type	c/n	Owner/Operator	Prev Regn

Reg	Yr	Type	c/n	Owner/Operator	Prev Regn
□ N226B	97	CitationJet	525-0200	B L Jennings Inc. Carson City, Nv.	N118A
□ N226G	89	Gulfstream 4	1122	W R Grace & Co. Boca Raton, Fl.	N40N
□ N226GA	71	Gulfstream 2	106	Gulfstream Aerospace Corp. Savannah, Ga.	N397LE
□ N226N	94	Citation V	560-0248	JFN Inc/Nordstrom Inc. Seattle, Wa.	N248CV
□ N226R	70	Falcon 20DC	226	Reliant Airlines Inc. Ypsilanti, Mi.	N21FE
□ N227GA	69	Gulfstream 2SP	76	KC Transportation Services, Ronkomkoma, NY.	N227G
□ N227GH	88	Gulfstream 4	1045	MDL Consulting Associates LLC. Nashua, NH.	N227G
□ N227HP	75	Citation Eagle	500-0227	Safari Motors SA. Santo Domingo, Dominican Republic.	C-GMMO
□ N227N	94	Learjet 60	60-049	Nordstrom Inc/NLC Inc. Seattle, Wa.	N50298
□ N227NL	91	Astra-1125SP	053	Stim-Air Inc. Hillsboro, Or.	N227N
□ N227R	70	Falcon 20DC	227	Reliant Airlines Inc. Ypsilanti, Mi. (status ?).	N24EV
□ N227TJ	68	Gulfstream 2	27	Hollywood Entertainment Inc. Wilsonville, Or.	N227TS
□ N228AJ	77	Citation 1/SP	501-0018	Arkansas Wholesale Lumber Co. Wilmington, NC.	N228AK
□ N228AK	80	Citation II/SP	551-0050	Tiftarea Air Charters Inc. Tifton, Ga. (was 550-0385).	N196HR
□ N228CC	80	Citation II	550-0133	Bettcher Industries Inc. Vermilion, Oh.	C-GRIO
□ N228N	94	Learjet 60	60-031	Nordstrom/NLC Inc. Seattle, Wa.	N4031L
□ N228S	74	Citation	500-0233	Great Western Aviation LLC. Beaverton, Or.	N223S
□ N228SW	77	Learjet 25D	25D-228	AirNet Systems Inc. Columbus, Oh.	
□ N229CJ	95	CitationJet	525-0129	B Four Flying Inc. Winfield, Ks.	
□ N229D	85	Westwind-Two	427	Diamond Management & Baker Management Inc. Tucson, Az.	N229N
□ N229R	70	Falcon 20DC	229	Reliant Airlines Inc. Ypsilanti, Mi.	N25EV
□ N229U	91	BAe 1000A	NA1002	United Technologies Corp. Hartford, Ct.	G-BTYN
□ N230R	77	Learjet 35A	35A-130	Hobart Corp. Dayton, Oh.	
□ N230RA	70	Falcon 20DC	230	Reliant Airlines Inc. Ypsilanti, Mi.	N26EV
□ N230TS	68	HS 125/3A-RA	NA700	Aero Toy Store Inc. Fort Lauderdale, Fl.	N946FS
□ N231JH	81	Falcon 10	176	Knoxville Aviation LLC. Knoxville, Tn.	N66HH
□ N231R	77	Learjet 35A	35A-128	Wilsonart International Inc. Temple, Tx.	
□ N232CC	81	Learjet 35A	35A-367	CCI Corp. Tulsa, Ok.	N97RJ
□ N232DM	79	Citation II	550-0079	Teterboro Aviation Inc. Teterboro, NJ.	N33RH
□ N232FX	86	Learjet 35A	35A-620	JABJ/Aviation Properties Inc. Winnetka, Il.	VR-BNI
□ N232K	94	Gulfstream 4SP	1232	Barbara Fasken Oil & Ranch, Midland, Tx.	N471GA
□ N232S	89	Astra-1125	032	Sherwin Williams Co. Cleveland, Oh.	N1125A
□ N233BC	95	Falcon 50	241	BASF Corp. Dover, De.	F-OKSI
□ N233CA	73	Learjet 25B	25B-133	Addison Aviation Services Inc. Dallas, Tx.	XA-RZY
□ N233CC	75	Learjet 35	35-031	CCI Corp. Tulsa, Ok.	N160AT
□ N233RS	78	Gulfstream 2TT	233	Rolling Stone-Ronkonkoma/Straight Arrow Publishers Inc. NY	N320TR
□ N233TW	70	Learjet 24B	24B-221	Sierra American Corp. Dallas, Tx.	N59JG
□ N234AQ	93	Citation V	560-0234	Aeroquip Corp. Swanton, Oh.	
□ N234AT	74	Citation	500-0240	Fox Hill Aviation Inc. Westwood, Ma.	N240CC
□ N234DB	85	Gulfstream 4	1000	Skybird Aviation, Van Nuys, Ca.	N404GA
□ N234G	65	Jet Commander	28	Aviation Business Corp. St Louis, Mo.	N77NR
□ N234JW	77	Citation 1/SP	501-0037	A & G Coal Corp. Wise, Va.	N19J
□ N234SV	77	Learjet 25D	25D-226	Valley Services, Sun Valley, Id.	N333SG
□ N234UM	73	Citation	500-0105	Marlin Air Inc. Detroit, Mi.	N32W
□ N234WS	95	CitationJet	525-0097	Four S's LLC. Greenville, NC.	N5153Z
□ N235AC	92	Learjet 35A	35A-676	Apache Corp. Houston, Tx.	N35LJ
□ N235CA	68	Falcon 20C	139	Jim Donaldson, Farmersville, Tx.	N900WB
□ N235DH	77	Learjet 35A	35A-134	JABJ/DHL Airways Inc. San Francisco, Ca.	N88EP
□ N235HR	84	Learjet 55	55-094	Hoffman La Roche Inc. Nutley, NJ.	
□ N235JS	78	Learjet 35A	35A-199	Fine Foligae International Inc. DeLeon Springs, Fl.	N444HC
□ N235KK	88	Citation III	650-0175	Kenneth Kirchman, Altamonte Springs, Fl.	N1820E
□ N235MC	80	Learjet 35A	35A-334	Smith's Food & Drug Centers Inc. Salt Lake City, Ut.	(N334AB)
□ N235SC	79	Learjet 35A	35A-275	McMillin Aircraft Inc. WinterPark, Fl.	(N65WH)
□ N235SV	93	Citation VI	650-0235	BI-GO Markets Inc/Supervalu Inc. Minneapolis, Mn.	N235CM
□ N236BN	79	HS 125/700A	NA0236	Naples Capital Development Corp. Naples, Fl.	N64HA
□ N236DJ	79	Falcon 10	138	Idaho Investment Inc. Idaho Falls, Id.	F-GGVR
□ N237AF	79	Learjet 35A	35A-262	Ameriflight Inc. Burbank, Ca.	N237GA
□ N237CJ	97	CitationJet	525-0237	Cessna Aircraft Co. Wichita, Ks.	
□ N237GA	88	Challenger 601-3A	5019	Orange County/Sunbird Aviation, Costa Mesa, Ca.	(N247GA)
□ N237TW	71	Learjet 24D	24D-237	Sierra American Corp. Dallas, Tx.	N825DM
□ N238CA	69	Learjet 25	25-040	Delaware DI Properties LLC. Dallas, Tx.	N23FN
□ N238RC	76	Learjet 35	35-061	Eagle Aviation Inc. West Columbia, SC.	N4246N
□ N239CA	74	Learjet 25B	25B-149	James Donaldson, Farmersville, Tx.	N149J
□ N239CD	81	Citation II	550-0223	B & A Aviation LLC. Jonesboro, Ar.	N81TJ
□ N240AA	66	Jet Commander	82	Darcor Holding, Denver, Co.	N103BW
□ N240AC	65	Sabre-40	282-41	Southwest Jet Inc. Belton, Mo.	(N116AC)
□ N240AK	82	Challenger 600S	1067	Oak Air Ltd. Akron, Oh.	N205EL
□ N240AT	70	Falcon 20E	240	Ciernes Overseas Inc. Geneva, Switzerland.	I-SNAG
□ N240B	97	Hawker 800XP	8358	The Budd Co. Troy, Mi.	
□ N240CF	74	Sabre-40A	282-132	Corporate Flight Inc. Detroit, Mi.	N70BC
□ N240CX	71	Gulfstream 2	101	Diamond Air Inc. Richmond Heights, Oh.	(N237LM)
□ N241CT	79	Westwind-1124	267	Cross Timbers Oil Co. Fort Worth, Tx.	N100SR
□ N241H	79	Sabre-65	465-5	Executive Aircraft Corp. Newton, Ks.	(N60CR)
□ N241JA	66	Learjet 24	24-131	Milam International Inc. Englewood, Co.	N11FH
□ N241LA	86	Citation S/II	S550-0091	Lambda Aviation Inc. Fayetteville, NC.	N595CM
□ N241RT	75	Learjet 35	35-024	Patron Aviation Inc. Atlanta, Ga.	N528EA

| Reg | Yr Type | c/n | Owner/Operator | Prev Regn |

Reg	Yr	Type	c/n	Owner/Operator	Prev Regn
☐ N242AC	92	Citation V	560-0177	AAA Cooper Transportation Inc. Dothan, AL.	D-CHHS
☐ N242DR	79	Learjet 35A	35A-242	Career Aviation Academy, Stockton, Ca.	VH-FSZ
☐ N242GS	78	Learjet 25D	25D-242	Air Fleet International Inc. Paramus, NJ.	(N242AF)
☐ N242LA	87	Citation S/II	S550-0153	Lambda Aviation Inc. Fayetteville, NC.	N153QS
☐ N242LJ	97	CitationJet	525-0242	Cessna Aircraft Co. Wichita, Ks.	
☐ N242MT	86	Learjet 35A	35A-621	Wal-Mart Stores Inc. Rogers, Ar.	PT-OTW
☐ N242WT	82	Citation II/SP	551-0066	Lawton Louisiana Inc. Lake Charles, La.	N6825X
☐ N243SH	75	Citation	500-0243	Trey Aviation LLC. Alamogordo, NM.	N53AJ
☐ N244A	86	Falcon 10	145	Archer Daniels Midland Co. Decatur, Il.	N209FJ
☐ N244AD	86	Falcon 50	162	Archer Daniels Midland Co. Decatur, Il.	C-GYPJ
☐ N244DM	68	Gulfstream 2	21	Joda Partnership, Town & Country, Mo.	N8PQ
☐ N244JM	88	BAe 125/800A	NA0428	Worthington Industries/JMAC Inc. Worthington, Oh.	N582BA
☐ N245BS	76	Learjet 25D	25D-214	Linrose Aviation Inc. Longview, Tx.	N214LJ
☐ N245CC	81	Citation II	550-0212	Eastern Air Center Inc & MaRiva Inc. Norwood, Ma.	(N6801V)
☐ N245SP	78	Falcon 10	135	Eagle Aviation of Conway Inc. Conway, Ar.	N707CX
☐ N245TT	86	Challenger 601-3A	5001	WOTAN America Inc. Fort Lauderdale, Fl.	C-GDDP
☐ N246AG	91	Falcon 900B	112	Cablevision Industries/Cableair Inc. Liberty, NY.	N473FJ
☐ N246CM	81	Learjet 35A	35A-395	Tri City Beverages Inc. Abilene, Tx.	N30GL
☐ N246NW	90	Citation V	560-0068	Northwestern Public Service Co. Huron, SD.	N712GF
☐ N246RR	74	Citation	500-0167	Producers Pipeline Corp. Houston, Tx.	N191AB
☐ N247EM	81	Falcon 50	48	Eugene Melnyk, Barbados.	N134AP
☐ N247VA	96	Vantage	001	VisionAire, Chesterfield, Mo. (Ff 16 Nov 96).	
☐ N248CJ	97	CitationJet	525-0248	Cessna Aircraft Co. Wichita, Ks.	
☐ N248TH	79	Gulfstream 2SP	248	Chelsea Aviation Inc. Dallas, Tx.	N510TL
☐ N249B	79	Learjet 35A	35A-240	The Budd Co. Troy, Mi.	N240B
☐ N249MW	66	HS 125/731	25115	The Jet Place Inc. & FirstPlus Financial Inc. Tulsa, Ok.	N249BW
☐ N249RA	72	Learjet 24D	24D-249	C C Medflight Inc. Lawrenceville, Ga.	XC-AA63
☐ N250AL	85	Citation S/II	S550-0042	Luhr Brothers Inc. Columbia, Il.	
☐ N250AS	87	Falcon 50	182	Boise Cascade Corp/Albertsons Inc. Boise, Id.	N182FJ
☐ N250CJ	97	CitationJet	525-0250	Cessna Aircraft Co. Wichita, Ks.	
☐ N250DH	69	HS 125/731	NA718	Arnoni Aviation Inc. Houston, Tx.	XA-SSV
☐ N250EC	72	Sabre-40A	282-110	Amjet Aircraft Corp. St Paul, Mn.	N477A
☐ N250J	90	Gulfstream 4	1144	Rikio Co.	N100PM
☐ N250JH	95	Citation V Ultra	560-0350	Cessna Aircraft Co. Wichita, Ks.	N645M
☐ N250JT	65	HS 125/1A	25053	Air 1st Aviation Companies Inc. Aiken, SC.	N254JT
☐ N250RA	84	Falcon 20F	481	Rite Aid Corp. Harrisburg, Pa.	N502F
☐ N250SP	93	Citation V	560-0211	Sonoco Products Co. Hartsville, SC.	
☐ N250VC	93	Gulfstream 4SP	1231	JABJ/Dacion Corp. Teterboro, NJ.	N470GA
☐ N250VP	81	Citation II	550-0250	Carolina Blue Air Corp. Fayetteville, NC.	N33GK
☐ N251AF	68	Learjet 25	25-004	Aero Freight Inc. El Paso, Tx.	N47MJ
☐ N251CT	79	Learjet 35A	35A-251	Beautiful Bird Brokers LLC. Las Vegas, Nv.	N27HF
☐ N251DS	77	Learjet 25D	25D-218	Air Star International Inc. Palm Springs, Ca.	N14NA
☐ N251JA	74	Learjet 25B	25B-150	Milam International Inc. Englewood, Co.	N888RB
☐ N251JS	79	Gulfstream 2	251	The Jet Place Inc. Dallas, Tx.	N36GS
☐ N251MD	82	Learjet 25D	25D-356	Superior Air Charter Inc. Medford, Or.	N25PT
☐ N251P	75	Citation	500-0250	Richard Brant Jr. Ridgeley, WV.	(N200BA)
☐ N251TJ	84	BAe 125/800A	8020	Tyler Jet LLC. Tyler, Tx.	N270HC
☐ N252BK	73	Learjet 25B	25B-107	Barken International Inc/Barbara Hepner, Salt Lake City, Ut	N25NB
☐ N252C	94	Gulfstream 4SP	1252	Cargill Inc. Minneapolis, Mn.	N433GA
☐ N252HS	85	Learjet 25G	25G-370	H S Aviation Inc. Newton Grove, NC.	N370LJ
☐ N252SC	68	Learjet 25	25-006	Hartford Holding Corp. Millersville, Md.	(N125SC)
☐ N252TJ	72	Learjet 24D	24D-252	Tyler Jet LLC. Tyler, Tx.	N157AG
☐ N253DV	87	B 737-39A	23800	RDV Sports/Orlando Magic & Orlando Solar Bears, Orlando, Fl.	N117DF
☐ N253L	80	Falcon 50	19	Broad River Aviation Inc. High Point, NC.	N63A
☐ N253S	83	Learjet 55	55-053	Shamrock Aviation Inc. NYC.	(N205EF)
☐ N253SC	78	Learjet 25D	25D-253	Dolphin Aviation Inc. Sarasota, Fl.	N253M
☐ N253W	81	Citation II	550-0221	Alumax Inc. Norcross, Ga.	N68026
☐ N254AM	79	Citation II	550-0081	Aeroservices International Inc. Herndon, Va.	I-FBCT
☐ N254CL	79	Learjet 25D	25D-275	Clay Lacy Aviation Inc. Van Nuys, Ca.	N211CD
☐ N254CR	76	Gulfstream 2	184	Richard & David Jacobs Group/Aviation Venture Inc. Cleveland	N220GA
☐ N254DV	81	Falcon 50	85	RDV Properties Inc. Grand Rapids, Mi.	N40TH
☐ N254JT	69	Learjet 24B	24B-181	John Travolta/ATLO Inc. Van Nuys, Ca.	N87CF
☐ N254NA	84	Falcon 50	148	Anschutz Corp. Denver, Co.	N28KB
☐ N254SC	82	Learjet 25B	25B-102	Dolphin Aviation Inc. Sarasota, Fl.	N64WH
☐ N255CC	95	Challenger 604	5302	Compaq Computer Corp. Houston, Tx.	N355CC
☐ N255CM	97	Falcon 50EX	255	Cargill Inc. Minneapolis, Mn.	F-WWHC
☐ N255DG	83	Diamond 1A	A056SA	Skyway Jet Service Inc. San Angelo, Tx.	I-FRTT
☐ N255DV	90	Learjet 31	31-030	RDV Corp. Grand Rapids, Mi.	N525AC
☐ N255JC	80	Learjet 35A	35A-326	JCIT-John Costanza Institute of Technology, Englewood, Co.	N155WL
☐ N255RB	81	Westwind-1124	336	Regal-Beloit Corp. Beloit, Wi.	N336SV
☐ N255RK	69	Falcon 20D	196	Purdey Aviation Corp. Raleigh, NC.	N79AE
☐ N255ST	82	Learjet 55	55-064	Southeast Toyota Distributors Inc. Deerfield Beach, Fl.	
☐ N255TC	90	Citation II	550-0638	Grey Bull Inc. Raleigh, NC.	N1717L
☐ N256A	86	Falcon 50	172	Ameritas Life Insurance Co/Bridgemark Assocs. Lincoln, Ne.	N9000F
☐ N256BC	94	Hawker 800	8256	Hawker LLC/Holiday Retirement Corp. Salem, Or.	N947H
☐ N256JC	84	Falcon 200	496	JCIT Institute of Technology Inc. Englewood, Co.	N227TA

Reg	Yr	Type	c/n	Owner/Operator	Prev Regn
□ N256M	95	Learjet 60	60-073	MAPCO Inc. Tulsa, Ok.	(N600LJ)
□ N256W	79	Falcon 10	151	Wendy's International Inc. Dublin, Oh.	N27AC
□ N257AJ	76	HS 125/700A	7001	T A Travel LLC. Westlake, Oh.	VH-LYG
□ N257CB	83	Diamond 1A	A050SA	Carlton-Bates Co. Little Rock, Ar.	N350DM
□ N257H	93	Gulfstream 4SP	1223	H J Heinz Co. Pittsburgh, Pa.	N935SH
□ N257W	78	Falcon 10	119	Wendy's International Inc. Dublin, Oh.	N191FJ
□ N258A	81	Falcon 20F-5	438	Daniel Ferguson, Naples, Fl.	N256A
□ N258G	81	Learjet 35A	35A-443	258G Corp. Dover, De.	N135RJ
□ N259DB	65	Learjet 23	23-064	Jet Investment Group Inc. Dover, De. (status ?).	ZS-MBR
□ N260GS	67	B 727-21	19261	Classic Designs of Tampa Bay, West Palm Beach, Fl.	N727RF
□ N260LF	90	Learjet 31	31-015	Landscapes Unlimited Inc. Lincoln, Ne.	N111TT
□ N260MB	73	Falcon 20F	274	Curtis Aviation Services Inc. Quincy, Il.	(D-CHEF)
□ N260QS	85	Citation S/II	S550-0060	EJI Inc/NetJets-The Small LLC. Columbus, Oh.	N314G
□ N261JP	93	Beechjet 400A	RK-76	Jefferson-Pilot Corp. Greensboro, NC.	N8166A
□ N261PC	95	Learjet 31A	31A-109	Park Corp. Cleveland, Oh.	N5029F
□ N261PG	80	Learjet 35A	35A-329	Kokoma Aviation/Aviation Charter Services, Indianapolis, In.	N261PC
□ N261WC	78	Learjet 25D	25D-261	Southeast Jet Leasing Inc. Deerfield Beach, Fl.	N24JK
□ N261WR	97	Citation V Ultra	560-0447	W R Meadows Inc. Hampshire, Il.	
□ N262WC	72	Westwind-1124	262	Westwind Charters Inc. Scottsdale, Az.	YV-393CP
□ N263PC	66	BAC 1-11/401AK	068	Park Corp. Portland, Or.	N111LP
□ N263PW	96	Falcon 900B	159	First Security Bank NA. Salt Lake City, Ut.	P4-NAN
□ N263S	95	Gulfstream 4SP	1263	Shamrock Aviation Inc. White Plains, NY.	N830CB
□ N264CL	78	Gulfstream 2SP	227	Clay Lacy Aviation Inc. Van Nuys, Ca.	N200LS
□ N264U	94	Citation V Ultra	560-0264	United Technologies Inc. Hartford, Ct.	N264CV
□ N265A	80	Sabre-65	465-47	AWI (Nevada) Inc. Las Vegas, Nv.	
□ N265C	80	Sabre-65	465-33	M D Lung Inc/Patrick Industries Inc. Elkhart, In.	N465SR
□ N265DP	75	Sabre-75A	380-30	O K Aviation Inc. Great Neck, NY.	N265CH
□ N265KC	76	Sabre-80A	380-49	Airmax LLC. East Pointe, Mi.	N221PH
□ N265M	80	Sabre-65	465-31	Aero Sales Inc. Chesterfield, Mo.	N65FC
□ N265MP	72	Falcon 20F	265	International Wire Group & Crain Industries, St Louis, Mo.	N606RP
□ N265SC	73	Sabre-40A	282-117	First Wing Inc. Carmel, In.	N3159U
□ N265SR	76	Sabre-60	306-120	Executive Aircraft Corp. Newton, Ks.	N265C
□ N265U	77	Sabre-60	306-132	Pre-Fab Transit Co. Farmer City, Il.	N60AG
□ N265WB		Sabreliner CT-39A	276-39	ILOC Inc. Ocala, Fl.	62-4486
□ N266TW	73	Learjet 24D	24D-266	Sierra American Corp. Dallas, Tx.	N266BS
□ N267TC	82	Citation II	550-0345	304 MC Ltd. Liverpool, NY.	N267TG
□ N267TG	88	Citation III	650-0159	S K U LP. Scottsdale, Az.	N683MB
□ N269CM	94	Citation V Ultra	560-0268	Columbia Management Co. Portland, Or.	N12012
□ N269HM	68	Gulfstream 2	13	LFP Inc/Flynt Aviation Inc. Van Nuys, Ca.	N269MH
□ N269MT	86	Citation S/II	S550-0080	Tibair Inc. Wilmington, De.	N581EA
□ N269RC	73	Citation	500-0078	Magnetic Land Inc. Las Vegas, Nv.	N110CK
□ N270AV	72	HS 125/F403B	25270	Intergraph Corp. Wilmington, De.	G-BKBA
□ N270AX	66	B 727-90C	19170	Omni Air Express Inc. Tulsa, Ok.	N798AS
□ N270BH	76	Citation	500-0330	EXIM Aviation Corp. Dover, De.	I-JESE
□ N270EX	97	Falcon 50EX	270	Dassault Falcon Jet Corp. Teterboro, NJ.	F-WWHU
□ N270KA	81	HS 125/700A	NA0307	Crusader Aviation Inc. Oxford, Ct.	N270MC
□ N270LC	79	Westwind-1124	245	Williamson-Dickie Manufacturing, Fort Worth, Tx.	N404CB
□ N270MC	83	Gulfstream 3	374	Key Air Inc/Parke Aviation Corp. Scarborough Manor, NY.	N24GA
□ N270NF	80	Citation 1/SP	501-0144	Super Food Services Inc. Dayton, Oh.	N270SF
□ N270PM	74	Citation	500-0196	Illinois Data Mart Inc. Geneva, Il.	(N711FW)
□ N270RA	97	Challenger 604	5337	Rite Aid Realty Corp. Harrisburg, Pa.	C-GLXD
□ N271AC	74	Citation	500-0218	Dominion Air Charter Inc. Trevilians, Va.	N4AC
□ N271CA	90	Citation V	560-0071	Bank IV Kansas NA. Wichita, Ks.	
□ N272BC	90	Beechjet 400A	RK-2	Bissell Inc. Grand Rapids, Mi.	N1902W
□ N272JP	72	Falcon 20F	272	Central Business Inc. Burnsville, Mn.	(N272FA)
□ N272JS	86	Gulfstream 3	489	Smurfit Packaging Corp. St Louis, Mo.	N328GA
□ N273LP	76	Gulfstream 2SP	192	Louisiana Pacific Corp. Hillsboro, Or.	HB-ITW
□ N273LR	76	Learjet 25	25-058	NKL Corp. El Paso, Tx.	N273LP
□ N273M	80	Learjet 25D	25D-315	American Jet International Corp. Houston, Tx.	N273KH
□ N273MC	85	Learjet 55	55-119	Meredith Corp. Des Moines, Ia.	N72613
□ N273RA	97	Astra-1125SPX	097	Galaxy Aerospace Corp. Fort Worth, Tx.	
□ N274CA	68	Sabre-60	306-31	Conrad Aviation Technologies Inc. Centerville, Oh.	N307D
□ N274FD	79	Learjet 35A	35A-274	Carlsbad Air Services Inc. Encinitas, Ca.	(N35WG)
□ N274HM	76	Westwind-1124	202	Shared Mail of Delaware Inc. Wilmington, De.	YV-297CP
□ N274K	79	Westwind-1124	274	Oklahoma Gas & Electric Co. Bethany, Ok.	N701W
□ N274QS	85	Citation S/II	S550-0074	EJI Inc/NetJets-Louiaiana Land & Exploration Co. Columbus.	N22EH
□ N275GC	88	Citation III	650-0162	York Transportation & Leasing Co. York, Pa.	N202RB
□ N275RA	97	Astra-1125SPX	098	Galaxy Aerospace Corp. Fort Worth, Tx.	
□ N277JM	81	Citation 1/SP	551-0035	Air Cruise/John Myers, Long Beach, Ca.	N277HM
□ N277RC	93	Citation V	560-0210	Richard & Mary Cree, Jackson, Wy.	N420DM
□ N277RP	87	Gulfstream 4	1026	Force Financial Corp/Petersen Aviation, Van Nuys, Ca.	N151A
□ N277SF	75	Falcon 10	44	Mandan Air Service LLC. Chicago, Il.	N62TJ
□ N277T	77	Gulfstream 2SP	209	E R Aviation Inc. Provo, Ut.	N806GA
□ N277TW	73	Learjet 24D	24D-277	Sierra American Corp. Wilmington, De.	N57BC
□ N279DM	79	Learjet 35A	35A-214	AirNet Systems Inc. Columbus, Oh.	
□ N279DP	87	Astra-1125	020	Monterey Airplane Co. Monterey, Ca.	4X-CUS

Reg	Yr	Type	c/n	Owner/Operator	Prev Regn

Reg	Yr	Type	c/n	Owner/Operator	Prev Regn
☐ N279DS	90	Astra-1125	040	Monterey Airplane Co. Monterey, Ca.	4X-CU.
☐ N279LE	73	Learjet 25B	25B-112	Air Ambulance Care Flight International Inc. Seminole, Fl.	N173J
☐ N279SP	81	Learjet 35A	35A-452	Southwestern Public Service Co. Amarillo, Tx.	N25MJ
☐ N280AZ	78	Westwind-1124	247	A & L Leasing LLC. Waterdown, SD.	N280LM
☐ N280BC	89	Falcon 900	71	Liberty Mutual Insurance Co. Boston, Ma.	PK-TRP
☐ N280BG	82	Falcon 50	109	M W Media Inc. Hillsboro, Or.	N280BC
☐ N280C	79	Learjet 25D	25D-280	Midtown Restaurants Corp. Mobile, Al.	(N510L)
☐ N280R	69	Learjet 24B	24B-188	Travel Lear Charter Service Inc. Oklahoma City, Ok.	N230R
☐ N280TA	80	Citation II	550-0206	Sesame Street Productions Inc. Boca Raton, Fl.	XA-SQW
☐ N281FP	73	Learjet 24D	24D-281	Valhi Inc. Dallas, Tx.	N23MJ
☐ N281RB	77	Gulfstream 2	200	International Group Inc. Horseheads, NY.	N17GG
☐ N282AC	67	Learjet 24	24-145	USA Jet Airlines Inc. Belleville, Mi.	(XA-LNA)
☐ N282Q	83	Gulfstream 3	379	North Carolina Air & Travel LLC. Cary, NC.	N379RH
☐ N282QS	85	Citation S/II	S550-0082	EJI Inc/NetJets, Columbus, Oh.	N9KH
☐ N282T	75	Falcon 10	42	CRST International Inc. Cedar Rapids, Ia.	N100UB
☐ N282WW	77	Sabre-60	306-134	IVEFA CA-Venezuela/Gadsden Holdings Inc. Miami, Fl.	N323EC
☐ N284AM	72	Citation	500-0028	Conquest Charters Corp. Dallas, Tx.	N103WV
☐ N284RJ	77	Citation 1/SP	501-0005	Warren Property Trust/Robert & Nani Warren, Portland, Or.	N143EP
☐ N284TJ	79	Learjet 25D	25D-284	Tyler Jet LLC. Tyler, Tx.	XC-CFM
☐ N285CP	81	Falcon 50	44	CITGO Petroleum Corp. Tulsa, Ok.	N44MK
☐ N285LM	78	JetStar 2	5224	Della One Inc. Pittsburgh, Pa.	N1924G
☐ N285TW	73	Falcon 20E	285	Sierra American Corp. Dallas, Tx.	N285AP
☐ N285XP	95	Hawker 800XP	8285	Fremont Administrative Services Corp. Santa Monica, Ca.	N808H
☐ N286GA	95	Gulfstream 4SP	1286	Electronic Data Systems Corp. Plano, Tx.	N480GA
☐ N286MC	97	Citation VII	650-7076	Maytag Corp. Newton, Ia.	
☐ N286PC	80	Citation 1/SP	501-0164	C L Swanson Corp. Sarasota, Fl.	N170JS
☐ N286WL	80	Learjet 35A	35A-286	Norwalk Aircraft Corp/Joseph Pagano, Aspen, Co.	PT-LSW
☐ N287MC	86	Citation S/II	S550-0102	Maytag Corp. Newton, Ia.	N1290Z
☐ N288DF	74	Learjet 24D	24D-288	Exploring the Electrical Content of the Air, Teterboro, NJ.	
☐ N288JP	79	Learjet 35A	35A-288	Prior Aviation Service Inc. Buffalo, NY.	N288JE
☐ N288QS	86	Citation S/II	S550-0088	EJI Inc/NetJets-Longaberger Co & co-owners, Columbus, Oh.	N825HL
☐ N288SJ	80	Westwind-Two	299	Sajalasi Corp. Carmel, In.	N600TC
☐ N288SP	75	Citation	500-0241	Aircraft Exchange, Carlsbad, Ca.	XB-EPN
☐ N288Z	88	Falcon 900	43	Chamarac Inc. NYC.	I-MTDE
☐ N289K	93	Challenger 601-3A	5132	Crawford Fitting Co. Cleveland, Oh.	N610DB
☐ N289SA	74	Learjet 24D	24D-289	Hartford Holding Corp. Millersville, Md.	N131MA
☐ N290EC	89	BAe 125/800A	NA0444	Ethyl Corp. Richmond, Va.	N596BA
☐ N290VP	79	Citation II	550-0090	Aviation Sales International Inc. Scottsdale, Az.	N410NA
☐ N291BC	89	Falcon 50	199	Boise Cascade Corp. Boise, Id.	N287FJ
☐ N291H	92	BAe 1000B	9016	Stevens Aviation Inc. Greenville, SC.	(5N-...)
☐ N291K	91	Learjet 35A	35A-665	Koch Industries Inc. Wichita, Ks.	
☐ N292GA	83	Challenger 601	3014	Republic Industries Inc. Fort Lauderdale, Fl.	N14PN
☐ N292ME	80	Learjet 35A	35A-292	Aircraft Specialists Inc. Sellersburg, In. (status ?).	N634H
☐ N292PC	84	Citation III	650-0058	Powers Construction Co. Florence, SC.	(N282PC)
☐ N293BC	83	Falcon 50	135	Boise Cascade Corp. Boise, Id.	N125FJ
☐ N293QS	95	Citation V Ultra	560-0293	EJI Inc/NetJets, Columbus, Oh.	N51575
☐ N293S	74	Citation	500-0193	Jet A Aviation Inc. Wilmington, De.	XA-SQZ
☐ N293SA	94	Learjet 31A	31A-101	Temple Inland Forest Products Corp. Diboll, Tx.	N5010J
☐ N294AW	89	Beechjet 400A	RK-1	Wornall 103 Partners, Kansas City, Mo.	N294FA
☐ N294M	69	Learjet 25	25-031	Southern Management Services Inc. Las Vegas, Nv.	N294NW
☐ N294S	97	Astra-1125SPX	094	Galaxy Aerospace Corp. Fort Worth, Tx.	
☐ N295DS	95	CitationJet	525-0091	DSW Development Corp. Cape Girardeau, Mo.	
☐ N295FA	93	Beechjet 400A	RK-68	Marc Fruchter Aviation Inc. Reading, Pa.	N8280J
☐ N295NW	74	Learjet 24XR	24XR-295	Northwestern Aircraft Capital Corp. Van Nuys, Ca.	N590CH
☐ N296FA	94	Beechjet 400A	RK-91	Marc Fruchter Aviation Inc. Reading, Pa.	N1545N
☐ N297GA	97	Astra-1125SPX	091	Galaxy Aerospace Corp. Fort Worth, Tx.	
☐ N297JD	70	HS 125/F403A	25235	West Cherry Sales LLC. Wilmington, De.	N227LA
☐ N297JS	87	Westwind-1124	435	VAS Aviation LLC. Bloomfield Hills, Mi.	(N669SB)
☐ N297S	74	Citation	500-0197	Aircare International Inc. Austin, Tx.	XA-SRB
☐ N297XP	96	Hawker 800XP	8297	Raytheon Aircraft Co. Wichita, Ks.	
☐ N298CM	80	Westwind-1124	298	CarpetMax Builder of Tennessee Inc. Kennesaw, Ga.	C-GRGE
☐ N298DR	79	Learjet 25D	25D-298	International Jet Aviation Inc. Van Nuys, Ca.	XA-ABH
☐ N298W	88	Falcon 900	45	Abex Inc. Portsmouth, NH.	N64BE
☐ N299BW	74	HS 125/F600A	6046	Grand Aviation Inc. Dallas, Tx.	XA-AGL
☐ N299MW	79	Learjet 25D	25D-299	Jacura Inc. Thomasville, Ga.	(N5B)
☐ N299QS	86	Citation S/II	S550-0099	EJI Inc/NetJets-General Electric Co. Columbus, Oh.	N44GT
☐ N299RP	78	Citation 1/SP	501-0073	Digital Communications Inc. Cedar Rapids, Ia.	N100SV
☐ N299SC	97	Learjet 60	60-112	Learjet Inc. Wichita, Ks.	
☐ N299SG	94	Learjet 60	60-025	Integrated Payment Systems Inc. Englewood, Co.	N299SC
☐ N299TB	75	Citation	500-0230	Illinois Data Mart Inc. Geneva, Il.	N24S
☐ N299TW	74	Learjet 24D	24D-299	Chaparral Leasing Inc. Newton, Ks.	XB-GJS
☐ N299XP	96	Hawker 800XP	8299	Raytheon Aircraft Credit Corp. Wichita, Ks.	(N32BC)
☐ N300A	81	Falcon 50	64	Management Air Svcs/Zeneca Inc & ICI Americas, New Castle.	N418S
☐ N300AA	83	Diamond 1A	A041SA	N300AA LLC/PJ Associates LLC. Phoenix, Az.	N83AE
☐ N300AK	97	Citation Bravo	550-0809	Gerald Mansbach, Ashland, Ky.	N800AK
☐ N300BS	79	HS 125/700A	NA0241	Tanara Inc/Aero-Dynamics, Dallas, Tx.	N25MK

Reg	Yr	Type	c/n	Owner/Operator	Prev Regn
☐ N300CR	91	Challenger 601-3A	5092	Crane Aerospace, White Plains, NY.	HB-IKV
☐ N300CV	74	Falcon 20F	322	Summit Seafood Suply Inc. Wilmington, De.	N464M
☐ N300DH	81	Diamond 1A	A010SA	David Hocker, Owensboro, Ky.	(N9FC)
☐ N300DL	69	Gulfstream 2	57	Solar Sportsystems Inc. Buffalo, NY.	N300DK
☐ N300ES	81	Falcon 50	70	Earth Star Inc. Burbank, Ca.	N699SC
☐ N300GB	66	HS 125/1A-522	25074	HEA Management Group Inc. Denton, Tx.	N400UW
☐ N300GN	86	BAe 125/800A	8057	Gannett Co Inc. Arlington, Va.	N362BA
☐ N300GX	91	Gulfstream 4	1164	Glaxo Wellcome Inc. Raleigh, NC.	N459GA
☐ N300K	95	Gulfstream 4SP	1266	Ford Motor Co. Detroit, Mi.	N412GA
☐ N300KC	89	Challenger 601-3A	5051	Ellis Aviation Inc. Dallas, Tx.	N190GG
☐ N300L	96	Gulfstream V	507	Annenberg Aviation, Philadelphia, Pa.	N507GV
☐ N300LS	87	BAe 125/800A	8098	Limited Stores/Southern Delaware Corp. Wilmington, De.	N536BA
☐ N300M	84	Gulfstream 3	417	Hunters Glen Inc. Dallas, Tx.	N1119C
☐ N300PM	90	BAe 125/800A	NA0457	Professional Funeral Associate, Lufkin, Tx.	N619BA
☐ N300PY	97	Citation Bravo	550-0806	Aurora Aircraft Co. Asuncion, Paraguay.	
☐ N300QW	86	Citation S/II	S550-0100	American Seafoods Co. Seattle, Wa.	(N616GB)
☐ N300R	81	Learjet 35A	35A-438	NASCAR Inc. Daytona Beach, Fl.	N12GJ
☐ N300RB	84	BAe 125/800B	8013	Bigelow Companies/Cosmos Air LLC. Las Vegas, Nv.	(N500RH)
☐ N300TC	78	Westwind-1124	241	Van Dyne-Crotty Co. Columbus, Oh.	N789TE
☐ N300TS	80	Diamond 1	A003SA	Pezold Air Charter LLC. Wilmington, De.	N300DM
☐ N300TW	79	Learjet 35A	35A-237	Wilson Holdings Inc. Montgomery, Al.	N237TJ
☐ N300WY	84	Gulfstream 3	427	Southwest Beta Services Inc. Dallas, Tx.	N87AC
☐ N300XL	80	Westwind-1124	276	Skyking Charter Co. Wilmington, De.	N800XL
☐ N301DM	81	Diamond 1A	A007SA	Diamond Air Inc & Margeaux Inc. Richmond Heights, Oh.	
☐ N301EL	80	Citation	500-0402	Navsink Corp. Newark, De.	XA-JCE
☐ N301K	95	Gulfstream 4SP	1267	Ford Motor Co. Detroit, Mi.	N417GA
☐ N301PC	82	Westwind-Two	377	Dude Inc. Fort Smith, Ar.	4X-CUJ
☐ N301PH	92	BAe 1000B	9031	Peabody Holding Co. St Louis, Mo.	G-HJCB
☐ N301QS	89	Citation V	560-0038	EJI Inc/NetJets-Stone Container Corp. Chicago, Il.	
☐ N301R	65	Falcon 20C	3	Reliant Airlines Inc. Ypsilanti, Mi.	N92MH
☐ N302PC	81	Learjet 25D	25D-351	Stamford Capital Aircraft Corp. Stamford, Ct.	N837CS
☐ N302QS	97	Citation V Ultra	560-0402	EJI Inc/NetJets-Allmac Inc. Columbus, Oh.	
☐ N303AJ	69	Jet Commander-B	149	Aviex Jet Inc. Houston, Tx.	(N303AJ)
☐ N303BC	96	Hawker 800XP	8324	NBC Air LLC. Las Vegas, Nv.	N324XP
☐ N303CB	77	Citation 1/SP	501-0051	Dominion Citation Group LC. Richmond, Va.	N150TJ
☐ N303GA	80	Gulfstream 3	303	Petersen Aviation/Airborne Charter Inc. Burbank, Ca.	N1761W
☐ N303JW	84	Falcon 50	140	Flying Wags Inc. Houston, Tx.	I-MMEA
☐ N303LC	94	CitationJet	525-0068	Lamar Air LLC. Baton Rouge, La.	N68CJ
☐ N303MW	74	HS 125/600B	6033	Metro West Ready Mix Inc. Salt Lake City, Ut.	XB-FMF
☐ N303P	83	Diamond 1A	A034SA	Motion Industries Inc. Birmingham, Al.	N318DM
☐ N303PC	86	Citation III	650-0110	J C Pace Holding Co & Kimball Inc. Fort Worth, Tx.	N76D
☐ N303PL	81	Falcon 10	187	Paul Lindsey LLC. Lebanon, Mo.	N81TJ
☐ N304KT	81	Citation II/SP	551-0304	Keen Transport Inc. Carlisle, Pa.	N702KH
☐ N304TS	80	Gulfstream 3	304	Tutor-Saliba Corp. Van Nuys, Ca.	VR-BSL
☐ N305AR	78	Falcon 20-5B	378	American Resources/Black Beauty Resources, Evansville, In.	(N6621)
☐ N305BB	78	Westwind-1124	228	Badgett Brown, Madisonville, Ky.	4X-CLZ
☐ N305CC	96	Challenger SE	7099	Carnival Cruise Lines/Funair, Fort Lauderdale, Fl.	N253SE
☐ N305CJ	95	CitationJet	525-0105	Aqua Sun Investments Inc. Ormond Beach, Fl.	(D-IAFD)
☐ N305FX	90	Challenger 601-3A	5070	FlexJets, Dallas, Tx.	N780HC
☐ N305PA	66	DC 9-15	45740	PharmAir Corp. Miami, Fl.	N911KM
☐ N305S	76	Citation	500-0301	Aircraft Marketing Inc. Albuquerque, NM.	OE-FNG
☐ N305TH	81	HS 125/700A	NA0305	Chelsea Aviation Inc. Dallas, Tx.	N730H
☐ N306CF	67	Sabre-60	306-13	National Flight Services Inc. Swanton, Oh.	N60EL
☐ N306FX	95	Challenger 601-3R	5175	FlexJets/First Delaware Financial Group Inc. Dallas, Tx.	N601FR
☐ N306JA	76	Learjet 24D	24D-306	E A S I Inc. Lighthouse Point, Fl.	XA-SAV
☐ N306P	81	Diamond 1A	A009SA	Chattem Inc. Chattanooga, Tn.	N909GA
☐ N307AJ	74	Learjet 25XR	25XR-175	Aviex Jet Inc. Houston, Tx.	(N96JJ)
☐ N307EW	76	Citation	500-0323	George A Mendenhall, Sunriver, Or.	(N268GM)
☐ N307FX	95	Challenger 601-3R	5179	FlexJets/Dell USA LP. Dallas, Tx.	N608CC
☐ N307QS	95	Citation V Ultra	560-0307	EJI Inc/NetJets-Barbara Hajim, Greenwich, Ct.	N5233J
☐ N308A	92	Citation II	550-0703	ARAMCO Associated Co. Dhahran, Saudi Arabia.	
☐ N308AJ	69	Learjet 25	25-039	Aviex Jet Inc. Houston, Tx.	N66NJ
☐ N308CK	79	Citation II	550-0106	Autocam Corp. Kentwood, Mi.	N37CR
☐ N308EL	69	Gulfstream 2	68	Eli Lilly & Co. Indianapolis, In.	
☐ N308FX	92	Challenger 601-3A	5110	FlexJets, Dallas, Tx.	N392PT
☐ N308HG	80	Gulfstream 3	308	Corporate Flight Inc. Wilmington, De.	N308GA
☐ N308SG	78	JetStar 2	5226	Peninsular Communications Inc. Los Angeles, Ca.	N815RC
☐ N308TS	80	Westwind-Two	308	Praegitzer Industries Inc. Dallas, Or.	N308JS
☐ N308TW	79	Citation II	550-0044	Jones Airways LLC. Cleveland, Tn. (was 551-0092)	N300TW
☐ N309AJ	69	Learjet 25	25-034	Aviex Jet/ Lear Jet Charter LC. Houston, Tx.	N19FN
☐ N309EL	79	Gulfstream 2	250	Eli Lilly & Co. Indianapolis, In.	N821GA
☐ N309FX	96	Challenger 604	5306	FlexJets/Aircraft Arrangentst Inc. Dallas, Tx.	C-G...
☐ N309G	87	BAe 125/800A	NA0408	GTE Southwest Inc. San Angelo, Tx.	N61CT
☐ N309GA	96	Gulfstream 4SP	1309	Electronic Data Systems Corp. Dallas, Tx.	N446GA
☐ N309WM	81	HS 125/700A	NA0309	William Morris/Atlantic Aviation Charters, Teterboro, NY.	N526DM
☐ N310AV	78	Citation II	550-0028	Avior Technologies Inc. Miami, Fl.	5Y-HAB

Reg	Yr	Type	c/n	Owner/Operator	Prev Regn
☐ N310FX	97	Challenger 604	5336	FlexJets/Aspenwood Aviation Inc. Dallas, Tx.	C-G...
☐ N310ME	80	Learjet 35A	35A-310	Aircraft Specialists Inc. Sellersburg, In.	N8280
☐ N310RG	81	Gulfstream 3	321	JABJ/Renco Group Inc. NYC.	N313RG
☐ N311AG	71	B 727-17	20512	Ann & Gordon Getty/Vallejo Co. San Francisco, Ca.	N767RV
☐ N311DB	77	Westwind-1124	208	Interplanetary Aviation Inc. Park City, Ut.	N324AJ
☐ N311DG	92	Citation V	560-0167	Woodlands Ltd. Manchester, UK.	N211DG
☐ N311EL	89	Gulfstream 4	1095	Eli Lilly & Co. Indianapolis, In.	N469GA
☐ N311FX	97	Challenger 604	5342	FlexJets/Renaissance Development Group, Dallas, Tx.	C-G...
☐ N311GX	87	Challenger 601-3A	5014	GTE Southwest Inc. San Angelo, Tx.	N311G
☐ N311JA	68	HS 125/731	NA714	ECA Inc/Act Two Inc. Southfield, Mi.	N176TS
☐ N311JD	78	HS 125/700B	7020	Eagle Hardware & Garden Inc. Renton, Wa.	N818
☐ N311QS	95	Citation V Ultra	560-0311	EJI Inc/NetJets, Columbus, Oh.	
☐ N311TP	81	Citation 1/SP	501-0196	Robert Rapaport, Palm Beach, Fl.	(N311TT)
☐ N312AM	96	Challenger 604	5312	Ameritech, Chicago, Il.	N604KC
☐ N312AT	96	Challenger 604	5313	Ameritech Corp. Chicago, Il.	N605KC
☐ N312EL	89	Gulfstream 4	1105	Eli Lilly & Co. Indianapolis, In.	N408GA
☐ N312K	75	Falcon 20F-5	324	Kiewit Engineering Co. Omaha, Ne.	N324TC
☐ N312NC	81	Citation II	550-0290	Progress Aviation Corp. Dallas, Tx.	N217LG
☐ N312QS	95	Citation V Ultra	560-0312	EJI Inc/NetJets, Columbus, Oh.	
☐ N313CC	85	BAe 125/800A	8043	Classic Services Inc. Wilmington, De.	N319AT
☐ N313CV	95	Citation V Ultra	560-0313	Florida Power & Light Co. Miami, Fl.	N5246Z
☐ N313QS	83	Citation III	650-0013	Excellent Aviation Rentals Inc. Houston, Tx.	(N13QS)
☐ N313RG	96	Gulfstream V	504	Renco Group Inc. NYC.	N504GV
☐ N314C	81	Learjet 35A	35A-412	JABJ/Carter-Wallace Inc. NYC.	(N31LM)
☐ N314EB	81	Citation 1/SP	501-0206	Village Leasing Inc. Dakota Dunes, SD.	OE-FBA
☐ N314MK	91	Learjet 31A	31A-040	Conex Freight Systems Inc. Portland, Or.	N340LJ
☐ N314QS	97	Citation V Ultra	560-0441	EJI Inc/NetJets-Mayo Foundation, Columbus, Oh.	
☐ N314RW	90	Citation V	560-0051	Red Wing Shoe Co. Red Wing, Mn.	N599SG
☐ N315AJ	66	Learjet 24	24-108	Millenium Jets Inc. Van Nuys, Ca.	N900JA
☐ N315CS	96	Citation V Ultra	560-0371	Pheasant Kay-Bee Toy Inc. Nashua, NH.	N371CV
☐ N315GA	96	Gulfstream 4SP	1315	EDS Corp. Dallas, Tx.	N413GA
☐ N315MR	97	CitationJet	525-0198	Guy Mabee, Midland, Tx.	
☐ N315QS	95	Citation V Ultra	560-0315	EJI Inc/NetJets-Annapurna Corp & ACK-SAF Corp. Columbus, Oh.	
☐ N315R	90	Beechjet 400A	RK-9	Arclar Co. Harrisburg, Il.	N8152H
☐ N315S	77	Citation 1/SP	501-0038	Bend Properties Inc. Irvine, Ca.	N36923
☐ N315SL	86	Challenger 601	3054	LPL Management Group Inc. Jupiter, Fl.	N375PK
☐ N315TR	81	Westwind-Two	315	Career Aviation Academy, Oakdale, Ca.	VH-NJW
☐ N316GS	93	Gulfstream 4SP	1225	Verochris Corp. White Plains, NY.	N459GA
☐ N317AF	75	Citation 2SP	168	Loyd's Aviation (USA) Corp. Wilmington, De.	N168JW
☐ N317CC	87	BAe 125/800A	8093	Marquette Electronics/Cozzens & Cudahy Air, Milwaukee, Wi.	N358LL
☐ N317JS	83	Westwind-Two	385	Jeffrey Bennett, Corona, Ca.	N962MV
☐ N317LJ	96	Learjet 31A	31A-117	Mid-South Management Group Inc. Bowling Green, Ky.	
☐ N317M	95	Citation VII	650-7055	MBNA Corp. Highland Heights, Oh.	N755CM
☐ N317MB	92	Citation VII	650-7012	MBNA Corp. Philadelphia, Pa.	N5144
☐ N317ML	85	Gulfstream 3	460	MBNA Corp. Highland Heights, Oh.	N2TF
☐ N317MZ	92	Citation VII	650-7010	MBNA Corp. Highland Heights, Oh.	N1902
☐ N317QS	95	Citation V Ultra	560-0317	EJI Inc/NetJets-Daniel C Searle, Columbus, Oh.	
☐ N317SM	80	Citation II/SP	551-0036	Spring Mountain Enterprises Inc. Costa Mesa, Ca.	N160D
☐ N317TC	72	HS 125/600A	6007	Ravenair LLC. NYC.	N3007
☐ N317TT	80	Learjet 35A	35A-317	Inter Island Yachts Inc. Zephyr Cove, Nv.	N98TE
☐ N317VP	76	Citation	500-0317	Delta Bravo Aviation Inc. Georgetown, De.	C-GPCO
☐ N318CT	90	Citation V	560-0081	Century Service Group Inc. Monroe, La.	N560HP
☐ N318MM	93	Citation V	560-0219	Schwan's Sales Enterprises Inc. Marshall, Mn.	N12580
☐ N318QS	97	Citation V Ultra	560-0418	EJI Inc/NetJets-Jacoboni Interest Inc. Columbus, Oh.	
☐ N319GP	74	Gulfstream 2	150	Corporate Wings/Columbus Bank & Trust Co. Columbus, Ga.	N613CK
☐ N320AF	76	HFB 320	1061	Kalitta Flying Service Inc. Morristown, Tn.	(D-CEDL)
☐ N320M	80	Learjet 35A	35A-320	Air Medical Leasing Inc. Wilmington, De.	N35FS
☐ N320MP	87	Westwind-1124	432	MedPartners Aviation Inc. Birmingham, Al.	(N317MT)
☐ N320QS	95	Citation V Ultra	560-0321	EJI Inc/NetJets-Jl Aviation Inc. Columbus, Oh.	N5262W
☐ N321AF	76	HFB 320	1060	Kalitta Flying Service Inc. Morristown, Tn.	(D-CCCH)
☐ N321AN	79	Learjet 35A	35A-272	Anderson Merchandisers Inc & co-owners, Knoxville, Tn.	N500EF
☐ N321GN	81	Citation II	550-0321	Woosie Aviation Inc/Ian Woosnam, Jersey, C.I.	TC-COY
☐ N322AF	75	HFB 320	1058	Kalitta Flying Service, Morristown, NJ.	16+21
☐ N322CP	93	Falcon 900B	134	Gallagher Enterprises LLC. Denver, Co.	N88YF
☐ N322K	94	Fokker 70	11521	Ford Motor Co. Detroit, Mi.	PH-MKS
☐ N322QS	97	Citation V Ultra	560-0421	EJI Inc/NetJets-Export Packaging Co. Columbus, Oh.	
☐ N322RG	96	Citation VII	650-7070	Greenhill Aviation LLC. NYC.	(N770VP)
☐ N322TP	68	HS 125/3A-RA	NA710	BTS Aviation LC. Provo, Ut.	C-FMKF
☐ N323AF	79	HFB 320	1062	Kalitta Flying Service, Morristown, Tn.	16+25
☐ N323LJ	96	Learjet 31A	31A-123	Richards Aviation Inc & M & S Aviation Inc. Memphis, Tn.	
☐ N323LM	97	CitationJet	525-0230	Cessna Aircraft Co. Wichita, Ks.	
☐ N323P	90	Astra-1125SP	049	Allease Inc. Tuscaloosa, Al.	N145AS
☐ N323QS	95	Citation V Ultra	560-0323	EJI Inc/NetJets-3JC Corp & Meyers Bakeries Inc. Columbus, Oh	N5097H
☐ N324AF	80	HFB 320	1064	Kalitta Flying Service Inc. Morristown, Tn.	16+27
☐ N324GA	97	Gulfstream 4SP	1330	Gulfstream Aerospace Corp. Savannah, Ga.	
☐ N324K	95	Fokker 70	11545	Ford Motor Co. Detroit, Mi.	PH-EZH
Reg	*Yr*	*Type*	*c/n*	*Owner/Operator*	*Prev Regn*

Reg	Yr	Type	c/n	Owner/Operator	Prev Regn
☐ N324L	81	Citation 1/SP	501-0197	Lee Lewis Construction Co. Lubbock, Tx.	N100SN
☐ N324QS	97	Citation V Ultra	560-0423	EJI Inc/NetJets-KN Interstate Gas Transmissio. Columbus, Oh.	
☐ N324SR	89	Challenger 601-3A	5038	Executive Aircraft Corp. Newton, Ks.	(N220LC)
☐ N324TW	76	Learjet 24D	24D-324	Chaparral Leasing Inc. Newton, Ks.	XA-SCY
☐ N325AF	80	HFB 320	1065	Kalitta Flying Service Inc. Morristown, Tn.	16+28
☐ N325AJ	75	Westwind-1124	181	Learjet Charter LC. Houston, Tx.	N107CF
☐ N325GA	97	Gulfstream 4SP	1326	Gulfstream Aerospace Corp. Savannah, Ga.	
☐ N325JL	77	Learjet 25D	25D-215	Universal Jet Aviation Inc. Boca Raton, Fl.	N44FE
☐ N325K	66	Sabre-40	282-63	Elm Tree Inc. Keene, NH.	
☐ N325LW	81	Westwind-Two	334	Longwood Industries/L W Aviation Inc. Florham Park, NJ.	N40MP
☐ N325NW	80	Learjet 35A	35A-325	Northwestern Aircraft Capital Corp. Van Nuys, Ca.	I-FFLY
☐ N325PJ	81	Learjet 25D	25D-342	Nuray Aircraft LLC. Hilton Head, SC.	N342GG
☐ N325QS	97	Citation V Ultra	560-0425	EJI Inc/NetJets-Rawlings Co. Columbus, Oh.	
☐ N326CB	69	JetStar-8	5143	Centurion Investments Inc. St Louis, Mo. (Wfu ?).	N620JB
☐ N326EW	79	Falcon 20F	392	Leco Corp. South Bend, In.	N328EW
☐ N326FB	81	Falcon 50	39	General Leasing Inc. Wilmington, De.	N754S
☐ N326KE	76	Learjet 24D	24D-326	Kauffman Engineering Inc. Lebanon, In.	(N400XB)
☐ N327GA	97	Gulfstream 4SP	1327	Gulfstream Aerospace Corp. Savannah, Ga.	
☐ N327GJ	76	Learjet 24D	24D-327	Richard Scruggs PA. Pascagoula, Ms.	F-GGPG
☐ N327QS	95	Citation V Ultra	560-0327	EJI Inc/NetJets-MB 1 Aviation Corp. Columbus, Oh.	
☐ N327SA	85	Westwind-Two	428	WEKEL SA. Bogota, Colombia.	(N92BE)
☐ N327TL	68	Gulfstream 2SP	33	Thomas H Lee Co. Bedford, Ma.	N217TL
☐ N328GA	97	Gulfstream 4SP	1328	Gulfstream Aerospace Corp. Savannah, Ga.	
☐ N328JK	69	Learjet 24B	24B-212	429K Corp. Bermuda Dunes, Ca.	N328TL
☐ N328NA	80	Citation 1/SP	501-0168	David Wyman, Seattle, Wa.	N601WT
☐ N328QS	97	Citation V Ultra	560-0428	EJI Inc/NetJets-Blue Water Airways, Columbus, Oh.	
☐ N329GA	97	Gulfstream 4SP	1329	Gulfstream Aerospace Corp. Savannah, Ga.	
☐ N329JS	77	JetStar 2	5206	Din Aero Inc. Troy, Mi.	XA-JML
☐ N329K	88	Falcon 900	46	Ford Motor Co. Detroit, Mi.	N434FJ
☐ N329TJ	76	Learjet 24E	24E-329	Tyler Jet LLC. Tyler, Tx.	XA-PFA
☐ N330G	66	HS 125/731	25087	Flight Enterprises LLC. Atlanta, ga.	N66AM
☐ N330K	88	Falcon 900	50	Ford Motor Co. Detroit, Mi.	N436FJ
☐ N330MC	97	Falcon 900EX	21	MCI Transcon Corp. Washington, DC.	F-WWFH
☐ N330QS	95	Citation V Ultra	560-0329	EJI Inc/NetJets-J T Co & co-owners, Columbus, Oh.	N5105F
☐ N330TP	94	Challenger 601-3R	5142	Washington Times Aviation Inc.Washington, DC.	VR-CJJ
☐ N330X	86	BAe 125/800A	8060	Texas Eastern Transmission Co. Houston, Tx.	N686CF
☐ N331CW	78	Westwind-1124	231	Paramount Aviation LLC. Cleveland, Oh.	(N27TA)
☐ N331DC	76	HS 125/F600A	6061	Interstate Distributor Co. Tacoma, Wa.	N169B
☐ N331GA	97	Gulfstream 4SP	1331	Gulfstream Aerospace Corp. Savannah, Ga.	
☐ N331MC	97	Falcon 900EX	22	MCI Transcon Corp. Washington, DC.	F-WWFQ
☐ N331N	90	Learjet 31	31-022	Nibco Inc. Elkhart, In.	
☐ N331P	68	Gulfstream 2	20	Imperial Palaceair Ltd. Las Vegas, Nv.	N4SP
☐ N331PR	97	Citation Bravo	550-0831	Poly-Resyn Inc. Dundee, Il.	
☐ N331QS	95	Citation V Ultra	560-0331	EJI Inc/NetNets-Baden Advisors Inc. Columbus, Oh.	N51072
☐ N331SJ	95	Learjet 31A	31A-113	Sterling Jets Ltd. Glasgow, UK.	N31LJ
☐ N331SK	92	Astra-1125SP	063	Kellett Aviation Inc. Boone, NC.	N74TJ
☐ N331TP	97	Challenger 604	5350	Washington Times Aviation Inc. Washington, DC.	C-GLXU
☐ N332DF	81	Westwind-Two	332	DFG Capital LLC. Minneapolis, Mn.	N24SR
☐ N332FG	80	Learjet 35A	35A-332	BLC/Fluor Daniel Inc. Greenville, SC.	N600LN
☐ N332GA	97	Gulfstream 4SP	1332	Gulfstream Aerospace Corp. Savannah, Ga.	
☐ N332LC	96	Citation V Ultra	560-0332	H C E Leasing Inc. Cleveland, Tn.	
☐ N332WE	82	HS 125/700A	NA0332	Double Eagle Aviation Inc. Wilmington, De.	XA-STX
☐ N333AV	66	Falcon 20C	28	Cegep Edouard-Montpetit=CEM Corp. Cincinnati, Oh.	C-GEAQ
☐ N333AX	88	Gulfstream 4	1063	ARR Inc/Richmor Aviation Inc. Hudson, NY.	N54SB
☐ N333CG	78	Learjet 25D	25D-262	D & D Aviation LC. Salt Lake City, Ut.	N440F
☐ N333CZ	96	Astra-1125SPX	080	Horizon Americas Inc. Buenos Aires, Argentina.	N333AJ
☐ N333DP	72	HS 125/731	NA775	Doane Products Co. Joplin, Mo.	N17HV
☐ N333EC	65	JetStar-731	5061	Flying Eagle Inc/Finch Air Ltd. Tortola, BVI.	N488EC
☐ N333GA	97	Gulfstream 4SP	1333	Gulfstream Aerospace Corp. Savannah, Ga.	
☐ N333GB	66	BAC 1-11/401AK	076	HM Industries Inc. Newark, De. (status ?).	VR-BHS
☐ N333GL	85	Challenger 601	3042	ESA Management Inc. Fort Lauderdale, Fl.	N900CC
☐ N333GM	72	Sabre-40A	282-106	Commercial Aviation Enterprises Inc. Delray Beach, Fl.	XA-RKG
☐ N333GZ	65	HS 125/1A-522	25070	Gonzo Zulu, Reno, Nv.	N470TS
☐ N333JH	76	Citation Eagle	500-0292	Nancy Harvey, Nashville, Ga.	N501HK
☐ N333MG	88	Challenger 601-3A	5035	Griffco Aviation Inc. NYC.	N606CC
☐ N333NM	65	Sabre-40	282-45	Commercial Aviation Enterprises Inc. Delray Beach, Fl.	N333GM
☐ N333PC	77	HS 125/700A	NA0205	Majestic Air Service/POLCO Inc. Landover, Md.	C-GYYZ
☐ N333PV	94	Gulfstream 4SP	1240	Peavey Electronics Corp. Meridian, Ms.	N486GA
☐ N333QS	95	Citation V Ultra	560-0333	EJI Inc/NetJets-Graham Land Co. Columbus, Oh.	
☐ N333RL	92	BAe 1000B	9027	Becker Group Inc. Warren, Mi.	N333RU
☐ N333TW	68	Learjet 24	24-168	Sierra American Corp. Dallas, Tx.	N155BT
☐ N333WM	96	Citation V Ultra	560-0385	Morris Communications Corp. Augusta, Ga.	N5109R
☐ N334GA	97	Gulfstream 4SP	1334	Gulfstream Aerospace Corp. Savannah, Ga.	
☐ N334H	85	Citation III	650-0071	Hillenbrand Industries Inc. Batesville, In.	(N1315B)
☐ N334JC	76	Citation	500-0334	Sierra Industries Inc. Uvalde, Tx. (status ?).	ZS-MPI
☐ N334MC	91	Falcon 900B	108	MCI Transcon Corp. Washington, DC.	N471FJ
Reg	*Yr*	*Type*	*c/n*	*Owner/Operator*	*Prev Regn*

Reg	Yr	Type	c/n	Owner/Operator	Prev Regn
☐ N334QS	97	Citation V Ultra	560-0434	EJI Inc/NetJets-Kojaian Management Corp. Columbus, Oh.	
☐ N334RC	73	Citation	500-0062	RealtiCorp Aviation LLC. Wilmington, De.	N4KH
☐ N335CC	83	Citation II	550-0480	Lone Star Jet Inc. Midland, Tx.	S5-BAC
☐ N335GA	97	Gulfstream 4SP	1335	Gulfstream Aerospace Corp. Savannah, Ga.	
☐ N335H	79	Gulfstream 2TT	238	Halliburton Co. Dallas, Tx.	N831GA
☐ N335JL	75	Learjet 35	35-015	Silver State Aviation Inc. Las Vegas, Nv.	N58CW
☐ N335JW	70	Learjet 24B	24B-226	C T Enterprises Inc/Jetwest International, Burbank, Ca.	(N335JR)
☐ N335K	81	Learjet 35A	35A-381	Koch Industries Inc. Wichita, Ks.	N300CM
☐ N335MC	95	Falcon 900B	150	MCI Transcon Corp. Washington, DC.	N150FJ
☐ N335QS	95	Citation V Ultra	560-0335	EJI Inc/NetJets-Patterson Aviation Inc. Columbus, Oh.	
☐ N336EA	80	Learjet 35A	35A-336	Eagle Aviation Inc. West Columbia, SC.	XA-BNO
☐ N336QS	95	Citation V Ultra	560-0336	EJI Inc/NetJets-Chesapeake Energy Corp. Columbus, Oh.	N5265B
☐ N336RJ	79	Sabre-65	465-10	Six Star Inc. Winter Park, Fl.	N77TC
☐ N337FP	90	Learjet 31	31-020	Florida Progress Corp. Clearwater, Fl.	N31LJ
☐ N337MC	95	Falcon 900B	152	MCI Transcon Corp. Washington, DC.	F-WWFJ
☐ N337QS	97	Citation V Ultra	560-0437	EJI/NetJets-Travelers Group Inc. Columbus, Oh.	
☐ N337RE	84	BAe 125/800A	8024	D & B Craft Corp. NYC.	N802D
☐ N337TV	74	Citation	500-0236	Emerald Aviation Inc. Fort Walton Beach, Fl.	N600SR
☐ N338AX	68	Gulfstream 2B	30	Whitehaven Aircraft Sales Co. Memphis, Tn.	N333AX
☐ N338FP	89	Learjet 55C	55C-138	Florida Progress Corp. St Petersburg, Fl.	N9LR
☐ N338MM	89	Gulfstream 4	1076	First Security Bank NA. Salt Lake City, Ut.	HZ-MNC
☐ N339A	81	Gulfstream 3	339	Bank of America/KaiserAir Inc. Oakland, Ca.	N522SB
☐ N339BA	78	Learjet 25D	25D-240	Bankair Inc. West Columbia, SC.	N33PT
☐ N339BC	82	Learjet 55	55-039	Chilton Aviation Inc. Wilmington, De.	VR-BQF
☐ N339H	74	Gulfstream 2	145	Halliburton Industries Corp. Houston, Tx.	N871E
☐ N339QS	95	Citation V Ultra	560-0339	EJI Inc/NetJets-William C Cox Jr. Columbus, Oh.	
☐ N339TG	77	Falcon 10	103	Galawayne LLC. Wilmington, De.	N103TJ
☐ N340DR	90	Citation V	560-0094	Stephens Inc. Little Rock, Ar.	N1823S
☐ N340GA	97	Gulfstream 4SP	1340	Gulfstream Aerospace Corp. Savannah, Ga.	
☐ N340GA	98	Gulfstream 4SP	1340	Gulfstream Aerospace Corp. Savannah, Ga.	
☐ N341AP	90	BAe 125/800A	NA0464	Air Products & Chemical Inc. Allentown, Pa.	N636BA
☐ N341GA	97	Gulfstream 4SP	1341	Gulfstream Aerospace Corp. Savannah, Ga.	
☐ N341QS	96	Citation V Ultra	560-0341	EJI Inc/NetJets-Helen of Troy Corp. Columbus, Oh.	
☐ N341TC	67	B 727-22	19148	Tracinda Corp. Las Vegas, Nv.	N7084U
☐ N342CC	82	Citation II	550-0414	Charter Services Inc. Mobile, Al.	(N414VP)
☐ N342GA	97	Gulfstream 4SP	1342	Gulfstream Aerospace Corp. Savannah, Ga.	
☐ N342K	76	Falcon 20F-5	357	Kiewit Engineering Co. Omaha, Ne.	N435TP
☐ N342TC	94	Challenger 601-3R	5155	Sun Microsystems Inc. Portland, Or.	C-GLWV
☐ N343CC	96	Citation V Ultra	560-0368	Westvaco Corp. NYC.	N5194J
☐ N343GA	97	Gulfstream 4SP	1343	Gulfstream Aerospace Corp. Savannah, Ga.	
☐ N343K	91	Challenger 601-3A	5086	Eastman Kodak Co. Rochester, NY.	N353K
☐ N343MG	90	Falcon 900	95	M G Transportation Inc. West Palm Beach, Fl.	N478A
☐ N343PJ	96	CitationJet	525-0166	PRJ Holdings Inc. San Francisco, Ca.	
☐ N343QS	97	Citation V Ultra	560-0444	EJI/NetJets-Olin Corp. Columbus, Oh.	
☐ N344BA	97	Challenger 604	5344	Boeing Aircraft Holding Co. Seattle, Wa.	C-GLXQ
☐ N344GA	97	Gulfstream 4SP	1344	Gulfstream Aerospace Corp. Savannah, Ga.	
☐ N344QS	95	Citation V Ultra	560-0344	EJI Inc/NetJets-Cyclo Corp. Columbus, Oh.	
☐ N345	80	Learjet 35A	35A-355	Gulfstream Aviation Enterprises Inc. Orlando, Fl.	N64RV
☐ N345AA	73	Gulfstream 2B	123	7-11 Air Corp. Teterboro, NJ.	N345CP
☐ N345BA	97	Challenger 604	5345	Boeing Aircraft Holding Co. Seattle, Wa.	C-G...
☐ N345BR	96	Hawker 800XP	8308	Willow Springs Aviation Corp. NYC.	(N11WC)
☐ N345DM	83	Diamond 1A	A060SA	Tri C Aviation Inc. Houston, Tx.	(N300SJ)
☐ N345GL	70	HS 125/400A	NA753	GMA Sales Corp. West Des Moines, Ia.	N840H
☐ N345MC	69	Learjet 25	25-046	MCOCO Inc. Houston, Tx.	N33PT
☐ N345N	81	Citation 1/SP	501-0204	Munoco LC. El Dorado, Ar.	(N123HP)
☐ N345QS	97	Citation V Ultra	560-0445	EJI Inc/NetJets, Columbus, Oh.	
☐ N345TR	81	Westwind-Two	345	Rollins Properties Inc. New Castle, De.	N534R
☐ N347GS	93	Learjet 60	60-026	Safari Aviation (UK) Ltd. Farnborough, UK. "Veliko Zlocesto"	(N700GS)
☐ N347HS	76	Falcon 20F	347	Henry I Siegel Co Inc. NYC.	N298CK
☐ N347JV	81	Learjet 25D	25D-347	Aviation Enterprises Inc. Wilmington, De.	N347AC
☐ N347K	73	Falcon 20F	281	Kiewit Engineering Co. Omaha, Ne.	N70830
☐ N348MC	95	Hawker 800XP	8290	Monsanto Co. Chesterfield, Mo.	N670H
☐ N348QS	95	Citation V Ultra	560-0348	EJI Inc/NetJets-Big Aviation Inc. Columbus, Oh.	
☐ N349K	87	Falcon 900	10	Peter Kiewit Sons Inc. Omaha, Ne.	N5MC
☐ N349MC	77	Westwind-1124	224	Tav Aero Inc. Houston, Tx.	N2756T
☐ N349MG	83	Falcon 200	479	Midland Financial Co. Oklahoma City, Ok.	(N200SA)
☐ N350AG	81	Learjet 25D	25D-350	Sky Enterprises Inc. Nashville, Tn.	(N428CH)
☐ N350DA	81	Learjet 35A	35A-366	Boston Chicken Inc. Golden, Co.	N119CP
☐ N350EF	81	Learjet 35A	35A-385	Executive Flight Inc. East Wenatchee, Wa.	N535MC
☐ N350JF	79	Learjet 35A	35A-219	Wilmington Trust Co. Wilmington, De.	N502G
☐ N350JH	75	Learjet 25B	25B-200	Frisco Aircraft Corp. Addison, Tx.	N680BC
☐ N350RD	89	Citation V	560-0026	Reynolds DeWitt & Co. Cincinnati, Oh.	N49MJ
☐ N350WB	82	Falcon 50	102	Wagner & Brown Inc. Midland, Tx.	(N50WB)
☐ N350WC	96	Citation V Ultra	560-0378	Williams Companies Inc. Tulsa, Ok.	
☐ N351AC	92	Learjet 31A	31A-051	Alpha Charlie Inc. Mountville, Pa.	N1905H
☐ N351AM	81	Learjet 35A	35A-409	Florida Jet Service Inc. Pembroke Pines, Fl.	N35FE

Reg	Yr	Type	c/n	Owner/Operator	Prev Regn
☐ N351AS	78	Learjet 35A	35A-146	Boise Cascade Corp/Albertsons Inc. Boise, Id.	N55AS
☐ N351C	79	Westwind-1124	264	Royal Charters Inc. Corpus Christi, Tx.	N88PV
☐ N351EF	77	Learjet 35A	35A-125	Executive Flight Inc. East Wenatchee, Wa.	N125GA
☐ N351GL	73	Learjet 35	35-001	Learjet Inc. Wichita, Ks. (experimental).	N731GA
☐ N351N	65	Learjet 23	23-054	RBS Aviation Group Inc. Fort Myers, Fl.	N351NR
☐ N351QS	97	Citation V Ultra	560-0451	EJI Inc/NetJets, Columbus, Oh.	
☐ N351SP	95	Hawker 800XP	8280	Sonoco Products Co. Hartsville, SC.	G-BWDC
☐ N351TC	81	Westwind-Two	351	C K E Aviation Inc. Los Angeles, Ca.	N106WT
☐ N351WC	95	Citation V Ultra	560-0330	Williams Companies Inc/TRANSCO, Tulsa, Ok.	
☐ N352AF	89	Challenger 601-3A	5041	Fayair Inc-Ali Al Fayed/Wayfarer Ketch Corp. White Plains	N641CL
☐ N352EF	91	Learjet 31A	31A-046	Executive Flight Inc. East Wenatchee, Wa.	N131PT
☐ N352MD	77	Learjet 24F	24F-352	CPA Aviation Inc. Tulsa, Ok.	(N449JS)
☐ N352QS	96	Citation V Ultra	560-0352	EJI Inc/NetJets-Bobby Rahal Inc. Columbus, Oh.	
☐ N352WC	92	Citation V	560-0194	Williams Companies Inc. Tulsa, Ok.	N194CV
☐ N353CA	69	Sabre-60	306-28	Charter America Inc. Miami, Fl.	N741RL
☐ N353CP	81	Falcon 20F	461	CITGO Petroleum Corp. Tulsa, Ok.	N747V
☐ N353EF	81	Learjet 35A	35A-364	Executive Flight Inc. E Wenatchee, Wa.	N490BC
☐ N353TC	89	Challenger 601-3A	5037	TCI Challenger Inc. Englewood, Co.	N707GG
☐ N353WC	96	Citation X	750-0008	Transcontinental Gas Pipeline Co. Houston, Tx.	
☐ N354CA	65	JetStar-731	5054	Chariot Air LLC. Dallas, Tx.	N20AP
☐ N354ME BOI	81	Learjet 35A	35A-378	Bean Ball Co. Seattle, Wa.	T-500/CX-
☐ N354QS	96	Citation V Ultra	560-0356	EJI Inc/NetJets-Fedders Corp. Columbus, Oh.	
☐ N354RZ	78	Learjet 35A	35A-170	Ronzoni Airplane Co. Redmond, Wa.	N100K
☐ N354TC	96	Challenger 601-3R	5192	Benco Inc. Las Vegas, Nv.	C-G...
☐ N354WC	97	Citation X	750-0027	Williams Companies Inc. Tulsa, Ok.	
☐ N355CD	81	Sabre-65	465-57	Sound Trak Inc. Omaha, Ne.	N903K
☐ N355DB	81	Learjet 55	55-006	Alamo Jet Inc. Fort Myers, Fl.	N126EL
☐ N355DF	82	Citation II	550-0355	Abaco Aviation Services Inc. Raleigh, NC.	N500BR
☐ N355JK	81	Westwind-Two	355	Kelley Aviation Associates Inc. Santa Fe, NM.	N355WW
☐ N355WC	97	Citation X	750-0030	Williams Companies Inc. Tulsa, Ok.	
☐ N356JW	90	Learjet 35A	35A-656	POLEAR Inc. Ventura, Ca.	N335SB
☐ N356WC	97	Citation V Ultra	560-0432	Texas Gas Transmission Corp. Owensboro, Ky.	
☐ N357BC	81	Westwind-1124	357	Automated Prescription Systems Inc. Pineville, La.	N66FG
☐ N357EC	94	Citation V Ultra	560-0269	Entergy Services Inc. Little Rock, Ar.	N331EC
☐ N357PR	83	Learjet 55	55-073	Full Gospel A M E Zion Church Inc. Temple Hills, Md.	N73WE
☐ N357RT	81	Challenger 600S	1033	Reading Jet Service Inc. Reading, Pa.	N101ST
☐ N357WC	90	Citation V	560-0069	Texas Gas Transmission Corp. Owensboro, Ky.	N70TG
☐ N358PG	80	Learjet 35A	35A-358	I F C Ltd. Sellersville, Pa.	N524HC
☐ N358QS	97	Citation V Ultra	560-0455	EJI Inc/NetJets, Columbus, Oh.	
☐ N359V	89	Astra-1125	039	Valmont Industries Inc. Omaha, Ne.	
☐ N359WJ	67	Sabre-60	306-1	Commercial Aviation Enterprises Inc. Delray Beach, Fl.	XA-REC
☐ N360JG	83	Learjet 25D	25D-360	First American National Bank, Nashville, Tn.	N618P
☐ N360M	84	Citation S/II	S550-0022	Murphy Oil Corp. El Dorado, Ar.	N258P
☐ N360MC	77	Citation 1/SP	501-0036	House of Raeford Farms Inc. Raeford, NC.	HI-493
☐ N360QS	97	Citation V Ultra	560-0460	EJI Inc/NetJets, Columbus, Oh.	
☐ N360X	78	HS 125/700A	NA0230	Panhandle Eastern Pipeline Co. Houston, Tx.	G-5-13
☐ N361EC	94	Citation V Ultra	560-0279	Entergy Services Inc. Jackson, Ms.	
☐ N361QS	95	Citation V Ultra	560-0361	EJI Inc/NetJets-Brown & Williamson Tobacco, Columbus, Oh.	
☐ N362CP	82	Citation II	550-0403	Colonial Pipeline Co. Atlanta, Ga. (was 551-0058).	N637EH
☐ N363FJ	77	Falcon 20F	363	Jetcraft Corp. Raleigh, NC.	HZ-DC2
☐ N363K	96	Beechjet 400A	RK-114	Spiral Inc. Chandler, Az.	N1084D
☐ N364CL	81	Learjet 35A	35A-383	Clay Lacy Aviation Inc. Van Nuys, Ca.	N66FE
☐ N364G	89	Gulfstream 4	1091	General Electric Co. White Plains, NY.	N467GA
☐ N365CX	84	Westwind-Two	425	CPX Charter Inc. Covington, Ky.	N600LE
☐ N365G	89	Gulfstream 4	1101	General Electric Co. White Plains, NY.	N404GA
☐ N365N	80	Learjet 35A	35A-300	ATC Air Inc. Lincoln, Ne.	
☐ N366F	88	Gulfstream 4	1041	Connell LP. Portsmouth, NH.	N433GA
☐ N366G	84	Citation III	650-0038	General Electric Co. White Plains, NY.	
☐ N366QS	97	Citation V Ultra	560-0466	EJI Inc/NetJets, Columbus, Oh.	
☐ N367G	84	Citation III	650-0053	General Electric Co. White Plains, NY.	
☐ N367JC	80	Citation II	550-0166	Walter Air Corp. Houston, Tx.	N166VP
☐ N367QS	96	Citation V Ultra	560-0367	EJI Inc/NetJets-Bryant Air Services Inc. Columbus, Oh.	
☐ N367TP	87	Falcon 50	181	Alumax Inc. Atlanta, Ga.	N345AP
☐ N367WW	82	Westwind-1124	367	Fred Hallmark, Warrior, Al.	N455S
☐ N368MD	82	Westwind-Two	368	FG Aviation Inc. Logan, Ut.	N28WW
☐ N368PU	83	Diamond 1A	A068SA	Purdue Research Foundation, W Lafayette, In.	N368DM
☐ N369BA	80	Learjet 35A	35A-312	Bankair Inc. West Columbia, SC.	LV-OFV
☐ N369BG	89	BAe 125/800A	NA0436	B/G Enterprises LLC. Carmel, In.	N354WG
☐ N369CA	77	Falcon 20F	359	Computer Associates International Inc. Islandia, NY.	(N508L)
☐ N369CS	83	Gulfstream 3	384	Wayfarer Ketch/Wesaire Inc. White Plains, NY.	N399BH
☐ N369JH	71	HS 125/400A	NA769	Aerohawk Leasing & Sales/James Fail, Dallas, Tx.	N42BL
☐ N370BH	71	Sabre-75	370-4	Gary Hall, Houston, Tx.	N37GF
☐ N370HF	77	Falcon 20F	370	Fuller Renting & Leasing Inc. Auburn, Al.	HL7234
☐ N370KP	82	Falcon 50	103	Comdisco Inc. Rosement, Il.	N83MP
☐ N370M	87	Citation S/II	S550-0128	Murphy Oil Corp. El Dorado, Ar.	N911BB
Reg	*Yr*	*Type*	*c/n*	*Owner/Operator*	*Prev Regn*

Reg	Yr	Type	c/n	Owner/Operator	Prev Regn
☐ N370RR	77	HS 125/700A	NA0213	Reynolds & Reynolds Co. Dayton, Oh.	N370M
☐ N370SL	73	Sabre-75	370-9	Centurion Investments Inc. St Louis, Mo.	XB-GJO
☐ N372BG	95	Gulfstream 4SP	1273	Berwind Corp. Philadelphia, Pa.	N372BC
☐ N372CM	88	Gulfstream 4	1049	Cordelia Scaife May, Pittsburgh, Pa.	N402GA
☐ N372G	83	Challenger 601	3006	Bombardier Business Jet Solutions, Dallas, Tx..	HB-IKX
☐ N372Q	67	Jet Commander	112	Legacy Aviation Inc. Dallas, Tx.	N773WB
☐ N372XP	98	Hawker 800XP	8372	Raytheon Aircraft Co. Wichita, Ks.	N1251K
☐ N373DH	65	HS 125/731	25066	373 Trading Corp. Dallas, Tx.	(N374DH)
☐ N373G	83	Challenger 601	3009	General Electric Co. White Plains, NY.	C-GLYO
☐ N373LP	79	Learjet 35A	35A-220	Louisiana Pacific Corp. Hillsboro, Or.	N873LR
☐ N373QS	96	Citation V Ultra	560-0373	EJI Inc/NetJets-Dow Chemical Co. Columbus, Oh.	
☐ N374G	97	Challenger 604	5351	General Electric Co. White Plains, NY.	C-GLXB
☐ N374QS	98	Citation V Ultra	560-0475	EJI Inc/NetJets, Columbus, Oh.	
☐ N375CM	95	Citation V Ultra	560-0365	Cessna Aircraft Co. Wichita, Ks.	N7547P
☐ N375G	84	Challenger 601	3019	General Electric Co. White Plains, NY.	C-G...
☐ N375NM	83	Gulfstream 3	375	Arizona Executive Air Service Inc. Scottsdale, Az.	VR-BOB
☐ N375QS	96	Citation V Ultra	560-0375	EJI Inc/NetJets-Burlington Northern Railroad, Columbus, Oh.	
☐ N375SC	95	Falcon 2000	23	Steelcase Inc. Grand Rapids, Mi.	N2036
☐ N376BE	83	Westwind-Two	376	BEF Aviation Co. Columbus, Oh.	N376WA
☐ N376D	75	Sabre-60A	306-101	Salyer Farms Airport, Corcoran, Ca.	N68NR
☐ N376QS	94	Citation V Ultra	560-0276	EJI Inc/NetJets, Columbus, Oh.	N183AJ
☐ N376SC	96	Falcon 2000	24	Steelcase Inc. Grand Rapids, Mi.	N2039
☐ N377BT	66	Falcon 20C	44	World Harvest Church Inc. Columbus, Oh.	N76TS
☐ N377GS	97	CitationJet	525-0179	James Jaeger, Cincinnati, Oh.	
☐ N377Q	78	Learjet 25D	25D-257	Airstar International Inc. Palm Springs, Ca.	N377C
☐ N377QS	96	Citation V Ultra	560-0377	Executive Jet Sales Inc. Columbus, Oh.	N450RA
☐ N378C	76	Falcon 10	73	Florida Crushed Stone Co. Leesburg, Fl.	VR-BNT
☐ N379RD		Gulfstream 2	...		
☐ N379XX	83	Gulfstream 3	394	Nexxus Products Inc. Santa Barbara, Ca. 'Jheri Reading'	N311GA
☐ N380BC	74	Sabre-80A	380-17	Bryce Corp/F C Leasing Inc. Memphis, Tn.	N1115
☐ N380DJ	75	Sabre-80A	380-32	American Air Inc. Farmington Hills, Mi.	(N86SH)
☐ N380SR	77	Sabre-75A	380-53	Executive Aircraft Corp. Newton, Ks.	XB-DVP
☐ N380TT	84	Gulfstream 3	437	Litton Industries Inc. Van Nuys, Ca.	
☐ N380X	94	Hawker 800	8269	Panhandle Eastern Corp. Houston, Tx.	N295H
☐ N381DA	67	Jet Commander	118	Jack Sisemore, Angel Fire, NM.	N716BB
☐ N382AA	66	Fan Commander	56	Wheeler Ridge Aviation Inc. Bakersfield, Ca.	(N53AA)
☐ N382E	78	Falcon 20F	382	U S West Inc. Englewood, Co.	N138E
☐ N382QS	96	Citation V Ultra	560-0382	EJI Inc/NetJets-Ultra Aviation LLC. Columbus, Oh.	
☐ N383MB	96	Learjet 60	60-083	M Bohlke Veneer Corp. Fairfield, Oh.	N683LJ
☐ N383SF	96	Astra-1125SPX	083	Smithfield Foods Inc. Norfolk, Va.	
☐ N384K	78	Falcon 20F	387	Quality Shipyards Inc. New Orleans, La.	N676DW
☐ N386CM	80	Learjet 35A	35A-283	Apple Jet Inc. Wilmington, De.	N205FL
☐ N386JM	68	Jet Commander-A	128	JMAR Inc. Stillwater, Ok.	N386MC
☐ N387HA	83	Citation II	550-0465	Highland Associates Inc. Birmingham, Al.	N784A
☐ N387PA	88	Astra-1125	025	GCH-General Cigar Holdings Transportation Inc. NYC	4X-CUH
☐ N387RE	88	Citation II	550-0575	Bell Aviation Inc. West Columbia, SC.	N337RE
☐ N388MA	79	Citation II	550-0093	Paul Sutton Aircraft Sales Inc. Columbia, SC.	N100SC
☐ N388MM	86	Gulfstream 3	490	Franklin Templeton Travel, San Mateo, Ca.	N28R
☐ N388PD	87	Learjet 35A	35A-630	Dooney & Bourke PR Inc. Yabucoa, PR.	N742P
☐ N388SB	81	Citation II	550-0245	Sierra Bravo Aviation LLC. Wichita, Ks.	N505GP
☐ N389AT	79	Learjet 25D	25D-297	Begley & Deaton Aviation LLC. Wilmington, De.	N24KW
☐ N389BG	80	Westwind-Two	304 • -	SouthTrust Bank NA. Birmingham, Al.	N369BG
☐ N389GA	80	Learjet 25D	25D-289	Starwood Industries Inc. Petersburg, Va.	RP-C400
☐ N389GS	95	Falcon 2000	20	The Travelers Inc. White Plains, NY.	F-WWME
☐ N389L	84	Citation S/II	S550-0013	Libby Owens Ford Co. Toledo, Oh.	N277AL
☐ N390GS	95	Falcon 2000	21	The Travelers Inc. White Plains, NY.	F-WWMA
☐ N390JP	81	Citation II	550-0326	King Leasing Inc. Asheville, NC.	N390AJ
☐ N391BC	80	Citation II/SP	551-0049	Best Chairs Inc. Ferdinand, In.	N155FL
☐ N391CV	96	Citation V Ultra	560-0391	Preformed Line Products Co. Mayfield, Oh.	N5092D
☐ N391QS	98	Citation V Ultra	560-0493	EJI Inc/NetJets, Columbus, Oh.	
☐ N391SC	97	Learjet 31A	31A-131	International Funeral Services Inc. Houston, Tx.	N31LR
☐ N392BS	92	Citation V	560-0164	Briggs & Stratton Corp. Milwaukee, Wi.	N164CV
☐ N392FV	84	Challenger 601	3032	Finova Capital Corp. Phoenix, Az.	N111GX
☐ N392QS	97	Citation V Ultra	560-0429	EJI Inc/NetJets-Albermarle Corp. Columbus, Oh.	
☐ N393BD	72	Gulfstream 2SP	120	Nineteen Forty CC Inc. Van Nuys, Ca.	N20FX
☐ N393E	85	Citation S/II	S550-0053	WCC/Business Properties Development Co. Irvine, Ca	N1223N
☐ N393QS	96	Citation V Ultra	560-0393	EJI Inc/NetJets-Dow Chemical Co. Columbus, Oh.	
☐ N393U	81	Gulfstream 3	325	Unisys Corp. Mercer County Airport, Trenton, NJ.	N89QA
☐ N394QS	96	Citation V Ultra	560-0394	EJI Inc/NetJets-Media General Inc. Columbus, Oh.	
☐ N395LJ	94	Learjet 31A	31A-095	Learjet Inc. Wichita, Ks.	OK-AJD
☐ N395QS	98	Citation V Ultra	560-0496	EJI Inc/NetJets, Columbus, Oh.	
☐ N395R	92	Citation V	560-0188	Ryan's Family Steak Houses Inc. Greer, SC.	N64PM
☐ N395RD	79	HS 125/700B	7064	Cheyenne Management Co. Auburn Hills, Mi.	VH-JFT
☐ N396M	82	Citation II	550-0362	West Coast Charters/Trishan Air Inc. Santa Barbara, Ca.	N12142
☐ N396QS	96	Citation V Ultra	560-0396	EJI Inc/JetNets-Witco Corp. Columbus, Oh.	
☐ N397AT	97	Beechjet 400A	RK-157	ITC Service Co & others, West Point, Ga.	
Reg	*Yr*	*Type*	*c/n*	*Owner/Operator*	*Prev Regn*

Reg	Yr	Type	c/n	Owner/Operator	Prev Regn
☐ N397BE	83	Challenger 600S	1053	BE Aerospace Inc. Wellington, Fl.	N32BQ
☐ N397CS	84	Citation III	650-0056	Consolidated Stores Corp. Columbus, Oh.	N760EW
☐ N397RD	68	Gulfstream 2	37	Cheyenne Management Co. Auburn Hills, Mi.	N994JD
☐ N397SL	82	Diamond 1A	A022SA	ARCH Air Medical Services, St Louis, Mo.	N811DJ
☐ N398AG	97	Astra C-38A	088	Tracor Inc. Austin, Tx.	
☐ N398RS	98	Citation Excel	560-5009	Cessna Aircraft Co. Wichita, Ks.	
☐ N399AG	97	Astra C-38A	090	Tracor Inc. Austin, Tx.	
☐ N399BA	81	Learjet 35A	35A-371	Bankair Inc. West Columbia, SC.	LV-ALF
☐ N399CB	95	Gulfstream 4SP	1261	Citiflight Inc. Newark, NJ.	N469GA
☐ N399CC	88	Gulfstream 4	1051	Citiflight Inc. Newark, NJ.	N403GA
☐ N399CF	90	Challenger 601-3A	5084	Citiflight Inc. Newark, NJ.	C-GLWT
☐ N399DM	81	Diamond 1A	A008SA	CorpWings Inc. Tyler, Tx.	(N56SK)
☐ N399FL	83	Challenger 600S	1083	FLD Corp/Flight Services Group, West Palm Beach, Fl.	N471SB
☐ N399G	96	CitationJet	525-0183	Dineair Corp. Holmdel, NJ.	N97CJ
☐ N399JC	97	Hawker 800XP	8334	Johnson Controls Inc. Milwaukee, Wi.	
☐ N399RP	82	Diamond 1A	A020SA	Diamond Air LLC. Tupelo, Ms.	
☐ N399SC	86	Gulfstream 3	488	SCI Texas Funeral Services Inc. Houston, Tx.	N401PJ
☐ N399SW	87	Challenger 601-3A	5009	Citiflight Inc. Newark, NJ.	C-G...
☐ N399W	94	Citation VII	650-7038	Williams International LLC. Pontiac, Mi.	N12642
☐ N400	96	Learjet 31A	31A-128	Indianapolis Motor Speedway Corp. Indianapolis, In.	N8082J
☐ N400AJ	97	Beechjet 400A	RK-137	Schuele German Technologies GmbH. Bad Radolfszell.	N1117Z
☐ N400CC	88	Gulfstream 4	1046	Wells Fargo & Co. San Francisco, Ca.	N119K
☐ N400CT	91	Citation V	560-0104	Aircraft Leasing & Sales Inc. Wichita, Ks.	N560CT
☐ N400D	70	Gulfstream 2	100	Maxim Financial Corp. Boulder, Co.	XB-FVL
☐ N400EP	70	Learjet 24XR	24XR-215	Charles McAdam Jr. Vero Beach, Fl.	(N57JR)
☐ N400FE	70	HS 125/731	NA747	Chastain Park Holdings Inc. Los Angeles, Ca.	N900EL
☐ N400FT	92	Beechjet 400A	RK-47	Featherlite Aviation Co. Cresco, Ia.	N8053V
☐ N400GK	82	Diamond 1	A019SA	Dalton Air Group Inc. Pittston, Pa.	N319DM
☐ N400GN	86	BAe 125/800A	8059	Gannett Co Inc. Arlington, Va.	N363BA
☐ N400GP	71	HS 125/731	NA762	Gladvest Inc. Oakland, Ca.	N523M
☐ N400HG	84	Diamond 1A	A091SA	Heartland Management Co. Topeka, Ks.	N611AG
☐ N400HH	82	Diamond 1	A025SA	H & H Aviation Ltd. Raytown, Mo.	N1843A
☐ N400J	86	Gulfstream 3	493	Johnson & Johnson, West Trenton, NJ.	N322GA
☐ N400JD	92	Gulfstream V	524	Deere & Co. Moline, Il.	N524GA
☐ N400JE	77	Learjet 35A	35A-120	AirNet Systems Inc. Columbus, Oh.	(N400RV)
☐ N400JH	77	Sabre-60	306-133	Wilton Investment Group Inc. Lyndeboro, NH.	N360CF
☐ N400JK	70	HS 125/731	NA756	KEVCO Inc. Atlanta, Ga.	N40Y
☐ N400KC	91	Challenger 601-3A	5090	Kimberly-Clark Corp. Irving, Tx.	N818TH
☐ N400KP	96	Beechjet 400A	RK-125	Kingston Aviation LLC. Idaho Falls, Id.	N1105U
☐ N400LX	83	Citation 1/SP	501-0251	Sacramento Jet Inc. Sacramento, Ca.	N945BC
☐ N400M	73	Gulfstream 2	132	Lorien Aviation Inc. Los Angeles, Ca.	N873GA
☐ N400MC	82	Learjet 35A	35A-487	Michael Cappy/MCAV Inc. Loveland, Co.	N206EC
☐ N400MP	78	JetStar 2	5228	TISMA Inc. Fairfax, Va.	XA-RMD
☐ N400MR	70	HS 125/400A	NA759	BTS Aviation LC. Provo, Ut.	N125CF
☐ N400NR	76	Sabre-75A	380-41	JHM Leasing Corp. Pittsfield, Ma.	N400N
☐ N400Q	92	Beechjet 400A	RK-55	Mid-American Waste Systems Inc. Canal Winchester, Oh.	
☐ N400RB	75	Learjet 35	35-011	Badgett Rogers Sr. Madisonville, Ky.	N3816G
☐ N400RG	67	B 727-22	19149	Reliance Group Holdings Inc. Newark, NJ.	N7085U
☐ N400SA	90	Gulfstream 4	1120	WEKEL SA. Bogota, Colombia.	VR-BOB
☐ N400SH	95	Beechjet 400A	RK-100	Oneok Inc. Tulsa, Ok.	N1570B
☐ N400TB	92	Challenger 601-3A	5120	Taco Bell Corp. Santa Ana, Ca.	(N500TB)
☐ N400TF	80	Westwind-1124	279	Tyson Foods Inc. Fayetteville, Ar.	N952HF
☐ N400VC	78	Learjet 25XR	25XR-235	N W Aircraft LLC. San Antonio, Tx.	N400JS
☐ N400VG	96	Beechjet 400A	RK-113	Vecellio & Grogan Inc. Beckley, WV.	N3263N
☐ N400VK	90	Beechjet 400A	RK-3	O Gara Aviation LLC. Las Vegas, Nv.	N400VG
☐ N400VP	96	Beechjet 400A	RK-110	Raytheon Aircraft Co. Wichita, Ks.	N1090X
☐ N400WP	81	HS 125/700A	NA0306	Wausau Paper Mills Co. Wausau, Wi.	C-FEXB
☐ N401AB	90	Beechjet 400A	RK-7	Gibraltar Aviation Ltd. Wethersfield, Ct.	(N631RR)
☐ N401AJ	74	Learjet 25B	25B-171	Travel Lear Charter Service Inc. Oklahoma City, Ok.	N888LR
☐ N401CG	88	Beechjet 400	RJ-43	George's Aviation LLC. Springdale, Ar.	N3143T
☐ N401EE	90	Beechjet 400A	RK-6	Raytheon Aircraft Co. Wichita, Ks.	N600CC
☐ N401G	83	Citation 1/SP	501-0320	The Young Family Trust, Portola Valley, Ca.	N70467
☐ N401JL	95	Gulfstream 4SP	1283	Colleen Corp. Philadelphia, Pa.	N472GA
☐ N401JR	69	HS 125/731	NA723	John & Mary Lou Kight, Las Vegas, Nv.	N723TS
☐ N401LG	96	CitationJet	525-0154	A Lakin & Sons Inc. Chicago, Il.	N51246
☐ N401LS	92	BAe 1000B	9032	Western (Delaware) Corp. Wilmington, De.	VP-CXX
☐ N401MS	68	Sabre-60	306-17	Holiday Air/Jacksonville Jet Center Inc. Jacksonville, Fl.	N13SL
☐ N401SK	66	BAC 1-11/401AK	073	Lukenbill Enterprises/Sky King Inc. Sacramento, Ca.	N5LC
☐ N402AC	66	HS 125/1A-522	25103	Liberty Aircraft Inc. Akron, Oh.	N60HU
☐ N402FB	85	Beechjet 400	RJ-2	Farm Bureau Life Insurance Co. West Des Moines, Ia.	N103AD
☐ N402JW	78	Falcon 10	120	Jetways Iowa LC. Minneapolis, Mn.	N100WG
☐ N402LM	83	Gulfstream 3	404	Lockheed-Martin Corp. Baltimore, Md.	N403LM
☐ N402ST	79	Citation II	550-0048	Shiloh Corp. Mansfield, Oh.	N558CB
☐ N402TJ	78	Citation II	551-0095	S & K Aviation Inc. Elk Grove, Il. (was 550-0049).	N400PC
☐ N403JW	67	Falcon 20C	102	General Aviation Inc/Jetways, Minneapolis, Mn.	N710WB
☐ N403W	84	Westwind-1124	403	Liberty Homes Inc. Goshen, In.	4X-CUH

Reg	Yr	Type	c/n	Owner/Operator	Prev Regn
☐ N404BF	79	Citation II	550-0071	Presidential Jet Inc. Somersworth, NH.	(N404BV)
☐ N404BS	96	Hawker 800XP	8294	BellSouth Telecommunications Inc. Atlanta, Ga.	N682H
☐ N404CE	96	Hawker 800XP	8293	BellSouth Telecommunications Inc. Atlanta, Ga.	N679H
☐ N404F	88	Falcon 900	41	WCF Aircraft Corp. Detroit, Mi.	N430FJ
☐ N404G	90	Citation V	560-0095	P H Glatfelter Co. Spring Grove, Pa.	N707CV
☐ N404JF	92	Citation VII	650-7001	R C Fisher, Palm Beach, Fl.	N111RF
☐ N404JW	76	Citation	500-0338	Jetways LLC. Minneapolis, Mn.	N73WC
☐ N404KA	81	Learjet 35A	35A-404	Palmer Aviation Leasing LLC. Germantown, Tn.	N404BB
☐ N404KS	81	Citation II	550-0334	4KS Aviation III LLC. Dallas, Tx.	N92LT
☐ N404LM	90	Gulfstream 4	1130	Lockheed-Martin Corp. Baltimore, Md.	N401MM
☐ N404M	89	Gulfstream 4	1110	Bristol Myers Squibb Co. NYC.	N415GA
☐ N404MA	73	Citation	500-0126	Software Customizers Inc. Las Vegas, Nv.	D-IDWH
☐ N404QS	96	Gulfstream 4SP	1304	Gulfstream Shares-FM Services Aviation LLC.Columbus, Oh.	N436GA
☐ N404R	88	Falcon 900	55	BellSouth Telecommunications Inc. Atlanta, Ga.	N495GA
☐ N404SB	82	Citation II	550-0426	BellSouth Telecommunications Inc. Atlanta, Ga.	N1218F
☐ N404VL	96	Falcon 900B	158	First Airplane LLC. Naples, Fl.	N900FJ
☐ N406LM	85	Citation III	650-0102	Lockheed Martin Corp. Baltimore, Md.	N406MM
☐ N406SS	82	Citation II	550-0337	Huntco Steel/HSI Aviation Inc. Springfield, Mo.	(N54HC)
☐ N407CA	84	Gulfstream 3	422	Chancellor Aviation Corp. West Palm Beach, Fl.	N750AC
☐ N407LM	85	Citation III	650-0103	Lockheed Martin Corp. Baltimore, Md.	N407MM
☐ N407W	84	Westwind-Two	407	Cross Creek Aviation Inc. Dallas, Tx.	4X-CUK
☐ N408JD	84	Citation III	650-0035	The Buckle Inc. Kearney, Ne.	N400JD
☐ N408QS	96	Gulfstream 4SP	1308	Gulfstream Shares-BT Southwest Inc. Columbus, Oh.	N446GA
☐ N408TR	64	Sabre-40	282-4	Eleanor Smith Corp. Ann Arbor, Mi.	N111MS
☐ N408W	85	Westwind-1124	408	Taughannock Aviation Corp. Ithaca, NY.	4X-CUB
☐ N409CC	82	Challenger 600S	1063	SS Platinum LLC/Stuller Settings Inc. Spring, Tx.	N8260D
☐ N409WT	63	Jet Commander	3	Flight Brothers Inc. Syosset, NY.	N400WT
☐ N410BT	91	BAe 125/800A	NA0468	Jackson National Life Insurance Co. Lansing, Mi.	N693C
☐ N410DM	92	Citation V	560-0184	Dibrell Brothers Tobacco Co. Danville, Va.	N873DB
☐ N410M	89	Gulfstream 4	1115	Bristol Myers Squibb Co. NYC.	N430GA
☐ N410NA	97	Citation V Ultra	560-0435	Samedan Oil Corp. Ardmore, Ok.	N52352
☐ N410PA	69	HS 125/731	NA728	Scope Leasing Inc. Columbus, Oh.	N320JJ
☐ N410QS	92	Gulfstream 4SP	1210	Gulfstream Shares, Columbus, Oh.	N9PC
☐ N410WW	92	Gulfstream 4SP	1203	W Wrigley-Zeno Air Inc. Wheeling, Il.	N434GA
☐ N411BB	96	Challenger 604	5316	Barnett Banks Inc. Jacksonville, Fl.	N604BB
☐ N411BW	85	Diamond-Two	A1008SA	Dodson International Parts inc. Ottawa, Ks. (status ?)	
☐ N411DS	77	Citation 1/SP	501-0033	Hartford Financial Corp. Smyrna, De.	(N101HC)
☐ N411PA	77	HS 125/700A	NA0211	PAS Inc. Wilmington, De.	N211WZ
☐ N411QS	96	Gulfstream 4SP	1311	Gulfstream Shares-Wilmington Trust Co. Columbus, Oh.	N449GA
☐ N411RA	90	BAe 125/800B	8177	Raytheon Aircraft Co. Wichita, Ks.	PT-OJC
☐ N411SC	75	Falcon 10	50	Medford Leasing Corp. Medford, Or.	PT-OHM
☐ N411SP	83	Diamond 1A	A049SA	Olympia Investments Inc. Pleasanton, Ca.	XA-SOD
☐ N411ST	96	Learjet 60	60-087	Benson Transportation Corp. Metairie, La.	N687LJ
☐ N411WW	89	Gulfstream 4	1121	W Wrigley-Zeno Air Inc. Wheeling, Il.	N7776
☐ N412AB	84	Falcon 200	492	Arkansas Best Corp. Fort Smith, Ar.	N803F
☐ N412LJ	97	Learjet 45	45-012	Learjet Inc. Wichita, Ks.	
☐ N412P	80	Citation II	550-0172	Pittston Coal Management Co. Lebanon, Va.	N78CS
☐ N412SE	81	Citation 1/SP	501-0225	Richard Carroll, Incline Village, Nv.	TG-KIT
☐ N412SP	74	Learjet 25B	25B-174	Jet Air Inc. Cincinnati, Oh.	N410SP
☐ N413CA	80	Citation II	550-0116	Alliance Air Inc. Miami, Fl.	PT-LGM
☐ N413CK	78	Citation II	550-0024	American Proteins Inc. Roswell, Ga.	N313CK
☐ N413LJ	97	Learjet 45	45-013	Learjet Inc. Wichita, Ks.	
☐ N413WF	69	Learjet 24B	24B-211	World of Faith International Christian Center, Redford, Mi.	N31LB
☐ N414BM	93	Gulfstream 4SP	1214	BMW Flight Service GmbH. Munich, Germany.	N405GA
☐ N414KL	84	Learjet 35A	35A-595	Krueger International & Shopko Stores Inc. Green Bay, Wi.	N85PM
☐ N414RF	79	HS 125/700A	NA0244	Alaska Eastern Inc. Anchorage, Ak.	N230DP
☐ N415EL	84	Westwind-Two	415	Marshall Industries Inc. Burbank, Ca.	N105BE
☐ N415FC	74	Citation	500-0145	Sierra Foxtrot Charlie Inc. Kansas City, Mo.	N145TA
☐ N415LJ	66	Learjet 23	23-092	AJM Airplane Co. Naples, Fl.	N344WC
☐ N415RD	80	HS 125/700A	7094	Rubloff Aviation LLC. Rockford, Il.	9M-STR
☐ N416AS	74	Falcon 10	16	Hines Jetcorp LLC. Muskegon, Mi.	F-GELA
☐ N416CC	82	Citation II	550-0416	Berry Leasing Co LP. Madera, Ca.	N12167
☐ N416F	79	Falcon 20F	416	U S West Inc. Englewood, Co.	N9VG
☐ N416K	68	Gulfstream 2SP	41	Peninsular Marine Services Inc. New Bern, NC.	N401GA
☐ N416KC	96	CitationJet	525-0130	Spenair LC. Salt Lake City, Ut.	N5093L
☐ N416QS	97	Gulfstream 4SP	1316	Gulfstream Shares, Columbus, Oh.	N427GA
☐ N416W	85	Westwind-1124	416	AFT Corp. New Castle, De.	4X-CUD
☐ N416WM	86	Gulfstream 3	487	ADI/Hi Flite Inc. Rochester Hills, Mi.	N618KM
☐ N417BA	79	Learjet 35A	35A-257	MCA-Midwest Corporate Aviation Lear 24 LP. Wichita, Ks.	N275DJ
☐ N417C	97	CitationJet	525-0174	Joubini Aviation Inc. Redwood City, Ca.	N817CJ
☐ N417CL	92	Challenger 601-3A	5107	Electronic Data Systems Corp. Dallas, Tx.	C-GLWZ
☐ N417LJ	98	Learjet 45	45-017	Learjet Inc. Wichita, Ks.	
☐ N417WW	68	Learjet 24	24-171	Westwood Industries Inc. Eugene, Or.	N737FN
☐ N418LJ	97	Learjet 45	45-018	Learjet Inc. Wichita, Ks.	
☐ N418MG	88	Beechjet 400	RJ-54	CoMar of Dayton Inc. Dayton, Oh.	XB-FDH
☐ N418R	94	Learjet 60	60-047	A S C-American Stores Co. Salt Lake City, Ut.	N5007P

Reg	Yr	Type	c/n	Owner/Operator	Prev Regn
N419MB	94	Beechjet 400A	RK-85	Applebee's International Inc.Overland Park, Ks.	N419MS
N419MK	92	Astra-1125SP	066	K & M Flight Corp. Chicago, Il.	N101NS
N419MS	96	Beechjet 400A	RK-121	Banking Consultants of America Inc. Memphis, Tn.	N419MB
N420CH	97	CitationJet	525-0066	C R H Investments, Fife, Wa.	N26495
N420DM	97	Citation V Ultra	560-0464	Dimon Inc. Danville, Va.	
N420DP	78	Falcon 20F	391	Dad's Products & Channllock Inc. Meadville, Pa.	N550PT
N420JM	70	Gulfstream 2	94	JM Aviation Inc/Executive Flight Management, Chicago, Il.	N623MW
N420JP	86	Falcon 50	168	Royal Jet Inc. San Diego, Ca.	N48GL
N420MP	86	Westwind-1124	418	MedPartners Aviation Inc. Birmingham, Al.	(N317MV)
N420PC	81	Falcon 10	186	Refreshment Services Inc. Springfield, Il.	N63TJ
N420QS	97	Gulfstream 4SP	1320	Gulfstream Shares, Columbus, Oh.	N437GA
N420SZ	88	Challenger 601-3A	5027	Eagle Flight Services Inc. Chicago, Il.	N421SZ
N420TJ	84	Westwind-Two	405	Ceridian Corp. Minneapolis, Mn.	N211DB
N421QL	82	Learjet 55	55-026	J L Leasing LLC. Valhalla, NY.	(N321GL)
N421SZ	87	Gulfstream 4	1025	Eagle Flight Services Inc. Chicago, Il.	N420SL
N422BC	80	Westwind-Two	302	Bradley Corp. Menomonee Falls, Wi.	N100AK
N422DV	68	Gulfstream 2	17	Otter Corp. Kirkland, Wa.	(N121PR)
N422MU	83	Falcon 200	484	Suiza Foods Corp. Dallas, Tx.	N357CL
N422QS	97	Gulfstream 4SP	1322	Gulfstream Shares, Columbus, Oh.	N445GA
N422X	79	HS 125/700A	NA0253	Carl Reichardt & Shorenstein Management Inc. Oakland, Ca.	(N831CJ)
N423JW	78	Learjet 36A	36A-043	Northwestern Aircraft Capital Corp. Mercer Island, Wa.	(N143JW)
N423SA	84	Gulfstream 3	429	Harbour Group Industries Inc. St Louis, Mo.	N429SA
N424BT	89	Beechjet 400	RJ-62	B J Tidwell Industries Inc.	N333RS
N424CV	97	Citation V Ultra	560-0424	Cx USA 9/97 to ?	(N324QS)
N424DA	72	Citation	500-0029	Bodmer Financing Co. Wilmington, De.	C-GDWN
N424GA	86	Gulfstream 4	1004	Aviation Enterprises Inc. Bedford, Tx.	
N424LB	85	Citation III	650-0076	Jet Flight Services Inc. Wilmington, De.	XA-SEP
N424R	74	Sabre-75A	380-15	Roblene Enterprises Inc. Ormond Beach, Fl.	N22JW
N425AS	80	Learjet 35A	35A-281	Miller Aviation, Johnson City, NY.	N425M
N425JF	77	Westwind-1124	210	Franks Petroleum Inc. Shreveport, La.	N337RE
N425JL	73	Learjet 25B	25B-127	Dolphin Aviation Inc. Sarasota, Fl.	N222AK
N425M	92	Learjet 31A	31A-055	GEICO Corp. Washington, DC.	N666RE
N425NA	68	Sabreliner CT-39E	282-95	NASA-GSFC, Wallops Flight Facility,	N4704N
N425TS	85	Astra-1125	004	Aero Toy Store Inc. Fort Lauderdale, Fl.	N96PC
N425WN	81	HS 125/700A	NA0311	Navellier Management Inc. Incline Village, Nv.	N311NW
N427CJ	75	Falcon 10	67	Superior Transportation Inc. Sioux City, Ia.	D-COME
N427LJ	66	Learjet 24A	24A-100	Dolphin Aviation Inc. Sarasota, Fl.	N616SC
N428CL	92	Challenger 601-3A	5108	Electronic Data Systems Corp. Farnborough, UK.	C-GLXD
N428DA	64	JetStar-6	5048	814K LLC. Pleasant View, Tn.	(N130LW)
N428FS	78	HS 125/700A	NA0219	Falstaff Air Corp. Dallas, Tx.	N428AS
N429DA	66	HS 125/1B-S522	25090	American Aircraft Sales International Inc. Sarasota, Fl.	N102TW
N429SA	96	Gulfstream 4SP	1314	USAA Special Services Co. San Antonio, Tx.	N461GA
N430A	85	Westwind-1124	430	COMM/NET Inc. Bellingham, Wa.	N821LG
N430HM	82	Learjet 55	55-043	OHM Corp. Findlay, Oh.	N30AF
N430JW	74	Learjet 24XR	24XR-285	Northwestern Aircraft Capital Corp. Van Nuys, Ca.	N995DR
N430MB	72	JetStar-731	5153	Leonard & Irma Fritz, Taylor, Mi.	N500PG
N430MP	73	Sabre-40A	282-113	Richard Hawkins, Nashua, NH.	N430MB
N430RG	79	Gulfstream 2TT	235	R & G Aviation LLC. Huntsville, Al.	N256M
N430SA	94	Citation VII	650-7041	USAA-United Services Auto Association, San Antonio, Tx.	(N449SA)
N431CB	85	Citation III	650-0084	C R Bard Inc. Murray Hill, NJ.	N85AW
N431CW	81	Learjet 35A	35A-431	Diamond Air Inc. Richmond Heights, Oh.	N34FD
N431NA	60	Sabreliner CT-39A	265-16	Des Moines Educational Resources, Des Moines, Ia.	USAF 60-3488
N431WM	87	Citation S/II	S550-0133	C W Ventures, Portland, Or.	N133VP
N432DG	77	Citation 1/SP	501-0039	TRB LLC. Portland, Or.	N800BH
N432EZ	81	Falcon 10	130	Texas Ezpawn Management Inc. Austin, Tx.	N921GS
N432QS	87	Gulfstream 4	1032	Gulfstream Shares, Columbus, Oh.	N888UE
N433CV	97	Citation V Ultra	560-0430	Beneto Inc. West Sacramento, Ca.	
N433DD	78	Learjet 35A	35A-161	D & D Aviation LC. Salt Lake City, Ut.	YV-65CP
N433GM	85	Westwind-Two	433	Glen Morgan/Park Street 1994 Investments, Dover, De.	N433WR
N434H	86	Citation III	650-0123	Hillenbrand Industries Inc. Batesville, In.	N624CC
N434JW	67	Gulfstream 2SP	2	JetWest International, Van Nuys, Ca.	N721SW
N435AS	76	Learjet 24E	24E-345	Comm Net Services Inc. Plano, Tx.	D-CFPD
N435JL	75	Learjet 35	35-018	Bellew Sky Inc. St Petersburg, Fl.	N696SC
N435MS	76	Learjet 35	35-054	Monte Sol Jet Inc. Santa Fe, NM.	N109MC
N435T	95	Falcon 2000	9	Tribune Co. Chicago, Il.	F-WWMG
N436BL	81	Learjet 35A	35A-436	Wilco Marine of Delaware Inc. Dallas, Tx.	PT-LDN
N436DM	81	Learjet 35A	35A-389	CVG Aviation/Comair Aviation, Cincinnati, Oh.	VR-BLU
N437H	79	Gulfstream 2	258	Halliburton Energy Services Inc. Dallas, Tx.	N929GV
N437SJ	86	Westwind-Two	437	Davison Transport Inc. Ruston, La.	N437WW
N438DM	78	Learjet 25D	25D-250	CVG Aviation Inc. Cincinnati. Oh.	N30LM
N439CL	92	Challenger 601-3A	5109	Electronic Data Systems Corp. Dallas, Tx.	C-GLXS
N439H	83	Citation III	650-0005	Honeywell Inc. Phoenix, Az.	N137S
N439WW	85	Westwind-Two	439	Commerce Leasing Co. Norcross, Ga.	4X-CUG
N440DM	81	Learjet 55	55-005	CVG Aviation Inc. Erlanger, Ky.	(N94TJ)
N440DS	90	Beechjet 400A	RK-8	State of Washington, Olympia, Wa.	

Reg	Yr	Type	c/n	Owner/Operator	Prev Regn
☐ N440MC	83	Learjet 35A	35A-495	McClane Company Inc. Temple, Tx.	
☐ N440RM	61	JetStar-6	5016	McKay Oil Corp. Roswell, NM.	N712GW
☐ N441BC	89	Astra-1125	033	International Bank of Commerce & others, Laredo, Tx.	N52KS
☐ N441PC	92	Learjet 35A	35A-668	Allmetal Inc. Wheeling, Il.	N9168Q
☐ N441TC	74	Citation	500-0140	Big Four D LLC. Brackettville, Tx.	N4ZK
☐ N442DM	81	Learjet 35A	35A-405	CVG Aviation Inc. Cincinnati, Oh.	N35FS
☐ N442ME	82	Citation II	550-0442	Midamerican Energy Co. Sioux City, Ia.	N442MR
☐ N442SW	97	Citation Bravo	550-0840	Cessna Aircraft Co. Wichita, Ks.	
☐ N442WT	93	Citation VII	650-7025	Wilson Trailer Co. Sioux City, Ia.	N1263B
☐ N444CW	84	Citation III	650-0064	Countrywide Home Loans Inc. Burbank, Ca.	N650TC
☐ N444ET	63	MS 760 Paris-2	101	Aviation Investments of Nevada Inc. Las Vegas, Nv.	N760PJ
☐ N444GB	80	Citation II	550-0159	National Air Lease Inc. Birmingham, Al.	C-GAPT
☐ N444GG	94	Citation V Ultra	560-0262	Diversified Health Group Inc. West Mifflin, Pa.	N262CV
☐ N444HC	93	Learjet 31A	31A-064	Florida Jet Center Inc. Fort Lauderdale, Fl.	N142GT
☐ N444MK	78	Learjet 25D	25D-252	Milam International Inc. Englewood, Co. 'Dream Chaser'	N4FH
☐ N444MV	77	Citation 1/SP	501-0034	Westfield Aviation Inc. Fairfax, Va.	N444MW
☐ N444MW	82	Westwind-1124	372	McWane Inc. Birmingham, Al.	N502BG
☐ N444PE	84	Falcon 50	143	P & E Properties, Teterboro, NJ.	N77CP
☐ N444QG	73	Gulfstream 2	133	Cedarhurst Air Charter Inc. Waukesha, Wi.	N44UP
☐ N444RH	91	CitationJet	525-0001	Jacket Express LLC. Wilmington, De.	N525CC
☐ N444TG	82	Learjet 35A	35A-469	Gaines Motor Lines Inc. Hickory, NC.	N660SA
☐ N444TJ	69	Jet Commander-B	146	Lawrence Wilkins, Venice, Fl.	N926JM
☐ N444TW	77	Learjet 24F	24F-348	Sierra American Corp. Dallas, Tx.	N8BG
☐ N444WB	88	Beechjet 400	RJ-42	Willbros USA Inc. Tulsa, Ok.	N735GA
☐ N444WC	65	Learjet 23	23-047	Graffitti Aviation Inc. Tampa, Fl.	N9260A
☐ N444WJ	75	Falcon 10	64	Western Jet Sales Inc. Arlington, Tx.	N718CA
☐ N444WW	79	Learjet 25D	25D-283	Alpha Aviation Inc. Dallas, Tx.	N312GK
☐ N445A	81	Westwind-1124	362	ARMCO Inc. West Mifflin, Pa.	4X-CUP
☐ N445BL	82	Westwind-1124	382	Jet Services Enterprises Inc. Bethany, Ok.	N999BL
☐ N445CC	93	Beechjet 400A	RK-45	C C Clark Inc. Starkville, Ms.	N445E
☐ N446D	81	Falcon 20F	446	U S West Inc. Englewood, Co.	N270RA
☐ N446U	85	Gulfstream 3	446	United Technologies Corp. Hartford, Ct.	N309GA
☐ N449ML	88	Challenger 601-3A	5022	JABJ/WFC Air Inc. Teterboro, NJ.	C-GLYK
☐ N450CL	81	Falcon 50	76	Werner Aire Inc. Omaha, Ne.	N411WW
☐ N450K	88	Falcon 50	186	Kimball International Transit Inc. Huntingburg, In.	N278FJ
☐ N450KK	81	Learjet 35A	35A-450	BYR 737/Beck & Yeaggy, Cincinnati, Oh.	
☐ N450KP	81	Falcon 50	82	Rock Island Air Inc. NYC.	N511GG
☐ N450MC	81	Learjet 35A	35A-368	McLane Co. Temple, Tx.	N99KW
☐ N451CS	69	Gulfstream 2B	70	5161 Corp. Burbank, Ca.	N165A
☐ N451DP	71	Falcon 20F	249	Dad's Products Co. Meadville, Pa.	N777JF
☐ N451GA	93	Gulfstream 4SP	1221	Dana Flight Operations Inc. Swanton, Oh.	
☐ N452LJ	96	Learjet 45	45-002	Learjet Inc. Wichita, Ks. (Ff 6 Apr 96).	N45LJ
☐ N453	76	Learjet 24D	24D-323	Omni Air Express Inc. Tulsa, Ok.	N104MC
☐ N453CV	97	Citation V Ultra	560-0453	Cessna Aircraft Co. Wichita, Ks.	
☐ N453GA	96	Gulfstream 4SP	1312	Gulfstream Aerospace Corp. Savannah, Ga.	
☐ N453JS	94	Falcon 900B	144	J B Scott & Kathryn Albertson, Boise, Id.	N144FJ
☐ N453LJ	96	Learjet 45	45-003	Learjet Inc. Wichita, Ks. (Ff 24 Apr 96).	
☐ N453S	82	Citation II/SP	550-0445	Selman Hangar, Monroe, La. (was 551-0445).	N1248K
☐ N453SB	74	Falcon 20F	308	Sierra Bravo Graco Inc. Lutherville, Md.	N81AJ
☐ N453TM	90	BAe 125/800A	8203	EMC Corp. Hopkinton, Ma.	N228G
☐ N454AC	77	Citation 1/SP	501-0015	McClain Enterprises Inc. Mountain Home, Ar.	N4446P
☐ N454DQ	81	Citation 1/SP	501-0174	Fornear Enterprises LLC. Marion, Ky.	N50JG
☐ N454GA	97	Gulfstream 4SP	1323	Gulfstream Aerospace Corp. Savannah, Ga.	
☐ N454JB	76	JetStar 2	5205	Shaw Group Inc. Baton Rouge, La.	N16BL
☐ N454LJ	96	Learjet 45	45-004	Learjet Inc. Wichita, Ks.	
☐ N455DW	87	Beechjet 400	RJ-20	Dudley Walker, Martinsville, Va.	N901P
☐ N455LJ	97	Learjet 45	45-005	Gainey Aircraft Corp. Grand Rapids, Mi.	
☐ N455RH	85	Learjet 55	55-110	Hunt Corp. Indianapolis, In.	N55GY
☐ N456AF	87	Citation III	650-0147	Motorola Inc. Farnborough, UK.	N141M
☐ N456CG	81	Learjet 25D	25D-343	L & W Leasing Inc. Pittsfield, Ma.	N3797L
☐ N456CL	81	Learjet 35A	35A-456	Jet Charter Inc. Van Nuys, Ca.	N711CD
☐ N456CM	83	Falcon 100	207	OrNda Health Corp. Nashville, Tn.	N207US
☐ N456CV	97	Citation V Ultra	560-0465	Cessna Aircraft Co. Wichita, Ks.	
☐ N456JW	94	Citation V Ultra	560-0266	Coca Cola Bottling Co. San Angelo, Tx.	N1295G
☐ N456LJ	97	Learjet 45	45-006	Learjet Inc. Wichita, Ks.	
☐ N456R	79	Citation 1/SP	501-0292	R Howard Strasbaugh Inc. San Luis Obispo, Ca. (was 500-0383)	YV-2295P
☐ N456SW	81	Gulfstream 3	337	Sentry Insurance a Mutual Co. Stevens Point, Wi.	
☐ N456WH	71	HS 125/400A	NA760	Woods Aviation Inc. Bozeman, Mt.	N70LY
☐ N457F	81	Falcon 20F	449	Southern Company Services Inc. Atlanta, Ga.	F-WLCT
☐ N457GA	98	Gulfstream 4SP	1324	Gulfstream Aerospace Corp. Savannah, Ga.	
☐ N457H	85	Gulfstream 3	457	GTE Southwest Inc. San Angelo, Tx.	N337GA
☐ N458N	79	Citation II	550-0061	DeJarnette Enterprises Inc. Lee's Summit, Mo.	N456N
☐ N458SW	66	Falcon 20-5	68	Sentry Insurance a Mutual Co. Stevens Point, Wi.	N577S
☐ N459GA	98	Gulfstream 4SP	1325	Gulfstream Aerospace Corp. Savannah, Ga.	
☐ N461AS	79	Falcon 10	146	Performance Consulting of N Florida Inc. Ponte Vedra, Fl.	F-GHVK
☐ N461GT	83	Gulfstream 3	411	Airmont Ltd-Deeside Trading Co. Reno, Nv.	N966H

| *Reg* | *Yr* | *Type* | *c/n* | *Owner/Operator* | *Prev Regn* |

Reg	Yr	Type	c/n	Owner/Operator	Prev Regn
☐ N461W	90	BAe 125/800A	NA0461	Buckeye Energy Co. Columbus, Oh.	N633BA
☐ N462B	89	Citation V	560-0016	James Koehler, Aberdeen, SD.	N68HQ
☐ N462QS	95	Gulfstream 4SP	1262	Gulfstream Shares-JI Aviation Inc. Columbus, Oh.	N496GA
☐ N464CL	66	Learjet 24A	24A-096	Clay Lacy Aviation Inc. Van Nuys, Ca.	N1972L
☐ N464EC	80	Westwind-Two	305	Barry Andrews & J Morgan Smith, Corpus Christi, Tx.	4X-CQY
☐ N464QS	95	Gulfstream 4SP	1264	Gulfstream Shares-RSL Investments Corp. Hilton Head, SC.	N499GA
☐ N465NW	82	Learjet 35A	35A-465	Bankair Inc. West Columbia, SC.	
☐ N465PM	80	Sabre-65	465-40	Flying Francis Inc. Houston, Tx.	N465RM
☐ N466CS	91	BAe 125/800A	NA0466	Central & South West Services Inc. Dallas, Tx.	N638BA
☐ N467H	63	Sabre-40	282-3	Hughes Aircraft Co. Van Nuys, Ca.	(N57QR)
☐ N467MW	81	Westwind-1124	325	Apache Rentals Inc. Phoenix, Az.	N525AJ
☐ N468GA	95	Gulfstream 4SP	1280	GRD Global Inc. Stamford, Ct.	
☐ N468KL	89	Challenger 601-3A	5036	LK Air Inc. Portland, Or.	N225N
☐ N469PW	76	Citation	500-0302	Jefferson County Racing Association, Montgomery, Al.	(N777QE)
☐ N470G	82	Falcon 20F	470	U S West Inc. Englewood, Co.	N607RP
☐ N472SW	93	CitationJet	525-0033	Jen Linz Ltd. Greenwich, Ct.	N526CA
☐ N472TS	91	Gulfstream 4	1172	El Paso Natural Gas Co. El Paso, Tx.	XA-SEC
☐ N473TC	69	Learjet 25	25-043	Tulsair Beechcraft Inc. Tulsa, Ok.	N234ND
☐ N475DJ	82	Gulfstream 3	358	Baymark Development Corp. Virginia Beach, Va.	(N1149E)
☐ N475QS	95	Gulfstream 4SP	1275	Gulfstream Shares-Wings Acquisition Corp. Hilton Head, SC.	N459GA
☐ N476BJ	97	Beechjet 400A	RK-176	Raytheon Aircraft Co. Wichita, Ks.	
☐ N477A	79	Citation II/SP	551-0020	Interwings Aircharter Corp. Wilmington, De.	YV-678CP
☐ N477DM	95	Challenger 601-3R	5174	CVG Aviation Inc/Jet Express, Cincinnati, Oh.	N605CC
☐ N477KM	82	Citation II	550-0353	King Management Inc. Billings, Mt. (was 551-0398).	N922RT
☐ N477MM	79	Learjet 25D	25D-291	Addison Insurance Marketing Inc. Dallas, Tx.	N530DC
☐ N477X	74	Sabre-60	306-78	Dillon Companies Inc. Hutchinson, Ks.	N140JA
☐ N479TS	84	Gulfstream 4	1079	Aero Toy Store Inc. Fort Lauderdale, Fl.	(N100WJ)
☐ N480CC	87	Citation S/II	S550-0129	Crounse Corp. Paducah, Ky.	N87TH
☐ N481FM	79	Learjet 35A	35A-218	Anderson Chemical Co. Macon, Ga.	N256TW
☐ N481JT	81	Challenger 600	1034	Inrots SA. Buenos Aires, Argentina.	(N209WE)
☐ N481MC	75	Westwind-1123	184	Deer Horn LC. Artesia, NM.	CC-CRK
☐ N481NS	82	Westwind-1124	378	Warren Manufacturing Inc. Birmingham, Al.	N84LA
☐ N481QS	96	Gulfstream 4SP	1281	Gulfstream Shares-KB Ventures Ltd. Hilton Head, SC.	N470GA
☐ N482CP	81	Learjet 25D	25D-331	Carmike Cinemas/Carl Patrick, Columbus, Ga.	N462B
☐ N482DM	84	Diamond 1A	A088SA	Mercantile Bank, St Louis, Mo.	
☐ N483DM	74	Learjet 24D	24D-291	Colvin Air Charter Inc. Athens, Ga.	N488DM
☐ N484CC	87	Beechjet 400	RJ-27	Corporate Wings Inc. Cleveland, Oh.	N3127R
☐ N484CS	80	Citation 1/SP	501-0153	Sampler Publications Inc. St Charles, Il.	N99CK
☐ N484KA	79	Citation Eagle	500-0387	Marshal Mize Ford Inc. Chattanooga, Tn.	N3206M
☐ N484TL	70	Gulfstream 2	93	Boxing Cat Productions Inc. Burbank, Ca.	N62K
☐ N485AC	82	Learjet 35A	35A-485	Sunward Corp. Billings, Mt.	N710WL
☐ N486DM	84	Diamond 1A	A086SA	Florida Rock Industries Inc. Jacksonville, Fl.	
☐ N486MJ	87	Beechjet 400	RJ-30	486MJ Inc. Spanish Fort, Al.	(N815BS)
☐ N487F	85	Falcon 50	160	A E Staley Manufacturing Co. Decatur, Il.	N48R
☐ N487GS	68	B 737-247	19600	Charlotte Hornets/Shinn Enterprises Inc. Charlotte, NC.	N307VA
☐ N487QS	95	Gulfstream 4SP	1287	Gulfstream Shares-QuarterGulf Inc. Hilton Head, SC.	N484GA
☐ N488DM	80	Sabre-65	465-26	Marshall Air LLC. Scottsdale, Az.	N31SJ
☐ N488GR	64	JetStar-6	5051	Sound Corp. Carson City, Nv.	N488JS
☐ N490CC	82	Citation II	550-0490	Scurlock Permian Corp. Houston, Tx. (was 551-0490)	N1254C
☐ N490ST	66	BAC 1-11/212AR	083	Pionus Corporation of Delaware, Amarillo, Tx.	N70611
☐ N491BT	73	Citation	500-0102	IBT=Industrial Bearing & Transmissions, Shawnee Mission, Ks.	N400K
☐ N491DB	82	Challenger 600S	1049	Riverhorse Investments Inc/Air Castle Inc. Santa Monica, Ca.	HB-VFW
☐ N491JB	89	Citation III	650-0182	J B Hunt Transport Inc. Lowell, Ar.	N682CC
☐ N491N	81	Citation II	550-0329	Indium Corp of America, NYC.	N6800C
☐ N492A	84	Gulfstream 3	425	Bank of America/KaiserAir Inc. Oakland, Ca.	N425SP
☐ N492BJ	98	Beechjet 400A	RK-192	Raytheon Aircraft Co. Wichita, Ks.	
☐ N492JT	70	Gulfstream 2SP	82	John Travolta/Atlo Inc. Van Nuys, Ca.	N728T
☐ N493QS	96	Gulfstream 4SP	1293	Gulfstream Shares-R Earl & King World Prod. Hilton Head, SC.	N415GA
☐ N494AT	87	BAe 125/800A	NA0404	Airtouch Communications, Oakland, Ca.	N916PT
☐ N495QS	96	Gulfstream 4SP	1295	Gulfstream Shares-Alberto Culver USA Inc. Hilton Head, SC.	N417GA
☐ N496EE	92	Beechjet 400A	RK-40	Progress Rail Services Corp. Albertville, Al.	N8252J
☐ N496SW	83	Learjet 35A	35A-496	Southwest Jet Aviation LLC. Scottsdale, Az.	N39TH
☐ N497DM	07	Challenger 604	5359	Bombardier Aerospace Corp. Hartford, Ct.	C-G...
☐ N497PT	80	HS 125/700A	NA0266	Patt Inc. Chaska, Mn.	N86MD
☐ N498CS	90	Citation III	650-0180	Pheasant Kay-Bee Toy Inc. Nashua, NH.	N768NB
☐ N499EH	78	Learjet 25D	25D-239	Eddie & Ercie Hill, Wichita Falls, Tx.	N45H
☐ N499NH	81	Sabre-65	465-56	Newman Racing, Lincolnshire, Il.	N65TL
☐ N499PB	65	JetStar-6	5063	Friendship Aviation Inc. Clearwater, Fl.	N420G
☐ N499QS	96	Gulfstream 4SP	1299	Gulfstream Shares-Dresser Industries Inc. Hilton Head, SC.	N423GA
☐ N499SC	94	Gulfstream 4SP	1238	Service Corp International Inc. Houston, Tx.	N483GA
☐ N500AD	72	Citation	500-0091	Kansas City Aviation Center Inc. Olathe, Ks.	(N1899)
☐ N500AF	86	Falcon 50	170	AFLAC Inc. Columbus, Ga.	N293K
☐ N500AG	68	JetStar-731	5119	A G Spanos Construction Co. Stockton, Ca.	N508TA
☐ N500AJ	94	Astra-1125SP	074	Cook Aircraft Leasing Inc. Bloomington, In.	
☐ N500AX	81	Westwind-Two	359	AVX Corp. Myrtle Beach, SC.	N500RR
☐ N500BG	67	Falcon 20C	121	United West Airlines/Tack 1 Inc. Pembroke Pines, Fl.	N25CP

Reg   *Yr Type*   *c/n*   *Owner/Operator*   *Prev Regn*

Reg	Yr	Type	c/n	Owner/Operator	Prev Regn
□ N500CD	97	Gulfstream 4SP	1321	Cendant Corp. NYC.	(N600CC)
□ N500CG	75	Learjet 24D	24D-304	Fox Gulf Inc. Pittsburgh, Pa.	N304LP
□ N500CM	86	Citation III	650-0111	Cargill Inc. Minneapolis, Mn.	(N13204)
□ N500CP	73	Citation	500-0087	Air Ambulance by Air Trek Inc. Punta Gorda, Fl.	(N911CJ)
□ N500CV	73	Citation	500-0076	Summit Seafood Supply Inc. Wilmington, De.	N90CC
□ N500CV	80	Citation II	550-0170	Fifth Third Bank of Kentucky Inc. Louisville, Ky.	N37BM
□ N500DL	69	Learjet 25	25-027	Able Jets Inc. Fort Pierce, Fl.	EC-EBM
□ N500DS	76	Learjet 35A	35A-079	American Jet International Corp. Houston, Tx.	N7777B
□ N500E	89	Gulfstream 4	1072	Exxon Center Services, Dallas, Tx.	N100A
□ N500ED	79	Learjet 35A	35A-241	Bankair Inc. West Columbia, SC.	N500EX
□ N500EL	74	Citation	500-0173	Advanced Drainage Systems Inc. Columbus, Oh.	N59CL
□ N500ES	66	JetStar-731	5075	Segerdahl Racing Corp. Wheeling, Il.	N1DB
□ N500ET	74	Citation	500-0180	Econoline Trailers Inc. Double Springs, Al.	N772C
□ N500EX	77	JetStar 2	5217	E C A Inc. Wilmington, De.	N814CE
□ N500FF	75	Falcon 10	58	Head Island Inc. Philadelphia, Pa.	(F-GHJL)
□ N500FK	90	Citation V	560-0047	Circuit City Stores Inc. Richmond, Va.	N2666A
□ N500FR	91	Citation VI	650-0208	Circuit City Stores Inc. Richmond, Va.	I-TALW
□ N500GA	81	Citation 1/SP	501-0260	Griffin Aircraft Industries, Cold Spring, Ky.	N224RP
□ N500GS	82	HS 125/700A	NA0312	General Signal Corp. Stamford, Ct.	N176RS
□ N500HG	79	Learjet 35A	35A-238	Harbour Group Industries Inc. St Louis, Mo.	N500CG
□ N500HH	74	Citation	500-0189	H R H Aviation Inc. Dover, De.	N189CC
□ N500J	91	BAe 125/800A	NA0469	Johnson & Johnson, New Brunswick, NJ.	N671BA
□ N500JE	94	Learjet 31A	31A-088	Eckerd Fleet Inc. Clearwater, Fl.	N31LJ
□ N500JS	68	Learjet 25	25-020	Kalitta Flying Service Inc. Ypsilanti, Mi.	N900CJ
□ N500JW	78	Gulfstream 2TT	234	Omicron Transportation Inc. Reading, Pa.	(N220GA)
□ N500KE	81	Westwind-1124	360	Trans-Exec Air Service/Kraco Enterprises Inc. Van Nuys, Ca.	N816S
□ N500KJ	89	Falcon 50	197	Sony Trading Corp. Tokyo, Japan.	N50FJ
□ N500LG	79	Learjet	28-005	Donald Allyn, Hanford, Ca.	N8LL
□ N500LJ	74	Citation	500-0195	Power Equity Inc. Henderson, Ky	N440EZ
□ N500LR	87	Challenger 601-3A	5012	Circuit City Stores Inc/Martinair, Sandston, Va.	N1868M
□ N500LS	69	B 727-31	20115	The Limited Inc. Columbus, Oh.	N505C
□ N500LW	77	Learjet 25D	25D-232	Manti Resources Inc. Covington, La.	N500EW
□ N500M	86	Astra-1125	018	Calcutta Aircraft Leasing Inc. Bloomington, In.	N1188A
□ N500MA	62	JetStar-731	5033	Mario Andretti/MA 500 Inc. Nazareth, Pa.	N50EC
□ N500MP	76	Learjet 25D	25D-208	King Leasing Co. Jackson, Mi.	N500PP
□ N500ND	80	Learjet 35A	35A-351	Mountain Aviation Inc. Laramie, Wy.	N500DD
□ N500NH	84	Learjet 55	55-100	Haas-Newman Enterprises LLC. Lincolnshire, Il.	N552SQ
□ N500PC	90	Challenger 601-3A	5071	Pepsico Inc. Westchester County Airport, NY.	C-G...
□ N500PE	86	Challenger 601	3065	Perkin-Elmer Corp. Westchester County Airport, NY.	(N128PE)
□ N500PG	85	Challenger 601	3039	Air Kelso LLC. NYC.	N639CL
□ N500PX	80	Citation II/SP	551-0215	Phoenix Investment Services Inc. Dover, De.	(N151PR)
□ N500R	82	Challenger 600S	1077	NASCAR Inc. Daytona Beach, Fl.	N71M
□ N500RE	82	Falcon 50	124	Ryan's Express Inc. Prior Lake, Mn.	N6666R
□ N500RH	70	Gulfstream 2	80	Hendrick Motosports Inc. Harrisburg, NC.	N85VT
□ N500RL	73	Gulfstream 2	122	Hawk Gathering System Inc. Longview, Tx.	N84A
□ N500RP	97	Citation X	750-0029	Omicron Transportation Inc. Reading, Pa.	N5090V
□ N500RR	83	Falcon 200	491	Sonic Financial Corp. Charlotte, NC.	N843MG
□ N500S	77	JetStar 2	5209	First Security Bank NA. Salt Lake City, Ut.	N5535L
□ N500SJ	74	Citation	500-0231	Wisconsin Aviation Inc. Watertown, Wi.	C-GMAT
□ N500SK	73	Citation	500-0129	Gulf Aire II Inc. Biloxi, Ms.	N8114G
□ N500SV	78	Learjet 36A	36A-040	Air Chariots Inc. Wilmington, De.	(N442SC)
□ N500SW	76	Learjet 24D	24D-325	Performance Aircraft Leasing Inc. Wilmington, De.	N416G
□ N500TL	78	Learjet 25D	25D-238	Labonte Racing Inc. Trinity, NC.	N238MP
□ N500TM	73	Citation Eagle	500-0112	Heizer Aviation Inc. St Louis, Mo.	N515DC
□ N500TS	77	Learjet 24E	24E-347	Cripple Creek Investments LLC. Mooresville, NC.	N500LL
□ N500TW	81	Citation 1/SP	501-0217	Eurosport Promotions/Heinz-Harald Frentzen, Nice, France.	N600SS
□ N500UB	90	Citation V	560-0052	Pomeroy Transport Inc. Stamford, Ct.	N500LE
□ N500WD	80	Sabre-65	465-48	Hotel Restaurant & Employees Union, Washington, DC.	XA-MLG
□ N500WJ	74	Citation	500-0202	K3C Inc. Uvalde, Tx.	N500VK
□ N500WN	69	JetStar-8	5135	Falcon General Inc. Plant City, Fl. (status ?).	(N500FG)
□ N500WR	91	Learjet 31A	31A-038	Diamond Aviation Inc. Mooresville, NC.	N131NA
□ N500WW	80	Gulfstream 3	318	Winged Wolf Inc. Burbank, Ca.	N150RK
□ N500XY	66	HS 125/731	25119	Oxy Air LLC. Temecula, Ca.	N213H
□ N500ZA	77	Learjet 24F	24F-350	Onager Co. Odessa, Tx.	N741GL
□ N501AF	80	Citation 1/SP	501-0139	Pacific Coast Aviation Inc. Brookings, Or.	N888BH
□ N501AT	74	Citation	500-0208	Bradley Flying Service Inc. Wethersfield, Ct.	(N508CC)
□ N501BA	78	Citation	500-0476	Bell Aviation Inc. West Columbia, SC.	F-GESZ
□ N501BB	79	Citation 1/SP	501-0087	Buck Air LLC. Carmel, Ca.	N501SJ
□ N501BE	77	Citation 1/SP	501-0263	Southwest Jet Inc. Kansas City, Mo. (was 500-0352)	(N501BF)
□ N501BG	90	Beechjet 400A	RK-5	Gaughan Leasing Corp. Las Vegas, Nv.	
□ N501BP	81	Citation 1/SP	501-0231	Apple Restaurants Management Co. Duluth, Ga.	N29HE
□ N501CC	70	Citation	701	Cessna Aircraft Co. Wichita, Ks. (was s/n 670).	
□ N501CD	78	Citation 1/SP	501-0066	Skagit Aviation LLC. Bow, Wa.	VP-CTB
□ N501CF	79	Citation 1/SP	501-0128	Delaroyal Corp. Wilmington, De.	N900MM
□ N501CG	79	Citation 1/SP	501-0102	Engineering & Technology Ltd. NYC. (stored GVA since 1/97).	I-AIRV
□ N501CW	75	Citation	500-0282	KC Air Inc. Wilmington, De.	N501SS

Reg	Yr	Type	c/n	Owner/Operator	Prev Regn
☐ N501D	79	Citation 1/SP	501-0298	Eagle Airways, Guernsey, Channel Islands.	VR-CMS
☐ N501DD	77	Citation 1/SP	501-0035	Del Webb Corp. Phoenix, Az.	N35JF
☐ N501DK	83	Citation II	550-0462	Niles Air Inc. Concord, NC.	XB-LTH
☐ N501DT	79	Westwind-1124	270	DT Industries Inc. Lebanon, Mo.	(N270DT)
☐ N501EM	77	Citation 1/SP	501-0273	Farah Manufacturing Co of NM Inc. El Paso. (was 500-0363).	N501JF
☐ N501F	95	Hawker 800XP	8286	U S Filter/Ionpure Inc. Beaverton, Or.	N807H
☐ N501FR	84	Citation 1/SP	501-0683	Fall River Group Inc. Mequon, Wi.	N501BK
☐ N501G	81	Citation 1/SP	501-0202	Michael & Joy Malcolm, Los Altos, Ca.	N477KM
☐ N501GG	79	Citation 1/SP	501-0141	Peterson Aviation LLC. Boise, Id.	N841MA
☐ N501GS	78	Citation 1/SP	501-0012	Silver Bay Logging Inc. Juneau, Ak.	N4196T
☐ N501GV	95	Gulfstream V	501	Gulfstream Aerospace Corp. Savannah, Ga. (Ff 28 Nov 95).	
☐ N501GW	81	Citation 1/SP	501-0177	Joe Morten & Son Inc. South Sioux City, Ne.	N501NC
☐ N501HS	79	Citation 1/SP	501-0096	Herbert Sutton/Southern Oregon Skyways Inc. Medford, Or.	N660KC
☐ N501JC	75	Citation	500-0252	International Jet Aviation Inc. Van Nuys, Ca.	N244WJ
☐ N501JE	83	Citation 1/SP	501-0253	NF Investments Inc. Atlanta, Ga.	N2650Y
☐ N501JM	81	Citation 1/SP	501-0226	Junior Miller Co. Hyrum, Ut.	N226VP
☐ N501KC	78	Citation II/SP	551-0077	Prompetta Ltd. Panama City. Panama. (was 550-0030).	(N200G)
☐ N501KG	75	Citation	500-0279	Eagle Mountain International Church, Newark, Tx.	N120S
☐ N501KK	81	Citation 1/SP	501-0181	Ross Brothers Construction Co. Ashland, Ky.	N250SR
☐ N501KR	78	Citation 1/SP	501-0071	King Ranch Oil & Gas Inc. Midland, Tx.	N501SR
☐ N501LC	80	Citation II	550-0146	ILC Enterprises Inc. San Antonio, Tx.	(N611RR)
☐ N501LE	83	Citation 1/SP	501-0323	Ed Bozarth Chevrolet Inc. Topeka, Ks.	N501RF
☐ N501LH	76	Citation	500-0342	Americraft Carton Group Inc. Kansas City, Mo.	N501DR
☐ N501LL	80	Citation 1/SP	501-0185	Craig Air Center Inc. Jacksonville, Fl.	N1CR
☐ N501LW	79	Citation 1/SP	501-0134	Citcon Inc/Concrete Systems Inc. Hudson, NH.	N2650S
☐ N501MB	79	Citation 1/SP	501-0122	Wistar Management Corp. Upper Darby, Pa.	(N501MD)
☐ N501MC	84	Citation II/SP	551-0500	Master Chemical Corp. Perrysburg, Oh.	N90RC
☐ N501MT	83	Citation 1/SP	501-0675	Conquest Services Inc. Corvallis, Mt.	N8900M
☐ N501NA	78	Citation	500-0375	Krapf C M Co. Wilmington, De.	LV-MMR
☐ N501PC	96	Gulfstream 4SP	1298	Pepsico Inc. Westchester County Airport, NY.	N422GA
☐ N501PV	77	Citation 1/SP	501-0028	Cytex Aviation LLC. Houston, Tx.	N1234X
☐ N501QS	89	Citation V	560-0024	EJI Inc/NetJets-Stone Container Corp. Chicago, Il.	(N12284)
☐ N501RB	73	Citation	500-0132	Richard Benes, Englewood, Co.	N132BP
☐ N501RC	80	Citation 1/SP	501-0165	New Hampshire Jet Holdings Inc. Nashua, NH.	N165NA
☐ N501RM	78	Citation 1/SP	501-0093	Capital Leasing Inc. Springfield, Il.	N623LB
☐ N501RS	79	Citation 1/SP	501-0097	C C Services of Nevada Inc. Reno, Nv.	N470JS
☐ N501SJ	78	Citation	501-0281	Alwyn LLC. Raleigh, NC. (was 500-0372).	LV-MGB
☐ N501SP	77	Citation 1/SP	501-0019	Fisher Sales Corp. St Charles, Il.	(N301MC)
☐ N501SS	78	Citation	500-0374	Silver State Aviation Inc. Las Vegas, Nv.	C-GRQA
☐ N501TB	84	Citation 1/SP	501-0685	S&P Aviation Services/Myrtle Blakley, Somerset, Ky.	(N243AB)
☐ N501TJ	77	Citation 1/SP	501-0013	VoiceNet Air Inc. Ivyland, Pa.	VR-BJK
☐ N501TK	74	Citation	500-0210	Word of Life Christian Center inc. Darrow, La.	YV-625CP
☐ N501TP	84	Citation 1/SP	501-0684	Stage Aviation Inc. Carmel, Ca.	N3683G
☐ N501VP	77	Citation 1/SP	501-0261	Jetair Inc. Wilmington, De. (was 500-0351).	N7NE
☐ N501WB	80	Citation 1/SP	501-0158	Bannen Enterprises/Davric Corp. Medford, Or.	N1MX
☐ N501WW	72	Citation	500-0044	N501WW Inc. Dover, De.	VR-CWW
☐ N501X	79	Citation 1/SP	501-0092	Ram Air Sales & Leasing LLC. Carefree, Az.	LV-WOI
☐ N502CC	79	Citation 1/SP	501-0113	Robert L Kimball, Edensburg, Pa.	N200ES
☐ N502E	93	Citation V	560-0232	Emerson Electric Co. St Louis, Mo.	
☐ N502GV	96	Gulfstream V	502	Gulfstream Aerospace Corp. Savannah, Ga.	
☐ N502PC	92	Challenger 601-3A	5121	Pepsico Inc. Westchester County Airport, NY.	C-G...
☐ N502RL	72	Citation	500-0043	Bayshore Air Inc. Miami, Fl.	N32DD
☐ N502RR	64	Sabre-40	282-13	Consorcio Aereo Inc. Wilmington, De.	XA-SMP
☐ N502T	96	Citation VII	650-7067	Tenneco Management Co. Houston, Tx.	
☐ N502TS	92	Citation V	560-0157	Turquoise Sky Inc. San Antonio, Tx.	N5704
☐ N503CC	72	Citation	500-0003	Ameriflight Aviation, Chesterfield, Mo.	
☐ N503GV	96	Gulfstream V	503	Gulfstream Aerospace Corp. Savannah, Ga.	
☐ N503PC	84	Challenger 601	3021	KFC West Inc. Westchester County Airport, NY.	N967L
☐ N503QS	90	BAe 1000B	9003	Executive Jet Sales Inc. Columbus, Oh.	G-ELRA
☐ N504T	94	Citation VII	650-7040	Tenneco Management Co. Houston, Tx.	N1265B
☐ N505AJ	67	Falcon 20C	89	Amermex Inc. Romulus, Mi.	N71CP
☐ N505CF	79	Citation 1/SP	501-0130	Coburn's Inc. St Cloud, Mn.	N102HS
☐ N505EE	83	Learjet 35A	35A-505	Commercial Plastics Co. Mundelein, Il.	N7259J
☐ N505HG	75	Learjet 36	36-009	General Transervice Inc. New Castle, De.	N505RA
☐ N505JC	76	Citation	500-0341	Teton Aviation Inc. Teton Village, Wy.	N2650
☐ N505JH	79	Citation 1/SP	501-0126	Jackson Hole Air Charter Inc. Jackson, Wy.	N505SP
☐ N505K	72	Citation	500-0004	Thomas Samuels MD. Decatur, Il.	N5005
☐ N505M	91	Challenger 601-3A	5100	Bristol-Myers Squibb Co. Evansville, In.	C-G...
☐ N505QS	91	BAe 1000A	9005	Executive Jet Sales Inc. Columbus, Oh. (was NA1000).	N410US
☐ N505RJ	77	Citation 1/SP	501-0009	John Rico Publishing Co. Vacaville, Ca.	N505BC
☐ N506E	93	Citation V	560-0236	Emerson Electric Co. St Louis, Mo.	
☐ N506T	86	Gulfstream 3	484	Tenneco Management Co. Houston, Tx.	N856W
☐ N506TF	77	Citation 1/SP	501-0001	Pelican Leasing Inc. Little Rock, Ar.	N51CJ
☐ N507HC	74	Challenger 600S	1026	Fort Mitchell Construction LLC. Fort Mitchell, Ky.	N507CC
☐ N507U	74	Sabre-60	306-93	Russ Darrow Leasing Co.	N200CX
☐ N508GP	81	Learjet 35A	35A-424	Frazier Group Corp. Mooresville, NC.	N2844
Reg	Yr	Type	c/n	Owner/Operator	Prev Regn

Reg	Yr	Type	c/n	Owner/Operator	Prev Regn
☐ N508HC	82	Challenger 600S	1057	Hoechst Marion Roussel, Kansas City, Mo.	N508CC
☐ N508R	80	Westwind-1124	280	Unitex Properties Inc. Marshall, Tx.	N500R
☐ N508T	78	Gulfstream 2	232	Tenneco Management Co. Houston, Tx.	N71WS
☐ N509GP	86	BAe 125/800A	8077	Georgia Pacific Corp. Atlanta, Ga.	N523BA
☐ N509QC	80	HS 125/700A	NA0277	Quikrete Companies, Atlanta, Ga.	(N799FL)
☐ N510GA	81	Citation 1/SP	501-0219	Sound Container Inc. Renton, Wa.	N18RN
☐ N510GP	82	Citation II	550-0421	Georgia Pacific Corp. Atlanta, Ga.	N67HW
☐ N510MS	69	Learjet 24B	24B-204	Surface Prep Inc. Richmond, Va.	N510ND
☐ N510MT	91	Citation V	560-0122	Biomet Inc. Warsaw, In.	N261WR
☐ N510PA	75	Learjet 24D	24D-305	Phoenix Air Group Inc. Cartersville, Ga.	(N43FN)
☐ N510SD	88	Citation III	650-0161	Marlin Air Inc. Detroit, Mi.	N500AE
☐ N510TP	82	Learjet 25D	25D-353	Ralph Owens & Owens Group Inc. Columbia, Tn.	XA-RLI
☐ N510TS	67	JetStar-731	5100	Airserv Inc. Brookfield, Wi.	XA-GZA
☐ N510US	78	Gulfstream 2SP	223	Lancaster Aviation Inc/Key Air Inc. Oxford, Ct.	N257H
☐ N510WS	87	Beechjet 400	RJ-24	World Savings & Loan Association, Oakland, Ca.	N3124M
☐ N511AB	81	Citation II	550-0299	Elizabeth Cardide Die Co. McKeesport, Pa. (was 551-0339).	HB-VIR
☐ N511AC	95	CitationJet	525-0098	Avis Industrial Corp. Upland, In.	N5156D
☐ N511AJ	70	Learjet 25	25-055	Able Jets Inc. Clearwater, Fl.	N65RC
☐ N511AT	76	Learjet 24E	24E-330	Sierra American Corp. Dallas, Tx.	
☐ N511DB	66	B 727-89	19139	Viscount Air Service Inc. Tucson, Az.	VR-CDB
☐ N511DR	80	Citation II	550-0185	Dan River Inc. Danville, Va.	N370AC
☐ N511JP	92	Beechjet 400A	RK-34	Windwalker Inc. Beckley, WV.	(N232BJ)
☐ N511KA	78	HS 125/700A	NA0237	Crusader Aviation Inc & Dubin Air Inc. Oxford, Ct.	N511GP
☐ N511PA	69	Gulfstream 2	49	Apex Oil Co/KB Aviation Inc. Chesterfield, Mo.	N830BH
☐ N511ST	94	Citation V Ultra	560-0281	SunTrust Banks Inc. Atlanta, Ga.	N.....
☐ N511TC	94	CitationJet	525-0074	Tom Chauncey, Scottsdale, Az.	N26581
☐ N511TD	69	JetStar-8	5145	Apex Oil Co & Fred Weber Inc. Chesterfield, Mo.	XA-JFE
☐ N511TS	67	JetStar-731	5101	Anthony Crane Rental LP. Beaumont, Tx.	N800AF
☐ N511WA	89	Astra-1125	034	SSW Jet Ltd. Cincinnati, Oh.	VR-CMG
☐ N511WD	90	BAe 125/800A	NA0459	CapX Transfer LLC. Wynnewood, Pa.	N511WM
☐ N511WS	82	Citation II	550-0439	Aquila Air LLC/Cin-Air LP. Cincinnati, Oh.	N550RS
☐ N511WV	91	Citation V	560-0138	Cin-Jet Inc. Cincinnati, Oh.	OY-FFV
☐ N512CC	72	Citation	500-0012	K3C Inc. Reno, Nv.	XC-FIU
☐ N513AG	66	Falcon 20C	64	Winter Haven Homes Inc. Atlanta, Ga.	N513AN
☐ N513MW	97	Gulfstream V	510	BMW Flight Service GmbH. Munich, Germany.	N598GA
☐ N513QS	91	BAe 1000A	NA1004	EJI Inc/NetJets, Columbus, Oh.	N125BA
☐ N514DS	97	CitationJet	525-0255	Cessna Aircraft Co. Wichita, Ks.	
☐ N514QS	91	BAe 1000A	NA1005	EJI Inc/NetJets, Columbus, Oh.	N125CJ
☐ N515AJ	66	JetStar-731	5078	CEMR Inc. Houston, Tx.	(N916RG)
☐ N515DB	87	Falcon 200	510	Songbird Inc. Livonia, Mi.	N79PM
☐ N515GP	95	Hawker 800XP	8289	Georgia-Pacific Corp. Atlanta, Ga.	N669H
☐ N515JT	76	Gulfstream 2B	189	J T Aviation Corp/Taughannock Aviation Corp. Ithaca, NY.	N512VB
☐ N515LG	76	Westwind-1124	193	Label Graphics Inc. Portland, Or.	N428JM
☐ N515QS	91	BAe 1000A	NA1006	EJI Inc/NetJets, Columbus, Oh.	N1000E
☐ N515TC	81	Learjet 25D	25D-354	Kress Enterprises Inc. Bartonville, Il.	N3795U
☐ N516CC	83	Westwind-Two	394	Churchill Industrial Leasing Inc. Minneapolis, Mn.	N352TC
☐ N516GP	96	Hawker 800XP	8316	Georgia-Pacific Corp. Atlanta, Ga.	N316XP
☐ N516LW	72	Sabre-40A	282-98	Longley Supply/Wright Corp. Wilmington, NC.	N516WP
☐ N516SM	80	Falcon 10	167	Alpha Charlie Inc. Round Rock, Tx.	N167AC
☐ N517AM	84	Learjet 55	55-108	Florida Jet Service Inc. Pembroke Pines, Fl.	(N551AM)
☐ N517MT	93	Citation VI	650-0232	ASASCO Aviation Co. Wilmington, De.	F-GJKS
☐ N518CL	95	Challenger 601-3R	5180	Bombardier Capital Inc. Colchester, Vt.	C-GLYO
☐ N518EJ	80	Falcon 50	18	EJI Inc. Columbus, Oh.	N82LP
☐ N518GA	97	Gulfstream V	518	Gulfstream Aerospace Corp. Savannah, Ga.	
☐ N518GS	73	Gulfstream 2	130	George Strait Productions Inc. San Antonio, Tx.	N872GA
☐ N518JG	81	Learjet 25D	25D-328	Joe Gibbs Racing Inc. Charlotte, NC.	N328JW
☐ N518JT	73	Gulfstream 2	135	J T Aviation Corp/Taughannock Aviation, Ithica, NY.	N113EV
☐ N520G	96	Citation V Ultra	560-0369	Oldenburg Group Inc. Milwaukee, Wi.	N5200
☐ N520MP	85	Westwind-1124	421	MedPartners Aviation Inc. Birmingham, Al.	N317MQ
☐ N520PA	75	Learjet 35	35-037	Phoenix Air Group Inc. Cartersville, Ga.	N600WT
☐ N520QS	92	BAe 1000A	NA1008	EJI Inc/NetJets-CK & BKS Aviation Inc. Columbus, Oh.	N676BA
☐ N520SC	83	Learjet 55	55-087	Stryker Corp. Kalamazoo, Mi.	N8564Z
☐ N520TJ	69	Falcon 20D	198	Worldwide Charters LLC. Stevensville, Md.	XA-SQS
☐ N521BH	93	Citation II	550-0727	Bradley Management Group Inc. Indianapolis, In.	N727CM
☐ N521TM	92	Citation II	550-0705	Idaho Power Co. Boise, Id.	
☐ N521WM	79	Citation II	550-0084	William McBride III, Santa Barbara, Ca.(was 551-0129).	(N467MW)
☐ N522AC	94	Falcon 900B	148	Amway Corp. Grand Rapids, Mi.	N148FJ
☐ N522CC	78	Citation II/SP	551-0027	Sherr & Co. Hickory, NC.	N552CC
☐ N522EE	92	Beechjet 400A	RK-38	Energy Education Inc. Wichita Falls, Tx.	N5685X
☐ N522JA	95	Citation V Ultra	560-0288	Seminole Aviation LLC. Seminole, Ok.	N5141F
☐ N522JS	81	Learjet 25D	25D-348	Stinair LLC. NYC.	N522GS
☐ N522PA	79	Learjet 35A	35A-254	Phoenix Air Group Inc. Cartersville, Ga.	N34FN
☐ N523AC	94	Falcon 900B	139	Amway Corp. Grand Rapids, Mi.	F-W...
☐ N523CC	72	Citation	500-0023	K3C Inc. Reno, Nv.	I-ALBS
☐ N523GA	97	Gulfstream V	523	Gulfstream Aerospace Corp. Savannah, Ga.	
☐ N523JM	92	Challenger 601-3A	5106	Worthington Industries/McAir Inc. Worthington, Oh.	(N601PR)

Reg	Yr	Type	c/n	Owner/Operator	Prev Regn

Reg	Yr	Type	c/n	Owner/Operator	Prev Regn
☐ N523KW	92	Citation V	560-0224	Capital Leasing Inc. Wichita, Ks.	N224CV
☐ N523QS	92	BAe 1000A	NA1010	EJI Inc/NetJets-Jetters Inc. Columbus,	N679BA
☐ N523WC	87	BAe 125/800A	8086	Towne Management Co. Youngstown, Oh.	N125BA
☐ N524DW	71	Learjet 25B	25B-081	Jet Charter Airlines Inc. Las Vegas, Nv.	N66TJ
☐ N524HC	95	Learjet 31A	31A-114	Harbert Corp. Birmingham, Al.	
☐ N524M	83	HS 125/700A	NA0343	Marathon Oil Co. Findlay, Oh.	G-5-15
☐ N524MA	78	Citation II	550-0029	Florida Custom Coach Inc. Leesburg, Fl.	N550TJ
☐ N524PA	75	Learjet 35	35-033	Phoenix Air Group, Cartersville, Ga.	N31FN
☐ N525AE	93	CitationJet	525-0017	JDP Aircraft II Inc. Raleigh, NC.	(N1328Q)
☐ N525AL	93	CitationJet	525-0011	Elmair Inc. New Castle, De.	(N1327N)
☐ N525AP	94	CitationJet	525-0045	Presco Inc. The Woodlands, Tx.	(N36854)
☐ N525AS	95	CitationJet	525-0092	All-State Industrial Rubber Co. West Des Moines, Ia.	
☐ N525BL	97	CitationJet	525-0191	Wenfield Farm Inc. Chesapeake, Va.	N525BF
☐ N525BT	96	CitationJet	525-0161	Dorand LLC. Portland, Or.	
☐ N525CH	94	CitationJet	525-0078	Cooper Hosiery Mills Inc. Fort Payne, Al.	
☐ N525CJ	91	CitationJet	702	Cessna Aircraft Co. Wichita, Ks.	
☐ N525CK	94	CitationJet	525-0058	Charles Key MD. Dallas, Tx.	
☐ N525DC	97	CitationJet	525-0195	Ilyas Chaudhary, Newport Beach, Ca.	
☐ N525DG	94	CitationJet	525-0057	Direct Charter LLC. Greensboro, NC.	
☐ N525EC	97	CitationJet	525-0246	Cessna Aircraft Co. Wichita, Ks.	
☐ N525FS	93	CitationJet	525-0026	G M Hock Construction Inc. Durham, NC.	(N1330G)
☐ N525GM	97	CitationJet	525-0240	Cessna Aircraft Co. Wichita, Ks.	
☐ N525HS	93	CitationJet	525-0035	HealthSouth Rehabilitation Corp. Birmingham, Al.	N1779E
☐ N525HV	97	CitationJet	525-0201	Hy-Vee Inc. Chariton, Ia.	
☐ N525J	97	CitationJet	525-0184	Jay Merten, Wilmington, De.	
☐ N525JH	95	CitationJet	525-0124	Medallion Exploration, Midvale, Ut.	
☐ N525JM	96	CitationJet	525-0134	John Mayes, Redwood City, Ca.	N234CJ
☐ N525JT	75	Gulfstream 2B	156	J T Aviation Corp/Eastway Aircraft Services, Ronkonkoma, NY.	N18NK
☐ N525KA	92	CitationJet	525-0019	Kal-Aero/PPAL Inc & Pilgrim Holdings Ltd. Kalamazoo, Mi.	(N525SP)
☐ N525KH	97	CitationJet	525-0197	Quaker Air Services Inc. Salem, Or.	
☐ N525KL	96	CitationJet	525-0136	C Kevin Landry, Lincoln, Ma.	
☐ N525KN	93	CitationJet	525-0007	Air South Inc/Westward Aviation, Long Beach, Ca.	(N1327E)
☐ N525MB	93	CitationJet	525-0036	Skylark East Inc. Great Neck, NY.	N1782E
☐ N525MC	93	CitationJet	525-0018	McCoy Corp. San Marcos, Tx.	(N1328X)
☐ N525NA	94	CitationJet	525-0048	DWT Aviation Inc. Baltimore, Md.	N500HC
☐ N525PE	95	CitationJet	525-0125	Plastics Engineering Co. Sheboygan, Wi.	
☐ N525PF	97	CitationJet	525-0232	Cessna Aircraft Co. Wichita, Ks.	
☐ N525PL	93	CitationJet	525-0043	Chandelle Investment Corp. Portland, Or.	(N3261M)
☐ N525PS	94	CitationJet	525-0061	Omyaviation Inc. Proctor, Vt.	N61CJ
☐ N525QS	92	BAe 1000A	9025	EJI Inc/NetJets-Marcus Cable Inc. Columbus, Oh.	N292H
☐ N525RC	97	CitationJet	525-0178	Rydell Co. Sioux Falls, SD.	
☐ N525RD	91	Citation V	560-0106	Western Leasing Inc. Phillips, Wi.	N60SH
☐ N525RF	93	CitationJet	525-0023	Robert Fiscella, Hampton, Va.	(N1329T)
☐ N525RM	97	CitationJet	525-0225	Molded Fiber Glass Companies, Ashtabula, Oh.	
☐ N525SC	95	CitationJet	525-0099	Falcon LLC. Carmel, In.	N5156V
☐ N525TF	94	CitationJet	525-0067	TFP Corp. Medina, Oh.	N594JB
☐ N525TW	68	Learjet 25	25-011	Sierra American Corp. Dallas, Tx.	N108GA
☐ N525WB	94	CitationJet	525-0079	William Barnet & Son Inc. Arcadia, SC.	N179CJ
☐ N525WC	95	CitationJet	525-0107	White Cap Industries Inc. Costa Mesa, Ca.	N5211F
☐ N525WM	97	CitationJet	525-0213	Wilkes & McHugh PA. Tampa, Fl.	
☐ N526AC	89	BAe 125/800A	8169	Amway Corp. Grand Rapids, Mi.	N47CG
☐ N526JP	94	CitationJet JPATS	705	Cessna Aircraft Co. Wichita, Ks.	
☐ N526JT	93	CitationJet JPATS	704	Model 526 JPATS CitationJet first flight 20 Dec 93.	
☐ N526M	85	BAe 125/800A	8032	Marathon Oil Co. Findlay, Oh.	
☐ N527AC	87	BAe 125/800A	NA0405	Amway Corp. Grand Rapids, Mi.	N542BA
☐ N527EW	83	Citation 1/SP	501-0322	Rockville Investments (Jersey) Ltd. Jersey, Channel Islands.	(N769EW)
☐ N527GA	98	Gulfstream V	527	Gulfstream Aerospace Corp. Savannah, Ga.	
☐ N527JG	97	Learjet 31A	31A-125	Gibbs International Inc. Spartanburg, SC.	N125LJ
☐ N527M	86	BAe 125/800A	8054	Marathon Oil Co. Findlay, Oh.	G-5-15
☐ N527PA	76	Learjet 36A	36A-019	Phoenix Air Group Inc. Cartersville, Ga.	N540PA
☐ N528AC	86	BAe 125/800A	8070	Amway Corp. Grand Rapids, Mi.	N520BA
☐ N528GA	98	Gulfstream V	528	Gulfstream Aerospace Corp. Savannah, Ga.	
☐ N528M	86	BAe 125/800A	8055	Marathon Oil Co. Findlay, Oh.	G-5-504
☐ N528RR	90	Astra-1125SP	042	Reese Rowling, Corpus Christi, Tx.	N588R
☐ N529AC	70	B 727-17	20327	Amway Corp. Grand Rapids, Mi.	N4002M
☐ N529DM	84	Challenger 601	3025	Trust Aviation Inc. Salem, NH.	N1620
☐ N529GA	98	Gulfstream V	529	Gulfstream Aerospace Corp. Savannah, Ga.	
☐ N529SC	70	Sabre-60	306-58	Columbus Properties LLC. New Orleans, La.	N1PN
☐ N530DL	80	Westwind-1124	287	Leprino Foods, Broomfield, Co.	N146BF
☐ N530G	67	JetStar-731	5096	Beta Aviation Inc. Houston, Tx.	N9252R
☐ N530KF	66	B 727-61	19176	DoJ-U S Marshals Service, Oklahoma City, Ok.	N2777
☐ N530P	81	Citation II	550-0310	Business Aircraft Group Inc. Cleveland, Oh.	N7798D
☐ N530QS	92	BAe 1000B	9030	EJI Inc/NetJets, Columbus, Oh.	G-DCCI
☐ N531AB	75	Sabre-60	306-98	Hubert Jet LLC. Pontiac, Mi.	N169AC
☐ N531AT	93	Learjet 31A	31A-085	Thompson Machinery Commerce Corp. LaVergne, Tn.	XA-PEN
☐ N531CC	92	Citation V	560-0178	Ruddick Corp/Teal Aviation, Charlotte, NC.	N1280A

Reg	Yr	Type	c/n	Owner/Operator	Prev Regn
☐ N531CM	85	Citation S/II	S550-0033	Freezer Services Inc. Omaha, Ne.	N550ST
☐ N531CW	77	Learjet 25D	25D-231	K A Leasing LLC. Reno, Nv.	N225HW
☐ N531GA	98	Gulfstream V	531	Computer Associates International Inc. Islandia, NY.	
☐ N531JF	86	Gulfstream 3	491	JetSet Inc. St Paul, Mn.	N998JB
☐ N531WB	87	Falcon 200	514	VBI Inc. Ventura, Ca.	PT-OQG
☐ N532CC	89	Citation III	650-0167	Charlotte Pipe & Foundry Co/Teal Aviation, Charlotte, NC.	N667CC
☐ N532GA	98	Gulfstream V	532	North Carolina Air & Travel LLC. Cary, NC.	
☐ N532JF	97	Citation VII	650-7077	JetSet Inc. St Paul, Mn.	N877CM
☐ N532MA	89	Citation V	560-0036	North American Jet Inc. Georgetown, Tx.	(N560EJ)
☐ N533CC	90	Citation III	650-0185	Ruddick Corp. Charlotte, NC.	N650HS
☐ N533GA	98	Gulfstream V	533	Gulfstream Aerospace Corp. Savannah, Ga.	
☐ N533JF	97	CitationJet	525-0247	Cessna Aircraft Co. Wichita, Ks.	
☐ N533MA	79	Citation II	550-0078	533MA LLC. Portland, Or.	(N277A)
☐ N533QS	92	BAe 1000A	9033	EJI Inc/NetJets, Columbus, Oh.	N850BL
☐ N534GA	98	Gulfstream V	534	Gulfstream Aerospace Corp. Savannah, Ga.	
☐ N534H	90	Citation III	650-0196	Hillenbrand Investment Advisors, Wilmington, De.	N896EC
☐ N534MA	80	Falcon 50	29	Wilmington Trust Co. Wilmington, De.	CS-TMF
☐ N535AF	78	Learjet 35A	35A-191	Ameriflight Inc. Burbank, Ca.	N35SE
☐ N535CS	85	Gulfstream 3	464	Campbell Soup Co. New Castle, De.	N340GA
☐ N535LR	95	CitationJet	525-0128	L R Green Co/Poster Display Co. Indianapolis, In.	N5092D
☐ N535MA	78	Citation II	550-0037	VEC Corp. Teaneck, NJ.	XA-SDV
☐ N535QS	93	BAe 1000A	9035	EJI Inc/NetJets-Jet Seven Inc. Columbus, Oh.	N160BA
☐ N536GA	98	Gulfstream V	536	Gulfstream Aerospace Corp. Savannah, Ga.	
☐ N537GA	98	Gulfstream V	537	Gulfstream Aerospace Corp. Savannah, Ga.	
☐ N538GA	98	Gulfstream V	538	Gulfstream Aerospace Corp. Savannah, Ga.	
☐ N539GA	98	Gulfstream V	539	Gulfstream Aerospace Corp. Savannah, Ga.	
☐ N539QS	93	BAe 1000A	9039	EJI Inc/NetJets-HFS Asset Co. Columbus, Oh.	N169BA
☐ N540B	79	HS 125/700A	NA0255	Virginia Binger/Jucamcyn Theaters Corp. Minneapolis, Mn.	N125AJ
☐ N540CH	96	Gulfstream 4SP	1306	Chase Manhattan Bank, NYC.	N441GA
☐ N540CL	65	Learjet 23	23-026	Visalia Community Bank, Visalia, Ca.	N404DB
☐ N540EA	75	Gulfstream 2SP	174	Jetmark Aviation LLC. NYC.	N900ES
☐ N540M	89	BAe 125/800A	NA0432	E I DuPont de Nemours & Co. Houston, Tx.	N586BA
☐ N540QS	93	BAe 1000A	9040	EJI Inc/NetJets-Nana Air Services Inc. Columbus, Oh.	N22UP
☐ N540W	95	Gulfstream 4SP	1265	Harpo Productions-Oprah Winfrey/Leap of Faith Inc. Chicago.	N465GA
☐ N541CW	80	Diamond 1	A004SA	Paramount Aviation LLC. Cleveland, Oh.	(N88TJ)
☐ N541PA	76	Learjet 35	35-053	Phoenix Air Group Inc. Cartersville, Ga.	N53FN
☐ N541QS	93	BAe 1000A	9041	EJI Inc/NetJets-Foremost Corp of America Inc. Columbus, Oh.	N936H
☐ N542PA	75	Learjet 35	35-030	Phoenix Air Group Inc. Cartersville, Ga.	N16FN
☐ N542QS	93	BAe 1000A	9042	EJI Inc/NetJets-Burlington Northern R R Inc. Columbus, Oh.	N941H
☐ N543PA	76	Learjet 35A	35A-070	Phoenix Air Group Inc. Cartersville, Ga.	N50FN
☐ N543SC	87	Citation S/II	S550-0144	Shea Aviation, Atlanta, Ga.	N6516V
☐ N544PA	79	Learjet 35A	35A-247	Phoenix Air Group Inc. Cartersville, Ga.	N523PA
☐ N544QS	94	Hawker 1000	9044	EJI Inc/NetJets-Cambridge Technology Partners, Columbus.	N956H
☐ N545GA	78	Citation II	550-0040	State of Georgia DoT, Atlanta, Ga. (was 551-0085).	(N551GA)
☐ N545JT	82	Gulfstream 3	347	J T Aviation Corp. NYC.	VR-BJE
☐ N545PA	77	Learjet 36A	36A-028	Phoenix Air Group Inc. Cartersville, Ga.	N75TD
☐ N545QS	94	Hawker 1000	9045	EJI Inc/NetJets-Warren Buffett & Platinum Air, Columbus, Oh.	G-5-801
☐ N545S	69	HS 125/731	NA719	Betman Inc. Wilmington, De.	G-AWXD
☐ N545TP	83	Diamond 1A	A045SA	S2 Yachts Inc. Holland, Mi.	N777DC
☐ N546BZ	92	Beechjet 400A	RK-41	Brazeway Inc. Adrian, Mi.	N920SA
☐ N546PA	80	Learjet 36A	36A-045	CFF Air Inc. Wilmington, De.	N13FN
☐ N546QS	94	Hawker 1000	9046	EJI Inc/NetJets-CGF Fleet Management Inc. Columbus, Oh.	N962H
☐ N547JG	78	Learjet 25D	25D-264	Goodman Aviation Inc. Las Vegas, Nv.	N502JC
☐ N547JL	79	Sabre-75A	380-69	Executive Aircraft Corp. Newton, Ks.	N111VX
☐ N547PA	75	Learjet 36	36-012	Phoenix Air Group Inc. Cartersville, Ga.	N712JE
☐ N547QS	94	Hawker 1000	9047	EJI Inc/NetJets-Starship 1 Inc. Columbus, Oh.	N296H
☐ N548PA	78	Learjet 36A	36A-038	Phoenix Air Group Inc. Cartersville, Ga.	N15PN
☐ N548QS	95	Hawker 1000	9048	EJI Inc/NetJets-Sara Lee Corp. Columbus, Oh.	N802H
☐ N549PA	77	Learjet 35A	35A-119	Phoenix Air Group Inc. Cartersville, Ga.	(N64DH)
☐ N549QS	96	Hawker 1000	9049	EJI Inc/NetJets-HK Air Ltd. Columbus, Oh.	G-5-837
☐ N550A	84	Citation S/II	S550-0009	Anguilla Properties Inc. Augusta, Ga.	N1256Z
☐ N550AJ	87	Citation S/II	S550-0141	Kiwi Aviation LLC. Newton, NC.	N26JJ
☐ N550AK	82	Learjet 55	55-045	Emax Oil Co. Charlottesville, Va.	(N123LC)
☐ N550AS	85	Citation S/II	S550-0020	A D Seenc Construction/Norsan Financial, Concord, Ca.	
☐ N550BB	95	Citation Bravo	550-0734	Cessna Aircraft Co. Wichita, Ks. (Ff 19 Apr 95).	
☐ N550BC	97	Citation Bravo	550-0804	Transcraft Corp. Anna, Il.	N804CB
☐ N550BD	89	Citation II	550-0619	Southeast Aviation Inc. Columbia, SC.	N170TC
☐ N550BG	93	Citation II	550-0719	Chabel Aviation Inc. Miami, Fl.	N12068
☐ N550CA	80	Citation II	550-0152	OIA Air Corp. Nashua, NH.	N88840
☐ N550CG	79	Citation II	550-0095	Lors Medical Corp. Roanake Rapids, NC.	N100UF
☐ N550CW	83	Challenger 600S	1084	Jayhawk Inc. Dallas, Tx.	N175ST
☐ N550DW	84	Citation II	550-0487	The Charles Machine Works Inc. Perry, Ok.	N444BL
☐ N550F	85	Citation S/II	S550-0056	FERSAN/Servicios Aereos Turisticos SA. Dominican Republic.	C-GERL
☐ N550FB	97	Citation Bravo	550-0803	Firebond Corp. Minden, La.	
☐ N550HB	81	Citation II	550-0213	KBH Corp. Clarksdale, Ms.	N421TX
☐ N550HC	86	Citation S/II	S550-0116	First Financial Resources Inc. Denton, Tx.	N125CG

Reg	Yr	Type	c/n	Owner/Operator	Prev Regn
☐ N550HF	81	Citation II	550-0251	Southwest Airplanes Inc. Georgetown, Tx.	PK-TRV
☐ N550HH	96	Citation Bravo	550-0802	Continental Resources Inc. Enid, Ok.	N802CB
☐ N550HT	86	Citation S/II	S550-0107	Havatampa Inc. Tampa, Fl.	N713DH
☐ N550J	82	Citation II	550-0418	Cardinal I G Co. Minnetonka, Mn.	N1217H
☐ N550JS	79	Citation II/SP	551-0149	Enhancement of Illinois Inc. Cerillos, NM. (was 550-0107).	N225FM
☐ N550K	67	Jet Commander-A	127	MRG Jet Commander Inc. Englewood, Co.	N100SR
☐ N550KA	79	Citation II	550-0070	VH Airplanes Inc. Georgetown, Tx.	N777FL
☐ N550KE	92	Citation II	550-0694	Capital Buyers Inc. Conway, Ar.	N694CM
☐ N550KH	98	Citation Bravo	550-0854	Cessna Aircraft Co. Wichita, Ks.	
☐ N550KL	97	Citation Bravo	550-0844	Cessna Aircraft Co. Wichita, Ks.	
☐ N550KM	85	Citation S/II	S550-0081	TEB Charter Services Inc. Seneca, NM.	N168HC
☐ N550LH	79	Citation II	550-0105	Heartland Aviation Inc, Eau Claire, Wi.	N105BA
☐ N550M	92	Astra-1125SP	061	Calcutta Aircraft Leasing Inc. Bloomington, In.	N60AJ
☐ N550MC	97	CitationJet	525-0205	Midwest Capital Services LLC. Indianapolis, In.	
☐ N550MH	83	Citation 1/SP	501-0321	MHC Transportation Inc. New London, Ct.	N321VP
☐ N550MT	81	Citation II/SP	551-0341	Moss & Rocovich, Roanoke, Va. (was 550-0306).	N303EC
☐ N550PA	80	Citation II	550-0191	JES Aircraft Services Inc. Columbia, Mo.	C-GWCJ
☐ N550PG	81	Citation II	550-0216	Eagle Helicopters Inc. Spokane, Wa.	(N911NJ)
☐ N550PL	87	Citation S/II	S550-0130	Questar Pipeline Co. Salt Lake City, Ut.	N302PC
☐ N550PR	81	Citation II	550-0258	Chaparral Aviation Inc. Hillsboro, Or.	N258JS
☐ N550QS	96	Hawker 1000	9050	EJI Inc/NetJets-Texaco Air Services Inc. Columbus, Oh.	G-5-846
☐ N550RB	80	Citation II	550-0117	Jet Service Corp. Winamac, In.	N150HE
☐ N550RM	92	Citation II	550-0698	Robert Moore, Norman, Ok.	YV-911CP
☐ N550RP	78	Citation II	550-0219	RDD Leasing Inc. Elgin, Il. (was 551-0008).	YV-606CP
☐ N550RV	84	Citation S/II	S550-0012	Guaranty Air Inc. Junction City, Or.	N550TB
☐ N550SA	81	Citation II	550-0248	Security Aviation Inc. Anchorage, Ak.	N6804C
☐ N550T	93	CitationJet	525-0013	Taft Broadcasting Co. Charlotte, SC.	N1328A
☐ N550TY	78	Citation II	550-0007	James Young, Beaverton, Or.	(N650WG)
☐ N550U	77	Citation Eagle	501-0047	Valcom Services LLC. Durham, NC.	N550T
☐ N550VW	81	Citation II	550-0307	Metal Industries Inc of California, Elizabethville, Pa.	N37BM
☐ N550WB	81	Citation II	550-0230	Southeast Equipment Co. Cambridge, Oh.	N550EK
☐ N550WW	92	Citation V	560-0180	Wasser & Winters Co. West Linn, Or.	N1280K
☐ N551AB	80	Citation II	550-0383	Air Trails LLC. Monterey, Ca.	(N98718)
☐ N551AS	83	Learjet 55	55-083	ASCO Transport Co. Denver, Co.	(N6780)
☐ N551BC	79	Citation II/SP	550-0242	Western Flight Inc. Carlsbad, Ca. (was 551-0012).	N1955E
☐ N551BE	86	Citation S/II	S550-0097	Beverly California Corp. Fort Smith, Ar.	N828WB
☐ N551BP	82	Citation II	550-0424	Kinnaird Properties, Grants Pass, Or.	N271CG
☐ N551EA	81	Citation II/SP	551-0360	Ernesto Ancira Jr. Wilmington, De. (was 550-0328).	N24CJ
☐ N551G	96	CitationJet	525-0153	Auto Glass Center of Kansas Inc. Lawrence, Ks.	
☐ N551HM	86	Learjet 35A	35A-612	Southeastern Jet Corp. Fort Lauderdale, Fl.	N551TW
☐ N551MC	79	Citation II	550-0371	Poly-Resyn Inc. Dundee, Il.	(N26621)
☐ N551QS	96	Hawker 1000	9051	EJI Inc/NetJets-Cincinnati Bell, Columbus, Oh.	G-5-859
☐ N551SC	81	Learjet 55	55-008	Hal E Dickson, San Angelo, Tx.	
☐ N551TK	81	Citation II/SP	551-0296	Preferred Acceptance Co. Wilmington, De. (was 550-0246).	N35BP
☐ N551TP	85	Westwind-Two	419	Tecumseh Products Co. Tecumseh, Mi.	N419W
☐ N551TT	80	Citation II/SP	551-0032	Stephens Pipe & Steel Transportation inc. Greensboro, NC.	YV-300CP
☐ N552BE	87	Citation S/II	S550-0138	Beverly Enterprises Inc. Fort Smith, Ar.	(N501BE)
☐ N552JH	62	JetStar-731	5037	Blake of P B Inc. North Palm Beach, Fl.	N6JL
☐ N552QS	97	Hawker 1000	9052	EJI Inc/NetJets-Unitech Automotive Systems, Columbus, Oh.	G-5-863
☐ N553AC	90	Citation III	650-0198	Adelphia Cable Communications Inc. Wellsville, NY.(status?).	N198CM
☐ N553DJ	81	Learjet 55	55-003	American Electric Co/SDI Operating Partners, Philadelphia.	N162GA
☐ N553M	85	BAe 125/800A	8027	Marmon Group Inc. Chicago, Il.	N800PM
☐ N553PF	91	Beechjet 400A	RK-32	IRC Realty Co. Burlington, Vt.	N998GP
☐ N553V	78	Learjet 35A	35A-141	Mayo Shaw LLC. Englewood, Co.	N553M
☐ N554CL	82	Learjet 55	55-040	Clay Lacy Aviation Inc. Van Nuys, Ca.	N55HK
☐ N554R	95	Citation V Ultra	560-0328	Rogers Group Inc. Nashville, Tn.	
☐ N555CB	78	HS 125/700A	NA0228	Cleveland Brothers Equipment Co. Wilmington, De.	G-5-11
☐ N555CJ	83	Learjet 55	55-089	King Aircraft Ltd. Riviera Beach, Fl.	N170VE
☐ N555CS	97	Gulfstream V	516	BankAmerica/KaiserAir Inc. Oakland, Ca.	N516GA
☐ N555DS	81	Citation II	550-0275	Delmar Systems Inc. West Linn, Or.	N550CP
☐ N555HD	73	Gulfstream 2	134	Huish Detergents Inc. Salt Lake City, Ut.	C-FROC
☐ N555KC	82	Gulfstream 3	366	Odin Aviation Inc/Avjet Corp. Burbank, Ca.	(N333KD)
☐ N555KK	94	Beechjet 400A	RK-92	MKT Investment Co. Fort Smith, Ar.	N3240J
☐ N555KT	82	Citation II/SP	551-0419	Ocean Equities Inc. Van Nuys, Ca.	I-AGSM
☐ N555LB	68	Learjet 24	24-177	American International Airways, Morristown, Tn.	(N524DW)
☐ N555MW	76	Gulfstream 2	188	McWane Inc. Birmingham, Al.	N662G
☐ N555MX	89	Learjet 55C	55C-142	Dr Max Schaldach, Nuremberg, Germany.	
☐ N555RS	67	Gulfstream 2	3	C C Services of Nevada Inc. Reno, Nv.	N311JJ
☐ N555RT	81	Citation II/SP	551-0323	Ret Butler Communications Corp. Steamboat Springs, Co.	N819Y
☐ N555SD	81	Learjet 25D	25D-333	Diamond Shamrock/Joe Brand Inc. Laredo, Tx.	N34MJ
☐ N555SG	66	JetStar-8	5090	United CCM Corp. Dallas, Tx. (status ?).	N55CJ
☐ N555SR	83	Falcon 20F-5	455	Ray Industries Inc/Sea Ray Boats Inc. Knoxville, Tn.	F-GKAL
☐ N555WD	97	Challenger 604	5355	WDW Aviation Inc. Jacksonville, Fl.	C-GLWV
☐ N555WH	78	Learjet 36A	36A-037	Rikuo Corp. Van Nuys, Ca.	RP-C5128
☐ N556GA	82	Learjet 55	55-028	Global Aviation P/L. Seletar, Singapore.	(N53HJ)
☐ N556WD	82	Challenger 600S	1047	WDW Aviation Inc/Executive Flight Service, Jacksonville, Fl.	N555WD
Reg	*Yr*	*Type*	*c/n*	*Owner/Operator*	*Prev Regn*

Reg	Yr	Type	c/n	Owner/Operator	Prev Regn
☐ N558AC	82	Learjet 55	55-048	Air Castle Corp. Santa Monica, Ca.	PT-OBS
☐ N559LC	74	Gulfstream 2	152	Little Caesar Enterprises Inc. Detroit, Mi.	N62WB
☐ N560AG	95	Citation V Ultra	560-0301	Aircraft Guaranty LLC. Houston, Tx. (rebuild).	VR-BQB
☐ N560BA	90	Citation V	560-0102	Bitz Aviation Inc. Baden, Pa.	VP-BUL
☐ N560BJ	95	Citation V Ultra	560-0303	Sweatmore Air Inc. Phoenix, Az.	(D-CAFB)
☐ N560BP	97	Citation V Ultra	560-0449	Cessna Aircraft Co. Wichita, ks.	
☐ N560CB	87	Citation II/SP	551-0555	Cimarron Aviation Inc. Bend, Or.	N93BA
☐ N560CC	77	Citation V	550-0001	Cessna Aircraft Co. Wichita, Ks.	N5050J
☐ N560CE	95	Citation V Ultra	560-0302	Hesco Parts Corp. Louisville, Ky.	I-NYSE
☐ N560DC	89	Citation V	560-0021	Dow Chemical Co. Freeland, Mi.	C-FDLT
☐ N560DM	91	Citation V	560-0101	SIAM Inc. Portland, Or.	N560EC
☐ N560EL	90	Citation V	560-0049	Dr Ernst Langler-Munich/Rent Air Inc. Cabin John, Md.	(N2672X)
☐ N560ER	89	Citation V	560-0003	RH Airplanes Inc. Georgetown, Tx.	N560BA
☐ N560FB	91	Citation V	560-0148	Firebond Corp. Minden, La.	N92HW
☐ N560GB	94	Citation V	560-0254	Accent Stripe Inc. Orchard Park, NY.	
☐ N560GL	90	Citation V	560-0079	Aviation Charter/Great Lakes Chemical Corp. W Lafayettte, In	(N2746C)
☐ N560H	89	Citation V	560-0017	Emerald Holding Co. Pittsburgh, Pa.	N89BM
☐ N560HC	89	Citation V	560-0020	Collins Brothers Corp. Las Vegas, Nv.	N520CV
☐ N560JR	89	Citation V	560-0027	Big Red Inc/Redman Investments Inc. Dallas, Tx.	(N12289)
☐ N560LC	95	Citation V Ultra	560-0296	Landmark Communications Inc. Norfolk, Va.	
☐ N560MC	65	Jet Commander	24	Morgan Merrill/Jeti Inc. Alexandria, Va. (status ?).	(N7GW)
☐ N560ME	89	Citation V	560-0012	Caguas Builders Corp. San Juan, PR.	(ZP-...)
☐ N560MH	91	Citation V	560-0105	MHC Transportation Inc. New London, Ct.	N147VG
☐ N560RA	66	Falcon 20C	56	Reliant Airlines Inc. Ypsilanti, Mi.	N388AJ
☐ N560RG	92	Citation V	560-0198	Rodger & Virginia Galland, Portland, Or.	N135BC
☐ N560RL	91	Citation V	560-0135	Royal Group Inc. Charlotte, NC.	N560BB
☐ N560RS	91	Citation V	560-0109A	C C Services of Nevada Inc. Reno, Nv.	N907SB
☐ N560TX	93	Citation V	560-0206	Textron Inc. Providence, RI.	N1284A
☐ N560WE	90	Citation V	560-0100	Eifel Aviation Inc/Aviation Beauport, Jersey, C.I.	PH-PBM
☐ N560WH	89	Citation V	560-0013	W W H Inc. Las Vegas, Nv.	(N1217P)
☐ N560XL	96	Citation Excel	706	Cessna Aircraft Co. Wichita, Ks. (Ff 29 Feb 96).	
☐ N561B	89	Citation V	560-0008	J G Boswell Co. Burbank, Ca.	
☐ N561BC	90	Citation V	560-0057	Southern Aircraft & Transport, Laredo, Tx.	N560BL
☐ N561D	75	Falcon 10	54	Lieblong Transport Inc. Conway, Ar.	VP-BFW
☐ N561EJ	89	Citation V	560-0035	Wings Aviation International Inc. Franklin Lakes, NJ.	N36H
☐ N561RP	72	HS 125/F600A	6001	Paige Charters Inc. Wilmington, De.	N709R
☐ N561ST	83	Gulfstream 3	388	Firstplus Financial Inc. Dallas, Tx.	N748T
☐ N561XL	96	Citation Excel	560-5001	Cessna Aircraft Co. Wichita, Ks.	
☐ N562MS	70	Sabre-60	306-44	Mount Nebo Aviation LLC. Wilsonville, Or.	HC-BQU
☐ N562R	69	Sabre-60	306-37	North American Royalties, Chattanooga, Tn.	N4SE
☐ N563C	92	Citation V	560-0174	Blue Cross & Blue Shield, Sandston, Va.	
☐ N564CL	70	Learjet 25	25-060	Clay Lacy Aviation Inc. Van Nuys, Ca.	N695LJ
☐ N565CC	73	Citation	500-0065	Lakeshore Air Corp. Holland, Mi.	(N565TW)
☐ N565JW	91	Citation V	560-0149	S C Johnson & Son Inc. Racine, Wi.	(N68786)
☐ N565KC	69	Gulfstream 2TT	46	KCS Aviation Corp/Taughannock Aviation Corp. Ithaca, NY.	N505JT
☐ N565NC	87	Citation II	550-0565	Idaho Potato Packers Corp. Blackfoot, Id.	N565JS
☐ N565SS	72	Citation	500-0017	Midwest Trophy Manufacturing Co. Del City, Ok.	N49EA
☐ N565VV	77	Citation 1/SP	501-0267	565VV Ltd. Union Hill, Il. (was 500-0355).	N944TG
☐ N566CC	80	Citation II	550-0279	GLGT Aviation Co. Detroit, Mi.	(N88838)
☐ N566NA	72	Learjet 25	25-064	NASA, Hampton, Va.	N266GL
☐ N566VP	89	Citation V	560-0006	Cessna Aircraft Co. Wichita, Ks.	N962JC
☐ N567CA	79	Citation II	550-0092	F T Aviation LLC. Richmond, Va.	F-GNLF
☐ N567EA	73	Citation	500-0067	P & L Aviation LLC. Birmingham, Al.	(F-GIHT)
☐ N567WB	79	Citation 1/SP	501-0109	Orion Aviation International LLC. Grant, Al.	N30RE
☐ N568PA	78	Learjet 35A	35A-205	CFF Air Inc/Phoenix Air, Cartersville, Ga.	N59FN
☐ N569BW	72	Falcon 20F	259	Choo Choo Aviation LLC. Chattanooga, Tn.	F-GIFP
☐ N570R	81	Sabre-65	465-75	Frontenac Properties Inc. Dover, De.	N2581E
☐ N570RC	85	Citation S/II	S550-0070	Rosewood Assets Inc. Dallas, Tx.	N570CC
☐ N570WD	88	Citation II	550-0570	W D 2 Associates, Houston, Tx.	N570VP
☐ N571BC	88	Citation II	550-0599	International Bank of Commerce, Laredo, Tx.	VR-BYE
☐ N571CH	79	HS 125/700A	NA0256	Chemed Corp & Federated Dept Stores Inc. Cincinnati, Oh.	N125AK
☐ N571MC	74	Westwind-1123	175	Deer Horn LC. Artesia, NM.	N523RB
☐ N572M	79	Westwind-1124	258	Carpau Corp. Tarpon Springs, Fl.	(N258CF)
☐ N573BB	85	Citation S/II	S550-0037	Blue Beacon International Inc. Salina, Ks.	N72WC
☐ N573F	92	Citation V	560-0171	Federated Corporate Services Inc. Cincinnati, Oh.	N5735
☐ N573L	78	Citation 1/SP	501-0060	Modern World Finance Corp. Deerfield Beach, Fl.	N5737
☐ N573LP	90	Learjet 35A	35A-658	Louisiana Pacific Corp. Hillsboro, Or.	N4290Y
☐ N573LR	78	Learjet 35A	35A-153	Louisiana Pacific Corp. Hillsboro, Or.	N573LP
☐ N574CF	84	Diamond 1A	A079SA	General American Enterprises Inc. Little Rock, Ar.	(N574U)
☐ N575CC	73	Citation	500-0075	Advanced Drainage Systems Inc. Columbus, Oh.	(N600TT)
☐ N575EW	87	Citation S/II	S550-0140	Century Air Inc. Dover, De.	C-GLCR
☐ N575GH	82	Learjet 55	55-042	American Jet International Corp. Houston, Tx.	D-CMTM
☐ N575MA	90	Challenger 601-3A	5061	BOC Group Inc. Morristown, NJ.	N661CL
☐ N575PC	93	Citation V	560-0223	PNC Bank Corp. West Mifflin, Pa.	
☐ N575SF	94	Gulfstream 4SP	1233	Chevron Corp. San Francisco, Ca.	N472GA
☐ N576CC	88	Citation II	550-0576	Cescite Corp. Wilmington, De.	(N1300G)

Reg	Yr	Type	c/n	Owner/Operator	Prev Regn
N577JT	78	Citation 1/SP	501-0057	Citation Aviation LLC. Birmingham, Al.	N577VM
N577SV	96	CitationJet	525-0131	Sky Vice Inc. Bensenville, Il.	N5039Y
N577T	89	BAe 125/800B	8149	612 Corp. Wilmington, De.	N155T
N577VM	89	Citation II	550-0615	Cleveland Newspapers Inc. Birmingham, Al.	N615EA
N578M	89	Citation II	550-0612	Menard Inc. Eau Claire, Wi.	(N534M)
N579L	88	Citation II	550-0579	Dana Flight Operations Inc. Swanton, Oh.	(N13001)
N579TG	84	Gulfstream 3	433	TIG Productions/Time Warner Entertainment Co. Burbank, Ca.	N399CB
N580R	73	Citation	500-0127	Mid Continent Corp. Sunset Hills, Mo.	N500R
N583BS	79	Learjet 35A	35A-258	Physician Sales & Service Inc. Jacksonville, Fl.	N583PS
N583CC	64	BAC 1-11/203AE	015	Gund Business Enterprises/Cleveland Cavaliers, Cleveland, Oh	N523AC
N583D	85	Gulfstream 3	471	E I DuPont de Nemours & Co. Houston, Tx.	N888WL
N583M	95	Citation V Ultra	560-0326	Menard Inc. Eau Claire, Wi.	N5103J
N583PS	97	Learjet 31A	31A-151	Physician Sales & Service Inc. Jacksonville, Fl.	
N584D	88	Gulfstream 4	1065	E I Dupont de Nemours & Co. New Castle, De.	N442GA
N584DB	65	HS 125/731	25023	Kendall Aircraft Sales Inc. Sarasota, Fl.	N284DB
N585D	94	Gulfstream 4SP	1258	E I DuPont de Nemours & Co. New Castle, De.	N400UP
N586C	85	Gulfstream 3	459	E I Dupont de Nemours & Co. New Castle, De.	N600B
N586CC	92	Citation V	560-0186	Runway Air LLC. Newport, Ky.	N583N
N586CS	97	Falcon 50EX	260	Valkyrie Aviation Corp. Seattle, Wa.	F-WWHK
N586RE	80	Citation II	550-0199	U S Customs Service, Oklahoma City, Ok.	N67983
N587S	90	Citation III	650-0188	Diamond Shamrock Refining, San Antonio, Tx.	N2617P
N588CA	79	Citation 1/SP	501-0089	Lone Eagle Inc. Roseville, Ca.	N106WV
N589HM	90	Gulfstream 4	1153	International Home Foods Inc. Dallas, Tx.	N110LE
N590A	89	Citation V	560-0029	Alascom Inc. Anchorage, Ak.	(N1229C)
N590AS	85	Citation III	650-0088	Stroobants International Inc. Forest, Va.	(N590AQ)
N590CH	76	Learjet 35	35-060	Chastain Park Holdings Inc/Elite Aviation Inc. Van Nuys, Ca.	N47BA
N591M	90	Citation V	560-0085	Modine Manufacturing Co. Racine, Wi.	(N2748F)
N592M	95	Citation V Ultra	560-0338	Modine Manufacturing Co. Racine, Wi.	
N592VP	90	Citation V	560-0092A	Bell Leasing, Albuquerque, NM.	(N713HH)
N593DC	81	Falcon 10	180	Dow Corning Corp. Freeland, Mi.	N245FJ
N593EM	92	Citation II	550-0714	Electro-Matic Products Inc. Farmington Hills, Mi.	N12035
N593LR	84	Learjet 35A	35A-593	L R Aviation Inc. Indianapolis, In.	I-FLYG
N593M	93	Citation V	560-0237	Modine Manufacturing Co. Racine, Wi.	
N593MD	90	Citation V	560-0074	Macklanburg-Duncan, Oklahoma City, Ok.	N174JS
N594G	83	Citation II	550-0482	Growmark Inc. Bloomington, Il.	N62WG
N594GA	97	Gulfstream V	525	First Security Bank NA. Salt Lake City, Ut.	
N595DC	84	Falcon 200	500	Dow Corning Corp. Freeland,	N214FJ
N595PC	97	Citation Bravo	550-0826	Presrite Corp. Cleveland, Oh.	N51072
N595PT	87	Beechjet 400	RJ-18	Harbor Air LC. Vero Beach, Fl.	ZS-NOD
N596A	96	Falcon 2000	28	Associates Information Services Inc. Irving, Tx.	F-WWMO
N596DA	72	Falcon 20F	273	Amerijet International Inc. Fort Lauderdale, Fl.	F-WQBK
N597FJ	86	Challenger 601	3061	James River Corp. Sandston, Va.	N999JR
N597GA	97	Gulfstream V	519	Gulfstream Aerospace Corp. Savannah, Ga.	
N597U	96	Citation X	750-0004	Diamond Shamrock Refining, San Antonio, Tx.	N754CX
N599EC	82	Learjet 55	55-018	Professional Funeral Associates Inc. Lufkin, Tx.	N39E
N599RR	75	Falcon 20F	325	Illinois Central Railroad Co. Chicago, Il.	N7WG
N599SC	97	Learjet 60	60-113	Learjet Inc. Wichita, Ks.	
N600AE	76	HS 125/F600A	6068	Air Eagle LLC. Detroit, Mi.	N501R
N600AS	97	Falcon 900EX	17	Allied Signal Inc. Morristown, NJ.	F-WWFB
N600AT	87	Citation II/SP	551-0551	ALLTEL Corp. Little Rock, Ar.	N487LD
N600AW	90	Learjet 31	31-017	Llama A W Corp. Fayetteville, Ar.	N17VG
N600BG	84	Gulfstream 3	430	Tenet Healthcare Corp. Van Nuys, Ca.	N23A
N600BP	79	Challenger 600S	1004	B P Air Inc. Tucson, Az.	N227CC
N600BW	90	Citation V	560-0087	Borg-Warner Automotive Inc. Chicago, Il.	N167WE
N600BZ	81	Challenger 600S	1028	Blanch Insurance Services Inc. San Antonio, Tx.	5B-CHX
N600C	82	Learjet 55	55-047	Fayez Sarofim & Co. Houston, Tx.	
N600CG	83	Diamond 1A	A055SA	CGT Inc. Wayne, Pa.	N600MS
N600CL	80	Challenger 600S	1005	Fifty Five Aviation Inc. Windermere, Fl.	D-BJET
N600DR	95	Challenger 601-3R	5176	Dominion Resources Inc. Richmond, Va.	ZS-CCT
N600DT	75	Learjet 35	35-017	ZZ Leasing Inc. Memphis, Tn.	N456MS
N600DW	91	Gulfstream 4	1169	Dreamworks LLC. Burbank, Ca.	N500DG
N600ES	81	Gulfstream 3	322	Earth Star Inc. NYC.	N555NT
N600G	76	HS 125/F600A	6066	Giant Industries Inc/Cairn Inc. Scottsdale, Az.	N800JP
N600GH	93	Citation II	550-0717	Jetcraft Corp. Raleigh, NC.	XA-TCM
N600GL	68	Sabre-60A	306-24	Six Hundred Golf Lima Inc. Dover, De.	(N995RD)
N600GM	79	Learjet 25D	25D-290	Milam International Inc. Englewood, Co.	N321RB
N600GP	79	Learjet 35A	35A-236	Ridge Partners LP. Northbrook, Il.	(N415RD)
N600HR	93	Citation II	525-0038	Hans Roeder & Partner GmbH. Miami, Fl.	N135MM
N600J	92	BAe 125/800A	NA0470	Johnson & Johnson, New Brunswick, NJ.	N672BA
N600JC	65	Falcon 20C	7	J Carroll Enterprises, Hollywood, Fl.	N93CP
N600JD	94	Citation VI	650-0236	Deere & Co. Moline, Il.	N1303M
N600K	69	Jet Commander-B	148	Kalitta Flying Service, Morristown, Tn.	(N22LL)
N600KC	91	BAe 125/800A	NA0467	Kimberly-Clark Corp. Oxford, UK.	N639BA
N600L	93	Learjet 60	60-020	Lincoln National Life Insurance Co. Fort Wayne, In.	
N600LC	93	Learjet 60	60-021	Lincoln National Life Insurance Co. Fort Wayne, In.	
N600LN	96	Learjet 60	60-082	Lincoln National Life Insurance Co. Fort Wayne, In.	N682LJ

Reg	Yr	Type	c/n	Owner/Operator	Prev Regn
☐ N600LS	87	BAe 125/800A	NA0411	Limited Stores/Eastern Corp II, Wilmington, De.	N556BA
☐ N600MS	85	Challenger 601	3041	Ross Investments Inc. Oakland, Ca.	N610MS
☐ N600PD	81	Challenger 600S	1020	Peter Dussmann/Pedus Service GmbH. Munich, Germany.	N600MG
☐ N600PM	94	Gulfstream 4SP	1255	Philip Morris Management Corp. White Plains, NY.	N437GA
☐ N600RA	77	Corvette	36	R L Riemenschneider Enterprises, Redmond, Or.	N601RC
☐ N600RE	83	Challenger 600	1079	Summit Jet Corp. Farmingdale, NY.	N888FW
☐ N600SV	68	HS 125/731	25159	Crown Delaware Investments Corp. Dallas, Tx.	N67TJ
☐ N600TE	82	Challenger 600S	1056	Triton Energy/Jet East Inc. Dallas, Tx.	N600CC
☐ N600VE	84	Citation S/II	S550-0019	Vulcan Engineering Co. Helena, Al.	N119EA
☐ N600WD	74	Falcon 20E	300	WRRCO Inc/Garrett Aviation Services, Phoenix, Az.	N300FJ
☐ N600WG	82	Falcon 50	98	Greenleaf Corp. Saegertown, Pa.	N50MK
☐ N600WJ	73	HS 125/600B	6017	World Jet Inc. Fort Lauderdale, Fl.	PK-PJD
☐ N601AB	91	Citation V	560-0158	AmSouth Bancorp. Birmingham, Al.	
☐ N601AE	85	Challenger 601	3050	Majestic Marketing Group/Pacific Flight Inc. Medford, Or.	N95SR
☐ N601AF	89	Challenger 601-3A	5045	Aviation Methods/Vulcan Northwest Inc. Bellevue, Wa.	N500GS
☐ N601BA	74	HS 125/600A	6040	Bayou Helicopters Inc. Houston, Tx.	N125GS
☐ N601BC	79	Citation II	550-0091	Hickory Springs Manufacturing Co. Hickory, NC.	N527AG
☐ N601BD	83	Challenger 601	3010	JJSA Aviation LLC. San Jose, Ca.	N601SQ
☐ N601BW	94	Challenger 601-3R	5150	Bindley Western Industries Inc. Indianapolis, In.	N602CC
☐ N601EG	87	Challenger 601-3A	5008	Caribbean Marine/South Seas Helicopter Co Inc. Carlsbad, Ca.	N601CC
☐ N601ER	93	Challenger 601-3R	5141	EMC Corp. Hopkinton, Ma.	N312AT
☐ N601FS	95	Challenger 601-3R	5172	TA Securities, Kuala Lumpur, Malaysia.	C-GLXD
☐ N601GT	86	Challenger 601	3062	International Game Technology, Reno, Nv.	N601HP
☐ N601JA	75	HS 125/600A	6051	Ayers Asset Management LLC. Parsons, Tn.	N95TS
☐ N601JJ	67	JetStar-8	5102	REN Aviation Inc. Henderson, NC.	N85CC
☐ N601JM	85	Challenger 601	3048	JMI Services Inc/John J Moores, Carlsbad, Ca.	N35FP
☐ N601KF	84	Challenger 601	3023	KFC National Management Corp. Louisville, Ky.	N524PC
☐ N601KJ	95	Challenger 601-3R	5187	Miranda International Aviation Inc/ACM, San Jose, Ca.	C-GLXK
☐ N601LJ	92	Learjet 60	60-001	Learjet Inc. Wichita, Ks.	
☐ N601RL	88	Challenger 601-3A	5028	Mylan Pharmaceuticals/Air Corps Services, Morgantown, WV.	N601TL
☐ N601RS	84	BAe 125/800A	8018	M I Schottenstein Homes Inc. Columbus, Oh.	N350WG
☐ N601S	86	Challenger 601	3060	Air Shamrock Inc. Burbank, Ca.	C-GLXY
☐ N601SR	93	Challenger 601-3A	5130	Knight Ridder Inc. Opa Locka, Fl.	VR-BQA
☐ N601ST	90	Challenger 601-3A	5081	John & June Rogers/Sky Trek Aviation, Modesto, Ca.	HL7202
☐ N601TJ	85	Challenger 601	3046	Tyler Jet LLC. Tyler, Tx.	N601HJ
☐ N601TP	94	Challenger 601-3R	5156	Big Sky Aviation LLC. Madison, Wi.	N255CC
☐ N601TX	83	Challenger 601	3005	Pegasus IV Inc/TXI Aviation Inc. Dallas, Tx.	C-FAAL
☐ N601UP	92	Challenger 601-3A	5123	Southern Pacific Rail Corp. Dallas, Tx.	C-G...
☐ N601VH	89	Challenger 601-3A	5043	R S V P Jet Inc. Van Nuys, Ca.	N601BH
☐ N601WM	88	Challenger 601-3A	5026	WMX Technologies Inc. Chicago, Il.	C-G...
☐ N601WW	82	Challenger 600S	1076	White Cloud Co. White Plains, NY.	I-BLSM
☐ N602AB	93	Citation V	560-0217	AmSouth Bancorporation, Birmingham, Al.	
☐ N602AS	82	Challenger 600S	1054	Sandler Management Group, Virginia Beach, Va.	N9008
☐ N602AT	89	Citation II	550-0606	ALLTEL Corp. Little Rock, Ar.	N770BB
☐ N602BC	74	Citation	500-0190	William Conner, Erie, Pa.	N500HK
☐ N602D	95	Challenger 601-3R	5181	Deere & Co. Moline, Il.	C-GLYH
☐ N602NC	76	Falcon 10	82	Newell Co. Freeport, Il.	N97MC
☐ N602SC	97	Learjet 60	60-095	Sundstrand Corp. Rockford, Il.	N5005X
☐ N602TJ	85	Challenger 601	3047	Tyler Jet LLC. Tyler, Tx.	N602HJ
☐ N603SC	97	Learjet 60	60-096	Great Planes Sales Inc. Tulsa, Ok.	N8086L
☐ N604AS	79	Learjet 25D	25D-292	Sandler Management Group, Virginia Beach, Va.	N711VK
☐ N604B	96	Challenger 604	5305	FMC Corp. Wheeling, Il.	C-G...
☐ N604CC	95	Challenger 604	5301	Itochu Aviation Inc. NYC.	C-FVUC
☐ N604CT	96	Challenger 604	5314	Cogen Technologies GP Capital Corp. Houston, Tx.	C-GLWZ
☐ N604CU	97	Challenger 604	5339	Citizens Consumer Services Inc. Stamford, Ct.	C-GBJA
☐ N604DS	96	Challenger 604	5323	GDSWA LLC/Snyder Communications Inc. Leesburg, Va.	C-GLXU
☐ N604FS	97	Challenger 604	5357	General Atlantic Inc. White Plains, NY.	C-GLYH
☐ N604HJ	88	Challenger 601-3A	5024	L & B Flight Corp. Chicago, Il.	B-4010
☐ N604JP	97	Challenger 604	5346	Jounaou & Parskevaides, Riyadh, Saudi Arabia.	C-G...
☐ N604KS	96	Challenger 604	5308	Palace Air Inc. Jeddah, Saudi Arabia.	C-G...
☐ N604M	90	Gulfstream 4	1132	J P Morgan Services Inc/K C Aviation Inc. Teterboro, NJ.	N60NY
☐ N604MC	87	Challenger 601-3A	5013	Atlas FlightLease Inc. Golden, Co.	I-CTPT
☐ N604PM	97	Challenger 604	5354	Philip Morris Management Corp. White Plains, NY.	C-GLYC
☐ N604WB	93	Challenger 601-3A	5125	Silver State Aviation Inc. Las Vegas, Nv.	N512BC
☐ N605AT	93	Citation V	560-0242	ALLTEL Corp. Little Rock, Ar.	N242CV
☐ N605BA	94	Challenger 601-3R	5152	Bombardier Capital Inc. Colchester, Vt.	VP-COJ
☐ N605GA	69	HFB 320	1038	Grand Aire Express Inc. Monroe, Mi.	N301AT
☐ N605PM	97	Challenger 604	5356	Philip Morris Management Corp. White Plains, NY.	C-GLYO
☐ N606AM	83	Falcon 100	205	Dart Container Corp. Leola, Pa.	N700DW
☐ N606AT	93	Citation VI	650-0225	ALLTEL Corp. Little Rock, Ar.	N1301Z
☐ N606JM	83	Diamond 1A	A044SA	Joe Morten & Son Inc. S Sioux City, Ne.	(N600GW)
☐ N606JR	97	CitationJet	525-0231	Cessna Aircraft Co. Wichita, Ks.	
☐ N606MM	95	CitationJet	525-0104	Mark Martin Enterprises Inc. Daytona Beach, Fl.	
☐ N606PM	97	Challenger 604	5360	Philip Morris Management Corp. White Plains, NY.	C-GLXX
☐ N607CF	76	Sabre-60	306-118	W R Fry, San Jose, Ca.	N607SR
☐ N607PM	97	Challenger 604	5362	Philip Morris Management Corp. White Plains, NY.	C-G...

Reg   *Yr Type*   *c/n*   *Owner/Operator*   *Prev Regn*

Reg	Yr	Type	c/n	Owner/Operator	Prev Regn
☐ N607RJ	96	Citation V Ultra	560-0370	James & Rhonda Harris, Houston, Tx.	N5262B
☐ N607RP	95	Challenger 601-3R	5184	Ralston-Purina Co. Chesterfield, Mo.	N605RP
☐ N608LB	85	Citation S/II	S550-0029	Interjet Inc. Addison, Tx.	N185SF
☐ N608RP	86	Challenger 601	3055	Ralston-Purina Co. Chesterfield, Mo.	N100HG
☐ N610AS	95	Falcon 2000	8	AlliedSignal Inc. Morristown, NJ.	F-WWMF
☐ N610CC	69	Gulfstream 2	56	Oxy Petroleum Inc. Bakersfield, Ca.	N805Y
☐ N610HC	81	Westwind-Two	346	H C Aviation Co. San Antonio, Tx.	N1124N
☐ N610JR	86	Learjet 55	55-125	Tele/Com Air Inc. Sugar Grove, Il.	
☐ N610MC	76	Gulfstream 2	196	Aviation Methods/May Department Stores Co. St. Louis, Mo.	N200BE
☐ N610TT	80	Citation 1/SP	501-0170	Westair Leasing Corp. La Jolla, Ca.	N170EA
☐ N611AT	79	Citation 1/SP	501-0103	Ozark Aircraft Sales Inc. Kansas City, Mo.	N49WC
☐ N611BA	90	BAe 125/800A	NA0449	Josephine Ford/JF Aircraft Corp. Dearborn, Mi.	G-5-658
☐ N611DB	75	Learjet 24D	24D-318	Phoenix Air Group Inc. Cartersville, Ga.	N114JT
☐ N611GS	90	Challenger 601-3A	5082	Gilman Securities/GSC Aviation Corp. Oxford, Ct.	N611NT
☐ N611JW	96	Falcon 900B	162	Tristram Inc. Boca Raton, Fl.	N162FJ
☐ N611MC	79	HS 125/700A	NA0257	May Department Stores Co. St. Louis, Mo.	N125HS
☐ N611PA	94	Beechjet 400A	RK-78	Premier Aviation LC. Dublin, Oh.	N8278Z
☐ N611SH	79	Learjet 35A	35A-253	Idaho Forest Industries Inc. Coeur d'Alene, Id.	N611CM
☐ N611ST	91	Citation V	560-0123	Sun Trust Banks Inc. Atlanta, Ga.	(N6806X)
☐ N612MC	82	HS 125/700A	NA0317	May Department Stores Co. St. Louis, Mo.	N700SS
☐ N613GA	67	Falcon 20C	77	Grand Aire Express Inc. Monroe, Mi.	F-GHSG
☐ N613MC	81	HS 125/700A	7151	May Department Stores Co. St Louis, Mo.	N161G
☐ N614DD	80	Citation 1/SP	501-0187	Stivers Midtown Lincoln-Mercury, Des Moines, Ia.	N70CG
☐ N614GA	82	Citation II	550-0366	N55FM LLC. Monroe, Mi.	N55FM
☐ N615AT	93	Citation V	560-0245	ALLTEL Corp. Little Rock, Ar.	
☐ N616AT	93	Citation VI	650-0230	ALLTEL Corp. Little Rock, Ar.	N1305N
☐ N616CC	94	Challenger 601-3R	5144	Boeing Aircraft Holding Co. Seattle, Wa.	C-GLYC
☐ N616MM	83	Diamond 1A	A062SA	Aquatic Innovations Inc & Kerry Murphy, Orinda, Ca.	N64EZ
☐ N616NA	79	Learjet 25	25-035	National Aeronautics, Cleveland, Oh.	N33TR
☐ N617GA	67	Falcon 20C	88	Grand Aire Express Inc. Monroe, Mi.	N41CD
☐ N618BR	65	Learjet 23	23-082A	Robert West Jr. Orinda, Ca.	(N118LS)
☐ N618DC	87	Challenger 601-3A	5004	KNC LLC. Newton, Ma.	N180KT
☐ N618GA	69	Falcon 20D	211	Grand Aire Express Inc. Monroe, Mi.	17101
☐ N618GH	70	Falcon 20F	236	Corporate Flight Management Inc. Wayne, Pa.	N128AP
☐ N618R	94	Learjet 60	60-044	ZB Industries Inc. Van Nuys, Ca.	(N618P)
☐ N619GA	69	Falcon 20D	215	Grand Aire Express Inc. Monroe, Mi.	17102
☐ N620A	79	Falcon 20F	412	AWI (Nevada) Inc. Las Vegas, Nv.	N12FU
☐ N620CC	77	Falcon 20F-5	373	Occidental Chemical Corp. White Plains, NY.	N610CC
☐ N620JF	95	Learjet 60	60-074	Shearwater Air inc. Saratoga, Ca.	N8074W
☐ N620JH	83	Gulfstream 3	387	Huntsman Chemical/Airstar Corp. Salt Lake City, Ut.	N621JA
☐ N620JM	78	Learjet 35A	35A-207	Aviation Charter Services/M-S Air Inc. Indianapolis, In.	N3PW
☐ N620K	92	Gulfstream 4	1193	Signet Leasing/Eastman Kodak Co. Rochester, NY.	N163M
☐ N620M	77	HS 125/700A	NA0203	Eldredge Air Inc. Ocala, Fl.	G-BERV
☐ N620S	81	Challenger 600S	1031	TADIC Inc. Kansas City, Mo.	C-GLXS
☐ N620TC	93	CitationJet	525-0014	Golden Eagle Aviation Inc. Southfield, Mi.	N70TP
☐ N621JH	95	Gulfstream 4SP	1272	Huntsman Petrochemical Corp/Airstar Corp. Salt Lake City, Ut	N454GA
☐ N621LJ	97	Learjet 60	60-121	Learjet Inc. Wichita, Ks.	
☐ N621MT	85	BAe 125/800A	8036	M T Leasing Co. Wilmington, De.	D-CFRC
☐ N621S	82	HS 125/700B	7178	SCS Investments Inc. Crystal Bay, Nv.	N700CJ
☐ N622LJ	97	Learjet 60	60-122	Learjet Inc. Wichita, Ks.	
☐ N622WG	86	Learjet 35A	35A-611	Zarego Inc. Mesquite, NM.	
☐ N623MS	82	Gulfstream 3	351	YONA Aviation/Interface Group Massachusetts Inc. Needham, Ma	N18TM
☐ N624BP	81	Gulfstream 3	320	Ziff Aircraft Services Inc. NYC.	N320WE
☐ N625AU	81	Learjet 25D	25D-340	Union Park Pontiac & GMC, Wilmington, De.	N980A
☐ N625CC	95	Citation VII	650-7058	Case Corp. Racine, Wi.	N52178
☐ N626BM	88	Learjet 35A	35A-634	Jet Air Holdings Inc. Wilmington, De.	I-EAMM
☐ N626CC	84	Citation III	650-0026	Damaged on production line 1984, now test airframe.	
☐ N626JS	81	Learjet 35A	35A-394	J W Childs Associates Inc. Boston, Ma.	N60DK
☐ N626KM	94	Learjet 60	60-012	Lear 60 Inc. Van Nuys, Ca.	N147CC
☐ N626TC	73	Gulfstream 2SP	129	Trillium Leasing LLC. Bellingham, Wa.	N83TE
☐ N627L	79	Citation 1/SP	501-0123	Lord Corp. Erie, Pa.	N513CC
☐ N627R	87	Citation III	650-0152	Camden Securities Publishing Co. Little Rock, Ar.	(N152VP)
☐ N627RP	97	Beechjet 400A	RK-155	Ring Power Corp. Jacksonville, Fl.	N2355T
☐ N628BS	72	Citation	500-0045	Eagle Jet LLC. Crossett, Ar.	N666BS
☐ N628ZG	78	Citation 1/SP	501-0070	Morton Transport LLC. Barrington Hills, Il.	(N277RW)
☐ N630M	97	Citation X	750-0021	Olin Corp. Norwalk, Ct.	N138A
☐ N630S	90	Astra-1125SP	046	Scores Inc. Fort Smith, Ar.	N140DR
☐ N631SF	93	Learjet 31A	31A-075	Sanderson Farms Inc. Laurel, Ms.	(N418RT)
☐ N632PB	91	Learjet 31	31-033	Modern Transportation Co. Owensboro, Ky.	(N131WS)
☐ N632PE	68	HS 125/1A	25058	located Freetown, Sierra Leone.	(EC-...)
☐ N633EE	85	Citation S/II	S550-0058	A K Guthrie, Big Spring, Tx.	N1271E
☐ N633L	91	Falcon 50	223	Kerr-McGee Corp. Oklahoma City, Ok.	N132FJ
☐ N633SL	69	Sabre-60	306-33	Centurion Investments Inc. St Louis, Mo.	CC-CGT
☐ N633WW	75	Falcon 10	59	Walker 33 Inc. Howell, Mi.	N302A
☐ N635AV	92	Gulfstream 4SP	1185	Avery Dennison Corp/Air Group Inc. Van Nuys, Ca.	N485GA
☐ N636GA	97	Gulfstream 4SP	1336	Gulfstream Aerospace Corp. Savannah, Ga.	

Reg	*Yr*	*Type*	*c/n*	*Owner/Operator*	*Prev Regn*

Reg	Yr	Type	c/n	Owner/Operator	Prev Regn
□ N636MF	97	Gulfstream V	512	Ropa Two Corp. Teterboro, NJ.	N512GA
□ N636N	78	Citation 1/SP	501-0069	Pelmont Aviation, NYC.	N501EF
□ N637GA	97	Gulfstream 4SP	1337	Gulfstream Aerospace Corp. Savannah, Ga.	
□ N637LJ	94	Learjet 60	60-037	Texas Instruments Inc. Dallas, Tx.	N4037A
□ N638GA	97	Gulfstream 4SP	1338	Gulfstream Aerospace Corp. Savannah, Ga.	
□ N638LJ	94	Learjet 60	60-038	Texas Instruments Inc. Dallas, Tx.	N4007J
□ N639GA	98	Gulfstream 4SP	1339	Gulfstream Aerospace Corp. Savannah, Ga.	
□ N640GA	98	Gulfstream V	540	Gulfstream Aerospace Corp. Savannah, Ga.	
□ N641GA	98	Gulfstream V	541	Gulfstream Aerospace Corp. Savannah, Ga.	
□ N642CC	78	Citation II	550-0038	Exchange Air Partners LLC. Oakhurst, NJ.	C-FLDO
□ N642GA	98	Gulfstream V	542	Gulfstream Aerospace Corp. Savannah, Ga.	
□ N642RP	69	Sabre-60	306-46	Puryear Aviation Inc. Spokane, Wa.	N100FN
□ N642TS	74	Sabre-75A	380-2	Downeast Networking Services Inc. Palm City, Fl.	N9GN
□ N643CR	82	Challenger 600	1055	Doubleplay Aviation LLC. Loveland, Co.	N217MB
□ N643GA	98	Gulfstream V	543	Gulfstream Aerospace Corp. Savannah, Ga.	
□ N643JL	69	HS 125/400A	NA737	Orcas Air Inc. Eastsound, Wa.	(N165AG)
□ N643MC	90	Citation II	550-0643	East Coast Jet Center, Fort Lauderdale, Fl.	PT-ODW
□ N643TD	82	Citation II	550-0438	Lynn Durham, Midland, Tx.	N437CC
□ N644GA	98	Gulfstream V	544	Gulfstream Aerospace Corp. Savannah, Ga.	
□ N644JW	78	JetStar 2	5223	JRW Aviation Inc. Dallas, Tx.	N1DB
□ N645G	76	Learjet 35	35-056	Gates Corp. Englewood, Co.	N106GL
□ N646G	82	Learjet 55	55-016	Gates Corp. Englewood, Co.	
□ N647JP	67	Falcon 20-5	120	Luiginos Inc. Billings, Mt.	N20AF
□ N648LJ	94	Learjet 60	60-048	Snap-On Aviation LLC. Kenosha, Wi.	N5008Z
□ N648WW	77	HS 125/700A	NA0218	Miller Aviation Inc. Johnson City, NY.	N746BC
□ N650	80	Citation VII	697	Cessna Aircraft Co Citation X testbed.	
□ N650AC	77	Citation 1/SP	501-0011	Continental Medical LLC. Lafayette, La.	N770MH
□ N650AF	86	Citation III	650-0125	Diamond Shamrock Refining & Marketing, San Antonio, Tx.	EC-EAP
□ N650AS	94	Citation VII	650-7053	AlliedSignal Inc. Morristown, NJ.	N344AS
□ N650CA	65	Learjet 24	24-050	Apex Flight Services LLC/C & B Aviation, Bloomington, Il.	N823M
□ N650CC	90	Citation III	650-0193	Gil Hodge Aviation Inc. Marietta, Ga.	N55HD
□ N650CD	84	Citation III	650-0066	Russell Corp. Alexander City, Al.	N138V
□ N650CE	86	Citation III	650-0106	Clark Transportation Co. Bethesda, Md.	N106CC
□ N650CG	84	Citation III	650-0023	C G Bretting Manufacturing Co. Ashland, Wi.	N38DD
□ N650CH	88	Citation III	650-0154	CHR Aviation Inc. Eden Prairie, Mn.	N696HC
□ N650CN	84	Citation III	650-0062	Conseco Investment Holding Co. Carmel, In.	N342HM
□ N650FC	87	Citation III	650-0146	Frank's Casing Crew & Rental Tools, Lafayette, La.	XA-PIP
□ N650G	78	Westwind-1124	233	Gulf Air Inc. DeRidder, La.	N650G
□ N650GE	93	Astra-1125SP	064	Gaylord Broadcasting Corp. Nashville, Tn.	4X-CUG
□ N650GH	84	Citation III	650-0034	Plains Air Inc. Amarillo, Tx.	(N45US)
□ N650GT	83	Citation III	650-0004	GTECH Corp. West Greenwich, RI.	N654GC
□ N650HC	86	Citation III	650-0124	Collins Bros/Terrible Herbst Inc. Las Vegas, Nv.	N7HV
□ N650J	83	Citation III	650-0022	Highlands Aviation Corp. Houston, Tx.	
□ N650KC	92	Citation VI	650-0215	Kansas City Southern Railway Co. Kansas City, Mo.	N215CM
□ N650L	91	Citation VI	650-0209	Lozier Corp. Omaha, Ne.	(N2634E)
□ N650LR	89	Learjet 35A	35A-650	Wal-Mart Stores Inc. Rogers, Ar.	N135MW
□ N650LW	83	Citation III	650-0010	DeBordieu/Dallas II Inc. Dallas, Tx.	OK-NKN
□ N650M	84	Citation III	650-0044	Moebel-Mann GmbH. Soellingen, Germany.	(N234HM)
□ N650MC	94	Citation VI	650-0237	Wilderness Investment Co. Silver Spring, Md.	N1306B
□ N650MM	84	Citation III	650-0048	Preco Inc. Boise, Id.	N986M
□ N650MP	86	Citation III	650-0107	Mosinee Papaer Corp. Mosinee, Wi.	N8000U
□ N650MW	81	Citation 1/SP	501-0180	XL Leasing Co. Alma, Mi.	N180VP
□ N650NY	84	Citation III	650-0027	Trans Marine Management Corp. Tampa, Fl.	N875SC
□ N650SB	83	Citation III	650-0018	Stockwood Inc. Morristown, NJ.	N275WN
□ N650SL	84	Citation III	650-0024	SEL Aviation LLC. Houston, Tx.	N95SR
□ N650SP	85	Citation III	650-0094	Jet 1 Inc. Naples, Fl.	5B-CSM
□ N650SS	87	Citation III	650-0140	Fleming Companies Inc.Oklahoma City, Ok.	N90CN
□ N650TC	84	Citation III	650-0061	Miami Valley CTC Inc. Dayton, Oh.	N650TP
□ N650TS	83	Citation III	650-0006	Aero Toy Store Inc. Fort Lauderdale, Fl.	N1TS
□ N650W	96	Citation VII	650-7065	The Wing Group, The Woodlands, Tx.	(N765W)
□ N650WE	84	Citation III	650-0040	Snowy Butte Aviation Inc. Eagle Point, Or.	HB-VIY
□ N650X	81	Falcon 50	69	Alumax Inc. Norcross, Ga.	N80FJ
□ N650Z	86	Citation III	650-0108	Dana Flight Operations Inc. Swanton, Oh.	(N1302U)
□ N651AF	86	Citation III	650-0114	Bendini Group Inc. Memphis, Tn.	N7000G
□ N651BH	84	Citation III	650-0051	Ty-Tex Exploration Inc. Tyler, Tx.	N910F
□ N651CC	82	Citation III	650-0001	United Foods Inc. Bells, Tn.	N654CC
□ N651JM	95	Citation VI	650-0241	Interstate Equipment Leasing Inc/Swift Air. Phoenix, Az.	N666JM
□ N651LJ	66	Learjet 24A	24A-125	Steven Lysdale, Bellevue, Wa.	
□ N651MK	81	Sabre-65	465-73	Mallinckrodt Group Inc. Clayton, Mo.	N64MQ
□ N651TC	85	Citation III	650-0090	Miami Valley CTC Inc. Dayton, Oh.	N1823S
□ N652CN	89	Challenger 601-3A	5040	Conseco Inc. Indianapolis, In.	C-G...
□ N652JM	86	Citation III	650-0113	J M Aviation Leasing LLC. Troutdale, Or.	(N650AF)
□ N652MK	86	Sabre-65	465-36	Mallinckrodt Group Inc. St Louis, Mo.	N651GL
□ N652ND	75	Citation	500-0277	Electrical Distributing Inc. Portland, Or.	N67MA
□ N652PC	95	Falcon 2000	10	Prince Transportation Inc. Holland, Mi.	F-WWMH
□ N654CN	93	Falcon 900B	127	Conseco Inc. Indianapolis, In.	N482FJ

Reg	Yr	Type	c/n	Owner/Operator	Prev Regn
N654E	68	Falcon 20C	164	MTI Vacations Inc. Downers Grove, Il.	N4367F
N654PC	78	Falcon 10	131	Prince Transportation Inc. Holland, Mi.	N196FJ
N655CN	90	Challenger 601-3A	5069	Conseco Inc. Carmel, In.	VR-BNF
N655DB	74	Falcon 10	28	U S Counseling Services Inc. Brooksfield, Wi.	N42EH
N656PS	78	Citation II	550-0009	California Oregon Broadcasting, Medford, Or.	N744DC
N657CC	88	Citation III	650-0157	Cable Jet Leasing Inc. Montvale, NJ.	N1327A
N657ER	93	Citation VII	650-7027	Arlington Aircraft LLC. Sun Lakes, Az.	N500
N657T	94	Citation VII	650-7042	Hershey Foods Corp. Middletown, Pa.	N1265K
N659AT	68	Learjet 24	24-157	I Q Air Services Inc. Miami, Fl.	XA-SNZ
N659HX	80	Learjet 25D	25D-300	Frederick Haddad/Hecks Inc. Charleston, WV.	(N46BA)
N660AH	93	Learjet 60	60-017	JABJ/American Home Products Corp. Madison, NJ.	N760AC
N660P	80	Falcon 20F-5	430	Phillips Petroleum Corp. Bartlesville, Ok.	N428F
N662P	97	Falcon 900EX	30	Phillips Petroleum Corp. Bartlesville, Ok.	F-W...
N663PD	73	Gulfstream 2B	139	Paul Davril Inc/Avjet Corp. Burbank, Ca.	N2UJ
N664CL	68	Learjet 24	24-167	Clay Lacy Aviation Inc. Van Nuys, Ca.	N888B
N664P	89	Falcon 50	200	Phillips Petroleum Corp. Bartlesville, Ok.	N288FJ
N665MC	90	Citation II	550-0665	Marsh Supermarkets Inc. Indianapolis, In.	N5348J
N665P	81	Falcon 20F-5	444	Phillips Petroleum Corp. Bartlesville, Ok.	N453F
N666JT	75	Gulfstream 2	162	Janus Transair Corp. Bedford, NY.	N74RV
N666LN	84	Citation S/II	S550-0005	Sky Eagle Corp. Reno, Nv.	(N1256G)
N666TW	73	Learjet 25B	25B-116	Sierra American Corp. Dallas, Tx.	(N818GY)
N667LC	96	Challenger 604	5324	Loews Corp/Clinton Court Corp. Teterboro, NJ.	N601CC
N667P	80	Falcon 20-5B	432	Phillips Petroleum Corp. Bartlesville, Ok.	N430F
N668S	76	Citation	500-0314	Nolan's RV Center Inc. Denver, Co.	N66ES
N668VP	97	CitationJet	525-0228	Value Plastics Inc. Fort Collins, Co.	
N669LJ	91	Learjet 35A	35A-669	D & R Investments LLC. Arlington, Va.	A6-FAJ
N669TW	66	Falcon 20DC	50	Kitty Hawk Air Cargo Inc. Dallas, Tx.	EC-EDO
N670AS	81	Sabre-65	465-58	AlliedSignal Avionics Inc. Olathe, Ks.	N65AM
N670C	72	Sabre-75A	370-7	Tyler Jet LLC. Tyler, Tx.	XB-ERU
N672AT	86	Beechjet 400	RJ-14	Raytheon Aircraft Co. Wichita, Ks.	N58AU
N673CA	88	Citation II	550-0590	Lone Star Equipment Inc. Corpus Christi, Tx.	C-GBCA
N673LR	80	Citation II	550-0179	Butler Air Inc. Butler, Pa.	N673LP
N673TM	85	BAe 125/800A	8033	Serra Investments/Team Management Inc. Grand Blanc, Mi.	N24RP
N674AC	82	Diamond 1A	A024SA	Northstar Aviation Inc. Grand Haven, Mi.	N450PC
N674G	82	Citation II	550-0435	Findlay Industries Inc. Findlay, Oh.	N390DA
N674JM	87	Citation S/II	S550-0127	Evan Morgan Massey, Richmond, Va.	N14UM
N675RW	98	Gulfstream V	526	Coca Cola Co. Atlanta, Ga.	N526GA
N676CC	84	Citation 1/SP	501-0676	Palo Alto Town & Country Village, Ca.	N76JY
N676CW	74	Citation	500-0169	RASCO Inc. Houston, Tx.	XA-SJW
N676PC	67	HS 125/731	25153	Rockwell-Ditzler Associates Inc. Pittsburgh, Pa.	N336MB
N676RW	94	Gulfstream 4SP	1253	Coca-Cola Co. Atlanta, Ga.	N435GA
N677RW	92	Gulfstream 4	1177	Coca-Cola Co. Atlanta, Ga. 'The Wind Ship'	(I-LADB)
N677SW	95	Gulfstream 4SP	1269	Joseph E Seagram & Sons, White Plains, NY.	N434GA
N678CG	81	Challenger 600S	1027	Marion Pepsi Cola Bottling Co. Marion, Il.	N420TX
N678JD	78	Citation 1/SP	501-0077	Jetcraft Corp & N Carolina Air & Travel LLC. Durham, NC.	N42HM
N678ML	81	Challenger 600S	1011	Midwest Aviation/Berkshire Hathaway Inc. Omaha, Ne.	N601JR
N679BC	88	Citation II	550-0589	Repoca Sales Co. Bristol, Va.	(N1301Z)
N679RW	90	Gulfstream 4	1131	Coca-Cola Co. Atlanta, Ga. 'The Wind Ship'	N437GA
N680BC	85	Citation III	650-0087	Quadion Corp. Minneapolis, Mn.	C-FQCY
N680FM	91	B 757-23A	24923	FM Services Aviation LLC. New Orleans, La.	N680EM
N681FM	83	Gulfstream 3	371	Freeport-McMoRan Inc. New Orleans, La.	(N8220F)
N682B	89	BAe 125/800A	NA0431	E I Dupont de Nemours & Co. New Castle, De.	N585BA
N682DC	84	Citation 1/SP	501-0682	Dement Construction Co. Jackson, Tn.	N55TK
N682G	66	B 727-76	19254	Occidental Petroleum Corp. Los Angeles, Ca.	N10XY
N682RW	66	BAC 1-11/401AK	061	Detroit Red Wings Inc. Detroit, Mi.	EI-BWR
N683E	87	BAe 125/800A	NA0410	E I Dupont de Nemours & Co. New Castle, De.	N555BA
N683EC	75	Gulfstream 2	157	Crown Aviation LLC. Holland, Mi.	N658PC
N684H	73	Citation	500-0113	Northern Air Inc. Grand Rapids, Mi.	N684HA
N684HA	77	Learjet 35A	35A-113	Anderson Management Corp. New Canaan, Ct.	N684LA
N685TA	86	Gulfstream 4	1003	JABJ/G IV Corp-American Home Products, Teterboro, NJ.	N986AH
N688CF	74	Citation	500-0147	Emerald Aviation Inc. Bay Springs, Ms.	N494G
N688GS	73	Learjet 25B	25B-123	Dickerson Associates, West Columbia, SC.	N906SU
N690EA	74	Citation	500-0201	Nicodis SA. Laval, France.	(F-GIHU)
N691HM	70	Gulfstream 2SP	92	Viasystems Inc. St Louis, Mo.	N589HM
N691RC	69	Gulfstream 2	43	Aero Toy Store Inc. Fort Lauderdale, Fl.	N33ME
N692BE	85	Citation III	650-0092	Bombardier Capital Inc. Colchester, Vt.	N692CC
N692FG	70	Learjet 25	25-052	EPC Transport Inc. Denver, Co.	N692FC
N692TV	83	Gulfstream 3	397	R Santulli/RTS-General Instrument Services Inc. Chicago, Il.	N978FL
N693M	85	Citation S/II	S550-0021	Hilltop Aviation Inc. Vandalia, Oh.	N593M
N693TJ	74	HS 125/F600B	6027	Diamond Leasing Inc. Knoxville, Tn.	OE-GIA
N696HC	94	Falcon 50	250	Henry Crown & Co. Wheeling, Il.	N250FJ
N696JH	76	Learjet 35A	35A-084	Dee Thomason Ford Co. Gladstone, Or.	N184TS
N696ST	97	CitationJet	525-0187	Dr Robert Knollenberg, Boulder, Co.	
N696TR	96	Beechjet 400A	RK-127	KNC LLC. Wellesley, Ma.	N1127U
N696US	80	Sabre-65	465-18	United Space Alliance LLC. Houston, Tx.	N4M
N697BJ	82	Gulfstream 3	370	S & H Fabricating & Engineering Inc. Fort Lauderdale, Fl.	N400K

Reg	Yr	Type	c/n	Owner/Operator	Prev Regn
☐ N697MC	85	Citation III	650-0097	Manor Care Aviation Inc. Silver Spring, Md.	(N1318Y)
☐ N699CC	96	Citation V Ultra	560-0351	Oliver Air & Co Ltd. Rochester, NH.	N5153K
☐ N699SC	94	Learjet 60	60-041	Service Corp International Inc. Houston, Tx.	N5004Y
☐ N700AC	81	HS 125/700A	NA0290	Lazy Lane Farms/Perpetual Corp. Washington, DC.	G-5-18
☐ N700AL	75	Falcon 10	55	Aerolease Inc. Itasca, Il.	N702NG
☐ N700BD	76	Falcon 10	81	Becton-Dickinson & Co. Teterboro Airport, NJ.	N162FJ
☐ N700BH	72	Gulfstream 2SP	115	Benco Inc/Hilton Aviation, Las Vegas, Nv.	N200BP
☐ N700BW	81	HS 125/700A	NA0288	Teton Aviation LLC. San Francisco, Ca.	N125AH
☐ N700CH	95	Learjet 60	60-056	Cardal Inc. Columbus, Oh.	N117RJ
☐ N700CL	81	Challenger 600S	1035	Sierra Aviation Inc. Kansas City, Mo.	VR-BLD
☐ N700CN	90	Gulfstream 4	1133	Copley Press Inc. Carlsbad, Ca.	N443GA
☐ N700DA	80	Learjet 25D	25D-302	DavisAir Inc. West Mifflin, Pa.	N740K
☐ N700FA	70	HS 125/731	NA752	Desert Air Charters Inc. Scottsdale, Az.	N700PL
☐ N700FC	72	Learjet 25B	25B-082	Air Response Inc. Fort Plain, NY.	N654
☐ N700FE	80	HS 125/700A	NA0278	Federated Investors Building Corp. Pittsburgh, Pa.	N86WC
☐ N700FH	79	Falcon 10	158	Flexair Ltd. Urbana, Il.	N790FH
☐ N700FS	82	Gulfstream 3	367	Jet Aviation-WPB/Flo-Sun Aircraft Inc. High Point, NC.	N367GA
☐ N700GD'	89	Gulfstream 4	1104	MC Group, Van Nuys, Ca.	N600ML
☐ N700HH	78	HS 125/700A	NA0240	Hilton Hotels Corp. Las Vegas, Nv.	N130BA
☐ N700JA	77	Citation 1/SP	501-0272	GordonAir Inc. Johnson City, Tn.	N700JR
☐ N700JC	81	Sabre-65	465-74	Oxley Petroleum Co. Tulsa, Ok.	
☐ N700JE	83	Learjet 55	55-091	Eckerd Fleet Inc. Clearwater, Fl.	(N69B)
☐ N700KC	88	Challenger 601-3A	5017	Kimberly-Clark Corp. Dallas, Tx.	C-GLWX
☐ N700LS	92	Gulfstream 4	1180	Limited Stores Inc. Columbus, Oh.	N472GA
☐ N700MD	77	Westwind-1124	212	Kal Kustom Mw Inc. Columbus, Oh.	N900CS
☐ N700MH	79	Sabre-60	306-142	Living Word Christian Center, Brooklyn Park, Mn.	N40KJ
☐ N700MP	74	Citation	500-0198	Contrails Inc/C&N Investments Inc. Kerrville, Tx.	N9UJ
☐ N700R	87	Learjet 55B	55B-133	NASCAR, Daytona Beach, Fl.	N55LF
☐ N700SB	81	Gulfstream 3	334	Heir to Air Inc. Van Nuys, Ca.	N41PG
☐ N700SJ	76	Learjet 35A	35A-082	AirNet Systems Inc. Columbus, Oh.	N700GB
☐ N700SW	85	Citation III	650-0096	Safeway Insurance Co. Westmont, Il.	N96VP
☐ N700TE	64	B 727-30	18365	Triton Energy Corp. Dallas, Tx.	N96B
☐ N700TF	89	Citation V	560-0011	Tyson Foods Inc. Fayetteville, Ar.	(N1217H)
☐ N700VA	80	Gulfstream 3	300	Varian Associates Inc. Palo Alto, Ca.	N300GA
☐ N700VC	72	Citation Eagle	500-0011	Jesse DuPlantis Ministries, New Orleans, La.	C-GJEM
☐ N700VT	81	HS 125/700A	7158	V T Inc. Phoenix, Az.	XB-DZN
☐ N701AS	76	Learjet 35A	35A-047	AirNet Systems Inc. Columbus, Oh.	N13MJ
☐ N701BR	74	Citation	500-0191	American Aircraft Leasing Inc. Raleigh, NC.	N155CA
☐ N701CF	79	HS 125/700A	NA0242	Aviation Enterprises Inc. Bedford, Tx.	N125BW
☐ N701DK	93	Citation V	560-0221	Keffer Co. Charlotte, NC.	N24UB
☐ N701FW	80	Sabre-65	465-21	Marine R Corp. Fort Myers, Fl.	(N265CA)
☐ N701JH	78	JetStar 2	5230	Bullfrog of Ohio County Inc. Owensboro, Ky.	N901EH
☐ N701LP	89	Beechjet 400	RJ-61	Leggett & Platt Inc. Carthage, Mo.	XA-RNE
☐ N701NW	77	HS 125/700A	NA0206	Miller Aviation Inc. Johnson City, NY.	XA-SNN
☐ N701QS	88	Gulfstream 4	1059	Stone Container Corp. Palwaukee, Il.	N415GA
☐ N701S	69	Gulfstream 2SP	69	Scores Inc. Carson City, Ca.	N440DR
☐ N701SC	71	Learjet 24XR	24XR-235	Lear 24 LLC. Wichita, Ks.	N51VL
☐ N701TA	79	HS 125/700A	7073	Telford Aviation/AAI Jet Inc. Waterville, Me.	(N59TJ)
☐ N701TS	77	HS 125/700A	NA0201	Aero Toy Store Inc. Fort Lauderdale, Fl.	G-IECL
☐ N701WC	97	Falcon 2000	48	Liberty Mutual Insurance Co. Bedford, Ma.	N2089
☐ N702DM	67	Falcon 20-5	74	D Q Automobiles Inc. Irving, Tx.	N800DC
☐ N702HC	73	HS 125/600A	6023	Hawkco Inc. Portland, Or.	N523MA
☐ N702JH	83	Diamond 1A	A035SA	Bullfrog of Ohio County Inc. Owensboro, Ky.	N135GA
☐ N702LP	94	Beechjet 400A	RK-87	Leggett & Platt Inc. Carthage, Mo.	N1567L
☐ N702NC	88	Falcon 100	220	Newell Co. Rockford, Il.	N326LW
☐ N703JP	81	HS 125/700A	NA0299	Patterson Dental Corp. Van Nuys, Ca.	N555RB
☐ N703JS	79	Falcon 10	157	U S FoodService Inc. Wilkes-Barre, Pa.	N64AM
☐ N703TS	78	HS 125/700B	7031	F Fifty Holdings Inc. Miami, Fl.	HB-VLA
☐ N705JH	78	HS 125/700A	NA0221	John H Harland Co. Decatur, Ga.	(N248JH)
☐ N705MA	86	Astra-1125	011	Pacific Diversified Investments Inc. Las Vegas, Nv.	N450BM
☐ N705NA	66	Learjet 24A	24A-102	NASA, Moffett Field, Ca.	N365EJ
☐ N705SP	85	Citation S/II	S550-0048	Sportsmans Market Inc. Batavia, Oh.	N999TJ
☐ N706TS	79	Gulfstream 2B	254	All American Communications Inc. Santa Monica, Ca.	N254AR
☐ N706VP	92	Citation VII	650-7006	Ramerica International Inc. NYC.	N966K
☐ N707AM	80	Falcon 10	159	AMC/Dart Container Corp. Sarasota, Fl.	N707PC
☐ N707BC	82	Westwind-1124	366	Westwind Acquisition LLC. McLean, Va.	N388GA
☐ N707FH	66	Sabre-40	282-74	Delta Fox Aviation, Birmingham, Al.	N707TG
☐ N707JC	75	Falcon 20F	335	Eagle Leasing Inc. Alpharetta, Ga.	N335AJ
☐ N707KS	69	B 707-321B	20025	Kalair USA Corp. London, UK.	N728Q
☐ N707MB	68	B 707-355C	19986	sold for military use 1/96.	N723GS
☐ N707PE	82	Citation II	550-0452	Blue Dolphin Corp/Ebensteiner Co. Van Nuys, Ca.	N707WF
☐ N707SC	65	Learjet 24	24-065	Dolphin Aviation Inc. Sarasota, Fl.	N957SC
☐ N707SG	97	Learjet 60	60-109	Valley Jet Corp. Portland, Or.	N109LJ
☐ N707SK	59	B 707-138B	17702	Prince Bandar/First City Texas Houston NA. Houston, Tx.	N707KS
☐ N707SQ	95	Learjet 60	60-062	Valley Jet Corp. Portland, Or.	N707SG
☐ N707TA	96	Hawker 800XP	8296	RTAC/Wabash Aviation Inc. Wichita, Ks.	(ZS-...)

| *Reg* | *Yr* | *Type* | *c/n* | *Owner/Operator* | *Prev Regn* |

Reg	Yr	Type	c/n	Owner/Operator	Prev Regn
☐ N707TF	71	Westwind-1123	155	Roever Evangelistic Association, Fort Worth, Tx.	N707TE
☐ N707W	79	Citation 1/SP	501-0085	Wellons Inc. Sherwood, Or.	N25DD
☐ N707WB	93	Falcon 900B	132	Home Depot Inc. Atlanta, Ga.	N132FJ
☐ N707WJ	70	B 707-358C	20301	WCJ Group Inc. Miami, Fl.	4X-ATY
☐ N707XX	64	B 707-138B	18740	Aviation Methods Inc. San Franscisco, Ca.	N108BN
☐ N708SP	97	Learjet 45	45-014	Sportsman's Market Inc. Batavia, Oh.	
☐ N708TA	98	Beechjet 400A	RK-178	Raytheon Travel Air Co. Wichita, Ks.	
☐ N709EL	92	Beechjet 400A	RK-52	G Kirkham/D F S Furniture Co. East Midlands, UK.	(N709EW)
☐ N709QS	98	Citation VII	650-7109	EJI Inc/NetJets, Columbus, Oh.	
☐ N709TA	98	Beechjet 400A	RK-180	Raytheon Travel Air Co. Wichita, Ks.	
☐ N709VP	92	Citation II	550-0709	Cessna Aircraft Co. Wichita, Ks.	N85KC
☐ N710A	88	BAe 125/800B	8110	ARAMCO Associated Co. Houston, Tx.	D-CMIR
☐ N710AG	82	HS 125/700A	NA0338	American General/Knickerbocker Corp. Houston, Tx.	N710BZ
☐ N710AT	96	Learjet 35A	35A-337	Global Aviation, Seletar, Singapore.	N337WC
☐ N710AW	97	Citation X	750-0033	Cessna Aircraft Co. Wichita, Ks.	N5093D
☐ N710EC	80	Gulfstream 3	315	Chouest Air Inc. Galliano, La.	N315GS
☐ N710K	62	MS 760 Paris	112	Edward G Martin, Orefield, Pa. (status ?).	N7277X
☐ N710MT	79	Citation II	550-0075	Biomet Inc/Air Warsaw Inc. Warsaw, In.	C-GSFA
☐ N710QS	98	Citation VII	650-7100	EJI Inc/NetJets, Columbus, Oh.	
☐ N710TA	98	Beechjet 400A	RK-183	Raytheon Travel Air Co. Wichita, Ks.	
☐ N710TV	68	Learjet 24	24-159	Hogan Air Inc. Middetwon, Oh.	(N269AL)
☐ N711AQ	68	HS 125/400A	NA711	RAK Air Inc. Coraopolis, Pa.	(N610HC)
☐ N711BP	70	HS 125/400A	NA744	Boles Parts Supply Inc. Atlanta, Ga.	(N382DA)
☐ N711CW	65	Learjet 24	24-055	Premier Jets Inc. Portland, Or.	N511WH
☐ N711EC	88	Beechjet 400	RJ-53	EFCO Corp. Monett, Mo.	N195KA
☐ N711FG	94	Learjet 31A	31A-092	Frontier Shoppes Inc. Las Vegas, Nv.	N50302
☐ N711FJ	79	Falcon 10	149	Embee Inc. Santa Ana, Ca.	(N830SR)
☐ N711GD	74	Sabre-80A	380-6	Corporate Flight Inc. Detroit, Mi.	N711GL
☐ N711GF	96	Citation VII	650-7075	IPD-Indeck Power Overseas LLC. Wheeling, Il.	N52613
☐ N711HL	70	HS 125/731	NA754	Point Zero Corp. Eugene, Or.	N60RE
☐ N711JC	75	Falcon 10	69	Object Development Corp. El Toro, Ca.	N7TJ
☐ N711JG	87	Astra-1125	017	Nirvana Air Inc. Minnetonka, Mn.	N996JP
☐ N711LT	81	Gulfstream 3	327	Townes Tele-Communications Inc. Paris, Tx.	N777RY
☐ N711MA	75	Learjet 35	35-032	McMillan Aircraft Inc. Dover, De.	N235JW
☐ N711MC	93	Gulfstream 4SP	1217	Marnell Corrao Associates, Las Vegas, Nv.	N981HC
☐ N711MD	85	Citation S/II	S550-0066	Marshall B Durbin Jr. Birmingham, Al.	N1272P
☐ N711NM	77	Learjet 25D	25D-224	Aero Charters Nashville Inc. Nashville, Tn.	(N32TJ)
☐ N711NV	87	Citation II/SP	551-0507	Nevada Department of Transport, Carson City. (was 550-0557).	N1298C
☐ N711R	75	Learjet 35	35-035	Cockrell Resources Inc. Houston, Tx.	
☐ N711RL	68	Gulfstream 2	25	JABJ/Polo Wings II Inc. Lyndhurst, NJ.	N527K
☐ N711RT	70	Falcon 20D	242	Keller Systems Inc/Priester Aviation Inc. Wheeling, Il.	N129AP
☐ N711SE	81	Westwind-1124	329	Southern Electric Supply Co. Meridian, Ms.	N30NS
☐ N711SX	83	Challenger 601	3007	Progressive Casualty Insurance, Richmond Heights, Oh.	N711SR
☐ N711TE	71	Gulfstream 2	105	Trans Exec Air Service Inc. Van Nuys, Ca.	N6060
☐ N711TF	75	Falcon 10	52	Contemporary Industries Leasing Corp. Omaha, Ne.	N860E
☐ N711VF	82	Citation 1/SP	501-0236	Gem Air Inc & Ameristar Casinos Inc. Las Vegas, Nv.	N26227
☐ N711VT	71	HS 125/731	25249	Automotive Investment Group. Phoenix, Az.	N200VT
☐ N711WD	79	Learjet 25D	25D-282	Transportation Management Inc. Rocky Mount, NC.	
☐ N711WM	80	HS 125/700A	NA0262	Coast Hotels & Casinos Inc. Las Vegas, Nv.	N350DH
☐ N711WV	80	Westwind-1124	313	Ross, Ross & Rector, Parkersburg, WV.	N711WU
☐ N711Z	82	Citation II	550-0436	Lockheed Martin Corp. Marietta, Ga.	N1219P
☐ N712CC	87	Gulfstream 4	1028	SAH Enterprises Inc/Bill Cosby, 'Camille'.	N712CW
☐ N712JB	88	Learjet 35A	35A-646	Laurel Aviation Enterprises Inc. Smithtown, NY.	N717JB
☐ N712ME	76	Falcon 20F-5	355	Liberty Mutual Insurance Co. Boston, Ma.	N200MK
☐ N712PC	69	B 707-323B	20176	Cx USA 2/96 to ?	N8437
☐ N712R	67	Learjet 24	24-156	Royal Cake Co. Winston Salem, NC.	N111RP
☐ N712TA	98	Beechjet 400A	RK-186	Raytheon Travel Air Co. Wichita, Ks.	
☐ N712TE	65	JetStar-731	5070	Liberty Aero Corp. Van Nuys, Ca.	N888WT
☐ N712VS	68	HS 125/400A	NA712	Williams Marketing Group International Inc. Guthrie, Ok.	N7777B
☐ N713HH	92	Citation V	560-0192	EGH & HMC Aviation Inc. Sanford, Fl.	D-CEWR
☐ N713QS	98	Citation VII	650-7103	EJI Inc/NetJets, Columbus, Oh.	
☐ N713SC	87	Astra-1125	013	Vulture One Corp. Portland, Or.	(N413SC)
☐ N714A	85	B 737-2K9	23405	ARAMCO Associated Co. Houston, Tx.	N701ML
☐ N715A	79	B 737-2S2C	21928	ARAMCO Associated Co. Houston, Tx.	N204FE
☐ N715AB	92	Citation II	550-0715	Gibraltar Aviation Ltd. Wethersfield, Ct.	PT-OTN
☐ N715CX	97	Citation X	750-0015	Cessna Aircraft Co. Wichita, Ks.	
☐ N715JS	72	Citation	500-0001	Jerry Savelle Ministries International, Crowley, Tx.	N501KG
☐ N715MH	73	Learjet 25B	25B-132	American Jet International Corp. Houston, Tx.	N715JF
☐ N715QS	97	Citation VII	650-7105	EJI Inc/NetJets, Columbus, Oh.	
☐ N715TA	98	Beechjet 400A	RK-189	Raytheon Travel Air Co. Wichita, Ks.	
☐ N716A	79	B 737-2S2C	21929	ARAMCO, Houston, Tx.	N205FE
☐ N716CB	72	Citation	500-0055	Town & Country Supermarkets/Cirrus Blue Inc. West Allis, Wi.	N999SF
☐ N716DB	86	Citation S/II	S550-0120	Western Leasing/Phillips Plastics Corp. Phillips, Wi.	N1283M
☐ N716HP	84	Challenger 601	3026	Joy Technologies Inc. Warrendale, Pa.	N80RP
☐ N716QS	98	Citation VII	650-7106	EJI Inc/NetJets, Columbus, Oh.	
☐ N716RD	89	Challenger 601-3A	5048	Readers Digest Sales & Service, White Plains, NY.	N2004G
Reg	*Yr*	*Type*	*c/n*	*Owner/Operator*	*Prev Regn*

Reg	Yr	Type	c/n	Owner/Operator	Prev Regn
☐ N716TE	72	Gulfstream 2	116	Trans-Exec Air Services Inc. Van Nuys, Ca.	(N410LR)
☐ N717AN	79	Learjet 25D	25D-272	Reagan Buick Inc. Omaha, Ne.	N747AN
☐ N717DD	91	Beechjet 400A	RK-18	Tony Downs Foods Co. St James, Mn.	N5598Q
☐ N717EP	78	Learjet 25D	25D-255	Apple Ten Aero LLC. McGee Tyson Airport, Tn.	N25GJ
☐ N717HB	83	Learjet 55	55-066	HBE Corp. Chesterfield, Mo.	N50DD
☐ N717JB	86	Learjet 55B	55B-128	Arbor Aviation Enterprises LLC. St James, NY.	N10BF
☐ N717MB	89	Citation V	560-0007	MPW Industrial Services Inc. Hebron, Oh.	N964JC
☐ N717PC	82	Citation II	550-0402	Pepsi Cola Bottling Co. La Crosse, Wi. (was 551-0057).	N700LB
☐ N717VL	91	Beechjet 400A	RK-21	Cactus Aircraft Inc. Glendive, Mt.	N1881W
☐ N718CK	82	Citation II	550-0368	Private Airways Inc. & Brenden Group Inc. Wilmington, De.	N94MF
☐ N718DW	81	Falcon 50	81	SmithKline-Beckman/Colleen Corp. Philadelphia, Pa.	N89FJ
☐ N718JS	69	Gulfstream 2	66	SEI-Sykes Enterprises inc. Clearwater, Fl.	N165U
☐ N718R	93	Challenger 601-3A	5127	A S C-American Stores Co. Salt Lake City, Ut.	N718P
☐ N718SA	81	Citation 1/SP	501-0179	Harris Air inc. Logas, Ut.	HB-VLD
☐ N719CC	80	Westwind-1124	290	A & C Air Inc/Executive Jet New York, Ronkonkoma, NY.	N800JJ
☐ N719JB	84	Learjet 35A	35A-166	Jet Partners LLC. St James, NY.	N831J
☐ N719L	94	CitationJet	525-0075	North American Air Charter Corp. Westboro, Ma.	
☐ N720A	79	B 737-2S2C	21926	ARAMCO Associated Co. Houston, Tx.	N201FE
☐ N720DC	66	B 727-77	19253	Santa Barbara Aerospace Inc. Santa Barbara, Ca.	N448DH
☐ N720DF	74	Falcon 10	26	Saturn Aviation Inc. Fort Wayne, In.	N707AM
☐ N720JR	62	B 720-047B	18451	J Raphael/J A R Aircraft Services Inc. Richardson, Tx.	N2143J
☐ N720JW	76	Gulfstream 2SP	178	Wing Aviation Inc. The Woodlands, Tx.	N42LC
☐ N720ML	94	Falcon 50	245	Northwestern Mutual Life Insurance Co. Milwaukee, Wi.	N240FJ
☐ N720SJ	96	Citation V Ultra	560-0386	Cessna Aircraft Co. Wichita, Ks.	N7274A
☐ N720TA	97	Hawker 800XP	8320	Raytheon Travel Air Co/First Finance Insurance, Wichita, Ks.	N2291X
☐ N720WC	92	Citation II	550-0708	Walbro Corp. Auburn Hills, Mi.	N12022
☐ N721BS	95	Falcon 2000	11	Golden Nugget Aviation Corp. Las Vegas, Nv.	N101NS
☐ N721CC	93	Citation II	550-0721	Information Technology Inc. Lincoln, Ne.	N1207B
☐ N721EW	89	MD-87	49767	Golden Nugget Aviation Corp. Las Vegas, Nv.	D-ALLI
☐ N721FF	84	Gulfstream 3	421	Frank Fertitta Enterprises Inc. Las Vegas, Nv.	N421GM
☐ N721LH	73	HS 125/600A	6025	Bradley Flying Service Inc. Wethersfield, Ct.	C-GTPC
☐ N721MF	81	B 727-2X8	22687	Wedge Group Europe, Paris, France.	N4523N
☐ N721RB	80	Gulfstream 3	311	Golden Nugget Aviation Corp. Las Vegas, Nv.	N311GA
☐ N721RL	72	Gulfstream 2SP	121	High Tech Aircraft Corp. Portsmouth, NH.	N507JC
☐ N722CC	84	BAe 125/800A	8008	Crown Central Petroleum Corp. Baltimore, Md.	G-5-11
☐ N722HP	82	Challenger 600S	1039	Richmor Aviation Inc. Carlsbad, Ca.	N1868S
☐ N722JB	95	Falcon 2000	13	Berry Investments Inc. Vandalia, Oh.	N2004
☐ N722Q	58	MS 760 Paris	9	Corey Garber, Carmel, Ca.	N334RK
☐ N722SG	94	CitationJet	525-0088	Southwest Gas Corp. Las Vegas, Nv.	N188CJ
☐ N722TA	97	Hawker 800XP	8322	Raytheon Travel Air Co. Wichita, Ks.	
☐ N723JR	81	Citation 1/SP	501-0190	Allsup's Convenience Stores Inc. Clovis, NM.	N40AW
☐ N723TA	97	Hawker 800XP	8349	Raytheon Travel Air Co. Wichita, Ks.	
☐ N724B	77	HS 125/700A	NA0204	BOY-CEN Air LLC. Boyne Falls, Mi.	G-BERX
☐ N724CL	66	B 727-51	19121	Clay Lacy Aviation Inc. Van Nuys, Ca.	N299LA
☐ N724DB	82	Gulfstream 3	372	Keystone Foods Corp. Philadelphia, Pa.	N500EX
☐ N724DS	76	Falcon 10	92	Limousin Air LLC. New Castle, De.	N95TJ
☐ N724TS	69	HS 125/731	NA724	Aero Toy Store Inc. Fort Lauderdale, Fl.	C-FSDH
☐ N725BA	82	Citation II	550-0369	B & A Aviation Corp. Dover, De.	N55MT
☐ N725DM	81	Falcon 10	184	Roppe Corp. Postoria, Oh.	N4AC
☐ N725L	96	CitationJet	525-0140	Brandt Interests Ltd. Oxford, Ms.	
☐ N725LB	96	Gulfstream 4SP	1296	Shoreline Aviation Inc/Mariner Health Group, East Haven, Ct.	N419GA
☐ N725RH	78	Citation II	550-0006	Ronson Aviation Inc. Trenton, NJ.	N152GA
☐ N726CC	66	HS 125/3A	25116	Astrakam International Corp. Glendale, Az.	N345CT
☐ N727AT	80	Westwind-1124	284	Thomas H Lee Co. Boston, Ma.	N217BL
☐ N727CS	80	Learjet 25D	25D-313	Milam International Inc. Englewood, Co.	N631CW
☐ N727GL	77	Learjet 35A	35A-127	Corporate Aviation Services Inc. Tulsa, Ok.	
☐ N727HC	67	B 727-35	19835	Clay Lacy Aviation, Van Nuys, Ca.	N900CH
☐ N727LA	67	B 727-21	19260	Carnival Cruise Lines/Fun Air Corp. Miami, Fl.	N727SG
☐ N727LM	80	Learjet 25D	25D-308	L & M Forwarding Inc. Laredo, Tx.	N102RR
☐ N727NA	85	Citation S/II	S550-0043	N L A Flight Inc. Birmingham, Al.	(N727AL)
☐ N727S	80	Falcon 50	17	Rainier Aviationc/727 Inc-Sloan Capital Companies, Seattle.	N349KS
☐ N727TA	77	HS 125/700A	NA0202	Q Air LLC. Ronkonkoma, NY.	N333ME
☐ N727TK	74	Citation	500-0141	K3C Inc/Sierra Industries Inc. Uvalde, Tx.	XB-EWQ
☐ N727TS	76	Falcon 10	76	Ajax Aviation Ltd. Federalsburg, Md.	F-BYCC
☐ N727WF	68	B 727-23	20045	Westfield Aviation Inc. Sydney, Australia.	N2913
☐ N727X	67	B 727-191	19394	DSTS Inc. Las Vegas, Nv.	N3964A
☐ N728A	69	DC 8-72	46081	ARAMCO Associated Co. Houston, Tx.	N8971U
☐ N728L	81	Westwind-Two	349	Kaman Corp. Windsor Locks, Ct.	N723L
☐ N728LB	81	Falcon 50	46	LB Aviation Inc. East Haven, Ct.	N725LB
☐ N728LW	78	Falcon 50	3	Laurence Carr, Anchorage, Ak.	N8805
☐ N728MC	79	Citation 1/SP	501-0115	Regal Corp. Knoxville, Tn.	N95RE
☐ N728PX	68	JetStar-731	5112	Paxson Communication Management Co. Clearwater, Fl.	N499PC
☐ N729PX	78	Citation 1/SP	501-0284	Paxson Communications Management Co. West Palm Beach, Fl.	N284PC
☐ N730CA	80	Westwind-Two	295	Cardinal American Corp/B K N Corp. Independence, Oh.	N555CW
☐ N730CP	72	Sabre-40A	282-103	Fenaire Corp. Toledo, Oh.	N730CA
☐ N731AS	77	Falcon 20F	344	7700 Properties LLC. Oklahoma City, Ok.	(N731F)

Reg	Yr	Type	c/n	Owner/Operator	Prev Regn
N731BW	66	HS 125/731	25075	First American Health Inc. St Simons Island, Ga.	N750GM
N731CW	73	Learjet 25B	25B-117	Johnson Aviation Corp. Columbus, In.	C-FMGM
N731L	66	JetStar-731	5095	AMC/Dart Container Corp. Sarasota, Fl.	N780RH
N731MS	70	HS 125/731	NA758	Select Air Corp. Mechanicsburg, Pa.	N6709
N732TS	69	HS 125/400A	NA732	Palm Beach Wings Inc. Lake Worth, Fl.	N70JC
N733A	92	Falcon 900B	126	HAC Inc/Humana Inc. Louisville, Ky.	N900FJ
N733AR	86	B 737-205	23466	ARCO Corp. Los Angeles, Ca.	LN-SUZ
N733CF	90	Challenger 601-3A	5057	C R Bard Inc. Morristown, NJ.	N830CD
N733EY	82	Learjet 55	55-057	JABJ-Ernst & Young/C B Applications LLC. Lyndhurst, NJ.	N10CR
N733H	91	Citation VI	650-0210	Humana Inc. Louisville, Ky.	N7059U
N733K	93	Citation VI	650-0222	Humana Inc. Louisville, Ky.	N222CD
N733M	94	Citation V	560-0249	HAC Inc/JAPC Inc. Louisville, Ky.	N10CN
N733MK	95	Beechjet 400A	RK-107	Merck & Co. Whitehouse Station, NJ.	N3227X
N733S	73	Falcon 20-5B	292	Shell Aviation Corp. Houston, Tx.	N4441F
N734S	74	Falcon 20-5B	316	Shell Aviation Corp. Houston, Tx.	N4451F
N735A	80	Learjet 35A	35A-323	National Collegiate Athletic Association, Kansas City, Mo.	
N736BP	86	B 737-205	23465	BP Exploration/ARCO Corp. Los Angeles, Ca.	LN-SUU
N737MM	88	Beechjet 400	RJ-35	Milgard Leasing LP. Tacoma, Wa.	N71GA
N737RJ	82	Citation 1/SP	501-0238	Roy Johnson Enterprises Inc. Clarksville, Ar.	N995PA
N737WH	81	B 737-2V6	22431	Huizenga Holdings Inc. Fort Lauderdale, Fl.	HB-IEH
N739CX	97	Citation X	750-0038	Cessna Aircraft Co. Wichita, Ks.	
N740E	97	Learjet 45	45-023	Eaton Corp. Cleveland, Oh.	
N740EJ	70	Learjet 24B	24B-222	Tango Juliet LLC. Dover, De.	N740F
N740F	92	Learjet 31A	31A-061	Eaton Corp. Cleveland, Oh.	N740E
N740R	94	Falcon 50	247	TRW Inc. Cleveland, Oh.	N247FJ
N741AM	79	JetStar 2	5236	Aircraft Management Co. Elkhart, In.	N34TR
N741C	80	Westwind-1124	292	Grover C Harned, Durango, Co.	N292JC
N741CC	97	CitationJet	525-0227	Cessna Aircraft Co. Wichita, Ks.	
N741E	97	Learjet 45	45-011	Eaton Corp. Cleveland, Oh.	
N741F	93	Learjet 31A	31A-070	Eaton Corp. Cleveland, Oh.	N741E
N742E	97	Learjet 45	45-025	Eaton Corp. Cleveland, Oh.	
N742F	93	Learjet 31A	31A-071	Eaton Corp. Cleveland, Oh.	N742E
N742R	95	Falcon 50	243	TRW Inc. Cleveland, Oh.	N243FJ
N743E	97	Learjet 45	45-016	Eaton Corp. Cleveland, Oh.	
N743F	93	Learjet 31A	31A-068	Eaton Corp. Cleveland, Oh.	N743E
N744E	93	Learjet 31A	31A-069	Eaton Corp. Cleveland, Oh.	N9173V
N744R	95	Citation V Ultra	560-0291	TRW Inc. Cleveland, Oh.	N5148B
N744X	81	Falcon 50	58	Shamrock Aviation/Pillsbury Co. Minneapolis, Mn.	N72FJ
N745E	97	Learjet 45	45-008	Eaton Corp. Cleveland, Oh.	
N745TA	97	Beechjet 400A	RK-145	Raytheon Travel Air Co. Wichita, Ks.	
N745TS	70	HS 125/731	NA745	Cherry One Aviation Inc. Westchester, Il.	N427DA
N746BR	94	Citation VII	650-7046	Flag Air LLC & Buffalo Rock Co. Birmingham, Al.	N746CM
N746UP	86	BAe 125/800A	8069	Union Pacific Railroad Co. Eppley Field, Ne.	N519BA
N747AC	97	CitationJet	525-0202	RABA LLC. Cleveland, Oh.	N202CJ
N747AN	86	Learjet 55	55-121	Worrell Investment Co. Charlottesville, Va.	N155SC
N747CP	83	Learjet 35A	35A-502	Bretford Manufacturing Inc. Waukegan, Il.	N8565X
N747CX	81	Falcon 20F	442	MorAir Inc. Conway, Ar.	I-SREG
N747GM	80	Learjet 35A	35A-308	Sage Well Services Inc. Encinal, Tx.	(N7LA)
N747LB	66	Jet Commander-B	55	Paisano Minerals Co. Corpus Christi, Tx.	N11MC
N747RL	76	Citation	500-0345	Sage Well Services Inc. Encinal, Tx.	N345TL
N747SC	66	Learjet 24	24-019	Florida Broadcast Management Inc. Bradenton, Fl.	N100EA
N747UP	86	BAe 125/800A	8072	Union Pacific Railroad Co. Eppley Field, Ne.	N522BA
N747Y	81	Falcon 50	50	F M C Corp. Wheeling, Il.	N747
N748FB	81	HS 125/700A	NA0292	Four Buoys LLC. Detroit, Mi.	N728JW
N748MN	78	Gulfstream 2	215	Merle Norman Cosmetics Inc. Van Nuys, Ca.	N816GA
N749CP	96	CitationJet	525-0158	C & P Aviation Services Inc. Blaine, Mn.	
N749TA	97	Beechjet 400A	RK-149	Raytheon Travel Air Co. Wichita, Ks.	N149TA
N750CC	80	Sabre-65	465-37	ANR Pipeline Co. Detroit, Mi.	N750CS
N750CX	93	Citation X	703	Model 750 Citation X first flight 21 December 1993.	
N750EC	96	Citation X	750-0007	SCB Enterprises inc. Incline Village, Nv.	
N750GT	83	Challenger 601	3002	GTECH Corp. Warwick, RI.	N601SR
N750LM	97	Citation X	750-0039	Northrop Grumman Aviation Inc. Hawthorne, Ca.	(N98TX)
N750PP	84	Citation 1/SP	501-0686	Peter Pfendler, Petaluma, Ca.	N6763M
N750RL	97	Citation X	750-0025	Russ Lyon Jr. Phoenix, Az. (2500th Citation built).	
N750SW	81	Gulfstream 3	338	Safeway Inc/Aviation Methods Inc. Oakland, Ca.	(N338RJ)
N750T	93	Beechjet 400A	RK-70	United Technologies Cortran Inc. Hartford, Ct.	C-FOPC
N751BH	98	Citation X	750-0059	Cessna Aircraft Co. Wichita, Ks.	(N570BH)
N751CA	73	Learjet 25B	25B-122	Critical Air Medicine Inc. San Diego, Ca.	N122WC
N751CX	94	Citation X	750-0001	Cessna Aircraft Co. Wichita, Ks. Ff 27 Sep 94.	
N751DB	82	Challenger 600	1075	Riverhorse Investments Inc/Air Castle Inc. Santa Monica, Ca.	N25SR
N751PJ	60	MS 760 Paris-1R	51	Airborne Turbine Inc. Ca.	No 51
N752CC	79	Citation II	550-0018	United States Customs Service, Washington, DC.	(N3225M)
N752EA	73	Learjet 25B	25B-137	Lyon Credit Corp. Stamford, Ct.	N752CA
N753CC	80	Citation II	550-0109	U S Customs Service, Oklahoma City, Ok.	N2665N
N754DB	68	Learjet 25	25-014	Air Methods Corp/David Beggrow-New Air, Durango, Co.	N14LJ
N754GL	78	Learjet 35A	35A-197	Ameriflight Inc. Burbank, Ca.	

Reg	*Yr Type*	*c/n*	*Owner/Operator*	*Prev Regn*

Reg	Yr	Type	c/n	Owner/Operator	Prev Regn
☐ N755WJ	70	HS 125/400A	NA755	Gulfstream Aviation Enterprises Inc. Orlando, Fl.	XB-AXP
☐ N756	78	Falcon 20F-5	388	Cleveland Cliffs Iron Co. Cleveland, Oh.	N731RG
☐ N756S	82	Gulfstream 3	348	Shell Aviation Corp. Houston, Tx.	
☐ N757AF	91	B 757-2J4	25155	Vulcan Northwest Inc. Seattle, Wa.	N115FS
☐ N757M	87	BAe 125/800A	NA0402	McCormick & Co. Sparks, Md.	N1125
☐ N757MA	97	B 757-200	28463		
☐ N757MC	95	Challenger 601-3R	5177	Frontliner Inc. San Diego, Ca.	N602MC
☐ N757T	97	Citation X	750-0014	Dayton Hudson Corp. Minneapolis, Mn.	
☐ N760AR	61	MS 760 Paris-2B	108	William Brunton & others, Las Vegas, Nv.	PH-MSX
☐ N760G	84	Gulfstream 3	428	Oxford Aviation Corp. Los Angeles, Ca. 'The Loan Ranger'	N760A
☐ N760J	58	MS 760 Paris	6	Don Hansen, Fort Worth, Tx. (wfu ?).	N84J
☐ N760R	61	MS 760 Paris-2B	104	Lyman & David Covell, Los Angeles, Ca.	N760P
☐ N760S	59	MS 760 Paris	43	B Air Inc. Alexandria, Va.	N760C
☐ N760X	59	MS 760 Paris	28	Aeronautical Systems Corp.	I-SNAI
☐ N761TA	97	Beechjet 400A	RK-161	Raytheon Travel Air Co-Zenith Drilling Corp. Wichita, Ks.	
☐ N765A	88	Gulfstream 4	1069	ARAMCO Associated Co. Dhahran, Saudi Arabia.	N459GA
☐ N765B	66	BAC 1-11/401AK	067	Calumet Inc. Lexington, Ky.	N109TH
☐ N766AF	81	Citation II	550-0203	Citation Holdings Corp. Wilmington, De.	N766AE
☐ N766CG	96	Citation VII	650-7066	CIT Leasing Corp. Livingstone, NJ.	N52630
☐ N766MH	83	Citation III	650-0015	Marketing Management Inc. Fort Worth, Tx.	N15QS
☐ N766NB	88	Citation S/II	S550-0156	Carl Panattoni & John Van Valkenburgh, Carmel, Ca.	(N400AJ)
☐ N767AC	81	Westwind-Two	356	Ames Construction Inc. Burnsville, Mn.	N861GS
☐ N767AG	76	Falcon 20F	349	J B Aviation Inc. Jacksonville, Fl.	N767AC
☐ N767FL	90	Gulfstream 4	1141	Gulfstream Aerospace Corp. Savannah, Ga.	N407GA
☐ N767JB		B 767-27C	27391	USAF, Wright-Patterson AFB. Oh.	
☐ N767SA	77	Learjet 25D	25D-216	RBS Aviation Group Inc. Dover, De.	N216SA
☐ N768J	89	Gulfstream 4	1119	Executive Skyfleet Inc. Fort Lauderdale, Fl.	N614HF
☐ N768TA	97	Beechjet 400A	RK-168	Raytheon Travel Air Co. Wichita, Ks.	N2168G
☐ N770BC	74	Learjet 24D	24D-308	TMJ Implants Inc. Golden, Co.	N39TT
☐ N770CC	92	Learjet 31A	31A-058	Carlisle Flight Services Inc. Syracuse, NY.	N26018
☐ N770E	80	Falcon 50	21	Jetcraft Corp. Raleigh, NC.	N77CE
☐ N770JC	82	Challenger 600S	1061	GMD Aviation LLC/Cash 4 Titles, Atlanta, Ga.	VH-MCG
☐ N770JM	79	Citation II	550-0072	McDonough Capital Co LLC. Lake Forest, Il.	N551SR
☐ N770MC	75	Falcon 20F	330	M C Aviation Inc. Tulsa, Ok.	C-GNTY
☐ N770MP	86	Citation III	650-0118	Air Travel Inc. Ronkonkoma, NY.	N118CD
☐ N770PC	84	Diamond 1A	A080SA	Bar-C Inc. Vienna, Ga.	N380CM
☐ N771AA	81	Citation II	550-0287	JFS Leasing Corp. Atlanta, Ga.	N550RL
☐ N771HR	74	Citation	500-0206	I H R Administrative Services Inc. Wichita Falls, Tx.	VH-LGL
☐ N771LD	67	Falcon 20C	59	Franklin Flight Corp. Bloomfield Hills, Mi.	N72BB
☐ N771ST	78	Citation II	550-0017	Executive Flight Inc. East Wenatchee, Wa.	P2-RDZ
☐ N771WW	81	Challenger 600S	1018	Wilkinson Flying Service Inc. Lexington, Ky.	N875PK
☐ N772AA	91	Citation V	560-0136	Island Outpost Aircraft Inc. Wilmington, De.	N501T
☐ N772AW	71	Westwind-1124	154	Avwest International LLC. Boulder, Co.	N919JH
☐ N772HP	81	Citation II	550-0226	Lanphere Enterprises Inc. Beaverton, Or.	N550RG
☐ N773AA	68	HS 125/731	NA713	Zebra Investments LC. Fort Lauderdale, Fl.	N272B
☐ N773LP	80	Learjet 35A	35A-362	Louisiana Pacific Corp. Hillsboro, Or.	N399KL
☐ N773W	74	Sabre-75A	380-20	U S D A Forest Service Aviation, Boise, Id.	N56
☐ N774CA	96	CitationJet	525-0141	Cyrk Inc. Gloucester, Ma.	
☐ N774TS	71	HS 125/731	NA774	Conanicut Aircraft Inc. NYC.	G-5-821
☐ N774W	75	Sabre-75A	380-37	U S D A Forest Service Aviation, Boise, Id.	N65
☐ N775M	92	Citation VII	650-7017	M & I Marshall & Ilsley Bank, Milwaukee, Wi.	
☐ N775US	98	Gulfstream V	535	Gulfstream Aerospace Corp. Savannah, Ga.	N593GA
☐ N776DF	95	CitationJet	525-0111	Delta Fox Inc. Zealand, Mi.	N52136
☐ N776MA	75	Gulfstream 2B	166	Deniston Enterprises Inc. Lutherville, Md.	XA-SWA
☐ N777AY	76	JetStar 2	5201	Space Master Investments Inc. Atlanta, Ga.	N745DM
☐ N777DC	67	Falcon 20C	91	Skyliner Inc. Omaha, Ne.	N8WN
☐ N777DM	80	Learjet 35A	35A-297	WYN Molded Plastics Inc. Rickenbacker, Oh.	N38US
☐ N777FC	86	Falcon 200	508	Ferro Corp/BP America Inc. Cleveland, Oh.	XA-RKE
☐ N777FH	90	Citation V	560-0076	Jetting Inc.Geneva, Il.	N777FE
☐ N777FL	90	Beechjet 400A	RK-4	James Warren, Sherwood, Or.	N400BE
☐ N777GA	68	HS 125/3A-RA	NA706	History Maker of Nevada Inc. Arlington, Tx.	(N899SA)
☐ N777GC	96	Beechjet 400A	RK-86	Grand Casinos Inc. Plymouth, Mn.	N1563V
☐ N777GG	79	Citation 1/SP	501-0108	Diversified Health Group Inc. Pittsburgh, Pa.	N777AJ
☐ N777GV	97	Gulfstream V	508	Steve Young International Ltd. Nassau, Bahamas.	N508GA
☐ N777HD	83	Westwind-1124	397	Gray & Co. Metairie, La.	N11CS
☐ N777JJ	74	Falcon 10	35	Falcon Enterprises LC. Virginia Beach, Va.	N54V
☐ N777KK	83	Challenger 600S	1082	Kohler Co. Sheboygan Falls, Wi.	I-PTCT
☐ N777KY	91	Citation V	560-0108	Mountain Enterprises Inc/Lenore Asphalt, Lenore, WV.	VH-NHJ
☐ N777LB	82	Learjet 35A	35A-476	Speedbird Inc. Cincinnati, Oh.	N476VC
☐ N777MC	83	Learjet 55	55-081	Meredith Corp. Des Moines, Ia.	N85631
☐ N777MR	67	Learjet 24	24-142	Air Response Inc. Nelliston, NY.	N200NR
☐ N777MW	86	Gulfstream 3	485	McWane Inc. Birmingham, Al.	N721CW
☐ N777NJ	74	Learjet 25XR	25XR-173	Dragon Leasing Corp. Naperville, Il.	N780AQ
☐ N777SG	66	JetStar-731	5074	Christian Advocates Serving Evangelists, Lawrenceville, Ga.	N168DB
☐ N777SK	80	Sabre-65	465-24	Sky King LLC. Wichita, Ks.	N741R
☐ N777SL	76	Citation	500-0307	K W Plastics, Troy, Al.	N2613
Reg	*Yr*	*Type*	*c/n*	*Owner/Operator*	*Prev Regn*

Reg	Yr	Type	c/n	Owner/Operator	Prev Regn
☐ N777SW	97	Gulfstream V	514	Joseph E Seagram & Sons Inc. NYC.	N514GA
☐ N777TK	73	HS 125/F600A	6015	Universal Jet Inc. Santa Monica, Ca.	N777SA
☐ N777TX	72	Learjet 25C	25C-084	Omimex Aviation Inc. Reno, Nv.	F-BYAL
☐ N777UE	90	Gulfstream 4	1146	Management Corp of America/MCA Inc. Burbank, Ca.	(N778W)
☐ N777WY	81	Citation II	550-0264	U S Energy Corp & Crested Corp. Riverton, Wy.	N550KC
☐ N779AZ	86	Citation III	650-0136	Federal Express Corp. Memphis, Tn.	N60AF
☐ N779QS	97	Citation VII	650-7079	EJI Inc/NetJets-Richard H Rogel, Columbus, Oh.	
☐ N780CF	78	Citation II	550-0014	CarFaye Inc. Eugene, Or.	N780GT
☐ N780E	91	Gulfstream 4	1165	IBM Corp. Dutchess County Airport, NY.	N460GA
☐ N780F	97	Gulfstream 5	530	Gulfstream Aerospace Corp. Savannah, Ga.	N530GA
☐ N780GT	89	Beechjet 400	RJ-55	Gordon Trucking Inc. Pacific, Wa.	N711FC
☐ N780RH	85	Gulfstream 3	472	HuffAir Inc & Huffco Group Inc. Houston, Tx.	N357H
☐ N782QS	97	Citation VII	650-7082	EJI Inc/NetJets, Dallas, Tx.	
☐ N785JM	90	Learjet 35A	35A-655	Yellow Brick Road LLC. Costa Mesa, Ca.	PT-MFR
☐ N785QS	98	Citation VII	650-7085	EJI Inc/NetJets, Dallas, Tx.	
☐ N787LP	91	Learjet 35A	35A-670	Coley Aircraft Sales & Leasing Inc. Memphis, Tn.	OY-CCO
☐ N787M	74	TriStar 100	1064	Operation Blessing/Medical Strike Force, Norfolk, Va.	C-GIES
☐ N787QS	98	Citation VII	650-7087	EJI Inc/NetJets, Columbus, Oh.	
☐ N787WB	77	JetStar 2	5210	Banair Inc. Houston, Tx.	N707WB
☐ N787WC	83	Citation II	550-0471	Beckett Enterprises/Lubrizol Corp. Wickcliffe, Oh.	N797WC
☐ N788MA	80	Westwind-1124	311	First Bank Systems Inc. St Paul, Mn.	N700MM
☐ N788NB	88	Citation III	650-0155	NationsBank NA. Charlotte, NC.	N13264
☐ N788R	97	Citation VII	650-7078	Burlington Resources Oil & Gas Co. Houston, Tx.	
☐ N789A	97	Astra-1125SPX	092	Galaxy Aerospace Corp. Fort Worth, Tx.	
☐ N789BR	78	Citation II	550-0036	Johnston Companies/JCI Transportation LLC. Chattanooga, Tn.	N711BP
☐ N789DD	76	Westwind-1124	187	United Aircraft Group Ltd. Dallas, Tx.	N516AC
☐ N789DJ	82	Diamond 1A	A015SA	Sepco Industries Inc. Houston, Tx.	N789DD
☐ N789DK	88	Gulfstream 4	1054	Platinum Air Charters LLC. Irving, Tx.	N745UR
☐ N789DR	82	Challenger 601	3001	GAR Aviation Ltd. Houston, Tx.	N601AG
☐ N789LB	93	BAe 125/800A	8248	Air Group/Superior Industries International Inc. Van Nuys.	N388H
☐ N789QS	98	Citation VII	650-7089	EJI Inc/NetJets, Columbus, Oh.	
☐ N789SR	93	Learjet 31A	31A-083	North Slope Borough Search & Rescue, Barrow, Ak.	N40363
☐ N789TP	83	Gulfstream 3	405	VCO Ltd. St Thomas, USVI.	(N9718P)
☐ N789TT	82	Citation II	550-0343	Tahoe-Teton Associates Inc. Menlo Park, Ca.	G-ORCE
☐ N790D	78	Citation II/SP	551-0071	AETEC International Inc. Tempe, Az.	N79CD
☐ N790FH	91	Astra-1125SP	056	Foundation Health California, Rancho Cordova, Ca.	N3175T
☐ N790JR	84	Westwind-Two	424	Journal Register Co. Trenton, NJ.	N424W
☐ N790L	95	Falcon 2000	15	IBM Corp. Dutchess County Airport, NY.	F-WWMO
☐ N790M	95	Falcon 2000	19	IBM Corp. Dutchess County Airport, NY.	F-WWMC
☐ N790QS	98	Citation VII	650-7090	EJI Inc/NetJets, Columbus, Oh.	
☐ N790US	76	Falcon 10	91	RVP Leasing Co. Grand Rapids, Mi.	D-CBAG
☐ N790Z	96	Falcon 2000	31	IBM Corp. Dutchess County Airport, NY.	N2032
☐ N791QS	98	Citation VII	650-7091	EJI Inc/NetJets, Columbus, Oh.	
☐ N791TA	95	Hawker 800XP	8291	Raytheon Travel Air Co. Wichita, Ks.	N291SJ
☐ N792MA	81	Citation II	550-0302	Matco Electric Corp. Vestal, NY.	N133BC
☐ N792QS	98	Citation VII	650-7092	EJI Inc/NetJets, Columbus, Oh.	
☐ N793BG	83	Westwind-Two	392	Boomerang Air Inc. Tampa, Fl.	N95WC
☐ N793CJ	93	CitationJet	525-0021	Cessna Aircraft Co. Wichita, Ks.	(N1329G)
☐ N793CT	94	Challenger 601-3R	5148	Caterpillar Inc. Peoria, Il.	C-G...
☐ N794GC	81	Learjet 35A	35A-446	First Security Bank, Salt Lake City, Ut.	N96CR
☐ N794QS	98	Citation VII	650-7094	EJI Inc/NetJets, Columbus, Oh.	
☐ N794TK	82	Westwind-Two	373	Lifestyle Aviation Inc. Coburg, Or.	N555DH
☐ N795A	81	HS 125/700B	7127	J S LLC. Indianapolis, In.	HB-VLC
☐ N795HE	81	HS 125/700A	NA0302	Harrah's Operating Co & Paramus Hotels Inc. Memphis, Tn.	N290PC
☐ N795HP	96	Astra-1125SPX	084	Hewlett-Packard Co. San Jose, Ca.	
☐ N795PH	88	BAe 125/800A	NA0426	Harrah's Operating Co & Promus Hotels Inc. Memphis, Tn.	N49VG
☐ N796HP	96	Astra-1125SPX	085	Hewlett-Packard Co. San Jose, Ca.	
☐ N796QS	98	Citation VII	650-7096	EJI Inc/NetJets, Columbus, Oh.	
☐ N796SF	76	Falcon 10	75	Farstad Oil Inc. Minot, ND.	N97TJ
☐ N797CB	96	Learjet 60	60-086	H E Butt Grocery Co. San Antonio, Tx.	N686LJ
☐ N797CW	81	Citation II	550-0232	U S Customs Service, Washington, DC.	N929DS
☐ N797QS	98	Citation VII	650-7097	EJI Inc/NetJets, Columbus, Oh.	
☐ N797SC	69	Learjet 25	25-042	Hartford Holding Corp. Millersville, Md.	(N25LG)
☐ N797T	90	Citation III	650-0197	Terra Industries Inc. Sioux City, Ia.	N197CC
☐ N797WC	77	JetStar 2	5216	Spears Manufacturing Co. Van Nuys, Ca.	N99E
☐ N799SC	95	Learjet 60	60-067	International Funeral Services Inc. Houston, Tx.	
☐ N800AB	73	Citation Eagle	500-0130	Nugget Oil Inc. Crestview, Fl.	ZS-MCP
☐ N800AJ	96	Astra-1125SPX	081	Asset Management Co. Palo Alto, Ca.	
☐ N800AR	82	Gulfstream 3	362	PJM/Riggs National Bank, Washington National, DC.	N408M
☐ N800AV	74	Citation	500-0209	Danis Heavy Construction Co. Dayton, Oh.	(N919AT)
☐ N800AW	78	Learjet 35A	35A-149	AirNet Systems Inc. Columbus, Oh.	(N40AN)
☐ N800BG	88	Gulfstream 4	1034	Tenet Healthcare Corp. Van Nuys, Ca.	N413GA
☐ N800CC	88	Gulfstream 4	1052	Chrysler Corp. Detroit, Mi.	N419GA
☐ N800CD	74	Sabre-75A	380-23	Perkndahl Management LLC. Minneapolis, Mn.	N102RD
☐ N800CH	81	Learjet 35A	35A-335	Cardal Inc. Dublin, Oh.	N800CD
☐ N800DA	66	HS 125/1A-522	25047	Dabia Corp. Banjul, Gambia.	(N717GF)
Reg	*Yr*	*Type*	*c/n*	*Owner/Operator*	*Prev Regn*

Reg	Yr	Type	c/n	Owner/Operator	Prev Regn
☐ N800DL	89	Citation V	560-0015	Carter Air/Home Interiors & Gifts Inc. Dallas, Tx.	N12171
☐ N800DM	71	DC 9-32	47466	Eagle Airlines Inc/Dallas Mavericks Inc. Dallas, Tx.	N1291L
☐ N800DR	90	BAe 125/800A	NA0462	Dominion Resources Inc. Richmond, Va.	N200GX
☐ N800DT	81	Citation 1/SP	501-0198	Hudson Holdings Corp/Spring Field Co LLC. Naples, Fl.	N198VP
☐ N800FJ	87	BAe 125/800A	8090	Fletcher Jones Management Group inc. Las Vegas, Nv.	N8090
☐ N800FL	84	BAe 125/800A	8005	Union Underwear Company. Bowling Green, Ky.	N601UU
☐ N800GE	69	HS 125/400A	NA735	Gan Eden Air Inc. Santa Fe, NM.	N165AG
☐ N800GG	68	Learjet 25	25-008	Grubbs Construction Co. Brooksville, Fl.	N88NJ
☐ N800GP	78	Learjet 35A	35A-158	Pal Waukee Aviation Inc. Wheeling, Il.	N158NE
☐ N800GT	94	Hawker 800XP	8266	GT-Green Tree Aircraft Corp. St Paul, Mn.	N800XP
☐ N800J	82	Gulfstream 3	359	Johnson & Johnson, West Trenton, NJ.	
☐ N800JH	80	Gulfstream 3	312	Law Offices of Gerald Hosier Ltd. Las Vegas, Nv.	N200JJ
☐ N800JT	73	HS 125/F400A	25272	John Tauber/J T Inc. Southfield, Mi.	N121VF
☐ N800KC	94	Challenger 601-3R	5157	Kimberly-Clark Corp. Dallas, Tx. (status ?).	N471SP
☐ N800LA	81	Citation II	550-0295	GFI Aviation Inc & co-owners, Augusta, Ga.	N483G
☐ N800M	80	Sabre-65	465-41	Fitness Management Corp. West Bloomfield, Mi.	N2556E
☐ N800MA	81	Westwind-Two	358	Old Mountain Air Inc. Waukegan, Il.	N13UR
☐ N800N	84	BAe 125/800A	8003	Emery Air Charter Inc. Rockford, Il.	N800BA
☐ N800NW	84	BAe 125/800A	8019	CYMI Investments Inc/IAMS Co. Dayton, Oh.	N799S
☐ N800PA	89	BAe 125/800A	NA0438	Polo Aviation Inc. St Paul, Mn.	N753G
☐ N800PC	98	Hawker 800XP	8369	Raytheon Aircraft Co. Wichita, Ks.	
☐ N800RF	79	Learjet 25D	25D-281	J-Fly LLC. Richmond, Va.	N555PG
☐ N800RT	69	Gulfstream 2	47	Robert E Torray & Co. Bethesda, Md.	N800FL
☐ N800RY	84	BAe 125/800B	8002	Raytheon Aircraft Co. Wichita, Ks.	(N1169D)
☐ N800TF	85	BAe 125/800A	8045	Tyson Foods Inc. Fayetteville, Ar.	N822AA
☐ N800VC	88	BAe 125/800A	NA0415	Crescent Trust + others, Bend, Or.	N560BA
☐ N800VF	96	Hawker 800XP	8300	Dwight Management LLC. Minneapolis, Mn.	N689H
☐ N800VJ	81	Citation II	550-0320	BOMAC LLC. Corvallis, Or.	N800EL
☐ N800WC	83	Gulfstream 3	392	Worldcom Inc. Jackson, Ms.	N9WN
☐ N800WG	89	BAe 125/800A	8152	Galtney Aerospace LC. Houston, Tx.	(N42US)
☐ N800WH	86	BAe 125/800A	8080	Whitman Corp. Waukegan, Il.	N800BP
☐ N800XL	68	Gulfstream 2	24	Cedar Enterprises LLC. Richardson, Tx.	(N224TS)
☐ N801	69	JetStar-8	5138	Essex Square Corp. Dania, Fl.	N31DK
☐ N801BB	96	Citation Bravo	550-0801	Carolina Mat Co. Plymouth, NC.	
☐ N801BC	74	HS 125/600A	6032	BudCo Ltd. San Antonio, Tx.	N921RD
☐ N801CC	94	Gulfstream 4SP	1254	Chrysler Corp. Detroit, Mi.	N436GA
☐ N801CE	94	BAe 125/800A	8253	Cummins Engine Co. Columbus, In.	N942H
☐ N801CH	83	BAe 125/800B	8001	Central Romana Corp. Santo Domingo, Dominican Republic.	OH-JOT
☐ N801FL	90	Challenger 601-3A	5063	Union Underwear Co. Bowling Green, Ky.	N811JW
☐ N801GC	86	Challenger 601	3052	G C R L Energy Ltd. Denver, Co.	(N601GF)
☐ N801JP	79	Citation II	550-0043	DavisAir Inc. Pittsburgh, Pa.	VR-CCI
☐ N801MB	86	BAe 125/800B	8067	Brass Aviation Inc. Fairfield, NJ.	N801MM
☐ N801P	91	Challenger 601-3A	5099	Sabrina Fisheries Corp. Bedford, Ma.	N504M
☐ N801SC	69	Falcon 20D	206	N801SC Corp/Michael Santaro, Syracuse, NY.	N632PB
☐ N801SM	80	Westwind-1124	297	Diversified Aircraft Holdings Ltd. Farmingdale, NY.	N51PD
☐ N801WB	95	Hawker 800XP	8287	Novell Inc. Orem, Ut.	(N287XP)
☐ N801WC	76	Gulfstream 2TT	183	Worldcom Inc. Jackson, Ms.	(N10NW)
☐ N802CC	76	Gulfstream 2	187	Chrysler Corp. Ypsilanti, Mi.	N202GA
☐ N802CE	94	Hawker 800	8270	Cummins Engine Co. Columbus, Oh.	N297H
☐ N802DC	94	Hawker 800	8257	David Clark, Naples, Fl.	N951H
☐ N802EC	82	Learjet 35A	35A-453	E A C Leasing Corp. Oakbrook, Il.	N802JW
☐ N802GA	82	Gulfstream 3	357	G 3 Charter Corp/Elite Aviation Inc. Van Nuys, Ca.	N340
☐ N802Q	80	Challenger 600S	1010	Rivett Group LLC/Quest Aviation. Aberdeen, SD.	N909MG
☐ N802X	88	BAe 125/800A	NA0418	Exxon Corp. Houston, Tx.	N563BA
☐ N803CC	83	Gulfstream 3	378	Chrysler Corp. Detroit, Mi.	N378HC
☐ N803CE	94	Hawker 800	8271	Cummins Engine Co. Columbus, Oh.	N298H
☐ N803E	86	Beechjet 400	RJ-16	Empak Airplane Co. Providence, RI.	N512WP
☐ N803JW	89	Astra-1125	038	Jeld-Wen Inc. Klamath Falls, Or.	
☐ N803L	69	Learjet 24B	24B-195	Frisco Aircraft Inc. Addison, Tx.	F-BUUV
☐ N803RA	76	Falcon 10	80	N803RA Inc. Houston, Tx.	N4RT
☐ N803X	88	BAe 125/800A	NA0420	Exxon Corp. Houston, Tx.	N565BA
☐ N804JT	97	Hawker 800XP	8311	DePuy Orthopaedics Inc. Warsaw, In.	
☐ N804JW	93	Astra-1125SP	069	Jeld-Wen Inc. Klamath Falls, Or.	
☐ N804PA	79	Sabre-65	465-4	Prime Aire Inc. Springfield, Mo.	N800TW
☐ N804X	88	BAe 125/800A	NA0421	Exxon Corp. Houston, Tx.	N566BA
☐ N805GT	91	Citation VI	650-0212	GTE Southwest, San Angelo, Tx.	
☐ N805JW	93	Astra-1125SP	070	Jeld-Wen Inc. Klamath Falls, Or.	N300AJ
☐ N805SM	69	Jet Commander-B	145	St Martin & Mahoney PLC. Houma, La.	N145AJ
☐ N805X	91	BAe 125/800A	NA0465	Exxon Corp. Houston, Tx.	N637BA
☐ N806LJ	65	Learjet 23	23-073	A Liner 8 Aviation, Livonia, Mi.	
☐ N807CC	77	Gulfstream 2TT	212	Chrysler Corp. Ypsilanti, Mi.	N551MD
☐ N808CC	72	HS 125/731	NA779	Christmas Cove Aircraft LLC. Portland, Me.	N408WT
☐ N808G	91	Challenger 601-3A	5098	The Gap Inc. San Francisco, Ca.	N812GS
☐ N808HS	94	CitationJet	525-0051	Spence Enterprises Inc. Crystal Lake, Il.	N800HS
☐ N808RP	75	HS 125/600A	6041	Rico-Perez Products Inc. Miami Beach, Fl.	N42TS
☐ N809F	81	Falcon 10	182	Admiral Beverage Corp. Worland, Wy.	N111MU

| Reg | Yr | Type | c/n | Owner/Operator | Prev Regn |

Reg	Yr	Type	c/n	Owner/Operator	Prev Regn
☐ N810BG	84	BAe 125/800A	8010	Basil Georges/Belchase Air Inc. Dallas, Tx.	N84A
☐ N810CR	79	HS 125/700A	NA0251	Omnicare Inc/Chemed Corp. Cincinnati, Oh.	N396U
☐ N810D	97	Challenger 604	5331	Jet Express Transit Corp. Short Hills, NJ.	C-G...
☐ N810GS	79	HS 125/700A	7061	Plain Vanilla Corp. East Hampton, NY.	N700SF
☐ N810MC	80	Citation II	550-0200	Central Business Jets, Burnsville, Mn.	N284
☐ N810MT	81	Challenger 600S	1024	Biomet Inc. Warsaw, In.	N326MM
☐ N810RA	67	Falcon 20C	81	Reliant Airlines Inc. Ypsilanti, Mi.	N93RS
☐ N810SS	96	CitationJet	525-0137	Super Service Inc. Somerset, Ky.	
☐ N811AA	68	Falcon 20D	187	USA Jet Airlines Inc. Belleville, Mi.	N750R
☐ N811BB	97	Challenger 604	5333	Barnett Banks Inc. Jacksonville, Fl.	N603CC
☐ N811BP	89	Challenger 601-3A	5039	N811BP Inc. Wilmington, De.	N811BB
☐ N811CC	94	Hawker 800	8267	Chrysler Corp. Detroit, Mi.	N294H
☐ N811DD	81	Learjet 35A	35A-384	Flynn Financial Corp. Boca Raton, Fl.	N811DF
☐ N811DF	79	Gulfstream 2	244	Flynn Financial Corp. Chicago, Il.	N509TT
☐ N811HL	79	Citation 1/SP	501-0114	Chariot Air LLC. Dallas, Tx.	(N725RH)
☐ N811JK	90	Gulfstream 4	1140	JBK Aviation/MacDonald's Hamburgers, Chicago, Il.	N405GA
☐ N811VC	81	Westwind-1124	331	Van Dyne Crotty Inc. Dayton, Oh.	N228L
☐ N811VG	79	Citation II/SP	551-0017	SBM Cleaning LLC. Corvallis, Or.	N811VC
☐ N812AA	66	Falcon 20C	57	USA Jet Airlines Inc. Belleville, Mi.	N711KG
☐ N812G	97	Challenger 604	5330	The GAP Inc/Aviation Methods Inc. San Francisco, Ca.	C-G...
☐ N812KC	81	Falcon 10	189	Clarence Scharbauer III, Midland, Tx.	N600PB
☐ N812RS	71	Gulfstream 2B	98	Rastar Holdings LLC/Avjet Corp. Burbank, Ca.	N198AV
☐ N812WN	80	Sabre-65	465-25	C R & J R Inc. Springfield, Mo.	N324ZR
☐ N813AA	66	Falcon 20C	25	USA Jet Airlines Inc. Belleville, Mi.	TG-GGA
☐ N813AS	78	Learjet 35A	35A-167	AirNet Systems Inc. Columbus, Oh.	N725P
☐ N813BR	67	Sabre-60	306-8	813BR LLC. Pleasant View, Tn.	N613BR
☐ N813JW	69	Learjet 25	25-038	Charter Airlines Inc. Las Vegas, Nv.	N400AJ
☐ N813TL	66	DC 9-15	45732	U S Department of Justice, Pineville, La.	N29
☐ N814CC	84	Citation S/II	S550-0018	Chrysler Corp. Ypsilanti, Mi.	N501NB
☐ N814CM	92	Citation V	560-0170	Hon Industries Inc. Muscatine, Ia.	N170CV
☐ N814D	71	HS 125/400A	NA761	813BR LLC. Pleasant View, Tn.	XA-RIL
☐ N814JR	69	Learjet 24B	24B-202	J & R Investments Inc. Wichita, Ks.	N814HH
☐ N814K	67	Jet Commander	106	814K LLC. Pleasant View, Tx.	N180TJ
☐ N814M	95	Falcon 900B	155	BP America Inc. Cleveland, Oh.	N730SA
☐ N814P	67	HS 125/3A-R	25148	N814P LLC. Chapmansboro, Tn.	XB-ERN
☐ N815A	96	Learjet 31A	31A-118	Clarcor Inc. Rockford, Il.	N318LJ
☐ N815AA	69	Falcon 20D	205	USA Jet Airlines Inc. Belleville, Mi.	(N426CC)
☐ N815CC	87	BAe 125/800A	NA0401	Chrysler Corp/NBD Transportation Co. Detroit, Mi.	N108CF
☐ N815CE	81	Citation II	550-0204	Restaurant Concepts Inc. Atlanta, Ga.	(N300PR)
☐ N815H	87	Citation S/II	S550-0146	T E Simpson, Winston Salem, NC.	N81SH
☐ N815L	77	Learjet 35A	35A-142	Starter Jet LLC. Atlanta, Ga.	N815A
☐ N815MC	96	CitationJet	525-0142	Kestrel Inc. Madison, In.	N5068R
☐ N816AA	73	Falcon 20E	290	USA Jet Airlines Inc. Belleville, Mi.	I-TIAL
☐ N816HB	89	Astra-1125	028	H P B Aviation Inc. NYC.	N11MZ
☐ N816M	82	Falcon 50	99	BP America Inc. Cleveland Hopkins Airport, Oh.	C-FMYB
☐ N816SQ	88	Challenger 601-3A	5030	Sprint/United Management Co. Kansas City, Ks.	N312CT
☐ N817AA	70	Falcon 20DC	233	USA Jet Airlines Inc. Belleville, Mi.	I-TIAG
☐ N817AM	83	Learjet 55	55-082	Florida Jet Service Inc. Pembroke Pines, Fl.	N68LP
☐ N817BD	66	JetStar-731	5083	Veronica Dawn Tisdale, Newport, Or.	N27FW
☐ N817CB	97	Citation Bravo	550-0817	Cessna Aircraft Co. Wichita, Ks.	
☐ N817JS	68	Falcon 20D	181	Bondstone Corp. Dallas, Tx.	N200GL
☐ N817M	80	Falcon 50	24	BP America Inc. Cleveland Hopkins Airport, Oh.	N51FJ
☐ N818AA	66	Falcon 20C	36	USA Jet Airlines Inc. Belleville, Mi.	OE-GUS
☐ N818CP	66	Falcon 20C	71	MLP 1 Ltd. Maitland, Fl.	(N293GT)
☐ N818LS	96	Challenger 604	5315	Falcon International Inc. Cleveland, Oh.	N604LS
☐ N818TH	82	Challenger 600S	1069	Arrow Aircraft Inc. Windsor Locks, Ct.	N788WG
☐ N819AA	66	Falcon 20C	26	USA Jet Airlines Inc. Belleville, Mi.	N11827
☐ N819GY	78	Sabre-75A	380-66	Sierra American Corp. Wilmington, De.	N943CC
☐ N819RC	76	Westwind-1124	192	Aero Advantage Inc. North Adams, Ma.	N319BG
☐ N820AA	67	Falcon 20C	118	USA Jet Airlines Inc. Belleville, Mi.	F-GGKE
☐ N820CB	97	Citation Bravo	550-0820	Bowman Gray School of Medicine, Winston Salem, NC.	
☐ N820FJ	90	Citation III	650-0183	Autocraft Industries Inc & Kerr McGee Corp. Oklahoma City,	EI-SNN
☐ N820L	65	Learjet 23	23-020	Avcon Industries Inc/Butler National Inc. Newton, Ks.	N388R
☐ N820MC	69	HS 125/731	NA739	Digital Communications Inc. Cedar Rapids, Ia.	N125DH
☐ N820RP	81	Learjet 35A	35A-410	Rich Aviation Inc. Lake Worth, Fl.	(N21WS)
☐ N821AA	69	Falcon 20D	203	USA Jet Airlines Inc. Belleville, Mi.	N36P
☐ N822AA	69	Falcon 20D	195	USA Jet Airlines Inc. Belleville, Mi.	N195MP
☐ N822BL	76	HS 125/600A	6067	Branch Law Firm Aviation Ltd. Albuquerque, NM.	XA-SKH
☐ N822CA	84	Learjet 35A	35A-591	ConAgra Inc. Omaha, Ne.	N500EX
☐ N822CB	97	Ciation Bravo	550-0822	Cessna Aircraft Co. Wichita, Ks.	N550TG
☐ N823AA	70	Falcon 20D	228	USA Jet Airlines Inc. Belleville, Mi.	OE-GRU
☐ N823CA	85	Learjet 35A	35A-600	ConAgra Inc. Omaha, Ne.	
☐ N823GA	86	Gulfstream 4	1005	Pratt Group, Melbourne, VIC. Australia.	VR-BJZ
☐ N824CA	86	Gulfstream 4	1010	ConAgra Inc. Omaha, Ne.	N444TJ
☐ N824CB	97	Citation Bravo	550-0824	Computer & Telecommunication Integration, Woburn, Ma.	
☐ N824CT	90	Citation II	550-0650	Cooper Tire & Rubber Co. Findlay, Oh.	N28RC

Reg	Yr	Type	c/n	Owner/Operator	Prev Regn
□ N824LJ	66	Learjet 23	23-083	Duncan Aviation Inc. Lincoln, Ne.	
□ N824MG	84	Learjet 55	55-106	Presidential Aviation Inc. Miami, Fl.	N318JH
□ N824R	82	Falcon 50	121	Sears Roebuck & Co. Chicago, Il.	N9311
□ N824TJ	69	HS 125/731	NA715	284DB Inc. Newtown Square, Pa.	N629P
□ N825CA	85	Learjet 35A	35A-605	ConAgra Inc. Omaha, Ne.	N35AS
□ N825CT	78	HS 125/700A	NA0229	Cooper Tire & Rubber Co. Findlay, Oh.	N700LS
□ N825D	79	Learjet 25D	25D-263	Carolina Industrial Products Inc.	N20DL
□ N825GA	93	CitationJet	525-0027	Parker Drilling Co. Tulsa, Ok.	(N1330N)
□ N825PS	81	Citation 1/SP	501-0224	Dawson Oil Co. LaCrosse, Wi.	N456CE
□ N825SB	68	Sabreliner 40	282-92	San Bernardino County Sheriff, Rialto, Ca.	158382
□ N825TC	66	Falcon 20C	52	Lyntex Inc. Wilmington, De.	N85DB
□ N826AA	67	Falcon 20C	67	USA Jet Airlines Inc. Detroit, Mi.	N821AA
□ N826CA	85	Learjet 35A	35A-596	ConAgra Inc. Omaha, Ne.	N850MM
□ N826CT	88	BAe 125/800A	NA0412	Cooper Tire & Rubber Co. Findlay, Oh.	N825PS
□ N826GA	94	Hawker 800	8263	PPG Industries Inc. Pittsburgh, Pa.	N961H
□ N826SU	93	BAe 125/800A	8249	Central Purchasing Inc. Camarillo, Ca.	N326SU
□ N827AA	74	Falcon 20E	298	USA Jet Airlines Inc. Belleville, Mi.	OE-GNN
□ N827CA	84	Learjet 35A	35A-590	ConAgra Inc. Omaha, Ne.	N969MC
□ N827GA	83	Gulfstream 3	398	PPG Industries Inc. Pittsburgh, Pa.	N88AE
□ N827JB	89	Citation II	550-0604	Aiglon Air LLC. Jackson, Tn.	N64VP
□ N828AA	66	Falcon 20C	31	Eagle and the Hawks Inc. Middletown, Oh.	N814AA
□ N828G	86	Citation III	650-0138	Publix Super Markets Inc. Lakeland, Fl.	(N1324R)
□ N829AA	79	Learjet 25D	25D-268	USA Jet Airlines Inc. Belleville, Mi.	N268WC
□ N829CA	82	Learjet 35A	35A-459	ConAgra Inc. Omaha, Ne.	N969MT
□ N829CB	97	Citation Bravo	550-0829	Cessna Aircraft Co. Wichita, Ks.	
□ N830EC	93	Gulfstream 4SP	1229	Entergy Services Inc. New Orleans, La.	N465GA
□ N830G	69	Gulfstream 2	44	E I Dupont de Nemours & Co. Ponca City, Ok.	N585A
□ N830KE	97	Citation Bravo	550-0830	Cessna Aircraft Co. Wichita, Ks.	
□ N830VL	82	Citation II	550-0412	SSI Properties Inc. Los Gatos, Ca.	N410CC
□ N831CB	81	Citation 1/SP	501-0237	Century II Management Inc. Wichita, Ks.	N237SC
□ N831CJ	89	Challenger 601-3A	5050	Marmon Group Inc. Chicago, Il.	N826JP
□ N831CW	81	Learjet 35A	35A-390	Noro Inc. Dover, De.	N508P
□ N831HG	74	Falcon 20E	310	HTG Corp. Atlanta, Ga.	N121AM
□ N831LH	78	Learjet 25D	25D-244	Chrysler Aviation Inc. Van Nuys, Ca.	N24EP
□ N831NW	70	HS 125/731	25231	Northwestern Aircraft Capital Corp. Mercer Island, Wa.	(N832MB)
□ N831S	93	CitationJet	525-0031	Insteel Industries Inc. Mount Aire, NC.	N31CJ
□ N831TJ	82	Diamond 1	A018SA	Tyler Jet LLC. Tyler, Tx.	N83BG
□ N832CB	92	Citation VII	650-7020	Siegel-Robert Inc. St Louis, Mo.	N700RR
□ N832MR	85	BAe 125/800B	8040	Maralo Inc & Lowe Petroleum Co. Houston, Tx.	N832MJ
□ N832QB	91	Citation V	560-0110	Technology Aviation Inc. Austin, Tx.	N832CB
□ N833JL	77	Citation 1/SP	501-0045	J Lewis Investments, Birmingham, Al.	N833
□ N833JP	85	BAe 125/800A	8044	Fairfax Realty Inc. Salt Lake City, Ut.	N72NP
□ N834CB	97	Citation Bravo	550-0834	Cessna Aircraft Co. Wichita, Ks.	
□ N834H	89	Citation III	650-0177	Hillenbrand Investment Advisory Corp. Wilmington, De.	N707HJ
□ N835CB	97	Citation Bravo	550-0835	Cessna Aircraft Co. Wichita, Ks.	
□ N835GA	76	Learjet 35A	35A-087	Parker Drilling Co. Tulsa, Ok.	N720GL
□ N835TS	85	BAe 125/800A	8035	Aero Toy Store Inc. Fort Lauderdale, Fl.	PT-WIA
□ N836ME	70	Gulfstream 2B	95	First Gibraltar (Parent) Holdings Inc. Dallas, Tx.	N836MF
□ N836MF	74	Gulfstream 2B	154	Ropa Two Corp. Teterboro, NJ.	N18JN
□ N837MA	73	Citation	500-0096	C & M Aero Inc. Wilmington, De.	(N187AP)
□ N838MF	87	Gulfstream 4	1012	Ropa Two Corp. Teterboro, NJ.	N636MF
□ N838QS	97	Hawker 800XP	8338	EJI Inc/NetJets, Columbus, Oh.	
□ N840QS	97	Hawker 800XP	8340	EJI Inc/NetJets, Columbus, Oh.	
□ N840SW	93	Learjet 31A	31A-084	Southwire Co. Carrollton, Ga.	N196HA
□ N841TT	81	Learjet 35A	35A-416	Salem Leasing Corp. Winston Salem, NC.	N40GG
□ N841WS	97	Citation Bravo	550-0841	Cessna Aircraft Co. Wichita, Ks.	
□ N843HS	86	Gulfstream 3	496	HealthSouth Aviation Inc. Birmingham, Al.	(N99SU)
□ N844F	82	Falcon 100	201	Semitool Inc. Kalispell, Mt.	N100NW
□ N844HS	96	Gulfstream 4SP	1289	HealthSouth Aviation Inc. Birmingham, Al.	N405GA
□ N844X	82	Falcon 50	93	Shamrock Aviation/Pillsbury Co. Minneapolis, Mn.	N98FJ
□ N845FW	81	Gulfstream 3	875	F-W Global Aviation Inc. Houston, Tx.	N728CP
□ N846HS	91	Citation II	550-0669	HealthSouth Aviation Inc. Birmingham, Al.	N6170C
□ N847C	84	Citation S/II	S550-0003	Transit Air Services Inc/Lynton Jet Centre, Morristown, NJ.	N847G
□ N847HS	90	Citation II	550-0661	HealthSouth Aviation Inc. Birmingham, Al.	N550RA
□ N847RH	93	BAe 125/800B	8224	First Security Bank NA. Salt Lake City, Ut.	N827RH
□ N848AB	77	JetStar 2	5214	Celeseixientos II Corp. Fort Lauderdale, Fl.	N106JL
□ N848C	89	Beechjet 400A	RJ-63	Midmark Corp. Versailles, Oh.	
□ N848G	88	Citation S/II	S550-0152	General Dynamics Corp. Washington, DC.	N26369
□ N848HS	93	Citation II	550-0720	HealthSouth Aviation Inc. Birmingham, Al.	N72WE
□ N849HS	81	Westwind-Two	344	HealthSouth Aviation Inc. Birmingham, Al.	N311BR
□ N850BM	88	BAe 125/800A	NA0430	Boston Chicken Inc. Golden, Co.	N1903P
□ N850CA	81	Falcon 50	75	Hunt-Wesson Inc. Ontario, Ca.	N45BE
□ N850CC	80	Sabre-65	465-38	Colorado Interstate Gas Co. Colorado Springs, Co.	N850CS
□ N850SM	86	BAe 125/800A	8074	Summit Aviation Inc. Newport Beach, Ca.	N300BW
□ N851BC	79	Citation 1/SP	501-0249	Chesnuts Investments Inc. Wilmington, De.	N249AS
□ N854TT	79	Citation 1/SP	501-0086	T & T Enterprises LLC. Santa Clara, Ca.	N11SQ

Reg	Yr	Type	c/n	Owner/Operator	Prev Regn
☐ N855CD	74	Sabre-60	306-85	B R Wings Inc. Dover, De.	N355CD
☐ N855DB	83	Learjet 55	55-062	Alamo Jet Inc. Fort Myers, Fl.	N292RC
☐ N855DH	83	Citation III	650-0014	Denstar Aviation, Minnetonka, Mn.	OE-GCN
☐ N856JB	65	Learjet 23	23-052	Sky Way Enterprises Inc/John Kowal, Washington, DC.	N360EJ
☐ N856MA	69	Sabre-60	306-41	Timothy Roman & Barry Baron, Hazleton, Pa.	(N62DW)
☐ N857BL	96	Citation V Ultra	560-0381	Cessna Aircraft Co. Wichita, Ks.	N2762J
☐ N860S	76	Learjet 35A	35A-086	Bullock Charter Inc. Princeton, Ma.	N86CS
☐ N860W	97	Citation VII	650-7086	Cessna Aircraft Co. Wichita, Ks.	
☐ N861CE	84	BAe 125/800A	8006	Coca-Cola Enterprises Inc. Atlanta, Ga.	N70SK
☐ N861MC		BAe 146-100	E1068	Moncrief Oil, Fort Worth, Tx.	G-BVUX
☐ N862CE	87	BAe 125/800A	8089	Coca-Cola Enterprises Inc. Atlanta, Ga.	N125JB
☐ N862G	81	Gulfstream 3	329	General Dynamics Corp. Washington, DC.	N301GA
☐ N863AB	76	Westwind-1124	196	Silverado Airways Inc. Raleigh, NC.	(N615DM)
☐ N863CE	80	Gulfstream 3	306	N14BK Inc. Beverly Hills, Ca.	N862CE
☐ N864CE	88	Gulfstream 4	1085	Coca-Cola Enterprises Inc. Atlanta, Ga.	N88GA
☐ N864EC	92	Citation VII	650-7014	Entergy Services Inc. New Orleans, La.	N1261A
☐ N865M	92	Citation V	560-0179	Central Trust Bank, Jefferson City, Mo.	N1280D
☐ N865VP	77	Falcon 20F	360	Volare Aviation Inc. Palwaukee, Il.	N165PA
☐ N866BB	95	Beechjet 400A	RK-98	Shamrock Aviation Charters Inc. Eden Prairie, Mn.	N400A
☐ N867CE	92	Gulfstream 4	1195	Coca-Cola Enterprises Inc. Atlanta, Ga.	N47HR
☐ N868CP	81	Westwind-Two	341	Charlie Papa Aviation Ltd. Van Nuys, Ca.	N728LM
☐ N868JT	95	Citation V Ultra	560-0310	Florida Power & Light Co. Miami, Fl.	N410CV
☐ N870MH	83	Citation II	550-0475	Oakwood Homes Corp. Greensboro, NC.	N1252B
☐ N870P	91	Beechjet 400A	RK-20	Pace Industries Inc. Fayetteville, Ar.	N82628
☐ N870PT	82	Citation II	550-0349	Dibb Air Inc. Wilmington, De.	N8FD
☐ N870WC	92	Citation II	550-0695	Walbro Corp. Case City, Mi.	N695VP
☐ N872AT	95	Hawker 800XP	8278	Action Digital Color Inc. Abbeville, SC.	G-BVZK
☐ N873LP	90	Learjet 35A	35A-659	Louisiana Pacific Corp. Hillsboro, Or.	
☐ N874A	96	Gulfstream 4SP	1285	Anardarko Petroleum Corp. Houston, Tx.	N477GA
☐ N874G	86	Citation III	650-0137	Lockheed Martin Corp. Fort Worth, Tx.	(N1324G)
☐ N874RA	82	Gulfstream 3	361	Anadarko Petroleum Corp. Houston, Tx.	(N875E)
☐ N875H	91	Challenger 601-3A	5093	Household International, Wheeling, Il.	N601CH
☐ N875HS	82	Westwind-1124	370	HealthSouth Aviation Inc. Birmingham, Al.	N471TM
☐ N875SC	87	BAe 125/800A	NA0409	The Jet Place, Tulsa, Ok.	N375SC
☐ N876CS	86	Learjet 35A	35A-616	COMM/SCOPE Inc. Hickory, NC.	N616LJ
☐ N876G	95	Citation VII	650-7062	General Dynamics Corp. Washington, DC.	N5262Z
☐ N876H	96	Hawker 800XP	8303	Household International Inc. Wheeling, Il.	N303XP
☐ N876MA	75	Falcon 10	63	Deniston Enterprises Inc. Baltimore, Md.	N70TS
☐ N876WB	77	Citation	500-0347	Schaefer Ambulance Service Inc. Van Nuys, Ca.	N500XY
☐ N877C	77	Citation 1/SP	501-0017	Paradise Developers Inc. New orleans, La.	N36869
☐ N877G	95	Citation VII	650-7063	General Dynamics Corp. Washington, DC.	N95CC
☐ N877J	93	Beechjet 400A	RK-69	Raytheon Aircraft Co. Wichita, Ks.	N877S
☐ N877RF	95	Citation V Ultra	560-0318	Reinhart Institutional Foods Inc. La Crosse, Wi.	N1273R
☐ N877RP	87	BAe 125/800A	8084	Petersen Publishing Co. Los Angeles, Ca.	N780A
☐ N877S	96	Hawker 800XP	8323	Sparks Companies Inc. Fort Morgan, Co.	N323XP
☐ N877SE	95	Challenger SE	7075	Fruit of the Loom Inc. Wheeling, Il.	C-FMLO
☐ N877Z	91	Beechjet 400A	RK-17	Aalfs Manufacturing Inc. Sioux City, Ia.	N877S
☐ N878SM	97	Gulfstream 4SP	1319	Gulfstream Aerospace Corp. Savannah, Ga.	N429GA
☐ N880CR	82	HS 125/700A	NA0324	Omnicare Management Co. Cincinnati, Oh.	N69SB
☐ N880DP	66	BAC 1-11/401AK	079	Detroit Pistons/Round Ball One Corp. Detroit, Mi.	N800DM
☐ N880GC	87	Gulfstream 4	1016	Guardian Industries Corp. Detroit, Mi.	N29GY
☐ N880RB	74	DC 9-32	47635	Detroit Pistons/Round Ball One Corp. Detroit, Mi.	PK-CNH
☐ N880RJ	75	Gulfstream 2	159	Fargo Aviation Inc/Aviation Methods, Minneapolis, Mn.	N800DJ
☐ N880SP	96	Hawker 800XP	8298	SP of Delaware Inc. Dover, De.	N298XP
☐ N880WD	78	Gulfstream 2	217	Guardian Industries Corp. Detrot, Mi.	N81728
☐ N880Z	77	Westwind-1124	203	Outback Steakhouse of Florida Inc. Tampa, Fl.	N124WW
☐ N881DM	74	Sabre-40A	282-137	Quarter M Farms Inc. Rose Hill, NC.	N870R
☐ N881G	91	Falcon 900B	104	Charles Schwab & Co. San Francisco, Ca.	N104FJ
☐ N881J	78	Falcon 20F-5	396	Westfield Aviation Inc. Fairfax, Va.	N179F
☐ N881M	82	Falcon 50	83	International Paper Co. White Plains, NY.	N88U
☐ N881P	94	Falcon 900B	146	International Paper Co. White Plains, NY.	N216FP
☐ N881TW	97	Challenger 604	5348	Trans West Air Services Inc. Salt lake City, Ut.	C-GLWR
☐ N882C	90	Challenger 601-3A	5065	Transit Air Services Inc. Morristown, NJ.	N601BF
☐ N882KB	85	Citation III	650-0095	F Korbel & Brothers Inc. Guerneville, Ca.	(N95VP)
☐ N882RB	85	Citation S/II	S550-0075	David Hutton, Killen, Al.	N882KB
☐ N883PF	86	Citation S/II	S550-0083	Prestage Farms Inc. Clinton, NC.	N511BR
☐ N886CS	75	Learjet 35	35-023	BYR737/Beck & Yeaggy, Cincinnati, Oh.	N886WC
☐ N886S	78	HS 125/700A	7025	Reptron Aviation Inc. Raleigh, NC.	N7782
☐ N888AZ	84	Challenger 601	3024	C & C Enterprises Inc. Ithaca, NY.	N93CR
☐ N888CJ	66	HS 125/1A	25084	Beeline Aviation Inc. Winchester, Ky.	N890RC
☐ N888CP	88	Learjet 31	31-003	William Bruggeman/Aero Plan Leasing, Minneapolis, Mn.	N331CC
☐ N888DE	75	Learjet 35	35-010	Pacific Coast Lease Corp. Las Vegas, Nv.	N888DH
☐ N888DH	86	BAe 125/800A	8051	AFG Industries/R D Hubbard Enterprises Inc. Palm Desert, Ca.	N889DH
☐ N888FA	72	Learjet 24D	24D-257	Creekwood Inc. Wilmington, De.	C-GHDP
☐ N888FL	77	Citation 1/SP	501-0014	L & L Aviation Inc. Vicksburg, Ms.	N22TP
☐ N888LK	89	Gulfstream 4	1125	Deerport Aviation Corp. Manila, Philippines.	N700WB

Reg	Yr	Type	c/n	Owner/Operator	Prev Regn
☐ N888MC	66	Learjet 24	24-106	Kempt Ville Plane Parts Ltd. Wilmington, De.	N103RB
☐ N888MJ	73	Citation 1/SP	501-0446	Hanover Air Service Inc. Ashland, Va. (was s/n 500-0097).	N63CF
☐ N888NA	93	Citation II	550-0723	Northern Automotive Systems LLC. West Salem, Wi.	N5NE
☐ N888PM	85	Gulfstream 3	435	Red White & Blue Pictures Inc. Van Nuys, Ca.	N435U
☐ N888RA	96	CitationJet	525-0135	Richard Auhll, Santa Barbara, Ca.	N5207A
☐ N888RT	81	Citation II	550-0254	Aladdin Inc. Oklahoma City, Ok.	N112SA
☐ N888SQ	96	Gulfstream 4SP	1305	Sprint/United Management Co. Kansas City, Ks.	N439GA
☐ N888TN	77	Learjet 36A	36A-026	Learstar LLC. Los Angeles, Ca.	N8U
☐ N888TW	74	Learjet 24D	24D-292	Sierra American Corp. Dallas, Tx.	N800PC
☐ N888TX	92	Citation VII	650-7003	AVCO Financial Services Management Co. Costa Mesa, Ca.	N17AP
☐ N888VS	85	Gulfstream 3	450	CPB Aviation Inc. Encino, Ca.	VR-CTG
☐ N888WJ	76	Falcon 10	89	Western Jet Sales Inc. Fort Worth, Tx. (maybe still TC-AND?)	TC-AND
☐ N888WS	94	Challenger 601-3R	5170	Williams-Sonoma Inc. San Francisco, Ca.	N166A
☐ N888XL	74	Citation	500-0177	Cedar Enterprises LLC. Richardson, Tx.	N192AB
☐ N889DW	92	Learjet 60	60-117	Learjet Inc. Wichita, Ks.	
☐ N890A	86	BAe 125/800A	8071	Aluminum Co of America, Pittsburgh, Pa.	N521BA
☐ N891CA	74	Citation	500-0168	Mike's Aviation Inc. Gainesville, Fl.	(N46JA)
☐ N892SB	78	Falcon 20F-5	379	Security Benefit Group Inc. Topeka, Ks.	N62570
☐ N893AC	88	Challenger 601-3A	5018	A C Nielsen Corp. White Plains, NY.	(N601GR)
☐ N893CA	77	Citation 1/SP	501-0270	Gulfstream Nautical, Key West, Fl.	N300WK
☐ N893CM	93	Citation V	560-0226	West Coast Charters Inc. Santa Ana, Ca.	
☐ N894TW	81	Westwind-1124	354	Panther Partners/JTA Air & RBW Enterprises, Yardley, Pa.	N124LS
☐ N895CC	79	HS 125/700A	NA0252	Blue Bear Air LLC. Bozeman, Mt.	N513GP
☐ N895LD	89	Citation V	560-0034	Casey Co. Long Beach, Ca.	N589LD
☐ N896MA	75	Citation	500-0290	Mississippi Air Service Inc. Jackson, Ms.	(N390S)
☐ N896R	96	Learjet 60	60-091	Botona Aviation Inc. Fort Lauderdale, Fl.	N8071L
☐ N897R	97	Learjet 60	60-097	Botona Aviation Inc. Fort Lauderdale, Fl.	N8067Y
☐ N898CB	91	Citation V	560-0097	H E Butt Grocery Co. San Antonio, Tx.	(N6790L)
☐ N898GF	91	Citation V	560-0121	Grede Transport Inc. Milwaukee, Wi.	PT-MTG
☐ N899DM	78	HS 125/700A	7028	Markin Aviation LLC. New Windsor, NY.	N7728
☐ N899SC	94	Learjet 60	60-040	SCI Texas Funeral Services Inc. Houston, Tx.	N399SC
☐ N899WA	76	Learjet 35	35-049	Wings Aviation of Nevada Inc. Reno, Nv.	N235JL
☐ N900AL	89	Gulfstream 4	1097	Abbott Laboratories/ABT Flight Inc. Waukegan, Il.	N402GA
☐ N900BF	77	Gulfstream 2	206	Gulfstream Wings LLC. Chesterfield, Mo.	N2PK
☐ N900BJ	77	Learjet 35A	35A-123	Mackey Communications Inc. Raleigh, NC.	N900JE
☐ N900CH	78	Falcon 20F-5	383	Cardal Inc. Dublin, Oh.	HB-VJS
☐ N900CL	92	Challenger 601-3A	5122	CIGNA Corp. Windsor Locks, Ct.	N7046J
☐ N900CM	83	Citation III	650-0017	Excel Corp/Cargill Leasing Corp. Minneapolis, Mn.	C-GHLM
☐ N900CR	62	JetStar-731	5036	Kenny & Marianne Rogers, Henderson, Nv.	N90KR
☐ N900DH	72	Gulfstream 2	111	Richmor Aviation/D & H Flying Corp. Olympic Valley, Ca.	N900BR
☐ N900DM	85	Citation S/II	S550-0067	Diemakers Inc. Monroe City, Mo.	N550HA
☐ N900DP	81	Challenger 600S	1036	BusAv/Del Inc. Aiken, SC.	N66MF
☐ N900EX	97	Falcon 900EX	12	Dassault Falcon Jet Corp. Teterboro, NJ.	(N900SB)
☐ N900FA	82	Learjet 55	55-024	Fabritek La Romana Inc. Columbus, Ms.	HB-VGZ
☐ N900FL	95	Citation VII	650-7049	Fleet Financial Group Inc. Providence, Rl.	N749CM
☐ N900FS	76	Westwind-1124	191	Executive Aviation of Louisiana Inc. Dallas, Tx.	N326AJ
☐ N900G	75	Citation	500-0268	Gulf Winds Inc. Wilmington, De.	A6-RKH
☐ N900GC	76	Citation	500-0298	Granite Construction Co. Watsonville, Ca.	N5298J
☐ N900GG	70	Learjet 24B	24B-216	North American Jet Charter Inc. Banning, Ca.	N777LB
☐ N900H	83	Westwind-1124	388	James M Hoak & Co. Dallas, Tx.	N1124K
☐ N900HC	89	Falcon 900	68	Robert M Bass Group Inc. Fort Worth, Tx.	N449FJ
☐ N900JB	81	Falcon 50	59	Brooks Telecommunications Corp. St Louis, Mo.	N31V
☐ N900JC	78	Learjet 35A	35A-178	U-Haul Company of Oregon, Phoenix, Az.	N35GG
☐ N900JD	91	Citation VI	650-0213	Deere & Co. Moline, Il.	(N6823L)
☐ N900LC	67	Falcon 20C-5	122	Mid Central Charter Corp. Chicago, Il.	N32PB
☐ N900LS	92	Gulfstream 4	1178	Limited Stores Inc. Columbus, Oh.	N470GA
☐ N900MA	88	Falcon 900	67	Danaher Corp. Washington, DC.	N448FJ
☐ N900MC	78	Citation 1/SP	501-0052	D Ray Booker, Tulsa, Ok.	N98715
☐ N900MJ	88	Falcon 900	48	ACM Aviation Inc/Apple Computers, San Jose, Ca.	N435FJ
☐ N900MP	71	Gulfstream 2SP	99	Berg Electronics Inc. St. Louis, Mo.	N900VL
☐ N900NA	66	Learjet 24A	24A-111	RBS Aviation Group Inc. Fort Myers, Fl.	N44WD
☐ N900NE	89	Falcon 900	83	Banc One Corp. Columbus, Oh.	N900WG
☐ N900P	97	CitationJet	525-0234	Cessna Aircraft Co. Wichita, Ks.	
☐ N900PA	83	Westwind-Two	400	Windsong Aviation LLC. Minneapolis, Mn.	N300LS
☐ N900Q	91	Falcon 900	93	Tarrant Partners LP. Fort Worth, Tx.	(N900P)
☐ N900R	82	Learjet 35A	35A-488	NASCAR Inc. Daytona Beach, Fl.	N30WY
☐ N900SB	97	Falcon 900EX	26	SBC Management Services Inc. San Antonio, Tx.	F-WWFX
☐ N900SE	79	Citation II/SP	551-0016	Ziegler Air Inc. Fort Worth, Tx. (was 550-0338).	N500QM
☐ N900SJ	87	Falcon 900	19	Sid R Bass Inc. Fort Worth, Tx.	N414FJ
☐ N900TW	80	Citation 1/SP	501-0167	Jetcraft Corp. Raleigh, NC.	N38RT
☐ N900VP	80	Westwind-1124	289	RVW Enterprises Inc. Williamsville, NY.	VR-CIL
☐ N900W	81	Falcon 50	60	BFI Transportation Inc. Houston, Tx.	JY-HZH
☐ N900WG	70	HS 125/731	NA757	Battelle Memorial Institute, Columbus, Oh.	N50NE
☐ N900WJ	82	Diamond 1A	A028SA	Matrix Aviation Corp. Denver, Co.	N331DC
☐ N900WK	88	Falcon 900	57	Kellogg Co. Battle Creek, Mi.	N441FJ
☐ N901C	77	JetStar 2	5218	Indianapolis Colts Inc. Indianapolis, In.	N816RD

Reg	Yr	Type	c/n	Owner/Operator	Prev Regn
☐ N901FH	81	Gulfstream 3	333	Coca-Cola Bottling Co Consolidated, Charlotte, NC.	N50PM
☐ N901JC	83	Learjet 55	55-088	Frederick Marine Corp. Wilmington, De.	N900JB
☐ N901K	97	Hawker 800XP	8329	Westinghouse Electric Corp. Pittsburgh, Pa.	N329XP
☐ N901NB	82	Citation 1/SP	501-0255	Nunes Company Inc. Salinas, Ca.	N400LX
☐ N901RM	91	Citation V	560-0116	Rubbermaid Inc. Wooster, Oh.	N68032
☐ N901SB	87	Falcon 900	33	SBC Management Services Inc. San Antonio, Tx.	N9138Y
☐ N901WG	73	Gulfstream 2	126	Gary, Williams et al, Stuart, Fl. 'Wings of Justice'	N416K
☐ N902	96	Gulfstream 4SP	1310	Owens-Illinois General Inc. Toledo, Oh.	(N2425)
☐ N902GA	68	Gulfstream 2	11	Tyler Jet LLC. Tyler, Tx.	N902
☐ N902K	89	Gulfstream 4	1113	Westinghouse Electric Corp. Allegheny County, Pa.	N423GA
☐ N902MP	79	Gulfstream 2SP	241	Viasystems Technologies Corp. St Louis, Mo.	N90MD
☐ N902QS	95	Citation X	750-0002	EJI Inc/NetJets, Columbus, Oh.	N752CX
☐ N902RM	93	Citation VII	650-7022	Wooster Rubber Co/Rubbermaid Inc. Wooster, Oh.	N722CM
☐ N902SB	92	Citation VII	650-7008	SBC Management Services Inc. San Antonio, Tx.	N901SB
☐ N902WC	68	B 737-247	19613	Nordam Group Inc. Tulsa, Ok.	N308VA
☐ N903G	75	Gulfstream 2	172	Owens-Illinois General Inc. Toledo, Oh.	N804GA
☐ N903HC	81	Learjet 35A	35A-440	Hydrol Conveyor Co. Jonesboro, Ar.	N101PK
☐ N903SB	95	Citation VII	650-7061	SBC Management Services Inc. San Antonio, Tx.	N5221Y
☐ N904SB	84	BAe 125/800A	8016	SBC Management Services Inc. San Antonio, Tx.	N415PT
☐ N905LC	88	Citation II	550-0581	Special Services Corp. Greenville, SC.	(N13007)
☐ N906SB	87	Falcon 900	14	SBC Management Services Inc. San Antonio, Tx.	N900SB
☐ N906WK	91	Falcon 900	102	Kellogg Co. Battle Creek, Mi.	N467FJ
☐ N908CL	88	Challenger 601-3A	5031	CIGNA Corp. Windsor Locks, Ct.	N900CL
☐ N908JE	74	Gulfstream 2B	151	Hyperion Air Inc. Wilmington, De.	N979GA
☐ N908R	92	Beechjet 400A	RK-44	Wolverine Tube Inc. Decatur, Al.	N404VP
☐ N908RF	75	Falcon 10	46	Jetways LLC & Wheel-Air Inc. Minneapolis, Mn.	N815LC
☐ N909QS	96	Citation X	750-0009	EJI Inc/NetJets-Dresser Industries Inc. Columbus, Oh.	N96TX
☐ N910A	82	Gulfstream 3	369	AMOCO Corp. Chicago, Il.	
☐ N910B	89	Gulfstream 4	1102	AMOCO Corp. Chicago, Il.	N405GA
☐ N910BH	77	Sabre-75A	380-54	USA Jet Airlines Inc. Belleville, Mi.	N380CF
☐ N910DS	88	Citation S/II	S550-0154	Turnberry Charters Inc. Opa Locka, Fl.	N500NS
☐ N910JW	87	Falcon 900	31	S C Johnson & Son Inc/Johnson's Wax, Racine, Wi.	N900FJ
☐ N910M	84	Citation III	650-0069	AMOCO Corp. Chicago, Il.	(N13142)
☐ N910MH	65	Jet Commander	45	College of Aeronautics, La Guardia, NY.	N121PG
☐ N910PC	94	Citation V Ultra	560-0273	August Busch III, St Peters, Mo.	
☐ N910RB	81	Citation II	550-0267	First Security Bank NA. Salt Lake City, Ut.	XA-AOC
☐ N910S	90	Gulfstream 4	1155	AMOCO Corp. Chicago, Il.	N1761B
☐ N910SH	93	Beechjet 400A	RK-72	Commander Airways Inc. Tampa, Fl.	(N72BJ)
☐ N910V	92	Citation V	560-0165	AMOCO Corp. Chicago, Il.	
☐ N911AE	77	Learjet 35A	35A-109	Keystone Group LLC. Portland, Or.	N506GP
☐ N911AJ	74	Learjet 25B	25B-163	AASP-Air Ambulance Support Programmes Inc. Houston, Tx.	(N65RC)
☐ N911AS	65	HS 125/1A	25039	ASAP Air Services/125TH LLC. Fort Lauderdale, Fl.	N125TB
☐ N911CB	90	Citation II	550-0662	Commerce Bancshares Inc. Kansas City, Mo.	N5294C
☐ N911CR	71	JetStar-731	5150	Juan Rodriguez, Sarasota, Fl.	N721CR
☐ N911DG	69	Falcon 20C	162	Aero Taxi Rockford Inc. Rockford, Il.	N162CT
☐ N911EM	80	Learjet 25D	25D-319	Image Air of Southwest Florida LC. Bloomington, Il.	N319EJ
☐ N911GM	72	Citation	500-0048	Miller & Schroeder/Air Lake Lines Inc. Minneapolis, Mn.	N67JR
☐ N911HB	85	Falcon 50	157	HBO & Co. Atlanta, Ga.	N341M
☐ N911KT	84	Gulfstream 3	438	Karthe Corp/Chrysler pentastar Aviation, Waterford, Mi.	N1841D
☐ N911ML	79	Learjet 35A	35A-256	Firstlear Leasing Inc. Wilmington, De.	N50DD
☐ N911MM	77	Citation Eagle	501-0030	Quad Fund Inc. Wilmington, De.	N96BA
☐ N911RF	87	Falcon 900	20	REFCO Inc. Chicago, Il.	N70FJ
☐ N911RG	68	Falcon 20C	144	Aero Taxi Rockford Inc. Rockford, Il.	N800KR
☐ N911SP	78	Westwind-1124	244	Chapman/Leonard Studio Equipment, N Hollywood, Ca.	N124PA
☐ N911TR	67	Learjet 24	24-134	San Juan Jet Charter Inc. Hato Rey, PR.	N7GN
☐ N912AS	67	HS 125/3A	25124	814D LLC. Fort Lauderdale, Fl.	N552N
☐ N912BD	88	Citation II	550-0580	Dillard Department Stores Inc. Little Rock, Ar.	(N13006)
☐ N912DA	69	Jet Commander-B	147	Agro Air Associates Inc. Miami, Fl.	(N888MP)
☐ N912SH	96	Beechjet 400A	RK-128	Commander Airways Inc. Tampa, Fl.	N1108Y
☐ N913JB	97	Challenger 604	5338	America Airfields Corp. Dallas/Candan Air, Lanseria, RSA.	N604PL
☐ N913MK	84	Gulfstream 3	407	Mary Kay Inc/MillionAir, Dallas, Tx.	N407GA
☐ N913SQ	92	Citation VII	650-7004	Sprint/United Management Co. Kansas City, Ks.	N708CT
☐ N913V	83	HS 125/700A	NA0345	Payless ShoeSource Inc. Topeka, Ks.	(N313VR)
☐ N913VL	77	Falcon 10	104	Diamond Shamrock Refining & Marketing Co. San Antonio, Tx.	N913V
☐ N914BD	89	Falcon 900	80	Dillard Department Stores Inc. Little Rock, Ar.	N459FJ
☐ N914CD	74	Citation	500-0150	C & D Investments Inc. Fort Smith, Ar.	N9V
☐ N914DZ	76	Gulfstream 2SP	190	DZ LLC. NYC.	N59JR
☐ N914J	97	Falcon 900EX	15	Metromedia Co. Teterboro, NJ.	N915EX
☐ N914X	95	Challenger 601-3R	5185	Xerox Corp. White Plains, NY.	N611CC
☐ N915BD	91	Challenger 601-3A	5091	Dillard Department Stores Inc. Little Rock, Ar.	C-GLXM
☐ N915RB	89	Learjet 35A	35A-647	FHC Flight Services Inc. Norfolk, Va.	N410RD
☐ N915US	76	Learjet 24B	24B-189	Gary Jet Center/Brian Easterman, Lake Forest, Il.	N711DX
☐ N916BD	94	Learjet 31A	31A-093	Dillard Department Stores Inc. Little Rock, Ar.	N4031K
☐ N916CS	96	Citation V Ultra	560-0400	Avlease Inc. Ann Arbor, Mi.	
☐ N916RC	80	Citation II	550-0164	Carik Services Inc. Denver, Co.	(N24PT)
☐ N917BD	94	Learjet 31A	31A-094	Dillard Department Stores Inc. Little Rock, Ar.	N31AX

| Reg | Yr | Type | c/n | Owner/Operator | Prev Regn |

Reg	Yr	Type	c/n	Owner/Operator	Prev Regn
☐ N917BE	80	Westwind-1124	291	Basler Electric Aviation Inc. Highland, Il.	N124WK
☐ N917W	91	Gulfstream 4	1158	Jet-Air Inc/Navair LLC. NYC.	N17582
☐ N918MK	97	Astra-1125SPX	089	K & M Flight Corp. Chicago, Il.	
☐ N919BT	85	Westwind-Two	434	Business Telecom Inc. Raleigh, NC.	N187EC
☐ N919TG	75	Gulfstream 2	160	TGA Ltd/Federal Aviation Services Corp. Dallas, Tx.	N900TP
☐ N920CC	80	Sabre-65	465-16	ANR Pipeline Co. Wilmington, De.	N112CF
☐ N920DS	69	Gulfstream 2B	73	Turnberry Charters Inc. Opa Locka, Fl.	N555CS
☐ N920DY	81	Sabre-65	465-50	Dixie Yarns Inc. Chattanooga, Tn.	N959C
☐ N920EA	70	Learjet 25	25-057	Eagle Aviation Inc. West Columbia, SC.	C-FTXT
☐ N920G	76	Falcon 20F-5	352	PAWS Inc. Albany, In.	N4466F
☐ N920MS	94	CitationJet	525-0089	S Enterprises Inc/Slipstream Aviation Inc. Birmingham, Al.	N189CJ
☐ N920PM	92	Citation V	560-0182	TPT Aviation Inc. Gastonia, SC.	N560RA
☐ N921CC	81	Sabre-65	465-67	Colorado Interstate Gas Co/ABCO Leasing Inc. Wilmington, De.	N65AR
☐ N921MB	78	Sabre-60	306-135	Metal Building Components Inc. Houston, Tx.	N60AM
☐ N921ML	67	Falcon 20C	99	AMI/Marion Merrell Dow-Marion Laboratories Inc. Kansas City.	N982F
☐ N922JW	87	Falcon 900	36	S C Johnson & Son Inc. Racine, Wi.	N91MK
☐ N922RR	69	HS 125/400A	NA726	Reschini Agency Inc. Indiana, Pa.	YV-141CP
☐ N923AR	94	CitationJet	525-0055	Union Planters Corp. Memphis, Tn.	
☐ N923ML	78	Gulfstream 2B	219	Hoechst Marion Roussel Inc. Kansas City, Mo.	N307AF
☐ N923QS	97	Citation X	750-0023	EJI Inc/NetJets-Chesapeake Energy Corp. Columbus, Oh.	
☐ N924BW	74	Learjet 25B	25B-158	AirNet Systems Inc. Columbus, Oh.	N71RB
☐ N924EJ	97	Citation X	750-0024	Interstate Equipment Leasing Inc/Swift Air, Phoenix, Az.	(N164M)
☐ N925AJ	94	Falcon 2000	4	Seneca Livestock Co. Eugene, Or.	F-WWMA
☐ N925DM	82	Learjet 35A	35A-486	Milam International Inc. Englewood, Co.	N810CC
☐ N925DW	76	Learjet 25D	25D-213	Leasco Aviation Inc. Reno, Nv.	(N803PF)
☐ N925EX	97	Falcon 900EX	25	Dassault Falcon Jet Corp. Teterboro, NJ.	F-WWFV
☐ N925WC	81	HS 125/700A	NA0294	Forest Hills Corp. Wilmington, De.	C-FIPG
☐ N926CB	83	Citation III	650-0008	Compass Bank, Birmingham, Al.	N84WU
☐ N926QS	97	Citation X	750-0026	EJI Inc/NetJets-Opportunity Aviation Ltd. Columbus, Oh.	
☐ N926TC	78	HS 125/700A	NA0214	PHH Corp. Baltimore, Md.	N926ZT
☐ N927AA	80	Sabre-65	465-22	Air Marty Co. Van Nuys, Ca.	VR-CEE
☐ N927EX	97	Falcon 900EX	27	Dassault Falcon Jet Corp. Teterboro, NJ.	F-WWFY
☐ N928CD	93	Learjet 60	60-010	A L Inc. Omaha, Ne.	N525CF
☐ N928RD	74	Citation	500-0204	Dodson Aviation Inc. Ottawa, Ks.	N204Y
☐ N929GV	72	Gulfstream 2B	119	Empire Airways/Empire Sanitary Landfill Inc. Taylor, Pa.	(N103EL)
☐ N929WG	83	Diamond 1A	A032SA	Fort Calumet Corp. Chicago, Il.	N320T
☐ N929WT	97	Gulfstream 4SP	1317	Gulfstream Aerospace Corp. Savannah, Ga.	N417GA
☐ N930RA	81	Sabre-65	465-68	Allen-Bradley Co. Milwaukee, Wi.	N165NA
☐ N930SD	71	Gulfstream 2SP	97	Longline Leasing Inc & Excel Armor Products, Fairfield, Oh.	N11AL
☐ N931CC	80	Falcon 50	31	EJI Inc. Columbus, Oh.	XA-AAS
☐ N931FD	96	Learjet 31A	31A-124	Family Dollar Stores Inc. Charlotte, NC.	
☐ N932QS	97	Citation X	750-0032	EJI Inc/NetJets-Times Mirror Co. Columbus, Oh.	
☐ N933JC	79	Sabre-75A	380-72	Jim Colbert Inc. Las Vegas, Nv.	N555JR
☐ N933NA	66	Learjet 23	23-049	Earth Resources Laboratory, NSTL Station, Ms.	(N933N)
☐ N933SH	83	Citation III	650-0009	Shaw Industries Inc. Chattanooga, Tn.	N933DB
☐ N934H	89	Citation III	650-0172	Hillenbrand Industries Inc. Batesville, In.	N672CC
☐ N934QS	97	Citation X	750-0034	EJI Inc/NetJets, Columbus, Oh.	
☐ N935H	92	BAe 125/800A	NA0475	Federal-Mogul Corp. Southfield, Mi.	N800CJ
☐ N936QS	97	Citation X	750-0036	EJI Inc/NetJets, Columbus, Oh.	
☐ N938H	93	BAe 125/800A	8252	Dudmaston Ltd. Hamilton, Bermuda.	G-5-788
☐ N940CC	65	Sabre-40	282-34	Coastal Corp/ANR Coal Co. Roanoke, Va.	N400CS
☐ N940HC	90	BAe 125/800A	NA0458	Oakwood Homes Corp. Greensboro, NC.	N100GX
☐ N940P	95	Learjet 60	60-071	Tri C Inc. Oklahoma City, Ok.	N60LJ
☐ N940SW	94	CitationJet	525-0071	Cottonaire Corp/TPT Aviation, Gastonia, NC.	N26509
☐ N941CE	80	HS 125/700A	NA0259	Corporate Eagle One LLC. Waterford, Mi.	N128CS
☐ N941CW	68	Gulfstream 2	29	Paramount Aviation LLC. Cleveland, Oh.	N71TJ
☐ N942BY	89	Learjet 31	31-005	ISC Aviation LLC. Fort Mill, SC.	XA-RFS
☐ N942CC	78	Sabre-75A	380-64	Sabre Transportation Inc. Huntington, WV.	N75Y
☐ N942DS	65	HS 125/731	25032	Storms Enterprises Inc. Pineville, NC.	N98TJ
☐ N942WN	66	HS 125/731	25079	YAGER/Internet Services Corp. Charlotte, NC.	N942Y
☐ N942Y	67	JetStar-731	5098	ISC Aviation LLC. Fort Mill, SC.	N792AA
☐ N943CE	81	HS 125/700A	NA0300	LCT Inc. Waterford, Mi.	N700SA
☐ N944AD	87	Falcon 900	17	Archer Daniels Midland Co. Decatur, Il.	N411FJ
☐ N944H	96	Citation X	750-0011	Honeywell Inc. Minneapolis, Mn.	
☐ N944KM	76	Learjet 24E	24E-334	Kelly Moss Aviation Inc. Madison, Wi.	N66MJ
☐ N944NA	74	Gulfstream 2	144	NASA Johnson Space Center, Houston, Tx.	HB-ITR
☐ N945CC	79	Sabre-65	465-13	ANR Pipeline Co. Detroit, Mi.	(N950CS)
☐ N945CE	81	HS 125/700A	NA0297	Corporate Eagle Capital LLC. Waterford, Mi.	N589UC
☐ N945MC	75	Falcon 10	37	Mountaire Corp. N Little Rock, Ar.	N72GW
☐ N945NA	72	Gulfstream 2	118	NASA Johnson Space Center, Houston, Tx. (NASA 650).	(N651NA)
☐ N946GM	77	Westwind-1124	215	ExecuJet Charter Services Inc. St Petersburg, Fl.	N500WH
☐ N946JR	68	Sabre-60	306-10	Aerotech GmbH-Saarbrucken/Aero Mobile Inc. Wilmington, De.	N125MC
☐ N946NA	74	Gulfstream 2	146	NASA Johnson Space Center, Houston, Tx.	N897GA
☐ N947ML	71	DC 9-32	47514	Homfeld II LLC. Eastpointe, Mi.	PH-MAX
☐ N947NA	74	Gulfstream 2	147	NASA Johnson Space Center, Houston, Tx.	N898GA
☐ N948NA	78	Gulfstream 2	222	NASA Johnson Space Center, Houston, Tx.	N5253A

Reg	Yr Type	c/n	Owner/Operator	Prev Regn

Reg	Yr	Type	c/n	Owner/Operator	Prev Regn
☐ N949L	66	DC 9-14	45844	International Airline Support Group, Miami, Fl.	N8963
☐ N949QS	98	Citation X	750-0049	EJI Inc/NetJets, Columbus, Oh.	
☐ N950F	88	Falcon 50	191	Russell Stover Candies Inc. Kansas City, Mo.	N282FJ
☐ N950FC	82	Citation II	550-0407	Cottonaire Corp. Gastonia, NC.	N758S
☐ N950G	77	Learjet 36A	36A-032	Winfair Aviation Ltd. Fort Lauderdale, Fl.	HB-VLK
☐ N950RA	67	Falcon 20C	95	Reliant Airlines Inc. Ypsilanti, Mi.	(OO-EEF)
☐ N950WA	90	Citation V	560-0082	Woodstock Aviation Inc. Santa Rosa, Ca.	(N2746U)
☐ N951DB	76	Westwind-1124	195	Interplanetary Aviation Inc. Culver City, Ca.	N195ML
☐ N951RK	76	Gulfstream 2SP	191	Kenair Inc. West Palm Beach, Fl.	N675RW
☐ N953C	92	Citation V	560-0163	JABJ/NYNEX Corp. White Plains, NY.	N529X
☐ N953F	89	Citation V	560-0005	Casto Plane LLC & Pizzuti Transportation Ltd. Columbus, Oh.	
☐ N955CC	69	Gulfstream 2B	54	Coastal Corp. Houston, Tx.	C-FNOR
☐ N955E	80	Falcon 50	14	JABJ/NYNEX Corp. White Plains, NY.	N9X
☐ N955H	88	Gulfstream 4	1081	Honeywell Inc. Minneapolis, Mn.	
☐ N955LS	81	Learjet 55	55-009	AV 8 Inc. Fort Lauderdale, Fl.	N955MD
☐ N956PP	83	Diamond 1A	A031SA	Printpack Inc. Atlanta, Ga.	N2220G
☐ N956S	72	Citation Eagle	500-0056	K3C Inc. Reno, Nv.	N360DA
☐ N957H	94	Hawker 800	8260	KORK Inc. Peoria, Il.	G-5-800
☐ N959H	94	Hawker 800	8262	Business Focus Sdn Bhd/Amin Shah, Kuala Lumpur.	G-5-803
☐ N959SA	76	Learjet 35A	35A-076	AirNet Systems Inc. Columbus, Ga.	
☐ N960CP	81	Citation II/SP	551-0345	Cornerstone Propane LP. Watsonville, Ca. (was 550-0313).	N246NW
☐ N960FA	81	Westwind-Two	348	Anne H Bass, Fort Worth, Tx.	N348SJ
☐ N960H	93	Learjet 60	60-015	Harris Corp. Melbourne, Fl.	
☐ N960TX	79	Falcon 20F	403	Texlon/PTC Airco Inc. Canton, Oh.	N100BG
☐ N961JC	86	BAe 125/800A	8062	Cooper Industries Inc. Houston, Tx.	N366BA
☐ N961MR	92	Learjet 60	60-003	AM & JB Inc. Cleveland, Oh.	N60LJ
☐ N962J	82	Citation II	550-0453	D & D Aviation, Grandview, Mo.	N962JC
☐ N964C	81	Sabre-65	465-66	Bail-Foster Aircraft Holding Corp. Wilmington, De.	
☐ N965CC	75	Gulfstream 2B	165	Coastal Corp. Houston, Tx.	N26L
☐ N965JC	95	Citation VII	650-7051	Cooper Industries Inc. Houston, Tx.	N95CC
☐ N965QS	98	Citation X	750-0065	EJI Inc/NetJets, Columbus, Oh.	
☐ N966H	96	Citation X	750-0012	Honeywell Inc. Minneapolis, Mn.	
☐ N966JM	93	Citation V	560-0240	Johnson Machinery Co. Riverside, Ca.	N91ME
☐ N966L	94	Hawker 800	8259	Pepsico Inc. Dallas, Tx.	N954H
☐ N966SW	94	Citation V Ultra	560-0284	Terry Coleman Inc. Manchester, UK.	N5108G
☐ N967L	94	Hawker 800	8273	Frito Lay Corp. Dallas, Tx.	N803H
☐ N968L	91	Challenger 601-3A	5089	KFC West, White Plains, NY.	C-GLXY
☐ N969SS	80	Learjet 25D	25D-317	Aviation Leasing Corp. Atlanta, Ga.	N821LM
☐ N970H	82	Learjet 55	55-055	Harris Corp. Melbourne, Fl.	N155LP
☐ N970SU	96	CitationJet	525-0173	SCCJ Inc. Stillwater, Ok.	
☐ N970WJ	80	Learjet 25D	25D-324	Jet Holding LLC. Nashville, Tn.	XA-POP
☐ N971AS	61	JetStar-731	5007	Douglas Matthews, West Palm Beach, Fl.	N72CT
☐ N971EC	82	Learjet 55	55-033	IES Diversified Inc. Cedar Rapids, Ia.	(N77JW)
☐ N971F	77	Learjet 35A	35A-095	Kelgen LP. Milwaukee, Wi.	N971H
☐ N971L	89	Gulfstream 4	1116	Interlease Aviation Corp/Air Group Inc. Van Nuys, Ca.	N431GA
☐ N971QS	98	Citation X	750-0071	EJI Inc/NetJets, Columbus, Oh.	
☐ N972H	76	Learjet 24D	24D-322	CITA Aviation Leasing Inc. Dover, De.	N105GL
☐ N972LM	80	HS 125/700A	NA0269	Execujet Charter Services Inc. St Petersburg, Fl.	N702NW
☐ N972TF	69	Jet Commander-B	138	Wisconsin Aviation Four Lakes Inc. Watertown, Wi.	N5BA
☐ N974JD	97	Beechjet 400A	RK-141	Wichita Air Services Inc. Wichita, Ks.	N1027S
☐ N975GR	84	Diamond 1A	A077SA	Ness, Motley, Loadholt, Richardson & Poole PA. Barnwell, SC.	N66PL
☐ N976BS	68	Learjet 25	25-016	Gerald Wingett, El Paso, Tx.	N8FF
☐ N977GA	70	Learjet 24B	24B-219	General Aviation Services Inc. Lake Zurich, Il.	F-GECI
☐ N977QS	98	Citation X	750-0077	EJI Inc/NetJets, Columbus, Oh.	
☐ N977TW	65	Falcon 20C	13	Sierra American Corp. Dallas, Tx.	F-BTCY
☐ N978E	77	Learjet 36A	36A-024	International Jet Aviation Inc. Van Nuys, Ca.	N38D
☐ N979C	94	Citation V Ultra	560-0263	The Colonial Co/REL Services Inc. Montgomery, Al.	N12945
☐ N979QS	98	Citation X	750-0079	EJI Inc/NetJets, Columbus, Oh.	
☐ N979RA	93	Gulfstream 4	1191	Ogden Corp. Westchester County, NY.	N404GA
☐ N979RF	81	Learjet 35A	35A-376	Fairbanks Communications Inc. West Palm Beach, Fl.	XA-SBF
☐ N980DM	77	Citation 1/SP	501-0062	Maggie Air LLC. Scottsdale, Az.	N900DM
☐ N980HC	94	Challenger 601-3R	5163	U S Health Aviation Corp. Philadelphia, Pa.	N709JM
☐ N980R	67	Falcon 20C	98	Reliant Airlines Inc. Ypsilanti, Mi.	N781AJ
☐ N981	97	Falcon 2000	53	Dassault Falcon Jet Corp. Teterboro, NJ.	F-WWMJ
☐ N981SW	86	Gulfstream 4	1001	Gulfstream Aerospace Corp. Savannah, Ga.	N31001
☐ N982HC	94	Gulfstream 4SP	1242	Penobscot Properties LLC. New Castle, De.	N490GA
☐ N983GT	80	HS 125/700A	NA0261	General Tranportation Services Inc. Liverpool, NY.	N500EF
☐ N984GC	97	Learjet 45	45-009	Learjet Inc. Wichita, Ks.	
☐ N986MA	93	Learjet 31A	31A-080	Mescalero Apache Tribe, Mescalero, NM.	N80LJ
☐ N987AR	91	Gulfstream 4	1156	Atlantic Richfield Co. Burbank, Ca.	N987AC
☐ N987QS	98	Citation X	750-0087	EJI Inc/NetJets, Columbus, Oh.	
☐ N988AA	74	Learjet 25B	25B-185	Applied Analytical Industries Inc. Wilmington, NC.	N988DB
☐ N988MT	62	Sabreliner CT-39A	276-32	Metro Tech Aviation Career Center, Oklahoma City, Ok.	62-4479
☐ N988QC	81	Learjet 35A	35A-455	Moyle Petroleum & Sodak Gaming Supplies Inc. Rapid City, SD.	N455NE
☐ N988RS	88	Citation II	550-0568	Estacada Lumber Co. Molawa, Or.	(N1299K)
☐ N988T	82	Falcon 50	130	Sears Roebuck & Co. Chicago, Il.	N630L

Reg	Yr	Type	c/n	Owner/Operator	Prev Regn

Reg	Yr	Type	c/n	Owner/Operator	Prev Regn
☐ N989QS	98	Citation X	750-0089	EJI Inc/NetJets, Columbus, Oh.	
☐ N989TL	68	Learjet 24	24-160	Booth Ranches/F Otis Booth Jr. Van Nuys, Ca.	N4791C
☐ N989TW	81	Citation II	550-0285	Wamberg Financial Corp. Wilmington, De.	N40PL
☐ N990AL	72	Citation	500-0033	Opercorp. Gainesville, Fl.	(N130AL)
☐ N990M	89	Citation II	550-0608	Menzil Enterprises Inc. Wexford, Pa.	N608AM
☐ N990MC	88	Falcon 900	65	MASCO Corp. Taylor, Mi.	N216FB
☐ N990PT	70	Learjet 24D	24D-236	Taylor Energy Co. New Orleans, La.	N93DD
☐ N990WC	95	Gulfstream 4SP	1268	Warnaco Aviation, Nashua, NH.	N427GA
☐ N991BM	79	Citation II	550-0114	N90BJ Inc. Warsaw, In.	(N900BM)
☐ N991PC	96	Citation V Ultra	560-0364	Iowa Packing Co. Des Moines, Ia.	N5260Y
☐ N991RF	86	Falcon 900	3	REFCO Inc. Chicago, Il.	N327F
☐ N991RV	74	Falcon 10	24	E I Aviation Inc/Eddie Irvine, Dublin, Ireland.	N301JJ
☐ N991TD	66	Learjet 24	24-124	Air Cargo Express Inc. Little Rock, Ar. (status ?).	XA-RTV
☐ N992	81	Falcon 50	77	Reynolds Metals Co. Richmond, Va.	N366F
☐ N992TD	65	Learjet 23	23-035	Air Cargo Express Inc. Little Rock, Ar.	(N10QX)
☐ N993	80	Falcon 50	38	Reynolds Metals Co. Richmond, Va.	N58FJ
☐ N993QS	98	Citation X	750-0094	EJI Inc/NetJets, Columbus, Oh.	
☐ N993TD	68	Learjet 24	24-166	Air Cargo Express Inc. Little Rock, Ar.	N124HF
☐ N994CT	94	Challenger 601-3R	5161	Caterpillar Inc. Peoria, Il.	C-GLWZ
☐ N994GC	69	Gulfstream 2	77	Guidant Corp. Indianapolis, In.	N125WM
☐ N995BC	84	Gulfstream 3	432	Rifton Enterprises Inc. Rifton, NY.	N713KM
☐ N995DC	65	Citation S/II	S550-0065	Delasystems LLC. El Dorado, Ar.	N612ST
☐ N995RD	74	Sabre-80A	380-9	Cheyenne Management Co. Waterford, Mi.	N383CF
☐ N995SK	78	HS 125/700A	NA0225	Tiburon Transportation Ltd. Healdsburg, Ca.	N486MJ
☐ N996JR	96	CitationJet	525-0147	Tango Corp. Minden, Nv.	
☐ N997BC	91	Gulfstream 4	1170	Rifton Enterprises Inc. Rifton, NY.	(N811SW)
☐ N997MX	83	Diamond 1	A036SA	Metal Exchange Corp. Matthews, NC.	N18BA
☐ N997TT	84	Falcon 20F-5	485	SRCG Holdings Inc/Richardson Aviation Inc. Fort Worth, Tx.	N161EU
☐ N998JR	85	Challenger 601	3045	James River Paper Co. Sandston, Va.	N3045
☐ N999AM	74	Citation	500-0232	JMK Holdings Inc & co-owners, Fort Worth, Tx.	N126R
☐ N999AU	93	Learjet 31A	31A-074	Piraya Aviation Inc. Fort Lauderdale, Fl.	(VP-B..)
☐ N999BL	86	Astra-1125	024	Brunswick Corp. Waukegan, Il.	N300JJ
☐ N999CB	85	Citation S/II	S550-0054	Charles Brewer Ltd. Phoenix, Az.	N57MB
☐ N999CV	74	Citation	500-0211	Classic Leasing Corp. Fort Smith, Ar.	N999CB
☐ N999EB	97	CitationJet	525-0210	Subaru of New England Inc. Norwood, Ma.	N210CJ
☐ N999EQ	72	Falcon 20E	275	Eagle Aircraft Services Ltd. Las Vegas, Nv.	SX-DKI
☐ N999F	74	Falcon 10	29	Trinity Industries Inc. Dallas, Tx.	N999F
☐ N999GH	84	Citation II/SP	551-0496	G M Hock Construction Inc. Durham, NC.	N8008F
☐ N999GP	92	Astra-1125SP	062	Gary Primm/Primadonna Resorts Inc. Las Vegas, Nv.	N1125
☐ N999HC	85	Citation S/II	S550-0030	Hach Co. Loveland, Co.	(N284L)
☐ N999JF	78	Learjet 35A	35A-188	J F Air Inc. Wilmington, De.	N88TJ
☐ N999LG	70	Sabre-60	306-53	Crown Air LLC. Palm Beach Gardens, Fl.	N999KG
☐ N999MF	69	Learjet 25	25-050	Rainbow Chaser LLC. Evergreen, Co.	N55FN
☐ N999PJ	61	MS 760 Paris-2	89	Stephen Griswold, San Francisco, Ca.	F-BJLY
☐ N999PM	93	Falcon 900B	128	Pacific Marine Leasing Inc. Lakewood, Wa.	N128FJ
☐ N999PW	80	Citation 1/SP	501-0160	2141 Corp. Atlanta, Ga.	C-GAAA
☐ N999TC	73	Citation	500-0120	United Vending & Food Services Inc. Sioux City, Ia.	N141DP
☐ N999TF	82	Challenger 600S	1042	Priester Aviation/Brinson Partners Inc. Chicago, Il.	N999SR
☐ N999TH	87	Falcon 200	512	Journal Publishing Co. Albuquerque, NM.	N45WH
☐ N999VT	65	Sabre-40	282-38	Video Lottery Technologies Inc. Bozeman, Mt.	(N68AA)
☐ N999WA	71	Learjet 24D	24D-242	Wings Aviation Inc. Shepherdsville, Ky.	N1972G
☐ N999WJ	87	Falcon 100	216	Western Jet Sales Inc. Fort Worth, Tx.	(N71M)
☐ N999WS	81	Citation 1/SP	501-0186	Weldon F Stump & Co Inc. Toledo, Oh.	N37HW
☐ N1000	77	Gulfstream 2	205	Swiflite Aircraft Corp/Oxy USA Inc. Bakersfield, Ca.	N25UG
☐ N1000E	94	CitationJet	525-0077	UTE Management Corp. Oklahoma City, Ok.	
☐ N1000W	92	Citation V	560-0204	Ashland Oil Inc. Worthington, Ky.	N1283Y
☐ N1040	92	Gulfstream 4SP	1206	Cox Enterprises Inc. Atlanta, Ga.	N437GA
☐ N1082A	88	Gulfstream 4	1082	Burlington Resources Oil & Gas Co. Houston, Tx.	
☐ N1083Z	96	Beechjet 400A	RK-131	Universal Health Management LLC. Detroit, Mi.	
☐ N1086	88	Gulfstream 4	1086	Stockwood V Inc. Morristown, NJ.	N23SY
☐ N1102U	97	Hawker 800XP	8343	Raytheon Aircraft Co. Wichita, Ks.	
☐ N1115G	97	Hawker 800XP	8347	Raytheon Aircraft Co. Wichita, Ks.	
☐ N1116R	96	Beechjet 400A	RK-116	Mid-South Milling Co. Naples, Fl.	
☐ N1119C	96	Beechjet 400A	RK-119	Gordon Management Inc. Des Moines, Ia.	
☐ N1121A	68	Jet Commander-A	123	KVA Resources Inc. Bellevue, Wa.	N155VW
☐ N1121E	65	Jet Commander	20	OK Turbines Inc. Hollister, Ca. (status ?).	N334LP
☐ N1121G	66	Jet Commander	67	Dodson International Parts Inc. Ottawa, Ks.	N650M
☐ N1121N	67	Jet Commander	110	Morvil & Clare Wilson, Destin, Fl.	N16GH
☐ N1121R	69	Jet Commander-A	125	Albert Lea Airport Inc. Albert Lea, Mn.	N30LS
☐ N1123H	73	Westwind-1123	167	Maitland Brothers Inc. Littleton, Pa.	N873EJ
☐ N1124F	80	Westwind-1124	281	Winner Aviation Corp. Sharon, Pa.	N4251H
☐ N1124G	78	Westwind-1124	216	Barken International Inc. Winston Salem, NC.	N216SC
☐ N1125	88	Astra-1125SP	076	Packard Bell Electronics Inc. Van Nuys, Ca.	4X-CUV
☐ N1125A	90	Astra-1125SP	051	Alabama River Pulp Co. Monroe County Airport, Al.	
☐ N1125E	91	Astra-1125SP	058	Atlantic Tele-Network Inc. St Croix, USVI.	
☐ N1125J	96	Astra-1125SP	078	Fred Meyer Inc. Portland, Or.	4X-C..

| *Reg* | *Yr Type* | | *c/n* | *Owner/Operator* | *Prev Regn* |

Reg	Yr	Type	c/n	Owner/Operator	Prev Regn
□ N1125K	89	Astra-1125	035	State Street Corp. Stratford, Ct.	
□ N1125L	94	Astra-1125SP	072	Terraire SA. Buenos Aires, Argentina.	
□ N1125M	82	Learjet 55	55-065	Bayer Corp fka Miles Laboratories Inc. Elkhart, In.	N555GL
□ N1125S	88	Astra-1125	021	PGA Tour Investments Inc. St Augustine, Fl.	N1125A
□ N1125Z	93	Astra-1125SP	068	GTN Inc/Superior Transportation Inc. Sioux City, Ia.	
□ N1126V	97	Beechjet 400A	RK-151	First Security Bank NA. Salt Lake City, Ut.	
□ N1128B	90	Citation III	650-0184	Florida Power & Light Co. Miami, Fl.	N95CC
□ N1129M	80	Learjet 35A	35A-360	Bayer/Miles Laboratories Inc. Elkhart, In.	N185FP
□ N1135X	97	Hawker 800XP	8345	Raytheon Aircraft Co. Wichita, Ks.	
□ N1136Q	97	Beechjet 400A	RK-136	Raytheon Aircraft Co. Wichita, Ks.	(9M-...)
□ N1140A	76	Learjet 35	35-045	AirNet Systems Inc. Columbus, Oh.	(N40AN)
□ N1200N	92	Citation II	550-0681	U S Customs Service, Oklahoma City, Ok.	N1200N
□ N1216N	97	CitationJet	525-0226	Cessna Aircraft Co. Wichita, Ks.	
□ N1217H	96	Citation V Ultra	560-0376	Horphag Research Ltd. Guernsey, Channel Islands.	(VR-BCY)
□ N1217V	87	Citation V	560-0001	Phillips Slick LLC & Earl Slick, Greensboro, NC.	N560CV
□ N1218S	82	Citation II/SP	551-0428	Valley Projects Inc. Robertson, Wy.	(N147RP)
□ N1223N	86	Citation III	650-0120	Ness, Motley & Partners, Barnwell, SC.	N650AF
□ N1249P	82	Citation II	550-0451	Flight Services Group/Greyhound Financial Corp. Paramus, NJ.	
□ N1252D	83	Citation II	550-0476	TVL Corp. Stateline, Nv.	
□ N1254X	84	Citation II	550-0494	U S Customs Service, Oklahoma City, Ok.	XC-JBR
□ N1255K	84	Citation II	550-0505	U S Customs Service, Oklahoma City, Ok.	XC-JAY
□ N1257B	84	Citation II	550-0497	U S Customs Service, Oklahoma City, Ok.	(XC-JBQ)
□ N1286C	93	Citation V	560-0222	Continental Ltd. Belleville, Mi.	
□ N1302A	93	Citation VI	650-0226	Standard Register Co. Dayton, Oh.	
□ N1326J	97	Beechjet 400A	RK-159	Raytheon Aircraft Co. Wichita, Ks.	N2159P
□ N1327G	93	CitationJet	525-0008	AirBob Corp. Freeport, Il.	
□ N1327J	93	CitationJet	525-0009	Air America Charter LLC. Englewood, Co.	
□ N1329D	93	CitationJet	525-0020	Glock Inc. Smyrna, Ga.	(OO-LFU)
□ N1329G	96	CitationJet	525-0146	Glock Inc. Smyrna, Ga.	
□ N1419J	86	Citation III	650-0115	AIRCOL Inc. NYC.	PT-LJC
□ N1454H	82	Gulfstream 3	350	Amerada Hess Corp. West Trenton, NJ.	N317GA
□ N1500	83	Challenger 600S	1078	JABJ/Wegmans Food Markets Inc. Rochester, NY.	N53SR
□ N1501	65	Falcon 20C	15	Mobile Telecommunications Technology, Jackson, Ms.	N1501
□ N1526R	83	Gulfstream 3	409	Stevens Express Leasing Inc. Memphis, Tn.	N1526M
□ N1540	88	Gulfstream 4	1044	Cox Aviation Inc. Honolulu, Hi.	N1040
□ N1547B	88	Beechjet 400	RJ-47	Euroflight Inc. Wichita, Ks.	
□ N1549J	88	Beechjet 400	RJ-49	Clarkson Construction Co. Kansas City, Mo.	
□ N1549W	94	Beechjet 400A	RK-88	Mueller Industries Inc. Memphis, Tn.	
□ N1565B	89	Beechjet 400	RJ-65	Ogden & Sons PLC. Leeds-Bradford, UK.	
□ N1621	79	Westwind-1124	275	Texaco Inc. Westchester County Airport, NY.	N1141G
□ N1622	90	Challenger 601-3A	5077	Texaco Inc. Westchester County Airport, NY.	C-G...
□ N1624	97	Gulfstream 4SP	1318	Texaco Inc. White Plains, NY.	N418GA
□ N1629	81	Westwind-1124	363	Texaco Inc. Englewood, Co.	N3320G
□ N1777T	66	Jet Commander	62	Westar Aviation Inc. Miami, Fl. (status ?).	C-GKFS
□ N1812C	87	Challenger 601-3A	5010	Citiflight Inc. Newark, NJ.	(N57HK)
□ N1818S	93	Falcon 900B	136	Stephens Group Inc. Little Rock, Ar.	N137FJ
□ N1823D	69	Gulfstream 2SP	59	Air Group Inc. Van Nuys, Ca.	N879GA
□ N1823S	93	Citation V	560-0225	Stephens Group Inc. Little Rock, Ar.	(N1865C)
□ N1824S	91	Citation V	560-0120	Stephens Group Inc. Little Rock, Ar.	N120CV
□ N1824T	84	Challenger 601	3029	JABJ/Chemical Banking Corp. Teterboro, NJ.	N5491V
□ N1828S	94	Citation VII	650-7047	Stephens Group Inc. Little Rock, Ar.	N647CM
□ N1847B	84	Falcon 200	493	Asia Air Charter P/L. Singapore.	N901SB
□ N1848U	91	Falcon 50	227	Colonial Companies Inc. Columbia, SC.	(N227FJ)
□ N1865M	85	Citation S/II	S550-0071	Milliken & Co. Greenville, SC.	N571CC
□ N1867M	96	Citation VII	650-7073	Mericos Aviation Ltd. South Pasadena, Ca.	
□ N1868M	96	Falcon 900B	157	Metropolitan Life Insurance Co. NYC.	N157FJ
□ N1871R	83	Gulfstream 3	381	Ingersoll-Rand Services Co. Woodcliff Lake, NJ.	N747G
□ N1873	96	Citation V Ultra	560-0353	George Koch Sons Inc. Evansville, In.	N353CV
□ N1879W	91	Citation II	550-0668	CUNA Mutual Life Insurance Co. Waverly, Ia.	N668CM
□ N1881Q	79	Falcon 20F	414	Taughannock Aviation Corp. Ithaca, NY.	N412F
□ N1883M	91	Citation II	550-0674	Meijer Inc. Grand Rapids, Mi.	N550FB
□ N1884	81	Challenger 600S	1032	Gulf States Paper Corp. Tuscaloosa, Al.	(N31DC)
□ N1886G	93	Citation II	550-0722	Meijer Inc. Grand Rapids, Mi.	XA-SMT
□ N1896F	82	Falcon 50	127	New York Times Co. NYC.	N1896T
□ N1897A	97	Hawker 800XP	8326	Raytheon Aircraft Co. Wichita, Ks.	N326XP
□ N1897S	77	Falcon 20-5B	376	J M Smucker Co. Orrville, Oh.	N1892S
□ N1900W	90	Gulfstream 4	1124	Whirlpool Corp. Benton Harbor, Mi.	N420GA
□ N1901M	88	Gulfstream 4	1039	Monsanto Co. St Louis, Mo.	N431GA
□ N1902P	93	Challenger 601-3R	5135	J C Penney Co. Dallas, Tx.	N1902J
□ N1903G	96	Challenger 604	5326	Tri C Inc/Gaylord Broadcasting, Oklahoma City, Ok.	N908G
□ N1904P	92	Challenger 601-3A	5116	J C Penney Co. Dallas, Tx.	N841PC
□ N1904S	97	Learjet 31A	31A-149	Learjet Inc. Wichita, Ks.	
□ N1904W	94	Gulfstream 4SP	1237	Whirlpool Corp. Benton Harbor, Mi.	N480GA
□ N1910A	90	BAe 125/800A	NA0454	Hallmark Cards Inc.Kansas City, Mo.	N616BA
□ N1910H	96	Hawker 800XP	8318	Hallmark Cards Inc. Kansas City, Mo.	N318XP
□ N1910J	84	BAe 125/800A	8023	RJW Brokerage Inc. Dallas, Tx.	N1910H
Reg	*Yr* *Type*		*c/n*	*Owner/Operator*	*Prev Regn*

Reg	Yr	Type	c/n	Owner/Operator	Prev Regn
☐ N1929P		Sabreliner CT-39A	276-48	Pittsburgh Institute of Aeronautics, Pittsburgh, Pa.	N6612S
☐ N1929Y	68	Gulfstream 2	19	Rokeby Farms, Washington, DC.	N839GA
☐ N1932P	94	Learjet 31A	31A-099	Marcair Inc. Roanoke, Va.	N5049J
☐ N1940	92	Learjet 60	60-002	Meadowlanders Inc. Rutherford, NJ.	N190AS
☐ N1944P	69	Jet Commander-C	142	Pittsburgh Institute of Aeronautics, Pittsburgh, Pa.	N51038
☐ N1955M	95	Gulfstream 4SP	1276	McDonalds Corp. West Chicago, Il.	N460GA
☐ N1956M	85	Gulfstream 3	469	Hill Air Corp. Dallas, Tx.	JY-HZH
☐ N1958N	85	Citation S/II	S550-0073	Milliken & Co. Greenville, SC.	(N1273E)
☐ N1962J	68	JetStar-731	5113	Eagle Mountain International Church, Fort Worth, Tx.	(N1967J)
☐ N1965L	65	Learjet 24A	24A-012	Clay Lacy Aviation Inc. Van Nuys, Ca.	N1969L
☐ N1968W	66	Learjet 23	23-089	GAR Inc. Cochranville, Pa.	N969B
☐ N1971R	84	Falcon 50	149	Motorola Inc. Farnborough, UK.	N149MD
☐ N1983Y	83	Learjet 55	55-079	Aviation Two LLC/ARAMARK Services Inc. Philadelphia, Pa.	(N2855)
☐ N2000	79	Sabre-65	465-7	Oxy USA Inc/Swiflite Aircraft Corp. Bakersfield, Ca.	N2800
☐ N2000A	97	Falcon 2000	44	Colorstrip Inc. Richmond, Ca.	N2074
☐ N2000M	91	Citation V	560-0146	Siebe PLC. Farnborough, UK.	(N6877Q)
☐ N2000X	91	Citation V	560-0144	Ashland Oil Inc. Worthington, Ky.	N.....
☐ N2002P	95	Gulfstream 4SP	1279	J C Penney Co. Dallas, Tx.	N466GA
☐ N2006	74	Sabre-40A	282-135	Flight Specialists Inc. North Canton, Oh.	(N67BK)
☐ N2015M	94	BAe 125/800A	8254	Monsanto Co. St Louis, Mo.	N943H
☐ N2056E	97	Beechjet 400A	RK-156	Raytheon Aircraft Co. Wichita, Ks.	
☐ N2094L	72	Learjet 25B	25B-095	Royal Air Freight Inc. Pontiac, Mi.	C-GRCO
☐ N2107Z	93	Gulfstream 4SP	1211	Richland Development Corp/Pennzoil Co. Houston, Tx.	N447GA
☐ N2114E	73	HS 125/600A	6022	Armoni Aviation Inc. Houston, Tx. (status ?).	XA-XET
☐ N2143H	59	B 707-123B	17644	Ess Jay Air Inc. Beaumont, Tx.	HZ-DAT
☐ N2158U	79	Citation 1/SP	501-0091	North American Jet Inc. Georgetown, Tx.	(N510NA)
☐ N2164Z	97	Beechjet 400A	RK-164	Raytheon Aircraft Co. Wichita, Ks.	
☐ N2200A	75	Sabre-75A	380-26	DoJ-U S Marshals Service, Oklahoma City, Ok.	N128MS
☐ N2213T	84	Learjet 25D	25D-369	DoJ-U S Immigration & Naturalization, Pineville, La.	N369MJ
☐ N2243	81	Citation 1/SP	501-0212	Smiley Investment Co. Little Rock, Ar.	N70AA
☐ N2267B	97	Beechjet 400A	RK-167	Raytheon Aircraft Co. Wichita, Ks.	
☐ N2273Z	97	Beechjet 400A	RK-173	Raytheon Aircraft Co. Wichita, Ks.	
☐ N2286L	97	Hawker 800XP	8336	Raytheon Aircraft Co. Wichita, Ks.	
☐ N2289B	97	Beechjet 400A	RK-170	Raytheon Aircraft Co. Wichita, Ks.	
☐ N2296S	81	Citation 1/SP	501-0214	Metal Transportation Systems Inc. Meredith, NH.	N3312T
☐ N2299T	97	Beechjet 400A	RK-166	Raytheon Aircraft Co. Wichita, Ks.	
☐ N2320J	97	Hawker 800XP	8342	Raytheon Aircraft Co. Wichita, Ks.	
☐ N2321S	98	Hawker 800XP	8352	Raytheon Aircraft Co. Wichita, Ks.	
☐ N2321V	98	Hawker 800XP	8353	Raytheon Aircraft Co. Wichita, Ks.	
☐ N2321Z	98	Hawker 800XP	8357	Raytheon Aircraft Co. Wichita, Ks.	
☐ N2329N	97	Beechjet 400A	RK-169	Raytheon Aircraft Co. Wichita, Ks.	
☐ N2351K	89	Citation II	550-0594	U S Customs Service, Oklahoma City, Ok.	N1302X
☐ N2358X	97	Beechjet 400A	RK-158	First Security Bank NA. Salt Lake City, Ut.	
☐ N2360F	97	Beechjet 400A	RK-160	Raytheon Aircraft Co. Wichita, Ks.	
☐ N2426	94	Hawker 800	8272	Owens Corning Fiberglas Corp. Swanton, Oh.	N299H
☐ N2428	95	Hawker 800	8274	Owens-Corning Fiberglas Corp. Swanton, Oh.	N804H
☐ N2440G	76	Sabre-75A	380-44	Southline Metal Products Co. Brenham, Tx.	N2116J
☐ N2600	89	Gulfstream 4	1088	Mobil Corp. Fairfax, Va.	N4UP
☐ N2604	83	Citation III	650-0021	Silver State Aviation Inc. Las Vegas, Nv.	N2624M
☐ N2605	87	Citation III	650-0144	Mobil Corp. Fairfax, Va.	N650KM
☐ N2606	90	Citation III	650-0194	Mobil Corp. Fairfax, Va.	N111VW
☐ N2610	89	Gulfstream 4	1094	Mobil Corp. Washington, DC.	
☐ N2617U	82	Citation 1/SP	501-0235	CAVU Inc. Saratoga, Ca.	(N31CF)
☐ N2648X	79	Citation 1/SP	501-0105	Casey's Services Co. Ankeny, Ia.	(N231LC)
☐ N2663Y	89	Citation II	550-0602	U S Customs Service, Oklahoma City, Ok.	XC-JAZ
☐ N2690M	80	Citation 1/SP	501-0151	Richards Aviation Inc & Richards SDA LLC. Memphis, Tn.	N269CM
☐ N3007	87	BAe 125/800A	8092	Empire Airlines Inc. Lebanon, Mo.	N2MG
☐ N3015F	97	Learjet 60	60-115	Learjet Inc. Wichita, Ks.	
☐ N3018C	97	Learjet 60	60-118	Learjet Inc. Wichita, Ks.	
☐ N3030C	80	Citation II	550-0180	Denison Poultry & Egg Co/Golden Distributing Co. Clinton, Mt	N77WU
☐ N3038V	94	Beechjet 400A	RK-93	The UND Aerospace Foundation, Grand Forks, ND.	
☐ N3113B	86	Beechjet 400	RJ-13	Mylan Pharmaceuticals Inc. Morgantown, WV.	
☐ N3121B	87	Beechjet 400	RJ-21	Sara Lee Corp. Chicago, Il.	
☐ N3123T	87	Beechjet 400	RJ-23	C C Medflight Inc. Lawrenceville, Ca.	
☐ N3141G	88	Beechjet 400	RJ-41	N916HC LLC. Leesburg, Va.	(N441EE)
☐ N3196N	94	Beechjet 400A	RK-96	The UND Aerospace Foundation, Grand Forks, ND.	
☐ N3197Q	95	Beechjet 400A	RK-97	Skyland Leasing/Logan Sarasota Realty Corp. Seffner, Fl.	
☐ N3218L	95	Beechjet 400A	RK-108	SMC Aviation Inc. Nashville, Tn.	
☐ N3232U	95	Beechjet 400A	RK-102	Transair USA Inc. Wichita, Ks.	
☐ N3235U	95	Beechjet 400A	RK-105	Hanna Steel Corp. Fairfield, Al.	(N8252J)
☐ N3240M	88	Beechjet 400	RJ-40	Leading Edge Charter, St Paul, Mn.	PK-ERA
☐ N3262M	91	Citation II	550-0652	U S Customs Service, Oklahoma City, Ok.	
☐ N3265A	96	Beechjet 400A	RK-115	HSI AirTravel LLC. Boulder, Co.	
☐ N3269A	96	Beechjet 400A	RK-109	SMC Aviation Inc. Nashville, Tn.	
☐ N3272L	96	Beechjet 400A	RK-112	Blue Bird Body Co. Macon, Ga.	
☐ N3278	69	Sabre-60	306-32	Executive Aircraft Corp. Newton, Ks.	N4743N
Reg	*Yr*	*Type*	*c/n*	*Owner/Operator*	*Prev Regn*

Reg	Yr	Type	c/n	Owner/Operator	Prev Regn
☐ N3280G	66	Sabre-40	282-70	United CCM Corp. San Antonio, Tx.	N34LP
☐ N3338	86	Gulfstream 4	1006	Maltaire Inc/Bucephalus Enterprises Inc. Ronkonkoma, NY.	N99GM
☐ N3399P	77	HS 125/700B	7010	Prado Aviation LLC. Atlanta, Ga.	N700ER
☐ N3444H	81	HS 125/700A	NA0287	Houston Industries Inc. Houston, Tx.	N299FB
☐ N3484U	78	Falcon 20F	380	Colonial Companies Inc. Columbia, SC.	N289MM
☐ N3490L	74	Citation	500-0128	Engineered Data Products Inc. Broomfield, Co.	N501AR
☐ N4110S	91	Citation V	560-0112	Samedan Oil Corp. Ardmore, Ok.	(N68027)
☐ N4200K	96	Citation V Ultra	560-0354	Kimball International Trabsit Inc. Huntingburg, In.	
☐ N4350M	84	Falcon 50	142	Mead Corp. Dayton, Oh.	N132FJ
☐ N4351M	82	Falcon 50	90	Mead Corp. Dayton, Oh.	(N650AS)
☐ N4358N	76	Learjet 35	35-065	AirNet Systems Inc. Columbus, Oh.	N425DN
☐ N4401	81	Learjet 35A	35A-434	Schwab Industries Inc. New Philadelphia, Oh.	
☐ N4402	90	BAe 125/800A	NA0460	GMRI fka General Mills Restaurants Inc. Orlando, Fl.	N632BA
☐ N4415D	86	Challenger 601	3051	JBC Aviation Inc/Prairie Aviation, Wichita, Ks.	N60MU
☐ N4415M	76	Learjet 35A	35A-072	L S Management Inc. Wichita, Ks.	N2015M
☐ N4415S	75	Learjet 35A	35A-021	L S Management Inc. Wichita, Ks.	N442JT
☐ N4415W	79	Learjet 35A	35A-229	L S Management Inc. Wichita, Ks.	N415LS
☐ N4444J	89	BAe 125/800A	NA0442	Aim High Enterprises, Bedford, NH.	(N442NW)
☐ N4447P	81	Learjet 25D	25D-338	U S Jets Inc. Athens, Ga.	XA-LOF
☐ N4500X	84	Gulfstream 3	416	Teratorn LLC. Seattle, Wa.	N883A
☐ N4545	98	Learjet 45	45-045	Learjet Inc. Wichita, Ks.	
☐ N4612Z	91	Citation	650-0203	Murphy Oil Corp. El Dorado, Ar.	(N4612S)
☐ N4614N	90	Citation II	550-0659	U S Customs Service, Oklahoma City, Ok.	XC-HGZ
☐ N5000C	83	Citation III	650-0002	BG Flight Services Inc. Naples, Fl.	N652CC
☐ N5060K	97	Citation Excel	...	Cessna Aircraft Co. Wichita, Ks.	
☐ N5073	90	Challenger 601-3A	5073	Mills Pride LP. West Palm Beach, Fl.	PK-HMK
☐ N5076K	97	Citation V Ultra	560-0440	Cessna Aircraft Co. Wichita, Ks.	
☐ N5100X	93	Regional Jet	7008	Xerox Corp. White Plains, NY.	C-G...
☐ N5103	84	Gulfstream 3	440	Go Jet Inc. Tulsa, Ok.	N304GA
☐ N5104	84	Gulfstream 3	443	General Motors Corp. Detroit, Mi.	N315GA
☐ N5105	84	Gulfstream 3	445	General Motors Corp. Detroit, Mi.	(N5103)
☐ N5109	86	Citation III	650-0135	General Mills Inc. Minneapolis, Mn.	(N1324B)
☐ N5112	96	Citation X	750-0010	General Motors Corp. Detroit, Mi.	N5225K
☐ N5113	97	Citation X	750-0013	General Motors Corp. Detroit, Mi.	
☐ N5114	97	Citation X	750-0017	General Motors Corp. Detroit, Mi.	
☐ N5115	96	Citation X	750-0018	General Motors Corp. Detroit, Mi.	N95CC
☐ N5116	97	Citation X	750-0019	General Motors Corp. Detroit, Mi.	N5109W
☐ N5117H	77	Gulfstream 2	197	Amerada Hess Corp. West Trenton, NJ.	N800GA
☐ N5118	92	Citation VII	650-7013	General Motors Corp. Detroit, Mi.	N5113
☐ N5119	92	Citation VII	650-7015	MedPartners Aviation Inc. Birmingham, Al.	(N317MX)
☐ N5175U	73	B 737/T-43A	20689	EG&G/Department of the Air Force, McCarran, Nv.	72-0282
☐ N5176Y	74	B 737/T-43A	20692	EG&G/Department of the Air Force, McCarran, Nv.	72-0285
☐ N5177C	74	B 737/T-43A	20693	EG&G/Department of the Air Force, McCarran, Nv.	72-0286
☐ N5207A	97	Citation V Ultra	560-0438	Cessna Aircraft Co. Wichita, Ks.	
☐ N5274U	66	HS 125/1A-522	25068	Jetlease/Finance Corp. Hollywood, Fl.	XB-SBC
☐ N5294E	74	B 737/T-43A	20691	EG&G/Department of the Air Force, McCarran, Nv.	72-0284
☐ N5294M	74	B 737/T-43A	20694	EG&G/Department of the Air Force, McCarran, Nv.	72-0287
☐ N5408G	91	Citation II	550-0666	U S Customs Service, Oklahoma City, Ok.	XC-JBS
☐ N5511A	80	HS 125/700A	NA0279	Amana Refrigeration Inc. Cedar Rapids, Ia.	N204N
☐ N5731	86	Falcon 900	8	Enron Corp & Enron Oil & Gas, Houston, Tx.	N406FJ
☐ N5732	91	Falcon 50	217	Enron Corp & Enron Oil & Gas, Houston, Tx.	N122FJ
☐ N5733	87	Falcon 900	39	Enron Corp & Enron Oil & Gas, Houston, Tx.	N181BS
☐ N5734	97	Hawker 800XP	8304	Enron Corp & Enron Oil & Gas, Houston, Tx.	N802JT
☐ N5735	97	Hawker 800XP	8309	Enron Corp & Enron Oil & Gas, Houston, Tx.	N803JT
☐ N5878	61	MS 760 Paris-2B	106	sale, Colorado Springs, Co. (11/94).	PH-MSV
☐ N5879	61	MS 760 Paris-2B	107	Larry Phillips, Fruitland Park, Fl.	PH-MSW
☐ N6100	97	Learjet 60	60-100	Learjet Inc. Wichita, Ks.	
☐ N6110	93	Citation VII	650-7023	General Mills Inc. Minneapolis, Mn.	N1262G
☐ N6177Y	67	Learjet 24	24-151	Executive Air Charters Inc. New Orleans, La.	N53GH
☐ N6262T	65	Learjet 23	23-071	Dodsons International Parts Inc. Ottawa, Ks. (status ?).	XA-RZC
☐ N6453	82	Gulfstream 3	349	Nike Inc. Hillsboro, Or.	N1KE
☐ N6513X	80	Gulfstream 3	310	Assembly Pointe Aviation Inc/Richmor Aviation, Hudson, NY.	(N373LP)
☐ N6555L	80	Falcon 20C	85	Threshold Ventures Inc. Wilmington, De.	VH-JSY
☐ N6581E	61	Sabreliner CT-39A	265-82	Spokane Community College, Spokane, Wa.	61-0679
☐ N6763L	91	Citation II	550-0673	U S Customs Service, Oklahoma City, Ok.	XC-HHA
☐ N6775C	91	Citation II	550-0677	U S Customs Service, Oklahoma City, Ok.	
☐ N6776T	92	Citation II	550-0680	U S Customs Service, Oklahoma City, Ok.	
☐ N6846T	90	Citation II	550-0625	Genesee Leroy Stone Corp. Pavillion, NY.	N625EA
☐ N6862Q	81	Citation II	550-0265	Prewitt Leasing Inc. Bedford, Tx.	(N265QS)
☐ N7000E	81	Falcon 20F-5	440	Ashland Oil Inc. Worthington, Ky.	N32TE
☐ N7005	94	Citation VII	650-7044	B F Goodrich Co & Geon Co. North Canton, Oh.	
☐ N7006	78	HS 125/700A	NA0217	K W Plastics Recycling Division, Troy, Al.	N7005
☐ N7008	94	Challenger 601-3R	5164	BLC Corp/B F Goodrich Co. North Canton, Oh.	N715BG
☐ N7050V	83	Diamond 1A	A058SA	Marvin Lumber & Cedar Co. Warroad, Mn.	VR-BKA
☐ N7070A	85	Citation S/II	S550-0068	Omega Air Inc. Washington, DC.	N4049
☐ N7074X	70	Learjet 24B	24B-223	Interlease Aviation Corp. Northfield, Il. (status ?).	D-CFVG

Reg	Yr	Type	c/n	Owner/Operator	Prev Regn
☐ N7092C	78	Learjet 35A	35A-184	Cefaratti International Inc. Coronado, Ca. (status ?).	HB-VFO
☐ N7117	82	Learjet 35A	35A-462	Business Jet Inc. Riverside, Ct.	N801K
☐ N7143N	61	Sabreliner CT-39A	265-70	Ascher Ward, Santa Monica, Ca.	61-0667
☐ N7148J	75	Sabre-75A	380-33	DOJ-U S Marshals Service, Oklahoma City, Ok.	N129MS
☐ N7158Q	69	HFB 320	1040	Kalitta Flying Service Inc. Morristown, Tn.	I-ITAL
☐ N7171	71	HS 125/F400A	25264	Condor Air Inc. Fort Lauderdale, Fl.	N264TS
☐ N7200K	66	Learjet 23	23-099	RBS Aviation Group Inc. Fort Myers, Fl.	
☐ N7228K	84	Falcon 50	146	First Bank System. Minneapolis, Mn.	N747
☐ N7270B	73	B 727-232	20641	ORCA BAY Aviation LLC.	N18786
☐ N7271P	65	B 727-30	18933	Imperial Palace Air Ltd. Las Vegas, Nv.	N727BE
☐ N7281Z	72	Citation	500-0047	Galvao & Arroyo Associates, Opa Locka, Fl.	PT-WAB
☐ N7301	97	Falcon 900EX	19	Samsung Semiconductor Inc. San Jose, Ca.	N919EX
☐ N7381	65	B 720-060B	18977	Hughes Aircraft Co. Los Angeles, Ca.	N440DS
☐ N7490A	82	HS 125/700A	NA0320	Oehmig Aviation LLC. Houston, Tx.	N700HW
☐ N7601	68	Gulfstream 2B	32	Unocal Corp. El Segundo, Ca.	N7602
☐ N7601R	60	MS 760 Paris-1R	60	Airborne Turbine Inc. Ca.	No 60
☐ N7602	94	Gulfstream 4SP	1245	Unocal Corp. El Segundo, Ca.	N101HC
☐ N7638S	69	Jet Commander	134	AAR Aircraft Group Inc. Wood Dale, Il.	4X-COP
☐ N7700L	92	Citation V	560-0203	Hutchens Industries inc/Tandem Air LLC. Springfield, Mo.	ZS-HNC
☐ N7700T	82	Citation 1/SP	501-0248	Distribution Air LLC. Springfield, Mo.	N2633N
☐ N7777B	83	Learjet 35A	35A-508	Bergstrom Pioneer Auto & Truck Leasing Inc. Appleton, Wi.	N741F
☐ N7788	83	Challenger 601	3011	Invemed Aviation Services Inc. NYC.	N700MK
☐ N7789	70	Gulfstream 2	90	Pal-Waukee Aviation Inc. Wheeling, Il.	N883GA
☐ N8000U	80	Falcon 20F-5	436	Ashland Oil Inc. Worthington, Ky.	N436MP
☐ N8005Y	73	Learjet 25B	25B-121	Delaware DI Properties LLC. Dallas, Tx.	XA-SAL
☐ N8040A	76	Learjet 35	35-048	AirNet Systems Inc. Columbus, Oh. (status ?).	F-GHMP
☐ N8073R	91	Beechjet 400A	RK-24	Raytheon Aircraft Co. Wichita, Ks.	
☐ N8080W	96	Learjet 60	60-080	Learjet Inc. Wichita, Ks.	
☐ N8083N	92	Beechjet 400A	RK-62	Horton Transportation Inc. Minneapolis, Mn.	
☐ N8085T	92	Beechjet 400A	RK-51	MARI LLC/Bloomington Aircraft Rentals Inc. Eden Prairie, In.	
☐ N8100E	96	Falcon 900EX	4	Emerson Electric Co. St Louis, Mo.	N204FJ
☐ N8167Y	93	Beechjet 400A	RK-67	Raytheon Aircraft Co. Wichita, Ks.	
☐ N8186	90	BAe 125/800B	8186	IAMS Co. Dayton, Oh.	G-5-683
☐ N8200E	87	Falcon 900	34	Emerson Electric Co. St Louis, Mo.	N8100E
☐ N8239E	92	Beechjet 400A	RK-46	Bunn-O-Matic Corp. Springfield, Il.	
☐ N8253Y	92	Beechjet 400A	RK-42	Sloan Financial Group Inc. Durham, NC.	
☐ N8270	95	Learjet 60	60-063	Boeing Aircraft Holding Co. Seattle, Wa.	N5003U
☐ N8271	95	Learjet 60	60-066	Boeing Aircraft Holding Co. Seattle, Wa.	N5006K
☐ N8281	79	Learjet 35A	35A-232	McDonnell-Douglas Corp. St. Louis, Mo.	
☐ N8288R	95	CitationJet	525-0090	Sixt Autovermietung GmbH. Taufkirchen, Germany.	
☐ N8300E	80	Falcon 50	33	Emerson Electric Co. St Louis, Mo.	N8100E
☐ N8400E	84	Falcon 50	150	Emerson Electric Co. St Louis, Mo.	N8200E
☐ N8484P	86	Astra-1125	014	Crabbe Huson Group Inc. Portland, Or.	N400JF
☐ N8534	68	Jet Commander	113	Anton Ptach, Poughkeepsie, NY.	4X-CPB
☐ N8550A	97	Falcon 50EX	263	Aluminum Co of America, Pittsburgh, Pa.	F-WWHN
☐ N8860	66	DC 9-15	45797	Richard Mellon Scaife, Pittsburgh, Pa.	(EC-BAX)
☐ N8940	97	CitationJet	525-0253	Cessna Aircraft Co. Wichita, Ks.	
☐ N9000F	93	Falcon 50	242	Ashland Inc. Worthington, Ky.	N599SC
☐ N9003	97	CitationJet	525-0257	Cessna Aircraft Co. Wichita, Ks.	
☐ N9035Y	61	MS 760 Paris-1A	86	North Pacific Aircraft Development Co. Helena, Mt.	F-BJLX
☐ N9108Z	75	Learjet 36	36-005	Stephen Meyer, Lansdale, Pa.	LV-LOG
☐ N9166Y		Sabreliner CT-39A	265-80	North Dakota University Aerospace Dept.	61-0677
☐ N9292X	97	Hawker 800XP	8315	Coast Services Inc. Santa Fe, NM.	N2169X
☐ N9300	87	Gulfstream 4	1020	Crown Cork & Seal Co. Philadelphia, Pa.	N600CS
☐ N9300C	82	Falcon 50	106	Crown Cork & Seal Co. Philadelphia, Pa.	N50BF
☐ N9550A	97	Falcon 50EX	265	Dassault Falcon Jet Corp. Teterboro, NJ.	F-WWHW
☐ N9700X	95	Challenger 601-3R	5186	Xerox Corp. White Plains, NY.	N612CC
☐ N10123	71	Gulfstream 2	107	HBD Industries Inc. Pittsburgh, Pa.	N5113H
☐ N10855	90	BAe 1000B	8159	Raytheon Aircraft Co. Wichita, Ks.	G-OPFC
☐ N10857	91	BAe 125/800B	8213	Raytheon Aircraft Co. Little Rock, Ar.	G-BURV
☐ N12549	84	Citation II	550-0501	U S Customs Service, Oklahoma City, Ok.	
☐ N12566	85	Citation T-47A	552-0012	Cessna Aircraft Co. Wichita, Ks.	(162766)
☐ N12568	85	Citation T-47A	552-0014	Cessna Aircraft Co. Wichita, Ks.	(162768)
☐ N12659	74	Sabre-75A	380-16	DoJ-U S Marshals Service, Oklahoma City, Ok.	N126MS
☐ N16251	87	Gulfstream 4	1013	Texaco Inc. White Plains, NY.	N1625
☐ N21092		Sabreliner CT-39A	265-42	Blackhawk Technical College, Janesville, Wi.	61-0639
☐ N23204	97	Hawker 800XP	8346	Raytheon Aircraft Co. Wichita, Ks.	
☐ N23207	98	Hawker 800XP	8350	Raytheon Aircraft Co. Wichita, Ks.	
☐ N23208	98	Hawker 800XP	8351	Raytheon Aircraft Co. Wichita, Ks.	
☐ N26494	89	Citation II	550-0605	U S Customs Service, Oklahoma City, Ok.	(N1304B)
☐ N26621	89	Citation II	550-0593	U S Customs Service, Oklahoma City, Ok.	(XC-JBT)
☐ N31403	62	Sabreliner CT-39A	276-4	Maple Woods Community College, Mo.	USAF 62-4451
☐ N31437	88	Beechjet 400	RJ-37	York Aviation Inc. York, Pa.	LV-RCT
☐ N32010	61	Sabreliner CT-39A	265-83	Central Missouri State University, Warrensburg, Mo..	USAF 61-0680

Reg	Yr	Type	c/n	Owner/Operator	Prev Regn

Reg	Yr	Type	c/n	Owner/Operator	Prev Regn
☐ N32508	61	Sabreliner T-39D	285-17	National Museum of Naval Aviation, Pensacola, Fl.	150985
☐ N37201	91	Citation II	550-0655	U S Customs Service, Oklahoma City, Ok.	
☐ N37971	82	Learjet 25D	25D-358	Caulkins Investment Co. Denver, Co.	
☐ N40593	65	Jet Commander	41	Sabin Air Charters Inc. Maputo, Mozambique.	ZP-...
☐ N41953	71	HS 125/F400A	25268	Arnoni Aviation Inc. Houston, Tx. (status ?).	(N268TS)
☐ N50207	94	Learjet 31A	31A-097	Parker Hannifin Corp/Travel 17325 Inc. Cleveland, Oh.	
☐ N61572	79	Citation 1/SP	501-0118	Aircraft Marketing Inc. Albuquerque, NM.	I-DECI
☐ N63537	80	Falcon 50	20	Cie IBM France, Paris-Le Bourget.	C6-BER
☐ N65618		Sabreliner CT-39A	276-42	Colorado Northwestern Community College, Rangely, Co.	62-4489
☐ N67983	81	Citation II/SP	551-0359	Scribner Equipment Co. Amory, Ms.	(N551SE)
☐ N68881	92	Citation V	560-0173	Mercantile Stores Co (NY), Fairfield, Oh.	
☐ N71460	74	Sabre-75A	380-5	DoJ-U S Marshals Service, Oklahoma City, Ok.	N223LP
☐ N71543	75	Sabre-75A	380-29	DoJ-U S Marshals Service, Oklahoma City, Ok.	N132MS
☐ N72028	74	Sabre-75A	380-14	Choctaw Nation of Oklahoma, Durant, Ok.	N53
☐ N72787	78	Westwind-1124	240	Sharper Image Corp. Oakland, Ca.	N400NE
☐ N77711	90	Citation V	560-0065	Par Avion Inc. Omaha, Ne.	N2721F
☐ N77794	94	CitationJet	525-0073	RSF Management Services Inc. Los Angeles, Ca.	
☐ N77797	97	Citation Bravo	550-0821	Miller's Professional Imaging, Pittsburgh, Ks.	
☐ N81366	67	JetStar-731	5097	B L Yachts Inc. Eaton Park, Il.	N77D
☐ N82400	93	Beechjet 400A	RK-75	Allmon Polis Partnership, Las Vegas, Nv.	
☐ N82679	81	Learjet 55	55-013	Atlanta Jet Inc. Lawrenceville, Ga.	(N155AJ)
☐ N90005	89	Gulfstream 4	1103	Siebe PLC. Farnborough, UK.	N433GA
☐ N92045	69	HFB 320	1041	Kalitta Flying Service Inc. Morristown, Tn.	D-CIRA
☐ N99114	73	Sabre-40A	282-128	Jett Aire Florida One Inc. West Palm Beach, Fl.	XA-LIX
Military					
☐ 01	85	Gulfstream 3	477	VC-20B, United States Coast Guard, Washington, DC.	86-0205
☐ 150542		Sabreliner T-39D	277-1	U. S. Navy. (stored China Lake, Ca.).	
☐ 150543		Sabreliner T-39D	277-2	U. S. Navy/to AMARC /?	
☐ 150544		Sabreliner T-39D	277-3	U. S. Navy/to AMARC 3/85 as 7T-006. Tt 5872.	
☐ 150545		Sabreliner T-39D	277-4	U. S. Navy.	
☐ 150546		Sabreliner T-39D	277-5	U. S. Navy/to AMARC 7/85 as 7T-014. Tt 9124.	
☐ 150547		Sabreliner T-39D	277-6	U. S. Navy/to AMARC 1/86 as 7T-021. Tt 9561.	
☐ 150548		Sabreliner T-39D	277-7	U. S. Navy/to AMARC 3/85 as 7T-008. Tt 8127.	
☐ 150549		Sabreliner T-39D	277-8	U. S. Navy/to AMARC 3/85 as 7T-011. Tt 7886.	
☐ 150550		Sabreliner T-39D	277-9	U. S. Navy.	
☐ 150551		Sabreliner T-39D	277-10	U. S. Marine Corp/to AMARC 10/01 as 7T-002.	
☐ 150969		Sabreliner T-39D	285-1	U. S. Navy/to AMARC /?	
☐ 150971		Sabreliner T-39D	285-3	U. S. Navy.	
☐ 150972		Sabreliner T-39D	285-4	U. S. Navy.	
☐ 150973		Sabreliner T-39D	285-5	U. S. Navy/to AMARC 4/85 as 7T-013. Tt 7155.	
☐ 150974		Sabreliner T-39D	285-6	U. S. Navy/to AMARC 7/85 as 7T-015. Tt 8737.	
☐ 150975		Sabreliner T-39D	285-7	U. S. Navy/to AMARC 3/85 as 7T-007. Tt 9524.	
☐ 150976		Sabreliner T-39D	285-8	U. S. Navy/to AMARC 7/85 as 7T-016. Tt 7943.	
☐ 150977		Sabreliner T-39D	285-9	U. S. Navy.	
☐ 150978		Sabreliner T-39D	285-10	U. S. Navy/to AMARC 3/85 as 7T-009. Tt 7576.	
☐ 150979		Sabreliner T-39D	285-11	U. S. Navy/to AMARC 7/85 as 7T-017. Tt 8631.	
☐ 150980		Sabreliner T-39D	285-12	U. S. Navy/to AMARC 3/85 as 7T-004.	
☐ 150981		Sabreliner T-39D	285-13	U. S. Navy/to AMARC 3/85 as 7T-012. Tt 8184.	
☐ 150982		Sabreliner T-39D	285-14	U. S. Navy/to AMARC 2/86 as 7T-022. Tt 8188.	
☐ 150983		Sabreliner T-39D	285-15	U. S. Navy/to AMARC 7/85 as 7T-023. Tt 9833.	
☐ 150984		Sabreliner T-39D	285-16	U. S. Navy/to AMARC 3/85 as 7T-010. Tt 8552.	
☐ 150986		Sabreliner T-39D	285-18	U. S. Navy/preserved Robins AFB. Ga.	
☐ 150987		Sabreliner T-39D	285-19	U.S. Navy/preserved at NAS Patuxent River, Md.	
☐ 150988		Sabreliner T-39D	285-20	U. S. Navy/to AMARC 3/85 as 7T-005. Tt 9208.	
☐ 150989		Sabreliner T-39D	285-21	U. S. Navy. (stored China Lake, Ca.).	
☐ 150990		Sabreliner T-39D	285-22	U. S. Navy/to AMARC 2/86 as 7T-024. Tt 9804.	
☐ 150991		Sabreliner T-39D	285-23	U. S. Navy/to AMARC 7/84 as 7T-003. Tt 8677.	
☐ 150992		Sabreliner T-39D	285-24	U. S. Navy, Naval Weapons Center, China Lake, Ca.	
☐ 151336		Sabreliner T-39D	285-25	U. 'S. Navy/to AMARC 2/86 as 7T-025. Tt 9822.	
☐ 151337		Sabreliner T-39D	285-26	U. S. Navy.	
☐ 151338		Sabreliner T-39D	285-27	U. S. Navy.	
☐ 151339		Sabreliner T-39D	285-28	U. S. Navy/preserved NAS Pensacola, Fl.	
☐ 151340		Sabreliner T-39D	285-29	U. S. Navy/to AMARC 1/86 as 7T-018. Tt 8563.	
☐ 151341		Sabreliner T-39D	285-30	U. S. Navy/to AMARC 1/86 as 7T-019. Tt 10118.	
☐ 151342		Sabreliner T-39D	285-31	U. S. Navy/to AMARC 1/86 as 7T-020. Tt 8146.	
☐ 151343		Sabreliner T-39D	285-32	U. S. Navy.	
☐ 157353	67	Sabreliner CT-39E	282-84	U. S. Navy/to AMARC /?	N2254B
☐ 157354	67	Sabreliner CT-39E	282-85	U. S. Navy, Code RW, VRC-30, North Island NAS, Ca.	N2255B
☐ 158381	68	Sabreliner CT-39E	282-93	U. S. Navy, VRC-50, Cubi Point, Phillipines. (status ?).	N4701N
☐ 158383	68	Sabreliner CT-39E	282-96	U. S. Navy, Code JK, VRC-40, Norfolk NAS, Va. (status ?).	N4705N
☐ 158843	71	Sabreliner CT-39G	306-52	U. S. Navy, Code JK, VRC-40, Norfolk NAS, Va.	N955R
☐ 158844	71	Sabreliner CT-39G	306-55	U. S. Navy, Department of the Navy, Andrews AFB. Md.	N5419
☐ 159361	73	Sabreliner CT-39G	306-65	U. S. Navy, Code 30, VR-24, NAF Sigonella, Italy.	N8364N
☐ 159362	73	Sabreliner CT-39G	306-66	U. S. Navy, Code 31, VR-24, NAF Sigonella, Italy.	N8365N
☐ 159363	73	Sabreliner CT-39G	306-67	U. S. Navy. (status ?).	
☐ 159364	74	Sabreliner CT-39G	306-69	U. S. Marine Corps, Iwakuni MCAS, Japan. (status ?).	

Reg	*Yr*	*Type*	*c/n*	*Owner/Operator*	*Prev Regn*

Reg	Yr	Type	c/n	Owner/Operator	Prev Regn
☐ 159365	74	Sabreliner CT-39G	306-70	U. S. Marine Corps, El Toro MCAS, Ca.	
☐ 160053	75	Sabreliner CT-39G	306-104	Commander Naval Reserve Forces, New Orleans NAS, La.	N65795
☐ 160054	75	Sabreliner CT-39G	306-105	Station Operations & Engineering Sqn. Cherry Point MCAS, NC.	N65796
☐ 160055	75	Sabreliner CT-39G	306-106	Station Operations & Engineering Sqn. Cherry Point MCAS, NC.	N65797
☐ 160056	75	Sabreliner CT-39G	306-107	Station Operations & Engineering Sqn. Cherry Point MCAS, NC.	N65798
☐ 163691	86	Gulfstream 3	480	C-20D, USN, CFLSW, Andrews AFB. Md.	N302GA
☐ 163692	86	Gulfstream 3	481	C-20D, USMC/USN, CFLSW, Andrews AFB. Md.	N304GA
☐ 165093	92	Gulfstream 4	1187	C-20G, USN, Andrews AFB. Md. 'City of Anapolis'	N481GA
☐ 165094	92	Gulfstream 4	1189	C-20G, USN, Andrews AFB. Md. 'City of Baltimore'	N402GA
☐ 165151	92	Gulfstream 4	1199	C-20G, U S Navy, Kaneohe. Hi.	N428GA
☐ 165152	92	Gulfstream 4SP	1201	C-20G, U S Navy, Kaneohe. Hi.	N431GA
☐ 165153	92	Gulfstream 4SP	1200	C-20G, U S Marine Corp. Washington, DC.	N430GA
☐ 165509	64	Sabre T-39N	282-9	U S Navy, Pensacola NAS, Fl.	N301NT
☐ 165510	66	Sabre T-39N	282-81	U S Navy, Pensacola NAS, Fl.	N302NT
☐ 165511	65	Sabre T-39N	282-29	U S Navy, Pensacola NAS, Fl.	N303NT
☐ 165512	63	Sabre T-39N	282-2	U S Navy, Pensacola NAS, Fl.	N304NT
☐ 165513	66	Sabre T-39N	282-66	U S Navy, Pensacola NAS, Fl.	N305NT
☐ 165514	65	Sabre T-39N	282-30	U S Navy, Pensacola NAS, Fl.	N306NT
☐ 165515	66	Sabre T-39N	282-72	U S Navy, Pensacola NAS, Fl.	N307NT
☐ 165516	67	Sabre T-39N	282-90	U S Navy, Pensacola NAS, Fl.	N308NT
☐ 165517	66	Sabre T-39N	282-61	U S Navy, Pensacola NAS, Fl.	N309NT
☐ 165518	66	Sabre T-39N	282-77	U S Navy, Pensacola NAS, Fl.	N310NT
☐ 165519	64	Sabre T-39N	282-19	U S Navy, Pensacola NAS, Fl.	N311NT
☐ 165520	65	Sabre-40	282-32	U S Navy, Pensacola NAS, Fl.	N312NT
☐ 165521	68	Sabre T-39N	282-94	U S Navy, Pensacola NAS, Fl.	N313NT
☐ 165522	65	Sabre T-39N	282-28	U S Navy, Pensacola NAS, Fl.	N314NT
☐ 165523	64	Sabre T-39N	282-20	U S Navy, Pensacola NAS, Fl.	N315NT
☐ 165524	66	Sabre T-39N	282-60	U S Navy, Pensacola NAS, Fl.	N316NT
☐ 165525	72	Sabre T-39N	282-100	U S Navy, Pensacola NAS, Fl.	N317NT
☐ 2101	77	Guardian HU-25B	374	USCG, Mobile, Al. (noted 2/94).	N1045F
☐ 2102	78	Guardian HU-25A	386	USCG, AMARC Davis-Monthan AFB. Park Code 410010.	N149F
☐ 2103	78	Guardian HU-25A	394	USCG, AR & SC Elizabeth City, NC. (noted 2/94).	N178F
☐ 2104	78	Guardian HU-25C	390	USCG, Miami, Fl. (noted 2/94).	N173F
☐ 2105	78	Guardian HU-25B	398	USCG, AMARC Davis-Monthan AFB. Park Code 410011.	N183F
☐ 2106	78	Guardian HU-25A	402	USCG, AMARC Davis-Monthan AFB. Park Code 410007.	N187F
☐ 2107	79	Guardian HU-25A	409	USCG, AR & SC Elizabeth City, NC. (noted 2/94).	N407F
☐ 2108	79	Guardian HU-25A	405	USCG, San Diego, Ca. (noted 2/94).	N405F
☐ 2109	79	Guardian HU-25A	407	USCG, AMARC Davis-Monthan AFB. Park Code 410012.	N406F
☐ 2110	79	Guardian HU-25A	411	USCG, noted stored Elizabeth City, NC. 9/95.	N408F
☐ 2111	79	Guardian HU-25B	413	USCG, Cape Cod, Ma. (noted 2/94).	N410F
☐ 2112	79	Guardian HU-25C	415	USCG, Carswell AFB. Tx. (noted 2/94).	N413F
☐ 2113	79	Guardian HU-25A	417	USCG, AMARC Davis-Monthan AFB. Park Code 410006.	N416FJ
☐ 2114	79	Guardian HU-25B	418	USCG, AMARC Davis-Monthan AFB. Park Code 410014.	N417F
☐ 2115	79	Guardian HU-25A	419	USCG, AR & SC Elizabeth City, NC. (noted 2/94).	N419F
☐ 2116	80	Guardian HU-25A	420	USCG, AMARC Davis-Monthan AFB. Park Code 410005.	N420F
☐ 2117	80	Guardian HU-25A	421	USCG, Astoria, Or. (noted 2/94).	N422F
☐ 2118	80	Guardian HU-25B	423	USCG, noted stored Elizabth City, NC. 9/95.	N423F
☐ 2119	80	Guardian HU-25A	424	USCG, AMARC Davis-Monthan AFB. Park Code 41002.	N424F
☐ 2120	80	Guardian HU-25A	425	USCG, Mobile, Al. (noted 2/94).	N425F
☐ 2121	80	Guardian HU-25A	431	USCG, San Diego, Ca. (noted 2/94).	N429F
☐ 2122	80	Guardian HU-25B	433	USCG, Mobile, Al. (noted 2/94).	N432F
☐ 2123	80	Guardian HU-25A	435	USCG, AMARC Davis-Monthan AFB. Park Code 41003.	N433F
☐ 2124	80	Guardian HU-25A	437	USCG, AR & SC Elizabeth City, NC. (noted 2/94).	N435F
☐ 2125	81	Guardian HU-25B	439	USCG, Cape Cod, Ma. (noted 2/94).	N443F
☐ 2126	81	Guardian HU-25A	441	USCG, Mobile, Al. (noted 2/94).	N445F
☐ 2127	81	Guardian HU-25A	443	USCG, AMARC Davis-Monthan AFB. Park Code 41001.	N447F
☐ 2128	81	Guardian HU-25A	445	USCG, AMARC Davis-Monthan AFB. Park Code 410013.	N449F
☐ 2129	81	Guardian HU-25A	447	USCG, Miami, Fl. (noted 2/94).	N455F
☐ 2130	81	Guardian HU-25A	450	USCG, AMARC Davis-Monthan AFB. Park Code 410004.	N458F
☐ 2131	81	Guardian HU-25A	452	USCG, Miami, Fl. (noted 2/94).	N459F
☐ 2132	81	Guardian HU-25A	454	USCG, ATTC Elizabeth City, NC. (noted 2/94).	N461F
☐ 2133	81	Guardian HU-25A	456	USCG, AR & SC Elizabeth City, NC. (noted 2/94).	N462F
☐ 2134	81	Guardian HU-25B	458	USCG, noted stored Elizabeth City, NC. 9/95.	N465F
☐ 2135	81	Guardian HU-25C	459	USCG, Miami, Fl. (noted 2/94).	N466F
☐ 2136	81	Guardian HU-25B	460	USCG, Miami, Fl. (noted 2/94).	N467F
☐ 2137	81	Guardian HU-25A	462	USCG, AMARC Davis-Monthan AFB. Park Code 410008.	N470F
☐ 2138	82	Guardian HU-25A	464	USCG, AMARC Davis-Monthan AFB. Park Code 410009.	N472F
☐ 2139	82	Guardian HU-25A	466	USCG, Miami, Fl. (noted 2/94).	N473F
☐ 2140	82	Guardian HU-25C	467	USCG, Miami, Fl. (noted 2/94).	N474F
☐ 2141	77	Guardian HU-25C	371	USCG, noted stored Elizabeth City, NC. 9/95.	N1039F
☐ 58-6971	59	C-137B	17926	USAF, SAM, 1st MAS/89th MAW, Andrews AFB. Md.	
☐ 58-6972	59	C-137B	17927	USAF, SAM, 1st MAS/89th MAW, Andrews AFB. Md.	
☐ 59-2868		Sabreliner CT-39A	265-1	USAF, (for Air Force Museum)?	(N2259V)
☐ 59-2869		Sabreliner CT-39A	265-2	USAF/to AMARC 10/84 as TG033. Tt 10744.	(N4999G)
☐ 59-2870		Sabreliner NT-39A	265-3	USAF, Edwards AFB. Ca.	
☐ 59-2872		Sabreliner CT-39A	265-5	USAF/to AMARC 6/84 as TG015. Tt 12728.	(N2296C)
Reg	Yr	Type	c/n	Owner/Operator	Prev Regn

Reg	Yr	Type	c/n	Owner/Operator	Prev Regn
☐ 59-2873		Sabreliner CT-39B	270-1	USAF/4950th Test Wing-ASD, Wright-Patterson AFB. Oh.	
☐ 59-2874		Sabreliner NT-39B	270-2	USAF, AMARC /96 as TG0103.	
☐ 59-5962	62	JetStar-6	5032	C-140A, USAF, 375th Aeromedical Airlift Wing, (status ?).	
☐ 60-0376		C-135E	18151	USAF, 55th SRW, Offutt AFB. Ne.	
☐ 60-0377		C-135A	18152	USAF, 4950 TW/ASD, Wright-Patterson AFB. Oh.	
☐ 60-0378		C-135E	18153	USAF, 55th SRW, Offutt AFB. Ne.	
☐ 60-3474		Sabreliner CT-39B	270-3	USAF/4950th Test Wing-ASD, Wright-Patterson AFB. Oh.	
☐ 60-3475		Sabreliner CT-39B	270-4	USAF to AMARC Davis Monthan 4/94 as TG098.	
☐ 60-3476		Sabreliner NT-39B	270-5	USAF, AMARC /96 as TG0102.	
☐ 60-3477		Sabreliner CT-39B	270-6	USAF, AMARC /96 as TG0101.	
☐ 60-3478		Sabreliner CT-39B	265-6	USAF, Edwards AFB. Ca.	
☐ 60-3479		Sabreliner CT-39A	265-7	USAF/to AMARC 8/85 as TG082. Tt 22494.	
☐ 60-3480		Sabreliner CT-39A	265-8	USAF/to AMARC 6/84 as TG013.	
☐ 60-3481		Sabreliner CT-39A	265-9	USAF/to AMARC 9/85 as TG085. Tt 9870.	
☐ 60-3483		Sabreliner CT-39A	265-11	USAF/preserved at Travis AFB. Ca.	
☐ 60-3485		Sabreliner CT-39A	265-13	USAF/to AMARC 11/83 as TG003. Tt 17269.	
☐ 60-3487		Sabreliner CT-39A	265-15	USAF/to AMARC 8/84 as TG021. Tt 20175.	(N510TD)
☐ 60-3489		Sabreliner CT-39A	265-17	USAF/to AMARC 3/85 as TG058. Tt 18158.	
☐ 60-3490		Sabreliner CT-39A	265-18	USAF/to AMARC 3/85 as TG062. Tt 19182.	
☐ 60-3491		Sabreliner CT-39A	265-19	USAF/to AMARC 5/84 as TG009. Tt 20372.	
☐ 60-3492		Sabreliner CT-39A	265-20	USAF/to AMARC 2/84 as TG007. Tt 17313.	
☐ 60-3493		Sabreliner CT-39A	265-21	USAF/to AMARC 2/85 as TG057. Tt 19313.	
☐ 60-3494		Sabreliner CT-39A	265-22	USAF/to AMARC 12/85 as TG094. Tt 18375.	
☐ 60-3495		Sabreliner CT-39A	265-23	USAF/preserved at Scott AFB. Il.	
☐ 60-3496		Sabreliner CT-39A	265-24	USAF/to AMARC 5/85 as TG072. Tt 19957.	
☐ 60-3497		Sabreliner CT-39A	265-25	USAF/to AMARC 3/85 as TG066. Tt 20311.	
☐ 60-3498		Sabreliner CT-39A	265-26	Williams Gateway, Chandler Municipal, Az. (noted 10/93).	
☐ 60-3499		Sabreliner CT-39A	265-27	USAF/to AMARC 10/84 as TG037. Tt 20506.	
☐ 60-3500		Sabreliner CT-39A	265-28	USAF/to AMARC 10/84 as TG030. Tt 19305.	
☐ 60-3501		Sabreliner CT-39A	265-29	USAF/to AMARC 12/85 as TG093. Tt 15817.	
☐ 60-3502		Sabreliner CT-39A	265-30	USAF/to AMARC 12/85 as TG095. Tt 21770.	
☐ 60-3505		Sabreliner CT-39A	265-33	USAF/preserved at Edwards Flight Test Museum, Ca.	
☐ 60-3507		Sabreliner CT-39A	265-35	USAF/to AMARC 3/85 as TG-061. Tt 19733.	
☐ 60-3508		Sabreliner CT-39A	265-36	USAF/to AMARC 12/84 as TG042. Tt 22378.	
☐ 61-0634		Sabreliner CT-39A	265-37	USAF/preserved at Dyess AFB. Tx.	
☐ 61-0635		Sabreliner CT-39A	265-38	Lafayette Regional Airport, La. (noted 10/92).	
☐ 61-0636		Sabreliner CT-39A	265-39	USAF/to AMARC 9/85 as TG-089. Tt 19841.	
☐ 61-0637		Sabreliner CT-39A	265-40	USAF/to AMARC 10/84 as TG035. Tt 21729.	
☐ 61-0638		Sabreliner CT-39A	265-41	USAF/to AMARC 9/85 as TG086.	
☐ 61-0641		Sabreliner CT-39A	265-44	USAF/to AMARC 10/84 as TG036. Tt 21184.	
☐ 61-0642		Sabreliner CT-39A	265-45	USAF/to AMARC 12/84 as TG045. Tt 17270.	
☐ 61-0643		Sabreliner CT-39A	265-46	USAF/to AMARC 9/84 as TG022. Tt 20288.	
☐ 61-0646		Sabreliner T-39A	265-49	USAF. (reported w/o 1978 ?).	
☐ 61-0647		Sabreliner CT-39A	265-50	USAF/to AMARC 6/85 as TG078. Tt 19781.	
☐ 61-0648		Sabreliner CT-39A	265-51	USAF/to AMARC 8/84 as TG017. Tt 20923.	
☐ 61-0649		Sabreliner T-39A	265-52	USAF/to AMARC 12/84 as TG047. Tt 5390.	(N1064)
☐ 61-0650		Sabreliner CT-39A	265-53	USAF/to AMARC 12/84 as TG043. Tt 22915.	
☐ 61-0651		Sabreliner CT-39A	265-54	USAF/to AMARC 12/84 as TG040. Tt 17587.	
☐ 61-0652		Sabreliner CT-39A	265-55	USAF/to AMARC 9/85 as TG087. Tt 16815.	(N4999H)
☐ 61-0653		Sabreliner CT-39A	265-56	USAF/to AMARC 5/85 as TG071. Tt 17913.	
☐ 61-0654		Sabreliner CT-39A	265-57	Embry-Riddle Aeronautical University, Daytona Beach, Fl.	
☐ 61-0655		Sabreliner CT-39A	265-58	USAF/to AMARC 12/83 as TG005. Tt 19975.	
☐ 61-0656		Sabreliner CT-39A	265-59	USAF/to AMARC 5/84 as TG010. Tt 21617.	
☐ 61-0657		Sabreliner CT-39A	265-60	USAF/to AMARC 9/84 as TG023.	
☐ 61-0658		Sabreliner CT-39A	265-61	USAF/to AMARC 10/84 as TG034. Tt 18138.	
☐ 61-0659		Sabreliner CT-39A	265-62	USAF/to AMARC 9/84 as TG026. Tt 19525.	
☐ 61-0660		Sabreliner CT-39A	265-63	USAF/preserved at McClellan AFB. Ca.	
☐ 61-0661		Sabreliner T-39A	265-64	USAF. (reported w/o 9 Feb 69 ?).	
☐ 61-0663		Sabreliner CT-39A	265-66	USAF/to AMARC 4/85 as TG067. Tt 21139.	
☐ 61-0664		Sabreliner CT-39A	265-67	USAF/to AMARC 3/85 as TG063. Tt 20916.	
☐ 61-0665		Sabreliner CT-39A	265-68	USAF/to AMARC 10/84 as TG028. Tt 19423.	
☐ 61-0666		Sabreliner CT-39A	265-69	USAF/to AMARC 6/84 as TG014. Tt 20336.	
☐ 61-0668		Sabreliner CT-39A	265-71	USAF/to AMARC 1/85 as TG051. Tt 21795.	
☐ 61-0669		Sabreliner CT-39A	265-72	USAF/to AMARC 5/85 as TG-075. Tt 22303.	
☐ 61-0670		Sabreliner T-39A	265-73	USAF. (status ?).	
☐ 61-0671		Sabreliner CT-39A	265-74	USAF/to AMARC 9/85 as TG090. Tt 22000.	
☐ 61-0673		Sabreliner CT-39A	265-76	USAF/to AMARC 11/84 as TG038. Tt 20181.	
☐ 61-0674		Sabreliner T-39A	265-77	USAF/preserved Norton AFB. Ca.	
☐ 61-0675		Sabreliner T-39A	265-78	USAF/preserved as 10475 - 475th ABW. Yokota AB. Japan.	
☐ 61-0676		Sabreliner CT-39A	265-79	USAF/to AMARC 12/84 as TG049. Tt 21692.	
☐ 61-0678		Sabreliner CT-39A	265-81	USAF/to AMARC 5/84 as TG012. Tt 14666.	
☐ 61-0681		Sabreliner CT-39A	265-84	USAF/located at Willow Run-Detroit Museum.	
☐ 61-0682		Sabreliner CT-39A	265-85	USAF/to AMARC 10/84 as TG031. Tt 19741.	
☐ 61-0684		Sabreliner T-39A	265-87	USAF. (status ?).	
☐ 61-0685		Sabreliner CT-39A	265-88	preserved U S Army Aviation Museum, Fort Rucker, Al.	
☐ 62-4125	62	C-135B	18465	USAF, 58th MAS, Ramstein AB. Germany.	

Reg	Yr	Type	c/n	Owner/Operator	Prev Regn

Reg	Yr	Type	c/n	Owner/Operator	Prev Regn
☐ 62-4126	62	C-135B	18466	USAF, 89th MAW, Andrews AFB. Md.	
☐ 62-4127	62	C-135B	18467	USAF, 89th MAW, Andrews AFB. Md.	
☐ 62-4129	62	C-135B	18469	USAF, 89th MAW, Andrews AFB. Md.	
☐ 62-4130	62	C-135B	18470	USAF, 89th MAW, Andrews AFB. Md.	
☐ 62-4449		Sabreliner CT-39A	276-2	USAF/preserved at Pima County Museum, Az.	
☐ 62-4450		Sabreliner CT-39A	276-3	USAF/to AMARC 1/84 as TG006. Tt 18151.	
☐ 62-4452		Sabreliner CT-39A	276-5	USAF/preserved at Travis AFB. Ca.	
☐ 62-4454		Sabreliner CT-39A	276-7	USAF/to AMARC 8/84 as TG018. Tt 20878.	
☐ 62-4455		Sabreliner CT-39A	276-8	USAF/to AMARC 3/85 as TG065. Tt 20973.	
☐ 62-4456		Sabreliner CT-39A	276-9	Northrop University, Los Angeles.	
☐ 62-4457		Sabreliner CT-39A	276-10	USAF/to AMARC 11/83 as TG002. Tt 19944.	
☐ 62-4459		Sabreliner CT-39A	276-12	USAF/to AMARC 12/84 as TG041. Tt 19376.	
☐ 62-4461		Sabreliner CT-39A	276-14	USAF/preserved at Robins AFB. Ga.	
☐ 62-4462		Sabreliner CT-39A	276-15	USAF/to AMARC 8/84 as TG046. Tt 16167.	
☐ 62-4463		Sabreliner CT-39A	276-16	USAF/to AMARC 10/94 as TG100.	
☐ 62-4464		Sabreliner CT-39A	276-17	USAF/to AMARC 12/83 as TG004. Tt 20825.	
☐ 62-4465		Sabreliner T-39A	276-18	USAF/preserved at March AFB. Ca.	
☐ 62-4466		Sabreliner CT-39A	276-19	USAF/to AMARC 8/84 as TG019. Tt 20689.	
☐ 62-4467		Sabreliner CT-39A	276-20	GTCC Aviation Center, Greensboro. (instructional airframe).	
☐ 62-4469		Sabreliner CT-39A	276-22	USAF/to AMARC 3/85 as TG064. Tt 20011.	
☐ 62-4470		Sabreliner T-39A	276-23	USAF. (status ?).	
☐ 62-4471		Sabreliner CT-39A	276-24	USAF/preserved at Ramstein AB. Germany.	
☐ 62-4473		Sabreliner CT-39A	276-26	USAF/to AMARC 10/84 as TG029. Tt 19320.	
☐ 62-4475		Sabreliner CT-39A	276-28	Milwaukee Technical College, Wi.	
☐ 62-4476		Sabreliner T-39A	276-29	USAF/to AMARC 10/94 as TG099.	
☐ 62-4477		Sabreliner CT-39A	276-30	USAF/to AMARC 12/84 as TG048. Tt 18763.	
☐ 62-4478		Sabreliner T-39A	276-31	USAF/preserved at Wright Patterson Museum, Oh.	
☐ 62-4482		Sabreliner T-39A	276-35	USAF/preserved at Kelly AFB. Tx.	
☐ 62-4483		Sabreliner CT-39A	276-36	USAF/to AMARC 1/85 as TG055. Tt 19812.	
☐ 62-4484		Sabreliner T-39A	276-37	USAF/preserved at Kadena AB. Okinawa, Japan.	
☐ 62-4485		Sabreliner T-39A`	276-38	USAF/fire dump Yokota AB. Japan.	
☐ 62-4487		Sabreliner T-39A	276-40	USAF/preserved Offutt AB. Ne.	
☐ 62-4488		Sabreliner CT-39A	276-41	USAF. (status ?).	
☐ 62-4490		Sabreliner CT-39A	276-43	USAF/to AMARC 6/85 as TG079. Tt 22319.	
☐ 62-4493		Sabreliner CT-39A	276-46	USAF/to AMARC 6/85 as TG076. Tt 20759.	
☐ 62-4494		Sabreliner CT-39A	276-47	USAF/preserved at Chanute AFB. Il.	
☐ 62-4497		Sabreliner CT-39A	276-50	USAF/to AMARC 1/85 as TG053. Tt 21247.	
☐ 62-4498		Sabreliner CT-39A	276-51	Salt Lake Community College, (noted 10/93).	
☐ 62-4499		Sabreliner T-39A	276-52	USAF. (status ?).	
☐ 62-4500		Sabreliner CT-39A	276-53	USAF. to AMARC 5/85 as TG070. Tt 21264.	
☐ 62-4501		Sabreliner CT-39A	276-54	USAF/to AMARC 5/85 as TG073. Tt 18175.	
☐ 62-4502		Sabreliner T-39A	276-55	USAF. (status ?).	
☐ 62-6000	62	C-137C	18461	Presidential Aircraft, 1st MAS/89th MAW, Andrews AFB. Md.	
☐ 71-0874	71	C-9A	47467	USAF, 374 TAW, Clark AB. Phillipines.	
☐ 71-0876	72	C-9A	47475	USAF, 435 TAW, SHAPE, Chievres, Belgium.	
☐ 71-0878	72	C-9A	47536	USAF, 435 TAW, Rhein-Main AB. Germany.	
☐ 71-0882	72	C-9A	47541	USAF, 435 TAW, Rhein-Main AB. Germany.	
☐ 72-0283	72	B 737-253	20690	USAF, T-43A, 58th MAS, Ramstein AB. West Germany.	
☐ 72-7000	72	C-137C	20630	USAF. 89 MAW, Andrews AFB. DC.	N8459
☐ 73-1681	75	C-9C	47668	USAF, SAM, 1st MAS/89th MAW, Andrews AFB. Md.	
☐ 73-1682	75	C-9C	47670	USAF, SAM, 1st MAS/89th MAW, Andrews AFB. Md.	
☐ 73-1683	75	C-9C	47671	USAF, SAM, 1st MAS/89th MAW, Andrews AFB. Md.	
☐ 82-8000	87	B 747-2G4B	23824	VC-25A, Presidential aircraft, 89th MAW, Andrews AFB. Md.	
☐ 83-0500	83	Gulfstream 3	382	C-20A, USAF, 58 MAS, Ramstein AB. Germany.	N305GA
☐ 83-0501	83	Gulfstream 3	383	C-20A, USAF, 58 MAS, Ramstein AB. Germany.	N308GA
☐ 83-0502	83	Gulfstream 3	389	C-20A, USAF, 58 MAS, Ramstein AB. Germany.	N310GA
☐ 84-0063	84	Learjet C-21A	35A-509	USAF, 375 FTS, Scott AFB. Il.	N7263C
☐ 84-0064	84	Learjet C-21A	35A-510	USAF, CS/21st SPW (APSPC), Peterson AFB. Co.	N7263D
☐ 84-0065	84	Learjet C-21A	35A-511	USAF, RA/12th FTW (ABTC), Randolph AFB. Tx.	N7263E
☐ 84-0066	84	Learjet C-21A	35A-512	USAF, 89th AW (AMC), Andrews AFB. Md.	N7263F
☐ 84-0067	84	Learjet C-21A	35A-513	USAF, 89th AW (AMC), Andrews AFB. Md.	N7263H
☐ 84-0068	84	Learjet C-21A	35A-514	USAF, 86th AW (USAFE), Stuttgart AB. Germany.	N7263K
☐ 84-0069	84	Learjet C-21A	35A-515	USAF, KS/81st TRW, Keesler AFB. Ms.	N7263L
☐ 84-0070	84	Learjet C-21A	35A-516	USAF, KS/81st TRW, Keesler AFB. Ms.	N7263N
☐ 84-0071	84	Learjet C-21A	35A-517	USAF, KS/81st TRW, Keesler AFB. Ms.	N7263R
☐ 84-0072	84	Learjet C-21A	35A-518	USAF, KS/81st TRW, Keesler AFB. Ms.	N7263X
☐ 84-0073	84	Learjet C-21A	35A-519	USAF, 89th AW (AMC), Andrews AFB. Md.	N400AD
☐ 84-0074	84	Learjet C-21A	35A-520	USAF, 89th AW (AMC), Andrews AFB. Md.	N400AK
☐ 84-0075	84	Learjet C-21A	35A-521	USAF, 89th AW (AMC), Andrews AFB. Md.	N400AN
☐ 84-0076	84	Learjet C-21A	35A-522	USAF, 89th AW (AMC), Andrews AFB. Md.	N400AP
☐ 84-0077	84	Learjet C-21A	35A-523	USAF, 89th AW (AMC), Andrews AFB. Md.	N400AQ
☐ 84-0078	84	Learjet C-21A	35A-524	USAF, 89th AW (AMC), Andrews AFB. Md.	N400AS
☐ 84-0079	84	Learjet C-21A	35A-525	USAF, 89th AW (AMC), Andrews AFB. Md.	N400AT
☐ 84-0080	84	Learjet C-21A	35A-526	USAF, 89th AW (AMC), Andrews AFB. Md.	N400AU
☐ 84-0081	84	Learjet C-21A	35A-527	USAF, 86th AW (USAFE), Stuttgart AB. Germany.	N400AX
☐ 84-0082	84	Learjet C-21A	35A-528	USAF, 86th AW (USAFE), Stuttgart AB. Germany.	N400AY

Reg	Yr	Type	c/n	Owner/Operator	Prev Regn

Reg	Yr	Type	c/n	Owner/Operator	Prev Regn
☐ 84-0083	84	Learjet C-21A	35A-529	USAF, 86th AW (USAFE), Stuttgart AB. Germany.	N400AZ
☐ 84-0084	84	Learjet C-21A	35A-530	USAF, 86th AW (USAFE), Ramstein AB. Germany.	N400BA
☐ 84-0085	84	Learjet C-21A	35A-531	USAF, 86th AW (USAFE), Ramstein AB. Germany.	N400FY
☐ 84-0086	84	Learjet C-21A	35A-532	USAF, 86th AW (USAFE), Ramstein AB. Germany.	N400BN
☐ 84-0087	84	Learjet C-21A	35A-533	USAF, 86th AW (USAFE), Ramstein AB. Germany.	N400BQ
☐ 84-0088	84	Learjet C-21A	35A-534	USAF, OF/55th WG (ACC), Offutt AFB. Ne.	N400BU
☐ 84-0089	84	Learjet C-21A	35A-535	USAF, OF/55th WG (ACC), Offutt AFB. Ne.	N400BY
☐ 84-0090	84	Learjet C-21A	35A-536	USAF, OF/55th WG (ACC), Offutt AFB. Ne.	N400BZ
☐ 84-0091	84	Learjet C-21A	35A-537	USAF, OF/55th WG (ACC), Offutt AFB. Ne.	N400CD
☐ 84-0092	85	Learjet C-21A	35A-538	USAF, OF/55th WG (ACC), Offutt AFB. Ne.	N400CG
☐ 84-0093	85	Learjet C-21A	35A-539	USAF, OF/55th WG (ACC), Offutt AFB. Ne.	N400CJ
☐ 84-0094	85	Learjet C-21A	35A-540	USAF, OH/88th ABW (AFWC), Wright-Patterson AFB. Oh.	N400CK
☐ 84-0095	85	Learjet C-21A	35A-541	USAF, OH/88th ABW (AFWC), Wright-Patterson AFB. Oh.	N400CQ
☐ 84-0096	85	Learjet C-21A	35A-542	USAF, OH/88th ABW (AFWC), Wright-Patterson AFB. Oh.	N400CR
☐ 84-0097	85	Learjet C-21A	35A-543	USAF, OH/88th ABW (AFWC), Wright-Patterson AFB. Oh.	N400CU
☐ 84-0098	85	Learjet C-21A	35A-544	USAF, OH/88th ABW (AFWC), Wright-Patterson AFB. Oh.	N400CV
☐ 84-0099	85	Learjet C-21A	35A-545	USAF, OH/88th ABW (AFWC), Wright-Patterson AFB. Oh.	N400CX
☐ 84-0100	85	Learjet C-21A	35A-546	USAF, OH/88th ABW (AFWC), Wright-Patterson AFB. Oh.	N400CY
☐ 84-0101	85	Learjet C-21A	35A-547	USAF, 374th AW (PACAF), Yokota AB. Japan.	N400CZ
☐ 84-0102	85	Learjet C-21A	35A-548	USAF, 374th AW (PACAF), Yokota AB. Japan.	N400DD
☐ 84-0103	85	Learjet C-21A	35A-549	USAF, CS/21st SPW (APSPC), Peterson AFB. Co.	N400DJ
☐ 84-0104	85	Learjet C-21A	35A-550	USAF, CS/21st SPW (APSPC), Peterson AFB. Co.	N400DL
☐ 84-0105	85	Learjet C-21A	35A-551	USAF, CS/21st SPW (APSPC), Peterson AFB. Co.	N400DN
☐ 84-0106	85	Learjet C-21A	35A-552	USAF, CS/21st SPW (APSPC), Peterson AFB. Co.	N400DQ
☐ 84-0107	85	Learjet C-21A	35A-553	USAF, CS/21st SPW (APSPC), Peterson AFB. Co.	N400DR
☐ 84-0108	85	Learjet C-21A	35A-554	USAF, 86th AW (USAFE), Ramstein AB. Germany.	N400DU
☐ 84-0109	85	Learjet C-21A	35A-555	USAF, 86th AW (USAFE), Ramstein AB. Germany.	N400DV
☐ 84-0110	85	Learjet C-21A	35A-556	USAF, 86th AW (USAFE), Ramstein AB. Germany.	N400DX
☐ 84-0111	85	Learjet C-21A	35A-557	USAF, 86th AW (USAFE), Ramstein AB. Germany.	N400DY
☐ 84-0112	85	Learjet C-21A	35A-558	USAF, 86th AW (USAFE), Ramstein AB. Germany.	N400DZ
☐ 84-0113	85	Learjet C-21A	35A-559	USAF, FF/1st FW (ACC), Langley AFB. Va.	N400EC
☐ 84-0114	85	Learjet C-21A	35A-560	USAF, FF/1st FW (ACC), Langley AFB. Va.	N400EE
☐ 84-0115	85	Learjet C-21A	35A-561	USAF, FF/1st FW (ACC), Langley AFB. Va.	N400EF
☐ 84-0116	85	Learjet C-21A	35A-562	USAF, FF/1st FW (ACC), Langley AFB. Va.	N400EG
☐ 84-0117	85	Learjet C-21A	35A-563	USAF, FF/1st FW (ACC), Langley AFB. Va.	N400EJ
☐ 84-0118	85	Learjet C-21A	35A-564	USAF, 375th AW (AMC), Scott AFB. Il.	N400EK
☐ 84-0119	85	Learjet C-21A	35A-565	USAF, 375th AW (AMC), Scott AFB. Il.	N400EL
☐ 84-0120	85	Learjet C-21A	35A-566	USAF, 375th AW (AMC), Scott AFB. Il.	N400EM
☐ 84-0122	85	Learjet C-21A	35A-568	USAF, AU/42nd ABW (AETC), Maxwell AFB. Al.	N400EQ
☐ 84-0123	85	Learjet C-21A	35A-569	USAF, AU/42nd ABW (AETC), Maxwell AFB. Al.	N400ER
☐ 84-0124	85	Learjet C-21A	35A-570	USAF, AU/42nd ABW (AETC), Maxwell AFB. Al.	N400ES
☐ 84-0125	85	Learjet C-21A	35A-571	USAF, AU/42nd ABW (AETC), Maxwell AFB. Al.	N400ET
☐ 84-0126	85	Learjet C-21A	35A-572	USAF, RA/12th FTW (ABTC), Randolph AFB. Tx.	N400EU
☐ 84-0127	85	Learjet C-21A	35A-573	USAF, 375th AW (AMC), Scott AFB. Il.	N400EV
☐ 84-0128	85	Learjet C-21A	35A-575	USAF, 375th AW (AMC), Scott AFB. Il.	N400EY
☐ 84-0129	85	Learjet C-21A	35A-576	USAF, 375th AW (AMC), Scott AFB. Il.	N400EZ
☐ 84-0130	85	Learjet C-21A	35A-577	USAF, 374th AW (PACAF), Yokota AB. Japan.	N400FE
☐ 84-0131	85	Learjet C-21A	35A-578	USAF, 374th AW (PACAF), Yokota AB. Japan.	N400FG
☐ 84-0132	85	Learjet C-21A	35A-579	USAF, 89th AW (AMC), Andrews AFB. Md.	N400FH
☐ 84-0133	85	Learjet C-21A	35A-580	USAF, CS/21st SPW (APSPC), Peterson AFB. Co.	N400FK
☐ 84-0134	85	Learjet C-21A	35A-581	USAF, RA/12th FTW (ABTC), Randolph AFB. Tx.	N400FM
☐ 84-0135	85	Learjet C-21A	35A-582	USAF, RA/12th FTW (ABTC), Randolph AFB. Tx.	N400FN
☐ 84-0137	85	Learjet C-21A	35A-585	USAF, RA/12th FTW (ABTC), Randolph AFB. Tx.	N400FR
☐ 84-0138	85	Learjet C-21A	35A-574	USAF, RA/12th FTW (ABTC), Randolph AFB. Tx.	N400EX
☐ 84-0139	85	Learjet C-21A	35A-587	USAF, HW/24th AW (ACC), Howard AFB. CZ.	N400FU
☐ 84-0140	85	Learjet C-21A	35A-588	USAF, 375th AW (AMC), Scott AFB. Il.	N400FV
☐ 84-0141	85	Learjet C-21A	35A-584	USAF, FF/1st FW (ACC), Langley AFB. Va.	N400FQ
☐ 84-0142	85	Learjet C-21A	35A-586	USAF, 375th AW (ACC), Scott AFB. Il.	N400FT
☐ 84-0193	74	B 727-30	18362	C-22A, USAF, 310th MAS, Howard AFB. Panama.	N78
☐ 85-0049	85	Gulfstream 3	456	C-20C, U. S. Army, 89th MAW, Andrews AFB. Md.	N336GA
☐ 85-0050	85	Gulfstream 3	458	C-20C, U. S. Army, 89th MAW, Andrews AFB. Md.	N338GA
☐ 86-0201	85	Gulfstream 3	470	USAF, C-20B, 89th MAW, Andrews AFB. Md.	N344GA
☐ 86-0202	85	Gulfstream 3	468	USAF, C-20B, 89th MAW, Andrews AFB. Md.	N342GA
☐ 86-0203	85	Gulfstream 3	475	USAF, C-20B, 89th MAW, Andrews AFB. Md.	N312GA
☐ 86-0204	85	Gulfstream 3	476	USAF, C-20B, 89th MAW, Andrews AFB. Md.	N314GA
☐ 86-0206	85	Gulfstream 3	478	USAF, C-20B, 59th MAW, Andrews AFB. Md.	N318GA
☐ 86-0374	87	Learjet C-21A	35A-624	USAF, 113th FW (DC ANG), Andrews AFB. Md.	
☐ 86-0375	87	Learjet C-21A	35A-625	USAF, 113th FW (DC ANG), Andrews AFB. Md.	
☐ 86-0376	87	Learjet C-21A	35A-628	USAF, 113th FW (DC ANG), Andrews AFB. Md.	N3801G
☐ 86-0377	87	Learjet C-21A	35A-629	USAF, 113th FW (DC ANG), Andrews AFB. Md.	
☐ 86-0403	85	Gulfstream 3	473	USAF, C-20D, 89th MAW, Andrews AFB. Md.	N326GA
☐ 87-0026	79	Learjet 35	35A-280	U S Army, OSAC PAT Det. Andrews AFB. Md.	YN-BVO
☐ 87-0139	86	Gulfstream 3	497	C-20E, U S Army, Western Command, Wheeler AFB. Hi.	N7096G
☐ 87-0140	86	Gulfstream 3	498	C-20E, U S Army, St Louis, Mo.	N7096E
☐ 89-0266	69	Gulfstream 2SP	45	VC-11A, U S Army, OSAC PAT det, Andrews AFB. Md.	N51741
☐ 89-0284	92	Jayhawk T-1A	TT-5	USAF, XL/47 FTW 86 FTS, Laughlin AFB. Tx.	N2876B

Reg   *Yr*   *Type*   *c/n*   *Owner/Operator*   *Prev Regn*

Reg	Yr	Type	c/n	Owner/Operator	Prev Regn
□ 90-0300	92	Gulfstream 4	1181	USAF, C-20G, Andrews AFB. Md.	N473GA
□ 90-0400	91	Jayhawk T-1A	TT-3	USAF, XL/47 FTW 86 FTS, Laughlin AFB. Tx.	N2892B
□ 90-0401	92	Jayhawk T-1A	TT-7	USAF, LB/64 FTW 52 FTS, Reese AFB. Tx.	N2869B
□ 90-0402	92	Jayhawk T-1A	TT-8	USAF, LB/64 FTW 52 FTS, Reese AFB. Tx.	N2868B
□ 90-0403	92	Jayhawk T-1A	TT-9	USAF, VN/71 FTW. Vance AFB. Ok.	
□ 90-0404	92	Jayhawk T-1A	TT-6	USAF, RA/12 FTW 99 FTS, Randolph AFB. Tx.	N2872B
□ 90-0405	92	Jayhawk T-1A	TT-4	USAF, RA/12 FTW 99 FTS, Randolph AFB. Tx.	
□ 90-0406	92	Jayhawk T-1A	TT-11	USAF, XL/47 FTW 86 FTS, Laughlin AFB. Tx.	
□ 90-0407	92	Jayhawk T-1A	TT-10	USAF, CB/14 FTW, Columbus AFB. Ms.	
□ 90-0408	92	Jayhawk T-1A	TT-12	USAF, LB/64 FTW 52 FTS, Reese AFB. Tx.	
□ 90-0409	92	Jayhawk T-1A	TT-13	USAF, LB/64 FTW 52 FTS, Reese AFB. Tx.	
□ 90-0410	92	Jayhawk T-1A	TT-14	USAF, VN/71 FTW 26 FTS, Vance AFB. Ok.	
□ 90-0411	92	Jayhawk T-1A	TT-15	USAF, VN/71 FTW 26 FTS, Vance AFB. Ok.	
□ 90-0412	91	Jayhawk T-1A	TT-2	USAF, LB/64 FTW 52 FTS, Reese AFB. Tx. (was RK-15).	N2887B
□ 90-0413	92	Jayhawk T-1A	TT-16	USAF, LB/64 FTW 52 FTS, Reese AFB. Tx.	
□ 91-0075	92	Jayhawk T-1A	TT-18	USAF, XL/47 FTW, Laughlin AFB. Tx.	
□ 91-0076	92	Jayhawk T-1A	TT-17	USAF, XL/47 FTW 86 FTS, Laughlin AFB. Tx.	
□ 91-0077	91	Jayhawk T-1A	TT-1	USAF, VN/71 FTW 26 FTS, Vance AFB. Ok.	N2886B
□ 91-0078	92	Jayhawk T-1A	TT-19	USAF, XL/47 FTW, Laughlin AFB. Tx.	
□ 91-0079	92	Jayhawk T-1A	TT-20	USAF, XL/47 FTW, Laughlin AFB. Tx.	
□ 91-0080	92	Jayhawk T-1A	TT-21	USAF, LB/64 FTW 52 FTS, Reese AFB. Tx.	
□ 91-0081	92	Jayhawk T-1A	TT-22	USAF, LB/ 64 FTW 52 FTS, Reese AFB. Tx.	
□ 91-0082	92	Jayhawk T-1A	TT-23	USAF, LB/64 FTW 52 FTS, Reese AFB. Tx.	
□ 91-0083	92	Jayhawk T-1A	TT-24	USAF, XL/47 FTW, Laughlin AFB. Tx.	
□ 91-0084	92	Jayhawk T-1A	TT-25	USAF, LB/64 FTW 52 FTS, Reese AFB. Tx.	
□ 91-0085	92	Jayhawk T-1A	TT-26	USAF, LB/64 FTW 52 FTS, Reese AFB. Tx.	
□ 91-0086	92	Jayhawk T-1A	TT-27	USAF, LB/64 FTW 52 FTS, Reese AFB. Tx.	
□ 91-0087	92	Jayhawk T-1A	TT-28	USAF, LB/64 FTW 52 FTS, Reese AFB. Tx.	
□ 91-0088	92	Jayhawk T-1A	TT-29	USAF, LB/64 FTW 52 FTS, Reese AFB. Tx.	
□ 91-0089	92	Jayhawk T-1A	TT-30	USAF, LB/64 FTW 52 FTS, Reese AFB. Tx.	
□ 91-0090	93	Jayhawk T-1A	TT-31	USAF, XL/47 FTW, Laughlin AFB. Tx.	
□ 91-0091	93	Jayhawk T-1A	TT-32	USAF, XL/47 FTW, Laughlin AFB. Tx.	
□ 91-0092	93	Jayhawk T-1A	TT-33	USAF, XL/47 FTW 86 FTS, Laughlin AFB. Tx.	
□ 91-0093	93	Jayhawk T-1A	TT-34	USAF, LB/64 FTW 52 FTS, Reese AFB. Tx.	
□ 91-0094	93	Jayhawk T-1A	TT-35	USAF, LB/64 FTW 52 FTS, Reese AFB. Tx.	
□ 91-0095	93	Jayhawk T-1A	TT-36	USAF, LB/64 FTW 52 FTS, Reese AFB. Tx.	
□ 91-0096	93	Jayhawk T-1A	TT-37	USAF, RA/12 FTW 99 FTS, Randolph AFB. Tx.	
□ 91-0097	93	Jayhawk T-1A	TT-38	USAF, LB/64 FTW 52 FTS, Reese AFB. Tx.	
□ 91-0098	93	Jayhawk T-1A	TT-39	USAF, LB/64 FTW 52 FTS, Reese AFB. Tx.	
□ 91-0099	93	Jayhawk T-1A	TT-40	USAF, RA/12 FTW 99 FTS, Randolph AFB. Tx.	
□ 91-0100	93	Jayhawk T-1A	TT-41	USAF, LB/64 FTW 52 FTS, Reese AFB. Tx.	
□ 91-0101	93	Jayhawk T-1A	TT-42	USAF, RA/12 FTW 99 FTS, Randolph AFB. Tx.	
□ 91-0102	93	Jayhawk T-1A	TT-43	USAF, RA/12 FTW 99 FTS, Randolph AFB. Tx.	
□ 91-0108	91	Gulfstream 4	1162	United States Army, St Louis, Mo.	N7096B
□ 92-0330	93	Jayhawk T-1A	TT-44	USAF, RA/12 FTW 99 FTS, Randolph AFB. Tx.	
□ 92-0331	93	Jayhawk T-1A	TT-45	USAF, RA/12 FTW 99 FTS, Randolph AFB. Tx.	
□ 92-0332	93	Jayhawk T-1A	TT-46	USAF, RA/12 FTW 99 FTS, Randolph AFB. Tx.	
□ 92-0333	93	Jayhawk T-1A	TT-47	USAF, RA/12 FTW 99 FTS, Randolph AFB. Tx.	
□ 92-0334	93	Jayhawk T-1A	TT-48	USAF, RA/12 FTW 99 FTS, Randolph AFB. Tx.	
□ 92-0335	93	Jayhawk T-1A	TT-49	USAF, RA/12 FTW 99 FTS, Randolph AFB. Tx.	
□ 92-0336	93	Jayhawk T-1A	TT-50	USAF, RA/12 FTW 99 FTS, Randolph AFB. Tx.	
□ 92-0337	93	Jayhawk T-1A	TT-51	USAF, RA/12 FTW 99 FTS, Randolph AFB. Tx.	
□ 92-0338	93	Jayhawk T-1A	TT-52	USAF, RA/12 FTW 99 FTS, Randolph AFB. Tx.	
□ 92-0339	93	Jayhawk T-1A	TT-53	USAF, RA/12 FTW 99 FTS, Randolph AFB. Tx.	
□ 92-0340	93	Jayhawk T-1A	TT-54	USAF, RA/12 FTW 99 FTS, Randolph AFB. Tx.	
□ 92-0341	93	Jayhawk T-1A	TT-55	USAF, LB/64 FTW 52 FTS, Reese AFB. Tx.	
□ 92-0342	93	Jayhawk T-1A	TT-56	USAF, XL/47 FTW 86 FTS, Laughlin AFB. Tx.	
□ 92-0343	93	Jayhawk T-1A	TT-57	USAF, XL/47 FTW 86 FTS, Laughlin AFB. Tx.	
□ 92-0344	93	Jayhawk T-1A	TT-58	USAF, XL/47 FTW 86 FTS, Laughlin AFB. Tx.	
□ 92-0345	93	Jayhawk T-1A	TT-59	USAF, XL/47 FTW 86 FTS, Laughlin AFB. Tx.	
□ 92-0346	93	Jayhawk T-1A	TT-60	USAF, XL/47 FTW 86 FTS, Laughlin AFB. Tx.	
□ 92-0347	93	Jayhawk T-1A	TT-61	USAF, XL/47 FTW 86 FTS, Laughlin AFB. Tx.	
□ 92-0348	93	Jayhawk T-1A	TT-62	USAF, XL/47 FTW 86 FTS, Laughlin AFB. Tx.	
□ 92-0349	93	Jayhawk T-1A	TT-63	USAF, XL/47 FTW 86 FTS, Laughlin AFB. Tx.	
□ 92-0350	93	Jayhawk T-1A	TT-64	USAF, XL/47 FTW 86 FTS, Laughlin AFB. Tx.	
□ 92-0351	93	Jayhawk T-1A	TT-65	USAF, XL/47 FTW 86 FTS, Laughlin AFB. Tx.	
□ 92-0352	93	Jayhawk T-1A	TT-66	USAF, XL/47 FTW 86 FTS, Laughlin AFB. Tx.	
□ 92-0353	93	Jayhawk T-1A	TT-67	USAF, XL/47 FTW 86 FTS, Laughlin AFB. Tx.	
□ 92-0354	93	Jayhawk T-1A	TT-68	USAF, XL/47 FTW 86 FTS, Laughlin AFB. Tx.	
□ 92-0355	93	Jayhawk T-1A	TT-69	USAF, XL/47 FTW 86 FTS, Laughlin AFB. Tx.	
□ 92-0356	94	Jayhawk T-1A	TT-70	USAF, XL/47 FTW 86 FTS, Laughlin AFB. Tx.	
□ 92-0357	94	Jayhawk T-1A	TT-71	USAF, XL/47 FTW 86 FTS, Laughlin AFB. Tx.	
□ 92-0358	94	Jayhawk T-1A	TT-72	USAF, XL/47 FTW 86 FTS, Laughlin AFB. Tx.	
□ 92-0359	94	Jayhawk T-1A	TT-73	USAF, XL/47 FTW 86 FTS, Laughlin AFB. Tx.	
□ 92-0360	94	Jayhawk T-1A	TT-74	USAF, XL/47 FTW 86 FTS, Laughlin AFB. Tx.	
□ 92-0361	94	Jayhawk T-1A	TT-75	USAF, XL/47 FTW 86 FTS, Laughlin AFB. Tx.	

Reg	Yr Type	c/n	Owner/Operator	Prev Regn
☐ 92-0362	94 Jayhawk T-1A	TT-76	USAF, XL/47 FTW 86 FTS, Laughlin AFB. Tx.	
☐ 92-0363	94 Jayhawk T-1A	TT-77	USAF, LB/64 FTW 52 FTS, Reese AFB. Tx.	
☐ 92-0375	94 Gulfstream 4SP	1256	C-20H, USAF, Andrews AFB. Md.	N438GA
☐ 92-9000	87 B 747-2G4B	23825	VC-25A, Presidential aircraft, 89th MAW, Andrews AFB. Md.	
☐ 93-0621	94 Jayhawk T-1A	TT-78	USAF, XL/47 FTW 86 FTS, Laughlin AFB. Tx.	
☐ 93-0622	94 Jayhawk T-1A	TT-79	USAF, XL/47 FTW 86 FTS, Laughlin AFB. Tx.	
☐ 93-0623	94 Jayhawk T-1A	TT-80	USAF, XL/47 FTW 86 FTS, Laughlin AFB. Tx.	
☐ 93-0624	94 Jayhawk T-1A	TT-81	USAF, XL/47 FTW 86 FTS, Laughlin AFB. Tx.	
☐ 93-0625	94 Jayhawk T-1A	TT-82	USAF, XL/47 FTW 86 FTS, Laughlin AFB. Tx.	
☐ 93-0626	94 Jayhawk T-1A	TT-83	USAF, XL/47 FTW 86 FTS, Laughlin AFB. Tx.	
☐ 93-0627	94 Jayhawk T-1A	TT-84	USAF, LB/64 FTW 52 FTS, Reese AFB. Tx.	
☐ 93-0628	94 Jayhawk T-1A	TT-85	USAF, LB/64 FTW 52 FTS, Reese AFB. Tx.	
☐ 93-0629	94 Jayhawk T-1A	TT-86	USAF, XL/47 FTW 86 FTS, Laughlin AFB. Tx.	
☐ 93-0630	94 Jayhawk T-1A	TT-87	USAF, LB/64 FTW 52 FTS, Reese AFB. Tx.	
☐ 93-0631	94 Jayhawk T-1A	TT-88	USAF, XL/47 FTW 86 FTS, Laughlin AFB. Tx.	
☐ 93-0632	94 Jayhawk T-1A	TT-89	USAF, XL/47 FTW 86 FTS, Laughlin AFB. Tx.	
☐ 93-0633	94 Jayhawk T-1A	TT-90	USAF, XL/47 FTW 86 FTS, Laughlin AFB. Tx.	
☐ 93-0634	94 Jayhawk T-1A	TT-91	USAF, XL/47 FTW 86 FTS, Laughlin AFB. Tx.	
☐ 93-0635	94 Jayhawk T-1A	TT-92	USAF, XL/47 FTW 86 FTS, Laughlin AFB. Tx.	
☐ 93-0636	94 Jayhawk T-1A	TT-93	USAF, XL/47 FTW 86 FTS, Laughlin AFB. Tx.	
☐ 93-0637	94 Jayhawk T-1A	TT-94	USAF, XL/47 FTW 86 FTS, Laughlin AFB. Tx.	
☐ 93-0638	94 Jayhawk T-1A	TT-95	USAF, VN/71 FTW 26 FTS, Vance AFB. Ok.	
☐ 93-0639	94 Jayhawk T-1A	TT-96	USAF, VN/71 FTW 26 FTS, Vance AFB. Ok.	
☐ 93-0640	94 Jayhawk T-1A	TT-97	USAF, LB/64 FTW 52 FTS, Reese AFB. Tx.	
☐ 93-0641	94 Jayhawk T-1A	TT-98	USAF, VN/71 FTW 26 FTS, Vance AFB. Ok.	
☐ 93-0642	94 Jayhawk T-1A	TT-99	USAF, VN/71 FTW 26 FTS, Vance AFB. Ok.	
☐ 93-0643	94 Jayhawk T-1A	TT-100	USAF, VN/71 FTW 26 FTS, Vance AFB. Ok.	
☐ 93-0644	95 Jayhawk T-1A	TT-101	USAF, VN/71 FTW 26 FTS, Vance AFB. Ok.	
☐ 93-0645	95 Jayhawk T-1A	TT-102	USAF, VN/71 FTW 26 FTS, Vance AFB. Ok.	
☐ 93-0646	95 Jayhawk T-1A	TT-103	USAF, VN/71 FTW 26 FTS, Vance AFB. Ok.	
☐ 93-0647	95 Jayhawk T-1A	TT-104	USAF, VN/71 FTW 26 FTS, Vance AFB. Ok.	
☐ 93-0648	95 Jayhawk T-1A	TT-105	USAF, VN/71 FTW 26 FTS, Vance AFB. Ok.	
☐ 93-0649	95 Jayhawk T-1A	TT-106	USAF, VN/71 FTW 26 FTS, Vance AFB. Ok.	
☐ 93-0650	95 Jayhawk T-1A	TT-107	USAF, VN/71 FTW 26 FTS, Vance AFB. Ok.	
☐ 93-0651	95 Jayhawk T-1A	TT-108	USAF, LB/64 FTW 52 FTS, Reese AFB. Tx.	
☐ 93-0652	95 Jayhawk T-1A	TT-109	USAF, LB/64 FTW 52 FTS, Reese AFB. Tx.	
☐ 93-0653	95 Jayhawk T-1A	TT-110	USAF, VN/71 FTW 26 FTS, Vance AFB. Ok.	
☐ 93-0654	95 Jayhawk T-1A	TT-111	USAF, VN/71 FTW 26 FTS, Vance AFB. Ok.	
☐ 93 0655	95 Jayhawk T-1A	TT-112	USAF, VN/71 FTW 26 FTS, Vance AFB. Ok.	
☐ 93-0656	95 Jayhawk T-1A	TT-113	USAF, VN/71 FTW 26 FTS, Vance AFB. Ok.	
☐ 94-0114	95 Jayhawk T-1A	TT-114	USAF, VN/71 FTW 26 FTS, Vance AFB, Ok.	
☐ 94-0115	95 Jayhawk T-1A	TT-115	USAF, VN/71 FTW 26 FTS, Vance AFB. Ok.	
☐ 94-0116	95 Jayhawk T-1A	TT-116	USAF, VN/71 FTW 26 FTS, Vance AFB. Ok.	
☐ 94-0117	95 Jayhawk T-1A	TT-117	USAF, VN/71 FTW 26 FTS, Vance AFB. Ok.	
☐ 94-0118	95 Jayhawk T-1A	TT-118	USAF, VN/71 FTW 26 FTS, Vance AFB. Ok.	
☐ 94-0119	95 Jayhawk T-1A	TT-119	USAF, VN/71 FTW 26 FTS, Vance AFB. Ok.	
☐ 94-0120	95 Jayhawk T-1A	TT-120	USAF, VN/71 FTW 26 FTS, Vance AFB. Ok.	
☐ 94-0121	95 Jayhawk T-1A	TT-121	USAF, VN/71 FTW 26 FTS, Vance AFB. Ok.	
☐ 94-0122	95 Jayhawk T-1A	TT-122	USAF, VN/71 FTW 26 FTS, Vance AFB. Ok.	
☐ 94-0123	95 Jayhawk T-1A	TT-123	USAF, VN/71 FTW 26 FTS, Vance AFB. Ok.	
☐ 94-0124	95 Jayhawk T-1A	TT-124	USAF, VN/71 FTW 26 FTS, Vance AFB. Ok.	
☐ 94-0125	95 Jayhawk T-1A	TT-125	USAF, VN/71 FTW 26 FTS, Vance AFB. Ok.	
☐ 94-0126	95 Jayhawk T-1A	TT-126	USAF, VN/71 FTW 26 FTS, Vance AFB. Ok.	
☐ 94-0127	95 Jayhawk T-1A	TT-127	USAF, VN/71 FTW 26 FTS, Vance AFB. Ok.	
☐ 94-0128	95 Jayhawk T-1A	TT-128	USAF, VN/71 FTW 26 FTS, Vance AFB. Ok.	
☐ 94-0129	95 Jayhawk T-1A	TT-129	USAF, VN/71 FTW 26 FTS, Vance AFB. Ok.	
☐ 94-0130	95 Jayhawk T-1A	TT-130	USAF, VN/71 FTW 26 FTS, Vance AFB. Ok.	
☐ 94-0131	96 Jayhawk T-1A	TT-131	USAF, CB/14 FTW, Columbus AFB. Ms.	
☐ 94-0132	96 Jayhawk T-1A	TT-132	USAF, VN/71 FTW 26 FTS, Vance AFB. Ok.	
☐ 94-0133	96 Jayhawk T-1A	TT-133	USAF, VN/71 FTW 26 FTS, Vance AFB. Ok.	
☐ 94-0134	96 Jayhawk T-1A	TT-134	USAF, CB/14 FTW, Columbus AFB. Ms.	
☐ 94-0135	96 Jayhawk T-1A	TT-135	USAF, CB/14 FTW, Columbus AFB. Ms.	
☐ 94-0136	96 Jayhawk T-1A	TT-136	USAF, CB/14 FTW, Columbus AFB. Ms.	
☐ 94-0137	96 Jayhawk T-1A	TT-137	USAF, CB/14 FTW, Columbus AFB. Ms.	
☐ 94-0138	96 Jayhawk T-1A	TT-138	USAF, CB/14 FTW, Columbus AFB. Ms.	
☐ 94-0139	96 Jayhawk T-1A	TT-139	USAF, CB/14 FTW, Columbus AFB. Ms.	
☐ 94-0140	96 Jayhawk T-1A	TT-140	USAF, CB/14 FTW, Columbus AFB. Ms.	
☐ 94-0141	96 Jayhawk T-1A	TT-141	USAF, CB/14 FTW, Columbus AFB. Ms.	
☐ 94-0142	96 Jayhawk T-1A	TT-142	USAF, CB/14 FTW, Columbus AFB. Ms.	
☐ 94-0143	96 Jayhawk T-1A	TT-143	USAF, CB/14 FTW, Columbus AFB. Ms.	
☐ 94-0144	96 Jayhawk T-1A	TT-144	USAF, CB/14 FTW, Columbus AFB. Ms.	
☐ 94-0145	96 Jayhawk T-1A	TT-145	USAF, CB/14 FTW, Columbus AFB. Ms.	
☐ 94-0146	96 Jayhawk T-1A	TT-146	USAF,	
☐ 94-0147	96 Jayhawk T-1A	TT-147	USAF,	
☐ 94-0148	96 Jayhawk T-1A	TT-148	USAF,	
☐ 95-0040	96 Jayhawk T-1A	TT-149	USAF,	

Reg	Yr Type	c/n	Owner/Operator	Prev Regn
☐ 95-0041	96 Jayhawk T-1A	TT-150	USAF,	
☐ 95-0042	96 Jayhawk T-1A	TT-151	USAF,	
☐ 95-0043	96 Jayhawk T-1A	TT-152	USAF,	
☐ 95-0044	96 Jayhawk T-1A	TT-153	USAF,	
☐ 95-0045	96 Jayhawk T-1A	TT-154	USAF,	
☐ 95-0046	96 Jayhawk T-1A	TT-155	USAF,	
☐ 95-0047	96 Jayhawk T-1A	TT-156	USAF, XL/47 FTW 86 FTS, Laughlin AFB. Tx.	
☐ 95-0048	96 Jayhawk T-1A	TT-157	USAF,	
☐ 95-0049	96 Jayhawk T-1A	TT-158	USAF,	
☐ 95-0050	96 Jayhawk T-1A	TT-159	USAF,	
☐ 95-0051	96 Jayhawk T-1A	TT-160	USAF,	
☐ 95-0052	96 Jayhawk T-1A	TT-161	USAF,	
☐ 95-0053	96 Jayhawk T-1A	TT-162	USAF,	
☐ 95-0054	96 Jayhawk T-1A	TT-163	USAF,	
☐ 95-0055	96 Jayhawk T-1A	TT-164	USAF,	
☐ 95-0056	96 Jayhawk T-1A	TT-165	USAF,	
☐ 95-0057	96 Jayhawk T-1A	TT-166	USAF,	
☐ 95-0058	96 Jayhawk T-1A	TT-167	USAF,	
☐ 95-0059	97 Jayhawk T-1A	TT-168	USAF,	
☐ 95-0060	97 Jayhawk T-1A	TT-169	USAF,	
☐ 95-0061	97 Jayhawk T-1A	TT-170	USAF,	
☐ 95-0062	97 Jayhawk T-1A	TT-171	USAF,	
☐ 95-0063	97 Jayhawk T-1A	TT-172	USAF,	
☐ 95-0064	97 Jayhawk T-1A	TT-173	USAF,	
☐ 95-0065	97 Jayhawk T-1A	TT-174	USAF,	
☐ 95-0066	97 Jayhawk T-1A	TT-175	USAF, XL/47 FTW 86 FTS, Laughlin AFB. Tx.	
☐ 95-0067	97 Jayhawk T-1A	TT-176	USAF,	
☐ 95-0068	97 Jayhawk T-1A	TT-177	USAF,	
☐ 95-0069	97 Jayhawk T-1A	TT-178	USAF, VN/71 FTW 26FTS, Vance AFB. Ok.	
☐ 95-0070	97 Jayhawk T-1A	TT-179	USAF, VN/71FTW 26FTS, Vance AFB. Ok.	
☐ 95-0071	97 Jayhawk T-1A	TT-180	USAF, VN/71FTW 26FTS, Vance AFB. Ok.	
☐ 95-0123	96 Citation UC-35A	560-0387	U S Army, 207th Aviation Co. Wiesbaden, Germany.	N5108G
☐ 95-0124	96 Citation UC-35A	560-0392	U S Army,	N5093L
☐ 96-0107	97 Citation UC-35A	560-0404	U S Army,	N5201M
☐ 96-0108	97 Citation UC-35A	560-0410	U S Army,	N52081
☐ 96-0109	97 Citation UC-35A	560-0415	U S Army,	N52457
☐ 96-0110	97 Citation V Ultra	560-0420	U S Army,	N51942
☐ 97-0400	98 Gulfstream C-37A	...	USAF,	
☐ 97-0401	98 Gulfstream C-37A	...	USAF,	

OB = Peru (Republic of Peru)

Civil

Reg	Yr Type	c/n	Owner/Operator	Prev Regn
☐ OB-1280	72 Citation Eagle	500-0019	ATSA=Aerotransportes SA. Lima.	OB-S-1280
☐ OB-1396	75 F 28-1000	11100	Government of Peru, Lima.	FAP-390
☐ OB-1429	74 Learjet 25B	25B-159	SAN-Servicio Aerofotografico Nacional, Lima.	FAP 522
☐ OB-1430	74 Learjet 25B	25B-164	SAN-Servicio Aerofotografico Nacional, Lima.	FAP 523
☐ OB-1431	83 Learjet 36A	36A-051	SAN-Servicio Aerofotografico Nacional, Lima.	FAP 524
☐ OB-1432	83 Learjet 36A	36A-052	SAN-Servicio Aerofotografico Nacional, Lima.	FAP 525
☐ OB-1433	80 Falcon 20F	434	SAN-Servicio Aerofotografico Nacional, Lima.	FAP 300
☐ OB-1626	91 Citation V	560-0124	Southern Peru Copper Corp. Lima.	N124VP

Military

Reg	Yr Type	c/n	Owner/Operator	Prev Regn
☐ FAP 370	69 DC 8-62CF	46078	FAP, Grupo Aereo 8, Las Palmas-Lima.	HB-IDK
☐ FAP 371	DC 8-62CF	45984	FAP, Grupo Aereo 8, Las Palmas-Lima.	HB-IDH
☐ PRP-001	95 B 737-528	27426	Government of Peru, Lima.	(F-GJNR)

OE = Austria (Republic of Austria)

Civil

Reg	Yr Type	c/n	Owner/Operator	Prev Regn
☐ OE-FAN	76 Citation	500-0289	Avanti Touristik GmbH. Vienna.	N939KS
☐ OE-FBS	87 Citation II/SP	551-0574	Airlink Luftverkehrs GmbH. Salzburg.	N60GF
☐ OE-FDM	79 Citation 1/SP	501-0140	Goldeck-Flug GmbH. Klagenfurt.	N96CF
☐ OE-FGN	75 Citation	500-0291	Airlink Luftverkehrs GmbH. Salzburg.	N291DS
☐ OE-FHH	82 Citation 1/SP	501-0246	Hoedlmayr GmbH/Hodi Air GmbH. Schwertberg.	N26LC
☐ OE-FHW	79 Citation 1/SP	501-0121	COMTEL Flug GmbH. Vienna.	D-IANO
☐ OE-FMS	79 Citation 1/SP	501-0239	Aircraft Innsbruck Luftfahrt GmbH. Innsbruck.	N164CB
☐ OE-FPA	87 Citation II/SP	551-0552	Airlink Luftverkehrs GmbH. Salzburg.	
☐ OE-FYC	81 Citation 1/SP	501-0207	Almeta Metallumschmelzwerke GmbH. Vienna.	(N207CF)
☐ OE-GAA	91 Citation V	560-0111	Tyrolean Jet Service, Innsbruck.	(N6802T)
☐ OE-GAV	78 Learjet 35A	35A-185	AVA-Bank GmbH/Grossman Air Service GmbH. Vienna.	TC-GEM
☐ OE-GBA	79 Citation II	550-0085	Bannert Air GmbH. Vienna.	N57AJ
☐ OE-GCC	91 Citation V	560-0125	Goldeck-Flug GmbH. Klagenfurt.	N6809V
☐ OE-GCF	88 Learjet 55C	55C-136	Schaffer GmbH. Vienna.	N155PS
☐ OE-GCI	78 Citation II	550-0041	Transair GmbH. Vienna.	N177HH
☐ OE-GCJ	72 Falcon 20C	184	City-Jet Business Airlines, Vienna.	F-GAPC
☐ OE-GCP	93 Citation V	560-0214	AVAG Air-Mercedez Austria/Automobilvertriebs AG. Salzburg.	N1285D
☐ OE-GCR	68 Falcon 20D-5	191	Lauda Air, Vienna.	N800CF

Reg	Yr Type	c/n	Owner/Operator	Prev Regn

Reg	Yr	Type	c/n	Owner/Operator	Prev Regn
☐ OE-GDA	92	Citation V	560-0200	Magna Air Luftfahrt GmbH. Vienna.	
☐ OE-GHS	87	BAe 125/800B	8078	Schaffer GmbH. Basle, Switzerland.	ZS-FSI
☐ OE-GIL	79	Citation II	550-0060	Air-Styria Luftfahrtunternehmen GmbH. Graz.	N315CK
☐ OE-GLZ	91	Citation II	550-0690	Taxiflug GmbH. Dornbirn.	
☐ OE-GMD	81	Learjet 36A	36A-047	Air MED Luftfahrzeug GmbH. Vienna.	N36SK
☐ OE-GMI	96	Citation V Ultra	560-0362	Magna Air Luftfahrt GmbH. Oberwaltersdorf.	N5183U
☐ OE-GNL	94	Learjet 60	60-032	Lauda Air, Vienna.	N5013D
☐ OE-GPA	90	Citation V	560-0099	Porsche Konstruktionen KG. Salzburg.	(N67905)
☐ OE-GRR	82	Learjet 55	55-059	Goldeck Flug GmbH. Klagenfurt.	N59LJ
☐ OE-GSC	78	Falcon 10	122	Tyrolean Jet Service GmbH/Heliair Helikopter, Innsbruck.	N312AT
☐ OE-GSW	90	Citation V	560-0088	Tyrolean Jet Service GmbH. Innsbruck.	
☐ OE-HET	83	Challenger 600S	1085	Grossmann Air Service GmbH. Vienna.	N600ST
☐ OE-HIT	91	Falcon 50	222	Friedrich Karl Flick, Vienna.	D-BELL
☐ OE-ILS	88	Falcon 900	58	Tyrolean Jet Service GmbH. Innsbruck.	F-WWFE
☐ OE-IMI	94	Falcon 900B	147	Magna Air Luftfahrt GmbH. Oberwatersdorf.	N901FJ

OH = Finland (Republic of Finland)

Civil

Reg	Yr	Type	c/n	Owner/Operator	Prev Regn
☐ OH-BAP	84	HS 125/700A	7212	Jetflite OY. Helsinki.	G-5-659
☐ OH-CAT	80	Citation II	550-0378	Jetflite OY. Helsinki.	N3999H
☐ OH-COC	74	Citation	500-0223	Lillbacka Jetair OY. Kauhava.	N400SA
☐ OH-FPC	76	Falcon 20F	345	Lillbacka Jetair OY. Kauhava.	N133AP
☐ OH-GLB	73	Learjet 24D	24D-262	Airdeal OY. Helsinki.	N110PS
☐ OH-IPP	83	Learjet 55	55-056	Cloudex OY. Kauniainen.	N270AS
☐ OH-JET	81	HS 125/700B	7136	Progressor OY/Jetflite OY. Helsinki.	G-5-545
☐ OH-KNE	81	Diamond 1	A014SA	Euro-Flite OY. Helsinki.	N339DM
☐ OH-RIF	93	Beechjet 400A	RK-79	Inter Flight Ltd OY. Helsinki.	(N30SF)
☐ OH-WIH	81	Challenger 600S	1029	Jetflite OY. Helsinki.	N205A

Military

Reg	Yr	Type	c/n	Owner/Operator	Prev Regn
☐ LJ-1	81	Learjet 35A	35A-430	Finnish Air Force, Utti.	N10870
☐ LJ-2	82	Learjet 35A	35A-451	Finnish Air Force, Utti.	N1462B
☐ LJ-3	82	Learjet 35A	35A-470	Finnish Air Force, Utti.	N3810G

OK = Czech Republic (Czech Republic)

Civil

Reg	Yr	Type	c/n	Owner/Operator	Prev Regn
☐ OK-BYA	92	Challenger 601-3A	5105	Czech Government, Prague.	C-FMVQ
☐ OK-UZI	89	Beechjet 400	RJ-56	Czech Government-Civil Aviation Inspectorate, Prague.	G-BSZP

OM = Slovak Republic

Civil

Reg	Yr	Type	c/n	Owner/Operator	Prev Regn
☐ OM-BYE	74	YAK 40	9440338	Government of Slovakia, Bratislava.	OK-BYE
☐ OM-BYL	96	YAK 40	9940560	Government of Slovakia, Bratislava.	OK-BYL
☐ OM-SKY	96	Hawker 800XP	8314	VSZ Steel/Cassovia Air AS. Bratislava.	N314XP

OO = Belgium (Kingdom of Belgium)

Civil

Reg	Yr	Type	c/n	Owner/Operator	Prev Regn
☐ OO-DCM	74	Citation	500-0182	Lambda Jet SA. Liege.	N13HJ
☐ OO-GBL	80	Learjet 35A	35A-284	Bank Lambert/GBL-Air, Brussels.	D-CCAX
☐ OO-IBI	75	Citation Eagle	500-0238	IBIS-International Business & Industrial Services, Antwerp.	N3QZ
☐ OO-LCM	72	Citation	500-0036	Lambda Jet SA. Liege.	N18HJ
☐ OO-LFR	80	Learjet 25D	25D-320	Abelag Aviation, Brussels.	N320EJ
☐ OO-LFT	81	Falcon 50	42	Abelag Aviation, Brussels.	OE-HCS
☐ OO-LFV	82	Learjet 35A	35A-481	Abelag Aviation, Brussels.	N27NR
☐ OO-LFY	78	Learjet 35A	35A-200	Abelag Aviation, Brussels.	D-CCAR
☐ OO-OSA	87	Citation S/II	S550-0147	Bosal International NV. Antwerp.	N1296N
☐ OO-PHI	95	CitationJet	525-0115	Avair BV. Gravenwezel.	N52141
☐ OO-RSE	81	Sabre-65	465-72	RST Aviation/R Steinmete & Sons, Antwerp.	N857W
☐ OO-SKS	79	Citation II	551-0117	Sky Service BV. Wevelgem. (was 550-0063).	N11AB

Military

Reg	Yr	Type	c/n	Owner/Operator	Prev Regn
☐ CA-01	85	Airbus A310-222	372	Belgian Air Force, 21 Sqn. 15TW, Melsbroek.	9V-STN
☐ CA-02	85	Airbus A310-222	367	Belgian Air Force, 21 Sqn. 15TW, Melbroek.	9V-STM
☐ CB-01	67	B 727-29C	19402	BAF 21, Belgian Air Force, 21 Sqn. 15TW, Melsbroek.	OO-STB
☐ CB-02	67	B 727-29C	19403	BAF 22, Belgian Air Force, 21 Sqn. 15TW, Melsbroek.	OO-STD
☐ CD-01	91	Falcon 900B	109	Belgian Air Force, 21 Sqn. 15TW, Melsbroek.	G-BTIB
☐ CM-01	73	Falcon 20E	276	BAF 31, Belgian Air Force, 21 Sqn. Melsbroek.	F-WNGL
☐ CM-02	73	Falcon 20E	278	BAF 32, Belgian Air Force, 21 Sqn. Melsbroek.	F-WNGM

OY = Denmark (Kingdom of Denmark)

Civil

Reg	Yr	Type	c/n	Owner/Operator	Prev Regn
☐ OY-APM	94	Challenger 601-3R	5153	Maersk Air A/S. Copenhagen.	N604BA
☐ OY-BDS	70	Falcon 20C	180	Danfoss Aviation/Air Alsie A/S. Sonderborg.	F-WMKF
☐ OY-BZT	81	Citation II	550-0259	Alebco Corp APS/Jetcopter-Air Center West, Stauning.	N810JT
☐ OY-CCG	83	Citation III	650-0003	Grundfos A/S. Bjerringbro/Alkair A/S. Naerum.	N92LA
☐ OY-CCJ	82	Learjet 35A	35A-468	Alkair Flight Operations A/S. Copenhagen. 'Skydreamer'	N486LM
Reg	*Yr*	*Type*	*c/n*	*Owner/Operator*	*Prev Regn*

Reg	Yr	Type	c/n	Owner/Operator	Prev Regn
☐ OY-CEV	75	Citation	500-0329	Falck Air A/S. Odense. 'Nancy'	N4999H
☐ OY-CKK	92	Falcon 900B	110	Kirkbi A/S-Lego Systems, Billund.	F-WWFH
☐ OY-CKT	90	Citation V	560-0078	Danfoss Aviation/Air Alsie A/S. Sonderborg.	SE-DLI
☐ OY-CPW	79	Citation 1/SP	501-0120	Aage Jensen Gruppen A/S-Danish Air Transport K/S. Vamdrup.	N487LS
☐ OY-CYV	82	Citation II	550-0440	Falck Air A/S. Odense.	N120TC
☐ OY-FFB	81	Citation	500-0406	Cimber Air, Sonderborg.	SE-DET
☐ OY-FLK	83	Learjet 55	55-050	Falck Air A/S. Odense.	N122JP
☐ OY-GKL	84	Citation III	650-0043	Kirkbi A/S-Lego Systems, Billund.	(N1310B)
☐ OY-INI	80	Citation 1/SP	501-0166	Jet Plane Corp/Sun Air of Scandinavia A/S. Billund.	I-CIPA
☐ OY-JEV	81	Citation II	550-0284	Weibel Scientific Inc. Copenhagen.	I-ARIB
☐ OY-JEY	81	Citation	500-0405	Air Alsie A/S-Jetair, Roskilde.	SE-DES
☐ OY-LIN	93	Falcon 50	230	Air Alsie A/S. Sonderborg.	3B-NSY
☐ OY-ONE	80	Citation 1/SP	501-0143	Magpie Aviation APS/Sun Air of Scandinavia A/S. Billund.	D-IGGK
☐ OY-PDN	82	Citation II/SP	551-0412	Nassau Doors A/S. Ringe.	N413VP
☐ OY-RAA	92	BAe 125/800B	8235	Armada Shipping A/S-Fredensborg/Air Alsie A/S. Sonderborg.	G-BWVA
☐ OY-RAC	97	Hawker 800XP	8335	Stein Erik Hagen AS. Oslo/Air Alsie A/S. Sonderborg.	N335XP
☐ OY-SBR	75	Corvette	23	North Flying A/S. Aalborg.	F-BVPF
☐ OY-SBT	76	Corvette	33	North Flying A/S. Aalborg.	F-BTTT
☐ OY-SVL	78	Citation 1/SP	501-0049	Sun-Air of Scandinavia A/S-Pharma Nord Aps. Vejle.	(N36WS)
☐ OY-TAM	74	Citation	500-0158	Torben & Marianne Andersen, Billund.	(N158TJ)
Military					
☐ F-249	79	Gulfstream 3	249	RDAF, ESK.721, Vaerlose-Copenhagen.	N901GA
☐ F-313	80	Gulfstream 3	313	RDAF, ESK.721, Vaerlose-Copenhagen.	

PH = Netherlands *(Kingdom of the Netherlands)*

Reg	Yr	Type	c/n	Owner/Operator	Prev Regn
Civil					
☐ PH-...	74	Falcon 20F	321	Berk Holding BV. Amersfoort.	N20FM
☐ PH-CTA	73	Citation	500-0088	Dynamic Air BV. Eindhoven.	G-HOLL
☐ PH-CTC	73	Citation	500-0098	Dynamic Air BV. Eindhoven.	G-BNVY
☐ PH-CTX	81	Citation II	550-0398	Dynamic Air BV. Eindhoven.	N398S
☐ PH-CTZ	79	Citation II	550-0052	Dynamic Air BV. Eindhoven.	N67TM
☐ PH-ERP	95	Falcon 900EX	1	Baan Investment BV. Amsterdam.	F-GREX
☐ PH-EZF	96	Fokker 70	11576	PT Caltex Pacific Indonesia/Pelita, Jakarta. 'Rokan'	
☐ PH-ILD	80	Falcon 50	23	N V Philips, Eindhoven.	D-BBWK
☐ PH-ILR	80	Falcon 50	15	N V Philips, Eindhoven.	F-WZHM
☐ PH-KBX	94	Fokker 70	11547	Dutch Royal Flight, Amsterdam.	PH-...
☐ PH-LAB	92	Citation II	550-0712	NLR-National Aerospace Laboratory, Amsterdam.	N12030
☐ PH-MDC	95	Citation V Ultra	560-0280	Cartier Europe BV. Amsterdam.	N1298G
☐ PH-MEX	97	Citation VI	650-0217	Martinair Holland NV. Amsterdam.	N217CM
☐ PH-MFX	94	Citation VI	650-0240	Martinair Holland NV. Amsterdam.	N51143
☐ PH-RMA	87	Citation S/II	S550-0145	Heerema Vliegbedrijf BV. Leiden.	(N1295Y)
☐ PH-VLG	94	Citation V Ultra	560-0271	Cartier Europe BV. Amsterdam.	
Military					
☐ V-11	86	Gulfstream 4	1009	RNAF, 334 Squadron, Eindhoven.	VR-BOY

PK = Indonesia *(Republic of Indonesia)*

Reg	Yr	Type	c/n	Owner/Operator	Prev Regn
Civil					
☐ PK-...	67	Falcon 20C	90		VH-CIR
☐ PK-CAG	79	Falcon 20F	408	Directorate of Civil Aviation, Jakarta. 'Dewa Rud'	F-WRQS
☐ PK-CAH	92	Learjet 31A	31A-066	Directorate of Civil Aviation, Jakarta.	N26006
☐ PK-CAJ	93	Learjet 31A	31A-077	Directorate of Civil Aviation, Jakarta.	N26002
☐ PK-CAP	80	Gulfstream 3	316	Directorate of Civil Aviation, Jakarta.	N26018
☐ PK-CTA	81	HS 125/700A	NA0313	TransIndo, Jakarta.	N18G
☐ PK-CTC	80	HS 125/700A	NA0270	TransIndo, Jakarta.	N621JA
☐ PK-CTP	84	Gulfstream 3	431	TransIndo/Mindo Petroleum Co. Jakarta.	(N259B)
☐ PK-DJW	68	HS 125/3B-RA	ʼ25147	Deraya Air Taxi, Jakarta. (status ?).	PK-PJR
☐ PK-HHS	79	B 737-2S9	21957	Gatari Air Services, Jakarta.	N80CC
☐ PK-HMG	74	HS 125/600B	6029	Gatari Air Services, Jakarta.	PK-PJE
☐ PK-KIG	87	Citation III	650-0151	KIG Air Transport Pte Ltd. Singapore.	N91D
☐ PK-NSP	88	Gulfstream 4	1077	P T Nugra Santana Air Service, Jakarta.	PK-MPZ
☐ PK-NZK	93	Gulfstream 4SP	1219	IPTN-Indonesian Aircraft Industries, Bandung.	PK-MPZ
☐ PK-OCF	68	B 737-247	19601	Airfast Indonesia PT/Freeport McMoRan Inc. New Orleans, La.	N466AC
☐ PK-OCG	70	B 737-293	20335	Airfast Indonesia PT/Freeport McMoRan Inc. New Orleans, La.	N469AC
☐ PK-OCI	70	B 737-230C	20255	Airfast-Indonesia PT/Freeport-McMoRan Inc. New Orleans, La.	N800WA
☐ PK-OCN	80	Gulfstream 3	305	Airfast Indonesia, Singapore.	N682FM
☐ PK-PFE	95	Fokker 70	11553	Pelita Air Service, Jakarta.	PH-MXN
☐ PK-PJA	83	Gulfstream 3	395	Pelita Air Service, Jakarta.	(N30GL)
☐ PK-PJF	66	BAC 1-11/401AK	065	Bouraq Airlines/Citra Group, Jakarta.	N117MR
☐ PK-PJJ	93	BAe 146/RJ-85	E2239	Pelita Air Service/Government of Indonesia, Jakarta.	G-5-239
☐ PK-PJK	83	F 28-4000	11192	Caltex/Pelita Air Service, Jakarta.	PH-EXW
☐ PK-PJL	76	F 28-4000	11111	Pelita Air Service, Jakarta.	PH-EXA
☐ PK-PJM	81	F 28-4000	11178	Pelita Air Service, Jakarta. 'Matak'	PH-EXW
☐ PK-PJN	90	Fokker 100	11288	Pelita Air Service, Jakarta. 'Lengguru'	PH-LMU
☐ PK-PJP	86	BAe 146 Statesman	E2050	Pelita Air Service/Government of Indonesia, Jakarta.	G-5-517
☐ PK-PJY	79	F 28-4000	11146	Pelita Air Service, Jakarta. 'Arun'	PH-EXN

Reg	Yr	Type	c/n	Owner/Operator	Prev Regn

Reg	Yr	Type	c/n	Owner/Operator	Prev Regn
PK-PJZ	68	Gulfstream 2	26	Pelita Air Service, Jakarta.	N202GA
PK-RJW	71	F 28-1000	11045	Rajawali Air Transport, Jakarta. 'Anugerah'	PH-PBX
PK-T..	68	BAC 1-11/423ET	118	Indonesia Air Transport, Jakarta.	G-BEJM
PK-TAL	78	BAC 1-11/488GH	259	Indonesian Air Transport, Jakarta.	G-BWES
PK-TIR	74	Falcon 20E	297	Indonesian Air Transport, Jakarta.	N121EU
PK-TRI	68	Falcon 20F	173	Indonesian Air Transport, Jakarta.	N729S
PK-TRS	67	BAC 1-11/422EQ	126	Indonesian Air Transport, Jakarta.	N51387
PK-TRU	80	BAC 1-11/492GM	262	Indonesian Air Transport, Jakarta.	G-BLDH
PK-VBA	66	B 727-25	18970	Bakrie Aviation Inc. Jakarta.	N680AM
PK-WSJ	88	BAe 125/800B	8106	EASTINDO-East Indonesia Air Charter Co. Jakarta.	G-5-580

Military

Reg	Yr	Type	c/n	Owner/Operator	Prev Regn
A-2801	71	F 28-1000	11042	TNI-AU/Government of Indonesia, Jakarta.	PK-PJT
A-7002	75	B 707-3M1C	21092	TNI-AU/Government of Indonesia, Jakarta.	PK-GAU
A-9446	64	JetStar-6	5046	TNI-AU/Government of Indonesia. 'Sapta Marga' (status ?).	T-9446

PP/PT = Brazil (Federative Republic of Brazil)

Civil

Reg	Yr	Type	c/n	Owner/Operator	Prev Regn
PT-...	67	HS 125/3A-RA	NA701		N125HS
PP-EIF	84	Citation 1/SP	501-0680	State Government of Parana, Curitiba, PR.	PT-LFR
PP-ERR	75	Learjet 35	35-008	State Government of Roraima, Boa Vista, RR.	PT-LFS
PP-ESC	89	Citation II	550-0618	Sec. Est. Casa Civil S Catarina, Florianopolis, SC.	PT-LXG
PP-JAA	84	Learjet 36A	36A-055	Learjet Latin American Sales Inc.	N365AS
PT-AAF	93	Falcon 50	234	Metro Taxi Aereo SA. Sao Paulo, SP.	N233FJ
PT-ALK	84	Gulfstream 3	418	Tendencia Corretora CVTM Ltda. Sao Paulo. SP.	JY-AMN
PT-CXJ	68	Learjet 24	24-176	Perfil Taxi Aereo Ltda. Curitiba, PR.	
PT-GAF	94	Hawker 800	8261	Banjet Taxi Aereo Ltda. Belo Horizonte, MG.	N958H
PT-GAP	84	Learjet 35A	35A-589	Imbralit Ltda. Criciuma, SC.	N8567K
PT-IIQ	72	Learjet 25C	25C-089	Pontual Leasing SA. Arrend. Mercantil, Barueri, SP.	
PT-ILJ	73	Citation	500-0057	TAM-Jatos Executivos, Sao Paulo, SP.	N557CC
PT-IQL	73	Citation	500-0069	Antonio Cabrera Mano, Gastao Vidigal, SP.	N569CC
PT-ISO	73	Learjet 25C	25C-115	Jose Carlos Ferreira Pimentel, Brasilia, DF.	
PT-JGU	73	Learjet 24D	24D-276	LRN Marketing e Promocoes Ltda. Sao Paulo, SP.	
PT-JKQ	74	Learjet 24D	24D-284	Proserc Proc. Dados SDVA Ltda. Curitiba, PR.	
PT-JMJ	73	Citation	500-0134	TAMIG-Mendes Junior SA. Belo Horizonte, MG.	N134CC
PT-JQM	93	Beechjet 400A	RK-63	First National Bank of Boston,	N82412
PT-KAP	74	Learjet 25C	25C-156	Banco Bandcirantes/Banjet Taxi Aereo Ltda. Belo Horizonte.	
PT-KBD	74	Learjet 25B	25B-166	Cia Real de Arrend. Mercantil, Barueri, SP.	
PT-KBR	74	Citation	500-0156	BCN Leasing Arrend. Mercantil SA. Barueri, SP.	
PT-KIR	73	Citation	500-0103	Alzira Delgado Garcete, Dourados, MS.	N103CC
PT-KPA	74	Citation	500-0181	Taxi Aereo Weston Ltda. Recife, PE.	N181CC
PT-KPB	74	Citation	500-0188	Heringer Taxi Aereo Ltda. Imperatriz, MA.	N5223J
PT-KPE	75	Learjet 24D	24D-315	KPE Empreendimentos Ltda. Sao Paulo, SP.	
PT-KQT	75	Learjet 36	36-011	Construtora Andrade Gutierrez SA. Belo Horizonte, MG.	
PT-KTU	75	Learjet 36A	36A-018	Hidroservice-Eng. Projetos Ltda. Sao Paulo, SP.	
PT-KZR	79	Learjet 35A	35A-252	Construtora Sa Cavalcante Ltda. Vitoria, ES.	N28CR
PT-LAA	80	Learjet 35A	35A-295	Taxi Aereo Weston Ltda. Recife, PE.	
PT-LAX	74	Citation	500-0194	TAM-Taxi Aereo Marilia SA. Sao Paulo, (SP).	N310U
PT-LBN	73	Citation	500-0079	Mario Celso Lopes, Andradina, SP.	N40JF
PT-LBW	70	Learjet 25XR	25XR-056	Fibra Leasing SA. Arrend Mercantil, S Caet. do Sul, SP.	N780A
PT-LBY	81	Learjet 35A	35A-411	Lojas Riachuelo SA. Sao Paulo, SP.	
PT-LCC	81	Citation Eagle	500-0413	Banco Rural SA. Belo Horizonte, MG.	(PT-LBZ)
PT-LCD	77	Learjet 35A	35A-103	Lider Taxi Aereo SA. Belo Horizonte, MG.	N50MJ
PT-LCR	80	Citation II	550-0142	Veloz Taxi Aerea Ltda. Sao Paulo.	N2648Z
PT-LCV	72	Learjet 24D	24D-254	Transamerica Taxi Aereo Ltda. Piracicaba, SP.	N13606
PT-LCW	81	Citation II	550-0333	Cresiful Comercio Exp. e Part. Ltda. Sao Paulo, SP.	N67990
PT-LDH	72	Citation	500-0049	Bradesco Leasing SA. Arrend. Mercantil, Osasco, SP.	PT-FXB
PT-LDI	77	Citation	500-0335	Net Aereo Fortaleza Ltda. Fortaleza, CE.	N2937L
PT-LDM	82	Learjet 35A	35A-494	Banestado Leasing SA. Arrend. Mercantil, Curitiba, PR.	
PT-LDR	88	Learjet 55B	55B-134	BRATA-Brasilia Taxi Aereo Ltda. Brasilia, DF.	N7261D
PT-LDY	79	Westwind-1124	251	TAMIG-Mendes Junior Engenharia SA. Belo Horizonte, MG.	CX-CMJ
PT-LEA	74	Learjet 25B	25B-155	Jet Service Taxi Aereo Ltda. Brasilia, DF.	N24TA
PT-LEB	82	Learjet 35A	35A-474	Taxi Aereo Weston Ltda. Recife, PE.	N37975
PT-LEM	73	Learjet 24D	24D-270	Manaca Taxi Aereo Ltda e Outros, Osasco, SP.	N3979P
PT-LEN	72	Learjet 25B	25B-093	Aeroexecutivos Taxi Aereo Ltda. Rio de Janeiro, RJ.	N33NM
PT-LET	83	Learjet 55	55-080	Mintage Corp. Sao Paulo, SP.	N85632
PT-LGI	85	Citation S/II	S550-0024	Veloz Taxi Aereo Ltda. Sao Paulo, SP.	(N12593)
PT-LGR	74	Learjet 35	35-009	Banco BMG SA. Sao Paulo, SP.	N263GL
PT-LGS	80	Learjet 35A	35A-299	Lider Taxi Aereo SA. Belo Horizonte, MG.	N244FC
PT-LGT	85	Citation III	650-0081	Lastro Operacoes Com. e Ind. Ltda. Sao Paulo, SP.	(N1316N)
PT-LGW	85	Learjet 35A	35A-598	Lider Taxi Aereo SA. Belo Horizonte, MG.	N8567T
PT-LHA	84	Citation III	650-0059	TAMIG-Mendes Junior SA. Belo Horizonte, MG.	N1313G
PT-LHB	85	BAe 125/800B	8031	CESP=Cia Energetica de Sao Paulo, Sao Paulo, SP.	PT-ZAA
PT-LHC	85	Citation III	650-0086	Interavia Taxi Aereo Ltda. Sao Paulo, SP.	(N1317X)
PT-LHD	85	Citation S/II	S550-0059	TAM-Taxi Aereo Marilia SA. Sao Paulo, SP.	(N1271N)
PT-LHK	69	HS 125/400B	25197	Macom Dist. de Petroleo Ltda. Sorocaba, SO.	PP-EEM
PT-LHR	82	Learjet 55	55-044	SOTAN-Soc de Taxi Aereo Nordeste Ltda. Maceio, AL.	N3797C

Reg	Yr	Type	c/n	Owner/Operator	Prev Regn
☐ PT-LHT	82	Learjet 35A	35A-479	Constructora Queiroz Galvao SA. Rio de Janeiro, RJ.	N30SA
☐ PT-LHY	82	Citation II	550-0427	Mario Branca Peres, Sao Paulo, SP.	N923RL
☐ PT-LIV	84	Citation II	550-0499	Antonio Gilberto Depieri, Sao Paulo, SP.	N550PT
☐ PT-LIX	74	Citation	500-0171	Aeroexecutivos Taxi Aereo Ltda. Rio de Janeiro, RJ.	N728US
☐ PT-LIY	75	Citation	500-0219	Bozano S Leasing SA. Rio de Janeiro, RJ.	N408CA
☐ PT-LIZ	82	Citation 1/SP	501-0234	BFB Leasing SA. Barueri, SP.	N711RP
☐ PT-LJF	81	Citation II/SP	551-0289	Itamarati Leasing Arrend Merc SA.Cuiaba, MT. (was 550-0244).	N551BW
☐ PT-LJI	86	Falcon 50	173	Morro Vermelho Taxi Aereo Ltda. Sao Paulo, SP.	N172FJ
☐ PT-LJJ	81	Citation II	550-0247	Antonio Luiz Fuchter, Sao Jose, SC.	N928DS
☐ PT-LJK	81	Learjet 35A	35A-372	Lider Taxi Aereo SA. Belo Horizonte, MG.	N372AS
☐ PT-LJL	86	Citation S/II	S550-0084	Banco Rural SA. Belo Horizonte, MG.	N1274N
☐ PT-LJQ	86	Citation S/II	S550-0113	Taxi Aereo Sinuelo Ltda. Vespasiano, MG.	N553CC
☐ PT-LKD	78	Learjet 24F	24F-356	Imperial Taxi Aereo Ltda. Sao Paulo, SP.	N113JS
☐ PT-LKR	82	Citation II	550-0344	Sa Rachid B. Saliba Ind. e Com. Alfenas, MG.	N532M
☐ PT-LKS	86	Citation S/II	S550-0114	Cia Itauleasing Arrend. Mercantil. Poa, SP.	N1292A
☐ PT-LLK	89	Learjet 31	31-010	Lider Taxi Aereo SA. Belo Horizonte, MG.	N446
☐ PT-LLN	74	Learjet 25C	25C-176	Pontual Leasing SA. Barueri, SP.	N28KV
☐ PT-LLQ	84	Citation II	550-0495	Arcel SA. Emp. e Particip. e Outros, Campinas, SP.	N505GL
☐ PT-LLS	80	Learjet 35A	35A-303	Bamerindus Leasing SA. Curitiba, PR.	N771A
☐ PT-LLT	81	Citation II	550-0327	Banorte Leasing SA. Barueri, SP.	N74JN
☐ PT-LLU	80	Citation II	550-0132	BFB Leasing SA. Barueri, SP.	(N330MG)
☐ PT-LME	80	Citation II/SP	551-0023	Itamarati Leasing SA. Cuiaba, MT.	N34DL
☐ PT-LMF	66	Learjet 24	24-120	Rico Taxi Aereo Ltda. Manaus, AM.	N44NJ
☐ PT-LML	78	Citation II	550-0013	TAM-Taxi Aereo Marilia SA. Sao Paulo, SP.	N21SV
☐ PT-LMM	80	Learjet 25D	25D-323	NTA-Nacional Taxi Aereo Ltda. Rio de Janeiro, RJ.	N6YY
☐ PT-LMO	75	Falcon 10	49	Flysul Aerotaxi Ltda. Porto Alegre, RS.	N700TT
☐ PT-LMS	74	Learjet 24D	24D-296	Transamerica Taxi Aereo Ltda. Sao Paulo, SP.	N500DJ
☐ PT-LMY	87	Learjet 35A	35A-627	Lider Taxi Aereo SA. Belo Horizonte, MG.	N7260T
☐ PT-LNC	81	Citation II	550-0222	Chapeco Taxi Aereo Ltda. Chapeco, SC.	N17RG
☐ PT-LND	81	Citation II	550-0227	Banco Agrimisa SA. Sao Paulo, SP.	N254CC
☐ PT-LNE	66	Learjet 24	24-114	Adhemar Goncalves Moreira Neto, Belo Horizonte, MG.	N99DM
☐ PT-LNN	83	Diamond 1A	A048SA	Taxi Aereo Pinhal Ltda. Sao Paulo, SP.	N335DM
☐ PT-LOE	81	Learjet 35A	35A-393	Wanair Taxi Aereo Ltda. Belo Horizonte, MG.	N700WJ
☐ PT-LOG	75	Citation	500-0284	Itamarati Leasing Arrend. Mercantil SA. Cuiaba, MT.	N37DW
☐ PT-LOJ	75	Learjet 24D	24D-303	EMS Industria Farmaceutica Ltda. S. B. do Campo, SP.	N303EJ
☐ PT-LOS	74	Citation	500-0239	TAM-Taxi Aereo Marilia SA. Sao Paulo, SP.	N6034F
☐ PT-LOT	77	Learjet 35A	35A-093	Antena um Radiodifusao Ltda. Rio de Janeiro, RJ.	N44PT
☐ PT-LPF	75	Citation	500-0249	Bahiatech Bahia Technologia Ltda. Ilheus, BA.	(N789DD)
☐ PT-LPH	73	Learjet 24D	24D-275	Nova Prospera Mineracao SA. Belo Horizonte, MG.	N216HB
☐ PT-LPK	78	Citation II	550-0010	TAM-Jatos Executivos, Sao Paulo, SP.	N806C
☐ PT-LPN	81	Citation II	550-0294	CBM Taxi Aereo Ltda. Belo Horizonte, MG.	N323CJ
☐ PT-LPP	80	Citation II	550-0218	EMS Industria Farmaceutica Ltda. S. B. de Campo, SP.	N45EP
☐ PT-LPZ	72	Citation	500-0015	TAM-Taxi Aereo Marilia SA. Sao Paulo, SP.	N14JL
☐ PT-LQG	75	Citation	500-0271	DM Construtura de Obras Ltda. Curitiba, PR.	N53FB
☐ PT-LQJ	88	Citation II	550-0578	Interavia Taxi Aereo Ltda. Sao Paulo, SP.	N1300N
☐ PT-LQK	76	Learjet 24E	24E-333	Safra Leasing SA. Rio de Janeiro, RJ.	N75GR
☐ PT-LQP	88	BAe 125/800B	8116	Metro Taxi Aereo Ltda. Sao Paulo, SP.	G-5-592
☐ PT-LQQ	79	Citation 1/SP	501-0129	Tectelcom Tecnia em Telecom Ltda. S J dos Campos, SP.	N70WP
☐ PT-LQR	75	Citation	500-0246	Radio e Televisao OM Ltda. Curitiba, PR.	N227VG
☐ PT-LSF	76	Citation	500-0328	Crasa Taxi Aereo Ltda. Curitiba, PR.	(N571K)
☐ PT-LSJ	78	Learjet 35A	35A-181	Aerotaxi Pampulha Ltda. Belo Horizonte, MG.	N5114G
☐ PT-LSN	84	Citation III	650-0049	Sersan Taxi Aereo Ltda. Brasilia, DF.	PJ-MAR
☐ PT-LSR	89	Citation II	550-0600	Indaia Taxi Aereo Ltda. Recife, PE.	N1303H
☐ PT-LTB	89	Citation III	650-0166	PONTAX-Ponta Grossa Taxi Aereo Ltda. Pres. Prudente, SP.	N1313J
☐ PT-LTI	75	Citation	500-0226	Helicentro Campinas SCRA Ltda. Campinas, SP.	N100AD
☐ PT-LTJ	81	Citation II	550-0225	Produtos Electricos Corona Ltda. Sao Paulo, SP.	N258CC
☐ PT-LUA	77	Citation	500-0346	Radio e Televisao Iguacu SA. Curitiba, PR.	N56DV
☐ PT-LUE	85	Citation III	650-0091	TAM-Jatos Executivos, Sao Paulo, SP.	N58HC
☐ PT-LUG	80	Learjet 35A	35A-356	LUG Taxi Aereo Ltda. Maceio, AL.	N800WJ
☐ PT-LUK	83	Learjet 55	55-086	Taxi Aereo Weston Ltda. Recife, PE.	N8227P
☐ PT-LUO	86	Citation III	650-0129	TAM-Jatos Executivos, Sao Paulo, SP.	(N309TA)
☐ PT-LUZ	81	Learjet 25D	25D-335	BCN Leasing SA. Rio de janeiro, RJ.	N27KG
☐ PT-LVB	81	Citation 1/SP	501-0205	Taxi Aereo Weston Ltda. Recife, PE.	N6784T
☐ PT-LVD	88	Falcon 100	223	Brasil Warrant Adm. Bens Empr. Ltda. Matao, SP.	N126FJ
☐ PT-LVF	89	Citation III	650-0171	TAM-Taxi Aereo Marilia SA. Sao Paulo, SP.	N1354G
☐ PT-LVO	88	Learjet 31	31-002	Itapemirim Taxi Aereo Ltda. Itapemirim, ES.	N7262Y
☐ PT-LVR	89	Learjet 31	31-013	Lojas Americanos SA. Rio de Janeiro, RJ.	
☐ PT-LXH	73	Citation	500-0133	Nacional Expresso Ltda. Uberlandia, MG.	N1270K
☐ PT-LXJ	89	Falcon 100	225	Taxi Aereo Par Ltda. Araras, SP.	N127FJ
☐ PT-LXO	88	Learjet 55C	55C-135	Aeropetrol Taxi Aereo SA. Curitiba, PR.	N1055C
☐ PT-LXS	73	Learjet 25B	25B-111	Bandeirantes SA. Belo Horizonte, MG.	N55ES
☐ PT-LXX	89	Learjet 31	31-007	Arbi Participacoes SA. Rio de Janeiro, RJ.	N3819G
☐ PT-LYA	89	Citation II	550-0620	Copasoja Taxi Aereo Ltda. Campo Grande, MS.	N1254G
☐ PT-LYE	77	Learjet 24F	24F-354	TAFETAL-Taxi Aereo Feijo e Tarauaca Ltda. Rio Branco, AC.	N678SP
☐ PT-LYS	89	Citation II	550-0624	Seguranca Taxi Aereo Ltda. Sao Jose do Rio Preto, SP.	N1255Y
☐ PT-LZO	80	Citation II	550-0249	CAVOK Taxi Aereo Ltda. Rio de Janeiro, RJ. (was 551-0236).	N201U

Reg	Yr Type	c/n	Owner/Operator	Prev Regn

Reg	Yr	Type	c/n	Owner/Operator	Prev Regn
PT-LZP	80	Learjet 35A	35A-339	Aerotaxi Pampulha Ltda. Belo Horizonte, MG.	N1500
PT-LZQ	89	Citation V	560-0045	Navegantes Taxi Aereo Ltda. Navegantes, SC.	N2665Y
PT-LZS	89	Learjet 55C	55C-139	Embrataxi Aereo Ltda. Belo Horizonte, MG.	N1039L
PT-MBZ	88	Astra-1125	022	Serv-Jet Ltda/Mercedes-Benz do Brazil, Sao Paulo, SP.	4X-CUT
PT-MCB	94	Learjet 31A	31A-100	Casa Bahia Comercial Ltda. Sao Paulo, SP.	N31LR
PT-MGS	92	Citation VII	650-7021	TAM-Taxi Aereo Marilia SA. Sao Paulo, SP.	N1262B
PT-MIL	95	CitationJet	525-0086	Viacao Barao de Maua Ltda. Sao Paulo, SP.	
PT-MJC	94	CitationJet	525-0085	Cremo Empreendimentos SA. Sao Paulo.	VR-CDN
PT-MML	97	Falcon 2000	43	Malharia Manz Ltda. Sao Paulo, SP.	N2077
PT-MMO	83	Citation II/SP	551-0455	Nor-Jet Particip. e Serv. Ltda. Rio de Janeiro, SP.	N90SF
PT-MPE	93	CitationJet	525-0015	TAM-Taxi Aereo Marilia SA. Sao Paulo, SP.	N115CJ
PT-MVI	93	Learjet 31A	31A-082	Indaia Taxi Aereo Ltda. Recife, PE.	N4022X
PT-OAA	90	Citation II	550-0635	TAM-Taxi Aereo Marilia SA. Sao Paulo, SP.	N1258H
PT-OAC	89	Citation II	550-0613	Equip Taxi Aereo Ltda. Curitiba, PR.	N1250P
PT-OAG	82	Citation II	550-0357	ZLC Intermed. de Negocios SC Ltda. Sao Paulo, SP.	N29FA
PT-OBD	71	Learjet 24B	24B-228	Angra Taxi Aereo SA. Sao Paulo, SP.	N150AB
PT-OBX	89	Citation III	650-0181	Banco Bradesco SA. Osasco, SP.	N181CC
PT-OCA	89	Learjet 55C	55C-140	Araucaria Aerotaxi Ltda. Curitiba, PR. 'Marcia Jose'	N72616
PT-OCZ	80	Learjet 35A	35A-361	BRATA-Brasilia Taxi Aereo Ltda. Brasilia, DF.	PT-FAT
PT-ODC	84	Citation 1/SP	501-0678	Martins Com. Imp. Exportacao Ltda. Uberlandia, MG.	N678CF
PT-ODL	90	Citation II	550-0640	Ezibras Imoveis e Represent. Ltda. Sao Paulo, SP.	N1308V
PT-ODZ	90	Citation II	550-0645	Banco Bandeirantes SA. Belo Horizonte, MG.	N1310C
PT-OER	82	Citation II	550-0356	Sudameris Arrend. Mercantil SA. Barueri, SP.	N6801Z
PT-OEX	90	Falcon 900	92	Safra Comercio e Servicos Ltda. Sao Paulo, SP.	N463FJ
PT-OHB	90	BAe 125/800B	8190	Jet Service Taxi Aereo Ltda. Brasilia, DF.	G-BTAE
PT-OHD	79	Learjet 25D	25D-296	Tupy Taxi Aereo Ltda. Sao Jose do Rio Preto, SP.	N55MJ
PT-OIG	72	Citation	500-0005	Aeroexecutivos Taxi Aereo Ltda. Sao Paulo, SP.	N815HC
PT-OJF	73	Citation	500-0131	Noronha Taxi Aereo Ltda. Teresina, PI.	N457CA
PT-OJG	91	Citation II	550-0676	Industrias Muller de Bebidas Ltda. Piarassununga, SP.	N67741
PT-OJO	91	Citation VI	650-0202	Refrigeracao Parana SA. Curitiba, PR.	N2627A
PT-OJT	87	Citation II	550-0562	Jose Claudio P. Goncalves e Outro, Belo Horizonte, MG.	N562CD
PT-OKP	83	Citation II	550-0460	Parana Jet Taxi Aereo Ltda. Curitiba, PR. (was 551-0460).	N6523A
PT-OMB	91	Citation II	550-0672	Distrib. Farmac. Panarello Ltda. Goiania, GO.	N6763C
PT-OMS	75	Citation	500-0251	ATA-Aerotaxi Abaete Ltda. Salvador, BA.	N790EA
PT-OMT	74	Citation	500-0179	ATA-Aerotaxi Abaete Ltda. Salvador, BA.	N179EA
PT-OMU	91	Citation	650-0205	TAM-Jatos Executivos, Sao Paulo, SP.	N2630N
PT-OOF	73	Citation	500-0074	Aeroexecutivos Taxi Aereo Ltda. Rio de Janeiro, RJ.	N8JG
PT-OOI	91	BAe 125/800B	8214	Grupo OK Benefica Cia Nacional de Pneus Ltda. Brasilia, DF.	VR-CCX
PT-OOK	72	Citation	500-0039	Taxi Aereo Taroba Ltda. Cascavel, PR.	N555CC
PT-OOL	73	Citation	500-0060	TTA-Teresina Taxi Aereo Ltda. Teresina, PI.	N712G
PT-OPJ	81	Learjet 35A	35A-396	Topjet Taxi Aereo Ltda. Uberlandia, MG.	N74JL
PT-OQD	75	Citation	500-0244	Superjet Taxi Aereo Ltda. Florianopolis, SC.	N516AB
PT-ORA	90	Learjet 55C	55C-146	Taxi Aereo Weston Ltda. Recife, PE.	N9125M
PT-ORC	92	Citation V	560-0195	TAM-Jatos Executivos, Sao Paulo, SP.	N1282N
PT-ORD	80	Citation II	550-0139	Taxi Aereo Marilia SA. Sao Paulo, SP.	N39FA
PT-ORE	91	Citation V	560-0131	TAM-Jatos Executivos, Sao Paulo, SP.	N131CV
PT-ORJ	81	HS 125/700A	NA0339		N125BJ
PT-ORO	81	Citation II	550-0271	Taxi Aereo Blumenau Ltda. Curitiba, PR.	N303J
PT-ORS	88	Falcon 100	219	Recrusul SA.	N2649
PT-OSA	90	Challenger 601-3A	5075	Centaurus Taxi Aereo Ltda. Rio de Janeiro, RJ.	VR-CCV
PT-OSD	75	Citation	500-0325	Vadao Transportes Ltda. Estr d'Oeste, SP.	N60MP
PT-OSM	88	Citation S/II	S550-0160	Empresa Baiana de Taxi Aereo Ltda. Salvador, BA.	N550GT
PT-OSW	90	BAe 125/800B	8184	Wanair Taxi Aereo Ltda. Belo Horizonte, MG.	G-5-678
PT-OTC	90	BAe 125/800B	8194	Centaurus Taxi Aereo Ltda. Rio de Janeiro, RJ.	G-5-692
PT-OTQ	72	Citation	500-0046	Diogenes Setti Sobreira, Rio de Janeiro, RJ.	N929RW
PT-OTS	93	Citation V	560-0213	Guaxupe Taxi Aereo Ltda. Guaxupe, MG.	N12845
PT-OTT	93	Citation V	560-0215	Citrosuco Paulista SA.	N1285G
PT-OVC	81	Learjet 35A	35A-399	Gobair Corp. Porto Alegre, RS.	N399AZ
PT-OVK	72	Citation	500-0027	TTA-Teresina Taxi Aereo Ltda. Teresina, PI.	N777AN
PT-OVU	93	Citation VII	650-7033	Banco Bradesco SA. Rio de Janeiro, RJ.	N1264B
PT-OVV	89	Citation II	550-0616	First National Bank of Boston, Nassau, Bahamas.	(D-IAFA)
PT-OVZ	91	Learjet 31A	31A-037	Araucaria Aerotaxi Ltda. Curitiba, PR.	N31TF
PT-OXT	83	Diamond 1A	A039SA	Construtora Borges Landeiro Ltda. Goiania, GO.	N399MJ
PT-OYA	73	Citation	500-0072	Interjet Aviation Inc.	N103AJ
PT-OYP	87	Citation II	550-0561	Solair Finance & Leasing Co. Curitiba; PR.	N916WJ
PT-OZB	94	Citation V	560-0258	Marcep International Trade Finance Ltd. Rio de Janeiro, RJ.	N1293G
PT-OZX	76	Citation Eagle	500-0299	BB Leasing Co Ltd. Presidente Prudente, SP.	YV-940CP
PT-POK	86	Learjet 35A	35A-619	OK Benefica Cia Nac. de Pneus Ltda. Brasilia, DF.	N8568V
PT-TOF	95	Learjet 31A	31A-103	Transbrasil Airlines Inc. Brasilia, DF.	N31AZ
PT-WAL	90	BAe 125/800B	8198	Construtora Cowan Ltda. Belo Horizonte, MG.	G-5-694
PT-WAR	79	Learjet 35A	35A-230	Wanair Taxi Aereo Ltda. Belo Horizonte, MG.	N714K
PT-WAU	88	BAe 125/800B	8133	EMSA Panama Inc. Goiania, GO.	HP-1262
PT-WBC	96	Astra-1125SPX	086	Companhia Cervejaria Brahma, Rio de Janeiro, RJ.	N793A
PT-WBV	84	Citation II	550-0485	Bamerindus Leasing SA. Rio de Janeiro, RJ.	N485A
PT-WBY	72	Citation	500-0008	Clipper Agencia de Viagens Ltda. Aracaju, SE.	ZP-TYP
PT-WEW	68	Learjet 24	24-158	Pedro Muffato e Cia Ltda. Cascavel, PR.	N220PM

Reg   *Yr* *Type*   *c/n*   *Owner/Operator*   *Prev Regn*

Reg	Yr	Type	c/n	Owner/Operator	Prev Regn
□ PT-WFD	95	Citation V Ultra	560-0308	Serveng Civilsan SA. Sao Paulo, SP.	N5235G
□ PT-WFT	74	Citation	500-0154	TASUL-Taxi Aereo Sul Ltda. Porto Alegre, RS.	N54MC
□ PT-WGD	95	CitationJet	525-0120	Banco Itamarati SA. Sao Paulo, SP.	N5264M
□ PT-WGF	80	Learjet 35A	35A-322	Learjet Latin American Sales Inc. Belo Horizonte, MG.	N305SC
□ PT-WGM	81	Learjet 36A	36A-048	Bougainville Part. e Repres. Ltda. Sao Paulo, SP.	N3NP
□ PT-WHB	93	Beechjet 400A	RK-73	Lider Taxi Aereo SA. Belo Horizonte, MG.	N8070Q
□ PT-WHC	93	Beechjet 400A	RK-58	Lider Taxi Aereo SA. Belo Horizonte, MG.	N56356
□ PT-WHD	93	Beechjet 400A	RK-77	Lider Taxi Aereo SA. Belo Horizonte, MG.	N8277Y
□ PT-WHE	93	Beechjet 400A	RK-81	Lider Taxi Aereo SA. Belo Horizonte, MG.	N8167G
□ PT-WHF	93	Beechjet 400A	RK-82	Lider Taxi Aereo SA. Belo Horizonte, MG.	N8282E
□ PT-WHG	92	Beechjet 400A	RK-54	Lider Taxi Aereo SA. Belo Horizonte, MG.	N80938
□ PT-WHH	95	Hawker 800XP	8282		N916H
□ PT-WHZ	75	Citation	500-0287	ASA Branca Taxi Aerea Ltda. Aracaju, SE.	N31LH
□ PT-WIB	87	Citation S/II	S550-0137	BB Leasing Co Ltda. Passo Fundo, RS.	N100TB
□ PT-WIV	93	Learjet 31A	31A-110	Du Valle Transporta Com. e Agro. Ltda. Vilhena, RO.	N40130
□ PT-WJS	96	Beechjet 400A	RK-122	Frigorifico Bertin Ltda. Sao Paulo. SP.	N1102B
□ PT-WJZ	81	Citation II	550-0318	TAM Jetos Exexutivos, Sao Paulo, SP.	PT-LJT
□ PT-WKL	74	Learjet 24D	24D-294	Transamerica Taxi Aereo Ltda. Sao Paulo, SP.	PP-EIW
□ PT-WKQ	91	Citation II	550-0675	BLD Leasing LLC. Miami, Fl. USA.	N275BD
□ PT-WKS	96	Citation V Ultra	560-0397	Transar Taxi Aereo SA. Sao Paulo. SP.	
□ PT-WLM	91	Beechjet 400A	RK-28	Confianca Factor. F. Merc. Ltda. Cuiaba, MT.	N411SK
□ PT-WLO	97	Learjet 31A	31A-122	Usiminas Adm Part S. Med, Belo Horizonte, MG.	N122LJ
□ PT-WLX	97	CitationJet	525-0176		N5161J
□ PT-WLY	96	Citation VII	650-7074	La Fonte Empresa de Shopping Centers SA. Sao Paulo, SP.	N5183V
□ PT-WLZ	95	Challenger 601-3R	5189	Globo Radio & TV Network, Rio de Janeiro, RJ.	C-FXCK
□ PT-WMA	96	Hawker 800XP	8301	Lider Taxi Aereo SA. Belo Horizonte, MG.	N1105Z
□ PT-WMD	96	Hawker 800XP	8312	Lider Taxi Aereo SA. Belo Horizonte, MG.	N312XP
□ PT-WMG	96	Hawker 800XP	8310	Lider Taxi Aereo SA. Belo Horizonte, MG.	N310XP
□ PT-WMO	96	Learjet 60	60-090	Globo Radio & Television Network, Rio de Janeiro, RJ..	N8090P
□ PT-WMQ	96	Citation V Ultra	560-0405		N5202D
□ PT-WMZ	97	Citation V Ultra	560-0406		N5203S
□ PT-WNE	97	Citation V Ultra	560-0411		N5226B
□ PT-WNF	97	Citation V Ultra	560-0412		N5228Z
□ PT-WNH	97	Citation Bravo	550-0814	Arcom Commercial Imp. Exp. Ltda.	
□ PT-WNH	97	Citation Bravo	550-0811	Morro Vermelho Taxi Aereo Ltda. Sao Paulo.	N5221Y
□ PT-WNO	95	Hawker 800XP	8284	UNIMED N N F C Trabajos Med. Ltda. Joao Pessoa.	N919H
□ PT-WOA	97	Citation V Ultra	560-0408		N5207A
□ PT-WOD	76	Citation	500-0340	Radio e TV Difus. do Maranho Ltda. Sao Luiz, MA.	N340DN
□ PT-WOM	92	Citation V	560-0176		N176VP
□ PT-WON	90	Citation II	550-0641	Cia Brasileiros de Metal e Mineracao, Sao Paulo, SP.	N1309A
□ PT-WPC	91	Citation V	560-0142	Executive Jet Leasing Co.	N7220L
□ PT-WRC	86	Gulfstream 3	492	Companhia Leco de Productos Alimenticios SA. Sao Paulo, SP.	N212AD
□ PT-WRR	96	Citation V Ultra	560-0389	R R Administracao e Comercio SA. Sao Paulo, SP.	N5090A
□ PT-WSB	97	Learjet 31A	31A-135	Bravespeed Trading Corp. Curitiba, PR.	N80645
□ PT-WSC	95	Falcon 50EX	253	Sucocitrico Cutrale Ltda. Sao Paulo.	N253EX
□ PT-WSF	80	Falcon 10	169	Flysul Aerotaxi Ltda. Porte Alegre, RS.	(N107AF)
□ PT-WSS	84	Learjet 55	55-102	Alcoa Aluminio SA. Sao Paulo, SP.	N44GA
□ PT-WVH	97	Citation V Ultra	560-0409		
□ PT-WXL	96	Challenger 604	5321		C-FZDY
□ PT-WZW	97	Citation V Ultra	560-0431		
Military					
□ 2401	68	B 707-345C	19840	FAB=Forca Aerea Brasileira.	PP-VJY
□ 2710	87	Learjet 35A	35A-631	Brazilian Air Force,	N3818G
□ 2711	87	Learjet 35A	35A-632	Brazilian Air Force,	N1461B
□ 2712	87	Learjet 35A	35A-633	Brazilian Air Force,	N39416
□ 2713	88	Learjet 35A	35A-636	Brazilian Air Force,	
□ 2714	88	Learjet 35A	35A-638	Brazilian Air Force,	
□ 2715	88	Learjet 35A	35A-639	Brazilian Air Force,	
□ 2716	88	Learjet 35A	35A-640	Brazilian Air Force.	N8568Y
□ 2717	88	Learjet 35A	35A-641	Brazilian Air Force,	N7261H
□ 2718	88	Learjet 35A	35A-642	Brazilian Air Force,	N7262X
□ EC93-2125	68	HS 125/3B-RC	25164	FAB=Forca Aerea Brasileira. Flight inspection.	
□ EU93-2119	73	HS 125/403B	25274	FAB, G.E.I.V., Rio-Santos Dumont. (flight calibration).	G-5-20
□ EU93-2121	68	HS 125/3B-RC	25165	FAB, G.E.I.V., Rio-Santos Dumont. (flight calibration).	VC93-2121
□ FAB6000	86	Learjet 35A	35A-613	FAB=Forca Aerea Brasileira.	N4289X
□ FAB6001	86	Learjet 35A	35A-615	FAB=Forca Aerea Brasileira.	N7260E
□ FAB6002	86	Learjet 35A	35A-617	FAB=Forca Aerea Brasileira.	N4289Z
□ VC93-2120	68	HS 125/3B-RC	25162	FAB, GTE=Grupo do Transporte Especiale, Brasilia.	
□ VC93-2123	68	HS 125/3B-RC	25167	FAB, GTE=Grupo do Transporte Especiale, Brasilia.	
□ VC93-2124	68	HS 125/3B-RC	25168	FAB, GTE=Grupo do Transporte Especiale, Brasilia.	
□ VC96-2115	76	B 737-2N3	21165	FAB, GTE=Grupo do Transporte Especiale, Brasilia.	
□ VC96-2116	76	B 737-2N5	21166	FAB, GTE=Grupo do Transporte Especiale, Brasilia.	
□ VU93-2114	70	HS 125/400A	NA740	FAB, GTE=Grupo do Transporte Especiale, Brasilia	N702P
□ VU93-2117	69	HS 125/400A	NA738	FAB, GTE=Grupo do Transporte Especiale, Brasilia.	N702D
□ VU93-2118	69	HS 125/400A	NA729	FAB, GTE=Grupo do Transporte Especiale, Brasilia.	N702SS
□ VU93-2126	73	HS 125/403B	25277	FAB, GTE=Grupo do Transporte Especiale, Brasilia.	
Reg	Yr	Type	c/n	Owner/Operator	Prev Regn

☐ VU93-2127	73	HS 125/403B	25288	FAB, GTE=Grupo do Transporte Especiale, Brasilia.	
☐ VU93-2128	73	HS 125/403B	25289	FAB, GTE=Grupo do Transporte Especiale, Brasilia.	G-5-16

P2 = Papua New Guinea *(Independent State of Papua New Guinea)*
Civil

☐ P2-MBN	80	Citation II	550-0145	MBA-Milne Bay Air P/L. Port Moresby.	VH-TFQ

P4 = Aruba
Civil

☐ P4-AFE	80	B 747SP-31	21962	Government of Brunei, Bandar Seri Begawan, Brunei.	UN-001
☐ P4-AMB	72	HS 125/400B	25252	Pro Air Ambulance, Oranjestad.	N48US
☐ P4-JLD	67	B 727-193	19620	Joylud Dist International/Government of Tatarstan.	VP-CWC
☐ P4-NSN	86	B 757-2M6	23454	Government of Kazakhstan, Almaty.	VR-CRK
☐ P4-TBN	74	B 707-3L6B	21049	TBN Aircraft Aruba Ltd.	A6-HPZ

RA = Russia
Civil

☐ RA-02800	77	HS 125/700B	7007	Tomsk Neft Gaz, Moscow.	G-5-721
☐ RA-02801	80	HS 125/700B	7097	Meridian Air/Magnitogorsk & Tjazhprom, Moscow.	G-5-810
☐ RA-02803	81	HS 125/700B	7139	Ward Business Inc-Panama/AVPROM-Moscow.	G-MHIH
☐ RA-02807	86	BAe 125/800B	8076	Meridian Air/Master Group. Moscow.	G-5-535
☐ RA-02809	79	HS 125/700A	7062	SIAT-Sibaviatrans/AVCOM, Moscow.	G-5-708
☐ RA-09000	92	Falcon 900B	118	Gazprom/Gazkomplektimpex, Moscow.	F-GNFI
☐ RA-09001	94	Falcon 900B	123	Gazprom/Gazkomplektimpex, Moscow.	F-WWFL
☐ RA-09003	68	Falcon 20D	183	AVCOM Aviation Commercial, Moscow.	EC-EFR
☐ RA-42424	95	YAK 142	...	LUKoil,	
☐ RA-86466	87	IL-62M	2749316	Government of Russia, Moscow.	
☐ RA-86559	91	IL-62M	...	Government of Russia, Moscow.	
☐ RA-86561	91	IL-62M	...	Government of Russia, Moscow.	
☐ RA-86712	86	IL-62M	4648339	Government of Russia, Moscow.	

Military

☐ 62 Black	69	Gulfstream 2	62	Russian Air Force, Chalovsky-Star City.	(N777TX)

RP = Philippines *(Republic of the Philippines)*
Civil

☐ RP-C57	79	Learjet 35A	35A-244	Philippines National Oil Co. Manila.	RP-57
☐ RP-C125	65	HS 125/1A-522	25033	Makar Properties Development Inc. Cebu.	N125LL
☐ RP-C235	81	HS 125/700B	7130	Philippine National Bank, Manila.	G-RJRI
☐ RP-C610	80	Learjet 35A	35A-338	United Coconut Planters Bank, Manila.	N610GE
☐ RP-C648	96	Learjet 60	60-093	Subic International Air Charter, Manila.	N80683
☐ RP-C653	80	Citation II	550-0181	Tagum Agricultural Development/House of Travel Inc. Manila.	N88826
☐ RP-C689	80	Citation II	550-0144	Philippine Overseas Telecom Corp. Manila.	N2649E
☐ RP-C848	65	Learjet 23	23-072	Goodyear Systems Philippines Inc. Quezon City.	N2SN
☐ RP-C1180	90	Citation II	550-0658		N550MZ
☐ RP-1250	79	F 28-3000	11153	Government of Philippines, Manila.	PH-ZBV
☐ RP-C1299	75	Citation	500-0259	Standard Construction & Heavy Equipment Rebuilding Corp.	JA8247
☐ RP-C1404	81	Learjet 35A	35A-441	Columbian Motors/Subic International Air Charter, Manila.	N404JS
☐ RP-C1426	81	Learjet 35A	35A-426	Columbian Motors/Subic International Air Charter, Manila.	N1128J
☐ RP-C1600	74	HS 125/600B	6037	L & L Corporate Jet Aviation Corp. Manila.	VH-NJA
☐ RP-C1747	73	Learjet 24XR	24XR-264	Subic International Air Charter Inc. Manila.	PI-C1747
☐ RP-C1911	80	Falcon 10	174	Associated Brokerage Systems Inc. Manila.	N402ES
☐ RP-C1926	92	BAe 125/800B	8226	A Soriano Aviation Inc/San Miguel Corp. Manila.	G-5-755
☐ RP-C1964	75	Citation	500-0242	Royal Star Aviation Inc. Manila.	N5242J
☐ RP-C2480	82	Westwind-Two	364	Airspan Corp. Manila.	N198HE
☐ RP-C4007	92	B 737-332	25996	Government of Philippines, Manila.	RP-C2000
☐ RP-C4654	92	Citation II	550-0707	Philippine Long Distance Telephone Co. Manila.	N1202T
☐ RP-C8008	91	BAe 125/800A	8212	Skyjet Inc/Gull Wing Center Inc. Manila.	D-CSRI
☐ RP-C8818	97	Citation V Ultra	560-0417		N1248B

SE = Sweden *(Kingdom of Sweden)*
Civil

☐ SE-DDY	80	Citation II	550-0115	Inter Air AB. Malmo-Sturup.	OY-CCU
☐ SE-DEG	75	Citation	500-0276	Golden Air AB. Lidkoping.	N473LR
☐ SE-DEY	77	Citation	500-0370	Inter Air AB. Malmo-Sturup.	N36897
☐ SE-DEZ	77	Citation 1/SP	501-0279	Business Jet Sweden AB. Vasteras. (was 500-0371).	N43BG
☐ SE-DHL	84	Citation III	650-0030	Stora Flight AB. Borlange.	N650SC
☐ SE-DHO	78	Learjet 35A	35A-195	Nyge-Aero AB. Nykoping.	N555JE
☐ SE-DHP	76	Learjet 35A	35A-075	Nyge-Aero AB. Nykoping.	N30FN
☐ SE-DKC	78	Falcon 10	123	Inter Air AB. Malmo-Sturup.	N25FF
☐ SE-DLB	81	Falcon 100	183	Andersson Business Jet AB. Stockholm.	N183SR
☐ SE-DLZ	82	Citation	500-0411	Jiv Air Marketing AB. Goteborg.	G-NCMT
☐ SE-DPG	90	Citation V	560-0086	Birgma Sweden AB. Ljungbyhed	(N2748V)
☐ SE-DPK	80	Falcon 10	152	Inter Air AB. Malmo-Sturup.	F-GDRN
☐ SE-DPZ	77	HS 125/700A	NA0210	Scanjet Sweden AB. Norrkoping.	N37P
Reg	*Yr*	*Type*	*c/n*	*Owner/Operator*	*Prev Regn*

Reg	Yr	Type	c/n	Owner/Operator	Prev Regn
SE-DRS	91	Beechjet 400A	RK-37	Stralfors AB. Ljungby-Ferringe.	N8014Q
SE-DRT	76	Citation	500-0311	Air Partner of Sweden AB. Aengelholm.	(OY-VIP)
SE-DRZ	76	Citation	500-0315	Jetair/Air Center West, Roskilde, Denmark.	N55SH
SE-DSA	75	Falcon 20F-5	339	SAABAIR AB. Linkoping.	N19MX
SE-DUZ	74	Citation	500-0143	Karlebo Aviation AB, Stockholm	N767BA
SE-DVA	80	Citation	500-0397	Blekinge Flyg AB. Ronneby.	OY-FFC
SE-DVB	76	Citation	500-0294	Team Jelbe Production On Air Television, Stockholm.	N924AS
SE-DVD	97	Hawker 800XP	8339	Birgma Sweden AB. Ljungbyhed.	N23395
SE-DVG	82	Falcon 50	104	Volvo AB. Goteborg.	N50VG
SE-DVK	94	Falcon 50	249	SAABAIR AB. Linkoping.	N663MN
SE-DVL	93	Falcon 50	238	SAABAIR AB. Linkoping.	N238DL
SE-DVP	89	Falcon 100	224		N135FJ
SE-DVS	70	HS 125/731	NA749	Scanjet Sweden AB/ABN Amro Bank AB. Stockholm.	(N45ND)
SE-DVT	90	Citation II	550-0634	Volvo AB. Goteborg.	N550SB
SE-DVY	92	Citation VII	650-7011	Inter Air AB, Malmo-Sturup	N700VP
SE-DVZ	97	Citation Bravo	550-0808	IFS AB.	N1299B
Military					
102001	87	Gulfstream 4/Tp 102	1014	Swedish Air Force, Stockholm.	N779SW
102002	93	Gulfstream 4/Tp 102B	1215	Swedish Air Force, Stockholm.	N426GA
102003	93	Gulfstream 4/Tp 102B	1216	Swedish Air Force, Stockholm.	N440GA
86001	65	Sabre-40/Tp 86	282-49	Defense Material Administration, Linkoping.	N905KB
86002	67	Sabre-40/Tp 86	282-91	Defence Material Administration, Linkoping.	N40NR

SP = Poland (Republic of Poland)

Civil

SP-FCP	68	Falcon 20C	136		HB-VBM

ST = Sudan (The Republic of the Sudan)

Civil

ST-DRS	75	B 707-368C	21104	Sudan Air/The Democratic Republic of the Sudan, Khartoum.	HZ-ACH
ST-PRS	77	Falcon 20F	372	Government of Sudan, Khartoum.	F-WRQV
ST-PSR	82	Falcon 50	114	Government of Sudan, Khartoum.	F-WPXM

SU = Egypt (Arab Republic of Egypt)

Civil

SU-AXJ	74	B 707-366C	20919	Government of Egypt, Cairo. (r/c Egyptian 01)	
SU-AXN	73	Falcon 20-5B	294	Government of Egypt, Cairo.	F-BVPM
SU-AYD	77	Falcon 20-5B	361	Government of Egypt, Cairo.	F-WMKF
SU-AZJ	76	Falcon 20-5B	358	Government of Egypt, Cairo.	F-WRQY
SU-DAF	62	JetStar-6	5025	ZAS Airline/Zarkani Aviation Services, Cairo. 'Dina'	11+01
SU-DAG	68	JetStar-8	5121	Zarkani Aviation Services, Cairo. 'Nadia'. (status ?).	11+02
SU-DAH	65	JetStar-6	5071	ZAS Airline/Zarkani Aviation Services, Cairo. 'Shereen'	11+03
SU-GGG	94	Airbus A340-211	061	Government of Egypt, Cairo.	F-W...
SU-OAE	68	Falcon 20D-5	175	Pyramid Airlines, Cairo.	HB-VJW
Military					
SU-BGM	88	Gulfstream 4	1048	Egyptian Air Force/Arab Republic of Egypt, Cairo.	N448GA
SU-BGU	85	Gulfstream 3	439	Egyptian Air Force/Arab Republic of Egypt, Cairo.	N17586
SU-BGV	85	Gulfstream 3	442	Egyptian Air Force/Arab Republic of Egypt, Cairo.	N17587

SX = Greece (Hellenic Republic)

Civil

SX-BNS	83	Learjet 55	55-072	Aegean Aviation Ltd. Athens.	N72ET
SX-BNT	79	Learjet 35A	35A-228	Aegean Aviation Ltd. Athens.	N4GB
SX-DCI	96	Citation V Ultra	560-0366	Interjet, Athens.	N52352
SX-ECH	87	Falcon 900	26	Olympic Airways/Government of Greece, Athens. 'King Minos'	HB-IAC

S5 = Slovenia

Civil

S5-BAA	86	Learjet 35A	35A-618	Government of Slovenia, Ljubljana.	S5-BAA
S5-BAB	75	Learjet 24D	24D-320	Government of Slovenia, Ljubljana.	SL-BAB

S9 = Sao Tome / Principe (Democratic Republic of Sao Tomé & Principe)

Civil

S9-NAD	65	JetStar-6	5065	Transafrik, Sao Tome. (status ?).	N1966G

TC = Turkey (Republic of Turkey)

Civil

Reg	Yr	Type	c/n	Owner/Operator	Prev Regn
TC-...	85	Citation S/II	S550-0072	Bon Air AS. Istanbul.	N686MC
TC-...	97	Falcon 900B	171		F-WWFW
TC-...	66	HFB 320	1023	Gokkusagi, Istanbul.	N103F
TC-AKH	94	Mach Air 1000	9043	Mach Air, Istanbul.	(N543QS)
TC-ANC	91	BAe 125/800B	8208	Mach Air/Olakoglu Metalurji A/S. Istanbul. 'Henza'	G-5-700
TC-ANT	93	Citation VI	650-0229	Emekli Ticaret/EMAIR, Ankara.	N1302X

Reg	Yr	Type	c/n	Owner/Operator	Prev Regn
☐ TC-ATA	88	Gulfstream 4	1043	Government of Turkey, Istanbul.	TC-ANA
☐ TC-ATC	94	Citation VII	650-7043	Rota Airlines, Istanbul.	N78DL
☐ TC-BAY	81	Citation II	550-0283	Bayindir Aviation, Ankara.	N316CC
☐ TC-BNY		Citation	...		
☐ TC-CAG	94	Falcon 900B	142	Nergis Air, Bursa. 'Caglar'	F-GSMF
☐ TC-CAO	84	Citation III	650-0060	Sonmez Airlines, Bursa.	N848US
☐ TC-CEN	75	Falcon 20E	326	Cen Ajans/Mach Air, Istanbul.	PH-ILY
☐ TC-CEY	91	Citation VI	650-0214	Ceylan Insaat, Ankara.	N95CM
☐ TC-CIN	96	Falcon 2000	26	Tayfunair/Demir Finansal Kirilama AS. Istanbul.	F-WQFL
☐ TC-CMY	87	Citation III	650-0141	Nergis Helikopter Servisi A/S. Bursa.	N110TM
☐ TC-COS	75	HS 125/600B	6048	Urayair/Uray Teknik AS. Istanbul.	N6567G
☐ TC-CRO	95	CitationJet	525-0102	Bilfer Madencilick AS. Ankara. 'Young Turks'	N202CJ
☐ TC-DEM	83	Falcon 200	489	Tayfunair/Demir Finansal Kiralama AS. Istanbul.	N7654F
☐ TC-DHB	91	Challenger 601-3A	5094	Dogus Air, Istanbul.	TC-OVA
☐ TC-EES	85	Citation III	650-0077	Kervansaray Turmal, Bursa.	(N20NA)
☐ TC-ELL	94	Learjet 60	60-030		N164PA
☐ TC-EMA	95	CitationJet	525-0121	Rubi Air, Istanbul.	N5264S
☐ TC-EYE	85	Falcon 50	161	Gokturk Air, Bursa.	N800BD
☐ TC-EZE	73	Falcon 20F	299	Bayair Havacilik AS. Istanbul.	F-GJSF
☐ TC-GAP	87	Gulfstream 4	1027	Government of Turkey, Istanbul.	N416GA
☐ TC-IHS	78	JetStar 2	5225	IHLAS Finance International, Istanbul.	N42KR
☐ TC-IOB		Citation	...	noted Istanbul 9/97.	
☐ TC-KAM	82	Falcon 50	95	Bayindir Insaat Turizm Ticaret Ve Sanayl AS. Ankara.	C-GSSS
☐ TC-KLS	90	Citation III	650-0191	Hamoglu Hava Tasimecilik Isletmecilik Tic AS. Istanbul.	N59B
☐ TC-LAA	93	Citation V	560-0212	DHMI Hava Taksi, Ankara.	N1284X
☐ TC-LAB	93	Citation V	560-0216	DHMI Hava Taksi, Ankara.	N1285N
☐ TC-LEY	69	HFB 320	1043	Genel Havacilik AS. Istanbul.	16+03
☐ TC-MDJ	96	Beechjet 400A	RK-120		N3261Y
☐ TC-MEK	93	Learjet 60	60-016	Cukorova Holding AS. Istanbul.	N50163
☐ TC-MSA	96	Beechjet 400A	RK-124	Sky Line Ulasim Ticaret AS. Ankara.	N1124Z
☐ TC-NEO	97	Beechjet 400A	RK-130	Sky Line Ulasim Ticaret AS. Ankara.	N1130B
☐ TC-NKB	79	Citation II	550-0053	Sonmez Airlines, Bursa.	N53KB
☐ TC-OMR	66	JetStar-731	5082		N82SR
☐ TC-ORM	74	Falcon 10	33	Albaraka Turk Ozel Finans Kurumu A/S. Istanbul.	N54WJ
☐ TC-RAM	89	Citation III	650-0178	KOC Holdings AS/Set Air, Istanbul.	N603AT
☐ TC-SBH	94	Citation VI	650-0234	Sabah Turizm/Mach Air, Istanbul.	N334CM
☐ TC-SMB	97	Beechjet 400A	RK-148		N1108T
☐ TC-TEK	92	BAe 125/800B	8229	Tefken Hava Tasimaciligi, Istanbul.	N229RY
☐ TC-TOP	85	Citation III	650-0083	Toprak Aviation, Istanbul.	N944CA
☐ TC-YSR	95	Falcon 50	246	Yasar Group/Bintur Air, Izmir.	N246FJ
☐ TC-YZB	82	Citation II	550-0323	Boronkay Hava Taksi Isletmesi, Istanbul.	TC-FMB
Military					
☐ 12-001	84	Citation II	550-0502	Turkish Air Force, 224 Filo, Etimesgut. (flight calibration)	N1255D
☐ 12-002	84	Citation II	550-0503	Turkish Air Force, 224 Filo, Etimesgut. (flight calibration)	N1255G
☐ 12-003	91	Gulfstream 4	1163	Turkish Air Force, 224 Filo, Etimesgut-Ankara.	N458GA
☐ ETI-024	93	Citation VII	650-7024	93-7024/Turkish Air Force, 224 Filo, Etimesgut-Ankara.	N1262Z
☐ ETI-026	93	Citation VII	650-7026	93-7026/Turkish Air Force, 224 Filo, Etimesgut-Ankara.	N1263G

TG = Guatemala (Republic of Guatemala)

Civil

Reg	Yr	Type	c/n	Owner/Operator	Prev Regn
☐ TG-...	94	Citation V Ultra	560-0274	International Coffee & Fertlizer Trading Co. Guatamala City.	N751CF
☐ TG-AIR	93	Learjet 31A	31A-067	Distribuidora Textil SA. Guatamala City.	N9173M
☐ TG-RIE	81	Citation 1/SP	501-0216	Trans RIF SA. Guatamala City.	TG-RIF
☐ TG-RIF	94	CitationJet	525-0072	Trans RIF SA. Guatamala City.	N.....
☐ TG-TJF	66	BAC 1-11/401AK	089	Tikal Jets, Guatamala City. 'Quirigua'	N97JF

TJ = Cameroon (Republic of Cameroon)

Civil

Reg	Yr	Type	c/n	Owner/Operator	Prev Regn
☐ TJ-AAM	78	B 727-2R1	21636	Government of Cameroun, Yaounde.	
☐ TJ-AAW	86	Gulfstream 3	486	Government of Cameroun, Yaounde.	N316GA
☐ TJ-AHR	75	Corvette	12	Air Affaires Afrique, Douala.	TR-LYM

TL = Central African Republic

Civil

Reg	Yr	Type	c/n	Owner/Operator	Prev Regn
☐ TL-FCA	60	Caravelle 3	42	Government of Central African Republic. (wfu ?).	TL-KAB

TN = Congo (People's Republic of Congo)

Civil

Reg	Yr	Type	c/n	Owner/Operator	Prev Regn
☐ TN-ADI	75	Corvette	9	Government of Congo Republic, Brazzaville. (status ?).	F-OCRN

TR = Gabon (Gabonese Republic)

Civil

Reg	Yr	Type	c/n	Owner/Operator	Prev Regn

	TR-KHC	81 Gulfstream 3	326	Government of Gabon, Libreville. 'Nyamga'	N17582
	TR-LCJ	87 Falcon 900	7	Government of Gabon, Libreville. 'Masuku II'	F-WWFG
	TR-LDB	90 BAe 125/800B	8192	Ste. Crossair, Zurich/Air Affaires Gabon, Libreville.	G-5-691
	TR-LTZ	69 DC 8-73CF	46053	Government of Gabon, Libreville.	N8638

Military

	TR-K..	97 Falcon 900EX	24	Government of Gabon, Libreville.	F-WWFU

TT = Chad *(Republic of Chad)*

Civil

	TT-AAI	79 Gulfstream 2	240	Government of Tchad, N'Djamena.	5A-DDR
	TT-DSZ	B 727	...	Government of Tchad, N'Djamena.	

TU = Ivory Coast *(Republic of the Ivory Coast)*

Civil

	TU-TIZ	76 F 28-1000C	11099	Air Ivoire, Abidjan.	PH-VAB

Military

	TU-VAA	87 Fokker 100	11245	Government of Ivory Coast, Abidjan.	PH-CDI
	TU-VAD	87 Gulfstream 4	1019	Government of Ivory Coast, Abidjan.	N17584
	TU-VAF	85 Gulfstream 3	462	Government of Ivory Coast, Abidjan.	N303GA
	TU-VAJ	F 28-4000VIP	11124	Government of Ivory Coast, Abidjan.	TU-VAZ

T9 = Bosnia Herzegovina *(Republic of Bosnia-Herzegovina)*

Civil

	T9-BIH	85 Citation S/II	S550-0045	Air Bosnia, Sarajevo.	BH-HIH

UR = Ukraine *(Ukraïna)*

Civil

	UR-CCA	71 Falcon 20F	256	CABI, Donetzk.	F-GKME
	UR-CCB	68 Falcon 20-5	141	CABI, Donetzk.	(UR-BCA)
	UR-CCC	93 Falcon 50	235	CABI, Donetzk.	(UR-ACA)
	UR-CCD	67 Falcon 20C	112	AVIS-Aviation Services, Donetzk.	CS-ATF
	UR-EFA	66 Falcon 20C	55	ALG Aeroleasing/Air Ukraine, Kiev.	CCCP-01100
	UR-EFB	67 Falcon 20C	75	ALG Aeroleasing/Air Ukraine, Kiev.	N77QM

VH = Australia *(Commonwealth of Australia)*

Civil

	VH-...	97 Beechjet 400A	RK-139	Hawker Pacific P/L. Yagoona, NSW.	N1099S
	VH-...	97 Beechjet 400A	RK-154	Hawker Pacific P/L. Yagoona, NSW.	N2354B
	VH-...	87 Falcon 900	37	Remorex P/L. Sydney.	N41SJ
	VH-AJJ	79 Westwind-1124	248	Pel-Air Aviation P/L. Sydney, NSW.	N25RE
	VH-AJK	79 Westwind-1124	256	Pel-Air Aviation P/L. Sydney, NSW.	4X-CNB
	VH-AJP	78 Westwind-1124	238	Pel-Air Aviation P/L. Sydney, NSW.	4X-CMJ
	VH-AJV	80 Westwind-1124	282	Pel-Air Aviation P/L. Sydney, NSW.	N186G
	VH-APU	83 Westwind-Two	395	Queensland Government Air Wing, Brisbane, QLD.	VH-SGY
	VH-ASM	89 Challenger 601-3A	5033	Associated Airlines P/L. Melbourne, VIC.	C-G...
	VH-ASQ	93 Gulfstream 4SP	1205	Associated Airlines P/L. Melbourne, VIC.	N439GA
	VH-BJD	92 Beechjet 400A	RK-35	Hawker Pacific P/L. Yagoona, NSW.	VH-LAW
	VH-BNK	81 Citation 1/SP	501-0171	SDS Holdings P/L. Perth, WA.	N67780
	VH-BRG	90 Challenger 601-3A	5064	Grollo Brothers/Grocon P/L. Melbourne, VIC.	C-FIOB
	VH-CCA	91 Gulfstream 4	1175	Coca-Cola Amatil Ltd. Sydney, NSW.	HB-ITJ
	VH-CCC	89 Gulfstream 4	1083	Crown Casinos Ltd. Melbourne, VIC.	HB-ITZ
	VH-CCO	89 Gulfstream 4	1107	Crown Casinos Ltd. Melbourne, VIC.	N11FX
	VH-CIT	95 CitationJet	525-0100	Capital Jet Charter P/L. Canberra, ACT.	N800HS
	VH-CPE	85 Falcon 200	504	Consolidated Press Holdings P/L. Sydney, NSW.	N702SB
	VH-ELJ	95 Hawker 800XP	8281	BHP Australia Coal Ltd. Brisbane, QLD.	N914H
	VH-EMO	85 Citation S/II	S550-0063	ESSO Australia Resources, Melbourne, VIC.	
	VH-EXM	81 Citation II	550-0228	Executive Airlines P/L. Melbourne, VIC.	N96CS
	VH-FHJ	94 Citation V Ultra	560-0278	Eagle Ltd/Tilling Associates, Jersey, C.I.	N2HJ
	VH-FIS	90 Astra-1125SP	045	Pearl Aviation Australia P/L. Brisbane, QLD.	D-CFIS
	VH-HEY	89 Citation V	560-0009	Heytesbury Aviation, Perth, WA.	N456FB
	VH-HFJ	74 Falcon 20E	306	Litmill P/L. Cairns, QLD.	N725P
	VH-HKX	72 Citation Eagle	500-0050	Metropolis City Promotions P/L. Melbourne, VIC.	N333PP
	VH-HVM	77 Citation Eagle	500-0349	Rosemount Wines, Denman, NSW.	N888GZ
	VH-ICX	72 Citation	500-0051	John Parker, Toowoomba, QLD.	N4646S
	VH-IMP	91 Beechjet 400A	RK-26	Impulse Transportation P/L. Sydney, NSW.	VH-BBJ
	VH-ING	80 Citation II	550-0141	Inghams Enterprises P/L. Sydney, NSW.	N26461
	VH-IWU	86 Citation S/II	S550-0118	Jewel's Food Barn, Sydney, NSW.	N820F
	VH-JCR	79 Learjet 35A	35A-231	Jet City P/L. Melbourne, VIC.	N62DK
	VH-JIG	81 Learjet 35A	35A-400	Quantum Resources/Shortstop Charter P/L. Melbourne, VIC.	VH-TPR
	VH-JPG	79 Citation II	550-0102	Pearl Aviation Australia P/L. Perth, WA.	VH-JCG
	VH-JPW	81 Westwind-1124	317	Jetcraft Aviation P/L. Brisbane, QLD.	(VH-NIJ)
	VH-KNS	81 Westwind-1124	323	Pel-Air Aviation P/L. Sydney, NSW.	N816H
	VH-LAW	96 Hawker 800XP	8295	McRoss Developments P/L. Sydney, NSW.	N683H
	VH-LLW	79 Westwind Sea Scan	253	Pearl Aviation Australia P/L. Perth, WA.	N253MD
Reg		*Yr Type*	*c/n*	*Owner/Operator*	*Prev Regn*

Reg	Yr	Type	c/n	Owner/Operator	Prev Regn
□ VH-LLX	79	Westwind Sea Scan	259	Pearl Aviation Australia P/L. Perth, WA.	N315JM
□ VH-LLY	79	Westwind Sea Scan	272	Pearl Aviation Australia P/L. Perth, WA.	N723R
□ VH-LMP	92	BAe 1000B	9022	National Jet Systems P/L-Santos Mining P/L. Adelaide, SA.	G-5-734
□ VH-MCX	78	Falcon 10	134	McCafferty's Air Charter P/L. Toowomba, QLD.	N509TC
□ VH-MGC	97	Citation Bravo	550-0810	Dick Smith Adventure P/L. Sydney, NSW.	N5218T
□ VH-MOJ	96	CitationJet	525-0138	Morgan & Co/Pearl Coast Airways P/L. Perth, WA.	N5211F
□ VH-MZL	80	Learjet 35A	35A-285	Westfield Aviation Inc. Sydney, NSW.	N34TB
□ VH-NCP	89	Gulfstream 4	1108	News Ltd. Sydney, NSW.	N522AC
□ VH-NGA	83	Westwind-Two	387	Pel-Air Aviation P/L. Sydney, NSW.	N97AL
□ VH-NJE	87	B 737-3Q8	23766	National Jet Systems P/L. Adelaide, SA.	N101GU
□ VH-NTH	89	Citation V	560-0041	NTH Aviation P/L. Melbourne, VIC.	N26643
□ VH-PFA	90	Learjet 35A	35A-661	Singapore Air Force/Pacific Flight Services P/L. Seletar.	N1268G
□ VH-PPF	88	Falcon 50	187	Paspaley Pearling Co P/L. Darwin, NT.	(N4QP)
□ VH-SCD	82	Citation II	550-0339	MacArthur Jet Charter, Werombi, NSW.	VH-KTK
□ VH-SGY	97	Hawker 800XP	8328	Queensland Government Air Wing, Brisbane, QLD.	N328XP
□ VH-SLD	78	Learjet 35A	35A-145	Pel Air Aviation P/L. Botany, NSW.	(N166AG)
□ VH-SLE	81	Learjet 35A	35A-428	Pel Air Aviation P/L. Botany, NSW.	N17LH
□ VH-SLF	81	Learjet 36A	36A-049	Pel Air Aviation P/L. Botany, NSW.	N136ST
□ VH-SLJ	75	Learjet 36	36-014	Pel Air Aviation P/L. Botany, NSW.	N200Y
□ VH-SOU	76	Citation	500-0333	Sydville P/L. Seven Hills, NSW.	N275AL
□ VH-WGJ	79	Citation II	550-0054	Pearl Aviation Australia P/L. Perth, WA.	N501AA
□ VH-WNZ	79	Citation II	550-0057	Queensland Police Department, Brisbane, QLD.	(N2661N)
□ VH-XDD	79	Citation II	550-0076	One Air, Vaucluse, NSW.	VH-TFY
□ VH-XMO	93	BAe 125/800B	8243	Southern Commander, Melbourne, VIC.	G-SHEB
□ VH-ZLE	82	Citation II	550-0347	China Southern WA Flying College, Jandakot, WA.	ZS-LEE
□ VH-ZMD	75	Citation	500-0263	Alistair Lamb, Melbourne, VIC.	VH-AQS
Military					
□ A20-261	76	B 707-368C	21261	RAAF, 33 Squadron, Richmond, NSW.	N7486B
□ A20-623	68	B 707-338C	19623	RAAF, 33 Squadron, Richmond, NSW. 'City of Sydney'	C-GRYN
□ A20-624	68	B 707-338C	19624	RAAF, 33 Squadron, Richmond, NSW. 'Richmond Town'	VH-EAD
□ A20-627	68	B 707-338C	19627	RAAF, 33 Squadron, Richmond, NSW. 'Windsor Town'	VH-EAG
□ A20-629	68	B 707-338C	19629	RAAF, 33 Squadron, Richmond, NSW.	C-GGAB
□ A26-070	89	Falcon 900	70	RAAF, 34 Squadron, Fairbairn, Canberra.	N450FJ
□ A26-073	89	Falcon 900	73	RAAF, 34 Squadron, Fairbairn, Canberra.	N452FJ
□ A26-074	89	Falcon 900	74	RAAF, 34 Squadron, Fairbairn, Canberra.	N453FJ
□ A26-076	89	Falcon 900	76	RAAF, 34 Squadron, Fairbairn, Canberra.	N454FJ
□ A26-077	89	Falcon 900	77	RAAF, 34 Squadron, Fairbairn, Canberra.	N455FJ

VP-B = Bermuda (UK-Colony of Bermuda)

Reg	Yr	Type	c/n	Owner/Operator	Prev Regn
Civil					
□ VP-BAC	96	Challenger 604	5309	Air Nauticus Ltd. Jeddah, Saudi Arabia.	VR-BAC
□ VP-BAT	79	B 747SP-21	21648	Government of Qatar, Doha.	VR-BAT
□ VP-BAW	92	BAe 125/800B	8237	Raytheon Aircraft Co. Wichita, Ks. USA.	9M-DRL
□ VP-BCC	94	Challenger 601-3R	5162	Consolidated Contractors Co. Athens, Greece.	VR-BCC
□ VP-BCF	84	HS 125/700B	7214	Costair Ltd/Viamax.	VR-BCF
□ VP-BCH	76	Falcon 10	70	Norasia Schiffahrts GmbH. Kiel, Germany.	VR-BCH
□ VP-BCI	96	Challenger 601-3R	5193	Consolidated Contractors Co. Athens, Greece.	VR-BCI
□ VP-BCP	77	JetStar 2	5222	Crescent Adventure Ltd.	C-GAZU
□ VP-BCT	80	Gulfstream 3	302	Canal Trading Associates Inc. Tortola, BVI.	(N561ST)
□ VP-BDJ	68	B 727-23	20046	Donald Trump/D J Aerospace (Bermuda) Ltd.	VR-BDJ
□ VP-BEF	97	Falcon 2000	49	E F Education Sweden/Executive Jet Charter, Farnborough, UK.	(PH-EFB)
□ VP-BEH	96	Falcon 900B	163	E F Education, Stockholm, Sweden.	(PH-EFA)
□ VP-BGT	77	Gulfstream 2	211	Ditco Air Ltd/Sheikh El Khereiji Group, Jeddah Saudi Arabia.	VR-BGT
□ VP-BGW	64	B 727-30	18366	Sigair Ltd/Sheikh El Khereiji, Saudi Arabia.	VR-BGW
□ VP-BHA	96	Challenger 604	5307	Air Nauticus Ltd. Jeddah, Saudi Arabia.	VR-BHA
□ VP-BHG	87	Gulfstream 4	1017	Sugra Ltd. Toronto, Canada.	(C-FNCG)
□ VP-BHI	82	Citation 1/SP	501-0258	International Prospect Investment Ltd.	VR-BHI
□ VP-BHJ	94	Falcon 900B	138	Frank Hackett-Jones/Eagle Ltd-Tilling Associates, Jersey.	VR-BHJ
□ VP-BHM	69	DC 8-62	46111	Brisair Ltd/Sheikh El Khereiji, Saudi Arabia.	VR-BHM
□ VP-BHW	83	HS 125/700B	7209	Pacific Fruit Co/Pan American Trading Corp.	VR-BHW
□ VP-BIA	63	DC-8-52	45658	JetAir Leasing Ltd. London, UK.	VR-BIA
□ VP-BIE	83	Challenger 601	3016	Inflite Executive Charter Ltd. Stansted, UK.	N601CL
□ VP-BJA	95	Falcon 900B	154	AVCON AG/KJJ Transportation, Zurich, Switzerland.	VR-BJA
□ VP-BJD	90	Gulfstream 4	1134	Transworld Oil America Inc. Newark, NJ. USA.	VR-BJD
□ VP-BJR	69	DC 8-72	46067	Al Nassar Ltd/Bruce J Rappaport, Geneva, Switzerland.	VR-BJR
□ VP-BJS	82	Learjet 35A	35A-464	Interfruit Co. Hamilton.	N464WL
□ VP-BJV	76	Gulfstream 2	186	Uniexpress Jet Services Ltd. UK.	VR-BJV
□ VP-BKC	71	B 727-1H2	20533	USAL Ltd. Geneva, Switzerland.	VR-BKC
□ VP-BKG	85	Falcon 50	147	Sioux Co Ltd. Cranfield, UK.	VR-BKG
□ VP-BKI	87	Gulfstream 4	1029	JABJ-Picton Ltd/Quetzal Inc. Geneva, Switzerland.	VR-BKI
□ VP-BKK	70	HS 125/731	25238	Air Man/Air 125/Business Real Estate Corp. Southampton, UK.	VR-BKK
□ VP-BKP	80	Citation 1/SP	501-0175	Star Aviation Ltd. Mannheim, Germany.	VR-BKP
□ VP-BKT	89	Gulfstream 4	1074	JABJ/Natascha Establishment, Vaduz, Liechtenstein.	VR-BKT
□ VP-BKY	68	HS 125/F3A	25150	Executive Aviation Services, Staverton, UK.	VR-BKY
□ VP-BLA	83	Challenger 601	3013	Sol Kerzner/Sun International Hotels, Fort Lauderdale, USA.	VR-BLA
□ VP-BLB	89	Falcon 900	49	Maritime Investment & Shipping/Stavros Niarchos, Athens.	VR-BLB
Reg	Yr	Type	c/n	Owner/Operator	Prev Regn

Reg	Yr	Type	c/n	Owner/Operator	Prev Regn
□ VP-BLF	84	Citation 1/SP	501-0679	Ekron Ltd/VCN,	VR-BLF
□ VP-BLG	69	DC 8-62H	46071	Consolidated Press Holdings, Sydney, NSW. Australia.	VR-BLG
□ VP-BLM	90	Falcon 900	72	Globus Travel/Aileron/Sen Montegazza, Lugano, Switzerland.	VR-BLM
□ VP-BLN	83	Gulfstream 3	402	Pegasus Aviation Ltd/JABJ-Jameel SAM, Monte Carlo.	VR-BLN
□ VP-BLV	76	Citation	500-0344	Santom Ltd	VR-BLV
□ VP-BMB	70	HS 125/400B	25240	Barton Ltd/Speedflight Ltd. Stansted, UK.	VR-BMB
□ VP-BMC	65	B 727-22	18323	U D Aviation Services Ltd.	VR-BMC
□ VP-BMD	83	HS 125/700B	7200	Airman Ltd. Hamilton.	VR-BMD
□ VP-BMF	90	Falcon 50	206	Sally Navigation Co. Farnborough, UK.	F-WWHB
□ VP-BMR	96	Beechjet 400A	RK-133	Capitol Aviation, Athens, Greece.	N1133T
□ VP-BMY	85	Gulfstream 3	463	Globair Ltd/JABJ, Zurich, Switzerland.	VR-BMY
□ VP-BNA	67	B 727-21	19262	Arapaho Ltd/Skyways International Inc. Houston, Tx. USA.	VR-BNA
□ VP-BND	77	Gulfstream 2B	199	ANB International, Antwerp, Belgium.	VR-BND
□ VP-BNF	97	Challenger 604	5332	Sural CA. Caracas, Venezuela.	C-FZRR
□ VP-BNG	93	Challenger 601-3A	5119	Alloy Aircraft Ltd. Istanbul, Turkey.	VR-BNG
□ VP-BNJ	93	Falcon 900	120	Silver Sand Air Services/Triair, Farnborough, UK.	VR-BNJ
□ VP-BNW	76	HS 125/600B	6057	Spectrum Aviation Corp. Riga, Latvia.	VR-BNM
□ VP-BNY	93	Gulfstream 4SP	1208	AMI-Stansted/Theberton USA Inc. Paris, France.	VR-BNY
□ VP-BNZ	85	Gulfstream 3	452	Dennis Vanguard International (Switchgear) Ltd. Coventry, UK	VR-BNZ
□ VP-BOA	92	Challenger 601-3A	5114	SAMCO Aviation, Riyadh, Saudi Arabia.	VR-BOA
□ VP-BOC	90	B 737-53A	24970	TAG Group SA/Theberton USA Inc. Paris, France.	VR-BOC
□ VP-BOK	83	Gulfstream 3	390	Pegasus Aviation Ltd/JABJ-Jameel SAM, Monte Carlo.	VR-BOK
□ VP-BOL	82	Learjet 55	55-022	New Pembroke Ltd. Hamilton.	VR-BOL
□ VP-BON	92	Astra-1125SP	060	Aerocentro de Servicios CA. Caracas, Venezuela.	VR-BON
□ VP-BOO	89	MD-87	49778	Ford Motor Co. Stansted, UK.	VR-BOO
□ VP-BOP	88	MD-87	49725	Ford Motor Co. Stansted, UK.	VR-BOP
□ VP-BOT	93	Gulfstream 4SP	1212	Geojet Ltd/International Aircraft Management, Teterboro, NJ.	VR-BOT
□ VP-BPE	78	HS 125/700B	7040	Swanmore Aviation Ltd/Sintez, Belarus, CIS.	VR-BPE
□ VP-BPI	95	Falcon 900B	149	Falconair/Rembrandt Tobacco Group, Capetown, RSA.	VR-BPI
□ VP-BPW	94	Falcon 900B	135	J N Somers/Tower House Consultants Ltd. Jersey, C.I.	VR-BPW
□ VP-BRF	87	Gulfstream 4	1015	Rashid Engineering, Riyadh/Eiger Jet Ltd. Geneva.	VR-BRF
□ VP-BRL	72	JetStar-731	5155	Rida Aviation Ltd. Bergamo, Italy.	VR-BRL
□ VP-BRO	97	Falcon 900EX	13	Rembrandt Tobacco Group, Capetown, RSA.	F-WWFK
□ VP-BSF	88	Gulfstream 4	1058	Seaflight Aviation Ltd. Athens, Greece.	N70PS
□ VP-BSH	68	JetStar-731	5117	Seaflight Aviation Ltd/George Livanos, Athens, Greece.	VR-BSH
□ VP-BSI	86	BAe 125/800B	8073	Group 4 Ltd. Staverton, UK.	VR-BSI
□ VP-BSK	94	Falcon 900B	125	Sky King International/Nusantara Satki Sdn Bhd.	VR-BSK
□ VP-BST	97	Falcon 50EX	258	Norasia Schiffahrts GmbH. Kiel, Germany. 'Samantha'	F-WQHU
□ VP-BTM	92	BAe 125/800B	8233	Tsakos Shipping & Trading SA. Athens, Greece.	VR-BTM
□ VP-BTR	91	Citation II	550-0693	Sumair Ltd. Geneva, Switzerland.	VR-BTR
□ VP-BTZ	80	HS 125/700B	7109	Palmyra Aviation Ltd.	VR-BTZ
□ VP-BUC	82	Falcon 50	107	Falconair/Rembrandt Tobacco Group, Capetown, RSA.	VR-BUC
□ VP-BUS	90	Gulfstream 4	1127	JABJ/Urs Schwarzenbach, Farnborough, UK.	VR-BUS
□ VP-BVV	81	Citation II	550-0272	Flash Aviation/Prince Aviation, Belgrade, Serbia. 'Vukica'	VR-BVV
□ VP-BWB	94	Challenger 601-3R	5151	Flight Management Services, Fort Myers, Fl. USA.	C-GBJA
□ VP-BWS	92	Falcon 900B	124	San Co/Dong Ah Construction Industrial Co. South Korea.	VR-BWS
□ VP-BZA	70	B 707-336C	20375	TBN Aircraft (Bermuda) Ltd/Diversified Aviation Services Ltd	VR-BZA
□ VP-BZE	84	Falcon 50	144	Flying Lion/Stork Ltd. Hamilton.	VR-BZE
□ VP-BZZ	98	CitationJet	525-0235	Fegotila Ltd. Staverton, UK.	

VP-C = Cayman Islands (UK-Colony of Cayman Islands)

Civil

Reg	Yr	Type	c/n	Owner/Operator	Prev Regn
□ VP-C..	97	Falcon 50EX	259	Volkswagenwerke AG. Wolfsburg, Germany.	F-WWHG
□ VP-CAN	97	Challenger 604	5335	United Dynasty Ltd. London.	C-FZVN
□ VP-CAS	90	BAe 125/800B	8167	Cavalier Aviation Ltd. London, UK.	VR-CAS
□ VP-CAT	81	Citation 1/SP	501-0232	Kestrel Aviation Intl. Biggin Hill, UK. 'Highland Flyer'	VR-CAT
□ VP-CAU	91	B 757-2J4	25220	Diamond International Aviation Corp. Ronkonkoma, NY. USA.	VR-CAU
□ VP-CBB	94	Gulfstream 4SP	1250	EAT-Zurich/Bugshan Construction Co. Jeddah, Saudi Arabia.	VR-CBB
□ VP-CBD	87	Falcon 900	16	Falcon Maintenance Corp. West Palm Beach, Fl.	N619BD
□ VP-CBE	79	Citation II	550-0108	Elliot Aviation Ltd/Eurojet Aviation Ltd. Birmingham.	(N65SA)
□ VP-CBM	94	Citation II	550-0729	Bernard Matthews PLC. Norwich, UK.	VR-CBM
□ VP-CBQ	78	B 727-212	21460	Mezel Air, Jersey, Channel Islands.	VR-CBQ
□ VP-CBT	97	Falcon 50EX	256	Volkswagenwerke AG. Wolfsburg, Germany.	F-WWHD
□ VP-CBW	89	Gulfstream 4	1096	Rolls Royce PLC. Farnborough, UK.	VR-CBW
□ VP-CBX	97	Gulfstream V	511	ARAVCO Ltd. Farnborough, UK.	N511GA
□ VP-CCC	93	CitationJet	525-0040	Gerhardt Schubert GmbH. Rothenburg ob der Tauber, Germany.	D-ISCH
□ VP-CCG	66	BAC 1-11/401AK	081	ATM Aviation, Jeddah, Saudi Arabia.	VR-CCG
□ VP-CCL	84	Falcon 200	482	Aerowest Flugcharter GmbH. Hanover, Germany.	VR-CCL
□ VP-CCR	90	Challenger 601-3A	5079	Bin Laden Aviation/Shoreditch Investments Ltd. Jeddah.	VR-CCR
□ VP-CCV	95	Citation V Ultra	560-0320	Multi Flight Ltd. Leeds, UK.	VR-CCV
□ VP-CDE	92	BAe 125/800B	8234	Arven Ltd/Saral Publications Inc C.A. Caracas, Venezuela.	VR-CDE
□ VP-CDL	79	B 727-2M7	21655		VR-CDL
□ VP-CDM	79	Citation 1/SP	501-0084	Grosvenor Estates, Hawarden-Chester, UK.	VR-CDM
□ VP-CEK	82	HS 125/700B	7175	AVCON AG. Zurich.	N770TJ
□ VP-CEZ	83	Falcon 50	138	IIR-Institute for International Research, Biggin Hill, UK.	N138NW
□ VP-CFG	81	Citation 1/SP	501-0176	Alpha Golf Aviation Ltd. Oxford, UK.	(VR-CIA)
Reg	*Yr*	*Type*	*c/n*	*Owner/Operator*	*Prev Regn*

Reg	Yr	Type	c/n	Owner/Operator	Prev Regn
☐ VP-CFL	96	Gulfstream 4SP	1282	Star Aircraft Leasing SA. Monaco.	VR-CFL
☐ VP-CGB	94	Falcon 900B	145	Volkswagen of America Inc. Waterford, Mi. USA.	VR-CGB
☐ VP-CHG	97	Falcon 50EX	262	Dassault Falcon Jet Corp. Teterboro, NJ. USA.	(N1896T)
☐ VP-CHH	73	Citation	500-0083	Aviation Leasing Group, Kansas City, Mo. USA.	VR-CHH
☐ VP-CHK	92	Challenger 601-3A	5102	Executive Air Transport Ltd. Riyadh, Saudi Arabia.	VR-CHK
☐ VP-CHT	71	Learjet 25B	25B-075	Island Air/Cayman Jet Holdings Ltd. Grand Cayman.	VR-CHT
☐ VP-CIC	87	Challenger 601-3A	5011	InvestCorp/TGC Aviation/Fakhar Ltd. Stansted, UK.	VR-CIC
☐ VP-CID	94	Falcon 900B	130	EAT-Zurich/Bugshan Construction Co. Jeddah, Saudi Arabia.	VR-CID
☐ VP-CJB	80	Citation 1/SP	501-0155	Brown Pestell Ltd. Biggin Hill, UK.	VR-CJB
☐ VP-CJR	82	Citation II	550-0354	Broome & Wellington (Aviation) Ltd.. Manchester, UK.	VR-CJR
☐ VP-CKA	71	B 727-82	20489	SAMCO-Saudi Aircraft Management & Operations Co. Jeddah.	VR-CKA
☐ VP-CKM	96	Citation V Ultra	560-0413	Gamston Aviation Ltd. Gamston, UK.	N413CV
☐ VP-CKO	68	DC 9-15	47151	K A K Aviation Ltd.	VR-CKE
☐ VP-CLD	84	Falcon 50	134	Volkswagenwerke AG. Braunschweig, Germany.	VR-CLD
☐ VP-CLN	95	Falcon 50EX	251	Volkswagenwerke AG. Braunschweig, Germany.	VR-CLN
☐ VP-CMC	89	Challenger 601-3A	5044	Michael Schumacher/M S Corp. Jersey, Channel Islands.	VR-CMC
☐ VP-CMF	88	Gulfstream 4	1062	Mohammed Fakhry/MSF Aviation, Heathrow, UK.	VR-CMF
☐ VP-CMI	69	BAC 1-11/212AR	183	Ashmawi Aviation/Kinyaa Ltd. Malaga, Spain.	VR-CMI
☐ VP-CMM	65	B 727-30	18368	MME Farms Maintenance Corp. Warrenton, Va. USA. "Amel"	VR-CMM
☐ VP-CMN	67	B 727-46	19282	IDG (Cayman) Ltd.	VR-CMN
☐ VP-CMO	73	Citation	500-0070	Tunstall Group, Leeds-Bradford, UK.	VR-CMO
☐ VP-CMZ	92	BAe 1000A	9021	EAT-Zurich/Aircraft Operations & Holding Ltd. Grand Cayman.	VR-CMZ
☐ VP-CNJ	84	Gulfstream 3	426	JABJ-Zurich/Starling Aviation, Grand Cayman.	VR-CNJ
☐ VP-COM	76	Citation	500-0318	Rapid 3864 Ltd/Colin McGill, Biggin Hill, UK.	VR-COM
☐ VP-CPA	77	Gulfstream 2SP	204	Global Aviation/Chief Harry Akande, Ibadan, Nigeria.	VR-CPA
☐ VP-CPO	94	Challenger 601-3R	5165	P & O Containers (Assets) Ltd. Stansted, UK.	VR-CPO
☐ VP-CPT	91	BAe 1000B	9004	Remo Investments, Biggin Hill, UK.	VR-CPT
☐ VP-CRT	82	Falcon 50	88	Volkswagenwerke AG. Braunschweig, Germany.	VR-CRT
☐ VP-CSC	97	Citation V Ultra	560-0439	Stadium City Ltd. Humberside.	(N39LX)
☐ VP-CSM	66	JetStar-731	5092	Ashmawi Aviation, Malaga, Spain.	VR-CSM
☐ VP-CSN	97	Citation V Ultra	560-0401	Scottish & Newcastle Breweries plc. Edinburgh.	N401CV
☐ VP-CSP	74	Citation	500-0165	SP Metal, Jersey, Channel Islands.	VR-CSP
☐ VP-CTE	92	Citation II	550-0716	Target Express Parcels Ltd. Chester, UK.	VR-CTE
☐ VP-CTJ	79	Citation II	550-0073	G I E DB Aviation, Le Havre.	F-GBTL
☐ VP-CTS	70	HS 125/F400B	25243	PCS Ltd. Grand Cayman.	N243TS
☐ VP-CTT	96	Falcon 900B	161	Walkfine Ltd. Manchester, UK.	F-GSAB
☐ VP-CUB	77	Gulfstream 2B	207	U B Ltd. Bangalore, India. "Sidharta"	(VT-UBG)
☐ VP-CVK	89	Challenger 601-3A	5049	Asprey Leasing/Falcon Maintenance Corp. W Palm Beach, Fl.	VR-CVK
☐ VP-CWI	85	Falcon 50	159	Global Wings/Executive Air Tansport AG. Zurich, Switzerland.	VR-CWI
☐ VP-CWM	91	Citation II	550-0667	Weber Managemen GmbH. Stuttgart, Germany.	VR-CWM
☐ VP-CYM	89	Gulfstream 4	1090	Jet Fly Corp. London, UK.	VR-CYM
☐ VP-CZY	79	B 727-2P1	21595	Dunview Ltd. Georgetown.	N727MJ

VT = India (Republic of India)

Civil

Reg	Yr	Type	c/n	Owner/Operator	Prev Regn
☐ VT-AAA	82	HS 125/700A	NA0323	Reliance Industries Ltd. Mumbai.	N2830
☐ VT-EAU	88	BAe 125/800B	8120	Khaltan Tea/Eastern Airways, Calcutta.	HB-VLI
☐ VT-EHS	82	Learjet 29	29-003	Aviation Research Centre/Government of India Agency, Delhi.	N289CA
☐ VT-EIH	84	Learjet 29	29-004	Aviation Research Centre/Government of India Agency, Delhi.	N294CA
☐ VT-EQZ	67	HS 125/3B	25133	India International Airways P/L. Delhi.	G-ILLS
☐ VT-ERO	65	Jet Commander	33	Aerocopter Services P/L-Bombay Dyeing & Manufacturing Co.	N104CJ
☐ VT-ETG	86	Citation S/II	S550-0089	Rusi H Modi/Tata Engineering Services Ltd. Calcutta.	VT-RHM
☐ VT-EUN	82	Citation II	550-0352	India International Airways P/L. Delhi.	N352AM
☐ VT-EUX	95	Citation V Ultra	560-0299	State Government of Tamil Nadu, Madras.	N5168F
☐ VT-KMB	87	Citation S/II	S550-0135	Grasim India Ltd. Bombay.	N2235
☐ VT-MPA	88	HS 125/700B	7172	Bombay Dyeing/Megapode Airlines, Bombay.	G-BKFS
☐ VT-OAM	88	Beechjet 400	RJ-38	Oswal Industries/Aerial Services P/L.	N447CC
☐ VT-OBE	84	HS 125/700B	7215	Oberoi Hotels Internationa/E India Hotel Ltd., New Delhi.	VH-HSP
☐ VT-OPJ	95	CitationJet	525-0112	Jindal Iron & Steel Co Ltd. Mumbai.	N1006F
☐ VT-SRR	82	HS 125/700A	NA0336	Essar Shipping/World Trade Ltd. Bombay.	N11TS
☐ VT-SWP	83	Learjet 25D	25D-367	Indian Steel & Wire Products, Jamshedpur.	N51DT
☐ VT-TEL	88	Beechjet 400	RJ-46	R P Goenka,	N146JB
☐ VT-TTA	87	Falcon 200	511	TELCO-Tata Engineering & Locomotive Co. Bombay.	F-OLET
☐ VT-UBG	71	HS 125/F400B	25254	United Breweries Group/UB Ltd. Bangalore. 'Sidhartha'	G-5-624
☐ VT-VPS	81	Citation II	550-0390	Raymond Woollen Mills, Bombay.	N136BC

Military

Reg	Yr	Type	c/n	Owner/Operator	Prev Regn
☐ K2412	83	B 737-2A8	23036	Indian Air Force, Delhi.	VT-EHW
☐ K2413	83	B 737-2A8	23037	Indian Air Force, Delhi.	VT-EHX
☐ K2899	68	B 707-337C	19988	Indian Air Force, Delhi.	VT-DXT
☐ K2900		B 707-337C	19248	Indian Air Force, Delhi.	VT-DVB
☐ K2980	84	Gulfstream SRA-1	420	Indian Air Force, New Delhi. (used VT-ENR 1987-1997).	N47449
☐ K2981	86	Gulfstream SRA-1	494	Indian Air Force, New Delhi.	N370GA
☐ K2982	86	Gulfstream SRA-1	495	Indian Air Force, New Delhi.	N371GA
☐ K3186	71	B 737-2A8	20484	Indian Air Force, Delhi.	VT-EAK
☐ K3187	70	B 737-2A8	20483	Indian Air Force, Delhi.	VT-EAJ

Reg   Yr Type   c/n   Owner/Operator   Prev Regn

VP-LA. = Anguilla (State of Antigua and Barbuda)

Civil

□ V2-LSF	76	HS 125/600A	6065	Stanford Financial Group Ltd. Houston, Tx. USA.	VR-CSF

V5 = Namibia (Republic of Namibia)

Civil

□ V5-CDM	91	Citation V	560-0151	Consolidated Diamond Mines,	ZS-NDU
□ V5-NAG	94	Learjet 31A	31A-091	Ministry of Works, Transport & Communications, Windhoek.	N5019Y
□ V5-NAM	91	Falcon 900B	103	Government of Namibia, Windhoek.	F-WWFJ
□ V5-NPC	97	Learjet 31A	31A-138	NamPower ,Windhoek.	N138LJ

V8 = Brunei

Civil

□ V8-...	98	B 747-430	28288	Government of Brunei, Bandar Seri Begawan.	
□ V8-...	90	Gulfstream 4	1109	Government of Brunei, Bandar Seri Begawan.	V8-007
□ V8-...	93	Gulfstream 4SP	1202	Government of Brunei, Bandar Seri Begawan.	V8-009
□ V8-007	97	Gulfstream V	515	Government of Brunei, Bandar Seri Begawan.	N599GA
□ V8-008	91	Gulfstream 4	1176	Government of Brunei, Bandar Seri Begawan.	N468GA
□ V8-009	97	Gulfstream V	509	Government of Brunei, Bandar Seri Begawan.	N509GA
□ V8-AC1	79	B 747SP-21	21649	Government of Brunei, Bandar Seri Begawan.	V8-JP1
□ V8-AL1	91	B 747-430	26426	Government of Brunei, Bandar Seri Begawan.	(D-ABVM)
□ V8-AM1	93	Airbus A340-211	009	Government of Brunei, Bandar Seri Begawan.	V8-JP1
□ V8-BKH	94	Airbus A340-212	046	Government of Brunei, Bandar Seri Begawan.	V8-PJB
□ V8-DPD	87	Airbus A310-304	431	Government of Brunei, Bandar Seri Begawan.	V8-HM1
□ V8-JBB	96	Airbus A340-213	151	Government of Brunei, Bandar Seri Begawan.	F-WWJY
□ V8-MJB	93	B 767-27G (ER)	25537	Government of Brunei, Bandar Seri Begawan.	
□ V8-PJB	94	Airbus A340-212	046	Government of Brunei, Bandar Seri Begawan.	AC9-HH
□ V8-SR1	90	Gulfstream 4	1150	Government of Brunei, Bandar Seri Begawan.	V8-009

XA/B/C = Mexico (United Mexican States)

Civil

□ XA-...	64	B 727-31	18752		N727PJ
□ XA-...	97	Beechjet 400A	RK-163	Aerolineas Ejecutivas SA. Toluca.	N2363A
□ XA-...	94	Citation II	550-0731		N550BP
□ XA-...	65	DC 9-14	45702	(still exec ?).	HB-IEF
□ XA-...	97	Falcon 900B	168	Telmex/Aero Frisco SA. Mexico City.	N167FJ
□ XA-...	93	Falcon 900B	131	Transportes Aereos de Monterrey SA. Monterrey, NL.	N158JA
□ XA-...	66	Learjet 24	24-109		N900DL
□ XA-...	62	Sabreliner CT-39A	276-21	Jett Paquetaria SA. San Luis Potosi.	N63611
□ XA-...	62	Sabreliner CT-39A	276-44	Jett Paquetaria SA. San Luis Potosi.	N63811
□ XA-...	60	Sabreliner CT-39A	265-14	Jett Paquetaria SA. Quetzalcoatl.	60-3484
□ XA-AAP	91	Learjet 31	31-032	Servicios Aereos Tribasa SA. Toluca.	N5010U
□ XA-ABA	73	Gulfstream 2SP	136	Aeromundo Ejecutivo SA. Toluca.	N26WB
□ XA-ACN	74	HS 125/600A	6038	Aero Central SA. Jurica, Quintana Roo.	N199SG
□ XA-ADC	66	BAC 1-11/211AH	084	Aliendros della Cruz/Aerotaxi Monse SA. Cuernavaca, Morelos.	S9-TAE
□ XA-AGA	79	Citation 1/SP	501-0095	Puerto Vallarta Taxi Aereo SA. Puerto Vallarta.	N612DS
□ XA-AGN	79	Citation II	550-0077	Antair SA. Toluca.	XA-PIJ
□ XA-AHM	75	Gulfstream 2	161	Antair SA/Grupo Acerero del Norte SA. Toluca.	XA-RUS
□ XA-AIS	89	Gulfstream 4	1098	Antair SA. Toluca.	N404CC
□ XA-ARA	70	Gulfstream 2SP	79	Aereo Transportes Comercial SA. Toluca.	XA-STO
□ XA-ARE	79	Sabre-60	306-146	Aviones ARE SA. Mexico City.	N360CH
□ XA-ATE	77	Sabre-60	306-123	Aero Transportes Empresariales SA. Toluca.	N97SC
□ XA-AVE	80	Falcon 50	22	Aerolineas Ejecutivas SA. Toluca	XA-SFP
□ XA-BAL	89	Gulfstream 4	1114	Aerovics SA. Toluca.	N555WL
□ XA-BCC	73	Falcon 20E	284	Servicios Aereos Grupo Cardena SA, Toluca.	N441FA
□ XA-BEB	69	JetStar-731	5132	TAESA. Toluca. 'Ishtar - Goddess of Luck'	XA-PSD
□ XA-BEG	91	Falcon 50	224	Servicios Ejecutivos Continental SA. Tampico.	XA-SDK
□ XA-BNG	87	Beechjet 400	RJ-33	Servicios Aereos Gana SA. San Luis Potosi.	XA-JJA
□ XA-BRE	76	Gulfstream 2	185	XABRE Aerolineas SA. Toluca.	N511WP
□ XA-BUX	78	Learjet 35A	35A-176	Transportes Aereos Tecnico Ejecutivo SA. Toluca.	(N67GA)
□ XA-CAG	92	Gulfstream 4	1197	Commander Mexicana SA. Toluca.	N150GX
□ XA-CEN	68	Sabre-60	306-26	Aerologic SA. Toluca.	(N377EM)
□ XA-CHP	68	Sabre-60	306-22	Aerotrans Privados SA/Seguros Chapultepec SA. Toluca.	(N450CE)
□ XA-CUR	77	Sabre-60	306-127	Peninter Aerea SA. Merida.	N60DD
□ XA-CYC	92	Citation V	560-0161	Aerozeus SA. Mexico City.	TR20-1
□ XA-CZG	78	Learjet 35A	35A-162	Aerojobeni SA. Toluca.	N222SL
□ XA-DAK	75	Learjet 25B	25B-190	TAESA-Transported Aereos Ejecutivos SA. Toluca.	
□ XA-DAN	65	Sabre-40	282-26	AeroDan SA. Sin Nombre.	XA-RED
□ XA-DET	76	Learjet 24F	24F-337	AEMSA/Gutsa Construcciones SA. Toluca.	XA-GEO
□ XA-DIJ	73	Learjet 24D	24D-269	Jet Rent SA. Toluca.	
□ XA-DPS	93	Gulfstream 4SP	1227	Servicios Aereos Trabasa SA. Toluca.	N463GA
□ XA-DSC	71	Sabre-60	306-56	TAESA-Transportes Ejecutivos SA. Toluca.	XA-RXP
□ XA-DUC	72	Falcon 20F	269	Aerolineas Ejecutivas SA. Toluca.	XA-NAY
□ XA-EKT	79	JetStar 2	5234	Aerotaxis Metropolitanos SA. Toluca.	(N234TS)

Reg	Yr Type		c/n	Owner/Operator	Prev Regn

Reg	Yr	Type	c/n	Owner/Operator	Prev Regn
☐ XA-EMO	69	JetStar-8	5140	TAESA-Transportes Aereos Ejecutivos SA. Toluca.	XA-JCG
☐ XA-FHR	79	JetStar 2	5231	TAESA-Transportes Aereos Ejecutivos SA. Toluca.	N112MX
☐ XA-FIR	93	Citation II	550-0718	Aero Taxi Autlan SA. Toluca.	N12060
☐ XA-FLM	77	Falcon 20F	364	Transpais Aereo SA. Del Norte, NL.	N285U
☐ XA-FMR	79	Learjet 25D	25D-274	Aero Copter SA. Toluca.	XA-RZE
☐ XA-FNY	75	Gulfstream 2	175	Naviera Mexicana SA. Toluca.	N770PA
☐ XA-FTN	66	Sabre-40	282-80	Taxi Aereo Turistico SA. Acapulco.	N40WH
☐ XA-FVK	80	Falcon 50	35	Aerolineas Ejecutivas SA. Toluca.	N350AF
☐ XA-GAC	74	Gulfstream 2B	155	Commander Mexicana SA. Toluca.	N308A
☐ XA-GAE	94	Falcon 900B	137	GAESA-Grupo de Aviacion Ejecutiva SA. Toluca.	N139FJ
☐ XA-GAN	92	Citation VI	650-0218	Antair SA. Toluca.	N6829X
☐ XA-GAP	79	Sabre-65	465-8	Commander Mexicana/El Heraldo de Mexico SA. Toluca.	N10581
☐ XA-GBA	72	Learjet 24D	24D-260	Taxi Aereos del Golfo SA. Monterrey.	XB-GMK
☐ XA-GDO	81	Learjet 35A	35A-449	Guja SA. Mexico City.	N449QS
☐ XA-GFB	86	BAe 125/800A	8075	Aerocer SA. Monterrey.	N533P
☐ XA-GHR	78	Sabre-75A	380-58	Taxirey SA. Monterrey.	XA-SEB
☐ XA-GIH	74	Sabre-60	306-72	Transportes Aereos de Xalapa SA. Jalapa.	N97SC
☐ XA-GME	93	Challenger 601-3A	5128	Industria Minera de Mexico/Mexico Transportes Aereos.	C-FPOX
☐ XA-GRB	94	Challenger 601-3R	5149	Aerolineas de Tehuacan, Tehuacan, Puebla.	N601GR
☐ XA-GTR	91	Falcon 900B	107	Aero Astra SA. Monterrey.	N470FJ
☐ XA-GYT	64	Sabre-40	282-6	Servicios Aereos Ilsa SA. Torreon.	XA-SBS
☐ XA-HEW	71	Falcon 20F	250	Commercial Aerea SA. Chihuahua.	N111AM
☐ XA-HFM	66	HS 125/1A-522	25107	Aeromedica SA. Toluca.	XA-GOC
☐ XA-HHF	74	Falcon 20F	327	Aerotaxi Grupo Tampico SA. Tamaulipas.	N25WG
☐ XA-HNY	73	JetStar-8	5162	Naviera Mexicana SA. Toluca.	N10JJ
☐ XA-HOS	80	Learjet 35A	35A-341	Aerolineas Ejecutivas SA. Toluca.	D-CARE
☐ XA-ICK	74	Sabre-60	306-86	Servicios Aereos del Centrol SA. Toluca.	N60TG
☐ XA-ICP	90	Citation II	550-0631	AVEMEX SA. Toluca.	XA-RUD
☐ XA-ILV	76	Gulfstream 2	195	Transpais Aereo SA. Del Norte, NL.	N71TP
☐ XA-INF	84	Citation S/II	S550-0010	Apoyo Aerea SA. Del Norte, NL.	N47MJ
☐ XA-JEX	79	Citation	500-0395	Aerotaxi Villa Rica SA. Veracruz.	N2651S
☐ XA-JFE	90	Challenger 601-3A	5059	Transporte Ejecutivo Aereo SA/El Universal SA. Toluca.	XA-GEO
☐ XA-JIQ	75	Learjet 24D	24D-317	Servicios de Alquiler Aereo SA. Toluca.	N45AJ
☐ XA-JJS	91	Challenger 601-3A	5097	Aero Lider SA/Arrendadora Banamex SA. Toluca.	C-FKVW
☐ XA-JLV	72	Citation	500-0021	Aerotaxis do Atizapan SA. Atizapan.	(N133N)
☐ XA-JMN	69	JetStar-731	5134	SACSA-Servicios Aereos del Centro SA. Toluca.	N136MA
☐ XA-JOC	80	Learjet 25D	25D-303	Jet Rent SA. Toluca.	
☐ XA-JRF	73	HS 125/600A	6018	Aerotaxi de Aguascalientes SA. Aguascalientes.	N288MW
☐ XA-JRH	85	Learjet 35A	35A-609	Aero Ermes SA. Del Norte, NL.	N788QC
☐ XA-JSO	66	Learjet 24	24-123	Servicios Aereos Poblanos SA. Puebla.	XA-JSC
☐ XA-JYO	91	Citation II	550-0689	Arrendadora Probursa SA/Aero Virel SA. Toluca.	XA-JPA
☐ XA-KAC	80	HS 125/700A	NA0271	SARSA=Servicios Aereos Regiomontanos SA. Monterrey.	G-5-15
☐ XA-KAJ	79	Learjet	28-004	SACSA-Servicios Aereos del Centro SA. Toluca.	N225MS
☐ XA-KCM	81	Learjet 35A	35A-418	Taxi Aereo de Mexico SA/Kimberly Clark SA. Toluca.	
☐ XA-KIM	83	Challenger 601	3015	Taxi Aereo de Mexico SA/Kimberly Clark SA. Toluca.	N374G
☐ XA-KMX	81	Citation II	550-0210	AVEMEX SA. Toluca.	XA-SDN
☐ XA-KOF	65	HS 125/1A	25065	Aerovias del Golfo SA. Veracruz.	N1YE
☐ XA-LAP	81	Learjet 25D	25D-336	Taxi Aereo de Veracruz SA. Jalapa.	
☐ XA-LEG	89	Beechjet 400	RJ-60	Corporacion Aerea Cencor SA.	N400FT
☐ XA-LIJ	80	Westwind-1124	285	Aerolineas Marcos SA. Queretaro.	VR-CBK
☐ XA-LIO	75	Falcon 10	40	Transpais Aereo SA. Del Norte, NL.	N15SJ
☐ XA-LML	73	Sabre-40A	282-115	Servicios Aereos Corporativos SA. Puerto Vallarta.	XA-GCH
☐ XA-LRJ	82	Learjet 25D	25D-359	Servicios Aereos del Centro SA. Toluca.	N116JR
☐ XA-LYH	85	Citation III	650-0101	Grupo IMSA SA. Del Norte, NL.	XB-GXV
☐ XA-MBM	65	HS 125/1A	25030	Aerotaxis Metropolitanos SA. Toluca.	XB-MBM
☐ XA-MDM	97	Learjet 60	60-089	Aeroservicios Dinamicos SA. Toluca.	N8089Y
☐ XA-MEA	79	Gulfstream 3	252	Aviones ARE SA/Manuel Espinosa Yglesias, Toluca.	(N301GA)
☐ XA-MHA	77	Learjet 25XR	25XR-222	Aerolineas Ejecutivas SA/Constructora Midas. Toluca.	XA-KEY
☐ XA-MIC	81	Gulfstream 3	323	Jet Ejecutivos SA/Aviones Televisa, Toluca.	
☐ XA-MII	89	Beechjet 400	RJ-58	Aeroservicios Ejecutivos de Occidente SA. Guadalajara.	XA-RNG
☐ XA-MIK	65	JetStar-731	5066	TAESA, Mexico City. (status ?).	XA-SAE
☐ XA-MJG	91	Learjet 31A	31A-044	Arrendora Bancomer SA. Toluca.	XA-HRM
☐ XA-MKI	94	Challenger 601-3R	5158	Servicios Aeronauticas Z SA. Chihuahua.	XA-ZTA
☐ XA-MMM	75	Falcon 10	36	Antair SA. Toluca.	N76AF
☐ XA-MVT	77	Sabre-75A	380-42	Servicios Aereos del Centro SA. Toluca.	N75AG
☐ XA-NGS	92	BAe 125/800A	8232	Aeroempresarial SA. Del Norte, NL.	N900KC
☐ XA-NLA	69	Learjet 24	24-180	Aerolineas Amanecer SA. Toluca.	XA-SBR
☐ XA-OAC	87	Beechjet 400	RJ-29	Corporacion Aerea Cencor SA.	(N597N)
☐ XA-OLI	71	JetStar-8	5148	TAESA-Transportes Aereos Ejecutivos SA. Toluca.	XA-ROK
☐ XA-OVR	94	Falcon 900B	141	Corporacion Aero Angeles SA. Toluca.	N141FJ
☐ XA-PAZ	80	Learjet 25D	25D-309	Transportes La PAZ SA. Mexico City.	XB-DKS
☐ XA-PFM	87	Falcon 200	515	Arrendadora Promex SA. Guadalajara.	F-GOBE
☐ XA-PIH	72	Sabre-40A	282-102	Verataxis SA. Vera Cruz.	(N157AT)
☐ XA-PIM	84	Learjet 25D	25D-368	Aeroservicios del Norte SA.	N8567J
☐ XA-PIU	80	Learjet 25D	25D-293	Servicios Aereos Estrella SA. Toluca.	N97JP
☐ XA-POG	72	Learjet 25B	25B-080	TAESA-Transportes Aereos Ejecutivos SA. Toluca.	N30AP
Reg	*Yr*	*Type*	*c/n*	*Owner/Operator*	*Prev Regn*

Reg	Yr	Type	c/n	Owner/Operator	Prev Regn
XA-PUF	72	Westwind-1123	153	Med Link/Aerolineas Marcos SA. Toluca.	N223WW
XA-PUR	67	Sabre-60	306-2	Aereo Fe SA. Monterrey.	N666BR
XA-RAP	74	Sabre-60	306-88	Aero Attie SA. Toluca.	N22CG
XA-RAR	87	Beechjet 400	RJ-32	SAI/Aerolineas Marcos SA. Monterrey.	N31432
XA-RAV	80	Learjet 35A	35A-290	Taxi Aereo de Veracruz SA. Jalapa.	N2022L
XA-RDM	95	Citation V Ultra	560-0342	Antair SA. Toluca.	(N14VF)
XA-REG	74	Sabre-40A	282-130	Ave Fenix Taxi Aereo SA. Mexico City.	XC-PGE
XA-REI	68	Sabre-60	306-20	Servicios Aereos Alfa SA. Mexico City.	N155EC
XA-REN	81	Citation II	550-0243	Taxi Aereo del Valle de Toluca SA. Toluca.	N1333Z
XA-RET	84	BAe 125/800A	8004	Aeroservicios Ejecutivos Corp SA. Toluca.	XA-SEH
XA-RFB	74	Sabre-60	306-87	Aero Quimmco SA. Monterrey.	N200CE
XA-RGB	87	Falcon 900	15	Transportes Aereos Tecnico Ejecutivo SA. Toluca.	HB-IAK
XA-RGC	65	Sabre-40R	282-48	Heliserv SA. Mexico City.	XA-CPQ
XA-RGS	90	Citation III	650-0189	Aeroempresarial SA/Vitro Corporativo SA. Monterrey.	N26174
XA-RIN	73	Learjet 25B	25B-104	Lineas Aereos Ejecutivos de Durango SA. Durango.	N392T
XA-RIR	69	Sabre-60	306-36	Servicios Aereos del Centro SA. Toluca.	N436CC
XA-RIZ	72	Westwind-1123	160	Servicios Aereos del Mar SA. Acapulco.	XA-MUI
XA-RKY	81	Learjet 35A	35A-370	Arrendadora Bancomer SA. Toluca.	N8216Q
XA-RLH	73	Sabre-40A	282-129	Aerotaxis Latino Americanos SA. Culiacan.	(N99FF)
XA-RLL	74	Sabre-60	306-83	Servindustria Aeronautica SA. Celaya.	N99FF
XA-RLS	70	Sabre-60	306-57	Organization de Servicios Ejecutivos Aereos SA. Toluca.	N465JH
XA-RMA	66	Falcon 20C	39	Aero Xtra SA. Toluca.	XB-EDU
XA-RMF	74	Learjet 24D	24D-290	Aerotonala SA. Guadalajara.	N24TK
XA-RMY	89	Citation III	650-0179	Alfa Corporativo/Transportes Aereos del Norte SA. Monterrey.	N679CC
XA-RNB	68	Falcon 20-5	142	Aerolineas Ejecutivas SA. Toluca.	(N205FJ)
XA-RNK	90	Learjet 31	31-021	Taxi Aereo del Noroeste SA. Cuidada Obregon, Sonora.	N3802G
XA-ROI	68	Gulfstream 2	10	Taxi Aereo T M M SA. Toluca.	N888CF
XA-RPS	66	Sabre-40	282-56	Servicios Aereos de Los Angeles SA. Puebla.	N85DA
XA-RPT	68	HS 125/3A-RA	NA708	Aerodan SA. Saltillo.	N75GN
XA-RQB	67	Learjet 24XR	24XR-150	Aerovanguardia SA. Monterrey.	N24XR
XA-RQP	68	Learjet 24	24-179	Taxi Aereo Nacional SA. Toluca.	XC-GII
XA-RRK	76	Learjet 24D	24D-307	Aerodin SA. Toluca.	N307BJ
XA-RSP	66	HS 125/1A	25091	Aerotaxi Monse SA. Morelos.	N65FC
XA-RSR	65	HS 125/1A-522	25017	Aerotaxi Monse SA. Morelos.	N333M
XA-RTM	65	Sabre-40R	282-39	Transportes Ejecutivos SA. Mexico City.	XA-RKQ
XA-RTP	76	Sabre-60	306-110	Aero Util SA. Toluca.	XA-DCO
XA-RTT	90	Citation V	560-0092	Casa de Bolsa Inverlat SA. Toluca.	N6784X
XA-RUQ	68	Sabre-60	306-15	Aerolineas del Oeste SA. Guadalajara.	N221PF
XA-RUU	89	Learjet 31	31-012	AVEMEX SA/Arrendadora Inverlat SA. Toluca.	N917MC
XA-RUY	96	Hawker 800XP	8302	Aerolineas Ejecutivas SA. Toluca.	N302XP
XA-RVT	78	Sabre-60	306-138	Aero Danta SA. Mexico City.	N702JR
XA-RVV	90	Falcon 50	213	Aerolineas Ejecutivas SA. Toluca.	N295FJ
XA-RXA	69	Learjet 24B	24B-197	Aero Rent SA. Toluca.	N710TJ
XA-RYE	73	Citation	500-0068	Cia Mexicana de Aeroplanos SA. Toluca.	XB-FDN
XA-RYH	81	Learjet 25D	25D-334	Aereo Ejecutiva Nieto SA. Capitan Castillo.	N334MD
XA-RYJ	71	Sabre-75	370-5	Servicios Aereos Corporativos SA. Mexico City.	N250BC
XA-RYN	68	Learjet 24	24-164	Aerotaxis Corporativos SA. Torreon.	N831RA
XA-RZD	91	Challenger 601-3A	5087	Aerotrans Privados/Consorcio Industrial Escorpion, Toluca.	N601CC
XA-SAG	73	Falcon 20E	287	Transpais Aereo SA. Monterrey.	XB-ALO
XA-SAH	78	Sabre-60	306-137	Aviones Ejecutivos JFA SA. Tuxtla Gutierrez Militar.	N18X
XA-SAI	74	HS 125/600A	6016	Omnirent Aviones SA. Toluca.	N99SC
XA-SAJ	91	Learjet 35A	35A-666	Servicios Aereos de Hidalgo SA.	N26002
XA-SAM	75	Citation	500-0255	Servicios Aereos Moritani SA. Guadalajara.	N885CA
XA-SAR	78	Falcon 10	125	Aerocer SA. Del Norte, NL.	N100CK
XA-SAU	78	HS 125/700A	NA0220	Aero Gisa SA. Saltillo.	N725CC
XA-SBV	75	Sabre-60A	306-109	Aero Rey SA. Del Norte, NL.	N602KB
XA-SBX	69	Sabre-60	306-40	Servicios Aereos Estrella SA. Toluca.	N997ME
XA-SCE	73	Learjet 24D	24D-271	Aerovias Castillo SA. Guadalajara.	N4305U
XA-SCL	68	Falcon 20C	130	Lineas Aereas Ejecutivas de Durango SA. Durango.	N514T
XA-SCR	81	Sabre-65	465-65	AESA/Arrendamiento Dinamico Serfin SA. Toluca.	N963WL
XA-SDI	77	Citation 1/SP	501-0031	Hotels Dinamicos SA. Guadalajara, Jalisco.	N395SC
XA-SDT	92	Citation V	560-0162	Stars de Mexico SA. Toluca.	N68864
XA-SDU	84	Citation III	650-0052	Servicios Aeronauticos Z SA. Chihuahua.	N20MW
XA-SEX	90	Citation II	550-0642	Aero Citro SA. Santa Ana Malopan.	XA-JRF
XA-SEY	81	Citation 1/SP	501-0228	Verataxis SA. Veracruz.	N228EA
XA-SFE	73	Citation	500-0125	Aeroservicio Corp de San Luis Potosi.	C-FCFP
XA-SFQ	72	HS 125/400A	NA768	Aero Cheyenne SA. Monterrey.	N2155P
XA-SGP	76	HS 125/3A	25114	Omni Flys SA/Omnitrition de Mexico SA. Guadalajara.	N25PM
XA-SGW	92	Falcon 900B	122	Aviacion Comercial America SA. Monterrey.	N479FJ
XA-SHZ	83	Challenger 601	3012	TAMSA-Transportes Aereos de Monterrey SA. Monterrey.	(N6165C)
XA-SIM	92	Falcon 900B	114	Aerolineas Mexicanos SA. Toluca.	N474FJ
XA-SIT	93	Citation V	560-0218	Servicios Aereonauticos de Oriente SA. Toluca.	N1285V
XA-SIV	90	BAe 125/800A	NA0452	SARSA-Servicios Aereos Regiomontanos SA. Monterrey.	(N207RC)
XA-SJC	92	Falcon 900B	560-0197	Taxi Aereos del Noroeste SA. Obregon.	(EI-DUN)
XA-SJN	83	Learjet 25D	25D-365	DGO-JET SA. Durango.	N365CM
XA-SJS	71	Learjet 25B	25B-076	SACSA-Servicios Aereos del Centro SA. Toluca.	N77KW

Reg	Yr	Type	c/n	Owner/Operator	Prev Regn
☐ XA-SJX	90	Falcon 900	97	Aero Personal SA/Banco Nacional de Mexico, Toluca.	EC-FFO
☐ XA-SKB	75	Sabre-60	306-111	Aereo Saba SA. Toluca.	N300RC
☐ XA-SKE	72	HS 125/F400A	25253	Hector Hugo Quezada Figueroa, Chetumal, Quintana Roo.	N610HC
☐ XA-SKX	91	Citation V	560-0118	Servicios Aeronauticos de Oriente SA. Toluca.	XA-RXO
☐ XA-SLA	93	Citation V	560-0228	Arrendadora Internacional SA. Toluca.	N228CV
☐ XA-SLB	93	Citation VI	650-0228	Gof-Air SA. Toluca.	N228CM
☐ XA-SLJ	77	Sabre-60	306-125	Organizacion de Transportes Aereos SA. Toluca.	N261T
☐ XA-SLP	72	HS 125/600A	6002	Aero Magar SA. Guadalajara.	N602MM
☐ XA-SLR	66	HS 125/3A	25112	Maria del Pilar Oliver Gaya Vd. Toluca.	XB-FFV
☐ XA-SMF	67	Sabre-60A	306-6	Aerotaxi Monse SA. Club de Vuelo Tepozotlan.	XA-ADC
☐ XA-SMQ	65	Sabre-40	282-50	Consorcio Aereo SA. Toluca.	N956CC
☐ XA-SMR	66	Sabre-40	282-71	Aerolineas Ejecutivas Tarascas SA. Morelia, Michoacan.	N957CC
☐ XA-SMU	72	Learjet 24D	24D-255	TAESA-Transportes Aereos Ejecutivos SA. Toluca.	XA-BBE
☐ XA-SND	67	Sabre-60	306-7	Aerovias Ejecutivas SA. Toluca.	N60CR
☐ XA-SNP	94	Beechjet 400A	RK-83	Aerolineas Ejecutivas SA. Toluca.	N8283C
☐ XA-SNX	93	Citation V	560-0229	Arrendadora Finac SA. Toluca.	N1287K
☐ XA-SOC	72	JetStar-8	5152	Servicios Aereos del Centro SA. Toluca. (status ?).	N113KH
☐ XA-SOK	94	Citation VII	650-7029	Aero Personal SA/Arrendadora GMB Atlantico SA. Toluca.	N95CM
☐ XA-SON	79	HS 125/700A	NA0268	Aero Sami SA. Del Norte, NL.	N501MM
☐ XA-SOR	94	Challenger 601-3R	5147	Aeropycsa SA. Toluca.	C-FRJX
☐ XA-SOY	69	JetStar-8	5142	Compania Ejecutiva SA. Toluca.	N23FE
☐ XA-SPM	80	Sabre-65	465-14	Aero Toluca International SA. Toluca.	N25SR
☐ XA-SPQ	93	Citation VII	650-7028	Aero Personal SA. Toluca.	N728CM
☐ XA-SPR	93	Learjet 31A	31A-076	Servicios Aereos del Centro SA. Toluca.	XA-PIC
☐ XA-SQA	73	Sabre-40A	282-125	Jett Paqueteria SA. Quetzalcoatl.	XB-NIB
☐ XA-SQV	80	Citation II	550-0198	Servicios Aereos Especializados Mexicanos SA. Del Norte.	XC-DOK
☐ XA-SSU	70	Learjet 24D	24D-230	Aero-Jet Express SA/Arrendadora Union SA. Guadalajara.	N32287
☐ XA-STI	74	Sabre-60	306-89	Multiservicios Aereos Queretanos SA. Queretaro.	N86RM
☐ XA-STT	76	Citation	500-0310	Transportes ABC SA. Costilla.	N820
☐ XA-SUN	79	Sabre-60	306-143	Horizontes Aereos SA. Toluca.	XA-SIM
☐ XA-SVX	75	Learjet 35	35-012	Aerotransportes de Toluca SA.	N97TJ
☐ XA-SWC	66	Falcon 20C	21	Aereo Escorpion SA./Multiva Arrendadora SA. Del Norte.	N91TS
☐ XA-SWK	73	HS 125/F600A	6026	Aero Sami SA. Del Norte, NL.	N818TP
☐ XA-SWM	93	Citation VII	650-7034	GAESA-Grupo de Aviacion Ejecutiva SA. Mexico City.	N1264E
☐ XΛ SWP	68	Gulfstream 2SP	14	TAESA-Transportes Aereos Ejecutivos SA. Toluca.	XA-RBS
☐ XA-SWX	83	Learjet 25D	25D-366	Transportes Aereos Santa Fe SA. Sin Nombre.	(ZS-LRI)
☐ XA-SXK	75	Sabre-75A	380-39	Aeroservicios Demex SA. Toluca.	N2093P
☐ XA-SYS	76	Sabre-60	306-121	Aerolineas Yasi SA. Toluca.	N880CK
☐ XA-SYY	73	Falcon 10	4	Aviacion Cemercial de America SA. Monterrey.	EC-FTV
☐ XA-TAB	85	Falcon 100	204	Aerovics SA. Toluca.	F-WGTG
☐ XA-TAQ	79	Learjet 25D	25D-286	Aerotransportes Internacionales de Torreon SA. Torreon.	XA-RVI
☐ XA-TAZ	67	JetStar-8	5103	Aerotaxis del Golfo SA. Acapulco, Guerrero.	N101AW
☐ XA-TBA	83	Citation III	650-0019	Servicios Aeronauticos de Oriente SA. Toluca.	(N833RL)
☐ XA-TBL	83	Falcon 100	203	VIP Servicios Aereos Ejecutivos SA. Chetumal.	N100CT
☐ XA-TBV	81	Learjet 25D	25D-325	Aviacion Ejecutiva de Hidalgo SA. Pachuca.	XA-RLG
☐ XA-TCA	70	Learjet 24B	24B-224	Servicios Aereos SAAR SA. Colonia Juarez Nuevo.	(N51GJ)
☐ XA-TCI	80	Learjet 35A	35A-349	Aerolineas Ejecutivas SA. Toluca.	N272T
☐ XA-TCO	83	Gulfstream 3	403	Servicios Aereos Norte SA. Villahermosa.	N403NW
☐ XA-TCR	80	HS 125/700A	NA0280	Aero Servicio Corporativo SA. Monterrey.	N810M
☐ XA-TCY	69	Learjet 25	25-048	Servicios Aereos del Golfo SA. Tampico. Tamaulipas.	N200G
☐ XA-TCZ	92	Citation VII	650-7019	AVEMEX SA. Toluca.	N12616
☐ XA-TDD	95	Falcon 50	252	Grupo Industrial Alfa SA. Toluca.	N50FJ
☐ XA-TDG	73	JetStar-8	5158	TAESA, Toluca. (status ?).	XA-FHR
☐ XA-TDK	72	Gulfstream 2	114	Omni Flys SA. Guadalajara.	N25BF
☐ XA-TDP	66	Learjet 24	24-128	Servicios Aerojal SA. Guadalajara.	N911KB
☐ XA-TDQ	76	Sabre-80A	380-50	Aero Jets Corporativos SA. Torreon.	XA-RLR
☐ XA-TDU	96	Falcon 2000	29	CEMEX SA. Monterrey.	N2028
☐ XA-TDX	62	Sabreliner CT-39A	276-27	Jett Paqueteria SA. Quetzalcoatl.	XB-GDV
☐ XA-TFD	60	Sabreliner CT-39A	265-10	Jett Paqueteria SA. San Luis Potosi.	(N510TA)
☐ XA-TFL	61	Sabreliner CT-39A	265-48	Jett Paqueteria SA. Quetzalcoatl.	N6CF
☐ XA-TGA	84	Citation III	650-0036	Servicios Aeronauticos de Oriente SA.	(N143RC)
☐ XA-TGO	62	Sabreliner T-39A	276-6	Jett Paqueteria SA. San Luis Potosi.	N6552R
☐ XA-THD	79	Learjet 35A	35A-243	Servicios de Taxi Aereo SA. Toluca.	N152TJ
☐ XA-THF	67	Jet Commander	109	Millenium Air Servicios Aereos Integrados SA. Guadalajara.	TG-VWA
☐ XA-THO	85	Citation S/II	S550-0035	Aero Zano SA. Del Norte, NL.	N711JN
☐ XA-TIE	83	Learjet 25D	25D-364	Aero Silza SA. Chihuahua.	N25TZ
☐ XA-TII	65	Learjet 23	23-070	Aerotonala SA. Guadalajara.	XA-RZM
☐ XA-TIP	74	Learjet 24D	24D-293	Aerolineas Ejecutivas/Aerovics SA. Toluca.	
☐ XA-TIW		Sabreliner	...	Jett Paqueteria SA. Quetzalcoatl.	
☐ XA-TMX	96	Citation VII	650-7069	Aero Frisco SA. Toluca.	N5117U
☐ XA-TRI	94	CitationJet	525-0060	Aero Taxi del Centro de Mexico SA. Mexico City.	XA-SOU
☐ XA-TZF	96	Learjet 60	60-088	Aero Silza SA. Chihuahua.	XA-TGE
☐ XA-VEL	69	Sabre-60	306-42	Arrendadora Financiera Dina SA. Merida.	N128JC
☐ XA-VTO	93	Falcon 900B	129	Aeroempresarial SA/Vitro Corporativo SA. Del Norte, NL.	N483FJ
☐ XA-XIS	93	Citation VII	650-7032	AVEMEX SA. Mexico City.	N12637
☐ XA-ZAP	77	Learjet 35A	35A-129	Aerolineas Ejecutivas SA. Toluca.	N229X
Reg	Yr	Type	c/n	Owner/Operator	Prev Regn

Reg	Yr	Type	c/n	Owner/Operator	Prev Regn
☐ XA-ZTH	89	Learjet 31	31-004	Servicios Especiales del Pacifico Jalisco, Leon.	
☐ XA-ZUM	80	Sabre-65	465-15	AESA/Farmaceuticos Lakeside SA. Mexico City.	N2513E
☐ XA-ZYZ	79	Learjet 25D	25D-287	Transportes Aereos Pegaso SA. Mixcoac.	N287MF
☐ XB-AMO	74	Citation	500-0152	Fabricas Orion SA. Monterrey, NL.	N2782D
☐ XB-BBL	73	Sabre-40A	282-116	Altos de Hornos SA. Mexico City.	N4PH
☐ XB-BON	90	Citation II	550-0654	Benjamin Trapero Bustamente, Los Mochis-Topolbampo.	XA-RZB
☐ XB-CUX	71	HS 125/400A	NA764	Aereo Fe SA. Monterrey, NL.	N59BH
☐ XB-DBS	73	JetStar-8	5159	Sindicato Petrolero Mexicano SA. Mexico City.	N520M
☐ XB-DLM	65	B 727-64	18912	UTAPEF, Mexico City.	XC-FAD
☐ XB-DVF	81	Citation	500-0408	Alonso Ayala Rodriguez, Del Monte, NL.	XA-LUD
☐ XB-DZD	77	Learjet 24F	24F-349	Impulsora Azucarera del Noroes, Culiacan.	XA-CAP
☐ XB-DZQ	81	Learjet 25D	25D-332	PRI, Mexico City.	XC-FIF
☐ XB-DZR	73	Learjet 24D	24D-273	Sindicato Nacional de Trabaja Education, Mexico City.	XC-DOP
☐ XB-EBI	70	Gulfstream 2	96	Vidreira Monterrey y Copropie Magallanes SA. Monterrey, NL.	XC-MEX
☐ XB-EQR	67	Sabre-40	282-82	Servicios Corporativos Jalisco SA. Guadalajara.	N366DA
☐ XB-ESS	73	Sabre-40A	282-123	Cafe Descafeinado de Chis. Cordoba.	XA-APD
☐ XB-ESX	70	Sabre-60A	306-47	Grupo Corporativo Cever SA. Toluca.	XA-ZOM
☐ XB-ETV	75	Sabre-60	306-96	Aceros San Luis SA. San Luis Potosi.	N315JM
☐ XB-FKT	90	Learjet 31	31-029	Zeferino Romero Bringas, Tehuacan.	N9173L
☐ XB-FKV	69	Jet Commander-B	137	Armando Alanis Rodriguez, Del Norte, NL.	N707TE
☐ XB-FMK	79	HS 125/700A	NA0249	Juan R Brittingham SA. Monterrey, NL..	N799SC
☐ XB-FNW	79	Learjet 35A	35A-255	Adame Barocio Alfonso, Toluca.	XB-LHS
☐ XB-FRP	69	HS 125/400A	NA720	Productos Rolmex SA. Monterrey, NL.	XA-RMN
☐ XB-FSZ	70	Sabre-60A	306-50	Miguel Kalifa Manzur/Multiva, Del Norte, NL.	XA-MUL
☐ XB-FXO	80	Citation 1/SP	501-0161	Righetti Bachthaler Heinz, Guadalajara.	N5UM
☐ XB-GBF	75	Citation	500-0273	Jaime M Benavides Pompa, Monterrey.	XA-RUR
☐ XB-GDJ	81	Citation	500-0412	Constructora y Pavimentadora SA. Leon, Guanjuato.	XA-LUV
☐ XB-GDW	61	Sabreliner CT-39A	265-86	Jett Paqueteria SA. Quetzalcoatl.	N510TB
☐ XB-GGK	66	HS 125/1A-522	25064	Transportes Aereos Corporativos SA. Guadalajara.	XA-TAL
☐ XB-GHO	67	Learjet 24	24-141	Jesus Arturo Armenta Castro, Del Norte, NL.	XB-FJW
☐ XB-GLZ	81	Citation II	550-0303	Lineas de Producciones SA y Perforadora Central SA. Toluca.	N450CC
☐ XB-GMD		Sabreliner	...	noted Little Rock 2/97.	
☐ XB-GNF	72	HS 125/403A	NA776	Desarrollo Inmobiliario SA. Monterrey, NL.	XA-SGM
☐ XB-GRN	80	Westwind-Two	301	Exportadora de Sal SA. Tijuana, Baja California Norte.	N815BC
☐ XB-GSP	77	Sabre-75A	380-55	Treviso Ballesteros Lillia, Del Norte, NL.	XB-RDB
☐ XB-GVY	77	Citation	500-0357	Maquinaria Diesel SA. Monterrey, NL.	N545G
☐ XB-HRA	67	Falcon 20C	127	Librada Angulo Navarrete,	XB-GCR
☐ XB-JHE	89	Beechjet 400	RJ-48	Industrial Patrona SA. Veracruz, Guadalajara.	N1548D
☐ XB-JMM	77	Sabre-60	306-130	Estado Mayor Presidencial, Mexico City.	XA-JIK
☐ XB-JMR	68	Sabre-60	306-35	Cia J M Romo SA. Aguascalientes.	N3456B
☐ XB-JTN	82	HS 125/700A	NA0331	Jamil Textil SA. Toluca.	N900BL
☐ XB-LAW	75	Sabre-60	306-100	Hunter's del Valle SA. Monterrey, NL.	XA-RLR
☐ XB-MCB	75	Sabre-75A	380-36	Aero Barloz SA. Tampico.	XA-MCB
☐ XB-MLC	82	HS 125/700A	NA0319	Cementos de Chihuahua SA. Chihuahua.	N319NW
☐ XB-MTS	89	Citation V	560-0030	Familia Sada Gonzales, Del Norte, NL.	N560W
☐ XB-MVG	73	Sabre-40A	282-134	Aceros San Luis SA. San Luis Potosi.	N66CD
☐ XB-OEM	88	Gulfstream 4	1055	Organization Editorial Mexicana, Toluca.	XB-EXJ
☐ XB-PGR	70	Gulfstream 2	81	Procurad General de la Republica, Mexico City.	N681AR
☐ XB-PUE	68	HS 125/3A-RA	25158	Latex Occidental SA. Guadalajara.	G-AVZK
☐ XB-PYC	94	Citation V Ultra	560-0261	Grupo Pycsa SA. Toluca.	N261CV
☐ XB-QND	68	Sabre-60	306-21	Aerotransportacion Eficaz SA. Toluca.	XB-LRD
☐ XB-RGO	73	Sabre-40A	282-114	Regio Empresas y Copropietarios, Del Norte, NL.	XB-RGS
☐ XB-RYO	81	Sabre-65	465-55	Industrial Perforadora de Campeche SA. Toluca.	XA-RYO
☐ XB-SHA	78	Sabre-75A	380-60	Rodolfo Junco de la Vega Gonzales, Del Monte, NL.	XB-RSG
☐ XB-SOL	82	Falcon 50	116	Banco Nacional de Mexico SA./CIFRA SA. Toluca.	N70AF
☐ XB-UAG	75	Citation	500-0278	Universidad Autonoma de Guadalajara, Jalisco.	XB-FQO
☐ XB-ZNP	80	Sabre-60	306-63	Aereo Representaciones y Servicios SA. Toluca.	XA-FNP
☐ XC-AA24	81	Learjet 36A	36A-050	TP-105, Procurad de la Republica, Mexico City.	XC-UJR
☐ XC-AA28	65	Learjet 23	23-037	Drug Enforcement Agency, Mexico City.	XC-UJP
☐ XC-AA60	80	Learjet 35A	35A-321	Drug Enforcement Agency, Mexico City.	XA-RVB
☐ XC-AA70	68	Gulfstream 2	18	Procurad General de la Republica, Mexico City.	XA-LZZ
☐ XC-AA73	72	Sabre-40A	282-105	Procurad General de la Republica, Mexico City.	XA-SCN
☐ XC-AA88	81	Learjet 25D	25D-330	Procurad General de la Republica, Mexico City.	XA-RCG
☐ XC-AA89	76	Sabre-75A	380-46	Procurad General de la Republica, Mexico City.	XC-HFY
☐ XC-AGR	80	Learjet 25D	25D-295	Agroasemex SA. Queretaro.	N45ES
☐ XC-BCS	81	Citation II	550-0241	Governor of the State of Baja California Sur, La Paz.	(N32TJ)
☐ XC-COL	69	Jet Commander-B	135	State Government of Colima.	N1121N
☐ XC-CUZ	79	Learjet 35A	35A-213	Fonseca Alvarez Guillermo, San Luis Potosi.	(N935NA)
☐ XC-DGA	72	Citation	500-0010	SCT CIAAC/TAF-Transporte Aereo Federal, Mexico City.	XC-FIT
☐ XC-DIP	73	Falcon 20E	282	Transporte Aereo Federal, Mexico City.	N282C
☐ XC-FEZ	81	Citation	500-0409	SCT/Dept de Verificaciones Aereo, Mexico City.	N67815
☐ XC-FIV	72	Citation	500-0013	SDN/Dept de Verificaciones Aereo, Mexico City.	N513CC
☐ XC-FVH	78	Falcon 20F-5	393	State Government of Sonora, Hermosillo.	XB-FVH
☐ XC-GAW	81	Citation	500-0410	State Government of Tamaulipas, Victoria.	N6780Z
☐ XC-GTO	80	Citation	500-0396	State Government of Guanajuato, Guanajuato.	(XA-JEW)
☐ XC-GUB	80	Learjet 25D	25D-306	SARH/Ministry of Agriculture, Mexico City.	XA-DUB

Reg	Yr	Type	c/n	Owner/Operator	Prev Regn
☐ XC-HGY	69	Sabre-60	306-38	Aerotaxis del Golfo SA. Acapulco.	XA-DCO
☐ XC-HHJ	81	Learjet 35A	35A-435	State Government of Chiapas, Tuxtla Gutierrez.	N435N
☐ XC-HIE	79	Learjet	29-002	SGDG/Director General of Security, Mexico City.	XC-DFS
☐ XC-HIS	80	Learjet 25D	25D-312	State Government of Chiapas, Tuxtla Gutierrez.	N94MJ
☐ XC-HIX	71	Falcon 20E	248	Government of Sinalao, Culiacan.	XB-VRM
☐ XC-HJA		Citation II	...		
☐ XC-IST	79	Learjet	29-001	ISSSTE/Institute of Security Social Services, Mexico City.	N929GL
☐ XC-JCC	65	JetStar-731	5053	Banco Nacional de Comercio Exterior SNC. Toluca.	XA-POU
☐ XC-JCV	89	Citation II	550-0595	Mexican Customs Service, Mexico City.	N2734K
☐ XC-JCW	89	Citation II	550-0607	Mexican Customs Service, Mexico City.	N26496
☐ XC-JCX	91	Citation II	550-0663	Mexican Customs Service, Mexico City.	N5314J
☐ XC-JCY	79	Citation II	550-0086	Mexican Customs Service, Mexico City.	N43SA
☐ XC-JCZ	91	Citation II	550-0670	Mexican Customs Service, Mexico City.	N6637G
☐ XC-JDA	80	Citation II	550-0169	Mexican Customs Service, Mexico City.	N6001L
☐ XC-JDC	79	Sabre-60	306-145	Governor of the State of Campeche.	XA-LOQ
☐ XC-NSP	75	Learjet 25B	25B-194	CONASUPO, Toluca.	XB-GPJ
☐ XC-OAH	74	Sabre-60	306-73	State Government of Coahuila, Ramos Arizpe, Coahuila.	N90EC
☐ XC-PGM	90	Citation II	550-0644	PGR/PJF-Policia Judicial Federal, Mexico City.	(N1310B)
☐ XC-PGN	89	Citation III	650-0165	Procuraduria General de la Republica, Mexico City.	XA-FCP
☐ XC-PGP	91	Citation II	550-0648	Procurad General/Mexican Customs Service, Mexico City.	N1260G
☐ XC-PVC		Citation	...	noted West Palm Beach 10/97,	
☐ XC-RPP	78	Learjet 25D	25D-236	Secretaria de Hacienda y Credito Publico, Aguascalientes.	N1466B
☐ XC-SCT	80	Citation II	550-0138	ASA, Mexico City. 'Mexico es Primero'.	N2646X
☐ XC-SEY	68	Falcon 20C	169	ASA-Aeropuertos y Servicios Auxiliares, Mexico City.	N4370F
☐ XC-SKI	68	JetStar-8	5124	Secretaria de la Reforma Agraria, Toluca.	XA-SKI
☐ XC-VSA	79	Learjet	28-002	State Government of Tabasco, Villahermosa. 'El Chontal'	N511DB
Military					
☐ ETE-1329	73	Citation	500-0090	Mexican Army, Mexico City.	XB-EFR
☐ JS 10201	69	JetStar-8	5144	DN-01, Ministry of Defence, Mexico City.	N5508L
☐ MTX-01	69	Sabre-60	306-34	Mexican Navy, Mexico City.	MTX-02
☐ MTX-02	75	Learjet 24D	24D-313	Mexican Navy, Mexico City.	MTX-01
☐ XC-FAY	65	B 727-14	18908	10503, FAM, Mexico City.	XA-SER
☐ XC-FAZ	65	B 727-14	18909	10504, FAM/SDN. Mexico City.	XA-SEU
☐ XC-IPP	75	Learjet 35	35-028	TP-104, TAF, Mexico City.	XC-IPP
☐ XC-UJA	66	B 727-51	19123	TP-05, Government of Mexico. 'Presidente Carranza'.	XC-UJA
☐ XC-UJB	89	B 737-33A	24095	TP-02, Estado Mayor Presidencial, Mexico City.	N731XL
☐ XC-UJD	78	Sabre-75A	380-68	TP-102, CGATP, Mexico City.	TP-104
☐ XC-UJE	78	Sabre-60	306-139	TP-103, CGATP, Mexico City.	TP-105
☐ XC-UJF	78	Sabre-60	306-144	Aeropuertos y Servicios Auxiliares, Mexico City.	XC-AA51
☐ XC-UJI	69	B 737-247	20127	Estado Mayor Presidencial, Mexico City.	B-12001
☐ XC-UJL	69	B 737-112	19772	TP-03, Government of Mexico. 'Emiliano Zapata'	TP-04/XC-UJJ
☐ XC-UJM	87	B 757-225	22690	TP-01, Government of Mexico. 'Presidente Juarez'	XC-CBD
☐ XC-UJN	82	Gulfstream 3	352	TP-06, Presidencia de la Republica, Chiapas.	HB-ITM
☐ XC-UJO	83	Gulfstream 3	386	TP-07, Government of Mexico, Mexico City.	N902KB

XT = Burkina Faso (People's Democratic Republic of Burkina Faso)

Civil

Reg	Yr	Type	c/n	Owner/Operator	Prev Regn
☐ XT-AOK	93	Citation II	550-0726	Entreprise El Hadj Oumorou Kanazoe, Ouagadougou.	N1209T
☐ XT-BBE	66	B 727-14	18990	Government of Burkina Faso, Ouagadougou.	N21UC

XU = Kampuchea (Kingdom of Cambodia)

Civil

Reg	Yr	Type	c/n	Owner/Operator	Prev Regn
☐ XU-001	69	F 28-1000	11012	Government of Cambodia, Phnom Penh.	F-GIAH
☐ XU-008	75	Falcon 20E	323	Government of Cambodia, Phnom Penh.	OE-GLF

XY = Myanmar (Union of Myanmar)

Military

Reg	Yr	Type	c/n	Owner/Operator	Prev Regn
☐ 4400	82	Citation II	550-0358	MoD/Burmese Air Force, Rangoon.	N6801Q

YI = Iraq (Republic of Iraq)

Civil

Reg	Yr	Type	c/n	Owner/Operator	Prev Regn
☐ YI-AHH	76	Falcon 20F	337	Government/Iraqi Airways, Baghdad. (photo-recce).	F-WRQR
☐ YI-AHJ	76	Falcon 20F	343	Government/Iraqi Airways, Baghdad. (photo recce).	F-WRQR
☐ YI-AKB	79	JetStar 2	5235	Government/Iraqi Airways, Baghdad.	N4055M
☐ YI-AKC	79	JetStar 2	5237	Government/Iraqi Airways, Baghdad.	N4058M
☐ YI-AKD	79	JetStar 2	5238	Government/Iraqi Airways, Baghdad.	N4062M
☐ YI-AKE	79	JetStar 2	5239	Government/Iraqi Airways, Baghdad.	N4063M
☐ YI-AKF	79	JetStar 2	5240	Government/Iraqi Airways, Baghdad.	N4065M

YK = Syria (Syrian Arab Republic)

Military

Reg	Yr	Type	c/n	Owner/Operator	Prev Regn

Reg	Yr Type	c/n	Owner/Operator	Prev Regn
☐ YK-ASA	75 Falcon 20F	328	Government of Syria, Damascus.	(N4459F)
☐ YK-ASB	75 Falcon 20F	331	Government of Syria, Damascus.	F-WRQS
☐ YK-ASC	91 Falcon 900	100	Government of Syria, Damascus.	F-WWFB

YL = Latvia *(Republic of Latvia)*
Civil
☐ YL-VIP	80 HS 125/700B	7103		VP-BOJ

YR = Romania *(Republic of Romania)*
Civil
☐ YR-DVA	73 HS 125/600B	6024	Grivco International Ltd. Bucharest-Baneasa.	G-OMGA

YU = Yugoslavia *(Federal Republic of Yugoslavia)*
Civil
☐ YU-BJG	75 Learjet 25B	25B-187	Government of Yugoslavia, Belgrade.	
☐ YU-BKR	77 Learjet 25D	25D-221	Government of Yugoslavia, Belgrade.	N3819G
☐ YU-BKZ	77 Citation	500-0373	Govt of Bosnia & Herzogovina/Air Service Sarajevo.	N98449
☐ YU-BML	80 Citation	500-0399	Air Service Sarajevo.	N2069A
☐ YU-BNA	81 Falcon 50	43	Government of Yugoslavia, Belgrade.	72102
☐ YU-BPY	78 Learjet 35A	35A-173	Air Service Podgorica, Montenegro.	(N83DM)
☐ YU-BPZ	80 Falcon 50	25	Government of Yugoslavia, Belgrade.	72101
☐ YU-BRA	76 Learjet 25B	25B-202	Government of Yugoslavia, Belgrade.	70401

Military
☐ 70402	76 Learjet 25B	25B-203	Government of Yugoslavia, Belgrade.	10402

YV = Venezuela *(Republic of Venezuela)*
Civil
☐ YV-....	80 Citation II/SP	551-0223	Inversiones Marafuera CA. Porlamar.	N754AA
☐ YV-03CP	67 JetStar-731	5106	Servicios Aerofacility SA. Caracas.	YV-03CP
☐ YV-05C	83 Citation II/SP	551-0463	Promociones Orizaba C.A. Caracas.	N131EL
☐ YV-05CP	78 Citation II/SP	551-0007	Pavimentadora Life C.A. Chua, Caracas.	YV-169CP
☐ YV-12CP	82 Learjet 55	55-031	Aero Servicios ALAS C.A. Caracas.	
☐ YV-19CP	77 Citation II/SP	551-0004	Servicios Aereos Rigres C.A.	N553CJ
☐ YV-21CP	73 Citation	500-0115	Transportes Inland, Caracas.	YV-TAFA
☐ YV-52CP	77 Citation	500-0367	Construcciones CADE.	N36906
☐ YV-55CP	74 Citation	500-0215	SABENPE, Caracas.	YV-T-OOO
☐ YV-58CP	74 Westwind-1123	172	West Wind Air C.A. Caracas.	N19EE
☐ YV-100CP	76 Learjet 35A	35A-083	Contivato C.A.	N121CL
☐ YV-125CP	86 Learjet 55	55-126	Transportes La Mona C.A. Caracas.	N7260J
☐ YV-162CP	81 Citation II	550-0300	Aero Servicios ALAS C.A. Caracas.	N68881
☐ YV-169CP	93 Citation V	560-0230		N.....
☐ YV-190CP	77 Westwind-1124	219	Macor C.A. Caracas.	4X-CLQ
☐ YV-203CP	70 Learjet 25C	25C-061	Tranarg C.A. Caracas.	N9CN
☐ YV-213CP	79 Citation II/SP	551-0015	Inversiones Lunfa C.A. Caracas.	(N26613)
☐ YV-278CP	65 Learjet 23	23-036	SIFINCA, Caracas.	N38DM
☐ YV-301P	79 Citation 1/SP	501-0131	S H Benacerraf,	N490WC
☐ YV-327CP	80 Learjet 35A	35A-344	Oficina Central Asesoria y Ayuda Tecnia C.A. Caracas.	N40149
☐ YV-332CP	81 Westwind-Two	330	Transporte 330 SA/Cia Tamesis SA. Caracas.	N723K
☐ YV-376CP	90 Citation II	550-0637	Pavimentadora Life C.A. Chua, Caracas.	N1258U
☐ YV-388CP	71 HFB 320	1057	Inversiones Guraica CA. Caracas.	VR-CYR
☐ YV-432CP	81 Learjet 35A	35A-437	Transporte Transilac C.A. Maracaibo.	N3803G
☐ YV-450CP	91 Falcon 50	219	Maraven S.A. Caracas.	N129FJ
☐ YV-452CP	79 Falcon 50	4	Maraven S.A. Caracas.	N50FJ
☐ YV-455CP	83 Falcon 50	136	Maraven S.A. Caracas.	N50HC
☐ YV-572CP	75 Corvette	17	Aerocolon S.A. Caracas. (status ?).	F-ODTM
☐ YV-604P	82 Citation II	550-0405	Consolid-Air S.A.	YV-276CP
☐ YV-646CP	72 Citation	500-0031	Executive Air C.A. Caracas.	N666SA
☐ YV-688CP	79 Citation 1/SP	501-0142	Derivados Plasticos CA. Valencia.	I-SATV
☐ YV-701CP	91 Citation II	550-0683	Luis Mendoza, Caracas.	(N6777X)
☐ YV-713CP	83 Citation II	550-0467		N64CM
☐ YV-738CP	85 Diamond-Two	RJ-6	Treol CA. Caracas. (was A1006SA).	YV-737CP
☐ YV-771CP	95 Astra-1125SP	077	Transporte Polar CA. Caracas.	N771CP
☐ YV-785CP	91 Astra-1125SP	057	Calle Guaicapuro, Caracas.	YV-2199P
☐ YV-811CP	91 Citation V	560-0134	Banco del Caribe S.A. Caracas.	N6812D
☐ YV-815CP	66 HS 125/731	25098		N29CR
☐ YV-824CP	68 Learjet 24	24-173	Editorial Roderick, Puerto Ordez.	N623RC
☐ YV-888CP	81 Citation II	550-0135	Inversiones Brancom S.A. Caracas.	N550BP
☐ YV-900CP	80 Citation II	550-0192	Inversiones Meyvi SA. Caracas. 'El Canguro'	N44ZP
☐ YV-901CP	72 Citation	500-0058		N6145Q
☐ YV-999P	69 HFB 320	1037	Gustavo Soto,	N555JM
☐ YV-1111CP	70 HS 125/731	NA748	Aeroservicio Ejecutivos Cocodrilos C.A. Caracas.	N728KA
☐ YV-2426P	81 Citation II/SP	551-0313	(was 550-0270).	N270CF
☐ YV-2454P	67 Jet Commander	96		(N2ES)
☐ YV-2477P	72 Citation	500-0052	Jesus A N Martinez,	YV-2267P

Reg	Yr Type	c/n	Owner/Operator	Prev Regn

☐ YV-2479P	72	Citation	500-0035		N10108
☐ YV-O-CVG		Westwind-Two	343	Corporacion Venezolana de Guyana, Puerto Ordaz.	YV-O-CVG-3
☐ YV-O-CVG-278		Citation II/SP	551-0006	Corporacion Venezolana de Guyana, Puerto Ordaz.	YV-06CP
Military					
☐ YV-2338P	82	Citation II	550-0449	FAV, MoD, Caracas.	(FAV1107)
☐ 0001		B 737-2N1	21167	Government of Venezuela, Caracas.	
☐ 0002	78	Citation II	550-0011	FAV, MoD, Caracas.	(N98876)
☐ 0004	73	Gulfstream 2	124	FAV, MoD, Caracas.	N203GA
☐ 0005	83	Gulfstream 3	400	FAV, MoD, Caracas.	N17585
☐ 0006	72	Learjet 24D	24D-250	FAV, MoD, Caracas.	N85CD
☐ 0013	79	Learjet 35A	35A-270	FAV, 41 Squadron, Grupo 4, La Carlotta.	YV-O-MRI-1
☐ 0222	73	Citation	500-0092	FAV, MoD, Caracas.	N592CC
☐ 0442	70	Falcon 20D	235	FAV, Palo Negro AB. Maracau.	(N442)
☐ 1650	83	Falcon 20F	476	FAV, MoD, Caracas.	F-ZJTD
☐ 2222	81	Citation II	550-0224	FAV, MoD, Caracas.	YV-OMTC20
☐ 5761	66	Falcon 20C	23	FAV, Palo Negro AB. Maracau.	(N582G)
☐ 5840	69	Falcon 20D	216	FAV, Palo Negro AB. Maracau.	N9FE

Z = Zimbabwe (Republic of Zimbabwe)
Civil
☐ Z-WPD	87	BAe 146 Statesman	E2065	Air Zimbabwe/Government of Zimbabwe,	G-5-065
☐ Z-WSY	77	Citation Eagle	501-0024	Executive Air/Skyline Charters P/L. Harare.	N711NR

ZK = New Zealand (Dominion of New Zealand)
Civil
☐ ZK-LJL	73	Citation Eagle	500-0123	PDL Industies/Laser Co-Op, Christchurch.	VH-LJL
☐ ZK-NLJ	87	Citation III	650-0133	Air Nelson Ltd. Nelson.	N133LH
Military					
☐ NZ7271	68	B 727-22C	19892	RNZAF, 40 Squadron, Whenuapai.	N7435U
☐ NZ7272	68	B 727-22C	19895	RNZAF, 40 Squadron, Whenuapai.	N7438U

ZP = Paraguay (Republic of Paraguay)
Civil
⊓ 7P-...	74	Citation	500-0185		N500AZ
☐ ZP-...	97	Citation Bravo	550-0819	B & R Enterprises Ltd. Asuncion.	N1259B
☐ ZP-AGD	71	Westwind-1123	151	Humberto Dominguez Dibb, Asuncion.	N88WP
☐ ZP-TYO	78	Citation II/SP	551-0084	Conempa SRL. Asuncion.	N78FA

ZS = South Africa (Republic of South Africa)
Civil
☐ ZS-...	97	Beechjet 400A	RK-162	Beechcraft Sales/NAC P/L. Rand.	N2362G
☐ ZS-...	68	Learjet 25XR	25XR-022		N111WB
☐ ZS-...	74	Westwind-1123	171	Eleven 23 Air Inc. Johannesburg.	N89XL
☐ ZS-AMB	73	Citation	500-0071	Lanseria Air Charter CC. Lanseria.	C-FBCM
☐ ZS-ARG	80	Citation II/SP	551-0163	Mineag Air Partnership P/L. Bryanston.	D-IGRC
☐ ZS-AVL	96	Challenger 604	5328	Anglo Vaal Air P/L. Saxonwold.	N712DG
☐ ZS-BCT	95	Astra-1125SP	075	Searay Aviation Charters Partnership, Capetown.	4X-CUW
☐ ZS-BPG	89	BAe 125/800B	8165	Billiton Aviation P/L. Lanseria.	VR-BPG
☐ ZS-CAL	69	HS 125/F3B-RA	25172	Directorate of Civil Aviation, Pretoria.	(G-5-506)
☐ ZS-CAR	86	Citation S/II	S550-0078	Director General of Transport, Pretoria.	N1273X
☐ ZS-CDS	97	Citation V Ultra	560-0414	K-Air Charter P/L. Rand.	N5235G
☐ ZS-EAG	97	Learjet 31A	31A-142	Ndizamtshina Leasing Ltd.	N142LJ
☐ ZS-EHL	78	Citation 1/SP	501-0059	Airworld P/L. Durbanville.	N13RC
☐ ZS-JBR	96	Learjet 45	45-007		N457LJ
☐ ZS-JRO	95	Beechjet 400A	RK-101	Maizecor Meulens P/L-Genfood, Wonderboom.	N3221T
☐ ZS-JWC	65	Learjet 23	23-030	Transair P/L. Wonderboom.	N431CA
☐ ZS-KJY	68	Learjet 24	24-165	AOC Surveys P/L. Lanseria.	V5-KJY
☐ ZS-LDV	82	Citation	500-0418	Roagsons International P/L. Geduld.	N2628B
☐ ZS-LHU	80	Citation II	550-0165	ICC Car Importers P/L. Cramerview.	3D-ACQ
☐ ZS-LNP	80	Citation II	550-0560	Barair P/L. Lanseria.	N1298J
☐ ZS-LTK	66	Learjet 24	24-103	Company for Research on Atmosphere Water Supply, Pretoria.	N90532
☐ ZS-LWU	69	Learjet 24B	24B-209	Interjet Maintenance P/L. Sun City. (status ?).	N14BC
☐ ZS-LXH	76	Learjet 25D	25D-206	Theron Airways P/L. Lanseria.	N206EQ
☐ ZS-LXT	81	Citation 1/SP	501-0215	Anerley Citation Partnership, Lanseria.	N50MM
☐ ZS-MAN	66	HS 125/1B	25067	M M Huisamen, North End. (status ?).	Z-TBX
☐ ZS-MCU	73	Citation	500-0137	Anomor Airways, Northcliff.	N12ME
☐ ZS-MDN	65	Learjet 23	23-081	Preftradar P/L. Randburg.	N418LJ
☐ ZS-MGJ	65	Learjet 24XR	24XR-207	Bonanza Hire, Nelspruit.	N457JA
☐ ZS-MGK	80	Learjet 35A	35A-357	Execujet Aviation P/L. Lanseria.	(N100L)
☐ ZS-MHN	89	Beechjet 400	RJ-59	South African Police, Pretoria.	N1559U
☐ ZS-MPN	77	Citation 1/SP	501-0275	Rand Air P/L. Wadwville. 'Eagles Flight II'	N31AJ
☐ ZS-MPT	90	Citation V	560-0089	Atair Executive Jet Charter P/L. Johannesburg.	N67830
☐ ZS-MTD	74	Learjet 25B	25B-160	Inter-Air P/L-Schwarz Jewellers P/L. Lanseria.	3D-AEZ
☐ ZS-MVX	93	CitationJet	525-0010	ELB Flying Services Ltd. Grand Central.	N210CJ

Reg	*Yr*	*Type*	*c/n*	*Owner/Operator*	*Prev Regn*

Reg	Yr	Type	c/n	Owner/Operator	Prev Regn
□ ZS-MVZ	90	Citation V	560-0064	Palabora Mining Co Ltd. Phalaborwa.	N2717X
□ ZS-NDT	92	Citation V	560-0160	Kersaf Investments P/L. Sandton.	N6886X
□ ZS-NDW	92	Citation V	560-0166	South African Breweries Ltd. Lanseria.	N68872
□ ZS-NDX	91	Citation V	560-0152	De Beers Consolidated Mines Ltd. Kleinzee.	N6882R
□ ZS-NFL	92	Citation II	550-0697	Foskor Ltd. Phalaborwa.	N6851C
□ ZS-NGG	73	Learjet 24XR	24XR-280	J R Brown/Blue Haze Air Charter, Lanseria.	N79RS
□ ZS-NGL	92	Citation V	560-0202	Sappi Manufacturing P/L. Braamfontein.	N1283V
□ ZS-NGM	92	Citation V	560-0201	Kindoc Airways P/L. Sandton.	N1283N
□ ZS-NGR	73	Citation	500-0080	Special Charter Aviation, Glenvista.	N10UP
□ ZS-NGS	93	Citation V	560-0241	Anglo American Corp of South Africa P/L. Jan Smuts.	N241CV
□ ZS-NHD	94	Citation V	560-0255	Alusaf Ltd. Richards Bay.	N255CV
□ ZS-NHF	75	Citation	500-0296	Shastean Investments (EDMS) Bpk. Hercules.	N882CA
□ ZS-NHO	81	Citation II	550-0237	Wilson Keller & Associates P/L. Capetown.	3D-ACT
□ ZS-NII	80	Citation II	550-0168	Vrede Textiles CC. Dassenberg.	N68GA
□ ZS-NKD	90	Challenger 601-3A	5060	Chartertech Partnership, Boksburg.	N630M
□ ZS-NMO	90	Gulfstream 4	1129	Anglo American Corp. Johannesburg.	EI-CAH
□ ZS-NNF	94	Falcon 2000	2	Pepkor Group P/L-Flicape P/L. Capetown.	F-WNEW
□ ZS-NNM	67	BAC 1-11/409AY	108	Nationwide Air Charter P/L. Lanseria.	G-BGTU
□ ZS-NTV	95	Learjet 60	60-052	Barlows Central Finance Corporation P/L. Lanseria.	N5022C
□ ZS-NUW	96	CitationJet	525-0150	Commercial Air Services (OFS), Lanseria.	N5090V
□ ZS-NUZ	96	Citation V Ultra	560-0398	Rembrandt Tobacco P/L. Capetown.	
□ ZS-NVV	95	Citation V Ultra	560-0322	Avmar Partnership 1996. Lanseria.	N5262X
□ ZS-NYG	73	Learjet 25C	25C-098	King Air Services Partnership, Melville.	N502MH
□ ZS-NYV	95	Learjet 31A	31A-115	Government of Kwazulu-Natal, Ulundi.	N31NR
□ ZS-NZO	92	Beechjet 400A	RK-57	Beechjet Charter Trust, Pinegowrie.	N8157H
□ ZS-OAM	73	Citation	500-0077	Absil Air Services P/L. Lanseria.	N869K
□ ZS-OCG	97	Beechjet 400A	RK-140	Beechcraft Sales/NAC P/L. Rand.	N1094N
□ ZS-OEA	73	Learjet 24XR	24XR-267	Mic-Dav Air P/L. Johannesburg.	N267MP
□ ZS-OFW	91	Learjet 31	31-031	F W Hangers CC. Lanseria.	N31HA
□ ZS-ONE	71	Citation Eagle	500-0002	Lanseria Air Charter CC. Lanseria.	C-FCPW
□ ZS-PTL	84	Learjet 35A	35A-594	Sandgate Investments P/L. Hennopsmeer.	N7007V
□ ZS-RCC	73	Citation	500-0106	Inter Air P/L-Schwartz Diamonds, Johannesburg.	N606CC
□ ZS-RKV	79	Citation II	550-0051	Red Cross Air Mercy Service Trust, Capetown.	N678CA
□ ZS-SEA	79	Falcon 10	156	ESCOM/Sapphire Executive Air, Halfway House.	SE-DEK
□ ZS-SMB	96	Citation V Ultra	560-0359	Mercedes-Benz of South Africa P/L. Lanseria.	N.....
□ ZS-TGG	67	Gulfstream 2SP	8	Gotvil Timber Africa P/L. Lanseria.	S9-GOT
□ ZS-TMG	74	Citation	500-0149	Transair Trust, Gillitts.	N149PJ
□ ZS-TOW	82	Learjet 35A	35A-475	Lonrho Management Services P/L. Lanseria.	3D-ADC
□ ZS-ZBB	94	Falcon 900B	143	Gencor/MacSteel P/L-Executive Wings P/L. Lanseria.	F-GNMR
Military					
□ ZS-CAQ	83	Falcon 50	133	South African Air Force, Pretoria.	HB-IEA
□ ZS-CAS	82	Falcon 50	91	South African Air Force, Pretoria.	ZS-BMB
□ ZS-JBA	71	HS 125/400B	25259	South African Air Force, Pretoria.	SAAF05
□ ZS-JIH	71	HS 125/400B	25260	South African Air Force, Pretoria.	SAAF06
□ ZS-LIG	83	Citation II	550-0474	South African Air Force, Pretoria.	N12514
□ ZS-LME	70	HS 125/403B	25242	South African Air Force, Pretoria.	3D-ABZ
□ ZS-LPE	69	HS 125/400B	25184	South African Air Force, Pretoria.	SAAF04
□ ZS-LPF	71	HS 125/400B	25269	South African Air Force, Pretoria.	SAAF07
□ ZS-MLN	81	Citation II/SP	551-0285	South African Air Force, Pretoria.	VDF-030
□ ZS-NAN	91	Falcon 900	99	South African Air Force, Pretoria.	F-WWFE

Z3 = Macedonia (Republic of Macedonia)

Civil

□ Z3-BAA	76	Learjet 25B	25B-205	Government of Macedonia, Skopje.	YU-BKJ

3A = Monaco (Principality of Monaco)

Civil

□ 3A-MGR	83	Falcon 20F	473	Prince Rainier of Monaco, Nice, France.	F-GEJR

3B = Mauritius (Republic of Mauritius)

Civil

□ 3B-GFI	81	Challenger 600S	1019	Generale Aviation et Finance Internaionale Ltd. Port Louis.	ZS-NER

4X = Israel (State of Israel)

Civil

□ 4X-...	85	Astra-1125	003	IAI third prototype, static and fatigue test airframe.	
□ 4X-CLK	77	Westwind-1124	213	Aereol Airways Ltd. Tel Aviv.	N30YM
□ 4X-COT	88	Challenger 601-3A	5032	Jet Link/ISCAR Ltd. Tel Aviv.	N667CC
□ 4X-COV	95	Hawker 800XP	8283	Jet Link/ISCAR Ltd. Tel Aviv.	(N283XP)
□ 4X-CTE	81	Westwind-Two	337	Jet Link/Control Centers Ltd. Tel Aviv.	N900NW
□ 4X-CZM	95	Hawker 800XP	8279	Jet Link/Control Centers Ltd. Tel Aviv.	(N817H)
□ 4X-IGA	97	Galaxy	003	Israel Aircraft Industries, Tel Aviv. (Ff 25 Dec 97).	
□ 4X-WIA	84	Astra-1125	002	Israel Aircraft Industries, Tel Aviv.	
Military					

Reg	Yr	Type	c/n	Owner/Operator	Prev Regn
☐ 4X-J..	75	B 707-3L6C	21096	..., ISDAF, Tel Aviv.	P4-MDJ
☐ 4X-JYB	72	B 707-3H7C	20629	255, ISDAF, Tel Aviv.	4X-BYR
☐ 4X-JYH	65	B 707-3J6B	20721	264, ISDAF, Tel Aviv.	B-2416
☐ 4X-JYJ	76	Westwind-1124N	185	927, ISDAF/195 Squadron, Tel Aviv.	N1123U
☐ 4X-JYO	76	Westwind-1124N	186	931, ISDAF/195 Squadron, Tel Aviv.	N1123R
☐ 4X-JYQ	69	B 707-344C	20110	242, ISDAF, Tel Aviv.	4X-BYQ
☐ 4X-JYR	71	Westwind-1124N	152	929, ISDAF/195 Squadron, Tel Aviv.	4X-CJC

5A = Libya (Socialist People's Libyan Arab Jamahiriya)

Civil

Reg	Yr	Type	c/n	Owner/Operator	Prev Regn
☐ 5A-DAG	68	Falcon 20C	143	Libyan Arab Airlines, Benghazi.	F-WMKH
☐ 5A-DAJ	69	JetStar-8	5136	Government of Libya, Tripoli.	LAAF001
☐ 5A-DAK	76	B 707-3L5C	21228	Government of Libya, Tripoli.	
☐ 5A-DCK	78	Corvette	38	Government of Libya, Tripoli.	F-ODIF
☐ 5A-DCM	81	Falcon 50	68	Government of Libya, Tripoli.	F-WZHQ
☐ 5A-DCO	70	Falcon 20SNA	190	Directorate of Civil Aviation, Tripoli.	LAAF002
☐ 5A-DDS	79	Gulfstream 2	242	Government of Libya, Tripoli.	

5B = Cyprus (Republic of Cyprus)

Civil

Reg	Yr	Type	c/n	Owner/Operator	Prev Regn
☐ 5B-...	66	HS 125/1A	25088	J A T Overseas Ltd. Nicosia.	N1230B
☐ 5B-CGP	68	JetStar-731	5128	Transair Ltd/Athenian Airlift Co. Nicosia.	N26S
☐ 5B-CHE	68	JetStar-731	5114	Medavia Ltd. Nicosia.	N26GL
☐ 5B-DBE	65	B 727-30	18371	A I M E S Co. Nicosia.	9M-SAS

5H = Tanzania (United Republic of Tanzania)

Civil

Reg	Yr	Type	c/n	Owner/Operator	Prev Regn
☐ 5H-CCM	78	F 28-3000	11137	Government of Tanzania, Dar es Salaam. 'Uhuru na Umoja'	PH-ZBS

5N = Nigeria (Federal Republic of Nigeria)

Civil

Reg	Yr	Type	c/n	Owner/Operator	Prev Regn
☐ 5N-...	70	BAC 1-11/414EG	160	(Executive Transport ?).	G-BFMC
☐ 5N-...	68	BAC 1-11/423ET	154	(Executive Transport ?).	G-BEJW
☐ 5N...	76	HS 125/600A	6069	Southern Airlines Ltd. Lagos.	N369TS
☐ 5N-...	69	Jet Commander-B	141	Triax Airlines Ltd. Enugu.	N163WS
☐ 5N-...	69	JetStar-8	5141	Oriental Airlines Ltd. Ikeja.	N747GB
☐ 5N-AGV	76	Gulfstream 2	177	Federal Government of Nigeria, Lagos.	N17587
☐ 5N-AGZ	89	BAe 125/800B	8143	Aero Contractors Ltd/Central Bank, Lagos.	5N-NPF
☐ 5N-ALH	66	HS 125/1B	25089	Aero Contractors Ltd. Lagos.	OO-SKJ
☐ 5N-AMY	70	HS 125/F403B	25227	Chief M K O Abiola/RCN, Lagos.	G-MKOA
☐ 5N-AOC	81	Learjet 25D	25D-322	AIC Co Ltd. Lagos.	
☐ 5N-AOL	75	HS 125/600B	6050	Okada Air, Benin City.	G-BLOI
☐ 5N-APN	75	Citation	500-0286	Nigerian Police Force, Lagos.	N286CC
☐ 5N-AVK	82	HS 125/700B	7160	Federal Civil Aviation Authority, Lagos.	G-5-19
☐ 5N-AVL	82	Citation 1/SP	501-0317	Federal Civil Aviation Authority, Lagos.	N2626Z
☐ 5N-AVM	82	Citation 1/SP	501-0233	Federal Civil Aviation Authority, Lagos.	N26264
☐ 5N-AVV	67	HS 125/3B	25138	Intercontinental Airlines Ltd. Lagos.	I-BOGI
☐ 5N-AVZ	67	HS 125/3B-RA	25113	Intercontinental Airlines Ltd. Lagos. (status ?).	G-AVDX
☐ 5N-AYA	90	Citation II	550-0632	Federal Government of Nigeria, Lagos.	N12570
☐ 5N-AYK	76	HS 125/600B	6060	Aero Contractors Ltd/Yakamata Air Services, Kano.	G-BFIC
☐ 5N-BBF	69	B 727-231	20049	Aviation Development Co/ADC Airlines, Lagos.	N44316
☐ 5N-BBG	69	B 727-231	20050	Aviation Development Co/ADC Airlines, Lagos.	N74317
☐ 5N-BCI	73	Citation	500-0085	AshakaCem PLC. Lagos.	(N64AJ)
☐ 5N-CCC	66	BAC 1-11/401AK	069	Kabo Air, Kano.	VR-CCS
☐ 5N-EAS	70	HS 125/403B	25217		G-OLFR
☐ 5N-FGE	90	Falcon 900	96	Federal Government of Nigeria, Lagos.	5N-OIL
☐ 5N-FGN	82	B 727-2N5	22825	Federal Government of Nigeria, Lagos.	5N-AGY
☐ 5N-FGO	88	Falcon 900	52	Federal Government of Nigeria, Lagos.	F-WWFC
☐ 5N-FGP	90	Gulfstream 4	1126	Federal Government of Nigeria, Lagos.	N426GA
☐ 5N-FGR	92	BAe 1000B	9018	Federal Government of Nigeria, Lagos.	G-5-741
☐ 5N-HHH	66	BAC 1-11/401AK	064	Government of Liberia, Monrovia.	HZ-NB2
☐ 5N-IMR	81	Gulfstream 3	344	Isiyako Rabiu & Sons Ltd. Kano.	N7000C
☐ 5N-MAY	76	HS 125/600B	6062	King Airlines & Travel, Lagos.	G-TMAS
☐ 5N-NPC	88	BAe 125/800B	8109	Aero Contractors Ltd/Nigerian National Petroleum Co. Lagos.	G-5-581
☐ 5N-NPF	80	Citation II	550-0125	Nigerian Police Force, Lagos.	N125RR
☐ 5N-NRC	67	BAC 1-11/217EA	124	Okada Air, Benin City. (executive transport ?).	5N-TOM
☐ 5N-SDP	67	BAC 1-11/217EA	125	Okada Air, Benin City. (executive transport ?).	A12-125
☐ 5N-WMA	68	HS 125/400B	25178	Pan African Airlines (Nigeria) Ltd. Lagos.	G-OOSP
☐ 5N-YET	73	HS 125/600A	6013	Associated Aviation Ltd. Ikeja.	VR-CDG
☐ 5N-YFS	75	HS 125/600B	6054	Aero Contractors Ltd/Yinka Folawiyo Petroleum Co. Lagos.	G-BCXF

5R = Madagascar (Democratic Republic of Madagascar)

Civil

Reg	Yr	Type	c/n	Owner/Operator	Prev Regn
☐ 5R-MHF	80	Citation II/SP	551-0171	Ste Henri Fraise Fils et Cie, Tananarive-Ivato.	ZS-PMC

5U = Niger (Republic of Niger)

Military

Reg	Yr	Type	c/n	Owner/Operator	Prev Regn
☐ 5U-BAG	78	B 737-2N9C	21499	Government of Niger Republic, Niamey. 'Monts Baghezan'	(5U-MAF)

5V = Togo (Togolese Republic)

Civil

Reg	Yr	Type	c/n	Owner/Operator	Prev Regn
☐ 5V-TAG	68	B 707-312B	19739	Government of Togo, Lome.	N600CS
☐ 5V-TAI	74	F 28-1000	11079	Government of Togo, Lome.	5V-MAB

5X = Uganda (Republic of Uganda)

Civil

Reg	Yr	Type	c/n	Owner/Operator	Prev Regn
☐ 5X-AMM	70	B 727-76	20371	Skyline International Ltd. Stansted, UK.	(VR-CAM)
☐ 5X-UOI	82	Gulfstream 3	345	Government of Uganda, Entebbe.	G-GIII

5Y = Kenya (Republic of Kenya)

Civil

Reg	Yr	Type	c/n	Owner/Operator	Prev Regn
☐ 5Y-...	88	Citation II	550-0569		G-OSNB

Military

Reg	Yr	Type	c/n	Owner/Operator	Prev Regn
☐ KAF308	95	Fokker 70	11557	Government of Kenya, Nairobi.	PH-MXM

6V = Senegal (Republic of Senegal)

Civil

Reg	Yr	Type	c/n	Owner/Operator	Prev Regn
☐ 6V-AEF	75	B 727-2M1	21091	Government of Senegal, Dakar.	N40104

7O = South Yemen (Republic of Yemen)

Civil

Reg	Yr	Type	c/n	Owner/Operator	Prev Regn
☐ 7O-ADC	85	BAe 125/800B	8037	Shaher Traders, Sana'a. 'Anisa IV'.	4W-ACN

7Q = Malawi (Republic of Malawi)

civil

Reg	Yr	Type	c/n	Owner/Operator	Prev Regn
☐ 7Q-YLF	92	Citation II	550-0706	Limbe Leaf Tobacco Co. Harare.	N.....

Military

Reg	Yr	Type	c/n	Owner/Operator	Prev Regn
☐ MAAW-J1	86	BAe 125/800B	8064	Government of Malawi, Zomba.	G-5-514

7T = Algeria (Democratic & Popular Republic of Algeria)

Civil

Reg	Yr	Type	c/n	Owner/Operator	Prev Regn
☐ 7T-VCW	82	HS 125/700B	7163	E.N. pour l'Exploitation Meteorlogique et Aeronautique.	G-5-12
☐ 7T-VHP	79	JetStar 2	5233	Air Algerie, Algiers.	YI-AKA

Military

Reg	Yr	Type	c/n	Owner/Operator	Prev Regn
☐ 7T-VPA	90	Falcon 900	81	Ministry of Defence, El Mouradia.	F-WWFL
☐ 7T-VPB	90	Falcon 900	82	Ministry of Defence, El Mouradia.	F-WWFM
☐ 7T-VPR	96	Gulfstream 4SP	1288	Government of Algeria, Algiers.	N403GA
☐ 7T-VPS	96	Gulfstream 4SP	1291	Government of Algeria, Algiers.	N412GA

8P = Barbados

Civil

Reg	Yr	Type	c/n	Owner/Operator	Prev Regn
☐ 8P-GAC	82	Gulfstream 3	355	Palm Beach Aircraft Management, Palm Beach, Fl. USA.	N876RW
☐ 8P-KAM	86	Citation III	650-0119	Petroleum Helicopters de Colombia SA. Bogota, Colombia.	N96AF
☐ 8P-MAK	92	Gulfstream 4	1186	Helicol SA. Bogota, Colombia.	N479GA

9A = Croatia (Republic of Croatia)

Civil

Reg	Yr	Type	c/n	Owner/Operator	Prev Regn
☐ 9A-BLY	78	Sabre-75A	380-65	Government of Croatia, Zagreb.	RC-BLY
☐ 9A-CAD	77	CitationJet	525-0199		N1216K
☐ 9A-CRO	96	Challenger 604	5322	Government of Croatia, Zagreb.	N604CL
☐ 9A-CRT	90	Challenger 601-3A	5067	Government of Croatia, Zagreb.	9A-CRO
☐ 9A-DVR	81	Citation 1/SP	501-0230	CRO Petrol,	HB-VLB

9G = Ghana (Republic of Ghana)

Civil

Reg	Yr	Type	c/n	Owner/Operator	Prev Regn
☐ 9G-MKC	64	DC 8-55F	45692	MK Air Cargo, Accra.	5V-TAF

Military

Reg	Yr	Type	c/n	Owner/Operator	Prev Regn
☐ G-530	77	F 28-3000	11125	Government/Ghana Air Force, Accra.	PH-ZBP

9H = Malta (Republic of Malta)

Civil

Reg	Yr	Type	c/n	Owner/Operator	Prev Regn
☐ 9H-ACR	78	Citation II	550-0025	Eurojet Ltd. Luqa.	N78PR
Reg	Yr	Type	c/n	Owner/Operator	Prev Regn

140

9J = Zambia (Republic of Zambia)
Civil

☐ 9J-RON	86	Challenger 601	3057	Roan Air/Mines Air Services Ltd. Lusaka.	N19J

9K = Kuwait (State of Kuwait)
Civil

☐ 9K-AGC	92	MD 83	49809	Government of Kuwait, Safat.	KAF 26
☐ 9K-AJA	91	Gulfstream 4	1157	Kuwait Airways Corp. Safat.	N17581
☐ 9K-AJB	91	Gulfstream 4	1159	Kuwait Airways Corp. Safat.	N17583
☐ 9K-AJC	91	Gulfstream 4	1161	Kuwait Airways Corp. Safat.	N17585
☐ 9K-ALD	93	Airbus A310-308	648	Government of Kuwait, Safat.	F-WWC.

Military

☐ KAF 321		DC 9-32	47690	Government/Kuwait Air Force, Safat.	160750

9M = Malaysia (Federation of Malaysia)
Civil

☐ 9M-...	83	Gulfstream 3	368	Karambunai Resorts Sdn Bhd. Kuala Lumpur.	N368TJ
☐ 9M-AZZ	92	BAe 125/800B	8219	Government of Sarawak/Hornbill Skyways, Kuching.	G-5-740
☐ 9M-BAB	92	Falcon 900B	121	Government of Malaysia, Kuala Lumpur.	N478FJ
☐ 9M-BAN	91	Falcon 900B	106	Government of Sarawak, Kuching.	F-GKDI
☐ 9M-BCR	66	Falcon 20C	35		N809P
☐ 9M-CAL	94	Learjet 60	60-034	Civil Aviation Directorate, Kuala Lumpur.	N5034Z
☐ 9M-CHG	95	B 737-33A	27456	Continental Heights Golf Resort, Kuala Lumpur.	9M-LKY
☐ 9M-DDW	86	BAe 125/800B	8079		SE-DRV
☐ 9M-FAZ	75	Citation	500-0245	Mofaz Air P/L. Kuala Lumpur.	N2019V
☐ 9M-FCL	95	Learjet 60	60-072	Department of Civil Aviation, Kuala Lumpur.	N5072L
☐ 9M-ISJ	89	Gulfstream 4	1106	Government of Johore, Johore Bahru.	N17608
☐ 9M-JJS	96	Falcon 900EX	5	Malaysian Plantations Sqn Bhd. Kuala Lumpur.	N205FJ
☐ 9M-NSA	91	Falcon 900B	550-0610	Nusantara Airlines, Kuala Lumpur.	9M-UEM
☐ 9M-NSK	94	Challenger 601-3R	5166	Nusantara Airlines, Kuala Lumpur.	N618CC
☐ 9M-SWG	92	Challenger 601-3A	5104		(N233SG)
☐ 9M-TAN	94	Challenger 601-3R	5154	Berjaya Air Sdn Bhd. Kuala Lumpur.	N602DP
☐ 9M TRI	93	Gulfstream 4SP	1230	Technology Resources Industry/MHS Aviation, Kuala Lumpur.	N466GA
☐ 9M-VVV	97	Hawker 800XP	8337		N337XP

Military

☐ M24-01	69	HS 125/400B	25189	9M-EDA, TUDM, 2 Bayan Squadron, Kuala Lumpur. (status ?).	FM1801
☐ M24-02	69	HS 125/400B	25209	9M-EDC, TUDM, 2 Bayan Squadron, Kuala Lumpur. (status ?).	FM1802
☐ M28-01	75	F 28-1000	11088	9M-EBE, TUDM, 2 Bayan Squadron, Kuala Lumpur.	FM2101
☐ M37-01	88	Falcon 900	64	9M-..., TUDM, 2 Bayan Squadron, Kuala Lumpur.	N446FJ
☐ M47-01	96	Challenger SE	7140	9M-..., TUDM, 2 Bayan Squadron, Kuala Lumpur.	N260SE

9Q = Dem Rep of Congo (Republic of Zaïre)
Civil

☐ 9Q-...	65	B 727-30	18369		N23AZ
☐ 9Q-CFW	74	HS 125/600B	6031	Gecamines, Lumbumbashi. (impounded Lanseria-RSA ?).	G-5-14
☐ 9Q-CGK	88	Falcon 50	177	ALG Aeroleasing/Gecamines, Kinshasa.	F-WPXF
☐ 9Q-CSN	71	HS 125/403B	25247	Shabair SPRL. Lumbumbashi.	G-AYRR
☐ 9Q-RDZ	66	B 727-30	18934	Government of Congo, Kinshasa. 'Hewa Bora'	VR-CHS

Military

☐ 9T-MSS	68	B 707-382B	19969	Government of Zaire, Kinshasa. 'Mount Hoyo'	CS-TBD

9U = Burundi (Republic of Burundi)
Civil

☐ 9U-BTB	81	Falcon 50	66	Government of Burundi, Bujumbura.	N4413N

9V = Singapore (Republic of Singapore)
Civil

☐ 9V-ATC	93	Learjet 31	31-033A	Singapore Airlines Ltd. Changi.	N2603S
☐ 9V-ATE	95	Learjet 31	31-033C	Singapore Airlines Ltd. Changi.	N5013L
☐ 9V-ATF	95	Learjet 31	31-033D	Singapore Airlines Ltd. Changi.	N5023D

9XR = Rwanda (Rwanda Republic)
Civil

☐ 9XR-CH		Caravelle 3	209	Government of Rwanda, Kigali.	F-BUFM

BUSINESS JETS
Withdrawn From Use and Written-Off

Civil

Regn	Yr	Type	c/n	Accident/Withdrawal details	Prev Regn
☐ A4O-AB		VC 10-1103	820	Wfu. Displayed at Brooklands Museum, Weybridge, UK.	G-ASIX
☐ B-98181	93	Learjet 35A	35A-675	W/o target towing Eastern Taiwan. 17 Sep 94.	N2602M
☐ C-FCFL	70	HS 125/400A	NA741	W/o Labrador, NF. Canada. 9 Dec 77.	G-AXTT
☐ C-FDTF	66	JetStar-6	5088	Wfu 9/86. Located Halifax, NS. Canada.	N9244R
☐ C-FDTX	61	JetStar-6	5018	Wfu 8/86. Located Rockcliffe Museum, Ottawa, ON. Canada.	N9287R
☐ C-FEYG	66	Jet Commander	81	W/o Winnipeg, MT. Canada. 26 May 78.	CF-KBI
☐ C-FHLL	65	HS 125/1A	25034	Wfu. Accident 4/83. Wing fitted to s/n 25027.	
☐ C-FMWW	82	Westwind-Two	380	W/o Meadowlake, SK. Canada. 27 Jan 94.	N380DA
☐ C-FRBC	73	JetStar-8	5160	Wfu as parted out 1988.	(N60EE)
☐ C-GBRW	73	Learjet 36	36-001	Wfu. Cx C- 4/97.	N26GL
☐ C-GBWA	73	Learjet 24D	24D-261	Wfu. located Bounty Aviation, Detroit-Willow Run, USA.	D-COOL
☐ C-GCGR-X	78	Challenger 600	1001	W/o Mojave, Ca. USA. 3 Apr 80.	
☐ C-GESZ	72	Citation	500-0022	Wfu.	N800JD
☐ C-GPUN	76	Learjet 35	35-058	W/o Queen Charlotte Island, BC. Canada. 11 Jan 95.	
☐ C-GWPB	67	Learjet 20C	92	W/o Labrador, BC. Canada. (ground instruction).	117503
☐ C-GXFZ	72	Citation	500-0032	W/o Orillia, ON. Canada. 26 Sep 84.	(N5364U)
☐ CF-BRL	72	Sabre-40A	282-107	W/o Frobisher Bay, NT. Canada. 27 Feb 74.	N40NR
☐ CF-CFL	69	HS 125/400A	NA725	W/o Labrador, Newfoundland, Canada. 11 Nov 69.	
☐ D-CARA	66	HFB 320	1021	Wfu. Cx D- register 3 Jul 84. Broken up for spares.	
☐ D-CARE	66	HFB 320	1022	Wfu Apr 72. Located at Finow Museum, Germany.	
☐ D-CARY	67	HFB 320	1026	Wfu & located Laatzen museum, Hanover Germany.	TC-FNS
☐ D-CASH	87	Citation II	550-0564	W/o Salzburg, Austria. 19 Feb 96.	N674CA
☐ D-CASY	68	HFB 320	1029	W/o Blackpool, UK. 29 Jun 72.	
☐ D-CATY	77	Learjet 35A	35A-114	W/o Moscow, Russia. 15 Dec 94. (N851L for Dodsons).	D-CATY
☐ D-CBUR	77	Falcon 10	98	W/o Friesenheim, Germany. 8 Aug 96.	F-WPXG
☐ D-CDFA	75	Learjet 36	36-006	W/o Libya, North Africa. 26 Mar 80.	D-CAFO
☐ D-CDPD	74	Learjet 25B	25B-177	W/o North Atlantic. 18 May 83.	N74SW
☐ D-CHFB	64	HFB 320	V1	W/o Torrejon, Spain. 12 May 65.	
☐ D-CHVB	89	Citation II	550-0629	W/o Allendorf, Germany 25 Jan 95.	N1256T
☐ D-CIRO	69	HFB 320	1044	W/o Texel Island, Netherlands. 18 Dec 70.	
☐ D-CLOU	64	HFB 320	V2	Wfu Sep 70. Located at Deutsches Museum-Munich, Germany.	
☐ D-COCO	82	Learjet 35A	35A-466	W/o Cologne-Bonn, Germany. 7 Jun 93. (parts at Dodson's, Ks)	N600WJ
☐ D-IAEC	81	Citation 1/SP	501-0203	W/o Blankensee Airport-Lubeck, Germany. 31 May 87.	N67830
☐ D-IHAQ	65	Learjet 23	23-007	W/o Zurich, Switzerland. 12 Dec 65.	N826L
☐ D-IHLZ	70	Learjet 24B	24B-225	W/o Mariensiel, Germany. 18 Jun 73.	N618R
☐ D-IJHM	80	Citation II/SP	551-0033	W/o Kassel, Germany. 19 May 82.	N88692
☐ EC-CGG	73	Citation	500-0108	W/o Barcelona, Spain. 20 Nov 74.	N108CC
☐ EC-CKR	75	Learjet 25B	25B-184	W/o Northolt, UK. 13 August 1996.	
☐ EC-DFA	78	Learjet 35A	35A-196	W/o Palma, Spain. 13 Aug 80.	HB-VFU
☐ EC-DQC	76	Corvette	24	Wfu and sold as spares 2/91,	F-BVPI
☐ EC-ECB	69	Falcon 20DC	210	W/o Las Palmas, Canary Islands. 30 Sep 87.	N66VG
☐ EC-EFI	68	Falcon 20D	189	W/o Nr Keflavik, Iceland. 11 Oct 87.	N444BF
☐ EC-EGS	74	HS 125/600A	6034	Wfu at Spirit of St Louis, USA 10/94.	EC-115
☐ EC-EGT	66	HS 125/1A-522	25080	Wfu at Dodson's, Ottawa, USA.	N23KL
☐ EC-EHF	73	HS 125/600A	6011	Wfu at Spirit of St Louis, USA 10/94.	N81D
☐ EP-AGX	73	Falcon 20E	283	W/o Kermanshah, Iran. 21 Nov 74.	F-WRQS
☐ F-BKMF	64	HS 125/1	25007	W/o Nice, France. 5 Jun 66.	HB-VAH
☐ F-BRNL	69	Learjet 24B	24B-183	W/o Toulouse, France. 18 Dec 85.	OY-AGZ
☐ F-BSQN	72	Falcon 10	03	Wfu. CoA expiry 4/81.	F-WSQN
☐ F-BSRL	69	Learjet 24B	24B-210	W/o Provins Nr Paris, France. 10 Jun 85.	ZS-LLG
☐ F-BSTM	65	AC 680V-TU	1540-6	Wfu. Cx F- 18 Nov 91. Museum exhibit Pelegry-Perpignan.	F-WSTM
☐ F-BTTU	77	Corvette	37	W/o St Yan, France. 31 Jul 90.	
☐ F-GAMA	65	Learjet 23	23-023	Wfu.	HB-VEL
☐ F-GAPY	65	Learjet 23	23-027	Wfu at White Industries, Bates City, Mo. USA.	(N108TW)
☐ F-GBTC	78	Falcon 10	124	W/o Chalons-Vatry nr Paris, France. 15 Jan 86.	F-WPUY
☐ F-GDAE	66	Learjet 24	24-105	Wfu.	TR-LYB
☐ F-GDAV	65	Learjet 23	23-017	W/o Lisbon, Portugal. 30 Jan 89.	F-GBTA
☐ F-GDHR	82	Learjet 55	55-070	W/o Nr Jakiri, Cameroun. 5 Feb 87.	
☐ F-GHLN	72	Falcon 20E	255	W/o Paris-Le Bourget, France. 20 Jan 95.	VH-MIQ
☐ F-GJGB	75	Falcon 10	47	W/o Besancon, France. 30 Sep 93.	N79PB
☐ F-GJHK	77	Falcon 10	108	W/o Brest, France. 26 Mar 92.	(F-GFJK)
☐ F-GKPP	61	MS 760 Paris-2	98	W/o Calvi, Corsica. Oct 91.	3A-MPP
☐ F-WAMD		Falcon 30	01	Wfu. Project shelved 1975. Wings to Falcon 20-486 F-GEFS.	
☐ F-WFAL	70	Falcon 10	01	W/o Romorantin-Loire Valley, France. 31 Oct 72.	
☐ F-WLKB	63	Falcon 20C	01	Wfu. Located Dassult Falcon Service, Paris-Le Bourget.	F-BLKB
☐ F-WMSH	65	Falcon 20C	1	Wfu. Donated to Rene Lemaire for Bordeaux-Merignac museum.	F-ZACV
☐ F-WNDB	76	Falcon 50	1	Wfu.	F-BNDB
☐ F-WRSN	70	Corvette	01	W/o Marseilles, France. 23 Mar 71.	(F-WSSE)
☐ F-WZIH	66	HFB 320	1024	Wfu at Musee de l'Air, Paris-Le Bourget. (back to 16+07 ?).	16+07

Regn	Yr	Type	c/n	Accident/Withdrawal details	Prev Regn
G-ARVF		VC 10-1101	808	Wfu Apr 83. Located Hermeskiel collection Nr Trier-W Germany	
G-ARYA	62	HS 125/1	25001	Wfu. Completely dismantled at Kelsterton College-Wales 1985.	
G-ARYB	62	HS 125/1	25002	Wfu. Located Midlands Air Museum, Coventry.	
G-ARYC	63	HS 125/1	25003	Wfu. Located Mosquito Museum-Hatfield.	
G-ASNU	64	HS 125/1	25005	Wfu & Cx G- 18 Nov 91. Wfu in Nigeria ?	D-COMA
G-ASSM	65	HS 125/1-522	25010	Wfu. Located Kensington Science Museum.	G-ASSM
G-ATPE	66	HS 125/1B-522	25092	Wfu & Cx G- 14 Mar 90.	
G-AVGW	67	HS 125/3B	25120	W/o Luton, UK. 23 Dec 67.	
G-AXPS	67	HS 125/3B	25135	W/o Edinburgh, Scotland. 20 Jul 70.	HB-VAY
G-AZCH	69	HS 125/3B-RA	25154	Wfu. Central fuselage in use at Luton as a mobile display.	EP-AHK
G-BBRT	74	HS 125/600B	6036	Wfu. Fuselage used in paint spraying trials, Chester.	
G-BCUX	74	HS 125/600B	6043	W/o Dunsfold, UK. 20 Nov 75.	
G-BOCB	66	HS 125/1B-522	25106	Wfu & Cx G- 22 Feb 95.	G-OMCA
G-BPCP	80	Citation	500-0403	W/o Jersey, Channel Islands. 1 Oct 80.	N1710E
G-DBAL	66	HS 125/3B	25117	Wfu.	G-BSAA
G-EXLR	90	BAe 1000B	8151	Wfu & Cx G- 25 Jun 97.	HZ-AA1
G-FANN	73	HS 125/600B	6019	Wfu.	
G-FIVE	63	HS 125/1	25004	Wfu. Wings used in rebuild of s/n 25008.	G-ASEC
G-JETB	81	Citation II	550-0288	W/o Southampton, UK. 26 May 1993.	G-MAMA
G-JSAX	69	HS 125/3B-RA	25157	Wfu. Broken up at Southampton Apr 86.	G-GGAE
G-LORI	71	HS 125/403B	25246	Wfu.	G-AYOJ
G-OBOB	66	HS 125/3B	25069	W/o Columbia, Mo. USA. 30 Jan 89. (parts at White Inds.).	G-BAXL
G-OHEA	67	HS 125/3B-RA	25144	Wfu.	G-AVRG
G-OMGB	74	HS 125/600B	6039	Wfu at Luton, UK 10/94. TT 6944. (parts at Hooks Airport-Tx)	EC-EAO
G-OMGC	75	HS 125/600B	6056	Wfu at Luton, UK 9/94. Tt 6404. (parts at Hooks Airport-Tx).	G-BKCD
G-TACE	70	HS 125/403B	25223	Wfu.	G-AYIZ
G-UESS	76	Citation	500-0326	W/o Stornaway, Scotland. 8 Dec 83.	N45LC
G-YUGO	66	HS 125/1B-522	25094	Wfu. Cx G- 3/93.	G-ATWH
HB-PAA	60	MS 760 Paris	69	Wfu & Cx HB- 25 Jun 84.	J-4117
HB-VAM	65	Learjet 23	23-044	W/o Innsbruck, Austria. 28 Aug 72.	N22B
HB-VAP	66	Falcon 20C	37	W/o Goose Bay, Newfoundland. 1 Oct 67.	(N11WA)
HB-VCG	70	Falcon 20D	231	W/o St Moritz, Switzerland. 20 Feb 72.	F-WPXE
HB-VFS	78	Learjet 36A	36A-042	W/o Zarzaitine, Algeria. 23 Sep 95.	N39391
HK-3885	73	Citation	500-0135	W/o Pereira, Colombia. 7 Mar 97.	YV-717CP
HZ-FNA	65	JetStar-8	5056	Wfu at Spirit of St Louis, Mo. USA.	HZ-FK1
HZ-GP5	75	Learjet 25XR	25XR-199	W/o Narssarssuag, Greenland. 11 Jan 82.	HZ-RI1
HZ-TAS	62	B 707-321B	18338	Wfu.	N98WS
SA-R-7		DH Comet 4C	6461	W/o Cuneo, Italy. 20 Mar 63.	
I-AIFA	76	Learjet 36A	36A-021	W/o Forli, Italy. 10 Dec 79.	N3524F
I-ALSU	90	Beechjet 400A	RK-11	W/o Parma, Italy. 20 Nov 91. (parts at Dodson's, Ks.).	N5680Z
I-AMME	75	Learjet 24D	24D-310	W/o Bari, Italy. 6 Feb 76.	HB-VDU
I-COTO	79	Learjet 25D	25D-285	Wfu. Broken up at Paris-Le Bourget 10/86.	N422G
I-KILO	81	Learjet 55	55-007	W/o Seville, Spain 4 Apr 94. (parts at Griffin, Ga. USA).	N41ES
I-MCSA	77	Learjet 35A	35A-099	W/o Palermo, Sicily. 22 Feb 78.	HB-VFC
I-NICK	64	Sabre-40	282-25	Wfu at Opa Locka, Fl. USA 12/95,	N40SJ
I-NLAE	68	Falcon 20C	134	W/o Kiel-Holtenau, West Germany. 25 Sep 91.	N897D
I-PIAI		PD 808	503	W/o San Sebastian, Spain. 18 Jun 68.	
I-RACE	64	HS 125/1	25006	Wfu.	HB-VAG
I-SNAF	68	HS 125/3B	25145	Wfu.	G-AVXL
I-SNAP	61	MS 760 Paris	99	W/o Milan, Italy. 27 Oct 62.	
JA8246	84	Diamond 1A	A092SA	W/o Sado Island, Japan. 23 Jul 86.	
JY-AEW	76	Learjet 35	35-052	W/o Riyadh, Saudi Arabia. 28 Apr 77.	
JY-AFC	76	Learjet 36A	36A-020	W/o Amman, Jordan. 21 Sep 77.	
LN-AAA	67	Falcon 20CC	73	Wfu as parted out by The Memphis Group 1/89.	LX-AAA
LN-AAB	65	Falcon 20C	12	Wfu. parted out at Memphis-Tn /89.	N51SF
LN-AAE	80	Citation II/SP	551-0245	W/o Bardufoss, Norway. 15 Nov 89.	N224CC
LN-FOE	80	Falcon 20C	62	W/o Norwich, England. 12 Dec 73.	(N17401)
LV-ALW	81	HS 125/700B	7133	W/o Salta, Argentina. 11 Apr 85.	LV-PMM
LV-MMV	78	Learjet 25D	25D-259	W/o Posadas, Argentina. 23 Sep 89.	LV-PAW
LV-RDB	65	Jet Commander	12	W/o Moron, Argentina. 1991.	N344DA
LV-TDF	82	Learjet 35A	35A-478	W/o Ushula, Argentina. 15 May 84.	N3815G
LV-WEN	69	Jet Commander-B	126	W/o Cordoba, Argentina. 29 Sep 94.	N87DL
LV-WLH	66	Falcon 20C	34	W/o Nr Salta, Argentina 7 Feb 97.	LV-PHV
LV-WMR	66	Learjet 24	24-135	W/o Pasadas, Argentina. 28 Aug 95.	N77LB
N1AH	81	Learjet 35A	35A-398	W/o Great Falls, Mt. USA. 16 May 97.	N3797A
N1AT	80	Citation 1/SP	501-0159	W/o Chub Cay, Berry Island, Bahamas. 11 Jun 97.	N308AT
N1DK	74	Citation	500-0175	W/o Allegheny County, Pa. USA. 6 Jan 98.	N175CC
N1EC	66	Jet Commander	51	Wfu at Dodson's, Ottawa, Ks. Cx USA 8/94.	N18JL
N1EM	66	JetStar-6	5077	W/o Chicago, Il. USA. 25 Mar 76.	N1924V
N1JR	75	Learjet 25B	25B-188	W/o Waterville, Me. USA. 28 Jul 84. (parts at White Inds.).	A40-AJ
N1JS	79	Westwind-1124	249	Wfu.	4X-CMU
N1JU	65	Jet Commander	13	Wfu at White Industries, Bates City, Mo.	XA-SFS
N1PT	67	Jet Commander	93	Wfu. sold as parts Cx USA 11/94.	(N999RA)
N1R	60	B 720-023B	18022	Wfu. To USAF for KC-135E spares 4/83.	N7536A
N1SJ	62	Sabreliner CT-39A	276-45	Wfu. Preserved at San Jose University, Ca.	62-4492
N2CA	79	Citation II/SP	551-0024	W/o Mountain View, Mo. USA. 18 Dec 82.	N26628

Regn	Yr	Type	c/n	Accident/Withdrawal details	Prev Regn
☐ N2TE	58	MS 760 Paris	5	W/o Nr John Wayne Airport, Ca. USA. 30 Nov 96.	(N760LB)
☐ N2WU	66	Jet Commander	72	Wfu at Don Torcuato, Argentina.	VR-CAU
☐ N3MF	66	HS 125/1A-522	25093	Wfu. Used in repair of s/n 25271.	N306L
☐ N3ZA	65	Learjet 23	23-024	Wfu at White Industries, Bates City, Mo. Cx USA 5/91.	N3ZA
☐ N4SX	66	JetStar-8	5081	Wfu as parted out 1987.	N4SP
☐ N5JR	66	Jet Commander	49	Wfu at White Industries, Bates City, Mo.	N430C
☐ N5NG	73	HS 125/600A	6020	Wfu at Monterrey-Mexico as XA-NTE.	XA-NTE
☐ N6CD	74	Citation	500-0151	Wfu at White Industries, Bates City, Mo.	N151CC
☐ N6ES	61	JetStar-6	5023	Wfu at White Industries, Bates City, Mo.	N2ES
☐ N6NF	68	Learjet 25	25-021	Wfu to Alabama Aviation & Technical College Foundation.	N40SN
☐ N7ES	69	HFB 320	1045	Wfu. located Opa Locka, Fl. USA.	N4ZA
☐ N7RC	78	Citation II	550-0019	W/o Walker's Cay, Bahamas. 26 Apr 95.	N900AF
☐ N8FE	69	Falcon 20DC	199	Wfu Aug 83. Located Paul E Gerber facilty, Md.	N4388F
☐ N8GE	66	Jet Commander	63	Wfu at White Industries, Bates City, Mo.	N8GA
☐ N8RA	67	Jet Commander	104	Wfu.	N87B
☐ N9FE	67	Falcon 20DC	84	Wfu 1986. Exhibited at FEDEX Corporate HQ. Memphis, Tn.	(N150FE)
☐ N10EA	65	Jet Commander	39	Wfu 15 May 82. Located at Copenhagen.	N16FP
☐ N10GE	77	Citation 1/SP	501-0022	W/o Nr Harrison Airport, Ar. USA. 21 May 85.	(N800WC)
☐ N10LN	68	HS 125/3A-RA	25156	Wfu.	N522M
☐ N10MB	72	Westwind-1123	157	Wfu at OK Aircraft, Gilroy, Ca.	(N820RT)
☐ N10YJ	75	Falcon 10	57	W/o White Plains, NY. USA 30 Jun 97.	(N50YJ)
☐ N11QM	66	Learjet 23	23-091	Wfu & Cx USA 12/89.	N110M
☐ N11UE	62	JetStar-6	5038	Wfu. Located at Van Nuys, Ca. USA.	(N44KF)
☐ N12MK	69	Learjet 24B	24B-192	W/o Palm Springs, Ca. USA. 6 Jan 77.	N1919W
☐ N13MJ	75	Learjet 24D	24D-314	W/o Elizabeth City, NC. USA. 6 Nov 82.	N501MH
☐ N14TX	77	Learjet 36A	36A-033	W/o Stephenville, NF. Canada. 6 Dec 96.	N762L
☐ N15NY	79	Citation 1/SP	501-0110	W/o Akron, Oh. USA. 2 Aug 79.	(N26481)
☐ N15TW	77	Learjet 35A	35A-106	W/o Minneapolis, Mn. USA. 8 Dec 85. (parts at Taylorville).	N101BG
☐ N15TX	74	Falcon 10	13	Wfu.	N777SN
☐ N16SK	67	Jet Commander	101	Wfu & Cx USA 5/88 to Norway as instructional airframe.	N16MA
☐ N17FN	70	Learjet 24B	24B-220	Wfu.	N248J
☐ N17SL	66	HS 125/1A-522	25082	Wfu at White Industries, Bates City, Mo.	N1MY
☐ N18CA	65	Jet Commander	5	Wfu at Dodson's, Ottawa, Ks. Cx USA 10/86.	C-GKFT
☐ N19BG	73	Sabre-40A	282-118	Wfu at Clarksville, Mo. Cx USA 8/95.	PT-JNJ
☐ N19LH	79	Learjet 35A	35A-279	W/o Avon Park, Fl. USA. 15 Jul 97.	
☐ N20EP	65	Learjet 23	23-008	Wfu. On display at White Industries, Bates City, Mo.	N20BD
☐ N20M	66	Learjet 23	23-094	W/o Detroit, Mi. USA. 15 Dec 72.	N417LJ
☐ N21AK	66	Jet Commander	59	Wfu.	N59JC
☐ N22FM	74	Citation	500-0229	W/o Wichita, Ks. USA. 26 Apr 83.	
☐ N22RB	67	JetStar-8	5093	Wfu as parted out 10/90.	N76EB
☐ N23AC	88	Gulfstream 4	1047	W/o Pal-Waukee Chicago, Il. USA. 30 Oct 96.	N461GA
☐ N23AJ	65	Learjet 23	23-053	Wfu as parted out 8/88.	F-BTQK
☐ N23ST	59	MS 760 Paris	50	W/o New Mexico, USA. 11 Sep 90.	N42BL
☐ N23TJ	65	Learjet 23	23-033	Wfu.	N60DH
☐ N24SA	65	Learjet 23	23-025	Wfu as parted out 6/89.	N508M
☐ N24VM	65	Learjet 24	24-051	Wfu at White Industries, Bates City, Mo. Cx USA 8/88.	N70JC
☐ N25BR	89	Beechjet 400	RJ-57	W/o Rome, Ga. USA. 11 Dec 91.	
☐ N25TA	75	Learjet 25B	25B-196	W/o New Mexico. USA. 11 Apr 80.	N711WD
☐ N26TL	65	HS 125/1A	25037	Wfu.	(N389DA)
☐ N27BD	66	Jet Commander	53	Wfu at White Industries, Bates City, Mo.	N925HB
☐ N27MD	67	Jet Commander	102	Wfu at White Industries, Bates City, Mo.	
☐ N27R	74	Falcon 20E	303	W/o Naples, Fl. USA. 12 Nov 76.	N4445F
☐ N28ST	65	Learjet 23	23-013	W/o Nr Guatamala City, Guatamala. 31 Jul 87.	N37BL
☐ N29FN	68	Learjet 25	25-018	Wfu at Brandis Aviation, Taylorville, Il.	(N23FN)
☐ N29LB	66	Jet Commander	61	W/o Many Airport, La. USA. 19 Dec 80.	N29LP
☐ N30AN	74	Westwind-1123	173	Wfu at OK Aircraft, Gilroy, Ca.	N30JM
☐ N30BE	74	Sabre-40	282-14	Wfu at Clarksville, Mo.	(N30PN)
☐ N30CC	74	Sabre-60A	306-81	Wfu at Clarksville, Mo. Cx USA 12/95.	N6ND
☐ N30EM	76	Learjet 24E	24E-338	Wfu at Brandis Aviation, Taylorville, Il. Cx USA 12/89.	N30LM
☐ N30W	64	Sabre-40	282-5	W/o Perryville, Mo. USA. 21 Dec 67.	
☐ N31SK	66	Learjet 24	24-118	W/o Vail, Co. USA. 27 Mar 87.	N1919W
☐ N32WE	73	Westwind-1123	164	Wfu at White Industries, Bates City, Mo.	N9114S
☐ N34NW	67	Jet Commander	117	Wfu at White Industries Inc. Bates City, Mo.	N54WC
☐ N34W	65	Sabre-40	282-47	W/o Midland, Tx. USA. 4 Jan 74.	N740R
☐ N36MK	66	HS 125/1A	25073	W/o Boise, Id. USA. 28 Dec 70.	N372GM
☐ N36PT	66	Jet Commander	79	Wfu. Cx USA 1/97.	N100LL
☐ N37CP	65	Learjet 23	23-028	Wfu at White Industries, Bates City, Mo.	N5QY
☐ N38B		King Air 200	BB-1	Wfu. Project shelved 1978.	
☐ N38DJ	75	Learjet 25B	25B-191	W/o Sheboygan, Wi. USA. 12 Jun 92.	N78BT
☐ N39DM	75	Learjet 35	35-040	W/o San Clemente Island, Ca. USA. 5 Mar 86.	C-GGYV
☐ N39Q	68	JetStar-8	5126	Wfu as spares 1983.	N39E
☐ N40BC	75	Learjet 25B	25B-128	W/o Pueblo, Co. USA. 6 Jul 79. (parts at White Industries).	N1MX
☐ N40BP	65	Sabre-40	282-40	Wfu at Clarksville, Mo.	N40BP
☐ N40LB	65	Learjet 23	25-009	W/o Omaha, Ne. USA. 25 Sep 73.	9Q-CHC
☐ N40PC	73	HS 125/600A	6010	W/o McLean, Va. USA. 28 Apr 77.	N23BH
☐ N40UA	66	Jet Commander	40	Wfu at White Industries, Bates City, Mo.	N40AJ
Regn	*Yr*	*Type*	*c/n*	*Accident/Withdrawal details*	*Prev Regn*

Regn	Yr	Type	c/n	Accident/Withdrawal details	Prev Regn
□ N41GS	64	Sabre-40	282-16	Wfu.	N40GP
□ N42QB	65	Jet Commander	6	Wfu at White Industries, Bates City, Mo.	N420P
□ N43CF	66	Sabre-40	282-59	Wfu. Cx USA 8/95.	N40SE
□ N44	69	Jet Commander-A	130	W/o Latrobe, Pa. USA. 2 Nov 88.	N84
□ N44CJ	67	Learjet 24	24-146	W/o Felt, Ok. USA. 1 Oct 81.	N235Z
□ N44GA	66	Learjet 24	24-129	W/o Santa Catlina, Ca. USA. 30 Jan 84.	C-GSAX
□ N46	64	B 727-30	18360	Wfu. Cx USA 8/94.	N97
□ N46TE	79	Gulfstream 2	243	W/o Little Rock, Ar. USA. 19 Jan 90.	N119RC
□ N48AJ	68	Learjet 24	24-172	Wfu at White Industries, Bates City, Mo.	N234WR
□ N48BA	67	Learjet 24	24-152	Wfu.	N9LM
□ N48CG	66	Sabre-40	282-75	Wfu at Clarksville, Mo. Cx USA 12/93.	(N48CE)
□ N49UC	67	JetStar-731	5110	Wfu as parted out 1991.	N788S
□ N50AS	66	HS 125/1A-522	25083	Wfu.	N538
□ N50BA	65	Learjet 24	24-043	Wfu & cx USA as N43AC on USCAR 8/89.	(N43AC)
□ N50CD	65	Sabre-40	282-42	Wfu at Elsberry, Mo. Cx USA 10/91.	N500RK
□ N50HH	65	HS 125/1A-522	25022	W/o Bedford, In. USA. 2 Aug 86. (parts at O K Aircraft, Ca.)	N100GB
□ N50JP	66	Jet Commander	69	Wfu at Dodson's, Ottawa, Ks. Cx USA 10/97.	N10SN
□ N50SK	80	Westwind-Two	309	W/o Bowie County, Tx. USA. 4 Apr 86.	N240S
□ N50TE	76	Falcon 10	86	Wfu at White Industries, Bates City, Mo.	N411WW
□ N50UD	61	JetStar-6	5019	Wfu.	N50UD
□ N51CA	69	Learjet 25	25-030	W/o Newark, NJ. USA. 30 Mar 83.	N45DM
□ N51DB	78	Learjet 25XR	25XR-246	W/o Nr Medina, Saudi Arabia. 24 Oct 86.	N40162
□ N51FN	76	Learjet 35	35-059	W/o Carlsbad, Ca. USA. 2 Apr 90. (parts at White Industries)	N221Z
□ N52	74	Sabre-75A	380-10	Wfu & Cx USA 9/95.	
□ N53CC	81	Citation II	550-0400	W/o Roxboro, NC. USA. 1 Oct 89.	N888EB
□ N55NC	65	JetStar-6	5060	Wfu & Cx USA 7/88.	N31F
□ N55NJ	68	Learjet 24	24-162	Wfu at Dodson's, Ottawa, Ks.	M835AG
□ N56B	65	BAC 1-11/401AK	055	Wfu and cx USA 8/91.	N1JR
□ N57TA	81	Learjet 55	55-010	W/o Waterkloof AFB. Pretoria, South Africa. 13 Nov 81.	
□ N59RD	63	B 707-441	17905	Wfu & Cx USA 7/89.	PP-VJA
□ N60	75	Sabre-75A	380-28	Wfu & cx USA 6/96.	
□ N60BC	68	JetStar-8	5116	Wfu at Spirit of St Louis, Mo.	N3HB
□ N60JN	67	Sabre-60	306-14	Wfu at Perryville, Mo.	N43GB
□ N60MB	74	Falcon 10	15	W/o Stapleton-Denver, Co. USA. 3 Apr 77.	N109FJ
□ N60XL	67	Learjet 60	55C-001	Wfu & Cx USA 9/96.	N551DF
□ N61TS	65	Learjet 23	23-029	Wfu as parted out 10/85.	N66AS
□ N64	75	Sabre-75A	380-35	W/o Liberal, Ks. USA. 29 Sep 86.	
□ N66HA	67	HS 125/3A	25126	W/o Houston, Tx. USA. 13 Aug 89.	N510X
□ N66JE	81	Westwind-1124	326	W/o ground fire Denver, Co. USA. 21 Feb 95. Cx USA 9/95.	(N88JE)
□ N66MP	61	JetStar-6	5015	Wfu.	N9046F
□ N67KM	74	Sabre-75A	380-7	W/o Watertown, SD. USA. 14 Jun 75.	
□ N69KB	65	Learjet 23	23-042	Wfu at White Industries, Bates City, Mo. Cx USA 12/91.	N701RZ
□ N70HC	72	Sabre-75	370-8	Wfu at Elsberry, Mo.	N3TE
□ N71JC	89	Learjet 31	31-008	W/o Amory, Ms. USA. 2 Sep 97.	
□ N77AP	65	Sabre-40	282-37	W/o New Orleans, La. USA. 7 Nov 77.	N265W
□ N77FV	65	Jet Commander	26	Wfu.	N10MC
□ N77NT	65	Jet Commander	7	Wfu.	N77KT
□ N77RS	73	Learjet 25C	25C-094	W/o Anchorage, Ak. USA. 4 Dec 78. (parts at White Inds.).	N97J
□ N77VK	65	HS 125/731	25051	Wfu. Parted out and cx USA register 1/87.	C6-BEY
□ N78JR	66	Falcon 20C	70	Wfu 20 Mar 89 as parted out.	(N400NL)
□ N79DD	75	Citation	500-0254	W/o San Luis Obispo, Ca. USA. 24 Sep 90.	N29991
□ N80DH	69	Learjet 24B	24B-191	Wfu at Dodson's, Ottawa, Ks. Cx USA 3/89.	(N44TL)
□ N81MC	77	Learjet 24F	24F-344	W/o St Thomas, Virgin Islands. 10 Nov 84.	
□ N82ML	67	Sabre-40	282-83	Wfu at Clarksville, Mo. Cx USA 3/95.	N160TC
□ N83LJ	65	Learjet 23	23-076	Wfu at Brandis Aviation, Taylorville, Il.	N50PJ
□ N84GP	83	Peregrine	551	Wfu. Oklahoma Air & Space Museum Exhibit.	N9881S
□ N85	68	Sabre-40	282-97	W/o 14 Jan 76.	N4706N
□ N85JM	76	Falcon 10	85	W/o Aurillac, France. 17 Feb 93. (parts at White Industries)	(N95DW)
□ N86	67	Sabre-40	282-86	Wfu & Cx USA 9/92.	N2255C
□ N86CC	65	Learjet 24	24-115	Wfu at Brandis Aviation, Taylorville, Il.	N591DL
□ N87CM	64	Sabre-40	282-21	Wfu at Clarksville, Mo.	N168D
□ N87DG	65	Jet Commander	14	Wfu at White Industries, Bates City, Mo. Cx USA 7/94.	N87DC
□ N88	67	Sabre-40	282-88	Wfu & Cx USA 12/93.	N2237C
□ N88CH	61	Convair 880	58	Wfu. Located Bonza Bay, East London, RSA.	VR-HGF
□ N88HA	90	Challenger 601-3A	5072	W/o Nr Bassett-Rock County Airport, Ne. USA. 20 Mar 94.	N609K
□ N88JF	66	Learjet 24A	24A-124	W/o Detroit, Mi. USA. Oct 86. (parts at Taylorville, Il.).	N35JF
□ N88JM	61	JetStar-731	5011	Wfu as parted out 1985.	N159B
□ N88MR	74	HS 125/1A	25013	Wfu at White Industries, Bates City, Mo. Cx USA 6/86.	N4646S
□ N88NB	63	BAC 1-11/201AC	005	Wfu. Cx USA 8/89.	N97KR
□ N89	67	Sabre-40	282-89	Wfu as parted out 10/91.	N2276C
□ N89MR	65	Jet Commander	9	Wfu at OK Aircraft, Gilroy, Ca.	(N98KK)
□ N90HM	73	Westwind-1123	170	Wfu & Cx USA 5/94.	N150HR
□ N90ME	65	JetStar-6	5057	Wfu & Cx USA 2/87.	N90U
□ N91MJ	79	Citation II	550-0101	W/o Marco Island, Fl. USA. 31 Dec 95.	(N42BM)
□ N93BE	65	Jet Commander	27	Wfu at White Industries, Bates City, Mo.	N93B
□ N93BR	70	Learjet 24D	24D-231	Wfu at Brandis Aviation, Taylorville, Il.	N37DH

| Regn | Yr | Type | c/n | Accident/Withdrawal details | Prev Regn |

Regn	Yr	Type	c/n	Accident/Withdrawal details	Prev Regn
☐ N95B	65	Jet Commander	19	Wfu to instructional airframe Norway 5/88.	
☐ N95GS	61	JetStar-6	5014	Wfu as parted out 1989.	N54BW
☐ N95TC	75	Learjet 35	35-020	W/o Waco, Tx. USA. 20 Dec 84. (parts at White Industries).	XA-BUK
☐ N96GS	65	JetStar-731	5068	W/o Miami, Fl. USA. 6 Jan 90.	N9231R
☐ N97DM	72	Learjet 24D	24D-253	W/o San Clemente Island, Ca. USA. 5 Mar 86.	N417JD
☐ N98KT	61	Caravelle 6R	102	Wfu. Located at Van Nuys, Ca. USA.	N2296N
☐ N99FT	65	JetStar-731	5055	Wfu at Chino, Ca. USA.	N304CK
☐ N99GS	65	Jet Commander	31	Wfu.	N399D
☐ N99KR	67	HS 125/3A-R	25149	Wfu & Cx USA 4/91.	N99GC
☐ N99S	81	Sabre-65	465-64	W/o Toronto, Canada. 11 Jan 83.	
☐ N99TC	66	Learjet 23	23-098	Wfu at Brandis Aircraft, Taylorville, Ml.	N711AE
☐ N100EP	78	Learjet 35A	35A-150	W/o Allegheny County Airport, Pa. USA. 11 May 87.	
☐ N100MK	68	Learjet 25	25-019	W/o Sandusky, Oh. USA. 21 Oct 78.	N88FP
☐ N100RC	66	Jet Commander	60	W/o Lexington, Ky. USA. 14 Nov 70.	N6545V
☐ N100TA	65	Learjet 23	23-045	W/o Savannah, Ga. USA. 6 May 82.	N711MR
☐ N100TR	66	Jet Commander	76	Wfu. parts at Hollister, Ca.	N100DR
☐ N100WM	66	Jet Commander	73	Wfu at Sarasota, Fl.	N100W
☐ N101AD	72	HS 125/731	NA777	Wfu as parted out 10/91.	(N425JF)
☐ N101LB	65	Jet Commander	8	Wfu.	N749MP
☐ N101PP	66	Learjet 23	23-085	W/o Windsor Locks, Ct. USA. 4 Jun 84.	N385J
☐ N102CJ	66	Jet Commander	78	Wfu at Opa Locka, Fl. 10/95.	N866DH
☐ N104SS	68	Sabre-60	306-30	Wfu. parts at White Industries, Bates City, Mo.	N1116A
☐ N107CJ	64	Sabre-40	282-12	Wfu. Cx USA 1/97.	N368DA
☐ N108PA	75	Learjet 25B	25B-195	Wfu.	OBM1004
☐ N111DC	68	HFB 320	1030	Wfu at Monroe, Mi. USA.	(N247GW)
☐ N111M	65	Falcon 20C	10	Wfu at Elsberry, Mo & Clarksville, Tn. Cx USA 8/87.	N810F
☐ N111TD	65	Jet Commander	11	Wfu.	N1172L
☐ N111YL	65	Jet Commander	42	Wfu.	N111Y
☐ N114GB	66	Learjet 23	23-022	Wfu at White Industries, Bates City, Mo.	N456SC
☐ N115DX	68	JetStar-8	5111	Wfu as parted out 8/91.	N115MR
☐ N118BA	66	JetStar-6	5091	Wfu at White Industries, Bates City, Mo. Cx USA 5/93.	N118B
☐ N120AR	66	JetStar-8	5089	Wfu.	(N85DL)
☐ N120ES	77	Citation 1/SP	501-0041	W/o San Salvador, El Salvador 24 Apr 95. (parts at Dodsons).	N173SK
☐ N120TA		BAC 1-11/520FN	236	Wfu.	PP-SDS
☐ N121AJ	66	Jet Commander	57	Wfu.	N770WL
☐ N121FJ	81	Falcon 100	192	W/o Rancho Murieta Airport, Ca. USA. 15 Oct 87.	N100FJ
☐ N121GW	65	Falcon 20C	4	W/o Memphis, Tn. USA. 18 May 78.	N116JD
☐ N121HM	65	Jet Commander	18	Wfu 17 Dec 78. Located at Copenhagen.	N1166Z
☐ N123AC	66	HS 125/3A	25122	Wfu at OK Aircraft, Gilroy, Ca.	N255CB
☐ N123CB	70	Learjet 24D	24D-232	W/o Butte, Mt. USA. 17 Apr 71.	
☐ N123DR	73	Westwind-1123	158	Wfu at Aeroparque Buenos Aires, Argentina.	N1123G
☐ N123RE	67	Learjet 24	24-154	W/o Lancaster, Ca. USA. 17 Oct 78.	N11AK
☐ N125AW	65	HS 125/1A	25057	Wfu. Located at American Flight & Technology Center, Pontiac	N188K
☐ N125CA	70	Falcon 20DC	208	W/o Catersville, Al. USA. 29 Jun 89.	N300JJ
☐ N125CM	71	HS 125/400A	NA767	Wfu at Spirit of St Louis, Mo. Oct 94. Cx USA 6/95.	N28GE
☐ N125E	66	HS 125/1A-522	25110	W/o Hobby-Houston, Tx. USA. 29 Jun 83.	N3125B
☐ N125NE	79	Learjet 25D	25D-271	W/o Gulf of Mexico. 21 May 80.	(N183AP)
☐ N127MS	74	Sabre-75A	380-18	Wfu at Perryville, Mo. Cx USA 6/95.	N55
☐ N127MW	67	HFB 320	1027	W/o Aberdeen, SD. USA. 5 Oct 84.	N905MW
☐ N128GA	66	BAC 1-11/401AK	058	Wfu.	N128TA
☐ N128SD	69	HFB 320	1035	Wfu.	PH-HFC
☐ N129K	66	Jet Commander	70	Wfu at White Industries, Bates City, Mo.	N1194Z
☐ N131MS	75	Sabre-75A	380-22	Wfu at Perryville, Mo. Cx USA 6/95.	N132MS
☐ N133ME	66	Jet Commander	50	Wfu.	N612JC
☐ N136DH	65	HS 125/1A	25036	Wfu. Cx USA 7/95.	C-FPQG
☐ N137GL	78	Learjet 25D	25D-237	W/o Detroit, Mi. USA. 19 Jan 79.	(N28BP)
☐ N140MM	64	Sabre-40	282-8	Wfu. Cx USA 9/96.	N369N
☐ N148E	65	Jet Commander	22	W/o Burbank, Ca. USA. 13 Sep 68.	N200M
☐ N148PE	60	JetStar-6	5002	Wfu.	N81JJ
☐ N158DP	61	JetStar-8	5013	Wfu at White Industries, Bates City, Mo.	(N5AX)
☐ N159DP	66	Jet Commander	52	Wfu at Aviation Museum, Darwin, Australia.	N159MP
☐ N165WC	68	Falcon 20C-PW300	140	Wfu at Detroit Willow Run, Mi. USA.	N314AE
☐ N169RF	69	Sabre-60	306-45	W/o Phoenix, Az. USA. 7 Nov 92.	N742K
☐ N171CC	68	JetStar-8	5127	Wfu.	N636
☐ N171MC	74	Falcon 10	30	Wfu, Parts at White Industries Inc. Bates City, Mo.	N191MC
☐ N172AC	63	Jet Commander	1	Wfu.	N112AC
☐ N181MA	79	Diamond-Two	A001SA	Wfu. Located Wichita-Beech Field, USA.	JQ8001
☐ N196KC	66	Jet Commander	68	W/o Fayetteville, Ar. USA. 1 Jul 68.	N619JC
☐ N200CK	62	JetStar-6	5039	Wfu at Spirit of St Louis, Mo. Cx USA 9/90.	N86HM
☐ N200LF	66	Jet Commander	47	Wfu at White Industries, Bates City, Mo.	N222HM
☐ N200RC	66	Jet Commander-B	140	W/o Tampa, Fl. USA. 25 Sep 73.	4X-CPG
☐ N203M	67	Jet Commander	120	Wfu.	N200M
☐ N204C	74	Gulfstream 2	143	W/o Kota Kinabalu, Borneo. 4 Sep 91.	N334
☐ N204RC	68	Gulfstream 2	34	W/o Caracas, Venezuela. 17 Jun 91.	N500JR
☐ N210RS	65	Falcon 20C	18	Wfu.	N9DM
☐ N211MB	70	Learjet 25	25-059	W/o Port au Prince, Haiti. 3 Aug 80.	N425JX

Regn	Yr	Type	c/n	Accident/Withdrawal details	Prev Regn
□ N212CW	66	Jet Commander	75	Wfu at White Industries, Bates City, Mo.	N1121R
□ N212NE	76	Learjet 25D	25D-212	Wfu as parted out 1989.	N911MG
□ N213AP	68	JetStar-8	5122	Wfu & Cx USA 3/89.	N1107M
□ N219TT	75	Sabre-75A	380-24	Wfu & Cx USA 2/95, to Tulsa Technology Center 5/96.	N58
□ N220RB	60	DC 8-21	45280	Wfu. Donated to Ministry of Health, Beijing, China.	N8003U
□ N221PH	66	Sabre-40	282-55	Wfu at Elsberry, Mo. Cx USA 8/86.	(N221PX)
□ N222KN	68	JetStar-8	5118	Wfu as parted out 1988.	N333KN
□ N222WL	81	Citation II	550-0208	Wfu.	N54RC
□ N232RA	70	Falcon 20DC	232	W/o Binghampton, NY. USA. 15 Feb 89.	N27EV
□ N234CM	70	Learjet 24B	24B-214	W/o Nr Monclova, Mexico. 16 Dec 88.	N42NF
□ N234F	65	Learjet 23	23-063	W/o Palm Springs, Ca. USA. 14 Nov 65.	
□ N234MR	66	Learjet 24	24-130	Wfu & Cx USA 13 Oct 87.	N330J
□ N235KC	66	HS 125/1A	25096	W/o Grand Bahama. 21 Nov 66.	G-ATNR
□ N235R	65	Learjet 23	23-032	W/o Clarendon, Tx. USA. 23 Apr 66.	
□ N236JP	67	Jet Commander	116	W/o Marion, Va. USA. 31 Oct 69.	N4743E
□ N250UA	67	Jet Commander-A	121	W/o Flatwood, La. USA. 27 Apr 78.	N1121R
□ N253K	74	Falcon 10	10	W/o Meigs-Chicago, Il. USA. 30 Jan 80. (parts at White Inds)	N105FJ
□ N265CM	66	Sabre-40	282-76	Wfu.	N8345K
□ N267L	65	JetStar-6	5067	W/o Luton, England. 29 Mar 81.	(N267AD)
□ N287W	69	Falcon 20D	194	Wfu at Elsberry, Mo & Clarksville, Tn.	N297W
□ N300HW	65	HS 125/1A	25021	Wfu.	N711WJ
□ N300JA	74	Learjet 24D	24D-282	W/o Dutch Harbour, Ak. USA. 2 Dec 79.	D-INKA
□ N300PL	78	Learjet 25D	25D-247	Wfu as parted out following accident 12/83.	
□ N301AJ	66	Jet Commander	48	W/o Cozumel, Mexico. 13 Aug 90.	N502U
□ N302EJ	75	Learjet 24D	24D-302	W/o Puerta Vallarta, Mexico. 14 Apr 83.	N39DM
□ N303AF	67	Learjet 24	24-144	Wfu at White Industries, Bates City, Mo.	N700C
□ N305AJ	67	Jet Commander	100	Wfu at OK Aircraft, Gilroy, Ca.	N11WP
□ N308WC	61	JetStar-6	5020	Wfu & Cx USA 8/88.	N300CR
□ N309CK	81	Westwind-Two	350	W/o Irvine, Ca. USA. 15 Dec 93.	N777LU
□ N316M	65	Learjet 23	23-061	W/o Lake Michigan, USA. 19 Mar 66.	
□ N320MC	68	HFB 320	1034	W/o Phoenix, Az. USA. 9 Mar 73.	N320J
□ N320MJ	69	B 707-321B	20028	W/o Marana, Az. USA. 20 Sep 90.	VR-CBN
□ N320W	65	Jet Commander	15	Wfu at White Industries, Bates City, Mo.	N125K
□ N329J	57	JetStar	1001	Wfu Aug 82. Located Pacific Vocational Institute-Vancouver.	
□ N331DP	65	Loarjet 23	23-067	W/o Dayton, Oh. USA. 18 Jan 90.	N720UA
□ N331DP	65	Learjet 23	23-059	Wfu at Willow Run Detroit, Mi. N331DP transferred to 23-067.	N3IDP
□ N332PC	65	Learjet 23	23-056	W/o Flint, Mi. USA. 6 Jan 77. (parts at White Inds).	N362EJ
□ N333BG	67	Jet Commander	98	Wfu. Cx USA 6/97.	N301L
□ N333SV	68	Jet Commander	114	Wfu. parts at Hollister, Ca.	N85MR
□ N349M	65	Jet Commander	23	Wfu at White Industries, Bates City, Mo.	N2100X
□ N360HK	73	Westwind-1123	166	Wfu.	C-GDOC
□ N366AA	74	Learjet 25B	25B-151	W/o Briggsdale, Co. USA. 31 Aug 74.	
□ N380AA	69	JetStar-8	5131	Wfu.	N212JW
□ N386G	65	Jet Commander	43	Wfu as parted out /89.	N121CS
□ N388LS	81	Learjet 35A	35A-388	W/o Lebanon, NH. USA. 24 Dec 96.	N388PD
□ N395BB	79	Falcon 20F	395	Wfu as OD-PAL at St Louis, Mo. USA.	OD-PAL
□ N397F	69	Gulfstream 2	72	W/o Burlington, Vt. USA. 22 Feb 76.	
□ N399P	67	Sabre-40	282-87	Wfu. donated Pittsburgh Institute of Aeronautics, Pa.	N36P
□ N400CP	65	Jet Commander	30	W/o Burlington, Vt. USA. 21 Jan 71.	N401V
□ N400M	61	JetStar-6	5008	W/o Saranac Lake, NY. USA. 27 Dec 72.	N500Z
□ N400PH	68	HS 125/400A	NA716	W/o Lexington, Ky. USA. 5 Dec 87.	N888CR
□ N401DE	67	Jet Commander	92	Wfu.	N33PS
□ N403M	69	Jet Commander	132	W/o Salt Lake City, Ut. USA. 16 Dec 69.	N200M
□ N409MA	70	Gulfstream 2	83	W/o Quito, Ecuador. 4 May 95.	(N48MS)
□ N425JA	66	Falcon 20C	51	Wfu. Cx USA 10/94.	N425JF
□ N428JX	73	Learjet 25B	25B-103	Wfu as parted out following accident 7/75.	
□ N431NA		Sabreliner T-39D	285-2	Wfu. Cx USA 12/91.	150970
□ N432EJ	65	Learjet 23	23-028A	W/o Muskegon, Mi. USA. 25 Oct 67.	N803LJ
□ N434AN	64	JetStar-731	5050	Wfu.	HZ-THZ
□ N434EJ	65	Learjet 23	23-046	W/o Pellston, Mi. USA. 9 May 70.	(N822LJ)
□ N440HM	80	Learjet 35A	35A-294	W/o Greenville, SC. USA. 27 Feb 97	N35VP
□ N442NE	81	Learjet 35A	35A-442	W/o Morristown, NJ. USA. 26 Jul 88. (parts at White Inds).	N35BK
□ N445	65	Jet Commander	37	Wfu.	N723JB
□ N448GG	65	Learjet 23	23-057	Wfu at Dodson's, Ottawa, Ks.	N448GC
□ N454RN	66	Learjet 24	24-121	W/o Atlanta, Ga. USA. 26 Feb 73.	N454GL
□ N455JA	74	Learjet 24XR	24XR-300	W/o Gulkana, Ak. USA. 30 Aug 85.	N300EJ
□ N456JA	73	Learjet 24XR	24XR-265	W/o Nr Juneau, Ak. USA. 24 Oct 85.	N32WL
□ N458J	73	Learjet 25XR	25XR-106	W/o Columbus, Oh. USA. 1 Jul 91.	N458JA
□ N460MC	67	Falcon 20C	105	Wfu at Clarksville, Mo.	N97FJ
□ N463LJ		Learjet 25	25-001	Wfu. Used in construction of s/n 25-002.	
□ N480LR	72	HFB 320	1054	Wfu & Cx USA 1/87.	N896HJ
□ N481DH	69	Jet Commander-B	139	Wfu at White Industries Inc 8/96.	N188G
□ N500AD	72	Citation	500-0006	Wfu.	(N500AH)
□ N500BF	65	Learjet 23	23-010	Wfu at Willow Run Detroit, Mi.	N400BF
□ N500CC	69	Citation	669	Citation 500 prototype. Ff 15 Sep 69. Wfu 10/76.	
□ N500CX	75	Citation	500-0300	W/o as OE-FAP Greece 6 Oct 84, to N500CX for spares 2/85.	OE-FAP

Regn	Yr	Type	c/n	Accident/Withdrawal details	Prev Regn
N500FM	66	Learjet 23	23-088	W/o Columbia, Tn. USA. 2 Jul 91. (parts at Detroit).	(N500LH)
N500J	69	Gulfstream 2	60	W/o Hot Springs, Va. USA. 26 Sep 76.	N892GA
N500JJ	59	B 707-138B	17699	Wfu.	G-AVZZ
N500JR	66	Jet Commander	65	W/o North Platte, SD. USA. 26 Sep 66.	
N500JW	64	Learjet 23	23-005	Wfu at Willow Run Detroit, Mi. Cx USA 8/87.	N15BE
N500MF	65	Jet Commander	34	Wfu at White Industries, Bates City, Mo.	TG-OMF
N500NL	74	Sabre-75A	380-8	W/o 23 Feb 75.	N5107
N500RW	78	Learjet 35A	35A-148	W/o Teterboro, NJ. USA. 24 May 88.	N333RP
N501AL	61	JetStar-6	5012	Wfu & Cx USA 1/94.	N500SJ
N501GP	72	Citation	500-0026	W/o Bluefield, WV. USA. 21 Jan 81.	N526CC
N501PS	74	Learjet 25B	25B-153	W/o Detroit, Mi. USA. 26 May 77. (parts at Dodson's, Ks.).	
N503U	66	Jet Commander	83	W/o nr Guatamala City, Guatamala. 19 Dec 95.	C-GHPR
N505PF	65	Learjet 23	23-006	Wfu. Kansas Aviation Museum exhibit Nov 96.	N111JD
N515VW	68	Learjet 25	25-013	W/o Delemont, Switzerland. 17 Apr 69.	
N520S	66	JetStar-731	5084	W/o Westchester, NY. USA. 11 Feb 81.	N901E
N521M	67	HS 125/3A	25129	W/o Findlay, Oh. USA. 12 Dec 72.	G-AVDM
N521PA	79	Learjet 35A	35A-239	W/o Fresno, Ca. USA. 14 Dec 94. Cx USA 5/95.	N239GJ
N535PC	80	Learjet 35A	35A-291	W/o Aspen, Co. USA. 13 Feb 91.	N7US
N545BF	70	JetStar-8	5146	Wfu as parted out 1987. Cx USA 2/97..	N499AS
N550CC	77	Citation	686	Wfu. (Citation II prototype).	
N555AJ	71	Citation	500-0007	W/o Denver, Co. USA. 19 Nov 79.	N500LF
N555DM	65	Jet Commander	25	Wfu at Dodson's, Ottawa, Ks.	
N555PB	64	JetStar-6	5047	Wfu & Cx USA 10/90.	N409MA
N555PT	66	Sabre-40	282-53	Wfu at Elsberry, MO. Cx USA 2/91.	N600BP
N556AT	72	Citation	500-0020	Wfu.	C-FBAX
N564MG	61	JetStar-6	5021	Wfu. Parted out and cx USA register 12/86.	C-FETN
N567DW	65	Sabre-40	282-35	Wfu at Clarksville, Mo.	N341AR
N604AN	75	Corvette	18	Wfu & Cx USA 12/90 to Spain for spares.	F-BTTO
N611JC	63	Jet Commander	2	Wfu. Test airframe for static fatigue.	
N617CC	81	Citation 1/SP	501-0211	Wfu & Cx USA 6/96.	N6785C
N621ST	64	HS 125/1A	25014	Wfu. Broken up for spares 3/85.	XA-JUZ
N627WS	74	Learjet 25B	25B-170	W/o Houston, Tx. USA. 13 Jan 98.	N98796
N630N	66	Sabre-40	282-73	Wfu at Clarksville, Mo. Cx USA 6/85.	N630M
N650CC	79	Citation	696	Wfu & Cx USA 11/89. Citation III prototype. Ff 30 May	
N650DH	66	BAC 1-11/2400	059	Wfu.	N700JA
N658TC	69	Learjet 25	25-044	W/o Victoria, Tx. USA. 18 Jan 72.	N962GA
N660A	68	Learjet 24	24-155	Wfu at OK Aircraft, Gilroy, Ca.	N210FP
N661LJ		Learjet 25	25-002	Wfu 7/72. AiResearch engine tests.	
N678BC	67	JetStar-731	5109	Wfu.	N968BN
N690LJ	65	Learjet 23	23-078	W/o Orlando, Fl. USA. 30 Nov 67.	
N700CW	74	Citation	500-0205	W/o Eagle Pass, Tx. USA. 1 Apr 83.	(N541NC)
N700DK	81	Falcon 10	191	W/o Pal-Waukee, Il. USA. 23 Sep 85.	N256FJ
N706A		Sabre-40	282-7	Wfu.	XA-STU
N707RZ	62	B 707-328	18375	Wfu. Broken up at Fort Lauderdale 4/85.	F-BHSU
N710JW	65	Jet Commander	35	Wfu.	N7HL
N710MB	84	Diamond 1A	A078SA	W/o Nr Goodland, Ks. USA. 15 Dec 93.	N378DM
N711AF	75	Learjet 35	35-029	W/o 100km South of Katab, Egypt. 11 Aug 79.	
N711JT	67	Jet Commander	91	W/o Tullahoma, Tn. USA. 13 Mar 75.	N73535
N711WM	82	Citation II/SP	551-0388	W/o 6 Nov 86.	
N711Z	60	JetStar	1002	Wfu. Located at Andrews AFB, Md.	N329K
N717JM	61	JetStar-6	5009	Wfu & Cx USA 8/84.	(HB-VET)
N720Q	69	Gulfstream 2	58	W/o Kline, SC. USA. 24 Jun 74.	N878GA
N723GL	77	Learjet 35A	35A-107	W/o College Station-Easterwood, Tx. USA. 12 Dec 85.	
N727RL	64	B 727-25	18253	Wfu.	EL-GOL
N727US	78	Sabre-75A	380-61	Wfu.	9L-LAW
N739R	66	Sabre-40	282-78	W/o Ventura, Ca. USA. 16 May 67.	
N743R	68	Sabre-60	306-11	W/o Montrose, Ca. USA. 13 Apr 73.	N723R
N745F	65	Learjet 23	23-077	W/o Perris, Ca. USA. 30 Jul 88.	(N611CA)
N747E	64	Sabre-40	282-22	W/o Buenos Aires, Argentina. 22 Dec 94.	N747
N750SB	68	HFB 320	1031	Wfu. located Opa Locka, Fl. USA.	N300SB
N751CR	66	Jet Commander	88	Wfu at OK Aircraft, Gilroy, Ca.	N70CS
N760FR	61	MS 760 Paris-1A	72	Wfu. parted out at Mojave, Ca. USA.	F-BJLV
N760M	59	MS 760 Paris	49	W/o Evadale, Tx. USA. 3 May 69.	
N760T	61	MS 760 Paris-2B	103	Wfu. located Mojave, Ca. USA.	N760N
N769K	74	Citation	500-0228	Wfu at White Industries, Bates City, Mo.	N6365C
N771WB	69	Sabre-60	306-29	Wfu. Destroyed by fire as N771WW. (parts at White Inds).	N771WW
N777EP	60	JetStar-6	5004	Wfu. Located at Gracelands, Memphis, Tn.	N69HM
N777PQ	70	HFB 320	1050	Wfu.	N777PZ
N779XX	83	Challenger 601	3018	W/o Milan, Italy. 7 Feb 85.	C-GBXW
N780PV	65	Jet Commander	36	Wfu & Cx USA 5/88 to Norway for instructional airframe.	N730PV
N784B	82	Falcon 50	118	W/o Teterboro, NJ. USA. 10 Nov 85.	(N183B)
N787PR	74	Sabre-60	306-77	Wfu.	N180AR
N793NA	59	B 707-138B	17700	Wfu. To USAF for KC-135E spares 4/84.	VP-BDE
N800CS	66	Sabre-40	282-64	Wfu & cx USA 2/90.	N9000S
N801L	63	Learjet 23	23-001	W/o Wichita, Ks. USA. 4 Jun 64.	
N802L	64	Learjet 23	23-002	Wfu. Located at Smithsonian Institute, Washington-DC.	
Regn	*Yr* *Type*		*c/n*	*Accident/Withdrawal details*	*Prev Regn*

Regn	Yr	Type	c/n	Accident/Withdrawal details	Prev Regn
☐ N804LJ	65	Learjet 23	23-015A	W/o Jackson, Mi. USA. 21 Oct 65.	
☐ N804LJ	64	Learjet 24	23-004	Wfu. Re-certificated s/n 23-015A. Subsequently W/o 21 Oct 65	
☐ N805C	81	Challenger 600	1037	W/o Sun Valley, Id. USA. 3 Jan 83.	C-GLYE
☐ N805F	66	Falcon 20C	60	W/o Boca Raton, Fl. USA. 5 Jul 71.	N885F
☐ N808JA	65	Learjet 23	23-050A	W/o in ground fire 23 May 82.	N808LJ
☐ N813M	78	Learjet 35A	35A-151	Wfu. Stolen ex Wichita 13 Apr 84. Cx USA 6/86.	N711L
☐ N814NA	60	JetStar-6	5003	Wfu. Stored at NASA Edwards AFB. Ca. Cx USA 12/89.	NASA14
☐ N821LG	80	Falcon 10	170	W/o Nr West Chester, Pa. USA. 2 Feb 86.	N236FJ
☐ N822LJ	65	Learjet 23	23-080	W/o Detroit, Mi. USA. 9 Dec 67.	
☐ N831LC	66	HS 125/1A-522	25095	W/o Nr San Diego, Ca. USA. 16 Mar 91.	N25AW
☐ N833NA	61	B 720-061	18066	W/o Edwards AFB. Ca. USA. 1 Dec 84.	N2697V
☐ N848C	66	Jet Commander	54	Wfu.	N6534V
☐ N864CL	70	Learjet 24B	24B-229	W/o San Francisco, Ca. USA. 9 Oct 84.	N551AS
☐ N866JS	65	Learjet 23	23-018	W/o Richmond, Va. USA. 6 May 80. (parts at White Inds.).	N866DB
☐ N873LP	77	Learjet 35A	35A-104	W/o Auburn, Al. USA. 32 May 85.	N87W
☐ N880A	66	Gulfstream 2	38	Wfu at Detroit Willow Run, Mi. Cx USA 4/95.	N80A
☐ N880EP	60	Convair 880	38	Wfu 6 Feb 84. Located at Elvis Presley's Graceland Estate.	N8809E
☐ N881FC	68	Learjet 24	24-175	Wfu & Cx USA 3/93.	N28BK
☐ N888AR	66	Falcon 20C	33	W/o Acapulco, Mexico. 7 Aug 76.	N369EJ
☐ N888DL	71	HFB 320	1051	Wfu & Cx USA 8/90.	N6ZA
☐ N888RW	62	JetStar-6	5040	Wfu & Cx USA 6/88.	N888RW
☐ N897WA	62	B 707-321B	18339	Wfu.	OE-IEB
☐ N900CD	66	HS 125/3A	25111	Wfu at White Industries, Bates City, Mo 10/94.	N177GP
☐ N920G	74	Sabre-60	306-74	W/o Lancaster, Pa. USA. 27 Dec 74.	
☐ N920KP	69	Jet Commander-C	144	Wfu at White Industries Inc. Bates City, Mo.	N20K
☐ N921FP	84	Learjet 55	55-103	W/o Rutland, Vt. USA. 6 Aug 86.	
☐ N925R	66	Jet Commander	80	Wfu.	N173AR
☐ N930GL	80	Learjet 35A	35A-330	Wfu and cx USA 7/91.	
☐ N957TH	66	Falcon 20C	38	Wfu at Elsberry, Mo.	N1107M
☐ N959SC	65	Learjet 23	23-045A	W/o Detroit City, Mi. USA. 23 Jul 91.	F-BSUX
☐ N960CC	59	B 707-123B	17634	Wfu.	N707AR
☐ N984HF	69	HS 125/731	NA717	W/o Sparta, Tn. USA. 7 Nov 85.	N100HF
☐ N984JD	87	Learjet 31	31-001	Wfu at Dodson's, Ottawa, Ks.	N311DF
☐ N990L	66	Falcon 20C	43	W/o Dallas NAS, Tx. USA. 8 Mar 75.	N872F
☐ N991PC	89	Citation V	560-0043	W/o Eagle River, Wi. USA. 30 Dec 95.	(N2665F)
☐ N999BH	80	Learjet 25D	25D-318	W/o nr Santa Fe, NM. USA. 5 Sep 93.	N522TA
☐ N999HG	74	Learjet 25B	25B-178	W/o Sanford, NC. USA. 8 Sep 77.	N999MV
☐ N1001U		Caravelle 6R	86	Wfu, destined for Pima Air Museum, Tucson, Az.	PT-DUW
☐ N1021B	66	Learjet 23	23-086	W/o Racine, Wi. USA. 6 Nov 69.	
☐ N1121F	69	Jet Commander-B	150	W/o La Carbonera, Zacatecas, Mexico. 21 May 97.	N121FM
☐ N1121M	67	Jet Commander	111	Wfu at OK Aircraft, Gilroy, Ca.	C-GDJW
☐ N1135K	65	HS 125/1A	25019	W/o Des Moines, Ia. USA. 24 Feb 66.	N1125G
☐ N1151K	68	JetStar-731	5115	Wfu. Cx USA 9/96.	N8300E
☐ N1181G	81	Falcon 50	72	W/o Lake Geneva, Wi. USA. 12 May 85. (parts at Clarksville).	N82FJ
☐ N1189A	69	JetStar-731	5139	Wfu & Cx USA 2/97.	XA-RVG
☐ N1846	66	Falcon 20C	47	W/o Parkersburg, WV. USA. 13 Mar 68.	N875F
☐ N1863T	66	Sabre-40	282-62	Wfu & Cx USA 8/87.	
☐ N1909D	66	Sabre-40	282-57	Wfu & Cx USA 4/97.	N1909R
☐ N1963A	66	Learjet 23	23-097	Wfu. parted out at Atlanta South Expressway /95.	N1968A
☐ N1966J	66	Jet Commander	66	Wfu at White Industries, Bates City, Mo.	
☐ N2120Q	67	Westwind-1123	107	Wfu.	CA-01
☐ N2265Z	76	Sabre-75A	380-43	Wfu at Clarksville, Mo.	N6NR
☐ N2286D	82	Learjet 35A	35A-482	W/o Straits of Malacca 14 Feb 93.	N482U
☐ N2579E	65	Jet Commander	21	Wfu at White Industries, Bates City, Mo.	CF-WOA
☐ N2627U	82	Citation 1/SP	501-0247	W/o Wichita, Ks. USA. 12 Nov 82.	(N24CH)
☐ N2652Z	80	Citation 1/SP	501-0145	W/o Lord Howe Island. 22 Apr 90. (parts at Dodson's, Ks.).	VH-LCL
☐ N2954T	66	Falcon 20C	58	Wfu and parted out 1987.	HB-VDG
☐ N3080	66	JetStar-6	5094	Wfu.	N3030
☐ N3118M	69	HS 125/400A	25199	Wfu at Dodson's, Ottawa, Ks.	(N905Y)
☐ N3274Q	66	HS 125/1A	25102	Wfu & Cx USA 12/95.	XB-AKW
☐ N3504		Sabreliner CT-39A	265-32	Wfu 7/86. Located Parks College Aviation School-St Louis, Mo	60-3504
☐ N3833L		B 720-047B	19523	Wfu. To USAF for KC-135E spares 9/83.	5V-TAD
☐ N4060K		Sabre-UTX	246-1	Wfu. Mock up until 1967, and subsequently broken up.	
☐ N4253A	73	HS 125/600B	6005	Wfu at White Industries Jun 93. Cx USA 6/95.	EC-EAC
☐ N4400E	65	HS 125/1A	25026	Wfu. Scrapped at St Louis-Mo 4/86.	(XA-...)
☐ N4550T	68	BAC 1-11/204AF	135	Wfu & Cx USA 12/93.	HZ-MO1
☐ N5038	59	B 707-123B	17652	Wfu.	N7525A
☐ N5075L	66	Sabre-60A	306-16	Wfu at Clarksville, Mo. Cx USA 5/95.	N38UT
☐ N5094B	68	Jet Commander	105	Wfu at White Industries, Bates City, Mo.	C-GWPV
☐ N5565	73	Sabre-40A	282-119	W/o Oklahoma City, Ok. USA. 15 Jan 74.	N8341N
☐ N5863		Convair 880	48	Wfu & Cx USA 10/86.	N58RD
☐ N6555C	67	Falcon 20C	78	Wfu & Cx USA 12/95.	VH-JSX
☐ N6887Y	81	Citation II	550-0293	W/o Billings, Mt. USA. 19 Dec 92.	
☐ N7028F	69	Jet Commander-A	131	Wfu at Fairmont State College, Fairmont, WV	N43
☐ N7145V	60	JetStar-731	5001	Wfu to Pratt Community College, Ks.	N11
☐ N7201U		B 720-022	17907	Wfu. Scrapped at Luton-UK 13 Jul 82.	

Regn   *Yr*   *Type*   *c/n*   *Accident/Withdrawal details*   *Prev Regn*

Regn	Yr	Type	c/n	Accident/Withdrawal details	Prev Regn
N7224U	62	B 720-022	18077	Wfu & Cx USA 2/87.	
N7572N	71	Sabre-75	370-1	Wfu. Used as parts in other test aircraft.	
N7775	66	JetStar-6	5073	Wfu & Cx USA 12/86.	
N7842M	66	Falcon 20C	42	W/o Fort Worth, Tx. USA. 16 Jan 74.	N1503
N8000Z	73	HS 125/600A	6012	Wfu & Cx USA 12/95. (parts at Hooks Airport-Tx).	EC-EOQ
N8070U	67	Jet Commander-B	124	Wfu. parts at O K Aviation Inc. Monterey, Ca.	XA-RQT
N8221M	84	Diamond 1A	A076SA	W/o Jasper Hinton, Alberta, Canada as C-GLIG. 1 Mar 95.	C-GLIG
N8733	69	B 707-331B	20062	Wfu. To USAF for KC-135E spares 5/86.	
N9023W	65	Jet Commander	10	Wfu to Norwegian Mechanics School 7/87.	N5BP
N9258U	78	Falcon 10	132	Wfu. parts at White Industries.	TC-ATI
N9503Z	64	Sabre-40	282-10	W/o Blaine, Mn. USA. 7 Mar 73.	N525N
N9739B	64	JetStar-6	5052	Wfu as parted out 1989.	C-FDTM
N12058	84	Citation T-47A	552-0004	W/o hangar fire Forbes Field, Topeka, Ks. USA. 20 Jul 93.	(162758)
N12065	85	Citation T-47A	552-0011	W/o hangar fire Forbes Field, Topeka, Ks. USA. 20 Jul 93.	(162765)
N12269	85	Citation T-47A	552-0015	W/o hangar fire Forbes Field, Topeka, Ks. USA. 20 Jul 93.	(162769)
N12557	84	Citation T-47A	552-0003	W/o hangar fire Forbes Field, Topeka, Ks. USA. 20 Jul 93.	(162757)
N12564	85	Citation T-47A	552-0010	W/o hangar fire Forbes Field, Topeka, Ks. USA. 20 Jul 93.	(162764)
N12660	85	Citation T-47A	552-0006	W/o hangar fire Forbes Field, Topeka, Ks. USA. 20 Jul 93.	(162760)
N12756	84	Citation T-47A	552-0002	W/o hangar fire Forbes Field, Topeka, Ks. USA. 20 Jul 93.	(162756)
N12761	84	Citation T-47A	552-0007	W/o hangar fire Forbes Field, Topeka, Ks. USA. 20 Jul 93.	(162761)
N12762	85	Citation T-47A	552-0008	W/o hangar fire Forbes Field, Topeka, Ks. USA. 20 Jul 93.	(162762)
N12763	85	Citation T-47A	552-0009	W/o hangar fire Forbes Field, Topeka, Ks. USA. 20 Jul 93.	(162763)
N12855	84	Citation T-47A	552-0001	W/o hangar fire Forbes Field, Topeka, Ks. USA. 20 Jul 93.	(167255)
N12859	84	Citation T-47A	552-0005	W/o hangar fire Forbes Field, Topeka, Ks. USA. 20 Jul 93.	(162759)
N12967	85	Citation T-47A	552-0013	W/o hangar fire Forbes Field, Topeka, Ks. USA. 20 Jul 93.	(162767)
N22265	61	JetStar-8	5005	Wfu.	XB-DLV
N29019	70	Sabre-75	370-6	Wfu at Clarksville, Mo. Cx USA 10/91.	(N30EV)
N29977	66	HS 125/1A	25028	Wfu at White Industries, Bates City, Mo. Cx USA 4/91.	XA-ESQ
N40180	76	Falcon 10	93	Wfu at White Industries, Bates City, Mo.	F-BYCV
N82197	69	Sabre-60	306-39	Wfu.	XA-SLH
N91669	65	Jet Commander	17	Wfu at OK Aircraft, Gilroy, Ca.	C-FSUA
N98386	65	Learjet 23	23-040	Wfu & Cx USA 8/82.	(N12HJ)
OB-1319	73	Sabre-40A	282-127	W/o Buenos Aires, Argentina. 3 Sep 93.	OB-T1919
OE-FFK	79	Citation 1/SP	501-0124	W/o Nr Salzburg, Austria. 26 Oct 88.	N95RE
OH-CAR	74	Citation	500-0144	W/o Nr Helsinki, Finland. 19 Nov 87.	N332H
OH-FFW	70	Falcon 20F	243	W/o Montreal, Canada. 1 Mar 72.	F-WMKH
OY-SBS		Corvette	21	W/o Nice, France. 3 Sep 79.	F-BVPE
PP-FMX	66	Learjet 23	23-090	W/o Rio de Janeiro, Brazil. 30 Aug 69.	
PP-SED	73	Sabre-40A	282-121	Wfu & Cx USA 8/97.	N8349N
PT-ASJ	76	Learjet 23	95	W/o Rio de Janeiro, Brazil. 17 Feb 89.	N173FJ
PT-CMY	74	Learjet 25C	25C-108	W/o Juiz de Fora, Brazil. 6 Apr 90.	
PT-CXK	66	Learjet 24	24-122	W/o Rio-Galeon, Brazil. 4 May 73.	N461LJ
PT-DVL	71	Learjet 25B	25B-077	W/o Sao Paulo, Brazil. 12 Nov 76.	
PT-DZU	71	Learjet 24D	24D-244	W/o Sao Paulo, Brazil. 23 Aug 79.	
PT-IBR	72	Learjet 25C	25C-072	W/o Sao Paulo, Brazil. 26 Sep 76.	N256GL
PT-ISN	73	Learjet 25C	25C-113	W/o Belo Horizonte, Brazil. 4 Nov 89. Cx PT- 1/90.	
PT-JBQ	73	Learjet 25B	25B-119	W/o Rio Branco, Brazil. 4 Sep 82.	N3810G
PT-JDX	73	Learjet 25C	25C-131	W/o Congonhas, Brazil. 26 Dec 78.	N3803G
PT-JXS	74	Citation	500-0162	W/o Belem, Brazil. 16 Mar 75.	
PT-KBC	74	Learjet 25C	25C-165	W/o Riberao Preto, Brazil. 4 Jun 96.	
PT-KIU	74	Citation	500-0172	W/o Aracatuba, Brazil. 12 Nov 76.	N172CC
PT-KKV	74	Learjet 25C	25C-172	W/o Macre, Brazil. 20 Feb 88, rebuilt, w/o again 11 Jan 91.	
PT-KYR	79	Learjet 25D	25D-266	W/o 1986.	
PT-KZY	76	Learjet 25B	25B-204	W/o Uberaba, Brazil. 16 May 82.	N472J
PT-LAU	71	Learjet 24D	24D-239	W/o Brasilia, Brazil. 10 Sep 94.	N83MJ
PT-LCN	74	Learjet 24D	24D-287	W/o Florianapolis, Brazil. 4 Apr 84.	N92565
PT-LGJ	85	Citation S/II	S550-0025	W/o Rio de Janeiro, Brazil. 6 Sep 88.	(N12596)
PT-LHU	73	Learjet 25C	25C-099	W/o Icuape, Brazil. 28 Jul 92. Cx PT- 10/94.	PT-FAF
PT-LIG	85	Learjet 55	55-111	W/o Guanabara Bay, Rio de Janeiro, Brazil. 9 Nov 94.	N7260G
PT-LIH	81	Learjet 35A	35A-433	W/o Uberlandia, Brazil. 15 Mar 91.	(N93RC)
PT-LKQ	65	Learjet 23	23-038	Wfu. Michigan Institute of Aeronautics, Detroit, USA.	N175BA
PT-LKT	86	Citation S/II	S550-0117	W/o Sao Paulo-Congonhas, Brazil. 1 Dec 92. Cx PT- 10/94.	N1292N
PT-LLL	78	Learjet 25D	25D-258	W/o nr Brasilia, Brazil. 18 Mar 91.	N258MD
PT-LMA	77	Learjet 24F	24F-353	W/o Macre, Brazil. 24 Feb 88.	N63BW
PT-LSD	78	Learjet 25D	25D-243	W/o Sierra de Cantareira, Sao Paulo, Brazil. 2 Mar 96.	N711JT
PT-OEF	77	Learjet 35A	35A-102	W/o Morelia, Mexico. 2 May 92.	N232R
PT-OMV	90	Citation VI	650-0200	W/o Nr Bogota, Colombia. 23 Mar 94.	N650CM
RP-C1500	75	Citation	500-0225	W/o between Cagayan & Butuan, Philippines. 1 Feb 97.	VH-OIL
RP-C1980	79	Falcon 20F	400	W/o Davao City, Philippines. 24 Apr 96.	F-WRQR
RP-C911		B 707-321	17606	Wfu 10/82. Located at Manila as 'Club 707' restaurant.	N728PA
SE-DCY	69	Jet Commander	136	W/o Stockholm, Sweden. 4 Dec 69.	N5044E
SE-DLK	76	Westwind-1124	197	W/o Umeaa, Sweden. 21 Sep 92.	N29CL
SX-ASO		Learjet 25B	25B-074	W/o Antibes, France. 18 Feb 72.	N251GL
TC-NSU	69	HFB 320	1046	Wfu.	16+04
TC-OMR	69	HFB 320	1047	Wfu.	16+05
TC-YIB	83	Diamond 1A	A051SA	Wfu at Dodson's, Ottawa, Ks.	D-CFGV

150

Regn	Yr	Type	c/n	Accident/Withdrawal details	Prev Regn
☐ TL-AAI		Caravelle 3	10	Wfu. Broken up at Paris-Orly 2/83.	F-BNGE
☐ TN-ADB	75	Corvette	22	W/o Nkayi, Congo Republic. 30 Mar 79.	F-ODFE
☐ TR-KHB	73	Gulfstream 2	127	W/o Ngaoundere, Cameroun. 6 Feb 80.	N17581
☐ TR-LZT	75	Corvette	20	Wfu.	(F-GKJB)
☐ TT-AAM		Caravelle 6R	100	Wfu.	(TT-AAD)
☐ TY-BBK	76	Corvette	29	W/o Lagos, Nigeria. 16 Nov 81.	F-OBZP
☐ TY-BBR	71	B 707-336B	20457	W/o Sebha, Libya. 13 Jun 85.	9G-ADB
☐ TY-BBW	62	B 707-321	18084	Wfu.	TY-AAM
☐ VH-AJS	77	Westwind-1124	221	W/o Alice Springs, Australia. 27 Apr 95.	(N969EG)
☐ VH-ANQ	75	Citation Eagle	500-0283	W/o Cairns, Australia. 11 May 90.	N18AF
☐ VH-CAO	65	HS 125/3B	25015	Wfu.	(9M-AYI)
☐ VH-ECE	66	HS 125/3B	25062	Wfu 1982. Located Camden Airport Museum, NSW.	
☐ VH-FSA	74	Citation	500-0237	W/o Prosperine, Queensland, Australia. 20 Feb 84.	N14TT
☐ VH-FWO	67	Falcon 20C	110	Wfu.	C-FWRA
☐ VH-IWJ	82	Westwind-1124	371	W/o Botany Bay, Australia. 10 Oct 85.	4X-CUH
☐ VR-BJB	70	Falcon 20F	244	W/o Lugano, Switzerland. 15 Jan 88. (parts at Dodson's, Ks.)	(OE-GCS)
☐ VR-BJI	71	JetStar-731	5149	Wfu as parted out 7/91.	N110MZ
☐ VR-BLJ	68	Gulfstream 2	40	W/o Jos, Nigeria. 20 Jun 96.	N1039
☐ VR-CAN	61	B 707-138B	18067	Wfu.	9Y-TDC
☐ VR-CCY	66	JetStar-6	5085	Wfu at Abu Dhabi Higher College of Technology-UAE 3/92.	S9-NAE
☐ XA-BBA	77	Learjet 25D	25D-223	W/o Washington-Dulles, USA. 18 Jun 94.	XA-RWH
☐ XA-COL	66	HS 125/1A	25086	W/o Acalpulco, Mexico. 12 Oct 73.	N3699T
☐ XA-CUZ	72	HS 125/400A	NA772	W/o Cancun, Mexico. 26 Dec 80.	N69BH
☐ XA-EEU	66	Sabre-40	282-54	W/o ground accident Mexico 1980.	N256CT
☐ XA-GBP	63	B 727-25	18252	Wfu.	XB-GBP
☐ XA-HOK	64	Sabre-40	282-17	Wfu at Clarksville, Mo.	N900CS
☐ XA-JLV	67	Learjet 24	24-136	Wfu.	N24LW
☐ XA-KEW	80	HS 125/700A	NA0276	W/o Norte-Monterrey, Mexico. 2 May 81.	G-5-14
☐ XA-KUT	74	HS 125/600A	6028	W/o Houston, Tx. USA. 18 Jan 88. (parts at Dodson's, Ks)	C-GDHW
☐ XA-LAN	79	Learjet 35A	35A-267	W/o Hermosillo, Mexico. 8 Jan 93.	N39418
☐ XA-NOG	81	Learjet 25D	25D-349	W/o Tijuana, Mexico. 2 Sep 93.	N20GT
☐ XA-POJ	72	Westwind-1123	161	Wfu at Dodson's, Ottawa, Ks.	N33WD
☐ XA-PUL	72	JetStar-8	5151	Wfu. TAESA ticket office at Wal-Mart S Mexico City.	N45K
☐ XA-RNR	70	Sabre-60	306-49	Wfu.	XA-POR
☐ XA-ROK	69	JetStar-8	5133	Wfu. TAESA ticket office at Wal-Mart E Mexico City.	HZ-WT1
☐ XA-RQI	69	Learjet 25	25-032	Wfu. Exhibit main terminal Mexico City 1994.	XA-ZYZ
☐ XA-RRC	72	Learjet 24D	24D-259	Wfu. Parts at White Industries, Bates City.	N22MH
☐ XA-SHA	66	Jet Commander	86	Wfu.	XA-RIW
☐ XA-SLQ	73	Citation	500-0111	W/o Ensenada, Mexico. 6 Feb 96.	XA-SHO
☐ XA-SMH	73	Citation	500-0084	W/o Aleman, Veracruz, Mexico. 25 March 1994.	XB-FPK
☐ XA-SWF	81	Learjet 35A	35A-391	W/o Nr Tepic Airport, Mexico. 23 Jun 95.	N888PT
☐ XA-TFC	60	Sabreliner CT-39A	265-12	W/o Monterrey, Mexico. 16 May 97.	XB-GDU
☐ XB-DUH	72	JetStar-8	5157	Wfu & preserved at Dodson's as N001D1.	N29WP
☐ XB-FJI	67	Jet Commander	115	Wfu at Monterrey, Mexico.	N500VF
☐ XB-JOY	73	Learjet 24D	24D-263	W/o Mexico City, Mexico. 29 Jun 76.	N3812G
☐ XC-AA26	68	Sabre-60A	306-12	Wfu.	XC-HHL
☐ XC-HAD	66	Jet Commander	85	Wfu.	N201S
☐ XC-PGR	82	Learjet 35A	35A-460	W/o Mexico City, Mexico. 1 Apr 94.	
☐ XC-TIJ	70	HFB 320	1049	W/o San Diego, Ca. USA. 6 Jun 84.	XC-DGA
☐ YI-AKH	82	HS 125/700B	7187	W/o as destroyed during Gulf War 2/91.	YI-AKH
☐ YI-AKI	83	Gulfstream 3	408	W/o as destroyed during Gulf war 2/91.	9K-AEG
☐ YI-AKJ	84	Gulfstream 3	419	W/o as destroyed during Gulf war 2/91.	9K-AEH
☐ YU-BJH	75	Learjet 25B	25B-186	W/o Sarajevo, Yugoslavia. 18 Jan 77.	
☐ YV-123CP	65	Jet Commander	16	Wfu.	N177A
☐ YV-160CP	77	Westwind-1124	211	W/o La Aurora, Guatamala City, Guatamala. 19 Feb 97.	4X-CLI
☐ YV-O-MAC1	76	Citation	500-0336	W/o Caracas, Venezuela. Jun 79.	N336CC
☐ 3D-ART	75	Falcon 10	61	W/o Magoebaskloof, Transvaal, RSA. 3 Oct 86.	F-BFDG
☐ 4X-AIP	78	Westwind-1124	243	W/o Rosh-Pina, Israel. 23 Jul 96.	N215SC
☐ 4X-COA	66	Jet Commander	71	Wfu. Located at tel Aviv.	N721GB
☐ 4X-COJ	68	Jet Commodore	29	W/o Tel Aviv, Israel. 21 Jan 70.	N615JC
☐ 4X-WIN	84	Astra-1125	001	Wfu 31 Aug 86.	
☐ 5A-DAD	65	Learjet 23	23-075	W/o Damascus, Syria. 5 Jun 67.	
☐ 5A-DAR	77	JetStar 2	5221	W/o en route Tripoli-Algeria, North Africa. 16 Jan 83.	N5547L
☐ 5N-AER	68	HS 125/1B-522	25099	Wfu. Located Zaria-Nigeria.	(N121AC)
☐ 5N-AMF	67	HFB 320	1028	W/o Abidjan, Ivory Coast. 25 Jul 77.	D-CASU
☐ 5N-AMR	78	Citation II	550-0045	W/o Bauchi, Nigeria. 21 May 91.	N4CR
☐ 5N-AOG	67	HS 125/3B-RA	25143	Wfu.	G-AVXK
☐ 5N-ASQ	81	Learjet 25D	25D-344	W/o Lagos, Nigeria. 22 Jul 83.	N37943
☐ 5N-ASZ	66	HS 125/1B	25063	Wfu. Broken up at Southampton-UK 11/86.	G-ONPN
☐ 5N-AWB	66	HS 125/1B	25025	Wfu.	(F-OCGK)
☐ 5N-AWD	64	HS 125/1	25008	Wfu.	G-ASSI
☐ 5N-AWS	75	HS 125/600B	6042	W/o Casablanca, Morocco. 31 Dec 86.	(G-5-505)
☐ 5N-AXO	83	HS 125/700B	7196	W/o Kano, Nigeria. 17 Jan 96.	G-5-766
☐ 5N-AXP	83	HS 125/700B	7203	W/o Kaduna, Nigeria. 31 Dec 85.	G-5-14
☐ 5T-RIM	61	Caravelle 6R	91	Wfu.	5T-MAL
☐ 5V-TAA	74	Gulfstream 2	149	W/o Lome, Togo, West Africa. 26 Dec 74.	N17586

Regn	Yr	Type	c/n	Accident/Withdrawal details	Prev Regn
☐ 6V-AAR	59	Caravelle 3	5	Wfu 1 Aug 85. Located Dakar-Senegal.	6V-ACP
☐ 7T-VHB	78	Gulfstream 2	230	W/o Nr Qotur, NW Iranian Border. 3 May 82.	N17586
☐ 7T-VRE	69	Falcon 20C	156	W/o 30 May 81.	F-WMKI
☐ 9M-HLG	72	HS 125/400B	25257	Wfu.	G-BATA
☐ 9Q-CBC	72	Learjet 24D	24D-248	W/o Kinshasa, Zaire. 18 Jan 94.	OO-LFA
☐ 9V-ATD	93	Learjet 31	31-033B	W/o between Phuket & Ranong, Thailand. 21 Jul 97.	N2600S
☐ 9XR-NN	79	Falcon 50	6	W/o Kigali, Rwanda. 6 Apr 94.	N815CA

Military

Regn	Yr	Type	c/n	Accident/Withdrawal details	Prev Regn
☐ 18351	61	B 720-051B	18351	Wfu. Preserved at Kangshan AB Museum, Taiwan.	N721US
☐ 144613	84	Challenger 601	3035	W/o Shearwater, NS. Canada. 24 Apr 95.	C-GCUN
☐ E-302	84	Citation III	650-0033	W/o Concepcion, Chile. 9 Jul 92.	CC-ECE
☐ 16+06	69	HFB 320	1048	Wfu. Located Luftwaffe Museum Gatow-Berlin, Germany.	D-CISI
☐ 16+22	76	HFB 320	1059	W/o Schwabmuenchen, Germany. 27 Nov 76.	98+25
☐ 16+26	79	HFB 320ECM	1063	Wfu. Located Luftwaffe Museum Gatow-Berlin, Germany.	D-CANU
☐ CA+102	62	JetStar-6	5035	W/o Bremen, Germany. 16 Jan 68.	(6212167)
☐ 236	71	HS 125/F600B	25256	W/o Casement-Dublin, Ireland. 27 Nov 79.	G-AYBH
☐ 5-3020	76	Falcon 20E	348	W/o Ardebil, Iran. 3 March 97.	5-4039
☐ 39	75	Falcon 10MER	39	W/o Toul-Rosieres, France. 30 Jan 80.	F-WPUX
☐ 141	63	Caravelle 3	141	Wfu 28 Mar 80, for preservation at Musee de l'Air-Le Bourget	F-BJTK
☐ 154	68	Falcon 20C	154	W/o Rambouillet, France. 22 Jan 76.	F-WLCV
☐ F-UGWP	75	Falcon 20E	309	W/o Nr Villacoublay-Paris, France. 2 Dec 1991.	F-RAFU
☐ XS710/O	65	Dominie T1	25012	Wfu. Instructional airframe at RAF Cosford, UK.	
☐ XS711/L	65	Dominie T1	25024	Wfu.	
☐ XS714/P	66	Dominie T1	25054	Wfu. Fire school RAF Manston.	
☐ XS726/T	66	Dominie T1	25044	Wfu. Instructional airframe at RAF Cosford, UK.	
☐ XS729/G	65	Dominie T1	25049	Wfu. Instructional airframe at RAF Cosford, UK.	
☐ XS732/B	66	Dominie T1	25056	Wfu.	
☐ XS733/Q	66	Dominie T1	25059	Wfu. Instructional airframe at RAF Cosford, UK.	
☐ XS734/N	66	Dominie T1	25061	Wfu. Instructional airframe at RAF Cosford, UK.	
☐ XS735/R	66	Dominie T2	25071	Wfu. Instructional airframe at RAF Cranwell, UK.	
☐ XS738/U	66	Dominie T1	25077	Wfu. Instructional airframe at RAF Cosford, UK.	
☐ XW930	66	HS 125/1	25009	Wfu at Jordan's scrapyard Portsmouth, UK.	G-ATPC
☐ AEE-402		Sabre-	...	W/o Quito, Ecuador. 10 Dec 92. (380-45 ?)	
☐ FAE-068	66	Sabre-40R	282-68	W/o Quito, Ecuador. 3 Jun 88.	N4469N
☐ 33-333	90	B 737-3Z6	24480	W/o Nr Khon Kaen, Thailand. 30 Mar 93.	
☐ MM61953		PD 808-TP	511	W/o Venice, Italy. 15 Sep 93.	
☐ 9203	90	Learjet U36A	36A-058	W/o Shikoku Island, Japan. 28 Feb 91.	N4290J
☐ T-24	80	Learjet 35A	35A-333	W/o Pebble Island, Falklands, South Atlantic. 7 Jun 82.	
☐ 157352	65	Sabreliner CT-39E	282-46	W/o Alameda AFB. Ca. USA. 21 Dec 75.	test
☐ 160057	75	Sabreliner CT-39G	306-108	W/o Glenview NAS, USA. 3 Mar 91.	N65799
☐ 58-6970	59	C-137B	17925	Wfu. Located Boeing Museum of Flight, Seattle, Wa. USA.	
☐ 59-2871		Sabreliner T-39A	265-4	W/o	
☐ 59-5958	61	JetStar-6	5010	Wfu at Travis AFB Museum, Ca. USA.	
☐ 59-5959	62	JetStar-6	5026	Wfu. C-140A, USAF at Scott AFB. Il.	
☐ 59-5960	62	JetStar-6	5028	Wfu. Located Greenville, Tx. USA.	
☐ 59-5961	62	JetStar-6	5030	W/o Robins AFB. Ga. USA. 7 Nov 62.	
☐ 60-3503		Sabreliner GCT-39A	265-31	Wfu. Located museum at Du Page Airport, Il. USA.	
☐ 60-3506		Sabreliner T-39A	265-34	W/o Colorado Springs, Co. USA. 9 Feb 74.	
☐ 61-0640		Sabreliner T-39A	265-43	W/o	
☐ 61-0644		Sabreliner T-39A	265-47	W/o	
☐ 61-0662	61	Sabreliner CT-39A	265-65	Wfu at Clarksville, Mo. USA.	
☐ 61-0672		Sabreliner T-39A	265-75	W/o Kunsong, Korea. 13 Mar 79.	
☐ 61-2488	61	JetStar-6	5017	Wfu. VC-140B, USAF/preserved Warner Robins AFB. Ga. USA.	N9286R
☐ 61-2489	61	JetStar-6	5022	VC-140B, USAF/AMARC as CL006. Tt 15637. Sub to Pima Museum.	
☐ 61-2490	61	JetStar-6	5024	VC-140B, USAF/to AMARC 4/87 as CL004. Tt 17701.	
☐ 61-2491	62	JetStar-6	5027	VC-140B, USAF, battle damage repair training Rhein-Main.	
☐ 61-2492	62	JetStar-6	5031	Wfu. VC-140B, USAF/preserved Wright-Patterson AFB. Oh.	
☐ 61-2493	62	JetStar-6	5034	VC-140B, USAF/to AMARC 2/84 as CL003. Tt 11732.	
☐ 62-4197	62	JetStar-6	5041	C-140B, USAF/to AMARC 7/87 as CL007. Tt 13646.	
☐ 62-4198	62	JetStar-6	5042	Wfu as broken up 1/92.	
☐ 62-4199	62	JetStar-6	5043	C-140B, USAF/to AMARC 2/84 as CL002. Tt 13462.	
☐ 62-4200	63	JetStar-6	5044	C-140B, USAF/to AMARC 6/87 as CL005. Tt 14082.	
☐ 62-4201	63	JetStar-6	5045	C-140B, USAF, preserved at Hill AFB. Ut.	
☐ 62-4448		Sabreliner T-39A	276-1	W/o Erfurt, East Germany. 28 Jan 64.	
☐ 62-4458		Sabreliner T-39A	276-11	W/o Clark AFB. Philippines. 25 Mar 65.	
☐ 62-4460		Sabreliner T-39A	276-13	W/o Torrejon, Spain. 28 Feb 70.	
☐ 62-4496		Sabreliner T-39A	276-49	W/o Scranton, Pa. USA. 20 Apr 85.	
☐ 84-0121	85	Learjet C-21A	35A-567	W/o Alabama, USA. 15 Jan 87.	N400EN
☐ 84-0136	85	Learjet C-21A	35A-583	W/o Thomas Russell Field, Alexander City, Al. USA. 17 Apr 95	N400FP
☐ F-330	81	Gulfstream 3	330	W/o Vagar, Faroe Islands. 3 Aug 96.	
☐ A-1645	65	JetStar-6	5059	Wfu. Museum exhibit at Yogjakarta-Adisutjipto, Indonesia.	T-1645
☐ VC93-2122	68	HS 125/3B-RC	25166	W/o Brasilia, Brazil. 18 Jun 79.	
☐ VU93-2129	73	HS 125/403B	25290	W/o Carajas, Brazil. 9 Sep 87.	
☐ A20-103	75	B 707-368C	21103	W/o Nr RAAF Richmond, Australia. 29 Oct 91.	HZ-ACG
Regn	*Yr*	*Type*	*c/n*	*Accident/Withdrawal details*	*Prev Regn*

152

Regn	Yr	Type	c/n	Accident/Withdrawal details	Prev Regn
XC-UJC	78	Sabre-75A	380-67	W/o Saltillo, Mexico. 26 Oct 89. (parts at Elsberry, Mo.).	TP-103
NZ7273	68	B 727-22C	19893	Wfu.	N7436U
SAAF01	69	HS 125/400B	25177	W/o Devil's Peak, South Africa. 26 May 71.	G-AWXN
SAAF02	69	HS 125/400B	25181	W/o Devil's Peak, South Africa. 26 May 71.	G-AXLU
SAAF03	69	HS 125/400B	25182	W/o Devil's Peak, South Africa. 26 May 71.	G-AXLV
KAF 320		DC 9-32	47691	W/o during Gulf War 2/91.	160749

BUSINESS JETS
Cross-Reference by Construction Number

Aerospatiale

Caravelle
42	TL-FCA
209	9XR-CH

Corvette
1	F-ZVMV
2	F-ZVMW
3	F-BUQN
4	F-BUQP
5	CN-TDE
6	F-BVPB
7	F-BVPK
8	F-GJAS
9	TN-ADI
10	F-ZVMX
11	F-GKGA
12	TJ-AHR
13	F-GFDH
14	F-GIRH
15	F-GNAF
16	F-BVPT
17	YV-572CP
23	OY-SBR
25	F-BVPG
26	EC-DQE
27	EC-DQG
28	F-GPLA
30	F-GLEC
31	F-GJAP
32	F-GILM
33	OY-SBT
34	F-GKGD
35	F-ODSR
36	N600RA
38	5A-DCK
39	F-GJLB
40	N200MT
50	F-GEPQ

Airbus
004	HZ-124
009	V8-AM1
026	A7-HHK
046	V8-BKH
046	V8-PJB
061	SU-GGG
151	V8-JBB
354	A6-SHZ
367	CA-02
372	CA-01
374	A6-PFD
421	F-RADA
422	F-RADB
425	15003
431	V8-DPD
434	10+24
441	15005
444	15004
446	15001
473	A7-...
482	15002
484	10+25
498	10+21
499	10+22
503	10+23
591	44-444
648	9K-ALD

Boeing

707
KC-135
17644	N2143H
17696	HZ-123
17697	N138SR
17702	N707SK
17926	58-6971
17927	58-6972
18151	60-0376
18152	60-0377
18153	60-0378
18334	CNA-NS
18461	62-6000
18465	62-4125
18466	62-4126
18467	62-4127
18469	62-4129
18470	62-4130
18586	HZ-SAK1
18740	N707XX
18757	T 17-2
18926	903
18928	N88ZL
19248	K2900
19374	904
19443	902
19498	N14AZ
19623	A20-623
19624	A20-624
19627	A20-627
19629	A20-629
19635	68-19635
19716	FAC-1201
19739	5V-TAG
19840	2401
19866	68-19866
19969	9T-MSS
19986	N707MB
19988	K2899
19997	10+01
19998	10+02
19999	10+03
20000	10+04
20025	N707KS
20060	T 17-1
20110	4X-JYQ
20176	N712PC
20301	N707WJ
20315	N108RA
20375	VP-BZA
20495	JY-ADP
20629	4X-JYB
20630	72-7000
20715	D2-TPR
20721	4X-JYH
20919	SU-AXJ
21049	P4-TBN
21081	HZ-HM2
21092	A-7002
21096	4X-J...
21104	ST-DRS
21228	5A-DAK
21261	A20-261
21334	A7-AAA
21367	T 17-3
21368	HZ-HM3
21396	EP-NHY
21956	CNA-NR

720
18024	C-FETB
18451	N720JR

18452	N92GS
18453	HZ-KA4
18977	N7381

727
18323	VP-BMC
18362	84-0193
18363	EP-PLN
18365	N700TE
18366	VP-BGW
18368	VP-CMM
18369	9Q-...
18370	N25AZ
18371	5B-DBE
18752	XA-...
18894	HP-500A
18908	XC-FAY
18909	XC-FAZ
18912	XB-DLM
18933	N7271P
18934	9Q-RDZ
18935	N113
18936	N18HH
18970	PK-VBA
18990	XT-BBE
18998	N111JL
19006	HZ-OCV
19121	N724CL
19123	XC-UJA
19124	HZ-DG1
19139	N511DB
19148	N341TC
19149	N400RG
19170	N270AX
19176	N530KF
19252	HZ-WBT
19253	N720DC
19254	N682G
19260	N727LA
19261	N260GS
19262	VP-BNA
19282	VP-CMN
19318	N44MD
19394	N727X
19399	2721
19402	CB-01
19403	CB-02
19520	2722
19535	N60FM
19620	P4-JLD
19818	2724
19835	N727HC
19854	N40
19859	C-FPXD
19892	NZ7271
19895	NZ7272
19987	HZ-HE4
20045	N727WF
20046	VP-BDJ
20049	5N-BBF
20050	5N-BBG
20111	2723
20115	N500LS
20228	JY-HS1
20327	N529AC
20371	5X-AMM
20489	VP-CKA
20512	N311AG
20533	VP-BKC
20641	N270B
21010	JY-HS2
21091	6V-AEF
21155	N109KM
21460	VP-CBQ
21595	VP-CZY
21636	TJ-AAM
21655	VP-CDL

21824	A9C-BA
21853	HZ-KA4
21948	N31TR
22362	HZ-AB3
22687	N721MF
22825	5N-FGN
22968	HZ-HR3

737
19600	N487GS
19601	PK-OCF
19605	N165W
19613	N902WC
19772	XC-UJL
20127	XC-UJI
20255	PK-OCI
20335	PK-OCG
20483	K3187
20484	K3186
20689	N5175U
20690	72-0283
20691	N5294E
20692	N5176Y
20693	N5177C
20694	N5294M
21165	VC96-2115
21166	VC96-2116
21167	0001
21317	EP-AGA
21499	5U-BAG
21613	N1PC
21926	N720A
21928	N715A
21929	N716A
21957	PK-HHS
22050	N5WM
22431	N737WH
22600	HZ-MIS
22628	A6-ESH
23036	K2412
23037	K2413
23059	22-222
23152	85101
23405	N714A
23465	N736BP
23466	N733AR
23468	HZ-TBA
23766	VH-NJE
23800	N253DV
23976	N37NY
24095	XC-UJB
24970	VP-BOC
25504	AP-BEH
25996	RP-C4007
26855	EZ-A001
27426	PRP-001
27456	9M-CHG
27906	55-555
28866	921

747
21648	VP-BAT
21649	V8-AC1
21652	HZ-HM1B
21785	A4O-SO
21786	A7-AHM
21961	A6-SMR
21962	P4-AFE
21963	A6-SMM
21992	A4O-SP
22750	HZ-AIJ
23070	HZ-HM1A
23610	A6-ZSN
23824	82-8000
23825	92-9000
24730	20-1101

24731	20-1102
26426	V8-AL1
28288	V8-...

757
22690	XC-UJM
23454	P4-NSN
24527	HB-IEE
24923	N680FM
25155	N757AF
25220	VP-CAU
25345	EZ-A010
25487	TP-01
25495	HZ-HMED
28463	N757MA

767
25537	V8-MJB
27391	N767JB

DC-8
45570	F-RAFE
45658	VP-BIA
45692	9G-MKC
45819	F-RAFC
45984	FAP 371
46013	F-RAFG
46043	F-RAFD
46053	TR-LTZ
46067	VP-BJR
46071	VP-BLG
46078	FAP 370
46081	N728A
46084	HZ-HM11
46111	VP-BHM
46130	F-RAFF

DC-9
MD80 series
45702	XA-...
45706	N13FE
45731	N120NE
45732	N813TL
45740	N305PA
45775	N40SH
45797	N8860
45826	N29AF
45844	N949L
47011	N79SL
47151	VP-CKO
47152	N66AF
47466	N800DM
47467	71-0874
47475	71-0876
47514	N947ML
47536	71-0878
47541	71-0882
47595	MM62012
47600	MM62013
47635	N880RB
47668	73-1681
47670	73-1682
47671	73-1683
47690	KAF 321
49670	N3H
49725	VP-BOP
49767	N721EW
49778	VP-BOO
49809	9K-AGC

DC-10
46501	N220AU

Mcdonnell
1	N4AZ

Bombardier

Challenger 600

Serial	Reg	Serial	Reg
1002	144612	1045	C-GBKB
1004	N600BP	1046	N46SR
1005	N600CL	1047	N556WD
1006	144603	1048	C-GDDR
1007	144604	1049	N491DB
1008	144605	1050	N82CW
1009	144606	1051	N91UC
1010	N802Q	1052	N110TD
1011	N678ML	1053	N397BE
1012	N121TW	1054	N602AS
1013	N129BA	1055	N643CR
1014	144607	1056	N600TE
1015	144608	1057	N508HC
1016	N16TS	1058	N60HJ
1017	144609	1059	N3HB
1018	N771WW	1060	N74JA
1019	3B-GFI	1061	N770JC
1020	N600PD	1062	N68SD
1021	N63HJ	1063	N409CC
1022	144610	1064	N100LR
1023	N90UC	1065	144602
1024	N810MT	1066	D-BSNA
1025	HB-ILH	1067	N240AK
1026	N507HC	1068	N160LC
1027	N678CG	1069	N818TH
1028	N600BZ	1070	D-BUSY
1029	OH-WIH	1071	N220LC
1030	144611	1072	N125AC
1031	N620S	1073	N125AN
1032	N1884	1074	HZ-WT2
1033	N357RT	1075	N751DB
1034	N481JT	1076	N601WW
1035	N700CL	1077	N500R
1036	N900DP	1078	N1500
1038	N65HJ	1079	N600RE
1039	N722HP	1080	N3JL
1040	144601	1081	N19HF
1041	N193DQ	1082	N777KK
1042	N999TF	1083	N399FL
1043	N100QR	1084	N550CW
1044	N55AR	1085	OE-HET

Challenger 601

Serial	Reg	Serial	Reg
3001	N789DR	3039	N500PG
3002	N750GT	3040	I2+02
3003	C-GESR	3041	N600MS
3004	N45PH	3042	N333GJ
3005	N601TX	3043	I2+03
3006	N372G	3044	N125N
3007	N711SX	3045	N998JR
3008	N38SW	3046	N601TJ
3009	N373G	3047	N602TJ
3010	N601BD	3048	N601JM
3011	N7788	3049	I2+04
3012	XA-SHZ	3050	N601AE
3013	VP-BLA	3051	N4415D
3014	N292GA	3052	N801GC
3015	XA-KIM	3053	I2+05
3016	VP-BIE	3054	N315SL
3017	C-GJPG	3055	N608RP
3019	N375G	3056	I2+06
3020	C-GCFI	3057	9J-RON
3021	N503PC	3058	N125PS
3022	C-GCFG	3059	I2+07
3023	N601KF	3060	N601S
3024	N888AZ	3061	N597FJ
3025	N529DM	3062	N601GT
3026	N716HP	3063	C-FURG
3027	N17CN	3064	N224F
3028	C-FBEL	3065	N500PE
3029	N1824T	3066	N105UP
3030	N39CD	3991	C-GCGT
3031	I2+01		
3032	N392FV		
3033	HB-ILK		
3034	N120MP		
3036	144614		
3037	144615		
3038	144616		

Challenger 601-3A

Serial	Reg	Serial	Reg
5001	N245TT	5050	N831CJ
5002	N43PR	5051	N300KC
5003	HB-IKT	5052	N4PG
5004	N618DC	5053	N5PG
5005	N14GD	5054	N3FE
5006	C-FLPC	5055	N46F
5007	N17TE	5056	N153NS
5008	N601EG	5057	N733CF
5009	N399SW	5058	N101SK
5010	N1812C	5059	XA-JFE
5011	VP-CIC	5060	ZS-NKD
5012	N500LR	5061	N575MA
5013	N604MC	5062	N142B
5014	N311GX	5063	N801FL
5015	N200DE	5064	VH-BRG
5016	N49UR	5065	N882C
5017	N700KC	5066	N3PC
5018	N893AC	5067	9A-CRT
5019	N237GA	5068	C-FNNT
5020	I-BEWW	5069	N655CN
5021	N48FU	5070	N305FX
5022	N449ML	5071	N500PC
5023	N175ST	5072	N5073
5024	N604HJ	5073	N23SB
5025	N1TK	5074	PT-OSA
5026	N601WM	5075	N5TM
5027	N420SZ	5076	N1622
5028	N601RL	5077	N53DF
5029	N83LC	5078	VP-CCR
5030	N816SQ	5079	N135BC
5031	N908CL	5080	N601ST
5032	4X-COT	5081	N611GS
5033	VH-ASM	5082	N189K
5034	C-GIOH	5083	N399CF
5035	N333MG	5084	D-ACTU
5036	N468KL	5085	N343K
5037	N353TC	5086	XA-RZD
5038	N324SR	5087	C-GPOT
5039	N811BP	5088	N968L
5040	N652CN	5089	N400KC
5041	N352AF	5091	N915BD
5042	HB-IKS	5092	N300CR
5043	N601VH	5093	N875H
5044	VP-CMC	5094	TC-DHB
5045	N601AF	5095	N2FE
5046	N6SG	5096	C-GSQI
5047	N140CH	5097	XA-JJS
5048	N716RD	5098	N808G
5049	VP-CVK	5099	N801P
5100	N505M	5138	N85
5101	N105BN	5139	N34CD
5102	VP-CHK	5140	N79AD
5103	N76CS	5141	N601ER
5104	9M-SWG	5142	N330TP
5105	OK-BYA	5144	N616CC
5106	N523JM	5145	C-GDPF
5107	N417CL	5146	LX-MMB
5108	N428CL	5147	XA-SOR
5109	N439CL	5148	N793CT
5110	N308FX	5149	XA-GRB
5111	N4SG	5150	N601BW
5112	N109NC	5151	VP-BWB
5113	N163MR	5152	N605BA
5114	VP-BOA	5153	OY-APM
5115	N25SB	5154	9M-TAN
5116	N1904P	5155	N342TC
5117	N80BF	5156	N601TP
5118	N24JK	5157	N800KC
5119	VP-BNG	5158	XA-MKI
5120	N400TB	5159	EI-SXT
5121	N502PC	5160	N94BA
5122	N900CL	5161	N994CT
5123	N601UP	5162	VP-BCC
5124	C-FBOM	5163	N980HC
5125	N604WB	5164	N7008
5126	N21NY	5165	VP-CPO
5127	N718R	5166	9M-NSK
5128	XA-GME	5167	N151CC
5129	N129RH	5168	C-GRPF
5130	N601SR	5169	N154NS
5131	N6JB	5170	N888WS
5132	N289K	5171	N213MC
5133	N121DF	5172	N601FS
5134	N43RK	5173	N181JC
5135	N1902P	5174	N477DM
5136	N20G	5175	N306FX
5137	N137CL	5176	N600DR
5177	N757MC	5186	N9700X
5178	CS-MAC	5187	N601KJ
5179	N307FX	5188	HS-JJA
5180	N518CL	5189	PT-WLZ
5181	N602D	5190	N190EK
5182	HL7577	5191	N191BE
5183	N55HF	5192	N354TC
5184	N607RP	5193	VP-BCI
5185	N914X	5194	EI-MAS

Challenger 604

Serial	Reg	Serial	Reg
5301	N604CC	5330	N812G
5302	N255CC	5331	N810D
5303	HL7522	5332	VP-BNF
5304	I-MILK	5333	N811BB
5305	N604B	5334	N43R
5306	N309FX	5335	VP-CAN
5307	VP-BHA	5336	N310FX
5308	N604KS	5337	N270RA
5309	VP-BAC	5338	N913JB
5310	C-GPGD	5339	N604SA
5311	N225LY	5340	N194WM
5312	N312AM	5341	C-GBJA
5313	N312AT	5342	N311FX
5314	N604CT	5343	C-GHKY
5315	N818LS	5344	N344BA
5316	N411BB	5345	N345BA
5317	D-AMIM	5346	N604JP
5318	C-FYYH	5347	C-GBDK
5319	N14R	5348	N881TW
5320	HZ-AFA2	5349	N.....
5321	PT-WXL	5350	N331TP
5322	9A-CRO	5351	N374G
5323	N604DS	5352	C-GBKE
5324	N667LC	5353	C-GRIO
5325	N60CT	5354	N604PM
5326	N1903G	5354	C-GLYC
5327	HB-IKJ	5355	N555WD
5328	ZS-AVL	5356	N605PM
5329	N1GC	5357	N604FS
		5358	C-GBRQ

5359 N497DM	E2239 PK-PJJ	500-0083 VP-CHH	500-0174 N16LG	500-0263 VH-ZMD
5360 N606PM		500-0085 5N-BCI	500-0176 N150TT	500-0264 G-OEJA
5362 N607PM	**Cessna**	500-0086 LX-YKH	500-0177 N888XL	500-0265 N38MH
5991 C-FTBZ		500-0087 N500CP	500-0178 HB-VKK	500-0266 N11MN
	500 Citation 1	500-0088 PH-CTA	500-0179 PT-OMT	500-0267 N41SH
Challenger SE	500-0001 N715JS	500-0089 N39LH	500-0180 N500ET	500-0268 N900G
7075 N877SE	500-0002 ZS-ONE	500-0090 ETE-1329	500-0181 PT-KPA	500-0269 D-ICCC
7099 N305CC	500-0003 N503CC	500-0091 N500AD	500-0182 OO-DCM	500-0270 G-OSCA
7136 HB-IDJ	500-0004 N505K	500-0092 0222	500-0183 N112CP	500-0271 PT-LQG
7138 B-....	500-0005 PT-OIG	500-0093 N62BR	500-0184 N67BF	500-0272 N30JN
7140 M47-01	500-0008 PT-WBY	500-0094 N96FB	500-0185 ZP-...	500-0273 XB-GBF
7149 B-....	500-0009 N147WS	500-0095 I-AMAW	500-0186 N186SC	500-0274 N225BC
7180 B-....	500-0010 XC-DGA	500-0096 N837MA	500-0187 N5FW	500-0275 N71HB
7189 B-....	500-0011 N700VC	500-0098 PH-CTC	500-0188 PT-KPB	500-0276 SE-DEG
7193 B-....	500-0012 N512CC	500-0099 N21CC	500-0189 N500HH	500-0277 N652ND
	500-0013 XC-FIV	500-0100 N58BT	500-0190 N602BC	500-0278 XB-UAG
Global Express	500-0014 N18FM	500-0101 N15CQ	500-0191 N701BR	500-0279 N501KG
9001 C-FBGX	500-0015 PT-LPZ	500-0102 N491BT	500-0192 I-AMCY	500-0280 N102AD
9002 C-FHGX	500-0016 N9AX	500-0103 PT-KIR	500-0193 N293S	500-0281 N144JP
9003 C-FJGX	500-0017 N565SS	500-0104 N40HP	500-0194 PT-LAX	500-0282 N501CW
9004 C-FKGX	500-0018 C-FSKC	500-0105 N234UM	500-0195 N500LJ	500-0284 PT-LOG
9005 C-GEGX	500-0019 OB-1280	500-0106 ZS-RCC	500-0196 N270PM	500-0285 N113SH
9006 C-....	500-0021 XA-JLV	500-0107 N79RS	500-0197 N297S	500-0286 5N-APN
	500-0023 N523CC	500-0109 I-AMCU	500-0198 N700MP	500-0287 PT-WHZ
RJ	500-0024 N94AJ	500-0110 N172MA	500-0199 N199SP	500-0288 N1DA
7008 N5100X	500-0025 N57LL	500-0112 N500TM	500-0200 CS-DBM	500-0289 OE-FAN
	500-0027 PT-OVK	500-0113 N684H	500-0201 N690EA	500-0290 N896MA
British Aerospace	500-0028 N284AM	500-0114 N65SA	500-0202 N500WJ	500-0291 OE-FGN
	500-0029 N424DA	500-0115 YV-21CP	500-0203 CC-PZM	500-0292 N333JH
BAC 1-11	500-0030 N52AN	500-0116 D-IATC	500-0204 N928RD	500-0294 SE-DVB
015 N583CC	500-0031 YV-646CP	500-0117 N161WC	500-0206 N771HR	500-0295 N10FG
054 N17MK	500-0033 N990AL	500-0118 N50MM	500-0207 N39J	500-0296 ZS-NHF
060 HZ-ABM2	500-0034 N11HJ	500-0119 N95Q	500-0208 N501AT	500-0297 N38SA
061 N682RW	500-0035 YV-2479P	500-0120 N999TC	500-0209 N800AV	500-0298 N900GC
064 5N-HHH	500-0036 OO-LCM	500-0121 N3QE	500-0210 N501TK	500-0299 PT-OZX
067 N765B	500-0037 N109AL	500-0122 N122AP	500-0211 N999CV	500-0301 N305S
068 N263PC	500-0038 N27L	500-0123 ZK-LJL	500-0212 N92B	500-0302 N469PW
069 5N-CCC	500-0039 PT-OOK	500-0124 N92SM	500-0213 N101HG	500-0303 N8DX
072 N119GA	500-0040 N98Q	500-0125 XA-SFE	500-0214 N214CA	500-0304 N10UH
073 N401SK	500-0041 N50AM	500-0126 N404MA	500-0215 YV-55CP	500-0305 N137WC
076 N333GB	500-0042 C-FBCL	500-0127 N580R	500-0216 N199CK	500-0306 N36SJ
078 N62WH	500-0043 N502RL	500-0128 N3490L	500-0217 N55GR	500-0307 N777SL
079 N880DP	500-0044 N501WW	500-0129 N500SK	500-0218 N271AC	500-0308 F-GSMC
080 HZ-BL1	500-0045 N628BS	500-0130 N800AB	500-0219 PT-LIY	500-0309 N668S
081 VP-CCG	500-0046 PT-OTQ	500-0131 PT-OJF	500-0220 G-ORHE	500-0310 XA-STT
083 N490ST	500-0047 N7281Z	500-0132 N501RB	500-0221 N24AJ	500-0311 SE-DRT
084 XA-ADC	500-0048 N911GM	500-0133 PT-LXH	500-0222 N52PM	500-0312 F-GJDG
086 HR-AMO	500-0049 PT-LDH	500-0134 PT-JMJ	500-0223 OH-COC	500-0313 HB-VLE
087 N162W	500-0050 VH-HKX	500-0136 N136SA	500-0224 N145CM	500-0314 N668S
088 HZ-MAJ	500-0051 VH-ICX	500-0137 ZS-MCU	500-0226 PT-LTI	500-0315 SE-DRZ
089 TG-TJF	500-0052 YV-2477P	500-0138 N138SA	500-0227 N227HP	500-0316 N97SK
090 N164W	500-0053 I-AEAL	500-0139 N15AW	500-0230 N299TB	500-0317 N317VP
108 ZS-NNM	500-0054 N98MB	500-0140 N441TC	500-0231 N500SJ	500-0318 VP-COM
111 EL-LIB	500-0055 N716CB	500-0141 N727TK	500-0232 N999AM	500-0319 F-GKID
118 PK-T..	500-0056 N956S	500-0142 N200GM	500-0233 N228S	500-0320 N70WA
119 N114MX	500-0057 PT-ILJ	500-0143 SE-DUZ	500-0234 N70CA	500-0321 JA8438
120 N87BL	500-0058 YV-901CP	500-0145 N415FC	500-0235 N12AM	500-0322 N108MC
124 5N-NRC	500-0059 N40PD	500-0146 N111ME	500-0236 N337TV	500-0323 N307EW
125 5N-SDP	500-0060 PT-OOL	500-0147 N688CF	500-0238 OO-IBI	500-0324 N52TC
126 PK-TRS	500-0061 N52AJ	500-0148 LV-WXJ	500-0239 PT-LOS	500-0325 PT-OSD
154 5N-...	500-0062 N334RC	500-0149 ZS-TMG	500-0240 N234AT	500-0327 HL7277
158 HZ-KB1	500-0063 N70MG	500-0150 N914CD	500-0241 N288SP	500-0328 PT-LSF
160 5N-...	500-0064 N27SF	500-0152 XB-AMO	500-0242 RP-C1964	500-0329 OY-CEV
163 N123H	500-0065 N565CC	500-0153 N153JP	500-0243 N243SH	500-0330 N270BH
183 VP-CMI	500-0066 N66CC	500-0154 PT-WFT	500-0244 PT-OQD	500-0331 G-LOFT
259 PK-TAL	500-0067 N567EA	500-0155 N88GJ	500-0245 9M-FAZ	500-0332 LV-LZR
260 HZ-KA7	500-0068 XA-RYE	500-0156 PT-KBR	500-0246 PT-LQR	500-0333 VH-SOU
262 PK-TRU	500-0069 PT-IQL	500-0157 CS-DCA	500-0247 C-GMAJ	500-0334 N334JC
263 ZH763	500-0070 VP-CMO	500-0158 OY-TAM	500-0248 N111BB	500-0335 PT-LDI
	500-0071 ZS-AMB	500-0159 N97DD	500-0249 PT-LPF	500-0337 N22MB
BAe 146	500-0072 PT-OYA	500-0160 C-GNSA	500-0250 N251P	500-0338 N404JW
E1006 G-OFOA	500-0073 C-FKMC	500-0161 N161CC	500-0251 PT-OMS	500-0339 G-JEAN
E1021 ZE700	500-0074 PT-OOF	500-0162 N192G	500-0252 N501JC	500-0340 PT-WOD
E1029 ZE701	500-0075 N575CC	500-0164 F-GEPL	500-0253 N8TG	500-0341 N505JC
E1068 N861MC	500-0076 N500CV	500-0165 VP-CSP	500-0255 XA-SAM	500-0342 N501LH
E1091 A6-SHK	500-0077 ZS-OAM	500-0166 N8DE	500-0256 N131SB	500-0343 N91DZ
E1124 ZE702	500-0078 N269RC	500-0167 N246RR	500-0257 N75GW	500-0344 VP-BLV
E2050 PK-PJP	500-0079 PT-LBN	500-0168 N891CA	500-0258 N125DS	500-0345 N747RL
E2065 Z-WPD	500-0080 ZS-NGR	500-0169 N676CW	500-0259 RP-C1299	500-0346 PT-LUA
	500-0081 I-PEGA	500-0170 N66LE	500-0260 N50RD	500-0347 N876WB
	500-0082 N173HH	500-0171 PT-LIX	500-0261 N58TC	500-0348 C-GCLQ
		500-0173 N500EL	500-0262 N110AB	500-0349 VH-HVM

500-0353 N27WW	501-0045 N833JL	501-0130 N505CF	501-0216 TG-RIE	501-0684 N501TP
500-0354 G-TJHI	501-0046 N5VP	501-0131 YV-301P	501-0217 N500TW	501-0685 N501TB
500-0356 AE-185	501-0047 N550U	501-0132 N91WZ	501-0218 N218JG	501-0686 N750PP
500-0357 XB-GVY	501-0048 I-OTEL	501-0133 N51WP	501-0219 N510GA	501-0687 N154CC
500-0358 I-UUNY	501-0049 OY-SVL	501-0134 N501LW	501-0220 N100LX	501-0688 D-IMRX
500-0361 F-GKIR	501-0050 N59MA	501-0135 N63CG	501-0221 N221EB	501-0689 N88MM
500-0364 G-ORJB	501-0051 N303CB	501-0136 N66AG	501-0222 N25HS	
500-0367 YV-52CP	501-0052 N900MC	501-0137 N46SC	501-0223 N18HC	**525 CitationJet**
500-0368 N53RG	501-0053 N52TL	501-0138 N74FH	501-0224 N825PS	525-0001 N444RH
500-0369 C-GVER	501-0054 N2BT	501-0139 N501AF	501-0225 N412SE	525-0002 N137AL
500-0370 SE-DEY	501-0055 N145DF	501-0140 OE-FDM	501-0226 N501JM	525-0003 N44FJ
500-0373 YU-BKZ	501-0056 N56WE	501-0141 N501GG	501-0227 N83DM	525-0004 N7CC
500-0374 N501SS	501-0057 N577JT	501-0142 YV-688CP	501-0228 XA-SEY	525-0005 N56K
500-0375 N501NA	501-0058 N211X	501-0143 OY-ONE	501-0229 N57MC	525-0006 N106CJ
500-0378 C-GBNE	501-0059 ZS-EHL	501-0144 N270NF	501-0230 9A-DVR	525-0007 N525KN
500-0381 I-ROST	501-0060 N573L	501-0146 N194RC	501-0231 N501BP	525-0008 N1327G
500-0386 LQ-MRM	501-0061 N202CF	501-0147 N27TS	501-0232 VP-CAT	525-0009 N1327J
500-0387 N484KA	501-0062 N980DM	501-0148 N148ED	501-0233 5N-AVM	525-0010 ZS-MVX
500-0389 HC-BVP	501-0064 N12WH	501-0149 N97FD	501-0234 PT-LIZ	525-0011 N525AL
500-0392 I-FLYB	501-0065 N33WW	501-0150 N73FW	501-0235 N2617U	525-0012 N12PA
500-0395 XA-JEX	501-0066 N501CD	501-0151 N2690M	501-0236 N711VF	525-0013 N550T
500-0396 XC-GTO	501-0067 HB-VJB	501-0152 N15CV	501-0237 N831CB	525-0014 N620TC
500-0397 SE-DVA	501-0068 N68EA	501-0153 N484CS	501-0238 N737RJ	525-0015 PT-MPE
500-0399 YU-BML	501-0069 N636N	501-0154 N154SC	501-0239 OE-FMS	525-0016 D-IKOP
500-0401 I-FARN	501-0070 N628ZG	501-0155 VP-CJB	501-0240 N77UB	525-0017 N525AE
500-0402 N301EL	501-0071 N501KR	501-0156 N44FM	501-0241 N101RR	525-0018 N525MC
500-0404 G-ZAPI	501-0072 N1HA	501-0157 N16VG	501-0242 N71L	525-0019 N525KA
500-0405 OY-JEY	501-0073 N299RP	501-0158 N501WB	501-0243 N43SP	525-0020 N1329D
500-0406 OY-FFB	501-0074 N28GC	501-0160 N999PW	501-0244 N176FB	525-0021 N793CJ
500-0408 XB-DVF	501-0075 N51ET	501-0161 XB-FXO	501-0245 A2-AGM	525-0022 G-BVCM
500-0409 XC-FEZ	501-0076 N150RM	501-0162 N44SW	501-0246 OE-FHH	525-0023 N525RF
500-0410 XC-GAW	501-0077 N678JD	501-0163 I-CIGB	501-0248 N7700T	525-0024 F-GNCJ
500-0411 SE-DLZ	501-0078 N13BT	501-0164 N286PC	501-0249 N851BC	525-0025 D-IOBO
500-0412 XB-GDJ	501-0079 N33CX	501-0165 N501RC	501-0250 N77PX	525-0026 N525FS
500-0413 PT-LCC	501-0080 N37LA	501-0166 OY-INI	501-0251 N400LX	525-0027 N825GA
500-0415 JA8474	501-0081 N12CV	501-0167 N900TW	501-0252 I-TOIO	525-0028 G-OICE
500-0418 ZS-LDV	501-0082 N5WF	501-0168 N328NA	501-0253 N501JE	525-0029 D-IWHL
500-0476 N501BA	501-0083 N101LD	501-0169 C-GFEE	501-0254 N84CF	525-0030 N177RE
	501-0084 VP-CDM	501-0170 N610TT	501-0255 N901NB	525-0031 N831S
500 Citation 1/SP	501-0085 N707W	501-0171 VH-BNK	501-0256 I-PAPE	525-0032 N95DJ
501-0001 N506TF	501-0086 N854TT	501-0172 HI-581SP	501-0257 N12NM	525-0033 N472SW
501-0002 N88TB	501-0087 N501BB	501-0173 N25GT	501-0258 VP-BHI	525-0034 N96G
501-0003 N81EB	501-0088 N23YZ	501-0174 N454DQ	501-0259 D-IEIR	525-0035 N525HS
501-0004 N142DA	501-0089 N588CA	501-0175 VP-BKP	501-0260 N500GA	525-0036 N525MB
501-0005 N284RJ	501-0090 N3GN	501-0176 VP-CFG	501-0261 N501VP	525-0037 HB-VKB
501-0006 N93TJ	501-0091 N2158U	501-0177 N501GW	501-0262 N58T	525-0038 N600HR
501-0007 N222WA	501-0092 N501X	501-0178 N83ND	501-0263 N501BE	525-0039 N39CJ
501-0008 N6HT	501-0093 N501RM	501-0179 N718SA	501-0267 N565VV	525-0040 VP-CCC
501-0009 N505RJ	501-0094 N159LC	501-0180 N650MW	501-0270 N893CA	525-0041 D-IAMM
501-0010 EC-EDN	501-0095 XA-AGA	501-0181 N501KK	501-0272 N700JA	525-0042 N96GD
501-0011 N650AC	501-0096 N501HS	501-0182 D-ISIS	501-0273 N501EM	525-0043 N525PL
501-0012 N501GS	501-0097 N501RS	501-0183 CS-AYY	501-0275 ZS-MPN	525-0044 N55DG
501-0013 N501TJ	501-0098 HB-VIC	501-0184 N90CF	501-0279 SE-DEZ	525-0045 N525AP
501-0014 N888FL	501-0099 I-FLYA	501-0185 N501LL	501-0281 N501SJ	525-0046 N123JN
501-0015 N454AC	501-0100 C-GSUM	501-0186 N999WS	501-0282 N82AJ	525-0047 N47FH
501-0016 N45TL	501-0101 N106EA	501-0187 N614DD	501-0283 N204CA	525-0048 N525NA
501-0017 N877C	501-0102 N501CG	501-0188 N61DT	501-0284 N729PX	525-0049 N49CJ
501-0018 N228AJ	501-0103 N611AT	501-0189 N80SF	501-0285 N13ST	525-0050 N70KW
501-0019 N501SP	501-0104 N33HC	501-0190 N723JR	501-0292 N456R	525-0051 N808HS
501-0020 N123EB	501-0105 N2648X	501-0191 N64RT	501-0294 N80SL	525-0052 N52PK
501-0021 N151SP	501-0106 D-IANE	501-0192 N190K	501-0297 N41GT	525-0053 N60ES
501-0023 N56MK	501-0107 EC-GJF	501-0193 N45MK	501-0298 N501D	525-0054 N54CJ
501-0024 Z-WSY	501-0108 N777GG	501-0194 N28JG	501-0314 N56LW	525-0055 N923AR
501-0025 N20RM	501-0109 N567WB	501-0195 N123KD	501-0317 5N-AVL	525-0056 JA8420
501-0026 N92CC	501-0111 N94SL	501-0196 N311TP	501-0319 N60EW	525-0057 N525DG
501-0027 JA8380	501-0112 N74PM	501-0197 N324L	501-0320 N401G	525-0058 N525CK
501-0028 N501PV	501-0113 N502CC	501-0198 N800DT	501-0321 N550MH	525-0059 N71GW
501-0029 N45AQ	501-0114 N811HL	501-0199 N62RG	501-0322 N527EW	525-0060 XA-TRI
501-0030 N911MM	501-0115 N728MC	501-0200 N47TL	501-0323 N501LE	525-0061 N525PS
501-0031 XA-SDI	501-0116 N7TK	501-0201 N55BM	501-0324 JA8493	525-0062 C-GINT
501-0032 N85WP	501-0117 N91AP	501-0202 N501G	501-0325 N64BH	525-0063 N55SK
501-0033 N411DS	501-0118 N61572	501-0204 N345N	501-0446 N888MJ	525-0064 D-IHEB
501-0034 N444MV	501-0119 N53EZ	501-0205 PT-LVB	501-0643 N54TS	525-0065 EC-FZP
501-0035 N501DD	501-0120 OY-CPW	501-0206 N314EB	501-0675 N501MT	525-0066 N420CH
501-0036 N360MC	501-0121 OE-FHW	501-0207 OE-FYC	501-0676 N676CC	525-0067 N525TF
501-0037 N234JW	501-0122 N501MB	501-0208 N82P	501-0677 N54CG	525-0068 N303LC
501-0038 N315S	501-0123 N627L	501-0209 N98RG	501-0678 PT-ODC	525-0069 N20FL
501-0039 N432DG	501-0124 N125EA	501-0210 N32FM	501-0679 VP-BLF	525-0070 D-ISGW
501-0040 I-KWYJ	501-0125 N505JH	501-0212 N2243	501-0680 PP-EIF	525-0071 N940SW
501-0042 N120DP	501-0126 N96SK	501-0213 I-AUNY	501-0681 N82LS	525-0072 TG-RIF
501-0043 N16NL	501-0127 N501CF	501-0214 N2296S	501-0682 N682DC	525-0073 N77794
501-0044 N50US	501-0129 PT-LQQ	501-0215 ZS-LXT	501-0683 N501FR	525-0074 N511TC

Reg		Reg		Reg		Reg		Reg	
525-0075	N719L	525-0157	N57HC	525-0242	N242LJ	550-0079	N232DM	550-0180	N3030C
525-0076	N80TF	525-0158	N749CP	525-0244	N66ES	550-0080	N45ME	550-0181	RP-C653
525-0077	N1000E	525-0159	N131RG	525-0246	N525EC	550-0081	N254AM	550-0182	N165MC
525-0078	N525CH	525-0160	N66AM	525-0247	N533JF	550-0082	N49U	550-0183	HB-VGS
525-0079	N525WB	525-0161	N525BT	525-0248	N248CJ	550-0083	N54CC	550-0184	N20TV
525-0080	N33DT	525-0162	N1XT	525-0250	N250CJ	550-0084	N521WM	550-0185	N511DR
525-0081	N181JT	525-0163	N51CD	525-0251	N50ET	550-0085	OE-GBA	550-0186	N80AW
525-0082	D-IHHS	525-0164	D-ICGT	525-0253	N8940	550-0086	XC-JCY	550-0187	C-GHOM
525-0083	N121CP	525-0165	D-IJYP	525-0255	N514DS	550-0089	N81CC	550-0188	HB-VIZ
525-0084	D-ITSV	525-0166	N343PJ	525-0256	N196HA	550-0090	N290VP	550-0189	D-CCCF
525-0085	PT-MJC	525-0167	N4EF	525-0257	N9003	550-0091	N601BC	550-0190	F-BTEL
525-0086	PT-MIL	525-0168	D-IRON	525-0258	N108CR	550-0092	N567CA	550-0191	N550PA
525-0087	N175PS	525-0169	N68CJ	525-0261	N31HD	550-0093	N388MA	550-0192	YV-900CP
525-0088	N722SG	525-0170	N170BG	525-0265	N198JH	550-0094	G-JETA	550-0193	N72FL
525-0089	N920MS	525-0171	N97VF			550-0095	N550CG	550-0194	N91B
525-0090	N8288R	525-0172	D-IAVB			550-0096	N87SF	550-0195	N61CK
525-0091	N295DS	525-0173	N970SU	**550 Citation II**		550-0097	N202CE	550-0196	HB-VLS
525-0092	N525AS	525-0174	N417C	550-0001	N560CC	550-0098	N212H	550-0197	HB-VIT
525-0093	I-IDAG	525-0175	N41PG	550-0004	F-GNCP	550-0099	N109JC	550-0198	XA-SGV
525-0094	N94MZ	525-0176	PT-WLX	550-0005	N77ND	550-0102	VH-JPG	550-0199	N586RE
525-0095	N61SH	525-0177	G-OSCB	550-0006	N725RH	550-0103	N90MA	550-0200	N810MC
525-0096	D-ICEE	525-0178	N525RC	550-0007	N550TY	550-0104	301	550-0201	N1GH
525-0097	N234WS	525-0179	N377GS	550-0008	N70X	550-0105	N550LH	550-0202	N175VB
525-0098	N511AC	525-0180	N123AV	550-0009	N656PS	550-0106	N308CK	550-0203	N766AF
525-0099	N525SC	525-0181	N88LD	550-0010	PT-LPK	550-0108	VP-CBE	550-0204	N815CE
525-0100	VH-CIT	525-0182	N177JB	550-0011	0002	550-0109	N753CC	550-0205	N30JD
525-0101	F-GPFC	525-0183	N399G	550-0012	N11FH	550-0110	N122G	550-0206	N280TA
525-0102	TC-CRO	525-0184	N525J	550-0013	PT-LML	550-0111	N123VP	550-0207	N196RJ
525-0103	D-IVHA	525-0185	N83TR	550-0014	N780CF	550-0112	N3FA	550-0209	N101KK
525-0104	N606MM	525-0186	N92ND	550-0016	HC-BTJ	550-0113	C-GDMF	550-0210	XA-KMX
525-0105	N305CJ	525-0187	N696ST	550-0017	N771ST	550-0114	N991BM	550-0211	N77PR
525-0106	N21VC	525-0188	D-IVIN	550-0018	N752CC	550-0115	SE-DDY	550-0212	N245CC
525-0107	N525WC	525-0189	N189CM	550-0021	N52RF	550-0116	N413CA	550-0213	N550HB
525-0108	N108CJ	525-0190	N41NK	550-0024	N413CK	550-0117	N550RB	550-0214	N75ZA
525-0109	N37DG	525-0191	N525BL	550-0025	9H-ACR	550-0118	N118EA	550-0215	N40MT
525-0110	N195ME	525-0192	N84FG	550-0026	N30AV	550-0121	N51FT	550-0216	N550PG
525-0111	N776DF	525-0193	D-I...	550-0027	G-BFRM	550-0122	C-GCUL	550-0217	N217SA
525-0112	VT-OPJ	525-0194	N81RA	550-0028	N310AV	550-0123	LN-NLA	550-0218	PT-LPP
525-0113	N111AM	525-0195	N525DC	550-0029	N524MA	550-0124	N109GA	550-0219	N550RP
525-0114	N96GM	525-0196	D-IURH	550-0031	N22GA	550-0125	5N-NPF	550-0220	N4CS
525-0115	OO-PHI	525-0197	N525KH	550-0032	N112JS	550-0127	G-GAUL	550-0221	N253W
525-0116	N41EB	525-0198	N315MR	550-0033	LX-GDL	550-0129	N122HM	550-0222	PT-LTJ
525-0117	N26CB	525-0199	9A-CAD	550-0034	N60CC	550-0130	N77RC	550-0223	N239CD
525-0118	D-IRWR	525-0200	N226B	550-0035	N74G	550-0131	C-GGSP	550-0224	2222
525-0119	N47TH	525-0201	N525HV	550-0036	N789BR	550-0132	PT-LLU	550-0225	PT-LTJ
525-0120	PT-WGD	525-0202	N747AC	550-0037	N535MA	550-0133	N228CC	550-0226	N772HP
525-0121	TC-EMA	525-0203	N33FW	550-0038	N642CC	550-0135	YV-888CP	550-0227	PT-LND
525-0122	N102AF	525-0204	B-....	550-0040	N545GA	550-0138	XC-SCT	550-0228	VH-EXM
525-0123	D-IRKE	525-0205	N550MC	550-0041	OE-GCI	550-0139	PT-ORD	550-0229	C-FGAT
525-0124	N525JH	525-0206	N17VB	550-0042	C-FPEL	550-0140	N55WL	550-0230	N550HB
525-0125	N525PE	525-0207	N31SG	550-0043	N801JP	550-0141	VH-ING	550-0231	N41SM
525-0126	D-IFUP	525-0208	N208JV	550-0044	N308TW	550-0142	PT-LCR	550-0232	N797CW
525-0127	N63LB	525-0209	D-ILAT	550-0046	C-GRHC	550-0143	N150RD	550-0234	C-FLPD
525-0128	N535LR	525-0210	N999EB	550-0047	N44AS	550-0144	RP-C689	550-0235	I-PNCA
525-0129	N229CJ	525-0211	D-IMMD	550-0048	N19ER	550-0145	P2-MBN	550-0236	LN-AAD
525-0130	N416KC	525-0212	N67VW	550-0050	F-GMCI	550-0146	N501LC	550-0237	ZS-NHO
525-0131	N577SW	525-0213	N525WM	550-0051	ZS-RKV	550-0147	N80GM	550-0238	N97S
525-0132	N132AH	525-0214	D-IFAN	550-0052	PH-CTZ	550-0149	N116K	550-0239	N66MC
525-0133	EC-GIE	525-0215	N18GA	550-0053	TC-NKB	550-0150	N1SV	550-0241	XC-BCS
525-0134	N525JM	525-0216	N28GA	550-0054	VH-WGJ	550-0151	N35HC	550-0242	N55SC
525-0135	N888RA	525-0217	D-IEWS	550-0055	N10EG	550-0152	N550CA	550-0243	XA-REN
525-0136	N525KL	525-0218	N64LF	550-0056	N89D	550-0153	N37MH	550-0245	N388SB
525-0137	N810SS	525-0219	N219CJ	550-0057	VH-WNZ	550-0154	G-JETJ	550-0247	PT-LJJ
525-0138	VH-MOJ	525-0220	N220CJ	550-0058	N100HB	550-0155	N155TJ	550-0248	N550SA
525-0139	N76AE	525-0221	D-IWIL	550-0060	OE-GIL	550-0156	N205SC	550-0249	PT-LZO
525-0140	N725L	525-0222	N111LR	550-0061	N458N	550-0157	N101BX	550-0250	N250VP
525-0141	N774CA	525-0223	D-IGAS	550-0062	C-GDLR	550-0158	N49HS	550-0251	N550HF
525-0142	N815MC	525-0224	N224CJ	550-0064	N64TF	550-0159	N444GB	550-0252	N6JL
525-0143	D-IALL	525-0225	N525RM	550-0065	N144GA	550-0162	N1UA	550-0253	N202TS
525-0144	D-IDAG	525-0226	N1216N	550-0066	N10JP	550-0164	N916RC	550-0254	N888RT
525-0145	N145CJ	525-0227	N741CC	550-0067	N74TC	550-0165	ZS-LHU	550-0255	I-JESO
525-0146	N1329G	525-0228	N668VP	550-0068	N402ST	550-0166	N367JC	550-0256	N75TP
525-0147	N996JR	525-0229	D-IHOL	550-0069	C-FFCL	550-0167	N100CJ	550-0257	N187TA
525-0148	N148CJ	525-0230	N323LM	550-0070	N550KA	550-0168	ZS-NII	550-0258	N550PR
525-0149	N67GH	525-0231	N606JR	550-0071	N404BF	550-0169	XC-JDA	550-0259	OY-BZT
525-0150	ZS-NUW	525-0232	N525PF	550-0072	N770JM	550-0170	N500CV	550-0260	N6HF
525-0151	N7EN	525-0233	N53CG	550-0073	VP-CTJ	550-0171	N19AJ	550-0261	N50AD
525-0152	N152KC	525-0234	N900P	550-0074	D-CIFA	550-0172	N412P	550-0263	N777WY
525-0153	N551G	525-0235	VP-BZZ	550-0075	N710MT	550-0174	N87PT	550-0264	N6862Q
525-0154	N401LG	525-0237	N237CJ	550-0076	VH-XDD	550-0175	N10JK	550-0266	N15NA
525-0155	I-EDEM	525-0240	N525GM	550-0077	XA-AGN	550-0176	N61MA	550-0267	N910RB
525-0156	N156ML	525-0241	N207BS	550-0078	N533MA	550-0179	N673LR	550-0269	N28RC

550-0271 PT-ORO	550-0383 N551AB	550-0489 N15RL	550-0630 N220AB	550-0712 PH-LAB	
550-0272 VP-BVV	550-0390 VT-VPS	550-0490 N490CC	550-0631 XA-ICP	550-0713 N95HE	
550-0273 N217FS	550-0393 I-FLYD	550-0491 I-AVRM	550-0632 5N-AYA	550-0714 N593EM	
550-0274 HB-VKT	550-0397 N114DS	550-0492 I-AVGM	550-0633 N7AB	550-0715 N715AB	
550-0275 N555DS	550-0398 PH-CTX	550-0493 N84GC	550-0634 SE-DVT	550-0716 VP-CTE	
550-0276 C-GGFW	550-0402 N717PC	550-0494 N1254X	550-0635 PT-OAA	550-0717 N600GH	
550-0277 N44LF	550-0403 N362CP	550-0495 PT-LLQ	550-0636 N50NF	550-0718 XA-FIR	
550-0279 N566CC	550-0405 YV-604P	550-0497 N1257B	550-0637 YV-376CP	550-0719 N550BG	
550-0280 N7SN	550-0406 C-FTAM	550-0498 N78FK	550-0638 N255TC	550-0720 N848HS	
550-0281 N33EK	550-0407 N950FC	550-0499 PT-LIV	550-0639 N100DS	550-0721 N721CC	
550-0282 G-JCFR	550-0408 N110WA	550-0501 N12549	550-0640 PT-ODL	550-0722 N1886G	
550-0283 TC-BAY	550-0409 N102HB	550-0502 12-001	550-0641 PT-WON	550-0723 N888NA	
550-0284 OY-JEV	550-0410 C-GNWM	550-0503 12-002	550-0642 XA-SEX	550-0724 LV-WEJ	
550-0285 N989TW	550-0411 C-FMPP	550-0504 N72SL	550-0643 N643MC	550-0725 N222FA	
550-0286 N2GG	550-0412 N830VL	550-0505 N1255K	550-0644 XC-PGM	550-0726 XT-AOK	
550-0287 N771AA	550-0414 N342CC	550-0550 HK-4128W	550-0645 PT-ODZ	550-0727 N521BH	
550-0289 N22HP	550-0415 F-GJYD	550-0553 N5XR	550-0646 N9VF	550-0728 LV-WJO	
550-0290 N312NC	550-0416 N416CC	550-0554 N40YC	550-0647 N205BE	550-0729 VP-CBM	
550-0291 N41C	550-0417 N17DM	550-0558 LV-WJN	550-0648 XC-PGP	550-0730 N2NT	
550-0292 C-FTOM	550-0418 N550J	550-0560 ZS-LNP	550-0649 N44LC	550-0731 XA-...	
550-0294 PT-LPN	550-0419 G-FJET	550-0561 PT-OYP	550-0650 N824CT	550-0732 N101AF	
550-0295 N800LA	550-0421 N510GP	550-0562 PT-OJT	550-0651 N24E	550-0733 C-GFCI	
550-0296 G-BJIR	550-0423 C-GUUU	550-0563 G-THCL	550-0652 N3262M	550-0734 N550BB	
550-0297 B-4104	550-0424 N551BP	550-0565 N565NC	550-0653 N30RL	550-0801 N801BB	
550-0299 N511AB	550-0425 U 20-1	550-0566 N15SP	550-0654 XB-BON	550-0802 N550HH	
550-0300 YV-162CP	550-0426 N404SB	550-0567 N41BH	550-0655 N37201	550-0803 N550FB	
550-0301 B-4103	550-0427 PT-LHY	550-0568 N988RS	550-0656 N30GR	550-0804 N550BC	
550-0302 N792MA	550-0428 N97BG	550-0569 5Y-...	550-0657 CC-DGA	550-0805 N108RF	
550-0303 XB-GLZ	550-0430 HB-VLY	550-0570 N570WD	550-0658 RP-C1180	550-0806 N300PY	
550-0304 N42PH	550-0432 I-ASAZ	550-0571 N90JJ	550-0659 N4614N	550-0807 C-FANS	
550-0305 B-4105	550-0433 N7ZU	550-0572 N193SS	550-0660 D-CILL	550-0808 SE-DVZ	
550-0307 N550VW	550-0434 N53FP	550-0573 N155AC	550-0661 N847HS	550-0809 N300AK	
550-0308 N15XM	550-0435 N674G	550-0575 N387RE	550-0662 N911CB	550-0810 VH-MGC	
550-0310 N530P	550-0436 N711Z	550-0576 N576CC	550-0663 XC-JCX	550-0811 PT-WNH	
550-0311 N2RC	550-0438 N643TD	550-0577 N120HC	550-0664 N70PC	550-0812 C-FJBO	
550-0312 N61CF	550-0439 N511WS	550-0578 PT-LQJ	550-0665 N665MC	550-0813 N100KU	
550 0315 N59GU	550-0440 OY-CYV	550-0579 N579L	550-0666 N5408G	550-0814 PT-WNH	
550-0316 N129TS	550-0441 HB-VKS	550-0580 N912BD	550-0667 VP-CWM	550-0815 N126TF	
550-0318 PT-WJZ	550-0442 N442ME	550-0581 N905LC	550-0668 N1879W	550-0816 C-FMCI	
550-0319 N78CK	550-0443 D-CGAS	550-0582 FAC-1211	550-0669 N846HS	550-0817 N817CB	
550-0320 N800VJ	550-0444 N47SW	550-0583 N12L	550-0670 XC-JCZ	550-0818 LV-WYH	
550-0321 N321GN	550-0445 N453S	550-0585 N79SE	550-0671 G-BWOM	550-0819 ZP-...	
550-0323 TC-YZB	550-0446 U 20-2	550-0586 F-GGGA	550-0672 PT-OMB	550-0820 N820CB	
550-0324 HB-VLQ	550-0447 HB-VIS	550-0587 N18HJ	550-0673 N763L	550-0821 N77797	
550-0326 N390JP	550-0448 N39HD	550-0588 N92BD	550-0674 N1883M	550-0822 N822CB	
550-0327 PT-LLT	550-0449 YV-2338P	550-0589 N679BC	550-0675 PT-WKQ	550-0823 N25FS	
550-0329 N491N	550-0450 N15EA	550-0590 N673CA	550-0676 PT-OJG	550-0824 N824CB	
550-0332 N120Q	550-0451 N1249P	550-0592 U 20-3	550-0677 N6775C	550-0825 N25HV	
550-0333 PT-LCW	550-0452 N707PE	550-0593 N26621	550-0678 EC-FES	550-0826 N595PC	
550-0334 N404KS	550-0453 N962J	550-0594 N2351K	550-0679 I-FJTO	550-0827 D-CCAB	
550-0335 N204AB	550-0454 N93BD	550-0595 XC-JCV	550-0680 N6776T	550-0828 N6FR	
550-0336 N90Z	550-0456 C-GMPQ	550-0596 D-CAWA	550-0681 N1200N	550-0829 N829CB	
550-0337 N406SS	550-0457 N63TM	550-0597 N213JS	550-0682 N90BL	550-0830 N830KE	
550-0339 VH-SCD	550-0458 N25MK	550-0598 N7WY	550-0683 YV-701CP	550-0831 N331PR	
550-0340 N38DD	550-0459 N15TV	550-0599 N571BC	550-0684 C-FJXN	550-0834 N834CB	
550-0341 C-FDYL	550-0460 PT-OKP	550-0600 PT-LSR	550-0685 C-FJWZ	550-0835 N835CB	
550-0343 N789TT	550-0461 N22FM	550-0601 G-OCDB	550-0686 C-FKCE	550-0840 N442SW	
550-0344 PT-LKR	550-0462 N501DK	550-0602 N2663Y	550-0687 C-FKDX	550-0841 N841WS	
550-0345 N267TC	550-0464 N117TA	550-0603 C-GMSM	550-0688 C-FKEB	550-0842 N86AJ	
550-0346 N106SP	550-0465 N387HA	550-0604 N827JB	550-0689 XA-JYO	550-0844 N550KL	
550-0347 VH-ZLE	550-0466 N10LY	550-0605 N26494	550-0690 OE-GLZ	550-0854 N550KH	
550-0348 C-GSCX	550-0467 YV-713CP	550-0606 N602AT	550-0691 C-GAPD		
550-0349 N870PT	550-0468 N120JP	550-0607 XC-JCW	550-0692 N75RJ	**S550 Citation II**	
550-0350 N86SG	550-0469 HB-VIP	550-0608 N990M	550-0693 VP-BTR	S550-0001 N86BA	
550-0351 I-ALKA	550-0470 N10RU	550-0609 F-GLTK	550-0694 N550KE	S550-0002 CC-CWW	
550-0352 VT-EUN	550-0471 N787WC	550-0610 9M-NSA	550-0695 N870WC	S550-0003 N847C	
550-0353 N477KM	550-0472 HZ-AFP	550-0611 F-GGGT	550-0696 N67PC	S550-0004 N72AM	
550-0354 VP-CJR	550-0473 HZ-AFQ	550-0612 N578M	550-0697 ZS-NFL	S550-0005 N666LN	
550-0355 N355DF	550-0474 ZS-LIG	550-0613 PT-OAC	550-0698 N550RM	S550-0006 N181G	
550-0356 PT-OER	550-0475 N870MH	550-0615 N577VM	550-0699 C-FKLB	S550-0007 N30CX	
550-0357 PT-OAG	550-0476 N1252D	550-0616 PT-OVV	550-0700 C-FJCZ	S550-0008 N40KM	
550-0358 4400	550-0477 N47TW	550-0618 PP-ESC	550-0701 C-FLZA	S550-0009 N550A	
550-0362 N396M	550-0478 N214RW	550-0619 N550BD	550-0702 C-FMFM	S550-0010 XA-INF	
550-0363 HK-3400	550-0479 N45NS	550-0620 PT-LYA	550-0703 N308A	S550-0011 N211QS	
550-0364 C-GLTG	550-0480 N335CC	550-0621 N99TK	550-0704 N197HF	S550-0012 N550RV	
550-0365 N100AY	550-0482 N594G	550-0622 HB-VKP	550-0705 N521TM	S550-0013 N389L	
550-0366 N614GA	550-0483 N89LS	550-0623 N89LS	550-0706 7Q-YLF	S550-0014 N214QS	
550-0367 N3MB	550-0484 N84EA	550-0624 PT-LYS	550-0707 RP-C4654	S550-0015 C-GMTV	
550-0368 N718CK	550-0485 PT-WBV	550-0625 N6846T	550-0708 N720WC	S550-0016 N85MP	
550-0369 N725BA	550-0486 A4O-SC	550-0626 LV-WOZ	550-0709 N709VP	S550-0017 N86PC	
550-0371 N551MC	550-0487 N550DW	550-0627 N17FL	550-0710 N90BJ	S550-0018 N814CC	
550-0378 OH-CAT	550-0488 N84EB	550-0628 IGM-628	550-0711 N58LC	S550-0019 N600VE	

S550-0020 N550AS	S550-0103 N103QS	551-0046 N81GD	560-0004 N189H	560-0086 SE-DPG
S550-0021 N693M	S550-0104 N224KC	551-0049 N391BC	560-0005 N953F	560-0087 N600BW
S550-0022 N360M	S550-0105 N105BG	551-0050 N228AK	560-0006 N566VP	560-0088 OE-GSW
S550-0023 N94RT	S550-0106 N106QS	551-0051 D-ICTA	560-0007 N717MB	560-0089 ZS-MPT
S550-0024 PT-LGI	S550-0107 N550HT	551-0053 N66MS	560-0008 N561B	560-0090 N30PC
S550-0025 N24PH	S550-0108 N108QS	551-0056 I-NIAR	560-0009 VH-HEY	560-0091 N3GT
S550-0026 N27FP	S550-0109 N38EC	551-0060 N59GB	560-0010 N205PC	560-0092 XA-RTT
S550-0027 HB-VHH	S550-0110 N45GP	551-0065 N99DE	560-0011 N700TF	560-0092A N592VP
S550-0028 N608LB	S550-0111 N111QS	551-0066 N242WT	560-0012 N560ME	560-0093 F-GKJL
S550-0029 N999HC	S550-0112 N112QS	551-0071 N790D	560-0013 N560WH	560-0094 N340DR
S550-0030 N50BK	S550-0113 PT-LJQ	551-0077 N501KC	560-0014 N12ST	560-0095 N404G
S550-0031 CS-DNA	S550-0114 PT-LKS	551-0084 ZP-TYO	560-0015 N800DL	560-0096 N10TD
S550-0032 N531CM	S550-0115 C-FDDD	551-0095 N402TJ	560-0016 N462B	560-0097 N898CB
S550-0033 N59EC	S550-0116 N550HC	551-0117 OO-SKS	560-0017 N560H	560-0098 N59DF
S550-0034 XA-THO	S550-0118 VH-IWU	551-0122 N10LR	560-0018 N114CP	560-0099 OE-GPA
S550-0035 N63JG	S550-0119 N11TS	551-0132 N78GA	560-0019 N61TW	560-0100 N560WE
S550-0036 N573BB	S550-0120 N716DB	551-0133 HB-V..	560-0020 N560HC	560-0101 N560DM
S550-0037 N3D	S550-0121 N23NM	551-0143 C-GLMK	560-0021 N560DC	560-0102 N560BA
S550-0038 N22UL	S550-0122 N122WS	551-0149 N550JS	560-0022 N211MA	560-0103 N98E
S550-0039 C-FEMA	S550-0123 N121CG	551-0163 ZS-ARG	560-0023 N31RC	560-0104 N400CT
S550-0040 N74BJ	S550-0124 N52CK	551-0169 N14RM	560-0024 N501QS	560-0105 N560MH
S550-0041 N250AL	S550-0125 N97CT	551-0171 5R-MHF	560-0025 CNA-NV	560-0106 N525RD
S550-0042 N727NA	S550-0126 N126QS	551-0174 EI-CIR	560-0026 N350RD	560-0107 N78NP
S550-0043 N92ME	S550-0127 N674JM	551-0179 N127BU	560-0027 N560JR	560-0108 N777KY
S550-0044 T9-BIH	S550-0128 N370M	551-0180 N166MA	560-0028 N6FE	560-0109 N109VP
S550-0045 N103VF	S550-0129 N480CC	551-0181 N29B	560-0029 N590A	560-0109A N560RS
S550-0046 N16RP	S550-0130 N550PL	551-0191 D-IVOB	560-0030 XB-MTS	560-0110 N832QB
S550-0047 N705SP	S550-0131 N87BA	551-0205 HB-VIO	560-0031 D-CHDE	560-0111 OE-GAA
S550-0048 B-4101	S550-0132 N91ML	551-0214 N9SS	560-0032 N96MT	560-0112 N4110S
S550-0049 B-4102	S550-0133 N431WM	551-0215 N500PX	560-0033 C-GAPC	560-0113 N4
S550-0050 CS-DNB	S550-0134 N134QS	551-0223 YV-.....	560-0034 N895LD	560-0114 D-CZAR
S550-0051 N57BJ	S550-0135 VT-KMB	551-0285 ZS-MLN	560-0035 N561EJ	560-0115 N91YC
S550-0052 N393E	S550-0137 PT-WIB	551-0289 PT-LJF	560-0036 N532MA	560-0116 N901RM
S550-0053 N999CB	S550-0138 N552BE	551-0296 N551TK	560-0037 N17LK	560-0117 D-CMEI
S550-0054 N87FL	S550-0139 N39TF	551-0304 N304KT	560-0038 N301QS	560-0118 XA-SKX
S550-0055 N550F	S550-0140 N575EW	551-0311 N38TT	560-0039 CNA-NW	560-0119 F-GLIM
S550-0056 N1UL	S550-0141 N550AJ	551-0313 YV-2426P	560-0040 N71NK	560-0120 N1824S
S550-0057 N633EE	S550-0142 C-FALI	551-0323 N555RT	560-0041 VH-NTH	560-0121 N898GF
S550-0058 PT-LHD	S550-0143 N1VA	551-0335 D-ILCC	560-0042 D-CAWU	560-0122 N510MT
S550-0059 N260QS	S550-0144 N543SC	551-0341 N550MT	560-0044 N111VP	560-0123 N611ST
S550-0060 N200LX	S550-0145 PH-RMA	551-0345 N960CP	560-0045 PT-LZQ	560-0124 OB-1626
S550-0061 I-AVVM	S550-0146 N815H	551-0351 N5TR	560-0046 G-CZAR	560-0125 OE-GCC
S550-0062 VH-EMO	S550-0147 OO-OSA	551-0355 I-ALPG	560-0047 N500FK	560-0126 LV-RED
S550-0063 N200CV	S550-0148 D-CSFD	551-0359 N67983	560-0048 N57CE	560-0127 N64HA
S550-0064 N995DC	S550-0149 N43RC	551-0360 N551EA	560-0049 N560EL	560-0128 N85KC
S550-0065 N711MD	S550-0150 N107RC	551-0361 LV-APL	560-0050 N208PC	560-0129 N22AF
S550-0066 N900DM	S550-0151 N151Q	551-0369 N68BK	560-0051 N314RW	560-0130 N14VF
S550-0067 N7070A	S550-0152 N848G	551-0378 F-OGVA	560-0052 N500UB	560-0131 PT-ORE
S550-0068 N43VS	S550-0153 N242LA	551-0393 N122SP	560-0053 C-GCUW	560-0132 HZ-SFA
S550-0069 N570RC	S550-0154 N910DS	551-0396 LV-WXD	560-0054 N100SC	560-0133 N88HF
S550-0070 N1865M	S550-0155 N155GB	551-0400 D-IMME	560-0055 N21JJ	560-0134 YV-811CP
S550-0071 TC-...	S550-0156 N766NB	551-0412 OY-PDN	560-0056 N78AM	560-0135 N560RL
S550-0072 N1958N	S550-0157 N157QS	551-0419 N555KT	560-0057 N561BC	560-0136 N772AA
S550-0073 N274QS	S550-0158 N66EH	551-0421 D-IAWA	560-0058 F-GKGL	560-0137 N193G
S550-0074 N882RB	S550-0159 N9GT	551-0428 N1218S	560-0059 F-GKHL	560-0138 N511WV
S550-0075 N89TD	S550-0160 PT-OSM	551-0431 N4MM	560-0060 N90MF	560-0139 N75F
S550-0076 CS-DNC		551-0436 N11SS	560-0061 D-CNCI	560-0140 N75G
S550-0077 ZS-CAR	**551 Citation II/SP**	551-0455 PT-MMO	560-0062 EC-GLM	560-0141 N141AQ
S550-0078 N97LB	551-0002 N39ML	551-0463 YV-05C	560-0063 N7FE	560-0142 PT-WPC
S550-0079 N269MT	551-0003 I-MESK	551-0481 CC-LLM	560-0064 ZS-MVZ	560-0143 N65HA
S550-0080 N550KM	551-0004 YV-19CP	551-0496 N999GH	560-0065 N77711	560-0144 N2000X
S550-0081 N282QS	551-0006 YV-O-CVG-2	551-0500 N501MC	560-0066 N60S	560-0145 N57MK
S550-0082 N282QS	551-0007 YV-05CP	551-0551 N600AT	560-0067 JA119N	560-0146 N2000M
S550-0083 PT-LJL	551-0009 D-IMTM	551-0552 OE-FPA	560-0068 N246NW	560-0147 N125RH
S550-0084 N143BP	551-0010 N321CA	551-0555 N560CB	560-0069 N357WC	560-0148 N560FB
S550-0085 N86QS	551-0015 YV-213CP	551-0556 N200GF	560-0070 F-GJXX	560-0149 N565JW
S550-0086 N21EG	551-0016 N900SE	551-0557 N711NV	560-0071 N271CA	560-0150 D-CTAN
S550-0087 N288QS	551-0017 N811VG	551-0559 N1AT	560-0071A N45RC	560-0151 V5-CDM
S550-0088 VT-ETG	551-0018 D-IEAR	551-0574 OE-FBS	560-0072 N72CT	560-0152 ZS-NDX
S550-0089 N97BP	551-0020 N477A	551-0584 N25QT	560-0073 N100WP	560-0153 N1SN
S550-0090 N241LA	551-0021 N55LS	551-0591 N1AT	560-0074 N593MD	560-0154 N154VP
S550-0091 N92QS	551-0022 N30EJ	551-0614 D-ILAN	560-0075 N75CV	560-0155 N40WP
S550-0092 N93QS	551-0023 PT-LME	551-0617 D-ILTC	560-0076 N777FH	560-0156 N75B
S550-0093 N6LL	551-0026 N25NH	552-0012 N12566	560-0077 HB-VLV	560-0157 N502TS
S550-0094 N200NK	551-0027 N522CC	552-0014 N12568	560-0078 OY-CKT	560-0158 N601AB
S550-0095 N29XA	551-0029 D-ICUR		560-0079 N560GL	560-0159 N68MA
S550-0096 N551BE	551-0031 N5T	**560**	560-0080 JA8576	560-0160 ZS-NDT
S550-0097 N98QS	551-0032 N551TT	**Citation V / Ultra**	560-0081 N318CT	560-0161 XA-CYC
S550-0098 N299QS	551-0035 N277JM	560-0001 N1217V	560-0082 N950WA	560-0162 XA-SDT
S550-0099 N300QW	551-0036 N317SM	560-0002 N90PG	560-0083 N22LP	560-0163 N953C
S550-0100 N101QS	551-0038 N103M	560-0003 N560ER	560-0084 C-GHEC	560-0164 N392BS
S550-0101 N287MC	551-0039 N71LP		560-0085 N591M	560-0165 N910V

RA-02801   British Aerospace HS 125/700B   Photo by: Harald Helbig

HB-VLT   British Aerospace BAe 125/800B   Photo by: Stefan Pfäffli

OM-SKY   Raytheon Hawker 800XP   Photo by: Wolfgang Zilske

C-GNAZ   British Aerospace HS 125/700A   Photo by: Jean-Luc Altherr

CN-TDE   Aerospatiale Corvette   Photo by: Jerôme Alché

TC-MSA   Raytheon Beechjet 400A   Photo by: Wolfgang Zilske

S5-BAC   Cessna Citation II   Photo by: Wolfgang Zilske

9H-ACR   Cessna Citation II   Photo by: Jean-Luc Altherr

HB-VLQ   Cessna Citation II   Photo by: Jean Luc Altherr

HB-VHH   Cessna Citation S/II   Photo by: Stefan Pfäffli

D-CZAR   Cessna Citation V   Photo by: Harald Helbig

N10JM   Cessna Citation X   Photo by: Nick Dean

D-ICEE   Cessna Citation Jet   Photo by: Stefan Pfäffli

OE-GAA   Cessna Citation V   Photo by: Steve Kinder

N444GG   Cessna Citation V Ultra   Photo by: Elliot Greenman

LX-TRG   Dassault Falcon 10   Photo by: Hansjörg Pfäffli

D2-ESV   Dassault Falcon 20F-5   Photo by: Frederic Vergneres

EP-FIF   Dassault Falcon 20E   Photo by: Wolfgang Zilske

XA-TDU   Dassault Falcon 2000   Photo by: Vicki Mills

HB-IBH   Dassault Falcon 2000   Photo by: Hansjörg Pfäffli

N33GQ   Dassault Falcon 50   Photo by: Vicki Mills

VP-CTT　　　Dassault Falcon 900B　　　Photo by: Jean-Luc Altherr

EC-GMO　　　Dassault Falcon 900EX　　　Photo by: Vicki Mills

N22CS　　　Dassault Falcon 900EX　　　Photo by: Hansjörg Pfäffli

A9C-BA   Boeing B727-2M7   Photo by: Manuel Negrerie

HZ-DGI   Boeing B727-51   Photo by: Manuel Negrerie

N511DB   Boeing B727-89   Photo by: Manuel Negrerie

HB-IEE   Boeing B757-23A   Photo by: Jean-Luc Altherr

N303NT   Rockwell Sabreliner T-39N   Photo by: Nick Dean

VP-BRL   Lockheed JetStar-731   Photo by: Jean-Luc Altherr

HZ-MSD   Gulfstream Aerospace G-2   Photo by: Wolfgang Zilske

N802GA   Gulfstream Aerospace G-3   Photo by: Harald Helbig

VP-BSS   Gulfstream Aerospace G-4   Photo by: Frederic Vergneres

N917W   Gulfstream Aerospace G-4   Photo by: Frederic Vergneres

165153   Gulfstream Aerospace C-20G   Photo by: Andreas Durr

N166RM   Israel Aircraft Industries Astra 1125-SP   Photo by: Jean-Luc Altherr

HZ-WT2   Canadair Challenger 600S   Photo by: Hansjörg Pfäffli

144611   Canadair Challenger CC144   Photo by: Andreas Durr

D-AMIM   Bombardier Challenger 604   Photo by: Anton Heumann

HZ-AFA2   Bombardier Challenger 604   Photo by: Jean-Luc Altherr

C-FZQP   Gates Learjet 35A   Photo by: Karel Jaeger

D-CION   Gates Learjet 55   Photo by: Harald Helbig

652   Beech King Air 200C   Photo by: Martin Pole

N8E   Grumman Gulfstream G-1   Photo by: Jean-Luc Altherr

D-ICIR   Beech King Air B200   Photo by: Harald Helbig

D-CSKY   Beech King Air 350   Photo by: Wolfgang Zilske

HC-DAC   Beech King Air E90   Photo by: Wolfgang Zilske

D-IHRA   Piaggio P-180 Avanti   Photo by: Wolfgang Zilske

560-0166 ZS-NDW	560-0248 N226N	560-0330 N351WC	560-0412 PT-WNF	650-0030 SE-DHL
560-0167 N311DG	560-0249 N733M	560-0331 N331QS	560-0413 VP-CKM	650-0031 N1ZC
560-0168 N168CV	560-0250 N205CM	560-0332 N332LC	560-0414 ZS-CDS	650-0034 N650GH
560-0169 N80AB	560-0251 LV-WGO	560-0333 N333QS	560-0415 96-0109	650-0035 N408JD
560-0170 N814CM	560-0252 N44GT	560-0334 N4TL	560-0416 N19PV	650-0036 XA-TGA
560-0171 N573F	560-0253 N46MT	560-0335 N335QS	560-0417 RP-C8818	650-0037 I-GASD
560-0172 N172CV	560-0254 N560GB	560-0336 N336QS	560-0418 N318QS	650-0038 N366G
560-0173 N68881	560-0255 ZS-NHD	560-0337 N108LJ	560-0419 EC-GOV	650-0039 N39WP
560-0174 N563C	560-0256 N22KW	560-0338 N592M	560-0420 96-0110	650-0040 N650WE
560-0175 N49LD	560-0257 N155PT	560-0339 N339QS	560-0421 N322QS	650-0041 N55BH
560-0176 PT-WOM	560-0258 PT-OZB	560-0340 N21CV	560-0422 N5XP	650-0042 C-GPOP
560-0177 N242AC	560-0259 N37WP	560-0341 N341QS	560-0423 N324QS	650-0043 OY-GKL
560-0178 N531CC	560-0260 C-GPAW	560-0342 XA-RDM	560-0424 N424CV	650-0044 N650M
560-0179 N865M	560-0261 XB-PYC	560-0343 N60AE	560-0425 N325QS	650-0045 N67SF
560-0180 N550WW	560-0262 N444GG	560-0344 N344QS	500-0427 N11LC	650-0046 N57TT
560-0181 N181SG	560-0263 N979C	560-0345 N75Z	560-0428 N328QS	650-0047 N33BQ
560-0182 N920PM	560-0264 N264U	560-0346 C-GFCL	560-0429 N392QS	650-0048 N650MM
560-0183 N83RR	560-0265 LV-WIJ	560-0347 N72FC	560-0430 N433CV	650-0049 PT-LSN
560-0184 N410DM	560-0266 N456JW	560-0348 N348QS	560-0431 PT-WZW	650-0050 N44M
560-0185 N29WE	560-0267 N96NB	560-0349 JA001A	560-0432 N356WC	650-0051 N651BH
560-0186 N586CC	560-0268 N269CM	560-0350 N250JH	560-0433 N33LX	650-0052 XA-SDU
560-0187 N60GL	560-0269 N357EC	560-0351 N699CC	560-0434 N334QS	650-0053 N367G
560-0188 N395R	560-0270 N68HC	560-0352 N352QS	560-0435 N410NA	650-0054 N17AN
560-0189 N62HA	560-0271 PH-VLG	560-0353 N1873	560-0436 N36LX	650-0055 N16AS
560-0190 LV-VFY	560-0272 N220JT	560-0354 N4200K	560-0437 N337QS	650-0056 N397CS
560-0191 N2JW	560-0273 N910PC	560-0355 N67GW	560-0438 N5207A	650-0057 N101YC
560-0192 N713HH	560-0274 TG-...	560-0356 N354QS	560-0439 VP-CSC	650-0058 N292PC
560-0193 TR 20-02	560-0275 D-CVHA	560-0357 N81SH	560-0440 N5076K	650-0059 PT-LHA
560-0194 N352WC	560-0276 N376QS	560-0358 N12TV	560-0441 N314QS	650-0060 TC-CAO
560-0195 PT-ORC	560-0277 D-CFOX	560-0359 ZS-SMB	560-0442 N23UD	650-0061 N650TC
560-0196 N4JS	560-0278 VH-FHJ	560-0360 N62WA	560-0443 N24UD	650-0062 N650CN
560-0197 XA-SJC	560-0279 N361EC	560-0361 N361QS	560-0444 N343QS	650-0063 N72LE
560-0198 N560RG	560-0280 PH-MDC	560-0362 OE-GMI	560-0445 N345QS	650-0064 N444CW
560-0199 N63HA	560-0281 N511ST	560-0363 N59KG	560-0446 HB-VLZ	650-0065 C-FIMO
560-0200 OE-GDA	560-0282 D-CBEN	560-0364 N991PC	560-0447 N261WR	650-0066 N650CD
560-0201 ZS-NGM	560-0283 N1CH	560-0365 N375CM	560-0448 C GSUN	650-0067 N210F
560-0202 ZS-NGL	560-0284 N966SW	560-0366 SX-DCI	560-0449 N560BP	650-0068 N9KL
560-0203 N7700L	560-0285 N147VC	560-0367 N367QS	560-0451 N351QS	650-0069 N910M
560-0204 N1000W	560-0286 N57MB	560-0368 N343CC	560-0453 N453CV	650-0070 LN-NLD
560-0205 F-GLYC	560-0287 N117CC	560-0369 N520G	560-0455 N358QS	650-0071 N334H
560-0206 N560TX	560-0288 N522JA	560-0370 N607RJ	560-0459 N79PM	650-0072 N72ST
560-0207 N52SN	560-0289 LV-WLS	560-0371 N315CS	560-0460 N360QS	650-0073 N85DA
560-0208 N88G	560-0290 N97BH	560-0372 N76CK	560-0464 N420DM	650-0074 N194DC
560-0209 N209CV	560-0291 N744R	560-0373 N373QS	560-0465 N456CV	650-0075 N16AJ
560-0210 N277RC	560-0292 HL7501	560-0374 N166KB	560-0466 N366QS	650-0076 N424LB
560-0211 N250SP	560-0293 N293QS	560-0375 N375QS	560-0470 N44FG	650-0077 TC-EES
560-0212 TC-LAA	560-0294 HL7502	560-0376 N1217H	560-0475 N374QS	650-0078 N50DS
560-0213 PT-OTS	560-0295 N61HA	560-0377 N377QS	560-0493 N391QS	650-0079 N217RR
560-0214 OE-GCP	560-0296 N560LC	560-0378 N350WC	560-0496 N395QS	650-0080 N69LD
560-0215 PT-OTT	560-0297 HL7503	560-0379 C-GWCR	560-5001 N561XL	650-0081 PT-LGT
560-0216 TC-LAB	560-0298 N200CK	560-0380 N190KL	560-5009 N398RS	650-0082 N4VF
560-0217 N602AB	560-0299 VT-EUX	560-0381 N857BL		650-0083 TC-TOP
560-0218 XA-SIT	560-0300 HL7504	560-0382 N382QS	**650 Citation III**	650-0084 N431CB
560-0219 N318MM	560-0301 N560AG	560-0383 N57ST	650-0001 N651CC	650-0085 I-CIST
560-0220 N73KH	560-0302 N560CE	560-0384 N196SA	650-0002 N5000C	650-0086 PT-LHC
560-0221 N701DK	560-0303 N560BJ	560-0385 N333WM	650-0003 OY-CCG	650-0087 N680BC
560-0222 N1286C	560-0304 N47VC	560-0386 N720SJ	650-0004 N650GT	650-0088 N590AS
560-0223 N575PC	560-0305 LV-WMT	560-0387 95-0123	650-0005 N439H	650-0089 N86WP
560-0224 N523KW	560-0306 N49MJ	560-0388 N92SS	650-0006 N650TS	650-0090 N651TC
560-0225 N1823S	560-0307 N307QS	560-0389 PT-WRR	650-0007 C-FLTL	650-0091 PT-LUE
560-0226 N893CM	560-0308 PT-WFD	560-0390 C-GMGB	650-0009 N926CB	650-0092 N692BE
560-0227 LV-WDR	560-0309 N212BD	560-0391 N391CV	650-0009 N933SH	650-0093 N222GT
560-0228 XA-SLA	560-0310 N868JT	560-0392 95-0124	650-0010 N650LW	650-0094 N650SP
560-0229 XA-SNX	560-0311 N311QS	560-0393 N393QS	650-0011 N17TN	650-0095 N882KB
560-0230 YV-169CP	560-0312 N312QS	560-0394 N394QS	650-0012 N15VF	650-0096 N700SW
560-0231 N12CQ	560-0313 N313CV	560-0395 N19MK	650-0013 N313QS	650-0097 N697MC
560-0232 N502E	560-0314 C-FYMM	560-0396 N396QS	650-0014 N855DH	650-0098 N54HC
560-0233 0233	560-0315 N315QS	560-0397 PT-WKS	650-0015 N766MH	650-0099 N26SD
560-0234 N234AQ	560-0316 N12RN	560-0398 ZS-NUZ	650-0016 N32MG	650-0100 N202JK
560-0235 N129PJ	560-0317 N317QS	560-0399 N97NB	650-0017 N900CM	650-0101 XA-LYH
560-0236 N506E	560-0318 N877RF	560-0400 N916CS	650-0018 N650SB	650-0102 N406LM
560-0237 N593M	560-0319 LV-WOE	560-0401 VP-CSN	650-0019 XA-TBA	650-0103 N407LM
560-0238 N194SA	560-0320 VP-CCV	560-0402 N302QS	650-0020 N10PN	650-0104 I-BETV
560-0239 N93CV	560-0321 N320QS	560-0403 JA01TM	650-0021 N2604	650-0105 N48TT
560-0240 N966JM	560-0322 ZS-NVV	560-0404 96-0107	650-0022 N650J	650-0106 N650CE
560-0241 ZS-NGS	560-0323 N323QS	560-0405 PT-WMQ	650-0023 N650CG	650-0107 N650MP
560-0242 N605AT	560-0324 N55LC	560-0406 PT-WMZ	650-0024 N650SL	650-0108 N650Z
560-0243 N39N	560-0325 N96AT	560-0407 F-OHRU	650-0025 N16SU	650-0109 N134MJ
560-0244 N60RD	560-0326 N583M	560-0408 PT-WOA	650-0026 N626CC	650-0110 N303PC
560-0245 N65AT	560-0327 N327QS	560-0409 PT-WVH	650-0027 N650NY	650-0111 N500CM
560-0246 LV-WGY	560-0328 N554R	560-0410 96-0108	650-0028 LN-NLC	650-0112 N93DK
560-0247 N94TX	560-0329 N330QS	560-0411 PT-WNE	650-0029 N70TT	

650-0113 N652JM	650-0195 C-GAPT	650-7032 XA-XIS	750-0007 N750EC	21 N40WJ
650-0114 N651AF	650-0196 N534H	650-7033 PT-OVU	750-0008 N353WC	22 F-GJLL
650-0115 N1419J	650-0197 N797T	650-7034 XA-SWM	750-0009 N909QS	23 N20WP
650-0116 C-FJJC	650-0198 N553AC	650-7035 N95RX	750-0010 N5112	24 N991RV
650-0117 F-GGAL	650-0199 N65KB	650-7036 N77HF	750-0011 N944H	25 N177BC
650-0118 N770MP		650-7037 N95TX	750-0012 N966H	26 N720DF

650 Citation VI

650-0119 8P-KAM	650-0201 N40PH	650-7038 N399W	750-0013 N5113	27 N38DA
650-0120 N1223N	650-0202 PT-OJO	650-7039 D-CACM	750-0014 N757T	28 N655DB
650-0121 N24VB	650-0203 N4612Z	650-7040 N504T	750-0015 N715CX	29 N999F
650-0122 N65WL	650-0204 N108WV	650-7041 N430SA	750-0016 N206PC	31 N27AJ
650-0123 N434H	650-0205 PT-OMU	650-7042 N657T	750-0017 N5114	32 32
650-0124 N650HC	650-0206 N39H	650-7043 TC-ATC	750-0018 N5115	33 TC-ORM
650-0125 N650AF	650-0207 N107CG	650-7044 N7005	750-0019 N5116	34 N220M
650-0126 N101PG	650-0208 N500FR	650-7045 CC-PGL	750-0020 N8JC	35 N777JJ
650-0127 N18PV	650-0209 N650L	650-7046 N746BR	750-0021 N630M	36 XA-MMM
650-0128 N125Q	650-0210 N733H	650-7047 N1828S	750-0022 N10JM	37 N945MC
650-0129 PT-LUO	650-0211 N211CC	650-7048 N18GB	750-0023 N923QS	38 F-GBRF
650-0130 N159MR	650-0212 N805GT	650-7049 N900FL	750-0024 N924EJ	40 XA-LIO
650-0131 303	650-0213 N900JD	650-7050 N33GK	750-0025 N750RL	41 N61TJ
650-0132 N24KT	650-0214 TC-CEY	650-7051 N965JC	750-0026 N926QS	42 N282T
650-0133 ZK-NLJ	650-0215 N650KC	650-7052 N24NB	750-0027 N354WC	43 N17TJ
650-0134 N123SL	650-0216 I-BLUB	650-7053 N650AS	750-0028 N100FF	44 N277SF
650-0135 N5109	650-0217 PH-MEX	650-7054 LV-WTN	750-0029 N500RP	45 C-FTEN
650-0136 N779AZ	650-0218 XA-GAN	650-7055 N317M	750-0030 N355WC	46 N908RF
650-0137 N874G	650-0219 G-HNRY	650-7056 N60PL	750-0031 N22RG	48 LX-EPA
650-0138 N828G	650-0220 B-4106	650-7057 N157CM	750-0032 N932QS	49 PT-LMO
650-0139 N96CP	650-0221 B-4107	650-7058 N625CC	750-0033 N710AW	50 N411SC
650-0140 N650SS	650-0222 N733K	650-7059 N4EG	750-0034 N934QS	51 N51BP
650-0141 TC-CMY	650-0223 N111Y	650-7060 N55SC	750-0035 N97DK	52 N711TF
650-0142 N20RD	650-0224 N1UP	650-7061 N903SB	750-0036 N936QS	53 N53WA
650-0143 N40FC	650-0225 N606AT	650-7062 N876G	750-0037 N75HS	54 N561D
650-0144 N2605	650-0226 N1302A	650-7063 N877G	750-0038 N739CX	55 N700AL
650-0145 N29AU	650-0227 N2UP	650-7064 HB-VLP	750-0039 N750LM	56 N56WJ
650-0146 N650FC	650-0228 XA-SLB	650-7065 N650W	750-0040 N68LP	58 N500FF
650-0147 N456AF	650-0229 TC-ANT	650-7066 N766CG	750-0041 N98TX	59 N633WW
650-0148 N50PH	650-0230 N616AT	650-7067 N502T	750-0042 N95CM	60 N69WJ
650-0149 N139N	650-0231 LV-WHY	650-7068 N111HZ	750-0045 N45BR	62 N6VG
650-0150 N150F	650-0232 N517MT	650-7069 XA-TMX	750-0049 N949QS	63 N876MA
650-0151 PK-KIG	650-0233 CC-DAC	650-7070 N322RG	750-0059 N751BH	64 N444WJ
650-0152 N627R	650-0234 TC-SBH	650-7071 HS-DCG	750-0065 N965QS	65 N66CF
650-0153 N47CM	650-0235 N235SV	650-7072 N35HS	750-0071 N971QS	66 N63TS
650-0154 N650CH	650-0236 N600JD	650-7073 N1867M	750-0077 N977QS	67 N427CJ
650-0155 N788NB	650-0237 N650MC	650-7074 PT-WLY	750-0079 N979QS	68 F-GFPF
650-0156 N74VF	650-0238 N19UC	650-7075 N711GF	750-0087 N987QS	69 N711JC
650-0157 N657CC	650-0239 N17UC	650-7076 N286MC	750-0089 N989QS	70 VP-BCH
650-0158 N135HC	650-0240 PH-MFX	650-7077 N532JF	750-0093 N993QS	71 N190H
650-0159 N267TG	650-0241 N651JM	650-7078 N788R		72 N50TY
650-0160 N33UL		650-7079 N779QS		73 N378C
650-0161 N510SD	**650 Citation VII**	650-7080 CS-DNF	**Chichester-Miles**	74 N5JY
650-0162 N275GC		650-7081 CS-DNG		75 N796SF
650-0163 N137MR	650-7001 N404JF	650-7082 N782QS	**Leopard**	76 N727TS
650-0164 N138MR	650-7002 N19SV	650-7085 N785QS	001 G-BKRL	77 N107TB
650-0165 XC-PGN	650-7003 N888TX	650-7086 N860W	002 G-BRNM	78 N83MD
650-0166 PT-LTB	650-7004 N913SQ	650-7087 N787QS		79 N73B
650-0168 N175J	650-7005 N200LH	650-7089 N789QS	**Convair**	80 N803RA
650-0169 N88JJ	650-7006 N706VP	650-7090 N790QS		81 N700BD
650-0170 N32JJ	650-7007 N28TX	650-7091 N791QS	55 N42	82 N602NC
650-0171 PT-LVF	650-7008 N902SB	650-7092 N792QS		83 N76MB
650-0172 N934H	650-7009 N93TX	650-7094 N794QS	**Falcon**	84 N106TW
650-0173 N173VP	650-7010 N317MZ	650-7096 N796QS		87 N99BL
650-0174 D-CLUE	650-7011 SE-DVY	650-7097 N797QS	**Falcon 10**	88 F-GHER
650-0175 N235KK	650-7012 N317MB	650-7100 N710QS	**Falcon 100**	89 N888WJ
650-0176 N2TF	650-7013 N5118	650-7103 N713QS	02 F-ZACB	90 N12TX
650-0177 N834H	650-7014 N864EC	650-7105 N715QS	1 F-BJLH	91 N790US
650-0178 TC-RAM	650-7015 N5119	650-7106 N716QS	2 C-GRIS	92 N724DS
650-0179 XA-RMY	650-7016 N68SK	650-7109 N709QS	3 N52TJ	94 N13BK
650-0180 N498CS	650-7017 N775M		4 XA-SYY	96 N174FJ
650-0181 PT-OBX	650-7018 N119RM	697 N650	5 F-BVPR	97 N175FJ
650-0182 N491JB	650-7019 XA-TCZ	701 N501CC	6 N59CC	99 F-GKBC
650-0183 N820FJ	650-7020 N832CB	702 N525CJ	7 HB-VKE	100 N100FJ
650-0184 N1128B	650-7021 PT-MGS	703 N750CX	8 N108KC	101 101
650-0185 N533CC	650-7022 N902RM	704 N526JT	9 N149TJ	102 N61BP
650-0186 N186VP	650-7023 N6110	705 N526JP	11 N211TJ	103 N339TG
650-0187 N78PT	650-7024 ETI-024	706 N560XL	12 N10F	104 N913VL
650-0188 N587S	650-7025 N442WT		14 N59TJ	105 N16DD
650-0189 XA-RGS	650-7026 ETI-026	**750 Citation X**	16 N416AS	106 N20CF
650-0190 N190JJ	650-7027 N657ER	750-0001 N751CX	17 F-GNDZ	107 N100T
650-0191 TC-KLS	650-7028 XA-SPQ	750-0002 N902QS	18 N1TJ	109 N89EC
650-0192 N15TT	650-7029 XA-SOK	750-0003 N1AP	19 LX-TRG	110 N104DD
650-0193 N650CC	650-7030 N95CC	750-0004 N597U	20 N42G	111 N10HE
650-0194 N2606	650-7031 N40N	750-0005 N99BB		112 N12MB
		750-0006 N76D		113 LX-DPA

114 N108GD
115 I-ITPR
116 F-GJMA
117 N18MX
118 N118AD
119 N257W
120 N402JW
121 F-GDLR
122 OE-GSC
123 SE-DKC
125 XA-SAR
126 N26WJ
127 N8GA
128 N99BC
129 129
130 N432EZ
131 N654PC
133 133
134 VH-MCX
135 N245SP
136 F-GFMD
137 C-GTVO
138 N236DJ
139 N110J
140 F-GHDX
141 N77SF
142 N174B
143 143
144 N144HE
145 N244A
146 N461AS
147 N212FJ
148 N79TJ
149 N711FJ
150 N99WA
151 N256W
152 SE-DPK
153 N81P
154 N149HP
155 D-CIEL
156 ZS-SEA
157 N703JS
158 N700FH
159 N707AM
160 LX-JCG
161 I-CREM
162 N47RK
163 N163AV
164 N222MU
165 N56LP
166 N211EC
167 N516SM
168 N43EC
169 PT-WSF
171 N42US
172 N10NC
173 N8LT
174 RP-C1911
175 N12EP
176 N231JH
177 F-GFGB
178 N87TH
179 F-GERO
180 N593DC
181 N138DM
182 N809F
183 SE-DLB
184 N725DM
185 185
186 N420PC
187 N303PL
188 N84TJ
189 N812KC
190 N190L
193 N3BY
194 F-GIPH
195 N95WJ
196 N125CA
197 N52N
198 N1PB
199 N96VR
200 N80BL

201 N844F
202 N80WJ
203 XA-TBL
204 XA-TAB
205 N606AM
206 N46MK
207 N456CM
208 N71M
209 HB-VKR
210 N85WN
211 F-GELT
212 CN-TNA
213 F-GKAE
214 N147G
215 F-GHPB
216 N999WJ
217 N68GT
218 F-GHSK
219 PT-ORS
220 N702NC
221 F-GPFD
222 N100CK
223 PT-LVD
224 SE-DVP
225 PT-LXJ
226 N121AT

Falcon 20
Falcon 200
Gardian
Guardian

2 F-BMSS
3 N301R
5 F-GJPR
7 N600JC
8 N190BD
9 LV-WMF
11 N216CA
13 N977TW
14 N41MH
15 N1501
16 N120AF
17 N55TH
19 C-GKHA
20 G-FRAJ
21 XA-SWC
22 F-ZACS
23 5761
24 N204JP
25 N813AA
26 N819AA
27 N174GA
28 N333AV
29 LV-WMM
30 F-GPIM
31 N828AA
32 F-GIVT
35 9M-BCR
36 N818AA
39 XA-RMA
40 C-GSKQ
41 041
44 N377BT
45 N175GA
46 EC-EHC
48 N91CV
49 F-RHFA
50 N669TW
52 N825TC
53 053
54 N205TS
55 UR-EFA
56 N560RA
57 N812AA
59 N771LD
61 N20NY
63 D-CBNA
64 N513AG
65 C-GSKN
66 N181RB
67 N826AA

68 N458SW
69 N31LT
71 N818CP
72 F-GJCC
74 N702DM
75 UR-EFB
76 F-GJDB
77 N613GA
79 F-ZACT
80 N24TW
81 N810RA
82 G-FRAS
83 N20PL
85 N6555L
86 F-ZACG
87 G-FRAT
88 N617GA
89 N505AJ
90 PK-...
91 N777DC
93 F-RAED
94 N.....
95 N950RA
96 F-ZACB
97 G-FRAU
98 N980R
99 N921ML
100 N179GA
101 N97WJ
102 N403JW
103 F-GPAA
104 F-ZACW
106 EC-EKK
107 N213LS
108 N108R
109 C-FIGD
111 N111BP
112 UR-CCD
113 N129JE
114 G-FRAW
115 F-UKJG
116 F-GLMM
117 F-GLMD
118 N820AA
119 N20FJ
120 N647JP
121 N500BG
122 N900LC
123 N45MR
124 F-ZACC
125 0125
126 N126R
127 XB-HRA
128 N70CK
129 N68TS
130 XA-SCL
131 F-ZACD
132 G-FFRA
133 N200JE
135 N194MC
136 SP-FCP
137 N200GT
138 F-ZACR
139 N235CA
141 UR-CCB
142 XA-RNB
143 5A-DAG
144 N911RG
145 F-ZACU
146 N182GA
147 N183GA
148 N148WC
149 EC-EQP
150 HC-BSS
151 G-FRAL
152 CNA-NN
153 N207CA
155 N68BC
156 C-GRSD
157 C-GRSD
158 D-CMAX
159 N96RT
160 N100UF

161 N10RZ
162 N911DG
163 N178GA
164 N654E
165 CNA-NM
166 N33D
167 F-RAEB
168 N112CT
169 XC-SEY
170 F-GHPA
171 N217AJ
172 I-LIAB
173 PK-TRI
174 HZ-NES
175 SU-OAE
176 F-GHDT
177 N41BP
178 G-FRBA
179 N17JT
180 OY-BDS
181 N817JS
182 F-UKJA
183 RA-09003
184 OE-GCJ
185 N147X
186 F-UKJE/463
187 N811AA
188 F-ZACX
190 5A-DCO
191 OE-GCR
192 N192R
193 N219CA
194 N822AA
195 N255RK
196 C-GTAK
197 N520TJ
200 HC-BUP
201 I-DRIB
202 N33L
203 N821AA
204 N204TW
205 N815AA
206 N801SC
207 G-FRAP
209 G-FRAR
211 N618GA
212 N212R
213 G-FRAK
214 G-FRAO
215 N619GA
216 5840
217 17103
218 N218CA
219 TM 11-3
220 N220CA
221 N221TW
222 TM 11-2
223 G-FRAH
224 G-FRAM
225 C-FONX
226 N226R
227 N227JP
228 N823AA
229 N229R
230 N230RA
233 N817AA
234 I-LIAC
235 0442
236 N618GH
237 HB-VJV
238 F-RAEE
239 I-AGEC
240 N240AT
241 I-FLYK
242 N711RT
245 EL-VDY
246 F-GLMT
247 N70PL
248 XC-HIX
249 N451DP
250 XA-HEW
251 EP-FIE

252 F-ZACA
253 T 11-1
254 F-GPAB
256 UR-CCA
257 HB-VKO
258 N20AE
259 N569BW
260 F-RAEA
261 N200WK
262 D2-ESV
263 F-ZACY
264 CS-ATG
265 N265MP
266 N184GA
267 I-REAL
268 F-RAEF
269 XA-DUC
270 G-FRAI
271 F-GKDB
272 N272JP
273 N596DA
274 N260MB
275 N999EQ
276 CM-01
277 J 753
278 CM-02
279 F-GROC
280 G-FRAE
281 N347K
282 XC-DIP
284 XA-BCC
285 N285TW
286 EP-AGY
287 XA-SAG
288 F-ZACV
289 N105TW
290 N816AA
291 F-RAEG
292 N733S
293 F-GOBZ
294 SU-AXN
295 G-FRAF
296 N20TX
297 PK-TIR
298 N827AA
299 TC-EZE
300 N600WD
301 EP-AKC
302 F-GOPM
304 G-FRAD
305 N34CW
306 VH-HFJ
307 F-GKIS
308 N453SB
310 N831HG
311 F-BVPN
312 N132AP
313 N183TS
314 F-GDLU
315 F-SEBI
316 N734S
317 HB-VEV
318 15-2533
319 N205K
320 EP-FIF
321 PH-...
322 N300CV
323 XU-008
324 N312K
325 N599RR
326 TC-CEN
327 XA-HHF
328 YK-ASA
329 D-CMET
330 N770MC
331 YK-ASB
332 TM 11-4
333 15-2235
334 EP-FIC
335 N707JC
336 5-2802
337 YI-AHH

338 EP-FID	423 2118	505 N221FJ	73 N15TW	156 N4MB
339 SE-DSA	424 2119	506 N147TA	74 N83FJ	157 N911HB
340 5-2803	425 2120	507 N50MG	75 N850CA	158 N54YR
341 F-GYSL	426 I-BAEL	508 N777FC	76 N450CL	159 VP-CWI
342 F-RAEC	427 I-ACTL	509 N200WY	77 N992	160 N487F
343 YI-AHJ	428 N148MC	510 N515DB	78 F-RAFJ	161 TC-EYE
344 N731AS	429 N149MC	511 VT-TTA	79 N60CN	162 N244AD
345 OH-FPC	430 N660P	512 N999TH	80 N50SJ	163 N5VH
346 5-2804	431 2121	513 N5UU	81 N718DW	164 N164MA
347 N347HS	432 N667P	514 N531WB	82 N450KP	165 LX-FMR
349 N767AG	433 2122	515 XA-PFM	83 N881M	166 N5VF
350 5-3021	434 OB-1433		84 T 16-1	167 N2T
351 5-9001	435 2123	**Falcon 50 / 50EX**	85 N254DV	168 N420JP
352 N920G	436 N8000U	2 F-GSER	86 N150UC	169 I-SNAB
353 5-9002	437 2124	3 N728LW	87 N55NT	170 N500AF
354 5-9003	438 N258A	4 YV-452CP	88 VP-CRT	171 N40AS
355 N712ME	439 2125	5 F-RAFI	89 N97BZ	172 N256A
356 N11UF	440 N7000G	7 N50HE	90 N4351M	173 PT-LJI
357 N342K	441 2126	8 N119PH	91 ZS-CAS	174 N79PF
358 SU-AZJ	442 N747CX	9 F-GGCP	92 N40CN	175 N200RT
359 N369CA	443 2127	10 N65B	93 N844X	176 C-FNNC
360 N865VP	444 N665P	11 F-GGVB	94 N212JP	177 9Q-CGK
361 SU-AYD	445 2128	12 CNA-NO	95 TC-KAM	178 N59PM
362 F-WDFJ	446 N446D	13 N150TX	96 C-FMFL	179 N222MC
363 N363FJ	447 2129	14 N955E	97 N33GQ	180 N50SF
364 XA-FLM	448 48	15 PH-ILR	98 N600WG	181 N367TP
365 N50BH	449 N457F	16 D-BFAR	99 N816M	182 N250AS
366 N100AC	450 2130	17 N727S	100 N14CG	183 I-CAFD
367 EP-SEA	451 F-UKJC	18 N518EJ	101 5-90..	184 N25MB
368 N23A	452 2131	19 N253L	102 N350WB	185 F-GKBZ
369 N138FJ	453 N189MM	20 N63537	103 N370KP	186 N450K
370 N370HF	454 2132	21 N770E	104 SE-DVG	187 VH-PPF
371 2141	455 N555SR	22 XA-AVE	105 N80CN	188 LV-WXV
372 ST-PRS	456 2133	23 PH-ILD	106 N9300C	189 N55SN
373 N620CC	457 N47LP	24 N817M	107 VP-BUC	190 CS-TMJ
374 2101	458 2134	25 YU-BPZ	108 N150K	191 N950F
375 F-ZACZ	459 2135	26 N190MC	109 N280BG	192 N96LT
376 N1897S	460 2136	27 F-RAFK	110 N77TE	193 MM62026
377 N30FT	461 N353CP	28 N47UF	111 N50AH	194 N95PH
378 N305AR	462 2137	29 N534MA	112 N144AD	195 17401
379 N892SB	463 N132EP	30 F-WQFZ	113 N35RZ	196 G-....
380 N3484U	464 2138	31 N931CC	114 ST-PSR	197 N500KJ
381 N20T	465 65	32 N80TR	115 N50TC	198 17402
382 N382E	466 2139	33 N8300E	116 XB-SOL	199 N291BC
383 N900CH	467 2140	34 F-RAFL	117 N124HM	200 N664P
384 N120TF	468 J 468	35 XA-FVK	119 N57DC	201 N54DA
385 N120WH	469 J 469	36 F-WWHZ	120 5-90..	202 N97LT
386 2102	470 N470G	37 I-SAME	121 N824R	203 C-GNCA
387 N384K	471 N44JC	38 N993	122 5-90..	204 EC-GIN
388 N756	472 72	39 N326FB	123 F-GPSA	205 N52JJ
389 I-CMUT	473 3A-MGR	40 N150JT	124 N500RE	206 VP-BMF
390 2104	474 N1HF	41 N76FD	125 I-DENR	207 N50CS
391 N420DP	475 T 11-5	42 OO-LFT	126 N52DC	208 N50HC
392 N326EW	476 1650	43 YU-BNA	127 N1896F	209 N96DS
393 XC-FVH	477 77	44 N285CP	128 N223DD	210 F-GICN
394 2103	478 N181WT	45 N9BX	129 N99JD	211 MM62029
396 N881J	479 N349MG	46 N728LB	130 N988T	212 N30TH
397 F-GBTM	480 80	47 N1BX	131 F-GOAL	213 XA-RVV
398 2105	481 N250RA	48 N247EM	132 I-EDIK	214 N55AS
399 N70U	482 VP-CCL	49 N43BE	133 ZS-CAQ	215 D-BOOK
401 301	483 F-UKJI	50 N747Y	134 VP-CLD	216 N56SN
402 2106	484 N422MU	51 F-GMGA	135 N293BC	217 N5732
403 N960TX	485 N997TT	52 N86AK	136 YV-455CP	218 D-BERT
404 N28C	486 N6VF	53 N90AM	137 C-GTPL	219 YV-450CP
405 2108	487 N137TA	54 LX-GED	138 VP-CEZ	220 N100RR
406 G-BGOP	488 N146CF	55 N96UH	139 N1S	221 17403
407 2109	489 TC-DEM	56 N84HP	140 N303JW	222 OE-HIT
408 PK-CAG	490 HC-BVH	57 N138F	141 N86MC	223 N633L
409 2107	491 N500RR	58 N744X	142 N4350M	224 XA-BEG
410 N200J	492 N412AB	59 N900JB	143 N444PE	225 N32TC
411 2110	493 N1847B	60 N900W	144 VP-BZE	226 N119AM
412 N620A	494 N49US	61 HB-IES	145 N50KD	227 N1848U
413 2111	495 C-GSCL	62 N50FH	146 N7228K	228 C-GAZU
414 N1881Q	496 N256JC	63 C-FKCI	147 VP-BKG	229 C-GMII
415 2112	497 N20CL	64 N300A	148 N254NA	230 OY-LIN
416 N416F	498 N69EC	65 F-GPPF	149 N1971R	231 N10PP
417 2113	499 N14CJ	66 9U-BTB	150 N8400E	232 N45NC
418 2114	500 N595DC	67 T-783	151 MM62020	233 N48HB
419 2115	501 F-GOJT	68 5A-DCM	152 N75WE	234 PT-AAK
420 2116	502 N64YP	69 N650X	153 N50HM	235 UR-CCC
421 2117	503 N50MW	70 N300ES	154 N154PA	236 N195SV
422 F-RAEH	504 VH-CPE	71 J2-KBA	155 MM62021	237 N94BJ

238 SE-DVL	49 VP-BLB	131 XA-...	8 N610AS	11557 KAF308
239 N200SG	50 N330K	132 N707WB	9 N435T	11576 PH-EZF
240 N33TY	51 N26LB	133 HZ-OFC3	10 N652PC	11992 FAC-0001
241 N233BC	52 5N-FGO	134 N322CP	11 N721BS	
242 N9000F	53 JA8570	135 VP-BPW	12 I-SNAW	

Gulfstream

243 N742R	54 I-FICV	136 N1818S	13 N722JB	**Gulfstream 2**
244 N95HC	55 N404R	137 XA-GAE	14 N70KS	1 N55RG
245 N720ML	56 JA8571	138 VP-BHJ	15 N790L	2 N434JW
246 TC-YSR	57 N900WK	139 N523AC	16 HB-IAW	3 N555RS
247 N740R	58 OE-ILS	140 N70HS	17 N77A	4 N36RR
248 N25UB	59 N32B	141 XA-OVR	18 EI-LJR	5 N65ST
249 SE-DVK	60 N91TH	142 TC-CAG	19 N790M	6 N122DU
250 N696HC	61 HZ-AFZ	143 ZS-ZBB	20 N389GS	7 N118NP
251 VP-CLN	62 N62FJ	144 N453JS	21 N390GS	8 ZS-TGG
252 XA-TDD	63 N127EM	145 VP-CGB	22 N200NE	9 N48EC
253 PT-WSC	64 M37-01	146 N881P	23 N375SC	10 XA-ROI
254 N50FJ	65 N990MC	147 OE-IMI	24 N376SC	11 N902GA
255 N255CM	66 CS-TMK	148 N522AC	25 N96FG	12 N212TJ
256 VP-CBT	67 N900MA	149 VP-BPI	26 TC-CIN	13 N269HM
257 F-OKSY	68 N900HC	150 N335MC	27 G-JCBI	14 XA-SWP
258 VP-BST	69 I-SNAX	151 G-OPWH	28 N596A	15 N125JJ
259 VP-C..	70 A26-070	152 N337MC	29 XA-TDU	16 N38GL
260 N586CS	71 N280BC	153 N57EL	30 HB-IAZ	17 N422DV
261 N140RT	72 VP-BLM	154 VP-BJA	31 N790Z	18 XC-AA70
262 VP-CHG	73 A26-073	155 N814M	32 N65SD	19 N1929Y
263 N8550A	74 A26-074	156 HL7301	33 HB-IAX	20 N331P
264 F-GVDN	75 HB-IAI	157 N1868M	34 HB-IAY	21 N244DM
265 N9550A	76 A26-076	158 N404VL	35 N27WP	22 N206MD
266 F-WWHQ	77 A26-077	159 N263PW	36 F-GSAA	23 N7TJ
267 F-WWHR	78 C-GSSS	160 N176CF	37 EC-GNK	24 N800XL
268 F-WWHS	79 N6BX	161 VP-CTT	38 N3BM	25 N711RL
269 F-WWHT	80 N914BD	162 N611JW	39 N151AE	26 PK-PJZ
270 N270EX	81 7T-VPA	163 VP-BEH	40 N1C	27 N227TJ
	82 7T-VPB	164 G-MLTI	41 N48CG	28 N68DM
Falcon 900	83 N900NE	165 G-EVES	42 HB-IBH	29 N941CW
1 F-GIDE	84 A6-AUH	166 F-GLI II	43 PT-MML	30 N338AX
2 F-RAFP	85 N74FS	167 F-GUEQ	44 N2000A	31 N105TB
3 N991RF	86 A6-UAE	168 XA-...	45 N45SC	32 N7601
4 F-RAFQ	87 N33GG	169 F-WWFP	46 F-GMCK	33 N327TL
5 F-GGRH	88 F-GNDA	170 F-WWFR	47 N220JM	35 N30PR
6 N80F	89 I-NUMI	171 TC-...	48 N701WC	36 N74A
7 TR-LCJ	90 T 18-2		49 VP-BEF	37 N397RD
8 N5731	91 A7-AAD	**Falcon 900EX**	50 D-BEST	39 N87TD
9 HB-IAB	92 PT-OEX	1 PH-ERP	51 N2AT	41 N416K
10 N349K	93 N900Q	2 N200L	52 N212T	42 N36PN
11 F-GLGY	94 A7-AAE	3 JA50TH	53 N981	43 N691RC
12 N77CE	95 N343MG	4 N8100E	54 F-WWML	44 N830G
13 N75V	96 5N-FGE	5 9M-JJS	55 F-WWMM	45 89-0266
14 N906SB	97 XA-SJX	6 EC-GMO		46 N565KC
15 XA-RGB	98 N59CF	7 N45SJ	**Fokker**	47 N800RT
16 VP-CBD	99 ZS-NAN	8 N30LB		48 N61WH
18 N72PS	100 YK-ASC	9 N70LF	**F 28**	49 N511PA
19 N900SJ	101 D-ALME	10 N22CS	11012 XU-061	50 N220JR
20 N911RF	102 N906WK	11 F-GOYA	11016 C-FHFP	51 N20H
21 HZ-AFT	103 V5-NAM	12 N900EX	11028 T-03	52 N211MT
22 N54DC	104 N881G	13 VP-BRO	11042 A-2801	53 N102AB
23 I-BEAU	105 CN-TFU	14 N72WS	11045 PK-RJW	54 N955CC
24 N93GR	106 9M-BAN	15 N914J	11048 T-02	55 N125DC
25 N70EW	107 XA-GTR	16 N67WB	11079 5V-TAI	56 N610CC
26 SX-ECH	108 N334MC	17 N600AS	11088 M28-01	57 N300DL
27 N90EW	109 CD-01	18 N18RF	11099 TU-TIZ	59 N1823D
28 N85D	110 OY-CKK	19 N7301	11100 OB-1396	61 N61LH
29 C-GTCP	111 N8BX	20 N158JA	11104 EP-PAZ	62 62 Black
30 I-DIES	112 N246AG	21 N330MC	11110 HB-AAS	63 N17ND
31 N910JW	113 HZ-SAB2	22 N331MC	11111 PK-PJL	64 N43RJ
32 N10MZ	114 XA-SIM	23 F-WWFS	11124 TU-VAJ	65 N58JF
33 N901SB	115 EC-FPI	24 TR-K..	11125 G-530	66 N718JS
34 N8200E	116 N82RP	25 N925EX	11137 5H-CCM	67 N67PR
35 HB-IAD	117 N70TH	26 N900SB	11145 5-T-20-0741	68 N308EL
36 N922JW	118 RA-09000	27 N927EX	11146 PK-PJY	69 N701S
37 VH-...	119 N22T	28 F-WWFZ	11147 5-T-10-0740	70 N451CS
38 T 18-1	120 VP-BNJ	30 N662P	11150 5-T-21-0742	71 N200AB
39 N5733	121 9M-BAB		11153 RP-1250	72 N920DS
40 N145W	122 XA-SGW	**Falcon 2000**	11178 PK-PJM	73 N74HH
41 N404F	123 RA-09001	1 F-WNAV	11192 PK-PJK	74 N74HH
42 N117TF	124 VP-BWS	2 ZS-NNF	11245 TU-VAA	75 N94TJ
43 N288Z	125 VP-BSK	3 N15AS	11288 PK-PJN	76 N227GA
44 HB-IBY	126 N733A	4 N925AJ	11521 N322K	77 N994GC
45 N298W	127 N654CN	5 N27R	11545 N324K	78 HP-1A
46 N329K	128 N999PM	6 N93GH	11547 PH-KBX	79 XA-ARA
47 A6-ZKM	129 XA-VTO	7 N28R	11553 PK-PFE	
48 N900MJ	130 VP-CID			

80 N500RH	167 N120GS	251 N251JS	372 N724DB	456 85-0049
81 XB-PGR	168 N317AF	252 XA-MEY	373 N162JC	457 N457H
82 N492JT	169 N169P	253 N154C	374 N270MC	458 85-0050
84 N27SL	170 N111GD	254 N706TS	375 N375NM	459 N586C
85 N93AT	171 HZ-AFH	255 N4NR	376 N70AG	460 N317ML
86 N179T	172 N903G	256 HZ-MSD	377 N40CH	461 N104AR
88 N80WD	173 N98FT	257 N56D	378 N803CC	462 TU-VAF
89 N203A	174 N540EA	258 N437H	379 N282Q	463 VP-BMY
90 N7789	175 XA-FNY	775 N6PC	380 N30WR	464 N535CS
91 N183SC	176 N15UC		381 N1871R	465 911
92 N691HM	177 5N-AGV	**Gulfstream 3**	382 83-0500	466 N37HE
93 N484TL	178 N720JW	300 N700VA	383 83-0501	467 JY-HAH
94 N420JM	179 HZ-PCA	301 N110BR	384 N369CS	468 86-0202
95 N836ME	180 N37WH	302 VP-BCT	385 HZ-MS3	469 N1956M
96 XB-EBI	181 N48CC	303 N303GA	386 XC-UJO	470 86-0201
97 N930SD	182 CNA-NL	304 N304TS	387 N620JH	471 N583D
98 N812RS	183 N801WC	305 PK-OCN	388 N561ST	472 N780RH
99 N900MP	184 N254CR	306 N863CE	389 83-0502	473 86-0403
100 N400D	185 XA-BRE	307 C-GGPM	390 VP-BOK	474 D2-ECB
101 N240CX	186 VP-BJV	308 N308HG	391 N14SY	475 86-0203
102 N102CX	187 N802CC	309 N1NA	392 N800WC	476 86-0204
103 HZ-MS4	188 N555MW	310 N6513X	393 A9C-BB	477 01
104 C-FHPM	189 N515JT	311 N721RB	394 N379XX	478 86-0206
105 N711TE	190 N914DZ	312 N800JH	395 PK-PJA	479 MM62025
106 N226GA	191 N951RK	313 F-313	396 N175BG	480 163691
107 N10123	192 N273LP	314 N93CX	397 N692TV	481 163692
108 N200GH	193 N54J	315 N710EC	398 N827GA	482 N164RJ
109 N73AW	194 N57HJ	316 PK-CAP	399 N188TJ	483 N66DD
110 N21AM	195 XA-ILV	317 HZ-DG2	400 0005	484 N506T
111 N900DH	196 N610MC	318 N500WW	401 N97AG	485 N777MW
112 N108DB	197 N5117H	319 N200SK	402 VP-BLN	486 TJ-AAW
113 N217JS	198 N91LA	320 N624BP	403 XA-TCO	487 N416WM
114 XA-TDK	199 VP-BND	321 N310RG	404 N402LM	488 N399SC
115 N700BH	200 N281RB	322 N600ES	405 N789TP	489 N272JS
116 N716TE	201 HZ-AFI	323 XA-MIC	406 N80L	490 N388MM
117 N75CC	202 A9C-BG	324 N67JR	407 N913MK	491 N531JF
118 N945NA	203 HZ-AFJ	325 N393U	409 N1526R	492 PT-WRC
119 N929GV	204 VP-CPA	326 TR-KHC	410 HZ-AFR	493 N400J
120 N393BD	205 N1000	327 N711LT	411 N461GT	494 K2981
121 N721RL	206 N900BF	328 N98RP	412 N105Y	495 K2982
122 N500RL	207 VP-CUB	329 N862G	413 N166WC	496 N843HS
123 N345AA	208 C-FNCG	331 HZ-RC3	414 N165G	497 87-0139
124 0004	209 N277T	332 N121JM	415 HZ-NR2	498 87-0140
125 N92LA	210 N30FW	333 N901FH	416 N4500X	875 N845FW
126 N901WG	211 VP-BGT	334 N700SB	417 N300M	
128 N128TS	212 N807CC	335 HB-IMX	418 PT-ALK	**Gulfstream 4**
129 N626TC	213 N96JA	336 N3PY	420 K2980	1000 N234DB
130 N518GS	214 N11NZ	337 N456SW	421 N721FF	1001 N981SW
131 N2JR	215 N748MN	338 N750SW	422 N407CA	1002 N168WC
132 N400M	216 HZ-HA1	339 N339A	423 N225SF	1003 N685TA
133 N444QG	217 N880WD	340 N4PC	424 N94FL	1004 N424GA
134 N555HD	218 N187PH	341 N1PR	425 N492A	1005 N823GA
135 N518JT	219 N923ML	342 N82AE	426 VP-CNJ	1006 N3338
136 XA-ABA	220 N117GL	343 N221CM	427 N300WY	1007 N59JR
137 N115MC	221 N2HF	344 5N-IMR	428 N760G	1008 N119R
138 N6JW	222 N948NA	345 5X-UOI	429 N423SA	1009 V-11
139 N663PD	223 N510US	346 HZ-HR2	430 N600BG	1010 N824CA
140 N159NB	224 N90CP	347 N545JT	431 PK-CTP	1011 A6-HHH
141 JA8431	225 N225TR	348 N756S	432 N995BC	1012 N838MF
142 N5RD	226 N5DL	349 N6453	433 N579TG	1013 N16251
144 N944NA	227 N264CL	350 N1454H	434 N23ET	1014 102001
145 N339H	228 N157LH	351 N623MS	435 N888PM	1015 VP-BRF
146 N946NA	229 N117FJ	352 XC-UJN	436 N10EH	1016 N880GC
147 N947NA	231 N47EC	353 HZ-108	437 N380TT	1017 VP-BHG
148 N180AR	232 N508T	354 N16NK	438 N911KT	1018 N43KS
150 N319GP	233 N233RS	355 8P-GAC	439 SU-BGU	1019 TU-VAD
151 N908JE	234 N500JW	356 A6-HEH	440 N5103	1020 N9300
152 N559LC	235 N430RG	357 N802GA	441 N80J	1021 N3M
153 N111VW	236 N211DH	358 N475DJ	442 SU-BGV	1022 N23M
154 N836MF	237 EC-FRV	359 N800J	443 N5104	1023 N85M
155 XA-GAC	238 N335H	360 N90LC	444 N110MT	1024 N96AE
156 N525JT	239 HZ-AFK	361 N874RA	445 N5105	1025 N421SZ
157 N683EC	240 TT-AAI	362 N800AR	446 N446U	1026 N277RP
158 N2S	241 N902MP	363 N77FK	447 N144PK	1027 TC-GAP
159 N880RJ	242 5A-DDS	364 HZ-AFN	448 I-MADU	1028 N712CC
160 N919TG	244 N811DF	365 CNA-NU	449 N85V	1029 VP-BKI
161 XA-AHM	245 N222NB	366 N555KC	450 N888VS	1030 N1WP
162 N666JT	246 N81RR	367 N700FS	451 MM62022	1031 HZ-AFU
163 N117JJ	247 N75MG	368 9M-...	452 VP-BNZ	1032 N432QS
164 N80AG	248 N248TH	369 N910A	453 HZ-103	1033 N1KE
165 N965CC	249 F-249	370 N697BJ	454 N111G	1034 N800BG
166 N776MA	250 N309EL	371 N681FM	455 N123CC	1035 HZ-AFV

1036 N152A	1119 N768J	1201 165152	1283 N401JL	518 N518GA
1037 HZ-ADC	1120 N400SA	1202 V8-...	1284 N1GN	519 N597GA
1038 HZ-AFW	1121 N411WW	1203 N410WW	1285 N874A	520 N17GV
1039 N1901M	1122 N226G	1204 N212AT	1286 N286GA	522 N158RA
1040 N74RP	1123 I-LUBI	1205 VH-ASQ	1287 N487QS	523 N523GA
1041 N366F	1124 N1900W	1206 N1040	1288 7T-VPR	524 N400JD
1042 N22	1125 N888LK	1207 N77SW	1289 N844HS	525 N594GA
1043 TC-ATA	1126 5N-FGP	1208 VP-BNY	1290 N6NB	526 N675RW
1044 N1540	1127 VP-BUS	1209 N157H	1291 7T-VPS	527 N527GA
1045 N227GH	1128 HZ-MFL	1210 N410QS	1292 N1GT	528 N528GA
1046 N400CC	1129 ZS-NMO	1211 N2107Z	1293 N493QS	529 N529GA
1048 SU-BGM	1130 N404LM	1212 VP-BOT	1294 HZ-KAA	530 N780F
1049 N372CM	1131 N679RW	1213 N56L	1295 N495QS	531 N531GA
1050 N153RA	1132 N604M	1214 N414BM	1296 N725LB	532 N532GA
1051 N399CC	1133 N700CN	1215 102002	1297 LV-WSS	533 N533GA
1052 N800CC	1134 VP-BJD	1216 102003	1298 N501PC	534 N534GA
1053 N165ST	1135 N100ES	1217 N711MC	1299 N499QS	535 N775US
1054 N789DK	1136 N27CD	1218 N5MC	1300 N1BN	536 N536GA
1055 XB-OEM	1137 N21CZ	1219 PK-NZK	1301 N92AE	537 N537GA
1056 N33M	1138 N200A	1220 N79RP	1302 N93AE	538 N538GA
1057 N43M	1139 N21KR	1221 N451GA	1303 75-3253	539 N539GA
1058 VP-BSF	1140 N811JK	1222 N71RP	1304 N404QS	540 N640GA
1059 N701QS	1141 N767FL	1223 N257H	1305 N888SQ	541 N641GA
1060 N1SF	1142 N222	1224 N18TM	1306 N540CH	542 N642GA
1061 F-GPAK	1143 HZ-AFX	1225 N316GS	1307 N94AE	543 N643GA
1062 VP-CMF	1144 N250J	1226 N41PR	1308 N408QS	544 N644GA
1063 N333AX	1145 N102MU	1227 XA-DPS	1309 N309GA	
1064 N7RP	1146 N777UE	1228 N18AN	1310 N902	
1065 N584D	1147 N200PM	1229 N830EC	1311 N411QS	
1066 N118R	1148 HB-IEJ	1230 9M-TRI	1312 N453GA	
1067 N200LC	1149 N152KB	1231 N250VC	1313 N94LT	
1068 N82A	1150 V8-SR1	1232 N232K	1314 N429SA	
1069 N765A	1151 N80AT	1233 N575SF	1315 N315GA	
1070 N107A	1152 N63M	1234 I-LXGR	1316 N416QS	
1071 N1	1153 N589HM	1235 N100A	1317 N929WT	
1072 N500E	1154 N186DS	1236 N100GN	1318 N1624	
1073 N75RP	1155 N010S	1237 N1904W	1319 N878SM	
1074 VP-BKT	1156 N987AR	1238 N499SC	1320 N420QS	
1075 N121JJ	1157 9K-AJA	1239 N1JN	1321 N500CD	
1076 N338MM	1158 N917W	1240 N333PV	1322 N422QS	
1077 PK-NSP	1159 9K-AJB	1241 N169CA	1323 N454GA	
1078 G-DNVT	1160 251	1242 N982HC	1324 N457GA	
1079 N479TS	1161 9K-AJC	1243 N39WH	1325 N459GA	
1080 N20XY	1162 91-0108	1244 JA002G	1326 N325GA	
1081 N955H	1163 12-003	1245 N7602	1327 N327GA	
1082 N1082A	1164 N300GX	1246 N49RF	1328 N328GA	
1083 VH-CCC	1165 N780E	1247 N14UH	1329 N329GA	
1084 HB-IMY	1166 HZ-AFY	1248 N62MS	1330 N324GA	
1085 N864CE	1167 N1SL	1249 N63HS	1331 N331GA	
1086 N1086	1168 A4O-AB	1250 VP-CBB	1332 N332GA	
1087 N1TM	1169 N600DW	1251 N60PT	1333 N333GA	
1088 N2600	1170 N997BC	1252 N252C	1334 N334GA	
1089 N53M	1171 N72RK	1253 N676RW	1335 N335GA	
1090 VP-CYM	1172 N472TS	1254 N801CC	1336 N636GA	
1091 N364G	1173 0K1	1255 N600PM	1337 N637GA	
1092 N3HX	1174 N174SJ	1256 92-0375	1338 N638GA	
1093 HB-ITX	1175 VH-CCA	1257 N4CP	1339 N639GA	
1094 N2610	1176 V8-008	1258 N585D	1340 N340GA	
1095 N311EL	1177 N677RW	1259 N1PG	1340 N340GA	
1096 VP-CBW	1178 N900LS	1260 N3PG	1341 N341GA	
1097 N900AL	1179 N41CP	1261 N399CB	1342 N342GA	
1098 XA-AIS	1180 N700LS	1262 N462QS	1343 N343GA	
1099 N199QS	1181 90-0300	1263 N263S	1344 N344GA	
1100 N100AR	1182 N200LS	1264 N464QS		
1101 N365G	1183 HB-IBX	1265 N540W		
1102 N910B	1184 N111NL	1266 N300K		
1103 N90005	1185 N635AV	1267 N301K		
1104 N700GD	1186 8P-MAK	1268 N990WC		
1105 N312EL	1187 165093	1269 N677SW		
1106 9M-ISJ	1188 HL7222	1270 75-3251		
1107 VH-CCO	1189 165094	1271 75-3252		
1108 VH-NCP	1190 JA001G	1272 N621JH		
1109 V8-...	1191 N979RA	1273 N372BG		
1110 N404M	1192 N212K	1274 LV-WOM		
1111 N111ZT	1193 N620K	1275 N475QS		
1112 N12U	1194 N77CP	1276 N1955M		
1113 N902K	1195 N867CE	1277 N5GF		
1114 XA-BAL	1196 A4O-AC	1278 N98LT		
1115 N410M	1197 XA-CAG	1279 N2002P		
1116 N971L	1198 N99GA	1280 N468GA		
1117 G-HARF	1199 165151	1281 N481QS		
1118 N2WL	1200 165153	1282 VP-CFL		

HFB

HFB 320

1023	TC-...
1025	16+08
1032	N132MW
1033	N130MW
1036	N136MW
1037	YV-999P
1038	N605GA
1039	N171GA
1040	N7158Q
1041	N92045
1042	N106TF
1043	TC-LEY
1052	N173GA
1053	N176GA
1055	N105TF
1056	D-COSA
1057	YV-388CP
1058	N322AF
1060	N321AF
1061	N320AF
1062	N323AF
1064	N324AF
1065	N325AF

Ilyushin

Il 62

2749316	RA-86466
4648339	RA-86712

Israeli Aircraft Industries

Jet Commander Westwind

3	N409WT
4	N72TQ
20	N1121E
24	N560MC
28	N234G
32	N98SC
33	VT-ERO
38	N37SJ
41	N40593
44	N60CD
45	N910MH
46	N99W
55	N747LB

Gulfstream 5

501	N501GV
502	N502GV
503	N503GV
504	N313RG
505	EI-WGV
506	N110LE
507	N300L
508	N777GV
509	V8-009
510	N513MW
511	VP-CBX
512	N636MF
513	HB-IVL
514	N777SW
515	V8-007
516	N555CS
517	HB-I..

56 N382AA	195 N951DB	282 VH-AJV	367 N367WW	014 N8484P
58 CP-2263	196 N863AB	283 N95JK	368 N368MD	015 N14SR
62 N1777T	198 N51MN	284 N727AT	369 N76ER	016 N221PA
64 N124VS	199 D-CHDL	285 XA-LIJ	370 N875HS	017 N711JG
67 N1121G	200 N4WG	286 N92FE	372 N444MW	018 N500M
74 N93RM	201 N29UF	287 N530DL	373 N794TK	019 N49MW
77 N177JC	202 N274HM	288 N48AH	374 N33MK	020 N279DP
82 N240AA	203 N880Z	289 N900VP	375 N66LX	021 N1125S
84 N16MK	204 N10UJ	290 N719CC	376 N376BE	022 PT-MBZ
87 N116KX	205 N205AJ	291 N917BE	377 N301PC	023 N23TJ
89 N163DC	206 N148H	292 N741C	378 N481NS	024 N999BL
90 N93SC	207 D-CHAL	293 N26T	379 N62ND	025 N387PA
94 N64AH	208 N311DB	294 N147A	381 N50FD	026 N9VL
95 CP-2259	209 N222LH	295 N730CA	382 N445BL	027 N199HF
96 YV-2454P	210 N425JF	296 N92WW	383 N84WU	028 N816HB
97 N34SW	212 N700MD	297 N801SM	384 N61RS	029 N154DD
99 N22RD	213 4X-CLK	298 N298CM	385 N317JS	030 N90UG
103 N77HH	214 N46BK	299 N288SJ	386 N68WW	031 N40AJ
106 N814K	215 N946GM	300 N10MV	387 VH-NGA	032 N232S
108 LV-WHZ	216 N1124G	301 XB-GRN	388 N900H	033 N441BC
109 XA-THF	217 N163WC	302 N422BC	389 N89AM	034 N511WA
110 N1121N	218 N74GR	303 N211ST	390 N122MP	035 N1125K
112 N372Q	219 YV-190CP	304 N389BG	391 C-GMPF	036 N195FC
113 N8534	220 N9RD	305 N464EC	392 N793BG	037 N100SR
118 N381DA	222 N3RC	306 HK-3971X	393 N53WW	038 N803JW
119 LV-...	223 N20KH	307 N97SM	394 N516CC	039 N359V
122 N122ST	224 N349MC	308 N308TS	395 VH-APU	040 N279DS
123 N1121A	225 N30MR	310 N78GJ	396 N37BE	041 N45MS
125 N1121R	226 D-CHBL	311 N788MA	397 N777HD	042 N528RR
127 N550K	227 N64FG	312 N97HW	398 N59AP	043 N56AG
128 N386JM	228 N305BB	313 N711WV	399 N48SD	044 N50AJ
129 N110ST	229 N40GG	314 N2HZ	400 N900PA	045 VH-FIS
133 N132LA	230 N1KT	315 N315TR	401 N84WW	046 N630S
134 N7638S	231 N331CW	316 N93KE	402 N51TV	047 N166RM
135 XC-COL	232 N4MH	317 VH-JPW	403 N403W	048 N88MF
137 XB-FKV	233 N650G	318 N38AE	404 N29CL	049 N323P
138 N972TF	234 N161X	319 N225N	405 N420TJ	050 N4EM
141 5N-...	235 N30AB	320 N204TM	406 N100CH	051 N1125A
142 N1944P	236 N22LZ	321 N93WW	407 N407W	052 N90AJ
143 N30AD	237 N24KL	322 N2AV	408 N408W	053 N227NL
145 N805SM	238 VH-AJP	323 VH-KNS	409 N26KL	054 N198MF
146 N444TJ	239 HK-2485W	324 N3VF	410 N26VF	055 N1MC
147 N912DA	240 N72787	325 N467MW	411 HC-BVX	056 N790FH
148 N600K	241 N300TC	327 N50M	412 N50HS	057 YV-785CP
149 N303AJ	242 N140DR	328 C-GPFC	413 N35LH	058 N1125E
151 ZP-AGD	244 N911SP	329 N711SE	414 N24MN	059 D-CCAT
152 4X-JYR	245 N270LC	330 YV-332CP	415 N415EL	060 VP-BON
153 XA-PUF	246 N101SV	331 N811VC	416 N416W	061 N550M
154 N772AW	247 N280AZ	332 N332DF	417 N115BP	062 N999GP
155 N707TF	248 VH-AJJ	333 HR-CEF	418 N420MP	063 N331SK
156 N35D	250 N60RV	334 N325LW	419 N551TP	064 N650GE
159 N96TS	251 PT-LDY	335 EC-GIB	420 N91SA	065 N50TG
160 XA-RIZ	252 N1WS	336 N255RB	421 N520MP	066 N419MK
162 N13GW	253 VH-LLW	337 4X-CTE	422 N87GS	067 N28NP
163 N163DL	254 N60AV	338 N50PL	423 N223WA	068 N1125Z
165 N102BW	255 N222MW	339 N90KC	424 N790JR	069 N804JW
167 N1123H	256 VH-AJK	340 N118MP	425 N365CX	070 N805JW
168 N111NF	257 N124UF	341 N868CP	426 N75BC	071 N60AJ
169 N44PR	258 N572M	342 N39RE	427 N229D	072 N1125L
171 ZS-...	259 VH-LLX	343 YV-O-CVG	428 N327SA	073 N173W
172 YV-58CP	260 N80FD	344 N849HS	429 C-FROY	074 N500AJ
174 N74TS	261 N39JN	345 N345TR	430 N430A	075 ZS-BCT
175 N571MC	262 N262WC	346 N610HC	431 C-FGGH	076 N1125
176 N27AT	263 N29PC	347 N178HH	432 N320MP	077 YV-771CP
177 N118AF	264 N351C	348 N960FA	433 N433GM	078 N1125J
178 N123CV	265 N7DJ	349 N728L	434 N919BT	079 C-FCFP
179 LV-WJU	266 N7HM	351 N351TC	435 N297JS	080 N333CZ
180 N3VL	267 N241CT	352 N117AH	436 N100AK	081 N800AJ
181 N325AJ	268 N56BP	353 C-GRGE	437 N437SJ	082 N121GV
182 LV-...	269 N50SL	354 N894TW	438 N100BC	083 N383SF
183 LV-WLR	270 N501DT	355 N355JK	439 N439WW	084 N795HP
184 N481MC	271 C-GWKF	356 N767AC	440 N127SA	085 N796HP
185 4X-JYJ	272 VH-LLY	357 N357BC	441 C-FZEI	086 PT-WBC
186 4X-JYO	273 N104RS	358 N800MA	442 N71WF	087 C-FRJZ
187 N789DD	274 N274K	359 N500AX		088 N398AG
188 C-GRDP	275 N1621	360 N500KE	**Astra**	089 N918MK
189 N42CM	276 N300XL	361 N3AV	002 4X-WIA	090 N399AG
190 N190WW	277 D-CHCL	362 N445A	003 4X-...	091 N297GA
191 N900FS	278 N10S	363 N1629	004 N425TS	092 N789A
192 N819RC	279 N400TF	364 RP-C2480	011 N705MA	093 N65TD
193 N515LG	280 N508R	365 N2BG	012 N27BH	094 N294S
194 N40TA	281 N1124F	366 N707BC	013 N713SC	095 N98AD

096 N66KG
097 N273RA
098 N275RA

Galaxy
003 4X-IGA

Learjet

Learjet 23
23-003 N3BL
23-009 N49CK
23-014 F-BXPT
23-016 N96CK
23-020 N820L
23-021 N133W
23-026 N540CL
23-030 ZS-JWC
23-034 N154AG
23-035 N992TD
23-036 YV-278CP
23-037 XC-AA28
23-039 N121CK
23-041 N77VJ
23-047 N444WC
23-048 N140RC
23-049 N933NA
23-052 N856JB
23-054 N351N
23-058 N153AG
23-062 N20TA
23-064 N259DB
23-065A N122M
23-066 N211TS
23-068 N73CE
23-069 N34TR
23-070 XA TII
23-071 N6262T
23-072 RP-C848
23-073 N806LJ
23-074 N83CE
23-079 N31CK
23-081 ZS-MDN
23-082 N7GP
23-082A N618BR
23-083 N824LJ
23-084 N119BA
23-089 N1968W
23-092 N415LJ
23-093 N7GF
23-095 N9RA
23-099 N7200K

Learjet 24
24-015 N88B
24-019 N747SC
24-050 N650CA
24-055 N711CW
24-060 N90J
24-065 N707SC
24-087 C-GEEN
24-101 N24WX
24-103 ZS-LTK
24-104 N45ED
24-106 N888MC
24-108 N315AJ
24-109 XA-...
24-112 N104GA
24-113 N100SQ
24-114 PT-LNE
24-119 N63CK
24-120 PT-LMF
24-123 XA-JSO
24-124 N991TD
24-126 N16HC
24-127 N124JL
24-128 XA-TDP
24-131 N241JA
24-132 N32CA
24-133 N133BL
24-134 N911TR

24-137 N151AG
24-138 N130RS
24-139 N96AA
24-140 N100VQ
24-141 XB-GHO
24-142 N777MR
24-143 N24WF
24-145 N282AC
24-147 N67CK
24-148 N41MP
24-149 N64HB
24-151 N6177Y
24-153 N120RA
24-156 N712R
24-157 N659AT
24-158 PT-WEW
24-159 N710TV
24-160 N989TL
24-161 N222TW
24-163 N68LU
24-164 XA-RYN
24-165 ZS-KJY
24-166 N993TD
24-167 N664CL
24-168 N333TW
24-169 N93BP
24-170 N151WW
24-171 N417WW
24-173 YV-824CP
24-174 N77WD
24-176 PT-CXJ
24-177 N555LB
24-178 N11AQ
24-179 XA-RQP
24-180 XA-NLA
24A-011 N24LG
24A-012 N1965L
24A-031 N175FS
24A-096 N464CL
24A-100 N427LJ
24A-102 N705NA
24A-107 N48L
24A-111 N900NA
24A-116 N105GA
24A-125 N651LJ
24B-181 N254JT
24B-182 N155J
24B-184 N58FN
24B-185 N44CP
24B-186 N196CF
24B-187 N129DM
24B-188 N280R
24B-189 N915US
24B-190 N190BP
24B-193 N83HC
24B-194 N62DM
24B-195 N803L
24B-196 N196TB
24B-197 XA-RXA
24B-198 N39KM
24B-199 N70TJ
24B-200 N119MA
24B-201 N100DL
24B-202 N814JR
24B-203 N203JL
24B-204 N510MS
24B-205 N64CE
24B-206 N116RM
24B-208 N14PT
24B-209 ZS-LWU
24B-211 N413WF
24B-212 N328JK
24B-213 N95AB
24B-216 N900GG
24B-217 C-FZHT
24B-218 N101VS
24B-219 N977GA
24B-221 N233TW
24B-222 N740EJ
24B-223 N7074X
24B-224 XA-TCA
24B-226 N335JW

24B-227 N27BJ
24B-228 PT-OBD
24D-230 XA-SSU
24D-234 LV-JTZ
24D-236 N990PT
24D-237 N237TW
24D-238 N48FN
24D-240 LV-JXA
24D-241 N63GA
24D-242 N999WA
24D-245 N44KB
24D-246 N24TE
24D-247 N42PG
24D-249 N249RA
24D-250 0006
24D-251 N39EL
24D-252 N252TJ
24D-254 PT-LCV
24D-255 XA-SMU
24D-256 C-GWFG
24D-257 N888FA
24D-258 N24CK
24D-260 XA-GBA
24D-262 OH-GLB
24D-266 N266TW
24D-268 N98WJ
24D-269 XA-DIJ
24D-270 PT-LEM
24D-271 XA-SCE
24D-272 N117K
24D-273 XB-DZR
24D-275 PT-LPH
24D-276 PT-JGU
24D-277 N277TW
24D-278 N202JS
24D-279 N101AR
24D-281 N281FP
24D-284 PT-JKQ
24D-288 N288DF
24D-289 N289SA
24D-290 XA-RMF
24D-291 N483DM
24D-292 N888TW
24D-293 XA-TIP
24D-294 PT-WKL
24D-296 PT-LMS
24D-297 N24S
24D-298 N169US
24D-299 N500JS
24D-301 N31BG
24D-303 PT-LOJ
24D-304 N500CG
24D-305 N510PA
24D-306 N306JA
24D-307 XA-RRK
24D-308 N770BC
24D-309 N80CK
24D-311 N10WJ
24D-312 N80AP
24D-313 MTX-02
24D-315 PT-KPE
24D-316 LV-LRC
24D-317 XA-JIQ
24D-318 N611DB
24D-320 S5-BAB
24D-321 N33TP
24D-322 N972H
24D-323 N453
24D-324 N324TW
24D-325 N500SW
24D-326 N326KE
24D-327 N327GJ
24D-328 D-CMMM
24E-329 N329TJ
24E-330 N511AT
24E-331 N32DD
24E-333 PT-LQK
24E-334 N944KM
24E-335 N8AE
24E-339 N52DD
24E-340 N106TJ
24E-341 N14DM

24E-343 N102C
24E-345 N435AS
24E-346 N117AE
24E-347 N500TS
24E-351 N77MR
24E-355 N165CM
24F-332 N56MM
24F-336 N9LD
24F-337 XA-DET
24F-342 N123DG
24F-348 N444TW
24F-349 XB-DZD
24F-350 N500ZA
24F-352 N352MD
24F-354 PT-LYE
24F-356 PT-LKD
24F-357 N129ME
24XR-117 N92DF
24XR-150 XA-RQB
24XR-207 ZS-MGJ
24XR-215 N400EP
24XR-233 N124TS
24XR-235 N701SC
24XR-243 N57FL
24XR-264 RP-C1747
24XR-267 ZS-OEA
24XR-274 N48CT
24XR-280 ZS-NGG
24XR-283 N47WT
24XR-285 N430JW
24XR-286 N56RD
24XR-295 N295NW
24XR-319 N173RD

Learjet 25
26 003 N97FN
25-004 N251AF
25-005 N39CK
25-006 N252SC
25-007 N52JA
25-008 N800GG
25-010 N121EL
25-011 N525TW
25-012 N102AR
25-014 N754DB
25-015 N25GJ
25-016 N976BS
25-017 N128JS
25-020 N500JS
25-023 N147TW
25-024 N20HJ
25-025 N111LM
25-026 N25EC
25-027 N500DL
25-028 N33PF
25-031 N294M
25-033 N77NJ
25-034 N309AJ
25-035 N616NA
25-036 N45BK
25-037 N155AG
25-038 N813JW
25-039 N308AJ
25-040 N238CA
25-041 N31AA
25-042 N797SC
25-043 N473TC
25-045 N28CK
25-046 N345MC
25-047 N222B
25-048 XA-TCY
25-049 N70HJ
25-050 N999MF
25-051 N12WW
25-052 N692FG
25-053 N153TW
25-054 N25MD
25-055 N511AJ
25-057 N920EA
25-058 N273LR
25-060 N564CL
25-062 N25ME

25-063 N25FM
25-064 N566NA
25B-075 VP-CHT
25B-076 XA-SJS
25B-078 N188BC
25B-080 XA-POG
25B-081 N524DW
25B-082 N700FC
25B-085 N8MQ
25B-086 N65WH
25B-088 N125JL
25B-090 C-GBOT
25B-091 N91PN
25B-092 N113ES
25B-093 PT-LEN
25B-095 N20NW
25B-096 N20NW
25B-100 N25TK
25B-101 N47MR
25B-102 N254SC
25B-104 XA-RIN
25B-105 N55PD
25B-107 N252BK
25B-109 C-GSAS
25B-110 N52SD
25B-111 PT-LXS
25B-112 N279LE
25B-114 N114HC
25B-116 N666TW
25B-117 N731CW
25B-118 N124MA
25B-120 N120SL
25B-121 N8005Y
25B-122 N751CA
25B-123 N688GS
25B-124 N33TW
25B-125 N10VG
25B-127 N425JL
25B-130 N26AT
25B-132 N715MH
25B-133 N233CA
25B-134 N65A
25B-135 N50RW
25B-136 N48WA
25B-137 N752EA
25B-138 N73LJ
25B-140 N140CA
25B-142 N49WA
25B-143 N143CK
25B-144 N44PA
25B-145 N145SH
25B-147 N147BP
25B-149 N239CA
25B-150 N251JA
25B-154 N82TS
25B-155 PT-LEA
25B-157 N50CK
25B-158 N924BW
25B-159 OB-1429
25B-160 ZS-MTD
25B-161 N61EW
25B-163 N911AJ
25B-164 OB-1430
25B-166 PT-KBD
25B-167 C-GBFP
25B-168 N88BY
25B-169 N59FL
25B-171 N401AJ
25B-174 N412SP
25B-180 N102VS
25B-182 N225JL
25B-183 N83CK
25B-185 N988AA
25B-189 YU-BJG
25B-190 XA-DAK
25B-191 FAB 008
25B-193 N125RM
25B-194 XC-NSP
25B-197 N197CF
25B-198 N29TS

25B-200 N350JH	25D-281 N800RF	25D-369 N2213T	31A-043 C-GLRJ	31A-125 N527JG
25B-201 N11TK	25D-282 N711WD	25D-372 N5NC	31A-044 XA-MJG	31A-126 N22SF
25B-202 YU-BRA	25D-283 N444WW	25D-373 EC-EGY	31A-045 N67SB	31A-127 HB-VLR
25B-203 70402	25D-284 N284TJ	25G-337 LV-WBP	31A-046 N352EF	31A-128 N400
25B-205 Z3-BAA	25D-286 XA-TAQ	25G-352 N25FN	31A-047 N39TW	31A-129 N115FX
25C-061 YV-203CP	25D-287 XA-ZYZ	25G-370 N252HS	31A-048 N43SF	31A-130 N31PV
25C-070 C-FZHU	25D-288 N100WN	25G-371 N1U	31A-049 D-CDEN	31A-131 N391SC
25C-071 N97AM	25D-289 N389GA	25XR-022 ZS-...	31A-050 N92UG	31A-132 N116FX
25C-083 N54FN	25D-290 N600GM	25XR-029 N107HF	31A-051 N351AC	31A-133 N117FX
25C-084 N777TX	25D-291 N477MM	25XR-056 PT-LBW	31A-052 N75MC	31A-134 N118FX
25C-087 N25TE	25D-292 N604AS	25XR-073 N45CP	31A-053 N44QG	31A-135 PT-WSB
25C-089 PT-IIQ	25D-293 XA-PIU	25XR-079 N85HR	31A-054 N82KK	31A-136 N119FX
25C-097 N22NJ	25D-294 N161RB	25XR-139 N111MP	31A-055 N425M	31A-137 N120FX
25C-098 ZS-NYG	25D-295 XC-AGR	25XR-141 N25HA	31A-056 N56LF	31A-138 V5-NPC
25C-115 PT-ISO	25D-296 PT-OHD	25XR-148 N98RS	31A-057 D-CSAP	31A-139 N131AR
25C-126 N14FN	25D-297 N389AT	25XR-152 N105BA	31A-058 N770CC	31A-140 N140FX
25C-129 N25MR	25D-298 N298DR	25XR-162 N150RS	31A-059 N31TK	31A-141 N121FX
25C-146 I-BMFE	25D-299 N299MW	25XR-173 N777NJ	31A-060 N156SC	31A-142 ZS-EAG
25C-156 PT-KAP	25D-300 N659HX	25XR-175 N307AJ	31A-061 N740F	31A-143 N122FX
25C-176 PT-LLN	25D-301 N25CZ	25XR-219 N55SL	31A-062 AP-BEK	31A-144 N144LJ
25C-181 N73TW	25D-302 N700DA	25XR-220 N99NJ	31A-063 N2	31A-145 N145LJ
25D-206 ZS-LXH	25D-303 XA-JOC	25XR-222 XA-MHA	31A-064 N444HC	31A-146 N218NB
25D-207 N207JC	25D-304 N25NY	25XR-235 N400VC	31A-065 N44SF	31A-147 N202LC
25D-208 N500MP	25D-305 N188R		31A-066 PK-CAH	31A-148 N148LJ
25D-209 N18NM	25D-306 XC-GUB	**Learjet 28**	31A-067 TG-AIR	31A-149 N1904S
25D-210 N75TJ	25D-307 LV-OEL	28-001 N3AS	31A-068 N743F	31A-150 N31NR
25D-211 FAB 010	25D-308 N727LM	28-002 XC-VSA	31A-069 N744E	31A-151 N583PS
25D-213 N925DW	25D-309 XA-PAZ	28-003 N28LR	31A-070 N741F	
25D-214 N245BS	25D-310 N211JC	28-004 XA-KAJ	31A-071 N742F	**Learjet 35**
25D-215 N325JL	25D-311 N199BT	28-005 N500LG	31A-072 N45UF	35-001 N351GL
25D-216 N767SA	25D-312 XC-HIS		31A-073 N46UF	35-002 C-GVVA
25D-217 N41H	25D-313 N727CS	**Learjet 29**	31A-074 N999AU	35-003 N111WB
25D-218 N251DS	25D-314 N95BP	29-001 XC-IST	31A-075 N631SF	35-004 C-GIRE
25D-221 YU-BKR	25D-315 N273M	29-002 XC-HIE	31A-076 XA-SPR	35-005 N178CP
25D-224 N711NM	25D-316 N17AH	29-003 VT-EHS	31A-077 PK-CAJ	35-006 N39FN
25D-225 N140GC	25D-317 N969SS	29-004 VT-EIH	31A-078 N112CM	35-007 N47JR
25D-226 N234SV	25D-319 N911EM		31A-079 N91DP	35-008 PP-ERR
25D-227 N25RE	25D-320 OO-LFR	**Learjet 31**	31A-080 N986MA	35-009 PT-LGR
25D-228 N228SW	25D-321 N25AM	31-002 PT-LVO	31A-081 N83WN	35-010 N888DE
25D-229 CX-ECO	25D-322 5N-AOC	31-003 N888CP	31A-082 PT-MVI	35-011 N400RB
25D-230 N161AC	25D-323 PT-LMM	31-004 XA-ZTH	31A-083 N789SR	35-012 XA-SVX
25D-231 N531CW	25D-324 N970WJ	31-005 N942BY	31A-084 N840SW	35-013 N35BN
25D-232 N500LW	25D-325 XA-TBV	31-006 N26LC	31A-085 N531AT	35-014 N190GC
25D-233 N75LM	25D-326 N25NB	31-007 PT-LXX	31A-086 N105FX	35-015 N335JL
25D-234 N88DJ	25D-327 N52DA	31-009 N173PS	31A-087 N106FX	35-016 N1SC
25D-236 XC-RPP	25D-328 N518JG	31-010 PT-LLK	31A-088 N500JE	35-017 N600DT
25D-238 N500TL	25D-329 N83TE	31-011 HB-VJI	31A-089 N77PH	35-018 N435JL
25D-239 N499EH	25D-330 XC-AA88	31-012 XA-RUU	31A-090 N78PH	35-019 N19NW
25D-240 N339BA	25D-331 N482CP	31-013 PT-LVR	31A-091 V5-NAG	35-022 N90WR
25D-241 N213CA	25D-332 XB-DZQ	31-014 N5VG	31A-092 N711FG	35-023 N886CS
25D-242 N242GS	25D-333 N555SD	31-015 N260LF	31A-093 N916BD	35-024 N241RT
25D-244 N831LH	25D-334 XA-RYH	31-016 N1DE	31A-094 N917BD	35-025 N185BA
25D-245 N60DK	25D-335 PT-LUZ	31-017 N600AW	31A-095 N395LJ	35-026 N89TC
25D-248 N95CK	25D-336 XA-LAP	31-018 N19TJ	31A-096 N30TK	35-027 N31WS
25D-249 N34TN	25D-338 N4447P	31-019 N19LT	31A-097 N50207	35-028 XC-IPP
25D-250 N438DM	25D-339 HK-2624P	31-020 N337FP	31A-098 N148C	35-030 N542PA
25D-251 N85TW	25D-340 N625AU	31-021 XA-RNK	31A-099 N1932P	35-031 N233CC
25D-252 N444MK	25D-341 N58HC	31-022 N331N	31A-100 PT-MCB	35-032 N711MA
25D-253 N253SC	25D-342 N329SS	31-023 N111VV	31A-101 N293SA	35-033 N524PA
25D-254 I-AVJE	25D-343 N456CG	31-024 N92EC	31A-102 N107FX	35-034 N37TA
25D-255 N717EP	25D-345 LV-WLG	31-025 I-AIRW	31A-103 PT-TOF	35-035 N711R
25D-256 N75CK	25D-346 N41TC	31-026 N45HG	31A-104 N108FX	35-036 N90AH
25D-257 N377Q	25D-347 N347JV	31-027 N2FU	31A-105 N109FX	35-037 N520PA
25D-260 N74RD	25D-348 N522JS	31-028 N90WA	31A-106 N29RE	35-038 C-FBFP
25D-261 N261WC	25D-350 N350AG	31-029 XB-FKT	31A-107 N31HY	35-039 N1HP
25D-262 N333CG	25D-351 N302PC	31-030 N255DV	31A-108 N110FX	35-041 N41PJ
25D-263 N825D	25D-353 N510TP	31-031 ZS-OFW	31A-109 N261PC	35-042 N73TJ
25D-264 N547JG	25D-354 N515TC	31-032 XA-AAP	31A-110 PT-WIV	35-043 C-GVCA
25D-265 N69GF	25D-355 LV-WRE	31-033 N632PB	31A-111 C-GRVJ	35-044 N130F
25D-267 N15ER	25D-356 N251MD	31-033A 9V-ATC	31A-112 LX-PCT	35-045 N1140A
25D-268 N829AA	25D-357 LV-WXY	31-033C 9V-ATE	31A-113 N331SJ	35-046 N58EM
25D-269 LV-WOC	25D-358 N37971	31-033D 9V-ATF	31A-114 N524HC	35-048 N8040A
25D-270 N123CG	25D-359 XA-LRJ	31-034 N45PK	31A-115 ZS-NYV	35-049 N899WA
25D-272 N717AN	25D-360 N360JG	31A-035 N3VJ	31A-116 N112FX	35-050 351
25D-273 N73DJ	25D-361 N218NR	31A-036 N127V	31A-117 N317LJ	35-051 N2BA
25D-274 XA-FMR	25D-362 N107RM	31A-037 PT-OVZ	31A-118 N815A	35-052 N541PA
25D-275 N254CL	25D-363 N197LS	31A-038 N500WR	31A-119 N114FX	35-053 N435MS
25D-276 N188TA	25D-364 XA-TIE	31A-039 N10ST	31A-120 I-TYKE	35-055 C-GJCD
25D-277 N81MW	25D-365 XA-SJN	31A-040 N314MK	31A-121 N121LJ	35-056 N645G
25D-278 N70JF	25D-366 XA-SWX	31A-041 N9CH	31A-122 PT-WLO	35-057 C-GTDE
25D-279 N417P	25D-367 VT-SWP	31A-042 D-CGGG	31A-123 N323LJ	35-060 N590CH
25D-280 N280C	25D-368 XA-PIM		31A-124 N931FD	35-061 N238RC

35-062	N31DP	35A-149	N800AW	35A-234	N35WR	35A-321	XC-AA60	35A-408	LV-AIT
35-063	N80PG	35A-152	FAB 009	35A-235	N166HL	35A-322	PT-WGF	35A-409	N351AM
35-064	N100GP	35A-153	N573LR	35A-236	N600GP	35A-323	N735A	35A-410	N820RP
35-065	N4358N	35A-154	N117RB	35A-237	N300TW	35A-324	G-JETN	35A-411	PT-LBY
35-066	352	35A-155	N110AE	35A-238	N500HG	35A-325	N325NW	35A-412	N314C
35A-021	N4415S	35A-156	N190DA	35A-240	N249B	35A-326	N255JC	35A-413	D-CFCF
35A-047	N701AS	35A-157	N26GP	35A-241	N500ED	35A-327	N32PE	35A-414	N196SD
35A-067	N135FA	35A-158	N800GP	35A-242	N242DR	35A-328	N35NX	35A-415	D-COSY
35A-068	T-781	35A-159	D-CAPO	35A-243	XA-THD	35A-329	N261PG	35A-416	N841TT
35A-069	N35NW	35A-160	D-CCCA	35A-244	RP-C57	35A-331	D-CFGC	35A-417	HC-BTN
35A-070	N543PA	35A-161	N433DD	35A-245	N30PA	35A-332	N332FG	35A-418	XA-KCM
35A-071	N199CJ	35A-162	XA-CZG	35A-246	N1DC	35A-334	N235MC	35A-419	N35SM
35A-072	N4415M	35A-163	N27BL	35A-247	N544PA	35A-335	N800CH	35A-420	N100KK
35A-073	N163A	35A-164	N50MJ	35A-248	N128CA	35A-336	N336EA	35A-421	I-VULC
35A-074	N198T	35A-165	N72CK	35A-249	C-FICU	35A-337	N710AT	35A-422	N45AE
35A-075	SE-DHP	35A-166	N719JB	35A-250	N63LF	35A-338	RP-C610	35A-423	D-CAVE
35A-076	N959SA	35A-167	N813AS	35A-251	N251CT	35A-339	PT-LZP	35A-424	N508GP
35A-077	N98LC	35A-168	C-FZQP	35A-252	PT-KZR	35A-340	N11AM	35A-425	N111KK
35A-078	N145AM	35A-169	N135ST	35A-253	N611SH	35A-341	XA-HOS	35A-426	RP-C1426
35A-079	N500DS	35A-170	N354RZ	35A-254	N522PA	35A-342	N56JA	35A-427	N42LL
35A-080	N10AZ	35A-171	N40DK	35A-255	XB-FNW	35A-343	N21NG	35A-428	VH-SLE
35A-081	N118DA	35A-172	N50AK	35A-256	N911ML	35A-344	YV-327CP	35A-429	G-ZENO
35A-082	N700SJ	35A-173	YU-BPY	35A-257	N417BA	35A-345	N30DK	35A-430	LJ-1
35A-083	YV-100CP	35A-174	N38AM	35A-258	N583BS	35A-346	I-DLON	35A-431	N431CW
35A-084	N696JH	35A-175	D-CDWN	35A-259	HK-3983X	35A-347	N85SV	35A-432	G-HUGG
35A-085	N15WH	35A-176	XA-BUX	35A-260	N40PK	35A-348	N35DL	35A-434	N4401
35A-086	N860S	35A-177	D-CITY	35A-261	N58MM	35A-349	XA-TCI	35A-435	XC-HHJ
35A-087	N835GA	35A-178	N900JC	35A-262	N237AF	35A-350	N35WB	35A-436	N436BL
35A-088	N72JF	35A-179	D-CFGA	35A-263	N37FN	35A-351	N500ND	35A-437	YV-432CP
35A-089	D-CCHB	35A-180	N44HG	35A-264	N64CP	35A-352	N35CZ	35A-438	N300R
35A-090	N88BG	35A-181	PT-LSJ	35A-265	G-LEAR	35A-353	C-GDJH	35A-439	N35LW
35A-091	N37FA	35A-182	N221SG	35A-266	N35GC	35A-354	N212GA	35A-440	N903HC
35A-092	N92NE	35A-183	N183FD	35A-268	D-CFGB	35A-355	N345	35A-441	RP-C1404
35A-093	PT-LOT	35A-184	N7092C	35A-269	N211WH	35A-356	PT-LUG	35A-443	N258G
35A-094	N94AF	35A-185	OE-GAV	35A-270	0013	35A-357	ZS-MGK	35A-444	N44SK
35A-095	N971F	35A-186	N96FN	35A-271	N40AN	35A-358	N358PG	35A-445	I-MOCO
35A-096	N94RL	35A-187	N32HM	35A-272	N321AN	35A-359	N136JP	35A-446	N794GC
35A-097	N108RB	35A-188	N999JF	35A-273	N103CL	35A-360	N1129M	35A-447	D-COKE
35A-098	N72DA	35A-189	I-AVJG	35A-274	N274FD	35A-361	PT-OCZ	35A-448	N222BG
35A-099	C-GRFO	35A-190	N202WR	35A-275	N235SC	35A-362	N773LP	35A-449	XA-GDO
35A-100	N109JR	35A-191	N535AF	35A-276	N69BH	35A-363	N19RP	35A-450	N450KK
35A-101	PT-LCD	35A-192	N49BE	35A-277	N42B	35A-364	N353EF	35A-451	LJ-2
35A-105	N18FN	35A-193	N9EE	35A-278	N12RP	35A-365	G-GJET	35A-452	N279SP
35A-108	D-CJPG	35A-194	N86BE	35A-280	87-0026	35A-366	N350DA	35A-453	N802EC
35A-109	N911AE	35A-195	SE-DHO	35A-281	N425AS	35A-367	N232CC	35A-454	N80AR
35A-110	N4J	35A-197	N754GL	35A-282	N62MB	35A-368	N400MC	35A-455	N988QC
35A-111	I-LIAD	35A-198	N198GJ	35A-283	N386CM	35A-369	VR-17	35A-456	N456CL
35A-112	D-CCAY	35A-199	N235JS	35A-284	OO-GBL	35A-370	XA-RKY	35A-457	N49WL
35A-113	N684HA	35A-200	OO-LFY	35A-285	VH-MZL	35A-371	N399BA	35A-458	N86RX
35A-114	T-21	35A-201	N35AZ	35A-286	N286WL	35A-372	PT-LJK	35A-459	N829CA
35A-116	N116AM	35A-202	D-CGPD	35A-287	N71HS	35A-373	N97AN	35A-461	N64CF
35A-117	N78MC	35A-203	N203RW	35A-288	N288JP	35A-374	HZ-106	35A-462	N7117
35A-118	N88JA	35A-204	D-CFTG	35A-289	N36TJ	35A-375	HZ-107	35A-463	N32HJ
35A-119	N549PA	35A-205	N568PA	35A-290	XA-RAV	35A-376	N979RF	35A-464	VP-BJS
35A-120	N400JE	35A-206	N46KB	35A-292	N292ME	35A-377	N18WE	35A-465	N465NW
35A-121	N43TJ	35A-207	N620JM	35A-293	N182K	35A-378	N354ME	35A-467	HZ-MS1
35A-122	N27TT	35A-208	N67PA	35A-295	PT-LAA	35A-379	N18LH	35A-468	OY-CCJ
35A-123	N900BJ	35A-209	N22MS	35A-296	N66NJ	35A-380	C-GAJS	35A-469	N444TG
35A-124	C-GTJL	35A-210	N210WL	35A-297	N777DM	35A-381	N335K	35A-470	LJ-3
35A-125	N351EF	35A-211	N44TT	35A-298	I-FLYC	35A-382	N60WL	35A-471	N110FT
35A-126	N15EH	35A-212	N180MC	35A-299	PT-LGS	35A-383	N364CL	35A-472	N35TN
35A-127	N727GL	35A-213	XC-CUZ	35A-300	N365N	35A-384	N811DD	35A-473	N44AB
35A-128	N231R	35A-214	N279DM	35A-301	N98AC	35A-385	N350EF	35A-474	PT-LEB
35A-129	XA-ZAP	35A-215	N35ED	35A-302	N51LC	35A-386	N13VG	35A-475	ZS-TOW
35A-130	N230R	35A-216	N142LG	35A-303	PT-LLS	35A-387	D-CARL	35A-476	N777LB
35A-131	N155AM	35A-217	N122JW	35A-304	N97QA	35A-389	N436DM	35A-477	N24JG
35A-132	N135AG	35A-218	N481FM	35A-305	N33NJ	35A-390	N831CW	35A-478	PT-LHT
35A-133	N133EJ	35A-219	N350JF	35A-306	N71E	35A-392	N1ED	35A-480	N39DK
35A-134	N235DH	35A-220	N373LP	35A-307	N120MB	35A-393	PT-LOE	35A-481	OO-LFV
35A-135	I-ZOOM	35A-221	N221TR	35A-308	N747GM	35A-394	N626JS	35A-482	N203AL
35A-136	T-22	35A-222	HB-VFZ	35A-309	D-CHPD	35A-395	N246CM	35A-483	VR-18
35A-137	N35TJ	35A-223	D-CGRC	35A-310	N310ME	35A-396	PT-OPJ	35A-484	N485AC
35A-138	N100MS	35A-224	N28MJ	35A-311	HC-BSZ	35A-397	N200TW	35A-485	N485DM
35A-139	D-CGFD	35A-225	N34TJ	35A-312	N369BA	35A-399	PT-OVC	35A-486	N925DM
35A-140	N40BD	35A-226	N30HJ	35A-313	N31WR	35A-400	VH-JIG	35A-487	N400MC
35A-141	N553V	35A-227	N85GW	35A-314	N35AK	35A-401	N177SB	35A-488	N900R
35A-142	N815L	35A-228	SX-BNT	35A-315	D-CCAA	35A-402	N35BG	35A-489	N222BE
35A-143	N20DK	35A-229	N4415W	35A-316	N18ST	35A-403	N101HW	35A-490	N64MP
35A-144	N56EM	35A-230	PT-WAR	35A-317	N317TT	35A-404	N404KA	35A-491	I-AGEN
35A-145	VH-SLD	35A-231	VH-JCR	35A-318	N103CF	35A-405	N442DM	35A-492	N37SY
35A-146	N351AS	35A-232	N8281	35A-319	T-23	35A-406	I-KELM	35A-493	I-FFRI
35A-147	N55F	35A-233	N35AW	35A-320	N320M	35A-407	C-GIWD	35A-494	PT-LDM
								35A-495	N440MC

c/n	Reg	c/n	Reg	c/n	Reg
35A-496	N496SW	35A-579	84-0132	35A-663	D-CCCB
35A-497	N15RH	35A-580	84-0133	35A-664	C-GRMJ
35A-498	I-FLYH	35A-581	84-0134	35A-665	N291K
35A-499	N1TS	35A-582	84-0135	35A-666	XA-SAJ
35A-500	N101US	35A-584	84-0141	35A-667	N135DE
35A-501	N35HW	35A-585	84-0137	35A-668	N441PC
35A-502	N747CP	35A-586	84-0142	35A-669	N669LJ
35A-503	N77NR	35A-587	84-0139	35A-670	N787LP
35A-504	G-RAFF	35A-588	84-0140	35A-671	LV-WPZ
35A-505	N505EE	35A-589	PT-GAP	35A-672	N45KK
35A-506	N10BD	35A-590	N827CA	35A-673	C-GPDO
35A-507	N42HP	35A-591	N822CA	35A-674	N22SN
35A-508	N7777B	35A-592	N93LE	35A-676	N235AC
35A-509	84-0063	35A-593	N593LR		
35A-510	84-0064	35A-594	ZS-PTL		

Learjet 36

c/n	Reg
36-002	N84FN
36-003	N55CJ
36-004	N180GC
36-005	N9108Z
36-007	N83FN
36-008	N101AJ
36-009	N505HG
36-010	N45FG
36-011	PT-KQT
36-012	N547PA
36-013	N71PG
36-014	VH-SLJ
36-015	N10FN
36-016	N12FN
36-017	N32JA
36A-018	PT-KTU
36A-019	N527PA
36A-022	N44EV
36A-023	N56PA
36A-024	N978E
36A-025	N32PA
36A-026	N888TN
36A-027	N27MJ
36A-028	N545PA
36A-029	N116MA
36A-030	N160GC
36A-031	N62PG
36A-032	N950G
36A-034	HY-985
36A-035	N71CK
36A-036	N36MJ
36A-037	N555WH
36A-038	N548PA
36A-039	N99RS
36A-040	N500SV
36A-041	N79SF
36A-043	N423JW
36A-044	N54JA
36A-045	N546PA
36A-046	N17A
36A-047	OE-GMD
36A-048	PT-WGM
36A-049	VH-SLF
36A-050	XC-AA24
36A-051	OB-1431
36A-052	OB-1432
36A-053	HY-984
36A-054	9201
36A-055	PP-JAA
36A-056	9202
36A-057	HB-VIF
36A-059	9204
36A-060	9205
36A-061	9206
36A-062	D-CGFE
36A-063	D-CGFF

Learjet 35A continued:

c/n	Reg	c/n	Reg
35A-511	84-0065	35A-595	N414KL
35A-512	84-0066	35A-596	N826CA
35A-513	84-0067	35A-597	N197JT
35A-514	84-0068	35A-598	PT-LGW
35A-515	84-0069	35A-599	N58GL
35A-516	84-0070	35A-600	N823CA
35A-517	84-0071	35A-601	HY-986
35A-518	84-0072	35A-602	HY-987
35A-519	84-0073	35A-603	HY-988
35A-520	84-0074	35A-604	N73LP
35A-521	84-0075	35A-605	N825CA
35A-522	84-0076	35A-606	N96GS
35A-523	84-0077	35A-607	N68MJ
35A-524	84-0078	35A-608	N14T
35A-525	84-0079	35A-609	XA-JRH
35A-526	84-0080	35A-610	N161MA
35A-527	84-0081	35A-611	N622WG
35A-528	84-0082	35A-612	N551HM
35A-529	84-0083	35A-613	FAB6000
35A-530	84-0084	35A-614	G-OCFR
35A-531	84-0085	35A-615	FAB6001
35A-532	84-0086	35A-616	N876CS
35A-533	84-0087	35A-617	FAB6002
35A-534	84-0088	35A-618	S5-BAA
35A-535	84-0089	35A-619	PT-POK
35A-536	84-0090	35A-620	N232FX
35A-537	84-0091	35A-621	N242MT
35A-538	84-0092	35A-622	N81MR
35A-539	84-0093	35A-623	60504
35A-540	84-0094	35A-624	86-0374
35A-541	84-0095	35A-625	86-0375
35A-542	84-0096	35A-626	C-GNPT
35A-543	84-0097	35A-627	PT-LMY
35A-544	84-0098	35A-628	86-0376
35A-545	84-0099	35A-629	86-0377
35A-546	84-0100	35A-630	N388PD
35A-547	84-0101	35A-631	2710
35A-548	84-0102	35A-632	2711
35A-549	84-0103	35A-633	2712
35A-550	84-0104	35A-634	N626BM
35A-551	84-0105	35A-635	60505
35A-552	84-0106	35A-636	2713
35A-553	84-0107	35A-637	2714
35A-554	84-0108	35A-638	2714
35A-555	84-0109	35A-639	2715
35A-556	84-0110	35A-640	2716
35A-557	84-0111	35A-641	2717
35A-558	84-0112	35A-642	2718
35A-559	84-0113	35A-643	G-LJET
35A-560	84-0114	35A-644	C-GMMY
35A-561	84-0115	35A-645	N43TR
35A-562	84-0116	35A-646	N712JB
35A-563	84-0117	35A-647	N915RB
35A-564	84-0118	35A-648	N97LE
35A-565	84-0119	35A-649	HB-VJJ
35A-566	84-0120	35A-650	N650LR
35A-568	84-0122	35A-651	HB-VJK
35A-569	84-0123	35A-652	D-CURE
35A-570	84-0124	35A-653	HB-VJL
35A-571	84-0125	35A-654	B-98183
35A-572	84-0126	35A-655	N785JM
35A-573	84-0127	35A-656	N356JW
35A-574	84-0138	35A-657	N10AH
35A-575	84-0128	35A-658	N573LP
35A-576	84-0129	35A-659	N873LP
35A-577	84-0130	35A-660	C-GLJQ
35A-578	84-0131	35A-661	VH-PFA
		35A-662	N35UK

Learjet 45

c/n	Reg	c/n	Reg
45-001	N45XL	45-009	N984GC
45-002	N452LJ	45-010	N41DP
45-003	N453LJ	45-011	N741E
45-004	N454LJ	45-012	N412LJ
45-005	N455LJ	45-013	N413LJ
45-006	N456LJ	45-014	N708SP
45-007	ZS-JBR	45-015	N31V
45 008	N745E	45-016	N743E
		45-017	N417LJ
		45-018	N418LJ
		45-019	N56WD
		45-022	N740E
		45-023	N740E
		45-024	N145ST
		45-025	N742E
		45-045	N4545

Learjet 55

c/n	Reg	c/n	Reg
55-003	N553DJ	55-069	N102ST
55-004	D-CLIP	55-071	N155JC
55-005	N440DM	55-072	SX-BNS
55-006	N355DB	55-073	N357PR
55-008	N551SC	55-074	N151PJ
55-009	N955LS	55-075	N90NE
55-011	N200BA	55-076	C-GKTM
55-012	N48HC	55-077	N85NC
55-013	N82679	55-078	I-ALPR
55-014	N155MP	55-079	N1983Y
55-015	D-CION	55-080	PT-LET
55-016	N646G	55-081	N777MC
55-017	D-CCGN	55-082	N817AM
55-018	N599EC	55-083	N551AS
55-019	N141SM	55-084	I-FLYJ
55-020	N55NY	55-085	N55NM
55-021	I-LOOK	55-086	PT-LUK
55-022	VP-BOL	55-087	N520SC
55-023	N110ET	55-088	N901JC
55-024	N900FA	55-089	N555CJ
55-025	N92MG	55-090	D-CGIN
55-026	N421QL	55-091	N700JE
55-027	B-3980	55-092	D-CLUB
55-028	N556GA	55-093	N32KJ
55-029	N100VA	55-094	N235HR
55-030	N55LJ	55-095	N55RT
55-031	YV-12CP	55-096	N126KD
55-032	N83SD	55-097	N20CR
55-033	N971EC	55-098	D-CCON
55-034	N123LC	55-099	N95WK
55-035	C-GPCS	55-100	N500NH
55-036	N155HM	55-101	C-FNRG
55-037	N53HJ	55-102	PT-WSS
55-038	N50AF	55-104	N18CG
55-039	N339BC	55-105	C-GQBR
55-040	N554CL	55-106	N824MG
55-041	HK-4016X	55-107	D-CWAY
55-042	N575GH	55-108	N517AM
55-043	N430HM	55-109	D-CVIP
55-044	PT-LHR	55-110	N455RH
55-045	N550AK	55-112	N7AU
55-046	N55HL	55-113	N57MH
55-047	N600C	55-114	N34GB
55-048	N558AC	55-115	N155BC
55-049	N150MS	55-116	N51V
55-050	OY-FLK	55-117	N155RB
55-051	D-CATL	55-118	C-FCLJ
55-052	D-COOL	55-119	N273MC
55-053	N253S	55-120	N55LK
55-054	N54NW	55-121	N747AN
55-055	N970H	55-122	C-FHJB
55-056	OH-IPP	55-123	N121US
55-057	N733EY	55-124	N58CG
55-058	N129SP	55-125	N610JR
55-059	OE-GRR	55-126	YV-125CP
55-060	N.....	55B-127	N73GP
55-061	D-CFUX	55B-128	N717JB
55-062	N855DB	55B-129	N75GP
55-063	N74RY	55B-130	N55VC
55-064	N255ST	55B-131	N52CT
55-065	N1125M	55B-132	N122SU
55-066	N717HB	55B-133	N700R
55-067	N127GT	55B-134	PT-LDR
55-068	N38D	55C-135	PT-LXO
		55C-136	OE-GCF
		55C-137	N155SP
		55C-138	N338FP
		55C-139	PT-LZS
		55C-139A	N55GM
		55C-140	PT-OCA
		55C-141	N155DB
		55C-142	N555MX
		55C-143	D-CMAD
		55C-144	N40CR
		55C-145	N66WM
		55C-146	PT-ORA
		55C-147	N111US

Learjet 60

c/n	Reg
60-001	N601LJ
60-002	N1940

60-003 N961MR	60-085 N99KW	5095 N731L	5233 7T-VHP	A006SA N200LP
60-004 N60UK	60-086 N797CB	5096 N530G	5234 XA-EKT	A007SA N301DM
60-005 N205FX	60-087 N411ST	5097 N81366	5235 YI-AKB	A008SA N399DM
60-006 N60VE	60-088 XA-TZF	5098 N942Y	5236 N741AM	A009SA N306P
60-007 N219FX	60-089 XA-MDM	5099 N18BH	5237 YI-AKC	A010SA N300DH
60-008 N222FX	60-090 PT-WMO	5100 N510TS	5238 YI-AKD	A011SA N211GA
60-009 N54	60-091 N896R	5101 N511TS	5239 YI-AKE	A012SA I-GIRL
60-010 N928CD	60-092 C-FBLJ	5102 N601JJ	5240 YI-AKF	A013SA I-VIGI

Tristar

60-011 N60T	60-093 RP-C648	5103 XA-TAZ	1064 N787M	A014SA OH-KNE
60-012 N626KM	60-094 A6-SMS	5104 N155AV	1067 N140SC	A015SA N789DJ
60-013 N55	60-095 N602SC	5105 HZ-MA1	1079 N125DT	A016SA N10NM
60-014 N7US	60-096 N603SC	5106 YV-03CP	1247 JY-HKJ	A017SA N33MM
60-015 N960H	60-097 N897R	5107 N69MT	1249 HZ-HM6	A018SA N831TJ
60-016 TC-MEK	60-098 N218FX	5108 N104CE	1250 HZ-HM5	A019SA N400GK
60-017 N660AH	60-099 N212FX	5112 N728PX		A020SA N399RP
60-018 N24G	60-100 N6100	5113 N1962J	**Morane**	A021SA N4LK
60-019 HB-VKI	60-101 N215FX	5114 5B-CHE		A022SA N393FX

MS760

60-020 N600L	60-102 LV-WXN	5117 VP-BSH		A023SA N22BN
60-021 N600LC	60-103 N216FX	5119 N500AG	01 F-BLKL	A024SA N674AC
60-022 N22G	60-104 N83WM	5120 HZ-TNA	2 N207MJ	A025SA N400HH
60-023 N60SB	60-105 N217FX	5121 SU-DAG	6 N760J	A026SA N140AK
60-024 LV-WFM	60-106 N140JC	5123 N57NP	8 N60GT	A027SA N7PW
60-025 N299SG	60-107 D-CFFB	5124 XC-SKI	9 N722Q	A028SA N900WJ
60-026 N347GS	60-108 N220FX	5125 N31BP	28 N760X	A029SA N100RS
60-027 N12FU	60-109 N707SG	5128 5B-CGP	39 F-BJET	A030SA N83SA
60-028 N206FX	60-110 N60LJ	5129 101	43 N760S	A031SA N956PP
60-029 N55KS	60-111 N221FX	5130 102	51 N751PJ	A032SA N929WG
60-030 TC-ELL	60-112 N299SC	5132 XA-BEB	53 N53PJ	A033SA N5EJ
60-031 N228N	60-113 N599SC	5134 XA-JMN	60 N7601R	A034SA N303P
60-032 OE-GNL	60-114 N199SC	5135 N500WN	81 N81PJ	A035SA N702JH
60-033 N56	60-115 N3015F	5136 5A-DAJ	86 N9035Y	A036SA N997MX
60-034 9M-CAL	60-116 N116LJ	5137 1004	89 N999PJ	A037SA N134RG
60-035 N116AS	60-117 N889DW	5138 N801	90 N69X	A038SA N212PA
60-036 N44EL	60-118 N3018C	5140 XA-EMO	97 N97PJ	A039SA PT-OXT
60-037 N637LJ	60-119 N119LJ	5141 5N-...	101 N444ET	A040SA N40GA
60-038 N638LJ	60-120 N120LJ	5142 XA-SOY	102 N20DA	A041SA N300AA
60-039 N57	60 121 N621LJ	5143 N326CB	104 N760R	A042SA N8LE
60-040 N899SC	60-122 N622LJ	5144 JS 10201	105 F-BXQL	A043SA N19R
60-041 N699SC	60-125 N60LR	5145 N511TD	106 N5878	A044SA N606JM

Lockheed

60-042 N90AG		5147 N212AP	107 N5879	A045SA N545TP
60-043 C-GHKY		5148 XA-OLI	108 N760AR	A046SA N109PW
60-044 N618R	**JetStar**	5150 N911CR	111 C6-BEV	A047SA N47PB
60-045 N60WM		5152 XA-SOC	112 N710K	A048SA PT-LNN
60-046 N214FX	5006 N6NE	5153 N430MB		A049SA N411SP
60-047 N418R	5007 N971AS	5154 N43AR	**Piaggio**	A050SA N257CB
60-048 N648LJ	5016 N440RM	5155 VP-BRL		A051SA N70XX
60-049 N227N	5025 SU-DAF	5156 N16AZ		A052SA JA30DA
60-050 N207FX	5029 N112TJ	5158 XA-TDG	**PD808**	A053SA N141H
60-051 N63BL	5032 59-5962	5159 XB-DBS	501 MM577	A054SA N600CG
60-052 ZS-NTV	5033 N500MA	5161 LY-AMB	502 MM578	A055SA N255DG
60-053 B-3981	5036 N900CR	5162 XA-HNY	504 I-PIAL	A056SA N119MH
60-054 N65BL	5037 N552JH	5201 N777AY	505 MM61958	A057SA N7050V
60-055 N1CA	5046 A-9446	5202 EC-FQX	506 MM61948	A058SA N126GA
60-056 N700CH	5048 N428DA	5203 1003	507 MM61949	A059SA N345DM
60-057 N58	5049 N96BB	5204 N202ES	508 MM61950	A060SA N61SA
60-058 N92BL	5051 N488GR	5205 N454JB	509 MM61951	A061SA N18T
60-059 N208FX	5053 XC-JCC	5206 N329JS	510 MM61952	A062SA N616MM
60-060 N209FX	5054 N354CA	5207 N34WR	512 MM61954	A063SA N51B
60-061 N98BL	5058 N200DW	5208 N38BG	513 MM61955	A064SA HI-646SP
60-062 N707SQ	5061 N333EC	5209 N500S	514 MM61956	A065SA N54RM
60-063 N8270	5062 EC-FGX	5210 N787WB	515 MM61957	A066SA N88ME
60-064 N210FX	5063 N499PB	5211 N118B	516 MM61959	A067SA N63DR
60-065 N30W	5064 N3QS	5212 N167G	517 MM61960	A068SA N368PU
60-066 N8271	5065 S9-NAD	5213 N60JM	518 MM61961	A069SA N197SL
60-067 N799SC	5066 XA-MIK	5214 N848AB	519 MM61962	A070SA N60EF
60-068 N95ZC	5069 N197JS	5215 N215DL	520 MM61963	A071SA N71GH
60-069 N60CE	5070 N712TE	5216 N797WC	521 MM62014	A072SA N174SA
60-070 N21AC	5071 SU-DAH	5217 N500EX	522 MM62015	A073SA N94LH
60-071 N940P	5072 N74AG	5218 N901C	523 MM62016	A074SA N32HP
60-072 9M-FCL	5074 N777SG	5219 N104BK	524 MM62017	A075SA N11WF
60-073 N256M	5075 N500ES	5220 N32KR		A077SA N975GR
60-074 N620JF	5076 N76HG	5222 VP-BCP	**Raytheon**	A079SA N574CF
60-075 N9CU	5078 N515AJ	5223 N644JW		A080SA N770PC
60-076 N211FX	5079 N58TS	5224 N285LM		A081SA N50EF
60-077 C-GLRS	5080 N77HW	5225 TC-IHS	**Diamond**	A082SA N62CH
60-078 N188TC	5082 TC-OMR	5226 N308SG	A002SA JA8248	A083SA N12WF
60-079 N95AG	5083 N817BD	5227 N171SG	A003SA N300TS	A084SA N160H
60-080 N8080W	5086 N27RC	5228 N400MP	A004SA N541CW	A085SA N70VT
60-081 N180CP	5087 N33SJ	5229 N222MF	A005SA N30HD	A086SA N486DM
60-082 N600LN	5090 N555SG	5230 N701JH		A087SA HB-VIA
60-083 N383MB	5092 VP-CSM	5231 XA-FHR		A088SA N482DM
60-084 N59FD		5232 N77C		A089SA N20PA
				A090SA C-GLIG

A091SA N400HG
A1008SA N411BW

Beechjet 400

RJ-1	N64VM
RJ-2	N402FB
RJ-3	N2023BA
RJ-4	N92RW
RJ-5	N77GA
RJ-6	YV-738CP
RJ-7	N85BN
RJ-9	N65RA
RJ-10	N131AP
RJ-11	N111BA
RJ-12	N106CG
RJ-13	N3113B
RJ-14	N672AT
RJ-15	N73BE
RJ-16	N803E
RJ-17	N84BJ
RJ-18	N595PT
RJ-19	N101CC
RJ-20	N455DW
RJ-21	N3121B
RJ-22	N48CK
RJ-23	N3123T
RJ-24	N510WS
RJ-25	N125RJ
RJ-26	N91MT
RJ-27	N484CC
RJ-28	N51EB
RJ-29	XA-OAC
RJ-30	N486MJ
RJ-31	I-ALSI
RJ-32	XA-RAR
RJ-33	XA-BNG
RJ-34	N96WW
RJ-35	N737MM
RJ-36	N52GA
RJ-37	N31437
RJ-38	VT-OAM
RJ-39	N48SR
RJ-40	N3240M
RJ-41	N3141G
RJ-42	N444WB
RJ-43	N401CG
RJ-44	I-GCFA
RJ-45	N58AU
RJ-46	VT-TEL
RJ-47	N1547N
RJ-48	XB-JHE
RJ-49	N1549J
RJ-50	N102MC
RJ-52	N196JH
RJ-53	N711EC
RJ-54	N418MG
RJ-55	N780GT
RJ-56	OK-UZI
RJ-58	XA-MII
RJ-59	ZS-MHN
RJ-60	XA-LEG
RJ-61	N701LP
RJ-62	N424BT
RJ-63	N848C
RJ-64	N195JH
RJ-65	N1565B

Beechjet 400A

RK-1	N294AW
RK-2	N272BC
RK-3	N400VK
RK-4	N777FL
RK-5	N501BG
RK-6	N401EE
RK-7	N401AB
RK-8	N440DS
RK-9	N315R
RK-10	D-CEIS
RK-13	N56BE
RK-14	N81TJ
RK-16	N46FE
RK-17	N877Z
RK-18	N717DD
RK-19	N41ME
RK-20	N870P
RK-21	N717VL
RK-22	N85CR
RK-23	N200BL
RK-24	N8073R
RK-25	D-CLBA
RK-26	VH-IMP
RK-27	N10FL
RK-28	PT-WLM
RK-29	I-IPIZ
RK-30	N205R
RK-31	N10J
RK-32	N553PF
RK-33	N197PF
RK-34	N511JP
RK-35	VH-BJD
RK-36	N57B
RK-37	SE-DRS
RK-38	N522EE
RK-39	N70BJ
RK-40	N496EE
RK-41	N546BZ
RK-42	N8253Y
RK-43	N45RK
RK-44	N908R
RK-45	N445CC
RK-46	N8239E
RK-47	N400FT
RK-48	N48SE
RK-49	N54HP
RK-50	N80KM
RK-51	N8085T
RK-52	N709EL
RK-53	N200GP
RK-54	PT-WHG
RK-55	N400Q
RK-56	N89KM
RK-57	ZS-NZO
RK-58	PT-WHC
RK-59	N50KH
RK-60	N61SM
RK-61	G-RAHL
RK-62	N8083N
RK-63	PT-JQM
RK-64	N53MS
RK-65	N81TT
RK-66	HB-VLM
RK-67	N8167Y
RK-68	N295FA
RK-69	N877J
RK-70	N750T
RK-71	N73BL
RK-72	N910SH
RK-73	PT-WHB
RK-74	N26JP
RK-75	N82400
RK-76	N261JP
RK-77	PT-WHD
RK-78	N611PA
RK-79	OH-RIF
RK-80	AP-BEX
RK-81	PT-WHE
RK-82	PT-WHF
RK-83	XA-SNP
RK-84	D-CHSW
RK-85	N419MB
RK-86	N777GC
RK-87	N702LP
RK-88	N1549W
RK-89	N94HE
RK-90	N165HB
RK-91	N296FA
RK-92	N555KK
RK-93	N3038V
RK-94	HB-VLN
RK-95	HS-UCM
RK-96	N3196N
RK-97	N3197Q
RK-98	N866BB
RK-99	N95PA
RK-100	N400SH
RK-101	ZS-JRO
RK-102	N3232U
RK-103	HB-VLW
RK-104	LV-WPE
RK-105	N3235U
RK-106	N1HS
RK-107	N733MK
RK-108	N3218L
RK-109	N3269A
RK-110	N400VP
RK-111	N42SK
RK-112	N3272L
RK-113	N400VG
RK-114	N363K
RK-115	N3265A
RK-116	N1116R
RK-117	N97FF
RK-118	LV-WTP
RK-119	N1119C
RK-120	TC-MDJ
RK-121	N419MS
RK-122	PT-WJS
RK-123	N110TG
RK-124	TC-MSA
RK-125	N400KP
RK-126	N197SD
RK-127	N696TR
RK-128	N912SH
RK-129	N129MC
RK-130	TC-NEO
RK-131	N1083Z
RK-132	N106KC
RK-133	VP-BMR
RK-134	N134BJ
RK-135	N135BJ
RK-136	N1136Q
RK-137	N400AJ
RK-138	N48PL
RK-139	VH-...
RK-140	ZS-OCG
RK-141	N974JD
RK-142	N142BJ
RK-143	N191NC
RK-144	N134CM
RK-145	N745TA
RK-146	N146TA
RK-147	N147BJ
RK-148	TC-SMB
RK-149	N749TA
RK-150	N100AG
RK-151	N1126V
RK-152	N97FB
RK-153	N153BJ
RK-154	VH-...
RK-155	N627RP
RK-156	N2056E
RK-157	N397AT
RK-158	N2358X
RK-159	N1326J
RK-160	N2360F
RK-161	N761TA
RK-162	ZS-...
RK-163	XA-...
RK-164	N2164Z
RK-165	N224MC
RK-166	N2299T
RK-167	N2267B
RK-168	N768TA
RK-169	N2329N
RK-170	N2289B
RK-173	N2273Z
RK-175	N175BJ
RK-176	N476BJ
RK-178	N708TA
RK-180	N709TA
RK-183	N710TA
RK-186	N712TA
RK-189	N715TA
RK-192	N492BJ
RK-194	N194BJ

Beechjet 400T

TX-1	41-5051
TX-2	41-5052
TX-3	41-5053
TX-4	41-5054
TX-5	41-5055
TX-6	51-5056
TX-7	51-5057
TX-8	51-5058
TX-9	71-5059

Jayhawk

TT-1	91-0077
TT-2	90-0412
TT-3	90-0400
TT-4	90-0405
TT-5	89-0284
TT-6	90-0404
TT-7	90-0401
TT-8	90-0402
TT-9	90-0403
TT-10	90-0407
TT-11	90-0406
TT-12	90-0408
TT-13	90-0409
TT-14	90-0410
TT-15	90-0411
TT-16	90-0413
TT-17	91-0076
TT-18	91-0075
TT-19	91-0078
TT-20	91-0079
TT-21	91-0080
TT-22	91-0081
TT-23	91-0082
TT-24	91-0083
TT-25	91-0084
TT-26	91-0085
TT-27	91-0086
TT-28	91-0087
TT-29	91-0088
TT-30	91-0089
TT-31	91-0090
TT-32	91-0091
TT-33	91-0092
TT-34	91-0093
TT-35	91-0094
TT-36	91-0095
TT-37	91-0096
TT-38	91-0097
TT-39	91-0098
TT-40	91-0099
TT-41	91-0100
TT-42	91-0101
TT-43	91-0102
TT-44	92-0330
TT-45	92-0331
TT-46	92-0332
TT-47	92-0333
TT-48	92-0334
TT-49	92-0335
TT-50	92-0336
TT-51	92-0337
TT-52	92-0338
TT-53	92-0339
TT-54	92-0340
TT-55	92-0341
TT-56	92-0342
TT-57	92-0343
TT-58	92-0344
TT-59	92-0345
TT-60	92-0346
TT-61	92-0347
TT-62	92-0348
TT-63	92-0349
TT-64	92-0350
TT-65	92-0351
TT-66	92-0352
TT-67	92-0353
TT-68	92-0354
TT-69	92-0355
TT-70	92-0356
TT-71	92-0357
TT-72	92-0358
TT-73	92-0359
TT-74	92-0360
TT-75	92-0361
TT-76	92-0362
TT-77	92-0363
TT-78	93-0621
TT-79	93-0622
TT-80	93-0623
TT-81	93-0624
TT-82	93-0625
TT-83	93-0626
TT-84	93-0627
TT-85	93-0628
TT-86	93-0629
TT-87	93-0630
TT-88	93-0631
TT-89	93-0632
TT-90	93-0633
TT-91	93-0634
TT-92	93-0635
TT-93	93-0636
TT-94	93-0637
TT-95	93-0638
TT-96	93-0639
TT-97	93-0640
TT-98	93-0641
TT-99	93-0642
TT-100	93-0643
TT-101	93-0644
TT-102	93-0645
TT-103	93-0646
TT-104	93-0647
TT-105	93-0648
TT-106	93-0649
TT-107	93-0650
TT-108	93-0651
TT-109	93-0652
TT-110	93-0653
TT-111	93-0654
TT-112	93-0655
TT-113	93-0656
TT-114	94-0114
TT-115	94-0115
TT-116	94-0116
TT-117	94-0117
TT-118	94-0118
TT-119	94-0119
TT-120	94-0120
TT-121	94-0121
TT-122	94-0122
TT-123	94-0123
TT-124	94-0124
TT-125	94-0125
TT-126	94-0126
TT-127	94-0127
TT-128	94-0128
TT-129	94-0129
TT-130	94-0130
TT-131	94-0131
TT-132	94-0132
TT-133	94-0133
TT-134	94-0134
TT-135	94-0135
TT-136	94-0136
TT-137	94-0137
TT-138	94-0138
TT-139	94-0139
TT-140	94-0140
TT-141	94-0141
TT-142	94-0142
TT-143	94-0143
TT-144	94-0144
TT-145	94-0145
TT-146	94-0146
TT-147	94-0147
TT-148	94-0148
TT-149	95-0040
TT-150	95-0041

TT-151 95-0042	NA0450 C-GAGU	25089 5N-ALH	25271 N70AP	7046 N55MT
TT-152 95-0043	NA0451 N63PM	25090 N429DA	25272 N800JT	7054 G-NCFR
TT-153 95-0044	NA0452 XA-SIV	25091 XA-RSP	25274 EU93-2119	7055 N46PL
TT-154 95-0045	NA0453 N60PM	25097 N67TS	25277 VU93-2126	7061 N810GS
TT-155 95-0046	NA0454 N1910A	25098 YV-815CP	25288 VU93-2127	7062 RA-02809
TT-156 95-0047	NA0455 N195KC	25100 N6SS	25289 VU93-2128	7064 N395RD
TT-157 95-0048	NA0456 N152NS	25101 N78AG		7067 HB-VKJ
TT-158 95-0049	NA0457 N300PM	25103 N402AC		7070 G-DEZC
TT-159 95-0050	NA0458 N940HC	25104 C-FMTC	**125-600**	7073 N701TA
TT-160 95-0051	NA0459 N511WD	25105 HZ-FMA	6001 N561RP	7076 N111ZS
TT-161 95-0052	NA0460 N4402	25107 XA-HFM	6002 XA-SLP	7082 N70HF
TT-162 95-0053	NA0461 N461W	25108 C-GTTS	6003 N91KH	7085 N10TN
TT-163 95-0054	NA0462 N800DR	25109 N4CR	6004 N4TS	7088 N222HL
TT-164 95-0055	NA0464 N341AP	25112 XA-SLR	6006 XX507	7091 G-OCAA
TT-165 95-0056	NA0465 N805X	25113 5N-AVZ	6007 N317TC	7094 N415RD
TT-166 95-0057	NA0466 N466CS	25114 XA-SGP	6008 XX508	7097 RA-02801
TT-167 95-0058	NA0467 N600KC	25115 N249MW	6009 N28TS	7100 G-BNFW
TT-168 95-0059	NA0468 N410BT	25116 N726CC	6013 5N-YET	7103 YL-VIP
TT-169 95-0060	NA0469 N500J	25118 N14HH	6014 N47HV	7107 N71MA
TT-170 95-0061	NA0470 N600J	25119 N500XY	6015 N777TK	7109 VP-BTZ
TT-171 95-0062	NA0471 N57PM	25121 N125TJ	6016 XA-SAI	7112 G-SVLB
TT-172 95-0063	NA0472 N58PM	25123 N125FD	6017 N600WJ	7115 G-BMIH
TT-173 95-0064	NA0473 N25W	25124 N912AS	6018 XA-JRF	7118 G-BWKL
TT-174 95-0065	NA0474 N44HH	25125 F-GFMP	6021 N37SG	7124 HZ-DA4
TT-175 95-0066	NA0475 N935H	25127 N125GK	6022 N2114E	7127 N795A
TT-176 95-0067	NA1001 N100U	25128 F-GECR	6023 N702HC	7130 RP-C235
TT-177 95-0068	NA1002 N229U	25130 F-BSIM	6024 YR-DVA	7136 OH-JET
TT-178 95-0069	NA1003 N208R	25131 F-GJDE	6025 N721LH	7139 RA-02803
TT-179 95-0070	NA1004 N513QS	25132 EI-WDC	6026 XA-SWK	7142 G-BJDJ
TT-180 95-0071	NA1005 N514QS	25133 VT-EQZ	6027 N693TJ	7151 N613MC
	NA1006 N515QS	25138 5N-AVV	6029 PK-HMG	7158 N700VT
HS125	NA1007 N2SG	25140 G-DJLW	6030 N217A	7160 5N-AVK
BAe125	NA1008 N520QS	25147 PK-DJW	6031 9Q-CFW	7163 7T-VCW
Dominie	NA1009 N52SM	25148 N814P	6032 N801BC	7166 F-BYFB
NA0401 N815CC	NA1010 N523QS	25150 VP-BKY	6033 N303MW	7169 B-HSS
NA0402 N757M		25151 N125F	6035 G-SUFC	7172 VT-MPA
NA0403 N89NC	25011 XS709/M	25152 N23CJ	6037 RP-C1600	7175 VP-CEK
NA0404 N494AT	25016 N222NG	25153 N676PC	6038 XA-ACN	7178 N621S
NA0405 N527AC	25017 XA-RSR	25155 N77BT	6040 N601BA	7181 ZD620
NA0406 C-FSCI	25018 N218TJ	25158 XB-PUE	6041 N808RP	7183 ZD703
NA0407 N70NE	25020 N55RF	25159 N600SV	6044 N116DD	7194 ZD704
NA0408 N309G	25023 N584DB	25162 VC93-2120	6045 G-BGYR	7184 G-OMGD
NA0409 N875SC	25027 N125BH	25164 EC93-2125	6046 N299BW	7189 N8KG
NA0410 N683E	25029 N10D	25165 EU93-2121	6047 N68GA	7190 ZD621
NA0411 N600LS	25030 XA-MBM	25167 VC93-2123	6048 TC-COS	7197 N3GL
NA0412 N826CT	25031 N105HS	25168 VC93-2124	6049 HZ-KA5	7200 VP-BMD
NA0413 N221RE	25032 N942DS	25169 N163AG	6050 5N-AOL	7205 ZE395
NA0414 C-FRPP	25033 RP-C125	25171 G-IFTC	6051 N601JA	7208 G-BLSM
NA0415 N800VC	25035 N151SG	25172 ZS-CAL	6052 G-BKBH	7209 VP-BHW
NA0416 C-GCGS	25038 N42CK	25178 5N-WMA	6053 N125WJ	7210 G-BLTP
NA0417 C-GCRP	25039 N911AS	25184 ZS-LPE	6054 5N-YFS	7211 ZE396
NA0418 N802X	25040 XS712/A	25189 M24-01	6055 N125GS	7212 OH-BAP
NA0419 HZ-BL2	25041 XS713/C	25194 G-AXDM	6057 VP-BNW	7213 N213C
NA0420 N803X	25042 N42FD	25197 PT-LHK	6058 N200XR	7214 VP-BCF
NA0421 N804X	25043 N65TS	25209 M24-02	6059 ZF130	7215 VT-OBE
NA0422 N222MS	25045 XS727/D	25215 D2-EXR	6060 5N-AYK	
NA0423 N204SM	25046 N125AD	25217 5N-EAS	6061 N331DC	**125-800**
NA0425 N110MH	25047 N800DA	25219 N219EC	6062 5N-MAY	8001 N801CR
NA0426 N795PH	25048 XS728/E	25227 5N-AMY	6063 N9AZ	8002 N800RY
NA0427 N45Y	25050 XS730/H	25231 N831NW	6064 N125HF	8003 N800N
NA0428 N244JM	25052 N125JR	25235 N297JD	6065 V2-LSF	8004 XA-RET
NA0429 N73WF	25053 N250JT	25238 VP-BKK	6066 N600G	8005 N800FL
NA0430 N850BM	25055 XS731/J	25240 VP-BMB	6067 N822BL	8006 N861CE
NA0431 N682B	25058 N632PE	25242 ZS-LME	6068 N600AE	8007 C-GYPH
NA0432 N540M	25060 N96SG	25243 VP-CTS	6069 5N-...	8008 N722CC
NA0433 N9UP	25064 XB-GGK	25247 9Q-CSN	6070 N83TJ	8009 N48Y
NA0434 N77W	25065 XA-KOF	25248 G-TCDI	6071 N171TS	8010 N810BG
NA0435 C-GMTR	25066 N373DH	25249 N711VT		8011 N186G
NA0436 N369BG	25067 ZS-MAN	25250 N7SJ	**125-700**	8012 N106JL
NA0437 C-FFTM	25068 N5274U	25251 LV-AXZ	7001 N257AJ	8013 N300RB
NA0438 N800PA	25070 N333GZ	25252 P4-AMB	7007 RA-02800	8014 N94WN
NA0439 N74PC	25072 XS736/S	25253 XA-SKE	7010 N3399P	8015 C-GWFM
NA0440 N74NP	25074 N300GB	25254 VT-UBG	7013 N101HF	8016 N904SB
NA0441 N75NP	25075 N731BW	25255 N4QB	7020 N311JD	8017 N217RM
NA0442 N4444J	25076 XS737/K	25258 G-OJPB	7022 N92RP	8018 N601RS
NA0443 C-FWCE	25078 N125NT	25259 ZS-JBA	7025 N886S	8019 N800NW
NA0444 N290EC	25079 N942WN	25260 ZS-JIH	7028 N899DM	8020 N251TJ
NA0445 N174A	25081 XS739/F	25264 N7171	7031 N703TS	8021 G-RCEJ
NA0447 N95AE	25084 N888CJ	25266 N125CK	7034 G-PLGI	8022 EC-ELK
NA0448 N60TC	25085 G-ATPD	25268 N41953	7037 G-IFTE	8023 N1910J
NA0449 N611BA	25087 N330G	25269 ZS-LPF	7040 VP-BPE	8024 N337RE
	25088 5B-...	25270 N270AV		8025 C-GMLR

8026 N6TM	8129 N94	8263 N826GA	8347 N1115G	NA0228 N555CB
8027 N553M	8130 G-ETOM	8264 HB-VLF	8349 N723TA	NA0229 N825CT
8028 G-TSAM	8131 N95	8265 HB-VLG	8350 N23207	NA0230 N360X
8029 N77LA	8133 PT-WAU	8266 N800GT	8351 N23208	NA0231 N225BJ
8030 N91CH	8134 N96	8267 N811CC	8352 N2321S	NA0232 N22KH
8031 PT-LHB	8136 I-SDFG	8268 52-3004	8353 N2321V	NA0233 C-GABX
8032 N526M	8143 5N-AGZ	8269 N380X	8354 N2G	NA0234 N205BS
8033 N673TM	8146 HZ-109	8270 N802CE	8357 N2321Z	NA0235 N10C
8034 N85DW	8148 HZ-110	8271 N803CE	8358 N240B	NA0236 N236BN
8035 N835TS	8149 N577T	8272 N2426	8359 N25WX	NA0237 N511KA
8036 N621MT	8152 N800WG	8273 N967L	8369 N800PC	NA0238 N120MH
8037 7O-ADC	8153 HB-VHV	8274 N2428	8372 N372XP	NA0239 C-GKPM
8038 D-CHEF	8154 N97	8275 N77TC		NA0240 N700HH
8039 N193TR	8155 D-CBMW	8276 N126KC	**125-1000**	NA0241 N300BS
8040 N832MR	8156 N98	8277 N97SH	9003 N503QS	NA0242 N701CF
8041 N71NP	8158 N99	8278 N872AT	9004 VP-CPT	NA0243 N104JG
8042 N112K	8159 N10855	8279 4X-CZM	9005 N505QS	NA0244 N414RF
8043 N313CC	8164 130	8280 N351SP	9007 G-BTSI	NA0245 N81QV
8044 N833JP	8165 ZS-BPG	8281 VH-ELJ	9008 N195L	NA0246 N79HC
8045 N800TF	8167 VP-CAS	8282 PT-WHH	9012 HZ-SJP2	NA0247 N120JC
8046 N125SB	8169 N526AC	8283 4X-COV	9016 N291H	NA0248 C-FBMG
8047 N84FA	8175 N204JC	8284 PT-WNO	9017 N204R	NA0249 XB-FMK
8048 C-GCIB	8176 HB-VJY	8285 N285XP	9018 5N-FGR	NA0250 N29GD
8049 N93CT	8177 N411RA	8286 N501F	9021 VP-CMZ	NA0251 N810CR
8050 G-ICFR	8180 G-XRMC	8287 N801WB	9022 VH-LMP	NA0252 N895CC
8051 N888DH	8182 N12F	8288 72-3005	9024 N5ES	NA0253 N422X
8052 N87EC	8184 PT-OSW	8289 N515GP	9025 N525QS	NA0254 N125XX
8053 N5G	8186 N8186	8290 N348MC	9026 G-GDEZ	NA0255 N540B
8054 N527M	8190 PT-OHB	8291 N791TA	9027 N333RL	NA0256 N571CH
8055 N528M	8192 TR-LDB	8292 N33BC	9028 D-CBWW	NA0257 N611MC
8056 G-JETI	8194 PT-OTC	8293 N404CE	9029 EZ-B021	NA0258 N193RC
8057 N300GN	8197 G-OMGE	8294 N404BS	9030 N530QS	NA0259 N941CE
8058 G-OMGG	8198 PT-WAL	8295 VH-LAW	9031 N301PH	NA0260 N184TB
8059 N400GN	8201 G-BWSY	8296 N707TA	9032 N401LS	NA0261 N983GT
8060 N330X	8203 N453TM	8297 N297XP	9033 N533QS	NA0262 N711WM
8061 N161MM	8208 TC-ANC	8298 N880SP	9034 N81HH	NA0263 N101FC
8062 N961JC	8210 G-RAAR	8299 N299XP	9035 N535QS	NA0264 N88MX
8063 N74ND	8211 N91DV	8300 N800VF	9036 N1AB	NA0265 N91CM
8064 MAAW-J1	8212 RP-C8008	8301 PT-WMA	9037 G-SHEC	NA0266 N497PT
8065 N77CS	8213 N10857	8302 XA-RUY	9038 N125GM	NA0267 N36GS
8066 N75CS	8214 PT-OOI	8303 N876H	9039 N539QS	NA0268 XA-SON
8067 N801MB	8215 29-3041	8304 N5734	9040 N540QS	NA0269 N972LM
8068 N68HR	8219 9M-AZZ	8305 72-3006	9041 N541QS	NA0270 PK-CTC
8069 N746UP	8222 G-VIPI	8306 N1103U	9042 N542QS	NA0271 XA-KAC
8070 N528AC	8224 N847RH	8307 N109TD	9043 TC-AKH	NA0272 N77D
8071 N890A	8226 RP-C1926	8308 N345BR	9044 N544QS	NA0273 N110EJ
8072 N747UP	8227 39-3042	8309 N5735	9045 N545QS	NA0274 HB-VLL
8073 VP-BSI	8228 HB-VKV	8310 PT-WMG	9046 N546QS	NA0275 N125SJ
8074 N850SM	8229 TC-TEK	8311 N804JT	9047 N547QS	NA0277 N509QC
8075 XA-GFB	8230 N71MT	8312 PT-WMD	9048 N548QS	NA0278 N700FE
8076 RA-02807	8231 N75MT	8313 N84BA	9049 N549QS	NA0279 N5511A
8077 N509GP	8232 XA-NGS	8314 OM-SKY	9050 N550QS	NA0280 XA-TCR
8078 OE-GHS	8233 VP-BTM	8315 N9292X	9051 N551QS	NA0281 N169TA
8079 9M-DDW	8234 VP-CDE	8316 N516GP	9052 N552QS	NA0282 N120YB
8080 N800WH	8235 OY-RAA	8317 N9NB		NA0283 N26SC
8081 N196MC	8236 N80PM	8318 N1910H	NA0201 N701TS	NA0284 N10UC
8082 N90ME	8237 VP-BAW	8319 C-FIPE	NA0202 N727TA	NA0285 N130YB
8083 N8UP	8238 N70PM	8320 N720TA	NA0203 N620M	NA0286 N150CA
8084 N877RP	8239 N62TC	8321 N32BC	NA0204 N724B	NA0287 N3444H
8085 G-WBPR	8240 HB-VLT	8322 N722TA	NA0205 N333PC	NA0288 N700BW
8086 N523WC	8241 N94NB	8323 N877S	NA0206 N701NW	NA0289 N27KL
8087 C-GAWH	8242 49-3043	8324 N303BC	NA0207 N33RH	NA0290 N700AC
8088 G-BTAB	8243 VH-XMO	8325 N1112N	NA0208 N38PA	NA0291 N45AF
8089 N862CE	8244 N95NB	8326 N1897A	NA0209 N74B	NA0292 N748FB
8090 N800FJ	8245 52-3001	8327 N111ZN	NA0210 SE-DPZ	NA0293 N7WG
8091 HB-VIK	8246 HB-VKW	8328 VH-SGY	NA0211 N411PA	NA0294 N925WC
8092 N3007	8247 52-3002	8329 N901K	NA0212 N125CS	NA0295 N89MD
8093 N317CC	8248 N789LB	8330 N139M	NA0213 N370RR	NA0296 N28GP
8094 D-CFAN	8249 N826SU	8331 N10NB	NA0214 N926TC	NA0297 N945CE
8095 C-FPCP	8250 52-3003	8332 N36H	NA0215 N215RS	NA0298 N80K
8096 N10TC	8251 N194JS	8334 N399JC	NA0216 N23SK	NA0299 N703JP
8097 HB-VIL	8252 N938H	8335 OY-RAC	NA0217 N7006	NA0300 N943CE
8098 N300LS	8253 N801CE	8336 N2286L	NA0218 N648WW	NA0301 N107LT
8099 N10YJ	8254 N2015M	8337 9M-VVV	NA0219 N428FS	NA0302 N795HE
8106 PK-WSJ	8255 N127KC	8338 N838QS	NA0220 XA-SAU	NA0303 N67PW
8109 5N-NPC	8256 N256BC	8339 SE-DVD	NA0221 N705JH	NA0304 N96PR
8110 N710A	8257 N802DC	8340 N840QS	NA0222 C-GNAZ	NA0305 N305TH
8112 N112NW	8258 N54SB	8342 N2320J	NA0223 N154JD	NA0306 N400WP
8115 G-TCAP	8259 N966L	8343 N1102U	NA0224 N200GY	NA0307 N270KA
8116 PT-LQP	8260 N957H	8344 N29GP	NA0225 N995SK	NA0308 N10CN
8118 105	8261 PT-GAF	8345 N1135X	NA0226 N7WC	NA0309 N309WM
8120 VT-EAU	8262 N959H	8346 N23204	NA0227 N81KA	NA0310 N18SH

NA0311 N425WN	NA752 N700FA	265-59 61-0656	**Sabreliner T-39B**	282-43 FAE-043
NA0312 N500GS	NA753 N345GL	265-60 61-0657	270-1 59-2873	282-44 N64MA
NA0313 PK-CTA	NA754 N711HL	265-61 61-0658	270-2 59-2874	282-45 N333NM
NA0314 N53GH	NA755 N755WJ	265-62 61-0659	270-3 60-3474	282-48 XA-RGC
NA0315 N26ME	NA756 N400JK	265-63 61-0660	270-4 60-3475	282-49 86001
NA0316 N18BA	NA757 N900WG	265-64 61-0661	270-5 60-3476	282-50 XA-SMQ
NA0317 N612MC	NA758 N731MS	265-66 61-0663	270-6 60-3477	282-51 N225LS
NA0318 N80CL	NA759 N400MR	265-67 61-0664		282-52 N64DH
NA0319 XB-MLC	NA760 N456WH	265-68 61-0665	**Sabreliner T-39D**	282-56 XA-RPS
NA0320 N7490A	NA761 N814D	265-69 61-0666	277-1 150542	282-58 N110FS
NA0321 C-GTLG	NA762 N400GP	265-70 N7143N	277-2 150543	282-60 165524
NA0322 N109G	NA763 N19H	265-71 61-0668	277-3 150544	282-61 165517
NA0323 VT-AAA	NA764 XB-CUX	265-72 61-0669	277-4 150545	282-63 N325K
NA0324 N880CR	NA765 N68CB	265-73 61-0670	277-5 150546	282-65 N145G
NA0325 C-GOCM	NA766 N150SA	265-74 61-0671	277-6 150547	282-66 165513
NA0326 N164WC	NA768 XA-SFQ	265-76 61-0673	277-7 150548	282-67 N104RF
NA0327 C-GAAA	NA769 N369JH	265-77 61-0674	277-8 150549	282-69 N49RJ
NA0328 N2HP	NA770 N38LB	265-78 61-0675	277-9 150550	282-70 N3280G
NA0329 N190WC	NA771 N4WC	265-79 61-0676	277-10 150551	282-71 XA-SMR
NA0331 XB-JTN	NA773 N32KB	265-80 N9166Y		282-72 165515
NA0332 N332WE	NA774 N774TS	265-81 61-0678	**Sabreliner T-39D**	282-74 N707FH
NA0333 N125AS	NA775 N333DP	265-82 N6581E	285-1 150969	282-77 165518
NA0334 N46WC	NA776 XB-GNF	265-83 N32010	285-3 150971	282-79 N35CC
NA0335 C-GJBJ	NA778 N89SR	265-84 61-0681	285-4 150972	282-80 XA-FTN
NA0336 VT-SRR	NA779 N808CC	265-85 61-0682	285-5 150973	282-81 165510
NA0337 N201PM	NA780 N65DL	265-86 XB-GDW	285-6 150974	282-82 XB-EQR
NA0338 N710AG		265-87 61-0684	285-7 150975	282-84 157353
NA0339 PT-ORJ	**Sabre**	265-88 61-0685	285-8 150976	282-85 157354
NA0340 N23BJ			285-9 150977	282-90 165516
NA0341 I-CIGH	**Sabreliner T-39A**	**Sabreliner T-39A**	285-10 150978	282-91 86002
NA0342 N93FR	265-1 59-2868	276-2 62-4449	285-11 150979	282-92 N825SB
NA0343 N524M	265-2 59-2869	276-3 62-4450	285-12 150980	282-93 158381
NA0344 C-GMBA	265-3 59-2870	276-4 N31403	285-13 150981	282-94 165521
NA0345 N913V	265-4 59-2871	276-5 62-4452	285-14 150982	282-95 N425NA
NA700 N230TS	265-5 59-2872	276-6 XA-TGO	285 15 150983	282-96 158383
NA701 PT-...	265-6 60-3478	276-7 62-4454	285-16 150984	282-98 N516LW
NA702 C-GMEA	265-7 60-3479	276-8 62-4455	285-17 N32508	282-99 N12BW
NA703 N140JS	265-8 60-3480	276-9 62-4456	285-18 150986	282-100 165525
NA704 C-GSKV	265-9 60-3481	276-10 62-4457	285-19 150987	282-101 N160W
NA705 N25MJ	265-10 XA-TFD	276-12 62-4459	285-20 150988	282-102 XA-PIH
NA706 N777GA	265-11 60-3483	276-14 62-4461	285-21 150989	282-103 N730CP
NA707 N160AG	265-13 60-3485	276-15 62-4462	285-22 150990	282-104 N104SL
NA708 XA-RPT	265-14 XA-...	276-16 62-4463	285-23 150991	282-105 XC-AA73
NA709 N55G	265-15 60-3487	276-17 62-4464	285-24 150992	282-106 N333GM
NA710 N322TP	265-16 N431NA	276-18 62-4465	285-25 151336	282-108 N85CC
NA711 N711AQ	265-17 60-3489	276-19 62-4466	285-26 151337	282-109 FAE-047
NA712 N712VS	265-18 60-3490	276-20 62-4467	285-27 151338	282-110 N250EC
NA713 N773AA	265-19 60-3491	276-21 XA-...	285-28 151339	282-111 N7KG
NA714 N311JA	265-20 60-3492	276-22 62-4469	285-29 151340	282-112 N40ZA
NA715 N824TJ	265-21 60-3493	276-23 62-4470	285-30 151341	282-113 N430MP
NA718 N250DH	265-22 60-3494	276-24 62-4471	285-31 151342	282-114 XB-RGO
NA719 N545S	265-23 60-3495	276-25 N35RG	285-32 151343	282-115 XA-LML
NA720 XB-FRP	265-24 60-3496	276-26 62-4473		282-116 XB-BBL
NA721 N12YS	265-25 60-3497	276-27 XA-TDX	**Sabre 40**	282-117 N265SC
NA722 N38TS	265-26 60-3498	276-28 62-4475	282-1 N116SC	282-120 N73DR
NA723 N401JR	265-27 60-3499	276-29 62-4476	282-2 165512	282-122 N188PS
NA724 N724TS	265-28 60-3500	276-30 62-4477	282-3 N467H	282-123 XB-ESS
NA726 N922RR	265-29 60-3501	276-31 62-4478	282-4 N408TR	282-124 N20ES
NA727 N100RH	265-30 60-3502	276-32 N988MT	282-6 XA-GYT	282-125 XA-SQA
NA728 N410PA	265-33 60-3505	276-33 N39FS	282-9 165509	282-126 N40GT
NA729 VU93-2118	265-35 60-3507	276-34 N33UT	282-11 N10SL	282-128 N99114
NA730 N101HS	265-36 60-3508	276-35 62-4482	282-13 N502RR	282-129 XA-RLH
NA731 N31TJ	265-37 61-0634	276-36 62-4483	282-15 N43W	282-130 XA-REG
NA732 N732TS	265-38 61-0635	276-37 62-4484	282-18 N131BH	282-131 LV-WND
NA733 N125RT	265-39 61-0636	276-38 62-4485	282-19 165519	282-132 N240CF
NA734 N99ST	265-40 61-0637	276-39 N265WB	282-20 165523	282-133 I-RELT
NA735 N800GE	265-41 61-0638	276-40 62-4487	282-23 N123CD	282-134 XB-MVG
NA736 N112M	265-42 N21092	276-41 62-4488	282-24 N8AF	282-135 N2006
NA737 N643JL	265-44 61-0641	276-42 N65618	282-26 XA-DAN	282-136 CP-2317
NA738 VU93-2117	265-45 61-0642	276-43 62-4490	282-27 N61RH	282-137 N881DM
NA739 N820MC	265-46 61-0643	276-44 XA-...	282-28 165522	
NA740 VU93-2114	265-48 XA-TFL	276-46 62-4493	282-29 165511	**Sabre 50**
NA742 N74RT	265-49 61-0646	276-47 62-4494	282-30 165514	287-1 N50CR
NA743 HK-3653	265-50 61-0647	276-48 N1929P	282-31 N34AM	
NA744 N711BP	265-51 61-0648	276-50 62-4497	282-32 165520	**Sabre 60**
NA745 N745TS	265-52 61-0649	276-51 62-4498	282-33 N168W	306-1 N359WJ
NA746 N103RR	265-53 61-0650	276-52 62-4499	282-34 N940CC	306-2 XA-PUR
NA747 N400FE	265-54 61-0651	276-53 62-4500	282-36 N40LB	306-3 LV-WPO
NA748 YV-1111CP	265-55 61-0652	276-54 62-4501	282-38 N999VT	306-4 N121JE
NA749 SE-DVS	265-56 61-0653	276-55 62-4502	282-39 XA-RTM	306-5 N161CM
NA750 N131LA	265-57 61-0654		282-41 N240AC	306-6 XA-SMF
NA751 HC-BTT	265-58 61-0655			

306-7 XA-SND	306-101 N376D	465-37 N750CC	380-41 N400NR
306-8 N813BR	306-102 N70HL	465-38 N850CC	380-42 XA-MVT
306-9 N4LG	306-103 N40TL	465-39 N203JK	380-44 N2440G
306-10 N946JR	306-104 160053	465-40 N465PM	380-45 FAE-045
306-13 N306CF	306-105 160054	465-41 N800M	380-46 XC-AA89
306-15 XA-RUQ	306-106 160055	465-42 N45NP	380-47 N25BX
306-17 N401MS	306-107 160056	465-43 N83TF	380-48 N100BP
306-18 N12PB	306-109 XA-SBV	465-44 N7NR	380-49 N265KC
306-19 N50DG	306-110 XA-RTP	465-45 N65TJ	380-50 XA-TDQ
306-20 XA-REI	306-111 XA-SKB	465-46 N65FF	380-51 N180NA
306-21 XB-QND	306-112 CC-CTC	465-47 N265A	380-52 N84NG
306-22 XA-CHP	306-113 N113T	465-48 N500WD	380-53 N380SR
306-23 N85HS	306-114 N65R	465-49 N82CR	380-54 N910BH
306-24 N600GL	306-115 FAB 001	465-50 N920DY	380-55 XB-GSP
306-25 LV-WOF	306-116 N39CB	465-51 N114LG	380-56 N22NB
306-26 XA-CEN	306-117 FAE-001A	465-52 N96RE	380-57 JY-AFH
306-27 N105DM	306-118 N607CF	465-53 N80R	380-58 XA-GHR
306-28 N353CA	306-119 N48MC	465-54 N65SR	380-59 N27LT
306-31 N274CA	306-120 N265SR	465-55 XB-RYO	380-60 XB-SHA
306-32 N3278	306-121 XA-SYS	465-56 N499NH	380-62 JY-AFP
306-33 N633SL	306-122 N56RN	465-57 N355CD	380-63 N75RS
306-34 MTX-01	306-123 XA-ATE	465-58 N670AS	380-64 N942CC
306-35 XB-JMR	306-124 N48WS	465-59 N61DF	380-65 9A-BLY
306-36 XA-RIR	306-125 XA-SLJ	465-60 N88BF	380-66 N819GY
306-37 N562R	306-126 N111F	465-61 N23BX	380-68 XC-UJD
306-38 XC-HGY	306-127 XA-CUR	465-62 N65AF	380-69 N547JL
306-40 XA-SBX	306-128 N117JL	465-63 N2N	380-70 N110AJ
306-41 N856MA	306-129 N95RC	465-65 XA-SCR	380-71 HZ-NR1
306-42 XA-VEL	306-130 XB-JMM	465-66 N964C	380-72 N933JC
306-43 N115CR	306-131 N131JR	465-67 N921CC	
306-44 N562MS	306-132 N265U	465-68 N930RA	
306-46 N642RP	306-133 N400JH	465-69 N25KL	
306-47 XB-ESX	306-134 N282WW	465-70 N58HT	
306-48 N4NT	306-135 N921MB	465-71 N75VC	
306-50 XB-FSZ	306-137 XA-SAH	465-72 OO-RSE	
306-51 N60JC	306-138 XA-RVT	465-73 N651MK	
306-52 158843	306-139 XC-UJE	465-74 N700JC	
306-53 N999LG	306-140 N26SQ	465-75 N570R	
306-54 N33TR	306-141 N8NR	465-76 N65L	
306-55 158844	306-142 N700MH		
306-56 XA-DSC	306-143 XA-SUN		
306-57 XA-RLS	306-144 XC-UJF		
306-58 N529SC	306-145 XC-JDC		
306-59 N10LX	306-146 XA-ARE		

Sabre 65

465-1 N117MB
465-2 N45H
465-3 N1CF
465-4 N804PA
465-5 N241H
465-6 N1CC
465-7 N2000
465-8 XA-GAP
465-9 N6GV
465-10 N336RJ
465-11 N57MQ
465-12 N112PV
465-13 N945CC
465-14 XA-SPM
465-15 XA-ZUM
465-16 N920CC
465-17 N74VC
465-18 N696US
465-19 N91BZ
465-20 N173A
465-21 N701FW
465-22 N927AA
465-23 N223LB
465-24 N777SK
465-25 N812WN
465-26 N488DM
465-27 N111AD
465-28 N24RF
465-29 N6NR
465-30 N65TC
465-31 N265M
465-32 HB-V..
465-33 N265C
465-34 N47SE
465-35 N65AK
465-36 N652MK

Sabre 75

370-2 N10M
370-3 N125NX
370-4 N370BH
370-5 XA-RYJ
370-7 N670C
370-9 N370SL

Sabre 75A

380-1 N100EJ
380-2 N642TS
380-3 T-10
380-4 LV-...
380-5 N71460
380-6 N711GD
380-9 N995RD
380-11 N151TB
380-12 N75BS
380-13 AE-175
380-14 N72028
380-15 N424R
380-16 N12659
380-17 N380BC
380-19 N80HG
380-20 N773W
380-21 N82AF
380-23 N800CD
380-25 N13NH
380-26 N2200A
380-27 N90GW
380-29 N71543
380-30 N265DP
380-31 N75CN
380-32 N380DJ
380-33 N7148J
380-34 N97SC
380-36 XB-MCB
380-37 N774W
380-38 N75AK
380-39 XA-SXK
380-40 N14TN

306-60 N15HF
306-61 N1JX
306-62 N162JB
306-63 XB-ZNP
306-64 N74BS
306-65 159361
306-66 159362
306-67 159363
306-68 FAE-049
306-69 159364
306-70 159365
306-71 N71CC
306-72 XA-GIH
306-73 XC-OAH
306-75 N11LX
306-76 N86CP
306-78 N477X
306-79 N43JG
306-80 N61FB
306-82 N59K
306-83 XA-RLL
306-84 N55ZM
306-85 N855CD
306-86 XA-ICK
306-87 XA-RFB
306-88 XA-RAP
306-89 XA-STI
306-90 N123FG
306-91 LV-WXX
306-92 N33JW
306-93 N507U
306-94 N217RN
306-95 N124DC
306-96 XB-ETV
306-97 N15DJ
306-98 N531AB
306-99 N66GE
306-100 XB-LAW

Swearingen

SJ 30
001 N30SJ

VFW

VFW614
G14 17+01
G17 D-ADAM
G18 17+02
G19 17+03

Visionaire

Vantage
001 N247VA

Yakovkev

9440338 OM-BYE
9940560 OM-BYL

BUSINESS TURBOPROPS
Country Index

BUSINESS TURBOPROPS
By Country

AP = Pakistan *(Islamic Republic of Pakistan)*
Civil

Regn	Yr	Type	c/n	Owner/Operator	Prev Regn
□ AP-...	91	Reims/Cessna F406	F406-0059	Aircraft Sales & Services P/L. Karachi.	N3122E
□ AP-BBR	90	DHC 6-300	782	Oil & Gas Development Corp.	C-GEVP
□ AP-BCQ	85	Conquest II	441-0352	Government of Punjab, Lahore.	N1213N
□ AP-BCW	85	Conquest II	441-0355	Government of Baluchistan, Quetta.	N1213Y
□ AP-BCY	85	Conquest II	441-0350	Government of Sind, Karachi.	N12127
□ AP-CAA	77	King Air 200	BB-278	Directorate of Civil Aviation, Karachi.	AP-CAD

Military

□ 11667	81	Gulfstream 840	11667	Pakistan Army,	N5919K
□ 11733	84	Gulfstream 840	11733	Pakistan Army,	N56GA
□ 927	82	King Air B200	BB-927	Pakistan Air Force, Rawalpindi.	N18262

A2 = Botswana *(Republic of Botswana)*
Civil

□ A2-...	82	King Air B200	BB-1046		ZS-MBN
□ A2-AEO	86	King Air B200	BB-1253	De Beers Botswana Mining Co P/L. Orapa.	N2614X
□ A2-AEZ	79	King Air 200	BB-421	Kalahari Air Services, Gaborone.	N4488L
□ A2-AGO	90	King Air B200	BB-1353	De Beers Botswana Mining Co P/L. Orapa.	N15599
□ A2-AHA	82	King Air B200	BB-961	Executive Charter P/L. Gaborone.	G-BNZH
□ A2-AHT	80	Conquest 1	425-0011	B L C Aviation Ltd.	N6161P
□ A2-AHV	83	King Air F90-1	LA-212	Sladden International P/L. Gaborone.	N6726P
□ A2-AHZ	76	King Air 200	BB-95	Executive Charter P/L. Gaborone.	ZS-JPD
□ A2-AIH	73	Rockwell 690A	11105	Executive Charter P/L. Gaborone.	ZS-MXE

A6 = United Arab Emirates
Civil

□ A6-MHH	96	King Air 350	FL-131	Government of UAE,	

Military

□ 825	96	King Air 350	FL-132	Dubai Air Force,	N3263Y

B = China *(People's Republic of China)*
Civil

□ B-...	94	King Air 350	FL-113		N8291Y
□ B-3551	85	King Air B200	BB-1204	China General Aviation Corp. Taiyuan.	N6927C
□ B-3552	85	King Air B200	BB-1205	China General Aviation Corp. Taiyuan.	N6927D
□ B-3553	85	King Air B200	BB-1206	Air China International Corp. Beijing.	N6927G
□ B-3621	90	PA-42 Cheyenne IIIA	5501051	CAAC Flying College, Guanghan.	(D-IOSF)
□ B-3622	90	PA-42 Cheyenne IIIA	5501052	CAAC Flying College, Guanghan.	(D-IOSG)
□ B-3623	90	PA-42 Cheyenne IIIA	5501054	CAAC Flying College, Guanghan.	N92409
□ B-3624	90	PA-42 Cheyenne IIIA	5501056	CAAC Flying College, Guanghan.	N9241D
□ B-3625	92	PA-42 Cheyenne IIIA	5501059	CAAC Flying College, Guanghan.	(OE-FAB)
□ B-3626	94	PA-42 Cheyenne IIIA	5501060	CAAC Flying College, Guanghan.	N9115X

B-H = Hong Kong *(Chinese Colony of Hong Kong)*
Civil

□ B-HZM	87	King Air B200C	BL-128	Government Flying Service, Kai Tak.	VR-HZM
□ B-HZN	88	King Air B200C	BL-130	Government Flying Service, Kai Tak.	VR-HZN

B = Taiwan *(Republic of China)*
Civil

□ B-.....	94	King Air 350	FL-111		N8139K
□ B-135	91	King Air 350	FL-52	Civil Aeronautical Administration, Taipei.	N81664
□ B-13152	79	King Air 200	BB-449	Government of Taiwan, Taipei.	N2068L
□ B-13153	93	King Air 350	FL-108	Department of Forests/Council of Agriculture, Taipei.	(VH-...)

C = Canada
Civil

Regn	Yr	Type	c/n	Owner/Operator	Prev Regn
□ C-....	81	Conquest 1	425-0080	785102 Ontario Inc.	N880EA
□ C-....	92	King Air 350	FL-87	Lawrence Construction Ltd. Beauport, PQ.	C-GSCL
□ C-....	80	King Air B100	BE-89		N737MG
□ C-....	82	King Air B100	BE-130	Thabet Aviation International Inc. Ste Foy, PQ.	N6241P
□ C-....	79	PA-31T Cheyenne II	7920069	Craig Evan Corp/CEC Flightexec, London, ON.	N250KA
□ C-....	76	PA-31T Cheyenne II	7620033	sold Canada 3/97 ?	N124AA
□ C-FAFD	70	King Air 100	B-42	Kenn Borek Air Ltd/United Nations, Islamabad, Pakistan.	LN-VIP
□ C-FAMO	69	HS 748-2A	1669	Aerial Recon Surveys Ltd. Whitecourt, AB.	
□ C-FAMU	73	King Air A100	B-166	Voyageur Airways Ltd. North Bay, ON.	N221SS
□ C-FANF	69	SA-26T Merlin 2A	T26-32	Keewatin Air Ltd. Rankin Inlet, NT.	N742G

Regn	Yr	Type	c/n	Owner/Operator	Prev Regn
☐ C-FAPP	73	King Air A100	B-169	Voyageur Airways Ltd. Sioux Lookout, ON.	N305TZ
☐ C-FASB	73	King Air A100	B-163	Thunder Airlines Ltd. Thunder Bay, ON.	SE-ING
☐ C-FATR	70	681 Turbo Commander	6020	Ministic Air Ltd. Winnipeg, MT.	N114MR
☐ C-FATW	76	King Air C90	LJ-685	Air Tindi Ltd. Yellowknife, NT.	N110SE
☐ C-FATX	75	King Air E90	LW-147	Pem-Air Ltd. Pembroke, ON.	N4RG
☐ C-FAWE	68	Gulfstream 1	188	Propair Inc. Rouyn-Noranda, PQ.	HB-LDT
☐ C-FAWG	63	Gulfstream 1	106	Airwave Transport/812971 Ontario Inc. Toronto, ON.	N64CG
☐ C-FAXN	66	680V Turbo Commander	1546-9	Aero Aviation Centre (1981) Ltd. Edmonton, AB.	N2549E
☐ C-FBCN	74	King Air 200	BB-7	Kenn Borek Air Ltd. Calgary, AB.	
☐ C-FBGS	74	King Air A100	B-204	Voyageur Airways Ltd. North Bay, ON.	N108JL
☐ C-FBPT	65	King Air 90	LJ-13	Toranda Leasing Inc. Calgary, AB.	N99W
☐ C-FCAV	81	PA-42 Cheyenne III	8001006	Carson Air/Kevin J Carson, Williams Lake, BC.	N131RC
☐ C-FCAW	70	SA-26AT Merlin 2B	T26-172E	Carson Air Ltd. Williams Lake, BC.	N135SR
☐ C-FCCD	69	Mitsubishi MU-2F	174	767070 Ontario Ltd. St Catherines, ON.	N889Q
☐ C-FCEC	81	Cheyenne II-XL	8166030	Craig Evan Corp (Flightexec), London, ON.	N76TW
☐ C-FCED	81	Cheyenne II-XL	8166013	CEC Flight Exec, London, ON.	N2501Y
☐ C-FCGB	75	King Air/Catpass 250	BB-24	Bar XH Air Ltd. Medicine Hat, AB.	N183MC
☐ C-FCGC	77	King Air/Catpass 250	BB-236	Northern Thundrbird Air Ltd. Prince George, BC.	N46KA
☐ C-FCGE	66	King Air A90	LJ-118	Transport Canada, Ottawa, ON.	
☐ C-FCGH	67	King Air A90	LJ-203	Transport Canada, Winnipeg, MT.	
☐ C-FCGI	67	King Air A90	LJ-220	Transport Canada, Toronto, ON.	
☐ C-FCGL	77	King Air/Catpass 250	BB-190	Central Mountain Air Ltd. Smithers, BC.	N190MD
☐ C-FCGM	77	King Air 200	BB-217	Northern Thunderbird Air Ltd. Prince George, BC.	N200CD
☐ C-FCGN	67	King Air A90	LJ-313	Transport Canada, Winnipeg, MT.	
☐ C-FCGT	76	King Air/Catpass 250	BB-159	Wasaya Airways Ltd. Pickle Lake, ON.	N47FH
☐ C-FCGU	78	King Air/Catpass 250	BB-301	Wasaya Airways Ltd. Pickle Lake, ON.	N611SW
☐ C-FCGW	77	King Air/Catpass 250	BB-207	Air Nunavut Ltd. Iqaluit, NT.	N111WH
☐ C-FCGX	77	King Air/Catpass 250	BB-250	Central Mountain Air Ltd. Smithers, BC.	N1008J
☐ C-FCOS	79	PA-31T Cheyenne 1	7904042	Skyservice FBO Inc. Dorval, PQ.	N23555
☐ C-FCSD	71	King Air 100	B-75	Voyageur Airways Ltd. North Bay, ON.	N24MK
☐ C-FDAM	69	King Air/Catpass 250	B-8	Pem-Air Ltd. Pembroke, ON.	N59T
☐ C-FDOR	72	King Air A100	B-103	Transport Canada, Ottawa, ON.	
☐ C-FDOS	72	King Air A100	B-106	Transport Canada, Montreal, PQ.	
☐ C-FDOU	72	King Air A100	B-112	Transport Canada, Edmonton, AB.	
☐ C-FDOV	72	King Air A100	B-117	Transport Canada, Edmonton, AB.	
☐ C-FDOY	72	King Air A100	B-120	Transport Canada, Montreal, PQ.	
☐ C-FEQB	79	PA-31T Cheyenne II	7920071	491549 Alberta Ltd. Calgary, AB.	SX-ABU
☐ C-FFFG	75	Mitsubishi MU-2L	662	878455 Ontario Ltd/Airmed Canada, Sudbury, ON.	N5191B
☐ C-FFSS	80	MU-2 Marquise	783SA	Air 500 Ltd. Toronto, ON.	N81604
☐ C-FGEM	80	MU-2 Solitaire	434SA	886875 Ontario Inc. St Catherines, ON.	N24MW
☐ C-FGFZ	78	King Air 200	BB-403	Provincial Airlines Ltd. St Johns, NF.	N147K
☐ C-FGIN	73	King Air A100	B-164	AeroPro, Ste Foy, PQ.	N164RA
☐ C-FGNL	74	King Air A100	B-184	Province of Newfoundland, St Johns, NF.	
☐ C-FGPK	79	Conquest II	441-0107	Crown Mail & Delivery Services Ltd. Brantford, ON.	N441SC
☐ C-FGSX	83	Cheyenne II-XL	8166048	160878 Canada/Aviation Commercial Aviation, Hearst, ON.	N600XL
☐ C-FGWA	79	PA-31T Cheyenne II	7920045	Craig Evan Corp/Flightexec, Toronto, ON.	N52LS
☐ C-FGWT	85	Gulfstream 900	15042	Grenfell Regional Health Services Board, St Anthony, NF.	N71GA
☐ C-FHBO	63	Gulfstream 1	104	Petro Canada Inc. Calgary, AB.	N719G
☐ C-FHGG	75	King Air A100	B-207	Wilderness Airline (1975) Ltd. Richmond, BC.	N727LE
☐ C-FHRV	84	PA-42 Cheyenne 400LS	5527010	Hughes Air Corp. Calgary, AB.	N100AK
☐ C-FHWI	67	King Air A90	LJ-309	Northern Dene Airways Ltd. Uranium City, SK.	N329H
☐ C-FIDN	69	King Air 100	B-3	Westair Aviation Inc. Fort St John, BC.	N128RC
☐ C-FIFE	76	Mitsubishi MU-2L	683	Nav Air Charter Inc. Victoria, BC.	OY-CEF
☐ C-FIFO	79	King Air 200	BB-527	Provincial Airlines Ltd. St Johns, NF.	N662L
☐ C-FJAK	81	Cheyenne II-XL	8166028	Centerline (Windsor) Ltd. Windsor, ON.	N355SS
☐ C-FJEL	78	Mitsubishi MU-2N	706SA	878455 Ontario Ltd/Airmed Canada, Sudbury, ON.	N866MA
☐ C-FJFH	73	King Air A100	B-171	Northern Thunderbird Air Ltd. Prince George, BC.	N888TB
☐ C-FJHP	68	King Air B90	LJ-325	Propair Inc. Rouyn-Noranda, PQ.	N900LD
☐ C-FJRT	76	King Air 200	BB-92	J T S Aviation Corp. Vancouver, BC.	145201
☐ C-FJVB	81	PA-31T Cheyenne II	8120012	Lochhead-Haggerty Engineering & Manufacturing Co. Delta, BC.	C-FZIH
☐ C-FKAL	96	Pilatus PC-XII	151	Kelner Airways Ltd. Goose Bay.	N151PB
☐ C-FKIJ	70	King Air 100	B-52	Pem-Air Ltd. Pembroke, ON.	N8NP
☐ C-FKIO	78	Mitsubishi MU-2N	725SA	La Ronge Aviation Services Ltd. La Ronge, SK.	N888RH
☐ C-FKJI	76	King Air 200	BB-105	Walsten Air Service (1986) Ltd. Kenora, ON.	N71TZ
☐ C-FLTS	73	King Air A100	B-149	Timberline Air Ltd. Chilliwack, BC.	N883CA
☐ C-FMAI	73	King Air A100	B-145	Myrand Aviation Inc. Ste-Foy, PQ.	N380W
☐ C-FMHB	76	PA-31T Cheyenne II	7620023	491549 Alberta Ltd. Calgary, AB.	SX-ABT
☐ C-FMHD	88	King Air 300	FA-151	9010-9091 Quebec Inc. St Georges, PQ.	N282PC
☐ C-FMKD	68	King Air B90	LJ-376	North Cariboo Flying Service Ltd. Fort St John, BC.	N300RV
☐ C-FMPA	96	Pilatus PC-XII	164	RCMP-GRC Air Services, Ottawa, ON.	HB-FRZ
☐ C-FMPE	80	King Air 200	BB-746	RCMP-GRC Air Services, Ottawa, ON.	N707BC
☐ C-FMPH	81	King Air 200	BB-757	RCMP-GRC Air Services, Ottawa, ON.	N72GA
☐ C-FMSK	80	Conquest II	441-0153	Anderson Air Ltd. Vancouver, BC.	N86CG
☐ C-FMSP	79	Conquest II	441-0072	Anderson Air Ltd. Vancouver, BC.	N441PL
☐ C-FMWM	70	King Air 100	B-59	Ministic Air Ltd. Winnipeg, MT.	N702JL
☐ C-FMXY	70	King Air 100	B-40	North Cariboo Flying Service Ltd. Fort St John, BC.	N923K

Regn	Yr	Type	c/n	Owner/Operator	Prev Regn
☐ C-FNAS	72	Rockwell 690	11003	First Nations Air Service Ltd. Deseronto, ON.	N72TT
☐ C-FNCB	78	King Air E90	LW-287	Pentastar Transportation Ltd. Spruce Grove, AB.	N23660
☐ C-FNCN	69	King Air B90	LJ-468	North Canada Corp Ltd. Yellowknife, NT.	N1FC
☐ C-FNED	76	King Air C90	LJ-680	Alberta Central Airways Ltd. Lac La Biche, AB.	N928RD
☐ C-FNGA	90	P-180 Avanti	1007	Union Gas Ltd. Chatham, ON.	I-RAII
☐ C-FNWC	81	Conquest II	441-0216	The North West Co. Winnipeg, Mt.	N441DM
☐ C-FOGY	76	King Air 200	BB-168	Propair Inc. Rouyn-Norana, PQ.	N10VW
☐ C-FOIL	97	King Air 350	FL-164	Stone Creek Resources Ltd. Calgary, AB.	N1104Y
☐ C-FPAJ	73	King Air A100	B-151	Propair Inc. Rouyn-Noranda, PQ.	N324B
☐ C-FPQQ	88	King Air B200	BB-1304	West Wind Aviation Inc. Winnipeg, MT.	(VT-...)
☐ C-FPWR	91	King Air 350	FL-62	Churchill Falls (Labrador) Corp. Churchill Falls, NF.	N82882
☐ C-FQOV	70	King Air 100	B-38	Little Red Air Service Ltd. Fort Vermilion, AB.	N931M
☐ C-FRJE	78	PA-31T Cheyenne II	7820002	Westair Aviation Inc. Kamloops, BC.	C-GCUL
☐ C-FROM	73	Mitsubishi MU-2J	601	Nav Air Charter Inc. Sidney, BC.	N308MA
☐ C-FROW	74	Mitsubishi MU-2J	628	Nav Air Charter Inc. Victoria, BC.	N4202M
☐ C-FRWK	81	MU-2 Marquise	1521SA	Samaritan Air Service Ltd. Toronto, ON.	N437MA
☐ C-FSEA	84	Conquest 1	425-0192	Sunlite Electric (St Paul) Ltd. St Paul, AB.	N1221T
☐ C-FSIK	78	King Air B100	BE-39	Max Aviation Inc. Three Rivers, PQ.	N129CP
☐ C-FSOZ	84	PA-42 Cheyenne IIIA	5501014	Huisson Aviation (1989) Ltd. Carp, ON.	N900MP
☐ C-FSVC	68	SA-26T Merlin 2A	T26-19	Keewatin Air Ltd. Rankin Inlet, NT.	N2JE
☐ C-FTIX	78	SA-226AT Merlin 4A	AT-066	Western Express Air Lines Inc. Vancouver, BC.	N5455M
☐ C-FTJC	76	SA-226AT Merlin 4A	AT-051	Bilingual Productions Ltd. Hamilton, ON. (was TC-226).	N4NY
☐ C-FTLB	87	King Air 300	FA-137	Chevron Canada Resources Ltd. Calgary, AB.	C-GMBA
☐ C-FTOO	72	Mitsubishi MU-2J	549	Nav Air Charter Inc. Victoria, BC.	N65198
☐ C-FTPE	93	King Air C90B	LJ-1342	Tatham Process Engineering Inc. Toronto, ON.	
☐ C-FTWO	75	Mitsubishi MU-2L	672	Nav Air Charter Inc. Victoria, BC.	N709US
☐ C-FUDY	83	Conquest II	441-0286	George Hogarth, Buttonville, ON.	N441WM
☐ C-FUFW	66	King Air 90	LJ-84	College d'Enseignement General et Professionel, Longueil, PQ	N619GS
☐ C-FVAX	83	Conquest II	425-0178	Venture Aviation Services Ltd. Richmond, BC.	(N90GM)
☐ C-FWAM	64	Gulfstream 1	148	Wabush Mines/Skyservice FBO Inc. Dorval, PQ.	N1701L
☐ C-FWCC	80	PA-42 Cheyenne III	8001001	Carson Air Ltd. Williams Lake, BC.	N124EL
☐ C-FWOL	71	King Air 100	B-84	Mission Aviation Fellowship of Canada, Angola.	N401TJ
☐ C-FWPG	71	King Air 100	B-67	Alberta Central Airways Ltd. Lac La Biche, AB.	N26KW
☐ C-FWPN	70	King Air 100	B-51	Alberta Central Airways Ltd. Lac La Biche, AB.	N16SW
☐ C-FWPT	83	Cheyenne II-XL	8166066	Sawridge Energy Ltd. Slave Lake, AB.	N9170C
☐ C-FWRL	78	Conquest II	441-0079	Skyways/565176 Ontario Ltd. Waterloo-Guelph, ON.	N441KR
☐ C-FWRM	72	King Air A100	B-125	Propair Inc. Rouyn-Noranda, PQ.	N89JM
☐ C-FWUT	76	PA-31T Cheyenne II	7620039	Ashe Aircraft Enterprises Ltd. Calgary, AB.	N82031
☐ C-FWWK	89	King Air 300	FA-182	Carrier Lumber Ltd. Prince George, BC.	N8840A
☐ C-FWYO	70	King Air 100	B-28	Airco Aircraft Salvage Ltd. Edmonton, AB.	N27JJ
☐ C-FXAJ	72	King Air A100	B-122	Air Mikisew Ltd/Contact Air, Fort McMurray, AB.	N8181Z
☐ C-FXNB	67	King Air A90	LJ-257	AeroPro, Ste Foy, PQ.	N711VP
☐ C-FYBV	75	PA-31T Cheyenne II	7520015	Ashe Aircraft Enterprises Ltd. Calgary, AB.	N11232
☐ C-FYTK	75	PA-31T Cheyenne II	7620008	Ashe Aircraft Enterprises Ltd. Calgary, AB.	F-BXSA
☐ C-FZIC	81	PA-31T Cheyenne II	8120104	McNutt Lumber Co. Fredrichton, NB. (was 8120069)	N9162Y
☐ C-FZPW	81	King Air B200	BB-940	Keewatin Air Ltd. Winnipeg, MT.	N519SA
☐ C-FZRQ	72	Rockwell 690	11025	Air Spray (1967) Ltd. Edmonton, AB.	N100LS
☐ C-FZVW	81	King Air 200	BB-787	Voyageur Airways Ltd. North Bay, ON.	N26G
☐ C-FZVX	77	King Air 200	BB-231	Voyageur Airways Ltd. North Bay, ON.	N200FH
☐ C-GAAL	73	Rockwell 690A	11104	Conair Aviation Ltd. Abbotsford, BC.	N690AZ
☐ C-GAFT	75	King Air/Catpass 250	BB-57	Alta Flights (Charters) Inc. Edmonton, AB.	N121DA
☐ C-GAGE	79	Conquest II	441-0086	Lake Central Airways, Toronto, ON.	N20BF
☐ C-GAMC	80	MU-2 Marquise	785SA	878455 Ontario Ltd/Airmed Canada, Sudbury, ON.	N273MA
☐ C-GAPK	74	King Air A100	B-198	Ministic Air Ltd. Winnipeg, MT.	N712AS
☐ C-GARO	72	King Air 200	BB-2	Pratt & Whitney Canada Inc. Longueil, PQ.	N200KP
☐ C-GASI	72	King Air A100	B-126	Thunder Airlines Ltd. Thunder Bay, ON.	N23BW
☐ C-GAST	73	King Air A100	B-173	Awood Air Ltd. Sidney, BC.	F-GECV
☐ C-GASW	72	King Air A100	B-108	Thunder Airlines Ltd. Thunder Bay, ON.	N110JJ
☐ C-GAVI	74	King Air A100	B-201	G-C Air North Inc. St Georges, PQ.	G-BBVM
☐ C-GBTI	74	King Air E90	LW-111	427112 Ontario Ltd/Fort Frances Air, Fort Frances, ON.	N11GE
☐ C-GBVB	79	MU-2 Solitaire	400SA	Highline Produce Ltd. Leamington, ON.	N666MA
☐ C-GBZM	69	SA-26AT Merlin 2B	T26-122	Air Dorval Ltee. Dorval, PQ.	N63SC
☐ C-GCAU	82	SA-227AT Merlin 4C	AT-469	Carson Air Ltd. Kelowna, BC.	N469GM
☐ C-GCFB	81	King Air C90	LJ-929	Transport Canada, Vancouver, BC.	N81DD
☐ C-GCFD	72	King Air A100	B-104	Transport Canada, Moncton.	N72X
☐ C-GCFF	79	King Air 200	BB-474	Transport Canada, Moncton.	
☐ C-GCFL	70	King Air B90	LJ-500	Transport Canada, Toronto, ON.	N20WC
☐ C-GCFM	80	King Air C90	LJ-886	Transport Canada, Vancouver, BC.	N15SL
☐ C-GCFZ	79	King Air C90	LJ-849	Transport Canada, Vancouver, BC.	N6647P
☐ C-GCTA	80	Conquest II	441-0140	Athabaska Airways Ltd. Prince Albert, SK.	(N26264)
☐ C-GDCL	74	Rockwell 690A	11192	Diamond Aviation Ltd. Fredericton, NB.	N57192
☐ C-GDEF	78	SA-226AT Merlin 4A	AT-069	Propair Inc. Dorval, PQ.	N63SC
☐ C-GDPB	82	King Air B200C	BL-44	Air Tindi Ltd. Yellowknife, NT.	N18379
☐ C-GDPI	73	King Air A100	B-156	Voyageur Airways Ltd. North Bay, ON.	N21RX
☐ C-GDSH	84	King Air B200	BB-1178	Derek Stimson Holdings Ltd. Lethbridge, AB.	N46CE
☐ C-GEAL	77	PA-31T Cheyenne II	7720062	Accent Aviation Services Inc. Calgary, AB.	SX-ABV

Regn	Yr	Type	c/n	Owner/Operator	Prev Regn
☐ C-GEAS	90	King Air 350	FL-17	Province of Saskatchewan, Regina, SK.	N56872
☐ C-GEBA	76	PA-31T Cheyenne II	7620029	Beaver Air Services Ltd. The Pas, MT.	N177JE
☐ C-GEOS	76	Rockwell 690A	11279	Geographic Air Survey Ltd. Edmonton, AB.	N57180
☐ C-GFAB	66	680V Turbo Comm	1601-43	Province of New Brunswick, Fredericton, NB.	N577RH
☐ C-GFDV	80	Conquest 1	425-0015	S'Wiches Ltd. Cambridge, ON.	N40RD
☐ C-GFFH	81	MU-2 Marquise	1522SA	Air 500 Ltd. Toronto, ON.	N902M
☐ C-GFLL	82	Cheyenne II-XL	8166041	Green Forest Lumber Ltd. Toronto, ON.	N61WA
☐ C-GFSA	97	King Air 350	FL-174	Province of Alberta, Edmonton, AB.	
☐ C-GFSB	75	King Air 200	BB-84	Province of Alberta, Edmonton, AB.	
☐ C-GFSG	75	King Air 200	BB-671	Province of Alberta, Edmonton, AB.	
☐ C-GFSH	81	King Air B200	BB-912	Province of Alberta, Edmonton, AB.	
☐ C-GGDC	80	MU-2 Marquise	796SA	Air 500 Ltd. Toronto, ON.	N700MA
☐ C-GGFL	78	PA-31T Cheyenne II	7820029	Ontario Secondary Teachers Federation, Buttonville, ON.	N783SW
☐ C-GGGL	65	King Air 90	LJ-38	Regionnair Inc. Chevery, PQ.	N1128B
☐ C-GGGQ	83	King Air B200	BB-1128	New Brunswick Telephone Co. St John, NB.	
☐ C-GGPS	78	PA-31T Cheyenne II	7820023	Province of Saskatchewan, Regina, SK.	
☐ C-GHOC	74	King Air A100	B-194	Kenn Borek Air Ltd. Calgary, AB.	N57237
☐ C-GHOP	76	King Air 200	BB-120	Toma Jetprop Ltd. Calgary, AB.	N6773S
☐ C-GHQG	77	Rockwell 690B	11353	Government of Canada, Ottawa, ON.	N4TX
☐ C-GHRM	76	PA-31T Cheyenne II	7620053	Crown Mail & Delivery Services Ltd. Brantford, ON.	(N717CB)
☐ C-GHSI	75	PA-31T Cheyenne II	7520025	Voyageur Airways Ltd. Kapuskasing, ON.	
☐ C-GHVR	68	King Air B90	LJ-337	Transport Canada, Ottawa, ON.	
☐ C-GIDC	80	PA-42 Cheyenne III	8001002	Alkan Air Ltd. Whitehorse, YU.	N61QR
☐ C-GILM	72	King Air A100	B-124	Voyageur Airways Ltd. North Bay, ON.	N100SJ
☐ C-GIND	81	King Air B200C	BL-42	Ministry of Health/Voyageur Airways Ltd. Sioux Lookout, ON.	N819CD
☐ C-GISH	73	King Air A100	B-152	Voyageur Airways Ltd. North Bay, ON.	(N67LG)
☐ C-GJAV	73	Mitsubishi MU-2J	593	Falls Aviation Ltd/Welland Aero Center, Welland, ON.	N400RX
☐ C-GJBV	72	King Air A100	B-100	Voyageur Airways Ltd. North Bay, ON.	N100S
☐ C-GJEI	81	Gulfstream 1000	96012	Irving Oil Transport Ltd. Saint John, NB.	N9931S
☐ C-GJHW	73	King Air A100	B-175	Bradley Air Services Ltd. Carp, ON.	N92DL
☐ C-GJJE	78	PA-31T Cheyenne II	7820053	TG Ventures Inc. Calgary, AB.	F-GJJF
☐ C-GJJF	72	King Air A100	B-123	Voyageur Airways Ltd. North Bay, ON.	N741EB
☐ C-GJLJ	77	King Air A100	B-235	Propair Inc. Rouyn-Noranda, PQ.	N23517
☐ C-GJLP	73	King Air A100	B-148	Propair Inc. Rouyn-Noranda, PQ.	N67V
☐ C-GJPT	75	PA-31T Cheyenne II	7520039	Province of Saskatchewan, Regina, SK.	
☐ C-GKAJ	77	King Air A100	B-232	Bearskin Lake Air Service Ltd. Thunder Bay, ON.	N400WH
☐ C-GKBB	73	King Air C90	LJ-607	Kenn Borek Air/Little Red Air Service Ltd. Ft Vermilion, AB.	N48DA
☐ C-GKBN	75	King Air 200	BB-29	Kenn Borek Air Ltd. Calgary, AB.	LN-ASG
☐ C-GKBP	79	King Air 200	BB-505	Kenn Borek Air Ltd. Calgary, AB.	C-GKBP
☐ C-GKBQ	70	King Air 100	B-62	Kenn Borek Air Ltd. Calgary, AB.	LN-NLB
☐ C-GKBZ	71	King Air 100	B-85	Kenn Borek Air Ltd. Calgary, AB.	LN-PAJ
☐ C-GKFG	59	Gulfstream 1	22	Kelowna Flightcraft Air Charter Ltd. Kelowna, BC.	N8BG
☐ C-GKMV	78	PA-31T Cheyenne II	7820003	Aries Aviation Services Corp. Calgary, AB.	N444ER
☐ C-GKMX	77	PA-31T Cheyenne II	7720058	491549 Alberta Ltd. Calgary, AB.	N7WS
☐ C-GKSC	81	King Air F90	LA-113	Sanjel Cementers Ltd. Calgary, AB.	N890CA
☐ C-GLAG	79	PA-31T Cheyenne II	7920027	Air Cascades Inc. Victoriaville, PQ.	N71QS
☐ C-GLBL	82	Conquest II	441-0250	Jo-Ad Industries Ltd. Toronto, ON.	N441MC
☐ C-GLKA	68	SA-26T Merlin 2A	T26-20	Keewatin Air Ltd. Rankin Inlet, NT.	N77WF
☐ C-GLMC	71	681 Turbo Commander	6044	Aero Photo (1961) Inc. Quebec City, PQ.	N4798M
☐ C-GLPG	73	King Air A100	B-159	AeroPro, Ste Foy, PQ.	N110KF
☐ C-GLRR	66	King Air A90	LJ-134	Little Red Air Service Ltd. High Level, AB.	N38LA
☐ C-GMDF	76	PA-31T Cheyenne II	7620019	Westside Logging Ltd. Quesnel, BC.	
☐ C-GMEH	92	King Air B200	BB-1433	Skyjet Aviation Inc. Halifax, NS.	N8043K
☐ C-GMFI	85	PA-42 Cheyenne 400LS	5527023	Malette Forest Inc. Calgary, AB.	N429BX
☐ C-GMGG	93	King Air B200	BB-1467	Lake Central Airways, Toronto, ON.	N124SC
☐ C-GMMO	82	Gulfstream 1000	96034	Wal-Mart Canada Inc. Toronto, ON.	N508AB
☐ C-GMOC	79	King Air/Catpass 250	BB-513	Alkan Air Ltd. Whitehorse, YT.	N513SA
☐ C-GMOU	68	SA-26AT Merlin 2B	T26-107	Earlton Airways Ltd. Earlton, ON.	N642RF
☐ C-GMPO	80	King Air 200	BB-667	RCMP-GRC Air Services, Ottawa, ON.	N183DW
☐ C-GMSL	83	Conquest II	441-0283	Anderson Air Ltd. Vancouver, BC.	N925WS
☐ C-GMTI	80	King Air F90	LA-65	1022978 Ontario Ltd/R S Lipic, Sudbury, ON.	C-GMIT
☐ C-GMWR	75	King Air 200	BB-68	Provincial Airlines Ltd. St Johns, NF.	N844N
☐ C-GMZG	69	SA-26AT Merlin 2B	T26-165	Earlton Airways Ltd. Earlton, ON.	N67HM
☐ C-GNAA	70	King Air 100	B-24	Air Mikisew Ltd/ Contact Air, Fort McMurray, AB.	N382WC
☐ C-GNAJ	72	King Air A100	B-107	Northern Air Charter (PR) Inc. Peace River, AB.	LN-AAH
☐ C-GNAK	65	Gulfstream 1	154	Business Flights/414660 Alberta Ltd, Calgary, AB.	G-BNKO
☐ C-GNAM	80	PA-31T Cheyenne II	8020065	Bar XH Air Inc. Medicine Hat, AB.	N118EL
☐ C-GNAR	74	King Air A100	B-190	Northern Air Charter (PR) Inc. Peace River, AB.	LN-AAG
☐ C-GNAX	70	King Air 100	B-47	Northern Air Charter (PR) Inc. Peace River, AB.	XA-PAK
☐ C-GNBB	79	King Air 200	BB-479	Pentastar Transportation Ltd. Spruce Grove, AB.	N200UQ
☐ C-GNDI	76	PA-31T Cheyenne II	7620036	Fast Air Ltd. Winnipeg, MT.	N73TB
☐ C-GNEX	75	King Air A100	B-211	Bearskin Lake Air Service Ltd. Thunder Bay, ON.	
☐ C-GNKP	75	PA-31T Cheyenne II	7520008	Province of Saskatchewan, Regina, SK.	
☐ C-GNVB	73	King Air A100	B-143	North Vancouver Airlines Ltd. Richmond, BC.	N151U
☐ C-GOGT	79	King Air 200	BB-535	Province of Ontario, Sault Ste Marie, ON.	
☐ C-GPCB	70	King Air 100	B-45	North Vancouner Airlines Ltd. Vancouver, BC.	N704S

Regn	Yr	Type	c/n	Owner/Operator	Prev Regn
□ C-GPCD	75	King Air 200	BB-76	Provincial Airlines Ltd. St Johns, NF.	N500DR
□ C-GPCL	74	SA-226AT Merlin 4	AT-017	Western Express Air Lines Inc. Vancouver, BC.	N5RT
□ C-GPCP	76	King Air 200	BB-140	Wilderness Airline (1975) Ltd. Richmond, BC.	
□ C-GPDX	76	Rockwell 690A	11319	O C Holdings '87 Inc. New Westminster, BC.	N22HP
□ C-GPPC	95	King Air 350	FL-127	Pancanadian Petroleum Ltd. Calgary, AB.	
□ C-GPRO	73	SA-226T Merlin 3	T-239	Southern Aviation Ltd. Regina, SK.	N833S
□ C-GPSP	78	Conquest II	441-0058	Hauts-Monts Inc. Beauport, PQ.	OY-BHM
□ C-GPTG	68	Gulfstream 1	189	Airwave Transport/812971 Ontario Inc. Mississauga, ON.	N776G
□ C-GPTN	62	Gulfstream 1C	88	Bradley Air Services Ltd. Carp, ON.	N857H
□ C-GQCC	75	PA-31T Cheyenne II	7620003	Odessey Aviation Ltd. Mississauga, ON.	LV-LZO
□ C-GQGA	81	King Air F90	LA-106	Simo Air Ltd. Mississauga, ON.	N19112
□ C-GQJG	77	King Air 200	BB-249	Avionair Inc. Montreal, PQ.	
□ C-GQNJ	77	King Air 200	BB-275	Province of Ontario, Sault Ste Marie, ON.	
□ C-GQXF	77	King Air 200	BB-285	Northern Lights Air Ltd. Smithers, BC.	
□ C-GRBF	70	SA-26AT Merlin 2B	T26-171E	Peace Air/Roberts Brothers Farming Ltd. Falher, AB.	N50AK
□ C-GRFN	82	King Air B200	BB-1054	Toma Jetprop Ltd. Calgary, AB.	N810V
□ C-GRJM	83	Conquest 1	425-0177	R J M Corp. Edmonton, AB.	N600TJ
□ C-GRSL	74	King Air C90	LJ-609	AeroPro, Ste Foy, PQ.	N38BA
□ C-GSAA	91	PA-42 Cheyenne IIIA	5501057	Province of Saskatchewan, Regina, SK.	N120GA
□ C-GSAX	76	King Air C90	LJ-697	Slave Air 1988 Ltd. Slave Lake, AB.	N90AW
□ C-GSBC	78	PA-31T Cheyenne II	7820014	Sommer Bros Contractors Ltd. Ponoka, AB.	N82216
□ C-GSFM	68	King Air B90	LJ-422	Kenn Borek Air Ltd. Calgary, AB.	N513SC
□ C-GSOC	74	SA-226AT Merlin 4A	AT-027	Schlumberger Canada Ltd. Calgary, AB.	N5339M
□ C-GSSA	79	King Air F90	LA-6	Silversat Aviation Inc. Westmount, PQ.	N7PB
□ C-GSVQ	66	680V Turbo Comm	1544-8	Northern Lights College, Dawson Creek, BC.	N146E
□ C-GSWF	82	King Air B100	BE-129	Sunwest International Aviation Services Ltd. Calgary, AB.	N2074M
□ C-GSWJ	69	SA-26AT Merlin 2B	T26-151	Western Express Airlines Inc. Vancouver, BC.	N396PS
□ C-GSYN	70	King Air 100	B-61	Adlair Aviation (1983) Ltd. Cambridge Bay, NT.	N418LA
□ C-GTFP	76	PA-31T Cheyenne II	7620016	491549 Alberta Ltd. Springbank, AB.	N57524
□ C-GTGA	80	King Air/Catpass 250	BB-728	Simo Air Ltd. Mississauga, ON.	(N124GA)
□ C-GTLA	73	King Air A100	B-165	Timberline Air Ltd. Chilliwack, BC.	N911CE
□ C-GTLF	70	King Air 100	B-72	Timberline Air Ltd. Chilliwack, BC.	N5476R
□ C-GTMA	68	King Air B90	LJ-348	Province of Alberta, Calgary, AB.	N805K
□ C-GTMW	70	SA-226AT Merlin 4	AT-002	Provincial Airlines Ltd. Halifax, NS.	N39RD
□ C-GTUC	77	King Air 200	BB-268	Conair Aviation Ltd. Abbotsford, BC.	N565RA
□ C-GTWW	75	King Air C90	LJ-657	Walsten Air Service (1986) Ltd. Kenora, ON.	N9030R
□ C-GUPP	73	King Air A100	B-157	Bearskin Lake Air Service Ltd. Sioux Lookout, ON.	N123CS
□ C-GURG	73	SA-226T Merlin 3	T-228	Air Charters Inc/Air Montreal, Doravl, PQ.	N188SC
□ C-GVKA	78	PA-31T Cheyenne II	7920008	Newfoundland Labrador Air Transport Ltd. Deer Lake, NF.	
□ C-GVKK	78	PA-31T Cheyenne II	7820038	2734141 Canada Inc/Knighthawk Air Express, Ottawa, ON.·	N679MM
□ C-GWKA	68	SA-26T Merlin 2A	T26-26	Keewatin Air Ltd. Rankin Inlet, NT.	N105EC
□ C-GWSL	75	SA-226AT Merlin 4A	AT-028	Max Aviation Inc. Three Rivers, PQ.	N5341M
□ C-GWSP	74	SA-226AT Merlin 4	AT-010	Avionair Inc. Dorval, PQ.	N603L
□ C-GWUY	75	King Air 200	BB-77	Alkan Air Ltd. Whitehorse, YT.	N300CP
□ C-GWWA	70	King Air 100	B-27	West Wind Aviation Inc. Saskatoon, SK.	G-BOFN
□ C-GWWN	74	King Air 200	BB-14	West Wind Aviation Inc. Saskatoon, SK.	N418CS
□ C-GWWQ	71	King Air 100	B-76	Accent Aviation Services Ltd. Calgary, AB.	N300DA
□ C-GXBF	76	PA-31T Cheyenne II	7620010	Province of New Brunswick, Fredericton, NB.	N54988
□ C-GXCD	76	PA-31T Cheyenne II	7620026	Ashe Aircraft Enterprises Ltd. Calgary, AB.	
□ C-GXHP	70	King Air A100	B-132	Bar XH Air Inc. Medicine Hat, AB.	XB-SLG
□ C-GXRX	70	King Air 100	B-36	Shuswap Flight Centre Ltd. Salmon Arm, BC.	N600CB
□ C-GXTC	77	PA-31T Cheyenne II	7720052	Diamond Aviation Ltd. Fredericton, NB.	(N82165)
□ C-GXVX	70	King Air 100	B-18	North American Airlines Ltd. Calgary, AB.	N7007
□ C-GXXD	92	TBM 700	75	Odessey Aviation Ltd. Toronto, ON.	F-OHBJ
□ C-GYQK	73	King Air A100	B-153	Bearskin Lake Air Service Ltd. Thunder Bay, ON.	N120AS
□ C-GZQD	75	PA-31T Cheyenne II	7520021	Newnes Transportation Ltd. Salmon Arm, BC.	N31PT
Military					
□ C-GMBC	92	King Air C90B	LJ-1300	DND, Portage La Prairie, MT.	
□ C-GMBD	92	King Air C90B	LJ-1301	DND, Portage La Prairie, MT.	
□ C-GMBG	92	King Air C90B	LJ-1304	DND, Portage La Prairie, MT.	
□ C-GMBH	92	King Air C90B	LJ-1309	DND, Portage La Prairie, MT.	
□ C-GMBW	92	King Air C90B	LJ-1310	DND, Portage La Prairie, MT.	
□ C-GMBX	92	King Air C90B	LJ-1313	DND, Portage La Prairie, MT.	
□ C-GMBY	92	King Air C90B	LJ-1317	DND, Portage La Prairie, MT.	
□ C-GMBZ	92	King Air C90B	LJ-1319	DND, Portage La Prairie, MT.	

CC = Chile *(Republic of Chile)*

Civil

Regn	Yr	Type	c/n	Owner/Operator	Prev Regn
□ CC-...	95	King Air C90B	LJ-1389		N3204K
□ CC-...	96	King Air C90B	LJ-1464		N1083K
□ CC-...	80	PA-31T Cheyenne II	8020053	Leasing Concepcion SA. Santiago.	(N200JH)
□ CC-...	79	PA-31T Cheyenne II	7920072	Servicios Aereos El Loa SA. Chuquicamata.	N620P
□ CC-CBD	83	PA-31T Cheyenne 1	8104070	Aerobenic SA. Santiago.	(D-IIHW)
□ CC-CLY	71	King Air 100	B-79	Maquinarias Pivcevic E Hijos Ltda/Aerovias DAP. Punta Arenas	CC-PIE
□ CC-CNH	80	PA-31T Cheyenne II	8020069	Linea de Aeroservicios SA. Santiago.	OB-1193
□ CC-CNT	81	PA-31T Cheyenne II	8120049	Empol Chile SA. Santiago.	N459CP
Regn	*Yr*	*Type*	*c/n*	*Owner/Operator*	*Prev Regn*

Regn	Yr	Type	c/n	Owner/Operator	Prev Regn
□ CC-COT	67	King Air A90	LJ-227	Maquinarias Pivcevic E Hijos Ltda/Aerovias DAP. Punta Arenas	CC-PIR
□ CC-CRE	74	Rockwell 690A	11155	Aero Lloyd Ltda. Santiago.	N536
□ CC-CRU	81	PA-31T Cheyenne 1	8104071	Soc Inversiones y Desarrollo Integral de Servicios, Santiago.	N122DM
□ CC-CWH	81	Cheyenne II-XL	8166016	Transportes Modern Air SA. Santiago.	N80DA
□ CC-DIV	81	King Air B200CT	BN-1	D.G.A.C. Santiago. (was BL-24).	CC-EAA
□ CC-DSN	75	King Air E90	LW-153	D.G.A.C. Santiago.	CC-EAB
□ CC-PBT	81	PA-31T Cheyenne 1	8104063	Metalurgica Ducasse Ltda. Santiago.	N2568Y
□ CC-PBZ	69	King Air B90	LJ-440		N181LL
□ CC-PCY	81	PA-31T Cheyenne II	8120028	Inversiones Antilco Ltda. Santiago.	N59KC
□ CC-PHM	79	PA-31T Cheyenne 1	7904054	Industria Metalurgica del Norte, Santiago.	CC-CPV
□ CC-PML	84	Cheyenne II-XL	1166002	Banco O Higgins/Rimac SA. Santiago. (was 8166074).	N2580Z
□ CC-PMZ	80	PA-31T Cheyenne II	8020090	Administradora Credisur Ltda. Santiago.	N925CA
□ CC-PNS	81	PA-31T Cheyenne 1	8104002	Constructora Raul del Rio SA. Santiago.	N780CA
□ CC-PTA	81	Cheyenne II-XL	8166070	Banco O Higgins y Forest Quinen, Santiago.	N333X
□ CC-PVE	81	Cheyenne II-XL	8166038	Soc de Servicios Latinoamericana Ltda. Santiago.	N161TC
Military					
□ ...	69	King Air B90	LJ-441	Fuerza Aerea de Chile, Santiago.	CC-ECF
□ 331	75	King Air A100	B-219	Fuerza Aerea de Chile, Santiago.	CC-ESA
□ 336	96	King Air B200	BB-1530	Fuerza Aerea de Chile, Santiago.	

CN = Morocco (Kingdom of Morocco)

Regn	Yr	Type	c/n	Owner/Operator	Prev Regn
Civil					
□ CN-CDF	80	King Air 200	BB-577	Royal Air Maroc, Casablanca.	
□ CN-CDN	80	King Air 200	BB-713	Royal Air Maroc, Casablanca.	N36741
□ CN-TAX	80	King Air C90	LJ-922	Omnium Nord Africain, Casablanca.	F-GCPN
□ CN-TOM		SA-227TT Merlin 3C	TT-480	Omnium Marocain de Peche, Casablanca.	N3021A
Military					
□ CNA-NB	74	King Air A100	B-181	Government of Morocco, Rabat.	
□ CNA-NC	74	King Air A100	B-182	Government of Morocco, Rabat.	
□ CNA-ND	74	King Air A100	B-183	Government of Morocco, Rabat.	
□ CNA-NE	74	King Air A100	B-186	Government of Morocco, Rabat.	
□ CNA-NF	74	King Air A100	B-187	Government of Morocco, Rabat.	
□ CNA-NG	82	King Air B200	BB-1072	Government of Morocco, Rabat.	
□ CNA-NH	82	King Air B200	BB-1073	Government of Morocco, Rabat.	
□ CNA-NI	82	King Air B200C	BL-57	Government of Morocco, Rabat.	
□ CNA-NX	89	King Air 300	FA-207	Government of Morocco, Rabat.	
□ CNA-NY	89	King Air 300	FA-208	Government of Morocco, Rabat.	

CP = Bolivia (Republic of Bolivia)

Regn	Yr	Type	c/n	Owner/Operator	Prev Regn
Civil					
□ CP-....	69	680W Turbo Commander	1835-40		N81LC
□ CP-....	80	Gulfstream 980	95004	Juan Carlos Vaca Artega, Santa Cruz.	N100TB
□ CP-....	73	King Air E90	LW-28	Aero Centro SA. La Paz.	N2XZ
□ CP-....	73	Rockwell 690	11055		N100MB
□ CP-....	77	Rockwell 690B	11424		N38LM
□ CP-1678	81	PA-31T Cheyenne II	8120017	Luis A Gomez,	EB-004
□ CP-1934		Rockwell 690	...		
□ CP-2042	70	681 Turbo Commander	6025		N10HG
□ CP-2050	81	Gulfstream 1000	96055	Aero Inca Ltda. La Paz.	N9975S
□ CP-2224	79	Rockwell 690B	11564	Omar Forti-Cons. F & L. Tarija.	N401SP
□ CP-2225	79	Rockwell 690B	11519	Aerosur, Santa Cruz.	N425DT
□ CP-2262	78	Rockwell 690B	11505	Aerojet SA. Cochabamba.	N28WR
□ CP-2266	77	Rockwell 690B	11395	Dr. Jorge Velasco Monasterio, Santa Cruz.	N816PC
Military					
□ EB-50		CASA 212-300	369	Ejercito de Bolivia, La Paz.	
□ FAB ...	74	King Air 200	BB-11	Fuerza Aerea Boliviana, La Paz.	FAB 001
□ FAB 018	81	King Air 200C	BL-28	Fuerza Aerea Boliviana, La Paz.	
□ FAB 023	79	Rockwell 690B	11562	Fuerza Aerea Boliviana, La Paz.	HK-2996P
□ FAB 026		King Air 90	...	Fuerza Aerea Boliviana, La Paz.	
□ FAB 028	73	Rockwell 690	11067	Fuerza Aerea Boliviana, La Paz.	CP-1076
□ FAB 030		Rockwell 690	...	Fuerza Aerea Boliviana, La Paz.	

CS = Portugal (Portuguese Republic)

Regn	Yr	Type	c/n	Owner/Operator	Prev Regn
Civil					
□ CS-ASG	77	Rockwell 690B	11452	OMNI Aviacao & Tecnologia Ltd. Cascais-Tires.	N115SB
□ CS-DBT	77	King Air C90	LJ-721	Aeromundo/Helisul, Cascais-Tires.	N757AL

CX = Uruguay (Republic of Uruguay)

Regn	Yr	Type	c/n	Owner/Operator	Prev Regn
Civil					
□ CX-...	85	Gulfstream 1000	96093		C-FALI
□ CX-...	82	PA-42 Cheyenne III	8001074	Banco de la Republica Oriental, Montevideo.	N82TD
□ CX-BRU	80	PA-31T Cheyenne 1	8004046	Banco de la Republica Oriental, Montevideo.	N811CM
Military					
□ 871	78	King Air 200T	BT-4	Uruguayan Navy. (was BB-408).	N2067D
Regn	Yr	Type	c/n	Owner/Operator	Prev Regn

C6 = Bahamas (Commonwealth of the Bahamas)

Civil

Regn	Yr	Type	c/n	Owner/Operator	Prev Regn
☐ C6-...	68	Mitsubishi MU-2F	122	Bahamas Customs Service, Nassau.	N98MA

C9 = Mozambique (Republic of Mozambique)

Civil

Regn	Yr	Type	c/n	Owner/Operator	Prev Regn
☐ C9-ASK	81	King Air C90	LJ-954	DNPCF & BMN-Brigada de Melhoramentos do Norte, Maputo.	F-GDCC
☐ C9-ASV	81	King Air 200C	BL-21	LAM-Linhas Aereas de Mocambique, Maputo.	N3831T
☐ C9-ASX	81	King Air 200C	BL-32	LAM-Linhas Aereas de Mocambique, Maputo.	N821CA
☐ C9-ATW	81	King Air B200	BB-937	U N WFP-World Food Programme - Skylink Canada,	N937SL
☐ C9-ENH	80	King Air 200	BB-626	ENH-Nacional Hidrocarbonetos de Mocambique, Maputo.	C9-ASS
☐ C9-ENT	79	PA-31T Cheyenne 1	7904034	TTA-Empresa Nacional de Transporte e Trabalho Aereo, Maputo.	N23HB
☐ C9-JTP	76	PA-31T Cheyenne II	7620012	Reliable Fork Lift & Truck Services, Alrode.	ZS-JTP
☐ C9-PMZ	82	King Air B200	BB-1076		ZS-MTW

D = Germany (Federal Republic of Germany)

Civil

Regn	Yr	Type	c/n	Owner/Operator	Prev Regn
☐ D-....	87	King Air B200	BB-1284	B200 Flug Charter Leer GmbH. Leer.	N6321V
☐ D-....	97	King Air B200	BB-1591	Beechcraft Vetrieb & Service GmbH. Augsburg.	N1819H
☐ D-....	96	King Air B200	BB-1538	Hanseatische Jet Charter GmbH. Bremen.	N3268L
☐ D-....	97	King Air C90B	LJ-1495	Beechcraft Vertrieb & Service GmbH. Augsburg.	N1135K
☐ D-CACB	83	King Air B200T	BT-27	Aerodata Flugmesstechnik GmbH. Braunschweig. (was BB-1105).	N7244R
☐ D-CADN	93	King Air 350	FL-101	R+R Flugzeughandels u Vermietungs GmbH. Kamenz.	N82311
☐ D-CAMM	91	King Air 350	FL-64	BASF AG. Ludwigshafen.	
☐ D-CASA	85	King Air 300	FA-76	Aero Dienst GmbH. Nuremberg.	N7247Y
☐ D-CBBB	94	King Air 350	FL-120	Hilde Schmitt-Woelfle, Augsburg.	N1512H
☐ D-CCBW	91	King Air 350	FL-46	R+R Flugzeughandels und Vermietungs Gmbh. Kamenz.	N81623
☐ D-CCCC	82	SA-227AT Merlin 4C	AT-511	Filder Air Service GmbH. Nuremberg.	N600N
☐ D-CFMA	92	King Air 350	FL-76	DFS-Deutsche Flugsicherung GmbH. Lechfeld AFB.	N8274U
☐ D-CFMB	93	King Air 350	FL-97	DFS-Deutsche Flugsicherung GmbH. Lechfeld AFB.	N8297L
☐ D-CINA	90	King Air 350	FL-7	ACH Hamburg Flug GmbH. Hamburg.	N5668F
☐ D-CKWM	94	King Air 350	FL-124	WEKA Firmengruppe GmbH. Augsburg.	N3198N
☐ D-COIL	88	King Air 300LW	FA-171	Fuchs Petroclub AG. Mannheim.	D-IKWM
☐ D-COLA	92	King Air 350	FL-75	AL Aviation Leasing GmbH. Dusseldorf.	HB-GJB
☐ D-CSKY	96	King Air 350	FL-130	Intro Verwaltungs GmbH/Aero-Dienst GmbH. Templehof.	
☐ D-FBFS	92	TBM 700	74	Brose Fahrzeugteile GmbH/BFS Flugservice GmbH. Coburg.	F-OHBK
☐ D-FCJA	97	Pilatus PC-XII	177		HB-FSM
☐ D-FGYY	91	TBM 700	7	Bruno Gantenbrink, Dortmund.	OE-EDB
☐ D-FTBM	90	TBM 700	1	Top Advice Consulting AG. Riehen, Switzerland.	F-GLBE
☐ D-IAAC	78	Conquest II	441-0073	Ruediger Kueckelhaus/CCF Manager Airline GmbH. Cologne.	N88834
☐ D-IAAD	91	Reims/Cessna F406	F406-0047	Arcus-Air Logistic GmbH. Mannheim.	N6589A
☐ D-IAAH	90	King Air C90A	LJ-1247	Dipl. Kfm. Steuerberater Dietmar Volkmann, Nuremberg.	N5651J
☐ D-IAAK	76	King Air 200	BB-134	Heitkamp u Thumann KG. Dusseldorf.	
☐ D-IAAV	80	Conquest II	441-0160	Aero Charter L U Bettermann GmbH. Menden.	(N2630B)
☐ D-IABB	90	King Air C90A	LJ-1235	B Boettcher GmbH-Bremerhaven/Aeroline GmbH. Westerland.	N1569N
☐ D-IACR	81	PA-31T Cheyenne 1	8104067	Cx D- 2/97 to ?	N52TW
☐ D-IADH	77	Rockwell 690B	11439	GRS Beteiligung/Flugdienst Carlos Pilars de Pilar,	(N81706)
☐ D-IAFF	90	King Air C90A	LJ-1229	Allkauf SB-Warenhaus GmbH. Moenchengladbach.	N422TW
☐ D-IAGA	86	Conquest II	441-0362	OFRA Systembau GmbH/Windrose Air Flugcharter GmbH. Berlin.	LX-ETB
☐ D-IAGB	80	King Air F90	LA-38	A & G Bumueller Grossbaeckerei - Konditorei, Stuttgart.	N888EM
☐ D-IAGT	81	Conquest 1	425-0061	AGIB-AG fur Grundbesitz u Industriebeteiligungen, Bielefeld.	N66QT
☐ D-IAHK	76	King Air 200	BB-149	AGV Mobilienvermietungs GmbH. Wiesbaden.	N123ST
☐ D-IAKK	87	King Air B200	BB-1265	Allkauf SB-Warenhaus GmbH. Moenchengladbach.	N550TF
☐ D-IAMB	81	King Air 200	BB-790	AERO-Flugcharter/Private Wings Flugcharter GmbH. Berlin.	F-GIAX
☐ D-IAMK	82	King Air B200	BB-956	Sylt Air Flugzeugbesitz/ACH Hamburg Flug GmbH. Hamburg.	N956WT
☐ D-IANA	95	King Air B200	BB-1517	KLW Mietflug-Verwaltungs GmbH. Paderborn-Lippstadt.	N3217W
☐ D-IAPA	81	PA-31T Cheyenne 1	8104032	Siemag Verwaltungs GmbH. Hilchenbach.	N5SL
☐ D-IAPW	81	Conquest II	441-0210	Hahn Kaffee GmbH. Munich.	(N24CJ)
☐ D-IARF	82	King Air C90-1	LJ-1034	Druckgusswerk Moessner GmbH. Munich.	(D-IFOC)
☐ D-IASW	81	PA-31T Cheyenne 1	8104101	Saarfeldspatwerke H Huppert GmbH. Saarbrucken.	N104MC
☐ D-IAWS	81	King Air B200	BB-933	Norbert Wrede/Air Evex Westfalia GmbH. Paderborn-Lippstadt.	N200LP
☐ D-IBAB	92	King Air 300LW	FA-225	OMG Objekt-Marketing GmbH. Donaueschingen.	N82396
☐ D-IBAD	85	King Air B200	BB-1229	Eheim/SFD Stuttgarter Flugdienst GmbH. Stuttgart.	
☐ D-IBAG	74	Rockwell 690A	11211	AGV GmbH-Wiesbaden/Manager Flugservice GmbH. Saarbrucken.	
☐ D-IBAR	87	King Air B200	BB-1280	Nordavia Flug GmbH. Hamburg.	
☐ D-IBDH	92	King Air C90B	LJ-1307	Peene-Werft GmbH. Peenemuende.	N8053U
☐ D-IBER	89	King Air 300LW	FA-184	Bertlesmann AG. Gutersloh.	
☐ D-IBFS	89	King Air B200	BB-1349	Heberger Bau GmbH. Schifferstadt.	N200KG
☐ D-IBFT	96	King Air B200	BB-1535	Brose Fahrzeugteile GmbH. Coburg.	N1135Z
☐ D-IBHK	78	King Air 200	BB-366	Euroflug Freiburg Gerhard Frenzel, Freiburg.	N1230
☐ D-IBIW	80	PA-31T Cheyenne 1	8004011	BIW-Beratung fuer Industrie u Wirtschaft GmbH. Weinstadt.	N76TG
☐ D-IBNK	89	King Air 300LW	FA-204	Gebr Knauf Westdeutsche Gipswerke KG. Iphofen.	N5662T
☐ D-IBSA	81	PA-31T Cheyenne II	8120033	Frau Amalie-Barbara Fuchs/Fuchs Mineraloelwerke, Mannheim.	N42TW
☐ D-IBSY	90	King Air B200	BB-1357	Dr Hanns-Guenther Seeg/GSW Charterflug GmbH. Baden-Baden.	N5551E
Regn	Yr	Type	c/n	Owner/Operator	Prev Regn

Regn	Yr	Type	c/n	Owner/Operator	Prev Regn
☐ D-ICBC	93	King Air 300	FA-227	Kronospan GmbH. Sandebeck.	N81418
☐ D-ICBH	81	PA-31T Cheyenne II	8120008	Berndt Heller GmbH. Stuttgart.	N812SW
☐ D-ICDU	84	Cheyenne II-XL	1166003	Quick Air Service GmbH/Karl Bittner, Cologne.	N2604R
☐ D-ICGA	81	Cheyenne II-XL	8166056	MFG-Milan Flug GmbH. Hanover.	N550TL
☐ D-ICGB	84	PA-42 Cheyenne IIIA	5501007	MFG-Milan Flug GmbH. Hanover.	N834CM
☐ D-ICHG	91	King Air B200	BB-1400	C H Scholz KG/Winair GmBH. Saarbrucken.	N8085D
☐ D-ICHS	85	Conquest 1	425-0233	Karl-Heinz Sengewald/Aerowest Braunschweig GmbH. Hannover.	N80938
☐ D-ICHT	90	King Air 300LW	FA-214	Sky Jet Flugservice GmbH/Medico Flugreisen Gmb. Baden-Baden	N5666S
☐ D-ICIR	83	King Air B200	BB-1051	Cirrus Luftfahrt GmbH. Saarbruecken-Ensheim..	(G-BJWG)
☐ D-ICKM	82	King Air B200	BB-1005	Kurt Marxer Anlagen u Maschinenbau GmbH. Friedberg.	OE-FKW
☐ D-ICLE	82	King Air C90	LJ-1002	Weber Management GmbH/Rieker Air Service GmbH. Stuttgart.	N643PU
☐ D-ICMC	81	PA-42 Cheyenne III	8001039	Heinz Oldenburg, Muehldorf.	F-GHAB
☐ D-ICMF	81	Conquest 1	425-0102	Erich Pohlmann Fenster u Tuerenwerk GmbH. Herne.	N151GA
☐ D-ICOA	82	King Air B200	BB-1065	Adolf Wuerth GmbH. Kuenzelsau.	
☐ D-ICOM	84	Conquest 1	425-0197	Comfort Air GmbH. Munich.	N406MA
☐ D-ICPA	81	PA-31T Cheyenne II	8120025	Hamburger Air Charter Erich Wagner GmbH. Hamburg.	N290CM
☐ D-ICRA	88	King Air B200	BB-1291	Guenter Daiss Versicherungsmakler GmbH. Freiberg.	
☐ D-IDBU	84	PA-42 Cheyenne IIIA	5501029	ACH Hamburg Flug GmbH. Hamburg.	N700CC
☐ D-IDEA	74	King Air E90	LW-103	Karl Ratzinger/MTM Aviation GmbH. Munich	C-FMGL
☐ D-IDIW	91	King Air C90A	LJ-1263	Peter Runge Rohrleitungsbau GmbH. Unterschleissheim.	N5680S
☐ D-IDSM	86	King Air B200	BB-1259	Diringer u Scheidel GmbH/EAS Executive Air Service, Mannheim	N734P
☐ D-IEAH	89	King Air C90A	LJ-1216	Fuchs & Partner GmbH. Augsburg.	N1562Z
☐ D-IEBE	91	King Air C90A	LJ-1267	Ebenhoehe Kies-u-Sandwerke GmbH/Isarflug GmbH. Munich.	A6-FAE
☐ D-IEEE	81	King Air F90	LA-104	Linde Airservice GmbH/Linde Planprojekt GmbH. Esslingen.	TC-NML
☐ D-IEFB	81	King Air B200	BB-897	Flugbereitschaft GmbH. Karlsruhe.	N200TM
☐ D-IEGA	81	Conquest II	441-0193	Partner Air KG/Windrose Air Flugcharter GmbH. Berlin.	OE-FRZ
☐ D-IEIS	79	PA-31T Cheyenne 1	7904002	Werner Moll-Gruibingen/Rieker Air Service GmbH. Stuttgart.	N52CA
☐ D-IEKG	79	King Air C90	LJ-867	Gilbert Alt/CONSETA Luftfahrt GmbH. Saarbruecken-Ensheim.	HB-GIH
☐ D-IEPA	84	PA-31T Cheyenne 1A	1104017	Lenoxhandels u Speditions GmbH. Hamburg.	N91204
☐ D-IERI	77	King Air B100	BE-29	Air Evex Westfalia GmbH. Paderborn-Lippstadt.	N7729B
☐ D-IFES	81	King Air 200	BB-827	MTM/Karl Ratzinger, Munich.	
☐ D-IFFB	93	King Air 300LW	FA-224	F & F Burda GmbH. Baden-Baden.	N56449
☐ D-IFHI	81	King Air C90	LJ-977	Ailana Vermoegens und Grundstuecksverwaltungs GmbH. Munich	N1813P
☐ D-IFHZ	84	PA-31T Cheyenne 1A	1104016	Zollern Flugdienste GmbH. Mengen.	N91201
☐ D-IFMI	85	King Air C90A	LJ-1101	Fritz Mueller, Ingelfingen.	N17EL
☐ D-IFUN	80	King Air C90	LJ-874	Air Connect Houfek GmbH. Hamburg.	N44486
☐ D-IGAF	73	Rockwell 690	11121	Air Evex Westfalia GmbH. Paderborn-Lippstadt.	N.....
☐ D-IGFD		SA-227TT Merlin 300	TT-536	Filder Air Service GmbH. Nuremberg.	OE-FLU
☐ D-IGKN	84	King Air C90A	LJ-1077	Georg Knuffmann, Munster-Osnabruck.	N4111U
☐ D-IGLB	78	Rockwell 690B	11456	Verwaltungsdienstleister Sozialer Einrichtungen GmbH. Langen	(N81750)
☐ D-IGLI	80	King Air C90	LJ-887	Glas Logistik International GmbH. Obernkirken.	D-IIWN
☐ D-IGOB	92	P-180 Avanti	1016	PBG GmbH-Mannheim/Manager Flugservice GmbH. Saarbruecken.	I-PJAT
☐ D-IHAH	94	King Air C90B	LJ-1370	Anton Haering, Bubsheim.	N1570C
☐ D-IHAN	94	King Air B200	BB-1478	Industrie Ott GmbH. Flugcharter KG. Stuttgart.	N8150N
☐ D-IHBP	95	King Air C90B	LJ-1424	Berger Bau GmbH/Comair Reise u Charter GmbH. Vilshofen.	N3252J
☐ D-IHDE	77	King Air C90	LJ-725	ATF-Air Transport Flug GmbH. Bayreuth.	
☐ D-IHHB	92	King Air 300LW	FA-223	Augusta Air GmbH. Augsburg.	N80775
☐ D-IHHE	93	King Air C90B	LJ-1327	FVG-Flugzeugvermietungs GmbH. Saarbrucken.	N8227P
☐ D-IHJK	81	PA-31T Cheyenne 1	8104057	BFC-Business Flight Charter GmbH. Velbert.	N62BW
☐ D-IHKH	79	Rockwell 690B	11541	WIP-Wirtschaft u Infrastruktur GmbH. Munich.	N9177N
☐ D-IHKM	87	King Air C90A	LJ-1158	Porta Flug GmbH. Porta Westfalica.	N38H
☐ D-IHLA	83	PA-42 Cheyenne IIIA	8301001	MFG-Milan Flug GmbH. Hanover.	C-GNRD
☐ D-IHMM	82	PA-31T Cheyenne 1	8104066	AFS Airway Flight Service GmbH/LBA, Braunschweig.	N82LC
☐ D-IHMO	91	P-180 Avanti	1009	Asko Deutsche Kaufhaus GmbH/MFG Flugdienst, Saarbruecken.	I-RAIH
☐ D-IHMS	81	PA-31T Cheyenne II	8120009	Hans-Martin Dopmann, Stuttgart.	N97MA
☐ D-IHMV	93	King Air C90B	LJ-1325	Wolfgang Osterloh, Ludwigsburg.	N8135M
☐ D-IHOT	91	P-180 Avanti	1011	Roy Foerster/Trend Air, Berlin.	N53MW
☐ D-IHRA	92	P-180 Avanti	1019	Hans Rabb GmbH. Quierscheid-Camphausen.	
☐ D-IHSI	80	Gulfstream 980	95039	AL Aviation Leasing GmbH. Dusseldorf.	N9790S
☐ D-IHSW	92	King Air C90B	LJ-1315	Beechcraft Vertieb & Service GmbH. Augsburg.	N8103E
☐ D-IHUT	97	King Air B200	BB-1590	Heitmann und Thumann GmbH. Dusseldorf.	
☐ D-IHVA	84	PA-42 Cheyenne IIIA	5501025	LGS Leasinggellscaft der Sparkasse GmbH. Bad Homberg.	
☐ D-IIAH	84	PA-31T Cheyenne 1A	1104015	Mietkauf GmbH. Mulheim.	N9168T
☐ D-IIHA	72	King Air C90	LJ-562	Dr Joachim von Meister, Bad Homburg.	(N333FJ)
☐ D-IIKM	85	King Air C90A	LJ-1120	Kurt Jaeger, Berlin.	N7237K
☐ D-IILG	85	King Air 300LW	FA-63	Eisenwerk Theodor Loos GmbH/Aero Dienst GmbH. Nuremberg.	N985GA
☐ D-IIRC	84	Cheyenne II-XL	1166008	DSF Flugdienst AG. Frankfurt. (was 1122002)	N362AB
☐ D-IIWB	93	King Air C90B	LJ-1340	GML-Gesellschaft fuer Mobilien-Leasing mbH. Weyhe.	N10799
☐ D-IIWB	80	PA-31T Cheyenne II	8020030	Sportproduction Diekkamp GmbH. Salzburg, Austria.	N37CA
☐ D-IJCL	92	P-180 Avanti	1017	Winair GmbH. Saarbruecken.	N14P
☐ D-IJET	81	PA-31T Cheyenne 1	8104042	LGS-Leasinggesellschaft der Sparkasse GmbH. Bad Homburg.	(N2590X)
☐ D-IJGW	84	Conquest 1	425-0193	Tobaccoland Grouehandels GmbH. Muenchengladbach.	N1221X
☐ D-IJLF	86	Conquest II	441-0358	Wilhelm Gronbach GmbH. Munich.	N85DJ
☐ D-IJOY	83	Conquest 1	425-0164	Berliner S-Bahn Werbung GmbH. Berlin.	N6887F
☐ D-IKBJ	85	King Air B200	BB-1209	KBA Kieler Business Air GmbH. Kiel-Holtenau.	N46GA
☐ D-IKES	81	King Air C90	LJ-942	Center Grundstuecksverwaltungs GmbH. Augberg.	

Regn	Yr	Type	c/n	Owner/Operator	Prev Regn
☐ D-IKET	80	PA-31T Cheyenne II	8020017	Karosserie Entwicklung Thurner, Munich.	N154CA
☐ D-IKEW	78	PA-31T Cheyenne II	7820066	Kress Elektrik GmbH. Bisingen.	N6108A
☐ D-IKIM	93	King Air C90B	LJ-1324	Rudolf Kimmerle Gewerbebau, Augsberg.	N82430
☐ D-IKIW	74	King Air C90	LJ-641	Beechcraft Vertrieb & Service GmbH. Augsburg.	N7128H
☐ D-IKKK	81	PA-31T Cheyenne II	8120004	Winair Winkler & Feycock GmbH. Saarbrucken.	N811SW
☐ D-IKKY	80	MU-2 Solitaire	420SA	Golden Europe Jet de Luxe Club E V. Stuttgart.	I-SOLT
☐ D-IKOB	81	King Air B200	BB-921	Hans Kobusch/Quick Air Service GmbH, Cologne.	N244JB
☐ D-ILCE	80	PA-31T Cheyenne 1	8004053	DSF-Deutsch Schweizer Flugdienst AG. Reichelsheim.	
☐ D-ILGA	84	PA-31T Cheyenne 1A	1104014	ElektroMstr. Willy Stuecker, Brokdorf.	N9382T
☐ D-ILGI	85	King Air C90A	LJ-1090	Georg Wissler, Grossostheim.	N7210H
☐ D-ILIN	79	King Air 200	BB-545	EAS-Executive Air Service Flug GmbH. Mannheim.	OY-CBY
☐ D-ILLF	97	King Air B200	BB-1568	Fischer Flug GmbH. Offenburg.	N1067V
☐ D-ILOH	82	King Air B200	BB-1080	Charterflug Rademacher GmbH/Kirberger Aviation, Burbach.	N200PL
☐ D-ILOR	79	PA-31T Cheyenne 1	7904001	Wensauer Automobile GmbH. Straubing.	N6183A
☐ D-ILPC	79	King Air 200	BB-524	Stahlbau Domesle GmbH. Straubing. (sold USA 12/97 ?).	N95PM
☐ D-IMAY	85	PA-42 Cheyenne 400LS	5527024	Dr Max Schaldach, Nuremberg.	(N400TM)
☐ D-IMLP	95	P-180 Avanti	1032		
☐ D-IMMF	88	King Air 300LW	FA-164	Multiflug GmbH. Troisdorf-Spich.	N1524H
☐ D-IMMM	85	King Air B200	BB-1201	IMM/AIL-Anlagen u Investitionsgueter Leasing GmbH. Munich.	N6815X
☐ D-IMWA	73	King Air E90	LW-59	Marco Polo GmbH. Karlsruhe.	VR-CGK
☐ D-IMWK	85	SA-227TT Merlin 300	TT-529	LTO-Lufttransport Osnabruck-Munster GmbH. Munster.	N3109S
☐ D-INGA	80	Conquest 1	425-0003	Centrum Bau, Bautrager fur Wohn-u Gewerbebau KG. Dusseldorf.	N98751
☐ D-INNN	81	PA-31T Cheyenne II	8120102	Dieter Eifler GmbH. Eckelhausen. (was 8120067).	N822SW
☐ D-INUS	90	Reims/Cessna F406	F406-0043	Flugdienst Fehlhaber GmbH. Cologne.	
☐ D-INWG	81	Conquest 1	425-0039	NWG-Nord West/RWL Luftfahrt GmbH. Moenchengladbach.	N81PE
☐ D-INWK	75	SA-226T Merlin 3A	T-255	Rudas Studios KG/Air Traffic GmbH. Dusseldorf.	N5349M
☐ D-IOAN	81	King Air B200	BB-872	Primair Executive Charter GmbH. Munich.	F-GHMN
☐ D-IOEB	91	King Air 300LW	FA-220	Dr August Oetker/Teuto Air Lufttaxi GmbH. Bielefeld.	
☐ D-IOHL	85	Conquest 1	441-0357	Ohlair Charterflug KG. Kiel.	PH-BMP
☐ D-IOMG	92	King Air C90B	LJ-1321	Disko Leasing KG. Dusseldorf.	N8232L
☐ D-IOSA	87	PA-42 Cheyenne IIIA	5501041	Lufthansa Flight Training GmbH. Bremen.	N9578N
☐ D-IOSB	87	PA-42 Cheyenne IIIA	5501042	Lufthansa Flight Training GmbH. Bremen.	
☐ D-IOSC	87	PA-42 Cheyenne IIIA	5501043	Lufthansa Flight Training GmbH. Bremen.	
☐ D-IOSD	87	PA-42 Cheyenne IIIA	5501044	Lufthansa Flight Training GmbH. Bremen.	
☐ D-IOTT	79	PA-31T Cheyenne II	7920010	Dr Bernd Koenes, Erkelenz.	N6196A
☐ D-IPAS	86	Conquest 1	425-0224	Peter Noecker-Dusseldorf/VHM Schul u Charterflug, Essen.	G-NORS
☐ D-IPIA	92	P-180 Avanti	1021	Primair Executive Charter GmbH. Munich.	
☐ D-IPOS	82	Conquest 1	425-0120	Dr Peters GmbH/RWL Luftfahrt GmbH. Moenchengladbach.	N425R
☐ D-IPWB	90	King Air B200	BB-1368	Flugdienst Baier GmbH. Marbach.	(ZS-...)
☐ D-IQAS	84	PA-42 Cheyenne 400LS	5527022	Quick Air Service GmbH. Cologne.	N322KW
☐ D-IRGW	83	Conquest 1	425-0165	Weber Vertrieb International BV. Amsterdam, Holland.	N425SP
☐ D-IRIS	85	King Air F90-1	LA-229	Frau Uta Ackermans-Meynen, Kerken.	N7209Z
☐ D-ISAR	85	Conquest 1	425-0214	WISAP-Wissenschaftlichen Apparatebau GmbH. Sauerlach.	N425JM
☐ D-ISAZ	82	King Air B200	BB-983	Aero-Dienst GmbH, Nuremberg.	(N983AJ)
☐ D-ISEM	89	King Air C90A	LJ-1207	EBM Werke GmbH/Elektrobau Mulfingen, Mulfingen.	N31559
☐ D-ISHY	88	Reims/Cessna F406	F406-0027	Flugdienst Fehlhaber GmbH. Cologne.	PH-FWH
☐ D-ISIG	81	PA-31T Cheyenne 1	8104055	Schindler Ingenieur GmbH. Aschaffenburg.	N123AT
☐ D-ISIX	94	King Air C90B	LJ-1355	Brinkmann Maschinenfabrik GmbH. Schloss Holte.	N995PA
☐ D-ITAB	83	King Air B200	BB-1166	Trajan-Bauregie GmbH/TAI Transair Luftreederei, Dusseldorf.	N717RM
☐ D-ITCH	86	King Air C90A	LJ-1138	Roland & Meckbach GmbH. Kassel-Calden.	N17KA
☐ D-ITLL	82	King Air F90	LA-192	Domesle Stahlverschalungs GmbH. Regensburg.	N17TS
☐ D-IUDE	93	King Air C90B	LJ-1323	Beechcraft Vetrieb & Service GmbH. Augsburg.	N90KA
☐ D-IVHM	90	King Air B200	BB-1369	VHM Schul und Charterflug GmbH. Essen-Muelheim.	N778HP
☐ D-IWAL	81	King Air F90	LA-100	Willi Waldhausen, Moenchen Gladbach.	HB-GHP
☐ D-IWKA	84	King Air F90-1	LA-218	Disko Leasing GmbH. Dusseldorf.	N137JP
☐ D-IWSH	93	King Air B200	BB-1462	Hilde Schmitt-Woelfle, Augsburg.	N82425
☐ D-IZZZ	85	King Air B200	BB-1235	M Graf von Krockow/LBG Luftfahrtdienst, Bonn-Hangelar	G-OAFB

D2 = Angola *(Republic of Angola)*

Civil

Regn	Yr	Type	c/n	Owner/Operator	Prev Regn
☐ D2-...	79	King Air 200	BB-603	Capricorn Systems Ltd. Lanseria, RSA.	N91TR
☐ D2-ALS	66	King Air 90	LJ-80		ZS-NED
☐ D2-EAA	73	Rockwell 690A	11132	DSAC-Directorate of Civil Aviation, Luanda.	CR-LAA
☐ D2-EBF	81	King Air 200	BB-836		S9-NAQ
☐ D2-ECN	86	Reims/Cessna F406	F406-0002	SAL-Sociedade de Aviacao Ligeira SA. Luanda.	PH-MNS
☐ D2-ECO	87	Reims/Cessna F406	F406-0011	SAL-Sociedade de Aviacao Ligeira SA. Luanda.	D-IDAA
☐ D2-ECP	87	Reims/Cessna F406	F406-0016	SAL-Sociedade de Aviacao Ligeira SA. Luana.	PH-LAS
☐ D2-ECQ	88	Reims/Cessna F406	F406-0019	SAL-Sociedade de Aviacao Ligeira SA. Luanda.	G-CVAN
☐ D2-ECX	90	King Air B200	BB-1362	SAL-Sociedade de Aviacao Ligeira SA. Luanda.	N1565F
☐ D2-ECY	89	King Air B200C	BL-135	SAL-Sociedade de Aviacao Ligeira SA. Luanda.	S9-NAP
☐ D2-ECZ	91	King Air 350	FL-59	SAL-Sociedade de Aviacao Ligeira SA. Luanda.	S9-NAY
☐ D2-EMX	79	King Air 200	BB-480	Intertransit, Luanda.	ZS-MJH
☐ D2-EOJ	90	King Air B200	BB-1371	Equator Leasing 14 Ltd. Luanda.	N56616
☐ D2-EQC	68	King Air B90	LJ-324	SONANGOL Aeronautica - Helipetrol, Luanda.	N892DF
☐ D2-ESO	86	King Air B200C	BL-127	SONANGOL Aeronautica - Helipetrol, Luanda.	
☐ D2-ESP	90	King Air B200	BB-1391	SONANGOL Aeronautica - Helipetrol, Luanda.	N8048W

Regn	Yr Type	c/n	Owner/Operator	Prev Regn

189

Regn	Yr	Type	c/n	Owner/Operator	Prev Regn
☐ D2-ESQ	91	King Air B200	BB-1407	SONANGOL Aeronautica - Helipetrol, Luanda. 'Luanda'	
☐ D2-EST	89	King Air B200	BB-1348	SONANGOL Aeronautica - Helipetrol, Luanda. 'Cobo'	S9-NAO
☐ D2-ETJ	82	King Air C90A	LJ-1193		PT-OJZ
☐ D2-EXD	64	Gulfstream 1	124	Absil Air Services P/L. Lanseria, RSA.	ZS-NKT
☐ D2-FMD	80	King Air/Catpass 250	BT-18	Capricorn Systems Ltd. Lanseria, RSA. (was BB-695).	N123PW

EC = Spain (Kingdom of Spain)

Civil

Regn	Yr	Type	c/n	Owner/Operator	Prev Regn
☐ EC-...	71	King Air C90	LJ-527		N55SG
☐ EC-CDI	73	King Air C90	LJ-603	D.G.A.C. Salamanca.	E22-1
☐ EC-CDJ	73	King Air C90	LJ-605	D.G.A.C. Salamanca.	E22-2
☐ EC-CDK	73	King Air C90	LJ-608	D.G.A.C. Salamanca.	E22-3
☐ EC-CHA	74	King Air C90	LJ-621	D.G.A.C. Salamanca.	E22-4
☐ EC-CHC	74	King Air C90	LJ-624	D.G.A.C. Salamanca.	E22-6
☐ EC-CHD	74	King Air A100	B-193	C N Air SA. Barcelona.	E23-1
☐ EC-CHE	74	King Air A100	B-195	D.G.A.C. Salamanca.	E23-2
☐ EC-COI	75	King Air C90	LJ-663	D.G.A.C. Salamanca. (status).	E22-7
☐ EC-COL	75	King Air C90	LJ-666	D.G.A.C. Salamanca.	E22-10
☐ EC-DHF	79	PA-31T Cheyenne II	7920073	Dominguez Toledo SA. Malaga.	N23699
☐ EC-DSA	66	680T Turbo Commander	1564-20	Ambulancias Insulares SA. Palma de Mallorca.	I-ARBO
☐ EC-DXA	76	Rockwell 690A	11328	Tur Air SA. Madrid. 'Don Mendo'	D-IHVB
☐ EC-DXG	67	680V Turbo Commander	1711-86	Ambulancias Insulares SA. Palma de Mallorca. (status ?).	N535SM
☐ EC-EAG	68	680W Turbo Commander	1776-14	Ambulancias Insulares SA. Palma de Mallorca.	N680W
☐ EC-EBG	78	Rockwell 690B	11433	Tur Air SA. Madrid.	N237SC
☐ EC-EFH	73	Rockwell 690A	11130	Tur Air SA. Madrid.	N111VS
☐ EC-EFS	72	Rockwell 690	11034	Ambulancias Insulares SA. Palma de Mallorca.	N400JJ
☐ EC-EIH	74	Rockwell 690A	11212	Ambulancias Insulares SA. Palma de Mallorca.	N690BT
☐ EC-EIL	72	Rockwell 690	11007	Ambulancias Insulares SA. Palma de Mallorca.	N171TT
☐ EC-EIM	79	PA-31T Cheyenne II	7904026	Jose Maria Caballe Horta, Alicante.	N458SC
☐ EC-ESV	81	King Air 200	BB-806	NAYSA Aerotaxis, Las Palmas.	N86GA
☐ EC-EVJ	60	Gulfstream 1	39	Drenair SA. Madrid. 'Francisco de Asis'	EC-376
☐ EC-EXB	65	Gulfstream 1	153	Drenair SA. Madrid. 'Vicante Ferrer'	EC-433
☐ EC-EXQ	64	Gulfstream 1	142	Drenair SA. Madrid. 'El Barranc' (status ?).	EC-461
☐ EC-EXS	61	Gulfstream 1	64	Drenair SA. Madrid. 'Maria' (status ?).	EC-460
☐ EC-EZO	60	Gulfstream 1	41	Drenair SA. Madrid. (status ?).	EC-494
☐ EC-FGK	77	SA-226AT Merlin 4A	AT-062	Flightline SL. Barcelona.	EC-125
☐ EC-FIO	60	Gulfstream 1	40	Drenair SA. Madrid.	EC-493
☐ EC-FOT	79	PA-31T Cheyenne II	7920036	Multiavionica SL. Madrid.	N525CA
☐ EC-FPF	91	TBM 700	12	Arturo Beltran SA. Zaragoza.	F-OHBD
☐ EC-FRR	74	Rockwell 690A	11185	Trabajos Aereos Espejo S L. Cordoba.	N53JJ
☐ EC-FUX	75	SA-226AT Merlin 4A	AT-038	Swiftair SA. Madrid.	EC-509
☐ EC-GBB	76	King Air 200	BB-182	PMS,	EC-727
☐ EC-GBI	75	SA-226AT Merlin 4A	AT-041	Swiftair SA. Madrid.	EC-867
☐ EC-GDR	80	SA-226AT Merlin 4A	AT-074	Ibertrans Aerea SL. Madrid.	EC-702
☐ EC-GDV	76	SA-226AT Merlin 4A	AT-043	Air Atlantic SL. Gran Canaria.	EC-975
☐ EC-GHZ	79	King Air 200	BB-555	Transported Aereos del Sur, Seville.	D-IFOR
☐ EC-GIJ	68	King Air B90	LJ-382	Helijet SA. Barcelona.	F-WQCC
☐ EC-GJZ	69	SA-26AT Merlin 2B	T26-149	Air Business SL. Madrid.	EC-202
☐ EC-GLU	74	Mitsubishi MU-2J	626	Aerofer, Alicante.	OY-ATZ

EI = Eire (Republic of Eire)

Civil

Regn	Yr	Type	c/n	Owner/Operator	Prev Regn
☐ EI-BHL	79	King Air E90	LW-321	Stewart Singlam Fabrics Ltd. Cork.	N60253
☐ EI-CRI	91	King Air 350	FL-66	Westair Aviation Ltd/Rothmans International, Luton, UK.	VR-CRI

Military

Regn	Yr	Type	c/n	Owner/Operator	Prev Regn
☐ 240	80	King Air 200	BB-672	Irish Air Corps. Casement-Dublin.	

EP = Iran (Islamic Republic of Iran)

Civil

Regn	Yr	Type	c/n	Owner/Operator	Prev Regn
☐ EP-AGU	70	681 Turbo Commander	6012	Government of Iran, Tehran.	N9061N
☐ EP-AGV	72	Rockwell 690	11045	Government of Iran, Tehran.	
☐ EP-AGW	72	Rockwell 690	11047	Government of Iran, Tehran.	
☐ EP-AHL	74	Rockwell 690A	11143	Sazemane Sanoye Nazami, Tehran.	N57142
☐ EP-AHM	74	Rockwell 690A	11182	State Television Company of Iran, Tehran.	
☐ EP-AKA	72	681B Turbo Commander	6065	National Oil Company of Iran, Tehran.	
☐ EP-AKB	72	681B Turbo Commander	6067	National Oil Company of Iran, Tehran.	
☐ EP-AKI	73	Rockwell 690	11075	Department of Forest Protection, Tehran.	
☐ EP-FIA	69	680W Turbo Commander	1849-45	Directorate of Civil Aviation, Tehran.	
☐ EP-FIB	69	680W Turbo Commander	1850-46	Directorate of Civil Aviation, Tehran.	
☐ EP-FSS	69	680W Turbo Commander	1848-44	Directorate of Civil Aviation, Tehran.	
☐ EP-KCD	75	Rockwell 690A	11256		

Military

Regn	Yr	Type	c/n	Owner/Operator	Prev Regn
☐ 501	73	Rockwell 690	11076	Iranian Navy, Mehrabad.	
☐ 4-901	73	Rockwell 690	11077	Iranian Army, Mehrabad.	
Regn	*Yr*	*Type*	*c/n*	*Owner/Operator*	*Prev Regn*

Regn	Yr	Type	c/n	Owner/Operator	Prev Regn
☐ 4-902	73	Rockwell 690	11078	Iranian Army, Mehrabad.	
☐ 4-903	73	Rockwell 690	11079	Iranian Army, Mehrabad.	
☐ 5-280	71	681B Turbo Commander	6062	Iranian Air Force, Tehran.	
☐ 5-281	72	681B Turbo Commander	6068	Iranian Air Force, Tehran.	
☐ 5-282	72	681B Turbo Commander	6072	Iranian Air Force, Tehran.	
☐ 5-2505	74	Rockwell 690A	11183	Iranian Navy, Mehrabad.	
☐ 5-4035	76	Rockwell 690A	11294	Iranian Navy, Mehrabad.	N9187N
☐ 5-4036	76	Rockwell 690A	11295	Iranian Navy, Mehrabad.	N81427
☐ 5-4037	76	Rockwell 690A	11333	Iranian Navy, Mehrabad.	N57196
☐ 5-4038	76	Rockwell 690A	11334	Iranian Navy, Mehrabad.	N81467
☐ 6-3201	74	Rockwell 690A	11181	Iranian Police Wing, Tehran.	N57196
☐ 6-3202	76	Rockwell 690A	11293	Iranian Police Wing, Tehran.	N81467

F = France *(French Republic)*

Civil

Regn	Yr	Type	c/n	Owner/Operator	Prev Regn
☐ F-....		Reims/Cessna F406	F406-0072		
☐ F-ASFA	73	King Air E90	LW-47	SFACT Operations Aeriennes, Melun-Villaroche.	
☐ F-BOSY	66	King Air A90	LJ-128	Altagna, Bastia-Poretta.	D-IMTW
☐ F-BRNO	69	King Air B90	LJ-482	DK Flight, Nancy.	HB-GEE
☐ F-BTEE	76	PA-31T Cheyenne II	7620045	Ste Bail Equipment, Nanterre.	F-ODEE
☐ F-BTQP	65	King Air 90	LJ-40	Air Transport Mediterranee, Ajaccio.	I-GNIS
☐ F-BUTS	73	King Air E90	LW-68	Ste Air Systems, Le Grand Luce.	
☐ F-BVET	75	King Air 200	BB-21	Air Provence International, Marseille-Marignane.	
☐ F-BVRS	74	King Air F90	LW-116	Ste Produits Usines Metallurgiques, Reims.	
☐ F-BVTB	73	King Air C90	LJ-579	Packair, Toussus-le-Noble.	D-INAF
☐ F-BXAP	71	King Air C90	LJ-522	Air Normandie, Le Havre-Octeville.	D-IHVB
☐ F-BXAS	75	Rockwell 690AT	11240	Ste Turbomeca, Pau.	F-WXAS
☐ F-BXON	75	King Air E90	LW-161	Regourd Aviation, Paris-Le Bourget.	
☐ F-BXPY	76	King Air C90	LJ-684	Atlantic-Ste Francaise de Developpement Thermique, La Roche	
☐ F-BXSI	76	King Air 200	BB-128	MIG Air, Paris-Le Bourget.	
☐ F-BXSK	76	PA-31T Cheyenne II	7620020	Air Bor, Dijon-Longvic.	
☐ F-BXSL	75	King Air C90	LJ-648	Sierra Lima SARL/Air Normandie, Le Havre-Octeville.	
☐ F-BXSN	76	King Air E90	LW-175	Air Picardie Investissement/Manag'Air, Amiens-Gilsy.	
☐ F-GABV	74	King Air E90	LW-102	Ste Soon, Toussus le Noble.	(F-GRCV)
☐ F-GALD	76	PA-31T Cheyenne II	7620032	Ste SITRAM Inox, St Benoit du Sault.	
☐ F-GALN	76	King Air 200T	BT-1	IGN France, Creil. (was BB-186).	
☐ F-GALP	77	King Air 200T	BT-2	IGN France, Creil. (was BB-203).	
☐ F-GALZ	76	King Air E90	LW-199	Ste Transport Air Centre, Roanne-Renaison.	
☐ F-GAMP	77	PA-31T Cheyenne II	7720029	Ste Air Service Affaires, Grenoble.	(N82122)
☐ F-GBLU	79	King Air C90	LJ-822	DIWAN, Rennes-St Jacques.	
☐ F-GBPB	66	King Air 90	LJ-98	Air ACG - Quick Speed, Lyon-Bron.	OY-ANP
☐ F-GBPZ	79	King Air C90	LJ-860	AVDEF-Aviation Defence Service, Nimes-Garons.	
☐ F-GCGA	80	King Air C90	LJ-894	Stylair, Colmar-Houssen.	
☐ F-GCJS	79	Mitsubishi MU-2P	406SA	Global Aero Finance, Cannes-Mandelieu.	N967MA
☐ F-GCLD	74	King Air C90	LJ-637	Air Bretagne Centrale, Vannes.	N95DD
☐ F-GCLH	80	PA-31T Cheyenne II	8020044	Alarme Service France/Locavions Aero Services, Pau.	N2330V
☐ F-GCTR	81	King Air F90	LA-115	BAICO-Business Aviation Compagnie, Annemasse.	
☐ F-GDAK	81	King Air F90	LA-141	Air Service Vosges, Epinal.	
☐ F-GDCS	82	King Air B200	BB-966	GIE Airsuc/EJA France, Paris-le Bourget.	
☐ F-GDJS	83	King Air B200	BB-1116	Aero Entreprise, Toussus le Noble.	(D-ILOC)
☐ F-GDLE	77	King Air 200	BB-230	Ste Aero Photo Europe Investigation, Moulins-Montbeugny.	G-BEHR
☐ F-GDMM	65	King Air 90	LJ-54	Ste TAT-Tours/CIPRA, Dinard. (status ?).	N66WC
☐ F-GEDV	66	King Air A90	LJ-150	SNC Air Armorique, St Brieuc.	D-ICPD
☐ F-GEFZ	81	Conquest 1	425-0059	Les Ailes Roussillonnaises, Montpellier.	LN-AFB
☐ F-GEJV	72	King Air A100	B-129	Air Normandie, Le Havre-Octeville.	N235B
☐ F-GEJY	79	King Air 200	BB-507	Tahiti Conquest Airlines/Ste Air West Indies, Raizet.	N600RM
☐ F-GELL	74	King Air E90	LW-88	Air Normandie, Le Havre-Octeville.	9Q-CTQ
☐ F-GEOU	81	King Air C90	LJ-941	Alsair SA. Colmar-Houssen.	N3804C
☐ F-GEPE	77	PA-31T Cheyenne II	7720031	Airlec Air Espace, Bordeaux-Merignac.	HB-LIW
☐ F-GEPY	81	King Air 200	BB-779	Nyco SA/DARTA Aero Charter, Paris-Le Bourget.	N811CB
☐ F-GEQM	80	MU-2 Marquise	790SA	Ste Financiere Delot et Compagnie, Orleans.	N279MA
☐ F-GERP	73	SA-226AT Merlin 4	AT-012	CEGEBAIL, Marcq en Baroeul.	N111MV
☐ F-GERS	80	King Air 200	BB-753	Aerobags, Paris-Le Bourget.	N3705B
☐ F-GESJ	74	King Air E90	LW-97	Transport Air Centre-Lyon/Ste Aerostock, Paris-Le Bourget.	F-WZIG
☐ F-GETI	80	King Air F90	LA-19	Air Poitiers, Poitiers-Biard.	N90NS
☐ F-GETJ	78	King Air E90	LW-296	Chalair, Caen-Carpiquet.	N207CP
☐ F-GEXK	68	King Air B90	LJ-331	Air Provence International, Marseille.	N886BD
☐ F-GEXL	76	King Air 200	BB-202	Aviasud Aerotaxi, Nice-Cote d'Azur.	N2425X
☐ F-GEXV	74	King Air A100	B-199	Flandre Air, Lille-Lesquin. (status ?).	N110TD
☐ F-GFBF	76	PA-31T Cheyenne II	7620054	Ste Soon, Toussus le Noble.	N39518
☐ F-GFCO	85	King Air C90A	LJ-1098	Plessis Air, Paris-Le Bourget.	N7218V
☐ F-GFCQ	64	Gulfstream 1	140	Air Provence International, Marseille-Marignane.	N92SA
☐ F-GFDJ	74	King Air E90	LW-86	Royal King Air SARL. Toussus-Le-Noble.	(N505N)
☐ F-GFEA	76	PA-31T Cheyenne II	7620011	France Europe Aviajet, Paris-Le Bourget.	N76PT
☐ F-GFEF	63	Gulfstream 1	122	DIWAN, Rennes-St Jacques.	N707MP
☐ F-GFGT	59	Gulfstream 1	5	Air Provence International, Marseille.'Le Provencal Club II'	N159AJ
Regn	*Yr*	*Type*	*c/n*	*Owner/Operator*	*Prev Regn*

Regn	Yr	Type	c/n	Owner/Operator	Prev Regn
☐ F-GFGV	60	Gulfstream 1	44	Air Provence International, Marseille. 'Le Provencal Club'	N717RD
☐ F-GFHC	77	King Air C90	LJ-717	Locavia France, Nantes.	N200BX
☐ F-GFHQ	68	King Air B90	LJ-347	Eolia SNC/Aerope 3S Aviation. Pointoise-Cormeilles.	N777SB
☐ F-GFIB	61	Gulfstream 1	71	DIWAN, Rennes-St Jacques.	N222EF
☐ F-GFIC	60	Gulfstream 1	49	Air Provence International, Marseille. 'Le Provencal Club 3'	N456
☐ F-GFIR	68	King Air B90	LJ-434	Heli Champagne Arden, Reims/Air ACG-Quick Speed, Lyon.	C-GRCN
☐ F-GFJR	81	Conquest 1	425-0032	Ste Vitreenne d'Abattage, Rennes.	N6773B
☐ F-GFLE	75	PA-31T Cheyenne II	7520002	Ste Prestavia SA. Coulommiers.	N90589
☐ F-GFMH	59	Gulfstream 1	20	Air Provence International, Marseille. 'Le Provencal Club 4'	N732US
☐ F-GFUV	77	PA-31T Cheyenne II	7720063	Roland Torterotot, Roanne-Renaison.	N3948A
☐ F-GFVN	82	King Air F90	LA-166	La Voix du Nord, Lille-Lesquin.	HB-GHM
☐ F-GFVO	79	PA-31T Cheyenne II	7920049	Air Service Vosges, Epinal-Mirecourt.	N500FC
☐ F-GGAM	65	King Air 90	LJ-32	SLIBAIL, Paris.	F-BTOK
☐ F-GGAT	78	PA-31T Cheyenne II	7820010	Ste Alpha Tango, Garches.	D-IFPD
☐ F-GGCH	81	PA-31T Cheyenne II	8120056	Lofinord/Bertrand Lehouck, Lille-Lesquin.	N51TW
☐ F-GGGY	61	Gulfstream 1	80	Air Provence International, Marseille. 'Le Francilien'	N200GJ
☐ F-GGLA	80	King Air 200	BB-744	Berry Flight, Chateauroux.	F-GGPJ
☐ F-GGLG	81	SA-227AT Merlin 4C	AT-493	Compagnie Aeronautique Europeene, Marseille-Marignane.	N121FA
☐ F-GGLN	79	King Air 200	BB-439	Air Net GIE. Montpellier.	N500JA
☐ F-GGLV	73	King Air A100	B-150	Chalair, Caen-Carpiquet.	N51BL
☐ F-GGMS	75	King Air 200	BB-80	ASE/Ste Aeronautique Auboise, Troyes.	N44TW
☐ F-GGMV	80	King Air 200	BB-616	Air Partner's International/Leadair-Unijet SA. Paris	SE-IUN
☐ F-GGPR	80	King Air 200	BB-681	CIP Transports/Aero Service Executive, Paris-Le Bourget.	LN-AXA
☐ F-GGRV	77	PA-31T Cheyenne II	7720036	Trans Helicoptere Service, Lyon-Bron.	N41RC
☐ F-GGTV	78	PA-42 Cheyenne III	7800002	EAS European Airlines, Perpignan.	N4494E
☐ F-GGVG	78	SA-226T Merlin 3B	T-293	Pan Europeenne Air Service, Chambery.	D-IBBB
☐ F-GHBB	71	King Air C90	LJ-510	Ste SNC Sun Air, Pointoise.	D-ILHB
☐ F-GHBD	72	King Air C90	LJ-545	Travelair Enterprises, Pointoise-Cormeiles.	D-ILHD
☐ F-GHCS	77	King Air/Catpass 250	BB-303	THS Helicopters, Lyon-Bron.	N18345
☐ F-GHDO	67	King Air A90	LJ-206	Ste Aero Stock, Toussus-le-Noble.	F-BTAK
☐ F-GHFE	72	King Air C90	LJ-544	Travelair Enterprises, Pointoise.	(F-GHFC)
☐ F-GHHV	72	King Air A100	B-91	Air Normandie, Le Havre-Octeville.	N9050V
☐ F-GHIV	80	King Air F90	LA-22	Air Poitiers/Iso Air, Vouille.	(N444EM)
☐ F-GHJV	77	PA-31T Cheyenne II	7720067	G F R, Annemasse.	N900SF
☐ F-GHLB	78	King Air 200	BB-349	Air Rhone Alpes, Lyon-Bron.	D-IEXD
☐ F-GHLD	77	King Air/Catpass 250	BB-233	Regourd Aviation, Paris-Le Bourget.	N50JD
☐ F-GHOC	78	King Air 200	BB-406	Air Ile de France, Paris-Le Bourget.	G-OEMS
☐ F-GHSV	80	King Air 200	BB-622	Lima Papa SARL/Air Normandie, Le Havre-Octeville.	N212BF
☐ F-GHTA	78	PA-31T Cheyenne II	7820015	Yankee Delta, Montpellier.	N107BK
☐ F-GHUV	78	King Air E90	LW-278	Fabris SA/Air Poitiers, Poitiers-Biard.	N700MA
☐ F-GHVF	82	SA-227AT Merlin 4C	AT-423	Compagnie Aeronautique Europeenne, Marseille-Marignane.	N10NB
☐ F-GHVV	80	King Air 200	BB-676	Chalair, Caen-Carpiquet.	N970AA
☐ F-GHYV	78	King Air 200	BB-364	Flandre Air, Lille-Lesquin.	N66171
☐ F-GIAL	81	King Air 200	BB-844	Ste Knauf la Rhenane/Alsair SA. Colmar-Houssen.	SE-IGV
☐ F-GICA	88	King Air 300	FA-146	Castorama Transports GIE. Lille-Lesquin.	N2650C
☐ F-GICE	68	King Air B90	LJ-363	Michel Bidoux/Air Transport Pyrenees SA. Pau.	N303WJ
☐ F-GIDV	80	King Air 200	BB-590	AVDEF-Aviation Defense Service, Nimes-Garons.	I-ALGH
☐ F-GIFB	69	King Air B90	LJ-453	Icare Franch Comte, Montebeliard/Air ACG-Qick Speed, Lyon..	(D-IAMX)
☐ F-GIFC	69	King Air B90	LJ-456	Aparcom/Air ACG - Quick Speed, Lyon-Bron.	D-ILTY
☐ F-GIFK	80	King Air F90	LA-62	OR Telematique, Tours.	D-ICBD
☐ F-GIII	80	PA-31T Cheyenne II	8020037	Ste Airways SA. Agen-La Garenne.	N805SW
☐ F-GIJB	74	King Air 200	BB-13	Chalair, Caen-Carpiquet.	N83MA
☐ F-GILB	79	King Air 200	BB-477	Ste MG Services, Marseilles.	OY-BPG
☐ F-GILH	78	King Air 200	BB-431	Compagnie Transair, Paris-Le Bourget.	9Q-CTF
☐ F-GILP	79	King Air 200	BB-542	Lima Papa SARL/Air Normandie. Le Havre-Octeville.	N3333X
☐ F-GIML	76	King Air E90	LW-180	Regourd Aviation, Paris-Le Bourget.	N2180L
☐ F-GIND	81	King Air 200	BB-822	Ste Air Norbert Dentressangles, Lyon-Bron.	N3FH
☐ F-GIPL	80	PA-31T Cheyenne II	8020091	Airco SNC. Brive.	N76SC
☐ F-GIZB	81	King Air C90	LJ-955	Atlantique Air Assistance, Nantes-Atlantique.	N768SB
☐ F-GJAD	72	King Air E90	LW-3	Azur Aviation Service, Cannes.	N888BH
☐ F-GJBS	83	King Air B200	BB-1181	Bongrain Service Air/Air Entreprise, Paris-Le Bourget.	N6725Y
☐ F-GJCD	84	King Air 300	FA-7	Ste J C Decaux, Toussus-Le-Noble.	N925AD
☐ F-GJCR	78	King Air E90	LW-251	Dassault Aviation, Istres.	N7ZU
☐ F-GJDK	77	PA-31T Cheyenne II	7720043	Aero Stock, Paris-Le Bourget.	(N80MA)
☐ F-GJFA	87	King Air B200	BB-1270	Air Normandie, Le Havre-Octeville.	N30391
☐ F-GJFC	89	King Air B200	BB-1347	SFACT Operations Aeriennes, St Cyr.	
☐ F-GJFD	90	King Air B200	BB-1379	SFACT Operations Aeriennes, St Cyr.	
☐ F-GJFE	91	King Air B200	BB-1399	SFACT Operations Aeriennes/ESMA, Montpellier.	
☐ F-GJHH	87	King Air 300	FA-112	Ste Accor/Ste de Transportes Internes, Evry.	N324NE
☐ F-GJHM	79	King Air E90	LW-316	Inter-France Air Exploitation, Paris-Le Bourget.	N38CR
☐ F-GJJJ	74	King Air A100	B-196	Chalair, Caen-Carpiquet.	N773SK
☐ F-GJLD	82	King Air C90-1	LJ-1035	Eiflavia, Ancenis.	G-NUIG
☐ F-GJMJ	82	King Air B200	BB-1032	Ste Knauf la Rhenane/Alsair SA. Colmar-Houssen.	I-CUVI
☐ F-GJPD	80	King Air E90	LW-328	Avialim, Limoges.	N551M
☐ F-GJPE	77	PA-31T Cheyenne II	7720042	Ste Air Mont Blanc, Sallanches.	YU-BKT
☐ F-GJPY	91	TBM 700	13	Fuz SA. Toussus-le-Noble.	(F-GKDJ)

Regn	Yr	Type	c/n	Owner/Operator	Prev Regn
☐ F-GJRD	67	King Air A90	LJ-217	Roland Duquairoux, Perpignan.	N601R
☐ F-GJSD	91	King Air C90A	LJ-1261	Week-End Aviation Location, Paris-Le Bourget.	
☐ F-GKCV	77	King Air 200	BB-251	Chalair, Caen-Carpiquet.	I-BMPE
☐ F-GKEL	76	King Air A100	B-228	Locavia France, Nantes.	ZS-LVL
☐ F-GKII	79	King Air 200	BB-515	Kyrn Air, Lyon-Bron.	N200HC
☐ F-GKJV	91	TBM 700	11	Jacques le Maigre du Breuil, Toussus-le-Noble.	
☐ F-GKKK	81	King Air F90	LA-129	Entreprise Philippe Lassarat, Le Havre.	OY-BEL
☐ F-GKRR	81	PA-31T Cheyenne II	8120015	Ste Airways SA. Agen-La Garenne.	N107TT
☐ F-GKSP	95	King Air C90B	LJ-1409	A F L SARL. Paris-Le Bourget/Chalair, Caen-Carpiquet.	F-GKDG
☐ F-GLBF	91	TBM 700	23	Airflo Europe,	F-WNGO
☐ F-GLBK	96	TBM 700	116	SOCATA, Tarbes.	
☐ F-GLBM	91	TBM 700	22	SOCATA, Tarbes.	F-OHBF
☐ F-GLJS	92	TBM 700	63	TBM SA. Toussus le Noble.	(F-GLJE)
☐ F-GLLL	95	TBM 700	107	Sofca Investissements SA. Bourges.	F-GLBG
☐ F-GLRZ	91	King Air C90A	LJ-1296	Entremont SNC. Annecy.	(F-GJDK)
☐ F-GMCP	92	P-180 Avanti	1022	Air Entreprise International, Paris-Le Bourget.	
☐ F-GMCR	79	King Air 200	BB-424	Ligier Jet Air SARL/Air Normandie, Le Havre-Octeville.	(HP-...)
☐ F-GMCS	80	King Air/Catpass 250	BB-688	Air Normandie, Le Havre-Octeville.	N250TR
☐ F-GMGB	90	King Air B200	BB-1390	IGN-Institut Geographique National, Creil.	(F-GLOP)
☐ F-GMLT	92	King Air B200T	BT-34	IGN France, Creil. (was BB-1426).	N56361
☐ F-GMPM	92	King Air C90B	LJ-1303	Ste O C I M Location et Cie, Toussus-le-Noble.	(F-GIAO)
☐ F-GMSM	94	Beech 1900D	UE-106	Group SM/SM Airlines, Bucharest, Romania.	N15574
☐ F-GMTO		SA-226AT Merlin 4A	AT-031	Etat Francais Meteo France, Bretigny.	N22KW
☐ F-GNAC	80	PA-31T Cheyenne II	8020076	Finist'Air, Brest-Guipavas.	N18AF
☐ F-GNAE	92	P-180 Avanti	1020	Air Entreprise International, Paris-Le Bourget.	(D-ISAP)
☐ F-GNBA	67	King Air A90	LJ-311	Carry Air, Lyon-Bron.	HB-GIN
☐ F-GNEE	93	King Air C90B	LJ-1328	G I E Avion Ecco, Lyon-Bron.	N90HB
☐ F-GNEG	90	King Air B200	BB-1377	Motorservice SA/Stylair, Colmar-Houssen.	HB-GIR
☐ F-GNGU	63	Gulfstream 1	101	Air Provence International, Marseille.	4X-ARV
☐ F-GNMA	92	King Air C90A	LJ-1299	Oyonnair SARL. Bourg-en-Bresse.	N8253D
☐ F-GNPD	77	King Air 200	BB-199	Aero Stock, Bamako, Mali.	LN-PAD
☐ F-GOON	81	PA-42 Cheyenne III	8001030	Ste Soon, Toussus Le Noble.	N21MS
☐ F-GOOO	89	King Air 300	FA-175	Sofaxis/Sofca Investissements, Bourges.	N175NJ
☐ F-GPAS	77	King Air 200	BB-209	Proteus Airlines, Dijon-Longvic..	D-IACS
☐ F-GRAN	78	King Air 200	BB-392	Regourd Aviation/EJA France, Paris-Le Bourget.	N400GM
☐ F-GSFA	86	King Air B200	BB-1244	SFACT Operations Aeriennes, St Cyr.	G-RIOO
☐ F-GSIN	77	King Air 200	BB-239	Sinair, Grenoble-St Geoirs.	N517JM
☐ F-O...	78	King Air C90	LJ-760	SNC Charlie Juliett, Point a Pitre-Le Raizet.	F-GHEM
☐ F-ODGS	77	PA-31T Cheyenne II	7720041	Franck Jean Gallo, Noumea, New Caledonia.	TS-LAZ
☐ F-ODGU	81	King Air F90	LA-88	New Caledonia High Commission, Noumea.	F-GCTB
☐ F-ODMM	80	PA-31T Cheyenne II	8020084	Ste Pacific Perles, Papeete, Tahiti.	F-WFLQ
☐ F-ODUJ	82	Conquest II	441-0264	Air Archipels, Papeete-Faaa.	G-EVNS
☐ F-ODYR	78	King Air 200	BB-716	Aviazur, Noumea-Magenta.	N200KW
☐ F-ODYZ	91	Reims/Cessna F406	F406-0057	Aviazur, Noumea-Magenta.	N31226
☐ F-OGOG	88	Reims/Cessna F406	F406-0026	Air Guyane, Cayenne-Rochambeau.	
☐ F-OGOX	75	King Air C90	LJ-668	Air Caraibes, Pointe a Pitre-Le Raizet.	F-GGMO
☐ F-OGPQ	77	King Air 200	BB-192	Air Caraibes, Pointe a Pitre-Le Raizet.	OY-CBV
☐ F-OGUY	90	King Air C90A	LJ-1250	Ste Tudy, St Martin Grand Case.	F-GKGT
☐ F-OGVS	91	Reims/Cessna F406	F406-0061		N3121X
☐ F-OHCP	81	King Air 200	BB-831	High Commissioner of French Polynesia, Papeete, Tahiti.	F-ODUA
☐ F-OHEV	92	TBM 700	52	SOCATA, Lisbon, Portugal.	VH-PTG
☐ F-OHJK	96	King Air B200	BB-1544	Air Archipels, Papeete-Faaa.	N1094S
☐ F-OHRT	93	King Air 300LW	FA-226	Wanair/Immeuble Tahiti Perles, Papeete.	N80907
☐ F-SEBK		ATR 42-320	264	CNET Operations Aeriennes, Lannion-Servel.	N99839
☐ F-WKDL	91	TBM 700	03	SOCATA/TBM SA. Tarbes.	
☐ F-WKPG	91	TBM 700	02	SOCATA/TBM SA. Tarbes.	
☐ F-WTBM		TBM 700	01	SOCATA/TBM SA. Tarbes.	
☐ F-WWRK	97	TBM 700	126	SOCATA, Paris-Le Bourget.	
☐ F-WWSR	96	Reims/Cessna F406	F406-0080	Reims Aviation,	
Military					
☐ 100 ABP	95	TBM 700	...	French Army, 3 GHL, Rennes.	
☐ F-MABM/0008		Reims/Cessna F406	F406-0008	French Army, Rennes.	
☐ F-MABN/0010		Reims/Cessna F406	F406-0010	French Army, Rennes.	
☐ F-MABQ/11596		TBM 700	115	French Army, 3 GHL, Rennes.	
☐ F-RAXA	92	TBM 700	33	33/65-XA, COTAM,	
☐ F-RAXB	92	TBM 700	35	35/65-XB, COTAM,	
☐ F-RAXC	93	TBM 700	70	70/65-XC, COTAM,	
☐ F-RAXD	93	TBM 700	77	77/65-XD, COTAM,	
☐ F-RAXE	93	TBM 700	78	78/65-XE, COTAM,	
☐ F-RAXF	93	TBM 700	80	80/65-XF, COTAM,	
☐ F-ZBAB		Reims/Cessna F406	F406-0025	French Customs,	
☐ F-ZBBB		Reims/Cessna F406	F406-0039	French Customs,	F-WZDS
☐ F-ZBBF	71	King Air C90	LJ-518	Securite Civile, Paris.	F-BTCA
☐ F-ZBCE		Reims/Cessna F406	F406-0042	French Customs,	F-WKRA
☐ F-ZBCF	95	Reims/Cessna F406	F406-0077	French Customs,	F-WZDZ
☐ F-ZBCG		Reims/Cessna F406	F406-0066	French Customs,	F-WZDT

| Regn | Yr | Type | c/n | Owner/Operator | Prev Regn |

Regn	Yr	Type	c/n	Owner/Operator	Prev Regn
□ F-ZBCH		Reims/Cessna F406	F406-0075	French Customs,	
□ F-ZBCI		Reims/Cessna F406	F406-0070	French Customs,	
□ F-ZBCJ		Reims/Cessna F406	F406-0074	French Customs,	
□ F-ZBEP		Reims/Cessna F406	F406-0006	French Customs,	
□ F-ZBES		Reims/Cessna F406	F406-0017	French Customs,	
□ F-ZBFA		Reims/Cessna F406	F406-01	French Customs,	F-GGRA
□ F-ZBFJ	83	King Air B200	BB-1102	GMA/Ministry of the Interior, Marseille-Marignane.	(F-GKDO)
□ F-ZBFK	81	King Air B200	BB-876	GMA/Ministry of the Interior, Marseille-Marignane.	F-GHSC
□ F-ZBFZ	81	PA-31T Cheyenne II	8120064	Customs & Excise, Paris-Le Bourget.	F-WFLQ
□ F-ZVMN	94	TBM 700	106	CEV-Centre d'Essais en Vol, Bretigny.	

G = United Kingdom (U.K. of Great Britain & Northern Ireland)

Civil

Regn	Yr	Type	c/n	Owner/Operator	Prev Regn
□ G-BGRE	79	King Air 200	BB-568	RM Aviation/Martin Baker (Engineering) Ltd. Chalgrove.	
□ G-BMKD	84	King Air C90A	LJ-1069	A E Bristow, Cranleigh.	N223CG
□ G-BNDY	85	Conquest 1	425-0236	Standard Aviation Ltd. Newcastle.	N1262T
□ G-BPPM	82	King Air B200	BB-1044	GAMA Aviation Ltd. Fairoaks.	N7061T
□ G-BVJT	94	Reims/Cessna F406	F406-0073	S F Herbert/NOR Leasing, London.	
□ G-BVMA	81	King Air 200	BB-797	Manhattan Air Ltd. Blackbushe.	G-VPLC
□ G-BVRS	69	King Air B90	LJ-481	The Eight Blew Ltd. Godalming.	G-KJET
□ G-BXMA	80	King Air 200	BB-726	Manhattan Air Ltd. Blackbushe.	N622JA
□ G-CEGR	78	King Air 200	BB-351	CEGA Aviation Ltd. Goodwood.	N68CP
□ G-DEXY	75	King Air E90	LW-136	Tornado Ltd. Guernsey, Channel Islands.	N750DC
□ G-DFLT	89	Reims/Cessna F406	F406-0036	Directflight Ltd. Norwich.	F-WZDZ
□ G-ECAV	79	King Air 200	BB-561	GEC-Marconi Avionics Ltd. Rochester.	N36GA
□ G-FCAL	83	Conquest II	441-0293	F R Aviation Ltd. Bournemouth-Hurn.	C-FMHD
□ G-FLTI	80	King Air F90	LA-59	Flightline Ltd. Southend.	N7P
□ G-FPLA	82	King Air B200	BB-944	F R Aviation Ltd. Teesside.	N31WL
□ G-FPLB	82	King Air B200	BB-1048	F R Aviation Ltd. Teesside.	N739MG
□ G-FPLC	81	Conquest II	441-0207	F R Aviation Ltd. Teesside.	G-FRAX
□ G-FRAZ	77	Conquest II	441-0035	F R Aviation Ltd. Bournemouth-Hurn.	SE-GYC
□ G-FRYI	76	King Air 200	BB-210	London Executive Aviation, London City.	G-OAVX
□ G-HAMA	75	King Air 200	BB-30	Bond Helicopters-Aberdeen/GAMA Aviation Ltd. Fairoaks.	N244JB
□ G-KMCD	89	King Air B200	BB-1325	Gamston Aviation Ltd. Gamston.	V5-BDL
□ G-LEAF	87	Reims/Cessna F406	F406-0018	Atlantic Air Transport Ltd. Coventry.	EI-CKY
□ G-LILI	81	Conquest 1	425-0054	Ortac Air, Guernsey, Channel Islands.	G-YOTT
□ G-NISR	75	Rockwell 690A	11243	Z I Bilbeisi, Stansted.	HB-GFS
□ G-OFHJ	82	Conquest II	441-0294	Tilling Associates Ltd/Eagle Airways Ltd. Guernsey, C.I.	G-HSON
□ G-OLDZ	81	King Air 200	BB-828	Gold Air, Biggin Hill.	G-MCEO
□ G-REBK	85	King Air B200	BB-1202	Planstable Enterprises Ltd/Gold Air, Biggin Hill.	D-IHAP
□ G-ROWN	80	King Air 200	BB-684	Holiday Chemical Holdings PLC/AJ Air Services, Leeds.	G-BHLC
□ G-SANB	78	King Air E90	LW-304	Maynard & Harris Holdings/Sterling Helicopters Ltd. Norwich.	G-BGNU
□ G-SBAS	82	King Air B200	BB-1007	Bond Helicopters Ltd. Aberdeen.	SE-IVZ
□ G-SFPA	91	Reims/Cessna F406	F406-0064	Fisheries Protection Agency, Edinburgh.	
□ G-SFPB	91	Reims/Cessna F406	F406-0065	Fisheries Protection Agency, Edinburgh.	
□ G-THUR	81	King Air 200	BB-782	Removed from register by CAA 11 Apr 97 (reason ?).	5Y-BIW
□ G-TINI	87	Reims/Cessna F406	F406-0015	Martini Airfreight Services Ltd.	PH-FWE
□ G-TURF	87	Reims/Cessna F406	F406-0020	Air Alba Ltd. Inverness.	EI-CND
□ G-VSBC	88	King Air B200	BB-1290	Vickers Shipbuilding & Engineering Ltd. Barrow-in-Furness.	N3185C
□ G-WELL	76	King Air E90	LW-198	Colt Transport Ltd/CEGA Aviation Ltd. Goodwood.	(N7PB)
□ G-WRCF	79	King Air B200	BB-472	Cx G- 19 Jan 98.	N4489A

HA = Hungary (Hungarian Republic)

Civil

Regn	Yr	Type	c/n	Owner/Operator	Prev Regn
□ HA-...	70	Mitsubishi MU-2F	191		SE-GHG
□ HA-ACS	78	King Air 200	BB-324	Fotex,	N900VG
□ HA-ACV	63	Gulfstream 1	102		N48CQ

HB = Switzerland (Swiss Confederation)

Civil

Regn	Yr	Type	c/n	Owner/Operator	Prev Regn
□ HB-FOA	91	Pilatus PC-XII	P-01	Pilatus Flugzeugwerke AG. Stans. (Ff 31 May 91).(status ?).	
□ HB-FOB	93	Pilatus PC-XII	P-02	Pilatus Flugzeugwerke AG. Stans.	
□ HB-FOE	95	Pilatus PC-XII	102	Pilatus Flugzeugwerke AG. Stans.	
□ HB-FOG	96	Pilatus PC-XII	134	Pilatus Flugzeugwerke AG. Stans.	
□ HB-FOI	96	Pilatus PC-XII	157	Share Plane AG/AVCON AG. Zurich.	
□ HB-FOJ	96	Pilatus PC-XII	158	Lutherworth AG. Zug/AVCON Ltd. Bruettisellen.	
□ HB-FQA	94	Pilatus PC-XII	104	Pilatus Flugzeugwerke AG. Stans. (was P-04).	JA8613
□ HB-FQB	97	Pilatus PC-XII	192	Pilatus Flugzeugwerke AG. Stans.	
□ HB-FQC	97	Pilatus PC-XII	193	Pilatus Flugzeugwerke AG. Stans.	
□ HB-FQD	97	Pilatus PC-XII	195	Pilatus Flugzeugwerke AG. Stans.	
□ HB-FSP	97	Pilatus PC-XII	180	Pilatus Flugzeugwerke AG. Stans.	
□ HB-FSV	97	Pilatus PC-XII	187	Pilatus Flugzeugwerke AG. Stans.	
□ HB-FSW	97	Pilatus PC-XII	188	Pilatus Flugzeugwerke AG. Stans.	
□ HB-FSX	97	Pilatus PC-XII	189	Pilatus Flugzeugwerke AG. Stans.	

Regn	Yr	Type	c/n	Owner/Operator	Prev Regn

Regn	Yr	Type	c/n	Owner/Operator	Prev Regn
☐ HB-FSY	97	Pilatus PC-XII	190	Pilatus Flugzeugwerke AG. Stans.	
☐ HB-FSZ	97	Pilatus PC-XII	191	Pilatus Flugzeugwerke AG. Stans.	
☐ HB-GDA	82	King Air C90-1	LJ-996	Girard-Perregaux SA. La Chaux de Fonds.	N600SB
☐ HB-GDL	82	King Air B200	BB-1079	STAC-Service des Transports Aeriens, Berne.	HB-GDI
☐ HB-GEV	67	King Air A90	LJ-215	FFA Air Travel, Altenrhein.	D-IEVW
☐ HB-GGU	79	King Air E90	LW-315	AISA-Automation Industrial SA. Vouvry.	
☐ HB-GHD	80	King Air F90	LA-50	Air Evasion SA. Lausanne.	F-GCLS
☐ HB-GHF	79	King Air 200	BB-417	Emoht Anstalt, Vaduz.	G-BFVZ
☐ HB-GHK	81	Gulfstream 1000	96023	Gofir SA Aerotaxi, Lugano.	ZS-KZV
☐ HB-GHN	85	King Air C90A	LJ-1093	Business Air AG. Altenrhein.	D-IDSR
☐ HB-GHO	81	King Air F90	LA-111	Luca Regusci, Castagnola.	G-BIEZ
☐ HB-GHS	82	King Air B200	BB-1039	Elf Huiles Minerales SA. Etagnieres.	N6572K
☐ HB-GHT	82	King Air C90	LJ-944	Air Glaciers SA. Sion.	(N135JA)
☐ HB-GHV	88	King Air 300	FA-170	Keller International AG. Wallisellen.	
☐ HB-GHW	85	King Air C90A	LJ-1089	Air Glaciers SA. Sion.	D-IEXT
☐ HB-GIE	81	King Air C90	LJ-931	Eco-Carrier Aviation AG/AVCON AG. Zurich.	I-FIRS
☐ HB-GII	93	King Air 350C	FN-1	BAMF Survey Flights, Duebendorf.	
☐ HB-GIL	77	King Air 200	BB-194	Air Glaciers SA. Sion.	N300PH
☐ HB-GIP	89	King Air 300	FA-202	Gert Kroll, Geuensee.	N1553V
☐ HB-GJA	81	King Air C90	LJ-992	Air Lemanic SA. Lausanne.	N992LJ
☐ HB-GJD	80	King Air 200C	BL-7	Zimex Aviation Ltd. Zurich.	F-GJBJ
☐ HB-GJE	95	King Air C90B	LJ-1391	Alain Prost, Yens/Centre Aeronautique Lemanique, Geneva.	
☐ HB-GJF	95	King Air C90B	LJ-1407	Weisstern AG/Gofir SA. Lugano 'Spirit of Santa Maria'	
☐ HB-GJH	81	King Air C90	LJ-972	Happy Lines SA. Geneva.	N18080
☐ HB-GJI	79	King Air 200	BB-451	Air Glaciers SA. Sion.	D-IBOW
☐ HB-GJL	97	King Air 350	FL-183	Breitling SA. Grenchen.	
☐ HB-KEI	91	TBM 700	3	Contaco Establishment, Vaduz.	F-GJTS
☐ HB-LLK	79	PA-31T Cheyenne 1	7904014	Bruno Keppeler, Conthey.	N401PT
☐ HB-LNL	80	PA-31T Cheyenne II	8020083	Daniel Knutti, Magglingen-Macolin.	N54JB
☐ HB-LNX	81	Cheyenne II-XL	8166050	AVCON AG/Transwing AG. Basel/Berne.	N700XL
☐ HB-LOE	81	Conquest 1	425-0016	Mecaplex AG. Grenchen.	ZS-KST
☐ HB-LOL	80	Gulfstream 840	11639	Alistar AG Mauren, Chiasso.	N63RB
☐ HB-LOZ	79	PA-31T Cheyenne II	7920039	Heli-Linth AG. Mollis.	N32745
☐ HB-LQA	81	Gulfstream 980	95069	Orchid Aviation Ltd. Zug.	VR-CBP
☐ HB-LQP	80	PA-31T Cheyenne 1	8004044	Ursella Holding AG. Berne.	OO-JMR
☐ HB-LRV	78	PA-31T Cheyenne II	7820017	Thomke AG. Stans.	N82222

HC = Ecuador *(Republic of Ecuador)*

Civil

Regn	Yr	Type	c/n	Owner/Operator	Prev Regn
☐ HC-...	80	Gulfstream 840	11615		N811EC
☐ HC-...	83	Gulfstream 900	15023		N88RC
☐ HC-...	80	Gulfstream 980	95026		N903L
☐ HC-BHU	80	Gulfstream 840	11634	TAESA. Quito. (status ?).	N5886K
☐ HC-BIF	81	PA-31T Cheyenne II	8120019	Ecuavia Ltda. Guayaquil.	N2425X
☐ HC-BNF	79	PA-31T Cheyenne 1	7904039	Banco del Pacifico, Quito.	N664RB
☐ HC-BPF	75	Rockwell 690A	11235	ICARO-Instituto Civil Aeronautico SA. Quito.	
☐ HC-BPX	74	Rockwell 690A	11187	Filanbanco SA.	N440CA
☐ HC-BPY	71	681B Turbo Commander	6049	Aviopacifico SA. Guayaquil.	N587KA
☐ HC-BSR	79	King Air/Catpass 250	BB-483	ICARO-Instituto Civil Aeronautico SA. Quito.	N27BG
☐ HC-BUD	81	Gulfstream 840	11669	SAEREO SA.	N844MA
☐ HC-BUS	81	Cheyenne II-XL	8166064	LAN Ecuador, Guayaquil.	N7194Y
☐ HC-BVA	78	PA-31T Cheyenne II	7820069	AXA 1002 Cia Ltda.	N25AB
☐ HC-DAC	76	King Air E90	LW-178	Directorate of Civil Aviation, Quito.	

Military

Regn	Yr	Type	c/n	Owner/Operator	Prev Regn
☐ AEE-101	81	King Air 200	BB-811	Ministry of National Defence, Quito.	AEE-001
☐ AN-231	81	King Air 200	BB-771	Aviacion Naval Ecuatoriana, Guayaquil.	N3831Q
☐ AN-232	85	King Air 300	FA-75	Aviacion Naval Ecuatoriana, Guayaquil.	N7247A
☐ AN-233	80	King Air 200	BB-580	Aviacion Naval Ecuatoriana, Guayaquil.	N48450
☐ AN-234	79	King Air 200	BB-458	Aviacion Naval Ecuatoriana, Guayaquil.	N169DB
☐ FAE-001	70	HS 748-2A	1684	Government of Ecuador, Quito.	HC-AUK
☐ IGM-240	78	King Air A100	B-242	Instituto Geografico Militar, Ejercito-army photo survey.	

HI = Dominican Republic

Civil

Regn	Yr	Type	c/n	Owner/Operator	Prev Regn
☐ HI-...	71	Mitsubishi MU-2F	194	Agroaplica Corp. Santo Domingo.	N282MS
☐ HI-586SP	90	King Air 350	FL-8	Empresas E Leon Jimenez, Santiago.	N5526V
☐ HI-678CA	69	Gulfstream 1	323	Caribe SA. Santo Domingo.	N980TT

HK = Colombia *(Republic of Colombia)*

Civil

Regn	Yr	Type	c/n	Owner/Operator	Prev Regn
☐ HK-....	95	DHC 8-202	391	BPX Colombia SA. Bogota.	C-GFBW
☐ HK-....	84	Gulfstream 1000	96073		N85DJ
☐ HK-....	80	Gulfstream 840	11602	Alicol Ltda.	N2647C
☐ HK-....	80	Gulfstream 840	11604	Carlos Alberto Cortez, Bogota.	N5855K
☐ HK-....	81	Gulfstream 980	95080		N999ST
Regn	*Yr*	*Type*	*c/n*	*Owner/Operator*	*Prev Regn*

Regn	Yr	Type	c/n	Owner/Operator	Prev Regn
□ HK-...	80	Gulfstream 980	95001	Gustavo Durango, Cali.	N980AA
□ HK-...	75	King Air 200	BB-60		N530JA
□ HK-...	86	King Air 300	FA-99	Amazonian Investment Inc. Bogota.	N6293V
□ HK-...	93	King Air 350	FL-98	Servicio Aereo del Choco Ltda. Bogota.	N8302N
□ HK-...	81	King Air C90-1	LJ-986		N18300
□ HK-...	74	King Air E90	LW-90		N114AT
□ HK-...	81	King Air F90	LA-84		N122GA
□ HK-...	80	PA-31T Cheyenne II	8020043	F de Jesus Ocampo de Sepulveda. (status ?).	
□ HK-...	81	PA-42 Cheyenne III	8001068		N4099Y
□ HK-...	77	Rockwell 690B	11386		N568H
□ HK-...	75	SA-226T Merlin 3A	T-253		N959M
□ HK-853W	69	King Air B90	LJ-458	Jorge Hernando Carvajal Administradores Ltda y Cia S En C.	HK-353P
□ HK-1770G	74	Rockwell 690A	11216	Instituto Geografico Agustin Codazzi, Bogota.	N57216
□ HK-1844P	73	Rockwell 690	11056	Jose Nelson Suarez,	HK-1844W
□ HK-1977	70	681 Turbo Commander	6035	SEARCA-Servicio Aereo de Capurgana Ltda.	HK-1977W
□ HK-1982	72	Rockwell 690	11014	TAS-Transporte Aereo de Santander,	HK-1982W
□ HK-2051W	77	Rockwell 690B	11350	Productos Alimenticios Doria SA.	HK-2051
□ HK-2055	72	Rockwell 690	11005	ARPA-Aerolineas Regionales de Paz de Ariporo Ltda.	HK-2055W
□ HK-2060W	72	Mitsubishi MU-2K	243	Agropecuaria Las Garzas y Cia Ltda.	HK-2060X
□ HK-2120P	74	Mitsubishi MU-2J	621	Carlos Alberto Aragon Orozco,	HK-2120
□ HK-2218W	78	Rockwell 690B	11453	Fabrica de Dulces Colombina Ltda.	N81736
□ HK-2281	72	Rockwell 690	11033	AVIEL-Aviones Ejecutivas Ltda. Bogota.	HK-2281P
□ HK-2282P	73	Rockwell 690A	11128	Gabriel Antonio Valencia Rendon,	HK-2282W
□ HK-2285P	67	680T Turbo Commander	1632-57	Jimenez Alfonso Mejia,	HK-2285
□ HK-2291X	77	Rockwell 690B	11364		HK-2291
□ HK-2347	79	PA-31T Cheyenne II	7920076	HELICONDOR-Helicopteros El Condor Ltda/Helicafe Ltda.	HK-2347P
□ HK-2376P	71	681 Turbo Commander	6043	Martha Lucy Hernandez de Dominguez,	HK-2376
□ HK-2390W	77	Conquest II	441-0016	Norman Gonzalez y Asociados Ltda.	N36938
□ HK-2403	76	Mitsubishi MU-2L	678	Jorge Alfonso Diaz Clavijo,	HK-2403W
□ HK-2414P	76	Rockwell 690A	11296	Nestor Manuel Gutierrez Sanchez,	HK-2414
□ HK-2451P	80	PA-31T Cheyenne II	8020040	Javier Dario Torres Tobon,	
□ HK-2455	80	PA-31T Cheyenne II	8020047	LIRA-Lineas Aereas de Uraba Ltda/Aero Atlantico.	HK-2455
□ HK-2479W	80	SA-226T Merlin 3B	T-320	AGROCASA Ltda/Inversiones Agropecuarias Casanare,	HK-2479
□ HK-2490P	77	Rockwell 690B	11354	Hugo Botero Correa,	HK-2490X
□ HK-2491	76	King Air E90	LW-183	Interamericana de Aviacion SA. Bogota.	YV-940P
□ HK-2495P	80	Gulfstream 840	11633	Dario Maja Velasquez,	N5885K
□ HK-2538P	80	Conquest II	441-0114	Pozo Hernando Buitrago,	HK-2538X
□ HK-2551P	78	Rockwell 690B	11489	Francisco Antonio Molina Laverde,	N7701L
□ HK-2585P	80	PA-31T Cheyenne II	8020077	Giraldo Eliecer Giraldo,	(N2519V)
□ HK-2595W	81	King Air C90	LJ-950	Almacen Mink Ltda.	HK-2595G
□ HK-2596	81	King Air C90	LJ-957	AEROTACA SA-Aerotransportes Casanare/Aerotaxi Casanare	HK-2596G
□ HK-2599P	80	Gulfstream 840	11642	Diego Paul Martinez Munoz,	HK-2599W
□ HK-2601	81	Gulfstream 840	11651	Lopez Montoya Cia S C S.	N840JP
□ HK-2631G	81	PA-31T Cheyenne II	8120032	Ministerio de Obras Publicas y Transporte, Bogota.	HK-2631X
□ HK-2642P	81	PA-31T Cheyenne II	8120037	Hector Gomez Alvarez,	
□ HK-2700	81	King Air 200C	BL-15	ALCOM-Aerolineas Comerciales del Meta Ltda.	HK-2700X
□ HK-2749P	81	Cheyenne II-XL	8166019	Adriana Gomez Larroche,	HK-2749X
□ HK-2772P	81	PA-42 Cheyenne III	8001059	Fernando Rafael Llanos Avila,	HK-2772X
□ HK-2873W	62	King Air C90	LJ-994	Inversiones Burgos Ciderautos S C A/Invers. M y Ciderautos,	HK-2873
□ HK-2888P	82	King Air F90	LA-181	Guillermo Ernesto Rengifo Arce,	HK-2888
□ HK-2909P	82	Gulfstream 1000	96045	Jorge Arturo Buelvas Marchena,	HK-2909
□ HK-2926P		Cheyenne T-1040	8275015	Felipe Sandoval Rincon,	HK-2926
□ HK-2951	82	Gulfstream 1000	96049	Aerotayrona Ltda.	HK-2951P
□ HK-2963	81	Cheyenne II-XL	8166053	Supertiendas y Droguerias Olimpicas SA/Char Hermanos Ltda.	N174CC
□ HK-3000G	81	PA-42 Cheyenne III	8001062	Unidad Administrativa Especial/Aeronautica Civil de Colombia	HK-3000X
□ HK-3009	83	Conquest II	441-0273	Helicopteros El Condor-Helicondor Ltda/Helicafe Ltda.	HK-3009P
□ HK-3025	81	PA-31T Cheyenne II	8120062	Aerovuelos y Servicios Ltda/Charter Ltda.	
□ HK-3036P	82	Conquest 1	425-0118	Pedro Antonio Pineda Duque,	N6881Q
□ HK-3043P	81	PA-31T Cheyenne II	8120051	Juan Climaco Prieto Cabrera,	(N2525Y)
□ HK-3060	82	Gulfstream 1000	96039	ATTA-Asociacion Tolimense de Transporte Aereo Ltda.	HK-3060W
□ HK-3074P	81	Cheyenne II-XL	8166035	Giraldo Arteaga Salvador Dario,	HK-3074
□ HK-3118W	83	King Air F90	LA-198	Leasing Boyaca SA/Mayaguez SA.	HK-3118X
□ HK-3147X	73	Rockwell 690A	11122		N9229Y
□ HK-3192W	85	Gulfstream 1000	96077	Aero Charter Ltda.	HK-3192X
□ HK-3193	85	Gulfstream 1000	96086	AEROES-Aerolineas Especiales/Taxi Aereo Aereos Ltda.	HK-3193X
□ HK-3214	81	King Air B200	BB-854	Servialas de Colombia Ltda	HK-3214X
□ HK-3218	83	Gulfstream 1000	96061	AVIEL-Aviones Ejecutivos Ltda. Bogota.	N184BB
□ HK-3236P	71	681B Turbo Commander	6046	Luis Miguel Astaiza Arias,	HK-3236X
□ HK-3239W	85	Gulfstream 1000	96010	Aero Antigua Ltda.	HK-3239X
□ HK-3253P	85	Gulfstream 1000	96074	Mario Montoya Gomez,	HK-3253X
□ HK-3275	85	Gulfstream 1000	96076	Hector Fabio Gonzalez Martinez,	HK-3275X
□ HK-3277P	80	King Air C90	LJ-892	Lazaro Estrada Ospina,	HK-3277W
□ HK-3279	84	Gulfstream 1000	96072	Aerocarbones del Cesar Ltda/Aerocarbon,	HK-3279W
□ HK-3283W	85	Gulfstream 1000	96099	Compania Exploradora del Pacifico Ltda.	HK-3293
□ HK-3284W	85	Gulfstream 1000	96080	Carlos Fernando Quiroz y Cia en S C.	HK-3284X
□ HK-3290P	81	Gulfstream 840	11653	Irene Leonor Clamote Lagnoo,	HK-3290

Regn *Yr Type* *c/n* *Owner/Operator* *Prev Regn*

196

Regn	Yr	Type	c/n	Owner/Operator	Prev Regn
☐ HK-3291P	85	Gulfstream 1000	96088	Diamar Ltda.	HK-3291W
☐ HK-3314W	75	Rockwell 690A	11255	SIRVE-Sociedad Importadora de Repuestos y Vehiculos Ltda.	HK-3314
☐ HK-3324P	85	Gulfstream 1000	96100	Adriana Lopez Cardona,	N111VP
☐ HK-3331	79	PA-31T Cheyenne II	7920088	Aeroexpreso Interamericano Ltda. Bogota.	N64CA
☐ HK-3337	85	Conquest II	441-0354	Minas de los Naranjos Ltda. Bogota. (status ?).	HK-3337W
☐ HK-3354W	79	Rockwell 690B	11518	Hilacol SA.	N57198
☐ HK-3366X	82	Gulfstream 1000	96033	LAPAZ Colombia, Paz de Ariporo.	G-IOOO
☐ HK-3367	82	Gulfstream 1000	96018	SERPA-Servicios Aereos del Pacifico,	YV-416CP
☐ HK-3379W	79	Rockwell 690B	11525	Velez y Valencia Compania S en C.	HK-3379
☐ HK-3381G	87	PA-42 Cheyenne IIIA	5501039	IDEMA-Instituto de Mercadeo Agropecuaria, Bogota.	HK-3381X
☐ HK-3385P	82	Gulfstream 840	11728	Hector Eduardo Tautna Cinturia,	HK-3385X
☐ HK-3389P	82	Gulfstream 1000	96037	Oscar Alberto Morales Salazar,	HK-3389
☐ HK-3390	82	Gulfstream 1000	96047	Air Caribe Ltda. Cartagena.	HK-3390X
☐ HK-3391W	82	Gulfstream 1000	96059	Aerominas de Colombia Ltda.	HK-3391
☐ HK-3397W	88	PA-42 Cheyenne 400LS	5527037	AGRONASUR-Agropecurias y Ganaderias del Sur Ltda.	N9561N
☐ HK-3401	78	Conquest II	441-0059	VIANA-Vias Aereas Nacionales Ltda. Arauca.	N441MS
☐ HK-3406	81	Gulfstream 980	95062	JET-Jet Express Ltda/Cia Sinuana de Transp. Aer Costa Ltda.	N92MT
☐ HK-3407	81	Gulfstream 980	95075	LIRA-Lineas Aereas de Uraba Ltda.	HK-3407X
☐ HK-3408	81	Gulfstream 980	95050	Servicios Aereos de la Capital Ltda. Bogota.	HK-3408X
☐ HK-3413	84	PA-42 Cheyenne 400LS	5527006	Aerotaxi de Valledupar Ltda.	HK-3413X
☐ HK-3418	84	Conquest II	441-0333	SADA-Sociedad Aeronautica de Armenia Ltda.	HK-3418P
☐ HK-3423	84	PA-42 Cheyenne 400LS	5527008	Taxi Aereo del Uraba Ltda.	N42MD
☐ HK-3424	80	Gulfstream 840	11611	APEL-Aerolineas Petroleras del Llano Express SA.	N35DR
☐ HK-3429W	79	King Air C90	LJ-831	NADAL Ltda/Asesores de Seguros y Servicios Generales,	HK-3429X
☐ HK-3432	85	King Air B200	BB-1227	SAVA-Servicios Aereos del Valle/Aviaco Ltda. Bogota.	N7237A
☐ HK-3433X	88	King Air 300	FA-159	Caja de Credito Agraria Industrial y Minero, Bogota.	N3083B
☐ HK-3439	82	Gulfstream 1000	96021	Aeroejecutivos Colombia SA. Bogota.	N132PR
☐ HK-3444	81	Gulfstream 980	95057	Servicios Aereos de la Capital Ltda. Bogota.	HK-3444X
☐ HK-3447P	82	Gulfstream 840	11722	Alvaro Jose Salgado Calle,	HK-3447
☐ HK-3450	81	Gulfstream 980	95083	RAA-Rutas Aereas Araucanas Ltda. Arauca.	HK-3450W
☐ HK-3451		PA-42 Cheyenne 400LS	5527033	SATURNO-Servicio Aereo Turistico del Nororiente Ltda.	N827PC
☐ HK-3455	81	Gulfstream 980	95065	RAA-Rutas Aereas Aruacanas Ltda. Arauca.	HK-3455X
☐ HK-3456W	83	Conquest II	441-0271	Aero Antigua Ltda.	HK-3456
☐ HK-3459P		PA-42 Cheyenne 400LS	5527038	Jesus Alberto Lugo Villafane,	HK-3459X
☐ HK-3461	81	Gulfstream 980	95043	ALICOI -Aerolineas Intercolombianos Ltda. Bogota.	HK-3461X
☐ HK-3463	85	King Air 300	FA-39	SAS-Servicio Aereo de Santander Ltda. Bogota.	G-SRES
☐ HK-3465P	73	Rockwell 690	11059	Sergio Jose Cardona Jimenez,	HK-3465X
☐ HK-3466	74	Rockwell 690A	11165	Helitaxi Ltda. Bogota.	N690LP
☐ HK-3470P	82	King Air B200	BB-960	Jose Alirio Valero Castellanos,	HK-3470
☐ HK-3473W	82	Gulfstream 900	15007	Anaconda Ltda.	HK-3473
☐ HK-3474	80	Gulfstream 980	95016	RAA-Rutas Aereas Aruacanas Ltda. Arauca.	N7050S
☐ HK-3481	81	Gulfstream 980	95079	Alfuturo Aereo Ltda/Aerocali Luego Alfuturo Air Ltda.	N9831S
☐ HK-3484	80	Gulfstream 980	95022	LAR-Lineas Aereas de Rionegro Ltda.	N20HG
☐ HK-3487X		Conquest II	...		
☐ HK-3495	85	King Air 300	FA-41	Lineas Aereas Paz de Ariporo Ltda.	N307ZC
☐ HK-3497	84	Conquest II	441-0335	LAR-Lineas Aereas de Rionegro Ltda.	HK-3497X
☐ HK-3504	82	King Air B200	BB-1063	Central Charter de Colombia SA/Aeroejecutivos Ltda. Bogota.	N256L
☐ HK-3505	84	King Air F90-1	LA-221	Aeropei Ltda.	HK-3505W
☐ HK-3507	82	King Air B200	BB-974	Lineas Aereas de Los Libertadores/Lineas Aereas Petroleras,	N1861B
☐ HK-3512	82	Conquest II	441-0261	AEROLLANO Ltda/Aerovias del Llano, Villavicencio (Meta).	N261DW
☐ HK-3514	74	Rockwell 690A	11221	Julio Cesar Gerena Diaz,	N132RD
☐ HK-3518	84	Conquest 1	425-0207	COARCO-Compania Aerea del Oriente Colombiano Ltda.	N1224B
☐ HK-3529	85	Conquest II	441-0348	VIANA-Vias Aereas Nacionales Ltda. Arauca.	HK-3529X
☐ HK-3532	85	Conquest II	441-0356	Aeroexpresso de la Sabana Ltda.	N441CS
☐ HK-3534	86	King Air 300	FA-96	SAETA-Servicios Aereos del Metay Territorios Nacionales Ltda	N2586E
☐ HK-3540E	83	Conquest II	441-0287	ARO-Aerovias Regionales del Oriente,	HK-3540
☐ HK-3547	85	King Air 300	FA-86	AEROATLANTICO-Aerovias del Atlantico Ltda.	N955AA
☐ HK-3549	84	Conquest II	441-0341	Taxi Aereo de Yopal Ltda.	OB-1337
☐ HK-3550	83	Conquest II	441-0320	SARPA-Servicios Aereos Panamericanos SA. Puerto Asis.	N189WS
☐ HK-3554	82	King Air B200	BB-1064	TAG-Trans Aereo de Girardot/Taxi Aereo Girardot Ltda.	N2157L
☐ HK-3556P	85	King Air 300	FA-47	Martha Lucia Obando Pena,	N961AA
☐ HK-3561	77	Rockwell 690B	11365	Helitaxi Ltda. Bogota.	LV-LZS
☐ HK-3562	82	Conquest II	441-0232	CALAMAR Ltda. Calamar (Vaupes).	N443WS
☐ HK-3568	83	Conquest II	441-0309	LARCO-Lineas Aereas de Corozal Ltda.	N441MT
☐ HK-3573	82	Conquest II	441-0251	SARPA-Servicios Aereos Panamericanos Ltda. Puerto Asis.	N711GF
☐ HK-3574	86	King Air 300	FA-108	Alfuturo Aereo Ltda/Aerocali Luego Alfuturo Air Ltda.	N654S ·
☐ HK-3580W	64	Gulfstream 1	145	Compania Rural de Occidente Ltda.	HK-3329X
☐ HK-3596	82	Conquest II	441-0274	Aerotaxi de Valledupar Ltda. Valledupar.	HK-3596X
☐ HK-3597	73	Rockwell 690A	11110	Aerotaxi de Valledupar Ltda. Valledupar.	HK-3597X
☐ HK-3603X	81	Cheyenne II-XL	8166034	HELIVALLE Ltda/Helicopteros del Valle Ltda. Palmira.	N730PC
☐ HK-3607X	79	Conquest II	...	Cie Interandina de Aviacion/Interandes SA.	
☐ HK-3611	84	Conquest II	441-0324	Taxi Aereo de Caldas Ltda.	N1209P
☐ HK-3613	81	Conquest II	441-0206	Aero Antigua Ltda.	N7049Y
☐ HK-3614	80	Conquest II	441-0178	Lineas Aereas del Sotavento/Rutas Aereas de la Guajira Ltda.	N446WS
☐ HK-3615	82	Conquest II	441-0239	Cx HK- 7/95 to ?	N45TF
☐ HK-3618W	93	PA-42 Cheyenne III	8001037	Cementos del Caribe,	N804CA
Regn	Yr	Type	c/n	Owner/Operator	Prev Regn

Regn	Yr	Type	c/n	Owner/Operator	Prev Regn
☐ HK-3620X	80	Conquest II	441-0152	Aerotaxi de Valledupar Ltda.	N2628X
☐ HK-3623X	69	Conquest II	...		
☐ HK-3632X		Gulfstream 840	...		
☐ HK-3648	86	King Air 300	FA-100	Aerolimosina Ltda. (status ?).	N2614C
☐ HK-3654	86	King Air 300LW	FA-101	Aerotaxi del Quindio Ltda.	G-BSTF
☐ HK-3656	76	Rockwell 690A	11314	SERCU-Servicios Aereos de Cucuta Ltda.	N14AD
☐ HK-3659	85	King Air 300	FA-60	APEL Express SA/Aerolineas Petroleras del Llano,	N285KA
☐ HK-3670	87	King Air 300	FA-123	Taxi Aereo de Ibaque Ltda.	N1844S
☐ HK-3680	80	Gulfstream 840	11620	Air Tropical Ltda. Bahia Solano.	G-BXYZ
☐ HK-3681	61	Gulfstream 1	78		N33CP
☐ HK-3693X	82	PA-42 Cheyenne III	8001075	HELIVALLE/Helicopteros del Valle Ltda.	N4998M
☐ HK-3699	77	King Air 200	BB-226	SERPA-Servicios Aereos del Pacifico Ltda.	YV-2359P
☐ HK-3700	82	Gulfstream 840	11721	APEL Express SA/Aerolineas Aereos del Pacifico,	XA-...
☐ HK-3703X	90	King Air/Catpass 250	BB-1360	BP Exploration Co Colombia Ltda. Bogota.	N35AR
☐ HK-3704X	90	King Air/Catpass 250	BB-1392	BP Exploration Co Colombia Ltda. Bogota.	N8026J
☐ HK-3705	75	King Air 200	BB-63	Aerotaxi del Quindio Ltda.	HK-3705X
☐ HK-3796	77	King Air/Catpass 250	BB-264	APSA-Aeroexpresso Bogota SA. Bogota.	C-FCGV
☐ HK-3819	81	Gulfstream 980	95059	SAHO-Servicios Aereos Horizonte Ltda.	N9811S
☐ HK-3822	77	King Air 200	BB-248	AERUBA-Aerotaxi de Uraba Ltda.	N123PM
☐ HK-3828	84	King Air 300	FA-10	Aerotaxi del Quindio Ltda.	N13PD
☐ HK-3854	76	King Air 200	BB-135	SERCU-Servicios Aereos de Cucuta Ltda.	N535E
☐ HK-3860	88	King Air 300	FA-169	SADA-Sociedad Aeronautica de Armenia Ltda.	XA-PUD
☐ HK-3894	82	King Air B200	BB-1049	SEARCA-Servicio Aereo de Capurgana Ltda.	F-GMLP
☐ HK-3907	73	King Air E90	LW-39	AERVAL-Aereos y Valores Ltda. Bogota.	OB-1466
☐ HK-3912	81	Gulfstream 840	11668	Taxi Aereo del Uraba Ltda.	YV-2413P
☐ HK-3918	88	King Air 300	FA-155	TASA-Transportes Aereos Sandoval Alvarez Ltda.	N3109N
☐ HK-3922P	78	King Air 200	BB-352	Oscar Alberto Monsalve Arbelaez,	YV-...
☐ HK-3923X	79	King Air 200	BB-450	JET-Jet EXpress/Cia Sinuana de Transportes Aer Costa Ltda.	C-GJCM
☐ HK-3935W	81	King Air F90	LA-151	Air Charter Ltda.	YV-445CP
☐ HK-3936	75	King Air 200	BB-75	TAXOR-Taxi Aereo de la Orinoquia Ltda.	N70LA
☐ HK-3954	89	King Air 300LW	FA-200	Transandina de Aviacion Ltda.	D-ICBW
☐ HK-3961X	84	Gulfstream 1000	96069	Danilo Botero Bustos, Bogota.	N6151T
☐ HK-3965X		King Air E90	...		
☐ HK-3975X	68	680W Turbo Commander	1820-34		N3SK
☐ HK-3988X	90	King Air 350	FL-18	Aeromor Ltd. Bogota.	N80CK
☐ HK-3990X	90	King Air 1300	BB-1376	APSA-Aeroexpresso Bogota SA. Bogota.	N914YW
☐ HK-3995X	77	King Air 200	BB-196		LN-FKF
☐ HK-4063W	67	Gulfstream 1000	...		
☐ HK-4095X		King Air 200	...		
Military					
☐ ARC-101		King Air 200	...		
☐ ARC-301		PA-31T Cheyenne	...		
☐ ARC-601		Rockwell 690	...		
☐ EJC-103		Gulfstream 1000	...	Colombian Army, Bogota.	
☐ EJC-108		King Air 200	...	Colombian Army, Bogota.	
☐ EJC-114	85	Gulfstream 1000	96083	Colombian Army, Bogota.	HK-3376
☐ EJC-115	81	Gulfstream 980	95066	Colombian Army, Bogota.	HK-2682P
☐ EJC-117		King Air 200	...	Colombian Army, Bogota.	
☐ FAC-542	66	680V Turbo Commander	1563-19	Fuerza Aerea Colombiana, Bogota-El Dorado.	HK-2539G
☐ FAC-5198	82	Gulfstream 1000	96030	Fuerza Aerea Colombiana, Bogota-El Dorado.	HK-...
☐ FAC-5553	81	Gulfstream 980	95055	Fuerza Aerea Colombiana, Bogota-El Dorado.	FAC-553
☐ FAC-5570	78	King Air C90	LJ-752	Fuerza Aerea Colombiana, Bogota-El Dorado.	FAC-570
☐ FAC-5739	75	PA-31T Cheyenne II	7520014	Fuerza Aerea Colombiana, Bogota-El Dorado.	N8TK
☐ FAC-5750		King Air 200	...	Fuerza Aerea Colombiana, Bogota-El Dorado.	
☐ PNC-203		King Air 200	...	Policia Nacional de Colombia, Bogota.	
☐ PNC-204	77	Conquest II	441-0031	Policia Nacional de Colombia, Bogota.	N918FE
☐ PNC-205		Gulfstream 1000	...	Policia Nacional de Colombia, Bogota.	
☐ PNC-208		King Air 300	...	Policia Nacional de Colombia, Bogota.	
☐ PNC-209		King Air 200	...	Policia Nacional de Colombia, Bogota.	
☐ PNC-210	77	King Air 200	...	Policia Nacional de Colombia, Bogota.	
☐ PNC-221	81	King Air 200	BB-833	Policia Nacional de Colombia, Bogota.	XA-RZH

HL = South Korea (Republic of Korea)

Civil

Regn	Yr	Type	c/n	Owner/Operator	Prev Regn
☐ HL5261	78	Rockwell 690B	11437	(status ?).	N9171S

HP = Panama (Republic of Panama)

Civil

Regn	Yr	Type	c/n	Owner/Operator	Prev Regn
☐ HP-...	84	Conquest II	441-0328	City Investments Ltd. Panama City.	N12093
☐ HP-...	79	Conquest II	441-0110	Cx 4/88 to HP-... ?	N84CH
☐ HP-...	82	Conquest II	441-0277	Revise SA.	N441HT
☐ HP-...	82	Gulfstream 1000	96003		N17CG
☐ HP-...	85	Gulfstream 1000	96081		N600BM
☐ HP-...	85	Gulfstream 1000	96082		N89GA
☐ HP-...	82	Gulfstream 900	15004		N5838N

Regn	*Yr*	*Type*	*c/n*	*Owner/Operator*	*Prev Regn*

Regn	Yr	Type	c/n	Owner/Operator	Prev Regn
□ HP-...	82	Gulfstream 900	15003		N900RH
□ HP-...	83	Gulfstream 900	15032		N112GA
□ HP-...	88	King Air B200T	BT-33	Chiriqui Land Co. Panama City. (was BB-1301).	N600RD
□ HP-...	77	Rockwell 690B	11368	Aviacion Panamena SA. Panama City.	N77PE
□ HP-...	79	Rockwell 690B	11565	TRANSPASA, Panama-Paitilla.	N81HK
□ HP-809	80	PA-31T Cheyenne II	8020015	Colony Trading SA.	N2353W
□ HP-960P	80	King Air 200	BB-617	Stowage SA.	HP-960
□ HP-976P	83	King Air C90-1	LJ-1051	Rales Internacional SA.	HP-976P
□ HP-1066	84	Cheyenne II-XL	1166001	Petroterminal de Panama SA. Panama City. (was 8166072).	HP-1066P
□ HP-1069		SA-226T Merlin 3A	T-288	Inversiones Humbolt SA. Panama City.	HP-1069P
□ HP-1118	80	King Air C90	LJ-876	Key West,	HP-1118P
□ HP-1136AP	68	SA-26AT Merlin 2B	T26-113	Raul Arias/Apair SA. Paitilla.	N878SC
□ HP-1149P	82	Gulfstream 1000	96031	World Interstate Corp.	C-FDGD
□ HP-1152	84	King Air C90A	LJ-1073	Kamura Holding SA. Panama City.	HP-1152P
□ HP-1180	83	King Air B200	BB-1131	Chiriqui Land Co.	N6642Z
□ HP-1182	90	King Air 300	FA-209	Motta Internacional SA.	N148M
□ HP-1189	84	Conquest II	441-0332	Marucci Holdings SA.	N888FL
□ HP-1203	91	King Air C90A	LJ-1272	Mays Zona Libre SA.	N8045T
□ HP-1211	91	King Air 350	FL-39	King Enterprises Holdings SA. Panama City.	N8040A
□ HP-1213	83	King Air B200	BB-1145	Aeronaves del Istmo SA.	(N71GA)
□ HP-1246	77	King Air E90	LW-218	Aeronaves del Atlantico SA.	N18243
□ HP-1264	91	King Air C90A	LJ-1285	Aldrin SA. Panama City.	N191SP
□ HP-1266	82	King Air F90	LA-160	Cockpit SA. Panama City.	N335R
□ HP-1298	86	King Air 300	FA-103	Aeronaves del Atlantico SA. El Dorado.	N827DL

HR = Honduras (Republic of Honduras)

Civil

□ HR-...	79	King Air 200	BB-481	Comercial y Invesriones Galatia SA.	N144K
□ HR-...	79	King Air E90	LW-308		N62CS
□ HR-...	67	Mitsubishi MU-2B	021		N22JZ
□ HR-AAJ	74	Rockwell 690A	11233	SETCO-Servicios Ejecutivos Turisticos Commander, Tegucigalpa	N9192N
□ HR-ANL		King Air 200	...		
□ HR-IAH	66	King Air A90	LJ-122	Islena Airlines, La Ceiba.	N860K
□ HR-JFA	81	PA-42 Cheyenne III	8001056		

Military

□ FAH-006	83	Gulfstream 1000	96060	Fuerza Aerea Honduras, Tegucigalpa.	HK-3194X

HS = Thailand (Kingdom of Thailand)

Civil

□ HS-...	97	King Air 350	FL-162		N1112Z
□ HS-DCB	76	King Air 200	BB-132	Department of Aviation, Bangkok.	HS-FFI
□ HS-DCF	88	King Air B200	BB-1315	Department of Aviation, Bangkok.	HS-AFI
□ HS-ITD	96	King Air 350	FL-151	Italian/Thai Development Co. Bangkok.	N10817
□ HS-PBA	91	TBM 700	4	P B Air Ltd. Bangkok.	F-OHBA
□ HS-PON	83	King Air B200	BB-1165		1165
□ HS-SLA	91	King Air 350	FL-53	Siam Land Flying Co. Bangkok.	HS-TFI
□ HS-SLB	90	King Air C90A	LJ-1243	Siam Land Flying Co. Bangkok.	HS-TFH
□ HS-TFG	78	Rockwell 690B	11482	Thai Flying Service Co. Bangkok.	N745T

Military

□ 00923	73	King Air E90	LW-26	Royal Thai Army, Don Muang.	01769
□ 0169		Beech 1900C-1	UC-169	Royal Thai Army, Don Muang.	N.....
□ 0170		Beech 1900C-1	UC-170	Royal Thai Army, Don Muang.	N.....
□ 0342	78	King Air 200	BB-342	Royal Thai Army, Don Muang.	794
□ 1112	95	Dornier 228-212	8226	Royal Thai Navy,	D-CBDF
□ 1113	95	Dornier 228-212	8227	Royal Thai Navy,	D-CCCP
□ 1114	96	Dornier 228-212	8228	Royal Thai Navy,	D-C...
□ 2011	96	King Air 350	FL-146	KASET-Thai Ministry of Agriculture, Bangkok.	N3268Z
□ 2012	96	King Air 350	FL-147	KASET-Thai Ministry of Agriculture, Bangkok.	N3269W
□ 41060	95	Jetstream 41	41060	Royal Thai Army, Don Muang.	G-BWGW
□ 41094	96	Jetstream 41	41094	Royal Thai Army, Don Muang.	G-BWTZ
□ 60501		SA-226AT Merlin 4A	AT-071	Royal Thai Air Force, Bangkok.	60301
□ 60502		SA-226AT Merlin 4A	AT-072	Royal Thai Air Force, Bangkok.	60302
□ 60503		SA-226AT Merlin 4A	AT-073	Royal Thai Air Force, Bangkok.	60303
□ 81491	76	Rockwell 690A	11340	Royal Thai Survey Department, Bangkok.	11340
□ 93303	92	King Air B200	BB-1436	Royal Thai Survey Dept. Bangkok.	N1564M
□ 93304	92	King Air B200	BB-1441	Royal Thai Survey Dept. Bangkok.	56385
□ 93305	92	King Air B200	BB-1443	Royal Thai Survey Dept. Bangkok.	56379

HZ = Saudi Arabia (Kingdom of Saudi Arabia)

Civil

□ HZ-AMA	80	MU-2 Marquise	793SA	Ashmawi Aviation, Malaga, Spain.	F-GDPH
□ HZ-PC2	87	Beech 1900C-1	UC-4	Directorate of Civil Aviation,	N3078C
□ HZ-SN6		SA-226AT Merlin 4	AT-007	DHL/SNAS WorldWide Express, Bahrain.	N2610
□ HZ-SN8	82	SA-227AT Merlin 4C	AT-434	DHL/SNAS WorldWide Express, Bahrain.	N3110F
□ HZ-SS1	72	Rockwell 690	11006	Saudi Arabian Photo Co. Riyadh.	N690SP

Regn	Yr	Type	c/n	Owner/Operator	Prev Regn

I = Italy (Italian Republic)

Civil

Regn	Yr	Type	c/n	Owner/Operator	Prev Regn
☐ I-....	78	PA-31T Cheyenne II	7820059	(status ?).	N11BC
☐ I-ACTC	92	P-180 Avanti	1014	Soc. Rinaldo Piaggio SpA. Genoa. (CoA expiry 11 Feb 93).	
☐ I-ALPN	93	P-180 Avanti	1015	Soc. ALPI Eagles SpA. Thiene (Vicenza).	
☐ I-AZME	86	King Air 300	FA-94	Delta Aerotaxi/Soc. Menarini Ind Farmaceutiche SRL. Firenze.	N25219
☐ I-BOMY	79	King Air C90	LJ-828	Albatros SRL. Baggiovara.	F-GFME
☐ I-CGAT	75	PA-31T Cheyenne II	7520033	Compagnia Generale Aeronautica SAS, Genoa.	N54964
☐ I-CGTT	78	PA-31T Cheyenne II	7820045	Soc. New Bristol SRL. Milan.	N82288
☐ I-FSAD	82	SA-227AT Merlin 4C	AT-440B	International Flying Services SRL. Bergamo.	N36JP
☐ I-HYDR	75	PA-31T Cheyenne II	7520034	Compagnia Generale Aeronautica, Genoa.	N54966
☐ I-LIAT	80	PA-31T Cheyenne 1	8004052	Soc. R F Celada SpA. Milan.	N2316X
☐ I-LIPO	74	King Air C90	LJ-616	Soc. Coronet SpA. Milan.	SE-GXD
☐ I-MKKK	68	Gulfstream 1	194	Sunline SRL. Cuneo-Levaldigi.	N81T
☐ I-NARB	74	SA-226AT Merlin 4	AT-020	BESIT SNC. Servizi Aerei, Olbia. 'Cises'	N747BD
☐ I-NARC	75	SA-226AT Merlin 4A	AT-035	BESIT SNC. Servizi Aerei, Olbia.	N90090
☐ I-NARW	77	SA-226AT Merlin 4A	AT-058	BESIT SNC. Servizi Aerei, Olbia.	D-ICFB
☐ I-PALS	75	PA-31T Cheyenne II	7620005	Soc. PAL SRL. Pollone (BI).	N54979
☐ I-PIAH	81	King Air 200	BB-777	Soc. Air Vallee SpA. Aosta.	
☐ I-PJAP	91	P-180 Avanti	1013	Soc. Rinaldo Piaggio SpA. Genoa.	N180AZ
☐ I-PJAR	87	P-180 Avanti	1002	Soc. Rinaldo Piaggio SpA. Genoa. (Ff 14 May 87).	
☐ I-PJAV	87	P-180 Avanti	1001	Soc. Rinaldo Piaggio SpA. Genoa. (Ff 23 Sep 86).	
☐ I-POMO	79	PA-31T Cheyenne 1	7904030	Soc. Adria Fly Shipping SAS. Udine.	N601PT
☐ I-RWWW	77	King Air E90	LW-220	Soc. Desio & Brianza Leasing SpA. Milan.	N17619
☐ I-SASA	80	PA-31T Cheyenne 1	8004024	Soc. Alibrixia Nord SRL. Brescia.	N2321V
☐ I-SERV	79	PA-31T Cheyenne 1	7904022	Soc. Aero Servizi Bresciana SRL. Montechiari-Brescia.	(D-IBWB)
☐ I-SWAA	76	SA-226T Merlin 3A	T-274	SERIB Wings, Turin.	N600TA
☐ I-TASA	76	Rockwell 690A	11273	Euraviation SRL. Milan.	(N892WA)
☐ I-TASC	3	Gulfstream 1	173	Sunline SRL. Cuneo-Levaldigi. (stored GVA since 4/94).	N49CB
☐ I-TASE	75	Rockwell 690A	11260	International Flying Services SRL. Bergamo.	N324BT
☐ I-TREP	87	PA-42 Cheyenne IIIA	5501045	Soc. Alitalia Flying School, Rome.	
☐ I-TREQ	87	PA-42 Cheyenne IIIA	5501046	Soc. Alitalia Flying School, Rome.	
☐ I-TRER	87	PA-42 Cheyenne IIIA	5501047	Soc. Alitalia Flying School, Rome.	

Military

Regn	Yr	Type	c/n	Owner/Operator	Prev Regn
☐ MM.....	92	P-180 Avanti	1027	ALE-Italian Army,	F-GMRM
☐ MM.....		P-180 Avanti	1031	ALE-Italian Army,	
☐ MM62159	94	P-180 Avanti	1024	AMI,	
☐ MM62160	94	P-180 Avanti	1025	AMI,	
☐ MM62161	94	P-180 Avanti	1026	ALE-Italian Army,	(D-IMED)
☐ MM62162	94	P-180 Avanti	1028	AMI,	
☐ MM62163	94	P-180 Avanti	1029	AMI,	
☐ MM62164	94	P-180 Avanti	1030	AMI,	
☐ MM62165	96	ATR 42-400MP	500	GF-13. AMI/Guardia di Finanza,	F-WWEW
☐ MM62166	96	ATR 42-400MP	502	GF-13. AMI/Guardia di Finanza,	F-WWEM

JA = Japan

Civil

Regn	Yr	Type	c/n	Owner/Operator	Prev Regn
☐ JA8598	93	King Air 350	FL-100	Noevir International Corp. Konan.	N8199W
☐ JA8599	95	Pilatus PC-XII	128	Auto Panther, Kagoshima.	HB-FQT
☐ JA8600	81	Gulfstream 980	95070	The Pilot & Co Ltd. Tokyo.	N35SA
☐ JA8604	81	Gulfstream 980	95044	Asia Air Survey/Mitsui Leasing Jigyo, Chofu.	N65664
☐ JA8614	94	King Air B200	BB-1491	C Itoh Aviation CO. Nagoya.	N15098
☐ JA8705	92	King Air B200	BB-1431	C Itoh Aviation Co. Nagoya.	N82696
☐ JA8724	90	PA-42 Cheyenne IIIA	5501055	Zen Nikku Shoji, Kumamoto.	N9198F
☐ JA8810	79	King Air 200T	BT-5	Japanese Maritime Safety Agency. (was BB-469).	N2071C
☐ JA8811	79	King Air 200T	BT-6	Japanese Maritime Safety Agency. (was BB-489).	N2071D
☐ JA8812	79	King Air 200T	BT-7	Japanese Maritime Safety Agency. (was BB-510).	N2071X
☐ JA8813	79	King Air 200T	BT-8	Japanese Maritime Safety Agency. (was BB-530).	N2071Y
☐ JA8814	79	King Air 200T	BT-9	Japanese Maritime Safety Agency. (was BB-551).	N2071Z
☐ JA8815	79	King Air 200T	BT-11	Japanese Maritime Safety Agency. (was BB-573).	N60576
☐ JA8816	80	King Air 200T	BT-12	Japanese Maritime Safety Agency. (was BB-591).	N60581
☐ JA8817	80	King Air 200T	BT-13	Japanese Maritime Safety Agency. (was BB-609).	N60587
☐ JA8818	80	King Air 200T	BT-14	Japanese Maritime Safety Agency. (was BB-627).	(9Q-CTZ)
☐ JA8819	80	King Air 200T	BT-15	Japanese Maritime Safety Agency. (was BB-647).	N6059D
☐ JA8820	80	King Air 200T	BT-16	Japanese Maritime Safety Agency. (was BB-665).	N60603
☐ JA8824	81	King Air 200T	BT-17	Japanese Maritime Safety Agency. (was BB-798).	N3718Q
☐ JA8826	81	Gulfstream 980	95078	Chunichi Press, Nagoya.	N9830S
☐ JA8828	74	SA-226AT Merlin 4	AT-016	Showa Aviation Co. Osaka-Yao.	N76MX
☐ JA8829	82	King Air 200T	BT-22	Japanese Maritime Safety Agency. (was BB-991).	N1841K
☐ JA8833	83	King Air B200T	BT-28	Japanese Maritime Safety Agency. (was BB-1117).	N1846M
☐ JA8838	85	King Air C90A	LJ-1097	Kawasaki Jitsugyo, Gifu.	N69261
☐ JA8840	86	King Air C90A	LJ-1139	Flying Training, Sendai.	N2785A
☐ JA8841	86	King Air C90A	LJ-1140	Flying Training, Sendai.	N30573
☐ JA8844	86	King Air C90A	LJ-1141	Flying Training, Sendai.	N2736D

Regn	Yr Type	c/n	Owner/Operator	Prev Regn

Regn	Yr	Type	c/n	Owner/Operator	Prev Regn
☐ JA8845	87	King Air C90A	LJ-1142	Flying Training, Sendai.	N2860A
☐ JA8846	87	King Air C90A	LJ-1143	Flying Training, Sendai.	N7239S
☐ JA8847	87	King Air C90A	LJ-1144	Flying Training, Sendai.	N7239U
☐ JA8848	87	King Air C90A	LJ-1145	Flying Training, Sendai.	N7239Y
☐ JA8849	87	King Air C90A	LJ-1148	Flying Training, Sendai.	N7240D
☐ JA8850	87	King Air C90A	LJ-1149	Flying Training, Sendai.	N7240E
☐ JA8851	87	King Air C90A	LJ-1150	Flying Training, Sendai.	N7240K
☐ JA8852	87	King Air C90A	LJ-1151	Flying Training, Sendai.	N7240L
☐ JA8853	85	PA-42 Cheyenne 400LS	5527026	Bell Hand Club, Sendai.	(N429BX)
☐ JA8854	87	King Air B200T	BT-31	Japanese Maritime Safety Agency. (was BB-1264).	N72392
☐ JA8855	85	Conquest 1	425-0235	Nozaki & Co Ltd. Tokyo.	N1262P
☐ JA8860	88	King Air B200T	BT-32	Japanese Maritime Safety Agency. (was BB-1289).	N3184A
☐ JA8867		PA-42 Cheyenne 400LS	5527035	Manichi Shimbun, Tokyo.	N9295A
☐ JA8870		PA-42 Cheyenne 400LS	5527040	Toyota Motors Inc.	N9219G
☐ JA8871	90	PA-42 Cheyenne IIIA	5501048	All Nippon Airways, Tokyo-Haneda.	N92264
☐ JA8872	90	PA-42 Cheyenne IIIA	5501049	All Nippon Airways, Tokyo-Haneda.	N92266
☐ JA8873	90	PA-42 Cheyenne IIIA	5501050	All Nippon Airways, Tokyo-Haneda.	N92275
☐ JA8874	91	PA-42 Cheyenne IIIA	5501058	All Nippon Airways, Tokyo-Haneda.	N9194X
☐ JA8879	91	King Air B200	BB-1401	C Itoh Aviation Co. Sendai.	N81525
☐ JA8880	91	King Air B200	BB-1406	Japan Air Systems,	N81536
☐ JA8881	91	King Air 300	FA-219	C Itoh Aviation Co. Sendai.	N82039
☐ JA8882	91	King Air C90A	LJ-1290	C Itoh Aviation Co. Sendai.	N81763
☐ JA8883	91	King Air C90A	LJ-1291	C Itoh Aviation Co. Sendai.	N8178W
☐ JA8884	91	King Air C90A	LJ-1292	C Itoh Aviation Co. Sendai.	N81826
☐ JA8894	92	TBM 700	38	TBM SA. Tokyo.	F-OHBG
☐ JA8951	96	SAAB 340B-SAR	385	JMSA-Japan Maritime Safety Agency,	SE-C85
☐ JA8952	97	SAAB 340B-SAR	405	JMSA-Japanese Maritime Safety Agency,	SE-C...
Military					
☐ 03-3208	70	Mitsubishi MU-2E	908	JASDF. (was s/n 186).	
☐ 03-3226	81	Mitsubishi MU-2E	926	JASDF. (was s/n 455).	
☐ 03-3227	85	Mitsubishi MU-2E	927	JASDF. (was s/n 464).	
☐ 13-3209	71	Mitsubishi MU-2E	909	JASDF. (was s/n 200).	
☐ 13-3210	71	Mitsubishi MU-2E	910	JASDF. (was s/n 201).	
☐ 13-3211	71	Mitsubishi MU-2E	911	JASDF. (was s/n 202).	
☐ 13-3212	71	Mitsubishi MU-2E	912	JASDF. (was s/n 204).	
☐ 22-001	67	Mitsubishi MU-2C	801	JGSDF. (was s/n 036). (status ?).	
☐ 22-003	72	Mitsubishi MU-2C	803	JGSDF. (was s/n 230).	
☐ 22-004	72	Mitsubishi MU-2C	804	JGSDF. (was s/n 236).	
☐ 22-005	73	Mitsubishi MU-2C	805	JGSDF. (was s/n 275).	
☐ 22-006	74	Mitsubishi MU-2C	806	JGSDF. (was s/n 317).	
☐ 22-007	76	Mitsubishi MU-2C	807	JGSDF. (was s/n 334).	
☐ 22-008	77	Mitsubishi MU-2C	808	JGSDF. (was s/n 359).	
☐ 22-009	78	Mitsubishi MU-2C	809	JGSDF. (was s/n 376).	
☐ 22-010	78	Mitsubishi MU-2C	810	JGSDF. (was s/n 394).	
☐ 22-012	80	Mitsubishi MU-2C	812	JGSDF. (was s/n 422).	
☐ 22-013	81	Mitsubishi MU-2C	813	JGSDF. (was s/n 442).	
☐ 22-014	81	Mitsubishi MU-2C	814	JGSDF. (was s/n 443).	
☐ 22-015	81	Mitsubishi MU-2C	815	JGSDF. (was s/n 444).	
☐ 22-016	83	Mitsubishi MU-2C	816	JGSDF. (was s/n 456).	
☐ 22-017	83	Mitsubishi MU-2C	817	JGSDF. (was s/n 457).	
☐ 22-018	85	Mitsubishi MU-2C	818	JGSDF. (was s/n 465).	
☐ 22-019	85	Mitsubishi MU-2C	819	JGSDF. (was s/n 463).	
☐ 22-020	85	Mitsubishi MU-2C	820	JGSDF. (was s/n 466).	
☐ 23-3213	72	Mitsubishi MU-2F	913	JASDF. (was s/n 225).	
☐ 23-3214	72	Mitsubishi MU-2E	914	JASDF. (was s/n 227).	
☐ 33-3215	72	Mitsubishi MU-2E	915	JASDF. (was s/n 234).	
☐ 33-3216	72	Mitsubishi MU-2E	916	JASDF. (was s/n 235).	
☐ 33-3217	73	Mitsubishi MU-2E	917	JASDF. (was s/n 278).	
☐ 43-3218	73	Mitsubishi MU-2E	918	JASDF. (was s/n 279).	
☐ 53-3219	74	Mitsubishi MU-2E	919	JASDF. (was s/n 318).	
☐ 53-3271	74	Mitsubishi MU-2J	951	JASDF. (was s/n 654).	
☐ 63-3220	76	Mitsubishi MU-2E	920	JASDF. (was s/n 335).	
☐ 63-3221	76	Mitsubishi MU-2E	921	JASDF. (was s/n 336).	
☐ 6801	73	King Air TC90	LJ-597	JMSDF-Japanese Maritime Self Defence Force.	N1845W
☐ 6802	73	King Air TC90	LJ-598	JMSDF-Japanese Maritime Self Defence Force.	N1846W
☐ 6803	73	King Air TC90	LJ-599	JMSDF-Japanese Maritime Self Defence Force.	N1847W
☐ 6804	75	King Air TC90	LJ-642	JMSDF-Japanese Maritime Self Defence Force.	N7312R
☐ 6805	75	King Air TC90	LJ-670	JMSDF-Japanese Maritime Self Defence Force.	N9395
☐ 6806	78	King Air TC90	LJ-778	JMSDF-Japanese Maritime Self Defence Force.	N23780
☐ 6807	79	King Air TC90	LJ-855	JMSDF-Japanese Maritime Self Defence Force.	N6062X
☐ 6808	80	King Air TC90	LJ-916	JMSDF-Japanese Maritime Self Defence Force.	N67233
☐ 6809	80	King Air TC90	LJ-917	JMSDF-Japanese Maritime Self Defence Force.	N6724D
☐ 6810	81	King Air TC90	LJ-976	JMSDF-Japanese Maritime Self Defence Force.	N3832G
☐ 6811	81	King Air TC90	LJ-980	JMSDF-Japanese Maritime Self Defence Force.	N3832K
☐ 6812	82	King Air TC90	LJ-1042	JMSDF-Japanese Maritime Self Defence Force.	N18460
☐ 6813	82	King Air TC90	LJ-1043	JMSDF-Japanese Maritime Self Defence Force.	N1846B

Regn   *Yr Type*   *c/n*   *Owner/Operator*   *Prev Regn*

Regn	Yr	Type	c/n	Owner/Operator	Prev Regn
☐ 6814	82	King Air TC90	LJ-1044	JMSDF-Japanese Maritime Self Defence Force.	N1846D
☐ 6815	82	King Air TC90	LJ-1047	JMSDF-Japanese Maritime Self Defence Force.	N1846F
☐ 6816	83	King Air TC90	LJ-1060	JMSDF-Japanese Maritime Self Defence Force.	N1875Z
☐ 6817	83	King Air TC90	LJ-1061	JMSDF-Japanese Maritime Self Defence Force.	N1876Z
☐ 6818	83	King Air TC90	LJ-1062	JMSDF-Japanese Maritime Self Defence Force.	N6886S
☐ 6819	84	King Air TC90	LJ-1083	JMSDF-Japanese Maritime Self Defence Force.	N6923Z
☐ 6820	84	King Air TC90	LJ-1084	JMSDF-Japanese Maritime Self Defence Force.	N69237
☐ 6821	85	King Air TC90	LJ-1110	JMSDF-Japanese Maritime Self Defence Force.	N7238J
☐ 6822	87	King Air TC90	LJ-1146	JMSDF-Japanese Maritime Self Defence Force.	N72400
☐ 6823	93	King Air TC90	LJ-1335	JMSDF-Japanese Maritime Self Defence Force.	N82323
☐ 6824	93	King Air TC90	LJ-1336	JMSDF-Japanese Maritime Self Defence Force.	N82326
☐ 6825	93	King Air C90B	LJ-1337	JMSDF-Japanese Maritime Self Defence Force.	N82349
☐ 6826	93	King Air C90B	LJ-1338	JMSDF-Japanese Maritime Self Defence Force.	N82366
☐ 6827	93	King Air C90B	LJ-1339	JMSDF-Japanese Maritime Self Defence Force.	N82376
☐ 73-3201	68	Mitsubishi MU-2E	901	JASDF. (was s/n 112).	
☐ 73-3202	68	Mitsubishi MU-2E	902	JASDF. (was s/n 128).	
☐ 73-3222	77	Mitsubishi MU-2E	922	JASDF. (was s/n 360).	
☐ 73-3272	78	Mitsubishi MU-2J	952	JASDF. (was s/n 715).	
☐ 83-3203	69	Mitsubishi MU-2E	903	JASDF. (was s/n 147).	
☐ 83-3204	69	Mitsubishi MU-2E	904	JASDF. (was s/n 148).	
☐ 83-3223	78	Mitsubishi MU-2E	923	JASDF. (was s/n 377).	
☐ 83-3224	78	Mitsubishi MU-2E	924	JASDF. (was s/n 378).	
☐ 83-3273	78	Mitsubishi MU-2J	953	JASDF. (was s/n 716).	
☐ 9102	82	King Air UC-90	LJ-1038	JMSDF-Japanese Maritime Self Defence Force.	N1839D
☐ 93-3206	69	Mitsubishi MU-2E	906	JASDF. (was s/n 176).	
☐ 93-3225	81	Mitsubishi MU-2E	925	JASDF. (was s/n 445).	
☐ 93-3274	78	Mitsubishi MU-2J	954	JASDF. (was s/n 717).	
☐ 9301	88	King Air LC-90	LJ-1182	JMSDF-Japanese Maritime Self Defence Force.	
☐ 9302	90	King Air LC-90	LJ-1248	JMSDF-Japanese Maritime Self Defence Force.	N56633
☐ 9303	90	King Air LC-90	LJ-1249	JMSDF-Japanese Maritime Self Defence Force.	N56638
☐ 9304	91	King Air LC-90	LJ-1281	JMSDF-Japanese Maritime Self Defence Force.	
☐ 9305	91	King Air LC-90	LJ-1282	JMSDF-Japanese Maritime Self Defence Force.	N8154G

LN = Norway *(Kingdom of Norway)*

Civil

Regn	Yr	Type	c/n	Owner/Operator	Prev Regn
☐ LN-FAH	77	Rockwell 690B	11367	Fjellanger-Wideroe A/S. Oslo.	OY-BEJ
☐ LN-FWA	81	Gulfstream 840	11681	Fjellanger-Wideroe A/S. Oslo.	SE-IUV
☐ LN-KCG	78	King Air C90	LJ-768	Master Aviation A/S. Oslo.	OY-SBU
☐ LN-MAA	79	King Air 200	BB-560	Flyvedlikehold A/S -Master Aviation A/S, Oslo.	N200NA
☐ LN-MOA	80	King Air 200	BB-582	Lufttransport A/S. Tromso..	N47PA
☐ LN-MOB	80	King Air 200	BB-584	Lufttransport A/S. Tromso.	N400WP
☐ LN-MOC	93	King Air B200	BB-1449	Lufttransport A/S. Tromso.	N200KA
☐ LN-MOD	93	King Air B200	BB-1459	Lufttransport A/S. Tromso.	N8163R
☐ LN-MOE	93	King Air B200	BB-1460	Lufttransport A/S. Tromso.	N8164G
☐ LN-MOF	93	King Air B200	BB-1461	Lufttransport A/S. Tromso.	N8261E
☐ LN-MOG	93	King Air B200	BB-1465	Lufttransport A/S. Tromso.	N8214T
☐ LN-MOH	93	King Air B200	BB-1466	Lufttransport A/S. Tromso.	N8216Z
☐ LN-MOI	93	King Air B200	BB-1470	Lufttransport A/S. Tromso.	N8225Z
☐ LN-NOA	81	King Air B200	BB-829	Airlift A/S. Forde-Bygstad.	N829AJ
☐ LN-PAG	76	King Air 200	BB-119	Nevi A/S. Oslo.	OY-AUJ
☐ LN-SFT	80	SA-226T Merlin 3B	T-342	Fjellanger-Wideroe A/S. Oslo.	N342NX
☐ LN-TWH	88	Reims/Cessna F406	F406-0032	Trans Wing AS. Gardemoen.	PH-CLE
☐ LN-VIU	77	King Air 200	BB-216	Ugland Air A/S. Kristiansand-Kjevik.	OY-AUZ
☐ LN-VIZ	83	King Air B200	BB-1136	Ugland Air A/S. Kristiansand-Kjevik.	D-IDOK

LQ/LV = Argentina *(Republic of Argentina)*

Civil

Regn	Yr	Type	c/n	Owner/Operator	Prev Regn
☐ LQ-BLU		PA-42 Cheyenne 400LS	...	Corte Suprema de Justica Nacional, Buenos Aires.	
☐ LV-APF	81	PA-42 Cheyenne III	8001026	TAC/Government Province of Catamarca, Catamarca.	LQ-APF
☐ LV-BNA	77	Rockwell 690B	11419	Government Province of Cordoba,	LV-PYH
☐ LV-JJW	69	King Air B90	LJ-449	E G L Corporacion SA. Buenos Aires.	LV-PIZ
☐ LV-JOJ	72	681 Turbo Commander	6007	CATA Linea Aerea S A C I F I., Buenos Aires.	N9057N
☐ LV-LEY	72	Rockwell 690	11019	CATA Linea Aerea S A C I F I., Buenos Aires.	LQ-LEY
☐ LV-LMU	74	Rockwell 690A	11176	Aero Servicios Omega SA. San Salvador, Jujuy.	LV-PTC
☐ LV-LRF	75	Rockwell 690A	11228	CATA Linea Aerea S A C I F I. Buenos Aires.	LV-PTJ
☐ LV-LRH	75	Rockwell 690A	11236	Odol SA. (status ?).	LV-PTT
☐ LV-LTB	75	Rockwell 690A	11238	Government Province of Jujuy, San Salvador.	LV-PUA
☐ LV-LTC	75	Rockwell 690A	11241	SAPSE-Servicios Aereos Patagonicas, Viedma.	LV-PUB
☐ LV-LTO	74	Rockwell 690A	11229	Government Province of San Luis.	LV-PTZ
☐ LV-LTU	75	Rockwell 690A	11261	SASA-Sudamericana de Aviacion SA. Buenos Aires.	LV-PUI
☐ LV-LTV	75	PA-31T Cheyenne II	7520022	Government Province of Corrientes.	LV-PTS
☐ LV-LTW	76	Rockwell 690A	11310	Moloco SA. Buenos Aires.	LV-PUF
☐ LV-LTX	75	Rockwell 690A	11258	SASA-Sudamericana de Aviacion SA. Buenos Aires.	LV-PUH
☐ LV-LTY	74	Rockwell 690A	11230	Government Province of Tucuman.	LV-PTY
☐ LV-LZL	75	Rockwell 690A	11246	Emepa SA. Buenos Aires.	LV-PUW

Regn	*Yr*	*Type*	*c/n*	*Owner/Operator*	*Prev Regn*

Regn	Yr	Type	c/n	Owner/Operator	Prev Regn
☐ LV-LZM	75	Rockwell 690A	11268	Banco Mercantil Argentino, Buenos Aires.	LV-PUX
☐ LV-MAG	77	Rockwell 690B	11392	Roemners S A C I F., Buenos Aires.	LV-PVH
☐ LV-MAU	77	Rockwell 690B	11394	TAN-Transportes Aereos Neuquen, Neuquen. `Alumine'	LV-PVL
☐ LV-MAW	77	Rockwell 690B	11398	TAN-Transportes Aereos Neuquen, Neuquen. `Collon Cora'	LV-PVN
☐ LV-MBY	77	Rockwell 690B	11412	Government Province of Chaco, Resistencia.	LV-PXN
☐ LV-MCV	77	Mitsubishi MU-2P	361SA	Industria Pescarmona Metalurgicas, Buenos Aires.	LV-PYI
☐ LV-MDG	77	PA-31T Cheyenne II	7720065	Lanin S A C I F., Buenos Aires.	LV-PYD
☐ LV-MDN	78	PA-31T Cheyenne II	11442	Government Province of Cordoba.	LV-PYT
☐ LV-MGC	77	Mitsubishi MU-2N	704SA	T A F T SRL. Buenos Aires.	LV-PZT
☐ LV-MGD	77	PA-31T Cheyenne II	7720059	Transportes Bragado SA. Buenos Aires.	LV-PXD
☐ LV-MHL	78	PA-31T Cheyenne II	7820020	Aguacil S A C I F., Buenos Aires.	LV-PZC
☐ LV-MLT	78	Mitsubishi MU-2P	385SA	Pioneer Confecciones SA. Buenos Aires.	LV-PAL
☐ LV-MMY	78	Conquest II	441-0075	Government Province of Buenos Aires.	LV-PBA
☐ LV-MNR	79	PA-31T Cheyenne II	7920017	Government Province of Rioja.	LV-P..
☐ LV-MNU	79	PA-31T Cheyenne II	7920003	Government Province of Formosa.	LV-DMO
☐ LV-MOD	79	PA-31T Cheyenne II	7920033	Government Province of Santa Cruz.	LV-P..
☐ LV-MOE	79	PA-31T Cheyenne II	7920050	TAC/Government Province of Catamarca, Catamarca.	LV-P..
☐ LV-MOO	79	Rockwell 690B	11543	Quequen Grande SA. Buenos Aires.	LV-PBS
☐ LV-MRN	79	King Air E90	LW-310	Cia Loma Negra SA. Buenos Aires.	LV-PAY
☐ LV-MRT	78	Conquest II	441-0082	Government Province of Buenos Aires.	LV-PAZ
☐ LV-MRU	78	Conquest II	441-0077	Government Province of Buenos Aires.	LV-PBB
☐ LV-MTU	79	PA-31T Cheyenne II	7920080	Junta Nacional de Carnes, Buenos Aires.	LV-P..
☐ LV-MYA	79	Rockwell 690B	11558	Government Province of Santa Fe.	LV-PDH
☐ LV-MYI	79	Rockwell 690B	11557	Arbol Solo S A G A C., Buenos Aires.	LV-PCC
☐ LV-MYX	79	PA-31T Cheyenne 1	7904045	Chincul S A C A I., Buenos Aires.	LV-P..
☐ LV-OAP	79	PA-31T Cheyenne II	7920065	Junta Nacional de Granos, Buenos Aires.	LV-PCJ
☐ LV-OBB	80	King Air E90	LW-330	Corrugadora Atuel SA. Buenos Aires.	LV-PFJ
☐ LV-ODZ	80	MU-2 Marquise	759SA	Vincente Robles S A M C I., Buenos Aires.	LV-PFD
☐ LV-OEI	80	Gulfstream 840	11612	Government Province of La Pampa, Santa Rosa.	LV-PGD
☐ LV-OFT	80	King Air 200	BB-699	Don Roberto S A A C I., Buenos Aires.	LV-PIF
☐ LV-OFX	67	680V Turbo Commander	1682-63	Empesur SA. Rio Gallegos.	AE-129
☐ LV-OGF	80	PA-31T Cheyenne II	8020013	Government Province of Chubut.	LV-P..
☐ LV-ONH	81	King Air 200	BB-764	Viajaire SA. Buenos Aires.	LV-PLN
☐ LV-OOO	81	PA-42 Cheyenne III	8001013	Government Province of San Juan.	LV-PGT
☐ LV-RAC	81	Mitsubishi MU-2P	386SA	Time Save Air SA. Buenos Aires.	N999BE
☐ LV-RAZ	69	Mitsubishi MU-2F	157	Sanatorio Panamericano SA. Munro.	LV-PAI
☐ LV-ROC	88	King Air C90A	LJ-1180	Peter Aero Servicios SRL. Buenos Aires.	LV-PAJ
☐ LV-RZB	69	SA-26AT Merlin 2B	T26-143	Coronaire SA. Quimes. (status ?).	LV-PFR
☐ LV-VFC	80	King Air E90	LW-339	Banco Macro SA. Buenos Aires.	N290TC
☐ LV-VGS	74	Rockwell 690A	11145	Aero Mak SA. Buenos Aires.	N86345
☐ LV-VHO	68	King Air B90	LJ-428	Pertrans SA.	N74GR
☐ LV-VHP	70	681 Turbo Commander	6014	Empresa Aerea Halcon SRL. Buenos Aires. (status ?).	N88RK
☐ LV-VHR	68	King Air B90	LJ-323	Aviajet SA. San Fernando.	N90SM
☐ LV-WDO	71	King Air 100	B-82	Navy Jet SA.	AE-100
☐ LV-WEW	81	King Air 200	BB-870	Industria Metalurgica Sud Americana, Don Torcuato.	N200EL
☐ LV-WFB	95	King Air C90B	LJ-1414	Express Post SA.	LV-P..
☐ LV-WFP	75	King Air E90	LW-129	Servicio Aerocomercial Essair, Venado Tuerto.	N423JD
☐ LV-WGP	79	King Air 200	BB-558	Minera del Altoplano SA. Don Torcuato.	N58JR
☐ LV-WHV	78	King Air E90	LW-259	Starman SA. Buenos Aires.	N269JB
☐ LV-WIO	80	King Air 200	BB-606	Cheyenne SA. Buenos Aires.	N29AJ
☐ LV-WIP	93	King Air 300LW	FA-229	GSP Empredimientos Agropecuario, Don Torcuato.	LV-PHI
☐ LV-WIR	73	SA-226T Merlin 3	T-232	Hawk Air SA. Buenos Aires.	N56TA
☐ LV-WIU	94	King Air C90B	LJ-1381	Loma Negra Cia Industrial, Don Torcuato.	LV-P..
☐ LV-WJE	94	King Air C90B	LJ-1354	Granbril SACIFIA, Lanus.	N8294Z
☐ LV-WJP	71	King Air C90	LJ-529	Air Service SA.	N78SE
☐ LV-WJY	74	Mitsubishi MU-2J	644	Parala SRL. Buenos Aires.	N494WC
☐ LV-WLT	92	King Air 300	FA-221	TIA SA. Don Torcuato.	N70FL
☐ LV-WLV	91	King Air C90A	LJ-1287	LAPA-Lineas Aereas Privadas Argentinas SA. Buenos Aires.	N8049R
☐ LV-WMA	93	King Air 300	FA-222	Banco Mercantil Argentino SA. Buenos Aires.	N8273L
☐ LV-WMD	70	King Air B90	LJ-493	Bravo Foxtro SRL. Don Torcuato.	N388AS
☐ LV-WME	69	Mitsubishi MU-2F	139	Aries del Sur SA. Don Torcuato.	N90SA
☐ LV-WMG	95	King Air C90B	LJ-1395	Aero Federal SA. Don Torcuato.	LV-PHZ
☐ LV-WNC	75	SA-226AT Merlin 4A	AT-036	Hawk Air SA. Don Torcuato.	N642TS
☐ LV-WNJ	95	King Air B200	BB-1507	Siderar SA. Buenos Aires.	N1567Z
☐ LV-WOR	96	King Air B200	BB-1521	Heli-Air SA. Don Torcuato.	N3241N
☐ LV-WOS	80	King Air 200	BB-639	Promas SA. La Rioja.	N550E
☐ LV-WPB	95	King Air C90B	LJ-1416	Government Province of Chubut.	N3254E
☐ LV-WPG	96	King Air 350	FL-138	Las Matildes SA.	
☐ LV-WPM	80	King Air/Catpass 250	BB-729	Government Province of La Pampa, Santa Rosa.	N743R
☐ LV-WSN	96	Pilatus PC-XII	135	Terraire SA. Buenos Aires.	N12CA
☐ LV-WXC	97	King Air C90B	LJ-1466	Express Post SA. Buenos Aires.	N1108K
Military					
☐ 4-F-43/0743	79	King Air 200	BB-460	FA1-EAN4/EAR4, Argentine Naval Base, Punta Indio.	4-G-43/0743
☐ 4-G-41/0697	75	King Air 200	BB-54	FA1/EAN4-EA4R, Argentine Naval Base, Punta Indio.	5-T-31/0697
☐ 4-G-42/0698	75	King Air 200	BB-71	FA1/EAN4-EA4R, Argentine Naval Base, Punta Indio.	5-T-32/0698
☐ 4-G-45/0745	79	King Air 200	BB-488	FA1-EAN4/EAR4, Argentine Naval Base, Punta Indio.	

Regn   *Yr*   *Type*   *c/n*   *Owner/Operator*   *Prev Regn*

Regn	Yr	Type	c/n	Owner/Operator	Prev Regn
☐ 4-G-47/0747	79	King Air 200	BB-546	FA1-EAN4/EA4R, Argentine Naval Base, Punta Indio.	
☐ 4-G-48/0748	79	King Air 200	BB-549	FA1-EAN4/EA4R, Argentine Naval Base, Punta Indio.	
☐ AE-176	77	SA-226T Merlin 3A	T-275	Argentine Army,	N5393M
☐ AE-178	77	SA-226T Merlin 3A	T-280	Argentine Army,	N5397M
☐ AE-179	77	SA-226T Merlin 3A	T-281	Argentine Army,	N5399M
☐ AE-180		SA-226AT Merlin 4A	AT-071E	Argentine Army. (was TC-286).	N5656M
☐ AE-181		SA-226AT Merlin 4A	AT-063	Argentine Army,	TS-01
☐ AE-182		SA-226AT Merlin 4A	AT-064	Argentine Army,	TS-02
☐ LV-RTC	79	King Air 200	BB-471	Argentine Navy, Buenos Aires.	4-G-44/0744

LX = Luxembourg *(Grand Duchy of Luxembourg)*

Civil

Regn	Yr	Type	c/n	Owner/Operator	Prev Regn
☐ LX-DAK	75	King Air C90	LJ-647	Air 7 SA. Luxembourg/Transavia GmbH. Worms, Germany.	N9075S
☐ LX-GDB	78	King Air 200	BB-397	Luxaviation SA. Luxembourg.	D-IAMW
☐ LX-LTX	78	King Air E90	LW-297	Sky-Service BV. Wevelgem, Belgium.	SE-IKD
☐ LX-RST	78	PA-31T Cheyenne II	7820027	Creditlease SA. Luxembourg.	F-GGPJ

LZ = Bulgaria *(Republic of Bulgaria)*

Civil

Regn	Yr	Type	c/n	Owner/Operator	Prev Regn
☐ LZ-FEO	95	King Air 200	...		
☐ LZ-PIA	90	P-180 Avanti	1008	Lada Air Ltd. Sofia.	EC-CFL
☐ LZ-RGP	75	King Air 200	BB-82	East West European, Sofia.	N788AA

N = USA *(United States of America)*

Civil

Regn	Yr	Type	c/n	Owner/Operator	Prev Regn
☐ N.....	85	Conquest II	441-0345	airframe used in development of Cessna 435 programme.	
☐ N.....	82	Gulfstream 900	15010	Turbine Group Inc. Melbourne, Fl.	HP-...
☐ N.....	95	King Air B200	BB-1512	Hawker Pacific Inc. Sun Valley, Ca.	HS-ITD
☐ N.....	97	King Air C90B	LJ-1497		
☐ N.....	97	Pilatus PC-XII	166		HB-F
☐ N1AQ	68	SA-26AT Merlin 2B	T26-114	Florence Enterprises Inc. Bogart, Ga.	N78WL
☐ N1BM	81	Conquest 1	425-0042	P & R Aviation Inc. Benton, Ky.	N67735
☐ N1BS	89	King Air 300	FA-172	Bank of Stockton, Portland, Or.	N871RC
☐ N1CB	81	King Air 200	BB-826	Garry Lewis, Baton Rouge, La.	N204EB
☐ N1CR	95	King Air B200	BB-1516	Roberts Properties Inc. Atlanta, Ga.	N3246S
☐ N1HE	79	King Air F90	LA-4	Southern Cross Aviation Inc. Fort Llauderdale, Fl.	N1DE
☐ N1HX	78	King Air 200	BB-361	Union Planters Corp. Memphis, Tn.	N334RR
☐ N1JG	78	Rockwell 690B	11510	Gay Company Inc. Austin, Tx.	N1NG
☐ N1KA	81	King Air B200	BB-899	CMH Homes Inc. Knoxville, Tn.	N18544
☐ N1MM	77	King Air 200	BB-274	Erin Air Inc. Seattle, Wa.	N18243
☐ N1MN		SA-227TT Merlin 3C	TT-486A	Devcon Construction Inc/Ceka Aviation, San Jose, Ca.	N86SH
☐ N1MT	79	King Air C90	LJ-824	Delta State University, Cleveland, Ms.	N724TD
☐ N1MW	76	King Air E90	LW-176	Michael Waltrip Inc. Sherrills Ford, NC.	N501MW
☐ N1NL	82	Conquest 1	425-0107	Mitchell Brothers Crane Co. Portland, Or.	N6850P
☐ N1NP	91	King Air C90A	LJ-1289	Nebraska Public Power District, Columbus, Ne.	N563AC
☐ N1PD	81	King Air B200	BB-908	JLJ Equipment Leasing Corp. Pottersville, NJ.	N88BA
☐ N1PN	78	King Air B100	BE-32	Superfos Construction (US) Inc. Dothan, Al.	N712MA
☐ N1SA	65	King Air 90	LJ-66	Schutts Land & Cattle Inc. Fort Worth, Tx.	N4PC
☐ N1TC	95	Pilatus PC-XII	130	Telecel International Ltd. Lanseria, RSA.	HB-FQU
☐ N1TP	87	King Air C90B	LJ-1490	N1TP Inc. Washington, DC.	N1127U
☐ N1TR	78	King Air A100	B-238	Trinity River Authority, Arlington, Tx.	N9100S
☐ N1TX	81	King Air 200	BB-800	State Aircraft Pooling Board, Austin, Tx.	N200GA
☐ N1UC	75	King Air E90	LW-140	Aristokraft Inc. Jasper, In.	N8069S
☐ N1UM	80	King Air C90	LJ-903	University of Mississippi, University, Ms.	N33GB
☐ N1UV	71	King Air C90	LJ-511	MAL Aircraft Sales Inc. Memphis, Tn.	N45BA
☐ N1VY	72	Mitsubishi MU-2J	567	Bankair/Dickerson Associates, West Columbia, SC.	N1VN
☐ N1WJ	65	King Air 90	LJ-60	Ridgeaire Inc. Wilmington, De.	N763K
☐ N1WV	86	King Air 300	FA-62	State of West Virginia, Charleston, WV.	N72345
☐ N1YC	69	SA-26AT Merlin 2B	T26-131	Van Lewis Inc. Henderson, Tx.	N4251R
☐ N1YS	79	King Air 200	BB-430	Excelsior Air Charter Inc. Palm Springs, Ca.	N1BS
☐ N2CJ	78	Mitsubishi MU-2N	726SA	Jones Air Corp. Wilmington, De. (status ?).	N898MA
☐ N2DB	72	Rockwell 690	11028	Keith Brown, Denver, Co.	N9228N
☐ N2DS	78	PA-31T Cheyenne II	7820081	Alpine Aviation Inc. White Lake, Mi.	N34CA
☐ N2ES	74	Rockwell 690A	11159	Indiana Aircraft Sales Inc. Indianapolis, In.	N655GG
☐ N2FJ	76	PA-31T Cheyenne II	7620050	Equity Air Services Inc. Hendersonville, NC.	(N39RP)
☐ N2GL	82	SA-227TT Merlin 3C	TT-426A	Giddings & Lewis Inc. Fond du Lac, Wi.	N4491E
☐ N2GZ	67	Mitsubishi MU-2B	025	Tarrant County Junior College, Fort Worth, Tx.	N2GT
☐ N2JB	96	Pilatus PC-XII	148	Breco International, Farnborough, UK.	N148PC
☐ N2MP	76	King Air C-12C	BC-32	State Highway Patrol, Jefferson City, Mo.	76-22556
☐ N2NA	62	Gulfstream 1	96	N A S A, OSF Johnson Space Centre, Houston, Tx.	N1NA
☐ N2NC	65	Gulfstream 1	158	Corporate Resources Inc. Bingham Farms, Mi. (status ?).	N2NR
☐ N2PX	86	King Air B200	BB-1260	Publix Super Markets Inc. Lakeland, Fl.	
☐ N2PY	77	King Air 200	BB-200	Changer Inc. Wilmington, De.	N2PX
☐ N2QE	81	King Air B100	BE-120	Q E Air Inc. Chesapeake, Va.	C-GBWB
☐ N2RA	77	Mitsubishi MU-2P	366SA	McCreary Enterprises Inc. Stuart, Fl.	N66UP
Regn	*Yr*	*Type*	*c/n*	*Owner/Operator*	*Prev Regn*

Regn	Yr	Type	c/n	Owner/Operator	Prev Regn
☐ N2S	90	King Air 350	FL-9	South Carolina Dept of Commerce, Columbia, SC.	N350S
☐ N2SC	82	King Air B200	BB-959	Elliott Aviation Aircraft Sales Inc. Des Moines, Ia.	N207SB
☐ N2TX	81	King Air B100	BE-103	Texland Aviation Co. Wilmington, De.	N3699T
☐ N2UV	69	King Air B90	LJ-480	Charles Greene, Fort Lauderdale, Fl.	N31SV
☐ N2UW	77	King Air 200T	BT-3	University of Wyoming, Laramie, Wy. (was BB-270).	
☐ N2VA	83	Gulfstream 1000	96062	Department of Aviation, Richmond, Va.	N120GA
☐ N2WC	76	King Air 200	BB-136	GT Aircraft Sales Inc. Youngstown, Oh.	(N2WQ)
☐ N3	65	Gulfstream 1	160	FAA Aircraft Maintenance Section, Washington, DC.	N752G
☐ N3AH	87	King Air 300	FA-130	American Aviation Inc. Lafayette, La.	N9UP
☐ N3AT	81	King Air C90	LJ-933	Appalachian Tire Products Inc. Charleston, WV.	N3709S
☐ N3AW	77	Conquest II	441-0036	Arrowhead Co. Burbank, Ok.	N36968
☐ N3CG	67	680V Turbo Commander	1680-62	J B I Co. New Albany, Ms.	N3CC
☐ N3CR	81	King Air 200	BB-850	Richard Childress Racing Enterprises Inc. Welcome, NC.	N2UH
☐ N3DE	78	King Air 200	BB-332	Champion Air Inc. Mooresville, NC.	N23807
☐ N3DF	65	King Air 90	LJ-36	Donald Bush, Warner Robins, Ga.	(N312DS)
☐ N3GC	73	King Air C90	LJ-576	AlliedSignal Avionics Inc. Everett, Wa.	
☐ N3KF	71	King Air C90	LJ-530	Scope Leasing Inc. Columbus, Oh.	N404VW
☐ N3LS	67	King Air E90	LW-245	Tecmetal/Chagrin Ventures Ltd. Cleveland, Oh.	N18DV
☐ N3NA	62	Gulfstream 1	92	N A S A, OSF Marshall Space Flight Centre, Huntsville, Al.	NASA3
☐ N3NC	82	Conquest II	441-0238	North Carolina Dept of Transport, Raleigh, NC.	N95TD
☐ N3PX	84	King Air B200	BB-1173	Publix Super Markets Inc. Lakeland, Fl.	N2842B
☐ N3RK	81	Cheyenne II-XL	8166022	Superior Distributing Co. Fostoria, Oh.	N81502
☐ N3TK	80	Conquest II	441-0158	Centaur Corp. Irvine, Ca.	N36EF
☐ N3U	81	Gulfstream 980	95041	Northwest Jet Sales & Leasing Inc. Bend, Or.	N400DW
☐ N3UN	78	Mitsubishi MU-2N	720SA	Gazelle Air Transport Ltd. Indianapolis, In.	VH-MIT
☐ N3WE	79	PA-31T Cheyenne 1	7904005	Chatham Chemical Co. Savannah, Ga.	N6181A
☐ N3WM	78	Conquest II	441-0068	D & M Aviation Corp. Houston, Tx.	N88827
☐ N3WU	76	Rockwell 690A	11336	Robert Huston, McMinnville, Or.	N16GL
☐ N3XY	80	Gulfstream 840	11638	Stanford Clinton Jr. Recluse, Wy.	N840SF
☐ N3ZC	85	King Air B200	BB-1207	H B Zachary Co. San Antonio, Tx.	(N321SF)
☐ N4AT	77	King Air 200	BB-281	Air Trails LLC. Monterey Peninsula Airport, Ca.	N315JW
☐ N4BC	70	SA-226T Merlin 3	T-205E	Caribbean Ventures Leasing Co. Orange Park, Fl. (status ?).	N1226S
☐ N4ER	69	SA-26AT Merlin 2B	T26-152	Vaero Inc. Blacksburg, Va.	(N939C)
☐ N4FB	67	680V Turbo Commander	1688-68	Fernando Rodrigo, Miami, Fl.	N4F
☐ N4NA	65	Gulfstream 1	151	N A S A, OSF J F Kennedy Space Center, Fl.	NASA4
⊓ N4PT	81	King Air B200	BB-879	Pope & Talbot Inc. Portland, Or.	
☐ N4PZ	75	Rockwell 690A	11269	Phillip Zeeck Inc. Odessa, Tx.	N690DM
☐ N4QG	80	PA-31T Cheyenne 1	8004023	Cedarhurst Air Charter Inc. Brookfield, Wi.	N180CA
☐ N4RP	84	PA-42 Cheyenne 400LS	5527021	M & M Airways Inc. Boston, Ma.	N500WE
☐ N4RX	79	PA-31T Cheyenne 1	7904012	Marvin Blount, Greenville, NC.	N331SW
☐ N4RY	65	King Air 90	LJ-8	R & J AG Manufacturing Inc. Ashland, Oh.	(N16CG)
☐ N4S	94	King Air 350	FL-107	Weyerhaeuser Co. Tacoma, Wa.	N8257V
☐ N4TA	78	Mitsubishi MU-2N	727SA	Xtal Technologies Ltd. Bay Shore, NY.	YV-717P
☐ N4TJ	75	King Air 200	BB-40	Jaco Oil Co. Bakersfield, SC.	N35TT
☐ N4UB	90	2000 Starship	NC-4	Raytheon Aircraft Credit Corp. Wichita, Ks.	N75WD
☐ N4WD	69	Mitsubishi MU-2F	179	Richardson Dunn Inc. Bellville, Tx.	N10FR
☐ N4WE	87	PA-42 Cheyenne 400LS	5527041	Shepherds Chapel Inc. Gravette, Ar.	N9518N
☐ N4YS	92	King Air 350	FL-82	W B Fry/CC2B Trust, Palo Alto, Ca.	N8182C
☐ N5AP	73	Mitsubishi MU-2J	604	Aero Taxi Rockford Inc. Rockford, Il.	N300WT
☐ N5CE	83	Conquest 1	425-0135	Whayne Supply Co. Louisville, Ky.	N6884D
☐ N5D	78	PA-31T Cheyenne II	7820080	William Packer, Orchard, Mi.	N6097A
☐ N5HT	91	TBM 700	9	Quik-Cook Inc. Rochester, NY.	N969RF
☐ N5LC	80	MU-2 Solitaire	433SA	Manitoba Leasing Corp. Lancaster, NY.	N209MA
☐ N5LE	77	King Air 200	BB-195	B-Air LLC. Nashville, Tn.	YV-92CP
☐ N5LJ	75	Mitsubishi MU-2P	321SA	Lane's Valley Forge Aviation Inc. Collegeville, Pa.	N513DQ
☐ N5LN	80	MU-2 Marquise	799SA	L W Aviation Inc. Wichita, Ks.	N928VF
☐ N5MK	79	King Air 200	BB-537	Mass Bay Kustom Leasing Inc. Braintree, Ma.	N700BX
☐ N5NA	64	Gulfstream 1	125	Central Missouri State University, Warrensburg, Mo.	N10NA
☐ N5NM	75	King Air E90	LW-123	Seven Bar Flying Service Inc. Albuquerque, NM.	N158D
☐ N5PA	81	PA-42 Cheyenne III	8001051	Aviation Plus Inc. Fargo, ND.	N810L
☐ N5PF	81	PA-42 Cheyenne III	8001003	Midlantic Flight Charter Inc. Dover, De.	N717MB
☐ N5PP	84	Conquest 1	425-0198	PPAL Inc. Kalamazoo, Mi.	N425PA
☐ N5PQ	82	MU-2 Marquise	1558SA	Corporate Aviation Services Inc. Tulsa, Ok.	N12WF
☐ N5PX	79	King Air 200	BB-554	Tim-Bar Corp/TB Container Corp. New Oxford, Pa.	N204BR
☐ N5RE	68	680W Turbo Commander	1818-32	George Fortier & LFS Inc. Gallatin Gateway, Mt.	N3RA
☐ N5ST	77	King Air 200	BB-289	Mayo Aviation Inc & R C Leasing LLC. Englewood, Co.	N7MB
☐ N5SY	81	PA-42 Cheyenne III	8001069	Dialysis Clinic Inc. Nashville, Tn.	N5SS
☐ N5TA	77	King Air C90	LJ-724	MacDonald Enterprises Inc. Spencerville, Oh.	N5TW
☐ N5TW	93	King Air B200	BB-1471	Warfield Inc. Santa Fe, NM.	
☐ N5UB	75	King Air 200	BB-69	Paw Enterprises Inc. Franklin, Tn.	N1US
☐ N5UN	80	King Air 200	BB-697	Eastway Aviation Inc. Ronkonkoma, NY.	N50N
☐ N5UV	80	King Air 200	BB-339	GordonAir Inc. Johnson City, Tn.	N93MF
☐ N5VK	93	King Air C90B	LJ-1334	K-Air Inc. Cato, Wi.	N8148N
☐ N5VX	59	Gulfstream 1	7	M I T Lincoln Laboratory, Lexington, Ma.	C-FLOO
☐ N5WC	79	PA-31T Cheyenne 1	7904028	Gunslinger Investment Corp. Telluride, Co.	N711D
☐ N5WG	67	King Air A90	LJ-289	K & K Aircraft Inc. Bridgewater, Va.	N35P

Regn	Yr	Type	c/n	Owner/Operator	Prev Regn
☐ N5WU	74	King Air C90	LJ-635	West Virginia University, Morgantown, WV.	N5GC
☐ N5XM	94	King Air 350	FL-115	XM Corp. Miami, Fl.	N35DT
☐ N5Y	67	King Air A90	LJ-272	B P Investments, Erie, Mi.	
☐ N6BZ	82	Gulfstream 900	15008	G Ponder & R Smith Jr/P & S Realty Service, Marshall, Tx.	ZS-KZT
☐ N6E	70	681 Turbo Commander	6002	Athol Chin, Miami, Fl.	N9052N
☐ N6EA	81	King Air B200	BB-890	Edwards & Assocs Inc. Bristol, Tn.	N1825H
☐ N6HU	78	King Air 200	BB-319	Harding University Inc. Searcy, Ar.	N300WJ
☐ N6JM	79	PA-31T Cheyenne 1	7904011	Gray Leasing Inc. Searcy, Ar.	N96MM
☐ N6KE	80	MU-2 Marquise	766SA	Kervick Enterprises Inc. Worcester, Ma.	N199MA
☐ N6KF	75	Mitsubishi MU-2L	659	Milam International Inc. Englewood, Co.	N5JE
☐ N6MF	95	2000A Starship	NC-53	Volante Corp. Coral Gables, Fl.	
☐ N6PE	81	King Air B200	BB-856	H A Braun, Burbank, Ca.	N1802H
☐ N6PW	85	King Air 300	FA-32	Bemis Manufacturing Co. Sheboygan Falls, Wi.	
☐ N6PX	77	King Air 200	BB-205	Tim-Bar Corp/Oxford Container Co. New Oxford, Pa.	N6PW
☐ N6SK	80	Conquest 1	425-0005	Knight Holdings Ltd. Belle Glade, Fl.	N98820
☐ N6SP	77	King Air C-12C	BC-42	Delaware State Police, Dover, De.	N116SP
☐ N6UD	82	King Air B200	BB-1055	Quality Wood Treating Co. Prairie du Chien, Wi.	N6LD
☐ N6VP	77	Conquest II	441-0019	Verama Inc. Wilmington, De.	(N55CC)
☐ N7BF	68	King Air B90	LJ-350	Anthony Crane Rental Inc. Beaumont, Tx.	(N579B)
☐ N7BQ	66	King Air 90	LJ-107	Lenora Co-Lithuania/Tesmo Don Co. Silver Springs, Md.	N7BF
☐ N7DD	74	Mitsubishi MU-2K	297	MGA Health Management Inc. Douglas, Ga.	N106GB
☐ N7EV	73	Rockwell 690A	11139	Deerfleet LLC. Miami, Fl.	N61WA
☐ N7FL	78	PA-31T Cheyenne II	7820082	D & M Aviation Inc & QuickFlite Corp. Chesterfield, Mo.	N6100A
☐ N7HN	84	MU-2 Marquise	1563SA	Robert & Judith Ann Hecht-Nielsen, Burbank, Ca.	(D-IOMX)
☐ N7KS	79	Rockwell 690B	11521	California Architectural Lighting, Portland, Or.	N81746
☐ N7NA	82	King Air/Catpass 250	BB-997	NASA HQ/OSSA,Van Nuys, Ca.	
☐ N7NW	80	Conquest II	441-0186	Aztlan Corp. Ashland, Or.	N2724S
☐ N7PA	92	King Air B200	BB-1444	Warren Spieker Jr. Soledad, Ca.	N663CS
☐ N7RW	75	King Air 200	BB-31	Rush Aviation Inc. Oxford, Al.	N61RR
☐ N7SL	59	Gulfstream 1	11	Sanders Lead Co. Troy, Al.	N100EL
☐ N7TD	78	King Air E90	LW-276	Twin Disc Inc. Racine, Wi.	N125L
☐ N7TW	69	King Air B90	LJ-478	Rush Aviation Inc. Oxford, Al.	N40WS
☐ N7VA	80	King Air 200	BB-670	Department of Aviation, Sandston, Va.	N1VA
☐ N7VR	78	PA-31T Cheyenne 1	7804010	Kelly Bruun, Portland, Or.	N93TW
☐ N7WF	82	Conquest 1	425-0119	Foxtrot Inc. Wichita Falls, Tx.	N17RC
☐ N7WU	75	King Air E90	LW-142	State of Washington, Olympia, Wa.	N7WS
☐ N7ZP	77	King Air E90	LW-227	Leasing of Mobile Inc. Camden, Ar.	N77P
☐ N7ZW	78	King Air E90	LW-262	H K Ventures LLC. Wallingford, Ct.	N77W
☐ N8CF	80	PA-31T Cheyenne II	8020062	Air East Charters of Ashoskie Inc. Ashoskie, NC.	N5452J
☐ N8DB	81	PA-31T Cheyenne II	8120014	J Baker, Hilton Head Island, SC.	N46CE
☐ N8E	62	Gulfstream 1	94	Saicam Air (US) Inc. Wonderboom, RSA.	N794G
☐ N8EE	83	Cheyenne II-XL	8166071	N6KH Corporation Inc. Winter Haven, Fl.	N44FH
☐ N8EF	80	King Air 200	BB-721	FBD Securities Inc. Steamboat Springs, Co.	N610HG
☐ N8EG	92	TBM 700	34	Malibu Leasing Corp. Aspen, Co.	
☐ N8FR	81	Conquest 1	425-0095	Frank Rhodes Jr. Newport Beach, Ca.	N68481
☐ N8GT	85	King Air C90A	LJ-1103	Business Aircraft Leasing Inc. Nashville, Tn.	N55LC
☐ N8KT	69	SA-26AT Merlin 2B	T26-158E	Augusta Aviation Inc. Daniel Field, Ga.	N256WC
☐ N8MG	78	King Air C90	LJ-747	Charter Services Inc. Mobile, Al.	N529JH
☐ N8NA	82	King Air B200	BB-950	NASA/OSSA, Wallops Island, Va.	
☐ N8PC	71	Mitsubishi MU-2F	193	A Lamas, Buenos Aires, Argentina.	N1210W
☐ N8PY	79	King Air 200	BB-487	Manxtrust, IOM/Eurooair Transport Ltd. Gatwick, UK.	VH-PIL
☐ N8RY	81	PA-42 Cheyenne III	8001036	Y T Timber Inc. Boise, Id.	N3JQ
☐ N8SV	79	King Air 200	BB-506	King Air IV Inc. Bellevue, Wa.	N945BV
☐ N8VF	85	King Air 300	FA-49	V F Corp. Wyomissing, Pa. (now TG-CPG ?).	N7224A
☐ N8VL	83	Gulfstream 840	11732	Laredo National Bank, Laredo, Tx.	N29DS
☐ N9AN	78	King Air C90	LJ-777	Ball Engineering Inc. Okolona, Ms.	
☐ N9HW	74	King Air B90	LJ-459	Michael Mullins, Germantown, Tn.	N2JR
☐ N9KG	77	Rockwell 690B	11426	Raffanti Enterprises Inc. Redding, Ca.	N81682
☐ N9LV	79	Rockwell 690B	11550	Smooth Propellor Co & Loy Hickman, Parkville, Mo.	N31GH
☐ N9MU	97	King Air C90B	LJ-1503	Raytheon Aircraft Co. Wichita, Ks.	
☐ N9NC	78	Mitsubishi MU-2N	728SA	Noland Co. Newport News, Va.	N900MA
☐ N9RU	82	King Air B200	BB-984	Power Ranch Inc. Santa Ynez, Ca.	N56JA
☐ N9TN	78	King Air C90	LJ-803	TNL Flight Services, Austin, Tx.	G-OLAF
☐ N9U	82	SA-227AT Merlin 4C	AT-495B	Northwest Jet Sales & Leasing Inc. Bend, Or.	N9UA
☐ N9VC	78	King Air C90	LJ-763	Eddie Wiggins Ford, Lincoln, Mercury Inc. Warner Robbins, Ga	N78JD
☐ N10AG	80	King Air B100	BE-100	Auto Glass Specialist Inc. Madison, Wi.	N3688F
☐ N10AU	83	King Air C90-1	LJ-1054	Auburn University, Auburn, Al.	N63908
☐ N10AY	66	King Air 90	LJ-88	A C McPhie, Bend, Or. (status ?).	(N90233)
☐ N10BQ	74	PA-31T Cheyenne II	7400005	West Tenessee Aviation Inc. Union City, Tn.	(N80BC)
☐ N10BY	80	King Air 200	BB-642	TPT Aviation, Gastonia, NC.	N30MK
☐ N10CW	79	King Air C90	LJ-832	South Holland Trust & Savings Corp. South Holland, Fl.	N91LE
☐ N10DR	81	King Air F90	LA-149	DeRoyal Industries Inc. Alcoa, Tn.	N707DR
☐ N10EC	85	King Air B200	BB-1211	State of Tennessee, Nashville, Tn.	N7220C
☐ N10HT	80	MU-2 Marquise	778SA	Epps Air Service Inc. Atlanta, Ga.	N264MA
☐ N10JE	65	King Air 90	LJ-10	Diversified Aviation Services Inc. Stuart, Fl.	N10J
☐ N10K	90	King Air 350	FL-4	Department of Central Services, Oklahoma City, Ok.	N97DL

Regn	Yr	Type	c/n	Owner/Operator	Prev Regn
□ N10MC	79	PA-31T Cheyenne II	7920028	Eastern Air Center Inc. Norwood, Ma.	N94TW
□ N10MD	78	King Air C90	LJ-750	Charles Rolle, Tucson, Az.	N214D
□ N10MR	77	Mitsubishi MU-2P	351SA	Greg & Lisa Krueger, Eagle, Id.	N10FR
□ N10PT	81	King Air B200C	BL-38	Purcell Tire & Rubber Co. Potosi, Mo.	ST-APW
□ N10SA	78	King Air E90	LW-273	McKnight LLC. Virginia Beach, Va.	N555TB
□ N10TB	62	Gulfstream 1	86	Tim Bar Corp. Hanover, Pa.	N86JK
□ N10TG	67	680V Turbo Commander	1684-65	AirAtlantic Airlines Inc. Centre Hall, Pa.	N1UT
□ N10TM	69	King Air B90	LJ-476	RLD Leasing Inc. Burr Ridge, Il.	N41AA
□ N10UT	76	Mitsubishi MU-2M	346	NFM Inc. Henderson, Nv.	N888RF
□ N10VU	81	MU-2 Solitaire	438SA	Chester Upham Jr. Mineral Wells, Tx. (status ?).	N99LC
□ N11CT	68	680W Turbo Commander	1773-11	Bob Jones University, Greenville, SC.	N4852E
□ N11DT	72	King Air E90	LW-11	David Turk, Honolulu, Hi.	N241B
□ N11FL	67	King Air A90	LJ-301	Core Investments Inc. Sunrise, Fl.	N800Q
□ N11HM	66	680V Turbo Commander	1620-51	R W Hawkins, Columbia, Il.	N680MH
□ N11HY	77	Rockwell 690B	11403	Harvey & Linda Peel, Tijeras, NM.	N29973
□ N11JJ	69	King Air 100	B-2	V E C Corp/Lease Air, Hackensack, NJ.	N3RC
□ N11MM	79	Conquest II	441-0105	Rogers Tool Works Inc. Rogers, Ar.	N4152G
□ N11SJ	74	Mitsubishi MU-2K	285	Anaconda Aviation Corp. Boca Raton, Fl.	N111JE
□ N11TE	80	King Air 200	BB-651	Tractor & Equipment Co Inc. Birmingham, Al.	N202HC
□ N11TN	68	King Air B90	LJ-419	M & C Aviation Inc. Bronx, NY.	N11TE
□ N12AC	72	King Air C90	LJ-554	Label Tech Inc. Somersworth, NH.	F-GESC
□ N12AQ	72	King Air A100	B-94	Golden Wings Aviation Inc. Uniontown, Pa.	N94CP
□ N12AU	78	King Air E90	LW-293	Auburn University, Auburn, Al.	
□ N12CF	79	King Air 200	BB-534	C A S Leasing LLC/Commonwealth Aviation, Richmond, Va.	
□ N12DE	78	Rockwell 690B	11501	Dealers Electrical Supply Co. Waco, Tx.	N25CL
□ N12FA	95	Pilatus PC-XII	117	Fisher Auto Parts Inc. Staunton, Va.	N117WF
□ N12GR	79	PA-31T Cheyenne II	7920058	Triple H Flying Co. Omaha, Ne.	N179SW
□ N12HF	78	PA-31T Cheyenne II	7820085	Central Aviation, Boise, Id.	N589GA
□ N12KA	73	King Air E90	LW-41	Aeronautic Services Inc. Albion, Il.	N2AS
□ N12LA	73	King Air E90	LW-49	Tulip City Air Services Inc. Holland, Mi.	N777AJ
□ N12LE	79	MU-2 Solitaire	403SA	Twins Aviation Inc. Liverpool, NY.	N7PW
□ N12MF	81	SA-226T Merlin 3B	T-417	Thriftway Marketing Corp. Farmington, NM.	N89RP
□ N12NG	80	King Air 200	BB-581	Northrop Grumman Aviation Inc. Hawthorne, Ca.	N14NA
□ N12NL	81	Conquest 1	425-0069	Ericksons Diversified Corp. Hudson, Wi.	N634M
□ N12RW	61	Gulfstream 1	82	Liberty Aviation I td. Greensboro, NC.	SE-LFV
□ N12SJ	76	Mitsubishi MU-2M	333	Air 1st Aviation Companies Inc. Aiken, SC.	N521MA
□ N12TW	77	PA-31T Cheyenne II	7720017	Executive Air Inc. New Port Richey, Fl.	N19AC
□ N12WC	69	SA-26AT Merlin 2B	T26-118	Whiting Wings, Alexander, Ar.	N1224S
□ N12WY	96	TBM 700	112	Barlu Inc. Dover, De.	F-GLBH
□ N13FH	84	Conquest 1	425-0209	Hertrich Aviation Inc. Seaford, De.	N1224K
□ N13NW	79	Conquest II	441-0090	North West Group Inc. Denver, Co.	N4061K
□ N13PF	78	Rockwell 690B	11455	Peco Foods Inc. Tuscaloosa, Al.	N690JL
□ N13RR	76	Mitsubishi MU-2L	682	Air Carriers Inc. Birmingham, Al.	N13PR
□ N13YS	77	Mitsubishi MU-2L	694	Youth Services International Inc. Owings Mills, Md.	N800PC
□ N14BM	81	Conquest 1	425-0076	Lanier & Assocs Consulting Engineers Inc. New Orleans, La.	N67735
□ N14CF	75	King Air A100	B-209	Beech Transportation Inc. Eden Prairie, Mn.	N100AN
□ N14CP	73	King Air C90	LJ-585	Metropolitan Dade County, Miami, Fl.	
□ N14CV	72	Rockwell 690	11020	Sheldon Mortenson/Imperial International, Inc. St Paul, Mn.	N340BP
□ N14EA	78	PA-31T Cheyenne II	7820006	Freight Systems Inc. Commerce, Ca.	F-GFPV
□ N14HB	96	King Air B200	BB-1533	King Air Leasing LC. Clewiston, Fl.	
□ N14HG	82	King Air B200	BB-1071	Fortuna Communications Corp. Fortune, Ca.	N82BS
□ N14NM	73	King Air E90	LW-35	Seven Bar Flying Service Inc. Albuquerque, NM.	N811JB
□ N14NW	80	Conquest II	441-0171	North West Group Inc. Denver, Co.	C-GKMA
□ N14SB	77	King Air E90	LW-214	Clayton Motors Inc. Knoxville, Tn.	N17603
□ N14TF	81	King Air 200	BB-810	State of Wisconsin, Madison, Wi.	N711AE
□ N14TP	72	SA-226T Merlin 3	T-220	Iowa City Flying Service Inc. Iowa City, Ia.	N905RK
□ N14V	68	King Air B90	LJ-411	Central Flying Service Inc. Little Rock, Ar.	N807K
□ N14VK	68	King Air B90	LJ-365	Godwin Aircraft Inc. Memphis, Tn.	N14V
□ N14VR	91	2000 Starship	NC-22	Raytheon Aircraft Services Inc. Wichita, Ks.	N14VP
□ N14YS	74	Mitsubishi MU-2K	290	YSI-Youth Services International, Owing Md.	N4186Y
□ N15	81	King Air F90	LA-138	FAA, Oklahoma City, Ok.	
□ N15CC	80	SA-226T Merlin 3B	T-339	Hudson International Inc. Charlote, NC.	N8167Z
□ N15CD		Rockwell 690A	11177	Lakin Law Firm, Wood River, Il.	N16TB
□ N15CT	67	King Air A90	LJ-192	Charles Thomas, Loves Park, Il.	N792K
□ N15DB	82	Conquest II	441-0278	Sierra Pacific Industries, Redding, Ca.	N98718
□ N15GA	75	King Air C90	LJ-656	Lane Aviation Inc. Rosenburg, Tx.	N600PC
□ N15HV	93	King Air B200	BB-1452	A Bar V Cattle & Commerce Corp. Prescott, Az.	N155V
□ N15JA	90	King Air B200	BB-1354	Raytheon Aircraft Receivables Corp. Wichita, Ks.	N161A
□ N15KA	80	King Air 200	BB-600	Joseph Hicks Pharmacy, Walnut Cove, NC.	
□ N15KW	81	Cheyenne II-XL	8166014	Alfons Ribitsch, Torrance, Ca.	(N31KP)
□ N15MV	81	MU-2 Marquise	1506SA	SamaritanAir Inc. Anchorage, Ak.	N450FA
□ N15NG	80	King Air 200	BB-666	Northrop Grumman Aviation Inc. Hawthorne, Ca.	N15NA
□ N15SF	79	Rockwell 690B	11528	Northstar Air Express Inc. Missoula, Mt.	N690SH
□ N15ST	83	Conquest 1	425-0140	Nicholas Street, Bristol, Tn.	N5829J
□ N15TP	96	Pilatus PC-XII	149	Atlantic Aero Inc & others, Greensboro, NC.	N464WF
□ N15TR	81	MU-2 Solitaire	446SA	F D P Inc. Hobbs, NM.	N81MW
Regn	*Yr*	*Type*	*c/n*	*Owner/Operator*	*Prev Regn*

Regn	Yr	Type	c/n	Owner/Operator	Prev Regn
N15VZ	72	Rockwell 690	11035	J L Smith Co. Yakima, Wa.	N882GS
N15WN	73	King Air E90	LW-78	G L Wilson Building Co. Statesville, NC.	N4378W
N15YS	73	Mitsubishi MU-2J	598	YSI-Youth Services International, Owing Mills, Md.	C-FMBL
N16	80	King Air C90	LJ-893	FAA, Washington, DC.	
N16CG	80	MU-2 Solitaire	418SA	Jerry Fambrough, Alamagordo, NM.	YV-11CP
N16GF	96	King Air B200	BB-1531	G F Air LLC. Charleston, SC.	N1081F
N16KM	81	King Air C90	LJ-961	Old NB Leasing Department, Evansville, In.	XA-ISL
N16LH	77	King Air E90	LW-217	Antlers Natural Gas Enterprises, Colorado Springs, Co.	
N16NG	78	DHC 6-300	596	Northrop Grumman Aviation Inc. Hawthorne, Ca.	N16NA
N16NM	73	King Air E90	LW-62	Seven Bar Flying Service Inc. Albuquerque, NM. (status ?).	N96DA
N16P	83	Conquest 1	425-0173	Stuart Irby Co. Jackson, Ms.	N30WF
N16SM	75	King Air A100	B-224	Stowers Machinery Corp. Knoxville, Tn.	N1524L
N16TB	89	King Air 300	FA-198	White Mountain Holdings & Haverford Utah LLC. Hanover, NH.	N5513E
N16TF	88	King Air B200	BB-1310	Friedkin International Aviation Service, Sugarland, Tx.	N310GA
N17	80	King Air C90	LJ-896	FAA, Washington, DC.	
N17AE	80	King Air F90	LA-80	Amprite Aviation Inc. Nashville, Tn.	N614RG
N17CK	86	Reims/Cessna F406	F406-0001	Clint Aero Inc. St Thomas, USVI.	PH-ALO
N17DW	80	King Air 200	BB-648	Darwal Inc. Harrisburg, NC.	N648MW
N17HF	73	Rockwell 690A	11127	Meyer Co. Garfield Heights, Oh.	N199WP
N17HG	81	MU-2 Marquise	1510SA	Glauser Construction/Harry Glauser III, Houston, Tx.	N417MA
N17HM	76	King Air 200	BB-94	Aeroland Co. JM Petroleum & HMTF Air Inc. Dallas, Tx.	N28AH
N17JG	68	680W Turbo Commander	1802-24	Tulsa County Vo-Tec School, Tulsa, Ok.	N102US
N17JJ	69	SA-26AT Merlin 2B	T26-166	Desert Aircraft Sales Inc. Bermuda Dunes, Ca.	N23X
N17JQ	78	Mitsubishi MU-2P	375SA	Jerry Woolf Inc. Bakersfield, Ca.	XA-TCS
N17NC	91	King Air 350	FL-42	Wachovia Corp. Winston Salem, NC.	N315P
N17NM	78	PA-31T Cheyenne II	7820035	Seven Bar Flying Service Inc. Albuquerque, NM.	N6NB
N17SA	66	King Air A90	LJ-164	Prestar Inc. Davis, Ca.	N8GF
N17SE	76	King Air E90	LW-169	Air Serv International, Maputo, Angola.	N600KC
N17TU	75	PA-31T Cheyenne II	7520035	Private Jets LLC. Oklahoma City, Ok.	N66LD
N17TW	89	King Air 300	FA-190	4 H Aviation Inc. South Bend, In.	N1538Q
N17ZD	82	Gulfstream 1000	96017	International Air Services Inc. (status ?).	N9937S
N18	81	King Air F90	LA-145	FAA, Oklahoma City, Ok.	
N18AF	81	PA-42 Cheyenne III	8001063	Allens Air Service Inc. Seaford, De.	N98LC
N18AH	72	King Air A100	B-118	Crestline Homes Inc. Laurinburg, NC.	N100MX
N18BF	69	Mitsubishi MU-2F	173	Snap Edge Corp. Gary, Il.	N887Q
N18BL	80	King Air F90	LA-13	Bobby Labonte Racing Inc. Trinity, NC.	N909K
N18CJ	77	King Air 200	BB-260	Apple Ten Aero LLC. Alcoa, Tn.	N1SC
N18CM	84	King Air F90-1	LA-219	Co-Mar of Dayton Inc. Dayton, Oh.	N8YK
N18DN	75	King Air 200	BB-81	Chaver Aviation Inc. Tulsa, Ok.	N697D
N18KA	78	King Air 200	BB-360	JamLeigh Corp. Telluride, Co.	N210PH
N18KK	71	681B Turbo Commander	6055	Karl Koster, Fremont, Ca.	N9127N
N18KP	79	PA-31T Cheyenne 1	7904013	Paddington Associates Inc. Wilmington, De.	N18KT
N18LP	67	King Air A90	LJ-278	David Verner, Vandalia, Il.	N741L
N18TF	60	Gulfstream 1	52	Eastway Aircraft Services Inc. Ronkonkoma, NY.	N3858H
N18U	70	King Air 100	B-57	United Methodist Committee on Relief, NYC.	N18X
N19BK	91	King Air C90A	LJ-1294	Dela Corp. Miami, Fl.	PT-OMO
N19DA	80	King Air B100	BE-87	Dominion Air Leasing Inc. Richmond, Va.	N980KA
N19GA	81	Mitsubishi MU-2P	454SA	Air 88 Inc/Crownair Charter Inc. San Diego, Ca.	I-FRTL
N19GR	93	King Air 350	FL-96	Gorman-Rupp Co. Mansfield, Oh.	N403P
N19GU	73	Mitsubishi MU-2J	574	Bush Field Aircraft Co. Augusta, Ga.	N243MA
N19LW	81	King Air C90	LJ-991	Daignault Aviation Inc. Towson, Md.	N504AB
N19NC	89	King Air 300	FA-183	Jack Henry & Associates Inc. Monett, Mo.	N1549D
N19NG	87	Beech 1900C-1	UC-2	Northrop Grumman Aviation Inc. Hawthorne, Ca.	N19NA
N19NM	93	PA-31T Cheyenne II	7520006	Seven Bar Flying Service Inc. Albuquerque, NM.	N33DT
N19RK	80	King Air F90	LA-12	RAPAC Inc & Ring Can Corp. Olive Branch, Ms.	N200E
N19UM	71	King Air C90	LJ-524	O & D Bird Aviation Inc. Wilmington, De.	N19CM
N20AU	76	King Air C-12C	BC-28	Auburn University, Auburn, Al.	76-22552
N20BL	66	King Air A90	LJ-163	Business Aircraft Sales Corp. Chattanooga, Tn.	N2RR
N20BM	67	680V Turbo Commander	1698-75	Western Granite Inc. Troutdale, Or.	N420J
N20BP	76	Rockwell 690A	11341	Sky-Tex Inc. Tifton, Ga.	N81493
N20DL	80	PA-31T Cheyenne 1	8004045	Central Iowa Energy Co-operative, Cedar Rapids, Ia.	N26DV
N20EF	69	SA-26AT Merlin 2B	T26-157	Air Cargo Carriers Inc. Milwaukee, Wi.	N20ER
N20EW	88	King Air 300LW	FA-152	Western Heritage Investments Corp. Missoula, Mt.	D-IMMB
N20FD	89	King Air C90A	LJ-1213	Meisterflieger Inc. Nashville, Tn.	N1559T
N20GC	70	SA-226T Merlin 3	T-205	Madan Construction Co. Michigan City, In.	N20JC
N20GT	82	Gulfstream 1000	96036	Alsois Aviation Corp. Silverside, De.	ZK-FRC
N20HC	77	Conquest II	441-0010	J W & Shirley Connolly, Sewickley, Pa.	N36930
N20KW	85	King Air 300	FA-44	K & W Restaurants Inc. Winston Salem, NC.	N48EB
N20LB	96	King Air B200	BB-1541	Bair Inc. Durham, NC.	N1003W
N20LH	74	King Air E90	LW-80	KEI Corp. Houston, Tx.	N3ZC
N20MA	78	Rockwell 690B	11514	First National Bank, Goodland, Ks.	N14BU
N20ME	78	Rockwell 690B	11440	Wrangler Aviation Corp. Norman, Ok.	N47CF
N20NL	85	King Air 300	FA-51	CNS Corp. Kansas City, Mo.	N7228Z
N20PJ	79	PA-31T Cheyenne II	7920021	Phillips & Jordan Inc. Knoxville, Tn.	N6188A
N20RE	81	King Air 200	BB-758	Helca Mining Co. Coeur d'Alene, Id.	
N20S	78	King Air E90	LW-267	Robert Schiffenhaus, Espoo, Finland.	N23726

Regn   Yr Type   c/n   Owner/Operator   Prev Regn

Regn	Yr	Type	c/n	Owner/Operator	Prev Regn
☐ N20SM	82	King Air B200	BB-1011	State of Mississippi, Jackson, Ms.	N22TZ
☐ N20TN	79	PA-31T Cheyenne II	7920019	Sierra Aviation Inc. Bishop, Ga.	N20TV
☐ N20VP	85	King Air B200	BB-1242	Valley Proteins Inc. Winchester, Va.	N72524
☐ N20W		Fairchild F-27F	97	Jet Support Systems Inc. Windsor Locks, Ct.	N2724R
☐ N20WE	81	PA-31T Cheyenne II	8120059	Kladstrup-Wetzel Associates Ltd. Englewood, Co.	
☐ N20WL	77	PA-31T Cheyenne II	7720011	Hamblen Aero Inc. Morristown, Tn.	N51BJ
☐ N20WS	73	King Air E90	LW-30	Southeast Airmotive Corp. Charlotte, NC.	(N24TF)
☐ N21CJ	80	MU-2 Marquise	789SA	Bankair Inc. West Columbia, SC.	N278MA
☐ N21DE	76	King Air 200	BB-117	Jones Air LLC. Mount Juliet, Tn.	N1DE
☐ N21FG	81	King Air 200	BB-839	Department of Fish & Game, Sacramento, Ca.	N3849B
☐ N21HA	80	King Air C90	LJ-918	Collins Pine Co. Portland, Or.	N499M
☐ N21HC	71	681B Turbo Commander	6054	General Atomic Aeronautical Society, San Diego, Ca.	N9124N
☐ N21HP	77	Mitsubishi MU-2P	357SA	Tele-Optics Inc. Kingsport, Tn.	N2HP
☐ N21JA	74	Mitsubishi MU-2J	614	Ridder Air Enterprises Inc. Whitman, Ma.	N998CA
☐ N21KN	76	PA-31T Cheyenne II	7620035	Jet Aircraft Acquisitions, Fernandina Beach, Fl.	N21AF
☐ N21MU	80	MU-2 Marquise	757SA	XTAL Technologies Ltd. Carle Place, NY.	YV-717P
☐ N21PS	89	King Air B200	BB-1344	PSC Inc. Wichita, Ks.	
☐ N21VF	88	King Air B200	BB-1297	Vanity Fair Intimates Inc. Monroeville, Al.	N100PL
☐ N22BB	67	King Air A90	LJ-241	Robert Spillman, Elyria, Oh.	N27UU
☐ N22BD	81	PA-42 Cheyenne III	8001055	Bill Davis Racing Inc. High Pint, NC.	N190AA
☐ N22BJ	72	King Air A100	B-133	Silverhawk Aviation Inc. Lincoln, Ne.	N1200Z
☐ N22BM	74	King Air C90	LJ-630	Alan Vester Management Corp. Littleton, NC.	XA-SGQ
☐ N22CG	80	Conquest II	441-0119	Griffith Aviation Co. Tulsa, Ok.	N26231
☐ N22CK	79	Rockwell 690B	11524	E Cohn, East St Louis, Il.	N81773
☐ N22DT	80	PA-31T Cheyenne II	8020057	Conyan Aviation Inc. Boise, Id.	N806CM
☐ N22DW		SA-226T Merlin 3B	T-317	MACI Leasing Corp. Edison, NJ.	N10143
☐ N22ER	72	King Air E90	LW-18	Walmer Leasing Co. Carrizo Springs, Tx.	N22EH
☐ N22ET	68	680W Turbo Commander	1793-23	R Con Aviation Inc. Billings, Mt.	N22RT
☐ N22F	82	King Air B200	BB-1025	Wesley West Interests Inc. Houston, Tx.	(N2227F)
☐ N22GW	70	SA-26AT Merlin 2B	T26-169	FEN Ltd. Sylvania, Oh.	N227CH
☐ N22RT	70	681 Turbo Commander	6001	J & J Service, Tulsa, Ok.	N66CP
☐ N22TL	80	King Air E90	LW-334	TX Aire Inc. Amarillo, Tx.	
☐ N22UC	78	PA-31T Cheyenne 1	7804007	Janis Lampedecchia, Clovis, Ca.	N244GC
☐ N22VF	81	Cheyenne II-XL	8166018	Cape Smythe Air Service Inc. Barrow, Ak.	N5UB
☐ N22WF	79	PA-31T Cheyenne II	7920007	Linsam Inc. McDonald, Pa.	N791SW
☐ N22YA	74	Mitsubishi MU-2J	623	Younkin Air Service Inc. Springdale, Ar.	N60GS
☐ N22YC	81	Cheyenne II-XL	8166020	W G Yates & Sons Construction Co. Philadelphia, Ms.	N11YC
☐ N23AE	73	SA-226T Merlin 3	T-241	C E I Aviation Inc. Midland, Tx.	N73542
☐ N23AW	81	Conquest 1	425-0023	Flying High Ventures Inc. Sun River, Or.	N6772S
☐ N23EA	80	Conquest II	441-0183	Samuel Mansbach, Greensboro, NC.	N120EA
☐ N23FH	97	King Air B200	BB-1565	Fox Hill Aviation Inc. Westwood, Ma.	N1094Y
☐ N23FL	93	2000A Starship	NC-32	Raytheon Aircraft Co. Wichita, Ks.	N23FH
☐ N23MV	77	PA-31T Cheyenne II	7720044	Navajo Inc. Onalaska, Wi.	N4015Y
☐ N23ST	78	King Air 200	BB-375	Senter Tool Service Inc. Portland, Or.	N89FC
☐ N23SV	91	P-180 Avanti	1010	Aviation Enterprises Inc. Wilmington, De.	I-ALPV
☐ N23TC	79	King Air 200	BB-453	DIHE Inc. Van Nuys, Ca.	N1284
☐ N23WP	80	PA-31T Cheyenne II	8020004	PFC Inc/Stingray Boat Co. Hartsville, SC.	N51RL
☐ N23WS	83	King Air B200	BB-1143	Hearne Aviation Inc. Hendersonville, NC.	N158LM
☐ N23YP	83	King Air B200	BB-1142	Yates Petroleum Corp. Artesia, NM.	N6659D
☐ N24A	80	Gulfstream 840	11627	Bushell Inc. NYC.	N41VY
☐ N24AR	81	King Air 200	BB-867	Arvin Industries Inc. Columbus, In.	N200RB
☐ N24BL	88	King Air 300	FA-153	Bennett Lumber Products Inc. Princeton, Id.	N21CY
☐ N24CC	74	Rockwell 690A	11201	Indiana Aircraft Sales Inc. Indianapolis, In.	N50ST
☐ N24CV	96	King Air B200	BB-1524	Calico Ventures LLC. Valley Center, Ca.	N1024A
☐ N24DD	79	PA-31T Cheyenne II	7920074	Summit Helicopters Inc. Cloverdale, Va.	N204EF
☐ N24FT	91	King Air 350	FL-72	Bee Line Inc. Portland, Or.	N284M
☐ N24GT	75	Rockwell 690A	11254	Universal Pacific Investments Corp. Bend, Or.	XA-RMZ
☐ N24QT	84	Conquest 1	425-0181	Quik Trip Corp. Tulsa, Ok.	N68732
☐ N24SA	79	King Air 200	BB-465	Hadley Auto Transport Inc/W Coast Charter, Santa Ana, Ca.	N2065P
☐ N24SM	77	King Air E90	LW-239	Cajun Air Inc. Monroe, La.	N89FN
☐ N24SP	74	King Air C-12L	BB-3	Kentucky State Police, Frankfort, Ky.	71-21058
☐ N24SX	79	King Air 200	BB-455	Lamco LP. Jonesboro, Ar.	N1122M
☐ N24TL	81	King Air F90	LA-89	Kennedy Rice Dryers Inc. Angel Fire, NM.	N871RC
☐ N24WC	78	Mitsubishi MU-2N	705SA	Mobile Instrument Service & Repair Inc. Bellefontaine, Oh.	N865MA
☐ N25AP	79	King Air E90	LW-309	Air Moore Inc. Pinehurst, NC.	N20GM
☐ N25BD	72	Rockwell 690	11009	Indiana Aircraft Sales Inc. Indianapolis, In.	N9209N
☐ N25BE	70	681 Turbo Commander	6041	Midwestern Jet Sales & Service Inc. Oklahoma City, Ok.	N9101N
☐ N25BL	75	King Air 200	BB-27	Velasco Aviation Inc. Lake Jackson, Tx.	N62MR
☐ N25CS	82	King Air B200	BB-948	All America Termite & Pest Contol Inc. Orlando, Fl.	N150RH
☐ N25CU	91	King Air 350	FL-41	Coca-Cola Bottling Co. Birmingham, Al.	N8051Q
☐ N25DL	77	King Air C90	LJ-716	APIO Inc. Guadalupe, Ca.	N25KW
☐ N25GA	91	King Air C90A	LJ-1284	Hansgrohe GmbH. Donaueschingen, Germany.	OY-GEF
☐ N25GE	97	King Air 350	FL-187	Raytheon Aircraft Co. Wichita, Ks.	
☐ N25GE	90	King Air B200	BB-1373	New York State Electric & Gas Corp. Binghampton, NY.	N5608J
☐ N25GM	79	MU-2 Solitaire	412SA	Rod Aero Aircraft Brokerage, Fort Lauderdale, Fl.	N814HH
☐ N25HB	78	King Air C90	LJ-761	Hempt Bros Inc. Camp Hill, Pa.	N9064S

Regn	Yr	Type	c/n	Owner/Operator	Prev Regn
☐ N25KA	81	King Air 200	BB-783	Barrows LC. Vero Beach, Fl.	N3714P
☐ N25KB	90	King Air 350	FL-15	F Kenneth Bailey Jr. Beaumont, Tx.	N75NC
☐ N25KW	76	King Air 200	BB-114	JRW Aviation Inc. Dallas, Tx.	N2114L
☐ N25LA	81	PA-42 Cheyenne III	8001012	Legal Air inc. Wilmington, De.	F-GCQY
☐ N25MG	79	PA-31T Cheyenne II	7920059	Global Applied Technologies Inc. Fort Worth, Tx.	N63SC
☐ N25SM	78	Rockwell 690B	11477	B A Mullican Lumber & Manufacturing Co. Maryville, Tn.	(N25SM)
☐ N25TG	79	PA-31T King Air E90	LW-238	Custom Air Services Inc. Chattanooga, Tn.	N707DC
☐ N26AJ	60	Gulfstream 1	54	Southwest Jet Inc. Kansas City, Mo.	C-FMUR
☐ N26AP	80	MU-2 Marquise	763SA	Corporate Aviation Services Inc. Tulsa, Ok.	N95BE
☐ N26BE	78	King Air 200	BB-388	Manchester Plastics Ltd. Troy, Mi.	N388MC
☐ N26BJ	79	Conquest II	441-0083	Gift Shop Group Inc. St Louis, Mo.	XA-TBM
☐ N26DV	83	PA-31T Cheyenne 1A	8304003	Vogel Paint & Wax Co. Orange City, Ia.	VH-SJJ
☐ N26JP	85	King Air 300	FA-35	Dick Broadcasting Co. Knoxville, Tn.	
☐ N26PJ	80	PA-31T Cheyenne II	8020087	Phillips & Jordan Inc. Knoxville, Tn.	N331BB
☐ N26PK	80	Conquest II	441-0143	Rocky Mountain Helicopters Inc. Provo, Ut.	(N66TR)
☐ N26RF	81	Conquest II	441-0194	C J L Enterprises Inc. Wilmington, De.	(N311HC)
☐ N26SJ	80	King Air 200	BB-592	Flying Dollar Air Inc. Wisconsin Rapids, Wi.	XA-JIM
☐ N26SL	79	PA-31T Cheyenne II	7920091	Sierra Aviation Inc. Bishop, Ca.	N19GA
☐ N26UT	81	King Air B200	BB-901	GAR Inc. Wilmington, De.	
☐ N27BF	81	PA-31T Cheyenne 1	8104038	Hilliard Corp. Hardinge Inc & co-owners, Elmira, NY.	N130CC
☐ N27CV	83	King Air B200	BB-1161	Valley International Properties, Stuart, Fl.	N77CV
☐ N27LJ	82	King Air B200	BB-1040	Pearlstine Distributors Inc. Ridgeland, SC.	N27LS
☐ N27PA	85	King Air F90-1	LA-231	Edward Helfrick, Elysberg, Pa.	N7220T
☐ N27RA	85	Beech 1900C	UB-37	Department of the Air Force, Layton, Ut.	
☐ N27TJ	81	MU-2 Marquise	1511SA	Frank Woodworth Inc. Pittsfield, Me.	N418MA
☐ N27VE	85	Gulfstream 1000B	96202	Edelbrock Corp. Torrance, Ca.	N695P
☐ N27VG	79	Rockwell 690B	11553	B & R Aviation LLC. Santa Monica, Ca.	N27VE
☐ N27WH	80	King Air F90	LA-23	Rebel Drilling Inc. Longview, Tx.	N14TT
☐ N28AB	66	King Air A90	LJ-151	H Richard Hunter Aviation, Brunswick, Oh.	N22HS
☐ N28BG	77	PA-31T Cheyenne II	7720005	Airlift Inc. Pensacola, Fl.	N444ET
☐ N28CG	94	Dornier 328-100	3024	Corning Inc. Corning, NY.	N95CG
☐ N28DA	82	PA-42 Cheyenne III	8001078	John J Gee Wax & Co. West Grove, Pa.	N22HD
☐ N28DC	68	Mitsubishi MU-2D	106	Eastern New Mexico University, Roswell, NM.	N518TQ
☐ N28J	78	King Air 200	BB-348	Sierra Pacific Industries Inc. Redding, Ca.	
☐ N28KC	78	King Air 200	BB-356	W A A M Inc. Columbus, Ne.	N9KA
☐ N28KP	79	King Air C90	LJ-845	State of Montana, Helena, Mt.	N28KC
☐ N28MS	74	King Air E90	LW-100	Rocky Mountain Holdings LLC. Provo, Ut.	N31FN
☐ N28RY	79	King Air 200	BB-519	Robert Yates Racing Inc. Charlotte, NC.	N979SR
☐ N28TC	78	Rockwell 690B	11449	Task Force Tips Inc. Valparaiso, In.	D-ICSM
☐ N28TM	82	King Air C90-1	LJ-1029	Frank Adams, Springfield, La.	N6567C
☐ N29CA	78	PA-31T Cheyenne II	7820076	Columbia Air Services Inc. Groton, Ct.	YV-133CP
☐ N29DE	67	680V Turbo Commander	1699-76	Gene Forsthofel, Palm Coast, Fl.	N7029P
☐ N29EB	80	King Air C90	LJ-871	BC & W Air Inc. Charleston, WV.	N25AW
☐ N29EC	85	King Air 300	FA-66	EC Aviation Services Inc. Zeeland, Mi.	N17ME
☐ N29GA	83	Gulfstream 900	15025	Powell Group, Baton Rouge, La.	
☐ N29GB	84	King Air F90-1	LA-224	Bianchi & Sons Packing Co. Merced, Ca.	N3855K
☐ N29HF	80	King Air 200	BB-685	Sherwood McGuigan, Midland, Tx.	N29CH
☐ N29TB	79	King Air C90	LJ-846	FFW Holdings LLC. Fripp Island, SC.	OY-MBA
☐ N29TF	84	PA-42 Cheyenne IIIA	5501003	Barden & Robeson Corp & Welliver McGuire Inc. Homer, NY.	N111BX
☐ N30BG	71	681B Turbo Commander	6059	Pacific Northwest Aviation Inc. Troutdale, Or.	N333RK
☐ N30CN	95	King Air C90B	LJ-1415	Champion International Corp. Stamford, Ct.	
☐ N30DU	77	PA-31T Cheyenne II	7720054	Air Saxton Leasing Inc. Waco, Tx.	(N98AS)
☐ N30EH	67	King Air A90	LJ-296	Ceasar JHA. Littleton, Co.	XB-FJF
☐ N30EM	82	King Air B200	BB-958	Oquossoc Corp. Oquossoc, Me.	N82AJ
☐ N30FJ	84	Conquest II	441-0325	Jay O'Call, Salem, Or.	N1209S
☐ N30GC	73	King Air A100	B-177	Federated Department Stores Inc. Miami, Fl.	(N31AS)
☐ N30GK	80	King Air C90	LJ-923	A N Rusche Distributing Co. Houston, Tx.	XA-REH
☐ N30GT	80	King Air F90	LA-37	Bi-Lo Inc. Mauldin, SC.	
☐ N30HF	71	King Air C90	LJ-506	Clifford & Irit Langness, Page, Az.	N63TH
☐ N30HS	69	SA-26AT Merlin 2B	T26-156	Dubois Aircraft LLC. Huntingburg, In.	N112A
☐ N30KC	77	King Air E90	LW-241	Kingsford Manufacturing Co. Louisville, Ky.	N17844
☐ N30MA	82	PA-42 Cheyenne III	8001101	Corporate Skyways Inc. Omaha, Ne.	F-GEHR
☐ N30MC	89	King Air 300	FA-199	Madden Contracting Co. Minden, La.	N42GA
☐ N30NH	80	King Air F90	LA-48	SQH Air Corp. Menlo Park, Ca.	C-FOMH
☐ N30PH	80	King Air 200	BB-635	Petroleum Helicopters Inc. Lafayette, La.	N30PM
☐ N30RR	73	Mitsubishi MU-2K	263	Joe Minard Construction Inc. Waterloo, Ia.	N10VU
☐ N30SE	78	King Air 200	BB-313	Terminal Aviation Inc. Wilmington, De.	N23313
☐ N30TF	69	SA-26AT Merlin 2B	T26-162	PWM Inc/Airspect Inc. Akron, Oh.	C-FTEL
☐ N30VP	97	King Air B200	BB-1560	Valley Proteins Inc. Winchester, Va.	N1103B
☐ N30XL	82	Cheyenne II-XL	8166040	Dalton Foundries Inc. Warsaw, In.	C-FTRX
☐ N30XY	78	King Air 200	BB-305	Occidental International Exploration, Bakersfield, Ca.	N50ZY
☐ N31A	78	King Air E90	LW-281	Stolle Corp. Sidney, Oh.	N502SC
☐ N31AD	81	Conquest 1	425-0049	Allaire Dupont, Chesapeake, Md.	N67741
☐ N31AT	77	SA-226AT Merlin 4A	AT-057	White Industries Inc. Bates City, Mo. (status ?).	N31264
☐ N31BR		CASA 212-200	294	1st Source Leasing, South Bend, In.	
☐ N31CR	80	Conquest II	441-0145	Quest Air Charters Inc. San Marcos, Ca.	(N711YK)

Regn	Yr	Type	c/n	Owner/Operator	Prev Regn
☐ N31FM	81	King Air 200	BB-869	HRO Inc. Danville, Ca.	N2135J
☐ N31GA	84	PA-42 Cheyenne IIIA	5501009	Four Star Inc. Medford, Wi.	N59KG
☐ N31GM	90	King Air C90A	LJ-1254	Garan Manufacturing Corp. Starkville, Ms.	N55486
☐ N31HL	81	PA-31T Cheyenne II	8120060	Hotel Lima Inc. New Orleans, La.	N816JA
☐ N31KF	85	PA-42 Cheyenne IIIA	5501026	Keating Fibre Inc. Fort Washington, Pa.	N300JQ
☐ N31MB	80	PA-31T Cheyenne 1	8004013	Gary Eelman, Pipersville, Pa.	
☐ N31SN	68	King Air B90	LJ-362	Scope Leasing Inc. Columbus, Oh.	N31SV
☐ N31SV	95	King Air B200	BB-1514	Sioux Valley Hospital Association, Sioux Falls, SD.	N214SE
☐ N31WB	72	King Air C90	LJ-536	Jim Kaseman Ministries Inc. Tulsa, Ok.	N21WF
☐ N31WE	79	PA-31T Cheyenne II	7920060	Worldwide Equipment Inc. Prestonburg, Ky.	N86MP
☐ N31WJ	80	King Air 200	BB-637	Spartan Chemical Co Inc. Toledo, Oh.	N961PS
☐ N31WM	81	PA-31T Cheyenne 1	8104020	Rayco Industries Inc. Wooster, Oh.	N2433X
☐ N31WP	74	King Air E90	LW-99	Hamman Aviation, Houston, Tx.	N31JJ
☐ N31XL	81	Cheyenne II-XL	8166003	Seven Wheels Inc. Morristown, Tn.	
☐ N32BA	69	King Air B90	LJ-475	CSPI-Computer Service Professionsls Inc. Jefferson City, Mo.	N500RK
☐ N32BG	82	King Air F90	LA-168	Kola Aviation Inc. Hazleton, Pa.	N6128P
☐ N32BW	79	Rockwell 690B	11545	B & B Transportation Inc. Paterson, NJ.	N25LS
☐ N32CK	73	Mitsubishi MU-2J	588	Kalitta Flying Service Inc. Morristown, Tn.	N292MA
☐ N32CM	80	King Air C90	LJ-881	Red Spot Paint & Varnish Co. Evansville, In.	N32SJ
☐ N32DF	66	680V Turbo Commander	1624-53	John Rutkosky, Flagler Beach, Fl. (status ?).	N1010M
☐ N32EC	77	Mitsubishi MU-2N	699SA	River City Aviation/McNeely Charter Service Inc. Memphis, Tn	N859MA
☐ N32FH	65	King Air 90	LJ-74	High Plains Ltd. Portland, Or.	XA-RMH
☐ N32HF	82	King Air F90	LA-171	Fuller Renting & Leasing Inc. Auburn, Al.	PT-LMC
☐ N32JP	80	PA-31T Cheyenne 1	8004018	Price Land Co. Monticello, Ar.	N4DF
☐ N32KC	77	King Air 200	BB-255	Aircraft Holdings Inc. Chattanooga, Tn.	N820DY
☐ N32KW	78	PA-31T Cheyenne II	7820090	John Brantigan, Santa Fe, NM.	D-IJPG
☐ N32LJ	82	King Air B200	BB-993	Jones Mechanical Inc. Salisbury, NC.	N32TJ
☐ N32NS	77	King Air E90	LW-209	Carpat LLC. Englewood, Co.	N700DC
☐ N32SJ	82	SA-227TT Merlin 3C	TT-453	Xpress Air Inc. Chattanooga, Tn.	N453SA
☐ N32SV	81	King Air 200	BB-865	Sioux Valley Hospital Assoc. Sioux Falls, SD.	N531BB
☐ N32WS	76	Rockwell 690A	11339	R & S Aviation LLC. Neenah, Wi.	ZS-JRL
☐ N33BK	91	King Air B200	BB-1403	BK Associates, Atlanta, Ga.	N8094Q
☐ N33DE	83	Conquest II	441-0308	State of North Carolina, Raleigh, NC.	N87494
☐ N33EW	81	MU-2 Marquise	1519SA	Florida Express Corp. Wilmington, De.	N331W
☐ N33FM	79	King Air 200	BB-536	Aircraft Inc. Southfield, Mi.	
☐ N33JA	95	Pilatus PC-XII	127	Jet Capital Inc. Vail, Co.	HB-FQS
☐ N33KM	79	King Air 200	BB-432	S & E Aviation LLC. Nashville, Tn.	N22NP
☐ N33LA	78	PA-31T Cheyenne II	7820004	NPP Inc. Valdosta, Ga.	N82197
☐ N33MS	81	PA-31T Cheyenne II	8120036	Maurice Shacket, West Bloomfield, Mi.	N25WA
☐ N33SB	67	King Air A90	LJ-252	David Harden, Washington, Pa.	N728K
☐ N33TG	79	King Air 200	BB-461	Glasgow Inc. Glenside, Pa.	N116PA
☐ N33WG	72	Rockwell 690	11038	Big Eye Helicopters, Agana, Guam.	(N38PR)
☐ N34AL	80	MU-2 Marquise	792SA	Corporate Aviation Services Inc. Tulsa, Ok.	N66LA
☐ N34BS	77	King Air E90	LW-242	Weaver Aero International Inc. Hesston, Ks.	N34WW
☐ N34CE	81	King Air C90	LJ-932	Collier Enterprises, Naples, Fl.	N37132
☐ N34GA	82	Gulfstream 1000	96057	Franklin Realty Group Inc. Blue Bell, Pa.	(N9977S)
☐ N34HA	77	King Air A90	LJ-315	Daryl Collins, Charleston, SC.	XB-FJM
☐ N34HM	77	PA-31T Cheyenne II	7720003	Pace Aviation Ltd. Reno, Nv.	N28FM
☐ N34LT	92	King Air B200	BB-1437	Family Aviation Inc. Fort Wayne, In.	N8059Y
☐ N34MF	76	King Air E90	LW-163	OMCO Equipment LLC. Wilmington, De.	(N112CM)
☐ N34RT	76	Rockwell 690A	11315	Air Skape LLC. Jackson, Ms.	N34SC
☐ N34S	91	P-180 Avanti	1006	Ozark Management Inc. Jefferson City, Mo.	I-PJAS
☐ N34SM	77	SA-226T Merlin 3A	T-279	Go Durango LLC. Durango, Co.	N5396M
☐ N34UA	69	SA-26AT Merlin 2B	T26-145	James Heth, Lindale, Tx.	
☐ N35	75	King Air 200	BB-88	FAA R&D Flight Program, Atlantic City, NJ.	N4
☐ N35CA	81	Cheyenne II-XL	8166008	Venture Air Inc. Jenkintown, Pa.	C-GKRS
☐ N35CM	73	King Air E90	LW-51	Warren Barse, Denver, Co.	N7LR
☐ N35GR	85	King Air B200	BB-1167	JNG Services Inc. Lexington, Ky.	N18CM
☐ N35RR	81	MU-2 Marquise	1525SA	Corporate Aviation Services Inc. Tulsa, Ok.	N442MA
☐ N35RT	78	PA-31T Cheyenne II	7820018	Townsend Engineering Co. Des Moines, Ia.	N353T
☐ N35SA	90	King Air B200	BB-1359	Raytheon Aircraft Credit Corp. Wichita, Ks.	ZS-MWB
☐ N35TK	81	Conquest 1	425-0088	Allied Resources Inc. Warwick, RI.	N425BW
☐ N35TV	73	King Air C90	LJ-572	Duncan Air Service Inc. Flint, Mi.	N46RF
☐ N36AG	83	Gulfstream 1000	96058	L C Leasing Co. Annapolis, Md.	HP-11GT
☐ N36AT	80	MU-2 Solitaire	426SA	Incast Inc. Gulf Breeze, Fl.	N181RS
☐ N36BA	74	Rockwell 690A	11167	GJF Enterprises Inc. Atlanta, Ga.	N85AB
☐ N36CP	76	King Air 200	BB-178	IFL-Industrial Freight Liquidators Group Inc. Waterford, Mi.	N355AF
☐ N36DD	69	King Air B90	LJ-464	Carbon Development Corp. Dallas, Tx.	N813K
☐ N36G	67	Mitsubishi MU-2B	035	Pearsons Enterprises Inc. Richmond, Va.	N601CT
☐ N36JF	78	Rockwell 690B	11478	Harkers Aircraft Corp. LeMars, La.	N81797
☐ N36JT	80	Gulfstream 980	95024	Aerographics Inc. Manassas, Va.	N66FP
☐ N36LC	81	SA-226T Merlin 3B	T-387	Liquid Charter Services Inc. San Diego, Ca.	N777JE
☐ N36PE	69	SA-26AT Merlin 2B	T26-164	Perry Equipment Corp. Mineral Wells, Tx.	N50MA
☐ N36R	90	King Air B200	BB-1367	J & B Nostrand LLC. Garden City, NY.	N367AJ
☐ N36TW	80	PA-31T Cheyenne II	8020075	L S S Properties LLC. Scottsdale, Az.	
☐ N36WH	77	King Air B100	BE-26	Western Industries Inc. Milwaukee, Wi.	N73JC
Regn	Yr	Type	c/n	Owner/Operator	Prev Regn

Regn	Yr	Type	c/n	Owner/Operator	Prev Regn
☐ N37AL	79	MU-2 Marquise	752SA	Corporate Aviation Services Inc. Tulsa, Ok.	N11WQ
☐ N37BW	73	Rockwell 690A	11129	Wrangler Aviation Corp. Norman, Ok.	(N500AL)
☐ N37DA	67	King Air A90	LJ-286	Red Carpet Helicopters Inc. New Smyrna Beach, Fl.	N935K
☐ N37GP	65	King Air 90	LJ-15	Anton Ptach, Poughkeepsie, NY.	N733KL
☐ N37HB	77	PA-31T Cheyenne II	7720020	R L Riemenschneider Enterprises Co. Redmond, Or.	N37RT
☐ N37HC	74	King Air E90	LW-108	Arthur Stevens, Fort Lauderdale, Fl.	N37X
☐ N37JT	85	Conquest 1	425-0201	Printer Inc. West Columbia, SC.	N777UP
☐ N37KW	89	PA-42 Cheyenne 400LS	5527039	Cumberland Management & Holding Inc. Dover, De.	C-FPQA
☐ N37NC	81	King Air 200	BB-781	National Commerce Bancorporation, Memphis, Tn.	N469JW
☐ N37PC	79	King Air B100	BE-66	Maple Point Aviation Inc. Columbus, Oh.	N6032E
☐ N37PJ	79	PA-31T Cheyenne 1	7904003	Rock Leasing, Las Vegas, Nv.	N800CM
☐ N37PS	81	Conquest 1	425-0050	Venture Trust, Henderson, Nv.	N425FB
☐ N37PT	77	King Air C90	LJ-731	Hayden Leasing LC. Scottsdale, Az.	N37PW
☐ N37RL	76	PA-31T Cheyenne II	7620049	Kansas Aircraft Corp. New Century, Ks.	N37TH
☐ N37RR	79	Rockwell 690B	11552	Direct Air LLC. Roswell, Ga.	N888SL
☐ N37SB	82	Gulfstream 840	11724	Stephen R Butter, Longview, Tx.	N5962K
☐ N37SR	82	Cheyenne II-XL	8166051	Seven Rivers Inc. Carlsbad, NM.	N162PM
☐ N37TW	81	PA-31T Cheyenne 1	8104006	Virga Corp/Virga Enviromental Corp. Loveland, Co.	(N816CM)
☐ N38AA	80	Conquest II	441-0163	Commercial Contracting Co. San Antonio, Tx.	(N38BR)
☐ N38AF	85	PA-42 Cheyenne 400LS	5527029	Airserve Inc. Pound Ridge, NY.	N689CA
☐ N38BE	71	Mitsubishi MU-2F	210	James Ewald, Daytona Beach, Fl.	N34DD
☐ N38CG	95	Dornier 328-100	3034	Corning Inc. Corning, NY.	D-CDXA
☐ N38GM	80	King Air 200	BB-737	Fairbanks Communications Inc. West Palm Beach, Fl.	F-GHLC
☐ N38GP	65	King Air 90	LJ-46	Anton Ptach, Poughkeepsie, NY.	C-GZIZ
☐ N38H	92	King Air 350	FL-71	HFL Leasing LP. Austin, Tx.	N56456
☐ N38P	77	King Air A100	B-231	Causey Aviation Service Inc. Liberty, NC.	N3EP
☐ N38RE	81	PA-31T Cheyenne 1	8104028	Javelin Air Inc. Blakely, Ga.	N241AC
☐ N38RP	79	King Air C90	LJ-722	The Card Place Inc. Birmingham, Al.	N35HP
☐ N38TA	81	SA-226T Merlin 3B	T-397	Armstrong International Inc. Stuart, Fl.	
☐ N38TW	81	PA-31T Cheyenne 1	8104008	Tango Whiskey inc. Wilmington, De.	
☐ N38V	91	King Air B200	BB-1412	El Paso Natural Gas Co. El Paso, Tx.	N8241J
☐ N38VT	80	PA-31T Cheyenne II	8020022	MKM Investment Co. Wilmington, De.	N38V
☐ N38WA	81	Cheyenne II-XL	8166006	Mini Max Warehouses Inc. Florence, SC.	
☐ N39A	83	Conquest 1	425-0138	Aircraft Owners & Pilots Association, Frederick, Md.	N6884Q
☐ N39AS	68	680W Turbo Commander	1721-1	Cx USA 2/97 to ?. (was 1721-92).	N954HE
☐ N39DT	92	King Air 350	FL-86	Raytheon Aircraft Co. Wichita, Ks.	M35DT
☐ N39FB	88	King Air C90A	LJ-1165	ICI Seeds/Farm Bureau Life Insurance Co. Des Moines, Ia.	N3081K
☐ N39MS	75	SA-226T Merlin 3A	T-257	Khan Aviation Inc. Elkhart, In.	N45BB
☐ N39PH	79	King Air 200C	BL-3	Petroleum Helicopters Inc. Lafayette, La.	N141GS
☐ N39TL	81	PA-31T Cheyenne 1	8104012	Panhandle Telecommunications & Systems, Guymon, Ok.	N39TW
☐ N39TU	91	2000 Starship	NC-23	Omni Solutions International Inc. Chesterfield, Mo.	N39TW
☐ N40AM	80	MU-2 Solitaire	427SA	Marvy Finger, Houston, Tx.	N166MA
☐ N40BA	69	King Air B90	LJ-444	San Juan Jet Charter Inc. Hato Rey, PR.	N5111C
☐ N40BC	84	King Air B200	BB-1169	MOCI Leasing Inc. Dover, De.	N90LP
☐ N40BR	80	King Air F90	LA-27	Helm Cost Corp & Sinmast of Pennsylvania Inc. York, Pa.	N44SR
☐ N40BT	80	PA-31T Cheyenne 1	8004047	Roller Die & Forming Companies, Louisville, Ky.	N2557V
☐ N40CK	68	King Air B90	LJ-358	Kenneth Fischer, Belleville, Il.	N603H
☐ N40FJ	86	Conquest II	441-0359	Flying J Inc. Brigham City, Ut.	N359BA
☐ N40HE	76	King Air 200	BB-569	Executive Air Charter Inc. Baton Rouge, La.	N39K
☐ N40JJ	78	Mitsubishi MU-2P	383SA	W-K Inc. Leewood, Ks.	N10JJ
☐ N40KC	75	Mitsubishi MU-2M	322	Keller Companies Inc. Manchester, NH.	N20KC
☐ N40MH	90	King Air C90A	LJ-1258	Brunner Drilling & Manufacturing Inc. Elroy, Wi.	N80CK
☐ N40NB	85	King Air 300	FA-55	Davison Transport Inc. Ruston, La.	N40NE
☐ N40PP	70	Mitsubishi MU-2G	515	Cavenaugh Aviation Inc. Conroe, Tx. (status ?).	N155WC
☐ N40RA	76	King Air 200	BB-104	Captex Land Corp. Austin, Tx.	
☐ N40RD	81	Conquest 1	425-0092	Air Quest Inc. Columbia, SC.	N89GC
☐ N40SM	79	Rockwell 690B	11559	K M Aviation Inc. Wilmington, De.	
☐ N40TD	87	King Air B200	BB-1276	Teledyne Industries Inc. San Diego, Ca.	N738P
☐ N40TT	75	PA-31T Cheyenne II	7520023	Ocean Aircraft Ltd. Chesapeake, Va.	N314GA
☐ N40WG	78	Rockwell 690B	11459	Golden Giant Inc. Kenton, Oh.	N333KD
☐ N41AD	77	Mitsubishi MU-2P	352SA	Cobarex International Inc. Miami, Fl.	N41WB
☐ N41AK	82	King Air F90	LA-188	Greenway Leasing Co. Elm Grove, Wi.	N41CK
☐ N41BA	73	SA-226T Merlin 3	T-248	Susquehanna Management Inc. New Cumberland, Pa.	P2-BCL
☐ N41BE	79	King Air A100	B-245	Beech Transportation Inc. Eden Prairie, Mn.	N41BP
☐ N41CV	80	King Air 200	BB-611	Chevron USA Inc. Houston, Tx.	N41C
☐ N41DZ	68	King Air B90	LJ-412	Prestar Inc. Davis, Ca.	N8093W
☐ N41HH	75	King Air E90	LW-146	Morning Star Aviation Service Inc. Lexington, Ky.	N141DA
☐ N41LH	65	Gulfstream 1	156	Larry Hedrick Motorsports Inc. Statesville, NC.	N22AS
☐ N41NS	78	PA-31T Cheyenne II	7820013	Bell Aviation Inc. West Columbia, SC.	N47NS
☐ N41PN	80	PA-31T Cheyenne 1	8004012	G T Aviation Inc. Columbus, Ga.	D-IIPN
☐ N41WC	68	King Air B90	LJ-430	Beech Transportation Inc. Eden Prairie, Mn.	N551SS
☐ N42AF	81	MU-2 Marquise	1539SA	Epps Air Service Inc. Atlanta, Ga.	ZS-MRJ
☐ N42AJ	87	King Air 300	FA-121	Cracker Barrel Old Country Store Inc. Lebanon, Tn.	HB-GIO
☐ N42CQ	67	King Air A90	LJ-245	Oklahoma State Department Vo-Tech, Stillwater, Ok.	(N221ML)
☐ N42DE	80	MU-2 Solitaire	417SA	Davis Electrical Constructors, Greenville, SC.	N131SA
☐ N42DT	82	Conquest II	441-0234	River Bend Aviation Inc. Idaho Falls, Id.	N41CR

Regn	Yr	Type	c/n	Owner/Operator	Prev Regn

Regn	Yr	Type	c/n	Owner/Operator	Prev Regn
☐ N42EL	96	King Air 350	FL-149	Recreational Enterprises Inc. Reno, Nv.	N350KA
☐ N42LJ	79	King Air 200	BB-564	Palmetto Mills Inc. Fort Payne, Al.	ZS-KHK
☐ N42ND	81	Cheyenne II-XL	8166033	University of Notre Dame, Notre Dame, In.	N59WA
☐ N42PC	74	King Air E90	LW-85	Schroeder Aircraft Leasing Inc. Indianapolis, In.	N20RF
☐ N42SC	96	King Air B200	BB-1550	CLC Leasing Inc. Dallas, Tx.	N1079Y
☐ N42SL	82	Gulfstream 900	15013	Seiple Lithograpg Co. N Canton, Oh.	N900AB
☐ N42SY	81	King Air F90	LA-131	Godwin Aircraft Inc. Memphis, Tn.	N42SC
☐ N42TD	82	King Air B200	BB-968	Shoney's Inc. Nashville, Tn.	N422TD
☐ N42TF	67	680V Turbo Commander	1687-67	Professional Aviation Sales, Millington, Tn.	N2TF
☐ N43CB	74	SA-226T Merlin 3	T-246	Bahamas Red Bird Inc. North Miami, Fl.	N43FG
☐ N43DC	83	MU-2 Marquise	1501SA	First Security Bank NA. Salt Lake City, Ut.	N405MA
☐ N43FC	77	King Air A100	B-237	Palmer Aviation Leasing LLC. Germantown, Tn.	
☐ N43MB	69	King Air B90	LJ-463	Way Point Inc. Wilmington, De.	(N300A)
☐ N43PC	78	King Air E90	LW-253	Prototype Equipment Corp. Lake Forest, Il.	N709DB
☐ N43PE	80	King Air 200	BB-585	Richard Petty Transportation LLC. Randleman, NC.	N438P
☐ N43SQ	75	PA-31T Cheyenne II	7520029	Ivey Mechanical Co. Kosciusko, Ms.	EC-...
☐ N43TA	92	King Air B200	BB-1432	Tiger Athletic Foundation, Baton Rouge, La.	N8037J
☐ N43TL	97	King Air B200	BB-1582	T-L Irrigation Co. Hastings, Ne.	N1130R
☐ N43TT	67	King Air A90	LJ-210	B S Aircraft Sales Inc. Norman, Ok.	N11UC
☐ N43WA	70	King Air B90	LJ-501	Western Aircraft Inc. Boise, Id.	N865MA
☐ N43WH	81	PA-31T Cheyenne II	8120068	WMJ Corp. Colorado Springs, Co.	D-IJOE
☐ N43WS	72	King Air E90	LW-16	Colemill Enterprises Inc. Nashville, Tn.	N80NC
☐ N44AX	80	MU-2 Solitaire	416SA	Lightning Blt Inc. South Kearny, NJ.	N44AB
☐ N44DT	78	PA-31T Cheyenne II	7820043	Sassafras Aviation LLC. Knoxville, Tn.	(N82267)
☐ N44EC	72	King Air E90	LW-4	Builders Gypsum Supply Co. Houston, Tx.	N112LS
☐ N44GK	78	King Air E90	LW-298	W W Aircraft Inc. Boise, Id.	N2029X
☐ N44GL		SA-226AT Merlin 4A	AT-032	W Lord, Lineville, Ia.	N44PB
☐ N44HP	77	King Air C90	LJ-702	Professional Office Services Inc. Waterloo, Ia.	
☐ N44HT	79	PA-31T Cheyenne II	7920061	MRL Industries Inc. Sonora, Ca.	N117FN
☐ N44KA	90	King Air 350	FL-19	International Union of Oper. Van Nuys, Ca.	N5532T
☐ N44KS	86	SAAB 340A	050	Goodyear Tire & Rubber Co. Akron, Oh.	N340SA
☐ N44KT	76	King Air 200	BB-154	Panther Partners, Wilmington, De.	N44KA
☐ N44KU	74	Mitsubishi MU-2J	647	Bankair Inc. West Columbia, SC.	(N110SS)
☐ N44MM	81	MU-2 Marquise	1526SA	Mike Moser Inc. Wilmington, De.	N443MA
☐ N44MR	81	King Air 200C	BL-27	Delta Life Insurance Co. Thomasville, Ga.	N3827Z
☐ N44MV	74	King Air E90	LW-107	Healdsburg Aviation Inc. Van Nuys, Ca.	N96TB
☐ N44NL	94	King Air B200	BB-1492	Pronto Enterprises Trust, Pittsburgh, Pa.	
☐ N44QM	82	King Air C90-1	LJ-1036	GLF Aviation Inc. Santa Barbara, Ca.	(N59WP)
☐ N44RG	68	King Air B90	LJ-417	Khan Aviation Inc. Elkhart, In.	N29TC
☐ N44SD	85	Gulfstream 1000	96096	Cannon Express Corp. Springdale, Ar.	N44SF
☐ N44SR	81	King Air B200	BB-853	Eastway Aviation Inc. Ronkonkoma, NY.	N3832B
☐ N44UF	75	King Air 200	BB-36	Maxair Inc. Appleton, Wi.	N816RB
☐ N44US	75	King Air 200	BB-56	Bill Elliott Aviation Inc. Dawsonville, Ga.	N55BN
☐ N44VP	80	King Air F90	LA-60	Vernon Peppard, Plano, Tx.	N233PT
☐ N44VR	80	MU-2 Solitaire	432SA	Samaritans Purse Inc. Boone, NC.	N15WF
☐ N44WV	96	King Air B200	BB-1554	WIV Air LLC. Calabasas, Ca.	N464A
☐ N45AR	92	Beech 1900D	UE-12	Range Systems Engineering Co. West Palm Beach, Fl.	N138MA
☐ N45AZ	77	Rockwell 690B	11383	State of Arizona, Phoenix, Az.	YV-149CP
☐ N45BS	72	Mitsubishi MU-2J	558	Internet Jet Sales Ltd. White Mills, Pa.	N43SR
☐ N45CE	74	Mitsubishi MU-2J	653	Precision Service M T R Inc. Addison, Il.	N350HE
☐ N45CF	80	King Air 200	BB-736	Gray Aircraft Accessories Inc. Louisville, Ky.	N577L
☐ N45MW	82	SA-227AT Merlin 4C	AT-452	Wilder Aviation Sales & Leasing, Clearwater, Fl.	N3010Q
☐ N45PE	79	King Air C90	LJ-830	Declan & Adrian Ryan, Anchorage, Ak.	N45PK
☐ N45PM	92	TBM 700	62	Time Tool Inc/Oakley Inc. Irvine, Ca.	
☐ N45PR	66	King Air A90	LJ-145	Lansing Community College, Lansing, Mi.	N89991
☐ N45QC	66	680T Turbo Commander	1542-7	Waggoner Aircraft Inc. Bethany, Ok. (status ?).	N45Q
☐ N45RL	72	King Air C90	LJ-565	Air Sal Leasing Inc. Miami, Fl.	N711MP
☐ N45ST	77	Rockwell 690B	11369	Turbine Group Inc. Melbourne, Fl.	N81550
☐ N45TP	83	Conquest II	441-0306	Construction Insurance Services Inc. Hilliard, Fl.	N886BC
☐ N45TT	67	King Air A90	LJ-312	A D Buffington, Gulf Shores, Al.	N675SB
☐ N45VT	73	Rockwell 690A	11134	Kolar Aviation Inc. Hudson, Oh.	N7RB
☐ N46AK	79	MU-2 Marquise	754SA	EPPS Air Service Inc. Atlanta, Ga.	XB-FBY
☐ N46AR	92	Beech 1900D	UE-27	Range Systems Engineering Co. West Palm Beach, Fl.	
☐ N46AX	67	King Air A90	LJ-317	Snip N Clip Haircut Shops Inc. Kansas City, Mo.	N46A
☐ N46AZ	76	Rockwell 690A	11307	State of Arizona, Phoenix, Az.	N48AZ
☐ N46BE	75	King Air A100	B-214	Beech Transportation Inc. Eden Prairie, Mn.	HZ-AFC
☐ N46BR	81	King Air 200	BB-852	South Coast Lumber Co. Brookings, Or.	
☐ N46DT	65	King Air 90	LJ-7	White Industries Inc. Bates City, Mo.	C-GJBK
☐ N46FL	95	King Air 350	FL-129	Texas Estrellas Inc. Tyler, Tx.	
☐ N46HA	77	Rockwell 690B	11381	William Greene, Elizabethton, Tn.	N81593
☐ N46KC	80	SA-226T Merlin 3B	T-332	John L Cox, Midland, Tx.	(N46KM)
☐ N46NA	70	SA-226T Merlin 3	T-206	Avtrend Inc. Miami, Fl.	N11RM
☐ N46RP	76	King Air E90	LW-193	McShares Inc. Salina, Ks.	N9057S
☐ N46SA	73	SA-226T Merlin 3	T-231	Corporate Aviation LLC. Lafayette, La.	N20QN
☐ N47AW	75	King Air E90	LW-130	Southern Securities Ltd. Greensboro, NC.	N75KC
☐ N47CA	79	PA-31T Cheyenne II	7920043	Power Rents, Tigard, Or.	

Regn   *Yr*   *Type*   *c/n*   *Owner/Operator*   *Prev Regn*

Regn	Yr	Type	c/n	Owner/Operator	Prev Regn
☐ N47CF	80	King Air 200	BB-640	Intermet Corp. Troy, Mi.	N581
☐ N47CK	71	681B Turbo Commander	6058	North Star Air Cargo Inc. Milwaukee, Wi.	N4ME
☐ N47HM	67	680V Turbo Commander	1691-70	Cochise Community College, Douglas, Az.	N4585E
☐ N47JF	80	PA-31T Cheyenne II	8020031	Minnetrista Corp. Muncie, In.	N102AR
☐ N47KS	79	King Air B100	BE-59	L J Associates Inc. Latrobe, Pa.	N777DQ
☐ N47LC	73	King Air E90	LW-64	Hoppman Corp. Chantilly, Va.	N213DS
☐ N47TT	80	Gulfstream 840	11600	Sterling Properties Inc. Sioux City, Ia.	N840RC
☐ N48A	68	King Air B90	LJ-381	K & K Aircraft Inc. Bridgewater, Va.	N91DT
☐ N48BA	81	Gulfstream 840	11665	Semitool Inc. Kalispell, Mt.	D-IWKW
☐ N48BS	80	Conquest II	441-0125	The Brothers Signal Co. Leesburg, Va.	N375K
☐ N48CR	80	King Air 200	BB-599	West Texas Gas Inc. Midland, Tx.	
☐ N48CS	86	King Air B200	BB-1247	General Equipment & Manufacturing Co. Louisville, Ky.	N392DF
☐ N48HH		SA-226T Merlin 3B	T-304	Charter Air Service Corp. Topeka, Ks.	F-GBOP
☐ N48N	82	King Air B200	BB-969	Airtex Division-Spartan Mills Inc. Greer, SC.	N188W
☐ N48NP	81	MU-2 Solitaire	447SA	Niles Chemical Paint Co. Niles, Mi.	(N348CP)
☐ N48PA	59	Gulfstream 1	18	Pal-Waukee Aviation Inc. Wheeling, Il.	N3UP
☐ N48Q	77	King Air 200	BB-263	Helicopters Inc. Cahokia, Il.	F-GCTP
☐ N48RA	78	PA-42 Cheyenne III	7801004	David Whitt, Texarkana, Tx.	XA-RLW
☐ N48TA	78	King Air E90	LW-283	Tidewater Aero Associates Inc. Norfolk, Va.	N845MC
☐ N48W	78	King Air E90	LW-254	Platina Investment Corp. Oklahoma City, Ok.	N22654
☐ N48XP	66	King Air 90	LJ-109	K & K Aircraft Inc. Bridgewater, Va.	N48N
☐ N49AC	80	PA-31T Cheyenne 1	8004021	James Cook III, Panama City, Fl.	N2317V
☐ N49B	81	PA-31T Cheyenne II	8120055	Acme Rocket Sleds Inc. Burbank, Ca.	N19BG
☐ N49CH	79	King Air F90	LA-2	Telford Aviation/Alternative Energy Inc. Bangor, Me.	N58AB
☐ N49E	78	King Air B100	BE-47	Mermaid Corp. Orange Beach, Al.	N4996M
☐ N49H	80	SA-226T Merlin 3B	T-310	Howell Instruments Inc. Fort Worth, Tx.	N400FF
☐ N49JG	81	King Air B200	BB-884	National Floor Products Co. Florence, Al.	(D-IKFB)
☐ N49K	94	King Air A90	LM-11	K & K Aircraft Inc. Bridgewater, Va.	N611ND
☐ N49KC	78	King Air 200	BB-318	Amsair Corp. Superior, Wi.	
☐ N49LL	92	King Air C90B	LJ-1316	Raytheon Aircraft Co. Wichita, Ks.	N8114P
☐ N49LM	96	Pilatus PC-XII	163	M L Morse LLC.	HB-FRY
☐ N49RM	79	PA-31T Cheyenne 1	7904051	Boldt Inc. Lubbock, Tx.	N779SW
☐ N49SS	78	King Air B100	BE-37	Furrow Services LLC. Knoxville, Tn.	N111XP
☐ N50AW	81	King Air F90	LA-142	Amwest Savings Association, Bryan, Tx.	N1815T
☐ N50DR	82	King Air B200	BB-1061	RWB Enterprises Inc/Raden Corp. Virginia Beach, Va.	N188H
☐ N50DX	74	Rockwell 690A	11227	Christopher Zupsic, Portland, Or.	N131JN
☐ N50EB	75	King Air E90	LW-128	The Durham Co. Lebanon, Mo.	N1CB
☐ N50JJ	67	King Air A90	LJ-290	Skydive Factory Inc. East Rochester, NH.	N50JP
☐ N50JN	85	Conquest 1	425-0232	G & W Air Inc. Middletown, De.	N50JG
☐ N50K	74	Mitsubishi MU-2K	305	Mid America Sprayers Inc. Johnson, Ks.	N465MA
☐ N50KK	65	King Air 90	LJ-55	Elite Air Inc. Cartersville, Ga.	N595MG
☐ N50KV	68	SA-26AT Merlin 2B	T26-115	Southland Leasing Co. Inman, SC.	N1223S
☐ N50KW	80	MU-2 Marquise	784SA	Bankair Inc. West Columbia, SC. (status ?).	N785MA
☐ N50MT	79	SA-226T Merlin 3B	T-308	Fly Away Inc. Atlanta, Ga.	N900TX
☐ N50RP	67	King Air A90	LJ-200	Big Chief Aviation Inc. Goshen, In.	N243D
☐ N50RV	80	King Air E90	LW-338	D W Davies Co. Racine, Wi.	N3682E
☐ N50VP	88	King Air C90A	LJ-1185	Southland Promotions Inc. Statesville, NC.	N44GP
☐ N50WF	81	Gulfstream 840	11701	Waupaca Foundry Inc. Waupaca, Wi.	N5953K
☐ N50YR	82	King Air B200	BB-1059	American General Finance Corp. Evansville, In.	
☐ N51CU	83	Conquest 1	425-0171	Mercy Healthcare North, Redding, Ca.	D-IBBP
☐ N51DM	82	Gulfstream 1000	96014	Donald & E L Mashburn, Oklahoma City, Ok.	N9934S
☐ N51GS	83	Cheyenne II-XL	8166076	General Shale Products Corp. Johnson City, Tn.	
☐ N51RD	78	PA-31T Cheyenne II	7820056	Weston Air Service Inc. Weston, Ma.	N4NH
☐ N51SG	72	King Air E90	LW-7	D-W Corp. Des Moines, Ia.	N107B
☐ N51WF	81	Gulfstream 840	11684	Waupaca Foundry Inc. Waupaca, Wi.	N5936K
☐ N52C	85	King Air 300	FA-40	Aristokraft Inc. Jasper, In.	N1MC
☐ N52EL	67	King Air A90	LJ-204	Allied Signal Corp. Fort Lauderdale, Fl.	N165U
☐ N52GP	81	King Air 200	BB-766	Global Industries Ltd. Lafayette, La.	N510H
☐ N52GT	78	KIng Air 200	BB-377	Hall & House Lumber Inc. Indianapolis, In.	N41GT
☐ N52KA	73	King Air E90	LW-42	Kenneth Rogers, Oklahoma City, Ok.	
☐ N52MW	90	King Air 300	FA-212	Warn Industries Inc. Milwaukee, Or.	N8PL
☐ N52PC	79	PA-31T Chetenne 1	7904023	Pasquinelli Produce Co & Glen Curtis Inc. Yuma, Az.	N23319
☐ N52PY	74	Rockwell 690A	11196	Phoenix Air Group Inc. Cartersville, Ga.	N52PB
☐ N52SF	75	King Air 200	BB-106	SAK Aviation Inc. Pittsburgh, Pa.	N383AS
☐ N52TT	79	PA-31T Cheyenne II	7920057	Minuteman Aviation Inc. Missoula, Mt.	N102E
☐ N52WA		PA-42 Cheyenne III	8001028	Haps Aerial Enterprises Inc. Sellersburg, In.	
☐ N53AM	79	PA-31T Cheyenne II	7920037	JMK Holdings Inc + others, Fort Worth, Tx.	N610MW
☐ N53BB	77	King Air E90	LW-224	Palm Aire Inc. St Thomas, USVI.	N21EH
☐ N53CE	76	King Air E90	LW-160	Engineering & Manufacturing Services Inc. Huntsville, Al.	N251SR
☐ N53CK	78	King Air 200	BB-329	Causey Aviation Service & Koury Aviation Inc. Liberty, NC.	N311GA
☐ N53DA	82	SA-227TT Merlin 3C	TT-438	BigHawk Corp. Wilmington, De.	(N312ST)
☐ N53G	82	King Air F90	LA-178	Liberty Self-Stor Ltd. Mentor, Oh.	N82HR
☐ N53GG	77	Conquest II	441-0022	G & G Trucking Inc. Franklin, Wi.	(N441GG)
☐ N53LB		SAAB 340A	022	Mellon Bank NA. Pittsburgh, Pa.	N19M
☐ N53MD	71	King Air 100	B-86	Bado Equipment Inc. Houston, Tx.	N500Y
☐ N53RF	74	Rockwell 690A	11153	U S Dept of Commerce, MacDill AFB. Fl.	N57074
Regn	*Yr*	*Type*	*c/n*	*Owner/Operator*	*Prev Regn*

Regn	Yr	Type	c/n	Owner/Operator	Prev Regn
☐ N53RT	81	King Air 200	BB-808	Robert Tobin, St Joseph, Mo.	N808EB
☐ N53TA	78	PA-31T Cheyenne II	7820083	Indiana Paging Network Inc. Michigan City, In.	N53TM
☐ N53TD	78	King Air B100	BE-53	Corporate Flight Inc. Detroit, Mi.	N2037C
☐ N53TJ	89	King Air C90A	LJ-1209	K K S Aviation Inc. Modesto, Ca.	N3030C
☐ N53TM	82	King Air B200	BB-973	E T Meredith III, Des Moines, Ia.	N6104A
☐ N53WA	81	Cheyenne II-XL	8166025	Aviasur del Caribe,	
☐ N54CF	68	King Air B90	LJ-374	Daniels Aircraft International Inc. Chestnut Hill, NJ.	N32SV
☐ N54CK	79	King Air B100	BE-73	Diamond Air Inc. Richmond Heights, Oh.	N730BR
☐ N54EZ	79	PA-31T Cheyenne 1	7904031	Jenco Inc & Dick Broadcasting Co. Greensboro, NC.	(N673BB)
☐ N54FB	85	King Air B200	BB-1212	Farm Bureau Life Insurance Co. Des Moines, Ia.	N7221Y
☐ N54GA	83	Gulfstream 1000	96066	Wal-Mart Stores Inc. Rogers, Ar.	
☐ N54GP	75	SA-226AT Merlin 4A	AT-034	Double EE Ranch Inc. Wilmington, De.	N717CC
☐ N54JW	70	King Air 100	B-60	Dodson Aviation Inc. Ottawa, Ks.	N2FA
☐ N54UM	80	Gulfstream 980	95025	P & P Enterprises Inc. Hollywood, Fl. (stolen 2/88 ?).	N9779S
☐ N54US	73	Mitsubishi MU-2J	590	Great Planes Sales Inc. Tulsa, Ok.	N550K
☐ N54WW	67	King Air A90	LJ-209	Concrete Structures of the Midwest, Chicago, Il.	N90BE
☐ N55AC	81	Conquest 1	425-0020	Rice Industries Inc. Wichita, Ks.	N200MT
☐ N55BK	96	Pilatus PC-XII	160	Sky Fun 1 Inc. Longmont, Co.	N160PC
☐ N55CE		SA-226AT Merlin 4	AT-004	U S Army Corps of Engineers, Omaha, Ne.	
☐ N55FG	76	King Air 200	BB-173	Flo-Con Systems Inc/Vesuvius USA, Champaign, Il.	N55AE
☐ N55FR	82	King Air B100	BE-123	Richard Morrison, Quanah, Tx.	N1838H
☐ N55FW	67	King Air A90	LJ-194	Far West Capital/Rey Del Aire Inc. Salt Lake City, Ut.	N1952L
☐ N55HC	75	King Air E90	LW-134	Tucson AeroService Center Corp. Marana, Az.	N663LS
☐ N55JS	74	Rockwell 690A	11195	Phase II Consulting/John & Calli Short, Salt Lake City, Ut.	N690DD
☐ N55KW	81	PA-31T Cheyenne II	8120007	C Keith West, Rock Springs, Wy.	N2369X
☐ N55LH	67	King Air A90	LJ-266	Hahn's Aircraft Co. Wichita, Ks.	N55GM
☐ N55MG	68	King Air B90	LJ-391	Northern Airways Inc/D & H Corp. South Burlington, Vt.	N22JJ
☐ N55MN	85	King Air C90	LJ-974	Minnesota Dept of Transport, St Paul, Mn.	N555RA
☐ N55MP	93	2000A Starship	NC-30	Maxon Corp. Muncie, In.	N8114Q
☐ N55MS	78	Conquest II	441-0061	Hangar One Inc. Spokane, Wa.	(N441DK)
☐ N55PC	84	King Air B200	BB-1170	Pizzagalli Construction Co. South Burlington, Vt.	N6930P
☐ N55TY	94	2000A Starship	NC-51	Tyco International Ltd. Exeter, NH.	N8286Q
☐ N55WJ	77	Rockwell 690B	11427	Ruidoso Connection Inc. Alto, NM.	N46802
☐ N55ZP	79	SA-226T Merlin 3B	T-299	Tom-Bob Inc. Pilot Mountain, NC.	N81QH
☐ N56AY	81	King Air B200	BB-882	A Y McDonald Industries Inc. Dubuque, Ia.	N9SB
☐ N56CC	77	King Air 200	BB-189	Rockwell-Ditzler Associates Inc. Pittsburgh, Pa.	N392K
☐ N56DA	79	King Air 200	BB-486	Belk Simpson Co. Greenville, SC.	N56RA
☐ N56DK	81	Cheyenne II-XL	8166039	Deka R & D Corp. Manchester, NH.	N65TJ
☐ N56DL	77	King Air C90	LJ-732	C & K Market Inc. Brookings, Or.	N473LP
☐ N56FL	79	King Air B100	BE-74	Flight Operations Leasing Inc. West Lake, Oh.	N488A
☐ N56HT	77	King Air E90	LW-215	Mobile Bay Air LLC. Mobile, Al.	I-MCCC
☐ N56KA	81	King Air 200	BB-763	S & S of Bowling Green Inc. Bowling Green, Ky.	N50PM
☐ N56MS	78	Conquest II	441-0007	Felts Field Aviation Inc. Spokane, Wa.	N441AK
☐ N56RT	81	King Air 200	BB-817	John L Rust Co. Albuquerque, NM.	
☐ N56WF	91	TBM 700	8	Skytech Inc. Baltimore, Md.	JA8892
☐ N57AF	81	Cheyenne II-XL	8166073	Alphaero Corp. Fort Lauderdale, Fl.	HK-3117W
☐ N57AG	68	King Air B90	LJ-343	State of Texas, Austin, Tx.	N880X
☐ N57EC	74	Rockwell 690A	11178	Eagle Construction & Enviromental, Eastland, Tx.	N124SB
☐ N57EM	67	King Air A90	LJ-295	Richard Harvey, Tyler, Tx.	N57FM
☐ N57HQ	92	TBM 700	90	Capital Partners Advisory Co. Baltimore, Md.	N57HC
☐ N57HT	76	King Air B100	BE-7	University of South Carolina, Columbia, SC.	N681PC
☐ N57KA	73	King Air C90	LJ-577	Aviation Sales Inc. Miamisburg, Oh.	
☐ N57MA	68	King Air B90	LJ-414	Thomas Howell Kiewit (USA) Inc. Miami, Fl. (status ?).	C-GVCC
☐ N57MR	76	PA-31T Cheyenne II	7620022	A1 Tech Speciality Steel/Great Circle Aviation, Dunkirk, NY	N57MK
☐ N57RS	74	Rockwell 690A	11149	Antonio Fernandez, Guaynabo, Peru.	N5KW
☐ N57SC	90	King Air 350	FL-34	HRB Systems Inc. State College, Pa.	N987HT
☐ N57SL	92	TBM 700	57	U S Financial Services Inc. Wilmington, De.	
☐ N57TJ	80	King Air B100	BE-102	San-Services LLC. Cleveland, Tn.	(N91L)
☐ N57TM	80	King Air F90	LA-34	B J's Limo Service Inc. Lakeland, Fl.	N43WS
☐ N57TS	97	King Air B200	BB-1557	Thunder Spring-Wareham LLC. Ketchum, Id.	
☐ N58AC	65	King Air 90	LJ-77	Sunny South Inc. Gainesville, Fl.	N56SQ
☐ N58AP	81	Cheyenne II-XL	8166062	Kansas Aircraft Corp. New Century, Ks.	N59AP
☐ N58AS	82	King Air B200C	BL-50	Aviation Specialities Inc. Manila, Philippines.	N54HF
☐ N58CA	69	Mitsubishi MU-2F	170	F J Grieser Co. Miami, Fl.	N239MA
☐ N58EZ	81	King Air F90	LA-97	Ed Snate Air LLC. Ketchum, Id.	N371CP
☐ N58GA	82	King Air B200	BB-1003	SP Farms Inc. Troup, Tx.	I-MEPE
☐ N58KA	65	King Air 90	LJ-58	Leonard Sallustro, Throgg Station, NY.	C-GXHD
☐ N58PL	81	Cheyenne II-XL	8166024	Coldwater Veneer Inc. Coldwater, Mi.	N2586Y
☐ N58WB	77	Rockwell 690B	11408	Elcor Corp. St Louis, Mo.	(N690SA)
☐ N59EK	80	King Air F90	LA-58	Van Dyke & Associates Inc. Brenham, Tx.	N67TM
☐ N59EZ	81	SA-226T Merlin 3B	T-394	Atlantic Aviation Leasing Inc. Jacksonville, Fl.	N1QL
☐ N59KA	73	King Air C90	LJ-589	Alumax Extrusions Inc. West Chicago, Il.	
☐ N59KS	75	Mitsubishi MU-2L	664	Kroksions Kontroll Honseri Inc. Dover, De.	N555BC
☐ N59MS	90	King Air C90B	LJ-1405	Snijder Air Service Inc. Antwerp, Belgium.	
☐ N59RW	73	Mitsubishi MU-2K	267	Power Play Inc. Kennewick, Wa.	N313MA
☐ N59TF	90	King Air 350	FL-26	Teleflex Inc. Plymouth Meeting, Pa.	N903M

Regn	Yr	Type	c/n	Owner/Operator	Prev Regn

Regn	Yr	Type	c/n	Owner/Operator	Prev Regn
☐ N59TP	69	SA-26AT Merlin 2B	T26-161	Jet Cap Aviation Corp. Lindner Regional Airport, Fl.	N101BE
☐ N60AZ	75	Mitsubishi MU-2M	329	Bush Field Aircraft Co. Augusta, Ga.	N35VS
☐ N60B	74	Rockwell 690A	11172	Interlease Aviation Corp. Northfield, Il.	(D-IIGI)
☐ N60BA	74	King Air E90	LW-79	Bemidji Aviation Services Inc. Bemidji, Mn.	N12AK
☐ N60BT	77	Mitsubishi MU-2P	358SA	Intercorpsa Inc. Miami, Fl.	C-GIRO
☐ N60CM	96	King Air 350	FL-139	Mewbourne Oil Co. Tyler, Tx.	
☐ N60DB	77	Rockwell 690B	11420	H T Haralambos, Portland, Or.	N813AW
☐ N60DR	77	PA-31T Cheyenne II	7720064	Hap's Aerial Enterprises Inc. Sellersburg, In.	N45LS
☐ N60FJ	78	Conquest II	441-0048	Warrington Development Corp. Birmingham, Al.	N441CT
☐ N60FL	81	MU-2 Marquise	1512SA	Flight Line Inc. Salt Lake City, Ut.	HB-LQB
☐ N60JW	80	PA-31T Cheyenne 1	8004031	Semford Management Co. Crestwood, Il.	N2349V
☐ N60KA	73	King Air C90	LJ-586	60KA Co LLC. Houston, Tx.	
☐ N60KC	80	MU-2 Marquise	781SA	Keller Companies Inc. Manchester, NH.	N7045X
☐ N60KG	70	Mitsubishi MU-2G	511	Corporate Aviation Services Inc. Tulsa, Ok.	N60KC
☐ N60KW	78	King Air C90	LJ-800	Added Communications Inc. Suwanee, Ga.	N2032N
☐ N60MH	78	King Air E90	LW-290	Commercial Aviation Service Inc. Miami, Fl.	N79NS
☐ N60NJ	76	Mitsubishi MU-2M	339	John Martins, Streamwood, Il.	N555DD
☐ N60PC	83	King Air B200	BB-1146	Southern Company Services Inc. Atlanta, Ga.	N6816T
☐ N60PD	75	King Air 200	BB-58	Mayo Aviation Inc/CWW Inc. Englewood, Co.	N60PC
☐ N60RJ	66	King Air 90	LJ-86	American Pioneer Life Insurance Co. Trumann, Ar.	N90FH
☐ N60SC	80	King Air 200	BB-693	Sun Co (R&M), Philadelphia, Pa.	
☐ N60TJ	77	King Air B100	BE-21	Andrea Leasing, Glendale, Az.	N17821
☐ N60VP	97	King Air C90B	LJ-1516	Raytheon Aircraft Co. Wichita, Ks.	
☐ N60VS	81	Gulfstream 840	11683	Integer Corp. Wilmington, De.	N840BM
☐ N61AP	84	King Air B200	BB-1192	James McMahan, Van Nuys, Ca.	N6743D
☐ N61BA	78	Mitsubishi MU-2N	729SA	Business Air Inc. Bennington, Vt.	C-FHXZ
☐ N61DP	79	MU-2 Solitaire	398SA	Thomas Garvey, Paris, Tn.	N222MS
☐ N61GN	95	King Air C90B	LJ-1421	King Air LLC. Fishers, In.	N3252B
☐ N61SB	63	Gulfstream 1	115	Apex Aviation Group, Southlake, Tx.	C-FASC
☐ N61SG	77	PA-31T Cheyenne II	7720057	Flightline Charter Services Inc. Gillette, Wy.	N610RG
☐ N62BL	78	King Air E90	LW-272	Robert Landry, Jonesboro, Ar.	(N301HC)
☐ N62DL	77	King Air 200	BB-208	Executive Airways Inc. Raleigh, NC.	N188WG
☐ N62E	80	PA-31T Cheyenne II	8020074	Gunther & Susan Balz, Highland Beach, Fl.	(N47DR)
☐ N62EA	80	King Air 200	BB-668	Eagle Aviation Inc. West Columbia, SC.	D-IKRA
☐ N62GA	80	King Air B200C	BL-16	Gantt Aviation Inc. Georgetown, Tx.	F-GFAA
☐ N62JT	94	Pilatus PC-XII	107	Serka Aviation Inc/Telecal International Ltd. New Canaan, Ct	HB-FQD
☐ N62KK	91	2000 Starship	NC-17	Raytheon Aircraft Co. Wichita, Ks.	N62KM
☐ N62LM	92	TBM 700	67	Omyaviation Inc. Proctor, Vt.	
☐ N62MA	74	Rockwell 690A	11168	Deerfleet LLC. Miami, Fl.	XA-LEY
☐ N62NC	73	King Air A100	B-158	Colnan Inc. Tampa, Fl.	N62KA
☐ N62SK	78	King Air C90	LJ-784	Little Laurel Aviation Co/Rader Aviation, Summersville, WV.	YV-254CP
☐ N63BV	78	King Air E90	LW-256	Mayo Aviation Inc. Englewood, Co.	N63BW
☐ N63CA	78	PA-31T Cheyenne II	7820033	Speciality Industries, Red Lion, Pa.	OO-DGS
☐ N63CM	80	PA-31T Cheyenne II	8020039	Conyan Aviation Inc. Boise, Id.	N2584W
☐ N63DL	80	Gulfstream 840	11601	Larry Lehmkuhl, Placida, Fl.	N840CM
☐ N63EC	76	King Air E90	LW-191	Flying Concepts Inc. Salt Lake City, Ut.	N300BJ
☐ N63GB	90	King Air 350	FL-27	Graham Brothers Construction Co. Dublin, Ga.	N16NE
☐ N63LP	84	King Air 300	FA-18	Marshall Management Co. Las Vegas, Nv.	N63LB
☐ N63SK	80	King Air 200	BB-747	Autohaus Ltd/Carousel Motors, Iowa City, Ia.	N117CM
☐ N64BA	84	MU-2 Marquise	1566SA	Horne Properties Inc. Knoxville, Tn.	N4CS
☐ N64DC	79	King Air 200	BB-492	King Air Inc. Madison Heights, Mi.	N200PD
☐ N64EZ	79	Rockwell 690B	11526	Coatesville Machine Inc. Coatesville, In.	XA-SCJ
☐ N64GG	94	2000A Starship	NC-47	Executive Air LLC. Indianapolis, In.	N8277Q
☐ N64KA	79	King Air C90	LJ-606	Farroll Equipment Co. Arroyo Grande, Ca.	
☐ N64LG	72	Mitsubishi MU-2K	240	International Business Aircraft Inc. Tulsa, Ok.	N222HL
☐ N64MD	82	MU-2 Marquise	1561SA	Anderson Aviation Inc. Anderson, In.	N486MA
☐ N64PS	81	Gulfstream 840	11702	Peter Schiff, Cookeville, Tn.	N5954K
☐ N64SS	84	King Air 300	FA-6	The Shoe Show Inc. Concord, NC.	N39FL
☐ N64WB	82	MU-2 Marquise	1550SA	Rivett Group LLC. Aberdeen, SD.	N476MA
☐ N65CE	60	Gulfstream 1	45	Corps of Engineers/Mississippi River Commission, Vicksburg.	N329CT
☐ N65CK	79	PA-31T Cheyenne 1	7904041	Air Alpha America Inc. Amarillo, Tx.	4X-CBS
☐ N65CR	91	King Air 350	FL-60	Rockwell Collins Inc. Cedar Rapids, Ia.	
☐ N65EB	84	King Air 200	BB-325	Morgan Aviation Inc. Garden City, Ga.	F-GGAK
☐ N65EZ	84	PA-31T Cheyenne 1A	1104013	Yates Aviation Inc. Texarkana, Tx.	N20MR
☐ N65GH	74	King Air C90	LJ-617	Godwin Aircraft Inc. Memphis, Tn.	C-FFAS
☐ N65JG	81	MU-2 Marquise	1523SA	T K Aviation Inc. Wilmington, De.	N439MA
☐ N65KA	74	King Air C90	LJ-611	L & N Leasing Co. Wilson, NC.	
☐ N65MM	81	King Air F90	LA-125	Apache Aviation LLC. Beaver Dam, Wi.	N3845S
☐ N65RT	76	King Air 200	BB-97	Professional Travel Inc. La Porte, In.	N7EG
☐ N65SF	80	King Air F90	LA-18	Leemore Air Inc. Fayetteville, Ar.	N22JW
☐ N65TA	72	King Air C90	LJ-538	S J C Inc & J Hindmarch & C Loepke, Gulfport, Ms.	VH-LLS
☐ N65TW	81	King Air B200	BB-902	Sanchez-O'Brien Oil & Gas Corp. Laredo, Tx.	N5TW
☐ N65WM	82	King Air B200	BB-1087	Pinos Produce Inc. San Diego, Ca.	N860MH
☐ N66	87	King Air 300	FF-1	FAA, Oklahoma City, Ok. (was FA-126).	
☐ N66AD	68	King Air B90	LJ-380	Hay Boy Farms, Moses Lake, Wa.	C-GHLA
☐ N66AW	74	Mitsubishi MU-2K	286	Anders Williams Aviation Inc. Norfolk, Va.	N666SP

Regn	Yr	Type	c/n	Owner/Operator	Prev Regn
☐ N66BS	80	King Air F90	LA-40	Simmons Inc of Virginia, Richlands, Va.	N6748P
☐ N66CL	68	Mitsubishi MU-2F	135	American International Airways, Morristown, Tn.	N44MA
☐ N66CN	75	King Air C90	LJ-644	NexAir Inc. Memphis, Tn.	N777SJ
☐ N66CY	72	Mitsubishi MU-2J	562	Magnolia River Development LLC. Foley, Al.	N29CY
☐ N66FV	67	680V Turbo Commander	1676-59	Gary Knostman, Fulton Beach, Tx.	N580M
☐ N66GA	82	SA-227AT Merlin 4C	AT-427	Thomas Ganley, Brecksville, Oh.	D-CAIR
☐ N66GS	67	King Air A90	LJ-237	Richard Whatley, Longmott, Tx.	N232A
☐ N66GW	74	Rockwell 690A	11174	Downtown Airpark Inc. Oklahoma City, Ok.	N6B
☐ N66JC	81	Cheyenne II-XL	8166029	Midwest Industries Inc. Ida Grove, Ia.	N346JC
☐ N66KA	73	King Air C90	LJ-582	Air Sal Leasing Inc. Miami, Fl.	
☐ N66LM	85	King Air C90A	LJ-1112	LMC Leasing Inc/Maximo Air Inc. Banner Elk, NC.	N737L
☐ N66LP	80	King Air F90	LA-21	Peterson Farms Inc. Decatur, Il.	
☐ N66MD	69	SA-26AT Merlin 2B	T26-159	Donald Paolucci/Mohican Air Service Inc. Fairview Park, Oh.	N14JK
☐ N66MT	81	Cheyenne II-XL	8166060	Manatts Inc. Brooklyn, Ia.	N88XL
☐ N66RE	67	King Air A90	LJ-307	Bodmer Financing Co. Wilmington, De.	N966CY
☐ N66TJ	75	King Air 200	BB-42	Tyler Jet Aircraft Sales Inc. Tyler, Tx.	N3GY
☐ N66TL	74	King Air C90	LJ-636	Bavarian Aircraft LLC. Deephaven, Mn.	N222BJ
☐ N66TS	80	King Air 200	BB-738	James Taubert, Wichita Falls, Tx.	
☐ N66TW	80	PA-31T Cheyenne 1	8004030	Corporate Air Service Inc. West Bountiful, Ut.	N2347V
☐ N66U	82	King Air B200	BB-996	Unison Industries Inc. Jacksonville, Fl.	N775M
☐ N66WJ	79	PA-31T Cheyenne II	7920041	David & Ami MacHugh, Pasco, Wa.	N302HA
☐ N67	87	King Air 300	FF-2	FAA, Oklahoma City, Ok. (was FA-129).	
☐ N67BA	81	Conquest 1	425-0007	IMIC Aviation Inc. Columbia, SC.	C-GJEB
☐ N67BS	81	King Air B100	BE-104	Turnberry Charters Inc. Aventura, Il.	N67KA
☐ N67CG	79	Rockwell 690B	11540	ESI Inc of Tennessee, Kennesaw, Ga.	N81697
☐ N67CL	87	King Air C90A	LJ-1154	Coastal Lumber Co. Weldon, NC.	N3076U
☐ N67FE	82	Gulfstream 840	11729	Furnas Electric Co. Sugar Grove, Il.	N5967K
☐ N67GA	76	King Air 200	BB-176	Z Z Leasing Inc. Memphis, Tn.	F-GFMJ
☐ N67JE	81	Conquest 1	425-0058	John & Colleen Gerken, Sandpoint, Id.	N67CA
☐ N67JG	81	PA-31T Cheyenne II	8120052	Hans-Juergen Guido, Regensburg, Germany.	D-IFGN
☐ N67LF	92	TBM 700	50	F Marcell Co. Berea, Oh.	N84HS
☐ N67MD	81	King Air 200	BB-834	Domar Leasing Corp. Dallas, Tx.	
☐ N67PL	72	King Air C90	LJ-566	Northwestern Aircraft Capital Corp. Mercer Island, Wa.	N67PC
☐ N67PS	74	King Air E90	LW-112	First Star Inc. Wilmington, De.	N3034W
☐ N67SM	70	Mitsubishi MU-2F	183	Turbine Aircraft Parts Inc. San Angelo, Tx.	N221KP
☐ N67TC	82	King Air B200	BB-1045	Jim Clark & Associates Inc. Oklahoma City, Ok.	N60SM
☐ N67TW	83	Cheyenne II-XL	8166046	Precision Concrete Construction Co. White Marsh, Md.	(N18KW)
☐ N67V	79	King Air E90	LW-306	Charles Hurd Jr. El Paso, Tx.	N90SR
☐ N68	87	King Air 300	FF-3	FAA, Oklahoma City, Ok.	
☐ N68AJ	84	King Air C90	LJ-1071	W Hampton Pitts, Nashville, Tn.	ZS-LOK
☐ N68BJ	79	PA-31T Cheyenne II	7920029	Johnson Enterprises LLC. Millbury, Ma.	TG-VDG
☐ N68CD	68	King Air B90	LJ-366	Tee Pee Contractors Inc. Casa Grande, Az.	N66CD
☐ N68CL	81	MU-2 Solitaire	448SA	Coastal Lumber Co. Weldon, NC.	N231LC
☐ N68DK	80	King Air F90	LA-56	Kelly Properties Inc. Aspen, Co.	N777AQ
☐ N68FA	82	King Air B200	BB-1088	First American National Bank, Nashville, Tn.	N868HC
☐ N68GK	81	King Air B200	BB-926	Terra Australis Corp. Wilmington, De.	N61CE
☐ N68HS	84	Conquest II	441-0331	Hastings Books, Music & Video, Amarillo, Tx.	N1210D
☐ N68KA	81	King Air B200	BB-793	Rockwell-Ditzler Associates Inc. Pittsburgh, Pa.	
☐ N68PC	82	King Air C90	LJ-1007	Kansas Aircraft Corp. New Century, Ks.	N61254
☐ N68PM	76	King Air E90	LW-188	Benchmark Aviation LLC. Baton Rouge, La.	N16TE
☐ N68TD	73	Rockwell 690	11066	Grover Harben & F Harrison Haynes, Gainesville, Ga.	N36WR
☐ N68TN	75	Mitsubishi MU-2L	675	Agamemenon Operating Inc/CirrusAir, Dallas, Tx.	N835MA
☐ N68TW	89	King Air C90A	LJ-1200	Davison Transport Inc. Ruston, La.	(N485JA)
☐ N69	88	King Air 300	FF-4	FAA, Oklahoma City, Ok.	
☐ N69AD	81	King Air F90	LA-143	Truman Arnold Distributors Co Inc. Texarkana, Tx.	N624LF
☐ N69BK	77	King Air 200	BB-271	Trical Inc. Hollister, Ca.	
☐ N69BS	91	TBM 700	10	Eagle Flight Center Inc. North Andover, Ma.	N19AP
☐ N69CD	76	King Air 200	BB-156	Galdy's Enterprises Inc. El Paso, Tx.	N69LD
☐ N69GA	84	Gulfstream 1000	96071	Duncan Enterprises Inc. Southern Pines, NC.	
☐ N69JH	82	King Air B200	BB-1089	American Executive Air Charter, Charleston, WV.	N87SA
☐ N69PC	81	PA-42 Cheyenne III	8001016	Southern Cross Aviation inc. Fort Lauderdale, Fl.	N300DK
☐ N69ST		SA-226AT Merlin 4A	AT-030	Camp Oil Co. Augusta, Ga.	N5440F
☐ N69VC	85	King Air B200	BB-1228	Executive Charters Inc. Stateline, Nv.	N6LD
☐ N70	88	King Air 300	FF-5	FAA, Oklahoma City, Ok.	
☐ N70AB	77	King Air C90	LJ-712	Turbo Air Inc. Boise, Id.	C-GUNG
☐ N70BA	78	King Air B100	BE-43	Anywhere Linkage Inc. Wilmington, De.	N32RT
☐ N70CU	68	King Air B90	LJ-369	Ambar Inc. Lafayette, La.	N700U
☐ N70FH	76	King Air C90	LJ-692	Frank Harrison, Lafayette, La.	N822TJ
☐ N70GM	82	Conquest 1	425-0108	Robert Powell, Van Buren, Ar.	D-IAAX
☐ N70GW	81	PA-31T Cheyenne 1	8104004	Soporific LLC. Reno, Nv.	N4WP
☐ N70HB	81	Conquest 1	425-0100	Abbott-Long Inc. Montrose, Co.	N55MS
☐ N70JL	71	King Air 100	B-87	Utah State University, Logan, Ut.	N125DB
☐ N70KC	80	MU-2 Marquise	775SA	Keller Companies Inc. Manchester, NH.	N93GN
☐ N70LF	79	King Air 200	BB-562	BK LLC. Jasper, In.	(N777JE)
☐ N70MD	74	Rockwell 690A	11210	Power Aviation, Wichita, Ks.	N1SS
☐ N70MN	93	King Air B200	BB-1447	State of Minnesota DoT Office of Aeronautics, St Paul, Mn.	N8138V

217

Regn	Yr	Type	c/n	Owner/Operator	Prev Regn
☐ N70MV	73	King Air E90	LW-48	Bank of Mountain View, Mountain View, Ar.	OK-DKH
☐ N70PQ	76	King Air 200	BB-164	Arkansas Aircraft Inc. Jonesboro, Ar.	N70PC
☐ N70QZ	71	King Air C90	LJ-521	John Freeman, Oklahoma City, Ok.	N700Z
☐ N70RF	70	681 Turbo Commander	6013	Modul Cast Construction Corp. Miami, Fl. (status ?)	N60BC
☐ N70RR	75	Rockwell 690A	11259	Nauta Inc/Heetco Jet Center, Quincy, Il.	SE-KYY
☐ N70SM	68	King Air B90	LJ-396	Turbo Air Inc. Boise, Id.	N783K
☐ N70SW	74	King Air E90	LW-101	Yankee Flying Ltd. Salinas, Ca.	N111PC
☐ N70TW	79	PA-31T Cheyenne 1	7904043	Cody Aircraft Leasing Inc. Wilmington, De.	
☐ N70UA	66	King Air A90	LJ-130	Murray Aviation Inc. Ypsilanti, Mi.	N70TG
☐ N70VM	67	King Air A90	LJ-300	Texas Biz Jet, Fort Worth, Tx.	N222MB
☐ N70VP	96	King Air C90B	LJ-1459	Goff Aloft Inc. Traverse City, Mi.	
☐ N71	88	King Air 300	FF-6	FAA, Oklahoma City, Ok.	
☐ N71BX	74	King Air E90	LW-74	Meredith Mallory Jr/Crown Air Service, San Antonio, Tx.	N713X
☐ N71CJ	63	Gulfstream 1	107	Pacific Gas & Electric Co. Oakland, Ca.	N7ZB
☐ N71DP	81	MU-2 Marquise	1502SA	Heetco Jet Center Inc. Quincy, Il.	VH-MVU
☐ N71EN	74	King Air C90	LJ-632	Beverly Fabrics Inc. Soquel, Ca.	N71FN
☐ N71FA	84	King Air 300	FA-17	First Alabama Bank, Birmingham, Al.	N6804M
☐ N71HE	81	Conquest 1	425-0085	Codding Investments Inc. Santa Rosa, Ca.	N5HE
☐ N71KA	73	King Air C90	LJ-578	David & Catherine Paine, Templeton, Ca.	N84JH
☐ N71MR	83	Gulfstream 1000	96054	Envirostat, Pompano Beach, Fl.	N8LB
☐ N71RD	69	Gulfstream 1	322	Saican Air (SA) Inc. Wonderboom, RSA.	N9QM
☐ N71SF	69	SA-26AT Merlin 2B	T26-154	Crumpets Inc. Comanche, Tx.	C-FGRA
☐ N71TB	81	King Air 200	BB-755	Reese Rowling, Corpus Christi, Tx.	N82SA
☐ N71VE	72	Rockwell 690	11043	Cooper Aerial Survey Ltd. Dover, De.	N71VT
☐ N71VG	91	King Air C90A	LJ-1279	Thomas & Vicki Griffin, Seattle, Wa.	N717A
☐ N71WB	75	King Air E90	LW-127	575 Flight Corp. Houston, Tx.	N711RQ
☐ N72	88	King Air 300	FF-7	FAA, Oklahoma City, Ok.	
☐ N72CF	69	SA-26AT Merlin 2B	T26-132	Jacksonville Jet Center Inc. Jacksonville, Fl.	N425MC
☐ N72DK	85	King Air 300	FA-42	Pacific Assembly Inc. Wilmington, De.	N791X
☐ N72GA	85	King Air 300	FA-72	George Miller II, Nashville, Tn.	ZS-MFW
☐ N72MM	79	King Air 200	BB-497	Northeastern Aviation Corp. Wilmington, De.	N73CA
☐ N72PK	96	King Air C90B	LJ-1449	Pre-Owned Electronics Inc. Bedford, Ma.	
☐ N72RL	79	King Air 200	BB-509	Createc Corp. Indianapolis, In.	N72DD
☐ N72SE	80	King Air 200	BB-596	Southern Energy Homes Inc. Addison, Al.	N918JN
☐ N72TA	74	King Air A100	B-202	Tellom Air Inc. Canal Fulton, Oh.	N5831A
☐ N72TB	80	Gulfstream 840	11619	FSB, Salt Lake City, Ut.	N16TG
☐ N72TJ	79	MU-2 Solitaire	414SA	JoRo LC. Waterloo, Ia.	XB-DVV
☐ N72VF	75	Rockwell 690A	11242	Yakima Theatres Inc. Yakima, Wa.	N72VT
☐ N73	88	King Air 300	FF-8	FAA, Oklahoma City, Ok.	
☐ N73BG	80	PA-31T Cheyenne 1	8004027	R W Air Services Inc. Hatfield, Pa.	(N701GP)
☐ N73DW	81	Conquest 1	425-0089	David R Webb Co. High Point, NC.	N74JW
☐ N73EF	80	Gulfstream 840	11617	Executive Flight Inc. East Wenatchee, Wa.	N840LC
☐ N73HC	69	SA-26AT Merlin 2B	T26-138	Hoosier Microfilm Inc/Midwest Air Charter, Washington, In.	N52L
☐ N73LC	90	King Air B200	BB-1393	Four Points Inc. Breckenridge, Tx.	N8064F
☐ N73MC	73	King Air C90	LJ-600	H B Aviation & Leasing Inc. Adrian, Mi.	N3PR
☐ N73MP	84	King Air B200	BB-1153	SAN LLC. Knoxville, Tn.	N7213B
☐ N73MW	75	King Air 200	BB-22	Royal Palm Airlease Inc. Deerfield Beach, Fl.	N7300R
☐ N73Q	67	King Air A90	LM-68	K K Aircraft Inc. Bridgewater, Va.	N117CP
☐ N74	88	King Air 300	FF-9	FAA, Oklahoma City, Ok.	
☐ N74BY	77	Conquest II	441-0039	Veneer Specialities/Michigan Veneer Ltd. Pewamo, Mi.	N36972
☐ N74CC	74	King Air C90	LJ-620	Dodson Aviation Inc. Ottawa, Ks.	
☐ N74ED	77	King Air 200	BB-228	Western Financial Services Inc. Oklahoma City, Ok.	N53G
☐ N74EF	80	Gulfstream 840	11614	Executive Flight Inc. East Wenatchee, Wa.	N74RF
☐ N74FB	81	PA-42 Cheyenne III	8001034	L24 Associates Inc. Exeter, NH.	N777FG
☐ N74GB	74	Rockwell 690A	11206	Chino Valley Commander Inc. Claremont, Ca.	N75RR
☐ N74GS	79	King Air 200	BB-429	GSC Enterprises Inc. Sulphur Springs, Tx.	N28BE
☐ N74HR	80	Conquest 1	425-0002	Southern Jet Management Inc. Houston, Tx.	N425CC
☐ N74LV	82	King Air B200	BB-1074	Franchise Enterprises Inc.	C-GTDX
☐ N74MA	69	King Air B90	LJ-479	Mayo Aviation Inc. Englewood, Co.	(N502M)
☐ N74ML	81	PA-31T Cheyenne 1	8104060	Kenneth Miller, Bakersfield, Ca.	N51WA
☐ N74RF	91	King Air B200	BB-1408	Richfood Holdings Inc. Mechanicsville, Va.	F-OGGK
☐ N74RG	82	King Air B200	BB-1000	Glass Enterprises Inc. Bentonville, Ar.	N51881
☐ N74RR	81	King Air B100	BE-122	Butler Air Inc. Butler, Pa.	XB-ESH
☐ N74TF	92	2000A Starship	NC-27	Tripefoods Inc. Buffalo, NY.	N8225Y
☐ N74TW	79	PA-31T Cheyenne II	7920067	MHE Inc. Bryan, Oh.	(N4301L)
☐ N74VR	75	King Air E90	LW-135	Greater Media Services Inc. East Brunswick, NJ.	N7275R
☐ N75	88	King Air 300	FF-10	FAA, Oklahoma City, Ok.	
☐ N75AH	80	King Air 200	BB-741	Air Holdings Inc. Helena, Mt.	N80GS
☐ N75AP	79	King Air B100	BE-57	Pentz Air Inc. Canton, Oh.	N57AK
☐ N75CF	77	King Air E90	LW-212	University of South Carolina, Columbia, SC.	
☐ N75FL	83	PA-42 Cheyenne IIIA	8301002	Frank Lill & Son Inc. Webster, NY.	(N830AM)
☐ N75GC	77	King Air C90	LJ-727	Global Air Rescue Inc. Clearwater, Fl.	N888BK
☐ N75GF	84	King Air B200	BB-1190	Eastern Holdings LLC. Wilmington, De.	N70AC
☐ N75GR	75	King Air A100	B-210	Silverhawk Aviation inc. Lincoln, Ne.	N75ZZ
☐ N75JP	76	King Air E90	LW-158	Prior Aviation Service Inc. Buffalo, NY.	N940SR
☐ N75LS	84	PA-42 Cheyenne IIIA	5501017	Freeman Decorating Co. Dallas, Tx.	N4118H

Regn	Yr	Type	c/n	Owner/Operator	Prev Regn
☐ N75LV	82	King Air B200	BB-1075	Cook-Fort Worth Children's Medical Center, Fort Worth, Tx.	C-GTDY
☐ N75LW	74	King Air E90	LW-75	Rocky Beach Properties Ltd. Montrose, Co.	XB-AEU
☐ N75ME	85	King Air 300	FA-57	Golden Rule Financial Corp. Indianapolis, In.	N75MC
☐ N75MS	83	King Air F90-1	LA-204	Soc King Air, Dizy, France.	XA-PEM
☐ N75MX	71	SA-226T Merlin 3	T-218	Flight Research, Hattiesburg, Ms.	N990M
☐ N75RM		SA-226T Merlin 3A	T-290	Riverside Manufacturing Co. Moultrie, Ga.	N5450M
☐ N75SC		SA-227TT Merlin 3C	TT-474	Stackpole Corp. Newton, Ma	
☐ N75SK	80	PA-31 Cheyenne 1	8004042	Montgomery Aviation Inc. Wilmington, De.	N2492V
☐ N75TF	78	PA-31 Cheyenne II	7820075	Crown Leasing Inc. Portland, Or.	N781CW
☐ N75TW	80	PA-31 Cheyenne 1	8004016	Trans-West Inc & Revere Corp. Missoula, Mt.	
☐ N75VF	77	King Air 200	BB-220	Hill Aircraft & Leasing Corp. Atlanta, Ga.	N70FH
☐ N75WD	91	King Air 350	FL-51	Wayne Densch Air Inc. Orlando, Fl.	N8055J
☐ N75WP	80	King Air 200	BB-574	Salt River Project Inc. Phoenix, Az.	
☐ N75X	82	SA-227TT Merlin 3C	TT-421	Blue Sky Air Inc. McAllen, Tx.	N1014B
☐ N75ZT	82	King Air B200	BB-967	Behlen Manufacturing Co. Columbus, Ne.	N75Z
☐ N76	88	King Air 300	FF-11	FAA, Oklahoma City, Ok.	
☐ N76CP	80	PA-31 Cheyenne 1	8004026	James Langley, Golden, Co.	N452DP
☐ N76EC	74	Rockwell 690A	11208	Century Capita/Ray & Nancy Gava, Lakeview, Or.	N74RR
☐ N76HC	82	King Air F90	LA-159	C P T Inc. Memphis, Tn.	N90GT
☐ N76HH	81	Gulfstream 980	95076	Eagle Air Inc. Memphis, Tn.	N9828S
☐ N76PM	82	King Air B200	BB-998	Pine Street Associates LLC. El Prado, NM.	N61474
☐ N76PW	74	King Air E90	LW-117	76PW LLC. Tulsa, Ok.	N31EE
☐ N76RJ	86	King Air B200	BB-1245	F Dietmar Seidler, Egelsbach, Germany.	N7275G
☐ N76SK	74	King Air E90	LW-115	ZZ Leasing Inc. Memphis, Tn.	N37MC
☐ N76WA	76	Rockwell 690A	11342	Promark Marketing & Sales Inc. Bentonville, Ar.	XA-RPD
☐ N77	88	King Air 300	FF-12	FAA, Oklahoma City, Ok.	
☐ N77CA	80	King Air 200	BB-717	Frances Christmann, Pinedale, NY.	
☐ N77CX	94	King Air B200	BB-1482	AMP Inc. Harrisburg, Pa.	N1559Y
☐ N77DA	66	King Air A90	LJ-146	Flight Safety International Inc. Dothan, Al.	N49EL
☐ N77DB	80	MU-2 Solitaire	429SA	Darrigo Brothers of California, Salinas, Ca.	N55DL
☐ N77DK	81	MU-2 Solitaire	449SA	Dennis Kranz, Portland, Or.	N22TG
☐ N77HD	91	King Air B200	BB-1397	Arkansas State Highway & Transportation Dept. Little Rock.	N8062J
☐ N77HE	81	King Air C90	LJ-969	Midwest Corporate Aviation Inc. Wichita, Ks.	9Q-CHE
☐ N77HN	91	King Air B200	BB-1380	Harold Nix & Cary Patterson, Texarkana, Tx.	N200KY
☐ N77HS	85	Gulfstream 900	15041	NBD Leasing Inc. Indianapolis, In.	N82BA
☐ N77JT	82	King Air B200	BB-1042	Modern Handcraft Inc. Fairway, Ks.	N400LM
☐ N77JX	73	King Air E90	LW-54	Barry Richardson, Marston, Mo.	N771HM
☐ N77M	85	SAAB 340A	036	AMP Inc. Harrisburg, Pa.	N260PM
☐ N77ML	81	Conquest II	441-0202	Martin & Evelyn Lutin Family Trust, Los Angeles, Ca.	N449DR
☐ N77NL	77	PA-31 Cheyenne II	7720038	New South Inc. Charleston, SC.	N82136
☐ N77PA	74	King Air E90	LW-89	Frank Godchaux III, Hamilton, Mt.	N22KF
☐ N77PF	70	King Air 100	B-70	Air Charter Service Inc. Washington, Pa.	N25JL
☐ N77PV	80	King Air F90	LA-68	Werk-Brau Co Inc. Findlay, Oh.	N99LM
☐ N77QX	80	King Air 200	BB-659	KFE Inc & SAROB Corp. San Diego, Ca.	N77CX
☐ N77R	83	Conquest II	441-0288	Rafter 7 Aviation Inc. Reno, Nv.	N352RT
☐ N77SS	67	King Air A90	LJ-230	K & K Aircraft Inc. Bridgewater, Va.	N93BA
☐ N77UA	77	Rockwell 690B	11422	University of Arkansas, Fayetteville, Ar.	N500MM
☐ N77UH	81	Cheyenne II-XL	8166026	U-Haul of Oregon, Phoenix, Az.	N31X
☐ N77VF	79	Rockwell 690B	11520	Leuchterm Inc. Jupiter, Fl.	N81785
☐ N77WM	86	King Air C90A	LJ-1133	Air Flite III Inc. Saginaw, Mi.	N300HH
☐ N77YP	82	Conquest 1	425-0111	Hayden Air Inc. Portland, Or.	N50FS
☐ N78	88	King Air 300	FF-13	FAA, Oklahoma City, Ok.	
☐ N78BA	74	Rockwell 690A	11166	Bob & Bud Inc. Wausau, Wi.	N77HS
☐ N78CA	79	PA-31 Cheyenne II	7920062	Air Alpha America Inc. Amarillo, Tx.	F-GIYV
☐ N78CH	70	681 Turbo Commander	6038	C & R Flying Service Inc. Baytown, Tx.	N46JC
☐ N78CP	74	SA-226AT Merlin 4A	AT-029	Superior Aviation Inc. Iron Mountain, Mi.	N294A
☐ N78CS	73	SA-226T Merlin 3A	T-229	Walleland Leasing LLC. San Antonio, Tx.	N105BB
☐ N78CT	81	King Air 200	BB-761	Springfield Aviation LLC. Santa Cruz, Ca.	N107TM
☐ N78DV	75	King Air 200	BB-67	First State Trucking Inc. Selbyville, De.	N400DB
☐ N78FS	79	MU-2 Solitaire	409SA	Steven Sandlin, Las Vegas, Nv.	N990MA
☐ N78GC	85	King Air B200	BB-1193	Golden Corral Corp/Investors Management Corp. Raleigh, NC.	N6728H
☐ N78MK	72	King Air A100	B-109	Joda Partnership, Town & Country, Mo.	N78CA
☐ N78NA	77	Rockwell 690B	11401	Nationair Insurance Agencies Ltd. Linoln, Ne.	N28SE
☐ N78TT	78	Rockwell 690B	11509	Tele-Tech Co. Lexington, Ky.	N3160G
☐ N78TW	79	PA-31 Cheyenne II	7920092	Richard & Linda Winkle, Canfield, Oh.	N73TW
☐ N78WD	78	Mitsubishi MU-2P	368SA	Pleasure Craft Marine Engines, Canal Winchester, Oh.	N16TQ
☐ N79	88	King Air 300	FF-14	FAA, Oklahoma City, Ok.	
☐ N79BJ	66	680V Turbo Commander	1610-46	Area Technical School, Atlanta, Ga. (status ?).	N79D
☐ N79CA	96	Pilatus PC-XII	136	H-S Air Inc. Englewood, NJ.	HB-FOH
☐ N79CF	79	King Air/Catpass 250	BB-441	Bering Air Inc. Nome, Ak.	
☐ N79CT	79	King Air E90	LW-303	Randall Stores Inc. Mitchell, SD.	N20351
☐ N79FB	79	PA-31 Cheyenne II	7920016	Flying Crown Inc. Wilmington, De.	N6180A
☐ N79GA	79	King Air 200	BB-556	Perdue Farms Inc. Salisbury, Md.	I-ARBX
☐ N79GS	81	King Air B200	BB-905	George Strait Productions Inc. San Antonio, Tx.	N80D
☐ N79HS	61	Gulfstream 1	79	Emery Worldwide, Palo Alto, Ca.	N190DM
☐ N79JS	80	PA-31 Cheyenne II	8020067	JDS Aviation Inc. Wilmington, De.	N301CH

Regn	Yr	Type	c/n	Owner/Operator	Prev Regn
☐ N79PG	82	King Air B200	BB-1043	Principal Mutual Life Insurance Co. Des Moines, Ia.	N62881
☐ N79PH	82	Gulfstream 1000	96029	Anodyne Corp.	N282AC
☐ N79SZ	83	Gulfstream 900	15024	Eagle Air Inc. Memphis, Tn.	N5911N
☐ N79TE	85	King Air 300	FA-67	Temple Inland Forest Products Inc. Diboll, Tx.	N55SC
☐ N79Z	91	TBM 700	19	Gabriel Enterprises Ltd. Staverton, UK.	
☐ N80	88	King Air 300	FF-15	FAA, Oklahoma City, Ok.	
☐ N80BC	97	King Air B200	BB-1571	Carter's Shooting Center Inc. Spring, Tx.	
☐ N80CP	79	PA-31T Cheyenne II	7920040	Danka Corp. Sparks, Nv.	
☐ N80DB	77	King Air B100	BE-31	DePuy Inc. Warsaw, In.	N18487
☐ N80DG	66	King Air A90	LJ-131	Pro Dealers Services, Visalia, Ca.	N700S
☐ N80GP	66	King Air A90	LJ-137	GPC Inc. Dover, De.	N24PR
☐ N80HH	79	MU-2 Marquise	732SA	Kenn-Air Leasing Corp. Gainesville, Fl.	(N30MA)
☐ N80M	81	King Air F90	LA-150	McLaughlin Body Co Inc. Moline, Il.	N700RL
☐ N80PA	77	King Air 200	BB-234	B & B Herricks & GPII Energy Inc. Kermit, Tx.	N200JC
☐ N80RT	78	King Air 200	BB-370	Irvan Aviation Inc. Concord, NC.	N117WD
☐ N80TB	76	Rockwell 690A	11300	Sportkraft Ltd. St College, Pa.	N471SC
☐ N80WM	81	King Air 200	BB-863	South Florida Water Management Division, West Palm Beach.	N44919
☐ N80WP	85	King Air F90-1	LA-228	Charles Putnam, Mills, Wy.	N7206N
☐ N80X	87	King Air 300	FA-117	Capital City Press Inc. Baton Rouge, La.	N29985
☐ N81	88	King Air 300	FF-16	FAA, Oklahoma City, Ok.	
☐ N81AT	70	King Air B90	LJ-490	Dallas Offset Inc. Dallas, Tx.	N881LT
☐ N81GC	80	King Air F90	LA-73	Guardian Corp. Rocky Mount, NC.	N369BR
☐ N81JN	81	Gulfstream 840	11660	Ozark Management Inc. Jefferson City, Mo.	(N140WJ)
☐ N81LT	81	King Air B200	BB-892	Vermdeco Aviation LPC. Wilmington, De.	(SE-IIS)
☐ N81MP	73	King Air E90	LW-72	Pegasus Aviation Inc. Cleveland, Oh.	N102MC
☐ N81PA	76	King Air 200	BB-158	Pennsylvania Department of Transportation, New Cumberland.	N8198M
☐ N81PG	68	King Air B90	LJ-395	Fayard Enterprises Inc. Louisburg, NC.	N81PA
☐ N81PN	77	Conquest II	441-0037	BA Leasing & Capital Corp. San Francisco, Ca.	N441TH
☐ N81PS	80	King Air F90	LA-74	Phillips Supply Co. Cincinnati, Oh.	
☐ N81RD	78	King Air 200	BB-385	Bird Space Technology, Sandpoint, Id.	
☐ N81SD	81	King Air F90	LA-130	Bi-Lo Inc. Maudlin, SC.	
☐ N81TF	81	King Air 200	BB-750	Thomas Fuqua Jr. Texarkana, Ar.	N200T
☐ N81WU	80	King Air 200	BB-576	Aero King Inc. Wilmington, De.	N505SC
☐ N82DD	81	King Air B200	BB-913	James Duncan, Big Springs, Tx.	N18371
☐ N82PC	80	PA-31T Cheyenne II	8020082	Automatic Business Products, New Smyrna Beach, Fl.	N2580V
☐ N82PG	82	PA-42 Cheyenne III	8001070	Basco Flying Service Inc. Pottstown, Pa.	N721CA
☐ N82TW	80	PA-31T Cheyenne 1	8004034	Dugan Funeral Services Inc. Fremont, Ne.	(N81TW)
☐ N82WC	78	Mitsubishi MU-2P	393SA	Wells Cargo Inc. Elkhart, In.	N9SR
☐ N83	88	King Air 300	FF-18	FAA, Oklahoma City, Ok.	
☐ N83A	78	Conquest II	441-0050	Lacks Industries Inc. Grand Rapids, Mi.	N55FC
☐ N83CH	81	Conquest 1	425-0025	Zond Systems Inc. Tehachapl, Ca.	N53MS
☐ N83FE	77	King Air E90	LW-219	Aviation Holdings Ltd. Houston, Tx.	N77AG
☐ N83FM	68	King Air B90	LJ-341	Wayne Muller & Robert Harris, Erie, Co.	N91NC
☐ N83G	73	Rockwell 690	11064	Don McCormack, Oklahoma City, Ok.	N830
☐ N83GA	79	King Air 200	BB-518	Beech Transportation Inc. Eden Prairie, Mn.	VH-MKR
☐ N83JN	79	King Air 200	BB-553	Noel Corp. Yakima, Wa.	N49KA
☐ N83KA	79	King Air 200	BB-436	Crusader Aviation Inc. Oxford, Ct.	N261GA
☐ N83LS	96	PWC PT6A-21	LJ-1461	Luck Stone, Richmond, Va.	N1076K
☐ N83MG	81	PA-31T Cheyenne 1	8104047	Jonathan Watkins, Vista, Ca.	N2420Y
☐ N83P	82	King Air C90-1	LJ-1027	Kyanite Mining Corp. Dillwyn, Va.	
☐ N83RH	82	King Air B200	BB-976	Richard Hormachea/RMH Co. Boise, Id.	N1845S
☐ N83TC	70	King Air B90	LJ-483	Unified Aero, Waukesha, Wi.	N210K
☐ N83TM	72	King Air A100	B-130	Tedesco Manufacturing Co. Connellsville, Pa.	N9UG
☐ N83WA	83	Gulfstream 1000	96063	Wal-Mart Stores Inc. Rogers, Ar.	N61508
☐ N83WE	78	King Air E90	LW-289	Welco Leasing Co. Aurora, Co.	N222MC
☐ N83XL	81	Cheyenne II-XL	8166075	North Shore Properties, Erie, Pa.	N2325W
☐ N84	88	King Air 300	FF-19	FAA, Oklahoma City, Ok.	
☐ N84CC	80	King Air 200	BB-691	Carolina Transco Co. Hickory, NC.	N40RL
☐ N84DT	72	Rockwell 690	11031	Ian Pringle, St Croix, USVI.	N1SS
☐ N84LJ	79	Conquest II	441-0084	Lovejoy Industries/L J Aircraft Inc. Oak Brook, Il.	N441GN
☐ N84LS	78	King Air 200	BB-346	Living Word International Inc. Midland, Mi.	(N15LS)
☐ N84P	83	King Air C90	LJ-1045	Sky-Lease Inc. Elkin, NC.	
☐ N84PA	81	King Air 200	BB-835	Pennsylvania DoT Bureau of Aviation, New Cumberland, Pa.	N81CT
☐ N84TP	80	King Air C90	LJ-911	Arrowhead Transfer Inc. Sitka, Ak.	N107K
☐ N85AJ	75	PA-31T Cheyenne II	7520037	Stevens Aviation Inc. Greenville, SC.	N79ND
☐ N85BC	81	King Air 200	BB-734	B & C Co. Columbus, Ga.	N3695B
☐ N85CM	80	PA-31T Cheyenne II	8020019	Trim Inc. Dothan, Al.	N71TW
☐ N85EM	81	Cheyenne II-XL	8166055	Stein Inc. Sandusky, Oh.	(CP-1699)
☐ N85HB	81	PA-31T Cheyenne II	8120021	Michael Spear, Columbia, Md.	N80WA
☐ N85LF	79	King Air B100	BE-62	B & P Aviation Inc. Tuscaloosa, Al.	N6044B
☐ N85TB	79	King Air C90	LJ-833	Fred Kajans, Carson City, Nv.	(N45AW)
☐ N85WA	85	Gulfstream 1000	96078	Wal-Mart Stores Inc. Bentonville, Ar.	N6151X
☐ N86BM	75	King Air A100	B-206	Silverhawk Aviation, Lincoln, Ne.	N86PA
☐ N86CR	84	PA-42 Cheyenne 400LS	5527011	Cumberland River Energies Inc. Brookside, Ky.	
☐ N86DD	84	King Air B200	BB-1171	DSD Aviation inc. Gorham, NH.	N100MS
☐ N86EA	81	Conquest 1	425-0082	Whirlwind Aviation Ltd. Anchorage, Ak.	N6846T
Regn	*Yr*	*Type*	*c/n*	*Owner/Operator*	*Prev Regn*

Regn	Yr	Type	c/n	Owner/Operator	Prev Regn
☐ N86FD	76	King Air B100	BE-14	Chattahoochee Aviation LLC. Columbus, Ga.	N4214S
☐ N86JG	84	King Air C90A	LJ-1076	Allen County Aviation/John Galvin, Lima, Oh.	N69275
☐ N86MG	67	King Air A90	LJ-264	Aviation Parts Exchange Inc. Southlake, Tx.	N86JR
☐ N86Q	77	King Air 200	BB-300	PowerFlame Inc. Parsons, Ks.	F-GHLV
☐ N86TR	77	King Air B100	BE-22	Laboratory Corp of America Holdings, Burlington, NC.	N50SS
☐ N86Y	78	King Air 200	BB-302	York International Holdings Ltd. Basildon, UK.	N300BW
☐ N87AG	79	King Air F90	LA-3	Alan & Diane Gaudenti, San Pedro, Ca.	N1MA
☐ N87BP	75	King Air 200	BB-34	Martinair Inc. Sandston, Va.	N72GC
☐ N87BT	67	680V Turbo Commander	1705-81	City of New York, Long Island City, NY.	N87D
☐ N87CA	78	King Air A100	B-240	James & Patti Latta, Fayetteville, Ar.	(N112CM)
☐ N87CH	72	King Air E90	LW-20	Mayo Aviation Inc. Englewood, Co.	N84LS
☐ N87GA	75	King Air 200	BB-87	Crusair LLC. Miami, Fl.	D-ITEC
☐ N87HB	90	King Air C90A	LJ-1251	Hill Brothers Leasing Co. Falkner, Ms.	N5607X
☐ N87JE	77	King Air B100	BE-27	SkyLife Aviation LLC. St Louis, Mo.	N777CR
☐ N87LP	78	King Air 200	BB-331	Labrador Leasing Corp. Las Vegas, Nv.	N111WA
☐ N87MM	68	King Air B90	LJ-415	Jack Wall Aircraft Sales Inc. Memphis, Tn.	N210X
☐ N87NW	78	King Air B100	BE-42	C & D Aerospace/C & D Plastics Inc. Santa Ana, Ca.	N700SB
☐ N87WS	77	Conquest II	441-0009	TMG Corp. NYC.	N777ED
☐ N87WZ	78	Rockwell 690B	11511	Ben Cart, St Simons Island, Ga.	N81795
☐ N87XX	81	King Air B100	BE-105	P C Design Inc. Manchester, NH.	N28PH
☐ N87YB	69	SA-26AT Merlin 2B	T26-130	Mark Thompson, Adelanto, Ca.	N87Y
☐ N87YP	81	PA-31T Cheyenne II	8120053	Yorktowne Paper Corp. Wilmington, De.	N53TW
☐ N88B	81	Cheyenne II-XL	8166023	F McLeskey, Virginia Beach, Va.	N2577Y
☐ N88CP	76	King Air 200	BB-145	Alcoa Fujikura Ltd. San Antonio, Tx.	N111MT
☐ N88EL	95	Pilatus PC-XII	131	E L 88 Corp. Seattle, Wa.	HB-FOF
☐ N88GC	73	King Air E90	LW-50	Chaparral Inc. Newton, Ks.	N10JQ
☐ N88GL	81	King Air C90	LJ-945	Austin Lopez, St John, USVI.	N979C
☐ N88GW	81	Conquest II	441-0187	Quest LLC. Las Vegas, Nv.	N405JB
☐ N88HL	82	MU-2 Marquise	1542SA	Rough Ryder Aircraft Corp. Wilmington, De.	N468MA
☐ N88HM	71	King Air C90	LJ-502	SQH Air Inc. Burlingame, Ca.	N881K
☐ N88MT	81	King Air 200	BB-830	Standard Commercial Tobacco Co. Wilson, NC.	N88P
☐ N88PD	90	King Air C90A	LJ-1242	Pipe Distributors Inc. Houston, Tx.	N15696
☐ N88QT	82	Conquest II	441-0228	Wyoming Department of Transportation, Cheyenne, Wy.	(N6854X)
☐ N88RP	70	King Air B90	LJ-491	American Logistics Aviation Inc. Longboat key, Fl.	N88RB
☐ N88SD	75	King Air C90	LJ-661	Richmor Aviation Inc. Hudson, NY.	N26CS
☐ N88SP	66	King Air A90	LJ-116A	PCS Computer Inc. Cincinnati, Oh.	N885K
☐ N88TR	73	King Air E90	LW-57	Chancellor Air Ltd. Scotia, NY.	N86TR
☐ N88TW	80	King Air F90	LA-29	Thomas Wheeler, Willoughby, Oh.	N667HE
☐ N89CR	72	Mitsubishi MU-2F	233	Corporate Aviation Services Inc. Tulsa, Ok.	N5NE
☐ N89DR	78	Mitsubishi MU-2N	713SA	R & M Foods Inc. Hattiseburg, Ms.	N874MA
☐ N89MP	76	King Air 200	BB-166	Clearwater Aviation Inc. West Bloomfield, Mi.	N19GB
☐ N89SC	81	MU-2 Marquise	1516SA	Nashville Air Flights Inc. Nashville, Tn.	N4SY
☐ N89TM	73	King Air C90	LJ-610	General Aviation Inc. Helena, Mt.	(N169WD)
☐ N89WA	96	King Air B200	BB-1540	Cloud Air Ltd. Tortola, BVI.	
☐ N90AF	73	King Air E90	LW-29	Astro Capital Lease & Finance, Dover, De.	N1739W
☐ N90AT	76	Rockwell 690A	11272	Paul Schmid, San Carlos, Ca.	N888PB
☐ N90BF	75	King Air E90	LW-124	Pio. Trans. Inc. Cleveland, Tn.	N90GK
☐ N90BP	77	King Air C90	LJ-718	Air Medical Leasing Inc. Lawrenceville, Ga.	
☐ N90CB	69	King Air B90	LJ-473	E & S Manufacturing Inc. Magnolia, Ar.	N9390C
☐ N90CH	96	King Air C90B	LJ-1445	Community Health Management Inc. Jasper, Al.	N209P
☐ N90CT	75	King Air C90	LJ-645	Russell Nelson, Logansport, In.	N9027R
☐ N90DA	72	King Air E90	LW-22	Walden Little, Tyler, Tx.	OB-1602
☐ N90DJ	86	King Air C90A	LJ-1130	Dur Jac Twenty Inc. Montgomery, Al.	N504EC
☐ N90DL	82	King Air C90	LJ-1003	Brandtjen & Kluge Inc. St Croix Falls, Wi.	
☐ N90DN	68	King Air B90	LJ-437	Tiforp Cprp. Dedham, Ma.	N12JG
☐ N90EL	79	Rockwell 690B	11548	Rocky Mountain Aviarion, Chino, Ca.	N29KG
☐ N90EP	85	King Air C90A	LJ-1124	Northern Vending Inc. Shelby, Mt.	D-IALL
☐ N90FA	66	King Air A90	LJ-133	Gateway Technical College, Kenosha, Wi.	N90SA
☐ N90FD	79	King Air F90	LA-5	McKinney Drilling Co. St Albans, WV.	
☐ N90FL	80	King Air F90	LA-33	E L P Aviation Inc. Bossier City, La.	N6726Z
☐ N90FS	80	PA-31T Cheyenne II	8020011	Fort Wayne Air Service Inc. Fort Wayne, In.	N73CA
☐ N90GB	69	King Air B90	LJ-469	C & S Asphalt Sealing Co. Alvin, Tx.	XA-COG
☐ N90GN	66	King Air A90	LJ-157	Little John Air Inc. Danville, Ar.	N288HH
☐ N90GT	83	SA-227TT Merlin 3C	TT-534	E C Menzies Electrical Inc. Wilmington, De.	N139F
☐ N90KA	97	King Air C90B	LJ-1493	Winfield Consumer Products Inc. Winfield, Ks.	
☐ N90KB	91	King Air C90B	LJ-1288	DDK Enterprises LLC. Fayetteville, NC.	N90KA
☐ N90LB	73	King Air C90	LJ-573	Falcon Aviation Ltd. Farmington, NM.	N707RW
☐ N90LF	79	King Air C90	LJ-852	Corporate Jets Inc. West Mifflin, Pa.	YV-250CP
☐ N90LG	68	King Air B90	LJ-351	Sunbird Planes Inc. Dover, De. (status ?).	N62BW
☐ N90MT	68	King Air B90	LJ-404	ATT Aviation Inc. Opelousas, La.	N835BG
☐ N90MU	76	King Air C90	LJ-679	L & W Leasing Inc. Dover, De.	N90ML
☐ N90MV	77	King Air C90	LJ-701	Holmes Leasing Inc. Des Moines, Ia.	N90LJ
☐ N90NB	80	SA-226T Merlin 3B	T-312	Westford Aviation Inc. Clevland, Oh.	
☐ N90PB	76	King Air 200	BB-125	Grant Aviation Inc. Anchorage, Ak.	TG-UGA
☐ N90PH	73	King Air E90	LW-60	HLC Hotels Inc. Savannah, Ga.	N999TB
☐ N90PR	96	King Air C90B	LJ-1437	Plumley Air Service Inc. Paris, Tn.	N1067L

Regn	Yr	Type	c/n	Owner/Operator	Prev Regn
☐ N90PU	83	King Air C90-1	LJ-1046	Purdue Research Foundation, West Lafayette, In.	N644PU
☐ N90PW	76	King Air C90	LJ-681	Philip & Linda Wencek, West Dundee, Il.	N1581L
☐ N90RT	81	King Air F90	LA-146	Save Mart Supermarkets, Modesto, Ca.	N90TX
☐ N90SB	81	King Air F90	LA-154	Activated Metals & Chemicals Corp. Sevierville, Tn.	(C-GFFY)
☐ N90SC	84	Cheyenne II-XL	1166004	Faust Distributing Co. Houston, Tx.	N6667A
☐ N90TD	82	King Air F90	LA-183	Permian Pegasus Inc. Midland, Tx.	N82WC
☐ N90TM	87	King Air B200	BB-1262	Tobin Surveys Inc. San Antonio, Tx.	N2997T
☐ N90TP	80	King Air F90	LA-66	Betteravia Farms, Santa Maria, Ca.	(YV-373CP)
☐ N90TW	84	PA-42 Cheyenne IIIA	5501013	Flying Winds Inc. Pueblo, Co.	N40833
☐ N90VM	81	PA-31T Cheyenne 1	8104031	Monaghan Management Corp. Commerce City, Co.	(N90FJ)
☐ N90VP	67	King Air A90	LJ-276	WRH Aircraft sales, Germantown, Tn.	N574M
☐ N90WE	81	Gulfstream 840	11687	Warren Administration, Midland, Tx.	XA-RWA
☐ N90WJ	71	King Air C90	LJ-525	Blackhawk Technical College, Janesville, Wi.	N70MT
☐ N90WL	69	King Air B90	LJ-461	W L Paris Enterprises Inc. Louisville, Ky.	N453SR
☐ N90WP	94	King Air C90B	LJ-1383	Alpha Aircraft Charter Ltd. Girard, Oh.	
☐ N90WT	73	King Air E90	LW-31	Aviacion Lider SA. Lima, Peru.	N17GD
☐ N90WW	84	PA-31T Cheyenne 1A	1104004	Williams Enterprises Inc. Carmi, Il.	N15AT
☐ N90YA	81	Conquest 1	425-0090	Hopkins Manufacturing Corp. Emporia, Ks.	N90GA
☐ N90ZH	73	King Air C90	LJ-594	Sky Harbor Air Services Inc. Omaha, Ne.	N3094W
☐ N91BM	97	TBM 700	128	Robert Felland, Three Lakes, Wi.	F-WWRP
☐ N91CD	91	King Air B200	BB-1402	Air Methods Corp/Deaconess Medical Center, Billings, Mt.	N5516Q
☐ N91CT	78	Rockwell 690B	11428	Clemson University Athletic Association, Clemson, SC.	N91CS
☐ N91HT	85	King Air B200	BB-1183	Jacob Stern & Sons Inc. Santa Barbara, Ca.	N44MH
☐ N91KA	90	King Air C90A	LJ-1232	Jack Henry & Associates Inc. Monett, Mo.	XA-RGL
☐ N91LW	82	King Air C90	LJ-1001	Midwest Corporate Aviation Inc. Wichita, Ks.	N88EL
☐ N91LY	91	King Air C90B	LJ-1295	Lazzara Yacht Corp. Tampa, Fl.	N91CR
☐ N91MM	71	Mitsubishi MU-2F	198	Anderson Aerial Spraying Service, Canby, Mn.	N71MF
☐ N91RK	76	King Air A100	B-226	X-Cell Speciality Corp. Odessa, Tx.	N9126S
☐ N91TJ	78	King Air C90	LJ-744	Steve Rayman Automotive Inc. Charlotte, NC.	N771PS
☐ N91TS	80	PA-31T Cheyenne II	8020050	Turbo South Inc. Perry, Ga.	C-FSRW
☐ N92CD	90	King Air C90A	LJ-1252	State of Utah, Salt Lake City, Ut.	N55008
☐ N92DE	88	King Air 300	FA-149	Desert Eagle Distrubuting of El Paso Inc. El Paso, Tx.	N87DC
☐ N92DV	78	King Air E90	LW-292	Flagstaff Medical Center, Flagstaff, Az.	N7MA
☐ N92JC		Conquest II	441-0223	The First W D Company Inc. Rockford, Il.	N52GA
☐ N92M	78	King Air 200	BB-382	Balmoral Central Contracts Inc. Durban, RSA.	(N700E)
☐ N92ST	74	Mitsubishi MU-2J	610	Air Carriers Inc. Birmingham, Al.	N342MA
☐ N92TC	81	King Air 200	BB-864	SWV Aviation LLC. Midland, Tx.	N12ST
☐ N92TW	90	King Air 350	FL-3	Noel Aviation Interests Ltd. Odessa, Tx.	(N350PC)
☐ N92V	77	King Air 200	BB-262	Pounds Aircraft Inc. Wilmington, De.	N18762
☐ N92WC	85	King Air 300	FA-69	Dalks Leasing Inc. Spartanburg, SC.	(N438P)
☐ N93A	73	King Air E90	LW-63	Hartford Holding Corp. Millersville, Md.	N808GA
☐ N93AH	73	Mitsubishi MU-2J	581	Trans Check Inc. Carlsbad, Ca.	N161WC
☐ N93BC	70	King Air 100	B-41	Badger Aircraft Leasing Inc. Marshfield, Wi.	N502CW
☐ N93CD	91	King Air B200	BB-1404	Deaconess Medical Center Inc. Billings, Mt.	N41TW
☐ N93CN	80	PA-31T Cheyenne 1	8004029	Jerry & Corrine Nothman, Portland, Or.	N49GG
☐ N93D	77	King Air B100	BE-23	Spectral Enterprises Inc. Grand Haven, Mi.	N13KA
☐ N93DA	81	Conquest 1	425-0013	Dreamer's Aviation Inc. Bryan, Tx.	(N117EE)
☐ N93EJ	68	King Air B90	LJ-388	Georgetown Jet Center Inc. Georgetown, Tx.	N93HA
☐ N93GA	86	King Air 300	FA-93	American Medical Society Inc. Green Bay, Wi.	XA-MIX
☐ N93HC	77	Conquest II	441-0011	Horizons Inc. Rapid City, SD.	N900JT
☐ N93KA	80	King Air F90	LA-24	CRST Inc/P S Air Inc. Cedar Rapids, Ia.	
☐ N93LP	80	King Air C90	LJ-901	Penn Warranty Corp. Taylor, Pa.	N23LF
☐ N93LV	76	King Air 200	BB-157	Lockheed Martin Vought Systems Corp. Grand Prairie, Tx.	N1200M
☐ N93MA	82	Gulfstream 1000	96035	Eagle Air Inc. Memphis, Tn. (now XC-HFZ ?).	D-IBER
☐ N93ME	89	King Air B200	BB-1345	Bullfrog of Ohio County Inc. Owensboro, Ky.	
☐ N93NM	82	Gulfstream 1000	96020	Ozark Management Inc. Jefferson City, Mo.	(N139WJ)
☐ N93NP	84	King Air B200	BB-1184	N Philadelphia Aviation Corp/Summit Aviation, Middletown, De	N27SE
☐ N93RA	68	680W Turbo Commander	1722-2	Cx USA 2/97 to ? (was 1722-93).	N93D
☐ N93RY	66	King Air A90	LJ-174	Ryan Aircraft LC. Wichita, Ks.	(N193SF)
☐ N93SF	76	King Air B100	BE-13	Winter Haven Homes Inc. Atlanta, Ga.	N4213S
☐ N93SH	79	PA-31T Cheyenne 1	7904029	Scope Leasing Inc. Columbus, Oh.	N29CA
☐ N93UM	71	Mitsubishi MU-2G	537	Bush Field Aircraft Co. Augusta, Ga. (status ?).	N161MA
☐ N93WT	80	King Air B100	BE-80	Wes-Tex Drilling Co. Abilene, Tx.	N13DR
☐ N93ZC	80	King Air 200	BB-654	Ziegler Inc. Minneapolis, Mn.	XB-BXU
☐ N94AC	78	Rockwell 690B	11486	Apollo Aviation Inc. Middletown, Ct.	N34EF
☐ N94AM	74	King Air C90	LJ-640	Chapparral Exploration Inc. College Station, Tx.	N4484W
☐ N94CD	81	King Air C90	LJ-939	Asset Investments Series A LLC. Wilmington, De.	N26803
☐ N94CS	84	PA-42 Cheyenne IIIA	5501016	Northern Air Aviation LLC. Batavia, NY.	N536CA
☐ N94EA	85	Gulfstream 1000	96094	Midtex Investments Inc. Austin, Tx.	N131GA
☐ N94EW	81	PA-42 Cheyenne III	8001054	Sea LLC. Grand Island, Ne.	N29HS
☐ N94FG	79	King Air 200	BB-433	Flex-Air Ltd. Urbana, Il.	N133GA
☐ N94HB	80	King Air C90	LJ-904	Central Flying Service Inc. Little Rock, Ar.	N90HB
☐ N94JK	80	MU-2 Marquise	761SA	Hartford Holding Corp. Millersville, Md.	LV-OAN
☐ N94KC	76	King Air 200	BB-172	Alliant Hospitals Inc. Louisville, Ky.	N68RR
☐ N94LC	85	King Air B200	BB-1224	Ranger Corp. Wilmington, De.	
☐ N94LL	95	King Air B200	BB-1506	Raytheon Aircraft Co. Wichita, Ks.	N3216L

Regn	Yr	Type	c/n	Owner/Operator	Prev Regn

Regn	Yr	Type	c/n	Owner/Operator	Prev Regn
☐ N94PA	82	Gulfstream 1000	96005	Quasar Aviation, Dover, De.	N815S
☐ N94QD	82	King Air B200	BB-1001	Burger Management South Bend 3 Inc. Mishawaka, In.	N94GA
☐ N94SA	65	Gulfstream 1	157	World Sales & Turbine Co. El Dorado, Panama.	N741G
☐ N94SR	89	King Air C90A	LJ-1221	Sylco Aviation Inc. Advance, NC.	N416P
☐ N94TK	94	King Air C90B	LJ-1358	Tem-Kil Co. Macogdoches, Tx.	N80927
☐ N94U	81	King Air F90	LA-124	Black & Co/Laurel Aviation LLC. Decatur, Il.	N13
☐ N95AC	81	SA-226T Merlin 3B	T-381	Reprinter (Delaware) Inc. Miami, Fl.	N80MJ
☐ N95BM	91	TBM 700	20	Robert Felland, Three Lakes, Wi.	(N700RF)
☐ N95GR	77	Rockwell 690B	11411	Aero Instruments & Avionics Inc. N Tonawanda, NY.	N81648
☐ N95JJ	77	King Air B100	BE-18	Fan 4 Inc. Fort Lauderdale, Fl.	N95CM
☐ N95JM	76	Rockwell 690A	11289	Valley Oil/Pond LLC. Salem, Or.	(N27JT)
☐ N95LB	73	King Air E90	LW-24	Bell-Carter Foods, Lafayette, Ca.	N95RB
☐ N95MJ	84	MU-2 Marquise	1564SA	Interlease Aviation Corp. Northfield, Il.	N5PC
☐ N95PC	85	King Air C90A	LJ-1109	J C Faw/Prism Companies, Wilkesboro, NC.	
☐ N95TT	81	King Air B200	BB-917	Texas Utilities Services Inc. Dallas, Tx.	N15TT
☐ N95UF	66	King Air 90	LJ-78	Anton Ptach, Poughkeepsie, NY.	N730K
☐ N95VR	81	PA-42 Cheyenne III	8001049	Blue Dolphin Aviation Inc. Largo, Fl.	C-FIZG
☐ N96AG	67	King Air A90	LJ-260	AG Trucking Inc. Goshen, In.	N303NH
☐ N96AH	75	King Air C90	LJ-643	AH Transmet Inc. Easton, Md.	N909HC
☐ N96AM	73	King Air A100	B-162	Texas A & M University, College Station, Tx.	N3DG
☐ N96DQ	79	King Air C90	LJ-814	King Air Aviation LLC. Ashland, Va.	N96DC
☐ N96GP	89	King Air B200	BB-1322	Caribbean Helicorp Inc. Hato Rey, PR.	N391L
☐ N96KA	90	King Air 350	FL-36	Cyprus Center Services Inc. Albuquerque, NM.	N4200K
☐ N96MA	78	PA-31T Cheyenne 1	7804005	Construction Concepts Inc. Oakland, Tn.	N6095A
☐ N96RL	68	SA-26AT Merlin 2B	T26-116	AVSAT Management Inc. San Antonio, Tx.	N100AW
☐ N96TT	80	King Air F90	LA-26	Katemo Corp. Emerald Isle, NC.	N6668C
☐ N96UB	77	King Air 200	BB-267	MOO Air Inc. Nashville, Tn.	N194WS
☐ N96WF	96	Pilatus PC-XII	139	David Arnold, Henniker, NH.	HB-FRA
☐ N96ZZ	82	King Air 200	BB-1035	American Equipment Racing Enterprise, Charlotte, NC.	N59TD
☐ N97AB	70	SA-226T Merlin 3	T-203	Reese Aircraft Inc. Indianapolis, In.	N5273M
☐ N97CV	97	King Air C90B	LJ-1492	Piney Branch Motors Inc/Allied Trailers, Savage, Md.	N2029Z
☐ N97DR	96	King Air 350	FL-133	Danchar Inc/Sun Aviation Inc. Vero Beach, Fl.	
☐ N97EB	86	King Air 300	FA-97	MAACO Enterprises Inc. King of Prussia, Pa.	N50NL
☐ N97KA	97	King Air C90B	LJ-1469	Crutchfield Corp. Charlottesville, Va.	
☐ N97PC	80	PA-31T Cheyenne II	8020034	King Forest Industries Inc. Wentworth, NH.	N4241Y
☐ N97SF	79	King Air C90	LJ-818	Seneca Foods Corp. Penn Yan, NY.	
☐ N97TW	79	PA-31T Cheyenne II	7920051	Grant Cox, Las Vegas, Nv.	
☐ N97WD	80	King Air B100	BE-97	Dash Multi-Corp Inc. St Louis, Mo.	N43RJ
☐ N98AJ	78	Rockwell 690B	11458	Cascadian Aviation LLC. Fircrest, Wa.	ZS-KEF
☐ N98B	66	King Air 90	LJ-87	Aerosports Inc. Beaverton, Or.	
☐ N98DA	97	King Air B200	BB-1555	First American Bank Texas SSB. Bryan, Tx.	
☐ N98DD	67	King Air A90	LJ-195	HC&Y Club/David J Ray, Asuncion, Paraguay.	N31CP
☐ N98EP	85	Conquest 1	425-0230	Ernest Pestana, Charleston, SC.	N1227V
☐ N98GF	80	PA-31T Cheyenne 1	8004009	Travel Express Aviation Ltd. St Charles, Il.	N2412W
☐ N98HB	67	King Air A90	LJ-285	Hermann Bodmer & Co. Zurich, Switzerland.	PH-IND
☐ N98MR	72	Rockwell 690	11022	Windmaster Manufacturing,	N14BH
☐ N98PC	75	King Air E90	LW-131	Alistar Beverages Corp. Latham, NY.	N9031R
☐ N98TB	78	PA-31T Cheyenne II	7820040	Jocund Inc. Knoxville, Tn.	N68MB
☐ N98TG	79	PA-31T Cheyenne 1	7904027	Scott Bell, Greeley, Co.	N98TW
☐ N98UC	81	SA-226T Merlin 3B	T-378	Southwest First Street Corp. Highlands, NC.	N1011P
☐ N99AC	75	King Air E90	LW-120	Design Technologies Inc. Memphis, Tn.	N9GC
☐ N99BT	73	Mitsubishi MU-2J	591	S & W Aircraft Sales Inc. Wilmington, De.	N295MA
☐ N99CD	73	King Air C90	LJ-601	B & E Leasing Inc. Jonesboro, Ar.	HP-918
☐ N99JW		SA-227TT Merlin 3C	TT-450	Bachman Co. Reading, Pa.	N53PC
☐ N99KF	79	PA-31T Cheyenne II	7920093	SST Corp. Klamath Falls, Or.	N2416R
☐ N99LL	82	King Air B200	BB-994	Alabama River Pulp Co. Purdue Hill, Al.	N25CN
☐ N99ML	96	King Air C90B	LJ-1460	Black River Construction Services, Black River Falls, Wi.	N1067K
☐ N99RK	70	SA-26AT Merlin 2B	T26-173	A G P L Inc & R L Bolin, Wichita Falls, Tx.	N49MZ
☐ N99SR	74	Mitsubishi MU-2K	315	Jetset Ltd. Cliffside Park, NJ.	N505MA
☐ N99VA	77	PA-31T Cheyenne II	7720007	Golden Nugget Aviation Corp. Las Vegas, Nv.	N721RP
☐ N99WC	76	Rockwell 690A	11308	David Lowe & Jim Poole, Dalhart, Tx.	N81419
☐ N100AM	77	Rockwell 690B	11361	Madonna Inns Inc. San Luis Obispo, Ca. (status ?).	(N81536)
☐ N100BG	90	King Air 350	FL-6	Brown Group Inc. St Louis, Mo.	N5643B
☐ N100BY	84	MU-2 Marquise	1565SA	Peter Luce, Englewood, Co.	N530DP
☐ N100CC	83	Conquest 1	425-0170	Riversville Aircraft Corp. Greenwich, Ct.	N68721
☐ N100CE	80	SA-226T Merlin 3B	T-325	Mi-Ka Aviation Inc. Boulder City, Nv.	HC-BUA
☐ N100CF	72	Mitsubishi MU-2F	229	Saxton Leasing inc. Plymouth, Mi.	N217MA
☐ N100CM	80	PA-31T Cheyenne II	8020073	Columbus Mills Inc. Columbua, Ga.	N2474V
☐ N100EC	75	King Air E90	LW-150	R & M Leasing Inc. New Castle, De.	N23AE
☐ N100FL	85	King Air 300	FA-34	Dept of Management Services, Tallahassee, Fl.	N865M
☐ N100GL	70	681 Turbo Commander	6028	Gala Air Taxi Inc/West Indies Sky Train, Grand Turk.	N9028N
☐ N100GV	72	King Air A100	B-116	King Air Corp. South Burlington, Vt.	N531DF
☐ N100HC	76	King Air A100	B-229	Bristol-American Holdings Inc. Tulsa, Ok.	N15LR
☐ N100HS	81	King Air C90	LJ-949	Executive Beechcraft Inc. Kansas City, Mo.	N8300C
☐ N100JD	72	King Air C90	LJ-555	J Hollingsworth on Wheels Inc. Greenville, SC.	N9498Q
☐ N100JF	67	King Air A90	LJ-292	BLM Aircraft Inc. Lake Mary, Fl.	N100HM

Regn	Yr	Type	c/n	Owner/Operator	Prev Regn
☐ N100KA	70	King Air 100	B-11	B & B Contractors Developer Inc. Youngstown, Oh.	
☐ N100KB	79	King Air C90	LJ-820	V F Management Co. Roanoke, Va.	
☐ N100KE	74	Mitsubishi MU-2P	313SA	Graben Family Trust, Rhome, Tx.	N100KP
☐ N100NP	80	MU-2 Solitaire	423SA	Retama Corp. Wilmington, De.	N156MA
☐ N100RU	96	King Air C90B	LJ-1433	Rental Uniform Services, Florence, SC.	N3254A
☐ N100SF	75	King Air 200	BB-32	Safe Flight Instrument Corp. White Plains, NY.	
☐ N100SM	77	King Air 200	BB-269	State of Missouri, Jefferson City, Mo.	N18269
☐ N100SN	69	SA-26AT Merlin 2B	T26-129	K & K Industries Inc. Montgomery, In.	N784AF
☐ N100TW	78	King Air B100	BE-51	Titan Wheel International Inc. Quincy, Il.	N1DA
☐ N100UE	66	King Air A90	LJ-138	Fayard Enterprises Inc. Louisberg, NC.	N100UF
☐ N100V	78	King Air C90	LJ-796	Legacy Aviation Inc. Monroe, La.	
☐ N100VK		PA-46 Turbo Malibu	1	V K Leasing Inc. St Petersburg, Fl.	
☐ N100WB	66	King Air A90	LJ-139	Brown Boy Aviation, Royal City, Wa.	N1515T
☐ N100WQ	78	PA-31T Cheyenne II	7820084	SBM Aviation Inc. Pascagoula, Ms.	N450DA
☐ N100WT	80	PA-31T Cheyenne 1	8004006	Windle Turley, Dallas, Tx.	N108SW
☐ N100Y	85	King Air 300	FA-68	Saratoga Inc. Lebanon, Or.	N100UF
☐ N100ZM	71	King Air 100	B-88	Lee Air LLC. Dover Plains, NY.	N100GM
☐ N101BS	68	King Air B90	LJ-375	Bruce Scheffer, Valparaiso, In.	N90RW
☐ N101BU	85	King Air C90A	LJ-1107	Baylor University, Waco, Tx.	N72260
☐ N101CA	82	Conquest 1	425-0142	State Aircraft Pooling Board, Austin, Tx.	OH-CIK
☐ N101GQ	79	King Air 200	BB-427	American Capital Corp. Chesterfield, Mo.	N101CC
☐ N101RG	78	Rockwell 690B	11463	Joe McShane III, Midland, Tx.	N93SA
☐ N101RW	79	Rockwell 690B	11517	Teamlease Inc. Fayetteville, NC.	N9054F
☐ N101SG	97	King Air C90B	LJ-1496	Andair Inc. McAllen, Tx.	
☐ N101SN	81	MU-2 Marquise	1505SA	Trace Air Inc. Orange Beach, Al.	(N81TJ)
☐ N101SQ	90	King Air C90A	LJ-1231	Fonvielle & Hinkle PA. Tallahassee, Fl.	N101SG
☐ N102AJ	75	King Air C90	LJ-649	GFSN, Wisconsin Dells, Wi.	OO-MCL
☐ N102BG	79	MU-2 Marquise	748SA	Bankair Inc. West Columbia, SC.	(N305PC)
☐ N102CS	94	King Air 350	FL-126	King Aire Inc. Van Nuys, Ca.	N3164C
☐ N102FG	72	King Air A100	B-131	University of Florida Athletics Assoc. Gainesville, Fl.	N90CC
☐ N102FL	81	King Air C90	LJ-982	State of Florida, Tallahassee, Fl.	N1820P
☐ N102JC	97	Jetcruzer 500	001	Advanced Aerodynamics & Structures Inc. Long Beach, Ca.	
☐ N102LF	70	King Air 100	B-65	Lynch Flying Service Inc. Billings, Mt.	N102RS
☐ N102PG	84	King Air 300	FA-19	The Pape' Group Inc. Coburg, Or.	N101PC
☐ N102WK	80	King Air F90	LA-36	Willis Knighton Medical Center, Shreveport, La.	N6736C
☐ N103AL	80	King Air 200	BB-730	Broad Street Aviation Inc. Charleston, SC.	N572AT
☐ N103BG	83	King Air B200	BB-1150	Boyd Central Region Inc. Memphis, Tn.	N18DN
☐ N103BN	77	King Air C-12C	BC-47	Mississippi Bureau of Narcotics, Meridian, Ms.	77-22936
☐ N103CB	81	King Air F90	LA-98	Charles E Brown Beverage Co. Lebanon, Mo.	N77JT
☐ N103FG	74	King Air C90	LJ-363	Universityof Florida Athletic Association, Gainesville, Fl.	N100VM
☐ N103FL	75	King Air C90	LJ-650	State of Florida, Tallahassee, Fl.	N9077S
☐ N103PA	70	SA-226T Merlin 3	T-202	Aircraft Guaranty Corp. Houston, Tx.	C-FDAC
☐ N103PM	78	King Air 200	BB-304	Prewett Hosiery Sales, Fort Payne, Al.	N451DB
☐ N103Q	73	Mitsubishi MU-2J	580	Roland Machinery Co. Springfield, Il.	N103BB
☐ N103RC	75	Mitsubishi MU-2L	673	River City Aviation Inc. West Memphis, Ar.	N4565E
☐ N103SP	80	PA-31T Cheyenne II	8020024	Indiana State Police, Indianapolis, In.	
☐ N103TF	80	King Air 200	BB-649	Jasper Engine Exchange Inc. Jasper, In.	N360CB
☐ N104AJ	88	King Air C90A	LJ-1164	Luwe Flug Inc. Bayreuth, Germany.	ZS-MIL
☐ N104BR	76	SA-226T Merlin 3A	T-269	Benchmark Research & Technology Inc. Midland, Tx.	N5039F
☐ N104JB	80	MU-2 Solitaire	435SA	John Broome Ranches, Oxnard, Ca.	N213MA
☐ N104LC	78	King Air C90	LJ-757	Occupational Consultants Inc. Memphis, Tn.	N23675
☐ N104LS	81	King Air B100	BE-115	Hickory Air LLC. Hickory, NC.	N126HU
☐ N104TA	77	SA-226T Merlin 3A	T-289	Berry Aviation Inc. San Marcos, Tx	(F-GBBD)
☐ N104Z	69	King Air B90	LJ-472	U S Department of Agriculture, Milwaukee, Wi..	
☐ N105CG	79	King Air C90	LJ-806	Hickory Tech Corp. Mankato, Mn.	N1MB
☐ N105FC	81	Conquest 1	425-0094	Fred & Christina Carroll, San Francisco, Ca.	N104HW
☐ N105K	66	King Air 90	LJ-113	Airlift Inc. Hollister, Ca.	N16CS
☐ N105RG	69	King Air B90	LJ-454	Private Air LLC. Oklahoma City, Ok.	N105RJ
☐ N105TC	84	King Air C90A	LJ-1086	Town & Country Food Stores Inc. San Angelo, Tx.	N18182
☐ N105WM	78	Mitsubishi MU-2N	709SA	LOU Corp. Paducah, Ky.	N105NM
☐ N106PA	79	King Air 200	BB-428	ALG Inc. Raleigh, NC.	G-BFWI
☐ N106TT	80	Gulfstream 840	11630	Skytel Aviation Inc.	D-INRO
☐ N106WA	94	Pilatus PC-XII	106	Western Aircraft Inc. Boise, Id.	N601BM
☐ N107AJ	82	King Air F90	LA-189	Ernest Spencer Inc. Meriden, Ks.	ZS-LFP
☐ N107FL	76	King Air 200	BB-150	State of Florida, Tallahassee, Fl.	N203SF
☐ N107GL	79	Rockwell 690B	11554	Gurley Leep Oldsmobile/Cadillac, Mishawaka, In.	N84WA
☐ N107MC	82	Cheyenne II-XL	8166032	W C McQuaide Inc/Penn Air Inc. Altoona, Pa.	N48CP
☐ N107MG	82	King Air B200	BB-924	BK Associates Inc. Atlanta, Ga.	N52YR
☐ N107SC	78	King Air C90	LJ-788	Southern Cross Aviation Inc. Fort Lauderdale, Fl.	N2030W
☐ N107Z	82	King Air B200C	BL-124	U S Department of Agriculture, Boise, Id.	
☐ N108BM	76	King Air 200	BB-108	Buddie Markum, Fort Worth, Tx.	RP-C1979
☐ N108EB	81	King Air B100	BE-108	Mill Steel Co/N108EB Inc/Rockford Constru. Grand Rapids, Mi.	N38005
☐ N108KU	73	King Air C90	LJ-568	Dodson International Parts Inc. Ottawa, Ks.	N100KU
☐ N108SA	77	Rockwell 690B	11416	Lund Assocs Inc. Rapid City, SD.	N81658
☐ N108SC	71	Mitsubishi MU-2G	545	William Whitehead, Llano, Tx. (status ?).	N169MA
☐ N108TJ	79	Conquest II	441-0108	Martin's Super Markets Inc. South Bend, In.	N104RD

Regn	Yr Type	c/n	Owner/Operator	Prev Regn

Regn	Yr	Type	c/n	Owner/Operator	Prev Regn
□ N108UC	80	PA-31T Cheyenne II	8020018	Eastgate Energy Services Ltd. Fairfield, Ia.	N802HC
□ N109DT	85	King Air C90A	LJ-1102	Supermarket Information Systems, Clearwater, Fl.	N682TA
□ N109P	63	Gulfstream 1	109	Phoenix Air Group Inc. Cartersville, Ga.	N307AT
□ N109TM	76	King Air 200	BB-124	M & F Air Inc. Greenville, NC.	N124AJ
□ N109TT	79	PA-31T Cheyenne II	7920082	Eastern Air Center Inc. Norwood, Ma.	N56HF
□ N110BM	76	King Air 200	BB-110	D & B Air Corp. Austin, Tx.	OE-FIM
□ N110EL	65	King Air 90	LJ-71	Avlease Inc. Wilmington, De.	N1100X
□ N110G	81	King Air 200	BB-792	Gared Graphics Inc. Camarillo, Ca.	
□ N110GC	77	Mitsubishi MU-2P	363SA	Engineered Products Co. San Juan, PR.	(N178GV)
□ N110MA	74	Mitsubishi MU-2J	616	Wings LLC. Pelham, Al.	SE-IUB
□ N110RB	64	Gulfstream 1	126	Rotec Industries Inc/R O Air Inc. Elmhurst, Il.	N63AU
□ N111AB	81	SA-226T Merlin 3B	T-407	Sunlight Corp. Dover, De.	
□ N111CT	76	PA-31T Cheyenne II	7620047	C L Thomas Inc. Victoria, Tx.	N82039
□ N111FL	74	Rockwell 690A	11163	Legal Air Flight Services Inc. Troutdale, Or. (status ?).	N9769S
□ N111FW	74	King Air E90	LW-105	Justin Industries Inc. Fort Worth, Tx.	N205CA
□ N111HF	71	Mitsubishi MU-2F	208	DBH Attachments Inc. Adamsville, Tn.	N288MA
□ N111JA	74	King Air E90	LW-84	Thorleifur Juliusson, El Paso, Tx.	TF-DCA
□ N111JW	81	King Air B200	BB-886	J A Whittenburg, Raton, NM.	
□ N111KU	86	Conquest 1	425-0226	G E C Precision Corp. Wellington, Ks.	N1226Z
□ N111KV	79	PA-31T Cheyenne II	7920035	American Medflight Inc. Reno, Nv.	N110MP
□ N111LP	80	King Air 200	BB-743	Agstar Inc/Lightnin' Truck Rentals, Lawrenceville, Ga.	N538AS
□ N111M	86	King Air 300	FA-90	Ingram Industries Inc. Nashville, Tn.	N77PA
□ N111MD	94	King Air C90SE	LJ-1367	Argosy Aircraft Inc. Reno, Nv.	N15599
□ N111NS	81	King Air 200C	BL-36	Columbia Helicopters Inc. Portland, Or.	
□ N111PV	81	King Air 200	BB-772	Peavey Electronics Corp. Meridian, Ms.	N111F
□ N111RC	81	Conquest 1	441-0188	N L Five Inc. Boca Raton, Fl.	N441GP
□ N111SK	67	680V Turbo Commander	1710-85	Bergman Photographic Services Inc. Portland, Or.	N111ST
□ N111SU	84	Conquest 1	425-0205	Pilgrims Holdings Ltd. Hickory Corners, Mi.	N105TC
□ N111UT	78	King Air 200	BB-374	University of Tennessee, Alcoa, Tn.	N7QR
□ N111YF	77	King Air B100	BE-30	Eagle Resources of Maine Inc. Mechanic Falls, Me.	N110EC
□ N112AB	85	King Air 300	FA-85	Arkansas Best Corp. Fort Smith, Ar.	N92GC
□ N112AF	95	Pilatus PC-XII	120	Alpha Flying/Rigi Inc. Nashua, NH.	HB-FQN
□ N112BB	81	PA-31T Cheyenne 1	8104021	Boyd Brothers Transportation, Clayton, Al.	N633AB
□ N112BC	81	PA-42 Cheyenne III	8001027	Williams Contraction Group, Stone Mountain, Ga.	
□ N112BL	77	PA-31T Cheyenne II	7720037	North Dakota Department of Transport, Bismarck, ND.	
□ N112CE	85	Gulfstream 1000	96097	Clark Transportation Co. Bethesda, Md.	N134GA
□ N112ED	80	PA-31T Cheyenne II	8020060	Darby Aviation Inc. Sheffield, Al.	N703CJ
□ N112EF	73	Rockwell 690A	11123	Executive Air Taxi Corp. Bismark, ND.	N122PG
□ N112EM	76	Rockwell 690A	11330	E Micah Aviation Inc. Cincinnati, Oh.	N567H
□ N112HF	87	King Air 300	FA-122	Fuller Renting & Leasing GmbH. Auburn, Al.	EC-FLX
□ N112MA	77	Mitsubishi MU-2N	689SA	International Business Aircraft Inc. Tulsa, Ok.	C-FTAD
□ N112SB	77	King Air E90	LW-232	Seven Bar Flying Service Inc. Albuquerque, NM.	N232CL
□ N113CT	70	681 Turbo Commander	6006	Bohlke International Airways, Kingshill, USVI.	N2725B
□ N113GW	79	King Air 200	BB-541	Emerald Aviation LLC. Bakersfield, Ca.	(N23WL)
□ N113RC	75	PA-31T Cheyenne II	7520009	Bob Lippner Pilot Services Inc. Alcoa, Tn.	N2654Z
□ N113SD	73	Mitsubishi MU-2J	600	Lindbergh Leasing Inc. St Louis, Mo.	N707TT
□ N113US	87	King Air B200	BB-1283	United Supermarkets Inc. Lubbock, Tx.	N200KA
□ N113WC	80	PA-42 Cheyenne III	8001005	Worsley Companies Inc. Wilmington, NC.	N13TT
□ N114CW	66	King Air A90	LJ-114	Colorado Well Service Inc. Rangely, Co.	N114KA
□ N114J	77	King Air C90	LJ-713	Robin Air Services Inc. Orlando, Fl.	N11TE
□ N114JF	77	King Air 200	BB-237	Charles Hodge & Air Investors Inc. Spartanburg, SC.	N717RM
□ N114JR	80	PA-31T Cheyenne 1	8004037	Rusk Cattle Co. Austin, Tx.	N900PC
□ N115AB	74	Rockwell 690A	11231	Allmand Brothers Inc. Holdrege, Ne.	N115BH
□ N115AP	68	Mitsubishi MU-2F	136	Earle Martin, Houston, Tx.	N75RJ
□ N115DT	75	King Air A100	B-212	Air Serv International Inc. Redlands, Ca.	N115D
□ N115GA	85	Gulfstream 1000B	96201	Horizon Aircraft Sales Inc. Miami, Fl.	
□ N115KU	82	King Air C90-1	LJ-1040	University of Kansas Medical Center, Kansas City, Ks.	N115MX
□ N115MX	89	King Air C90A	LJ-1203	Smithway Motor Xpress Inc. Fort Dodge, Ia.	N505EB
□ N115PA	66	King Air A90	LJ-117	Angus Hastings, Fort McCoy, Fl.	N10430
□ N115PC	78	PA-31T Cheyenne II	7820001	Saratoga Air Service Inc. Panama City, Fl.	N711FN
□ N116AF	96	Pilatus PC-XII	137	CML Air Inc & Rigi Inc/Keyson Airways, Nashua, NH.	HB-FQZ
□ N116DG	81	King Air B100	BE-116	Directors Air Corp. Reno, Nv.	N116AC
□ N116JP	81	PA-42 Cheyenne III	8001050	Turbines Inc. Terre Haute, In.	N4098K
□ N117CA	81	King Air 200	BB-873	Casino America Inc. Biloxi, Ms.	N48BA
□ N117FH	76	King Air E90	LW-194	Buck Inc. Fuquay Marina, NC.	N194KA
□ N117H	79	MU-2 Marquise	751SA	Berlyn Industries Inc. New Castle, NH.	N948MA
□ N117MF	78	King Air C90	LJ-779	Mercy Flights Inc. Medford, Or.	(N107RE)
□ N117TJ	80	King Air F90	LA-124	RaCon Inc. Tuscaloosa, Al.	N19EG
□ N117W	80	King Air F90	LA-20	Wallace Oil Co. Smryna, Ga.	N754D
□ N118AF	97	Pilatus PC-XII	175	Kinder Properties Inc & co-owners,	HB-FSK
□ N118BW	80	PA-31T Cheyenne 1	8004001	Aircraft Guaranty LLC. Houston, Tx.	D-IHJJ
□ N118CA	81	King Air C-12C	BC-57	California Highway Patrol, Sacramento, Ca.	77-22946
□ N118CR	76	Rockwell 690A	11276	Hokie Airco Inc. Roanoke, Va.	N57172
□ N118GW	87	King Air 300	FA-119	Super King Air LLC. Taft, Ca.	N688RL
□ N118P	74	Mitsubishi MU-2J	646	Tulip Aviation Inc. Granville, Il.	N113P
□ N118WC	80	PA-31T Cheyenne II	8020020	Halair Inc. Wilmington, De.	N113WC

Regn	Yr	Type	c/n	Owner/Operator	Prev Regn
N119EB	77	PA-31T Cheyenne II	7720012	Michael Harper, Duncanville, Tx.	N79ML
N119MC	79	King Air 200	BB-572	Gene & Mary McAuley, Proctor, Ar.	N972LL
N119WM	80	King Air 200	BB-662	Coast Hotels & Casinos Inc. Las Vegas, Nv. (status ?).	N117WM
N120EK	83	Gulfstream 900	15026	GECC, Wilton, Ct.	N28BF
N120FA	82	SA-227AT Merlin 4C	AT-461	Transair America Inc. Buffalo Grove, Il.	F-GGLF
N120JM		SA-227AT Merlin 4C	AT-577	Path Corp. Rehoboth Beach, De.	N31136
N120P	81	King Air 200	BB-786	Duckwall-ALCO Stores Inc. Abilene, Ks.	
N120PR	83	King Air B200T	BT-29	Commonwealth of Puerto Rico, San Juan, PR.(was BB-1097).	N150BA
N120RC	81	King Air F90	LA-117	R R Cassidy Inc. Baton Rouge, La.	N7775
N120RJ	79	King Air 200	BB-423	Burt Aviation Inc. Rochester Hills, Mi.	N300KC
N120SC	78	SA-226AT Merlin 4A	AT-067	River City Aviation Inc. West Memphis, Ar.	C-FJTL
N120TM	81	SA-226T Merlin 3B	T-405	Rosemary Medders, Wichita Falls, Tx.	N189CC
N121B	72	King Air E90	LW-21	Goodland Regional Medical Center, Goodland, Ks.	
N121BA	73	Mitsubishi MU-2J	599	Priority Air Inc. New Orleans, La.	N304MA
N121BE	80	PA-31T Cheyenne 1	8004036	Bruce Erickson, Great Falls, Mt.	N657DC
N121CH	73	Mitsubishi MU-2K	265	Air 1st Aviation Companies Inc. Aiken, SC.	SE-IOY
N121EB	94	King Air C90B	LJ-1372	Designer Programs Inc. Raleigh, NC.	N151JL
N121EG	73	King Air E90	LW-71	Transfur Air Inc. Wilmington, De.	
N121HC	68	King Air B90	LJ-392	Chapparal Inc. Newton, Ks.	N392CA
N121KB	68	SA-26T Merlin 2A	T26-33	Eastern Aviation & Marine Underwriters, Towson, Md.	N1215S
N121LB	79	King Air 200	BB-475	Arkansas Aircraft Inc. Jonesboro, Ar.	
N121P	81	King Air C90	LJ-970	Sundance Aviation Inc. Galesburg, Il.	
N121RF	95	Pilatus PC-XII	114	Richard A Foreman Associates Inc. Wilmington, De.	N114SV
N121RH	85	King Air 300	FA-84	Hunter Ltd. Santa Barbara, Ca.	N7251K
N122CK	78	Mitsubishi MU-2P	374SA	Air 1st Aviation Companies Inc. Aiken, SC.	N9HA
N122H	90	King Air B200	BB-1370	Fox Hill Inc. Cleveland, Oh.	N75LC
N122K	77	King Air C90	LJ-707	Executive Leasing Inc. Shelby, NC.	(N32HF)
N122LC	94	King Air B200	BB-1490	Lowe's Companies Inc. North Wilkesboro, NC.	N1563N
N122MA	71	Mitsubishi MU-2F	207	ZMP Corp. Wilmington, De.	
N122NC	75	King Air E90	LW-125	State of North Carolina, Raleigh, NC.	
N122RF	76	King Air 200	BB-122	Que' Pass Aviation 1 Ltd. Houston, Tx.	N122TJ
N122RG	78	King Air C90	LJ-746	Granberry Supply Corp. Phoenix, Az.	(N122RG)
N122SC	82	King Air F90	LA-158	Sunrise Community Inc. Miami, Fl.	N123GM
N122U	70	King Air 100	B-32	H R Air Inc. Mentor, Oh.	N122H
N123AF	85	King Air 300	FA-46	MacMillan Bloedel Inc. Montgomery, Al.	N7222U
N123AG	81	PA-42 Cheyenne III	8001031	Aqua Glass Corp. Adamsville, Tn.	N123CN
N123BL	80	King Air B100	BE-83	Gary Mathews Motors Inc. Clarksville, Tn.	N412FC
N123CH	80	King Air F90	LA-32	Jet Sales & Services Inc. Park City, Ut.	N969MC
N123GM		SA-227TT Merlin 3C	TT-512A	Camelot Aviation Ventures Inc. Coral Gables, Fl.	N927DC
N123LH	82	SA-227TT Merlin 3C	TT-433	North Park Aviation LLC. San Marcos, Tx.	C-GFCE
N123LL	80	King Air C90	LJ-885	ZZ Leasing Inc. Memphis, Tn.	N1WB
N123LN	78	King Air E90	LW-271	Robert Ingraham Homes Inc. Charlotte, NC.	N23798
N123ME	88	King Air B200	BB-1299	Elliott Aviation Aircraft Sales Inc. Des Moines, Ia.	N123ML
N123MH	74	King Air E90	LW-104	Mark Osborn, Birmingham, Al.	N133PL
N123ML	97	King Air B200	BB-1587	George Beitzel, Chappaqua, NY.	N3270Q
N123NA	82	King Air C90	LJ-1006	Allianz Life Insurance Co. Minneapolis, Mn.	N61188
N123SC	82	Conquest 1	425-0117	Salvatore Cangiano Ltd. Sioux Falls, SD.	N6881L
N123V	67	King Air A90	LJ-258	Milwaukee Area Technical College, Milwaukee, Wi.	N915BD
N123WH	82	King Air B100	BE-126	Hilty Quarries Inc. Clinton, Mo.	D-IALT
N124BB	84	King Air 300	FA-2	South Beach Co. Palm Beach, Fl.	N32323
N124BW	95	Pilatus PC-XII	124	RAW Inc. Raleigh, NC.	HB-FQP
N124CM	84	King Air 300	FA-24	Carter Machinery Co Inc. Salem, Va.	N300KA
N124DP	84	PA-42 Cheyenne 400LS	5527018	Amistad Aviation Inc. Amarillo, Tx.	N4112Z
N124JS	83	King Air B200C	BL-64	J R Simplot Co. Boise, Id.	N6912F
N124MB	85	King Air C90A	LJ-1088	AND Inc. Stamford, Ct.	N71WW
N124PS	69	SA-26AT Merlin 2B	T26-135	Montana Pacific Aviation Inc. Santa Barbara, Ca.	N175P
N125A	68	King Air B90	LJ-360	Aerolease Inc. Carrollton, Tx.	N1250B
N125D	81	King Air B100	BE-114	James Naples, Steamboat Springs, Co.	
N125MM	80	Gulfstream 840	11605	Malibu Aviation Co. Concord, NH.	HB-GPB
N125NC	80	King Air B200	BB-1023	N Carolina Department of Commerce, Raleigh, NC.	N62546
N125PG	82	Conquest 1	425-0125	Valero Management Co. San Antonio, Tx.	N125VE
N125SC	83	Conquest 1	425-0136	James E Simon Co. Cheyenne, Wy.	N6884G
N125TS	92	King Air B200	BB-1422	Tri-State Generation & Transmission Association, Denver, Co.	N422BW
N125VH	77	King Air B100	BE-25	Valley Hope Association, Norton, Ks.	N125U
N126AP	84	King Air B200	BB-1157	ConAero Inc. Houston, Tx.	HB-GIG
N126AT	67	King Air A90	LJ-265	Apple Ten Aero LLC. Alcoa, Tn.	N10YP
N126DS	90	King Air 350	FL-31	William Holmes, Macon, Ga.	N8013R
N126M	80	Gulfstream 980	95033	Kasey & Associates Inc. Opa Locka, Fl.	N8774P
N126RD	97	King Air B200	BB-1597	Davis Aviation Inc.	
N126RD	77	King Air E90	LW-230	DeClark Aviation Inc. Newport Beach, Ca.	N345MB
N127AT	81	PA-31T Cheyenne II	8120022	Ducap Inc. West Point, Ne.	HK-2607X
N127BB	89	King Air 300	FA-196	Air Pegasus Inc. Longview, Tx.	N1557R
N127DC	80	Conquest 1	425-0012	43701 Corp. Zanesville, Oh.	D-IAAS
N127EC	78	King Air E90	LW-299	Skyline Investments LLC. Forest, Va.	N127BB
N127GA	77	King Air 200	BB-312	Beech Transportation Inc. Eden Prairie, Mn.	F-GEBC
N127MC	85	Conquest 1	425-0231	McCoy Corp. San Marcos, Tx.	N12270

Regn	Yr	Type	c/n	Owner/Operator	Prev Regn
☐ N127SD	83	King Air B200	BB-1099	Wachovia Leasing Corp. Winston Salem, NC.	N349C
☐ N127TA	80	King Air 200	BB-636	Rierba Leasing Inc. Santo Domingo, Dominican Republic.	N806TC
☐ N127Z	74	King Air A100	B-179	USDA Forest Service, Boise, Id.	N20EG
☐ N128FL	95	King Air 350	FL-128	Raytheon Aircraft Co. Wichita, Ks.	D-CDDD
☐ N128V	92	King Air B200	BB-1442	El Paso Natural Gas Co. El Paso, Tx.	
☐ N129AF	64	Gulfstream 1	129	Aerial Films Inc. Morristown, NJ.	N113GA
☐ N129C	73	King Air E90	LW-61	Sunfresh Inc. Royal City, Wa.	N120GR
☐ N129D	82	King Air B200	BB-1064	Canned Foods Inc. Berkeley, Ca.	N129DP
☐ N129DH	96	Pilatus PC-XII	140	Virtual Village Aircraft Inc. Plano, Tx.	HB-FRB
☐ N129LA	66	King Air A90	LJ-129	Liberty Air Management Inc. Hagerstown, Md.	N7202L
☐ N129TB	81	Gulfstream 840	11676	Semitool Inc. Stansted, UK.	N5928K
☐ N130AT	80	King Air B100	BE-88	Roche Biomedcal Lab Inc. Burlington, NC.	
☐ N130DM	68	King Air B90	LJ-385	Kenneth Ingram Jr. Alexander City, Al.	N33AS
☐ N130MA	84	King Air C90A	LJ-1075	Steven Anderson, Victoria, Tx.	N68474
☐ N130MS		MU-2 Marquise	750SA	Arkansas Bolt Co. Little Rock, Ar.	N947MA
☐ N130PA	76	King Air 200	BB-330	Mesa Air Group Inc. Farmington, NM.	N200CE
☐ N130TT	78	Rockwell 690B	11495	T & T Pump Co. Fairmont, WV.	N76EC
☐ N131AF	78	PA-31T Cheyenne II	7820005	White Industries Inc. Bates City, Mo. (status ?).	N5MC
☐ N131CL	90	King Air C90A	LJ-1268	John Landrum/Springlake Farms, Wilmington, De.	N636
☐ N131DF	79	PA-31T Cheyenne 1	7904015	KIF Aviation Inc. Fort Wayne, In.	N23235
☐ N131HF	87	King Air 300	FA-131	Fuller Renting & Leasing Inc. Auburn, Al.	EC-ETM
☐ N131MP	78	PA-31T Cheyenne II	7820039	Weather Modification Inc. Fargo, ND.	N900JM
☐ N131PA	76	King Air 200	BB-161	Mesa Air Group Inc. Farmington, NM.	C-GTIM
☐ N131PC	78	PA-31T Cheyenne II	7820009	Rombauer Vineyards Inc. St Helena, Ca.	
☐ N131SA	93	King Air C90B	LJ-1318	Comflight LLC. Duncan, Ok.	CC-CDL
☐ N131SJ	76	King Air 200	BB-131	C J Aviation LLC. Sunriver, Or.	TC-TAB
☐ N131SP	83	King Air F90-1	LA-206	Lucky Star Industries Inc. Baldwyn, Ms.	D-IIBS
☐ N131XL	81	Cheyenne II-XL	8166015	Leconte Aviation Inc. Alcoa, Tn.	C-GSGJ
☐ N132B	72	King Air E90	LW-8	Hull Properties Inc. Augusta, Ga.	
☐ N132BK	81	MU-2 Marquise	1529SA	Barken International Inc. Salt Lake City, Ut.	N818R
☐ N132CC	84	King Air 300	FA-11	Club Car Inc. Augusta, Ga.	N732P
☐ N132JH	73	Rockwell 690A	11126	Tom Davis PC. Austin, Tx.	(N345SP)
☐ N132K	81	King Air B200	BB-938	Dan & Bruce Harrison, Houston, Tx.	
☐ N132MC	92	King Air B200	BB-1395	Mississippi Chemical Corp. Yazoo, Ms.	N8049H
☐ N133CZ	95	Pilatus PC-XII	133	Horizon Americas Inc. Buenos Aires, Argentina.	HB-FQW
☐ N133DL	80	Gulfstream 840	11616	Safeguards International Inc. Charlotte, NC.	D-IGEL
☐ N133GA	89	King Air B200	BB-1321	Superior Industries International, Van Nuys, Ca.	F-GILI
☐ N133K	94	King Air 200	BB-1487	Raytheon Aircraft Co. Wichita, Ks.	N1548S
☐ N133LC	93	King Air B200	BB-1464	Lowe's Companies Inc. North Wilkesboro, NC.	N200TW
☐ N134AM	87	King Air 300	FA-134	Apparelsoft Inc. Seneca, SC.	EC-EGF
☐ N134MA	70	Mitsubishi MU-2G	507	AJM Airplane Co. Naples, Fl.	
☐ N134W	65	King Air 90	LJ-52	Lucaya Air Enterprise Inc. Opa Locka, Fl.	N90BL
☐ N135MA	80	PA-31T Cheyenne II	8020025	P-Baron Leasing LLC. La Crosse, Wi.	N2407W
☐ N135MK	84	King Air 300	FA-3	Manitowoc Co. Manitowoc, Wi.	N67683
☐ N135SP	68	SA-26AT Merlin 2B	T26-111	Opex Aviation Inc. Santa Paula, Ca.	VH-CAH
☐ N136JH	70	King Air 100	B-25	Bullfrog Inc. Owensboro, Ky.	N352NR
☐ N137BW	78	Rockwell 690B	11446	B & B Transportation Inc. Paterson, NJ.	N117SA
☐ N137C	62	Gulfstream 1	93	Phoenix Air Group Inc. Cartersville, Ga.	XB-...
☐ N137CP	69	SA-26AT Merlin 2B	T26-136	Helitech Inc. Hampton, Va.	N51LF
☐ N137CW	79	PA-31T Cheyenne 1	7904052	Lawrence Smith, Austin, Tx.	N72TW
☐ N137D	82	King Air B100	BE-128	AAA Cooper Transportation Corp. Dothan, Al.	
☐ N138BC	76	King Air C-12C	BC-22	Marion County Sheriffs Office, Ocala, Fl.	76-22546
☐ N139B	72	King Air C90	LJ-563	Aircraft Management Inc. Elyria, Oh.	
☐ N139SC	79	King Air C90	LJ-868	Elliott Aviation Aircraft Sales inc. Des Moines, Ia.	F-GERL
☐ N140CM	70	Mitsubishi MU-2F	190	Cacciamani Architectural Corp. Wilmington, De.	N111MA
☐ N140CP	77	Mitsubishi MU-2P	362SA	Decathlon Performance Inc. Wilmington, De.	N801L
☐ N140MP	80	Conquest II	441-0165	Montana Power Co. Butte, Mt.	N27214
☐ N140MT		King Air C-12C	BD-3	Marion County Sheriffs Office, Ocala, Fl.	73-1207
☐ N140PA	67	King Air A90	LJ-297	Shasta Livestock Auction Yard, Cottonwood, Ca.	N839K
☐ N141DR	93	King Air 350	FL-92	E'OLA Products Inc. St George, Ut.	N8138E
☐ N141DT	79	PA-31T Cheyenne II	7920002	Eric Reimers, Missoula, Mt.	N400PB
☐ N141GA	80	PA-31T Cheyenne 1	8004022	W M Jackson & Co. Marion, Il.	G-BHTP
☐ N141TC	84	PA-42 Cheyenne IIIA	5501021	Joseph Meyer, Edwardsville, Il.	N200JH
☐ N142AF	81	PA-42 Cheyenne III	8001066	Air 1st Aviation Companies Inc. Aiken, SC.	VH-NMA
☐ N142BK	79	MU-2 Marquise	733SA	Corporate Aviation Services Inc. Tulsa, Ok.	N533MA
☐ N142JC	80	King Air B100	BE-93	Valley Plating Inc. Green Bay, Wi.	XC-IMC
☐ N142NR	73	SA-226T Merlin 3	T-227	Navajo Refining Co. Artesia, NM.	N2746Z
☐ N142SR	78	King Air 200	BB-380	Nemco Motor Sports Inc. Mooresville, NC.	(N380GW)
☐ N143JA	75	Mitsubishi MU-2M	324	BAR Realty Corp. San Juan, PR.	N512MA
☐ N143KB	68	King Air B90	LJ-339	Flying W Diamond Ranch Inc. Albuquerque, NM.	N120LG
☐ N143LG	79	King Air 200C	BL-5	Life Guard Alaska/Era Aviation Inc. Anchorage, Ak.	N390AC
☐ N143RJ	77	PA-31T Cheyenne II	7720027	Apollo Air Charter Inc. Spencer, Oh.	N82120
☐ N143Z		DHC 6-300	437	USDA Forest Service Aviation, Boise, Id.	
☐ N144GP		Beech 1900C-1	UC-44	Kiti International/GP Express Airlines, Grand Island, Ne.	N31261
☐ N144JB	85	Gulfstream 900	15039	Skytel Inc. Fort Lauderdale, Fl.	N61GA
☐ N144PC	96	Pilatus PC-XII	154	Altaire Services Corp.	HB-FRP

Regn	Yr	Type	c/n	Owner/Operator	Prev Regn
☐ N144PL	84	PA-42 Cheyenne 400LS	5527014	Four-Four Inc. Farmington, NM.	N400SN
☐ N145FS	81	MU-2 Solitaire	437SA	Picacho Aviation LLC. Las Cruces, NM.	N666AM
☐ N146BC	81	King Air B200	BB-885	TR Builder Corp. Newport Beach, Ca.	N38340
☐ N146BT	78	King Air B100	BE-46	James River Coal Service Co. London, Ky.	N45BT
☐ N146EA	80	Conquest II	441-0146	Kinetic Concepts Inc. San Antonio, Tx.	D-ICRI
☐ N146GA	81	Conquest 1	425-0074	Aircraft Guaranty Corp. Houston, Tx.	HB-LPU
☐ N146GW	83	Conquest 1	425-0146	Cimarron Aviation Inc. Bend, Or.	N146FW
☐ N148CA	81	PA-42 Cheyenne III	8001046	Andrew & Williamson Sales Co. San Diego, Ca.	N148CA
☐ N149CC	79	PA-31T Cheyenne 1I	7904036	Cianbro Corp. Pittsfield, Me.	N55EU
☐ N150BA	78	Mitsubishi MU-2N	718SA	Heart of Texas Dodge Inc. Austin, Tx.	C-GJSD
☐ N150PB	96	Pilatus PC-XII	150	Pilatus Business Aircraft Ltd. Broomfield, Co.	HB-FRL
☐ N150TJ	76	King Air B100	BE-3	Pennington Seed Inc. Madison, Ga.	N9103S
☐ N150VE	76	King Air C90	LJ-678	Pride Air Inc. Lake Village, Ar.	N9091S
☐ N150YA	82	King Air B100	BE-124	Hospital Investors Management Corp. Englewood Cliffs, NJ.	(N150YR)
☐ N150YR	82	King Air B100	BE-132	Central Control Inc. Alexandria, La.	
☐ N151BU	66	King Air A90	LJ-183	Baylor University, Waco, Tx.	C-FPCB
☐ N151CF	96	King Air B200	BB-1551	Country Fresh Inc. Grand Rapids, Mi.	N969TS
☐ N151E	78	King Air 200	BB-371	Waterloo Industries Inc. Waterloo, Ia.	N4660M
☐ N151FB	81	Cheyenne II-XL	8166061	Texas Farm Bureau Building, Waco, Tx.	N151TC
☐ N151JB	68	Mitsubishi MU-2D	104	Smyrna Air Center Inc. Smyrna, Tn.	N177DM
☐ N151MP	81	Cheyenne II-XL	8166063	Minnkota Power Co-Operative, Grand Forks, ND.	N776AK
☐ N151PB	96	Pilatus PC-XII	151	Pilatus Business Aircraft Ltd. Broomfield, Co.	HB-FRM
☐ N152BK	82	MU-2 Marquise	1537SA	Barken International Inc. Salt Lake City, Ut.	OY-BHY
☐ N152C	83	King Air B200	BB-1132	Consol Inc. Pittsburgh, Pa.	N66480
☐ N152D	75	King Air E90	LW-119	State of Texas, Austin, Tx.	N102CU
☐ N152WE	81	King Air F90	LA-152	Hannay Reels Inc. Westerlo, NY.	HP-1215
☐ N152WW	75	King Air C90	LJ-654	Han Aero Services Inc. Wilmington, De.	N917K
☐ N152X	81	Gulfstream 840	11692	Trans World Oil & Gas Corp. Billings, Mt.	(N5944K)
☐ N153JM	83	Conquest 1	425-0153	Mark Jacobson & Associates Inc. Bingham Farms, Mi.	C-GAGE
☐ N153ML	75	King Air 200	BB-23	K & E Aircraft Leasing Inc. South Burlington, Vt.	N814KA
☐ N153PB	96	Pilatus PC-XII	153	E L Thompson & Son Inc.	HB-FRO
☐ N153PB	96	Pilatus PC-XII	153	Pilatus Business Aircraft Ltd. Broomfield, Co.	HB-FRO
☐ N153TC	79	King Air A100	B-247	Jim Clark & Associates Inc. Oklahoma City, Ok.	F-GJPA
☐ N154CA	80	PA-31T Cheyenne II	8020085	Sunday River Transportation Inc. Bethel, Me.	D-IIPA
☐ N154L	77	SA-226T Merlin 3A	T-284	Owners Jet Services Ltd. Las Vegas, Nv.	(N28TA)
☐ N154PC	85	King Air 300	FA-82	Pontiac Coil Inc. Waterford, Mi.	N488WG
☐ N154RH	64	Gulfstream 1	132	B J Management Inc. Greensboro, NC.	N154NS
☐ N154TC	78	King Air A100	B-239	Jim Clark & Associates Inc. Oklahoma City, Ok.	F-GELR
☐ N154WC	70	Mitsubishi MU-2G	509	Schuybroeck Aviation Inc. Dover, De.	C-GBOX
☐ N155A	91	King Air C90A	LJ-1257	ATC Air Inc. Lincoln, Ne.	N8255A
☐ N155AS	73	Mitsubishi MU-2J	589	DHB Aviation Inc. Buffalo, NY.	N293MA
☐ N155BA	73	Mitsubishi MU-2J	582	Corporate Aviation Services Inc. Tulsa, Ok.	N286MA
☐ N155CA	78	PA-31T Cheyenne II	7820024	Reno Flying Service Inc. Reno, Nv.	YV-...
☐ N155CG	73	King Air E90	LW-40	Gunn Air Charter Inc. San Antonio, Tx.	N61NA
☐ N155DS	80	PA-31T Cheyenne II	8020026	AWH Inc. Sevierville, Tn.	N89CA
☐ N155LS	78	King Air E90	LW-286	V F Management Co. Roanoke, Va.	N18754
☐ N155MA	67	Mitsubishi MU-2B	014	Kirtland Community College, Roscommon, Mi.	N4WD
☐ N155QS	75	King Air 200	BB-79	Winco Industries Inc. Tipp City, Oh.	N155RJ
☐ N155S	75	King Air 90	LJ-14	Sam Investment Co. Little Rock, Ar.	D-ILTI
☐ N156GA	83	Conquest 1	425-0134	Gene Messer Ford Inc. Lubbock, Tx.	HB-LNT
☐ N156SB	96	Pilatus PC-XII	156	Ranger Corp. Woodinville, Wa.	HB-FRR
☐ N157CB	78	King Air C90	LJ-758	Peachtree Aviation Inc. Wilmington, De.	N900SC
☐ N157GA	81	MU-2 Marquise	1536SA	Carter County Bank & Summers-Taylor Inc. Elizabethton, Tn.	RP-C585
☐ N157JB	91	TBM 700	6	JAD Aviation Inc & Commercial Bag Co. Bloomington, Il.	N700ZL
☐ N157MA	80	MU-2 Solitaire	424SA	International Paper Box Machine Co. Nashua, NH.	
☐ N157TA	73	Rockwell 690A	11102	Michael Hayde, Newport Beach, Ca.	C-GADI
☐ N158EF	83	King Air B200	BB-1158	East Ford Inc. Jackson, Ms.	N158TJ
☐ N158TJ	81	PA-42 Cheyenne III	8001057	Central California Aviation Inc. Fresno, Ca.	N303PL
☐ N159DG	79	PA-31T Cheyenne II	7920024	David Garbocauskas, Stonington, Ct.	N111HT
☐ N159PB	96	Pilatus PC-XII	159	Pilatus Business Aircraft Ltd. Broomfield, Co.	HB-FRU
☐ N160AB	91	King Air 300	FA-216	Shenandoah Aviation Ltd. Stephen's City, Va.	N160AC
☐ N160AC	96	King Air 350	FL-154	Elkhorn Aviation Inc. San Jose, Ca.	N1063F
☐ N160AD	81	King Air B200	BB-919	Aircraft Management Co/Dart Container Corp. Leola, Pa.	N160AC
☐ N160NA	77	PA-31T Cheyenne II	7720060	S & H Sales LP. Kalispell, Mt.	XB-JIC
☐ N161RC	90	King Air B200	BB-1356	Rieke Corp. Auburn, In.	N5598N
☐ N162PA	77	King Air 200	BB-232	Aero Lima/John Air Services Inc. Miami, Fl. (status ?).	YV-112CP
☐ N162PB	96	Pilatus PC-XII	162	Gary Miller.	HB-FRX
☐ N162Q	83	King Air B200	BB-1104	Cirrus Corp. Jackson, Ms.	N1628
☐ N162RB	71	681B Turbo Commander	6051	Barnstormers Ltd. Savery, Wy.	N681FV
☐ N163SA	79	PA-31T Cheyenne II	7920025	White Industries Inc. Bates City, Mo.	N6195A
☐ N164TC	80	King Air B200	BB-939	Greater Ozarks Aviation Co. Springfield, Mo.	N175WW
☐ N165BC	80	Gulfstream 840	11646	Ian Herzog A Professional Corp. Santa Monica, Ca.	N65Y
☐ N165FA	88	King Air 300LW	FA-165	ConAgra Inc. Gilroy, Ca.	VH-KDV
☐ N165PD	96	Pilatus PC-XII	165	Pilatus Business Aircraft Ltd. Broomfield, Co.	HB-FSA
☐ N166K	93	Beech 1900D	UE-63	Raytheon Aircraft Co. Wichita, Ks.	
☐ N168WA	97	Pilatus PC-XII	168	Western Aircraft Inc. Boise, Id.	HB-FSD

Regn	Yr	Type	c/n	Owner/Operator	Prev Regn
☐ N169DC	80	PA-31T Cheyenne 1	8004048	Dakota Aircraft Inc. Pierre, SD.	N14PT
☐ N169DR	89	King Air C90A	LJ-1205	New South Communications Inc. Meridian, Ms.	N1552F
☐ N169MM	69	King Air 100	B-6	Jim Clark & Associates Inc. Oklahoma City, Ok.	N203AJ
☐ N169RA	71	King Air 100	B-89	Truman Arnold Companies, Texarkana, Tx.	N4209L
☐ N169TC	81	PA-42 Cheyenne III	8001045	Jim Clark & Associates Inc. Oklahoma City, Ok.	C-FCJP
☐ N170AS	72	King Air A100	B-127	Samoa Aviation Inc. Pago Pago, American Samoa.	C-GJVK
☐ N170PD	97	Pilatus PC-XII	170	Pilatus Business Aircraft Ltd. Broomfield, Co.	HB-FSF
☐ N170S	96	King Air B200	BB-1527	Shanair Inc. Visalia, Ca.	N1027Y
☐ N170SP	76	King Air 200	BB-139	Bill's Dollar Stores Inc. Jackson, Ms.	
☐ N171AF	79	PA-31T Cheyenne II	7920089	Maresco Services Inc. Annondale, NJ.	N101AF
☐ N171AT	66	680V Turbo Commander	1616-49	H E Wright, Anchorage, Ak.	N311PD
☐ N171CP	80	Gulfstream 980	95006	Inter City Investments Inc. Dallas, Tx.	N4468F
☐ N171JP		PA-31T Cheyenne II	7720028	James Parker, Forest, Va.	N29JM
☐ N171MA	80	MU-2 Solitaire	431SA	Drug & Laboratory Disposal Inc. Plainwell, Mi.	
☐ N171PC	68	SA-26AT Merlin 2B	T26-103	Acme Aircraft Sales Inc. Wilmington, De.	N136SP
☐ N171RD	79	King Air 200	BB-610	LJSEGT Inc. North Las Vegas, Nv.	5Y-CDO
☐ N172JS	97	Pilatus PC-XII	171	J B Scott, Boise, Id.	N171PD
☐ N172PB	97	Pilatus PC-XII	172	Pilatus Business Aircraft Ltd. Broomfield, Co.	HB-FSH
☐ N172RD	66	Gulfstream 1	172	Dodson International Parts Inc. Ottawa, Ks.	N11NY
☐ N173DB	78	Rockwell 690B	11485	DMB Packing Corp. Newman, Ca.	N46NH
☐ N173KS	97	Pilatus PC-XII	173	Sutton Development Co. Naples, Fl.	HB-FSI
☐ N174MA	79	MU-2 Marquise	753SA	Bankair Inc. West Columbia, SC.	N100BY
☐ N174PC	97	Pilatus PC-XII	174	Pilatus Business Aircraft Ltd. Broomfield, Co.	HB-FSJ
☐ N174PW	75	King Air 200	BB-64	Spitfire Sales & Leasing Inc. Goldsboro, NC.	N96QM
☐ N174WB	81	King Air 200	BB-804	BACE International Inc. Charlotte, NC.	N3824P
☐ N175AA	81	PA-42 Cheyenne III	8001017	CLB Corp/Alpine Air, Provo, Ut.	XA-SEK
☐ N175BC	83	King Air B200	BB-1126	Interstate Equipment Inc. Bozeman, Mt.	N13PR
☐ N175BM	79	King Air 200	BB-501	Butch Mock Motorsport Inc. Mooresville, NC.	N9UN
☐ N175CA	79	MU-2 Marquise	736SA	Corporate Aviation Services Inc. Tulsa, Ok.	N711PD
☐ N175DB		680V Turbo Commander	1519-95	Richard Rohrman, Shawnee Mission, Ks. (status ?).	N1161Z
☐ N175SA	76	King Air 200	BB-183	Spitfire Sales & Leasing Inc. Goldsboro, NC.	N418GA
☐ N176BJ	69	Mitsubishi MU-2F	144	Intercept Systems Inc. Norcross, Ga.	N11KR
☐ N176CS	80	PA-31T Cheyenne 1	8004054	The Sky's The Limit LLC. Dyersburg, Tn.	N176FE
☐ N176M	82	King Air B200	BB-1029	APACO Inc. Ooltewah, Tn.	N95BD
☐ N176RS	82	Cheyenne II-XL	8166054	Brindlee Mountain Telephone Co. Arab, Al.	N1234L
☐ N176TW	74	King Air E90	LW-76	Sierra American Corp. Dallas, Tx.	ZS-LJF
⊓ N177CN	84	King Air B200	BB-1191	AMP Inc. Harrisburg, Pa.	N.....
☐ N177EM	78	Rockwell 690B	11471	Wal-Mart Stores Inc. Rogers, Ar.	N70ES
☐ N177JW	85	King Air 300	FA-77	Air Everett Inc. Tampa, Fl.	N678SB
☐ N177MK	75	King Air E90	LW-149	Trans Equipment Services Inc. Greenville, SC.	N177KA
☐ N177SA	69	Mitsubishi MU-2F	177	Lab of Path PA. Searcy, Ar.	N891Q
☐ N178CD	81	PA-31T Cheyenne II	8120046	Ralph Betts, Cary, NC.	N101DH
☐ N178NC	77	King Air B100	BE-17	Meisner Aircraft Inc. Lake in the Hills, Il.	(N1981B)
☐ N178PC	97	Pilatus PC-XII	178	Pilatus Business Aircraft Ltd. Broomfield, Co.	HB-FSN
☐ N178RC	78	King Air C90	LJ-762	Beech Travel Services Inc. Palm Beach, Fl.	N24176
☐ N178WM	70	King Air 100	B-35	Harry Gotschall, Austin, Tx.	N77WM
☐ N179D	85	King Air C90A	LJ-1091	Wyoming Machinery Co. Casper, Wy.	
☐ N180BP	90	P-180 Avanti	1004	Planes of Fame/Robert J Pond, Eden Prairie, Mn.	I-RAIH
☐ N180CA	96	King Air B200	BB-1547	Ruand Inc. Osage Beach, Mo.	N1117N
☐ N180TE	92	P-180 Avanti	1012	Keith Graham, Austin, Tx.	(D-ITSV)
☐ N181CG	77	King Air E90	LW-225	Urethane Management Council Inc. Lancaster, Pa.	N101CG
☐ N181DC	75	PA-31T Cheyenne II	7520026	Choptank Air Inc. Easton, Md.	N180JS
☐ N181NK	66	King Air A90	LJ-142	Jack Wall Aircraft Sales Inc. Memphis, Tn.	N101NK
☐ N181SW	81	PA-31T Cheyenne 1	8104001	MacFarlane Co USA LLC. El Dorado, Ar.	
☐ N181Z	73	King Air E90	LW-52	USDA/Forest Service Fire & Aviation, Boise, Id.	N74171
☐ N182CA	81	King Air F90	LA-121	Talbot Aviation Inc. Snyder, Tx.	N82DD
☐ N182ME	78	PA-31T Cheyenne II	7820021	Minnetrista Corp. Muncie, In.	N102FL
☐ N182PE	97	Pilatus PC-XII	182	Pilatus Business Aircraft Ltd. Broomfield, Co.	HB-FSQ
☐ N182TC	81	PA-31T Cheyenne II	8120103	Chief Industries Inc. Grand Island, Ne. (was 8120065).	
☐ N183MA	72	Mitsubishi MU-2F	217	Corporate Aviation Services Inc. Tulsa, Ok.	
☐ N183PC	97	Pilatus PC-XII	183	Aviation Sales Inc. Englewood, Co.	HB-FSR
☐ N183SA	73	King Air C90	LJ-571	Skyward Aviation, Washington, Pa.	D-ILKB
☐ N184JS	86	King Air B200	BB-1256	J R Simplot Co. Boise, Id.	N2676M
☐ N184K	61	Gulfstream 1	84	Indiana University Foundation, Bloomington, In.	N362GP
☐ N185GA	78	Conquest II	441-0066	Rocky Mountain Helicopters Inc. Provo, Ut.	D-INKA
☐ N185MA	72	Mitsubishi MU-2F	219	Ginger Leaven, Burr Ridge, Il.	
☐ N185MV	82	King Air B200	BB-1034	Midwest Aviation, Marshall, Mn.	N185MC
☐ N185PB	97	Pilatus PC-XII	185	Roger & Gayle Block,	HB-FST
☐ N185XP	82	King Air B200	BB-952	U S Department of Energy, Las Vegas, Nv.	N1852B
☐ N186DD	81	King Air F90	LA-83	Shelburne Limestone Corp. Essex Junction, Vt.	N771JB
☐ N186E	79	Rockwell 690B	11566	Winner Aviation Corp. Sharon, Pa.	N186EC
☐ N186EB	84	King Air B200	BB-1186	Executive Business Aviation Inc. Nashville, Tn.	ZS-NBA
☐ N186GA	78	PA-31T Cheyenne II	7820086	Stephen Chevrolet-Oldsmobile Inc. Granby, Ct.	N44CS
☐ N186MQ	80	King Air 200	BB-720	Highwoods Services Inc. Raleigh, NC.	N186MC
☐ N186WF	97	Pilatus PC-XII	186	Pilatus Business Aircraft Ltd. Broomfield, Co.	HB-FSU
☐ N187J	77	King Air B100	BE-24	Nocar Leasing Corp. Raleigh, NC.	N117EP

Regn	Yr	Type	c/n	Owner/Operator	Prev Regn
☐ N187MQ	80	King Air 200	BB-689	Elliott Aviation Aircraft Sales Inc. Des Moines, Ia.	N187MC
☐ N187SB	72	Mitsubishi MU-2F	226	Internet Jet Sales Inc. White Mills, Pa.	N31SK
☐ N187SC	79	PA-31T Cheyenne 1	7904020	Aviation Parts Exchange Inc. Bedford, Tx.	N800MP
☐ N187Z	68	SA-26T Merlin 2A	T26-29	U S Department of Agriculture, Washington, DC.	N22CE
☐ N188BF	81	King Air F90	LA-105	Bruce Foods Corp. Wilson, NC.	D-IAWK
☐ N188CA	79	PA-31T Cheyenne 1	7904032	R K Turnipseed, Sumner, Ms.	N20JS
☐ N188CP	84	Conquest 1	425-0188	Carolina Packers Inc. Smithfield, NC.	N188EA
☐ N188MC	93	King Air 350	FL-93	MASCO Corp. Romulus, Mi.	N82678
☐ N188RB	84	Conquest 1	425-0220	Irby Construction Co. Jackson, Ms.	N22HS
☐ N188SC	77	SA-226T Merlin 3B	T-276	Kosoy Investments Inc. Palm Beach, Fl.	N262PC
☐ N189JR	80	King Air F90	LA-61	Interjet Inc. Addison, Tx.	N444RS
☐ N189MC	96	King Air 350	FL-136	MASCO Corp. Romulus, Mi.	N35017
☐ N190BT	65	King Air 90	LJ-59	Tiger Pass Properties Inc. Gretna, La.	N1909R
☐ N190FD	80	King Air F90	LA-42	Air Limo, Liege, Belgium.	N47PE
☐ N190JL	65	King Air 90	LJ-69	Anton Ptach, Poughkeepsie, NY.	N120DP
☐ N190JS	90	King Air C90A	LJ-1228	Jewett Scott Truck Line Inc. Mangum, Ok.	N90BJ
☐ N190PA	68	Gulfstream 1	195	Phoenix Air Group Inc. Cartersville, Ga.	N1900W
☐ N190RF	67	King Air A90	LJ-238	C L Frates & Co. Oklahoma City, Ok.	N7HD
☐ N190RM	72	King Air E90	LW-1	R & M Aviation Inc. Dereham, Norfolk, UK.	N64RJ
☐ N191A	78	King Air C90	LJ-765	Integrated Control Systems Inc. Punta Gorda, Fl.	5N-WNL
☐ N191FL	76	King Air 200	BB-107	South Florida Aviation Inc. Boca Raton, Fl.	N585FL
☐ N191MA	81	PA-31T Cheyenne 1	8104019	Insurance Management Associates, Wichita, Ks.	N817QT
☐ N192PA	65	Gulfstream 1	149	Phoenix Air Group Inc. Cartersville, Ga.	N684FM
☐ N193A	71	King Air C90	LJ-503	First Air Inc. Booneville, Ar.	
☐ N193AA	79	MU-2 Marquise	741SA	FMS Flight Management LLC. Smyrna, Tn.	VH-SMZ
☐ N193CS	79	SA-226T Merlin 3B	T-300	Jetways of Iowa LC. Minneapolis, Mn.	N210SW
☐ N194DB	90	2000 Starship	NC-8	Dorsey Bryan Hardeman, Albuquerque, NM.	OY-GEA
☐ N194LJ	86	King Air 300	FA-89	Long John Silver's Restaurants Inc. Lexington, Ky.	N1948J
☐ N194MA	81	PA-31T Cheyenne II	8120050	C C Leasing Inc. Tiffin, Oh.	N234SE
☐ N194PC	97	Pilatus PC-XII	194	Pilatus Flugzeugwerke AG. Stans.	HB-FQC
☐ N195AL	86	King Air 300	FA-102	LightAir Ltd. Twinsburg, Oh.	C-FPCC
☐ N195DP	66	King Air 90	LJ-89	Pegasus Air Inc. Laurel, De.	N195DR
☐ N195FW	81	Conquest II	441-0195	Beach Manufacturing Co. Donnelsville, Oh.	(N2626B)
☐ N195KQ	76	King Air 200	LJ-688	Forest Siding Supply, Oklahoma City, Ok.	N195KC
☐ N195MA	70	King Air 100	B-20	Alpha Aviation Inc. Dallas, Tx.	N49GW
☐ N196HA	96	King Air C90B	LJ-1436	Richard Henson, Naples, Fl.	N8236B
☐ N196MA	80	MU-2 Marquise	764SA	Carl Short, Tulsa, Ok.	
☐ N196MP	79	King Air 200	BB-523	Manchester Plastics Inc. Troy, Mi.	N240RL
☐ N196PA	64	Gulfstream 1	139	Phoenix Air Group Inc. Cartersville, Ga.	C-FRTU
☐ N196WC	82	King Air F90	LA-196	Plane 1 Leasing Co. Green Bay, Wi.	N196JP
☐ N197RM	68	Gulfstream 1	197	Revival Ministries International, Tampa, Fl.	N20HE
☐ N197SC	78	King Air C90	LJ-783	Jabil Circuit Inc. St Petersburg, Fl.	(N32DF)
☐ N198AA	80	PA-31T Cheyenne II	8020041	Air Alpha America Inc. Amarillo, Tx.	PT-OHS
☐ N198KA	66	King Air A90	LJ-162	National Lease Services Inc. Elizabthtown, NC.	N198T
☐ N198MA	72	Mitsubishi MU-2J	563	Aero Taxi Rockford Inc. Rockford, Il.	
☐ N199BC	80	King Air F90	LA-30	Skyway Air Inc. Wilmington, NC.	N6690C
☐ N199MA	81	PA-31T Cheyenne 1	8104005	Network Investments LLC. Brookings, Or.	N944JG
☐ N199TT	76	King Air E90	LW-157	RAM Air Sales & Leasing LLC. Carefree, Az.	N81AS
☐ N200AE	66	Gulfstream 1	169	American International Television, Marietta, Ga.	N400WF
☐ N200AF	75	King Air 200	BB-102	Quebec Enterprises Inc. Burbank, Ca.	N997MA
☐ N200AJ	78	King Air 200	BB-378	Aviex Jet Inc. Houston, Tx.	N482SW
☐ N200AP	77	King Air 200	BB-277	American Precision Industries/AF Aviation Inc. Buffalo, NY.	N946BF
☐ N200BC	75	King Air 200	BB-55	Western Precooling Systems, Fremont, Ca.	G-OAKM
☐ N200BE	81	King Air 200	BB-832	Land Use Corp. Summerville, WV.	N45BR
☐ N200BH	80	King Air 200	BB-587	Giradi, Keese & Crane, Van Nuys, Ca.	
☐ N200BM	80	King Air 200	BB-712	Buddie Markham Ranches, Fort Worth, Tx.	N77WM
☐ N200BT	77	King Air 200	BB-293	Mondo Flight Management LLC. Smyrna, Tn.	N500CP
☐ N200CJ	76	King Air 200	BB-143	Curt Joa Inc. Boynton Beach, Fl.	N1PC
☐ N200CP	86	King Air B200	BB-1246	Cary Patterson, Daingerfield, Tx.	N7257T
☐ N200DK	82	Gulfstream 1000	96019	Coast Aircraft Brokerage Inc. Tarzana, Ca.	(N9939S)
☐ N200DT	68	680W Turbo Commander	1763-9	Commander Services Inc. Hayward, Ca.	N97011
☐ N200EW	83	King Air B200	BB-1163	Edmund Littlefield, San Francisco, Ca.	N200DB
☐ N200EZ	74	King Air 200	BB-9	Decatur Aviation Inc. Decatur, Il.	N5PC
☐ N200FB	79	PA-31T Cheyenne 1	7904037	University of Wyoming, Laramie, Wy.	N2332R
☐ N200FV	77	King Air 200	BB-299	AeroGenesis Inc. Mason, Mi.	N6111
☐ N200GS	85	King Air B200	BB-1214	Navajo Nation, Window Rock, Az.	N7225D
☐ N200HD	82	King Air B200	BB-987	Carport Inc. Tuscaloosa, Al.	N358ST
☐ N200HF	83	King Air B200	BB-1149	Fuller Renting & Leasing Inc. Auburn, Al.	TC-SAY
☐ N200HV	91	King Air C90A	LJ-1262	Hy-Vee Food Stores Inc. West Des Moines, Ia. (status ?).	N15527
☐ N200HW	82	King Air B200	BB-957	H S Williams Co. Marion, Va.	N7CT
☐ N200JL	76	King Air 200	BB-127	John Lyddon, Las Vegas, Nv.	N2127L
☐ N200JQ	74	Rockwell 690A	11224	Corona Aircraft Inc. Corona, Ca.	N200JN
☐ N200JR	87	King Air B200	BB-1285	JRN Chicken Stores Inc. Columbia, Tn.	N36809
☐ N200KK	80	King Air 200	BB-677	Prince Manufacturing Corp. Sioux City, Ia.	YV-93CP
☐ N200LU	84	King Air B200	BB-1177	Rosfam Airplane Co. Birmingham, Mi.	N200LV
☐ N200MG	81	King Air B200	BB-923	Besser Co. Alpena, Mi.	N666PC

Regn	Yr	Type	c/n	Owner/Operator	Prev Regn
□ N200MH	69	SA-26AT Merlin 2B	T26-148	CPA Leasing Inc. Williamsport, Pa.	N540MC
□ N200MJ	82	King Air B200	BB-1012	Marks-Joy Corp. Buffalo, NY.	N14BW
□ N200MP	84	King Air B200	BB-1174	MAC Papers Inc. Jacksonville, Fl.	N168AC
□ N200NB	77	PA-31T Cheyenne II	7720035	National By Products Inc. Des Moines, Ia.	
□ N200NY	76	King Air C-12C	BC-25	New York State Police, Latham, NY.	N1172J
□ N200PG	79	PA-31T Cheyenne II	7920015	General Aviation Services Inc. Lake Zurich, Il.	N200GH
□ N200PH	82	King Air B200	BB-1017	Pioneer Hi-Bred International Inc. Des Moines, Ia.	
□ N200PL	78	King Air 200	BB-410	CAL Leasing Inc. Prairie Village, Ks.	N275Z
□ N200PU	94	King Air B200	BB-1477	Purdue University, Lafayette, In.	N8301D
□ N200PY	80	King Air 200	BB-700	K A Aviation Inc. Wilmington, De.	N101TS
□ N200QS	83	King Air B200	BB-1130	EJI Inc. Hilton Head, SC.	N16KK
□ N200RE	76	King Air E90	LW-164	Metz Baking Co. Sioux City, Ia.	
□ N200RM	73	King Air E90	LW-45	Phoenix Partners Leasing LLC. University Place, Wa.	N777SS
□ N200RR	87	King Air B200	BB-1282	State Aircraft Pooling Board, Austin, Tx.	N3082C
□ N200RW	77	King Air 200	BB-242	200RW Inc. St Louis, Mo.	N24201
□ N200RX	72	Mitsubishi MU-2J	548	Internet Jet Sales Ltd. Wilmington, De.	N68DA
□ N200SC	89	King Air C90A	LJ-1227	560 Inc. Chattanooga, Tn.	N1556G
□ N200SE	85	King Air B200	BB-1208	King Flyers Partners LP. Magnolia, Tx.	N7213K
□ N200SL	88	King Air C90A	LJ-1189	Sutherland Lumber Co. Tulsa, Ok.	
□ N200SN	80	SA-226T Merlin 3B	T-354	Medical Review Consultants, Independence, Mo.	N656PS
□ N200SR	95	Cirrus SR-20	001	Cirrus Design Corp. Duluth, Mn. (Ff 31 Mar 95).	
□ N200SV	84	King Air B200	BB-1187	Cutter Aviation Inc. Phoenix, Az.	N610SW
□ N200TB	67	680V Turbo Commander	1708-83	East Coast Aero Technical School, Holden, Ma.	N500HY
□ N200TG	73	King Air C90	LJ-583	Gemi Trucking Inc. Eatonton, Ga.	N21WB
□ N200TK	85	King Air B200	BB-1240	Fairview Air Inc. Wilmington, De.	N90BL
□ N200TP	77	King Air 200	BB-191	The November Group LLC. Aspen, Co.	N1PC
□ N200TR	84	King Air C90A	LJ-1067	Taylor Ramsey Enterprises Inc. Lynchburg, Va.	N224P
□ N200TT	81	Gulfstream 980	95073	Arizona Wholesale Supply Co. Phoenix, Az.	N9825S
□ N200V	97	King Air B200	BB-1599	Raytheon Aircraft Co. Wichita, Ks.	
□ N200VA	77	King Air 200	BB-246	Vermont American Corp. Louisville, Ky.	N46WD
□ N200WZ	76	King Air 200	BB-89	Elliott Aviation Aircraft Sales Inc. Des Moines, Ia.	N16BF
□ N200XL	81	Cheyenne II-XL	8166010	Thortech International Inc. Wilmington, De.	
□ N200YB	85	King Air B200	BB-1219	Century Equipment Co. Salt Lake City, Ut.	N123D
□ N200ZC	75	King Air 200	BB-41	AlliedSignal Avionics Inc. Olathe, Ks.	
□ N201CH	76	King Air 200	BB-103	Richard Fessler, Horseshoe Bay, Tx.	
□ N201SM	67	SA-26T Merlin 2A	T26-5	J Sills, Lakeland, Fl.	N400P
□ N201UV	76	Mitsubishi MU-2L	680	Royal Air Freight, Waterford, Mi.	N201U
□ N202AJ	79	King Air 200	BB-511	Alpha Aviation Services Ltd. Bloomfield, Ct.	N200LW
□ N202BB	84	King Air 200	BB-624	Bogey Bird Inc. Jefferson, In.	N123SR
□ N202HC	78	PA-31T Cheyenne II	7820036	Jent Aviation LLC. Marion, Ky.	N82251
□ N202KA	77	King Air 200	BB-272	CullieAir Inc. Springfield, Mo.	N2000X
□ N202MC	78	Mitsubishi MU-2P	369SA	Embers Asset Managers Inc. Indianapolis, In.	N755MA
□ N202PV	91	King Air 350	FL-50	Owen Industries Inc. Omaha, Ne.	N8092F
□ N202VT	80	King Air 200	BB-709	Virginia Polytechnic Institute & State University.	N202NC
□ N202WS		SA-226AT Merlin 4A	AT-037	GAS Wilson Inc. Lake Zurich, Il.	N216GA
□ N203BS	79	King Air 200	BB-476	Bill Smith, Oklahoma City, Ok.	N6040T
□ N203GA	73	Mitsubishi MU-2J	573	Joseph Gibson III, Pompano Beach, Fl.	N3929L
□ N203HC	90	King Air 350	FL-28	Hytrol Conveyor Co. Jonesboro, Ar.	N5655W
□ N203PC	78	King Air E90	LW-258	Flying Swan Inc. Wilmington, De.	N20316
□ N204AJ	70	King Air 100	B-10	Aviex Jet Inc. Houston, Tx. (status ?).	N400BE
□ N204WB	84	King Air B200	BB-1179	Copper & Brass Sales Inc. Eastpointe, Mi.	N444WB
□ N205BN	77	Rockwell 690B	11404	Boone Newspapers Inc. Tuscaloosa, Al.	N100MF
□ N205P	97	King Air C90B	LJ-1474	Thermal Engineering Corp. Columbia, SC.	
□ N205SP	81	King Air C90	LJ-965	Colorado State Patrol, Denver, Co.	N62BB
□ N206K	85	King Air 300	FA-36	Diamond 1A Inc. Divide, Co.	
□ N206P	79	King Air 200	BB-466	Piedmont Aviation Services Inc. Winston Salem, NC.	N200TW
□ N206R	91	2000 Starship	NC-21	Raytheon Corp. Lexington, Ma.	
□ N207CM	80	King Air 200	BB-621	Charlotte Memorial Hospital & Medical Center, Charlotte, NC.	N6757M
□ N207DB	81	King Air 200	BB-862	Burns Investment Co. Redmond, Wa.	N260F
□ N207P	95	King Air C90B	LJ-1406	First South Development & Investment Corp. Greensboro, NC.	
□ N207R	90	King Air 350	FL-14	Raytheon Corp. Lexington, Ma.	
□ N207SB	71	King Air 100	B-69	B & C Aviation Co Inc. Nashville, Tn.	N2SC
□ N207SS	80	Conquest II	441-0136	Grady White Boats Inc. Greenville, NC.	N441HC
□ N208AJ	76	King Air 200	BB-126	Grand Casino Coushatta, Kinder, La.	N388CC
□ N208CL	76	Rockwell 690A	11297	Beecher Securities Inc. Houston, Tx.	N514GP
□ N208F	81	King Air 200	BB-851	Air Transportation LLC. Ridgeland, Ms.	N200E
□ N208RC	80	King Air F90	LA-46	King Air Corp. Miami, Fl.	ZP-TZF
□ N209CM	85	King Air B200	BB-1232	Charlotte Macklenburg Hospital, Charlotte, NC.	G-BLYB
□ N210AC	85	King Air B200	BB-1216	Woodland Aircraft Leasing Inc. Plymouth, Mn.	N55MV
□ N210EC	84	King Air C90A	LJ-1070	State of Tennessee, Nashville, Tn.	N67554
□ N210SU	79	King Air 200	BB-420	Ohio State University, Columbus, Oh.	N2920A
□ N211AD	85	Gulfstream 1000	96092	Allison Development Corp. Charlotte, NC.	N127GA
□ N211BE	74	Mitsubishi MU-2J	641	Write Stuff Aviation Inc. Wilmington, De.	N114MA
□ N211CP	81	King Air 200	BB-843	AeroGenesis Inc. Mason, Mi.	N3850K
□ N211NA	82	King Air F90	LA-199	Charles Cook Jr. Center Harbour, NH.	N196HA
□ N211RV	70	Mitsubishi MU-2G	502	Jack Tilka, Hammond, In. (was s/n 153).	C-FAXP
Regn	*Yr*	*Type*	*c/n*	*Owner/Operator*	*Prev Regn*

Regn	Yr	Type	c/n	Owner/Operator	Prev Regn
☐ N211X	67	King Air A90	LJ-279	Aviation Parts Exchange Inc. Bedford, Tx. (status ?).	
☐ N212BA	73	Mitsubishi MU-2J	587	Air Cargo Masters Inc. Brandon, SD.	N102BH
☐ N212CC	71	Mitsubishi MU-2F	209	Bobby Hamilton Inc. Mount Juliet, Tn.	N100VC
☐ N212D	67	King Air A90	LJ-234	Central Missouri State University, Warrenburg, Mo.	N800BP
☐ N212DM	82	King Air B200	BB-1053	David & Diane Miller, Bandon, Or.	N132N
☐ N212GM	81	PA-31T Cheyenne 1	8104013	212 Aviation Corp. Des Moines, Ia.	N200HV
☐ N212HH	79	PA-31T Cheyenne 1	7904010	Fred Netterville Lumber Co. Woodville, Ms.	N25BF
☐ N212JB	85	King Air B200	BB-1238	Beech 212JB LP. El Paso, Tx.	(N86GA)
☐ N212WP	65	King Air 90	LJ-62	Volunteer Machinery Sales Co. Portland, Tn.	N512WP
☐ N213DB	93	King Air B200	BB-1450	Marshall Air Systems Inc. Charlotte, NC.	N8059Q
☐ N213RW	76	King Air E90	LW-167	Polaris Industries LP. Minneapolis, Mn.	N214RW
☐ N214CK	74	King Air A100	B-191	Causey Aviation Service Inc. Liberty, NC.	N4391W
☐ N214EC	85	King Air 300	FA-50	Shamrock Aviation Inc. Warren, Mi.	N85GA
☐ N214JB	90	2000 Starship	NC-14	Raytheon Aircraft Credit Corp. Wichita, Ks.	N5674B
☐ N214P	91	King Air C90A	LJ-1293	Navajo Nation, Window Rock, Az.	
☐ N214SC	74	King Air E90	LW-96	Allied Signal Avionics Inc. San Francisco, Ca.	N35KA
☐ N215BA	74	Rockwell 690A	11215	Lauren Engineering & Constructors Corp. Abile, Tx.	LN-LMF
☐ N215HC	78	King Air E90	LW-266	Plane Lease Co. Minneapolis, Mn.	N525KA
☐ N215P	90	King Air B200	BB-1381	SET AIR LLC. Rocky Mount, NC.	
☐ N215PA	77	King Air 200	BB-224	Leonard Hudson Drilling Co. Pampa, Tx.	OE-FAY
☐ N216AJ	67	King Air A90	LJ-190	CEMR Inc. Houston, Tx.	N700US
☐ N216CD	75	Mitsubishi MU-2K	323	Camile Landry, Baton Rouge, La.	N511MA
☐ N216RP	82	King Air B200	BB-1015	Primore Inc. Adrian, Mi.	N901EB
☐ N217CP	79	King Air 200	BB-490	AeroSmith/Penny Inc. Houston, Tx.	F-ODZE
☐ N217SB	73	Mitsubishi MU-2J	586	Geiger Aviation Leasing Inc. Huntington, In.	N21AU
☐ N219MA	81	MU-2 Solitaire	440SA	Solitaire Inc. Portland, Or.	
☐ N220CB	79	King Air 200	BB-426	Mesquite Aviation Co. Houston, Tx.	N11HY
☐ N220DK	82	King Air B200	BB-1058	DeKalb Genetics Corp. DeKalb, Il.	N88SR
☐ N220HC	76	Rockwell 690A	11320	Brady Industries Inc. Las Vegas, Nv.	XA-PAC
☐ N220JB	80	King Air 200	BB-638	Memphis Aereo Co SA. Memphis, Tn.	N8DX
☐ N220KW	73	King Air A100	B-185	Columbia Pacific Aviation Inc. Moses Lake, Wa.	N818AS
☐ N220PB	76	King Air E90	LW-168	J B Investments, Mesa, Az.	N711HA
☐ N220RJ	82	King Air B200	BB-1022	P & M Aviation Inc. Old Forge, Pa.	N404AU
☐ N220SC	81	PA-31T Cheyenne II	8120041	Entrechato Inc. Wilmington, De.	N8361T
☐ N220TA	78	King Air 200	BB-409	Abilene Aero Inc. Abilene, Tx.	N2044D
☐ N220TB	82	King Air B200	BB-1057	Rocky Mountain Holdings LLC. Provo, Ut.	F-GILY
☐ N220TT	79	King Air 200	BB-462	Mel Farr Air Inc. Detroit, Mi.	G-MOAT
☐ N220TW	92	P-180 Avanti	1018	Tom Walkinshaw Racing, Oxford, UK.	LZ-VPA
☐ N220W	81	MU-2 Solitaire	450SA	JCL Corp. Bentonville, Ar.	N28PP
☐ N221K	81	Gulfstream 980	95077	Nayko Inc. Portland, Or.	N9829S
☐ N221MA	80	MU-2 Marquise	771SA	Aerographics Inc. Manassas, Va.	N248NW
☐ N221MM	88	King Air 300	FA-161	Franklin/Templeton Travel Inc. San Mateo, Ca.	N3113A
☐ N221NC	68	King Air B90	LJ-393	Wheel-Air Inc. Minneapolis, Mn.	N800CT
☐ N221RC	81	PA-31T Cheyenne 1	8104046	Sports Aero Ltd. Wilmington, De.	N478PC
☐ N221SP	84	King Air B200	BB-1180	Addington Enterprises Inc. Ashland, Ky.	VR-BNN
☐ N221TC	83	King Air B100	BE-136	Texas Christian University, Fort Worth, Tx.	N18348
☐ N221TM	75	King Air C90	LJ-672	J & G Industries Inc. Toledo, Oh.	(N11AB)
☐ N222BC	81	PA-31T Cheyenne 1	8104007	Bradenton Auto Sales Inc. Gallipolis, Oh.	N281SW
☐ N222CM	95	Pilatus PC-XII	111	Planes Unlimited Inc. Dallas, Tx.	HB-FQE
☐ N222GW	79	PA-31T Cheyenne 1	7904050	Dodson International Parts Inc. Ottawa, Ks.	N2512R
☐ N222HH	85	MU-2 Marquise	1569SA	Hoover Inc. Nashville, Tn.	N502MA
☐ N222KA	75	King Air 200	BB-49	State of Washington, Olympia, Wa.	
☐ N222LP	80	King Air B100	BE-85	Line Power Manufacturing Co. Bristol, Va.	N6723T
☐ N222LR	72	Mitsubishi MU-2K	239	Anaconda Aviation Corp. Boca Raton, Fl.	N221MA
☐ N222ME	76	Rockwell 690A	11338	Stormy Petrel LLC. Atlanta, Ga.	N46906
☐ N222MV	71	SA-226T Merlin 3	T-212	TRI-CON Aviation Inc. New Knoxville, Oh.	VR-BHQ
☐ N222WJ	85	King Air C90A	LJ-1106	Jen Rob Aviation, Mine Hill, NJ.	
☐ N223CH	68	King Air B90	LJ-321	CMH Investments Inc. Springdale, Ar.	N269RR
☐ N223HC	81	King Air/Catpass 250	BB-855	Cape Smythe Air Service Inc. Barrow, Ak.	
☐ N223JR	85	King Air 300	FA-45	Delta Coals Inc. Nashville, Tn.	N7220L
☐ N223JS	82	MU-2 Marquise	1544SA	Richland LLC. Columbia, SC.	N222JM
☐ N224BB	84	King Air 300	FA-5	B B I Inc. Salina, Ks.	N444HK
☐ N224BH	76	King Air C90	LJ-676	Race Air LLC. Wilmington, De.	N90MM
☐ N224CC	89	King Air C90A	LJ-1218	James Harris, Houston, Tx.	N15628
☐ N224EC	79	PA-31T Cheyenne II	7920020	EyeCorp Inc. Memphis, Tn.	N866RA
☐ N224HR	71	SA-226T Merlin 3	T-217	Suncoast Custom Builders Inc. Pinellas Park, Fl.	N199Z
☐ N224P	85	King Air B200	BB-1230	Branch Banking & Trust Co. Elm City, NC.	
☐ N225CA	81	PA-31T Cheyenne 1	8104004	Windward Aviation Corp. Katonah, NY.	SE-IUG
☐ N225LH	76	King Air C-12C	BC-33	Department of Pubic Safety, Montgomery, Al.	76-22557
☐ N225MM	78	Rockwell 690B	11462	Export Air Corp. Miami, Fl.	N81764
☐ N225WL	85	King Air B200	BB-1226	Lantis Enterprises Inc. Spearfish, SD.	(N346EA)
☐ N226DD	78	SA-226T Merlin 3A	T-286	Hurley Aircraft Corp. Titusville, Fl.	XB-ZAO
☐ N226FC		SA-226T Merlin 3B	T-306	Rosebriar Transportation Inc. Dallas, Tx.	N5661M
☐ N226HA	69	SA-26AT Merlin 2B	T26-142	T G Air Inc. Long Beach, Ca.	N991CB
☐ N226JW	68	King Air B90	LJ-406	Para Drop, Las Vegas, Nv.	N438
☐ N226SR	76	SA-226T Merlin 3A	T-270	Falcon Flight, Telluride, Co.	N953AE

Regn	Yr	Type	c/n	Owner/Operator	Prev Regn
N227	70	Mitsubishi MU-2G	510	L Concepcion & K Weichinger, Downey, Ca.	N137MA
N227BC	76	King Air A100	B-227	J R Wood Inc. Atwater, Ca.	N77715
N227JW	84	SA-227AT Merlin 4C	AT-585	Hasta Manana Inc/Ceka Aviation Inc. San Jose, Ca.	(F-....)
N227KM	92	King Air 350	FL-91	Stuart Investment Co. Wilmington, De.	N350KA
N227TT	82	SA-227TT Merlin 3C	TT-444	Transport Co. Ruston, La.	VH-UZA
N228CX	93	TBM 700	84	Turbine Aviation Inc. Dover, De.	
N228RA	76	King Air E90	LW-197	Flagstaff Medical Center, Flagstaff, Az.	N222JD
N228WP	72	Mitsubishi MU-2F	228	Wayne Perry Construction Inc. Buena Park, Ca.	N963MA
N229EM	77	King Air C-12C	BC-59	Department of Public Safety, Montgomery, Al.	77-22948
N230GK	81	King Air B200	BB-894	AGES Holding Corp. Boca Raton, Fl.	N18VG
N231PT	74	PA-31T Cheyenne II	7400003	CCI Corp/Safetran Systems Corp. Tulsa, Ok.	(N177PT)
N231SW	79	PA-31T Cheyenne 1	7904007	RidgeAire Inc. Jacksonville, Tx.	
N232DH	79	PA-31T Cheyenne 1	7904021	Coatue Inc. Dover, De.	N23263
N233RC	81	Conquest II	441-0263	Rico Marketing Corp. Flint, Mi.	N815MC
N234CC	79	Conquest II	441-0100	Robert Smith, San Marcos, Ca.	N388DR
N234CW	85	King Air F90	LA-236	Charles Weyerhaeuser, Big Fork, Mt.	PT-OZE
N234K	84	PA-31T Cheyenne II	7520001	Weather Modification Inc. Fargo, ND.	N668DH
N234PC	84	PA-31T Cheyenne 1A	1104010	Aircraft Guaranty Corp. Houston, Tx.	N817CT
N234SB	81	PA-31T Cheyenne I	8104023	Paul Wojnowich, Fort Mill, SC.	N124DF
N235GA	85	Gulfstream 1000B	96209	GAC/U S Government,	
N236CP	76	King Air B100	BE-9	Medic Air, Reno, Nv.	(N360BA)
N238GA	85	Gulfstream 1000B	96210	GAC/U S Government,	
N239C	77	King Air C90	LJ-738	Comair Aviation Academy, Sanford, Fl.	N666PC
N240DH	85	SA-227AT Merlin 4C	AT-602B	Ameriflight Inc/DHL Airways Inc. Cincinnati, Oh.	N3117P
N240K	68	King Air B90	LJ-340	Merlin Associates Inc. Princeton, NJ.	N925B
N240NM	73	SA-226T Merlin 3	T-240	Navajo Refining Co. Artesia, NM.	N76U
N240RE	73	King Air C90	LJ-570	Alfred Tria Jr. Princeton, NJ.	N240RL
N240S	73	King Air 300	FA-116	Four Corners Aviation Inc. Farmington, NM.	N2998X
N241DH	85	SA-227AT Merlin 4C	AT-607B	Ameriflight Inc/DHL Airways Inc. Cincinnati, Oh.	N3118A
N242DH	85	SA-227AT Merlin 4C	AT-608B	Ameriflight Inc. Burbank, Ca.	N3118G
N242DM	81	King Air 200	BB-754	Mallard Aviation LLC. Memphis, Tn.	N20GZ
N242LC	81	King Air 200	BB-765	Hemisphere Airways Ltd. Louisville, Ky.	N36GA
N242RA	80	PA-42 Cheyenne III	8001003	Ross Aviation Inc. Cortland, Oh.	N8EA
N242TC	74	Rockwell 690A	11219	Executive Aircraft Inc. Kincheloe, Mi.	N50MP
N243DH	85	SA-227AT Merlin 4C	AT-609B	Ameriflight Inc/DHL Airways Inc. Cincinnati, Oh.	N3118H
N244AB	80	PA-31T Cheyenne 1	8004015	Westmoreland Mechanical Testing & Research, Youngstown, Pa.	N2442W
N244CH	81	King Air 200	BB-801	CEH Inc. Drummond Island, Mi.	N123WN
N244DH	85	SA-227AT Merlin 4C	AT-618B	Ameriflight Inc/DHL Airways Inc. Cincinnati, Oh.	
N244J	80	King Air F90	LA-44	Ozark Management Inc. Columbia, Mo.	V5-KLT
N244JP	76	King Air 200	BB-109	Jerrys Enterprises Inc. Minneapolis, Mn.	N908R
N244MP	79	Rockwell 690B	11531	GOR Commercial Enterprises Inc. Miami, Fl.	N101RF
N245CF	76	Rockwell 690A	11313	Howell Instruments Inc. Fort Worth, Tx.	N245CT
N245CT	82	King Air B200	BB-1016	Continental Telephone Co. Victorville, Ca.	N88TS
N245DH	85	SA-227AT Merlin 4C	AT-624B	Ameriflight Inc/DHL Airways Inc. Cincinnati, Oh.	
N245DL	83	Conquest II	441-0298	Magna Lomason Corp. Farmington Hills, Mi.	N16EF
N246DH	85	SA-227AT Merlin 4C	AT-625B	Ameriflight Inc/DHL Airways Inc. Cincinnati, Oh.	
N246MC	78	Rockwell 690B	11448	Olen Spurgeon, Washington, Il.	XC-COL
N246SP	82	Conquest II	441-0246	C Vincent Phillips MD. Bakersfield, Ca.	D-IMAX
N246W	82	MU-2 Marquise	1552SA	Pamida Inc. Omaha, Ne.	N478MA
N247B	73	King Air A100	B-139	Kokomo Aviation/Aviation Charter Services, Indianapolis, In.	
N247DH	85	SA-227AT Merlin 4C	AT-626B	Ameriflight Inc/DHL Airways Inc. Cincinnati, Oh.	
N248DH	85	SA-227AT Merlin 4C	AT-630B	Ameriflight Inc/DHL Airways Inc. Cincinnati, Oh.	
N249DH	85	SA-227AT Merlin 4C	AT-631B	Ameriflight Inc/DHL Airways Inc. Cincinnati, Oh.	
N249PA	65	King Air 90	LJ-21	Loan Guarantee Insurance Agency Inc. Irvine, Ca.	N5Y
N249RL	74	SA-226T Merlin 3A	T-249	GFI International Inc. Stuart, Fl.	D-IFWZ
N249WM	75	King Air E90	LW-139	Richards Aviation & Meadows Consulting Inc. Memphis, Tn.	N345KA
N250DL	81	King Air 200	BB-799	D & L Equipment Inc. Van Nuys, Ca.	N666RL
N250PD	81	King Air 300	FA-33	Falcone Aero Services Inc. Syracuse, NY.	N85RR
N250TJ	81	PA-42 Cheyenne III	8001024	CLB Corp. Orem, Ut. (status ?).	C-GRCY
N250TT	78	PA-31T Cheyenne II	7820050	DMG Investments Inc. Fargo, ND.	F-GGPV
N250U	67	King Air A90	LJ-213	Dr G William Woods, New Ulm, Tx.	N417CS
N251ES	75	Rockwell 690A	11251	Schwertner Farms Inc. Schwertner, Tx.	N690RC
N252AF	89	King Air 1300	BB-1339	Arizona Airways Inc. Tucson, Az.	N393YV
N253TA	91	2000 Starship	NC-10	Elliott Flying Service Inc. Moline, Il.	N1563Z
N254PW	76	Rockwell 690A	11275	Bay Air Inc. Annapolis, Md. (status ?).	N952HE
N255C	71	Mitsubishi MU-2G	536	Sorelivo Investment Inc. Miami, Fl.	N58BC
N256AF	89	King Air 1300	BB-1340	Raytheon Aircraft Credit Corp. Wichita, Ks.	(N132AZ)
N256DD	82	Conquest II	441-0256	Cal West Land & Livestock Co. Coalinga, Ca.	N256DP
N256TA	67	King Air A90	LJ-256	Flanagan Enterprises Inc. Stateline, Nv.	C-FOLR
N257CG	95	King Air C90B	LJ-1419	FS Corsair Inc. NYC.	
N257YA	97	King Air B200	BB-1585	Rolling Green Enterprises LLC. Edina, Mn.	N1104X
N258VB	82	Conquest II	441-0258	High Plains Pizza Inc. Great Falls, Mt.	(N6859Y)
N259SC	72	King Air E90	LW-17	Health Care Administration Co. San Antonio, Tx.	LV-WEY
N260G	96	King Air B200	BB-1525	260G Corp/GPI Aviation Inc. Toms River, NJ.	N1015X
N260WE	79	Rockwell 690B	11536	Tumac Industries Inc. Grand Junction, Co.	N81707
N261AC	78	King Air 200	BB-321	CLT Enterprises LC. Fort Worth, Tx.	N38JL

Regn	Yr	Type	c/n	Owner/Operator	Prev Regn
☐ N261MA	80	MU-2 Marquise	758SA	L C Leasing Co. Annapolis, Md.	N777FL
☐ N261WB	75	Mitsubishi MU-2M	330	Roy Kinsey Jr. Pensacola, Fl.	N261WR
☐ N262SA	78	SA-226T Merlin 3B	T-292	Joda Partnership, Town & Country, Mo.	LV-MRL
☐ N263AL	95	Pilatus PC-XII	112	Lost Tree Aviation Inc. Indianapolis, In.	HB-FQG
☐ N263CM	72	SA-226T Merlin 3	T-223	Hunter Aviation Inc. Fort Worth, Tx.	N2630M
☐ N263ND	77	Mitsubishi MU-2N	698SA	Philip Fagan Jr PC. Burbank, Ca.	N858MA
☐ N263SW	87	King Air B200	BB-1263	Dan A Hughes Co. Beeville, Tx.	N2997W
☐ N264B	72	SA-226T Merlin 3	T-221	Hoosier Microfilm Inc/Midwest Air Charter, Washington, In.	N2649
☐ N264KW	73	Mitsubishi MU-2K	264	Kenneth Wolf MD. Lewiston, Me.	N50TT
☐ N264PA	80	King Air B100	BE-86	Air East Inc. Johnstown, Pa.	N85KA
☐ N265EB	76	King Air 200	BB-185	EBCO Manufacturing Co. Columbus, Oh.	N185DA
☐ N265EX	80	Gulfstream 980	95032	Benchmark Ventures Ltd. Wilmington, De.	N9785S
☐ N265JH	81	Gulfstream 840	11689	Clouds on High Inc. Jackson. 'Spirit of Wyoming'	N5941K
☐ N265K	71	King Air 100	B-66	Jera Investments, Houston, Tx.	C-GBFD
☐ N266F	93	King Air C90B	LJ-1330	Eclipse Manufacturing Co. Sheboygan, Wi.	N8064A
☐ N266M	82	SA-227TT Merlin 3C	TT-424	Air Charter Ltd. Wausau, Wi.	N900AK
☐ N266RH	81	King Air 200	BB-778	Robinson Humphrey Co. Atlanta, Ga.	N36CP
☐ N269BW	80	King Air 200	BB-618	Big Country Charters LLC. Fayetteville, Ar.	HB-GID
☐ N269JR	94	Pilatus PC-XII	105	ROI Development Corp/Newmar, Newport Beach, Ca.	HB-FQB
☐ N269M	85	Gulfstream 1000	96098	w/o Boca Raton, Fl 22 Jan 98.	N699GN
☐ N269TA	81	King Air 200	BB-749	Truman Arnold Companies Inc. Texarkana, Tx.	N12KW
☐ N270CS	81	King Air B200	BB-911	Head Island Inc. Philadelphia, Pa.	N73LX
☐ N270M	67	King Air A90	LJ-288	Clark Brothers Felt, Waco, Tx.	
☐ N270TC	79	King Air C90	LJ-368	Execjet LLC. Ronkonkoma, NY.	F-GHFS
☐ N271BC	85	King Air 300	FA-48	Robert & Colleen Haas, San Francisco, Ca.	N7221N
☐ N271DC	81	SA-226T Merlin 3B	T-414	Southern Air & Sea Inc. Tampa, Fl.	N1013T
☐ N271WN	67	King Air A90	LJ-226	Carole Broderick, Pinellas Park, Fl.	N14TG
☐ N273AZ	79	Conquest II	441-0104	JMS Automotive Rebuilders Inc. Yuma, Az.	N8194Z
☐ N273TA	92	King Air 350	FL-69	GECC. Emeryville, Ca.	N8230Q
☐ N274GC	85	Conquest II	441-0349	Pro Aviation II LLC. Ogden, Ut.	N12125
☐ N274KA	78	King Air E90	LW-274	Martin Aviation Corp. NYC.	C-GTIO
☐ N275CA	78	PA-31T Cheyenne II	7820089	Twin Otter International Ltd. Las Vegas, Nv.	XA-JGS
☐ N275DP	65	King Air 90	LJ-34	I Fly South Inc. Wilmington, De.	N200SW
☐ N275LE	68	King Air B90	LJ-373	Rockford Motors Inc. Rockford, Il.	N99KA
☐ N275X	79	King Air 200	BB-502	K A 200 Inc. Houston, Tx.	N77PA
☐ N276VM	66	King Air 90	LJ-81	Nuway Inc. Wilmington, De.	N950K
☐ N277GA	90	King Air B200	BB-1389	Glass Equipment Development Inc. Twinsburg, Oh.	N389SA
☐ N279CA	77	King Air 200	BB-279	Corporate Air, Billings, Mt.	4X-ARD
☐ N279DD	77	Rockwell 690B	11373	Monterey Airplane Co. Monterey, Ca.	N911AC
☐ N280RA	75	King Air 200	BB-53	Cor Aviation Inc. Orlando, Fl.	N6ES
☐ N280SW	80	PA-31T Cheyenne 1	8004057	Portage Electrical Products Inc. North Canton, Oh.	
☐ N280TT	77	King Air 200	BB-280	Sunquest Executive Air Charter Inc. Van Nuys, Ca.	F-GJPD
☐ N280YR	82	King Air B200	BB-943	Roby Inc. Marysville, Oh.	N250YR
☐ N281JH	79	King Air 200	BB-516	Hermitage Aircraft Inc. Nashville, Tn.	N231JH
☐ N281MA	73	Mitsubishi MU-2K	261	MU-2 Inc. Huntsville, Al.	
☐ N282CT	82	King Air B200	BB-1002	Warren Transportation Inc. Birmingham, Al.	N6169S
☐ N282SJ	81	King Air 200	BB-925	Greenlaw Grupe Jr Operating Co. Stockton, Ca.	N115D
☐ N282TA	83	King Air B200	BB-1123	Tanimura & Antle Inc. Yuma, Az.	N272TA
☐ N283B	76	King Air C-12C	BC-35	A H Charters Inc. Belle Chasse, La.	N7067B
☐ N283JP	77	King Air 200	BB-283	Cogdell Investors (BB283) LLC. Matthews, NC.	(N68AA)
☐ N283PM	70	King Air 100	B-46	Planemasters Ltd. W Chicago, Il.	N678RM
☐ N284BB	84	PA-31T Cheyenne 1A	1104012	Wells Dairy Inc. Le Mars, Ia.	N284BL
☐ N284K	80	King Air 200	BB-608	Indiana University Foundation, Bloomington, In.	N16AS
☐ N284KW	82	King Air B200	BB-1084	Kladstrup-Wetzel Associates Ltd. Englewood, Co.	JA8831
☐ N284PM	77	King Air C90	LJ-734	Planemasters Ltd. West Chicago, Il.	(N700BN)
☐ N285KW	84	Cheyenne II-XL	1166007	Chabel Aviation Inc. Key Biscayne, Fl. (was 1122001)	D-IAQA
☐ N288CB	96	King Air 350	FL-135	W C Bradley Co & Columbus Bank & Trust Co. Columbus, Ga.	N3263M
☐ N288CR	78	King Air E90	LW-269	B & K Airplanes LLC. Amarillo, Tx.	N288CB
☐ N288DC	79	PA-31T Cheyenne II	7920030	Brannan Leasing Inc. Wichita, Ks.	N65EL
☐ N288RA	69	King Air 100	B-5	MacAviation LLC. Jonesboro, Ar.	N280RA
☐ N290AS	65	Gulfstream 1	164	Stroobants International, Forest, Va.	N590AQ
☐ N290CC	66	King Air A90	LJ-132	Southern Pride Aviation Inc. Fort Lauderdale, Fl.	ZP-TYF
☐ N290DK	81	King Air F90	LA-90	DeKalb Genetics Corp. DeKalb, Il.	N3735D
☐ N290GC	74	Mitsubishi MU-2K	310	Twins Aviation Inc. Liverpool, NY.	VH-MUK
☐ N290JS	97	King Air C90B	LJ-1500	Jewett Scott Truck Line Inc. Mangum, Ok.	N18XJ
☐ N290K	80	King Air E90	LW-337	June & John Rogers, Hughson, Ca.	
☐ N290PF	76	Rockwell 690A	11290	Pacific Flights Inc. Medford, Or.	N100JJ
☐ N290RS	79	PA-31T Cheyenne 1	7904033	JSA Aviation LLC. Pineville, La.	N627NB
☐ N290SJ	77	King Air 200	BB-290	Legacy Leasing of N Carolina, Salt Lake City, Ut.	YV-161CP
☐ N291CC	77	King Air C90	LJ-728	Chapel Creek Air Inc. Wilmington, De.	N275L
☐ N291MB	74	Mitsubishi MU-2K	291	International Business Aircraft Inc. Tulsa, Ok.	CP-1147
☐ N291MM	73	King Air E90	LW-32	H McMurrey, Houston, Tx.	N123PP
☐ N291RB	74	Rockwell 690A	11161	Ron Blackwell Ford Inc. Bentonville, Ar.	XA-RZS
☐ N292A	66	King Air 90	LJ-99	Goldco Energy Corp. Shreveport, La.	N73679
☐ N292RG	91	TBM 700	14	Chemrite Industries Inc. Lannon, Wi. (status ?).	N711GH
☐ N292Z	66	680T Turbo Commander	1566-22	Pasco Inc. Memphis, Tn.	OB-T-924

Regn	Yr Type	c/n	Owner/Operator	Prev Regn

Regn	Yr	Type	c/n	Owner/Operator	Prev Regn
N294WT	88	King Air B200	BB-1294	Century Equipment Co. Salt Lake City, Ut.	N294CS
N295CP	88	King Air B200	BB-1295	Central & South West Services Inc. Dallas, Tx.	N95MW
N296A	67	King Air A90	LJ-208	Sowela Regional Technical Institute, Lake Charles, La.	N29SA
N298MA	73	Mitsubishi MU-2J	594	Executive Cash Services Inc. Mays Landing,, NJ.	
N300AE	94	TBM 700	97	Yankee Bravo Inc. Ponte Verde, Fl.	
N300AJ	82	King Air B200	BB-965	S I Aircraft & Sales Inc. Arlington, Tx.	N184MQ
N300AL	80	SA-226T Merlin 3B	T-326	M & C Leasing Inc. New Bern, NC.	N19J
N300BA	75	King Air E90	LW-143	Mid South Title Agency Inc. Ridgeland, Ms.	N30MD
N300CH	75	King Air E90	LW-148	U S Helicopters Inc & Diamond Aviation Inc. Marshville, NC.	N300KD
N300CW	87	King Air 300	FA-115	CWB Inc & Winburn Stewart Jr. Macon, Ga.	XC-AA20
N300DG	76	King Air B100	BE-15	Dowdle Butane Gas Co. Columbus, Ms.	N300LE
N300DK	76	King Air 200	BB-167	U S Motorsports Inc. Statesville, NC.	N167KA
N300DM	82	King Air F90	LA-165	Addco Manufacturing Inc/3N Properties, St Paul, Mn.	N686LD
N300FL	84	King Air C90A	LJ-1066	Fleet Financial Group Inc. Providence, RI.	N6727M
N300GM	78	Mitsubishi MU-2N	724SA	Scope Leasing Inc. Columbus, Oh.	N44MX
N300HB	79	King Air 200	BB-566	Gantt Aviation inc. Georgetown, Tx.	N200ER
N300HM	79	King Air 200	BB-514	C C Medflight Inc. Lawrenceville, Ga.	
N300HP	80	PA-31T Cheyenne II	8020038	Weather Modification Inc. Fargo, ND.	N804CM
N300JC	90	King Air 350	FL-13	Cashman Equipment Co. Las Vegas, Nv.	N56862
N300JK	79	King Air 200	BB-532	Dot Foods Inc. Mount Sterling, Il.	N300JD
N300MC	86	King Air 300	FA-91	Ridgeway Enterprises Inc/Rose Packing Co. Barrington, Il.	N990PT
N300MP	78	King Air B100	BE-44	Minnesota Power & Light Co. Duluth, Mn.	
N300MY	89	King Air 300	FA-187	Mayo Yarns Inc. Madison, NC.	N300VY
N300NK	83	Conquest II	441-0318	Enkay Corp. Shreveport, La.	N1208M
N300PE	84	King Air 300	FA-27	Iron Eagle Inc. Fayetteville, Ar.	N205K
N300PH	89	King Air 300	FA-201	Pioneer Hi-Bred International Inc. Des Moines, Ia.	
N300PK	84	King Air 300	FA-8	Welder Exploration Inc. Victoria, Tx.	
N300PT	73	SA-226T Merlin 3	T-233	Toy Air Inc. Southfield, Mi.	N84LC
N300PV	81	PA-31T Cheyenne II	8120038	Lucian McElroy, Metamora, Mi.	N806CA
N300RX	73	Mitsubishi MU-2J	572	Diamond Aviation Inc. Statesboro, Ga.	N986MA
N300SP	76	King Air E90	LW-166	Flagstaff Medical Center, Flagstaff, Az. (status ?).	N26902
N300SR	81	PA-31T Cheyenne 1	8104022	Seastar Inc. Carmel, In.	N77NH
N300TA	80	King Air F90	LA-9	Michael Williams, LaPorte, In.	N200FA
N300TJ	84	King Air 300	FA-13	Galaxy Systems Management Inc. Sikeston, Mo.	N6669M
N300TM	88	King Air 300	FA-168	Taylor & Martin Inc - Auctioneers, Fremont, Ne.	N1517R
N300TP	87	King Air B200	BB-1279	Otter Tail Power Co. Fergus Falls, Mn.	N58AB
N300TR	74	King Air 200	BB-6	ELOW Inc. Birmingham, Al.	N301TT
N300US	77	King Air 200	BB-243	Hagair Inc. Dallas, Tx.	N234AM
N300VA	80	King Air C90	LJ-877	Virginia Student Aid Founndation, Charlottesville, Va.	N66877
N300WC	92	TBM 700	82	Woodcraft Industries Inc. St Cloud, Mn.	
N300WP	79	PA-31T Cheyenne 1	7904006	Spectrum Jet Center Inc. Hayden, Co.	N6182
N301DK	68	King Air B90	LJ-372	Sky-Med Inc. Honolulu, Hi.	N800KA
N301ER	91	King Air C90A	LJ-1286	C & E Holdings Inc. Madison, Md.	N56534
N301GM	77	Mitsubishi MU-2L	696	Columbus Airways Inc. Columbus, In.	N856MA
N301K		Convair 580	387	IFL Group Inc. Waterford, Mi.	N321K
N301KS	85	King Air 300	FA-61	State of Kansas, Topeka, Ks.	N7234E
N301PS	80	King Air 200	BB-657	International Muffler Co. Schulenberg, Tx.	
N301PT	78	PA-31T Cheyenne 1	7804002	Kankakee Aeronautics Inc. Kankakee, Il.	
N301TS	79	King Air B100	BE-76	Bika Air Inc. Solon, Oh.	N812WJ
N302BA	76	Rockwell 690A	11302	San Pedro Aviation Inc. Dover, De.	XA-PIR
N302NC	84	King Air 300	FA-9	Newell Co. Freeport, Il.	N930G
N303CA	86	King Air C90A	LJ-1134	Dunlap & Kyle Co. Batesville, Ms.	D-IEVO
N303D	83	King Air C90-1	LJ-1055	Alan Binette, Medford, Or.	
N303DK	79	King Air 200	BB-578	CLG Associates Inc. Charlotte, NC.	N578G
N303G		Rockwell 690B	11355	Clemson Univesrity, Clemson, SC.	N303G
N303GM	82	Gulfstream 1000	96042	South Pacific Air Transport, Houston, Tx.	N9962S
N303KL	78	PA-31T Cheyenne II	7820061	All Star Airlines, Lebanon, Mo.	(N794PL)
N303QC	78	King Air C90	LJ-754	Chaparral Inc. Newton, Ks.	N3030C
N304HC	76	Rockwell 690A	11304	Barry Aviation Inc. San Marcos, Tx.	EC-EAQ
N304JS	88	King Air 300	FA-142	Juliet Sierra Aviation Inc. Newport Beach, Ca.	N27TB
N305CW	75	Mitsubishi MU-2L	667	Murray Aviation Inc. Ypsilanti, Mi.	N300CW
N305RL	90	King Air 350	FL-10	Royal Insurance Co. Charlotte, NC.	N305P
N307CL	78	Rockwell 690B	11508	Champion Laboratories Inc. West Salem, Il.	N81861
N309L	69	King Air E90	LJ-436	Charles Collins Aviation Inc. Carlsbad, Ca.	
N309M	91	King Air 350	FL-48	Bristol-Myers Squibb Co. White Plains, NY.	N8068R
N310MA	69	Mitsubishi MU-2F	167	Jackson Tennessee Leasing Co. Jackson, Tn.	N549LK
N310VE	86	King Air 300	FA-104	Love Box Co. Wichita, Ks.	N215GA
N311AC	81	PA-31T Cheyenne II	8120016	Pacific Service Co. Upland, In.	N241PS
N311AV	78	King Air 200	BB-336	Avior Technologies Inc. Miami, Fl.	N50TW
N311CM	80	King Air B100	BE-101	LJ Associates Inc. Latrobe, Pa.	N3695A
N311DS	80	King Air F90	LA-41	Ridgaire Inc. Jacksonville, Tx.	N37JT
N311JB	74	Mitsubishi MU-2K	311	Kalitta Flying Services Inc. Morristown, Tn.	N501MA
N311LM	77	PA-31T Cheyenne II	7720016	Freefall Express inc. Keene, NH.	C-FACM
N311MP	83	King Air B200	BB-1112	Med-Pro Leasing, Oklahoma City, Ok.	(N7090U)
N311RV	75	SA-226T Merlin 3A	T-250	Banner Associates Inc. Lake Forest, Il.	N70X
N311SC	82	Cheyenne T-1040	8275006	SouthCentral Air Inc. Kenai, Ak.	N9176Y

Regn	Yr	Type	c/n	Owner/Operator	Prev Regn
☐ N312AC	0	SA-226T Merlin 3B	T-322	Airvac Inc. Rochester, In.	N70312
☐ N312BC	94	Pilatus PC-XII	101	Cheyenne Air Inc. Wilmington, De.	HB-FOC
☐ N312KJ	94	2000A Starship	NC-46	JBF Systems Inc. Crystal Falls, Mi.	N8170Q
☐ N312MA	73	Mitsubishi MU-2K	266	Raymond Kinney & Finley Ledbetter, Gainesville, Tx.	
☐ N312PC	96	Pilatus PC-XII	144	Eastern Seaboard Packaging Inc. Holliston, Ma.	HB-FRF
☐ N312VF	67	King Air A90	LJ-299	Suns Products Inc. Eldon, Mo.	N312RF
☐ N313BA	89	King Air 300	FA-178	Beta Aire Ltd. Toledo, Oh.	ZS-MNG
☐ N313D	82	SA-227AT Merlin 4C	AT-464	Del Airways of Delaware Inc. Wilmington, De.	N30364
☐ N313DS	83	Conquest II	441-0313	Christian Business Mens Committee, Chattanooga, Tn.	(N12072)
☐ N313DW	96	King Air C90B	LJ-1434	Durair Inc. Sedona, Az.	
☐ N313EE	80	King Air 200	BB-708	Denison Aviation LLC. Indianapolis, In.	N313EL
☐ N313ES	88	King Air B200	BB-1300	Lewis Webb, Irvine, Ca.	N875EC
☐ N313KY	86	King Air 300	FA-110	Jobe Ski Corp. Redmond, Wa.	ZS-NPL
☐ N313MP	84	King Air 300	FA-23	Rudd Performance MotorSports Inc. Moresville, NC.	N313MB
☐ N314DD	80	PA-31T Cheyenne 1	8004025	North Wind Aviation Inc. Fort Wayne, In.	N2323V
☐ N314P	82	King Air F90	LA-170	Makena Hawaii Inc. Kailua, Hi.	N605CC
☐ N314SC	82	Cheyenne T-1040	8275007	Southcentral Air Inc. Kenai, Ak.	N9180Y
☐ N314TD	80	PA-31T Cheyenne II	8020056	Park Services Inc. Salem, Or.	N144RL
☐ N315DB	80	SA-226T Merlin 3B	T-315	Bannex Aircraft Corp. Wilmington, De.	N318AK
☐ N316EN		Mitsubishi MU-2M	349SA	Jaguar Oil & Gas Corp. Prestonsburg, Ky.	N734MA
☐ N317P	92	King Air 350	FL-74	Washington Asphalt Inc & Red Samm Mining Co. Bellevue, Wa.	
☐ N318MA	76	PA-31T Cheyenne II	7620018	Weather Modification Inc. Fargo, ND.	N941SS
☐ N318W	78	King Air 200	BB-340	USDA Forest Service, Boise, Id.	N400QK
☐ N319BF	81	Gulfstream 840	11675	Jose Manuel Gardilla,	N39SA
☐ N320E	68	King Air B90	LJ-378	CounselAir LLC. Syracuse, NY.	N595AF
☐ N320JS	82	King Air B200C	BL-53	Aviation Specialities Inc. Washington, DC.	N494AC
☐ N321DB	74	Rockwell 690A	11218	Deane Beman Gold Enterprises Inc. Ponte Verde, Fl.	N690VM
☐ N321DH	95	Pilatus PC-XII	116	Virtual Village Aircraft Inc. Wilmington, De.	HB-FQJ
☐ N321DM	77	King Air E90	LW-250	Aerostar Executive Aviation Inc. San Antonio, Tx.	N600EF
☐ N321LB	81	Cheyenne II-XL	8166012	Coosa Air Inc. Wilmington, De.	N888PT
☐ N321LH	84	PA-42 Cheyenne 400LS	5527012	Northeast Air Charter Inc. Greenwich, Ct.	C-GRGE
☐ N321TP	80	MU-2 Marquise	780SA	Ted Price, Ventura, Ca.	N16HA
☐ N322AN	80	Conquest II	441-0127	Anderson News Corp. Amarillo, Tx.	N2625D
☐ N322CP	85	King Air 300	FA-65	CITGO Petroleum Corp. Tulsa, Ok.	N7232Z
☐ N322GB	80	MU-2 Solitaire	419SA	Flying Services Inc. Corpus Christi, Tx.	N153MA
☐ N322GK	80	King Air F90	LA-64	Milford Bostik, Waco, Tx.	N3698H
☐ N322P	83	Conquest II	441-0274	Herbert & Marigrace Boyer, Novato, Ca.	N98468
☐ N322TA	80	MU-2 Marquise	760SA	Indiana State Medical Association, Indianapolis, In.	N321RC
☐ N322TC	92	King Air C90B	LJ-1302	KJV Aviation Inc. Petersburg, WV.	N5670D
☐ N323HA	79	King Air E90	LW-323	Dale Aviation Inc. Rapid City, SD.	N323KA
☐ N325KW	85	PA-42 Cheyenne 400LS	5527025	MCS Transport Inc. Quincy, Ma.	N552AC
☐ N325MA	75	Mitsubishi MU-2M	325	Hahn Transportation Ltd. Decatur, Il.	N440RB
☐ N325MM	82	Gulfstream 1000	96015	Interselect Inc. Alto, NM.	XB-DSF
☐ N325WP	80	King Air F90	LA-10	Blue Ridge King Inc. Martinsville, Va.	N90BN
☐ N326PS	89	King Air B200	BB-1326	Langston Air LLC. Greenville, SC.	N1544G
☐ N326RS	83	Conquest II	441-0326	R S & I Oregon Inc. Roseburg, Or.	N171SP
☐ N327ML		Lear Fan	E003	EAA Aviation Foundation Inc. Oshkosh, Wi.	N21LF
☐ N327RK	78	King Air 200	BB-335	Deffenbaugh Industries Inc. Shawnee, Ks.	N371TA
☐ N328AJ	84	SA-227TT Merlin 300	TT-483A	Columbus Fair Auto Auction Inc. Columbus, Oh.	N300CV
☐ N328CA	63	Gulfstream 1C	116	Chrysler Pentastar Aviation, Ypsilanti, Mi.	N159AN
☐ N328DC	94	Dornier 328-100	3019	Pacific Gas & Electric Co. San Francisco, Ca.	
☐ N328VP	83	Conquest II	441-0301	Way International, New Knoxville, Oh.	N6853A
☐ N329HS	80	SA-226T Merlin 3B	T-323	Harold's Air Trans Inc. Norman, Ok.	(N329HP)
☐ N330BR	76	King Air 200	BB-115	Bahari Racing Inc. Charlotte, NC.	N80XC
☐ N330CB	65	King Air 90	LJ-27	S Butler, Gainesville, Fl.	N400PQ
☐ N330CS	89	King Air B200	BB-1337	Concord Service Corp/Cable Services, Williamsport, Pa.	N1550U
☐ N330DR	97	King Air B200	BB-1578	Robertson Properties, Big Bear Lake, Ca.	N1095G
☐ N330V	79	King Air C90	LJ-811	Truck Body Aviation Inc. Lynchburg, Va.	
☐ N330VP	85	King Air F90-1	LA-227	Leitch Inc. Chesapeake, Va.	N7206Z
☐ N331AM	66	King Air A90	LJ-144	PWA Aircraft Inc. Ponca City, Ok.	(N3DG)
☐ N331GB	70	King Air B100	BE-16	Chris Aviation Inc. Southfield, Mi.	
☐ N331GM	80	SA-226T Merlin 3B	T-331	B & B Aviation LLC. New Orleans, La.	N331TB
☐ N331H	61	Gulfstream 1	70	Dodson International Parts Inc. Ottawa, Ks.	N770G
☐ N331J	79	MU-2 Marquise	734SA	Whitehall Corp. Dallas, Tx.	N988S
☐ N331MA	80	MU-2 Marquise	768SA	Aerographics Inc. Manassas, Va.	N36MF
☐ N331RC	73	Rockwell 690A	11137	Ace Air Inc. Lafayette, La.	N876MC
☐ N332DE	76	King Air C90	LJ-674	PAM Aviation Enterprises LLC. Sioux Falls, SD.	N9074S
☐ N332DM	80	Conquest II	441-0167	Defiance Metal Products Co. Defiance, Oh.	N312KJ
☐ N332SA	81	PA-42 Cheyenne III	8001020	BSA Air Inc. Dothan, Al.	N328AJ
☐ N333AP	83	King Air B200	BB-1147	Andy Petree Racing Inc. East Flat Rock, NC.	N133LJ
☐ N333BM	97	King Air C90B	LJ-1504	Raytheon Aircraft Co. Wichita, Ks.	
☐ N333CA	73	Rockwell 690A	11136	Hamburg Aero Leasing LP. Hamburg, NY.	
☐ N333DV	88	King Air 300	FA-144	Vincent Aviation Inc. Van Nuys, Ca.	N950JM
☐ N333EB	81	King Air F90	LA-139	Subaru of New England, Norwood, Ma.	A2-AFI
☐ N333G	66	King Air 90	LJ-101	Bramco Aircraft Leasing, Miami, Fl.	N42B
☐ N333HC	74	Rockwell 690A	11150	Robert Holt, Midland, Tx.	N47150

Regn	Yr	Type	c/n	Owner/Operator	Prev Regn
N333LE	77	King Air E90	LW-223	Melvin Merton Kaftan, Southfield, Mi.	N7UM
N333P	80	PA-31T Cheyenne II	8020002	Outdoor World Corp. Bushkill, Pa.	(N4LH)
N333PA	79	Rockwell 690B	11523	K Bryant Leasing Inc. Portsmouth, NH.	C-GMDD
N333RD	81	King Air B200	BB-915	ETCO Inc. Warwick, RI.	ZS-LFT
N333RK	78	Mitsubishi MU-2P	380SA	Intercontinental Jet Inc. Tulsa, Ok.	ZS-NIY
N333TL	82	King Air C90	LJ-999	AKSM Equipment LLC. Columbus, Oh.	N333TP
N333TN	80	PA-31T Cheyenne II	8020089	Betnr Engineering & Construction Corp. Pittsfield, Ma.	D-IGEM
N333TP	88	King Air B200	BB-1292	Textile Printing Co. Chattanooga, Tn.	N3109Y
N333TS	79	King Air/Catpass 250	BB-512	HWV Aviation, Piney Flats, Tn.	F-GHCU
N333UR	72	Rockwell 690	11016	Eoff Electric Co. Salem, Or.	N333UP
N333WF	78	Mitsubishi MU-2P	387SA	Jack Jaax Gerhard, Calexico, Ca.	N8TB
N333XX	79	PA-31T Cheyenne II	7920038	Robert Arnot, NYC.	N333CE
N334EB	73	Mitsubishi MU-2J	568	Lakeside Aviation Inc. Terra Alta, WV.	N99SL
N334FP	83	PA-42 Cheyenne III	8001104	N W Regional Inc. Irving, Tx.	N712CA
N335S	77	King Air 200	BB-227	Thomas Malone PC. Atlanta, Ga.	N90BR
N337C	80	Conquest II	441-0126	Chapman Corp. Washington, Pa.	N98630
N337KC	84	Conquest II	441-0337	Kinsley Construction, York, Pa.	XA-PEH
N338CM	77	Mitsubishi MU-2N	703SA	W W & D Ltd. Hinsdale, Il.	N333GM
N339W	92	TBM 700	39	Stage Aviation Inc. Carmel, Ca.	
N340RE	97	Pilatus PC-XII	179	Ringhaver Equipment Co. Riverview, Fl.	HB-FSO
N344KL	73	Mitsubishi MU-2K	257	Flight Service Co. Munhall, Pa.	N344K
N345CM	77	Rockwell 690B	11402	RORAN Investments Inc. San Francisco, Ca.	N81638
N345DF	78	PA-31T Cheyenne II	7820088	State of Nevada, Carson City, Nv.	N4491C
N345MB	83	King Air B200	BB-1148	Galen Aviation Corp. Beverly, Ma.	N211DQ
N345RD	92	TBM 700	76	Ray Dolby, San Francisco, Ca.	
N345SA	72	Mitsubishi MU-2J	557	European Aircraft Leasing Inc. Dover, De.	OH-MIC
N345T	68	SA-26AT Merlin 2B	T-26-105	Boyd Air Corp. Dallas, Tx.	
N345TT	68	680W Turbo Commander	1791-21	Thunder Commander Ltd. Van Nuys, Ca.	N77JL
N345V	72	King Air E90	LW-23	Fred Fiore, Pittsburgh, Pa.	
N345WK	97	King Air B200	BB-1580	AviaServ LLC. Los Angeles, Ca.	N200KA
N346CM	75	King Air 200	BB-18	Tri City Beverage Corp. Midland, Tx.	N211JB
N346VL	72	Mitsubishi MU-2F	231	Diamond, Lakee & Pearson Resources, Austin, Tx.	N666MA
N347AR	79	Conquest II	441-0085	Gilliland Construction Co. Scranton, Ar.	C-GSJJ
N347D	85	King Air B200	BB-1197	Oil Well Perforators Inc. Glendive, Il.	
N348AC	67	King Air A90	LJ-196	St Louis Public Schools Gateway Institute of Technology, Mo.	N300DD
N348KN	83	SA-227TT Merlin 3C	TT-541	N348KN Inc. Fremont, Mi.	N541SA
N349AC	72	Rockwell 690	11032	Marko Air Service Inc. Fort Lauderdale, Fl.	N8LD
N349MA	74	Mitsubishi MU-2J	615	International Business Aircraft Inc. Tulsa, Ok.	YV-108CP
N350AC	78	King Air 200	BB-405	Aero Charter Inc. Chesterfield, Mo.	N83GB
N350AT	73	King Air E90	LW-58	Executive Wings Inc. Wilmington, De.	N250HP
N350BD	94	King Air 350	FL-123	Alfa Financial Corp. Montgomery, Al.	N30YR
N350CD	89	King Air 300	FA-177	Russell Corp. Alexander City, Al.	(N25CN)
N350CS	91	King Air 350	FL-61	Central & South West Services Inc. Dallas, Tx.	N8140F
N350DW	88	King Air 300	FA-157	Crow Air Ltd. Wheatland, Wy.	N552M
N350EA	94	King Air 350	FL-116	European Aircraft Corp. Miramar, Fl.	D-CAAA
N350EB	92	King Air 350	FL-49	Aviacorp Corp.Miami, Fl.	N27BH
N350FT	91	King Air 350	FL-45	Featherlite Aviation Co. Cresco, Ia.	N350CA
N350GA	90	King Air 350	FL-16	Gasser Air Co. Youngstown, Oh.	N48HP
N350KA	97	King Air 350	FL-173	Raytheon Aircraft Co. Wichita, Ks.	
N350LL	97	King Air 350	FL-157	Leavitt Leasing Co. Los Angeles, Ca.	N1093Q
N350MC	81	SA-226T Merlin 3B	T-400	Sanair Corp. Boston, Ma.	N10126
N350MS	93	King Air 350	FL-99	Mississippi State University Foundation Inc. Starkville, Ms.	N8299L
N350RG	80	MU-2 Marquise	773SA	International Amusement Ltd. Durand, Mi.	N259MA
N350TJ	82	King Air B100	BE-125	Dauphin Enterprises Inc. Wilmington, De.	N314EB
N350TR	91	King Air 350	FL-67	Thompson Realty Co. Shoal Creek, Al.	N8213Q
N351EB	94	King Air 350	FL-121	Leggett & Platt Inc. Carthage, Mo.	D-CSAG
N351GR	79	King Air E90	LW-324	Golden Rule Financial Corp. Indianapolis, In.	N2043C
N352GR	74	King Air E90	LW-93	Travel Air Aircraft Co. Goddard, Ks.	N655F
N355ES	83	Conquest 1	425-0158	Edward & Patricia Shorma, Wahpeton, ND.	N25QL
N355TW	79	King Air 200	BB-521	Quincy Newspapers Inc. Quincy, Il.	N220GK
N356GA	79	King Air 200	BB-447	NorAm Energy Corp. Houston, Tx.	N300CT
N356M	91	TBM 700	32	Stephen Robertson, Jackson, Wy.	
N356WG	82	King Air B200	BB-1066	Texas Gas Transmission Co. Owensboro, Ky.	N356WC
N357CC	82	King Air F90	LA-180	Conklin Instrument Corp. Pleasant Valley, NY.	N999KK
N357HP	82	King Air C90-1	LJ-1030	Department of Public Safety, Jackson, Ms.	N6504H
N357JR	65	King Air 90	LJ-5	EMPC LLC. Middlefield, Oh.	N702K
N358HF	79	Rockwell 690B	11551	B-Air LC. Sikeston, Mo.	XA-RPY
N359JT	86	King Air C90A	LJ-1136	Alan Silvestri Music Inc. Carmel Highlands, Ca.	N90RK
N359K	78	King Air 200	BB-359	Koch Industries Inc. Wichita, Ks.	N780W
N360C	70	King Air 100	B-21	Cutter Flying Service Inc. Albuquerque, NM.	N11AG
N360D	67	King Air A90	LJ-240	SCS Group Ltd/Springdale Air Service Inc. Springdale, Ar.	N36030
N360JK	70	Mitsubishi MU-2G	525	Hartford Holding Corp. Millersville, Md.	YV-O-CDM-1
N360MP	84	King Air C90A	LJ-1085	Lakeshore Village, Atlanta, Ga.	N720CT
N360RA	79	MU-2 Marquise	740SA	Hirschfield Steel Co. Lynchburg, Va.	XA-EFU
N360SC	82	King Air B200	BB-1030	Food City Aviation LLC & W-L Aviation LLC. Abingdon, Va.	N6348H
N361DB	95	Pilatus PC-XII	132	Radio Flyer II Inc. Wilmington, De.	HB-FQV

Regn	Yr	Type	c/n	Owner/Operator	Prev Regn
N361EA	83	King Air B200	BB-1103	Hill Aircraft & Leasing Corp. Atlanta, Ga.	N57SC
N361JA	76	Mitsubishi MU-2L	681	International Business Aircraft Inc. Tulsa, Ok.	C-GJWM
N362D	78	King Air E90	LW-265	BBKR LLC. Las Vegas, Nv.	
N362EA	77	King Air 200	BB-257	Metz Baking Co. Sioux City, Ia.	N27DA
N363D	95	King Air B200	BB-1503	Penn Aire Aviation Inc. Franklin, Pa.	N363K
N363EA	83	King Air B100	BE-134	Elliott Aviation Flight Services, Omaha, Ne.	N888RK
N364BC	78	King Air B100	BE-35	Kerrville Bus Co. Kerrville, Tx.	N564BC
N364C	86	King Air 300	FA-109	Cutter Aviation Inc. Phoenix, Az.	LV-WGJ
N364D	79	King Air C90	LJ-837	Aviation Charter Services/Lafayette Aviation Inc. In.	
N364EA	80	King Air 200	BB-664	Elliott Aviation Aircraft Sales Inc. Des Moines, Ia.	N280SC
N364SB	79	King Air 200	BB-445	Beartooth Communications Co. Helena, Mt.	N79KA
N366GW	79	King Air E90	LW-320	Joe Morten & Son Inc. Sioux City, Ne.	N366JM
N366SP	80	King Air F90	LA-53	Machavia Inc. Manhattan Beach, Ca.	N3667U
N369B	89	King Air C90A	LJ-1226	Brian Bemis Inc. Sycamore, Il.	(N702RA)
N369CD	76	King Air E90	LW-162	Commander Inc. Friars Point, Ms.	N36GS
N369GA	81	King Air C90	LJ-934	Eastway Aviation Inc. Ronkonkoma, NY.	F-GCTA
N369MK	89	King Air 300	FA-203	Aileron Air Inc. Forest Park, Il.	LV-RBM
N369PC	85	Conquest II	441-0351	Peter Carras, Midland, Mi.	N118TS
N369TA	81	King Air 200	BB-820	Truman Arnold Companies/Tac Air, Texarkana, Ar.	N935SJ
N370AE	87	SA-227AT Merlin 4C	AT-506	First Bank NA. Minneapolis, Mn.	N87FM
N370K	66	680V Turbo Commander	1570-25	Western Granite Inc. Troutdale, Or.	N222JK
N370MA	78	Mitsubishi MU-2P	370SA	Robert Nass Inc. Asheville Regional, Pa.	N370AC
N370TC	77	King Air 200	BB-311	Berquist Co. Edina, Mn.	F-GILE
N370X	69	SA-26AT Merlin 2B	T26-106	Twins Charter Aircraft Inc. Miami Beach, Fl.	
N373LP	81	Conquest 1	425-0141	Aero Property Management, Sacramento, Ca.	D-IIGA
N373Q	81	PA-42 Cheyenne III	8001014	Sullivan Leasing Co. Durant, Ok.	C-GAKA
N375AA	74	Rockwell 690A	11179	Miroslav Liska, Bend, Or.	N57179
N375AC	75	Mitsubishi MU-2M	327	Celco Constantine Engineering Laboratories, Wilmington, De.	N515MA
N375CA	74	Mitsubishi MU-2J	643	Corporate Air Leasing Inc. Fort Wayne, In.	N881DT
N376TC	73	Rockwell 690	11057	Mississippi State University, Mississippi State, Ms.	N37TB
N376WS	80	PA-31T Cheyenne II	8020064	Polar Star Co. Hodges, SC.	(N32WS)
N377CA	83	Conquest II	441-0084	Arturo Saldana, Coppell, Tx.	N88638
N379VM	73	King Air E90	LW-27	JBM Enterprises LLC. Laramie, Wy.	N999ES
N380AA	80	King Air C90	LJ-883	DEGP Ltd. Columbus, Oh.	D-IAXX
N380M	67	King Air A90	LJ-218	Baker Aviation Inc. Kotzebue, Ak.	N360M
N380SC	79	King Air C90	LJ-862	M & M Aircraft Inc. Wilmington, De.	F-GMJP
N381MG	97	King Air 350	FL-182	Raytheon Aircraft Co. Wichita, Ks.	
N381SW	81	PA-31T Cheyenne 1	8104015	Morrill Motors Inc. Fort Wayne, In.	
N382MB	81	PA-31T Cheyenne II	8120057	Manfred Bohlke Veneer Corp. Fairfield, Oh.	N9087Y
N382TW	75	King Air E90	LW-141	Bud Antle Inc. Salinas, Ca.	(N167JR)
N383AA	72	King Air E90	LW-13	Semrau Aircraft Co. Wilmington, De.	N21DJ
N383DA	73	King Air C90	LJ-592	Executive Aviation Logistics Inc. Chino, Ca.	N41CV
N383JC	67	King Air A90	LJ-207	Rob & Simon Diver, Georgetown, Tx.	N7177
N383SS	83	Conquest 1	425-0139	Joseph Skilken & Co. Columbus, Oh.	N6884R
N384DB	78	King Air 200	BB-383	RDB Aviation Inc. Oklahoma City, Ok.	N384JD
N384H	78	King Air C90	LJ-745	Charles Nicholson Jr. Concord, NC.	N92BE
N386CP	85	King Air C90A	LJ-1119	Tri Counties Bank, Chico, Ca.	N72369
N386GA	78	King Air C90	LJ-775	Tamaqua Corp. Milford, De.	HB-GIW
N387CC	80	SA-226T Merlin 3B	T-360	Trigon Inds/Controls Corp of America, Virginia Beach, Va.	N118BR
N387GA	77	King Air C90	LJ-726	General Aviation Service Inc. Lake Zurich, Il.	ST-AGZ
N388MC	68	King Air B90	LJ-383	B & J of Destin Inc. Dover, De.	N63GB
N388NC	81	MU-2 Solitaire	452SA	Paul Connolly, Nashua, NH.	N230MA
N390K	73	Mitsubishi MU-2K	277	International Business Aircraft Inc. Tulsa, Ok.	C-GMUK
N390PS	78	King Air E90	LW-279	Pro Air Inc. Columbus, Ne.	N390MT
N391BT	81	King Air C90	LJ-983	La Tour Air Inc. New Baltimore, Mi.	N491BT
N391GM	80	SA-226T Merlin 3B	T-391	Fun Times Boat Inc. Slidell, La.	F-GCTC
N391RR	90	King Air 350	FL-23	Rogers Enterprises LLC. Sun Valley, Id.	(F-GKIZ)
N392P	78	MU-2 Solitaire	392SA	Aero Classics Inc. Daytona Beach, Fl.	D-IFMU
N393CE	85	King Air F90	LA-230	J & J Black Diamond LLC. Greenwich, Ct.	N393CF
N393CF	97	King Air 350	FL-165	Hiddenbrook Air, Salt Point, NY.	N1135L
N393JW	77	King Air 200	BB-292	Pelican Air Ltd. Memphis, Tn.	N393K
N394AL	68	King Air B90	LJ-394	Alberta Aircraft Leasing Inc. Las Vegas, Nv.	C-GASR
N394G	83	Conquest II	441-0272	Kir-Nie Aviation Inc. Quincy, Il.	N594G
N395AM	83	King Air B200	BB-1101	Amcast Industrial Financial Services Inc. Dayton, Oh.	N48CE
N395CA	84	Cheyenne II-XL	1166005	Anderson Hay & Grain Co. Ellensburg, Wa.	HB-LRM
N395DR	81	PA-42 Cheyenne III	8001065	CLB Corp/Alpine Air, Provo, Ut.	N742RB
N396FW	80	PA-42 Cheyenne III	8001021	Zephyr Aviation LLC. Houston, Tx.	C-FWAB
N397WM	88	King Air 300	FA-156	West Michigan Aviation Services, Wyoming, Mi.	N642BL
N398CA	95	Pilatus PC-XII	115	Wilba Inc. Wilmington, De.	N74BE
N398HM	78	King Air 200	BB-398	Apparelsoft Inc. Seneca, SC.	N627BM
N399BM	78	King Air 200	BB-399	Woodland Aviation Inc. Woodland, Ca.	F-GIRM
N399LA	83	King Air B200	BB-1135	Netwave Systems Inc. Monroe, La.	(N171SP)
N400AC	76	King Air B100	BE-12	Midlantic Jet Charters Inc. Egg Harbor Town, NJ.	
N400AL	89	King Air 350	FL-2	Abbott Laboratories Inc. Waukegan, Il.	N350KR
N400BX	76	King Air C90	LJ-686	Robert Browning, Austin, Tx.	N100BT
N400CM	76	PA-31T Cheyenne II	7620040	Robert Brocker, Youngstown, Oh.	

Regn	Yr	Type	c/n	Owner/Operator	Prev Regn
☐ N400CP	80	PA-31T Cheyenne II	8020071	Raymond & Celesta Shefland, Bemidji, Mn.	N2471V
☐ N400DC	69	SA-26AT Merlin 2B	T26-167	Parts & Turbines Inc. St Simons Island, Ga.	N5353M
☐ N400DG	68	SA-26T Merlin 2A	T26-9	Hodges Packing Co. Palestine, Tx.	N22EK
☐ N400DS	78	Rockwell 690B	11512	Commander LLC. Newport Beach, Ca.	N81872
☐ N400ES	78	Mitsubishi MU-2P	389SA	Northwind Corp. Miami, Fl.	(N787X)
☐ N400JH	78	Mitsubishi MU-2P	381SA	AJ Inc. Dover, De.	N771MA
☐ N400LP	67	680V Turbo Commander	1697-74	Country Club Investment & Development, Fairmont, WV.	N400LR
☐ N400PS	79	MU-2 Solitaire	411SA	West Texas Executive Leasing Inc. San Angelo, Tx.	N8LC
☐ N400PT		PA-42 Cheyenne 400LS	5527001	Piper Aircraft Corp. Vero Beach, Fl. (was 8427001).	
☐ N400RK	78	King Air B100	BE-49	Elliott Flying Services Inc. Des Moines, Ia.	N98D
☐ N400RV	79	King Air C90	LJ-853	Dawson International Aviation Inc. Wilmington, De.	N6LD
☐ N400SF	77	King Air E90	LW-221	Hughston Orthopaedic Clinic, Columbus, Ga.	
☐ N400SG	74	Mitsubishi MU-2J	634	Pan American Importing Corp. Wilmington, De.	C-GODE
☐ N400ST	91	TBM 700	16	Howard Patrick, Kirkland, Wa.	
☐ N400TJ	81	King Air B100	BE-111	Galaxy Air Services Inc. Sikeston, Mo.	N622RP
☐ N400TR	69	Mitsubishi MU-2F	161	Corporate Aviation Services Inc. Tulsa, Ok.	N11LQ
☐ N400VB		PA-42 Cheyenne 400LS	5527002	Kelleher Corp. San Rafael, Ca. (was 8427002).	N400PS
☐ N400WS	81	Cheyenne II-XL	8166044	Aspen Aero Inc. Albuquerque, NM.	N31KW
☐ N401NS	85	King Air 300	FA-28	Naylor Inc. Atlanta, Ga.	N481NS
☐ N401TS	97	King Air C90B	LJ-1501	Raytheon Aircraft Co. Wichita, Ks.	N2029Z
☐ N402G	70	King Air 100	B-14	ISO Aero Service Inc. Kinston, NC.	N40TG
☐ N402KA	77	King Air 200	BB-296	Kokomo Aviation/Aviation Charter Services, Indianapolis, In.	N402CJ
☐ N402NC	85	King Air 300	FA-80	Newell Co. Freeport, Il.	N51UH
☐ N402NG	81	Conquest 1	425-0035	Darrell & Dorothy Sawyer, Paradise Valley, Az.	N402NC
☐ N403R	78	King Air 200	BB-396	Western Repair Inc. Houston, Tx.	N203R
☐ N404EW	89	King Air 300	FA-186	E W Marine, Elkhart, In. (status ?)	(N700FT)
☐ N404FA	82	King Air B200	BB-981	Falk Air Corp. Norfolk, Va.	N481BC
☐ N404JP	82	King Air C90-1	LJ-1039	Hurd Millwork Co. Medford, Wi.	N6746S
☐ N404SC	79	King Air C90	LJ-843	Jay Hall Jr. Cheshire, Oh.	N843CP
☐ N404SS		Beech 1900C	UB-39	BellSouth Telecommunications Inc. Atlanta, Ga.	N7227E
☐ N405BC	69	King Air B90	LJ-457	Roush Corp/Roush Racing, Livonia, Mi.	N696MM
☐ N405MA	83	Conquest II	441-0316	Preformed Line Products Co. Mayfield, Oh.	G-LOVX
☐ N406CP	81	Gulfstream 840	11655	Palumbo Aircraft Sales Inc. Uniontown, Oh.	YV-406CP
☐ N406RS	79	PA-31 Cheyenne II	7920048	Santiago Management Co. Santa Ana, Ca.	N404AF
☐ N407MA	81	MU-2 Marquise	1503SA	H & F Executive Aviation Inc. Wilmington, De.	
☐ N408G	85	King Air 300	FA-73	M A Inc. Oshkosh, Wi.	N403G
☐ N409GA	78	King Air 200	BB-394	Data Supplies Inc. Duluth, Ga.	IAC234
☐ N410CA	81	Cheyenne II-XL	8166005	Teerling Aviation Inc. Lockport, Il.	N127GP
☐ N410JA	95	Jetstream 41	41030	AlliedSignal Inc. Morristown, NJ.	G-4-...
☐ N410MA	82	Conquest II	441-0104	Miller Aviation Inc. Johnson City, NY.	(D-ITTW)
☐ N410SP	73	King Air A100	B-146	Jet Air Inc. Cincinnati, Oh.	N208SB
☐ N410TH	68	680W Turbo Commander	1790-20	Charles Bella, Chaparral, NM.	N5061E
☐ N410TW	84	PA-42 Cheyenne 400LS	5527017	Reuben Setliff III MD. North Platte, Ne.	N4118Y
☐ N410VE	81	Conquest 1	425-0097	Miami Air Corp. Wilmington, De.	N6849D
☐ N410WA	84	Conquest 1	425-0221	Future Vision Inc. Tempe, Az.	N12256
☐ N411BG		PA-42 Cheyenne 400LS	5527004	Robert Gottsch, Elkhorn, Ne.	N4119B
☐ N411FT	69	King Air B90	LJ-443	James Taylor, Albany, Ga.	N796K
☐ N411LM	77	PA-31T Cheyenne II	7720069	Mansfield Industrial Coatings Inc. Pensacola, Fl.	N45TX
☐ N411RS	66	King Air 90	LJ-106	Sundance Aviation Inc. Yukon, Ok.	N271MB
☐ N412SR	72	King Air E90	LW-2	Thomas Flyers Inc. Lubbock, Tx.	N710TK
☐ N413PC	82	Conquest II	441-0240	Paul Chiapparone, Plano, Tx.	N10FJ
☐ N414GN	75	King Air E90	LW-156	Newberg Flying Enterprises Inc. Chicago, Il.	(N5PC)
☐ N415CA	59	Gulfstream 1C	27	Phoenix Air, Cartersville, Ga.	N114GA
☐ N415P	93	King Air C90B	LJ-1345	S & R Leasing Co. Hickory, NC.	
☐ N415RB	83	King Air F90-1	LA-210	Barnhill Contracting Co. Tarboro, NC.	N715JT
☐ N416BK	79	King Air C90	LJ-816	Airflight Travel Inc. Hollandale, Ms.	N40PS
☐ N416DY	89	King Air 300	FA-197	Brookwood Enterprises LLC. Atlanta, Ga.	N59AH
☐ N416LF	67	King Air A90	LJ-267	Consolidated Investment Group, Wichita, Ks.	N5963H
☐ N416P	80	King Air F90	LA-67	CRLI Aviation Inc. Bannockburn, Il.	N614ME
☐ N417MC	96	King Air B200	BB-1526	Mayo Foundation for Education & Research, Rochester, Mn.	N3258P
☐ N417VN	89	King Air C90A	LJ-1202	Corn Construction Inc. Albuquerque, NM.	HB-GIZ
☐ N418DY	80	King Air A90	LA-177	Broodwood Enterprises LLC. Atlanta, Ga.	N416DY
☐ N419R	78	PA-31T Cheyenne II	7820034	Reno Flying Service Inc. Reno, Nv.	N82249
☐ N420MA	82	Conquest 1	425-0116	Jubilee Airways Inc. Glasgow, Scotland.	N45TP
☐ N420PA	78	PA-42 Cheyenne III	7800001	Piper Aircraft Corp. Vero Beach, Fl.	
☐ N421HV	91	King Air B200	LJ-1266	Hy Vee Food Stores Inc. West Des Moines, Ia.	
☐ N422HV	81	PA-31T Cheyenne 1	8104052	DGA Inc. Nebraska City, Ne.	N422DM
☐ N422RJ	92	King Air C90B	LJ-1306	Coastal Lumber Co. Weldon, NC.	N8290T
☐ N422Z	81	King Air F90	LA-135	All Star Advertising Agency Inc. Baton Rouge, La.	
☐ N423JB	82	Cheyenne II-XL	8166052	G L Wilson & Sioux City Truck Sales Inc. Sioux City, Ia.	D-IOKA
☐ N423LW	97	King Air B200	BB-1561	Linweld Inc. Waverly, Ne.	N1069F
☐ N423TG	81	Conquest II	441-0200	Dodson International Parts Inc. Ottawa, Ks. (status ?)	N313GA
☐ N424BS	76	King Air 200	BB-179	Straub Enterprises, Coto de Caza, Ca.	N630DB
☐ N424CM	80	PA-31T Cheyenne 1	8004002	H G Woten, Winter Haven, Fl.	(N2499R)
☐ N424CP	82	King Air F90	LA-182	Finhoff & Associates Inc & others,. Boulder, Co.	N125TS
☐ N424CR	78	King Air 200	BB-347	Hauck Casualty II LLC. Cincinnati, Oh.	ZS-KCB

Regn	Yr	Type	c/n	Owner/Operator	Prev Regn

Regn	Yr	Type	c/n	Owner/Operator	Prev Regn
☐ N424PP	69	680W Turbo Commander	1821-35	Corona Skyways Inc. Corona, Ca.	TG-DAM
☐ N424RA	97	King Air B200	BB-1583	Edward Rorer & Co. Philadelphia, Pa.	
☐ N424SW	71	King Air 100	B-80	Windham Aviation Inc. Williamantic, Ct.	N99KA
☐ N424TV	83	King Air C90-1	LJ-1057	Cannan Communications, Wichita Falls, Tx.	
☐ N425AT	80	Conquest 1	425-0004	Anderson Trucking Service/RBJ Industries Inc. St Cloud, Mn.	N425TF
☐ N425BA	81	Conquest 1	425-0046	Bell Aviation Inc. West Columbia, SC.	ZS-NES
☐ N425BZ	82	Conquest 1	425-0104	ZD Air Inc. Menlo Park, Ca.	N425NT
☐ N425CF	81	Conquest 1	425-0018	Catawba Management Corp. Alexandria, Va.	N425EE
☐ N425CL	84	Conquest 1	425-0206	Case Leasing & Rental Inc. Celina, Oh.	N12238
☐ N425DC	83	Conquest 1	425-0185	Kent Business Systems Inc. Dallas, Tx.	N6874L
☐ N425DD	81	Conquest 1	425-0083	NorCal Beverage Co. West Sacramento, Ca.	N6846X
☐ N425DM	81	Conquest 1	425-0098	Marco Equipment Inc. Jacksonville, Fl.	N6849L
☐ N425E	81	Conquest 1	425-0096	Western Pneumatics Inc. Eugene, Or.	N425EA
☐ N425EZ	81	Conquest 1	425-0099	Eli's Bread (Eli Zabar) Inc. NYC.	N404EW
☐ N425FG	83	Conquest 1	425-0186	Glover Feed Mills Inc. Mount Pleasant, Tx.	N444AK
☐ N425GM	81	Conquest 1	425-0033	Brian Shore, Oyster Bay, NY.	D-IAJA
☐ N425HB	81	Conquest 1	425-0073	Conquest Aviation/Howard Baker, Huntsville, Tn.	N550SC
☐ N425HS	81	Conquest 1	425-0044	B Van Milders NV. Antwerp, Belgium.	N555BE
☐ N425JB	81	Conquest 1	425-0043	John Macguire, Santa Teresa, NM.	N550SC
☐ N425JP	81	Conquest 1	425-0038	James Pugh Jr. Winter Park, Fl.	N3FC
☐ N425K	68	King Air B90	LJ-318	K & K Aircraft Inc. Bridgewater, Va.	
☐ N425KC	83	Conquest 1	425-0174	Linjo Corp. Rancho Palos Verde, Ca.	N384MA
☐ N425KD	85	Conquest 1	425-0203	John Legat Inc. Barrington Hills, Il.	N45TP
☐ N425LD	83	Conquest 1	425-0149	State Aircraft Pooling Board, Austin, Tx.	N425EJ
☐ N425MM	81	Gulfstream 840	11699	Aardex Corp. Lakewood, Co.	XB-DYZ
☐ N425NW	81	Conquest 1	425-0070	New West Foods & Fruit Corp. Watsonville, Ca.	N6854P
☐ N425PG	85	Conquest 1	425-0200	John Fleischhacker, St Paul, Mn.	N17HM
☐ N425PJ	83	Conquest 1	425-0157	Pleasantville Aviation Inc. Pleasantville, NJ.	(N68860)
☐ N425PL	81	Conquest 1	425-0010	Burlington Aviation LLC. Bow, Wa.	N813JL
☐ N425RM	83	Conquest 1	425-0180	Corbi Aircraft Sales Inc. Salem, Oh.	N68731
☐ N425SF	81	Conquest 1	425-0037	Presbyterian Medical Services, Santa Fe, NM.	N6773L
☐ N425SG	81	Conquest 1	425-0052	Accutronics Inc. Englewood, Co.	N377SC
☐ N425SP	83	Conquest 1	425-0184	Jet 1 Inc. Naples, Fl.	VH-JER
☐ N425SR	83	Conquest 1	425-0133	Lowcountry Air Inc. Wilmington, De.	N425TS
☐ N425SX	82	Conquest 1	425-0106	Callis Papa Hale Jensen etc Law Firm PC. Granite City, Il.	N425GA
☐ N425TB	81	Conquest 1	425-0072	Sierra Bells Inc. Grand Junction, Co.	N425ET
☐ N425TC	80	Conquest 1	425-0014	High Times Inc. Bentonville, Ar.	(N6770W)
☐ N425TM	84	Conquest 1	425-0217	Six L's Packing Co. Clinton, NC.	N1225T
☐ N425TV	83	Conquest 1	425-0176	AIRMOD Inc. Carlsbad, Ca. (status ?).	ZS-LDR
☐ N425TW	83	Conquest 1	425-0161	Teton West Construction, Rexburg, Id.	N707NY
☐ N425TY	81	Conquest 1	425-0063	James Young, Bellevue, Wa.	(N20YC)
☐ N425WB	81	Conquest 1	425-0028	Bobbie Devlin, Forest Hills, Ca. (status ?).	N67224
☐ N425WL	83	Conquest 1	425-0166	Law Offices of Gerald Hosier Ltd. Las Vegas, Nv.	N6872D
☐ N425WT	83	Conquest 1	425-0175	Grant General Contractors Inc. Carlsbad, Ca.	N74LD
☐ N426PC	90	PA-42 Cheyenne IIIA	5501053	Southern Cross Aviation Inc. Fort Lauderdale, Fl.	D-IOSF
☐ N428A	86	King Air B200	BB-1243	I A Leasing Co. Greensboro, NC.	N804CT
☐ N428P	80	King Air 200	BB-745	LVA Management & Consulting Inc. Las Vegas, Nv.	N3698S
☐ N429DM	79	King Air C90	LJ-804	310 Enterprises Inc. Wilmington, De.	HK-3276
☐ N429E	77	King Air 200	BB-287	Interstate Power Co. Dubuque, Ia.	
☐ N429K	76	Rockwell 690A	11282	Downtown Airpark Inc. Oklahoma City, Ok.	N57193
☐ N429LC	71	SA-226T Merlin 3	T-208	A Z Aviation, Berryville, Ar.	N969EE
☐ N429WM	81	MU-2 Marquise	1520SA	Southern Cities fasteners & Supply Co. Muscle Shoals, Al.	N777MJ
☐ N430DA	73	Mitsubishi MU-2K	253	Bush Field Aircraft, Augusta, Ga.	
☐ N430MC	81	King Air B200	BB-904	John DeNault/KJB Air, Fullerton, Ca.	(N202KC)
☐ N431AC	75	PA-31T Cheyenne II	7520024	Inter-Tel Inc. Salem, Or.	N431LS
☐ N431CF	81	Cheyenne II-XL	8166002	Michael O'Neil, Jupiter, Fl.	C-FKEY
☐ N431G	60	Gulfstream 1	29	Comtran International Inc. San Antonio, Tx.	N222SE
☐ N431GW	80	PA-31T Cheyenne II	8020088	M C Aviation Inc. Barrington, Il.	N58GG
☐ N431MC	85	PA-42 Cheyenne 400LS	5527031	Lou Four Inc. Wheeling, Il.	N41199
☐ N431R	71	King Air 100	B-71	C & R Air Inc. Rayne, La.	N7771R
☐ N431S	82	SA-227TT Merlin 3C	TT-431	Sirius Aviation Inc. Indianapolis, In.	N431SA
☐ N432CV	95	Pilatus PC-XII	119	Avex Inc. Van Nuys, Ca.	HB-FQM
☐ N432R	81	PA-42 Cheyenne III	8001032	Dan Williams/Diversified Interior Co. El Paso, Tx.	C-GLMO
☐ N434AE	79	Conquest II	441-0097	Samaritan Health System, Phoenix, Az.	N30RP
☐ N434CA	82	CASA 212-200	286	Fayard Enterprises Inc. Louisburg, NC.	
☐ N435A	67	King Air A90	LJ-229	Flight Safety International Inc. Dothan, Al.	
☐ N437CF	66	King Air A90	LJ-140	J V Ventures LLC. Brentwood, Tn.	N50GH
☐ N438BM	79	King Air 200	BB-438	American Merchandising Inc. Duncanville, Tx.	F-GGPT
☐ N438SP	85	King Air C90A	LJ-1108	Southern Pipe & Supply Co. Meridian, Ms.	N7277R
☐ N439BA	81	MU-2 Solitaire	439SA	Brit International Aircraft Sales, Conroe, Tx.	D-IBBB
☐ N439WA	77	King Air E90	LW-216	World Acceptance Corp. Greenville, SC.	N190DB
☐ N440CA	85	PA-42 Cheyenne IIIA	5501027	Spirit Wing Aviation Ltd. Edmond, Ok.	D-IHGO
☐ N440CC	74	Rockwell 690A	11191	C F P & P Inc. Fort Vernon, In.	N616SD
☐ N440CE	80	King Air 200	BB-698	Charles Prothro, Wichita Falls, Tx.	
☐ N440D	67	King Air A90	LJ-235	Flight Safety International Inc. Dothan, Al.	N7SD
☐ N440KF	80	King Air C90	LJ-878	Marine Lumber Co. Nantucket, Ma.	N440KC

Regn	Yr	Type	c/n	Owner/Operator	Prev Regn
☐ N440TP	65	King Air 90	LJ-17	Apex Aviation Group Inc. Bedford, Tx.	(N110AS)
☐ N441AB	83	Conquest II	441-0284	Wright Brand Foods Inc. Vernon, Tx.	N6888C
☐ N441AG	84	Conquest II	441-0327	Superior Air Charter Inc. Medford, Or.	(N1209X)
☐ N441AW	81	Conquest II	441-0199	Alexander Leasing LLC. Farmingdale, NJ.	N441JG
☐ N441BB	81	Conquest II	441-0217	Head Inc. Fayetteville, NC.	N441WM
☐ N441BL	78	Conquest II	441-0080	Wilbur & Orville Inc. Grand Rapids, Mi.	N441DT
☐ N441BW	77	Conquest II	441-0034	Premier Air Center Inc. East Alton, Il.	N441CM
☐ N441CA	78	Conquest II	441-0046	G G Trucking Inc. Franklin, Wi.	(N36989)
☐ N441CC	78	Conquest II	441-0008	Country Coach Inc. Junction City, Or.	N441CP
☐ N441CD	82	Conquest II	441-0259	Seminis Vegetable Seeds Inc. Saticoy, Ca.	N87HT
☐ N441CJ	80	Conquest II	441-0117	PFD Supply Corp. Carlinville, Il.	(N2623Z)
☐ N441CX	83	Conquest II	441-0305	Blue Ridge Cellular Inc. Christianburg, Va.	N441FB
☐ N441DB	83	Conquest II	441-0300	Charolais Corp. Madisonville, Ky.	N68439
☐ N441DD	85	Conquest II	441-0342	Dyson, Dyson & Dunn Inc. Winnetka, Il.	N986PC
☐ N441DK	80	Conquest II	441-0176	Felts Field Aviation Inc. Spokane, Wa.	N139ML
☐ N441DS	77	Conquest II	441-0028	Delmar Systems Inc. West Linn, Or.	N36955
☐ N441DZ	79	Conquest II	441-0089	Express Shippers Inc. Fort Myers, Fl.	N90GC
☐ N441EB	78	Conquest II	441-0049	Plane Space Inc. West Palm Beach, Fl.	N441FW
☐ N441EL	82	Conquest II	441-0245	Flying Fox Inc. Rancho Santa Fe, Ca.	N441WF
☐ N441EW	81	Conquest II	441-0214	Kenneth & Carol Lindstrom, Morton, Il.	N837RE
☐ N441FC	78	Conquest II	441-0078	Thunderskies Inc. Dover, De.	G-HOSP
☐ N441FS	78	Conquest II	441-0041	Rosbottom Aviation LLC. Shreveport, La.	N555GB
☐ N441G	81	Conquest II	441-0230	Robinson Manufacturing Co. Dayton, Tn.	N14PN
☐ N441GE	80	Conquest II	441-0116	Roger Ferguson/Golden Eagle Ranch, Rexburg, Id.	(N444WS)
☐ N441HA	82	Conquest II	441-0243	Conquest Investors, Reno, Nv.	(N6856X)
☐ N441HF	83	Conquest II	441-0315	Hawkes Brazos Inc. Omaha, Ne.	CC-CWZ
☐ N441JA	80	Conquest II	441-0137	Cheyenne Aviation Inc. Wilmington, De.	N301KB
☐ N441JA	78	Conquest II	441-0045	Poly Aero Inc. Asheville, NC.	N441TA
☐ N441JD	84	Conquest II	441-0336	Banks Lumber Co. Elkhart, In.	N1210U
☐ N441JK	81	Conquest II	441-0197	J I Kislak Mortgage Corp. Miami Lakes, Fl.	N8108Z
☐ N441JR	83	Conquest II	441-0303	SR Aviation Inc. Wilmington, De.	N6860A
☐ N441K	79	Conquest II	441-0093	Ocean Mist Farms, Castroville, Ca.	N441P
☐ N441KP	78	Conquest II	441-0062	IDC Jet Express Inc. Wichita, Ks.	N76DA
☐ N441LL	80	Conquest II	441-0139	Peninsular Airways Inc. Anchorage, Ak.	N10CC
☐ N441M	77	Conquest II	441-0032	Boyd Aircraft/Data Research Associates Inc. St Louis, Mo.	N30WW
☐ N441MD	79	Conquest II	441-0103	Keystone Aerial Surveys Inc. Philadelphia, Pa.	N441CE
☐ N441MJ	78	Conquest II	441-0053	KFTJ Inc. Carson City, Nv.	N40KM
☐ N441MS	78	Conquest II	441-0056	W/o Lakeland Linder Airport, Fl. USA. 2 Jan 97.	C-GDBB
☐ N441ND	80	Conquest II	441-0123	Aurora Aviation Inc. Batavia, Il.	(N441JH)
☐ N441PJ	84	Conquest II	441-0321	Mobile Crane Services Inc. Ottumwa, Ia.	C-FNNC
☐ N441RA	78	Conquest II	441-0040	Keeton-Riemenschneider Partnership, Sisters, Or.	N12DT
☐ N441RB	77	Conquest II	441-0029	John DeJoria. (status ?).	(N36956)
☐ N441RC	78	Conquest II	441-0076	Rocky Mountain Helicopters Inc. Provo, Ut.	(N88838)
☐ N441RK	78	Conquest II	441-0074	Mighty Horn Ministries Inc. Cleveland, Tn.	N1962J
☐ N441RW	86	Conquest II	441-0360	R B W Enterprises, West Conshohocken, Pa.	N441AR
☐ N441S	83	Conquest II	441-0323	Simmons Foods Inc. Siloam Springs, Ar.	N323JG
☐ N441SA	80	Conquest II	441-0172	Security Aviation/Michael & Marilyn O'Neill, Anchorage, Ak.	N441VP
☐ N441SC	77	Conquest II	441-0012	Exports Inc. Sweet Grass, Mt.	N36932
☐ N441SM	80	Conquest II	441-0134	D & H Airways LLC. Fort Collins, Co.	N441DE
☐ N441ST	77	Conquest II	441-0014	RAL Enterprises Inc.	N441AA
☐ N441SX	82	Conquest II	441-0248	Pratt Bradford & Tobin PC. East Alton, Il.	N441JV
☐ N441TA	81	Conquest II	441-0204	Thomas Appleton, Hidden Hills, Ca.	N2727L
☐ N441TM	80	Conquest II	441-0128	Step 2 Corp. Streetsboro, Oh.	N2625H
☐ N441VH	81	Conquest II	441-0225	Satellite Aero Inc. Jackson, Wy.	N6854D
☐ N441W	80	Conquest II	441-0181	Warrington Development Corp. Birmingham, Al.	N81WS
☐ N441WP	79	Conquest II	441-0098	Aero Transporte Nationale, Brownsville, Tx.	(N35DD)
☐ N441WT	82	Conquest II	441-0242	ITEC Inc. Huntsville, Al.	N3127R
☐ N441YA	81	Conquest II	441-0231	Paxton Media Group Inc. Paducah, Ky.	HK-3647X
☐ N441Z	81	Conquest II	441-0203	Air Tijerina Inc. Brownsville, Tx.	N93BA
☐ N442JA	77	Conquest II	441-0013	Pegasus & Crew Inc. Fort Lauderdale, Fl.	N441BG
☐ N442KA	79	King Air 200	BB-442	Nielsons Inc. Cortez, Co.	
☐ N443CL	79	King Air E90	LW-318	Ventana Aviation LLC. Morgan Hill, Ca.	N943CL
☐ N443TC	77	King Air B100	BE-20	Hanley Co. Knoxville, Tn.	N525WE
☐ N444AD	80	King Air 200	BB-733	Dept of Public Safety, Anchorage, Ak.	
☐ N444AK	84	Conquest II	441-0334	Black Hills Aviation Inc. Alamogordo, NM.	N334FW
☐ N444BK	89	King Air B200	BB-1332	Koop Tjuchem BV. Groningen, Holland.	D-ICSM
☐ N444CM	96	Pilatus PC-XII	152	Sun Aero II Inc. Billings, Mt.	HB-FRN
☐ N444KA	83	King Air B200	BB-1160	Energy Equipment Resource Inc. Dallas, Tx.	N388CP
☐ N444KF	83	Conquest 1	425-0191	GEM Aviation Inc. Weddington, NC.	N1221N
☐ N444KK	83	King Air F90	LA-209	Darryl Unruh, Asheville, NC.	N6727U
☐ N444LB	82	SA-227TT Merlin 3C	TT-428A	M T Group Inc. Nashville, Tn.	
☐ N444MF	80	King Air F90	LA-81	Phoenix Agro-Invest Inc/Martori Farms Inc. Scottsdale, Az.	N423DA
☐ N444NC	79	Rockwell 690B	11549	Cosgrove Aircraft Service Inc. Hauppauge, NY.	N81622
☐ N444PC	79	PA-31T Cheyenne II	7920066	Business Aircraft Center Inc. Danbury, Ct.	N111AM
☐ N444PS	74	King Air C90	LJ-615	Northwest Boring Co. Woodinville, Wa.	
☐ N444RC	75	PA-31T Cheyenne II	7520030	Four R C Corp. Baton Rouge, La.	N54959

Regn	Yr	Type	c/n	Owner/Operator	Prev Regn
□ N444RK	83	King Air B100	BE-137	McWorth Management Co. Duluth, Ga.	N65187
□ N444RR	83	Conquest 1	425-0128	Sasnak Management Corp. Wichita, Ks.	N444RH
□ N444RU	83	Conquest 1	425-0190	Louisiana Pacific Corp. Hillsboro, Or.	N444RH
□ N444TH	84	Conquest 1	425-0219	NPC International Inc. Pittsburgh, Ks.	N1225Y
□ N444WE	82	MU-2 Marquise	1551SA	Big Rapids Air LLC. Wilmington, De.	N444WF
□ N445AE		Conquest II	441-0043	BLC Corp/Samaritan Airevac, Phoenix, Az.	N445WS
□ N445CR	79	King Air C90	LJ-838	Claire Rice, Palm Desert, Ca.	N445DR
□ N447WS	81	Conquest II	441-0198	Charles Harris/Space Cabotage Navigation Co. San Francisco.	C-FCWZ
□ N448CA	86	PA-42 Cheyenne 400LS	5527034	Columbia Aircraft Sales Inc. Groton, Ct.	JA8878
□ N448T	89	King Air B200	BB-1336	Terra Industries Inc. Sioux City, Ia.	N448M
□ N449CA	84	PA-42 Cheyenne IIIA	5501023	Moyer Aviation Inc. Ulysses, Ks.	D6-ECA
□ N450HC	81	Cheyenne II-XL	8166021	Cessford Construction Co. Le Grand, Ia.	N300XL
□ N450LM	90	King Air 350	FL-5	Shelter Enterprises Inc. Columbia, Mo.	N5634E
□ N450MW	84	PA-42 Cheyenne 400LS	5527013	RASair LLC. East Hamstead, NH.	N32KK
□ N451A	93	King Air C90B	LJ-1348	Airfoil Corp. Oxnard, Ca.	
□ N451WS	79	Conquest II	441-0096	Hoff Ford Inc/Tony Copeland Ford, Lewiston, Id.	N8977N
□ N452MA	81	MU-2 Marquise	1533SA	C B Anderson, Jacksonville, Tx.	
□ N452TT	69	King Air B90	LJ-452	Tyler Turbine Ltd. Tyler, Tx.	D-ILKA
□ N454CA	81	PA-31T Cheyenne 1	8104053	Frederick Weissberg, Elk Grove, Il.	G-JVAJ
□ N454EA	78	Conquest II	441-0054	CTF Air Inc. The Dalles, Or.	C-GRSL
□ N454MA	81	MU-2 Marquise	1535SA	Rocky Mountain Holdings LLC. Provo, Ut.	
□ N455JW	81	PA-42 Cheyenne III	8001040	Gas'N Shop Inc. Lincoln, Ne.	N456JW
□ N456CD	79	King Air 200	BB-456	Christian Duplications International Inc. Orlando, Fl.	N86LD
□ N456CS	76	King Air 200	BB-177	Golden Triangle Constructors Inc. Port Arthur, Tx.	
□ N456FC	83	Conquest 1	425-0127	Lamar Leasing Co. Lamar, In.	N6882V
□ N456PS	76	Mitsubishi MU-2M	338	Southport Aviation Inc. Dover, De.	N79AC
□ N456Q	89	King Air 300	FA-181	Quinn Co. Fresno, Ca.	N5547K
□ N457CP	67	King Air A90	LJ-275	Texas State Technical College, Waco, Tx.	N457SR
□ N457G	69	SA-26AT Merlin 2B	T26-150	Royal Sons Inc. Clearwater, Fl.	N4257X
□ N457TC	80	PA-31T Cheyenne II	8020059	Brittany Aviation Services Inc. Dallas, Tx.	SE-IDM
□ N458BB	85	MU-2 Solitaire	458SA	Beeline Flight Inc. Wilmington, De.	XB-FQM
□ N459SA	85	MU-2 Solitaire	459SA	Solitaire Leasing LLC. Wilmington, De.	OY-CGW
□ N460LC	76	PA-31T Cheyenne II	7620014	Sunrise Air Inc. Pinehurst, Tx.	N98AT
□ N461LM	96	TBM 700	122	Smarter Charter Inc. Tampa, Fl.	F-GLBJ
□ N461MA	74	Mitsubishi MU-2K	301	Jack Goodale, Grand Rapids, Mi.	N10T
□ N462MA	74	Mitsubishi MU-2K	302	BAM International Inc. Wilmington, De.	
□ N463DC	76	SA-226T Merlin 3A	T-273	Jaime Hernandez, Brownsville, Tx.	N5390M
□ N464AB	67	King Air A90	LJ-224	T & G Aviation Inc. Chandler, Az.	N464AL
□ N464G	67	King Air A90	LJ-202	H & Z Corp. San Antonio, Tx.	(N474DP)
□ N464WF	96	Pilatus PC-XII	149	Atlantic Aero Inc. Greensboro, NC.	HB-FR.
□ N465MC		King Air C-12C	BC-18	Monroe County Sheriffs Department, Key West, Fl.	73-22267
□ N466MW	87	King Air B200	BB-1273	Free Skate Co - BHTC, Boston, Ma.	N30486
□ N469BL	69	SA-26AT Merlin 2B	T26-139	Blue Angel Holding Co. Bossier City, La.	C-FCAR
□ N469MA	82	MU-2 Marquise	1543SA	Trans Costa Rica, San Jose, Costa Rica.	TF-FHM
□ N470MM	97	King Air 350	FL-156	Schwan's Sales Enterprises Inc. Marshall, Mn.	N1093A
□ N471CD		SA-227AT Merlin 4C	AT-549	Ashley Aire Inc. Upper Marlboro, Md.	89-1471
□ N471SC	83	Gulfstream 900	15031	Aaron Rents Inc & Shepherd Construction Co. Atlanta, Ga.	N615JB
□ N475CA	75	Mitsubishi MU-2M	319	Southeast Air Inc. Farmingdale, NY.	N6SG
□ N475JA	87	King Air C90A	LJ-1147	DeBruce Grain Inc. N Kansas City, Mo.	
□ N475MG	77	SA-226T Merlin 3A	T-278	Mitchell Air Inc. Wilmington, De.	N117PB
□ N477B	78	Conquest II	441-0055	Glory Charters Inc. Bellevue, Wa.	HB-LFF
□ N477JA	88	King Air C90A	LJ-1162	Special Investments Inc. Encino, Ca.	
□ N477SJ		Conquest II	441-0162	Greenleaf Air Inc. Kalamazoo, Mi.	N66CP
□ N478JA	88	King Air C90A	LJ-1163	Special Investments Inc. Encino, Ca.	
□ N479JA	88	King Air C90A	LJ-1177	Japan Air Lines, Los Angeles, Ca.	
□ N479MA	82	MU-2 Marquise	1553SA	Rocky Mountain Holdings LLC. Provo, Ut.	
□ N479SW	79	PA-31T Cheyenne 1	7904047	Plymouth Foam Products Inc. Plymouth, Wi.	
□ N480BC	68	SA-26AT Merlin 2B	T26-108	Utah Valley State College, Orem, Ut.	N122NK
□ N480CA	80	PA-31T Cheyenne 1	8004051	BB&T Corp. Winston-Salem, NC.	
□ N480JA	88	King Air C90A	LJ-1178	Japan Air Lines, Los Angeles, Ca.	
□ N480MA	82	MU-2 Marquise	1554SA	Monroe Aviation Inc. Green Bay, Wi.	
□ N480TC	97	King Air B200	BB-1600	Raytheon Aircraft Co. Wichita, Ks.	
□ N481JA	88	King Air C90A	LJ-1179	Japan Air Lines, Los Angeles, Ca.	
□ N481SW	81	PA-31T Cheyenne 1	8104016	Charles White Construction Co. Clarksdale, Ms.	
□ N482JA	88	King Air C90A	LJ-1183	Japan Air Lines, Los Angeles, Ca.	
□ N482TC	83	PA-31T Cheyenne 1	8104072	Pierce-Western Inc. San Angelo, Tx.	
□ N483JA	88	King Air C90A	LJ-1184	Japan Air Lines, Los Angeles, Ca.	
□ N483JM	66	King Air A90	LJ-147	McDonald Sales & Leasing Inc. Charlotte, NC.	N102RC
□ N484JA	89	King Air C90A	LJ-1196	Japan Air Lines, Los Angeles, Ca.	
□ N484SC	78	PA-31T Cheyenne II	7820022	Carl Bolander & Sons Co. St Paul, Mn.	N9DK
□ N485JD	89	King Air C90A	LJ-1197	Japan Air Lines, Los Angeles, Ca.	N485JA
□ N485K	79	King Air 200	BB-485	Koch Industries Inc. Wichita, Ks.	N120K
□ N486JD	89	King Air C90A	LJ-1198	Japan Air Lines, Los Angeles, Ca.	N486JA
□ N488AD	81	King Air B200	BB-860	Phoenix Color Corp. Hagerstown, Md.	N488CA
□ N488FT	81	King Air F90	LA-137	Jet 1 Inc. Naples, Fl.	N200MW
□ N488GA	87	King Air 300	FA-125	T & L Inc/Timber Products Co. Springfield, Or.	N3042K

Regn	Yr Type	c/n	Owner/Operator	Prev Regn

Regn	Yr	Type	c/n	Owner/Operator	Prev Regn
N488GB	68	Mitsubishi MU-2F	133	Scope Leasing Inc. Columbus, Oh.	
N488JR	92	King Air C90B	LJ-1305	Japan Air Lines, Los Angeles, Ca.	N488JD
N488LL	97	King Air C90B	LJ-1458	Conex International Inc. Beaumont, Tx.	
N489GA	79	Rockwell 690B	11530	VF Aircraft Holdings Inc/VF Air, Collinsville, Il.	N81734
N489JS	92	King Air C90B	LJ-1311	Japan Air Lines, El Segundo, Ca.	N489JD
N489SC		Gulfstream 840	11635	S E Cone Jr. Hobbs, NM.	N331SC
N489WC	82	Conquest II	441-0262	Calcutta Aircraft Leasing Inc. Bloomington, In.	N489ST
N490J	92	King Air C90B	LJ-1312	Fort Wayne Air Service Inc. Fort Wayne, In.	N490JT
N490MA	74	Mitsubishi MU-2J	640	Lawrence Shory, Birmingham, Al.	
N490TN	92	King Air 350	FL-89	TennOhio Transportation Co. Columbus, Oh.	N8112F
N491JV	93	King Air C90B	LJ-1349	Japan Air Lines, El Segundo, Ca.	
N491MB	81	Cheyenne II-XL	8166007	Midwest Organ Bank Inc. Westwood, Ks.	N187GA
N492JW	93	King Air C90B	LJ-1350	Japan Air Lines, El Segundo, Ca.	
N493JX	93	King Air C90B	LJ-1351	Japan Air Lines, El Segundo, Ca.	
N494JY	93	King Air C90B	LJ-1352	Japan Air Lines, El Segundo, Ca.	
N495CA	91	PA-42 Cheyenne 400LS	5527044	LB Aviation Inc. East Haven, Ct.	HL5205
N495MA	74	Mitsubishi MU-2K	306	Carl Air Inc. Bayard, WV.	
N495NM	78	King Air C90	LJ-731	State of New Mexico, Santa Fe, NM.	N4953M
N496MA	82	Gulfstream 1000	96006	Midwest Aviation Inc. Omaha, Ne.	N7031J
N497P	76	King Air C90	LJ-698	Rialto Riverside Corp. Corona, Ca.	N9872C
N497P	85	King Air C90A	LJ-1126	Piedmont Aviation Services Inc. Winston-Salem, NC.	N300CK
N497SL	79	King Air B100	BE-64	Skylife Aviation LLC. St Louis, Mo.	N4490M
N499WC	78	Rockwell 690B	11499	AeroMitchell Inc. City of Industry, Ca.	N81832
N500CE	77	Conquest II	441-0005	Ziegler Investments Inc. Nauvoo, Il.	N9917G
N500CR	80	King Air 200	BB-714	Central Rock Mineral Co. Lexington, Ky.	N7CC
N500CS	81	King Air 200	BB-773	Hall Auto World Inc. Virginia Beach, Va.	N83JE
N500CY	81	King Air B200	BB-935	Cale Yarborough Motorsports Inc. Florence, SC.	N19GB
N500DE	78	King Air 200	BB-357	Champion Air Inc. Mooresville, NC.	N200JV
N500EA	77	King Air E90	LW-240	Clement Auto & Truck Inc. Fort Dodge, Ia.	N80WP
N500EQ	94	King Air C90B	LJ-1387	Dearborn Associates, Panorama City, Ca.	N500ED
N500FC	87	King Air 300	FA-124	Mechanics Laundry Supply Inc. Indianapolis, In.	N1845C
N500GK	77	Mitsubishi MU-2P	348SA	Anaconda Aviation Corp. Boca Raton, Fl.	D-IEMU
N500GM	81	Conquest II	441-0222	International Auto Brokers Inc. Scottsdale, Az.	D-IDRJ
N500GN	75	King Air 200	BB-62	Springfield Newspapers Inc. Springfield, Mo.	N300GN
N500HY	78	King Air 200	BB-306	NACCO Materials Handling Group Inc. Greenville, NC.	N274K
N500KA	81	King Air 200	BB-819	Joseph Hicks Pharmacy, Walnut Cove, NC.	N425P
N500KD	80	King Air B100	BE-79	Winner Aviation Corp. Vienna, Oh.	N500JE
N500KR	77	King Air C90	LJ-708	NC Stste Bureau of Investigation, Raleigh, NC.	N500KS
N500KS	75	King Air 200	BB-59	Ken Schrader Racing Inc. Concord, NH.	N504WR
N500LM	84	PA-42 Cheyenne 400LS	5527016	James Smith, Longmont, Co.	
N500LP	83	King Air B200	BB-1141	Quail Aero Service of Syracuse inc. Schenectady, NY.	N66549
N500MC	83	Conquest 1	425-0147	E A Martin Co. Springfield, Mo.	N425WM
N500MS	74	King Air C90	LJ-626	Mike Skinner Enterprises Inc. Sophia, NC.	N9CR
N500MT	81	King Air C90	LJ-958	Spevco Inc. Pfafftown, NC.	N458P
N500MY	78	PA-31T Cheyenne 1	7804004	Surefoot LC. Park City, Ut.	N500MT
N500N	81	King Air B100	BE-110	Competitor Liaison Bureau Inc. Daytona Beach, Fl.	N300R
N500NA	68	King Air B90	LJ-401	K & K Aircraft Inc. Bridgewater, Va.	N8NM
N500PB	81	Cheyenne II-XL	8166027	Broyhill Industries Inc. Lenoir, NC.	N2602Y
N500PH	81	King Air 200C	BL-29	Petroleum Helicopters Inc. Lafayette, La.	N3847H
N500PJ	75	Mitsubishi MU-2L	668	Premier Jets Inc. Portland, Or.	N500RM
N500PS	72	Mitsubishi MU-2F	224	Corporate Aviation Services Inc. Tulsa, Ok.	N190MA
N500QT	92	King Air C-90B	LJ-1322	Medical Foundation of Inc. Chapel Hill, NC.	N8121C
N500RJ	83	Conquest II	441-0302	ARM Investments Inc. Lemont, Il.	N68599
N500RN	62	Gulfstream 1	95	Mission Air Support Inc. Roanoke, Va.	N500RL
N500SN	68	SA-26AT Merlin 2B	T26-100	Ossid Corp. Rocky Mount, NC.	N333F
N500SX	80	SA-226T Merlin 3B	T-366	Merlin Aviation Inc. Indianapolis, In.	N911JZ
N500UW	77	Conquest II	441-0017	Ultra Wheel Co. Buena Park, Ca.	(N711RD)
N500WF	81	King Air B200	BB-947	William Flowers, Dothan, Al.	G-FOOD
N500X	68	Mitsubishi MU-2D	114	Bush Field Aircraft Co. Augusta, Ga.	N908108
N500XX	73	Mitsubishi MU-2K	250	Techno Aero Corp. Wilmington, De.	N740FN
N501EB	79	King Air 200	BB-422	Heritage Imports Inc. Pella, Ia.	N551E
N501FS	69	SA-26AT Merlin 2B	T26-146	F S Air Service Inc. Anchorage, Ak. (status ?).	N7WY
N501PT	79	PA-31T Cheyenne 1	7904048	Flournoy Development Co. Columbus, Ga.	
N501RH	81	King Air 200	BB-805	Hendrick Motorsports Inc. Harrisburg, NC.	N3812S
N501TD	81	King Air F90	LA-82	Transdigm Inc. Richmond Heights, Oh.	N531DS
N502BR	80	Conquest II	441-0135	U S Leasing & Financial Inc. Wilmington, De.	N93RK
N502MM	82	PA-31T Cheyenne 1	8104064	Kenmar Aircraft Inc. Wilmington, De.	N2571Y
N502NC	81	Conquest 1	425-0056	D & D Aviation Inc. Grandview, Mo.	N56DA
N502RH	80	King Air 200	BB-673	Hendrick Motorsports Inc. Harrisburg, NC.	N673YV
N502SE	77	King Air C90	LJ-740	Aviation Resources Ltd. Fargo, ND.	
N502W	67	King Air A90	LJ-287	Phoenix Air Inc. Frostproof, Fl.	N823K
N503AA	74	Mitsubishi MU-2J	633	Air Carriers Inc. Birmingham, Al.	(N56JS)
N503M	66	King Air A90	LJ-158	Robert Spillman, Elyria, Oh.	(N900WM)
N503WR	79	PA-31T Cheyenne 1	7904016	Diamond Aviation II Inc. Wilmington, De.	N93CV
N504TQ	79	PA-31T Cheyenne 1	7904019	Tucker Inc. Asheboro, NC.	N504TC
N505BG	79	King Air E90	LW-312	Gunn Oil Co. Wichita Falls, Tx.	N20509
Regn	*Yr*	*Type*	*c/n*	*Owner/Operator*	*Prev Regn*

Regn	Yr	Type	c/n	Owner/Operator	Prev Regn
N505FS		SA-227AT Merlin 4C	AT-591	F S Air Service Inc. Anchorage, Ak.	N176SW
N505HC	82	Conquest II	441-0257	Cardio Equipment IV LC. Albuquerque, NM.	N100YA
N505MW	75	King Air E90	LW-155	Mike Wallace Aviation LLC. Concord, NC.	N505GA
N505RH		Beech 1900C	UB-56	Hendrick Motorsports Inc. Harrisburg, NC.	
N505RT	68	680W Turbo Commander	1752-5	Twin City Leasing Corp. Kansas City, Mo.	N5051E
N505WR	84	King Air 300	FA-15	J M Air Inc. Pinedale, Wy.	
N506AB	78	King Air 200	BB-362	Daybreak Properties Inc. Lebanon, Mo.	G-VSEL
N506EB	88	King Air B200	BB-1312	Quest II LLC.	TC-KOC
N506GT	80	King Air 200	BB-612	Foster Poultry Farms Inc. Livingston, Ca.	
N507EB	82	King Air 350	FL-81	Wachovia Leasing Corp. Winston-Salem, NC.	N8053R
N507EF	95	King Air B200	BB-1511	Cape Clear Capital Corp. Portsmouth, NH.	N3231F
N507W	67	King Air A90	LJ-269	Sierra Nevada Helicopters Inc. Reno, Nv.	N788K
N508BM	95	King Air B200	BB-1497	TOMCO of Delaware Inc. Wilmington, De.	N123SA
N508JA	80	King Air 200	BB-735	Aviation Anonymous Inc/JetSun Aviation Centre, Sioux City.	N500GC
N508MV	81	King Air B200	BB-877	Ziff Air Services Inc. NYC.	N711BU
N510CB	79	King Air 200	BB-473	Curt Bean Lumber Co. Amity, Ar.	
N510WP	94	King Air 350	FL-109	Westpoint Stevens Inc. Westpoint, Ga.	N8203C
N510WR	86	King Air 300	FA-111	B & W Aviation LLC. Houston, Tx.	N510WP
N511BF	66	King Air A90	LJ-179	Executive Air Charters Inc. New Orleans, La.	N711CF
N511D	97	King Air 350	FL-172	Bunn-O-Matic Corp. Springfield, Il.	
N511SC	81	Cheyenne II-XL	8166069	Westside Timber Inc. Stayton, Or.	N9190Y
N512G	65	King Air 90	LJ-64	American Aircraft Inc. Newton, Ks.	N512Q
N512JD	66	680T Turbo Commander	1584-36	Chisholm Construction Co.	N512JC
N513DC	81	MU-2 Marquise	1513SA	Air Hi Ho Inc. Wilmington, De. (status ?).	N275CA
N513DM	82	MU-2 Marquise	1560SA	Incoe Corp. Troy, Mi.	N486MA
N513KL	81	King Air F90	LA-123	Learner Financial Corp. Walnut Creek, Ca.	N211EC
N514GP	94	Jetstream 41	41038	Georgia Pacific Corp. Atlanta, Ga.	N438JX
N514M	81	PA-31T Cheyenne 1	8104045	Menard Inc. Eau Claire, Wi.	N150BC
N514MC	83	Conquest II	441-0312	J Cunningham, Phoenix, Az.	(N312SA)
N514NA	68	680W Turbo Commander	1772-10	NASA, Langley,	83-24126
N515AC	90	2000 Starship	NC-16	Stevens Aviation Inc. Greenville, SC.	N80KK
N515AM	80	Gulfstream 980	95008	Allen Myland Inc. Broomall, Pa.	N9761S
N515BA	81	PA-31T Cheyenne II	8120044	Quaker Air Services Inc. Salem, Oh.	N444PD
N515BC	75	King Air E90	LW-121	R & M Aviation Inc. DeKalb, Il.	N500TR
N515CR	84	King Air B200	BB-1115	Travel King Inc. Atlanta, Ga.	SE-IXA
N515GA	82	PA-31T Cheyenne 1	8104050	Ingram Aviation LLC. San Braunfels, Tx.	N49WA
N515JS	95	2000A Starship	NC-52	Osborn Energy LLC. Plano, Tx.	N1564Q
N515M	82	PA-42 Cheyenne III	8001079	Air 1 Inc. Albany, Mn.	N515DW
N515RC	82	PA-42 Cheyenne IIIA	5501018	Ruan Inc. Des Moines, Ia.	N516GA
N516WB	65	King Air 90	LJ-70	University of Alaska, Fairbanks, Ak.	N516W
N518DM	67	King Air A90	LJ-251	L & L Sandblasting Inc. Eunice, La.	N516DM
N518NA	67	King Air A90	LM-80	Lake Technical Institute, Watertown, SD.	66-18080
N519CC	83	Gulfstream 1000	96068	Eagle Air Inc. Memphis, Tn.	N62GA
N519HB	80	Gulfstream 980	95013	L C Leasing Co. Herndon, Va.	N500TH
N520CS	71	681B Turbo Commander	6061	BRS Services Inc. Visalia, Ca.	(N22FF)
N520D	82	King Air B200	BB-989	Aviation Resources Inc. East Hampton, NY.	
N520MC	75	King Air 200	BB-43	Mayo Aviation Inc. Englewood, Co.	N500UR
N520RM	83	Conquest 1	425-0129	EMI Flight Services Inc. Denver, Co.	N68822
N520WS	81	King Air 200	BB-751	World Savings & Loan Association, Oakland, Ca.	N2000C
N521LB	77	King Air E90	LW-249	Blue Aero Leasing Inc. Dallas, Tx.	N68CC
N521M	87	Beech 1900C	UB-68	Menard Inc. Eau Claire, Wi.	N30CY
N522CF	79	King Air B100	BE-67	Lake Erie Airlines Inc. Cleveland, Oh.	N6035H
N523PD	79	PA-31T Cheyenne 1	7904044	Zephyr Corp. Chicago, Il.	N23569
N524BA	80	King Air B100	BE-99	Midwest Aviation Services Inc. West Paducah, Ky.	N3699P
N524SC	70	King Air 100	B-37	Star Care V, Lincoln, Ne.	N627L
N525JK	67	King Air A90	LJ-305	Fayard Enterprises Inc. Louisburg, NC.	N732NM
N526RR	67	King Air A90	LJ-263	Flight Safety International Inc. Dothan, Al.	N526BT
N527JC	81	PA-31T Cheyenne 1	8104043	John & Cynthia Romito, Leawood, Ks.	N45TW
N528AM	85	King Air 300	FA-59	Franklin Johnson, Palo Alto, Ca.	N7233U
N528DS	79	PA-31T Cheyenne 1	7904008	Danny Tay Smith, Tuscaloosa, Al.	N38HG
N528WG	76	King Air 200	BB-151	Tigress Air III LLC. California, Md.	N98CM
N529NA	83	King Air B200	BB-1091	NASA Langley Research Center, Hampton, Va.	N9NA
N531CB	83	King Air B100	BE-135	Lake Aero Enterprises Inc. Valley, Ne.	N531CM
N531LP	73	SA-226T Merlin 3A	T-234	Dr William Reed Jr MD. Leawood, Ks.	N5316M
N531MC	84	PA-42 Cheyenne 400LS	5527007	Pittman Flight Inc. NYC.	N400SL
N531SW	79	PA-31T Cheyenne 1	7904020	Central Arkansas Nursing Centers, Fort Smith, Ar.	
N533DM	74	Mitsubishi MU-2N	652SA	Orval Yarger, Normal, Il.	N533MA
N533P	87	King Air B200	BB-1278	C W Parker Aviation LLC. Cary, NC.	D-IAIR
N535WM	75	Mitsubishi MU-2L	655	Systec 2000 Inc. Pell City, Al.	N535MA
N536MA	83	Conquest II	441-0322	Miller Aviation Inc. Johnson City, NY.	D-IEWA
N538AS	86	King Air/Catpass 250	FA-107	Akins Aviation Inc. Morresville, NC.	N90GA
N538EA	81	MU-2 Marquise	1538SA	Corporate Aviation Services Inc. Tulsa, Ok.	N538MC
N539MA	72	Mitsubishi MU-2J	552	International Business Aircraft, Tulsa, Ok.	N246W
N540CB	87	King Air 300	FA-135	Pepco Inc. Rocky Mount, NC.	HB-GHY
N540GA	81	Conquest 1	425-0026	Dept of Transportation, State of Georgia, Atlanta, Ga.	N17PL
N540SP		King Air C-12C	BC-5	Oregon State Police, Salem, Or.	73-22254

Regn	Yr	Type	c/n	Owner/Operator	Prev Regn
☐ N541MM	80	King Air F90	LA-71	U I S Inc. Fairfield, Il.	N45WL
☐ N541SC	87	King Air 300	FA-138	Inland Steel Co. Chicago, Il.	N30757
☐ N543GA	81	Conquest 1	425-0101	Department of Transportation, State of Georgia, Atlanta, Ga.	N425AF
☐ N543S	66	680T Turbo Commander	1556-14	Mark Lundell, Paradise Valley, Az.	
☐ N544AL	80	Conquest II	441-0120	Peninsula Airways Inc. Anchorage, Ak.	N441DM
☐ N544GA	82	Gulfstream 900	15015	State of Georgia, Atlanta, Ga.	N27MW
☐ N544UP	83	SA-227AT Merlin 4C	AT-544	Ameriflight Inc. Burbank, Ca.	N68TA
☐ N547GA	81	Conquest 1	425-0084	Department of Transportation, State of Georgia, Atlanta, Ga.	N1918W
☐ N548GQ	70	681 Turbo Commander	6027	DPG Corp. Cincinnati, Oh.	N681AS
☐ N548UP	83	SA-227AT Merlin 4C	AT-548	Ameriflight Inc. Burbank, Ca.	N548SA
☐ N549BR	79	King Air C90	LJ-809	Booby Ross Group-BRG Transportation Inc. Austin, Tx.	N922DT
☐ N549GA	78	Conquest II	441-0044	State of Georgia Department of Transportation, Atlanta, Ga.	N441AD
☐ N550BE	73	SA-226T Merlin 3	T-224	Lorraine Aviation Inc. Cartersville, Ga.	N5307M
☐ N550JC	78	King Air 200	BB-354	Goody's Family Clothing Inc. Alcoa, Tn.	N198GH
☐ N550MM	81	Cheyenne II-XL	8166017	Lima Alpha Corp. Allentown, Pa.	N2536Y
☐ N550NP	75	Rockwell 690A	11234	National Propane Corp. Cedar Rapids, Ia.	N680RH
☐ N550P	78	King Air E90	LW-301	Public Service Co. Denver, Co.	
☐ N550SP	73	King Air C-12C	BC-6	Oregon State Police, Salem, Or.	73-22255
☐ N550TP	90	King Air 350	FL-35	Tecumseh Products Co. Tecumseh, Mi.	N8035H
☐ N551AT	68	King Air B90	LJ-399	Addison King Air Inc. Dallas, Tx.	N551F
☐ N551ES	92	King Air 350	FL-78	IES Industries Inc. Cedar Rapids, Ia.	N8291K
☐ N551JL	81	King Air 200	BB-788	Midwest Fastener Corp. South Holland, Il.	
☐ N551MS	79	King Air E90	LW-311	Michael Simpson, Bridgeport, Tx.	N83RH
☐ N552GA	75	King Air A100	B-215	Penny Klotz, St Charles, Il.	OY-CCA
☐ N552R	78	King Air C90	LJ-749	Lynch Flying Service Inc. Billings, Mt.	N93BA
☐ N552TB	83	King Air B200	BB-1159	Roy Goodart, Red Lodge, Mt.	(D-IRUS)
☐ N552TP	89	King Air 350	FL-1	Tecumseh Products Co. Tecumseh, Mi.	N120SK
☐ N553HC	81	King Air B200	BB-928	Electric Systems Inc. Chattanooga, Tn.	N555WF
☐ N553R	80	King Air 200	BB-589	Financial Business Systems Inc. Fargo, ND.	N202BE
☐ N554CF	73	King Air E90	LW-66	Destin Airways LLC. Little Rock, Ar.	(N106TB)
☐ N555AT	81	PA-31T Cheyenne 1	8104054	MARKAY Enterprises Ltd. Wilmington, De.	N818SW
☐ N555C	68	Mitsubishi MU-2F	129	ISD, Rivervalls, Mn.	N555S
☐ N555CK	76	King Air C90	LJ-677	Oustalet Ford Inc. Jennings, La.	N555WF
☐ N555FT	80	Conquest II	441-0124	S & J Operating Co. Wichita Falls, Tx.	N2624N
☐ N555FW	74	King Air E90	LW-248	Justin Industries Inc. Fort Worth, Tx.	N9FC
☐ N555GB	82	SA-227AT Merlin 4C	AT-439	Merlin Express/B A Leasing & Capital Corp. San Francisco.	N41BP
☐ N555GD	79	Conquest II	441-0094	Davis Moore Oldsmobile Inc. Wichita, Ks.	N555JE
☐ N555HP	95	TBM 700	108	STS Inc. Dover, De.	
☐ N555JJ	77	Conquest II	441-0004	Rip Griffin Truck Service Center, Lubbock, Tx.	N555JK
☐ N555JP	76	Mitsubishi MU-2M	332	Ericksen Properties Inc. Naples, Fl.	N555LL
☐ N555KG	93	2000A Starship	NC-36	Starship Industries Inc. Boca Raton, Fl.	N555KK
☐ N555MS	80	King Air C90	LJ-872	Craven Sports Properties Inc. Concord, NC.	N6668U
☐ N555PM	76	PA-31T Cheyenne II	7620028	Elite Holdings Inc. Wilmington, De.	N531PT
☐ N555TB	68	King Air B90	LJ-364	Ozark Management Inc. Jefferson City, Mo. (status ?).	(N110SL)
☐ N555TT	82	Conquest 1	425-0110	Personnel Management Inc. Framington Hills, Mi.	N6851A
☐ N555TT	76	King Air E90	LW-170	R E L Leasing Inc. Wichita Falls, Tx.	N500TL
☐ N555VK	67	Mitsubishi MU-2B	023	David McCredie, Flint, Mi.	N3555X
☐ N555WA	81	Conquest II	441-0019	Anderson Merchandisers Inc. Amarillo, Tx.	(N441CX)
☐ N555WF	85	King Air 300	FA-58	Ring Can Corp. Oakland, Tn.	N282HC
☐ N556UP	83	SA-227AT Merlin 4C	AT-556	Ameriflight Inc. Burbank, Ca.	N3113B
☐ N560UP	83	SA-227AT Merlin 4C	AT-560	Ameriflight Inc. Burbank, Ca.	N3113A
☐ N561SS	79	King Air 200	BB-464	General Manufactured Housing Inc. Waycross, Ga.	N79SE
☐ N561UP	83	SA-227AT Merlin 4C	AT-561	Ameriflight Inc. Burbank, Ca.	N3113F
☐ N562GA	95	Pilatus PC-XII	113	General Aviation Services Inc.	JA8204
☐ N562P	67	King Air A90	LJ-248	David Verner, Vandalia, Il.	N202RW
☐ N563GA	89	Reims/Cessna F406	F406-0041	General Aviation Services Inc. Lake Zurich, Il.	9H-ACI
☐ N563MC	68	King Air B90	LJ-384	Fly Boys LLC. Olive Branch, Ms.	N736K
☐ N564BC	83	King Air B200	BB-1107	Bright Coop Inc. Nacogdoches, Tx.	N737US
☐ N564CA	79	King Air B100	BE-58	Carver Aero Inc. Davenport, Ia.	N100P
☐ N564GA	93	King Air B200	BB-1463	General Aviation Services Inc. Lake Zurich, Il.	JA8784
☐ N565M	88	Beech 1900C	UB-43	Menard Inc. Eau Claire, Wi.	N34GT
☐ N566CA	67	King Air A90	LJ-184	Carver Aero Inc. Muscatine, Ia.	N858K
☐ N566UP	83	SA-227AT Merlin 4C	AT-566	Ameriflight Inc. Burbank, Ca.	N3113N
☐ N567GJ	74	King Air E90	LW-95	Fegaras Brothers Aircraft Sales, Michigan City, In.	OY-CFO
☐ N567JD	82	King Air B200	BB-949	Appalachian Aviation Services Inc. Pikeville, Ky.	N61415
☐ N567R	73	Rockwell 690	11051	Jorge Gonzalez, Hialeah, Fl.	N500R
☐ N568K	81	King Air B100	BE-106	Speciality Retailers Inc. Houston, Tx.	(N59SS)
☐ N569GB	96	King Air C90B	LJ-1447	Fun Bike Center Inc. San Diego, Ca.	
☐ N569H	72	Rockwell 690	11029	Phoenix Air Group Inc. (status ?).	
☐ N569UP	83	SA-227AT Merlin 4C	AT-569	Ameriflight Inc. Burbank, Ca.	N31134
☐ N570AB	78	PA-31T Cheyenne II	7820026	J B Air Corp. Newtown, Pa.	N36JM
☐ N570GB	74	Rockwell 690A	11209	True Speed Enterprises Inc. Indianapolis, In.	N570WA
☐ N573G	82	SA-227AT Merlin 4C	AT-446	Ameriflight Inc. Burbank, Ca.	N3008L
☐ N573MA	69	Mitsubishi MU-2F	162	Dragon Leasing Corp. Naperville, Il.	N875Q
☐ N575C	71	King Air C90	LJ-532	David Lilly, Alentown, Pa.	N57SC
☐ N575CA	80	MU-2 Solitaire	425SA	G A C F Corp. Wilmington, De.	N111GP

Regn	Yr	Type	c/n	Owner/Operator	Prev Regn

Regn	Yr	Type	c/n	Owner/Operator	Prev Regn
☐ N575T	90	King Air 1300	BB-1384	Frontier Flying Service Inc. Fairbanks, Ak.	N912YW
☐ N577D	70	King Air 100	B-22	Donald Fanetti Trucking, Franklin, Wi.	N577L
☐ N577RW	76	King Air B100	BE-5	Abilene Aero Inc. Abilene, Tx.	N90EM
☐ N578BM	79	King Air 200	BB-588	Mullins Enviromental Testing Co. Dallas, Tx.	N132GA
☐ N580AC	95	King Air C90B	LJ-1388	Anderson Columbia Co. Lake City, Fl.	N83KK
☐ N580BC	61	Gulfstream 1	63	Mission Air Support Inc. Roanoke, Va.	N144NK
☐ N580RA	73	King Air C90	LJ-580	Green Tree Financial Corp. St Paul, Mn.	(F-GFYI)
☐ N580S	79	King Air B100	BE-77	DeJarnette's Enterprises Inc. Lee's Summit, Mo.	N55US
☐ N581B	84	King Air C90A	LJ-1170	J G Boswell Co. Burbank, Ca.	N4131S
☐ N581RJ	80	King Air F90	LA-14	Geoffrey Miller MD. Woodland Hills, Ca.	N205BC
☐ N583AT	92	King Air B200	BB-1439	ITC Holding Co & others, West Point, Ga.	N200KA
☐ N586BC	85	King Air B200	BB-1223	Blue Cross & Blue Shield, Des Moines, Ia.	N7231M
☐ N586TC	75	King Air C90	LJ-653	Focus Enterprises Inc. Valparaiso, In.	N585TC
☐ N586UC	83	King Air B200	BB-1118	Trans UCU Inc. Kansas City, Mo.	N913PG
☐ N587M	68	King Air B90	LJ-361	Fayard Enterprises Inc. Louisburg, NC.	N530M
☐ N587PB	95	King Air C90B	LJ-1408	Quik Way Aviation of Nevada LLC. Oklahoma City, Ok.	N749RN
☐ N588FM	80	SA-226T Merlin 3B	T-330	Target Aviation Holdings Inc. Fort Lauderdale, Fl.	N10074
☐ N590DL	75	PA-31T Cheyenne II	7520031	Octavius Hunt Inc.	SE-GLA
☐ N592G	85	Conquest II	441-0361	Growmark Inc. Bloomington, Il.	N1283B
☐ N595RC	81	King Air C90	LJ-964	John Bunn III, Paradise Valley, Az.	N595PC
☐ N597DM	85	King Air C90A	LJ-1111	DK Airways Inc. Concord, NC.	D-IOPL
☐ N600AC	76	King Air E90	LW-185	Seagate Aviation Corp. Frankfort, Mi.	
☐ N600AL	70	King Air B90	LJ-495	Aerolease Inc. Itasca, Il.	N395DA
☐ N600AM	78	King Air 200	BB-337	Flightcraft Inc. Birmingham, Al.	
☐ N600BF	67	King Air A90	LJ-193	James Vecchio, Grand Prairie, Tx.	N600BW
☐ N600BM	84	Gulfstream 840	11734	Butler Machinery Co. Fargo, ND.	N1931S
☐ N600BS	81	Cheyenne II-XL	8166045	Polycom Products Inc. Winchester, Ky.	N9138Y
☐ N600BV	77	King Air 200	BB-254	AGES Holding Corp. Boca Raton, Fl.	N600BW
☐ N600CB	91	King Air 350	FL-38	Bashas Inc & Chapman Automotive, Chandler, Az.	N350SR
☐ N600CM	83	Gulfstream 900	15020	City Aviation Services Inc. Detroit, Mi.	N102VF
☐ N600CX	73	King Air E90	LW-43	Whiteco Industries, Merrillville, In.	D-INAC
☐ N600DJ	72	King Air C90	LJ-549	Profile Aviation Center Inc. Hickory, NC.	N9449Q
☐ N600DK	72	King Air A100	B-138	U S Motorsports Inc. Statesville, NC.	N611CC
☐ N600DL	75	SA-226T Merlin 3A	T-252	Lady Val Aviation Inc. Pensacola, Fl.	YV-500CP
☐ N600DM	78	King Air 200	BB-414	Jack Adams Co. Olive Branch, Ms.	N5VG
☐ N600FC	79	King Air E90	LW-322	Dusan Inc.	N6050F
☐ N600FL	81	King Air C90	LJ-935	Fleet Financial Group Inc. Providence, RI.	N3709W
☐ N600KA	88	King Air 300	FA-143	R & L Transfer Inc. Wilmington, Oh.	N300KA
☐ N600MM	80	Gulfstream 980	95029	Bush Field Aircraft Co. Augusta, Ga.	(N700MM)
☐ N600P	68	SA-26T Merlin 2A	T26-6	Bell Leasing Inc. Mesquite, Tx.	
☐ N600RM	79	PA-31T Cheyenne II	7920081	RKM Inc. Rockbridge Baths, Va.	(N701RM)
☐ N600SF	82	King Air F90	LA-186	Dan Alfaro, Corpus Christi, Tx.	N928RS
☐ N600TA	73	SA-226AT Merlin 4	AT-018	Fifth Third Bank, Columbus, Oh.	N316MW
☐ N600VT	80	Conquest II	441-0149	Clausen Investments Inc. Bryan, Tx.	N441DW
☐ N600WM	80	King Air F90	LA-75	STG Consultants Inc. Atlanta, Ga.	N3686B
☐ N601BM	97	Pilatus PC-XII	181	Butler Machinery Co. Fargo, ND.	HB-FSB
☐ N601CF	75	King Air 200	BB-25	Epps Air Service Inc. Atlanta, Ga.	N1555N
☐ N601DM	79	King Air C90	LJ-825	Executive Beechcraft STL Inc. Chesterfield, Mo.	N11LS
☐ N601JT	80	SA-226T Merlin 3B	T-319	Day Star Aviation LLC. Aberdeen, SD.	N808LB
☐ N601SC	78	King Air C90	LJ-753	Duncan Aviation Inc. Lincoln, Ne.	
☐ N601TA	66	King Air A90	LJ-120	Turbine Power Inc. Oklahoma City, Ok.	N601T
☐ N603PA	68	King Air B90	LJ-403	Blue Water Aircraft Inc. Wilmington, De.	N2000X
☐ N605W	79	King Air E90	LW-305	Liberty Aviation Inc. Baton Rouge, La.	N625W
☐ N606AJ	86	King Air B200	BB-1257	Coley Aircraft Sales & Leasing Inc. Memphis, Tn.	G-IJJB
☐ N606PS	79	SA-226T Merlin 3B	T-307	James Mason, DeKalb, Il.	N300MT
☐ N607KW	82	King Air B200	BB-977	Books-A-Million Inc + co-owners, Birmingham, Al.	N733NM
☐ N610CA	80	MU-2 Marquise	788SA	Dickerson Associates Inc. Columbia, SC.	N277MA
☐ N610P	75	PA-31T Cheyenne II	7620002	Forrest Wood, Flippin, Ar.	N54977
☐ N611	82	Gulfstream 900	15018	Department of the Interior, Boise, Id.	N5886N
☐ N611LM	80	PA-31T Cheyenne II	8020023	Larry Schlasinger/Executive Sales & Leasing, Minnetonka, Mn.	N457RS
☐ N611R	77	King Air E90	LW-244	City Markets Inc. Grand Junction, Co.	OY-ASU
☐ N612BB	81	Cheyenne II-XL	8166068	Wells Dairy Inc. Lemars, Ia.	N9185C
☐ N612PC	96	Pilatus PC-XII	146	Kensington Transcom Inc.	HB-FRH
☐ N613BR	65	King Air 90	LJ-9	MMS Air Inc. Vidalia, Ga.	N649MC
☐ N613CS	80	King Air 200	BB-613	AvFuel Corp. Ann Arbor, Mi.	N2TX
☐ N613HC	77	PA-31T Cheyenne II	7720053	Vader Air Corp. Saratoga Springs, NY.	(N779SW)
☐ N614ML	92	King Air 350	FL-84	Oliver Aviation Inc. Wichita, Ks.	N8084J
☐ N615	76	Rockwell 690A	11317	Canadian Valley Vo-Tech, El Reno, Ok.	
☐ N615AA	67	King Air A90	LJ-298	Alberta Aircraft Leasing Inc. Las Vegas, Nv.	N788W
☐ N615C	59	Gulfstream 1	16	B F Aircraft Leasing LLC/Execair LLC. Loveland, Co.	N202HA
☐ N615GA	73	SA-226AT Merlin 4	AT-009	Arrow Trading Inc. Fort Lauderdale, Fl.	C-FJTC
☐ N616AS	66	King Air A90	LJ-160	Fayard Enterprises Inc. Louisberg, NC. (status ?).	N42CG
☐ N616CK	94	King Air 350	FL-114	Collins & Aikman Products Co. Charlotte, NC.	N8288Q
☐ N616CP	80	King Air F90	LA-25	Coperco Inc. Houston, Tx.	N616BH
☐ N616DR	85	King Air B200	BB-1194	Bryan Automotive Group Inc. Sanford, Fl.	N743E
☐ N616E	67	680V Turbo Commander	1707-82	Church of Bible Understanding, Philadelphia, Pa.	N616MC

Regn	Yr	Type	c/n	Owner/Operator	Prev Regn
N616F	66	King Air A90	LJ-165	Alberta Aircraft Leasing Inc. Las Vegas, Nv.	N6HF
N616GB	81	King Air 200	BB-752	King Air LLC. Sioux Falls, SD.	N300QW
N616PS	80	SA-226T Merlin 3B	T-316	West One Bank/Canyon Springs Aviation LLC. Sun Valley, Id.	N51DA
N616SC	89	King Air C90A	LJ-1192	Webb Road Development Inc & K A Corp. Wichita, Ks.	D-ISAG
N617BB		Mitsubishi MU-2G	522	AIR-HI-O Corp. Washington, Oh.	N147MA
N617KM	85	King Air C90A	LJ-1095	Refrigeration Supplies Distributor, Monterey Park, Ca.	N850CE
N617LM	82	King Air C90-1	LJ-1012	BandyCo/Cumberland Petroleum Transport Inc. Stanford, Ky.	N61797
N618	90	King Air B200	BB-1378	U S Department of the Interior, Boise, Id.	N5637Y
N618SW	81	Cheyenne II-XL	8166036	Mon Air Inc & Corporate Air Management Inc. West Mifflin, Pa	
N619JB	81	PA-42 Cheyenne III	8001011	Bailey Marketing Inc. Wilmington, De.	N795KW
N620AD	81	Cheyenne II-XL	8166059	Aviation Business Management Inc. Charlotte, NC.	HK-2907
N620MW	83	PA-42 Cheyenne III	8001105	B C Air LC. Moundridge, Ks.	N4109W
N620WE	78	King Air C90	LJ-743	W H Ebert Corp. San Jose, Ca.	N161AC
N621SC	78	Conquest II	441-0070	J W & Shirley Connolly, Sewickley, Pa.	N910EA
N621TA	73	Mitsubishi MU-2J	605	Metro Enviromental Associates Inc. Washington, Ga.	XB-ARE
N621TB	68	King Air B90	LJ-334	Thomas Bryan, Atlanta, Ga.	N722TS
N621WP	89	King Air C90A	LJ-1208	W J Page Inc. Falls Church, Va.	(N485JA)
N622DC	79	King Air 200	BB-491	Spirit Aviation Inc. Van Nuys, Ca.	RP-C243
N623AW	67	King Air A90	LJ-282	Flight Safety International Inc. Dothan, Al.	N853K
N623BB	67	King Air A90	LJ-277	Double B Foods Inc. Weimar, Tx.	N228CF
N623R	66	King Air A90	LJ-173	Global Air Service Inc. Nashville, Tn.	XB-SFS
N623VG	81	King Air 200	BB-769	K G Beverage Inc. Jacksonville, Fl.	N769AJ
N623WA	83	King Air B200	BB-1134	Waukesha-Pearce Industries Inc. Houston, Tx.	N66460
N624CB	80	Conquest II	441-0166	Industrials International Inc. St Thomas, USVI.	I-GEFI
N625MD	81	King Air B200	BB-942	Ultrmar Inc. Hanford, Ca.	N525BC
N625W	91	King Air B200	BB-1394	National Bank of Commerce, Lincoln, Ne.	N8047Y
N626PS	80	SA-226T Merlin 3B	T-329	Vacair Co/Paul Songer Sr. Washington, Pa.	N300KF
N627KW	84	PA-42 Cheyenne IIIA	5501010	Repimex Inc. Dover, De.	D-IFRC
N629TM	74	Mitsubishi MU-2J	631	Sanmat Aviation Inc. Wilmington, De.	N58BC
N630DB	76	King Air 200	BB-187	Air Lex-Sprite Flite Jets/R R Dawson Bridge Co. Lexington.	
N630MW	81	Cheyenne II-XL	8166011	Marvin Lumber & Cedar Co. Warroad, Mn.	N218SW
N631DS	94	2000A Starship	NC-44	Dartswift Inc. Philadelphia, Pa.	N1863Q
N631PC	81	PA-42 Cheyenne III	8001060	Reinbeck Motors Inc. Reinbeck, Ia.	(N623KW)
N631WF	76	PA-31T Cheyenne II	7620051	Wilcox Family Farms Inc. Roy, Wa.	N631MC
N632DS	88	King Air 300	FA-141	Dartswift Inc. Rosement, Pa.	N3080F
N633ST	73	SA-226T Merlin 3	T-237	State Tool & Manufacturing, Benton Harbor, Mi.	C-FCPH
N633WC	81	PA-31T Cheyenne 1	8104036	Werner Construction Co. Hastings, NE.	N75BR
N634TT	86	King Air B200	BB-1252	Ted Taylor, Birmingham, Al.	N125GA
N635AF	77	King Air C90	LJ-736	Auburn Foundry Inc. Auburn, In.	N19HT
N636JM	80	King Air E90	LW-332	J Morten & Son Inc. Sioux City, Ne.	
N636SP	77	SA-226T Merlin 3A	T-285	Worldwide Aircraft Services Inc. Springfield, Mo.	N65P
N637KC	85	PA-42 Cheyenne IIIA	5501030	Air One Corp. Rye, Co.	N627KW
N637WG	74	Mitsubishi MU-2J	637	Bankair Inc. West Columbia, SC.	N951MS
N638D	68	King Air B90	LJ-424	Turbine Power Inc. Bridgewater, Va.	
N640MW		Beech 1900C-1	UC-1	Marvin Lumber & Cedar Co. Warroad, Mn.	N3114B
N641PE	81	King Air F90	LA-91	Platt Electric Supply, Beaverton, Or.	N90GS
N641TC	80	King Air 200	BB-641	Bon Aero Inc. Las Vegas, Nv.	F-GGVV
N642DH	68	King Air B90	LJ-420	Diamond Airways Inc. Columbus, Oh.	N759KX
N642JL	85	King Air 300	FA-30	Pizza Hut Inc/J Larry Fugate Enterprises Inc. Wichita, Ks.	N70CR
N642TD	78	King Air C90	LJ-766	J M Associates Inc. Little Rock, Ar.	N44VC
N642TF	78	King Air 200	BB-363	Felber Motivation inc. Colorado Springs, Co.	N96GA
N644SP	76	King Air C90	LJ-696	Legacy Aviation Inc. Monroe, La.	(N525PC)
N646BM	80	King Air 200	BB-646	Kennecott Energy Co. Gillette, Wy.	N53GA
N647JM	83	King Air B200	BB-1147	Mathis Aviation Inc. Lexington, SC.	N6627V
N650TJ	79	King Air B100	BE-60	Jetman LLC. Midland, Tx.	(N331BB)
N652L	80	King Air E90	LW-329	Laurence Aviation Inc. Dover, De.	F-GJBG
N652MC	85	Conquest 1	425-0225	McCoy Corp. San Marcos, Tx.	N6522T
N653PC	94	Dornier 328-100	3027	Prince Transportation Inc. Holland, Mi.	D-CDHM
N654BA	82	King Air B200C	BL-54	Department of the Air Force, Layton, Ut.	N6563C
N654C	65	King Air 90	LJ-12	Turbine Power Inc. Bridgewater, Va.	N90MR
N655BA	80	King Air 200	BB-655	Dana Kirk, Houston, Tx.	G-OGAT
N655JG	92	King Air B200	BB-1440	Ariel Corp. Mount Vernon, Oh.	N8046H
N655PC	80	Gulfstream 980	95015	Arvest Bank Group Inc. Bentonville, Ar.	N9768S
N657P	65	Gulfstream 1	165	Executive Aircraft Leasing LLC. Lake Zurich, Il.	N657PC
N659PC	69	Gulfstream 1	196	Chariot Air LLC. Dallas, Tx.	N134PA
N660JM	75	King Air C90	LJ-673	J Morten & Son Inc. Sioux City, Ne.	N666JM
N660MW	93	King Air/Catpass 250	BB-1428	Marvin Lumber & Cedar Co. Warroad, Mn.	N8265V
N660PB	90	King Air B200	BB-1364	William Wilson & Associates, San Mateo, Ca.	N5510Y
N661BA	83	King Air B200C	BL-61	Department of the Air Force, McCarran, Nv.	N6564C
N661DP	80	MU-2 Marquise	798SA	Great River Finance Co. La Grange, Mo.	VH-MIU
N661DW	92	TBM 700	61	Danwell Corp. Springfield, Il.	
N662BA	83	King Air B200C	BL-62	Department of the Air Force, McCarran, Nv.	N6566C
N663AC	85	King Air 300	FA-54	AG-CHEM Equipment Co. Minnetonka, Mn.	N1HS
N665JK	75	King Air C90	LJ-565	OK Aviation Inc. Monterey, Ca.	EC-COK
N665TM	81	Conquest II	441-0215	Abigail Kawananakoa, Honolulu, Hi.	ZS-KPB
N666DC	73	King Air E90	LW-44	Blue Ash Industrial Supply Co. Cincinnati, Oh.	N166A

Regn	Yr	Type	c/n	Owner/Operator	Prev Regn
□ N666FG	66	King Air A90	LJ-121	Robert Smith, Neustadt an der Aisch, Germany.	F-GERH
□ N666HC	93	Conquest II	441-0292	Haskell Co. Jacksonville, Fl.	N88716
□ N666JL	79	PA-31T Cheyenne II	7920084	Bean Enterprises Inc. Dover, De.	N700JR
□ N666MN	66	680T Turbo Commander	1568-24	Regional Missouri Bank, Marceline, Mo.	N2637M
□ N666PC	84	King Air C90A	LJ-1068	666PC Inc. Shreveport, La.	N96MR
□ N666RH	90	King Air 350	FL-21	Airbank Inc. Brentwood, Tn.	
□ N666RK	68	Mitsubishi MU-2D	110	Antroguz Corp. Miami, Fl. (status ?).	N857Q
□ N666RL	73	Mitsubishi MU-2K	272	Air 1st Aviation Companies, Aiken, SC.	(N500BE)
□ N667AM	81	MU-2 Marquise	1534SA	Earle Martin, Houston, Tx.	HB-LQS
□ N667JJ	95	Pilatus PC-XII	108	Lynton Group Inc. Morristown, NJ.	HB-...
□ N669CA	81	PA-42 Cheyenne III	8001041	OMTAE Financial & Concept Products Inc. Springfield, Pa.	G-BWTX
□ N669SP	68	SA-26AT Merlin 2B	T26-101	James & Sherry Purdy, Meridian, Ms.	N11PM
□ N671L	73	King Air A100	B-174	Airjet Inc. Tulsa, Ok.	N763K
□ N671LL	66	King Air A90	LJ-148	B & L Air Service Inc. Cincinnati, Oh.	N671L
□ N675J	80	King Air E90	LW-336	T L James & Co. Ruston, La.	N6759P
□ N675PC	97	King Air 350	FL-178	Pacific Coast Building Products Inc. Sacramento, Ca.	N2054Q
□ N675PG	80	King Air 200	BB-680	Raytheon Aircraft Co. Wichita, Ks.	N675PC
□ N676BB	95	King Air B200	BB-1501	Becker Trading Co. Fort Pierce, Fl.	N3180S
□ N676DM	71	681B Turbo Commander	6048	Graves & Graves Construction Co. Parsons, Tn.	N911JM
□ N676J	76	King Air E90	LW-179	T L James & Co. Ruston, La.	N211MH
□ N677J	78	King Air E90	LW-294	T L James & Co. Ruston, La.	N9DF
□ N677KA	67	680V Turbo Commander	1677-60	H F Payne Construction Co. Monrovia, Md.	N2UL
□ N678EB	83	King Air B200	BB-1124	Executive Beechcraft STL inc. Chesterfield, Mo.	ZS-MMO
□ N678SS	82	King Air B200	BB-1021	Southern Systems Inc. Memphis, Tn.	N195KA
□ N679BK	76	Mitsubishi MU-2L	679	Intercontinental Business Aircraft Inc. Tulsa, Ok.	C-FYBN
□ N680CA	74	Mitsubishi MU-2J	642	Thor Enterprises Inc. Beloit, Wi.	N492MA
□ N680CB	82	King Air B200	BB-1006	C C Medflight Inc. Lawrenceville, Ga.	(N46CE)
□ N680JD	68	680W Turbo Commander	1792-22	Michael Knudsen, Anaheim, Ca.	N43WL
□ N680KM	68	680W Turbo Commander	1812-29	Great American Financial Services, Portland, Or.	N500NR
□ N680WA	81	Gulfstream 840	11680	Nevada Department of Transportation, Carson City, Nv.	XA-JYM
□ N681PC	89	King Air B200	BB-1330	U S Aircraft Leasing Inc. Dover, De.	N773M
□ N681SW	81	PA-31T Cheyenne 1	8104027	Sure Alloy Steel Corp. Warren, Mi.	
□ N682DR	76	King Air 200	BB-130	Air Services Brokerage LLC. Traverse City, Mi.	N323MB
□ N682KA	76	King Air C90	LJ-682	Pezold Air Charter LLC. Columbus, Ga.	N3980D
□ N686AC	82	King Air B100	BE-127	The G W Van Keppel Co. Kansas City, Ks.	N666AC
□ N686GW	84	King Air C90A	LJ-1082	Joe Morten & Son Inc/Cargill Leasing Corp. Minnetonka, Mn.	N666GW
□ N687AE	79	Conquest II	441-0087	Samaritan Health Service, Phoenix, Az.	N441DW
□ N687HB	76	Mitsubishi MU-2L	687	Golden Eagle Aviation Inc. Portsmouth, Va.	XB-ARF
□ N687L	66	680V Turbo Commander	1560-17	RAF Air Cargo Inc. Wilmington, De.	N419S
□ N688AA	76	King Air 200	BB-434	Adventure Aviation Inc. Bluefield, WV.	N107CT
□ N688CA	78	PA-31T Cheyenne II	7820046	MCR Aviation Inc. Lancaster, Oh.	N777LM
□ N688DS	76	King Air B100	BE-8	Delta Star Inc. Lynchburg, Va.	N27WT
□ N688RA	76	Mitsubishi MU-2L	688	Royal Air Freight Inc. Waterford, Mi.	N688MA
□ N688SH	66	680T Turbo Commander	1597-42	Bohlke International Airways, St Croix, USVI.	N676MB
□ N689AE	83	Conquest II	441-0281	BLC Corp/Samaritan Airevac, Phoenix, Az.	N6832C
□ N689BV	76	King Air 200	BB-338	Richmor Aviation Inc. Hudson, NY.	VR-BGN
□ N689EB	76	King Air C90	LJ-689	M I T C Aviation LLC. Shawnee Mission, Ks.	(N45FE)
□ N690AC	79	Rockwell 690B	11527	Tomahawk Aviation LLC. Bay City, Mi.	N81765
□ N690AH	73	Rockwell 690A	11119	Therma-Tron-X Inc. Sturgeon Bay, Wi.	XA-RTQ
□ N690AT	74	Rockwell 690A	11202	Central Air Southwest, Kansas City, Mo.	N600PB
□ N690BA	78	Rockwell 690B	11513	SKG Inc. California, NJ.	N74WA
□ N690BE	79	Rockwell 690B	11544	Guaranty Development Co. Livingston, Mt.	I-ACLR
□ N690CA	72	King Air C90	LJ-540	SHH Inc. Lakeland, Fl.	N123SK
□ N690CB	77	Rockwell 690B	11387	Simone Aviation Inc. Wilmington, De.	(N790CB)
□ N690CC	77	Rockwell 690B	11379	Edgar Cruft, Tijeras, NM.	
□ N690CE	73	Rockwell 690A	11103	Dynamic Engineering Inc. Fort Smith, Ar.	N690MF
□ N690CH	79	Rockwell 690B	11542	Crabbe Huson Co. Portland, Or.	N76DT
□ N690CM	72	Rockwell 690	11040	Iguana Enterprises Inc. Fort Worth, Tx.	N701CM
□ N690CP	78	Rockwell 690B	11451	JW & JA Inc. Indianapolis, In.	N71MA
□ N690DB	74	Rockwell 690A	11226	Indiana Aircraft Sales Inc. Indianapolis, In.	N717AP
□ N690DE	76	Rockwell 690A	11316	Com-Net Construction Services Inc. Johnson City, Tn.	N666K
□ N690DT	77	Rockwell 690B	11413	DT Industries Inc. Lebanon, Mo.	N528BE
□ N690EH	76	Rockwell 690A	11309	Surdex Corp. Chesterfield, Mo.	N2NQ
□ N690EM	73	Rockwell 690A	11125	Myler Co. Crawfordsville, In.	N690AE
□ N690ES	77	Rockwell 690B	11388	Equipment Supply Co. Burlington, NJ.	N532
□ N690FD	77	Rockwell 690B	11393	Stegall Motor Sports LLC. Greenville, SC.	N699GN
□ N690FR	78	Rockwell 690B	11436	Opequon Air LLC. Winchester, Va.	D-IFAB
□ N690G	73	King Air E90	LW-34	American Jet International Corp. Houston, Tx.	N387SC
□ N690GH	77	Rockwell 690B	11357	Haas Chemical Co. Mobile, Al.	N690MH
□ N690GS	77	Rockwell 690B	11363	Great Southern Wood Preserving Inc. Abbeville, Al.	N690JB
□ N690HB	74	Rockwell 690A	11205	J Brooks LLC. Louisville, Ky.	N9007
□ N690HF	76	Rockwell 690A	11298	Nicolas Cagnoni, Springfield, Il.	(N80LE)
□ N690HS	78	Rockwell 690B	11431	Townley Manufacturing Co. Candler, Fl.	N72RF
□ N690HT	78	Rockwell 690B	11467	Extraordinaire Inc. Washington, DC.	N690PG
□ N690JC	78	Rockwell 690B	11479	Aircraft Sales International Inc. Fort Lauderdale.(status?).	N51MF
□ N690JH	79	Rockwell 690B	11529	Commercial Aviation Inc. Kerrville, Tx.	N226BP

Regn	Yr	Type	c/n	Owner/Operator	Prev Regn
☐ N690JJ	74	Rockwell 690A	11171	Aero Leasing Inc. Indianapolis, In.	N690EM
☐ N690JP	76	King Air C90	LJ-690	Cibolo Air/Southwestern Holdings Aviation Inc. Houston, Tx.	N80TB
☐ N690KC	77	Rockwell 690B	11384	Dunning-White Air Inc. Dothan, Al.	N295NM
☐ N690L	84	PA-42 Cheyenne IIIA	5501020	Offshore Logistics Inc. Lafayette, La.	N623KW
☐ N690LH	78	Rockwell 690B	11487	Litehawk Aviation LLC. Sandpoint, Id.	N690PR
☐ N690NA	77	Rockwell 690B	11359	Sencon International Inc. Fort Lauderdale, Fl.	XB-OCI
☐ N690PJ	73	Rockwell 690	11058	American Aerial Surveys Inc. Sacramento, Ca.	N2069B
☐ N690PT	75	Rockwell 690A	11252	N690PT LLC. Aiken, SC.	(N690AR)
☐ N690RA	72	Rockwell 690	11010	Nevada Exchange LLC. Carson City, Nv.	XA-RYF
☐ N690RK	73	Rockwell 690A	11138	RBK Aviation Inc. Wilson, Wy.	N690EC
☐ N690RP	78	Rockwell 690B	11493	Business Air LLC. Taylor, Mi.	N1HR
☐ N690SG	74	Rockwell 690A	11146	Amber Aviation Inc. Beaver Creek, Oh.	F-OHJE
☐ N690SM	76	Rockwell 690A	11337	Pascomar Inc. Chicago, Il.	N1547A
☐ N690TC	77	Rockwell 690B	11385	Wal-Mart Stores Inc. Rogers, Ar.	VH-EXT
☐ N690TG	77	Rockwell 690B	11444	Gramer Air II LLC. Birmingham, Mi.	N1KC
☐ N690WC	72	Rockwell 690	11017	West Coast Airlines/Wadell Engineering Corp. Burlingame, Ca.	N101RQ
☐ N690WM	77	Rockwell 690B	11399	Wal-Mart Stores Inc. Rogers, Ar.	N81633
☐ N690WP	79	Rockwell 690B	11561	Bob Mitchell, Sallisaw, Ok.	N9196Q
☐ N691AS	90	King Air C90A	LJ-1240	Warn Industries Inc. Milwaukee, Or.	N1565X
☐ N691CP	78	Rockwell 690B	11457	JW & JA Inc. Indianapolis, In.	N900R
☐ N691SM	78	Rockwell 690B	11492	Steven Myers & Associates Inc. Newport Beach, Ca.	N42MS
☐ N691WM	79	Rockwell 690B	11522	Wal-Mart Stores Inc. Rogers, Ar.	N50MS
☐ N692T	79	Rockwell 690B	11555	Timberline Aviation LLC. Grand Junction, Co.	N150SP
☐ N694AB	75	King Air 200	BB-78	J-Hawk Servicing Corp. Waco, Tx.	N783DY
☐ N695JJ	76	King Air C90	LJ-695	J & J Sporting Goods/Joseph Gagliardi, Campbell, Ca.	N93BB
☐ N695NC	82	Gulfstream 1000	96032	Thomas Norton, Madras, Or.	VH-GAB
☐ N695RC	85	Gulfstream 1000	96087	Montana Flying Machine LLC. Ennis, Mt.	N722SG
☐ N695WF	95	Pilatus PC-XII	125	George & Dianna Archuleta, Lakeside, Mt.	HB-FQQ
☐ N696AB	72	King Air A100	B-101	R & L Aviation Inc. West Homestead, Pa.	N98TR
☐ N696JB	70	King Air 100	B-26	Dolph Briscoe Jr. Uvalde, Tx.	N610KR
☐ N696RA	81	King Air F90	LA-87	United River Aviation Inc. Hickory, NC.	N200SC
☐ N696TS	97	Pilatus PC-XII	169	Virtual One LLC. San Mateo, Ca.	HB-FSE
☐ N696WW	94	King Air C90B	LJ-1371	GECC, Atlanta, Ga.	C-FSGZ
☐ N697P	85	King Air B200	BB-1217	Guthy-Renker Aviation LLC. Palm Desert, Ca.	D-IMOL
☐ N698GN	66	680V Turbo Commander	1589-40	Michael Franzblau, Los Angeles, Ca.	N699GN
☐ N698X	69	SA-26AT Merlin 2B	T26-137	Dr. George Swanson MD. Port Arthur, Tx.	(N192GK)
☐ N699KM	75	SA-226T Merlin 3A	T-256	OFC Holding Co. Wilmington, De.	N311GM
☐ N700AT	73	King Air A100	B-136	Air Eagle LLC/Talon Inc. Detroit, Mi.	(N45PL)
☐ N700BA	80	King Air B100	BE-84	Byerly Aviation Inc. Peoria, Il.	N300TN
☐ N700BE	79	King Air 200	BB-528	SE Aircraft Sales Inc. Jacksonville, Fl.	N203BC
☐ N700BF	92	TBM 700	53	William Foster, Sherborn, Ma.	
☐ N700BK	81	King Air F90	LA-107	Bank of Mississippi, Tupelo, Ms.	N4237M
☐ N700BS	80	PA-31T Cheyenne 1	8004020	Stevens Racing Consultants Inc. Mooresville, NC.	N120ET
☐ N700CB	70	Rockwell 690	11001	T O G Ltd. Vero Beach, Fl. (was s/n 6031).	N9100N
☐ N700CC	96	TBM 700	113	Jeanette Symons, San Francisco, Ca.	F-OHBU
☐ N700CP	76	King Air C90	LJ-700	Consolidated Pipe & Supply Corp. Birmingham, Al.	N700FC
☐ N700CS	95	TBM 700	109	Clive Samuels & Associates Inc. Princeton, NJ.	
☐ N700DD	78	King Air E90	LW-288	Sammann Corp. Michigan City, In.	N717US
☐ N700DH	74	King Air E90	LW-114	Aero Ventures, Woodbury, NJ.	N89L
☐ N700EF	91	TBM 700	21	Capital Technologies Inc. Newton Centre, Ma.	
☐ N700GJ	95	TBM 700	102	N700GJ Inc. Gaithersburg, Md.	F-GNHP
☐ N700GM	91	King Air C90A	LJ-1283	International Auto Brokers Inc. Scottsdale, Az.	D-IHMW
☐ N700HK	93	TBM 700	60	WIC Inc. Rio Rancho, NM.	(N95DW)
☐ N700HM	79	King Air 200	BB-448	Hollywood Marine Inc. Houston, Tx.	N865W
☐ N700JG	82	Conquest II	441-0237	Laidlaw Enviromental Services Inc. Columbia, SC.	N700JE
☐ N700JJ	91	TBM 700	2	JCT Aviation Inc. Dover, De.	F-GLBA
☐ N700KL	92	TBM 700	88	C Kevin Landry, Lincoln, Ma.	
☐ N700LL	93	TBM 700	86	Spring Brook Marina Inc. Seneca, Il.	
☐ N700LT	79	PA-31T Cheyenne II	7920046	Teufel Nursery Inc. Portland, Or.	N700TF
☐ N700LW	75	Mitsubishi MU-2M	326	Living Word Christian Center, Brooklyn Park, Mn.	XA-RLZ
☐ N700MB	67	King Air A90	LJ-233	Flight Safety International Inc. Dothan, Al.	N100HT
☐ N700PC	77	Rockwell 690B	11389	James Lindsey Jr Family Trust, Bakersfield, Ca.	N90CH
☐ N700PP	92	TBM 700	59	Power Service Products Inc. Weatherford, Tx.	
☐ N700PT	78	PA-31T Cheyenne II	7820032	Quaker Air Services Inc. Salem, Oh.	N43CW
☐ N700PU	91	TBM 700	15	John Montgomery, San Francisco, Ca.	
☐ N700PW	91	TBM 700	29	TBM North America Inc. Grand Prairie, Tx.	
☐ N700RG	78	PA-31T Cheyenne II	7820042	OCAAS LLC. Amarillo, Tx.	N400RT
☐ N700RX	72	Mitsubishi MU-2K	241	Internet Jet Sales Ltd. Wilmington, De.	C-GJAV
☐ N700SF	91	TBM 700	26	Air Frantz Inc. Wilmington, De.	
☐ N700SP	94	TBM 700	98	Jim Jaeger, Cincinnati, Oh.	
☐ N700SR	74	Rockwell 690A	11164	GEN-COR Inc. Plymouth, In.	N57152
☐ N700TB	96	TBM 700	123	Fred Beans Ford Inc. Doylestown, Pa.	
☐ N700TJ	91	TBM 700	25	Sweatmore Air Inc. Phoenix, Az.	N303WB
☐ N700U	81	King Air B200	BB-888	Ohio University, Athens, Oh.	N922CR
☐ N700VF	81	PA-42 Cheyenne III	8001053	Flying A Flight Service Inc. Silver City, NM.	N543FM
☐ N700VM	92	TBM 700	72	Rentco/Idaho, Boise, Id.	

Regn   *Yr Type*   *c/n*   *Owner/Operator*   *Prev Regn*

Regn	Yr	Type	c/n	Owner/Operator	Prev Regn
N700VX	96	TBM 700	118	Paul Pennell Sr. Cramerton, NC.	F-WNGF
N700WD	91	TBM 700	27	Broadreach Sales & Service Co. Stuart, Fl.	N700GB
N700WJ	81	Conquest 1	425-0036	Bighorn Airways Inc. Sheridan, Wy.	(N32TJ)
N700WP	88	King Air B200	BB-1313	Bullfrog of Ohio County Inc. Owensboro, Ky.	8P-BAR
N700WT	93	TBM 700	91	Jerry's Enterprises Inc. Edina, Mn.	
N700Z	79	King Air 200	BB-443	Bramco Inc. Louisville, Ky.	
N701BN	61	Gulfstream 1	74	Battelle Corp. Pasco, Wa.	N5619D
N701CR	83	Conquest 1	425-0148	Aon Operations/Globe Auto Leasing Inc. Wheeling, Il.	N425EK
N701K	79	MU-2 Solitaire	410SA	Lucky Landing's Charter Inc. Placeda, Fl.	N329WM
N701NA	83	King Air B200	BB-1164	NASA/OAST, Ames, Moffett Field, Ca.	
N701NC	80	King Air F90	LA-55	Buchanan LLC. Fargo, ND.	N311DB
N701PF	91	TBM 700	31	P F Flyers Inc. Lakeland, Fl.	N64TW
N701PT	80	PA-31T Cheyenne 1	8004008	Natoli Engineering Co. Chesterfield, Mo.	(N701DH)
N701RJ	70	King Air 100	B-23	Precision Products Racing Inc. Denver, NC.	N711AU
N701X	76	King Air E90	LW-165	SouperKruser Inc. Fort Lauderdale, Fl.	N700DH
N702MA	82	King Air B200	BB-1010	Michigan Aeronautics Commission, Lansing, Mi.	N62FC
N702TA	97	King Air B200	BB-1573	Raytheon Travel Air Co. Wichita, Ks.	
N703DM	69	Mitsubishi MU-2F	138	Bush Field Aircraft Co. Aiken, SC.	N252DC
N703JT	80	King Air F90	LA-39	King Management Corp. Santa Rosa, Ca.	YV-342CP
N703KH	80	King Air 200	BB-703	B & C Aviation Inc. Ridgefield, Ct.	(OY-AUY)
N703TA	97	King Air B200	BB-1574	Raytheon Travel Air Co. Wichita, Ks.	
N705TA	97	King Air B200	BB-1575	Raytheon Air Travel Co-St Joe Corp. Wichita, Ks.	
N706DG	80	King Air 200	BB-548	Davis-Garvin Agency Inc. West Columbia, SC.	N78SC
N706DM	67	Mitsubishi MU-2B	038	Bush Field Aircraft Co. Augusta, Ga.	N8BL
N706KC	76	Rockwell 690A	11343	BRS Services Inc. Reno, Nv.	F-GBGL
N706TA	98	King Air B200	BB-1598	Raytheon Travel Air Co. Wichita, Ks.	
N707FN	73	Mitsubishi MU-2K	280	International Business Aircraft Inc. Tulsa, Ok.	N707DM
N707MA	83	Conquest II	441-0285	Caribbean Air & Marine Service, Memphis, Tn.	D-IIAA
N707ML	75	PA-31T Cheyenne II	7520017	Maritime Sales & Leasing Inc. Newnan, Ga.	(N5RZ)
N707MP	74	Rockwell 690A	11198	American Companies Inc. Las Vegas, Nv.	N700MP
N707PK	76	SA-226T Merlin 3B	T-271	William Cook Agency/River City Aviation, West Memphis, Ar.	N707DB
N707SS	71	King Air 100	B-81	Triangle Copter Inc-Wichita/Alpha Aviation Inc. Dallas, Tx.	N858B
N708DG	71	King Air C90	LJ-508	Dark Light Inc. Rockford, Il.	N706DG
N708DM	73	Mitsubishi MU-2K	271	Bush Field Aircraft Co. Augusta, Ga.	N23HR
N709FN	74	Mitsubishi MU-2K	308	Bush Field Aircraft Co. Augusta, Ga.	N709DM
N710G	78	Mitsubishi MU-2N	710SA	Anaconda Aviation Corp. Boca Raton, Fl.	I-MLST
N711BL	82	King Air B200	BB-1013	United Rotary Brush Corp. Lenexa, ks.	N144C
N711BN	73	King Air C90	LJ-588	Koosharem Corp. Santa Barbara, Ca.	N711BL
N711BP	77	Rockwell 690B	11356	Ohio Kentucky Oil Corp. Lexington, Ky.	N653PC
N711CR	74	King Air 200	BB-16	Florida Executive Air Charter Inc. Orlando, Fl.	N700CP
N711ER	81	PA-31T Cheyenne 1	8104056	Aviation Inc. Fulton, Ms.	(D-IAHM)
N711FN	81	Cheyenne II-XL	8166043	Romeo Hotel Air Service Inc. Montgomery, Al.	N820SW
N711GE	81	Conquest II	441-0209	Haggen Inc & Allsop Inc. Bellingham, Wa.	N2728G
N711GM	70	King Air 100	B-31	T & W Leasing, Brookesmith, Tx.	N38HB
N711HV	77	King Air E90	LW-246	Manitowoc Co. Manitowoc, Wi.	N711NV
N711KW	81	King Air F90	LA-99	Del-R Air LLC. Holland, Mi.	N3741U
N711L	84	King Air F90-1	LA-222	U S Investments Ltd. Crystal Bay, Nv.	N69283
N711LD	81	PA-31T Cheyenne 1	8104040	Ideal Electric Co. Mansfield, Oh.	N981SR
N711LV	79	PA-31T Cheyenne II	7920042	Anthony Bianco, Barrington, Il.	
N711MB	79	King Air 200	BB-437	HealthCorp of N Carolina Inc. Morehead City, NC.	N711MD
N711MZ	80	King Air 200	BB-849	Champion Air Inc. Mooresville, NC.	N74F
N711RD	79	PA-31T Cheyenne II	7920068	J E L Farms Ltd. Hale Center, Tx.	N40JC
N711RE	77	King Air A100	B-233	Tulip City Air Service Inc. Holland, Mi.	N99HE
N711TZ	77	King Air E90	LW-226	The Grefenson Clinic, Twin Falls, Id.	N976
N711VH	79	King Air 200	BB-544	Frontenac Aviation Inc/Air Fleet International, Teterboro.	N548WB
N712GA	78	King Air C-12C	BC-63	Georgia Dept of Public Safety, Kennesaw, Ga.	78-23127
N712JC	90	King Air C90A	LJ-1259	Keffer Co. Charlotte, NC.	
N713FP	97	King Air B200	BB-1595	Raytheon Aircraft Co. Wichita, Ks.	
N713TA	98	King Air B200	BB-1610	Raytheon Travel Air Co. Wichita, Ks.	
N714BX	78	King Air E90	LW-257	R & F Wings Inc. Liverpool, NY.	N711BX
N714F	77	King Air E90	LW-237	Seven Bar Flying Service Inc. Albuquerque, NM.	N23707
N715CA	81	PA-31T Cheyenne II	8120063	Giumarra Brothers Fruit Co. Los Angeles, Ca.	
N715CG	92	King Air 200	BB-1421	Carl Gregory Cars/J J & C Air Inc. Columbus, Ga.	N8230Z
N715MA	78	King Air 200	BB-322	DeWitt Enterprises/Corporate Aircraft Inc. Kinoleloe, Mi.	N202LJ
N715MC	91	TBM 700	30	Socata Aircraft, Grand Prairie, Tx.	
N715RD	80	King Air 200	BB-707	A T Massey Coal Co. Richmond, Va.	(D-IEHD)
N715WA	73	King Air A100	B-168	Colemill Enterprises Inc. Nashville, Tn.	F-GFVM
N716GA	76	King Air C-12C	BC-24	Georgia Dept of Public Safety, Kennesaw, Ga.	76-22548
N716TA	95	King Air B200	BB-1509	Raytheon Travel Air Co. Wichita, Ks.	N109NT
N716WA	80	PA-31T Cheyenne II	8020042	Jon Word, Albuquerque, NM.	N300TB
N717D	80	King Air B100	BE-91	DeCol Aviation Inc. Altoona, Pa.	N6740D
N717DC	87	King Air B200	BB-1261	Danella Companies Inc. Plymouth Meeting, Pa.	N2997N
N717ES	80	PA-42 Cheyenne III	8001004	J Bar B Transportation Inc. Waelder, Tx.	N42WZ
N717HT	74	King Air 200	BB-133	Dr Daniel Capen, Westlake Village, Ca.	N113RL
N717LW	81	Conquest 1	425-0086	James Stewart Jr. El Paso, Tx.	N911RD
N717PD	70	SA-26AT Merlin 2B	T26-178	Pennsylvania Dutch Co. Mount Holly Springs, Pa.	N61775
Regn	Yr	Type	c/n	Owner/Operator	Prev Regn

Regn	Yr	Type	c/n	Owner/Operator	Prev Regn
☐ N717PS	76	Mitsubishi MU-2L	686	Royal Air Freight Inc. Waterford, Mi.	N23RA
☐ N717RA	65	Gulfstream 1	167	Orco Aviation/Riverside Air Service, Houston, Tx.	C-GDWM
☐ N717RD	89	King Air C90A	LJ-1199	J & S Yacht Services Inc. Wilmington, De.	N487JD
☐ N717RS	60	Gulfstream 1	38	Orco Aviation/Riverside Air Service, Riverside, Ca.	N333AH
☐ N717X	73	King Air C90	LJ-581	Mid Valley Air, Bakersfield, Ca.	(N581VP)
☐ N717Y	91	TBM 700	17	J H Van Zant II, Albany, Tx.	
☐ N718GL	69	SA-26AT Merlin 2B	T26-140	The Cole 1st Family LP. Amarillo, Tx.	N699F
☐ N718JP	96	Pilatus PC-XII	155	Pezold Air Charters LLC. Wilmington, De.	N361FB
☐ N718K	68	King Air B90	LJ-371	B90 King Group LLC. Austin, Tx.	
☐ N719HC	82	King Air B200	BB-962	Harbert Corp. Birmingham, Al.	N721HC
☐ N719RA	60	Gulfstream 1	28	Dodson International Parts Inc. Ottawa, Ks.	N118X
☐ N720JK	69	Mitsubishi MU-2F	154	Aircraft Management Inc. Poughkeepsie, NY.	(N81LJ)
☐ N720K	65	King Air 90	LJ-37	Anton Ptach, Poughkeepsie, NY.	
☐ N721FC	68	Mitsubishi MU-2D	111	East Mississippi Community Co. Mayhew, Ms.	N858Q
☐ N721MR	82	Gulfstream 900	15002	Fulghum Fibres Inc. Augusta, Ga.	N131KS
☐ N721SG	85	PA-42 Cheyenne 400LS	5527027	Douglas Reichardt, Henderson, Nv.	N4119V
☐ N721TB	77	Rockwell 690B	11352	David C Poole Co. Greenville, SC.	N46JC
☐ N721VB	84	Conquest 1	425-0210	Van Beruden Insurance Services, Kingsbury, Ca.	N1224N
☐ N722DR	81	King Air F90	LA-147	Del Rio Flying Service Inc. Del Rio, Tx.	XB-ROE
☐ N722KR	84	King Air C90A	LJ-1065	Reilly Enterprises LLC. Baton Rouge, La.	N900LE
☐ N722LJ	66	680V Turbo Commander	1538-5	R & M Leasing, Fremont, Ne.	N100BP
☐ N722M	72	King Air E90	LW-6	Teixeira Farms Inc. Santa Maria, Ca.	(N60DC)
☐ N722MU	78	Mitsubishi MU-2N	722SA	Southern Cross Aviation Inc. Fort Lauderdale, Fl.	F-GFHB
☐ N722PT	85	King Air F90-1	LA-235	Penkhus Motor Co. Colorado Springs, Co.	N901GS
☐ N722SR	92	TBM 700	49	122 Ventures Inc. Wilmington, De.	N567T
☐ N723AC	76	Rockwell 690A	11249	American National Bank, Albert Lea, Mn.	N161JB
☐ N723JP	79	PA-31T Cheyenne II	7920009	Missouri Forge Inc. Doniphan, Mo.	N723JR
☐ N723W	76	King Air A100	B-225	Optimair Inc. New London, NH.	N696RH
☐ N724A	96	DHC 8-202	440	ARAMCO Associated Co. Houston, Tx.	C-GFBW
☐ N724KW	96	King Air C90B	LJ-1426	MidAmerican Holding Co. Kansas City, Mo.	
☐ N725A	96	DHC 8-202	441	ARAMCO Associated Co. Houston, Tx.	C-GFCF
☐ N725FN	73	Mitsubishi MU-2K	248	Bush Field Aircraft Co. Augusta, Ga.	N725DM
☐ N725MC	76	King Air 200	BB-169	N725MC Inc. Lafayette, Ca.	N200KA
☐ N725SV	78	Conquest II	441-0051	Sunview Air Inc. Delano, Ca.	N986SG
☐ N726FN	74	Mitsubishi MU-2K	284	Aviation Systems & Manufacturing, Tuskegee, Al.	N726DM
☐ N727CC	82	Gulfstream 1000	96027	United States Aviation Underwriters, NYC.	N9947S
☐ N727DD	81	King Air 200	BB-861	Southern Aviation & Marine Corp. Belleair Bluffs, Fl.	N270L
☐ N727DP		SA-226AT Merlin 4A	AT-039	Glendal Inc/Glendale Aviation, Sunburg, Mn.	N439BW
☐ N727PC	81	PA-42 Cheyenne III	8001025	Hangar Nine Inc. Wilmington, De.	(D-IBTV)
☐ N727RS	81	King Air B100	BE-112	Modern Technologies Corp & Soin Aviation Inc. Dayton, Oh.	N990SV
☐ N727SM	78	PA-31T Cheyenne II	7820019	Graves Aircraft Inc. El Reno, Ok.	N9113Y
☐ N727TP	81	MU-2 Marquise	1517SA	Ted Price Sr. Ventura, Ca.	(N321TP)
☐ N728DS	78	King Air C90	LJ-773	TumiAir Inc. Woodside, Ca.	N8TZ
☐ N728F	67	Mitsubishi MU-2B	019	Internet Jet Sales Ltd. Wilmington, De.	N3551X
☐ N729CC	79	Rockwell 690B	11539	Structural Steel Services Inc. Meridian, Ms.	N81699
☐ N729MS	76	King Air B100	BE-2	MCS Leasing Inc. Walnut Creek, Ca.	N43KA
☐ N730CE	79	King Air 200	BB-452	Gantt Aviation Inc. Georgetown, Tx.	
☐ N730PT	77	PA-31T Cheyenne II	7720008	Twin Otter International Ltd. Las Vegas, Nv.	XA-HAY
☐ N730SS	77	Rockwell 690B	11415	DLZ Aircraft Inc. Columbus, Oh.	N700SS
☐ N730WB	77	King Air C90	LJ-730	Klintek Inc. Fletcher, NC.	ZS-KAM
☐ N731PC		PA-31T Cheyenne II	7920053	Alpine Engineered Products Inc. Pompano Beach, Fl.	
☐ N731RJ	82	King Air C90	LJ-1023	Weekend Air Charter Services Inc. Oklahoma City, Ok.	PT-OKW
☐ N733NM	93	King Air 350	FL-104	Niagara Mohawk Power Corp. Syracuse, NY.	N8207D
☐ N734A	88	King Air B200	BB-1298	Raytheon Aircraft Receivables Corp. Wichita, Ks.	HK-3440W
☐ N735DB	73	King Air A100	B-172	New Air, Bend, Or.	N16FA
☐ N735K	65	King Air 90	LJ-35	South Seattle Community College, Seattle, Wa.	
☐ N736P	85	King Air 300	FA-53	CSL 340 Corp. Truro, Ma.	
☐ N737E	74	Rockwell 690A	11189	R A Swick & Associates Inc. Toledo, Oh.	(N300CG)
☐ N737WB	76	PA-31T Cheyenne II	7620052	Lule Corp. Wilmington, De.	N65MD
☐ N738MA	77	Mitsubishi MU-2P	353SA	Aircraft & Engines Ltd. Grand Turk, Turks & Caicos Island.	
☐ N738R	71	King Air C90	LJ-517	Pine Tree Aviation LLC. Boulder, Co.	N738RH
☐ N739K	71	King Air C90	LJ-533	Kobill Airways Ltd. Farmingdale, NY.	
☐ N739P	86	King Air 300	FA-98	Cottonaire Corp. Gastonia, NC.	
☐ N740GL	80	King Air 200	BB-650	Air Serv International, Redlands, Ca.	N33TJ
☐ N740P	85	King Air B200	BB-1218	Navajo Nation, Window Rock, Az.	
☐ N740PB	75	Mitsubishi MU-2L	657	Systec 2000 Inc. Pell City, Al.	N740PC
☐ N740PC	85	King Air 300	FA-78	Oglethorpe Power Corp. Tucker, Ga.	N72479
☐ N741FN	75	Mitsubishi MU-2L	658	Agamemnon Operating Inc/CirrusAir, Dallas, Tx.	N740DM
☐ N741K	68	King Air B90	LJ-390	Cessna Air Inc. Wilmington, De.	YV-94P
☐ N741MA	77	Mitsubishi MU-2P	355SA	Bio Vim Inc. Naples, Fl.	N54EC
☐ N742FN	75	Mitsubishi MU-2L	670	Flight International Inc. Jacksonville, Fl. (status ?).	N742DM
☐ N742GR	70	SA-26AT Merlin 2B	T26-172	Neuhoff Aviation Inc. Dallas, Tx.	N120FS
☐ N743JA	73	King Air E90	LW-73	Condor Aviation Inc. Rio Raucho, NM.	
☐ N744CH	76	Rockwell 690B	11470	CH Aviation Inc. Wilmington, De.	N86MP
☐ N744JD	73	Rockwell 690A	11135	Accessories Inc. Allison Park, Pa.	N160G
☐ N744WP	80	PA-31T Cheyenne 1	8004040	Gregory Aviation Co. Des Moines, Ia.	N420DW

Regn	Yr	Type	c/n	Owner/Operator	Prev Regn
□ N745R	83	King Air B200	BB-1113	TRW Inc. Cleveland, Oh.	
□ N747HN	80	King Air 200	BB-658	HAN Airways International Ltd. Bedford, NH.	N127AP
□ N747JB	84	Conquest II	441-0330	N747JB inc. Wilmington, De.	N55CH
□ N747P		SAAB 340A	029	AMP Inc. Harrisburg, Pa.	N77A
□ N747RE	77	PA-31T Cheyenne II	7720032	Weather Modification Inc. Fargo, ND.	(N748RL)
□ N747SY	82	MU-2 Marquise	1556SA	Smith Young Partnership, Liverpool, UK.	N277JR
□ N749MA	77	Mitsubishi MU-2P	364SA	Charles Anderson, Greensboro, NC.	
□ N749RH	91	King Air B300C	FM-2	Bayport Broadcasting Group. Jackson, Wy.	N350TW
□ N750CA	79	MU-2 Solitaire	407SA	Mid-Missouri Cellular Inc. Pilot Grove, Mo.	N979MA
□ N750FC	70	King Air 100	B-58	Carlina Inc. Durham, NC.	C-FJLJ
□ N750G	69	Gulfstream 1	200	ITT Industries Inc. Allentown, Pa.	
□ N750Q	67	Mitsubishi MU-2B	017	McClure Coal & Oil Co. Marion, In.	N269AA
□ N753D	80	King Air B100	BE-98	L E Bell Construction Co. Heflin, Al.	
□ N755DM	94	TBM 700	101	News Flight Inc. NYC.	
□ N755MA	72	Mitsubishi MU-2J	553	International Business Aircraft Inc. Tulsa, Ok.	N7034K
□ N755Q	68	Mitsubishi MU-2F	131	Tobul Accumulator Inc. Wexford, Pa.	C6-BFA
□ N756Q	68	Mitsubishi MU-2F	132	Heetco Inc/Interair P/L. Melbourne, VIC. Australia.	
□ N757H	81	Conquest 1	425-0006	Silverwings Charter Inc. Silver City, NM.	C-GMSV
□ N759A	96	DHC 8-202	435	ARAMCO Associated Co. Houston, Tx.	C-G...
□ N760NE	80	King Air 200	BB-594	Ardall Services Inc. Conway, SC.	N760NB
□ N760NP	75	King Air 200	BB-46	ZZ Leasing Inc. Memphis, Tn.	N760NB
□ N761K	68	King Air B90	LJ-426	Dr John Manchin II & Anker Energy Corp. Morgantown, WV.	
□ N762JC	80	MU-2 Marquise	762SA	Smith & Sons Aircraft LC. Paso Robles, Ca.	OY-SUH
□ N762NB	81	King Air B200	BB-893	Sunrise Air Inc. Miami, Fl.	
□ N764K	74	King Air A100	B-178	United Air Services Co. Cobham, Va.	N3078W
□ N764NB	83	King Air B200	BB-1115	Books-A-Million & others, Birmingham, Al.	N6679E
□ N766MA	78	Mitsubishi MU-2P	373SA	Frank Sanderson, Hampton, Va.	
□ N766RB	82	King Air F90	LA-176	Robert Hansen Trucking Inc. Rockford, Il.	N6344H
□ N767CW	94	TBM 700	96	High Sierra Inc. Los Angeles, Ca.	
□ N767DM	81	Cheyenne II-XL	8166042	D & M Aviation Inc & QuickFlite Corp. Chesterfield, Mo.	N500XL
□ N767MC	73	King Air C90	LJ-595	Gulf States Systems Inc. Gulfport, Ms.	N19UW
□ N767MD	75	Mitsubishi MU-2M	328	International Business Aircraft Inc. Tulsa, Ok.	N500HA
□ N767WF	78	King Air 200	BB-314	Wheat Service & Equipment Co. Richmond, Va.	N101BP
□ N768MB	80	PA-31T Cheyenne 1	8004041	Hitch Enterprises Inc. Guymon, Ok.	N769MB
□ N769	77	King Air 200	BB-291	Loftin Constructors Inc. Brandon, Ms.	N286TC
□ N769D	80	King Air F90	LA-52	Charles Putman, Mills, Wy.	
□ N769MB	79	King Air 200	BB-571	American Health Centers Inc. Parsons, Tn.	N702AS
□ N770D	80	King Air B100	BE-90	Roberts Publishing Co. Andrews, Tx.	N67460
□ N770GX	91	King Air 300	FA-218	Flexsteel Industries Inc. Dubuque, Ia.	(D-IHIT)
□ N770M	96	King Air B200	BB-1546	Marshall & Ilsley Bank, Milwaukee, Wi.	
□ N770SD	80	King Air F90	LA-72	Holiday Management Co. Evansville, In.	N90SK
□ N771BA	78	Rockwell 690B	11429	Indiana Aircraft Sales Inc. Indianapolis, In.	(D-ILAT)
□ N771CW	78	King Air B100	BE-52	Air Charter Service of W N Y Inc. Niagara Falls, NY.	N771S
□ N771FF	82	Gulfstream 900	15014	S E C Inc. Athens, Ga.	N5869N
□ N771HC	76	King Air 200	BB-147	Southern Sky Aviation Inc. Corpus Christi, Tx.	N777FL
□ N771HM	82	King Air B200	BB-1078	HMHC Inc. Plymouth Meeting, Pa.	
□ N771SC	96	King Air C90B	LJ-1463	Coller Real Estate, Bloomington, In.	N1099K
□ N771SG	95	King Air C90B	LJ-1410	Raytheon Aircraft Co. Wichita, Ks.	N771SC
□ N772DA	80	MU-2 Marquise	772SA	Epps Air Service Inc. Atlanta, Ga.	I-MPLT
□ N772MA	78	Mitsubishi MU-2P	382SA	Jetprop Inc. Waterloo, Ia.	
□ N773	83	King Air F90-1	LA-214	Jack Wall Aircraft Sales Inc. Memphis, Tn.	N77P
□ N773CA	77	Rockwell 690B	11418	CYRK Inc. Gloucester, Ma.	N555MT
□ N773S	67	King Air A90	LJ-283	N M Aire Inc. Searcy, Ar.	
□ N774A	97	King Air 350	FL-184	Raytheon Aircraft Co. Wichita, Ks.	
□ N774DK	96	Pilatus PC-XII	167	Franklin M Orr Jr Trust, Stanford, Ca.	HB-FSC
□ N774KV	81	Cheyenne II-XL	8166057	State of Nebraska, Lincoln, Ne.	N9168Y
□ N774MA	78	Mitsubishi MU-2P	384SA	R M Equipment Inc. Miami, Fl.	(N127RM)
□ N775CA	83	PA-42 Cheyenne IIIA	5501004	Northern Aircraft Sales, Bismarck, ND.	D-IEEF
□ N775DM	78	King Air C90	LJ-764	Dynamic International Inc. Hoffman Estates, Il.	N201KA
□ N776CC	72	Mitsubishi MU-2J	560	Artis James Jr & P Holcomb Hector/Sparc Co. Diamondhead, Ms.	N195MA
□ N776DC	77	King Air E90	LW-235	Aerovida SA. Buenos Aires, Argentina.	C-FBCS
□ N776L	70	King Air 100	B-54	MacAviation LLC. Jonesboro, Ar.	N1776L
□ N777AG	90	King Air B200	BB-1366	General Council of Assemblies of God, Springfield, Mo.	N11EQ
□ N777AQ	80	King Air 200	BB-583	Raceco Inc. Baytown, Tx.	N777AG
□ N777AS	83	King Air F90-1	LA-202	Aerospace Safety Technologies Inc. Stratford, Ct.	D-IIBB
□ N777AT	66	King Air A90	LJ-166	Gray Aviation LLC. Wilmington, De.	N28J
□ N777EB	79	King Air C90	LJ-863	Baird Kurtz & Dobson, Springfield, Mo.	N200SC
□ N777EL	79	Rockwell 690B	11538	Westak of Oregon Inc. Forest Grove, Or.	N81701
□ N777G	81	PA-31T Cheyenne 1	8104065	Hawkeye Aviation Inc. Alden, Ia.	N2580Y
□ N777GF	72	King Air C90	LJ-564	Stevens Aviation Inc. Greenville, SC.	N246DA
□ N777GS	79	King Air A100	B-241	Corporate Jets Inc. West Mifflin, Pa.	N777GF
□ N777HE	76	Rockwell 690A	11287	Sedgwick County, Wichita, Ks.	N777HC
□ N777HF	83	King Air C90-1	LJ-1048	C & E Holdings Inc. Madison, Md.	N46CR
□ N777KA	78	King Air E90	LW-285	Limine Air Inc. Little Rock, Ar.	
□ N777KU	68	King Air B90	LJ-377	Jimmy & Ivy Ballard, Wichita, Ks.	N8473N
□ N777LP	78	Mitsubishi MU-2N	719SA	Lincoln Leasing Co. Lincoln, Ne.	N911JE

Regn	Yr Type	c/n	Owner/Operator	Prev Regn

Regn	Yr	Type	c/n	Owner/Operator	Prev Regn
□ N777NP	65	King Air 90	LJ-67	Anton Ptach, Poughkeepsie, NY.	(N21AM)
□ N777NW	72	King Air C90	LJ-561	Al Ted Co. Parkesburg, Pa.	N851MK
□ N777PR	78	King Air 200	BB-404	Mathews Trucking Co. Ruston, La.	N921DT
□ N777SS	80	King Air 200	BB-661	Seven S Air Inc. Oklahoma City, Ok.	N111JW
□ N777SW	77	PA-31T Cheyenne II	7720002	Legalair Inc. Wilmington, De.	N505GP
□ N777TE	74	Rockwell 690A	11162	Tex Edwards Co. Pensacola, Fl.	N124HQ
□ N777VN	81	King Air C90	LJ-979	Fox Valley Aircraft Corp. Elgin, Il.	N777NQ
□ N777WM	79	MU-2 Solitaire	397SA	Straubel Investments, Davenport, Ia.	N666SP
□ N777YP	84	PA-42 Cheyenne III	5501008	Yellow Pages/L M Berry & Co. Dayton, Oh.	N41117
□ N779DD	92	King Air C90A	LJ-1297	Basha's - Chapman Group, Chandler, Az.	N8239Q
□ N779M	80	SA-226T Merlin 3B	T-363	Star Aeroservices Inc. Santa Barbara, Ca.	(N155AM)
□ N780BF	85	King Air 300	FA-70	PHH-CFC Leasing Inc. Hunt Valley, Md.	N72451
□ N780BP	73	Rockwell 690	11052	Long Star Air Service Inc. Miami, Fl.	N1230D
□ N780RC	81	King Air 200	BB-780	D W Machine Products Inc. Highland, Il.	F-GIHK
□ N781SW	78	PA-31T Cheyenne II	7820012	LADAS Properties LLC. Hobbs, NM.	
□ N782MA	78	MU-2 Solitaire	390SA	Kern Helicopters Inc. Bakersfield, Ca.	
□ N783MA	78	Mitsubishi MU-2P	391SA	Mitchell Slayman, Bakersfield, Ca.	
□ N783MC	73	King Air C-12C	BC-2	MCA Construction Co. Albuquerque, NM.	N7064B
□ N783ST	80	Conquest II	441-0170	Silver Eagle Air Enterprises Inc. Roaring Springs, Pa.	N720JM
□ N784K	68	King Air B90	LJ-427	Engineering & Manufacturing Services Inc.	
□ N786CB	76	King Air B100	BE-4	Aeronaut Ltd. Palm Beach Gardens, Fl.	N9104S
□ N787K	66	King Air 90	LJ-102	Ameron-Enmar Finishes Division, Little Rock, Ar.	
□ N787SW	78	PA-31T Cheyenne II	7820074	Frost Flying Inc. Marianna, Ar.	(N6062A)
□ N788SC	84	King Air 300	FA-4	Southern Comm Aviation Inc. Fort Lauderdale, Fl.	N111SS
□ N788SW	79	King Air E90	LW-327	Sugar Whiskey LLC. Nashville, Tn.	G-JGAL
□ N789CH	79	PA-31T Cheyenne II	7920079	American Medflight Inc. Reno, Nv.	N555RC
□ N789DS	79	King Air 200	BB-478	Petroleum Helicopters Inc. Lafayette, La.	N789BT
□ N789G	62	Gulfstream 1	89	Cummins Engine Co. Columbus, In.	
□ N790A	82	King Air C90-1	LJ-1016	Robert Ash, Los Angeles, Ca.	N6199P
□ N790CA	73	Mitsubishi MU-2K	260	Internet Jet Sales Ltd. White Mills, Pa.	N457SC
□ N790SD	79	PA-31T Cheyenne II	7920004	Silverwings Air Ambulance Ltd.. Silver City, NM.	N790SW
□ N792JM	88	King Air 300	FA-140	Jamestown Metal Marine Sales Inc. Boca Raton, Fl.	N117DR
□ N793EM	94	King Air 350	FL-106	Empire Southwest Co. Mesa, Az.	N8211K
□ N794A	84	PA-42 Cheyenne IIIA	5501015	First Security Bank NA. Salt Lake City, Ut.	D-ILSW
□ N794B	81	Conquest 1	425-0017	J P Air Charter, Rancho Cacamonga, Ca.	N58CH
□ N794CA	81	PA-31T Cheyenne II	8120018	Robert Calhoun/Ram Oil Well Service Inc. Hobbs, NM.	N406TM
□ N794MA	80	MU-2 Marquise	794SA	West Texas Executive Leasing, San Angelo, Tx.	N54CE
□ N794WB	79	King Air B100	BE-69	Central Virginia Aircraft Sales, Lynchburg, Va.	N710KC
□ N795CA	79	King Air 200	BB-559	Plane South Leasing Inc. Dunwoody, Ga.	N559BM
□ N795GB	82	King Air B200	BB-1069	Walkabout Inc. Tampa, Fl.	N87DR
□ N795K	66	King Air 90	LJ-95	College of Aeronautics, La Guardia, NY.	
□ N795PA	78	King Air 200	BB-328	Van Aviation Inc. Jackson, Ms.	(D-IDOL)
□ N797CF	78	King Air C90	LJ-797	D S Aviation Inc. Irving, Tx.	N61GA
□ N798K	66	King Air A90	LJ-178A	Fayard Enterprises Inc. Louisberg, NC.	
□ N798TW	66	King Air A90	LJ-167	Truck-A-Way Inc. Vacaville, Ca.	C-FIYA
□ N799DD	72	King Air A100	B-102	Chris Finkbeiner, Little Rock, Ar.	N777SD
□ N799MA	74	Mitsubishi MU-2J	612	International Business Aircraft Inc. Tulsa, Ok.	VH-UUJ
□ N800BF	74	King Air E90	LW-87	Wardaire Inc. Atlanta, Ga.	N107TB
□ N800BM	81	Gulfstream 840	11691	Sunridge Nurseries Inc.. Bakersfield, Ca.	C-GIIT
□ N800BW	81	King Air 200	BB-818	Borg Warner Automotive Inc. Ithaca, NY.	N3828E
□ N800BY	72	Mitsubishi MU-2F	221	Bluevue Corporation Leasing, Wilmington, De.	N800BR
□ N800GS	91	TBM 700	5	Sutliff Chevrolet Co. Harrisburg, Pa.	N107BP
□ N800JF	97	King Air C90B	LJ-1505	Raytheon Aircraft Co. Wichita, Ks.	
□ N800JR	82	King Air B200	BB-982	Home Interiors & Gifts Inc. Dallas, Tx.	N703MD
□ N800KT	99	King Air B200	BB-1346	TNS Mills Inc. Greenville, SC.	
□ N800LD	78	SA-226T Merlin 3A	T-291	Interlata Aviation Inc. San Antonio, Tx.	N19SD
□ N800LS	79	King Air 200	BB-457	Lone Star Steel Co. Dallas, Tx.	(N141BB)
□ N800MM	81	SA-226T Merlin 3B	T-375	Great Western Hotels Corp. La Habra, Ca.	(N98FT)
□ N800PP	81	King Air 200	BB-776	American Racing Series Inc. Jackson, Mi.	N800KC
□ N800PW	72	King Air C90	LJ-548	Crystal Air Charters Inc. Wilmington, De.	N937K
□ N800RP	74	King Air C90	LJ-628	Perkndahl Management LLC. Minneapolis, Mn.	F-GNUV
□ N800YM	83	Conquest II	441-0266	Bill Elliott Racing Enterprises, Blairsville, Ga.	N8881N
□ N801AR	90	King Air 300	FA-191	Albritton Communication Co. Washington, DC.	N1541Q
□ N801BT	68	SA-26T Merlin 2A	T26-18	Hoyt Axton, Tahoe City, Ca.	N12NA
□ N801CA	81	PA-31T Cheyenne 1	8104018	La Stella, Pueblo, Co.	
□ N801EB	73	Rockwell 690A	11111	Sterling Marlin Enterprises Inc. New London, NC.	XA-...
□ N801HL	80	PA-31T Cheyenne II	8020016	Harrigan Lumber Co. Monroeville, Al.	N2355W
□ N801JW	82	PA-42 Cheyenne III	8001072	B E Air Inc. Huntingdon Valley, Pa.	N63WA
□ N801ST	80	PA-31T Cheyenne II	8020014	Corporate Skies Inc. Elizabethton, Tn.	N801GC
□ N802DG	79	King Air C90	LJ-807	D-G Airways Inc. Huntington, WV.	N877WL
□ N802DJ	69	SA-26AT Merlin 2B	T26-117	Spartan Fleet Management Inc. Tryon, NC.	C-GYLP
□ N802HS	95	Pilatus PC-XII	118	Keystone Builders Resource Group Inc. Richmond, Va.	HB-FQL
□ N802ME	78	SA-226T Merlin 3B	T-294	SkyKnight Air Services inc. Fort Lauderdale, Fl.	N800TA
□ N802MW	82	PA-42 Cheyenne III	8001081	MIC-PLANE Inc. Baltimore, Md.	N881AM
□ N802RD	80	King Air B100	BE-96	Matrix Aero Corp. Fenton, Mo.	N55TJ
□ N802SM	81	MU-2 Marquise	1515SA	Aberg Ltd. Rockford, Il.	N910DA

Regn	Yr	Type	c/n	Owner/Operator	Prev Regn
N803CA	81	PA-31T Cheyenne 1	8104029	Aircraft Guaranty Corp. Salzburg, Austria.	(N90WA)
N803DJ	68	SA-26AT Merlin 2B	T26-102	Tri State Aviation Services Inc. Berryville, Ar.	C-FBWU
N804	80	King Air 200	BB-724	Sause Brothers Ocean Towing Co. Coos Bay, Or.	N604
N804C	84	Cheyenne II-XL	1166006	D & M Performance Aero, Dumas, Tx. (was s/n 1122001).	N80BC
N806JW	96	King Air 350	FL-140	Jeld-Wen Inc. Klamath Falls, Or. (5000th King Air).	N1070D
N806W	61	Gulfstream 1	67	U S Immigration & Naturalization Service, Pineline, La.	N5241Z
N807M	82	SA-227AT Merlin 4C	AT-454	Ameriflight Inc. Burbank, Ca.	N3013T
N808DS	82	King Air F90	LA-172	D S Air Inc. Chesapeake, Va.	N101ET
N808GU	66	680V Turbo Commander	1579-32	Kenneth Erickson, San Diego, Ca.	N6536V
N808NC	85	Gulfstream 1000	96085	Equipment Supply Co & others, Burlington, NJ.	N205AB
N808NT	81	Gulfstream 980	95081	Eagle Air Inc. Memphis, Tn.	N888NT
N808W	85	King Air C90A	LJ-1114	Danny Souders, Boulder, Co.	N7232U
N809E	84	PA-42 Cheyenne IIIA	5501006	Cheyenne Aircraft LLC. Searcy, Ar.	N942PC
N809SW	80	PA-31T Cheyenne II	8020080	Demolition Services Ltd. Leeds, UK.	(F-ODMM)
N810CM	86	King Air 300	FA-106	Noranda Aluminium Corp. New Madrid, Mo.	C-GPKP
N810EC	81	Gulfstream 980	95071		N9823S
N811DA	81	PA-42 Cheyenne III	8001029	Duncan Aviation Inc. Lincoln, Ne.	N40PT
N811FA	80	King Air 200	BB-678	Marc Fruchter Aviation Inc. Reading, Pa.	N811VG
N811GA	79	King Air C90	LJ-810	Marshall Air Service Inc. Chesterton, In.	ZS-MZG
N811LC	89	King Air 300	FA-194	Charles Ruppman Ltd. Peoria, Il.	N1568T
N811MM	81	King Air E90	LW-345	First National Bank Group Inc. Edinburg, Tx.	N911MM
N811R	86	King Air C90A	LJ-1131	Western Oil Fields Supply Co. Bakersfield, Ca.	N72508
N812BJ	84	PA-42 Cheyenne 400LS	5527020	Skyview Unlimited LC. Hillsborough, Or.	N402TW
N812CM	80	PA-31T Cheyenne 1	8004050	TRP Services Corp. Austin, Tx.	
N812P	65	King Air 90	LJ-2	Cascade Air Services Inc. Issaquah, Wa.	N812Q
N813BL	79	King Air B100	BE-55	Transport Industries Corp. Joplin, Mo.	N546BZ
N813Q		SA-227TT Merlin 3C	TT-507	TENN-GA Truck Equipment & Sales, Chattanooga, Tn.	N212Q
N813ZM	83	Conquest 1	425-0159	Greenwood Chevrolet Inc. Youngstown, Oh.	N425BB
N814G	66	King Air 90	LJ-104	814G LLC. Pleasant View, Tn.	N37PP
N815BC	83	Gulfstream 840	11731	Flyaway Inc. Odessa, Tx.	N840DA
N815D	89	King Air 300	FA-188	Valley Products Corp.	N423MK
N815K		King Air A90	LJ-123	Thomas Longley, Sisters, Or.	
N816CM	81	PA-31T Cheyenne 1	8104009	Municipal Energy Agency, Lincoln, Ne.	
N816RL	76	King Air E90	LW-187	Peter Earp, Staverton, UK.	N66BP
N817F	86	King Air C90A	LJ-1129	U S Filter Distribution Group, Waco, Tx.	N717DW
N818L	66	680V Turbo Commander	1558-16	Turcotte Enterprises Inc. Houston, Tx.	N818EC
N818MS	65	King Air 90	LJ-53	Wildflower Inc. Pueblo, Co.	N123PP
N818PL	82	Conquest 1	425-0109	Kansas Power & Light Co/Western Resources Inc. Topeka, Ks.	N66LL
N819MH	77	King Air C90	LJ-735	Windward Charter Ltd. Indianapolis, In.	N190TT
N819SW	81	PA-31T Cheyenne 1	8104059	Blue Sky Aviation LLC. Wichita, Ks.	
N821CT	79	King Air C90	LJ-821	Wheel Air Inc & Jetways LLC. Blaine, Mn.	C-GKCA
N821MA	75	Mitsubishi MU-2N	661SA	Win Team Insurance Services Inc. Irvine, Ca.	
N821U	65	King Air 90	LJ-19	Vibrators Inc. Wilmington, De.	
N822BA	82	King Air F90	LA-191	Coit Services Inc. Burlingame, Ca.	(F-GRLN)
N822CM	81	PA-31T Cheyenne 1	8104037	Builders Supply Corp. Janesville, Wi.	(N822WC)
N823MA	75	Mitsubishi MU-2L	663	Bankair Inc/Dickerson Associates, West Columbia, SC.	YV-409P
N823SB	66	King Air A90	LM-16	San Bernardino County, Rialto, Ca.	66-18015
N824TT	80	King Air 200	BB-824	Elliott Aviation Aircraft Sales Inc. Des Moines, Ia.	F-GJCF
N824VA	81	PA-31T Cheyenne 1	8104039	Joseph Ramirez, Guymon, Ok.	N824CM
N825B	82	Conquest 1	425-0123	Certified Leasing Co. Bakersfield, Ca.	N6882L
N825K	66	King Air 90	LJ-91	Spirit Fighters Inc. Wilmington, De.	N2085W
N825KA	81	King Air 200	BB-825	B & A Aviation Corp. Shawnee Mission, Ks.	XC-FUS
N825ST	89	King Air B200	BB-1320	June & John Rogers, Modesto, Ca.	N300LX
N825SW	82	Cheyenne II-XL	8166058	Chaparral Boats Inc. Nashville, Ga.	
N826CM	82	PA-31T Cheyenne 1	8104048	Mike Vaughn Custom Sports Inc. Lake Orion, Mi.	
N827CC	85	King Air 300	FA-43	Calcot Ltd. Bakersfield, Ca.	N85TH
N827DP	80	King Air F90	LA-77	Drug Plastics & Glass Co. Boyertown, Pa.	N3704S
N827T	66	King Air 90	LJ-83	Quest Aviation Inc. Fort White, Fl.	(YV-980P)
N828CA	85	King Air 300LW	FA-79	ConAgra Inc. Omaha, Ne.	N427KW
N828CM	84	SA-226T Merlin 3A	T-258	Fly By Night Leasing LLC. Nevada, Mo.	N15GS
N828JB	81	King Air 200	BB-795	Walco International Inc. Porterville, Ca.	N502EB
N829JC	80	Conquest II	441-0156	C C Industries Inc. Wheeling, Il.	N7RC
N830CM	81	PA-42 Cheyenne III	8001048	W & M Air Inc. Somerset, Pa.	N830CB
N830EM	70	King Air 100	B-29	White Industries Inc. Bates City, Mo.	N8300E
N832AD	83	Conquest II	441-0311	Donald & Alice Fehrenbach, Chandler, Az.	XA-OAC
N833BK	85	King Air B200	BB-1239	Jamm Aviation Corp. Columbus, Oh.	N7250V
N835CC	83	Gulfstream 840	11730	State of New Mexico, Santa Fe, NM.	N28GA
N835MA	85	Conquest 1	441-0343	Calcutta Aircraft Leasing Inc. Bloomington, In.	N12114
N836MA	78	SA-226AT Merlin 4A	AT-068	Fly Cargo SA. Buenos Aires, Argentina.	F-GFPR
N838RA	92	TBM 700	71	Chester Prior, Echo, Or.	N888RA
N839AB	81	PA-31T Cheyenne 1	8104011	Kent Kelly, Montgomery, Al.	(N839CH)
N840AA	80	Gulfstream 840	11610	Duke Realty Services LP. Indianapolis, In.	YV-609CP
N840CF	80	Gulfstream 840	11624	Michael Francis, Crowley, La.	N840XL
N840DC	80	Gulfstream 840	11661	Dempsey Construction Corp. Mammoth Lakes, Ca.	N840TC
N840JC	80	Gulfstream 840	11643	Semitool Inc. Kalispell, Mt.	(N5895K)
N840JW	81	Gulfstream 840	11658	H G Westerman, Richardson, Tx.	N840FK

Regn	Yr Type	c/n	Owner/Operator	Prev Regn

Regn	Yr	Type	c/n	Owner/Operator	Prev Regn
☐ N840LC	80	Gulfstream 840	11647	Indiana Aircraft Sales Inc. Indianapolis, In.	HK-3460P
☐ N840LE	81	Gulfstream 840	11709	O Henriksen/Crystal Ltd. Guernsey, Channel Islands.	N690BA
☐ N840MD	81	Gulfstream 840	11693	Wal-Mart Stores Inc. Rogers, Ar.	N79SA
☐ N840NB	81	Gulfstream 840	11663	N B Harty General Contractors, Dexter, Mo.	(N690NB)
☐ N840SM	81	Gulfstream 840	11700	B & S Industries Inc. Eugene, Or.	(N83SA)
☐ N840V	82	Gulfstream 840	11727	K V Oil & Gas Inc. Versailles, Ky.	N5965K
☐ N840VB	81	Gulfstream 840	11679	Bering Marine Corp. Seattle, Wa.	N5931K
☐ N840VM	80	Gulfstream 840	11607	Commercial Aviation Services Inc. Miami, Fl.	N840EA
☐ N841K	81	King Air 200	BB-841	Tharaldson Equipment Inc. Helena, Mt.	F-GHLG
☐ N842DS	79	King Air A100	B-244	Windham Aviation Inc. Williamantic, Ct.	N942DS
☐ N844C	79	King Air C90	LJ-866	U S Silica Co. Berkeley Springs, WV.	N6664P
☐ N844MP	84	King Air B200	BB-1168	Mass Bay Kustom Leasing Inc. Quincy, Ma.	N179MC
☐ N844S	92	TBM 700	46	Thionville Laboratories Inc. New Orleans, La.	
☐ N844TS	90	King Air 350	FL-33	Fowler Foods Inc. Jonesboro, Ar.	(N180CA)
☐ N846BB	69	SA-26AT Merlin 2B	T26-149E	Bastante II Inc. San Antonio, Tx.	N200BC
☐ N846BE	84	King Air 300	FA-16	MD Airways Inc. Derby, Ks.	
☐ N846YT	81	Conquest II	441-0218	True Oil Co. Casper, Wy.	N41TA
☐ N847YT	82	Conquest 1	425-0114	True Oil Co. Casper, Wy.	N68803
☐ N848CE	76	Rockwell 690A	11303	CTS Aviation Inc. Climax, Mi.	N81409
☐ N848NA	81	King Air 200	BB-848	Nebrig & Associates Inc. Dallas, Tx.	RP-C267
☐ N848PF	85	King Air B200	BB-1225	PFP One Inc. Panama City, Fl.	N40FQ
☐ N849KM	80	PA-31T Cheyenne 1	8004055	Wyoming Aviation, Huntsville, Al.	N46HM
☐ N850C	80	King Air 200	BB-710	Cutter Aviation Inc. Phoenix, Az.	N3669Z
☐ N850D	92	King Air 350	FL-70	Moeller Transport Leasing Inc. Spencerville, Oh.	N350KC
☐ N850DB	70	King Air 100	B-53	Westchester Air Inc. White Plains, NY.	N879K
☐ N851KA	79	King Air C90	LJ-851	Beech Transportation Inc. Eden Prairie, Mn.	N4B
☐ N851MK	80	King Air 200	BB-674	Bates & Associates Inc. Bainbridge, Ga.	N351MK
☐ N855MA	76	Rockwell 690A	11299	Hageland Aviation Services, Anchorage, Ak.	N856MA
☐ N856JC	80	MU-2 Solitaire	430SA	Bankair Inc. West Columbia, SC.	N170MA
☐ N857C	83	Conquest 1	425-0156	Conquest Charters Inc. San Francisco, Ca.	N425BX
☐ N857MA	77	Mitsubishi MU-2N	697SA	Daniel Knopper, Inkster, Mi.	
☐ N859CC	78	King Air 200	BB-389	E & B Leasing Corp. Brookfield, Wi.	(OY-FCT)
☐ N859Q	68	Mitsubishi MU-2D	113	Leon County School Board, Tallahassee, Fl.	
☐ N860CC	88	King Air 300	FA-147	Citation Corp. Birmingham, Al.	N3085Z
☐ N860H	82	King Air B200	BB-1067	State of Ohio, Columbus, Oh.	N47MM
☐ N860MA	77	MU-2 Marquise	700SA	NJTS Corp. Murfreesboro, Tn.	
☐ N860MH	77	King Air E90	LW-210	Oakwood Homes Corp. Greensboro, NC.	N717TM
☐ N860SM	80	MU-2 Solitaire	415SA	Select Homes Inc. Asheboro, NC.	F-ODRZ
☐ N861CC	94	King Air 350	FL-94	Citation Corp. Birmingham, Al.	N8194Q
☐ N861E	75	Mitsubishi MU-2L	674	Agnes Aviation Inc. Corpus Christi, Tx.	N834MA
☐ N862DD	77	King Air 200	BB-298	Drake & Drake Inc. Gravois Mills, Mo.	N5110
☐ N866A	67	King Air A90	LJ-201	Fayard Enterprises Inc. Louisburg, NC.	(D-ILNY)
☐ N866D	75	Mitsubishi MU-2L	656	Internet Jet Sales Ltd. Wilmington, De.	N666D
☐ N868MA	78	Mitsubishi MU-2N	708SA	Jaax Flying Service, Colexico, Ca.	N15UD
☐ N869	76	King Air 200	BB-174	Jaymar Ruby Inc. Michigan City, In.	N69DD
☐ N869AM	80	King Air 200	BB-625	TAMS Air Inc. East Brunswick, NJ.	N8SZ
☐ N869D	71	Mitsubishi MU-2G	540	Turbine Aircraft Parts Inc. San Angelo, Tx.	(PT-...)
☐ N869MA	76	King Air/Catpass 250	BB-170	Aero Condor SA. Lima, Peru.	C-FIWH
☐ N869P	77	Mitsubishi MU-2L	692	Premier Jets Inc. Portland, Or.	N623DC
☐ N870CA	83	King Air B200C	BL-65	Capital Aerolease, Bethesda, Md.	N9701Y
☐ N871KS	91	King Air C90A	LJ-1277	Kansas State University, Salina, Ks.	N8012U
☐ N872CA	83	King Air B200	BB-1114	Capital Aerolease, Bethesda, Md.	N9768S
☐ N872S	69	SA-26AT Merlin 2B	T26-123	Skyway Enterprises Inc. Kissimmee, Fl. (status ?)	N872D
☐ N873K	68	King Air B90	LJ-344	Jose Gonzalez, Dorado, PR.	
☐ N874RJ	93	TBM 700	87	Mediajet Inc. Beverly, Ma.	
☐ N877AQ	65	King Air 90	LJ-41	Dodson International Parts Inc. Ottawa, Ks.	N877AG
☐ N877JE	85	King Air B200	BB-1203	Jasper Engines & Transmissions Inc. Huntingburg, In.	TC-CHS
☐ N877RC	82	King Air B200	BB-978	Kelleher Construction Inc. Burnsville, Mn.	N877RF
☐ N877W	92	King Air 350	FL-68	Georgia Crown Distributing Co. Columbus, Ga.	N8131E
☐ N878K	70	King Air B90	LJ-496	Wehco Equipment Corp. Little Rock, Ar.	F-GFLQ
☐ N878MS	78	Rockwell 690B	11435	Jack Hilton MD. Herrin, Il.	N338UP
☐ N879C	86	King Air 200	BB-1249	McElroy Metal Mill Inc. Bossier City, La.	N2512R
☐ N879PC	81	King Air F90	LA-120	Tewel Corp. Hubbard, Oh.	N80CK
☐ N879SW	79	PA-31T Cheyenne 1	7904056	Cartex Production Inc. Paso Robles, Ca.	
☐ N880AC	82	MU-2 Marquise	1559SA	Regal Air Inc. San Juan, PR.	HB-LQN
☐ N880CA	81	PA-31T Cheyenne II	8120001	303 Air Inc.	
☐ N880H	73	King Air C90	LJ-596	State of Ohio, Columbus, Oh.	N2896W
☐ N880SW	80	PA-31T Cheyenne II	8020086	Dixon Brothers Inc. Newcastle, Wy.	
☐ N881CD	81	Conquest II	441-0221	CDT Corp. New Milford, Ct.	N881GB
☐ N881CS	81	King Air B200	BB-881	Cook Sales Inc. Anna, Il.	VH-USD
☐ N881SW	81	PA-31T Cheyenne 1	8104033	Alpine Air Leasing Inc. Fort Lauderdale, Fl.	
☐ N882AC	77	Rockwell 690B	11375	Restrepo Aircraft Corp. Pompano Beach, Fl.	N412AC
☐ N883CA	92	TBM 700	83	JK Moving & Storage Inc. Sterling, Va.	N783DJ
☐ N883SW	93	King Air B200	BB-1458	Palmer Air LP. Sarasota, Fl.	N8258V
☐ N884CA	78	PA-31T Cheyenne II	7820011	Tool Crib Aero Inc. Grand Forks, ND.	(N8RS)
☐ N884D	81	Gulfstream 840	11696	Aerosales & Services Inc. Austin, Tx.	N88PD
Regn	*Yr*	*Type*	*c/n*	*Owner/Operator*	*Prev Regn*

Regn	Yr	Type	c/n	Owner/Operator	Prev Regn
☐ N884PG	76	King Air 200	BB-91	Midway Air Group Inc. Springfield, Il.	N9CJ
☐ N885HT	79	King Air 200	BB-446	Mid-South Engineering Co. Hot Springs, Ar.	N773KA
☐ N885RA	72	Rockwell 690	11012	Sooner Air Charter Inc. Norman, Ok.	N921HB
☐ N886AT	97	King Air C90B	LJ-1485	Albert & Patricia Nichols, Aspen, Co.	
☐ N887KU	68	King Air B90	LJ-357	Jimmy & Iva Ballard, Wichita, Ks.	N887K
☐ N887T	85	King Air B200	BB-1233	Terra Industries Inc, Sioux City, Ia.	N4445T
☐ N888AH	79	PA-31T Cheyenne 1	7904038	Corporate Helicopters Inc. Gulfport, Ms.	(N528KC)
☐ N888AS	87	King Air 300LW	FA-136	Harris Farms Inc. Coalinga, Ca.	N600CB
☐ N888CS	88	King Air B200	BB-1311	FB Air Holding Inc. Naples, Fl.	N1553P
☐ N888DC	79	King Air 200	BB-454	Drummond Ltd. Birmingham, Al.	
☐ N888DS	69	Mitsubishi MU-2F	159	Dixie Continental Charter Group, Dover, De.	N30MA
☐ N888EM	77	King Air 200	BB-240	Express Marine Inc. Camden, NJ.	N605EE
☐ N888ET	77	King Air 200	BB-258	JBQ Inc. Las Vegas, Nv.	N40PS
☐ N888FW	84	PA-42 Cheyenne IIIA	5501011	Din Aero Inc. Troy, Mi.	N288FA
☐ N888GN	75	King Air C90	LJ-667	Nebrig & Associates Inc. Dallas, Tx.	G-BMZD
☐ N888HG	81	King Air B200	BB-891	H S G Aviation Inc. Dover, De.	(N18VG)
☐ N888JM	77	PA-31T Cheyenne II	7720066	Brazos Roofing Inc. Rapid City, SD.	N499EH
☐ N888JS	85	Conquest 1	425-0215	J Von Schaesburg, Beek, Holland.	N1225D
☐ N888LB	81	PA-31T Cheyenne II	8120035	Speedbird Inc. Wilmington, De.	D-IIRR
☐ N888PH	79	MU-2 Marquise	731SA	Filmar Aviation Inc. Nashville, Tn.	(N81FR)
☐ N888RH	79	MU-2 Marquise	737SA	Cape Central Airways/River City Aviation Inc. Marion, Ar.	N315MA
☐ N888SE	82	MU-2 Marquise	1549SA	Epps Air Service Inc. Atlanta, Ga.	N475MA
☐ N888SF	80	Gulfstream 980	95023	Steiner Film Inc. Williamstown, Ma.	D-IMKO
☐ N888TR	75	King Air 200	BB-50	PAC Truck & Equipment Ltd. Manchester, Jamaica.	N500FE
☐ N888WW	80	MU-2 Marquise	791SA	Air 1st Aviation Companies Inc. Aiken, SC.	N111LG
☐ N888ZX	83	King Air B200	BB-1140	Dwight Stuart Air Group Inc. Van Nuys, Ca.	N42KA
☐ N894FL	83	Conquest II	441-0314	Flying Lion Inc. Nashville, Tn.	N186EC
☐ N895FK	78	King Air C90	LJ-759	Macomb Air LLC. Macomb, Il.	N133E
☐ N895MA	78	Mitsubishi MU-2N	723SA	Value Outlet Stores Inc. NYC.	
☐ N896SB	73	King Air A100	B-160	Rogers Helicopters Inc. Clovis, Ca.	OY-CCS
☐ N899D	68	King Air B90	LJ-386	Galbraith Aviation Inc. St Petersburg, Fl.	N610K
☐ N899GP	76	King Air E90	LW-184	Transportation Travel Services Inc. Lakeside, Mt.	N555LW
☐ N899HC	90	King Air B200	BB-1374	Warren Manufacturing Inc. Birmingham, Al.	N999HC
☐ N900AC	78	King Air E90	LW-282	Spencer Eccles, Bellevue, Id.	N50MB
☐ N900BE	80	King Air C90	LJ-907	Chandler Aviation inc. Richmond, Va.	N900TJ
☐ N900CK	66	King Air 90	LJ-85	Cosco Aviation Services, Crestview, Fl.	N900CF
☐ N900CP	74	King Air 200	BB-17	Allied Services Inc. Norfolk, Va.	(N17TJ)
☐ N900DG	74	King Air C90	LJ-618	All Day Inc. Little Rock, Ar.	N92AM
☐ N900DN	73	King Air A100	B-170	Lynch Flying Service Inc. Billings, Mt.	N900DH
☐ N900DS	84	Gulfstream 900	15035	Aspen Base Operations Inc. Aspen, Co.	OE-FGS
☐ N900EC	82	Gulfstream 1000	96050	JNS Aviation Inc. Nara Vista, NM.	ZS-KZX
☐ N900EC	82	Gulfstream 900	15001	VF Aircraft Holdings Inc/VF Air, Collinsville, Il.	N14072
☐ N900ET	84	Gulfstream 900	15037	West Coast Air Charter/Burlingame Industries Inc. Rialto, Ca	C-FAWG
☐ N900FD	83	KIng Air B200	BB-1109	Family Dollar Stores Inc. Charlotte, NC.	N6580B
☐ N900GB	87	King Air 300	FA-133	Platt Electric Supply Inc. Beaverton, Or.	N2203Z
☐ N900HA	78	Conquest II	441-0027	Auerbach Aeronautics Assocs Inc. Dover, De.	(N29HA)
☐ N900HM	81	KIng Air F90	LA-119	Retro Aviation Inc. Omaha, Ne.	N100PH
☐ N900M	82	MU-2 Marquise	1545SA	Heritage Air Inc. Cape Canaveral, Fl.	N471MA
☐ N900MS	81	Conquest 1	425-0079	Tristate Electrical Supply Co. Hagerstown, Md.	N6846K
☐ N900PS	80	Conquest II	441-0129	K Eaton LLC. Morrisville, NC.	C-FETE
☐ N900RB	89	King Air 300	FA-192	Becker Holding Corp. Fort Pierce, Fl.	N1552K
☐ N900RD	77	King Air B100	BE-33	Chandler Aircraft Marketing Inc. Chandler, Tx.	(N232EB)
☐ N900RH	81	King Air 200	BB-816	Rite-Hite Corp. Milwaukee, Wi.	N510G
☐ N900RJ	66	680V Turbo Commander	1572-27	Raja Mohammed, South Gate, Ca.	(N713SP)
☐ N900TV	69	Mitsubishi MU-2F	140	Ham Air Inc. Cincinnati, Oh.	N333RK
☐ N900YH	73	Mitsubishi MU-2J	584	Styles Aviation Inc. NY.	N791MA
☐ N901AJ	79	King Air C90	LJ-829	Aero Advantage inc. North Adams, Ma.	D-IFCL
☐ N901BR	75	King Air 200	BB-65	Braxton Ranch Corp. Mills, Wy.	N90ML
☐ N901FD	82	Conquest 1	425-0145	Family Dollar Stores Inc. Charlotte, NC.	N6885T
☐ N901JA	76	King Air C90	LJ-694	Emerald Aviation Inc. London, Ky.	N712K
☐ N901MC	81	SA-226T Merlin 3B	T-369	McBee Co. Dallas, Tx.	N1009Y
☐ N901TP		AP68TP-600	9001	Aeromanagement Inc. Herndon, Va.	I-BAML
☐ N901WL	68	King Air B90	LJ-410	W L Paris Enterprises Inc. Louisville, Ky.	N33CS
☐ N902LT	97	King Air C90B	LJ-1480	BLT Enterprises Inc. Oxnard, Ca.	
☐ N903FH		BAe Jetstream 31	605	Coca-Cola Bottling Co. Chattanooga, Tn.	N331NY
☐ N903GP	79	King Air C90	LJ-861	Myers Medical Equities Inc. Paterson, NJ.	N905GP
☐ N903SE	95	King Air C90B	LJ-1403	SP of Delaware Inc. Dover, De.	
☐ N904DG	84	King Air B200	BB-1176	Delta Health Group Inc. Pensacola, Fl.	N81LC
☐ N904FH	83	BAe Jetstream 31	613	Coca-Cola Bottling Co. Chattanooga, Tn.	N331BA
☐ N904HB	90	King Air C90A	LJ-1256	Midwest Surgical Services Inc. Bloomington, Mn.	
☐ N904JP	85	King Air C90A	LJ-1121	McLane Trailer Sales Inc. Poplar Bluff, Mo.	(N20LK)
☐ N904MC	91	King Air 350	FL-44	Malco Leasing Corp. Danbury, Ct.	N90PR
☐ N905GP	81	King Air 200	BB-789	Godwin Pumps of America Inc. Bridgeport, NJ.	N70KM
☐ N905LC	84	PA-42 Cheyenne IIIA	5501022	reported stolen 1988. (status ?).	N4115K
☐ N906F	64	Gulfstream 1	146	F B Air Inc. Opa Locka, Fl.	4X-JUD
☐ N908K	71	King Air C90	LJ-504	Midway Ford Inc. Wilmington, De.	(N479SJ)

Regn	Yr	Type	c/n	Owner/Operator	Prev Regn
N909J	97	King Air B200	BB-1579	Carpenter Technology Corp. Reading, Pa.	
N909ST	87	King Air B200	BB-1286	Star Delivery & Transfer Inc. Morton, Il.	N1525C
N910AJ	81	King Air B200	BB-910	L'Eagle Air II Inc. Key West, Fl.	ZS-LJA
N910EC	81	Gulfstream 840	11682	State of Tennessee, Nashville, Tn.	N5934K
N910HM	74	PA-31T Cheyenne II	7400009	MarkFlite LLC. Wilmington, De.	N66838
N910P	77	King Air 200	BB-212	Arawak Inc. St Thomas, USVI.	(N190P)
N911AZ	78	King Air E90	LW-300	State of Arizona, Phoenix, Az.	N2035C
N911ER	82	Conquest II	441-0249	Acadian Ambulance Services Inc. Lafayette, La.	N800BN
N911JW	79	Rockwell 690B	11537	Moxy Trucks of America Inc. Cincinnati, Oh.	N700PQ
N911KA	67	King Air A90	LJ-254	Global Air Services Inc. Nashville, Tn.	N781JT
N911MN	77	King Air 200	BB-229	Presentation Sisters Inc/McKennan Hospital, Sioux Falls, SD.	N904CM
N911ND	78	King Air C90	LJ-774	Aviation Resources Ltd. Fargo, ND.	(N713PA)
N911RB	91	King Air 300	FA-217	Becker Holdings Corp. Fort Pierce, Fl.	N8115M
N911RL	70	King Air 100	B-55	Dale Aviation Inc. Rapid City, SD.	N4167P
N911SR	81	King Air B200	BB-898	Lynch Flying Service Inc. Billings, Mt.	N98GA
N911UM	78	King Air C90	LJ-769	C C Medflight Inc. Lawrenceville, Ga.	N2KQ
N913CR	78	PA-31T Cheyenne 1	7804003	W Carl Reynolds PC. Cashiers, NC.	N21JA
N913DM	68	SA-26AT Merlin 2B	T26-144	BLT Aviation Inc. Odessa, Tx.	N599MS
N913YW	90	King Air 1300	BB-1383	Northeast Airlines Inc. Portland, Me.	N383YV
N914CT	90	King Air B200	BB-1351	NGF Corp. New Castle, De.	N1562F
N915RF	76	Mitsubishi MU-2L	677	Thor Enterprises Inc. Beloit, Wi.	N2ND
N915YW	89	King Air 1300	BB-1338	Raytheon Aircraft Services Inc. Wichita, Ks.	N338YV
N916HC	81	King Air B200	BB-916	Texas Drydock Inc. Orange, Tx.	N18261
N917BH	85	King Air C90A	LJ-1123	Samedan Oil Corp. Ardmore, Ok.	N7245Z
N917CT	91	King Air 350	FL-56	Langa Inc. Dover, De.	N717CT
N917F	81	PA-31T Cheyenne 1	8104030	Hughes Venture Group Inc. Wilmington, De.	N71LA
N918VS	76	King Air E90	LW-300	San Tomo Partners, Chiloquin, Or.	N913VS
N919AG	68	King Air B90	LJ-432	Aircraft Services Group LC. Little Rock, Ar.	N345LL
N919WM	73	King Air A100	B-154	Westling Manufacturing Co. Princeton, Mn.	N70MN
N920C	79	King Air 200	BB-565	Air Travel Services Inc. Nashville, Tn.	N335GA
N920S	71	Mitsubishi MU-2G	534	Bankair Inc. West Columbia, SC. (status ?).	(N78V)
N920WJ	80	Gulfstream 840	11621	Gulfstream Aviation Enterprises Inc. Orlando, Fl.	(N23TX)
N921AZ	88	King Air B200	BB-1287	Department of Public Safety, Phoenix, Az.	N713DH
N921S	78	King Air 200	BB-307	Central Flying Service Inc. Little Rock, Ar.	N23687
N921ST	74	Rockwell 690A	11200	Sun Pacific International Inc. Tucson, Az.	
N922FM	72	Mitsubishi MU-2F	216	Oakton International Corp. Laporte, In.	N922ST
N923CR	84	King Air C90A	LJ-1074	JMS Partners LLC. Jackson, Mi.	N6583K
N923S	87	King Air 300	FA-127	Air King Aviation LLC. Moorehead City, NC.	N92SS
N924PC	72	Rockwell 690	11041	United CCM Corp. San Antonio, Tx. (status ?).	(N54MH)
N924RM	79	King Air B100	BE-63	Executive Leasing Inc. Shelby, NC.	N924WS
N925B	82	King Air B200	BB-1050	Deluxe Aviation Inc. Indianapolis, In.	(N19KA)
N925BC	80	King Air C90	LJ-925	Medico Rentals & Leasing Co. Ridgeland, Ms.	N925GS
N925GC	65	Gulfstream 1	161	Guidant Corp.San Jose, Ca.	N307EL
N925RM	89	PA-42 Cheyenne IIIA	5501040	RJM Aviation Associates Inc. Berlin, Ct.	N561GA
N925X	64	King Air 90	LJ-1	Medair Leasing Ltd. NYC. (status ?)	N26CH
N926ES	81	Conquest 1	425-0029	Fullers White Mountain Motors Inc. Show Low, Az.	N926FS
N926FS	81	Conquest 1	425-0093	First Security Bank of Idaha, Boise, Id.	N6848R
N926HS	79	King Air B100	BE-70	Mayflower Aviation LLC. Raleigh, NC.	C-GAPT
N926LD	78	PA-31T Cheyenne II	7820047	North Dakota State University, Fargo, ND.	N290T
N926S	69	King Air B90	LJ-447	Oahu North Shore Aviation Ltd. Honolulu, Hi.	C-GSUN
N926SC	80	Gulfstream 840	11622	S E Cone Jr. Hobbs, NM.	N49BB
N928VF	81	MU-2 Solitaire	436SA	Vincent Forschan, D O Medical Corp. Rancho Mirage, Ca.	N62CP
N930K	67	King Air A90	LJ-294	DATS Aviation Inc. Truckee, Ca.	
N930SP	81	King Air B200	BB-930	Arkansas State Police, Little Rock, Ar.	F-GHMY
N930SU	93	TBM 700	89	John Miller Enterprises LLC. Park City, Ut.	
N932BF	81	PA-42 Cheyenne III	8001064	Tricon Metals & Services Inc. Birmingham, Al.	N42SJ
N932G	81	King Air F90	LA-96	Flat Top Ranch/Louis Beecherl III, Walnut Springs, Tx.	N37390
N933DG	83	PA-42 Cheyenne III	8001100	Genter Airways LLC. Los Angeles, Ca.	N42KA
N933RT	82	King Air B200	BB-955	Phelps-Tointon Inc. Greeley, Co.	G-WILK
N934DC	77	King Air E90	LW-202	Dixie Capital Corp. Richmond, Va.	N127
N934SH	77	King Air 200	BB-252	Shaw Industries Inc. Dalton, Ga.	N475U
N935CA	80	PA-31T Cheyenne 1	8004033	JIB Inc. East Haddam, Ct.	XB-FDE
N939K	68	King Air B90	LJ-349	Ballard Aviation Inc. Wichita, Ks.	
N940AC	80	Gulfstream 840	11629	Conley, Lott & Nichols Machinery Co. Dallas, Tx.	VH-NCM
N940U	69	680W Turbo Commander	1843-42	University of Oklahoma, Norman, Ok.	(YV-723P)
N941MA	79	MU-2 Marquise	744SA	Epps Air Service Inc. Atlanta, Ga.	
N941S	79	MU-2 Marquise	738SA	Stokely Affiliated Financial Corp. Knoxville, Tn.	N916MA
N942CE	79	King Air 200	BB-494	Corporate Eagle II Inc. Southfield, Mi.	N30FL
N942ST	79	MU-2 Marquise	745SA	Bankair Inc. West Columbia, SC.	N942MA
N944CC	80	King Air 200	BB-604	ANR Pipeline Co. Detroit, Mi.	N75C
N944CE	78	King Air 200	BB-326	Corporate Eagle Four LLC. Waterford, Mi.	N40BL
N944K	69	King Air B90	LJ-467	Carmine Labriola Inc. Scarsdale, NY.	
N944RS	76	King Air E90	LW-177	Troy Fraser, Big Spring, Tx.	N2177L
N946CE	79	King Air 200	BB-540	Corporate Eagle Capital LLC. Waterford, Mi.	N65NL
N946WA	70	King Air B90	LJ-492	Waterside Aircraft Sales LLC. Guilford, Ct.	LN-HAC
N948CC	77	King Air E90	LW-236	Regent Air Service Inc. Kelseyville, Ca.	N48V

Regn	Yr	Type	c/n	Owner/Operator	Prev Regn
☐ N949SW	78	King Air B100	BE-34	Truck Components Inc. Rockford, Il.	N203KA
☐ N950M	74	Rockwell 690A	11173	State of Indiana, Indianapolis, In.	N501MQ
☐ N950TJ	82	Gulfstream 900	15016	Northwest Arkansas Paper Co. Springdale, Ar.	N27KG
☐ N951K	70	King Air 100	B-17	Air Activities Inc. Tyler, Tx.	
☐ N953HF	74	Rockwell 690A	11194	Grover Woolston, Minneapolis, Mn.	N101UC
☐ N953L	82	King Air B200	BB-953	Ignacio Lozano/West Coast Charter, Newport Beach, Ca.	N10EH
☐ N955FC	78	King Air B100	BE-50	MB Transportation LLC. Morehead City, NC.	N188JB
☐ N955RA	83	King Air F90	LA-201	Apollo Leasing Co. Ocean Springs, Ms.	N653LP
☐ N956DS	77	SA-226T Merlin 3A	T-282	Allen Associates, Newburgh, In.	N51RX
☐ N960V	80	King Air F90	LA-31	Silver Dollar City Inc. Branson, Mo.	N6675W
☐ N961LL	83	King Air B200	BB-1139	State of Illinois, Springfield, Il.	N6661C
☐ N961PC	96	Pilatus PC-XII	129	Gregory Aviation Co. Des Moines, Ia.	HB-FQX
☐ N962R	70	King Air 100	B-44	Air Transport Services LLC. Athens, Ga.	(N101PF)
☐ N963DC	80	SA-226T Merlin 3B	T-314	Tyler Jet LLC. Tyler, Tx.	HZ-OCE
☐ N965J	79	King Air C90	LJ-856	MMM Ltd. Barrington, Il.	N27RF
☐ N965LC	77	King Air E90	LW-204	Lane Construction Corp. Windsor, Ct.	N4204S
☐ N966MA	79	MU-2 Solitaire	405SA	South Coast Tumor Institute, San Diego, Ca.	N711TF
☐ N968T	79	King Air 200	BB-570	Avastar Jet Charter & Management Service, Waterford, Mi.	
☐ N970P	97	King Air C90B	LJ-1487	Piedmont Aviation Services Inc. Winston-Salem, NC.	
☐ N971CF	81	King Air C90	LJ-971	Execuflight Inc. Monroeville, Al.	OY-BEB
☐ N971LL	85	King Air B200	BB-1234	State of Illinois, Springfield, Il.	N991LL
☐ N973BB	81	MU-2 Marquise	1509SA	Elias Rodriguez, Fairfax, Va.	N973MA
☐ N974DC	81	PA-31T Cheyenne II	8120027	OCAAS LLC. Amarillo, Tx.	N815SW
☐ N976JT	76	King Air C90	LJ-699	Tri-State Aero Inc. Evansville, In.	
☐ N976KC	80	King Air 200	BB-601	Central Aviation Inc/Corsicana Co. Dallas, Tx.	N6687H
☐ N977LX	76	King Air 200	BB-141	Specchio Cablevision Co. Rantoul, Il.	
☐ N977MP	83	Conquest II	441-0310	Lloyd & Lois Martin, Chambersburg, Pa.	N444KE
☐ N977SB	72	King Air E90	LW-10	Hatfield Lumber Co. Hatfield, Ar.	N14MW
☐ N977XT	81	PA-42 Cheyenne III	8001008	R Y Timber Inc. Boise, Id.	N808CA
☐ N978BC	78	PA-31T Cheyenne II	7820025	Jamatt Aviation Inc. Sarasota, Fl.	N82228
☐ N980AK	80	Gulfstream 840	11636	Universal Pacific Investments Corp. Stateline, Nv.	N7649J
☐ N980BH	80	Gulfstream 980	95002	Semitool Inc. Kalispell, Mt.	N8LN
☐ N980CA	81	PA-31T Cheyenne 1	8104010	TLH Enterprises Inc. Corsicana, Tx.	
☐ N980DB	74	King Air 200	BB-10	Radio Flyer Inc/King Air BB10 Inc. Bloomington, Il.	N660M
☐ N980DT	81	Gulfstream 980	95048	Detroit Tool & Engineering Co. Lebanon, Mo.	N9800S
☐ N980EA	80	Gulfstream 980	95031	Eagle Air Inc. Memphis, Tn. (now XC-HFW ?).	HP-1132P
☐ N980EC	80	Gulfstream 980	95011	Amstar Aviation, Dover, De.	ZP-...
☐ N980GR	81	Gulfstream 980	95049	Gulf Resources Corp. San Antonio, Tx.	N3982C
☐ N980GZ	81	Gulfstream 980	95063	Alken-Ziegler Inc. Kalkaska, Mi.	N980GM
☐ N980MD	80	Gulfstream 980	95030	Air Operations International, Charlotte, NC.	N980AB
☐ N980SA	80	Gulfstream 980	95012	Amstar Aviation, Dover, De.	YV-129P
☐ N980WM	80	Gulfstream 980	95034	Wal-Mart Stores Inc. Rogers, Ar.	(N200JN)
☐ N981LL	80	King Air 200	BB-602	State of Illinois, Springfield, Il.	N6735T
☐ N982BA	82	King Air C90-1	LJ-1031	Raslan Air Services Flying Club Corp. Tampa, Fl.	SU-PAC
☐ N983K	66	King Air A90	LJ-169	Silver Lining Service Inc. Montgomery, NY.	N903K
☐ N984AA	68	King Air B90	LJ-429	USA Jet Airlines Inc. Belleville, MI.	N811AA
☐ N984RE	80	MU-2 Marquise	787SA	Epps Air Service Inc. Atlanta, Ga.	N267PC
☐ N985AA	67	King Air A90	LJ-214	Murray Aviation Inc. Ypsilanti, Mi.	N813AA
☐ N986MC	82	Conquest II	441-0280	SBC Co LLC. Rock Island, Il.	N6831N
☐ N987B	80	King Air B100	BE-81	Mirage Aviation Inc. Jackson, Tn.	(N3737G)
☐ N987GM	73	King Air E90	LW-65	Bank One Arizona Leasing Corp. Phoenix, Az.	N3065W
☐ N987MA	68	Mitsubishi MU-2F	124	Fast Air Brokerage Inc. Miami, Fl.	N18UT
☐ N988AE	80	Conquest II	441-0175	Samaritan Health Services, Phoenix, Az.	N2723A
☐ N988ME	92	King Air 350	FL-77	Flight Investors LP. Hagerstown, Md.	N4000K
☐ N988P	97	King Air C90B	LJ-1506	Piedmont Aviation Services Inc. Winston Salem, NC.	
☐ N988SC	78	King Air 200	BB-310	Alex Air Ltd. Cincinnati, Oh.	N100KM
☐ N988SL	69	King Air B90	LJ-438	Image Tecrhnology Solutions Inc. Wood Dale, Il.	N900LS
☐ N989GM	74	King Air E90	LW-109	Flagstaff Medical Center Inc. Flagstaff, Az.	N388SC
☐ N990BM	80	King Air F90	LA-70	N90BJ Inc. Warsaw, In.	N3687S
☐ N990CB	94	King Air C90B	LJ-1362	Springfield Flying Service, Springfield, Mo.	
☐ N990F	82	King Air F90	LA-164	Air Twerps Inc. Wilmington, De.	(F-GFDM)
☐ N990SA	67	King Air A90	LJ-261	Summit Aviation Inc. Dallas, Tx.	N5115D
☐ N991GA	81	King Air 200	BB-812	Avex Inc. Santa Paula, Ca.	5Y-PLM
☐ N991LL	96	King Air 350	FL-142	State of Illinois, Springfield, Il.	N1072S
☐ N992C	85	King Air C90	LJ-1122	Illinois Data Mart Inc. Geneva, Il.	N7244J
☐ N992TJ	82	King Air B200	BB-992	Safeguard Technologies Inc/Marc Fruchter, Reading, Pa.	N340TT
☐ N992TT	80	PA-31T Cheyenne 1	8004019	Air Alpha America Inc. Amarillo, Tx.	D-ICGD
☐ N993M	81	King Air F90	LA-134	California Oregon Television, Medford, Or.	N2420M
☐ N994PE	78	Mitsubishi MU-2N	714SA	Private Enterprise Investment Corp. Madison, NH.	OY-NIH
☐ N995MS	81	King Air B200	BB-931	Morton International Inc. Waukegan, Il.	N6789
☐ N995TA	96	King Air C90B	LJ-1441	Transair USA Inc. Wichita, Ks.	ZS-NUE
☐ N996AB	77	Rockwell 690B	11425	Natural Gas Processing Co. Worland, Wy.	ZP-TWV
☐ N996TT	95	King Air C90B	LJ-1398	StarKing Inc. Fort Worth, Tx.	N3238K
☐ N997ME	82	King Air B100	BE-135	S Robert Davis, Dublin, Oh.	N135AR
☐ N999BE	80	Conquest II	441-0147	Erickson Air Crane Co. Central Point, Or.	N999PP
☐ N999BT	78	King Air 200	BB-400	Sanford Farms Inc. East Haddam, Ct.	N164AB

Regn	Yr	Type	c/n	Owner/Operator	Prev Regn
☐ N999CY	77	King Air 200	BB-253	Iowa State University, Ames, Ia.	N999HC
☐ N999DF	80	Conquest II	441-0185	National Imperial Corp. Deer Park, NY.	N999DB
☐ N999DT	76	King Air 200	BB-138	Davison Transport Inc. Ruston, La.	N925BQ
☐ N999FG	79	Rockwell 690B	11535	HealthSouth Aviation Inc. Birmingham, Al.	N81709
☐ N999G	73	King Air A100	B-144	Bud Holding Co. Greensboro, NC.	N999TB
☐ N999GA	81	King Air B200	BB-929	Air Taxi Aviation & Badger Airlines Inc. Milwaukee, Wi.	N81AJ
☐ N999MC	91	King Air 350	FL-65	Mason Corp. Birmingham, Al.	N40FM
☐ N999MM	80	SA-226T Merlin 3B	T-309	Planned Residential Communities Financial, Long Branch, NJ.	N5668M
☐ N999MX		SA-227AT Merlin 4C	AT-501	Fair Weather Aviation Inc. Rochester Hills, Mi.	N3051H
☐ N999NP	81	PA-31T Cheyenne 1	8104058	N & B Cheyenne LLC. Las Cruses, NM.	N278HM
☐ N999RC	83	King Air F90-1	LA-208	Carpenter Inc. Naples, Fl.	N901SA
☐ N999RF	90	2000 Starship	NC-9	Raytheon Aircraft Services Inc. Wichita, Ks.	N2009W
☐ N999SE	80	King Air E90	LW-344	Great Northern Charter Co. Syracuse, NY.	N888RT
☐ N999SF	72	King Air E90	LW-5	J Jacques Mistrot, Raleigh, NC.	N999SE
☐ N999TA	70	Mitsubishi MU-2G	514	Air Cargo Express Inc. Little Rock, Ar.	N514WG
☐ N999UP	82	MU-2 Marquise	1557SA	Universal Bow Transport Inc. Bow, NH.	N988RR
☐ N999VB	80	King Air 200	BB-645	Vaughan & Bushnell Manufacturing Co. Hebron, Il.	N80GB
☐ N999WW	73	Rockwell 690	11073	Triple Nine Whiskey Whiskey Corp. Westlake Village, Ca.	N9140N
☐ N1007B	80	SA-226T Merlin 3B	T-327	Swearingen Aviation Corp. San Antonio, Tx.	
☐ N1008U	80	SA-226T Merlin 3B	T-351	Planeholder Inc. Miami, Fl.	
☐ N1010V	81	SA-226T Merlin 3B	T-372	Aviation Business Machines, Hillsboro, Or.	
☐ N1014V	80	SA-226T Merlin 3B	T-321	Aarque Steel Corp. Jamestown, NY.	
☐ N1031Y	96	King Air C90B	LJ-1431	Majo Trading Corp. Miami, Fl.	
☐ N1039Y	70	SA-26AT Merlin 2B	T26-180E	Sheila DeForest, Pitt Meadows, Canada.	A2-KAM
☐ N1057L	96	King Air C90B	LJ-1457	Sam Shapiro, Atlanta, Ga.	
☐ N1057Q	97	King Air 350	FL-163	Golfo International Air Services USA Inc. Wichita, Ks.	
☐ N1068K	96	King Air C90B	LJ-1448	McClatchy Newspapers Inc. Sacramento, Ca.	
☐ N1069F	97	King Air C90B	LJ-1489	Raytheon Aircraft Co. Wichita, Ks.	
☐ N1069S	97	King Air B200	BB-1549	JJB Sports PLC/Westair Flying Services Ltd. Blackpool, UK.	
☐ N1070E	96	King Air B200	BB-1545	Raytheon Aircraft Co. Wichita, Ks.	
☐ N1070F	96	King Air C90B	LJ-1440	Jim Causley Aviation Inc. Grosse Pointe, Mi.	
☐ N1071S	97	King Air 350	FL-175	HPP Aviation Services Inc. Sun Valley, Ca.	
☐ N1074G	96	King Air B200	BB-1534	FGC Inc. Hialeah, Fl.	
☐ N1078	80	King Air C90	LJ-908	Info Systems of North Carolina Inc. Charlotte, NC.	N91HM
☐ N1079D	96	King Air C90B	LJ-1462	Cowherd Aviation Inc. Phoenix, Az.	
☐ N1083N	96	King Air 350	FL-153	Calabasas B C D Inc. Agoura Hills, Ca.	
☐ N1083S	96	King Air C90B	LJ-1443	Hunt Aviation Inc. Ruston, La.	
☐ N1084N	96	King Air C90B	LJ-1444	Carriage Leasing Inc. Wilmington, Oh.	
☐ N1086Z	96	King Air C90B	LJ-1446	JAK Aircraft Inc. Las Vegas, Nv.	
☐ N1089L	96	King Air C90B	LJ-1439	The Outback LLC. Eden Prairie, Mn.	
☐ N1089V	97	King Air B200	BB-1577	Martin Aviation Inc. Deephaven, Mn.	
☐ N1090X	97	King Air C90B	LJ-1473	Aeroservices of Central Florida Inc. Fern Park, Fl.	
☐ N1092G	96	King Air C90B	LJ-1453	Raytheon Aircraft Credit Corp. Wichita, Ks.	
☐ N1092H	96	King Air C90B	LJ-1454	Raytheon Aircraft Credit Corp. Wichita, Ks.	
☐ N1092N	96	King Air B200	BB-1556	Cigris Ventures Ltd. Tortola, BVI.	
☐ N1095M	96	King Air C90B	LJ-1435	Maverick Tube Corp. Chesterfield, Mo.	
☐ N1095Q	97	King Air 350	FL-158	Universal Avionics Systems Corp. Tucson, Az.	
☐ N1095W	96	King Air C90B	LJ-1456	Robert Watson, San Antonio, Tx.	
☐ N1097B	96	King Air C90B	LJ-1452	Perryman Enterprises, Houston, Pa.	
☐ N1097S	73	King Air C90	LJ-574	Ken Air Aviation Corp. Gainesville, Fl.	N318F
☐ N1099D	97	King Air C90B	LJ-1471	Metal Processors Inc. Stevensville, Ky.	
☐ N1099L	97	King Air C90B	LJ-1472	Stevens Aviation Inc. Greer, SC.	
☐ N1099Z	97	King Air C90B	LJ-1470	Elliott Aviation Aircraft Sales, Des Moines, Ia.	
☐ N1100M	73	King Air E90	LW-25	BETCO Inc. Statesville, NC.	N4406W
☐ N1100W	91	King Air 350	FL-47	WestOne Bancorp. Boise, Id.	N8069F
☐ N1101U	97	King Air C90B	LJ-1481	Mueller Transportation inc. Springfield, Mo.	
☐ N1102K	97	King Air C90B	LJ-1465	Montpelier Partners LLC. Darien, Ct.	
☐ N1103G	97	King Air C90B	LJ-1475	Kenworth Inc. Birmingham, Al.	
☐ N1105X	97	King Air 350	FL-155	Image Air LLC & others, Warsaw, In.	
☐ N1106J	97	King Air B200	BB-1566	Raytheon Aircraft Co. Wichita, Ks.	
☐ N1106M	97	King Air C90B	LJ-1476	Seahatch Laboratory NV. Oranjestad, Aruba.	
☐ N1107W	97	King Air C90B	LJ-1477	Raytheon Aircraft Co. Wichita, Ks.	
☐ N1108M	97	King Air C90B	LJ-1478	Elliott Aviation Aircraft Sales Inc. Des Moines, Ia..	
☐ N1110K	97	King Air C90B	LJ-1486	Peach Air Corp. Coral Gable, Fl.	
☐ N1114K	97	King Air B200	BB-1559	RPR LLC. Tacoma, Wa.	
☐ N1118G	97	King Air B200	BB-1576	Southern Farm Bureau Casualty Insurance Co. Ridgeland, Ms.	
☐ N1118W	97	King Air 350	FL-168	Perkins Transportation Co. Atlanta, Ga.	
☐ N1119Z	97	King Air B200	BB-1581	The Scotts Co. Marysville, Oh.	N12MG
☐ N1120Z	97	King Air B200	BB-1570	Mallen Industries Inc. Norcross, Ga.	
☐ N1126J	97	King Air C90B	LJ-1483	Carlisle Mechanical & Welding Inc. Carlisle, Pa.	
☐ N1128M	90	Beech 1900C-1	UC-139	Bayer Corp. Elkhart, In.	
☐ N1130J	97	King Air C90B	LJ-1467	B/W Air, Elkhart Lake, WI.	
☐ N1134D	97	King Air C90B	LJ-1484	David Wolfe, Tempe, Az.	
☐ N1134G	97	King Air C90B	LJ-1468	Woodland Aviation Inc. Woodland, Ca.	
☐ N1135G	97	King Air C90B	LJ-1488	Executive Air LLC. Indianapolis, In.	
☐ N1154S	66	King Air 90	LJ-108	Mocaan Inc. Miami, Fl.	

Regn	Yr	Type	c/n	Owner/Operator	Prev Regn

Regn	Yr	Type	c/n	Owner/Operator	Prev Regn
☐ N1162V	80	King Air 200	BB-711	Harland & Shirley Stonecipher, Centrahoma, Ok.	F-GILJ
☐ N1164F	82	MU-2 Marquise	1562SA	Epps Air Service Inc. Atlanta, Ga.	D-ICDG
☐ N1183G	79	PA-31T Cheyenne II	7920086	Naumes Inc. Medford, Or.	OE-FMR
☐ N1184U	76	King Air C-12C	BD-16	World Wide Aviation Services Corp. Fayetteville, NC.	76-0159
☐ N1191K	89	King Air 300	FA-185	Institute of Nuclear Power, Atlanta, Ga.	
☐ N1194C	80	King Air 200	BB-679	Cardinal IG Co. Minnétonka, Mn.	N256PL
☐ N1194V	83	Cheyenne T-1040	8375001	Vegas Jet LLC. North Las Vegas, Nv.	5Y-JJB
☐ N1198S	68	SA-26T Merlin 2A	T26-12	Osage Drilling Inc. Chanute, Ks. (status ?).	C-GRDT
☐ N1205S	79	King Air E90	LW-319	Randall Stores Inc. Mitchell, SD.	N2065K
☐ N1207S	68	SA-26T Merlin 2A	T26-24	Henry Wurst Inc. Apex, NC.	
☐ N1208S	68	SA-26T Merlin 2A	T26-25	South Jersey Airways Inc. Atlantic City, NJ.	
☐ N1210S	68	SA-26T Merlin 2A	T26-27	Wepco Inc/Ridgley Aviation, Tyler, Tx.	
☐ N1210Z	85	Conquest II	441-0339	RBM Aviation Inc. Dover, De.	
☐ N1211C	84	Conquest II	441-0340	Southern Rainbow Air Inc. Miami, Fl.	
☐ N1212C	85	Conquest II	441-0346	Bil Mar Foods Inc. Muskegon, Mi.	
☐ N1212K	85	Conquest II	441-0347	MTW Aviation Inc. Wilmington, De..	
☐ N1220W	78	Conquest II	441-0065	Overland West Investments LLC. Billings, Mt.	N67DM
☐ N1222B	82	Conquest 1	425-0060	Reid Dennis, Woodside, Ca.	N68436
☐ N1222G	84	Conquest 1	425-0196	Ford Steel Co. Hillsboro, Or.	
☐ N1223C	76	King Air 200	BB-152	Chaparral Inc. Newton, Ks.	N48HF
☐ N1223P	84	Conquest 1	425-0204	Gottschalks Inc. Fresno, Ca.	
☐ N1224J	84	Conquest 1	425-0208	Lark Aviation Inc. Laguna Hills, Ca.	
☐ N1224S	84	Conquest 1	425-0211	Inductotherm Corp. Rancocas, NJ. (status ?).	
☐ N1224T	84	Conquest 1	425-0212	Grass Valley Group Inc. Grass Valley, Ca.	
☐ N1225V	84	Conquest 1	425-0218	Babco Inc. Omaha. Ne.	
☐ N1226B	84	Conquest 1	425-0222	Samuel Heffner, Phoenix, Md.	
☐ N1227J	85	Conquest 1	425-0229	Cessna Aircraft Co. Wichita, Ks.	
☐ N1250	86	King Air 300	FA-88	Raytheon Appliances Commercial Laundry, Ripon, Wi.	N7255X
☐ N1262K	85	Conquest 1	425-0234	Costa Aircraft Corp. Wilmington, De.	
☐ N1290A	65	King Air 90	LJ-30	Turbine Power Inc. Bridgewater, Va.	N538M
☐ N1347Z	72	King Air A100	B-114	Westernair of Albuquerque Inc. Albuquerque, NM.	XC-FIX
☐ N1362N	76	King Air B100	B-230	Will-Flite Aviation Ltd. Marshall, Tx.	D-IKUL
☐ N1500X	69	SA-26AT Merlin 2B	T26-127	F S Air Service Inc. Anchorage, Ak.	
☐ N1509G	88	King Air B200	BB-1308	New York State Power Authority, NYC.	
☐ N1515H	84	PA-42 Cheyenne 400LS	5527019	Corporate Aviation Services Inc. Pepper Pike, Oh.	
☐ N1517K	88	King Air 300	FA-167	World Color Press Inc. NYC.	
☐ N1525C	92	King Air 350	FL-88	The Cretex Companies Inc. Elk River, Mn.	N8288W
☐ N1543H	89	King Air 300	FA-174	Lykes Services Co. Tampa, Fl.	
☐ N1543Z	89	King Air B200	BB-1327	Barefoot Air Service LLC. N Myrtle Beach, SC.	
☐ N1544V	94	King Air C90B	LJ-1366	Graham NB, First Coleman NB , Farmers NB. Norman, Ok.	
☐ N1546	77	King Air C-12C	BC-58	U S Customs Service, Oklahoma City, Ok.	77-22947
☐ N1547	77	King Air C-12C	BC-50	U S Customs Service, Oklahoma City, Ok.	77-22939
☐ N1547V	88	King Air B200	BB-1307	State of South Dakota, Pierre, SD.	
☐ N1548B	89	King Air B200	BB-1319	Mutual Protective Insurance Co. Omaha, Ne.	
☐ N1548K	89	King Air 300	FA-179	Cutter Aviation Inc. Phoenix, Az.	
☐ N1549		King Air C-12C	BC-45	U S Customs Service, Oklahoma City, Ok.	77-22934
☐ N1550S	90	2000 Starship	NC-5	Raytheon Aircraft Co. Wichita, Ks.	N42SR
☐ N1551	76	King Air C-12C	BC-39	U S Customs Service, Oklahoma City, Ok.	76-22562
☐ N1551A	90	King Air 350	FL-24	B & C Co. Columbus, Ga.	
☐ N1551C	94	King Air C90B	LJ-1365	Raytheon Aircraft Co. Wichita, Ks.	
☐ N1551H	89	King Air C90A	LJ-1211	GECC. Plymouth Meeting, Pa.	
☐ N1551T	90	King Air 350	FL-25	P & P Producing Inc. Midland, Tx.	
☐ N1552D	89	King Air C90A	LJ-1204	O'Gara Aviation LLC. Marietta, Ga.	
☐ N1552G	89	King Air C90A	LJ-1206	Tee Lok Corp. Edenton, NC.	
☐ N1552S	90	2000 Starship	NC-12	Robert Erwin, Dallas, Tx.	
☐ N1553	76	King Air C-12C	BC-30	U S Customs Service, Oklahoma City, Ok.	76-22254
☐ N1553D	89	King Air C90A	LJ-1210	Wesley Medical Center, Wichita, Ks.	
☐ N1553E	88	King Air B200	BB-1306	Versa Technologies Inc. Racine, Wi.	
☐ N1553G	89	King Air C90A	LJ-1214	Cargill Detroit Corp. Clawson, Mi.	
☐ N1553M	89	King Air B200	BB-1328	Texas A & M University, College Station, Tx.	
☐ N1553N	90	King Air C90A	LJ-1238	Quik Way Aviation of Nevada LLC. Oklahoma City, Ok.	
☐ N1553S	90	2000 Starship	NC-13	Cutter Aviation Inc. Phoenix, Az.	
☐ N1553V	94	King Air 350	FL-117	Green Tree Financial Corp. St Paul, Mn.	
☐ N1553Y	92	2000A Starship	NC-25	Max Flight Inc. Wilmington, De.	
☐ N1554	76	King Air JC-12C	BC-21	U S Customs Service, Oklahoma City, Ok.	76-22545
☐ N1554U	89	King Air 300	FA-195	Gateway 2000 Aviation Inc. North Sioux City, SD.	
☐ N1556S	90	2000 Starship	NC-6	Raytheon Aircraft Credit Corp. Wichita, Ks.	
☐ N1558	73	King Air C-12C	BC-20	U S Customs Service, Oklahoma City, Ok.	73-22269
☐ N1558H	94	King Air 350	FL-119	Blue Cross & Blue Shield Inc. Jacksonville, Fl.	
☐ N1558S	93	2000A Starship	NC-31	Liberty Aviation Inc. Philadelphia, Pa.	
☐ N1559	73	King Air C-12C	BC-16	U S Customs Service, Oklahoma City, Ok.	73-22265
☐ N1559G	94	King Air B200	BB-1480	Horton Transportation Inc. St Paul, Mn.	
☐ N1559W	94	King Air B200	BB-1481	Triple AAA, Bastanchury & Yosemite Water Cos. Fullerton, Ca	
☐ N1559Z	94	King Air B200	BB-1483	CX USA 10/96 to ?	
☐ N1560T	94	King Air C90B	LJ-1357	West Side Leasing Inc. Helena, Mt.	
☐ N1564P	90	King Air C90A	LJ-1230	Six-Four Papa Inc. Little Rock, Ar.	

Regn	Yr	Type	c/n	Owner/Operator	Prev Regn
☐ N1564W	90	King Air C90A	LJ-1234	V J Coleman & Son, Ackerly, Tx.	
☐ N1565L	77	King Air C90	LJ-704	JSW Investments Inc. Martinsburg, WV.	
☐ N1567F	89	King Air B200	BB-1335	Woodland Aircraft Leasing Inc. Plymouth, Mn.	
☐ N1567G	89	King Air C90A	LJ-1217	Southland Oil Corp & co-owners, Jackson, Ms.	
☐ N1568D		Beech 1900C-1	UC-76	Dow Chemical Co.Freeland, Mi.	
☐ N1568E	94	King Air B200	BB-1488	Roquette America Inc. Keokuk, Ia.	
☐ N1568X	94	King Air C90B	LJ-1368	Kookaburra Air LLC. Morristown, NJ.	
☐ N1569S	90	2000 Starship	NC-11	Raytheon Aircraft Credit Corp. Wichita, Ks.	
☐ N1570F	90	King Air B200	BB-1350	John Crump Automotive Inc. Jasper, Al.	(N147VC)
☐ N1573L	76	King Air E90	LW-173	B & H Aircraft Ltd. Wilmington, De.	
☐ N1660W	78	King Air 200	BB-390	Cutter World Inc. Atlanta, Ga.	N1000W
☐ N1727S	81	MU-2 Marquise	1504SA	Garratt Family Trust of 1989, San Mateo, Ca.	N541NC
☐ N1728S	81	MU-2 Solitaire	451SA	Bernt Hamarback, Eugene, Or.	N229MA
☐ N1790M	80	MU-2 Marquise	756SA	Corporate Aviation Services Inc. Tulsa, Ok.	N179CM
☐ N1801B	74	King Air C90	LJ-634	Banana Express Inc. McKinney, Tx.	N19R
☐ N1803P	81	King Air F90	LA-133	Progress Printing Co. Lynchburg, Va.	
☐ N1804T	81	King Air B200	BB-880	Pair Corp. Wilmington, De.	
☐ N1807H	81	King Air B100	BE-119	Brut Air Inc. Worthington, Oh.	
☐ N1811S	81	King Air B200	BB-896	Forum Communications Co. Fargo, ND.	(N360EA)
☐ N1836H	81	King Air C90	LJ-990	Pacjets Ltd, Baraboo, Wi.	
☐ N1837F	81	King Air C90	LJ-993	CEBU Air Inc. Manila, Philippines.	
☐ N1840S	78	King Air E90	LW-263	Paul Flowers Jr. Dothan, Al.	N184JS
☐ N1843S	83	Conquest II	441-0317	Imperial Holly Corp. Sugar Land, Tx.	N1208J
☐ N1845	87	King Air 300	FA-132	National City Corp. Columbus, Oh.	N7241K
☐ N1848S	96	King Air C90B	LJ-1455	Stephens Group Inc. Little Rock, Ar.	N1085V
☐ N1848T	82	Conquest 1	425-0105	Knapheide Manufacturing Co. Quincy, Il.	N425FT
☐ N1850X	82	King Air B200	BB-946	BioMed Research Inc. Lake San Marcos, Ca.	
☐ N1853T	82	King Air C90	LJ-997	CTB Inc. Goshen, In.	
☐ N1857F	82	King Air F90	LA-167	Cariva Inc. Perryton, Tx.	
☐ N1860N	81	King Air B200	BB-907	Carolyn Vance Cook, McAllen, Tx.	(N440KC)
☐ N1865A	93	King Air 350	FL-103	Wachovia Bank of Georgia NA. Atlanta, Ga.	N8192M
☐ N1869	71	King Air C90	LJ-520	Air Sal Leasing Inc. Miami, Fl.	N880M
☐ N1870S	83	King Air B200	BB-1106	Skywater Lodge, Glenbrook, Nv. (status ?).	N63882
☐ N1879D	80	PA-31T Cheyenne II	8020046	Gibson's Discount Centers Inc. Dodge City, Ks.	N1879D
☐ N1888M	81	King Air B100	BE-113	BWOC Air Services Inc. Columbus, Ga.	(N88TL)
☐ N1906K	81	Conquest 1	425-0068	Griffith Aviation Co. Tulsa, Ok.	(N425PF)
☐ N1907W	70	SA-26AT Merlin 2B	T26-176	Minter Corp. St Simons Island, Ga.	N777PE
☐ N1911L	76	King Air B100	BE-11	Dominion Air Leasing Inc. Richmond, Va.	
☐ N1925P	83	King Air B200	BB-1094	Rolscreen Co. Pelia, Ia.	N35EC
☐ N1926A	81	King Air B200	BB-922	Capital Management Systems inc. Philadelphia, Pa.	N80BT
☐ N1928H	68	680W Turbo Commander	1789-19	Reed Kent Nixon, Aurora, Or.	(N260RC)
☐ N1932H	88	King Air B200	BB-1303	Fleeman Aviation Inc. Rogers, Ar.	N1932P
☐ N1955E	80	King Air 200	BB-727	Troutman Investment, Eugene, Or.	N55CC
☐ N1969C	78	King Air 200	BB-391	Pine Telephone Co. Broken Bow, Ok.	N526AP
☐ N1975G	78	King Air E90	LW-252	Ed's Flying Service Inc. Walnut Cove, NC.	N505RG
☐ N1976J	75	PA-31T Cheyenne II	7520005	Lock Haven Aircraft Sales Inc. Wilmington, De.	
☐ N1978P	71	Mitsubishi MU-2F	212	International Aircraft Recovery, Fort Pierce, Fl.	YV-1978P
☐ N1981S	81	Gulfstream 980	95047	Air Operations International, Charlotte, NC.	TG-...
☐ N1999G	68	King Air B90	LJ-319	Fayard Enterprises Inc. Louisberg, NC.	N845K
☐ N2000E	66	King Air A90	LJ-172	Soldwedel LPC. Yuma, Az.	
☐ N2000S	93	2000A Starship	NC-43	Hawker Pacific Inc. Seletar, Singapore.	
☐ N2015G	97	King Air 350	FL-170	Euroflight Inc. Wichita, Ks.	
☐ N2025M	78	King Air 200	BB-384	Water Soft Inc. Saxonburg, Pa.	
☐ N2025S	78	King Air B100	BE-54	Diamond Air Inc. Richmod Heights, Oh.	
☐ N2029Z	97	King Air 350	FL-177	HPP Aviation Services Inc. Sun Valley, Ca.	
☐ N2030P	78	King Air 200	BB-412	UCG Energy Corp. Brentwood, Tn.	
☐ N2047V	97	King Air 350	FL-179	First Security Bank NA. Salt Lake City, Ut.	
☐ N2050A	79	King Air C90	LJ-813	Beech Transportation Inc. Eden Prairie, Mn.	N517PC
☐ N2057C	79	King Air C90	LJ-827	Wiley Elick, Lemoore, Ca.	
☐ N2057N	79	King Air C90	LJ-815	McLinch Aviation Corp. Stamford, Ct.	
☐ N2057S	79	King Air 200	BB-425	Whayne Supply Co. Louisville, Ky.	
☐ N2061B	79	King Air 200	BB-413	State Aircraft Pooling Board, Austin, Tx.	
☐ N2062A	79	King Air E90	LW-317	Dupps Co. Germantown, Oh.	
☐ N2063A	79	King Air C90	LJ-819	Executive Aircraft Corp. Newton, Ks.	
☐ N2100T	70	Mitsubishi MU-2F	182	Ranger Aviation Enterprises, Sonora, Tx.	N105MA
☐ N2132W	97	King Air B200	BB-1594	Gentry Brothers, Lubbock, Tx.	N2132W
☐ N2141B	78	Rockwell 690B	11484	690 Inc. Dallas, Tx.	G-BLPT
☐ N2155B	72	Rockwell 690	11046	University of South Alabama, Mobile, Al.	
☐ N2164L	80	King Air F90	LA-79	Transport Aircraft Inc. Trenton, NJ.	XB-DQP
☐ N2176D	67	Mitsubishi MU-2B	028	Thief River Falls Technical College, Mn.	CP-1962
☐ N2186L	76	King Air E90	LW-186	San Francisco Welding & Fabricating Co. Alamo, Ca.	
☐ N2192L	76	King Air E90	LW-192	Siller Brothers Inc. Yuba City, Ca.	
☐ N2247R	84	PA-42 Cheyenne IIIA	5501005	DoJ/U S Marshals Service, Oklahoma City, Ok.	N5381X
☐ N2267U	82	Gulfstream 1000	96007	Apex Aviation Inc. Haslet, Tx.	N9456T
☐ N2270T	82	Gulfstream 1000	96011	Omega Air Inc. Long Beach, Ca.	N8159G
☐ N2272H	97	King Air B200	BB-1572	Elk River Aviation Inc. Newland, NC.	

Regn	Yr	Type	c/n	Owner/Operator	Prev Regn
☐ N2274L	77	King Air E90	LW-213	Carlton Forge Works, Paramount, Ca.	
☐ N2288B	97	King Air B200	BB-1589	Citicorp Leasing International Inc. New Castle, De.	
☐ N2297C	97	King Air C90B	LJ-1499	North American Off Road Adventures, Escondido, Ca.	
☐ N2303P	69	SA-26AT Merlin 2B	T26-167E	New Vistas Inc. Wilmington, De.	D-IBMD
☐ N2325V	80	PA-31T Cheyenne 1	8004028	National Group Protection Inc. Kitty Hawk, SC.	
☐ N2338V	80	PA-31T Cheyenne II	8020049	Malco Leasing Corp. N Palm Beach, Fl.	
☐ N2348W	79	PA-31T Cheyenne 1	7904057	Air Kor Inc. Wilmington, De.	
☐ N2349R	79	PA-31T Cheyenne II	7920090	5K Corp. Columbia, Ms.	
☐ N2356X	81	PA-31T Cheyenne 1	8104003	Hartco Contractors Inc. Paducah, Ky.	
☐ N2366X	81	PA-31T Cheyenne II	8120006	Eckerd Family Youth Alternatives Inc. Clearwater, Fl.	(N101TR)
☐ N2369V	80	PA-31T Cheyenne 1	8004035	Elsea Inc. Circleville, Oh.	
☐ N2403X	81	PA-31T Cheyenne 1	8104014	William Schmidt & Associates, Evansville, In.	
☐ N2415W	80	PA-31T Cheyenne 1	8004010	Aegon USA/Investors Warranty of America, Cedar Rapids, Ia.	
☐ N2427W	84	PA-31T Cheyenne 1A	1104005	Hap's Aerial Enterprises Inc. Sellersburg, In.	
☐ N2434V	84	PA-31T Cheyenne 1A	1104011	Lauren Manufacturing Co. New Philadelphia, Oh.	
☐ N2434W	84	PA-31T Cheyenne 1A	1104006	Davison Terminal Service Inc. Ruston, La.	
☐ N2435Y	81	PA-31T Cheyenne 1	8104049	Evans Equipment Inc. Concordia, Mo.	
☐ N2458W	80	PA-31T Cheyenne 1	8004017	Miller Welding & Iron Works Inc. Washington, Il.	
☐ N2467V	80	PA-31T Cheyenne II	8020068	Harold Levy, NYC.	
☐ N2467X	81	PA-31T Cheyenne 1	8104025	Quarter M Farms Inc. Rose Hill, NC.	
☐ N2469V	80	PA-31T Cheyenne II	8020070	Casey-Fogli Concrete Contractors Inc. Belmont, Ca.	
☐ N2480X	81	PA-31T Cheyenne 1	8104026	Seward Motor Freight Inc. Seward, Ne.	
☐ N2483W	80	PA-31T Cheyenne II	8020027	Mid-States Development Inc. Fargo, ND.	(N248WW)
☐ N2488Y	81	Cheyenne II-XL	8166009	Breit, Bosch, Levin & Coppola PC. Englewood, Co.	
☐ N2489Y	82	Cheyenne T-1040	8275001	Vegas Jet LLC. North Las Vegas, Nv.	
☐ N2519X	81	PA-31T Cheyenne 1	8104035	G & B Oil Co. Elkin, NC.	
☐ N2519Y	86	King Air B200	BB-1250	McKenzie Tanklines Inc. Tallahassee, Fl.	
☐ N2522V	80	PA-31T Cheyenne II	8020078	Regency Acquisitions Inc. Vero Beach, Fl.	
☐ N2522Z	83	PA-31T Cheyenne 1A	8304002	Donald Farris, Olive Branch, Ms.	N2522Z
☐ N2552Y	82	PA-31T Cheyenne 1	8104061	Continental Fire Sprinkler Co. Omaha, Ne.	
☐ N2556R	80	PA-31T Cheyenne 1	8004003	Air Casino Inc/Gas Supply Inc. Minneapolis, Mn.	
☐ N2587R	80	PA-31T Cheyenne 1	8004004	Baron Flight Inc. Portsmouth, NH.	
☐ N2624D	80	Conquest II	441-0122	Wing Dancer Inc. Louisville, Ky.	
☐ N2628M	80	Conquest II	441-0151	Conquest Air Corp. Jacksonville, Fl.	
☐ N2643B	71	681B Turbo Commander	6060	Duane Egli, Wevelgem, Belgium.	9Q-CGE
☐ N2709Z	87	SA-227AT Merlin 4C	AT-695B	Robert Wright Aircraft Inc. Tampa, Fl.	
☐ N2721D	80	Conquest II	441-0161	DKV Investments Inc. Denver, Co.	
☐ N2722D	80	Conquest 1	441-0168	USA Gasoline Corp. Agoura, Ca.	N2722D
☐ N2722Y	79	Conquest II	441-0173	AIA Inc/Reed Taylor, Lewiston, Id.	
☐ N2723X	80	Conquest II	441-0180	Eagle Aviation Inc. West Columbia, SC.	(N33AR)
☐ N2725N	80	Conquest II	441-0190	Thumb Energy Inc. Bad Axe, Mi.	
☐ N2725U	80	Conquest II	441-0192	Beall Lessors Inc. Portland, Or.	
☐ N2748X	86	King Air B200	BB-1258	Booneville Power Administration, Portland, Or.	
☐ N2755B	68	680W Turbo Commander	1762-8	O'Kelly Air Inc. Oakland Park, Fl.	
☐ N2830S	78	King Air B100	BE-48	Florida Air Charters Inc. Tampa, Fl.	N5009M
☐ N2877K	80	King Air 200	BB-683	Annett Holdings Inc/TMC Transportation Inc. Des Moines, Ia.	OY-ASS
☐ N2883	76	King Air 200	BB-144	Maxair Inc. Appleton, Wi.	N28S
☐ N3002S	66	King Air 90	LJ-103	Allen Godwin, Memphis, Tn.	
☐ N3015Q	95	King Air B200	BB-1493	Venus Airlines, Athens, Greece.	N3015Q
☐ N3019W	74	King Air C90	LJ-639	AeroCenter Inc. Zephyr Hills, Fl.	
☐ N3025Z	87	King Air 300	FA-120	Armstrong Telephone Co of Maryland, Butler, Pa.	
☐ N3026H	94	King Air B200	BB-1494	Moorman Manufacturing Property Co. Quincy, Il.	
☐ N3030G	85	King Air C90A	LJ-1117	Pierre Melcher, Austin, Tx.	N3030C
☐ N3035T	97	King Air C90B	LJ-1482	Tulsair Beechcraft Inc. Tulsa, Ok.	
☐ N3051K	94	King Air B200	BB-1495	Eagle Leasing Inc. Alpharetta, Ga.	
☐ N3066V		SA-227TT Merlin 3C	TT-489	Oilfield Aviation LC. Houston, Tx.	(N108TA)
☐ N3072Y		SA-227TT Merlin 3C	TT-518	Rockwell-Ditzler Associates Inc. Pittsburgh, Pa.	
☐ N3076W	73	King Air A100	B-176	Lowell Dunn Co. Hialeah, Fl.	YV-O-CVG4
☐ N3077Y	87	King Air C90A	LJ-1155	Dock Foundry Co. Three Rivers, Mi.	
☐ N3083K	94	King Air C90B	LJ-1378	Special Devices Inc. Van Nuys, Ca.	
☐ N3084K	88	King Air 300	FA-160	Southwestern Energy Co. Fayetteville, Ar.	
☐ N3086D	88	King Air 300	FA-148	Fresh Express Inc. Salinas, Ca.	
☐ N3090A	87	King Air C90A	LJ-1152	Executive Aircraft Corp. Newton, Ks.	
☐ N3092S	94	King Air B200	BB-1498	First Security Bank, Salt Lake City, Ut.	
☐ N3100K	69	King Air 100	B-1	Bushnell Aviation Inc. Baton Rouge, La.	N3400K
☐ N3107W	88	King Air 300	FA-150	Hangar One Inc. Atlanta, Ga. (status ?).	
☐ N3112W	74	King Air C90	LJ-612	Welch Aviation Inc. Alpena, Mi.	
☐ N3120U	94	King Air C90B	LJ-1382	C & C Management Inc. Elsa, Tx.	
☐ N3122Z	88	King Air 300	FA-162	Anheuser-Busch Companies Inc. St Louis, Mo.	
☐ N3125J	82	King Air B200C	BL-46	Aviation Specialities Inc. Washington, DC.	XA-MIN
☐ N3127K	88	King Air B200	BB-1293	J H Kelly Inc. Longview, Wa.	
☐ N3181Q	97	King Air B200	BB-1586	Raytheon Aircraft Co. Wichita, Ks.	
☐ N3190S	88	King Air C90A	LJ-1190	RLI Aviation Inc. Peoria, Il.	
☐ N3196K	94	King Air C90B	LJ-1384	Kevin Smith, Coral Gables, Fl.	
☐ N3199A	95	King Air B200	BB-1499	Texas Smokehouse Foods Inc. Lufkin, Tx.	
☐ N3203P	70	681 Turbo Commander	6019	Marsh Aviation International, Mesa, Az.	C-GFAE
Regn	*Yr*	*Type*	*c/n*	*Owner/Operator*	*Prev Regn*

Regn	Yr	Type	c/n	Owner/Operator	Prev Regn
□ N3208T	95	King Air B200	BB-1508	Five Rs-Richards Group, Goleta, Ca.Ks.	
□ N3213G	95	King Air B200	BB-1510	R J Transportation, Rupert, Id.	
□ N3216K	95	King Air C90B	LJ-1392	Raytheon Aircraft Co. Wichita, Ks.	
□ N3216U	95	King Air C90B	LJ-1397	Cemco Inc. Albuquerque, NM.	
□ N3217N	95	King Air B200	BB-1513	Utopia Services Inc. Naples, Fl.	
□ N3217X	95	King Air C90B	LJ-1417	Aero Maquila Inc. Del Rio, Tx.	
□ N3220L	95	King Air C90B	LJ-1420	Cavalier Industries Inc. Addison, Al.	
□ N3223H	95	King Air B200	LJ-1425	Mark Aviation Inc. Orlando, Fl.	
□ N3230X	95	King Air B200	BB-1520	W A C Charter Inc. Portland, Or.	
□ N3234S	87	2000 Starship	NC-3	Raytheon Aircraft Co. Wichita, ks.	
□ N3237K	95	Beech C90B	LJ-1390	City Transportation LLC. Birmingham, Al.	
□ N3237S	88	King Air 300	FA-163	Cheds Investments Inc. Charlotte, NC.	
□ N3242L	95	King Air C90B	LJ-1413	Norfolk Iron & Metal Co. Norfolk, Ne.	(N90KA)
□ N3242Z	96	King Air C90B	LJ-1428	Grouper LLC. Bozeman, Mt.	
□ N3250V	96	King Air B200	BB-1523	Standard Air News Inc. Fort Stockton, Tx.	
□ N3251Q	96	King Air C90B	LJ-1429	Jetflight Aviation Inc. Lugano, Switzerland.	
□ N3252V	96	King Air 350	FL-134	Executive Transportation Inc. Los Angeles, Ca.	
□ N3252X	96	King Air B200	BB-1532	Hawker Pacific Inc. Sun Valley, Ca.	
□ N3263C	96	King Air C90B	LJ-1432	R K Petroleum Corp. Midland, Tx.	
□ N3263X	97	King Air 350	FL-143	Euroflight Inc. Wichita, Ks.	
□ N3265K	96	King Air C90B	LJ-1450	JMR Oil Inc. Wilmington, De.	
□ N3268M	96	King Air C90B	LJ-1438	American Speciality Insurance Services Inc. Roanoke, Va.	
□ N3270K	95	King Air C90B	LJ-1401	Raytheon Aircraft Credit Corp. Wichita, Ks.	
□ N3270V	96	King Air C90B	LJ-1451	C S Flight Service Inc. Rock Island, Il.	
□ N3330K	72	Mitsubishi MU-2J	551	Air Response Inc. Nelliston, NY.	N111WN
□ N3330S	81	Conquest II	441-0205	Lario Oil & Gas Co. Wichita, Ks.	(N2727X)
□ N3500E	71	King Air 100	B-74	Vernon Sorenson MD. Bakersfield, Ca.	N3500P
□ N3606T	70	King Air 100	B-30	Sunshine Aero Industries Inc. Crestview, Fl.	N360BT
□ N3663B	80	King Air B100	BE-94	Central Crude Inc & Jordan Oil Co. Lake Charles, La.	
□ N3663M	80	King Air 200	BB-686	Siro Partners, Evansville, In.	
□ N3688P	80	King Air C90	LJ-915	Hawthorne Services Inc. Charleston, SC.	
□ N3690F	80	King Air C90	LJ-921	Elliott Aviation Aircraft Sales Inc. Des Moines, Ia.	
□ N3695W	80	King Air C90	LJ-924	Maryland State Police, Baltimore, Md.	
□ N3697F	80	King Air 200C	BL-14	Columbia Helicopters Inc. Aurora, Or.	
□ N3699B	81	King Air B100	BE-107	Sair Aviation Inc. Syracuse, NY.	
□ N3700M	80	King Air E90	LW-340	McDonnell Douglas Corp. St Louis, Mo.	
□ N3710A	81	King Air 200	BB-760	Tulsair Beechcraft Inc. Tulsa, Ok.	
□ N3722Y	80	King Air 200	BB-719	Benson Chrysler Aviation Dept. Greer, SC.	YV-384CP
□ N3738B	81	King Air 200	BB-774	Hertel Aviation LLC. Minneapolis, Mn.	
□ N3739C	90	King Air 350	FL-11	Bass Enterprises Production Co. Fort Worth, Tx.	
□ N3741M	81	King Air E90	LW-342	LJT Leasing LLC. Albertville, Mn.	
□ N3805E	81	King Air C90	LJ-943	Champ Resources Inc. Houston, Tx.	
□ N3809C	81	King Air F90	LA-112	Bruce Foods Corp. New Iberia, La.	
□ N3811F	81	King Air B100	BE-109	Maddox Petroleum Inc. Wichita Falls, Tx.	
□ N3813C	76	King Air E90	LW-196	For The Birds Inc. San Diego, Ca.	N2269L
□ N3817H	81	King Air C90	LJ-938	Penn State University, University Park, Pa.	
□ N3821S	81	King Air E90	LW-346	D K Leasing, Dallas, Tx.	
□ N3824V	81	King Air F90	LA-110	Deutsch Co Electronic Components, Santa Monica, Ca.	
□ N3825E	81	King Air F90	LA-116	Galco Leasing Co. Houston, Tx.	
□ N3867N	70	681 Turbo Commander	6010	Sierra American Corp. Dallas, Tx.	XB-PAO
□ N3929G	73	King Air E90	LW-55	Gale Investments Ltd. San Antonio, Tx.	XB-IEI
□ N3951F	75	PA-31T Cheyenne II	7520010	Barron Thomas Aviation Inc. Las Vegas, NM.	XA-JOF
□ N3998Y	80	PA-31T Cheyenne II	8020055	Prestige Aero Ltd. Wilmington, De.	PH-TAX
□ N4000	77	King Air 200	BB-247	MCM Construction Inc. North Highlands, Ca.	N18347
□ N4019	73	SA-226AT Merlin 4	AT-014	Career Aviation Academy, Oakdale, Ca.	C-GSDR
□ N4042J	81	King Air B200	BB-874	Stevens Express Leasing Inc. Memphis, Tn.	N200GK
□ N4051X	69	SA-26AT Merlin 2B	T26-124	Rick Fowler/Western Aviatiors, Grand Junction, Co.	
□ N4065D	75	Mitsubishi MU-2L	660	International Business Aircraft, Tulsa, Ok.	TF-FHL
□ N4095S	75	King Air C90	LJ-655	Pinney Leasing LLC. Tulsa, Ok.	
□ N4098T	81	PA-42 Cheyenne III	8001058	Citrus World Inc. Lake Wales, Fl.	
□ N4100B	68	680W Turbo Commander	1803-25	Mid-State Drainage Products Inc. Port Gibson, Ms.	N5079E
□ N4100L	81	PA-42 Cheyenne III	8001061	Executive Leasing Inc. Shelby, NC.	
□ N4114A	82	PA-42 Cheyenne III	8001102	Mawson & Mawson Inc. Langhorne, Pa.	
□ N4116Q	84	PA-42 Cheyenne IIIA	5501012	Ned Good, Pasadena, Ca.	
□ N4116W	75	PA-31T Cheyenne II	7520032	Pace Aviation Ltd. Reno, Nv.	F-BXLC
□ N4119X	85	PA-42 Cheyenne 400LS	5527028	Rowe Aircraft Inc. Millington, Tn.	
□ N4200A	70	King Air 100	B-64	Jet Plane Inc. Pasadena, Tx.	
□ N4203C	75	Mitsubishi MU-2L	671	Epps Air Service Inc. Atlanta, Ga.	EC-EDE
□ N4209S	74	SA-226T Merlin 3	T-245	Jones Aviation Inc. Lexington, Ky.	RP-C1261
□ N4216S	77	King Air E90	LW-211	Tica Investment Corp. San Diego, Ca.	N88RG
□ N4260X	69	SA-26AT Merlin 2B	T26-158	James Benham, Baton Rouge, La.	
□ N4262X	69	SA-26AT Merlin 2B	T26-153	Latham Aviation Inc. Pelham, Al.	
□ N4273X	76	SA-226T Merlin 3A	T-268	Arens Industries Inc. Northfield, Il.	N19SD
□ N4276Z	81	Conquest 1	425-0103	Interforest Corp. Greensboro, NC.	C-GINT
□ N4277C	88	King Air 1300	BB-1309	Frontier Flying Service Inc. Fairbanks, Ak.	N309YV
□ N4277E	88	King Air 1300	BB-1314	Raytheon Aircraft Co. Wichita, Ks.	N314YV

Regn	Yr	Type	c/n	Owner/Operator	Prev Regn
☐ N4288S	76	King Air 200	BB-188	Porta Kamp/P K Air Charter, Bellville, Tx.	
☐ N4298S	77	King Air 200	BB-198	Corporate Air, Billings, Mt.	
☐ N4392W	74	King Air A100	B-192	Air Sal Leasing Inc. Miami, Fl.	
☐ N4420F	82	Conquest 1	425-0053	Nybo Aviation Inc. Skakopee, Mn.	(D-IFLY)
☐ N4430V	78	King Air 200	BB-323	Air Laurel Inc. Jackson, Ms.	YV-74CP
☐ N4441T	80	Conquest II	441-0133	Tennyson Enterprises Inc. Ottumwa, Ia.	N332S
☐ N4447W	74	King Air C90	LJ-627	ILPI Inc. Coral Gables, Fl.	
☐ N4449Q	80	King Air C90	LJ-895	Warren Oil Co. Dunn, NC.	
☐ N4488L	95	King Air B200	LJ-1423	Kidentair Management Systems Inc. Van Nuys, Ca.	N3253Q
☐ N4495N	72	King Air E90	LW-14	Horizon Aviation/Woven Electronics Corp. Greenville, SC.	YV-72CP
☐ N4495U	81	King Air C90	LJ-948	McGriff Seibels & Williams, Birmingham, Al.	YV-381CP
☐ N4600K	84	King Air 300	FA-21	Blue Sky Aviation Inc. Portland, Or.	N4000K
☐ N4622E	68	680W Turbo Commander	1723-3	Cx USA 2/97 to ? (was 1723-94).	
☐ N4679K		SA-226AT Merlin 4	AT-006	Hurley Aircraft, Yukon. Ok.	XC-
UTF/TP207					
☐ N4679M	78	King Air 200	BB-343	Raytheon Aircraft Co. Wichita, Ks.	
☐ N4682E	66	680V Turbo Commander	1630-56	Cincinnati Technical College, Cincinnati, Oh. (status ?).	N4682E
☐ N4700K	81	King Air B100	BE-118	Mechanical Tool & Engineering Co. Rockford, Il.	N3866B
☐ N4715W	85	Gulfstream 1000B	96204	U S Department of Justice, Fort Worth, Tx.	N224GA
☐ N4717V	74	Rockwell 690A	11220	International Jet Aviation Inc. Van Nuys, Ca.	YV-236P
☐ N4725M	78	King Air E90	LW-275	James Hall & James Branton, San Antonio, Tx.	
☐ N4764A	88	King Air C90A	LJ-1161	M & I First National Leasing Corp. Milwaukee, Wi.	N476JA
☐ N4774M	78	King Air C90	LJ-771	State Highway Commission, Jefferson City, Mo.	
☐ N4776M	78	King Air C90	LJ-776	The Pickens Co. Dallas, Tx.	
☐ N4799M	78	King Air 200	BB-373	Security Investments Inc. Oshkosh, Wi.	
☐ N4820M	78	King Air E90	LW-280	Flight Specialists Inc. North Canton, Oh.	
☐ N4947M	78	King Air C90	LJ-780	THF Aircraft LLC. St Louis, Mo.	
☐ N4948W	65	King Air 90	LJ-31	Stewart-Davis International Inc. N Hollywood, Ca.	(N505M)
☐ N4950C	74	King Air C90	LJ-629	G-Star Corp. Memphis, Tn.	YV-39CP
☐ N4954S	77	King Air E90	LW-231	Four Corners Aviation Inc. Farmington, NM.	YV-27P
☐ N5007	76	Rockwell 690A	11271	Cornerstone Propane GP Inc. Lebanon, Mo.	N690JT
☐ N5095K	81	King Air C90	LJ-928	M Cameron Head Inc. Camilla, Ga.	HI-366CT
☐ N5111U	66	King Air A90	LJ-154	Aircraft Aloft LP/R Zadow & C Lingenfelser, Las Vegas, Nv.	N5111
☐ N5123	95	SAAB 2000	020	General Motors Corp. Detroit, Mi.	N5123L
☐ N5124	95	SAAB 2000	027	General Motors Corp. Detroit, Mi.	SE-027
☐ N5125	95	SAAB 2000	030	General Motors Corp. Detroit, Mi.	SE-030
☐ N5136V	76	PA-31T Cheyenne II	7620037	Scenic Airlines Inc. Las Vegas, Nv.	C-GFIN
☐ N5160S	91	2000A Starship	NC-24	Raytheon Aircraft Co. Wichita, Ks.	
☐ N5245F	66	King Air 90	LJ-92	McBee Leasing Co. Tulsa, Ok.	XC-FUR
☐ N5317M	73	SA-226T Merlin 3	T-236	Michael Mass, San Ysidro, Ca.	
☐ N5356M	80	Gulfstream 980	95036	Mavax Ltd. Chesapeake, Va.	D-IOEB
☐ N5371	80	King Air C90	LJ-906	Extru-Tech Inc. Sabetha, Ks.	
☐ N5441M	77	SA-226T Merlin 3A	T-283	Worldwide Aircraft Services Inc. Springfield, Mo.	SE-GXV
☐ N5450J	82	Gulfstream 1000	96024	Gallo Air Inc. Dover, De.	YV-484CP
☐ N5462G	73	King Air E90	LW-69	Northwest Aircraft Leasing Corp. Wilmington, De.	N769AM
☐ N5521T	94	King Air C90B	LJ-1361	George Inc/Balkema Inc. Kalamazoo, Mi.	
☐ N5530H	89	King Air B200	BB-1331	Teton Equity Investments Corp. Jackson, Wy.	
☐ N5549B	92	2000 Starship	NC-15	Raytheon Aircraft Services Inc. Wichita, Ks.	
☐ N5552U	90	King Air B200	BB-1361	Mission of Nevada Inc. Las Vegas, Nv.	
☐ N5559X	90	King Air B200	BB-1372	Gold Run Aviation Inc. Menlo Park, Ca.	
☐ N5568V	90	King Air B200	BB-1352	Circle L Aviation Corp. Key Largo, Fl.	
☐ N5598L	90	King Air C90A	LJ-1236	Interlease Aviation Corp. Augsburg, Germany.	
☐ N5626Y	91	King Air 350	FL-43	Bank One Indiana NA. Indianapolis, In.	
☐ N5639K	90	King Air C90A	LJ-1239	Flightcraft Inc. Portland, Or.	
☐ N5641X	90	King Air C90A	LJ-1241	Southwestern Energy Co. Fayetteville, Ar.	
☐ N5644E	90	King Air C90A	LJ-1244	T-L Irrigation Co. Hastings, Ne.	
☐ N5647Q	90	King Air 300	FA-213	Bank One Arizona NA. Phoenix, Az.	
☐ N5655K	90	King Air 350	FL-12	Crescent Electric Supply Co. East Dubuque, Il.	
☐ N5682P	90	King Air B200	BB-1358	Design Homes Inc. Prairie du Chien, Wi.	
☐ N5727	70	King Air 100	B-48	Kenosha Beef International Inc. Kenosha, Wi.	N572
☐ N5801D	84	King Air C-12F	BL-74	USAF, 1400th Military Airlift Squadron, Nellis AFB. Nv.	84-0144
☐ N5803F	84	King Air C-12F	BL-105	USAF, 1400th Military Airlift Squadron, Nellis AFB. Nv.	84-0175
☐ N5805	80	SA-226T Merlin 3B	T-324	Department of Justice, El Paso, Tx.	N77UU
☐ N5852K	85	Gulfstream 1000B	96205	Freelance Air Inc. Atlanta, Ga.	N226GA
☐ N5878K	80	Gulfstream 840	11626	Jack Canine, Crawfordsville, In.	
☐ N5900K	80	Gulfstream 840	11648	Erwin Industries Inc. Hillsboro, Or.	
☐ N5911P	73	King Air E90	LW-70	Aerocraft International Inc. Coconut Grove, Fl.	N133K
☐ N5914K	81	Gulfstream 840	11662	Safeguards International Inc. Charlotte, NC.	OY-SVG
☐ N5955K	81	Gulfstream 840	11703	Security Group Inc. Fort Pierce, Fl.	
☐ N5956K	81	Gulfstream 840	11719	Leo Sullivan, O'Fallon, Il.	PT-LRQ
☐ N6026K	79	King Air C90	LJ-826	Irv Guinn Construction Co. Bakersfield, Ca.	(N20GT)
☐ N6045S	79	King Air B100	BE-65	Duty Free Aviation Inc. Glen Burnie, Md.	
☐ N6051C	79	King Air 200	BB-499	Marc Fruchter/Arrow International Inc. Reading, Pa.	N302PC
☐ N6080A	78	PA-31T Cheyenne II	7820077	Trendwest-Windmill Inns of America, Klamath Falls, Or.	
☐ N6107A	78	PA-31T Cheyenne II	7820087	Mid-South Industries Inc. Gadsden, Al.	
☐ N6120C	82	King Air B200	BB-979	Hershey Foods Corp. Hershey, Pa.	
Regn	*Yr*	*Type*	*c/n*	*Owner/Operator*	*Prev Regn*

Regn	Yr	Type	c/n	Owner/Operator	Prev Regn
N6125A	78	PA-31T Cheyenne 1	7804009	Steve Locklear, Tuscaloosa, Al.	
N6134A	78	PA-31T Cheyenne 1	7804006	Marion McCann, Tazewell, Va.	
N6137	65	King Air 90	LJ-47	Turbine Power Inc. Bridgewater, Va.	N613M
N6166A	79	PA-31T Cheyenne II	7920005	Over & Out Inc. Clayton, NM.	
N6173C	82	King Air C90-1	LJ-1014	73 Charlie Co. Burnsville, Mn.	
N6175A	79	PA-31T Cheyenne II	7920011	David Roeberg, Wilmington, De.	C-GHXG
N6182A	94	King Air B200	BB-1484	Colleen Corp. Philadelphia, Pa.	N200KA
N6192A	79	PA-31T Cheyenne 1	7904009	Cianbro Corp. Pittsfield, Me.	
N6200B	67	King Air A90	LJ-250	Clark Marine Corp. Mechanicsburg, Pa.	(N66FS)
N6207F	82	King Air C90-1	LJ-1017	Roche Fruit Co. Yakima, Wa.	
N6228Q	67	King Air A90	LJ-280	Cumberland Board of Vocational Education, Bridgeton, NJ.	N46G
N6238N	66	King Air A90	LJ-124	Kerr-Muter Inc. Athens, Oh.	F-BINE
N6271C	82	King Air B200	BB-1036	University of Texas, Austin, Tx.	
N6280E	82	King Air C90-1	LJ-1015	Beech Transportation Inc. Eden Prairie, Mn.	
N6281R	81	Conquest II	441-0226	Holli Aero, Hollister, Ca.	HK-3456
N6308F	82	King Air B200	BB-1014	State of Texas, Austin, Tx.	
N6335F	82	King Air F90	LA-190	Prestige Care Inc. Portland, Or.	
N6354H	82	King Air B100	BE-131	Duty Free Aviation Inc. Glen Burnie, Md.	
N6356C	83	King Air C90-1	LJ-1052	Medical University of South Carolina, Charleston, SC.	
N6406S	77	King Air E90	LW-206	NLF Inc. Wheelersburg, Oh.	
N6451D	82	King Air B200	BB-1009	U S Department of Energy, Las Vegas, Nv.	
N6492C	83	King Air C90-1	LJ-1050	Midwest Aviation Services Inc. West Paducah, Ky.	
N6507B	79	King Air 200	BB-498	U S Customs Service, Washington, DC.	N23707
N6509F	79	King Air 200	BB-493	HAB Enterprises, Oklahoma City, Ok.	N6040W
N6531N	82	King Air B200	BB-1081	SEPCO Industries Inc. Houston, Tx.	
N6540V	66	680V Turbo Commander	1571-26	Portland Community College, Portland, Or. (status ?).	
N6563K	82	King Air C90-1	LJ-1032	Austra Corp. Austin, Mn.	
N6569L	74	Mitsubishi MU-2L	645	XCorp. Tulsa, Ok.	YV-11CP
N6571S	76	King Air E90	LW-171	Indianapolis Aviation, Fishers, In.	
N6590Y	90	Reims/Cessna F406	F406-0052	Cessna Aircraft Co. Wichita, Ks.	
N6591L	90	Reims/Cessna F406	F406-0053	Cessna Aircraft Co. Wichita, Ks.	
N6591R	90	Reims/Cessna F406	F406-0054	Cessna Aircraft Co. Wichita, Ks.	
N6604L	83	King Air B200	BB-1121	Cubic Corp. San Diego, Ca.	
N6606R	83	King Air B200	BB-1122	Dalton Woods, Shreveport, La.	
N6609K	83	King Air B200	BB-1120	Farmland Industries Inc. Kansas City, Mo.	
N6642B	84	King Air 350	FA-1	Raytheon Aircraft Co. Wichita, Ks.	
N6644J	82	King Air B200	BB-1031	Pellerin Milnor Corp. Kenner, La.	
N6646R	79	King Air C90	LJ-836	El Aero Services Inc. Elko, Nv.	
N6656D	80	King Air C90	LJ-879	Jim's Supply Co. Bakersfield, Ca.	
N6662D	79	King Air C90	LJ-869	Associated Vintage Aviation, Napa, Ca.	
N6663A	79	King Air C90	LJ-864	Penn State University, University Park, Pa.	
N6672N	80	King Air C90	LJ-875	Corporate Flight inc/Lawrence Printing Co. Greenwood, Ms.	
N6681S	79	King Air C90	LJ-850	Paper Chemicals Inc. Texarkana, Tx.	
N6683W	83	King Air B200	BB-1154	Red Tail LLC. Lake Osweego, Or.	
N6684B	80	King Air 200	BB-631	KMR Aviation Inc. Ontario, Ca.	
N6685P	82	King Air F90	LA-194	John Sutter, Charleston, WV.	
N6689D	80	King Air 200	BB-623	Utility Air Inc + co-owners, Moberly, Mo.	
N6690L	85	King Air F90-1	LA-226	Hudbro Aviation Inc. Wilmington, De.	
N6690R	83	King Air F90-1	LA-205	S S Air Inc. Vidalia, Ga.	
N6692D	84	King Air C90A	LJ-1072	William Coleman, Spokane, Wa.	
N6723Y	82	King Air C90-1	LJ-1013	NA Degerstrom Inc. Spokane, Wa.	N2872B
N6732V	80	King Air C90	LJ-899	Commercial Plastics Co. Mundelein, Il.	
N6739P	80	King Air 200	BB-628	Moorman Pontiac Inc & Shawnco Inc. Datyon, Oh.	
N6749E	80	King Air F90	LA-43	Metrolimas Green Houses/T & S Aircraft Inc. Huntersville, NC	
N6754H	80	King Air C90	LJ-891	Commander Properties Inc. Wilmington, De.	
N6756P	80	King Air B100	BE-92	Passport Leasing Corp. Fort Lauderdale, Fl.	
N6763K	84	King Air C90A	LJ-1064	Tri Star Aviation Inc. Fulton, Ms.	
N6767M	83	Gulfstream 1000	96084	P A Bergner & Co. Peoria, Il.	OY-BPF
N6772P	81	Conquest 1	425-0022	Western Slope Auto Co. Grand Junction, Co.	
N6774Z	81	Conquest 1	425-0048	Ricarda Corp. Beatrice, Ne.	(N425E)
N6786S	97	King Air 300	FL-166	Hawker Pacific Inc. Sun Valley, Ca.	
N6789	90	King Air B200	BB-1382	Olivebus Corp. Wilmington, De.	N5669B
N6812W	85	King Air 300	FA-38	Diversified Energy Inc. Knoxville, Tn.	
N6816A	85	King Air B200	BB-1200	Fiber Materials Inc. Biddeford, Me.	
N6832M	83	Conquest II	441-0282	Firemans Fund Insurance Companies,	
N6837R	83	Conquest II	441-0295	Montana Dakota Utilities Co. Bismarck, ND.	
N6838T	83	Conquest II	441-0297	Griffith Aviation Co. Tulsa, Ok.	C-GBFM
N6840T	83	Conquest II	441-0299	Citation Oil & Gas Corp. Houston, Tx.	
N6844D	81	Conquest 1	425-0062	SSF Aviation Corp. Charlotte, NC.	
N6844S	81	Conquest 1	425-0065	Edward Austin Jr. Hebbronville, Tx.	N81TR
N6844T	81	Conquest 1	425-0066	McDevitt & Street Co. Charlotte, NC.	
N6845R	81	Conquest 1	425-0071	Woods Equipment Co. Oregon, Il.	
N6846S	81	Conquest 1	425-0081	William Wheeler, Las Vegas, Nv.	
N6847P	81	Conquest 1	425-0087	W P Hobby, Houston, Tx.	
N6851G	81	Conquest 1	425-0112	Jack Bowles, Oklahoma City, Ok.	
N6851L	82	Conquest 1	425-0113	Master Craft Engineering Co. Tifton, Ga.	

Regn	Yr	Type	c/n	Owner/Operator	Prev Regn
☐ N6851T	81	Conquest II	441-0211	Ward Leasing Co. Stamford, Ct.	
☐ N6851X	81	Conquest II	441-0212	Superior Aviation, Iron Mountain, Mi.	
☐ N6851Y	81	Conquest II	441-0213	Pulice Construction Inc. Phoenix, Az.	
☐ N6853T	81	Conquest II	441-0220	Escape Air Service Inc. Wilmington, De.	
☐ N6855P	82	Conquest II	441-0233	Bear Claw Aviation Inc. Los Angeles, Ca.	
☐ N6857T	82	Conquest II	441-0247	Wyoming Aeronautics Commission, Cheyenne, Wy.	
☐ N6860C	83	Conquest II	441-0304	Calibrated Forms Co. Columbus, Ks.	YV-34CP
☐ N6872L	83	Conquest 1	425-0167	Vinlytech Corp. Phoenix, Az.	
☐ N6881S	69	King Air B90	LJ-450	Dodson International Parts Inc. Ottawa, Ks.	N979LX
☐ N6882C	82	Conquest 1	425-0121	Roxine Stone, Idaho Falls, Id.	
☐ N6882D	81	Conquest 1	425-0122	JLS Air Inc. Claremont, NC.	G-BJYA
☐ N6883R	83	Conquest 1	425-0132	Fisher Sand & Gravel Co. Dickinson, ND.	
☐ N6885P	82	Conquest 1	425-0143	Manor Care Inc. Silver Spring, Md.	
☐ N6885S	82	Conquest 1	425-0144	State Aircraft Pooling Board, Austin, Tx.	
☐ N6886V	83	Conquest 1	425-0154	Tomarce Inc. Danville, Il.	
☐ N7000B	66	King Air A90	LM-1	Turbine Power Inc. Bridgewater, Va.	66-18000
☐ N7007G	66	King Air A90	LM-2	Turbine Power Inc. Bridgewater, Va.	66-18001
☐ N7007Q	66	King Air A90	LM-5	Turbine Power Inc. Bridgewater, Va.	66-18004
☐ N7010L	66	King Air A90	LM-7	Turbine Power Inc. Bridgewater, Va.	66-18006
☐ N7018F	66	King Air A90	LM-13	Turbine Power Inc. Bridgewater, Va.	66-18013
☐ N7026H	66	King Air A90	LM-15	Dynamic Avlease Inc. Bridgewater, Va.	66-18014
☐ N7031K	82	Gulfstream 1000	96009	Eagle Air Inc. Memphis, Tn.	HK-3271
☐ N7034K	66	King Air A90	LM-22	Dynamic Avlease Inc. Bridgewater, Va.	66-18021
☐ N7040J	96	King Air A90	LM-31	Dynamic Avlease Inc. Bridgewater, Va.	66-18030
☐ N7057A	81	Gulfstream 840	11664	MALECO, Salem, Or.	G-RNCO
☐ N7066D	76	King Air C-12C	BC-40	U S Customs Service, San Diego, Ca.	76-22563
☐ N7069A	77	King Air C-12A	BC-54	U S Customs Service, Washington, DC.	77-22943
☐ N7071N	66	King Air A90	LM-59	Dynamic Avlease Inc. Bridgewater, Va.	66-18058
☐ N7074G	73	King Air C-12C	BC-17	AVPAC Inc. Oklahoma City, Ok.	73-22266
☐ N7087N	89	King Air B200	BB-1324	Chiara Aeroservices Inc. Dover, De.	G-JHAN
☐ N7101L	80	Gulfstream 980	95027	Southwest Jet Inc. Belton, Mo.	N83SA
☐ N7139B	81	PA-42 Cheyenne III	8001042	Justin Meyer & Raymond Duncan, Oakville, Ca.	N54568
☐ N7155P	96	King Air A90	LM-107	Turbine Power Inc. Bridgewater, Va.	66-18111
☐ N7157K	66	King Air A90	LM-115	Turbine Power Inc. Bridgewater, Va.	66-18119
☐ N7166P	79	King Air 200	BB-482	Manker Aerial Mapping Inc. Oklahoma City, Ok.	N6017
☐ N7202Y	84	King Air 300	FA-12	McDonnell Douglas Corp. St Louis, Mo.	
☐ N7203R	85	King Air B200	BB-1199	Lufkin Industries Inc. Lufkin, Tx.	
☐ N7206E	85	King Air F90-1	LA-234	Stroeh Corporate Ventures Ltd. Incline Village, Nv.	
☐ N7218Y	85	King Air 300	FA-37	Baker Hughes Oilfield Operations, Houston, Tx.	
☐ N7223X	85	King Air C90A	LJ-1104	David Kieffaire, Laramie, Wy.	
☐ N7228T	84	King Air 300	FA-20	Lonnie Pilgrim, Pittsburg, Tx.	
☐ N7230H	85	King Air C90A	LJ-1113	Skybird Ltd. Sanford, NC.	
☐ N7231P	85	King Air 300	FA-56	Chuck Collins & Associates Inc. Carlsbad, Ca.	
☐ N7233R	83	King Air B200C	BL-69	U S Department of Energy, Albuquerque, NM.	N2811B
☐ N7247Y	73	King Air C-12C	BC-3	U S Customs Service, Oklahoma City, Ok.	73-22252
☐ N7250L	85	King Air 300	FA-81	Wheless Industries Inc. Shreveport, La.	
☐ N7250T	85	King Air B200	BB-1237	Puget Sound Power & Light Co. Bellevue, Wa.	
☐ N7252S	84	King Air C90A	LJ-1087	Spencers Inc. Mount Airy, NC.	
☐ N7256K	85	King Air B200	BB-1241	Carlton Co. Albany, Ga.	
☐ N7377	66	King Air A90	LJ-115	Godwin Aircraft Inc. Memphis, Tn.	N30KS
☐ N7388K	90	2000 Starship	NC-7	Raytheon Aerospace Co. Madison, Ms.	(N4NV)
☐ N7400V	75	King Air E90	LW-152	Santa Rosa de Aviacion SRL. Pucallpa, Peru.	N7400V
☐ N7586Z	85	King Air 300	FA-71	Northport Air Inc. Tuscaloosa, Al.	ZS-LOI
☐ N7603	68	SA-26AT Merlin 2B	T26-112	TRH Inc/Robo Aviation, Fayetteville, Tn.	N1222S
☐ N7610U		680V Turbo Commander	1548-10	Joseph Fink Jr. New Orleans, La.	C-FHAP
☐ N7644R	68	King Air B90	LJ-335	Dynamic Avlease Inc. Bridgewater, Va.	
☐ N7703L	78	PA-31T Cheyenne II	7820073	814G LLC. Chapmansboro, Tn.	C-GBVO
☐ N7812	82	Gulfstream 1000	96089	Omega Air Inc. Long Beach, Ca.	N20TX
☐ N7896G	84	Gulfstream 1000	96070	Lindsey Aviation Services, Haslet, Tx.	N67GA
☐ N7931D	83	King Air C90-1	LJ-1049	Safety Seal Piston Ring Co. Marshall, Tx.	
☐ N8000Q	92	2000A Starship	NC-26	Plastic Ingenuity Inc. Cross Plains, Wi.	
☐ N8001V	91	King Air C90A	LJ-1265	Ocala Flight Line Inc. Summerfield, Fl.	
☐ N8002J	92	King Air B200	BB-1425	Bohemia Inc. Portland, Or.	
☐ N8017M	92	King Air B200	BB-1438	A R Wings, Concord, NH.	
☐ N8021P	91	King Air C90A	LJ-1269	Contractors Material Co. Jackson, Ms.	
☐ N8022Q	92	King Air C90B	LJ-1314	Harden Manufacturing Corp. Haleyville, Al.	
☐ N8025L	91	2000 Starship	NC-19	Piedmont Aviation Services Inc. Wichita, Ks.	
☐ N8048U	91	King Air 350	FL-40	Archer Daniels Midland Co. Decatur, Il.	
☐ N8074S	93	2000A Starship	NC-33	Raytheon Aircraft Co. Wichita, Ks.	
☐ N8080C	92	King Air 350	FL-85	RDK Charter Services LLC. Goshen, In.	
☐ N8080Q	91	King Air 350	FL-54	Eagle Wings Inc. Blowing Rock, NC.	
☐ N8083A	79	MU-2 Marquise	739SA	Epps Air Service Inc. Atlanta, Ga.	N707EZ
☐ N8093W	91	King Air B200	BB-1396	Utah Department of Transportation, Salt Lake City, Ut.	
☐ N8096U	93	King Air C90B	LJ-1326	Dennis Smith, Salem, Or.	
☐ N8099Q	91	King Air C90A	LJ-1274	Dodson Aviation Inc. Rantoul, Ks.	
☐ N8100M	80	PA-31T Cheyenne II	8020072	Performance Aviation Inc. Hurst, Tx.	N810CM
Regn	*Yr*	*Type*	*c/n*	*Owner/Operator*	*Prev Regn*

Regn	Yr	Type	c/n	Owner/Operator	Prev Regn
N8116N	91	King Air 350	FL-58	Hardy Boys Motorsports Inc. Dallas, Ga.	
N8118R	75	King Air E90	LW-118	Kopp Clay Co. Malvern, Oh.	
N8119N	92	King Air C-90B	LJ-1320	O'Reilly II Aviation, Springfield, Mo.	
N8119S	93	2000A Starship	NC-34	Bob Jones University, Greenville, SC.	
N8131F	76	PA-31T Cheyenne II	7620042	Scenic Airlines Inc. Las Vegas, Nv.	F-GAJC
N8140P	92	King Air B200	BB-1417	Hoyt Air Inc. Wilmington, De.	
N8145E	93	King Air 350	FL-95	Gilman Aircraft Corp. St Marys, Ga.	
N8148F	91	King Air 350	FL-57	Specialized Aircraft Services Inc. Wichita, Ks.	N319P
N8149S	93	2000A Starship	NC-35	Starship Enterprises Inc. Los Angeles, Ca.	
N8156Z	93	King Air C90B	LJ-1333	SICO-Saudi Investment Co SA. Geneva, Switzerland.	
N8158X	94	2000A Starship	NC-42	Architectural Air LLC. Fort Worth, Tx.	
N8168S	91	2000 Starship	NC-20	Elliott Aviation Inc. Moline, Il.	
N8194S	94	2000A Starship	NC-37	Starship Enterprises Inc. Los Angeles, Ca.	
N8196Q	94	2000A Starship	NC-48	Raytheon Aircraft Co. Wichita, Ks.	
N8208C	93	King Air C90B	LJ-1343	B B Kats Inc. Finchville, Ky.	
N8210C	93	King Air C90B	LJ-1347	Oklahoma Rig & Supply Co. Muskogee, Ok.	
N8215Q	94	2000A Starship	NC-45	FINSA Industrial Park Inc. Brownsville, Tx.	
N8220V	93	King Air C90B	LJ-1344	PCI LLC. Park Ridge, Il.	
N8224Q	95	2000A Starship	NC-49	H L Aircraft Leasing LLC. Atlanta, Ga.	
N8236K	92	King Air B200	BB-1424	Deposit Guaranty Corp. Jackson, Ms.	
N8244L	92	2000A Starship	NC-29	Starship Enterprise Leasing LLC. Las Vegas, Nv.	(N121GV)
N8246S	91	2000 Starship	NC-18	Raytheon Aircraft Co. Wichita, Ks.	
N8259Q	93	King Air C90B	LJ-1332	Jerry's Inc. West Palm Beach, Fl.	
N8275P	92	King Air B300C	FM-4	Stevens Express Leasing Inc. Memphis, Tn.	
N8280S	94	2000A Starship	NC-38	RCO Corp. Wilmington, De.	
N8282S	94	2000A Starship	NC-39	Four Eight Romeo Alpha Inc. Jacksonville, Tx.	
N8283S	94	2000A Starship	NC-41	Continental Datalabel Inc. Elgin, Il.	
N8285Q	94	2000A Starship	NC-50	Principal Edge Group, Las Vegas, Nv.	
N8287E	94	King Air C90B	LJ-1356	Vincent Zaninovich & Sons Inc. Richgrove, Ca.	
N8300S	94	2000A Starship	NC-40	P M B Aviation Inc. Van Nuys, Ca.	
N8484T	74	Mitsubishi MU-2J	617	Cal-Air Charter, Carson City, Nv.	EI-AWY
N8514B	76	King Air B100	BE-6	Rebel Express LLC/Central Coast Aviation Inc. Salinas, Ca.	HB-GEP
N8520L	66	King Air A90	LJ-156	Federal Bureau of Investigation,	N22
N8534W	77	King Air 200	BB-225	Barken International Inc. Salt Lake City, Ut,	I-PIAO
N8535	73	Rockwell 690A	11131	State of Alabama, Montgomery, Al.	N57099
N8617K	83	Conquest II	441-0307	North Carolina Dept of Transportation, Raleigh, NC.	
N8838T	77	Conquest II	441-0003	F M Roberts, Miami, Fl.	N2899P
N8897Y	81	SA-227AT Merlin 4C	AT-492	Career Aviation Academy, Oakdale, Ca.	C-FJTA
N8970N	79	Conquest II	441-0092	Rocky Mountain Holdings LLC. Provo, Ut.	
N9029R	75	King Air E90	LW-132	Dairyland Power Corp. La Crosse, Wi.	
N9032H	68	SA-26T Merlin 2A	T26-7	Napier Air Service Inc. Dothan, Al.	C-GGFJ
N9052Y	79	MU-2 Solitaire	399SA	Superior Builders, St Joseph, Mi.	YV-70CP
N9058N	70	681 Turbo Commander	6008	Broward Community College, Pembroke Pines, Fl.	YV-06CP
N9059S	76	King Air E90	LW-159	State of New Mexico, Santa Fe, NM.	
N9076S	77	King Air C90	LJ-715	Cadogan Properties Inc. Baton Rouge, La.	
N9081R	79	King Air C90	LJ-859	Boulais Aviation Inc. Glendale, Az.	HR-AHJ
N9085U	85	PA-42 Cheyenne IIIA	5501034	U S Customs Service, Oklahoma City, Ok.	
N9091J	85	PA-42 Cheyenne IIIA	5501035	U S Customs Service, Oklahoma City, Ok.	
N9092S	78	King Air C90	LJ-767	Imokolee Transportation Inc. Greenville, De.	
N9097N	70	681 Turbo Commander	6040	A & L Aviation LLC. Trussville, Al.	
N9116Q	86	PA-42 Cheyenne IIIA	5501037	U S Customs Service, Oklahoma City, Ok.	
N9142B	86	PA-42 Cheyenne IIIA	5501038	U S Customs Service, Oklahoma City, Ok.	
N9150R	76	King Air 200	BB-181	White Industries Inc. Bates City, Mo. (status ?).	C-GPKK
N9150T	84	PA-42 Cheyenne IIIA	5501024	U S Customs Service, Washington, DC.	N41182
N9159Y	84	PA-42 Cheyenne IIIA	5501028	U S Customs Service, Washington, DC.	
N9168N	73	Rockwell 690	11068	REL Aviation & Marine Inc. Arlington, Tx.	
N9175N	73	Rockwell 690A	11071	Western Marine Electronics, Woodinville, Wa.	
N9180K	83	Conquest 1	425-0179	Westark Leasing Co. Fort Smith, Ar.	N6873Z
N9183C	83	PA-31T Cheyenne 1A	8304001	Tennessee Turbine Aircraft Sales Inc. Alcoa, Tn.	
N9223N	72	Rockwell 690	11023	BRS Services Inc.	
N9233T	85	PA-42 Cheyenne IIIA	5501032	U S Customs Service, Washington, DC.	
N9266Y	84	PA-31T Cheyenne 1A	1104008	Douglas Olson, Excelsior, Mn.	
N9268Y	84	PA-31T Cheyenne 1A	1104009	Great Southern Trucking Co. Durham, NC.	
N9279A	86	PA-42 Cheyenne IIIA	5501036	U S Customs Service, Oklahoma City, Ok.	
N9426	68	King Air B90	LJ-421	Orange Grove Holdings Inc. Wilmington, De.	
N9442Q	72	King Air C90	LJ-542	Flight Review Inc. Scottsdale, Az.	
N9450Q	72	King Air C90	LJ-550	Jay Breaux, Junction City, Or.	XB-FRW
N9683N	82	Conquest II	441-0255	Crownair Charter inc. San Diego, Ca.	N68587
N9711B	68	King Air B90	LJ-367	Soldwedel LP. Yuma, Az.	N72WE
N9756S	80	Gulfstream 980	95003	Eagle Air Inc. Memphis, Tn.	N501NB
N9767S	80	Gulfstream 980	95014	3MD Charter, Portland, Or.	(N321MD)
N9812S	81	Gulfstream 980	95060	Aerographics, Manassas, Va.	
N9838Z	68	King Air B90	LJ-435	K & K Aircraft Inc. Bridgewater, Va.	D-IHCH
N9898	87	King Air B200	BB-1269	Grede Foundries Inc. Milwaukee, Wi.	N877WA
N9900	76	SA-226T Merlin 3A	T-266	Liquid Charter Services Inc. San Diego, Ca.	(N97FT)
N9901	66	King Air 90	LJ-93	Great Oaks Joint Vocational School District, Cincinnati, Oh.	N48A

Regn	Yr	Type	c/n	Owner/Operator	Prev Regn
☐ N9902S	82	Gulfstream 1000	96002	Hodge Electronics Inc. South Miami, Fl.	
☐ N9933S	79	King Air B100	BE-56	Hiawatha Aviation of Rochester Inc. Rochester, Mn.	N251DA
☐ N9942S	82	Gulfstream 1000	96022	Sioux Aviation Inc. Fort Lauderdale, Fl.	
☐ N9945S	82	Gulfstream 1000	96025	Zuleta Services & Trading Co. Boca Raton, Fl.	
☐ N9948S	82	Gulfstream 1000	96028	Air Operations International, Charlotte, NC.	
☐ N9966S	82	Gulfstream 1000	96046	Franks Petroleum Inc.	
☐ N9968S	82	Gulfstream 1000	96048	Eagle Air Inc. Memphis, Tn.	
☐ N9973S	82	Gulfstream 1000	96053	Ram-Air International Inc. Wilmington, De.	
☐ N10024	96	King Air B300C	FM-9	Stevens Express Leasing Inc. Memphis, Tn.	
☐ N10127	67	King Air A90	LJ-222	A & R Transportation Inc. Omaha, Ne.	N7LH
☐ N10436	96	King Air B200	BB-1536	Hawker Pacific P/L. Seletar, Singapore.	
☐ N11135X	97	King Air C90B	LJ-1494	Raytheon Aircraft Co. Wichita, Ks.	
☐ N11692	78	King Air C90	LJ-772	Four Corners Aviation Inc. Farmington, NM.	F-GFBO
☐ N12099	83	Conquest II	441-0329	Security Aviation/Michael & Marilyn O'Neill, Anchorage, Ak.	
☐ N12201	83	Conquest 1	425-0187	Sawyer Aviation, Phoenix, Az.	(N6874Z)
☐ N12214	84	Conquest 1	425-0194	L T Garner Jr. Wilmington, NC.	
☐ N12244	84	Conquest 1	425-0213	D L S Inc. Fort Payne, Al.	
☐ N12268	85	Conquest 1	425-0227	Sharpline Converting Inc. Wichita, Ks.	
☐ N13622	78	Rockwell 690B	11469	Wal-Mart Stores Inc. Rogers, Ar.	F-GCMJ
☐ N15234	89	King Air C90A	LJ-1194	Erickson Oil Products Inc. Hudson, Wi.	
☐ N15613	89	King Air 300	FA-193	Spencer Aviation Inc. Rockford, Il.	
☐ N17573	77	King Air C90	LJ-714	Elliot Farms, Visalia, Ca.	
☐ N17792	78	King Air B100	BE-41	D P Bolin, W K Altman & Bellevue Properties, Wichita Falls.	
☐ N18260	81	King Air B200	BB-900	Middletown Corporate Aviation Inc. Franklin, Oh.	
☐ N18264	87	King Air C90A	LJ-1156	Aeronautics Commission, Lansing, Mi.	
☐ N18343	77	King Air E90	LW-243	Reuben Richards, Far Hills, NJ.	
☐ N18383	77	King Air C90	LJ-733	Global Mountain Air Inc. Pierre, SD.	
☐ N18471	82	King Air F90	LA-161	Polygon Air Corp. White Plains, NY.	
☐ N20564	79	King Air E90	LW-314	Mario Saikhon Co Trustee, Holtville, Ca.	
☐ N20880	70	King Air 100	B-63	Tony Frost & Gregory Behrens, Red Rock, Az.	F-GGFE
☐ N21868	97	King Air C90B	LJ-1498	First Security Bank NA. Salt Lake City, Ut.	
☐ N22071	76	King Air 200	BB-111	Skylark Aviation Inc. Horseshoe Bay, Tx.	N633EB
☐ N22522	74	Mitsubishi MU-2J	625	Diamond Aviation Inc. Statesboro, Ga.	C-GSKM
☐ N23139	79	PA-31T Cheyenne II	7920032	Kladstrup-Wetzel Associates Ltd. Englewood, Co.	
☐ N23243	79	PA-31T Cheyenne 1	7904017	Stark Oldsmobile Inc. Menomonee Falls, Wi.	
☐ N23250	79	PA-31T Cheyenne 1	7904018	Phoenix Financial LLC. Spokane, Wa.	
☐ N23334	79	PA-31T Cheyenne 1	7904024	Flight One Inc. Dunn, NC.	TG-LIA
☐ N23404	77	King Air A100	B-234	Bradley Flying Service Inc. Wethersfield, Ct.	
☐ N23407	79	PA-31T Cheyenne II	7920055	Mid-South Agricultural Products Inc. Lafayette, La.	
☐ N23426	79	PA-31T Cheyenne 1	7904035	Selland Pontiac-GMC Inc. Moorhead, Mn.	
☐ N23605	77	King Air A100	B-236	Windham Aviation Inc. Willimantic, Ct.	
☐ N23646	79	PA-31T Cheyenne II	7920078	D A Davidson & Co. Great Falls, Mt.	(PH-BAM)
☐ N23658	79	PA-31T Cheyenne II	7904046	Hanks Holding Co. Monroe, Tx.	
☐ N23718	79	PA-31T Cheyenne II	7920087	Bell Aviation Inc. West Columbia, SC.	
☐ N24203	78	King Air B100	BE-40	Montgomery Aviation Corp. Montgomery, Al.	
☐ N25677	75	SA-226T Merlin 3A	T-254	Dept of Forestry & Fire Protection, Mather, Ca..	N58018
☐ N26540	67	King Air A90	LJ-270	Graves Aircraft inc. El Reno, Ok.	MT-231
☐ N27856	76	PA-31T Cheyenne II	7620041	Blowouts & Failures Inc. Baytown, Tx.	D-IEEA
☐ N29997	87	King Air B200	BB-1268	Wallace Westwind Inc. Las Cruces, NM.	
☐ N30234	87	King Air B200	BB-1274	Weis Markets Inc. Sunbury, Pa.	
☐ N30833	88	King Air C90A	LJ-1172	Saddle Creek Corp. Lakeland, Fl.	
☐ N30844	88	King Air C90A	LJ-1169	McElhaney Cattle Co. Welton, Az.	
☐ N31210		Beech 1900C-1	UC-43	Kiti International Corp.	JA8863
☐ N31447	88	King Air C90A	LJ-1187	Lufkin-Conroe Telephone Exchange Inc. Lufkin, Tx.	
☐ N32229	65	King Air 90	LJ-49	Airlift Inc. Hollister, Ca.	C-GNUX
☐ N36805	87	King Air C90A	LJ-1157	Dillon Companies Inc. Hutchinson, Ks.	
☐ N37392	81	King Air 200	BB-768	Marshall Tobins, Weston, Ma.	
☐ N37990	81	King Air F90	LA-101	III B LLC. Oskaloosa, Ia.	
☐ N38280	81	King Air C90	LJ-953	Air Services of Springfield LLC. Springfield, Mo.	
☐ N38381	81	King Air B200	BB-934	Wheels of Africa, Rand, RSA.	
☐ N38473	81	King Air 200	BB-842	National Gypsum Co. Charlotte, NC.	
☐ N38920	77	Mitsubishi MU-2B	031	Per Hellum, Bellevue, Wa.	N28C
☐ N41054	83	Conquest 1	425-0172	Fred Hibbert Jr. The Dalles, Or.	N425LS
☐ N41198	85	PA-42 Cheyenne 400LS	5527030	Aircraft Guaranty Corp. Houston, Tx.	
☐ N44776	80	Conquest II	441-0121	P M Inc. Plano, Tx.	N44776
☐ N44882	81	King Air F90	LA-102	Spirol International Corp. Danielson, Ct.	YV-399CP
☐ N45818 UTE/TP206	73	SA-226T Merlin 3	T-235	U S Department of Justice, El Paso, Tx.	XC-
☐ N50525	66	King Air A90	LJ-159	Dynamic Aviation Group Inc. Bridgewater, Va.	86-0092
☐ N50655	67	680V Turbo Commander	1714-88	Park Leasing Co. Charlotte, NC.	XB-CED
☐ N53474		Reims/Cessna F406	F406-0023	Del-Air Management Inc. Dover, De.	F-WZIJ
☐ N54026	85	DHC 7-103	106	U S Department of Energy, Albuquerque, NM.	
☐ N54163	68	680W Turbo Commander	1774-12	North American Container Corp. Mableton, Ga.	N5416
☐ N55495	90	King Air C90A	LJ-1255	Jerome Sacks, Friday Harbor, Wa.	
☐ N55796	70	King Air 100	B-9	Jesus Alive Ministries, Rand, South Africa.	CF-QDI
☐ N55947	92	King Air B300C	FM-3	Freelance Air Inc. Atlanta, Ga.	

Regn	Yr	Type	c/n	Owner/Operator	Prev Regn
N56016	92	King Air B300C	FM-5	Los Amigos Aircraft Co. Sacramento, Ca.	
N57092	74	Rockwell 690A	11160	Nationwide Marketing Inc. Oklahoma City, Ok.	
N57096	73	Rockwell 690A	11120	HHR Corp/McOco Inc. Houston, Tx.	
N57104	75	Rockwell 690A	11262	Title Financial Group, Blackfoot, Id.	
N57106	73	Rockwell 690A	11106	VEC Corp of Delaware, Teaneck, NJ.	
N57109	73	Rockwell 690A	11109	Pioneer Aviation LLC. Eugene, Or.	
N57112	75	Rockwell 690A	11263	Hogan Manufacturing Inc. Escalon, Ca.	
N57113	73	Rockwell 690A	11113	Ken Muldrow, Brownfield, Tx.	
N57118	76	Rockwell 690A	11311	William Latham, Manassas, Va.	
N57133	73	Rockwell 690A	11133	Corporate Air, Billings, Mt.	(N47EC)
N57154	74	Rockwell 690A	11154	Prairie Wood Products, Riddle, Or.	
N57175	72	Rockwell 690	11004	Aero-Metric Engineering Inc. Sheboygan, Wi.	N37546
N57237	75	Rockwell 690A	11237	Links Air Inc. Jupiter, Fl.	C-GHMD
N57292	75	Rockwell 690A	11270	United CCM Corp. Dallas, Tx.	
N58280	75	King Air 200	BB-61	TIPS Air Inc. San Antonio, Tx.	C-GSEP
N60049	75	PA-31T Cheyenne II	7520036	Arkansas Aircraft inc. Jonesboro, Ar.	C-GPCW
N61228	82	King Air F90	LA-169	DA/PRO Rubber Inc. Tulsa, Ok.	
N61383	82	King Air C90A	LJ-1009	Northland Aluminium Products Co. Minneapolis, Mn.	
N62366	73	Rockwell 690A	11141	Ron Air Inc. Las Vegas, Nv.	HB-GFQ
N62525	76	King Air C90	LJ-691	G W W Partnership, Lamper, Ca.	JA8839
N62526	73	King Air A100	B-161	Aviles Pernia V Inc. Pembroke Pines, Fl.	PT-LJM
N62569	82	King Air B200	BB-1028	Department of Human Resources, Austin, Tx.	
N63200	81	King Air C90	LJ-989	Sunland Fire Protection Inc. High Point, NC.	(YV-442CP)
N63593	79	King Air 200	BB-552	Airtrade Inc. Wilmington, De.	G-BGRD
N63686	82	King Air B200	BB-1085	Terry Johnson, Longmont, Co.	
N63791	83	King Air B200	BB-1100	U S Department of Energy, Portland, Or.	
N66404	83	King Air B200	BB-1129	Executive Beechcraft STL Inc. Chesterfield, Mo.	
N66804	80	King Air B100	BE-82	VC Cattle Co. Palestine, Tx.	(N200JL)
N66820	79	King Air B100	BE-68	Bayport Air Co. Mobile, Al.	N6052C
N66825	83	King Air B200	BB-1156	Commerce Leasing Co. Norcross, Ga.	
N66837	74	PA-31T Cheyenne II	7400008	Booth Enterprises Inc. Roseburg, Or.	
N67262	84	King Air B200	BB-1182	Mark Schmaltz, Van Nuys, Ca.	
N67511	80	King Air C90	LJ-888	Marc Fruchter Aviation Inc. Reading, Pa.	(N111GF)
N67726	81	Conquest 1	425-0030	J & L Oil Inc. Vernon Hills, Il.	
N68734	84	Conquest 1	425-0182	Seanaire Inc. Midland Park, NJ.	
N68823	83	Conquest 1	425-0130	James Thomas Hamm, Micanopy, Fl.	
N68865	83	Conquest 1	425-0160	State Aircraft Pooling Board, Austin, Tx.	
N69084	82	King Air F90	LA-157	Elmo Greer & Sons Construction Co. London, Ky.	
N69301	84	King Air C90A	LJ-1079	Peoples Gas System Inc. Tampa, Fl.	
N70088	66	King Air A90	LM-6	Turbine Power Inc. Bridgewater, Va.	66-18005
N70135	66	King Air A90	LM-9	Turbine Power Inc. Bridgewater, Va.	66-18008
N72069	84	King Air 300	FA-22	Cx USA 6/96 to ? HK- ?	N72069
N72146	85	King Air 300	FA-29	Precise Airplane Co. Clayton, Mo.	
N72470	73	King Air C-12C	BC-4	U S Customs Service, Oklahoma City, Ok.	73-22253
N72472	73	King Air C-12C	BC-14	Los Angeles Police Department, Los Angeles, Ca.	73-22263
N72476	76	King Air C-12C	BC-26	U S Customs Service, Oklahoma City, Ok.	76-22550
N75368	65	King Air 90	LJ-75	Bruffey Flying Inc. Rome, Ga.	C-GJBE
N75465	77	PA-31T Cheyenne II	7720019	Aircraft Restoration Co. Provo, Ut.	C-GSKR
N80605	92	King Air B300C	FM-6	Kasey & Assocs Inc. Opa Locka, Fl.	
N80904	91	King Air C90A	LJ-1271	Robert Garriott, Austin, Tx.	
N81432	85	Gulfstream 1000B	96207	Kasey & Associates Inc. Opa Locka, Fl.	N229GA
N81448	76	Rockwell 690A	11327	Robert Burchette Jr & D P Riggins & Assocs Inc. Linville, NC	
N81601	73	Mitsubishi MU-2J	577	Gale Force Corp. Janesville, Wi.	C-FBAN
N81674	77	Rockwell 690B	11423	Wal-Mart Stores Inc. Rogers, Ar.	
N81703	78	Rockwell 690B	11438	Frasca International Inc. Champaign, Il.	
N81737	78	Rockwell 690B	11454	Tepuy Corp. Coral Gables, Fl.	YV-182CP
N81799	78	Rockwell 690B	11480	Indiana Aircraft Sales Inc. Indianapolis, In.	
N81831	78	Rockwell 690B	11497	Alpar Resources Inc. Perryton, Tx.	
N82094	77	PA-31T Cheyenne II	7720014	Jack Wall Aircraft Sales Inc. Memphis, Tn.	
N82156	77	PA-31T Cheyenne II	7720045	Alfred & Pearl Lion, Fresno, Ca.	
N82161	77	PA-31T Cheyenne II	7720049	Varsity Carpet Services Inc. Dalton, Ga.	
N82163	77	PA-31T Cheyenne II	7720048	Krebes Cattle Co. Scott City, Ks.	
N82232	78	PA-31T Cheyenne II	7820028	Country Health Associates Inc. Marathon, Fl.	
N82290	78	PA-31T Cheyenne II	7820058	Larson Aircraft Sales, North Glenn, Co.	
N82307	71	King Air C90	LJ-519	International Yachting Information Center, Antwerp, Belgium.	XA-RUI
N82324	92	King Air B300C	FM-7	Lindsey Aviation Services, Haslet, Tx.	
N82428	92	2000A Starship	NC-28	Raytheon Aircraft Credit Corp. Wichita, Ks.	
N87699	81	King Air B200	BB-887	Gulf Air Inc. Wilmington, De.	YV-2222P
N88598	78	Conquest II	441-0060	Federated Rural Electric Inc. Madison, Wi.	N441CF
N88692	83	Conquest II	441-0290	Cattlemens Industries Inc. Miami, Fl.	
N88727	82	Conquest II	441-0267	M-E-C Co. Neodesha, Ks.	
N88798	82	Conquest II	441-0265	John & Betty Van Denburgh, Phoenix, Az.	
N88823	78	Conquest II	441-0064	William Paganetti & Natale Carasali, Reno, Nv.	
N88834	82	Conquest II	441-0266	Pridgeon & Clay Inc. Grand Rapids, Mi.	N88834
N90806	75	King Air 200	BB-33	Corporate Jets Inc. West Mifflin, Pa.	ZS-AAA
N91384	73	Rockwell 690A	11118	Cooper Aerial Surveys Ltd. Dover, De.	SE-FLN

269

Regn	Yr	Type	c/n	Owner/Operator	Prev Regn
☐ N92427	79	SA-226T Merlin 3B	T-297	Apex Aviation Inc. Haslet, Tx.	N839KA
☐ N96954	80	SA-226T Merlin 3B	T-311	Lindsey Aviation Services, Haslet, Tx.	N27563
☐ N97315	85	Gulfstream 1000B	96206	U S Department of Justice, Fort Worth, Tx.	N227GA
☐ N97696	81	Gulfstream 840	11697	Los Amigos Aircraft Co. Sacramento, Ca.	HK-....
Military					
☐ 02	62	Gulfstream 1	91	United States Coast Guard, Opa Locka, Fl.	USCG1380
☐ 155724	67	Gulfstream TC-4C	180	U S Navy/to AMARC 10/94 as 4G002.	N786G
☐ 155725	67	Gulfstream TC-4C	182	U S Navy, NAS Whidbey Island, Wa.	N762G
☐ 155726	67	Gulfstream TC-4C	183	U S Navy, NAS Whidbey Island, Wa.	N766G
☐ 155727	67	Gulfstream TC-4C	184	U S Navy/to AMARC 10/94 as 4G004.	
☐ 155728	67	Gulfstream TC-4C	185	U S Navy/to AMARC 10/94 as 4G003.	
☐ 155729	67	Gulfstream TC-4C	186	U S Navy/to AMARC 10/94 as 4G001.	
☐ 155730	67	Gulfstream TC-4C	187	U S Navy/to AMARC 4/95 as 4G005.	
☐ 161185		King Air UC-12B	BJ-1	U.S. Navy, Test Pilot School, NAS Patuxent River, Md.	
☐ 161186		King Air UC-12B	BJ-2	U.S. Navy, Code 7G, NAS Whidbey Island, Wa.	
☐ 161187		King Air UC-12B	BJ-3	U.S. Marine Corps. MCAS Cherry Point, NC.	
☐ 161188		King Air UC-12B	BJ-4	U.S. Navy, Code 7W, COMMIDEASTFOR, Bahrain.	
☐ 161190		King Air UC-12B	BJ-6	U.S. Navy, Code 7K, NAS Memphis, Tn.	
☐ 161191		King Air UC-12B	BJ-7	U.S. Navy, Code 7E, NAS Jacksonville, Fl.	
☐ 161192		King Air UC-12B	BJ-8	U.S. Marine Corps. Code 5Y, MCAS Yuma, Az.	
☐ 161193		King Air UC-12B	BJ-9	U.S. Navy, Code 7M, NAS North Island, Ca.	
☐ 161194		King Air UC-12B	BJ-10	U.S. Marine Corps. Code EZ, MWHS-4, NAS New Orleans, La.	
☐ 161195		King Air UC-12B	BJ-11	U.S. Navy, Code 8F, NAS Guantanamo Bay, Cuba.	
☐ 161196		King Air UC-12B	BJ-12	U.S. Navy, Code 8N, NAS El Centro, Ca.	
☐ 161197		King Air UC-12B	BJ-13	U.S. Navy, Code RW, VRC-30, NAS North Island, Ca.	
☐ 161198		King Air UC-12B	BJ-14	U.S. Navy, Code 7x, NAS New Orleans, La.	
☐ 161199		King Air UC-12B	BJ-15	U.S.M.C., Code 5A, MAG-49 Detachment A, Andrews AFB. Md.	
☐ 161200		King Air UC-12B	BJ-16	U.S. Marine Corps. Code 5Y, MCAS Yuma, Az.	
☐ 161201		King Air UC-12B	BJ-17	U.S. Marine Corps. Code 5T, MCAS El Toro, Ca.	
☐ 161202		King Air UC-12B	BJ-18	U.S. Navy, Code 7L, NAS Point Mugu, Ca.	
☐ 161203		King Air UC-12B	BJ-19	U.S. Navy, Code 7C, NAS Norfolk, Va.	
☐ 161204		King Air UC-12B	BJ-20	U.S. Navy, Code 8N, NAF El Centro, Ca.	
☐ 161205		King Air UC-12B	BJ-21	U.S. Navy, Code 8K, COMMIDEASTOR, Bahrain.	
☐ 161206		King Air UC-12B	BJ-22	U.S. Navy, Code 7G, NAS Whidbey Island, Wa.	
☐ 161306		King Air UC-12B	BJ-23	U.S. Navy, Code 5B, COMMIDEASTFOR, Bahrain.	
☐ 161307		King Air UC-12B	BJ-24	U.S. Navy, Code 7F, NAS Brunswick, Me.	
☐ 161308		King Air UC-12B	BJ-25	U.S. Navy, Code 7J, NAS Alameda, Ca.	
☐ 161309		King Air UC-12B	BJ-26	U.S. Navy, Code RW, VRC-30, NAS North Island, Ca.	
☐ 161310		King Air UC-12B	BJ-27	U.S. Navy, Code 8E, NAS Roosevelt Roads, PR.	
☐ 161311		King Air UC-12B	BJ-28	U.S. Navy, Code 7M, NAS North Island, Ca.	
☐ 161312		King Air UC-12B	BJ-29	U.S. Navy, Code 7J, NAS Alameda, Ca.	
☐ 161313		King Air UC-12B	BJ-30	U.S. Navy, Code 7G, NAS Whidbey Island, Wa.	
☐ 161314		King Air UC-12B	BJ-31	U.S.M.C., Code 5A, MAG-49 Detachment A, Andrews AFB. Md.	
☐ 161315		King Air UC-12B	BJ-32	U.S. Navy, Code F, Training Wing 6, NAS Pensacola, Fl.	
☐ 161316		King Air UC-12B	BJ-33	U.S. Navy, Code G, Training Wing 4, NAS Corpus Christi, Tx.	
☐ 161317		King Air UC-12B	BJ-34	U.S. Marine Corps. MCAS Cherry Point, NC.	
☐ 161318		King Air UC-12B	BJ-35	U.S. Navy, Code 7Q, NAS Key West, Fl.	
☐ 161319		King Air UC-12B	BJ-36	U.S. Navy, Code 7M, NAS North Island, Ca.	
☐ 161320		King Air UC-12B	BJ-37	U.S. Navy, Code RW, VRC-30, NAS North Island, Ca.	
☐ 161321		King Air UC-12B	BJ-38	U.S. Navy, Code 7C, NAS Norfolk, Va.	
☐ 161322		King Air UC-12B	BJ-39	U.S. Navy, Code 7X, NAS New Orleans, La.	
☐ 161323		King Air UC-12B	BJ-40	U.S. Navy, Code RW, VRC-30, NAS North Island, Ca.	
☐ 161324		King Air UC-12B	BJ-41	U.S. Navy, Code 7S, NAS Lemoore, Ca.	
☐ 161325		King Air UC-12B	BJ-42	U.S. Marine Corps. Code 5D, MCAS New River, NC.	
☐ 161326		King Air UC-12B	BJ-43	U.S. Army, Code 7G, NAS Whidbey Island, Wa.	
☐ 161327		King Air UC-12B	BJ-44	U.S. Navy, Code 8J, COMMIDEASTFOR, Bahrain.	
☐ 161497		King Air UC-12B	BJ-45	U.S. Navy, Code 7K, NAS Memphis, Tn.	
☐ 161498		King Air UC-12B	BJ-46	U.S. Navy, Code A, Training Wing 1, NAS Meridian, Ms.	
☐ 161499		King Air UC-12B	BJ-47	U.S. Navy, Code 7Z, NAS South Weymouth, Ma.	
☐ 161500		King Air UC-12B	BJ-48	U.S. Navy, Code 7E, NAS Jacksonville, Fl.	
☐ 161501		King Air UC-12B	BJ-49	U.S. Navy, Code 8U, NAS Mayport, Fl.	
☐ 161502		King Air UC-12B	BJ-50	U.S. Navy, Code 8E, NAS Roosevelt Roads, PR.	
☐ 161503		King Air UC-12B	BJ-51	U.S. Navy, Code 7D, NAF Fort Worth, Fort Worth JRB. Tx.	
☐ 161504		King Air UC-12B	BJ-52	U.S. Navy, Code 7S, NAS Lemoore, Ca.	
☐ 161505		King Air UC-12B	BJ-53	U.S. Navy, Code F, Training Wing 6, NAS Pensacola, Fl.	
☐ 161506		King Air UC-12B	BJ-54	U.S. Navy, Code 7N, NAF Washington, Andrews AFB. Md.	
☐ 161507		King Air UC-12B	BJ-55	U.S. Marine Corps. Code 5D, MCAS New River, NC.	
☐ 161508		King Air UC-12B	BJ-56	U.S. Navy, Code G, NAS Corpus Christi, Tx.	
☐ 161509		King Air UC-12B	BJ-57	U.S. Navy, Code 8F, NAS Guantanamo Bay, Cuba.	
☐ 161510		King Air UC-12B	BJ-58	U.S. Navy, Code 7B, NAS Atlanta, Ga.	
☐ 161511		King Air UC-12B	BJ-59	U.S. Navy, Code 7V, NAS Glenview, Il.	
☐ 161512		King Air UC-12B	BJ-60	U.S. Navy, Code .., NAF Fort Worth, Fort Worth JRB, Tx.	
☐ 161513		King Air UC-12B	BJ-61	U.S. Navy, Code 7R, NAS Oceana, Va.	
☐ 161514		King Air UC-12B	BJ-62	U.S.M.C., Code EZ, MWHS-4, NAS New Orleans, La.	
☐ 161515		King Air UC-12B	BJ-63	U.S. Navy, Code 5B, MCAS Beaufort, SC.	
☐ 161516		King Air UC-12B	BJ-64	U.S. Navy, Code 7W, NAS Willow Grove, Pa.	

Regn   *Yr Type*   *c/n*   *Owner/Operator*   *Prev Regn*

Regn	Yr	Type	c/n	Owner/Operator	Prev Regn
☐ 161517		King Air UC-12B	BJ-65	U.S. Navy, Code 7U, NAS Cecil Field, Fl.	
☐ 161518		King Air UC-12B	BJ-66	U.S. Marine Corps. Code 5T, MCAS El Toro, Ca.	
☐ 163553		King Air UC-12F	BU-1	U.S. Navy, Code 8M, NAS Misawa, Japan.	
☐ 163554		King Air UC-12F	BU-2	U.S. Navy, Code 8H, NAF Atsugi, Japan.	
☐ 163555		King Air UC-12F	BU-3	U.S. Navy, Code RW, NAF Atsugi, Japan.	
☐ 163556		King Air UC-12F	BU-4	U.S. Navy, NAF Atsugi, Japan.	
☐ 163557		King Air UC-12F	BU-5	U.S. Navy, Code 8M, NAF Misawa, Japan.	
☐ 163558		King Air UC-12F	BU-6	U.S. Marine Corps, MCAS Iwakuni, Japan.	
☐ 163559		King Air UC-12F	BU-7	U.S. Marine Corps, MCAS Iwakuni, Japan.	
☐ 163560		King Air UC-12F	BU-8	U.S. Marine Corps. MCAS Futenma, Okinawa, Japan.	
☐ 163561		King Air UC-12F	BU-9	U.S. Marine Corps. MCAS Futenma, Okinawa, Japan.	
☐ 163562		King Air UC-12F	BU-10	U.S. Navy, Code RW, VRC-30, NAS North Island, Ca.	
☐ 163563		King Air UC-12F	BU-11	U.S. Navy, Pacific Missile Range Facility, Barking Sands, Hi	
☐ 163564		King Air UC-12F	BU-12	U.S. Navy, Pacific Missile Range Facility, Barking Sands, Hi	
☐ 163836		King Air UC-12M	BV-1	U.S. Navy, Code 7C, NAS Norfolk, Va.	
☐ 163837		King Air UC-12M	BV-2	U.S. Navy, Code 8G, NAF Mildenhall, England.	
☐ 163838		King Air UC-12M	BV-3	U.S. Navy, NAS Sigonella, Sicily.	
☐ 163839		King Air UC-12M	BV-4	U.S. Navy, NS Rota, Spain.	
☐ 163840		King Air UC-12M	BV-5	U.S. Navy, Code 8G, NAF Mildenhall, England.	
☐ 163841		King Air UC-12M	BV-6	U.S. Navy, NAS Sigonella, Sicily.	
☐ 163842		King Air UC-12M	BV-7	U.S. Navy, NS Rota, Spain.	
☐ 163843		King Air UC-12M	BV-8	U.S. Navy, Code 8G, NAF Mildenhall, England.	
☐ 163844		King Air UC-12M	BV-9	U.S. Navy, NAS Sigonella, Sicily.	
☐ 163845		King Air UC-12M	BV-10	U.S. Navy, Code 7C, NAS Norfolk, Va.	
☐ 163846		King Air UC-12M	BV-11	U.S. Navy, Code 8E, NAS Roosevelt Roads, PR.	
☐ 163847		King Air UC-12M	BV-12	U.S. Navy, Naval Air Weapons Center, NAS Point Mugu, Ca.	
☐ 60112	97	Beech 1900D	UE-256	U S Army,	N10931
☐ 66-15361	66	King Air A90 N901R	LJ-153	VC-6A, U S Army, White Sands Missile Range, Holloman AFB. NM	
☐ 70-15908	72	King Air U-21J	B-95	U S Army,	
☐ 70-15909	72	King Air U-21J	B-96	U S Army	
☐ 70-15910	72	King Air U-21J	B-97	U S Army,	
☐ 70-15911	72	King Air U-21J	B-98	U S Army,	
☐ 70-15912	72	King Air U-21J	B-99	U S Army,	
☐ 71-21059	74	King Air C-12L	BB-4	U S Army, Military District of Washington, Davison AAF, Va.	
☐ 71-21060	74	King Air C-12L	BB-5	U S Army, Military District of Washington, Davison AAF, Va.	
☐ 73-1205		King Air C-12C	BD-1	USAF, U S Embassy Flight, Islamabad, Pakistan.	
☐ 73-1206		King Air C-12C	BD-2	USAF, U S Embassy Flight, Ankara, Turkey.	
☐ 73-1208		King Air C-12C	BD-4	USAF, U S Embassy Flight, Kinshasa, Zaire.	
☐ 73-1209		King Air C-12C	BD-5	U S Army, South Dakota State Area Command, Rapid City, SD.	
☐ 73-1210		King Air C-12C	BD-6	USAF, U S Embassy Flight, Abidjan, Ivory Coast.	
☐ 73-1212		King Air C-12C	BD-8	USAF, tail code KS, 81st Training Wing, Keesler AFB. Ms.	
☐ 73-1213		King Air C-12C	BD-9	USAF, tail code KS, 81st Training Wing, Keesler AFB. Ms.	
☐ 73-1214		King Air C-12C	BD-10	USAF, U S Embassy Flight, Bangkok, Thailand.	
☐ 73-1215		King Air C-12C	BD-11	USAF, U S Embassy Flight, Bangkok, Thailand.	
☐ 73-1216		King Air C-12C	BD-12	USAF, U S Embassy Flight, Ankara, Turkey.	
☐ 73-1217		King Air C-12C	BD-13	USAF, U S Embassy Flight, Manila, Philippines.	
☐ 73-1218		King Air C-12C	BD-14	USAF, U S Embassy Flight, Athens, Greece.	
☐ 73-22250		King Air C-12C	BC-1	U S Army, OSAC, Virginia Flight Detachment, Langley AFB. Va.	
☐ 73-22256		King Air C-12C	BC-7	U S Army, Operational Support Airlift Command,	
☐ 73-22257		King Air C-12C	BC-8	U S Army, OSAC HQ. Davison AAF, Va.	
☐ 73-22258	73	King Air C-12C	BC-9	U S Army,	N381PD
☐ 73-22259		King Air C-12C	BC-10	U S Army, Camp Zama, Japan.	
☐ 73-22260	73	King Air C-12C	BC-11	U S Customs Service, Oklahoma City, Ok.	73-22260
☐ 73-22261		King Air C-12C	BC-12	U S Army, 5th Btn. 158th Aviation Regiment, Wiesbaden,	
☐ 73-22262		King Air C-12C	BC-13	U S Army, OSAC, N Carolina Flight Det. Simmons AAF, NC.	
☐ 73-22264	73	King Air C-12C	BC-15	Los Angeles Police Department, Los Angeles, Ca.	73-22264
☐ 73-22268		King Air C-12C	BC-19	U S Army, OSAC HQ. Davison AAF, Va.	
☐ 76-0158		King Air C-12C	BD-15	USAF, U S Embassy Flight, Islamabad, Pakistan.	
☐ 76-0160		King Air C-12C	BD-17	USAF, U S Embassy Flight, Riyadh, Saudi Arabia.	
☐ 76-0161		King Air C-12C	BD-18	USAF, U S Embassy Flight, Bogota, Colombia.	
☐ 76-0162		King Air C-12C	BD-19	deleted from inventory	
☐ 76-0163		King Air C-12C	BD-20	USAF, U S Embassy Flight, Canberra, Australia.	
☐ 76-0164		King Air C-12C	BD-21	USAF, U S Embassy Flight, Dhahran, Saudi Arabia.	
☐ 76-0165		King Air C-12C	BD-22	USAF, U S Embassy Flight, Tegucigalpa, Honduras.	
☐ 76-0166		King Air C-12C	BD-23	USAF, U S Embassy Flight, Cairo, Egypt.	
☐ 76-0168		King Air C-12C	BD-25	USAF, 89th Airlift Wing, Andrews AFB. Md.	
☐ 76-0169		King Air C-12C	BD-26	U S Army,	
☐ 76-0170		King Air C-12C	BD-27	USAF, U S Military Training Mission, Dhahran, Saudi Arabia.	
☐ 76-0171		King Air C-12C	BD-28	USAF, U S Embassy Flight, Ankara, Turkey.	
☐ 76-0172		King Air C-12C	BD-29	U S Army,	
☐ 76-0173		King Air C-12C	BD-30	USAF, U S Embassy Flight, Ankara, Turkey.	
☐ 76-22547		King Air C-12C	BC-23	U S Army, OSAC, N Carolina Flight Det. Simmons AAF, NC.	
☐ 76-22551		King Air C-12C	BC-27	U S Army, OSAC, Texas Flight Detachment, Robert Gray AAF, Tx	
☐ 76-22553		King Air C-12C	BC-29	U S Army, OSAC HQ. Davison AAF, Va.	
☐ 76-22555		King Air C-12C	BC-31	U S Army, Special Operations Command, MacDill AFB. Fl.	
Regn	Yr	Type	c/n	Owner/Operator	Prev Regn

Regn	Yr Type	c/n	Owner/Operator	Prev Regn
☐ 76-22559	King Air C-12C	BC-36	U S Army, Operational Support Airlift Command,	
☐ 76-22560	King Air C-12C	BC-37	U S Army, OSAC, Georgia Flight Detachment, Fulton County, Ga	
☐ 76-22561	King Air C-12C	BC-38	U S Army, OSAC, Virginia Flight Detachment, Langley AFB. Va.	
☐ 76-22564	King Air C-12C	BC-41	U S Army, OSAC, N Carolina Flight Det. Simmons AAF, NC.	
☐ 76-3239	King Air C-12C	BD-24	USAF, U S Embassy Flight, Jakarta, Indonesia.	
☐ 77-22932	King Air C-12C	BC-43	U S Army, 6th Aviation Detachment, SETAF, Vicenza, Italy.	
☐ 77-22933	King Air C-12C	BC-44	U S Army, OSAC HQ. Davison AAF, Va.	
☐ 77-22935	King Air C-12C	BC-46	U S Army, Operational Support Airlift Command,	
☐ 77-22937	King Air C-12C	BC-48	U S Army, Operational Support Airlift Command,	
☐ 77-22938	King Air C-12C	BC-49	U S Army, Operational Support Airlift Command,	
☐ 77-22940	King Air C-12A	BC-51	U S Army, Operational Support Airlift Command,	
☐ 77-22941	King Air C-12A	BC-52	U S Army, OSAC, N Carolina Flight Det. Simmons AAF, NC.	
☐ 77-22942	King Air C-12C	BC-53	U S Army, OSAC HQ. Davison AAF, Va.	
☐ 77-22944	King Air C-12C	BC-55	U S Army, OSAC, N Carolina Flight Det. Simmons AAF, NC.	
☐ 77-22945	King Air C-12C	BC-56	U S Army, Special Operations Command, MacDill AFB. Fl.	
☐ 77-22949	King Air C-12C	BC-60	U S Army, OSAC, Texas Flight Detachment, Robert Gray AAF, Tx	
☐ 77-22950	King Air C-12C	BC-61	U S Army, 207th Aviation Company, Heidelberg AAF, Germany.	
☐ 78-23126	King Air C-12C	BC-62	U S Army, 207th Aviation Company, Heidelberg AAF, Germany.	
☐ 78-23128	King Air C-12C	BC-64	U S Army, 207th Aviation Company, Heidelberg AAF, Germany.	
☐ 78-23129	King Air C-12C	BC-65	U S Army, OSAC HQ, Davison AAF, Va.	
☐ 78-23130	King Air C-12C	BC-66	U S Army, OSAC HQ, Davison AAF, Va.	
☐ 78-23131	King Air C-12C	BC-67	U S Army, OSAC, Utah Flight Detachment, Salt Lake City, Ut.	
☐ 78-23132	King Air C-12C	BC-68	U S Army, Army Aviation Element, Izmir, Turkey.	
☐ 78-23133	King Air JC-12C	BC-69	U S Army, Army Aviation Center, Cairns AAF, Al.	
☐ 78-23134	King Air C-12C	BC-70	U S Army, 1st Btn. 501st Aviation Regiment, Seoul, S Korea.	
☐ 78-23135	King Air C-12C	BC-71	U S Army, Operational Support Airlift Command,	
☐ 78-23136	King Air C-12C	BC-72	U S Army, Operational Support Airlift Command,	
☐ 78-23137	King Air C-12C	BC-73	U S Army, Operational Support Airlift Command,	
☐ 78-23138	King Air C-12C	BC-74	U S Army, Operational Support Airlift Command,	
☐ 78-23139	King Air C-12C	BC-75	U S Army, OSAC, N Carolina Flight Det. Simmons AAF, NC.	
☐ 78-23140	King Air JC-12D	BP-1	U S Army, Army Aviation Center, Cairns AAF, Al.	
☐ 78-23141	King Air RC-12D	GR-6	U S Army, Comms & Electronics Command, NAS Lakehurst, NJ.	
☐ 78-23142	King Air RC-12D	GR-7	U S Army, 15th Mil Intelligence Btn. Robert Gray AAF, Tx.	
☐ 78-23143	King Air RC-12D	GR-8	U S Army, 15th Mil Intelligence Btn. Robert Gray AAF, Tx.	
☐ 78-23144	King Air RC-12D	GR-9	U S Army, 3rd Mil Intelligence Btn. Camp Humphries, S Korea.	
☐ 78-23145	King Air RC-12D	GR-10	U S Army, Military Intelligence Btn. (LI), Orlando, Fl.	
☐ 80-23371	King Air RC-12D	GR-2	U S Army, Comms & Electronics Command, NAS Lakehurst, NJ.	
☐ 80-23372	King Air RC-12G	FC-3	U S Army, Military Intelligence Btn. (LI), Orlando, Fl.	
☐ 80-23373	King Air RC-12D	GR-4	U S Army,	
☐ 80-23374	King Air RC-12D	GR-12	U S Army, 304th Mil Intelligence Btn. Fort Huachuca, Az.	
☐ 80-23375	King Air RC-12D	GR-5	U S Army, 15th Mil Intelligence Btn. Robert Gray AAF, Tx.	
☐ 80-23376	King Air RC-12D	GR-11	U S Army, 15th Mil Intelligence Btn. Robert Gray AAF, Tx.	
☐ 80-23377	King Air RC-12D	GR-3	U S Army, 15th Mil Intelligence Btn. Robert Gray AAF, Tx.	
☐ 80-23378	King Air RC-12D	GR-13	U S Army, Army Aviation Center, Cairns AAF, Al.	
☐ 80-23379	King Air RC-12G	FC-1	U S Army, Military Intelligence Btn. (LI), Orlando, Fl.	
☐ 80-23380	King Air RC-12G	FC-2	U S Army, Military Intelligence Btn. (LI), Orlando, Fl.	
☐ 81-23541	King Air C-12D	BP-22	U S Army, OSAC, Utah Flight Detachment, Salt Lake City, Ut.	
☐ 81-23542	King Air RC-12D	GR-1	U S Army, 15th Military Intelligence Btn.Robert Gray AAF, Tx	
☐ 81-23543	King Air C-12D	BP-24	U S Army, Fort Wainwright, Ak.	
☐ 81-23544	King Air C-12D	BP-25	U S Army, Western Command, Fort Schafter, Wheeler AFB. Hi.	
☐ 81-23545	King Air C-12D	BP-26	U S Army, Special Operations Command, MacDill AFB. Fl.	
☐ 81-23546	King Air C-12D	BP-27	U S Army, Fort Wainwright, Ak.	
☐ 82-23780	King Air C-12D	BP-28	U S Army, Kansas State Area Command, ARNG, Topeka-Forbes, Ks	
☐ 82-23781	King Air C-12D	BP-29	U S Army, Aviation Training Site, ARNG, Clarksburg, WV.	
☐ 82-23782	King Air C-12D	BP-30	U S Army, Montana State Area Command, ARNG, Helena, Mt.	
☐ 82-23783	King Air C-12D	BP-31	U S Army, Arizona State Area Command, ARNG, Papago AAF, Az.	
☐ 82-23784	King Air C-12D	BP-32	U S Army, Louisiana State Area Command, New Orleans, La.	
☐ 82-23785	King Air C-12D	BP-33	U S Army, Missouri State Area Command, St Joseph-Rosecrans.	
☐ 83-0494	King Air C-12D	BP-40	USAF, U S Embassy Flight, Abidjan, Ivory Coast.	
☐ 83-0495	King Air C-12D	BP-41	USAF, U S Embassy Flight, Budapest, Hungary.	
☐ 83-0496	King Air C-12D	BP-42	USAF, 89th Airlift Wing, Andrews AFB. Md.	
☐ 83-0497	King Air C-12D	BP-43	USAF, U S Embassy Flight, Buenos Aires, Argentina.	
☐ 83-0498	King Air C-12D	BP-44	USAF, U S Embassy Flight, Mexico City, Mexico.	
☐ 83-0499	King Air C-12D	BP-45	USAF, U S Embassy Flight, La Paz, Bolivia.	
☐ 83-24145	King Air C-12D	BP-34	U S Army, North Dakota State Area Command, ARNG, Bismarck.	
☐ 83-24146	King Air C-12D	BP-35	U S Army, Michigan State Area Command, Grand Ledge-Abrams.	
☐ 83-24147	King Air C-12D	BP-36	U S Army, New Mexico State Area Command, ARNG, Santa Fe, NM	
☐ 83-24148	King Air C-12D	BP-37	U S Army, Washington State Area Command, Tacoma-Camp Murray.	
☐ 83-24149	King Air C-12D	BP-38	U S Army, Oklahoma State Area Command, ARNG, Lexington, Ok.	
☐ 83-24150	King Air C-12D	BP-39	U S Army, Mississippi State Area Command, Jackson, Ms.	
☐ 83-24313	King Air RC-12H	GR-14	U S Army, 3rd Military Intelligence Btn. South Korea.	
☐ 83-24314	King Air RC-12H	GR-15	U S Army, 3rd Military Intelligence Btn. South Korea.	
☐ 83-24315	King Air RC-12H	GR-16	U S Army, 3rd Military Intelligence Btn. South Korea.	
☐ 83-24316	King Air RC-12D	GR-17	U S Army, 3rd Military Intelligence Btn. South Korea.	
☐ 83-24317	King Air RC-12H	GR-18	U S Army, 3rd Military Intelligence Btn. South Korea.	
☐ 83-24318	King Air RC-12H	GR-19	U S Army, 3rd Military Intelligence Btn. South Korea.	

Regn	Yr	Type	c/n	Owner/Operator	Prev Regn
☐ 84-0143		King Air C-12F	BL-73	U S Army,	
☐ 84-0144		King Air C-12F	BL-74	U S Army,	
☐ 84-0145		King Air C-12F	BL-75	U S Army,	
☐ 84-0146		King Air C-12F	BL-76	U S Army,	
☐ 84-0147		King Air C-12F	BL-77	USAF, tail code AK,517th Airlift Squadron, Elmendorf AFB. Ak	
☐ 84-0148		King Air C-12F	BL-78	U S Army,	
☐ 84-0149		King Air C-12F	BL-79	U S Army,	
☐ 84-0150		King Air C-12F	BL-80	U S Army,	
☐ 84-0151		King Air C-12F	BL-81	U S Army,	
☐ 84-0152		King Air C-12F	BL-82	U S Army,	
☐ 84-0153		King Air C-12F	BL-83	U S Army,	
☐ 84-0154		King Air C-12F	BL-84	U S Army,	
☐ 84-0155		King Air C-12F	BL-85	U S Army,	
☐ 84-0156		King Air C-12F	BL-86	U S Army,	
☐ 84-0157		King Air C-12F	BL-87	U S Army, 5-158th Aviation, Wiesbaden, Germany.	
☐ 84-0158		King Air C-12F	BL-88	U S Army, HQ USEUCOM, Stuttgart, Germany.	
☐ 84-0159		King Air C-12F	BL-89	U S Army, HQ USEUCOM, Stuttgart, Germany.	
☐ 84-0160		King Air C-12F	BL-90	U S Army, HQ USEUCOM, Stuttgart, Germany.	
☐ 84-0161		King Air C-12F	BL-91	U S Army,	
☐ 84-0162		King Air C-12F	BL-92	U S Army,	
☐ 84-0163		King Air C-12F	BL-93	U S Army,	
☐ 84-0164		King Air C-12F	BL-94	U S Army,	
☐ 84-0165		King Air C-12F	BL-95	U S Army,	
☐ 84-0166		King Air C-12F	BL-96	U S Army,	
☐ 84-0167		King Air C-12F	BL-97	U S Army,	
☐ 84-0168		King Air C-12F	BL-98	U S Army,	
☐ 84-0169		King Air C-12F	BL-99	U S Army,	
☐ 84-0170		King Air C-12F	BL-100	U S Army,	
☐ 84-0171		King Air C-12F	BL-101	U S Army,	
☐ 84-0172		King Air C-12F	BL-102	U S Army,	
☐ 84-0173		King Air C-12F	BL-103	U S Army,	
☐ 84-0174		King Air C-12F	BL-104	U S Army,	
☐ 84-0175		King Air C-12F	BL-105	U S Army,	
☐ 84-0176		King Air C-12F	BL-106	U S Army,	N5819T
☐ 84-0177		King Air C-12F	BL-107	U S Army,	
☐ 84-0178		King Air C-12F	BL-108	U S Army,	
☐ 84-0179		King Air C-12F	BL-109	U S Army,	
☐ 84-0180		King Air C-12F	BL-110	U S Army,	
☐ 84-0181		King Air C-12F	BL-111	U S Army,	
☐ 84-0182		King Air C-12F	BL-112	U S Army,	
☐ 84-0484		King Air C-12F	BL-118	U S Army,	
☐ 84-0485		King Air C-12F	BL-119	U S Army,	
☐ 84-0486		King Air C-12F	BL-120	U S Army,	
☐ 84-0487		King Air C-12F	BL-121	U S Army,	
☐ 84-0488		King Air C-12F	BL-122	U S Army,	
☐ 84-0489		King Air C-12F	BL-123	U S Army,	
☐ 84-24375		King Air C-12D	BP-46	U S Army, Operational Support Airlift Command,	
☐ 84-24376		King Air C-12D	BP-47	U S Army, 1st Btn. 501st Aviation Regiment, Seoul, S Korea.	
☐ 84-24377		King Air C-12D	BP-48	U S Army, OSAC, Dhahran, Saudi Arabia.	
☐ 84-24378		King Air C-12D	BP-49	U S Army, OSAC, Texas Flight Detachment, Robert Gray AAF, Tx	
☐ 84-24379		King Air C-12D	BP-50	U S Army, Special Operations Command, MacDill AFB. Fl.	
☐ 84-24380		King Air C-12D	BP-51	U S Army, 207th Aviation Company, Heidelberg AAF, Germany.	
☐ 85-0147		King Air RC-12K	FE-1	U S Army, 1st Military Intelligence Btn. Wiesbaden, Germany.	
☐ 85-0148		King Air RC-12K	FE-2	U S Army, 1st Military Intelligence Btn. Wiesbaden, Germany.	
☐ 85-0149		King Air RC-12N	FE-3	U S Army, Aviation Systems Command, Wichita, Ks.	
☐ 85-0150		King Air RC-12K	FE-4	U S Army, 1st Military Intelligence Btn. Wiesbaden, Germany.	
☐ 85-0151		King Air RC-12K	FE-5	U S Army, 1st Military Intelligence Btn. Wiesbaden, Germany.	
☐ 85-0152		King Air RC-12K	FE-6	U S Army, 1st Military Intelligence Btn. Wiesbaden, Germany.	
☐ 85-0153		King Air RC-12K	FE-7	U S Army, 1st Military Intelligence Btn. Wiesbaden, Germany.	
☐ 85-0154		King Air RC-12K	FE-8	U S Army, 1st Military Intelligence Btn. Wiesbaden, Germany.	
☐ 85-0155		King Air RC-12K	FE-9	U S Army, 1st MIlitary Intelligence Btn. Wiesbaden, Germany.	
☐ 85-1262		King Air C-12F	BP-53	U S Army, Tennessee State Area Command, (ARNG), Smyrna, Tn.	
☐ 85-1263		King Air C-12F	BP-54	U S Army, Puerto Rico State Area Command, Isla Grande, PR.	
☐ 85-1264		King Air C-12F	BP-55	U S Army, Information Systems Command, Sinop AAF, Turkey.	
☐ 85-1265		King Air C-12F	BP-56	U S Army, 1st Btn. 228th Aviation Regiment, Howard AFB. CZ.	
☐ 85-1266		King Air C-12F	BP-57	U S Army, Special Operations Command, MacDill AFB. Fl.	
☐ 85-1267		King Air C-12F	BP-58	U S Army, Army Safety Center, Fort Rucker, Cairns AAF, Al.	
☐ 85-1268		King Air C-12F	BP-59	U S Army, Military Intelligence Btn (LI), Howard AFB. Panama	
☐ 85-1270		King Air C-12F	BP-61	U S Army, District of Columbia, Davison AAF, Va.	
☐ 85-1271		King Air C-12F	BP-62	U S Army, Utah State Area Command (ARNG), Salt Lake No 2, Ut	
☐ 85-1272		King Air C-12F	BP-63	U S Army, Idaho State Area Command (ARNG), Boise, Id.	
☐ 85-1609	77	PA-31T Cheyenne II	7720051	U S Army, H & H C, 1-58th Aviation Det. Simmons AAF, NC.	N27KM
☐ 86-0078		King Air C-12J	UD-1	USAF, 51st Fighter Wing, Osan AB. South Korea.	
☐ 86-0079		King Air C-12J	UD-2	USAF, 3rd Wing, Elmendorf AFB. Ak.	
☐ 86-0080		King Air C-12J	UD-3	USAF, 46th Test Wing, Eglin AFB, Fl.	
☐ 86-0081		King Air C-12J	UD-4	USAF, 3rd Wing, Elmendorf AFB, Ak.	

☐ 86-0082	King Air C-12J	UD-5	USAF, 3rd Wing, Elmendorf AFB. Ak.		
☐ 86-0083	King Air C-12J	UD-6	USAF, 51st Fighter Wing, Osan AB. South Korea.		
☐ 86-0084	King Air C-12F	BP-64	U S Army, Virginia State Area Command (ARNG), Richmond, Va.		
☐ 86-0085	King Air C-12F	BP-65	U S Army, New York State Area Command (ARNG), Albany, NY.		
☐ 86-0086	King Air C-12F	BP-66	U S Army, California State Area Command, Mather AFB, Ca.		
☐ 86-0087	King Air C-12F	BP-67	U S Army, Pennsylvania State Area Command, Muir AAF, Pa.		
☐ 86-0088	King Air C-12F	BP-68	U S Army, Florida State Area Command (ARNG), Craig, Fl.		
☐ 86-0089	King Air C-12F	BP-69	U S Army, Illinois State Area Command (ARNG), Decatur, Il.		
☐ 87-0160	King Air C-12F	BP-70	U S Army, Texas State Area Command (ARNG), Austin-Mueller.		
☐ 87-0161	King Air C-12F	BP-71	U S Army, Alabama State Area Command (ARNG), Montgomery, Al.		
☐ 88-0325	King Air RC-12N	FE-10	U S Army, 224th Military Intelligence Btn. Hunter AAF, Ga.		
☐ 88-0326	King Air RC-12N	FE-11	U S Army, 224th Military Intelligence Btn. Hunter AAF, Ga.		
☐ 88-0327	King Air RC-12N	FE-12	U S Army, 224th Military Intelligence Btn. Hunter AAF, Ga.		
☐ 89-0267	King Air RC-12N	FE-13	U S Army, 224th Military Intelligence Btn. Hunter AAF, Ga.		
☐ 89-0268	King Air RC-12N	FE-14	U S Army, 224th Military Intelliegence Btn. Hunter AAF, Ga.		
☐ 89-0269	King Air RC-12N	FE-15	U S Army, 224th Military Intelligence Btn. Hunter AAF, Ga.		
☐ 89-0270	King Air RC-12N	FE-16	U S Army, 224th Military Intelligence Btn. Hunter AAF, Ga.		
☐ 89-0271	King Air RC-12N	FE-17	U S Army, 224th Military Intelliegnce Btn. Hunter AAF, Ga.		
☐ 89-0272	King Air RC-12N	FE-18	U S Army, 224th Military Intelligence Btn. Hunter AAF, Ga.		
☐ 89-0273	King Air RC-12N	FE-19	U S Army, 304th Military Intelliegnce Btn. Libby AAF, Az.		
☐ 89-0274	King Air RC-12N	FE-20	U S Army, 304th Military Intelligence Btn. Libby AAF, Az.		
☐ 89-0275	King Air RC-12N	FE-21	U S Army, 224th Military Intelligence Btn. Hunter AAF, Ga.		
☐ 89-0276	King Air RC-12N	FE-22	U S Army,		
☐ 91-0516	King Air RC-12K	FE-23	U S Army,		
☐ 91-0517	King Air RC-12K	FE-24	U S Army,		
☐ 91-0518	King Air RC-12K	FE-25	U S Army,		
☐ 91-0519	King Air RC-12K	FE-26	U S Army,		
☐ 92-13120	King Air RC-12K	FE-27	U S Army,		
☐ 92-13121	King Air RC-12K	FE-28	U S Army,		
☐ 92-13122	King Air RC-12K	FE-29	U S Army,		
☐ 92-13123	King Air RC-12K	FE-30	U S Army,		
☐ 92-13124	King Air RC-12K	FE-31	U S Army,		
☐ 92-13125	King Air RC-12K	FE-32	U S Army,		
☐ 92-3327	King Air C-12F	BW-1	U S Army, 2nd Btn 228th Aviation Regiment, NAS Willow Grove	N2843B
☐ 92-3328	King Air C-12F	BW-2	U S Army, 2nd Btn 228th Aviation Regiment, NAS Willow Grove.	N2844B
☐ 92-3329	King Air C-12F	BW-3	U S Army, 2nd Btn 228th Aviation Regiment, NAS Willow Grove.	N2845B
☐ 93-0698	King Air RC-12P	FE-33	U S Army,		
☐ 93-0699	95 King Air RC-12P	FE-34	U S Army,		
☐ 93-0700	95 King Air RC-12P	FE-35	U S Army,		
☐ 93-0701	95 King Air RC-12P	FE-36	U S Army,		
☐ 94-0315	King Air C-12R	BW-4	U S Army,		
☐ 94-0316	King Air C-12R	BW-5	U S Army,		
☐ 94-0317	King Air C-12R	BW-6	U S Army,		
☐ 94-0318	King Air C-12R	BW-7	U S Army,		
☐ 94-0319	King Air C-12R	BW-8	U S Army,		
☐ 94-0320	King Air C-12R	BW-9	U S Army,		
☐ 94-0321	King Air C-12R	BW-10	U S Army,		
☐ 94-0322	King Air C-12R	BW-11	U S Army,		
☐ 94-0323	King Air C-12R	BW-12	U S Army,		
☐ 94-0324	King Air C-12R	BW-13	U S Army,		
☐ 94-0325	King Air C-12R	BW-14	U S Army,		
☐ 94-0326	King Air C-12R	BW-15	U S Army,		
☐ N58009	King Air C-12F	BL-83	USAF, 1400th Military Airlift Squadron, Nellis AFB. Nv.	84-0153

OB = Peru (Republic of Peru)

Civil

Regn	Yr Type	c/n	Owner/Operator	Prev Regn
☐ OB-....	77 King Air/Catpass 250	BB-214	Aero Condor SA. Lima.	N26LE
☐ OB-....	81 PA-42 Cheyenne III	8001009		(N401MD)
☐ OB-1146	SA-226AT Merlin 4A	AT-064E		OB-M-1146
☐ OB-1228	81 PA-31T Cheyenne II	8120048	Aerovias SA. Satipo.	OB-M-1228
☐ OB-1234	81 PA-42 Cheyenne III	8001019	Aero Transporte SA. Lima. (status ?).	OB-S-1234
☐ OB-1284	74 Mitsubishi MU-2K	282	(status ?).	N3MP
☐ OB-1297	68 King Air B90	LJ-326	Aero Condor SA. Lima.	OB-T-1297
☐ OB-1308	79 PA-31T Cheyenne II	7920075	Videma SA-Compania de Taxi Aereo, Trujillo.	OB-S-1308
☐ OB-1330	88 King Air 300	FA-145	Centro Internacional de la Papa, Lima.	OB-M-1330
☐ OB-1364	68 King Air B90	LJ-330	Aero Condor SA. Lima.	N66MS
☐ OB-1365	81 PA-42 Cheyenne III	8001018	Aero Transporte SA. Lima.	N19CD
☐ OB-1420	74 King Air E90	LW-106	Aerolineas Lambayecanas EIRL. Lima.	SE-IIU
☐ OB-1457	66 King Air B90	LJ-180	TAAPSA-Taxi Aereo Alas del Palcazu SA. Izcozacin.	N610W
☐ OB-1468	77 King Air 200	BB-193	TAS-Taxi Aereo Selva SRL. Lima, Peru.	N131MB
☐ OB-1495	68 King Air B90	LJ-333	Peruana de Aviacion SA. Lima.	N891K
☐ OB-1509	75 King Air 200	BB-20	TAMSA-Transportes Aereos Maranon SA. Tarapoto.	N9023R
☐ OB-1558	68 King Air B90	LJ-405	Aero Selva SA. Pucallpa.	N68RT
☐ OB-1567	67 King Air A90	LJ-228	Air Atlantic SRL. Lima.	N946K
☐ OB-1593	69 King Air B90	LJ-477	Aero Condor SA. Lima.	N7777
☐ OB-1594	68 King Air B90	LJ-322	Aero Condor SA. Lima.	N45SC

Regn	Yr	Type	c/n	Owner/Operator	Prev Regn
□ OB-1595	68	King Air B90	LJ-400	TATSA-Transporte Aereo Taxi SA. Lima.	N501PP
□ OB-1629	81	PA-42 Cheyenne III	8001067	Aero Transporte CA. Lima.	N183CC
□ OB-1630	81	PA-42 Cheyenne III	8001022	Aero Transporte CA. Lima.	N145CA
□ OB-1631	78	PA-42 Cheyenne III	7801003	Aero Transporte CA. Lima.	N134KM
□ OB-932	69	King Air B90	LJ-465	Aero Condor SA. Santa Fe, Argentina.	OB-T-932

Military

Regn	Yr	Type	c/n	Owner/Operator	Prev Regn
□ AE-571	83	King Air B200CT	BN-2	Peruvian Navy. (was BL-58).	N6904Q
□ AE-572	83	King Air B200CT	BN-3	Peruvian Navy. (was BL-59).	N2856B
□ AE-573	83	King Air B200CT	BN-4	Peruvian Navy. (was BL-60).	N2790B
□ AE-574	83	King Air B200T	BT-25	Peruvian Navy. (was BB-1096).	N2795B
□ AE-575	83	King Air B200T	BT-26	Peruvian Navy. (was BB-1098).	N2826B
□ PNP-230	73	King Air E90	LW-36	Policia Nacional del Peru, Lima.	OB-1598

OE = Austria (Republic of Austria)

Civil

Regn	Yr	Type	c/n	Owner/Operator	Prev Regn
□ OE-BAZ	59	Gulfstream 1	23	BAZ-Bundesamt fuer Zivilluftfahrt, Vienna.	N1929B
□ OE-BBB	79	King Air 200	BB-526	Federal Ministry for the Interior Air Police, Vienna.	
□ OE-EHG	91	TBM 700	28	Flugtaxi GmbH. Ried-im-Innkreis.	
□ OE-EKD	96	Pilatus PC-XII	142	Klaus Durst GmbH. Vienna.	HB-FRD
□ OE-FAK	85	King Air C90A	LJ-1125	Alpla Air Charter GmbH. Hard.	N7208L
□ OE-FAM	83	Conquest 1	425-0131	Automobilvertriebs AG. Salzburg.	OE-FIB
□ OE-FBO	78	PA-31T Cheyenne II	7820051	Goldeck-Flug GmbH. Klagenfurt.	D-IGAK
□ OE-FDS	77	PA-31T Cheyenne II	7720056	Businessair Flugzeuglieh GmbH. Vienna.	(N82169)
□ OE-FHL	85	King Air C90A	LJ-1115	Airlink Luftverkehrs GmbH. Salzburg.	D-IBPE
□ OE-FJB	85	King Air B200	BB-1231	Bertsch Aviation GmbH. Bludenz.	D-IBAC
□ OE-FKG	80	PA-31T Cheyenne II	8020036	Airlink Luftverkehrs GmbH. Salzburg.	N30DJ
□ OE-FME	93	King Air 300LW	FA-228	Airlink Luftverkehrs GmbH. Salzburg.	(D-CAIR)
□ OE-FMO	81	PA-31T Cheyenne II	8120058	Dipl Ing Alexander Moser, Vienna.	N3GF
□ OE-FOW	70	SA-226T Merlin 3B	T-318	Charter-Air GmbH. Vienna.	D-IBBD
□ OE-FSO	91	King Air 300	FA-215	Wilhelm Schwarzmueller GmbH. Schaerding.	N8017G
□ OE-GBB	96	Dornier 328-100	3078	Tyrolean Jet Service GmbH. Vienna.	D-C...
□ OE-LEA	79	King Air 1300	BB-468	Eagle Airlines GmbH. Klagenfurt.	TF-ELI

OH = Finland (Republic of Finland)

Civil

Regn	Yr	Type	c/n	Owner/Operator	Prev Regn
□ OH-...	81	King Air C90	LJ-978		N725KR
□ OH-ACN	76	Rockwell 690A	11301	Maanmittaushallitus/Jetflite OY. Helsinki.	(N81405)
□ OH-ADA	74	SA-226T Merlin 3	T-248	VIP Leasing OY/Airdeal OY. Helsinki.	N120TT
□ OH-BAX	81	King Air C90	LJ-984	Aerial OY. Helsinki.	LN-FOD
□ OH-BCX	78	King Air C90	LJ-770	Metro Jet OY. Helsinki.	N88CG
□ OH-BIF	81	King Air 200	BB-847	Global Flight Service Ltd/Inter Flight Ltd OY. Helsinki.	N635GA
□ OH-BKA	70	King Air 100	B-39	Airwings OY. Tampere.	HB-GEN
□ OH-BSA	89	King Air 300	FA-205	Finnair Training OY. Helsinki.	N5672A
□ OH-BSB	89	King Air 300	FA-206	Finnair Training OY. Helsinki.	N5672J
□ OH-PHA	75	PA-31T Cheyenne II	7620001	Savair OY/Keski-Suomen Liikelentopalvelu OY. Tikkakoski.	OY-BSB
□ OH-PNT	75	PA-31T Cheyenne II	7520007	Jetflite OY. Helsinki.	
□ OH-PYE	79	PA-31T Cheyenne II	7920094	Juris Economica OY/Airecon OY. Helsinki.	SE-ICS
□ OH-UTI	74	Rockwell 690A	11204	Utin Lento OY. Utti.	SE-GSR

OK = Czech Republic

Civil

Regn	Yr	Type	c/n	Owner/Operator	Prev Regn
□ OK-BKS	96	King Air C90B	LJ-1430	Charouz Holding/Autoklub Bohemia Assistance, Prague.	N3251V
□ OK-YES	93	Beech 1900D	UE-44	Topair Ltd. Prague.	D-CBSG

OM = Slovak Republic

Civil

Regn	Yr	Type	c/n	Owner/Operator	Prev Regn
□ OM-VKE	89	King Air C90A	LJ-1222	Cassovia Air AS. Kosice.	OK-VKE

OO = Belgium (Kingdom of Belgium)

Civil

Regn	Yr	Type	c/n	Owner/Operator	Prev Regn
□ OO-KNM	85	Conquest II	441-0344	Begimmo NV. Antwerp.	N413MA
□ OO-LFL	80	Conquest II	441-0184	Abelag Aviation, Brussels.	(N2724M)
□ OO-ROB	77	Rockwell 690B	11409	Robsa International NV. Essen.	N81646
□ OO-SNA	75	King Air A100	B-217	Etat Belge Admin Aeronautique, Brussels.	
□ OO-SXB	81	Xingu 121A	040	Airventure BV. Antwerp.	PT-MBH

Military

Regn	Yr	Type	c/n	Owner/Operator	Prev Regn
□ CF-01	76	SA-226T Merlin 3A	T-259	Belgian Air Force, Melsbroek.	N5374M
□ CF-02	76	SA-226T Merlin 3A	T-260	Belgian Air Force, Melsbroek.	N5373M
□ CF-04	76	SA-226T Merlin 3A	T-264	Belgian Air Force, Melsbroek.	N5378M
□ CF-05	76	SA-226T Merlin 3A	T-265	Belgian Air Force, Melsbroek.	N5381M
□ CF-06	76	SA-226T Merlin 3A	T-267	Belgian Air Force, Melsbroek.	N5382M

OY = Denmark *(Kingdom of Denmark)*

Civil

Regn	Yr	Type	c/n	Owner/Operator	Prev Regn
☐ OY-ARV	74	Mitsubishi MU-2J	635	Jetair Aircraft Sales A/S. Roskilde.	LN-MTU
☐ OY-BHU	79	PA-31T Cheyenne 1	7904004	A/S Kongedybet, Copenhagen.	N131SW
☐ OY-BPM	81	SA-226T Merlin 3B	T-410	North Flying A/S. Aalborg.	(D-IFLY)
☐ OY-BVB	79	King Air 200	BB-419	Aviation Assistance A/S-United Nations, Zagreb.	N256EN
☐ OY-BVS	68	King Air B90	LJ-418	Danish Air Transport K/S. Copenhagen.	(SE-LEN)
☐ OY-BVW	80	King Air 200	BB-705	Skandinavisk Motor A/S-Volkswagen, Brondby.	D-IBAB
☐ OY-CBP	77	King Air 200	BB-235	Air Service Co. Kiev, Ukraine.	(N777MW)
☐ OY-CVC	83	King Air F90-1	LA-203	CODAN Invest A/S. Copenhagen.	N72KA
☐ OY-DLM	70	Mitsubishi MU-2F	187	Sun Air of Scandinavia A/S. Billund.	(N508MA)
☐ OY-EEF	97	King Air B200	BB-1548	Thrane & Thrane A/S. Roskilde.	
☐ OY-GEB	81	King Air 200C	BL-40	Aviation Assistance A/S-UNO, Nairobi, Kenya.	VH-NSR
☐ OY-GIG	97	King Air 350	FL-167	Aviation Assistance A/S. Roskilde.	
☐ OY-GRB	81	King Air 200	BB-845	Air Alpha A/S. Odense.	N486DC
☐ OY-IFH	96	King Air C90B	LJ-1422	Institut fur Funktionsanalyse Hospitalprojektering, Roskilde	
☐ OY-JAB	89	King Air C90A	LJ-1223	Flyjet A/S. Copenhagen.	N5522X
☐ OY-JAR	80	King Air 200C	BL-13	Air Alpha A/S-Aviation Assistance A/S (Red Cross), Roskilde.	PH-ILG
☐ OY-JRO	68	King Air B90	LJ-327	Danish Air Transport K/S. Vamdrup.	(N507M)
☐ OY-LEL	97	King Air 350	FL-161	Kirkbi A/S-Lego Systems A/S. Billund.	N11176
☐ OY-NUK	80	King Air 200	BB-634	Groenlandsfly A/S. Nuuk.	N101CP
☐ OY-PEB	78	King Air 200	BB-309	Faxe-Jydske Breweries, Skive.	(OY-FTC)
☐ OY-PEU	89	Reims/Cessna F406	F406-0045	NAC/Danish Air Transport K/S. Copenhagen.	5Y-LAN
☐ OY-YES	81	PA-42 Cheyenne III	8001043	Air Alpha A/S. Odense.	N809AA

PH = Netherlands *(Kingdom of the Netherlands)*

Civil

Regn	Yr	Type	c/n	Owner/Operator	Prev Regn
☐ PH-...	74	Mitsubishi MU-2J	630	Hezemans Air BV. Eindhoven.	OY-BIS
☐ PH-ACZ	85	King Air B200	BB-1215	ACE Air Charters, Groningen-Eelde.	D-IEEE
☐ PH-AJS	92	TBM 700	24	Airflo Europe NV.	F-GLBD
☐ PH-ATM	76	King Air/Catpass 250	BB-123	Tulip Air BV. Rotterdam.	N120DA
☐ PH-BOA	81	MU-2 Marquise	1507SA	Quick Airways BV. Groningen-Eelde.	N888FS
☐ PH-BRN	93	King Air 350	FL-80	Trans Travel Airlines BV. Lelystad.	N8275D
☐ PH-DDB	77	King Air/Catpass 250	BB-221	Tulip Air BV. Rotterdam.	SE-KYL
☐ PH-DUS	88	King Air 1300	BB-1296	Trans Travel Airlines BV. Lelystad.	N296YV
☐ PH-FWC		Reims/Cessna F406	F406-0007	Duijvestijn Aviation BV. Lelystad.	EC-ESF
☐ PH-FWD		Reims/Cessna F406	F406-0014	Duijvestijn Aviation BV. Lelystad.	F-WZDS
☐ PH-FWM	82	MU-2 Marquise	1548SA	Rijnmond Air Services BV. Rotterdam.	N474MA
☐ PH-GPX	90	Reims/Cessna F406	F406-0058	Duijvestijn Aviation BV. Lelystad.	F-OGPX
☐ PH-GUG	91	Reims/Cessna F406	F406-0060	Duijvestijn Aviation BV. Lelystad.	F-OGUG
☐ PH-GUI	91	Reims/Cessna F406	F406-0063	Duijvestijn Aviation BV. Lelystad.	F-OGUI
☐ PH-HUB	97	TBM 700	127	SOCATA, Eindhoven.	F-OHBV
☐ PH-JOE	83	Conquest 1	425-0168	Westerheide Management & Consultancy BV. Arnhem.	G-BJYC
☐ PH-MNZ	92	Dorner 228-212	8206	Kustwacht-Dutch Coastguard,	D-CDIV
☐ PH-PHO	88	Reims/Cessna F406	F406-0034	Duijvestijn Aviation BV. Lelystad.	AP-BFB
☐ PH-SBK	76	King Air 200	BB-180	Nationale Luchtvaartschool BV. Maastricht.	G-BHVX
☐ PH-SKP	80	King Air 200C	BL-11	Schreiner Airways BV. Rotterdam.	F-GIMD
☐ PH-TBD	92	TBM 700	85	TOVO BV.	N300PW
☐ PH-VMP	96	King Air B200	BB-1538	Pon Holdings BV.	N3268L

Military

Regn	Yr	Type	c/n	Owner/Operator	Prev Regn
☐ U-01	81	Fokker 60UTA-N	20321	RNAF, 334 Squadron, Eindhoven. 'Marinus Van Meel'	PH-UTL
☐ U-02	95	Fokker 60UTA-N	20324	RNAF, 334 Squadron, Eindhoven. 'Willem Versteegh'	PH-UTN
☐ U-03	95	Fokker 60UTA-N	20327	RNAF, 334 Squadron, Eindhoven. 'Jan Borghouts'	PH-UTP
☐ U-04	95	Fokker 60UTA-N	20329	RNAF, 334 Squadron, Eindhoven. 'Jules Zegers'	PH-UTR
☐ U-05	96	Fokker 50	20253	RNAF, 334 Squadron, Eindhoven. 'Fons Aler'	PH-KXO
☐ U-06	96	Fokker 50	20287	RNAF, 334 Squadron, Eindhoven. 'Robbie Wijting'	PH-MXI

PJ = Netherlands Antilles

Civil

Regn	Yr	Type	c/n	Owner/Operator	Prev Regn
☐ PJ-...	83	Gulfstream 1000	96065	Zunoca Freezone NV. Aruba.	N46GA
☐ PJ-CEB	76	Rockwell 690A	11292	CEB Investments BV.	HB-GEH
☐ PJ-NAF	82	Gulfstream 1000	96008		N9917S

PK = Indonesia *(Republic of Indonesia)*

Civil

Regn	Yr	Type	c/n	Owner/Operator	Prev Regn
☐ PK-...	96	TBM 700	119	SOCATA, Tangerong.	F-OHBQ
☐ PK-...	96	TBM 700	120	SOCATA, Tangerong.	F-OHBR
☐ PK-CAE	75	King Air A100	B-222	Directorate of Civil Aviation, Jakarta.	
☐ PK-CAK	93	King Air B200C	BL-140	Directorate of Civil Aviation, Jakarta.	N82410
☐ PK-CAL	96	TBM 700	114	Directorate of Civil Aviation, Jakarta.	F-OHBS
☐ PK-CAM	96	TBM 700	121	Directorate of Civil Aviation, Jakarta.	F-OBHT
☐ PK-CTE	67	Gulfstream 1	177	TransIndo, Jakarta.	G-BRWN
☐ PK-DYR	78	PA-31T Cheyenne II	7820054	Deraya Air Taxi PT. Jakarta.	VH-MWT
☐ PK-FKL	80	Conquest II	441-0179	Kayan River Timber Products PT. Jakarta.	N2723S

Regn	*Yr*	*Type*		*c/n*	*Owner/Operator*	*Prev Regn*

Regn	Yr	Type	c/n	Owner/Operator	Prev Regn
□ PK-HPH	87	King Air B200	BB-1281	Gatari Air Service PT. Jakarta.	(N521FA)
□ PK-HSN	89	King Air B200C	BL-134	Gatari Air Service PT. Jakarta.	PK-YPR
□ PK-HTI	86	King Air B200	BB-1255	Roma Avasindo PT/Gatari Air Services PT. Jakarta.	N125CU
□ PK-NSI	90	King Air 350	FL-30	Nugra Santana Air Service PT. Jakarta.	
□ PK-ODR	80	Gulfstream 980	95019	(status ?) (A-2022 ?).	N9772S
□ PK-PJH	76	PA-31T Cheyenne II	7620015	Pelita Air Service, Jakarta. (status ?)	9V-BHF
□ PK-RJA	68	Gulfstream 1	191	Rajawali Air Transport, Singapore.	VH-JPJ
□ PK-TRA	76	King Air 200	BB-113	Indonesian Air Transport, Jakarta.	
□ PK-TRL	60	Gulfstream 1	60	Indonesia Air Transport, Jakarta.	C-FIOM
□ PK-TRM	60	Gulfstream 1	57	Indonesia Air Transport, Jakarta.	N66JD
□ PK-TRN	68	Gulfstream 1	193	Indonesia Air Transport, Jakarta.	N754G
□ PK-TRO	64	Gulfstream 1	130	Indonesia Air Transport, Jakarta.	N3416
□ PK-TRW	96	Beech 1900D	UE-177	Mobil Oil/Indonesian Air Transport, Jakarta.	N3237H
□ PK-TRX	96	Beech 1900D	UE-186	Mobil Oil/Indonesian Air Transport, Jakarta.	N3233J
□ PK-VKA	80	King Air 200	BB-732	PENAS-Perum Survai Udara, Jakarta.	N3716D
□ PK-VKB	81	King Air 200	BB-794	PENAS-Perum Survai Udara, Jakarta.	N3720U
□ PK-VKY	67	King Air A90	LJ-197	PENAS-Perum Survai Udara, Jakarta.	N2510L
□ PK-VKZ	67	King Air A90	LJ-189	PENAS-Perum Survai Udara, Jakarta.	N123KA
□ PK-YPS	81	King Air B200	BB-920	P T Trigana Air Service, Jakarta.	N83TJ

PP/PT = Brazil (Federative Republic of Brazil)

Civil

Regn	Yr	Type	c/n	Owner/Operator	Prev Regn
□ PT-...	67	680V Turbo Commander	1703-79	(status ?)	N161XX
□ PT-...	81	Cheyenne II-XL	8166049		N500XL
□ PT-...	80	Gulfstream 980	95038	(status ?).	N91SA
□ PT-...	97	King Air 350	FL-169		N1099E
□ PT-...	65	King Air 90	LJ-16	Carcara Taxi Aereo Ltda. (status ?).	N51KA
□ PT-...	84	King Air B200	BB-1062	Rio Real Empreendimentos Ltda. Belo Horizonte, MG.	N985GA
□ PT-...	97	King Air C90B	LJ-1491		N2316H
□ PT-...	77	Rockwell 690B	11391	EXBRA Impotacao e Exportacao Ltda. Sao Paulo, SP. (status ?)	N73MA
□ PP-EFC	72	King Air E90	LW-15	State Government of Goias, Goiania, GO.	
□ PP-EHE	74	King Air C90	LJ-638	State Government of Goias, Goiania, GO.	PT-KFV
□ PP-EJG	91	King Air B200	BB-1410	State Government of Goias, Goiania, GO.	
□ PP-EOP	76	King Air 200	BB-137	State Government of Roraima, Boa Vista, RR.	PP-IKN
□ PP-EPB	81	PA-42 Cheyenne III	8001035	State Government of Paraiba (Sec. Adm.), Joao Pessoa, PB.	PT-OSX
□ PP-EPD	86	King Air 300	FA-92	State Government of Amazonas, Manaus, AM.	PT-OSZ
□ PP-EUE	68	King Air B90	LJ-409	State Government of Parana, Curitiba, PR.	
□ PP-FOY	73	King Air A100	B-142	Departamento de Policia Federal, Brasilia, DF.	
□ PP-FPP	73	King Air E90	LW-56	Departamento de Policia Federal MJ, Brasilia, DF.	PT-FGA
□ PT-ASN	85	King Air F90-1	LA-232	Ayrton Senna Prom. e Empreend. Ltda. Sao Paulo, SP.	D-IWPF
□ PT-BOY	69	Mitsubishi MU-2F	145	Oeste Redes. Aer. SA/ORA Taxi Aereo, Varzea Grande, MT.	N769Q
□ PT-BPY	69	Mitsubishi MU-2F	158	Antonio Nunes galvao, Ribeiro Preto, SP.	N871Q
□ PT-BZW	69	Mitsubishi MU-2F	175	CITEP-Com. e Imp. T. Posses Ltda. Sao Caet. do Sul, SP.	N890Q
□ PT-BZY	70	Mitsubishi MU-2F	188	Franca Taxi Aereo Ltda. Sao Luiz, MA.	N109MA
□ PT-DEU	68	King Air B90	LJ-355	Madeireira Juary Ltda. Redencao, PA.	
□ PT-DIQ	68	King Air B90	LJ-398	Helio Abrao Iunes Trad. Goiania, GO.	
□ PT-DKV	70	King Air 100	B-43	Saenge Eng. de Saneam. e Edif Ltda. Sao Paulo, SP.	
□ PT-DNP	70	King Air 100	B-56	Banco Brasileiro Comercial SA. Goiania, GO.	
□ PT-DQX	70	681 Turbo Commander	6018	Ernesto Mahle, Sao Paulo, SP.	N9068N
□ PT-DTL	71	Mitsubishi MU-2F	196	Sete Taxi Aereo Ltda. Goiania, GO.	N116MA
□ PT-IBE	71	King Air C90	LJ-531	Translima Taxi Aereo Ltda. Belo Horizonte, MG. (status ?).	
□ PT-ICD	72	Mitsubishi MU-2F	215	Sete Taxi Aereo Ltda. Goiania, GO.	N181MA
□ PT-ICP	72	King Air C90	LJ-558	Voar Taxi Aereo Ltda, Goiania, GO.	
□ PT-IEC	72	681B Turbo Commander	6069	BCN Leasing Arrend. Mercantil SA. Barueri, SP.	
□ PT-IED	72	681B Turbo Commander	6070	Geral do Comercio Arrend. Mercantil, Sao Paulo, SP.	
□ PT-IEE	72	681B Turbo Commander	6071	Banestado Leasing SA. Arrend. Mercantil, Curitiba, PR.	
□ PT-IGD	72	King Air E90	LW-9	Marcos Paixao de Araujo, Belo Horizonte, MG.	
□ PT-JGA	73	Mitsubishi MU-2K	268	Cia de Cimento Portland Maringa, Sao Paulo, SP.	N314MA
□ PT-JUB	93	King Air B200	BB-1455		N8105Q
□ PT-KGV	67	King Air A90	LJ-221	Confeccoes Dujor C E Repres. Ltda. Nova Friburgo, RJ.	N38V
□ PT-KME	75	PA-31T Cheyenne II	7520012	TTA-Teresina Taxi Aereo Ltda. Teresina, PI.	
□ PT-KYF	61	Gulfstream 1	75	Goiazem Goiania de Armazens Gerais Ltda. Goiania, GO.	N304K
□ PT-KYY	69	680W Turbo Commander	1833-38	Francisco Celso Tinoco Chagas Fo. Fortaleza, CE.	PT-FNK
□ PT-LBZ	66	King Air A90	LJ-181	CAESGO-Cia Agricola of the State of Goias, Goiania, GO.	N223KD
□ PT-LCE	81	King Air E90	LW-347	Corazza Partic. e Agropec. Ltda. Piracicaba, SP.	N3841V
□ PT-LDA	72	Rockwell 690	11036	Santa Barbara Taxi Aereo Ltda. Maringa, PR.	PT-FRC
□ PT-LDL	72	Rockwell 690	11037	Transar Taxi Aereo SA. Sao Paulo, SP.	PT-FRD
□ PT-LER	81	King Air F90	LA-148	Cia Paulista de Forca e Luz, Campinas, SP.	N1826P
□ PT-LEW	72	Mitsubishi MU-2K	244	TAM-Taxi Aereo Marilia SA. Sao Paulo, SP.	N400SM
□ PT-LFX	74	Mitsubishi MU-2J	650	Aires Moreira de Melo, Santarem, PA.	N990M
□ PT-LHH	81	MU-2 Marquise	1508SA	AB Promocoes E P. Artist S/C Ltda. Anapolis, GO.	N618RT
□ PT-LHJ	82	King Air C90	LJ-1010	Granasa Granitos Nacionais Ltda. Belo Horizonte, MG.	N6135Z
□ PT-LHM	66	King Air 90	LJ-105	Aluisio Gregorio Motta Jr. Gurupi, TO.	PP-ENF
□ PT-LHV	77	Rockwell 690B	11376	Transar Taxi Aereo SA. Sao Paulo, SP.	N81567
□ PT-LHZ	75	King Air E90	LW-133	BMG Leasing Arrend. Mercantil, Contagem, MG.	(N52CB)
□ PT-LIF	84	King Air F90-1	LA-223	Wanair Taxi Aereo Ltda. Belo Horizonte, MG.	N83KA

277

Regn	Yr	Type	c/n	Owner/Operator	Prev Regn
☐ PT-LIK	82	MU-2 Marquise	1546SA	Soc. Com. Triangulo Ltda. Sao Paulo, SP.	N472MA
☐ PT-LIR	80	MU-2 Solitaire	428SA	Unicos Comercio e Adm. Ltda. Sao Jose do Rio Preto, SP.	N124AX
☐ PT-LIS	79	MU-2 Marquise	749SA	Agropastoril Sapucaia Ltda. Babaculandia, GO.	N980MA
☐ PT-LJN	72	King Air A100	B-121	Cia Real de Arrend. Mercantil, Barueri, SP.	PP-EGK
☐ PT-LJS	85	MU-2 Marquise	1568SA	Uniair Taxi Aereo Ltda. Vitoria, ES.	N501MA
☐ PT-LLG	80	PA-31T Cheyenne II	8020054	Agropecuaria N.S. do Carmo Ltda. Ariranha, SP.	LV-OGB
☐ PT-LLO	89	King Air C90A	LJ-1225	Fatty Taxi Aereo Ltda. Sao Paulo, SP.	N1564M
☐ PT-LLP	79	King Air F90	LA-7	Aratu Taxi Aereo Ltda. Salvador, BA.	N67RP
☐ PT-LLR	81	King Air C90	LJ-946	Francisco Jose Rodrigues Filho, Brasilia, DF.	N3236T
☐ PT-LLV	80	King Air C90	LJ-897	Embra Taxi Aereo SA. Belo Horizonte, MG.	N758D
☐ PT-LMD	67	Mitsubishi MU-2B	026	Taxi Aereo Marilia SA. Sao Paulo, SP.	N482G
☐ PT-LMI	80	King Air C90	LJ-913	Agropecuaria Grao Para SA. S do Araguaia, PA.	N715AT
☐ PT-LNG	81	PA-31T Cheyenne II	8120061	Extremo Sul Taxi Aereo Ltda. Pelotas, RS.	N831CM
☐ PT-LNJ	85	King Air 300	FA-31	CEMIG-Cia Energetica de Minas Gerais, Belo Horizonte, MG.	HK-3689
☐ PT-LOH	68	Mitsubishi MU-2F	126	Belair Taxi Aereo Ltda. Belo Horizonte, MG.	N3917J
☐ PT-LPB	85	MU-2 Marquise	1567SA	Sebastiao Gilberto Tavares e Outros, Sao Jose do Rio Preto.	N499MA
☐ PT-LPD	88	King Air C90A	LJ-1173	CESP-Cia Energetica de Sao Paulo, Sao Paulo, SP.	
☐ PT-LPG	87	King Air B200	BB-1271	A.C. Agro Mercantil SA. Paracatu, MG.	N3048U
☐ PT-LPJ	82	King Air C90-1	LJ-1026	Carlos Roberto Alves, Belo Horizonte, MG.	N6364H
☐ PT-LPL	80	King Air F90	LA-28	Construtora Jalk SA. Belo Horizonte, MG.	N90LL
☐ PT-LPS	79	King Air C90	LJ-817	Prominas Taxi Aereo Ltda. Belo Horizonte, MG.	N3981Y
☐ PT-LQC	81	King Air F90	LA-132	Construtora Cowan SA. Belo Horizonte, MG.	N38649
☐ PT-LQD	79	King Air C90	LJ-844	Economico SA. Arrend. Mercantil, S. Caet. do Sul, SP.	N707CV
☐ PT-LQE	83	King Air C90-1	LJ-1056	Itapoan Taxi Aereo Ltda. Salvador, BA.	N90GH
☐ PT-LQS	81	King Air C90	LJ-966	Fazendas Reunidas Ligacao, Ananindeua, PA.	N181JH
☐ PT-LRT	81	PA-31T Cheyenne II	8120040	Aerobert Emp. e Participacoes Ltda. Carapicuisa, SP.	N44TW
☐ PT-LSE	84	King Air C90A	LJ-1063	TERCAM-Terraplen. Const. Inc. Ltda. Belo Horizonte, MG.	N76DS
☐ PT-LSH	81	King Air F90	LA-94	Gebepar SA Participao Investimento, Goiania, GO.	N3735W
☐ PT-LSO	78	King Air C90	LJ-794	Java Consultoria e Comercio Ltda. Sao Paulo, SP.	N57JB
☐ PT-LSP	82	King Air F90	LA-197	Cia Cacique de Cafe Soluvel, Sao Paulo, SP.	YV-494CP
☐ PT-LSQ	81	MU-2 Marquise	1530SA	Complemento Taxi Aereo Ltda. Sao Paulo, SP.	N449MA
☐ PT-LTC	74	Mitsubishi MU-2K	314	Agropaecuaria Vale Novo Ltda. Goiana, GO.	N982MA
☐ PT-LTF	71	King Air C90	LJ-543	Jose Paulino Pires, Belo Horizonte, MG.	N29791
☐ PT-LTO	81	King Air F90	LA-156	Emp. Aerotaxi e Man. Pampulha Ltda. Belo Horizonte, MG.	N1827F
☐ PT-LTT	81	King Air F90	LA-103	Encalso Construcoes Ltda. Sao Jose do Rio Preto, SP.	N3802F
☐ PT-LUF	75	King Air C90	LJ-651	Anibal Camillo Togni, Pocos de Caldas, MG.	N7300N
☐ PT-LUJ	77	PA-31T Cheyenne II	7720039	Transjunior Transp. Comercio Ltda. Imperatriz, MA.	N1144Z
☐ PT-LUT	84	King Air F90-1	LA-215	Locadora Brasal Ltda. Brasilia, DF.	N6730S
☐ PT-LVI	79	King Air C90	LJ-834	UPSI Informatica Ltda. Belo Horizonte, MG.	N42QC
☐ PT-LVK	89	King Air C90A	LJ-1201	Itapemirim Taxi Aereo Ltda. Itapemirim, ES.	(N486JA)
☐ PT-LXI	80	King Air F90	LA-11	Belfair Participacoes S/C Ltda. Rio de Janeiro, RJ.	N18EH
☐ PT-LXY	82	King Air F90	LA-195	Gal Air Taxi Aereo Ltda. Recife, PE.	N70132
☐ PT-LYI	71	Mitsubishi MU-2F	213	Heringer Taxi Aereo Ltda. Imperatriz, MA. (CoA expired ?)	N100BR
☐ PT-LYK	88	King Air C90A	LJ-1188	Usina Santa Adelia SA. Jaboticabal, SP.	N1537H
☐ PT-LYM	82	King Air F90	LA-185	Inter Air Taxi Aereo Ltda. Rio de Janeiro, RJ.	N61DH
☐ PT-LYP	81	King Air F90	LA-126	Acucar e Alcool Osw. Rib. Mend. Ltda. Guaira, SP.	N3848V
☐ PT-LYT	82	King Air C90-1	LJ-1037	BFB Leasing SA. Barueri, SP.	N283DP
☐ PT-LYW	73	King Air C90	LJ-584	Transpel Aerotaxi Ltda. Sao Carlos, SP.	N243TC
☐ PT-LYZ	81	King Air F90	LA-109	Novadata Sist. e Computadores SA. Brasilia, DF.	N3806U
☐ PT-LZA	74	King Air A100	B-200	Jet Sul Taxi Aereo Ltda. Curitiba, PR.	PT-FOB
☐ PT-LZB	79	PA-31T Cheyenne II	7920063	Aluminal Quimica do Nordeste Ltda. Simoes Filho, BA.	N23KF
☐ PT-LZD	81	PA-42 Cheyenne III	8001038	Destilaria de Alcool Goio Ere Ltda. Moreira Sales, PR.	LV-ONL
☐ PT-LZH	79	King Air C90	LJ-808	BMG Leasing SA. Arrend. Mercantil, Contagem, MG.	(N711WT)
☐ PT-LZR	79	PA-31T Cheyenne II	7920083	Nortox Agro Quimica SA. Arapongas, PR.	N31DC
☐ PT-LZT	84	King Air F90-1	LA-216	Paliber Participacoes Ltda. Sao Paulo, SP.	N390D
☐ PT-MCM	85	King Air 300	FA-52	Conserva de Estradas Ltda. Belo Horizonte, MG.	N50KA
☐ PT-MFL	82	PA-42 Cheyenne III	8001080	Fabio O Luchesi Advoc. Terras S/C. Sao Paulo, SP.	N882SW
☐ PT-MFW	83	Cheyenne II-XL	8166067	Gobair Corp. Porto Alegre, RS.	(N67ER)
☐ PT-MMB	82	King Air B200	BB-971	Malharia Diana Ltda. Timbo, SC.	N503RH
☐ PT-MMC	87	King Air 300	FA-113	MMC Automotores do Brazil Ltda. Curitiba, PR.	N299GS
☐ PT-MPC	76	King Air C90	LJ-683	INCOSPAL-Ind. Concreto de Sao Paulo SA. Vitoria, ES.	N1583L
☐ PT-MPN	78	Rockwell 690B	11465	Nome & Cia Ltda. Sarandi, PR.	CS-ASA
☐ PT-MPR	81	King Air B200	BB-840	MPE-Montagens e Projetos Especiais SA. Rio.	N36QS
☐ PT-MTD	89	King Air C90A	LJ-1219	Solair Leasing & Financial Co . Miami, Fl. USA.	N61HB
☐ PT-OAB	82	King Air C90-1	LJ-1022	CMS Taxi Aereo Ltda. Belo Horizonte, MG.	N6280P
☐ PT-OAJ	80	PA-31T Cheyenne 1	8004005	Taxi Aereo Texas Ltda. Sao Paulo, SP.	N2594R
☐ PT-OAM	80	PA-31T Cheyenne II	8020028	Leasing Bank of Boston SA. Arrend. Mercantil, Barueri, SP.	LV-OGG
☐ PT-OBF	89	King Air C90A	LJ-1224	Rima Taxi Aereo Ltda. Belo Horizonte, MG.	N1556Z
☐ PT-OBW	68	King Air B90	LJ-353	Aluminal Quimica do Nordeste Ltda. Simoes Filho, BA.	PT-FOA
☐ PT-OCC	81	King Air C90	LJ-960	Constructora Eferco Ltda, Belo Horizonte, MG.	N3861H
☐ PT-OCE	84	King Air F90-1	LA-217	BMG Leasing SA. Arrend. Mercantil, Belo Horizonte, MG.	N6756L
☐ PT-OCI	82	King Air C90	LJ-998	Rodoban Transportes Terrest. e Aereos Ltda. Uberlandia, MG.	N17EN
☐ PT-OCL	80	PA-31T Cheyenne II	8020033	CIKEL-Comercio e Industria Keila SA. Curitiba, PR.	N11WC
☐ PT-OCT	72	King Air C90	LJ-567	Polux Taxi Aereo Ltda. Belo Horizonte, MG.	PP-IAF
☐ PT-OCY	79	King Air C90	LJ-847	Aparte Taxi Aereo Ltda. Sao Paulo, SP.	N317EC
☐ PT-ODA	69	King Air B90	LJ-466	Antonio Leme Nunes Galvao, Bonfin Paulista, SP.	PP-IAG

Regn   *Yr Type*   *c/n*   *Owner/Operator*   *Prev Regn*

Regn	Yr	Type	c/n	Owner/Operator	Prev Regn
☐ PT-ODE	87	King Air C90A	LJ-1159	Soylent Ltda.	N3078D
☐ PT-ODH	86	King Air C90A	LJ-1128	Concordia Taxi Aereo Ltda. Concordia, SC.	N7248G
☐ PT-ODM	81	PA-31T Cheyenne II	8120042	Eucatur Taxi Aereo Ltda. Cascavel, PR.	N131CC
☐ PT-ODN	81	King Air F90	LA-85	NTA-Nacional Taxi Aereo Ltda. Brasilia, DF.	N3697P
☐ PT-ODO	83	King Air F90-1	LA-213	NTA-Nacional Taxi Aereo Ltda. Brasilia, DF.	N77M
☐ PT-ODR	80	PA-31T Cheyenne II	8020079	Co-operative Centr. Oeste Catarinense, Chapeco, SC.	LV-OEU
☐ PT-OED	80	PA-31T Cheyenne II	8020029	Alberto Soares e Outro, Campo Grande, MS.	(N661AE)
☐ PT-OEH	80	King Air C90	LJ-914	Lumar Taxi Aereo Ltda. Belem, PA.	N6RU
☐ PT-OEP	82	King Air C90-1	LJ-1019	BMG Leasing SA. Arrend. Mercantil, Belo Horizonte, MG.	N25AJ
☐ PT-OFB	83	King Air F90	LA-200	Construtora Andrade Gutierrez SA. Belo Horizonte, MG.	N6685H
☐ PT-OFC	72	King Air C90	LJ-534	Lamar Engenharia e Comercio Ltda. Belo Horizonte, MG.	N120JJ
☐ PT-OFD	81	King Air F90	LA-118	BCN Leasing Arrend. Mercantil SA. Barueri, SP.	N715GW
☐ PT-OFF	91	King Air C90A	LJ-1264	Aerotaxi Aracaju Ltda. Recife, PE.	N5680Z
☐ PT-OFG	76	Rockwell 690A	11274	Safra Leasing SA. Arrend. Mercantil, Barueri, SP.	N4432W
☐ PT-OFH	79	PA-31T Cheyenne II	7920034	Japi Taxi Aereo Ltda. Sao Paulo, SP.	N29KR
☐ PT-OFS	84	King Air F90-1	LA-225	Umuarama Constr. Terrap. e Pav. Ltda. Araguiana, TO.	N713DB
☐ PT-OFV	74	Mitsubishi MU-2K	298	Germano Franzoni, Sao Paulo, SP.	N188RM
☐ PT-OFY	85	King Air C90A	LJ-1094	Empresa Brasileira de Taxi Aereo SA. Belo Horizonte, MG.	N7215L
☐ PT-OFZ	79	King Air A100	B-243	Frimar Frigorifico Araguaina SA. Araguaina, TO.	N72EH
☐ PT-OHH	81	King Air C90	LJ-975	Joao Cesar Presotto, Guapore, RS.	N94SC
☐ PT-OHK	80	MU-2 Marquise	774SA	Economico SA. Arren. Mercantil, S Caet do Sul, SP.	N15ZM
☐ PT-OHX	83	King Air F90-1	LA-211	Berneck & Cia. Curitiba, PR.	(JA8865)
☐ PT-OHZ	82	King Air F90	LA-173	Colorado Auto Pecas Ltda.Goiania, GO.	N56TW
☐ PT-OIF	80	King Air F90	LA-49	SOTAN-Soc. Taxi Aereo Nordeste Ltda. Maceio, AL.	N200BM
☐ PT-OIP	77	Mitsubishi MU-2P	354SA	Rodoviario Sao Lucas Ltda. Manaus, AM.	N739MA
☐ PT-OIU	71	King Air C90	LJ-515	TRANSPEN-Transp. Col. e Encom. Ltda. Itarare, SP.	N953K
☐ PT-OIV	82	King Air F90	LA-163	Quarup Taxi Aereo Ltda. Uberlandia, MG.	N90FP
☐ PT-OIY	81	MU-2 Solitaire	453SA	BFB Leasing SA. Arrend. Mercantil, Barueri, SP.	N24FJ
☐ PT-OIZ	88	King Air C90A	LJ-1174	ICAL-Industria de Calcinacao Ltda. Sao Jose de Lapa, MG.	N31398
☐ PT-OJA	81	King Air C90	LJ-952	Banco Digibanco SA. Sao Paulo, SP.	N4490L
☐ PT-OJE	81	PA-31T Cheyenne II	8120031	Matosul Com. Imp. e Exportacao Ltda. Campo Grande, MS.	N628DE
☐ PT-OJI	79	King Air C90	LJ-812	Blue Sky Taxi Aereo Ltda. Brasilia, DF.	N627KP
☐ PT-OJM	81	PA-31T Cheyenne II	8120070	EPAGRI SA. Florianopolis, SC.	N826SW
☐ PT-OJQ	88	King Air 300	FA-154	Marina Taxi Aereo Ltda. Belo Horizonte, MG.	N1563K
☐ PT-OJU	80	King Air C90	LJ-900	Felisberto Moutinho Rodrigues Jr. Osasco, SP.	N415MA
☐ PT-OKL	82	PA-42 Cheyenne III	8001103	Sococo SA Industrias Alimenticias, Maceio, AL.	N4114D
☐ PT-OKQ	89	King Air C90A	LJ-1195	BTA-Braganca Taxi Aereo Ltda. Arapongas, PR.	N70PA
☐ PT-OKT	81	PA-31T Cheyenne 1	8104041	Sementes Maggi Ltda. Rondonopolis, MT.	N805CA
☐ PT-OLF	80	PA-31T Cheyenne 1	8004039	Eucatur Taxi Aereo Ltda. Cascavel, PR.	N500AQ
☐ PT-OLI	84	King Air 300	FA-26	Citrosuco Paulista SA, Congonhas, SP.	N984CF
☐ PT-OLM	81	King Air F90	LA-86	Amper Construcoes Electricas Ltda. Cuiaba, Mt.	(D-ILLL)
☐ PT-OLP	84	King Air F90	LA-220	Maeda Taxi Aereo Ltda. Ituverava, SP.	N6837C
☐ PT-OLQ	80	King Air C90	LJ-884	Quintal Taxi Aereo Ltda. Limeira, SP.	N88RB
☐ PT-OLW	81	King Air C90	LJ-985	Jose Francisco da Cunha e Outros, Cambui Campinas, SP.	N409ND
☐ PT-OLX	81	King Air C90	LJ-963	Obradek Erce Armazens Gerais Ltda. Nova Eguacu, RJ.	N38589
☐ PT-OLZ	80	PA-31T Cheyenne II	8120005	Silvio Name e Outro, Curitiba, PR.	N57656
☐ PT-OMZ	89	King Air C90A	LJ-1220	Companhia Ferroligas Minas Gerais Ltda. Contagem, MG.	N5520X
☐ PT-ONE	81	King Air F90	LA-144	Rio das Pedras Empreendimentos Ltda. Brasilia, DF.	(N300BF)
☐ PT-ONJ	84	King Air C90	LJ-1078	NTA-Nacional Taxi Aereo Ltda. Goiania, GO.	N78SR
☐ PT-ONO	81	King Air F90	LA-92	Montplas Ind. Mont. Mec. Plast. Ltda. Curitiba, PR.	N3715T
☐ PT-ONQ	82	King Air C90-1	LJ-1018	Geraldo Vilela Couto, Brasilia, DF.	N501LA
☐ PT-ONU	81	King Air F90	LA-128	Jubran Taxi Aereo Ltda. Presidente Prudente, SP.	N3867A
☐ PT-OOD	82	King Air C90	LJ-1000	Banco Bamerindus do Brasil SA. NYC. USA.	N6623D
☐ PT-OOS	78	Mitsubishi MU-2P	388SA	Meta Taxi Aereo/Sada Transp. e Armazenagens Ltda. Betim, MG.	N91CM
☐ PT-OOT	82	King Air C90	LJ-995	Fausto Jorge, Vera Cruz, SP.	N1855H
☐ PT-OOX	82	King Air F90	LA-162	SICOM Ltda. Sao Carlos, SP.	N90BL
☐ PT-OOY	80	King Air C90	LJ-882	Wortham Inc. Tortola-BVI/Arias Fabrega,	N181GA
☐ PT-OPC	81	PA-31T Cheyenne II	8120010	Alberto Youssef, Londrina, PR.	(D-IIKW)
☐ PT-OPD	81	King Air C90	LJ-920	Gama Indl. e Com. de Sec. e Molh. Ltda.Apar de Goiania, GO.	N42KA
☐ PT-OPE	81	King Air C90	LJ-940	APEC-Assoc. Prudent. Educ. e Cultura, Pres. Prudente, SP.	N82P
☐ PT-OPF	80	PA-31T Cheyenne 1	8004038	Canaa Agropastoril Ltda. Recife, PE.	N977CP
☐ PT-OPH	76	PA-31T Cheyenne II	7620044	ATR-Travessia Taxi Aereo SA. Recife, PE.	N92FC
☐ PT-OPQ	80	PA-31T Cheyenne 1	8004007	Taxi Aereo Taroba Ltda. Cascavel, PR.	N2379W
☐ PT-OPR	80	King Air C90	LJ-870	Luiz Carlos Ferreira, Belo Horizonte, MG.	N500MB
☐ PT-OQH	72	Rockwell 690	11011	Taxi Aereo Florianopolis Ltda. Florianopolis, SC.	N9211N
☐ PT-OQP	69	King Air 100	B-7	Rubens Correia Coimbra, Penapolis, SP.	N800MD
☐ PT-OQQ	70	681 Turbo Commander	6021	Tania Kuhnen, Florianapolis, SC.	N10RN
☐ PT-OQS	82	King Air C90	LJ-1005	Bandeirantes SA Arrend. Mercantil, Barueri, SP.	N6661J
☐ PT-OQY	85	Gulfstream 900	15038	Banco Bamerindus do Brasil SA. NYC. USA.	N77PK
☐ PT-ORB	92	King Air B200	BB-1435	INCOBRASA, Porto Alegre, RS.	N8050X
☐ PT-ORG	94	King Air C90B	LJ-1308	Frigorifico Bertim Ltda. Sao Paulo, SP.	
☐ PT-ORW	82	King Air C90	LJ-1004	Banco Bamerindos do Brasil SA.	N45US
☐ PT-ORZ	90	King Air C90A	LJ-1233	Usina Alto. Alegre SA. Asucar Alcool, Pres. Prudente, SP.	N113TP
☐ PT-OSI	81	King Air C90	LJ-936	Solair Financial & Leasing Co. Miami, Fl. USA.	N49FA
☐ PT-OSN	90	King Air C90A	LJ-1260	Unileasing Financing Ltd.	N5618Z
☐ PT-OSO	81	King Air C90	LJ-927	Tamandare Taxi Aereo Ltda. Teresina, PI.	N4492D

Regn	Yr	Type	c/n	Owner/Operator	Prev Regn

Regn	Yr	Type	c/n	Owner/Operator	Prev Regn
☐ PT-OSR	81	King Air 200	BB-784	Sergio Reis Prod. e Prom. Art. Ltda. Sao Paulo, SP.	N789H
☐ PT-OTA	82	King Air F90	LA-187	Ourivio Taxi Aereo Ltda. Belo Horizonte, MG.	N6416P
☐ PT-OTF	79	PA-31T Cheyenne II	7920044	Clipper Agencia de Viagens Ltda. Sao Paulo, SP.	N163MC
☐ PT-OTG	85	King Air C90A	LJ-1096	Empresa Gontijo de Transportes Ltda. Belo Horizonte, MG.	N7216H
☐ PT-OTI	90	King Air C90A	LJ-1237	King Aircraft Corp.	N338DR
☐ PT-OTO	80	King Air 200	BB-694	Aluminal Quimica do Nordeste Ltda. Simoes Filho, BA.	N90FQ
☐ PT-OTV	81	PA-31T Cheyenne 1	8104017	Impres. Cia Bras. de Impr.e Prop. Sao Paulo, SP.	N4494U
☐ PT-OUF	81	King Air E90	LW-343	Macauba Citros Ltda. Sao Paulo, SP.	N3710Y
☐ PT-OUJ	81	King Air F90	LA-155	First National Bank of Boston,	N155GA
☐ PT-OUL	66	King Air A90	LJ-125	Aluminal Quimica do Nordeste Ltda. Simoes Filho, BA.	N120JM
☐ PT-OUO	70	King Air B90	LJ-499	Ricardo Vicente Baptista, Barueri, SP.	N44454
☐ PT-OUX	81	King Air C90	LJ-937	Christiana Arcangeli, Sao Paulo, SP.	OY-BEK
☐ PT-OVB	81	PA-31T Cheyenne 1	8104051	Frigorifico Vale do Itajai Ltda. Itajai, SC.	N47TW
☐ PT-OVD	84	PA-42 Cheyenne 400LS	5527005	BNY Leasing Corp.	N25HE
☐ PT-OVE	80	PA-31T Cheyenne 1	8004014	BB Leasing Co. Grand Cayman.	N234PC
☐ PT-OVN	82	King Air C90-1	LJ-1041	Brickell Finance & Trade Corp. Aruba, Netherlands Antilles.	N106AJ
☐ PT-OVP	66	King Air A90	LJ-152	Paulo Panarello Neto, Goiania, GO.	N8180
☐ PT-OVQ	72	King Air A100	B-137	Safra Leasing SA. Arrend. Mercantil, Barueri, SP.	N95GA
☐ PT-OVW	77	Mitsubishi MU-2P	350SA	Allied Leasing & Finance Corp. Brasilia, DF.	N58MA
☐ PT-OVY	79	King Air C90	LJ-835	Nacional Leasing SA. Arrend. Mercantil, Corumba, MS.	N414AF
☐ PT-OXG	94	King Air B200	BB-1473	Sucocitrico Cutrale Ltda. Araraquara, SP.	N8064Q
☐ PT-OXH	85	King Air C90A	LJ-1127	Marcep International Trade Finance Ltd. Sao Paulo, SP.	N7254B
☐ PT-OXU	72	King Air C90	LJ-535	Pif Paf SA. Industria e Comercio, Belo Horizonte, MG.	N794K
☐ PT-OYD	69	King Air B90	LJ-455	Arnaldo Machado Diniz, Brasilia, DF.	N444WG
☐ PT-OYN	84	King Air C90A	LJ-1081	Ecen Engenharia Ltda. Goiania, GO.	N60CW
☐ PT-OZJ	81	King Air C90	LJ-951	Construmil Const. e Terrapl. Ltda. Goiania, GO.	N511D
☐ PT-OZK	75	King Air 200	BB-45	Constructora Lima Araujo Ltda. Maceio, AL.	N46JK
☐ PT-OZL	93	King Air C90B	LJ-1341	Azucar Zillor-Lorenzetti,	
☐ PT-OZN	80	PA-31T Cheyenne II	8020061	Lucio Christovan Furtado de Miranda, Ponta Grossa, PR.	N711DH
☐ PT-OZP	82	King Air F90	LA-175	First National Bank of Boston, Recife, PE.	N415GN
☐ PT-OZR	83	King Air C90-1	LJ-1059	Atlantic Corp. Belo Horizonte, MG.	N6581B
☐ PT-OZY	78	PA-31T Cheyenne II	7820030	Marialdo Rangel dos Santos, Sao Paulo, SP.	N700TR
☐ PT-WAE	67	King Air A90	LJ-191	Joao Carlos Soares de Matos, Florianapolis, SC.	N737K
☐ PT-WAG	75	King Air E90	LW-138	Luis Alexandre Igayara, Rio de Janeiro, RJ.	N90GD
☐ PT-WAH	90	King Air C90A	LJ-1245	Henrique Duarte Prata, Sao Paulo-Marte, SP.	N5654E
☐ PT-WBQ	69	King Air B90	LJ-460	Taza Com. Veic. Pec. Aces. I. Exp. Ltda. Sao Paulo, SP.	N113TT
☐ PT-WCB	92	King Air B200	BB-1419	Mouran Taxi Aereo Ltda. Sao Paulo, SP.	N8248W
☐ PT-WCS	94	King Air C90B	LJ-1377	Beech Acceptance Corp.	N3042K
☐ PT-WDU	78	King Air C90	LJ-791	MPE-Montagens e Projetos Especiais SA. Rio de Janeiro, RJ.	N791RC
☐ PT-WEF	81	PA-31T Cheyenne 1	8104034	Concremax Concreto Eng. e San. Ltda. Cuiaba, MT.	N100GY
☐ PT-WEG	81	King Air B200	BB-875	EGESA-Empreend. Gerais de Eng. SA. Belo Horizonte, MG.	N313SC
☐ PT-WET	80	King Air F90	LA-78	Auguri Constructora Participacoes Ltda. Sao Paulo, SP..	N90MH
☐ PT-WFB	80	PA-31T Cheyenne II	8020048	Procomp Agropecuaria e Exportadora Ltda. Ponta Grossa, PR.	N42EJ
☐ PT-WFN	93	King Air C90B	LJ-1346	Gianni Franco Samaja, Cuiaba, MT.	LV-WDP
☐ PT-WFQ	78	PA-31T Cheyenne II	7820049	Aguil Algodeira Gulmaraes Ltda. Mir D'Oeste, MT.	N689AC
☐ PT-WGH	73	SA-226AT Merlin 4	AT-011	Comask Ind. e Com. Ltda e Outra, Sorocaba, SP.	N750AA
☐ PT-WGJ	81	PA-31T Cheyenne II	8120101	Delta N Bank of New York, Curitiba, PR. (was 8120066).	N152CC
☐ PT-WGS	93	King Air B200	BB-1446	BB Leasing Co Ltda. Belo Horizonte, MG.	N5685X
☐ PT-WGU	94	King Air C90B	LJ-1363	Citicorp Leasing International Inc. Porto Alegre, RS.	N1534T
☐ PT-WHA	90	King Air C90A	LJ-1253	CONTERPAV-Const. Terrap. e Pav. Ltda. Goiania, GO.	N309P
☐ PT-WHI	79	PA-31T Cheyenne II	7920077	Istoril SA. Joinville, SC.	LV-MZA
☐ PT-WHN	81	PA-31T Cheyenne 1	8104073	Irmaos Muffato e Cia Ltda. Cascavel, PR.	N9185Y
☐ PT-WHP	89	King Air C90A	LJ-1212	Leasing bank of Boston SA. Goiania, GO.	N1551J
☐ PT-WIC	80	Gulfstream 840	11625	Banco Itamarati SA. Sao Paulo, SP.	N690HC
☐ PT-WIH	95	King Air C90B	LJ-1396	Coffee Holdings Inc. Tortola, BVI.	N3218K
☐ PT-WIT	95	King Air C90SE	LJ-1394		N3217K
☐ PT-WIX	72	Mitsubishi MU-2F	232	Luiz Antonio Portelinha Bueno, Bauru, SP.	N800HR
☐ PT-WJD	96	King Air C90B	LJ-1427	Frigorifico Independencia Ltda. Cajamar.	N3251E
☐ PT-WJF	94	King Air C90B	LJ-1386	Marcep International Trade Finance Ltd. Uberaba, MG.	N3165M
☐ PT-WKF	96	Pilatus PC-XII	141	Flamingo Unimed AirTaxi Aereo Ltda. Sao Paulo.	N141BL
☐ PT-WLD	83	Gulfstream 900	15027	Braulino Basilio Maia Filho, Sao Paulo, SP.	N900ST
☐ PT-WLF	96	King Air B200	BB-1528	Fundacao de E E T de Alfenas, Alefenas, MG.	N3261L
☐ PT-WLJ	81	PA-31T Cheyenne II	8120011	Francisco Preto Ribeiro, Sao Paulo, SP.	N31FR
☐ PT-WLK	96	King Air B200	BB-1543	CEMIG-Cia Energetica de Minas Gerais, Belo Horizonte, MG.	N1082S
☐ PT-WLT	87	King Air B200	BB-1275	Itaquere Participacoes Ltda. Sao Paulo, SP.	N3215K
☐ PT-WME	80	PA-31T Cheyenne II	8020058	IFS Flugzeug Leasing GmbH. Rio de Janeiro, RJ.	D-IGKG
☐ PT-WMT	81	King Air C90	LJ-956	Industria de Madeiras Tozzo Ltda. Chapeco.	N225AT
☐ PT-WMU	80	PA-31T Cheyenne 1	8004043		N53WM
☐ PT-WMX	81	PA-31T Cheyenne II	8104062	Fabrica de Pecas Elet. Delmar, Tatui, SP.	N2560Y
☐ PT-WNC	81	PA-31T Cheyenne II	8120020	F da Silva Machado, Sao Paulo.	N56MC
☐ PT-WND	96	King Air 350	FL-141		N1061Q
☐ PT-WNI	96	King Air C90B	LJ-1442	Lider Taxi Aereo SA. Belo Horizonte, MG.	N1072G
☐ PT-WNL	97	King Air 350	FL-159	Sementes Maggi Ltda. Rondonopolis.	N1100N
☐ PT-WNN	97	King Air B200	BB-1568	UNIMED N N F C Trabajos Med. Ltda. Joao Pessoa.	N1108A
☐ PT-WNW	85	King Air C90A	LJ-1092		N11755
☐ PT-WOF	82	King Air B200	BB-986	Fotop an Materials Fotograficos, Sao Paulo, SP.	N986TJ
Regn	Yr	Type	c/n	Owner/Operator	Prev Regn

Regn	Yr	Type	c/n	Owner/Operator	Prev Regn
☐ PT-WOR	81	PA-31T Cheyenne II	8120030	Campo Alegre Locacion Veiculos Maquineros Equipmentos Ltda.	N199RC
☐ PT-WPV	78	King Air B100	BE-45	Futura Taxi Aereo Ltda. Araraquara, SP.	N263DC
☐ PT-WRA	94	King Air C90B	LJ-1385	Marcep International Trade Finance Ltd. Rio de Janeiro, RJ.	N3198K
☐ PT-WSJ	96	King Air 350	FL-152		N1092S
☐ PT-WST	78	Mitsubishi MU-2N	711SA	Sete Taxi Aereo Ltda. Goiana, GO.	N171CA
☐ PT-WSX	87	King Air B200	BB-1266	Londrina, PR.	N204MS
☐ PT-WTN	68	King Air B90	LJ-346	C Nacife Jr & Partner, Belo Horizonte, MG.	PP-EOC
☐ PT-WVI	93	King Air C90B	LJ-1331	Citicorp Leasing Internationa Inc. Recife, PE.	N8089J
☐ PT-WYY	88	King Air 1300	BB-1302	MM Aerotaxi Ltda. Sao Paulo, SP.	N302YV

P2 = Papua New Guinea (Independent State of Papua New Guinea)

Civil

Regn	Yr	Type	c/n	Owner/Operator	Prev Regn
☐ P2-CAA	79	King Air 200	BB-415	Department of Civil Aviation, Port Moresby.	P2-PNH
☐ P2-HCN	81	King Air 200C	BL-22	Hevi Lift (PNG) P/L. Mount Hagen.	P2-PJV
☐ P2-IAH	77	King Air 200	BB-297	Islands Helicopter Services P/L. Rabaul. (status ?).	VH-IBD
☐ P2-MBH	92	King Air B200	BB-1423	MBA P/L. POrt Moresby.	VH-OXL
☐ P2-MBZ	92	King Air B200	BB-1420	MBA P/L. Port Moresby.	VH-OXA
☐ P2-PNG	92	King Air 350	FL-79	Government of Papua New Guinea, Port Moresby.	N8246Q
☐ P2-VIC	82	King Air B200	BB-990	Hevi Lift (PNG) P/L. Mount Hagen.	N61369

P4 = Aruba

Civil

Regn	Yr	Type	c/n	Owner/Operator	Prev Regn
☐ P4-JML	61	Gulfstream 1	76		G-BRAL

RP = Philippines (Republic of the Philippines)

Civil

Regn	Yr	Type	c/n	Owner/Operator	Prev Regn
☐ RP-C...	82	Gulfstream 1000	96038	Rayfront Corp. Makati.	N707TS
☐ RP-C...	96	King Air 350	FL-144	Invest Building Corp. Manila.	N1084W
☐ RP-C...	96	King Air B200	BB-1529		N3261E
☐ RP-C201	75	King Air E90	LW-126	Philippine National Bank, Manila.	(G-BCKE)
☐ RP-C203	74	SA-226T Merlin 3	T-244	A L Po. Manila.	N5330M
☐ RP-C223	75	King Air 200	BB-66		(N219DM)
☐ RP-C264	80	King Air 200	BB-692	Development Bank of the Philippines, Manila.	RP-C5139
☐ RP-C282	71	King Air 100	B-78	A Soriano Aviation Inc/Philippine Cotton Corp. Manila.	PI-C282
☐ RP-C289	72	King Air E90	LW-12	Far East Bank & Trust Co. Manila.	N2KA
☐ RP-C290	79	King Air C90	LJ-857	National Irrigation Administration, Manila.	N6064A
☐ RP-C291	79	King Air E90	LW-325	Lepanto Consolidated Mining Co.	N60575
☐ RP-C292	78	King Air E90	LW-277	Marcopper Mining Corp.	N4977M
☐ RP-C298	79	King Air E90	LW-302	Allied Banking Corp. Manila.	(N209DM)
☐ RP-C304	79	King Air 200	BB-503	Jaka Air Transport Corp. Manila.	N400GC
☐ RP-C319	80	King Air E90	LW-331	National Sugar Administration, Manila.	N67262
☐ RP-C323	80	SA-226T Merlin 3B	T-313	Filipino Amusement & Casino Gaming Corp. Manila.	N5673M
☐ RP-C367	82	King Air B200	BB-963	Aboitiz Inc. Manila.	N37GA
☐ RP-C410	81	King Air F90	LA-136		(N215DM)
☐ RP-C415	76	King Air E90	LW-190	Benguet Corp.	
☐ RP-C549	80	Conquest II	441-0115	Rio Tuba Nickel Mining Corp.	N2623Q
☐ RP-C574	80	Conquest 1	425-0008	Semirara Coal Corp.	N98876
☐ RP-C582	81	King Air 200	BB-785	Commercial & Industrial Bank, Manila.	(N88DA)
☐ RP-C704	80	King Air 200	BB-615	A Soriano Aviation Inc. Manila.	N16PM
☐ RP-C711	75	King Air 200	BB-83	Soriano Aviation, Manila.	
☐ RP-C755	82	King Air B200	BB-975	United Coconut Planters Bank, Manila.	(N208DM)
☐ RP-C775	72	681B Turbo Commander	6066	Coconut Industrial Development Corp. Manila.	N2NR
☐ RP-C879	75	King Air E90	LW-145	Philippine Airlines, Manila.	N122HC
☐ RP-C969	82	King Air B200	BB-951		VH-AGB
☐ RP-C990	67	King Air A90	LJ-247	Lepanto Consolidated Mining Co.	
☐ RP-C1260	80	King Air 200	BB-595		N6672V
☐ RP-C1341	66	King Air	...	Transglobal Aviation, Manila.	
☐ RP-C1502	94	King Air B200	BB-1500	Royal Duty Free Shop/AAI Island Hopper, Manila.	N3199B
☐ RP-C1515	82	King Air B200	BB-1004	Benpres Corp.	F-GKAN
☐ RP-C1577	82	King Air/Catpass 250	BB-945	A Soriano Aviation Inc/Island Aviation Inc. Manila.	N3000R
☐ RP-C1587	90	King Air 350	FL-20	Philippine American Life Co. Manila.	N1552Q
☐ RP-C1728	94	King Air 350	FL-118	Philippine Long Distance Telephone Co. Manila.	N1555E
☐ RP-C1807	88	King Air C90A	LJ-1181	A Soriano Aviation Inc/Rikio Co. Manila.	N90PE
☐ RP-C1890	85	King Air 300	FA-64	A Soriano Aviation Inc/San Miguel Corp. Manila.	N7232L
☐ RP-C1956	73	Rockwell 690	11072	JMB Aircraft Corp. Manila.	N70WA
☐ RP-C1977	70	681 Turbo Commander	6036	Planters Products Inc. Manila.	N9089N
☐ RP-C1978	71	King Air 100	B-77	Development Bank of the Philippines, Manila.	PI-C1978
☐ RP-C1990	80	King Air E90	LW-341	CRC Concrete Mix Corp/AAI Island Hopper, Manila.	N3722G
☐ RP-C1995	92	King Air B200	BB-1429	Southstar Aviation Co. Davao City.	N8008A
☐ RP-C2100	91	King Air B200	BB-1405	Air Transportation Office, Manila.	N8129A
☐ RP-C2340	65	King Air 90	LJ-62	Benigno Toda Jr/Philippine Airlines, Manila.	N2340M
☐ RP-C2638	96	King Air 350	FL-137	United Laboratories Inc.	N1067S
☐ RP-C2850	96	King Air 350	FL-145	National Power Corp. Manila.	N1075G
☐ RP-C3650	75	King Air C90	LJ-662	Orient Leaf Tobacco Co.	
☐ RP-C4650	81	King Air 200	BB-762	Philippine Long Distance Telephone Co.	N37225
Regn	Yr	Type	c/n	Owner/Operator	Prev Regn

□ RP-C5129	78	King Air 200	BB-358	National Steel Corp. Manila.	
□ RP-C7188	82	King Air B200	BB-1077	Fil-Estate Properties Inc. Manila.	N258L
□ RP-C8300	92	King Air 350	FL-83		OY-CVL
Military					
□ 11250	75	Rockwell 690A	11250	Government of Philippines, Manila.	N44WV

SE = Sweden *(Kingdom of Sweden)*

Civil
□ SE-FGO	68	Mitsubishi MU-2D	102	Nyge-Aero AB. Nykoping.	HB-LED
□ SE-GHA	74	Mitsubishi MU-2K	283	Nyge-Aero AB. Nykoping.	N327MA
□ SE-GHB	74	Mitsubishi MU-2K	287	Nyge-Aero AB. Nykoping.	N331MA
□ SE-GHC	74	Mitsubishi MU-2K	289	Nyge-Aero AB. Nykoping.	N334MA
□ SE-GHD	74	Mitsubishi MU-2K	293	Nyge-Aero AB. Nykoping.	N453MA
□ SE-GHE	74	Mitsubishi MU-2K	294	Nyge-Aero AB. Nykoping.	N454MA
□ SE-GHF	74	Mitsubishi MU-2K	299	Nyge-Aero AB. Nykoping.	N459MA
□ SE-GHH	72	Mitsubishi MU-2F	222	Nyge-Aero AB. Nykoping.	N9PN
□ SE-GSS	80	Gulfstream 840	11613	V-Air/Vaermlandsflyg AB. Torsby.	D-IBOB
□ SE-IIB	77	King Air C90	LJ-723	Varmforzinkning AB. Smilandsstehar.	OY-ASI
□ SE-INI	80	King Air 200	BB-687	SOS Flygambulans AB. Goteborg.	EI-BIP
□ SE-IOV	75	Mitsubishi MU-2M	337	Nyge-Aero AB. Nykoping.	N522MA
□ SE-IOZ	75	Mitsubishi MU-2M	320	Nyge-Aero AB. Nykoping.	N641KE
□ SE-IUA	76	Mitsubishi MU-2M	345	Nyge-Aero AB. Nykoping.	N730MP
□ SE-IUX	80	King Air 200	BB-675	SOS Flygambulans AB. Goteborg.	N26AD
□ SE-IVA	75	Mitsubishi MU-2L	666	International Business Air AB. Karlshamn.	N826RC
□ SE-IXC	85	King Air B200	BB-1210	SOS Flygambulans AB. Goteborg.	N7213J
□ SE-KBX	73	Mitsubishi MU-2K	247	Air Service i Falkoping AB. Falkoping. (status ?).	N5TQ
□ SE-KDK	81	King Air B200C	BB-909	Joh Sjoe AB/Air Express AB. Norrkoping.	N171M
□ SE-KFP	88	King Air B200C	BL-132	SOS Flygambulans AB. Goteborg.	
□ SE-KGO	80	MU-2 Marquise	755SA	Billingen Flyg AB. Skovde.	OY-CGN
□ SE-KOL	89	King Air 300LW	FA-189	Pioneer Electronic Svenska AB/Air Express AB. Norrkoping.	N8208B
□ SE-LCB	93	King Air B200C	BL-139	SOS Flygambulans AB. Goteborg.	N82431
□ SE-LCE	90	King Air B200	BB-1355	Stora Flight AB. Borlange.	N404SK
□ SE-LCT	84	King Air 300	FA-25	V-Air/Vaermlandsflyg AB. Torsby.	N147CA
□ SE-LDL	75	King Air A100	B-213	Stanson Air AB. Stockholm-Bromma.	F-GFEV
Military					
□ 100001	89	SAAB 340B/Tp 100	170	Swedish Air Force, Stockholm.	SE-F70
□ 101002	79	King Air 200	BB-459	Swedish Air Force, Stockholm. (Code 012 of F21).	OY-BVC
□ 101003	80	King Air 200	BB-619	Swedish Air Force, Stockholm.	LN-MOD
□ 101004	81	King Air B200	BB-932	Swedish Air Force, Stockholm. (Code 014 of F7).	SE-KKM

ST = Sudan *(The Republic of the Sudan)*

Civil
□ ST-AFO	75	King Air C90	LJ-669	Coptrade Air Transport, Khartoum.	N9397S
□ ST-ANH	79	King Air C90	LJ-823	Sudan Airways Co Ltd. Khartoum.	N580C
□ ST-SFS	79	King Air 200	BB-539	Sudan Airways Co Ltd. Khartoum.	N555SK

SU = Egypt *(Arab Republic of Egypt)*

Civil
□ SU-...	96	King Air B200	BB-1518	Orca Air, Sharm El Sheikh.	N3218V
□ SU-...	82	King Air B200	BB-1024	Raslan Air Service, Cairo.	N300DK
□ SU-BAX	78	King Air 200	BB-353	Government of Egypt, Cairo.	SU-AYD
□ SU-UAA	95	King Air C90B	LJ-1418	Orca Air, Sharm El Sheikh.	N3218X
□ SU-ZAA	94	King Air C90B	LJ-1353	Nuclear Materials Authority, Cairo.	N8292Y

SX = Greece *(Hellenic Republic)*

Civil
□ SX-APJ	78	King Air 200	BB-401	Aviator Ltd. Athens.	OY-JAO
□ SX-ECG	78	King Air 200	BB-372	Civil Aviation Authority, Athens.	N4937M
Military					
□ 401		King Air C-12C	BC-34	Aeroporias Stratu-Army Aviation, Megara.	

S2 = Bangladesh *(People's Republic of Bangladesh)*

Military
□ S3-BHN	84	PA-31T Cheyenne 1A	1104007	Government of Bangladesh, Dacca.	N2436W

S5 = Slovenia

Civil
□ S5-CAE	80	Conquest II	441-0150	Iskra Investment Services/Adria Airways, Ljubljana.	SL-CAE
□ S5-CMO	94	King Air C90B	LJ-1360	Kondorair, Portoroz.	N1560U

S7 = Seychelles *(Republic of Seychelles)*

□ S7-006		Reims/Cessna F406	F406-0035	Government of Seychelles, Mahe.	S7-AAM
□ S7-AAI		Reims/Cessna F406	F406-0051	Government of Seychelles, Mahe.	(9V-...)
Regn	*Yr*	*Type*	*c/n*	*Owner/Operator*	*Prev Regn*

S9 = Sao Tome / Principe (Democratic Republic of Sao Tomé & Principe)

Civil

Regn	Yr	Type	c/n	Owner/Operator	Prev Regn
☐ S9-IHD	91	Reims/Cessna F406	F406-0048		5Y-MIM
☐ S9-TAP	93	King Air 350	FL-102	SAL-Sociedade de Aviacao Ligeira SA. Luanda, Angola.	D2-ECW

TC = Turkey (Republic of Turkey)

Civil

Regn	Yr	Type	c/n	Owner/Operator	Prev Regn
☐ TC-...	82	King Air B200	BB-1082		N801BC
☐ TC-...	95	King Air C90B	LJ-1412	Altay Kollectif Sirketi/Dardanel Ltd. Ankara.	N3106Y
☐ TC-ACN	81	King Air 200	BB-791	Kalyon Air, Ankara.	N54LG
☐ TC-AEM	91	Beech C90A	LJ-1275	Rubi Air, Istanbul.	N8108E
☐ TC-AUT	74	King Air C90	LJ-622	Anadolu University Air Taxi, Eskisehir.	N104TT
☐ TC-AUV	73	King Air C90	LJ-587	Anadolu University Air Taxi, Eskisehir.	N61KA
☐ TC-AUY	78	King Air 200	BB-333	Anadolu University Air Taxi, Eskisehir.	F-GHLH
☐ TC-CSA	78	King Air C90	LJ-801	Metro Air, Istanbul.	N18BG
☐ TC-DBY	81	King Air 200	BB-821	Top Air, Istanbul.	N144TM
☐ TC-DBZ	77	King Air C90	LJ-703	Top Air, Istanbul.	(F-GKSR)
☐ TC-DHA	90	King Air 350	FL-37	DOGUS Air, Istanbul.	N4937M
☐ TC-DHC	91	King Air 350	FL-63	DOGUS Air, Istanbul.	TC-SAB
☐ TC-EEE	88	PA-42 Cheyenne 400LS	5527036	Gokturk Air, Bursa.	TC-SCM
☐ TC-FAH	85	PA-42 Cheyenne IIIA	5501033	THK-Turk Hava Kurumu, Ankara.	
☐ TC-FRT	80	King Air C90	LJ-910	Firat Aviation, Istanbul. (sold USA 12/97 ?).	N217GA
☐ TC-LMK	84	King Air C90A	LJ-1080	Menekse Air, Ankara.	N6931W
☐ TC-MAZ	95	King Air	...		
☐ TC-MCK	81	King Air C90	LJ-962	KOC Holdings AS/Set Air, Istanbul.	N1213P
☐ TC-MDE	96	King Air B200	BB-1539	Guven Air, Istanbul.	N1089S
☐ TC-MSS	91	King Air C90A	LJ-1276	Sky Line Ulasim Ticaret AS. Ankara.	N8065R
☐ TC-NAZ	78	King Air C90	LJ-787	Top Air, Istanbul.	N170DB
☐ TC-OZD	94	King Air B200	BB-1496	Overseas Aviation Resources Inc. Istanbul.	N3047L
☐ TC-SKO	89	King Air B200	BB-1334	Mach Air, Istanbul.	N5545B
☐ TC-THK	85	PA-42 Cheyenne IIIA	5501031	THK-Turk Hava Kurumu, Ankara.	TC-FAG
☐ TC-UPS	75	SA-226AT Merlin 4A	AT-044	Uensped Paket Servisi, Istanbul. `Beril'	C-GGPT
☐ TC-YPI	81	King Air B200	BB-883	Cukurova Holding AS. Istanbul.	N940WT
☐ TC-YSM	82	King Air B200	BB-1086	AND Air, Bursa.	TC-SDR

Military

Regn	Yr	Type	c/n	Owner/Operator	Prev Regn
☐ 10010	91	King Air B200	BB-1409	Turkish Air Force,	
☐ 10011	91	King Air B200	BB-1411	Turkish Air Force,	
☐ 10012	91	King Air B200	BB-1413	Turkish Air Force,	
☐ 10013	91	King Air B200	BB-1414	Turkish Air Force,	
☐ 10014	91	King Air B200	BB-1415	Turkish Air Force,	
☐ 4005	92	King Air B200	BB-1434	Turkish Air Force,	N81148
☐ 4006	90	King Air B200	BB-1375	Turkish Air Force,	M-1375
☐ J-11265	75	Rockwell 690A	11265	Turkish Jandarma	UN-77

TF = Iceland (Republic of Iceland)

Civil

Regn	Yr	Type	c/n	Owner/Operator	Prev Regn
☐ TF-ELT	77	King Air 200	BB-276	Islandflug HF. Reykjavik.	(N198SC)
☐ TF-FMS	85	King Air B200	BB-1221	Directorate of Civil Aviation, Reykyavik.	TF-UUU

TG = Guatemala (Republic of Guatemala)

Civil

Regn	Yr	Type	c/n	Owner/Operator	Prev Regn
☐ TG-...	94	King Air B200	BB-1479	Valores Turisticos SA. Guatamala City.	N8155L
☐ TG-...	95	King Air C90B	LJ-1399		N3212Y
☐ TG-...	95	King Air C90B	LJ-1411		N3218P
☐ TG-...	81	PA-31T Cheyenne II	8120003		N57KW
☐ TG-...	80	PA-31T Cheyenne II	8020032	Caimo Real, Guatamala City.	N803SW
☐ TG-...	80	PA-31T Cheyenne II	8020045	Compania de Jarabes y Bebidas La Mariposo SA. Guatamala City	N75CA
☐ TG-CCA	94	King Air C90B	LJ-1364		
☐ TG-GMI	70	681 Turbo Commander	6032	Multillantas SA. Guatamala City.	N75GM
☐ TG-LIA	81	PA-31T Cheyenne II	8120043	Distribudora Piper SA. Guatamala City.	N40H
☐ TG-MEE	78	Rockwell 690B	11472	Fumasa, Guatamala City.	(N81775)
☐ TG-OIL	80	PA-31T Cheyenne II	8020003	Distribuidora Piper SA. Guatamala City.	N985CA
☐ TG-RWC	94	King Air C90B	LJ-1373		
☐ TG-UME	77	Conquest II	441-0024	COCESNA. Guatamala City.	N169CA
☐ TG-VAL	81	PA-31T Cheyenne II	8120045	Jose Maria Valdes,	N2441Y
☐ TG-VAS	78	King Air C90	LJ-782		
☐ TG-WIZ	70	681 Turbo Commander	6022		C-GBIT

Military

Regn	Yr	Type	c/n	Owner/Operator	Prev Regn
☐ TG-MDN	86	King Air 300	FA-105	FAG/Guatamala Air Force, Guatamala City.	HK-3628

TI = Costa Rica (Republic of Costa Rica)

Civil

Regn	Yr	Type	c/n	Owner/Operator	Prev Regn
☐ TI-...	81	Gulfstream 980	95056		N980BM

Regn	Yr	Type	c/n	Owner/Operator	Prev Regn
☐ TI-...	80	King Air F90	LA-76	Compania Palma Tica SA. San Jose, Costa Rica.	N781VC
☐ TI-....	75	PA-31T Cheyenne II	7520043		N29KL
☐ TI-GEV	78	King Air E90	LW-268	Aires de Pavas SA.	N23681

TJ = Cameroon (Republic of Cameroon)

Civil
☐ TJ-AHZ	77	Conquest II	441-0001	SEBC=Soc d'Exploitation des Bois du Cameroun, Douala.	N983SM

TN = Congo (People's Republic of Congo)

Civil
☐ TN-AFG	79	King Air E90	LW-326	Congolaise Industrielle des Bois, Brazzaville.	D-IHCE

TR = Gabon (Gabonese Republic)

Civil
☐ TR-...	80	King Air 200	BB-620		F-GILU
☐ TR-LBP	83	King Air B200C	BL-67	Ste Crossair, Zurich/Air Affaires Gabon, Libreville.	(XA-...)
☐ TR-LDM	85	King Air B200	BB-1220	COMILOG-Cie Miniere de l'Ogoue, Moanda.	N93GA
☐ TR-LDU	83	King Air B200	BB-1110	COMUF, Mounana.	N110GA

Military
☐ TR-KJD	95	ATR 42	...	Government of Gabon, Libreville.	

TU = Ivory Coast (Republic of the Ivory Coast)

Civil
☐ TU-TJL	77	PA-31T Cheyenne II	7720033	Air Inter Ivoire, Abidjan.	N82152

Military
☐ TU-VBB	77	King Air 200	BB-295	Government of Ivory Coast, Abidjan. (status ?).	F-GAPV

VH = Australia (Commonwealth of Australia)

Civil
Regn	Yr	Type	c/n	Owner/Operator	Prev Regn
☐ VH-...	96	King Air 350	FL-148	Hawker Pacific P/L. Yagoona, NSW.	N3268H
☐ VH-...	97	King Air 350	FL-171	Hawker Pacific P/L. Yagoona, NSW.	N2315A
☐ VH-...	97	King Air 350	FL-181	Hawker Pacific P/L. Yagoona, NSW.	N2281S
☐ VH-...	82	King Air B200T	BT-23	Airborne Research Australia, Adelaide, SA.	N312D
☐ VH-...	74	Mitsubishi MU-2K	303	DGH P/L.Brunswick, VIC.	N728FN
☐ VH-AAG	73	Rockwell 690A	11101	Transair, Brisbane, QLD.	N57101
☐ VH-AKT	79	King Air 200	BB-579	Southern Cross Air P/L. Cairns, QLD.	N6064B
☐ VH-AMB	89	King Air B200C	BL-131	Ambulance Service of NSW/Pearl Aviation Australia P/L. Perth	N3228X
☐ VH-AMM	85	King Air B200C	BL-125	Ambulance Service of NSW/Pearl Aviation Australia P/L. Perth	N72385
☐ VH-AMR	85	King Air B200C	BL-126	Ambulance Service of NSW/Pearl Aviation Australia P/L. Perth	N72381
☐ VH-AMS	90	King Air B200C	BL-133	Ambulance Service of NSW/Pearl Aviation Australia P/L. Perth	N15588
☐ VH-ATF	74	Rockwell 690A	11158	Ringflow P/L. Brisbane, QLD.	N57158
☐ VH-AWU	79	SA-226T Merlin 3B	T-298	McKinlay Air Charter P/L. McKinlay, QLD.	N5495M
☐ VH-AZB	80	Conquest II	441-0182	Skippers Transport P/L. Perth, WA.	N983GA
☐ VH-AZW	77	Conquest II	441-0026	Skippers Transport P/L. Perth, WA.	VH-FWA
☐ VH-BUW	81	PA-42 Cheyenne III	8001047	Specialised Container Transport, S Melbourne, VIC.	VH-ISW
☐ VH-CFD	80	Conquest II	441-0141	Royal Flying Doctor Service, Sydney, NSW.	N26267
☐ VH-CWO	84	King Air B200C	BL-72	Royal Flying Doctor Service, Port Hedland.	N43CE
☐ VH-DLK	76	Rockwell 690A	11321	Department of Lands, Adelaide, SA.	N81430
☐ VH-DRV	78	PA-31T Cheyenne II	7820079	Christmas Island Resort P/L. Perth, WA.	N6109A
☐ VH-DYN	67	King Air A90	LJ-281	Paul Ng. Rochedale, QLD.	N5451U
☐ VH-EEN	83	SA-227AT Merlin 4C	AT-563	Pel-Air Express P/L. Brisbane, QLD.	N563UP
☐ VH-EEO	83	SA-227AT Merlin 4C	AT-564	Pel-Air Express P/L. Brisbane, QLD.	N546UP
☐ VH-EEP	83	SA-227AT Merlin 4C	AT-567	Pel-Air Express P/L. Brisbane, QLD.	N565UP
☐ VH-EGC	70	SA-226T Merlin 3	T-204	Alturas Airways, Sydney, NSW.	N224SB
☐ VH-EGQ	84	Conquest 1	425-0202	Royal Flying Doctor Service, Jandakot, WA. 'Keith Parry'	VH-ICO
☐ VH-EGR	83	Conquest 1	425-0195	Royal Flying Doctor Service, Jandakot. 'Womens Auxiliary'	(N195RB)
☐ VH-EGS	83	Conquest 1	425-0183	Royal Flying Doctor Service, Jandakot, WA. 'Laverton'	VH-JEC
☐ VH-EGT	84	Conquest 1	425-0216	Royal Flying Doctor Service, Jandakot, WA. 'Kookynie'	N1225J
☐ VH-EVP	79	Conquest II	441-0088	Central Air P/L. Perth, WA.	N8936N
☐ VH-FAM	96	Pilatus PC-XII	161	Farnsway Australia P/L. Samford, QLD.	HB-FOK
☐ VH-FDA	82	King Air B200C	BL-55	Royal Flying Doctor Service, Brisbane, QLD.	VH-NSD
☐ VH-FDB	81	King Air 200C	BL-26	Royal Flying Doctor Service, Sydney. 'Alan Earnshaw'	VH-WLH
☐ VH-FDG	84	King Air B200	BB-1172	Royal Flying Doctor Service, Jandakot. 'Alec McLaughlan'	G-OJGA
☐ VH-FDI	82	Beech B200	BB-1037	Royal Flying Doctor Service, Brisbane, QLD.	VH-DAX
☐ VH-FDM	82	King Air C90-1	LJ-1024	Royal Flying Doctor Service, Sydney. 'Tim O'Leary'	N618DB
☐ VH-FDO	82	King Air B200	BB-1056	Royal Flying Doctor Service, Brisbane, QLD.	VH-RFX
☐ VH-FDP	81	King Air C90	LJ-968	Royal Flying Doctor Service, Brisbane. 'John Flynn'	N102EP
☐ VH-FDR	81	King Air 200C	BL-39	Royal Flying Doctor Service, Norwood, SA.	N75WR
☐ VH-FDS	83	King Air 200C	BL-68	Royal Flying Doctor Service, Sydney. 'Marjorie Loveday'	N83GA
☐ VH-FDT	79	King Air C90	LJ-842	Royal Flying Doctor Service, Jandakot 'Sir Robert Law-Smith'	N6052F
☐ VH-FDW	82	King Air C90-1	LJ-1011	Royal Flying Doctor Service, Sydney. 'Allan Vickers'	N6139U
☐ VH-FDZ	82	King Air C90-1	LJ-1021	Royal Flying Doctor Service, Sydney. 'Alf Traeger'	N117D
☐ VH-FIX	93	King Air 350	FL-90	Pearl Aviation Australia P/L. Perth, WA.	D-C...
☐ VH-FMC	95	Pilatus PC-XII	109	Royal Flying Doctor Service, Norwood, SA.	
Regn	Yr	Type	c/n	Owner/Operator	Prev Regn

Regn	Yr	Type	c/n	Owner/Operator	Prev Regn
☐ VH-FMF	95	Pilatus PC-XII	110	Royal Flying Doctor Service, Norwood, SA.	
☐ VH-FMN	82	King Air 200C	BL-47	Royal Flying Doctor Service, Norwood, SA.	N6334F
☐ VH-FMP	95	Pilatus PC-XII	122	Royal Flying Doctor Service, Parafield, SA.	HB-FQ.
☐ VH-FMQ	79	Conquest II	441-0109	Skippers Transport P/L. Perth, WA.	(N26226)
☐ VH-FMW	95	Pilatus PC-XII	123	Royal Flying Doctor Service, Parafield, SA.	HB-FQ.
☐ VH-FMZ	96	Pilatus PC-XII	138	Royal Flying Doctor Service, Kent Town, SA.	HB-F..
☐ VH-HEO	81	King Air 200C	BL-41	Royal Flying Doctor Service, Norwood, SA.	N100QR
☐ VH-ICO	93	TBM 700	69	Austcom P/L. Melbourne, VIC.	F-OHBL
☐ VH-IJQ	80	Conquest II	441-0174	Corporate Air, Canberra, ACT.	N441HW
☐ VH-ITA	78	King Air 200	BB-344	Interair P/L. Melbourne, VIC.	P2-IAG
☐ VH-ITH	79	King Air 200	BB-463	Interair P/L. Melbourne, VIC.	P2-NAT
☐ VH-JES	70	Mitsubishi MU-2G	516	Interair P/L. Melbourne, VIC.	N881DP
☐ VH-JFD	79	Conquest II	441-0095	Royal Flying Doctor Service, Sydney, NSW.	VH-JEB
☐ VH-JJR	82	King Air B200	BB-1019	CPC Energy P/L. Kirribilli, NSW.	VH-ARZ
☐ VH-JMZ	72	Mitsubishi MU-2J	561	Pel-Air Express, Sandgate, QLD.	N22FL
☐ VH-JWO	70	681 Turbo Commander	6039	GAM Air Services P/L. Melbourne, VIC.	N420MA
☐ VH-KDN	79	Conquest II	441-0130	O'Connor Airlines, Mount Gambier, SA.	N2625N
☐ VH-KFN	81	King Air 200C	BL-31	Royal Flying Doctor Service, Jandakot, WA. 'John Uhrig'	N200LG
☐ VH-KFT	93	TBM 700	92	KFT Services, Lilydale, VIC.	F-OBHP
☐ VH-KJD	95	King Air 350	FL-125	Executive Air, Brisbane, QLD.	N3026K
☐ VH-KOF	71	Mitsubishi MU-2G	544	Adventure Air Service P/L. Coffs Harbour Jetty, NSW	VH-WMW
☐ VH-KOH	70	Mitsubishi MU-2G	521	Adventure Air Service P/L. Coffs Harbour Jetty, NSW.	VH-WYY
☐ VH-KUZ	80	Conquest II	441-0138	Anindilyakwa Air P/L. Darwin, NT.	N311RR
☐ VH-KZL	80	King Air 200C	BL-9	N T Aerial Medical Service/Pearl Aviation Australia P/L.	VH-NSG
☐ VH-LBA	78	Conquest II	441-0042	Skippers Transport P/L. Perth, WA.	N46MR
☐ VH-LBC	82	Conquest II	441-0236	Skippers Transport P/L. Perth, WA.	VH-TFG
☐ VH-LBD	83	Conquest II	441-0296	Skippers Transport P/L. Perth, WA.	N6838K
☐ VH-LBX	78	Conquest II	441-0091	Skippers Transport P/L. Perth, WA.	VH-AZY
☐ VH-LBY	77	Conquest II	441-0023	Skippers Transport P/L. Perth, WA.	VH-TFW
☐ VH-LBZ	77	Conquest II	441-0038	Skippers Transport P/L. Perth, WA.	VH-HWD
☐ VH-LEM	79	Conquest II	441-0081	Kevron P/L-Kevron Aerial Survey, Perth, WA.	N4490C
☐ VH-LFD	80	Conquest II	441-0164	Royal Flying Doctor Service, Sydney. 'W B (Bill) Blown'	VH-LBC
☐ VH-LJG	82	King Air C90-1	LJ-1020	Vonly P/L. Hughesdale, VIC.	N71SL
☐ VH-LKB	77	King Air 200	BB-259	Skippers Transport P/L. Perth, WA.	VH-APA
☐ VH-LKF	80	King Air 200	BB-660	Skippers Transport P/L. Perth, WA.	N200TK
☐ VH-LTM	85	Gulfstream 1000	96208	Air Services Australia, Melbourne, VIC.	N230GA
☐ VH-MKA	94	King Air 350	FL-110	Westrac Aviation, Midland, WA.	N8279P
☐ VH-MNU	70	Mitsubishi MU-2G	527	Brook Armstrong Aviation Services P/L. Sydney, NSW.	VH-UZN
☐ VH-MSH	91	King Air B200	BB-1416	Royal Flying Doctor Service, Sydney, NSW.	N8254H
☐ VH-MSM	92	King Air B200	BB-1430	Royal Flying Doctor Service, Sydney, NSW.	N773AM
☐ VH-MSO	80	Conquest II	441-0132	Maroomba Air Services, Perth, WA.	VH-ANJ
☐ VH-MSU	82	King Air 200C	BL-48	Royal Flying Doctor Service, Sydney. 'Philip H Bushell'	N1860B
☐ VH-MSZ	81	King Air B200	BB-866	Royal Flying Doctor Service, Sydney. 'Fred McKay'	ZK-FBG
☐ VH-MVL	89	King Air B200	BB-1333	Royal Flying Doctor Service, Sydney, NSW.	N1101W
☐ VH-MXK	80	King Air 200	BB-653	Pearl Aviation Australia P/L. Perth, WA.	N67224
☐ VH-MYO	94	King Air B200	BB-1472	Airking P/L. Bankstown, NSW.	
☐ VH-NEY	73	Rockwell 690	11062	Tamair P/L. Tamworth, NSW.	VH-BLH
☐ VH-NFD	80	Conquest II	441-0159	Royal Flying Doctor Service, Sydney.	N45FM
☐ VH-NIA	79	King Air 200	BB-470	Great Western Aviation P/L. Perth, WA.	N3018C
☐ VH-NMU	78	Mitsubishi MU-2N	707SA	Pel-Air Express, Sandgate, QLD.	N15CN
☐ VH-NYA	74	Rockwell 690A	11152	W J Aviation, Launceston.	N42MM
☐ VH-NYC	72	Rockwell 690	11026	Transair/Lessbrook P/L. Brisbane, QLD.	N9226N
☐ VH-OAA	79	Conquest II	441-0102	O'Connor Airlines, Mount Gambier, SA.	N4246Z
☐ VH-OCS	77	Conquest II	441-0030	Air Charter Australia P/L. Adelaide, SA.	N441MM
☐ VH-OWN	81	King Air B200	BB-936	Five Star Aviation P/L. Coolangatta, QLD.	N200NS
☐ VH-OXF	94	King Air 350	FL-122	Five Star Aviation P/L. Coolangatta, QLD.	VH-BTL
☐ VH-OYA	78	King Air 200	BB-365	Pearl Aviation Australia P/L. Perth, WA.	P2-SML
☐ VH-OYD	82	King Air B200	BB-1041	Pearl Aviation Australia P/L. Perth, WA.	N200BK
☐ VH-PCV	76	Rockwell 690A	11283	Asia-Pacific Airlines/Commander Land & Air Invs. Sydney.	N57228
☐ VH-PJC	78	Rockwell 690B	11475	Pasdonnay P/L. Bayswater, WA.	VH-NPT
☐ VH-PSK	90	King Air 350	FL-29	Police Air Wing, Brisbane, QLD.	ZS-NAV
☐ VH-RSW	81	Cheyenne II-XL	8166001	Statewide Air Charter, Brisbane, QLD.	N450CB
☐ VH-SBM	82	King Air B200	BB-964	Saint Barbara Mines, Meekatharra, WA.	VH-HTU
☐ VH-SGQ	96	King Air 350	FL-150	Queensland Emergency Services, Brisbane.	N10691
☐ VH-SGT	75	King Air 200	BB-73	Central Air P/L. Cloverdale, WA.	
☐ VH-SGV	80	King Air 200	BB-718	Great Western Aviation P/L. Hamilton, QLD.	N6728N
☐ VH-SKC	75	King Air 200	BB-47	Flight West Airlines P/L. Brisbane, QLD. 'Capt Jack Treacy'	RP-C200
☐ VH-SKN	80	King Air 200	BB-690	Flight West Airlines P/L. Brisbane. 'Sir Hudson Fysh KBE'	ZK-WNL
☐ VH-SMB	78	King Air 200	BB-355	Pearl Aviation Australia P/L. Perth, WA.	P2-SMB
☐ VH-SMT	76	King Air 200	BB-162	Maroomba Air Service, Perth, WA.	RP-C22
☐ VH-SMZ	83	King Air B200	BB-1155	Maroomba Air Service, Perth, WA.	VH-HPA
☐ VH-SSD	71	SA-226T Merlin 3	T-213	Winrye Aviation P/L. Sydney.	N174SP
☐ VH-SSL	71	SA-226T Merlin 3	T-207	Winrye Aviation P/L. Sydney.	N173SP
☐ VH-SWO	80	King Air 200C	BL-12	N T Aerial Medical Service/Pearl Aviation Australia P/L.	F-GILF
☐ VH-SWP	79	King Air 200	BB-529	N T Aerial Medical Service/Pearl Aviation Australia P/L.	F-GIQV
☐ VH-TFB	82	Conquest II	441-0260	Transair/Lessbrook P/L.Brisbane, QLD.	N68597

Regn	Yr	Type		c/n	Owner/Operator		Prev Regn

Regn	Yr	Type	c/n	Owner/Operator	Prev Regn
□ VH-TLX	79	King Air 200	BB-550	Pearl Aviation Australia P/L. Perth, WA.	P2-MBM
□ VH-TNP	79	PA-31T Cheyenne II	7920026	Lampion P/L. Windsor.	VH-LJK
□ VH-TNQ	81	King Air 200C	BL-30	N T Aerial Medical Service/Pearl Aviation Australia P/L.	N3723Y
□ VH-TTD	80	PA-31T Cheyenne II	8020005	Australasian Jet P/L. Melbourne, VIC.	VH-DXI
□ VH-UJN	71	681B Turbo Commander	6047	GAM Air Services P/L. Melbourne, VIC.	VH-NYE
□ VH-ULX	82	Conquest 1	425-0124	Abril Properties P/L. Sunnybank, QLD.	N6882M
□ VH-UUA	82	SA-227AT Merlin 4C	AT-502	Jetcraft Aviation P/L. Brisbane, QLD.	OY-CHH
□ VH-UZB	70	Mitsubishi MU-2G	528	Brook Armstrong Aviation Services P/L. Sydney.	ZK-ESM
□ VH-UZC	70	Mitsubishi MU-2G	519	(status ?).	ZK-EKZ
□ VH-UZI	83	SA-227AT Merlin 4C	AT-570	JetCraft Aviation P/L. Brisbane, QLD.	N570UP
□ VH-WCE	81	PA-42 Cheyenne III	8001033	Whim Creek Consolidated P/L. Perth, WA.	N582SW
□ VH-WLO	72	Rockwell 690	11030	National Parks & Wildlife Service, Bankstown, NSW.	VH-WLS
□ VH-WNH	76	King Air 200	BB-148	Pearl Aviation Australia P/L. Perth, WA.	C-GBGW
□ VH-WNT	72	King Air C90	LJ-552	Knispel Fruit Juices, Regency Park, SA.	VH-AMH
□ VH-XDB	79	King Air 200	BB-533	Flight West Airlines P/L. Brisbane, QLD. 'Lores Bonney MBE'	N87RK
□ VH-XMD	77	Conquest II	441-0025	Ross Aviation P/L. Adelaide, SA.	N441HD
□ VH-XMJ	80	Conquest II	441-0113	Rossair Charter, Adelaide, SA.	N990AR
□ VH-XRF	76	King Air 200	BB-165	Flight West Airlines P/L. Brisbane, QLD. 'Arthur Affleck'	N76MB
□ VH-XRP	78	King Air 200	BB-327	Flight West Airlines P/L. Brisbane, QLD. 'Bob Norman OBE'	N771HA
□ VH-YFD	80	Conquest II	441-0157	Royal Flying Doctor Service, Sydney. 'James Hardie'	N51LR
□ VH-YNE	80	King Air 200	BB-605	Yartayne P/L-Corporate Air, Canberra, ACT.	N71GA
□ VH-YOL	79	Conquest II	441-0106	Comserv P/L. Southport, QLD.	VH-HXM
□ VH-ZZE	95	Reims/Cessna F406	F406-0076	Surveillance Australia P/L. Cairns, QLD.	F-WZDX
□ VH-ZZF	95	Reims/Cessna F406	F406-0078	Surveillance Australia P/L. Adelaide, SA.	VH-BPH
□ VH-ZZG	96	Reims/Cessna F406	F406-0079	Surveillance Australia P/L. Adelaide, SA.	F-WZDZ
Military					
□ VH-HPP	90	King Air B200C	BL-137	Australian Army, Oakey, QLD.	ZS-NSD
□ VH-HPW	95	King Air B200	BB-1504	Australian Army, Oakey, QLD.	N3214D
□ VH-HPX	95	King Air B200	BB-1505	Australian Army, Oakey, QLD.	N3197L
□ VH-HPZ	92	King Air B200C	BL-138	Australian Army, Oakey, QLD.	VH-AJM
□ VH-KBH	84	King Air B200	BB-1189	RAAF, 34 ANS, East Sale, VIC.	N843CK
□ VH-KCH	83	King Air B200	BB-1125	RAAF, 34 ANS, East Sale, VIC.	N12LD

VN = Vietnam *(The Socialist Republic of Vietnam)*

Civil

Regn	Yr	Type	c/n	Owner/Operator	Prev Regn
□ VN-B594	89	King Air B200	BB-1329	VASCO-Vietnam Air Services Co.	VH-SWC

VP-B = Bermuda *(UK-Colony of Bermuda)*

Civil

Regn	Yr	Type	c/n	Owner/Operator	Prev Regn
□ VP-BBK	96	King Air B200	BB-1519	Video Vision Broadcasting, Ronaldsway, IOM.	VR-BBK
□ VP-BDR	84	Conquest 1	425-0199	Inter-City Air Ltd.	VR-BDR
□ VP-BJT	81	Conquest 1	425-0027	Rig Design Services Group Ltd. Fairoaks, UK.	VR-BNM
□ VP-BKW	79	King Air C90	LJ-805	David J Sewell, Ibadan, Nigeria.	VR-BKW
□ VP-BLK	81	Gulfstream 840	11672	Control Techniques (Bermuda) Ltd. Welshpool, UK.	VR-BLK
□ VP-BLS	97	Pilatus PC-XII	176	Bruno Schroder, Fairoaks, UK.	(N176BS)
□ VP-BMZ	84	Gulfstream 900	15033	Aviatica Trading Co/Marlborough Fine Arts Ltd. Fairoaks, UK.	VR-BMZ
□ VP-BPH	82	Conquest II	441-0268	Interfruit Co. San Jose, Costa Rica.	VR-BPH

VP-C = Cayman Islands *(UK-Colony of Cayman Islands)*

Civil

Regn	Yr	Type	c/n	Owner/Operator	Prev Regn
□ VP-CAY	82	Gulfstream 900	15011		VR-CAY
□ VP-CCT	82	King Air C90A	LJ-1028	Corgi Investments Ltd. Guernsey, Channel Islands.	VR-CCT
□ VP-CHE	97	King Air B200	BB-1569		N20505
□ VP-CMA	97	King Air B200	BB-1564	Rangemile, Coventry/Moffat Aviation Ltd. Carlisle, UK.	N205JT

VT = India *(Republic of India)*

Civil

Regn	Yr	Type	c/n	Owner/Operator	Prev Regn
□ VT-ASB	89	Reims/Cessna F406	F406-0031	Bharatair, Calcutta. (for VH- ?).	D-IBOM
□ VT-AVB	81	King Air B100	BE-121	Hindustan Aluminium Corp. Bombay.	VT-EGQ
□ VT-BAL	97	King Air B200	BB-1563	Indamer Co Pte Ltd. Mumbai.	N204JT
□ VT-BSA	94	King Air B200	BB-1485	Border Security Force, Delhi.	N1509X
□ VT-CIL	93	King Air B200	BB-1469	Coal India Ltd.	N82378
□ VT-CSK	97	King Air B200	BB-1567	Kirloskar Group, Bangalore.	N1107F
□ VT-DEJ	95	King Air C90B	LJ-1404	Lakshmi Machine Works Ltd. Coimbatore.	N3106P
□ VT-DXU	67	King Air A90	LJ-308	Mahindra & Mahindra Ltd. Delhi.	N7010N
□ VT-EBB	94	King Air B200	BB-1486	National Remote Sensing Agency, Hyderabad.	N1542Z
□ VT-ECA	77	King Air C90	LJ-537	Kirloskar Oil Engine Co. Poona.	N9030V
□ VT-EFB	77	King Air C90	LJ-706	Tata Iron & Steel Co. Jamshedpur.	N23856
□ VT-EFE	77	King Air C90	LJ-711	United Breweries/UB Air, Bangalore.	N23875
□ VT-EFF	77	King Air C90	LJ-705	Baja Auto Co. Poona.	N2075L
□ VT-EFG	77	King Air C90	LJ-719	Government of Bihar, Patna.	N23917
□ VT-EFP	77	King Air C90	LJ-720	Government of Madhya Pradesh, Bhopal.	N23929
□ VT-EFZ	78	King Air C90	LJ-790	Indamer Co Pte Ltd. Bombay.	N4908M
□ VT-EGR	81	King Air C90	LJ-967	Government of Maharashtra, Bombay.	N3832X

Regn	Yr	Type	c/n	Owner/Operator	Prev Regn

Regn	Yr	Type	c/n	Owner/Operator	Prev Regn
☐ VT-EHB	82	King Air B200	BB-972	Government of Orissa, Cuttack.	N18409
☐ VT-EHY	82	King Air C90	LJ-1008	Government of Punjab, Chandigarh.	N1842A
☐ VT-EID	82	King Air B200C	BL-56	Government of Madhya Pradesh, Bhopal.	N1844C
☐ VT-EIE	83	King Air B200C	BL-63	Government of Uttar Pradesh, Lucknow.	N6921D
☐ VT-EJZ	85	King Air C90A	LJ-1100	Government of Haryana, Ambala.	N7219K
☐ VT-ELZ	85	King Air F90-1	LA-233	Indian Steel Authority,	N72224
☐ VT-EMI	86	King Air C90A	LJ-1135	Indira Gandhi Rashtriya Uran Academy,	N2602M
☐ VT-EMJ	86	King Air C90A	LJ-1137	Indira Gandhi Rashtriya Uran Academy,	N6690N
☐ VT-ENL	86	King Air B200	BB-1248	Aviation Research Centre, Cuttack, Orissa.	N7256G
☐ VT-ENM	85	King Air B200	BB-1236	Aviation Research Centre, Cuttack, Orissa.	N72473
☐ VT-EOA	87	King Air B200C	BL-129	Border Security Force, Delhi.	N72401
☐ VT-EPA	86	King Air B200	BB-1254	Aviation Research Centre, Cuttack, Orissa. (status ?).	N2646K
☐ VT-EPY	87	King Air B200	BB-1277	Government of Maharashtra, Bombay.	N7241L
☐ VT-EQD	87	King Air B200	BB-1272	Bokaro Steel Plant, Bokaro.	N3043W
☐ VT-EQK	88	King Air B200	BB-1288	National Remote Sensing Agency, Hyderabad.	N30850
☐ VT-EQN	88	King Air C90A	LJ-1160	Government of Rajasthan, Jaipur.	N31174
☐ VT-EQO	87	King Air C90A	LJ-1153	Government of Uttar Pradesh, Lucknow.	N70491
☐ VT-ESR	80	King Air 200	BB-739	Avtar Singh Sandhu/Bechtel Overseas,	N28TL
☐ VT-ETI	87	King Air C90A	LJ-1160	Finolax Cables Ltd.	N909GA
☐ VT-HYA	94	King Air C90B	LJ-1376	Government of Haryana, Nissar.	N15542
☐ VT-IRC	94	King Air 350	FL-112	Indian Railways Corp. Delhi.	N82112
☐ VT-JKC	85	King Air B200	BB-1198	JK Corp. New Delhi.	N9MS
☐ VT-JNK	97	King Air 350	FL-160	Government of Jammu & Kashmir.	N1100A
☐ VT-LNT	93	King Air B200	BB-1468	Larsen & Toubro Ltd. Bombay. (status ?).	N8230E
☐ VT-MNM	94	King Air 350	FL-105	Mahindra & Mahindra Ltd.	N8215W
☐ VT-MPG	93	King Air B200	BB-1445	Government of Madhya Pradesh, Bhopal.	N8121M
☐ VT-NEF	80	King Air C90	LJ-890	National Energy Processing Co/NEPC Airlines, Madras.	ZS-LOL
☐ VT-NEI	85	King Air C90A	LJ-1116	National Energy Processing Co/NEPC Airlines, Madras.	N25AE
☐ VT-NKF	95	King Air C90B	LJ-1402	Bajaj Tempo Ltd. Akurdi.	N3234K
☐ VT-RLK	91	King Air C90A	LJ-1278	Kirloskar Oil Engine Co. Juhu.	N8141K
☐ VT-RLL	94	King Air C90B	LJ-1369	Ranbaxy Laboratories Ltd. Delhi.	N3222K
☐ VT-RSB	89	King Air B200	BB-1317	Hiadalco Industries Ltd/Birla Co. Bombay.	N591EB
☐ VT-SAA	90	Reims/Cessna F406	F406-0050	Spanair Aviation (I) Ltd. Pune.	(9V-...)
☐ VT-SAB	88	King Air 1300	BB-1305	Spanair Aviation (I) Ltd. Pune.	N789GA
☐ VT-SAC	90	Reims/Cessna F406	F406-0033	Spanair Aviation (I) Ltd. Pune.	(9V-...)
☐ VT-SAD	89	King Air 1300	BB-1341	Spanair Aviation (I) Ltd. Pune.	N41AV
☐ VT-SAE	89	King Air 1300	BB-1342	Spanair Aviation (I) Ltd. Pune.	N99DX
☐ VT-SAF	89	King Air 1300	BB-1343	Spanair Aviation (I) Ltd. Pune.	N98DX
☐ VT-SLK	91	King Air C90A	LJ-1270	Kirloskar Oil Engine Co. Juhu.	N324AB
☐ VT-SLS	80	King Air 200	BB-575	Inlac Gnavston Ltd. Bangalore.	OY-PAL
☐ VT-TIS	95	King Air C90B	LJ-1393	Tata Iron & Steel Co.	N3217M
☐ VT-UPA	94	King Air 300LW	FA-230	Government of Uttar Pradesh, Lucknow.	N80679
☐ VT-UPZ	95	King Air C90B	LJ-1400	Government of Uttar Pradesh, Lucknow.	N3239K
☐ VT-VHL	87	King Air B200	BB-1267	Venkateshwara Hatcheries Group, Pun, Maharashtra.	N67GA
☐ VT-VIL	94	King Air C90B	LJ-1374	Videcon India Ltd.	N15116

V5 = Namibia (Republic of Namibia)

Civil

Regn	Yr	Type	c/n	Owner/Operator	Prev Regn
☐ V5-CIC	75	King Air 200	BB-85	National Airlines Joint Venture, Randburg, RSA.	ZS-NTL
☐ V5-EEZ	71	Reims/Cessna F406	F406-0004	Government of Namibia, Windhoek.	F-WIVD
☐ V5-INN	77	King Air C90	LJ-523	G T Joubert,	ZS-INN
☐ V5-LYZ	81	Conquest 1	425-0021	Eerste Nas Ontwkoop of South West Africa,	ZS-LYZ
☐ V5-MJW	81	Conquest 1	425-0077	Eros Air P/L. Windhoek.	ZS-MJW
☐ V5-RTZ	92	King Air 350C	FM-1		N1564D

XA/B/C = Mexico (United Mexican States)

Civil

Regn	Yr	Type	c/n	Owner/Operator	Prev Regn
☐ XB-...		680W Turbo Commander	1834-39		N200QT
☐ XA-...	85	Gulfstream 1000	96075		N6151W
☐ XB-...	68	King Air B90	LJ-425		N41LH
☐ XA-...	97	King Air C90B	LJ-1479	Aerolineas Ejecutivas SA. Toluca.	N1119U
☐ XA-...	76	PA-31T Cheyenne II	7620034	Aereo Servicios Empresariales SA. Nueva Vizcaya.	N7276C
☐ XA-...	77	Rockwell 690B	11370	Aerotaxi del Cabo SA.	N200TE
☐ XA-ALT	89	King Air 300	FA-176	Aerovics SA. Toluca.	N1570H
☐ XA-CHM	85	Gulfstream 900	15040	Chilchota Taxi Aereo SA. Coahuila. (status ?).	N4NT
☐ XA-DER	73	Rockwell 690	11060	Aerotaxis del Itsmo SA. Cuernavaca, Morelos.	N.....
☐ XA-EOC		Rockwell 690A	11225	Salvador Vargas/Aero Centro SA. Celaya, Guanajuato.	XB-EXC
☐ XA-EYZ	84	PA-42 Cheyenne 400LS	5527009	Aviacion Comercial de America SA. Monterrey.	XB-EYZ
☐ XA-GEM	75	Rockwell 690A	11267	Transportes Aereos Virva SA. Mexico City.	N57035
☐ XA-KUU	72	Rockwell 690	11042	Commander Mexicana SA. Mexico City.	XC-FUJ
☐ XA-LEK	81	Gulfstream 980	95042	Aerotux SA.	N9794S
☐ XA-LGT	80	King Air F90	LA-45	Aero Rentas de Coahuila SA. Coahuila.	N748GM
☐ XA-MIM	68	King Air B90	LJ-342	Aerodiplomatic SA. Toluca.	XB-PEZ
☐ XA-MOB	83	King Air B200	BB-1151	Aero Barloz SA. Monterrey.	XA-MCB
☐ XA-MUR	65	King Air 90	LJ-63	Aerorent SA. Mexico City.	XB-ZAA

Regn	Yr	Type	c/n	Owner/Operator	Prev Regn

Regn	Yr	Type	c/n	Owner/Operator	Prev Regn
☐ XA-PEU	75	King Air 200	BB-48	Aerojalisco SA. Guadalajara.	XB-EDZ
☐ XA-POY	84	King Air 300	FA-14	Deluxe Mexico Hotels International, San Diego, Ca.	XA-VID
☐ XA-PUY	72	Rockwell 690	11018	Aerovallarta SA. Quetzalcoatl.	N63PG
☐ XA-RFN	90	King Air C90A	LJ-1246	Aerotransportes Rafilher SA. San Luis Potosi.	
☐ XA-RJB	65	Gulfstream 1	159	Aero Guadalajara SA/Aerolineas Damojh SA.	G-BNKN
☐ XA-RLB	83	PA-31T Cheyenne 1	8104068	Aereos Servicios de Nuevo Laredo SA. Quetzalcoatl.	XB-WOE
☐ XA-RLK	64	Gulfstream 1	138	Aero Guadalajara SA/Aerolineas Damojh SA. Guadalajara.	N126K
☐ XA-RMT	78	PA-31T Cheyenne II	7820060	Aero Cheyenne SA. Del Norte, NL.	N89TW
☐ XA-RMX	81	King Air 200	BB-814	Aerogisa SA. Saltillo.	N6VW
☐ XA-RNL	80	MU-2 Marquise	777SA	Servicios Aereos Interestatales SA. Del Norte, NL.	XB-DJX
☐ XA-RVH	90	King Air B200	BB-1365	Antair SA. Toluca.	N56562
☐ XA-RVJ	82	King Air B200	BB-995	La Valenciana Taxi Aereo SA. Torreon, Coahuila.	N11RM
☐ XA-RWJ	70	SA-26AT Merlin 2B	T26-175	Servicios Aereos Especiales de Jalisco SA. Guadalajara.	N60Y
☐ XA-RWR	79	PA-31T Cheyenne II	7920047	Aerolineas Comerciales SA. Michoacan.	N79CA
☐ XA-RXT	91	King Air C90A	LJ-1280	Aereo Servicios Saltillo SA. Saltillo.	
☐ XA-SAW	81	King Air F90	LA-95	Aero Quimmco SA. Monterrey.	XB-CGP
☐ XA-SFD	79	Rockwell 690B	11534	Aerolineas Villaverde SA. Cordoba.	XB-BGH
☐ XA-SHP	71	681B Turbo Commander	6063	GBM Aereo SA. Cuidad Acuna, Coahuila.	XB-HUL
☐ XA-SPW	74	Rockwell 690A	11175	Aeropremier de Mexico SA. Merida.	N276H
☐ XA-STD	77	Conquest II	441-0015	Servicios Aereos del Norte SA.	XA-REF
☐ XA-SWE	70	King Air 100	B-34	Lineas Aereas Comerciales SA. Plan de Guadalupe.	N61MR
☐ XA-TBT	64	Gulfstream 1	136	Industrial Minera Mexico SA.	XB-GAW
☐ XA-TDJ	64	Gulfstream 1	127	Aerolineas Bonanza SA.	N717JP
☐ XA-TDW	94	King Air C90B	LJ-1375	Aeroservicios de Nuevo Leon SA. Monterrey.	N3194K
☐ XA-TFQ	76	Rockwell 690A	11332	Servicios Aereos de Chihuahua SA. Chihuahua.	N72AB
☐ XA-THE	77	Rockwell 690B	11400	Aeroejecutivos del Centro SA. Guadalajara.	XB-GQO
☐ XA-TIL	82	PA-42 Cheyenne III	8001071	Servicios Aereos Especializados Mexicanos SA. Del Norte, NL.	N65MC
☐ XB-ACO	81	Gulfstream 980	95051	Procurad General de la Republica, Mexico City.	XC-HGG
☐ XB-BED	74	Rockwell 690A	11188	Estacion Cuauhtemoc SPR.	XA-FIZ
☐ XB-BRB	78	King Air 200	BB-395	Tomas Gonzales Sada y Pablo GL. Garza Garcia, NL.	XA-IIY
☐ XB-BZQ	67	King Air A90	LJ-316	Fideicomiso Lazaro Cardenas, Michoacan.	YV-79CP
☐ XB-CIO	68	King Air B90	LJ-387	Representaciones Cristaleria y Peltre de Zamora, Zamora.	N824K
☐ XB-CYR	86	PA-42 Cheyenne 400LS	5527032	Commander Mexicana SA.Toluca.	N4120G
☐ XB-DJN	70	681 Turbo Commander	6023	Gonzalez Trejo Alejandro, Mexico City.	XA-CAG
☐ XB-DMT	77	Rockwell 690B	11360	Carrillo Caraza Rene, Culiacan, Sinalao.	N100WC
☐ XB-DTD	67	680W Turbo Commander	1788-18	Fernando Nunez Chavez, Durango.	XA-DII
☐ XB-DTW	80	Gulfstream 840	11650	Dulce V Gonzalez Garduno, Mexico City.	XA-JPA
☐ XB-DZP	77	Rockwell 690B	11410	P R I. Toluca.	XC-SPP
☐ XB-ECT	82	Gulfstream 1000	96004	Francisco Lira Ortega, Ensenada, Baja California.	XA-LUV
☐ XB-EFZ	80	SA-226T Merlin 3B	T-328	Dist Int Productos Agricolas SA. Celaya.	N23X
☐ XB-EGZ	77	PA-31T Cheyenne II	7720055	Arrendadora Financiera del Norte, Toluca.	XA-IEQ
☐ XB-EIH	74	Rockwell 690A	11214	Compras y Comisiones SA. Leon.	N57214
☐ XB-EWO	82	Conquest II	441-0235	Parson Meikle Joseph Richard, Nva Casas Grandes, Chihuahua.	N333GE
☐ XB-FKC	77	Rockwell 690B	11406	Constructadora y Edificado Comal. Toluca.	XC-SPI
☐ XB-FLF	74	Rockwell 690A	11169	Aida Lucia Rodriguez, Mexico City.	XC-GIM
☐ XB-FLL	60	Gulfstream 1	58	Celanese Mexicana SA. Toluca.	N47TE
☐ XB-FMS	82	Conquest 1	425-0051	Arturo Armendariz Chaparr, Chihuahua.	XA-RLN
☐ XB-FMV	67	Mitsubishi MU-2B	034	Christian Esquino,	N28HR
☐ XB-FND	80	Conquest II	441-0155	Doroteo Couret Nolasco, Los Mochis, Sinalao.	XB-BON
☐ XB-FQC	74	Rockwell 690A	11144	Lin. de Produc. SA. Toluca.	XA-RAO
☐ XB-FSG	83	Cheyenne II-XL	8166047	Grupo Fralvar de Mexico SA. Mexico City.	N457SR
☐ XB-FVK	80	Conquest II	441-0148	Agrupacion Muralla SA.	N148GA
☐ XB-FXK		Conquest 1	425-0091	Jose Luis Franco Alvarez, Mexico City.	N68478
☐ XB-FXS	73	Rockwell 690A	11107	Movimiento de Maquinaria SA. Guadalajara.	N115CW
☐ XB-FXU	79	PA-31T Cheyenne 1	7904053	Krum SA. Del Norte, NL.	N2379R
☐ XB-FYD	73	Rockwell 690	11065	Gabriela Ortega Larraury, Del Norte, NL.	XB-CZX
☐ XB-GBN	77	Conquest II	441-0021	Malva Inmobiliara SA. Queretaro.	XA-SKJ
☐ XB-GCU	78	Rockwell 690B	11460	Fausto Bermudez Hernandez, Toluca.	N42DK
☐ XB-GDS	77	Rockwell 690B	11371	Pascual Oyarvide Sanchez, San Luis Potosi.	XA-RMG
☐ XB-GGZ	77	PA-31T Cheyenne II	7720068	Miguel Angel Cagnasso Cantu, Del Norte, NL.	XA-SBW
☐ XB-GJL	73	Rockwell 690	11074	Manuel Vargas Ramirez, San Luis Rio Col, Baja California.	XB-AEL
☐ XB-GJQ	78	Rockwell 690B	11503	Jauregui Garcia Marcelino, Mexicali.	XB-DKQ
☐ XB-GMT	81	Gulfstream 840	11677	Vazquez Garcia Miguel Angel, Puebla, Chihuahua.	XB-GCV
☐ XB-GQI	80	PA-31T Cheyenne 1	8004056	Rocha Fuentes Jorge Miguel, Del Norte, NL.	N8TH
☐ XB-GQU	76	Rockwell 690A	11277	Rodriguez L Agustin, Toluca.	XB-EBD
☐ XB-GVI	80	PA-31T Cheyenne II	8020008	Rodriguez Gracia Juan Gerardo, Del Norte, NL.	N165CA
☐ XB-HNA	79	PA-31T Cheyenne II	7920012	Constructora Brisa SA. Jalapa.	XB-RTG
☐ XB-HOV	69	Mitsubishi MU-2F	156	Armando Alanis Rodriguez, Del Norte, NL.	N869Q
☐ XB-JLA	80	King Air E90	LW-333	Jose Lorca Avalos, San Luis Potosi.	N90BE
☐ XB-KLY	79	Rockwell 690B	11563	Eduardo de la Vega Canelo, Culiacan.	N140CA
☐ XB-LIJ	76	Mitsubishi MU-2M	341	Dirigir SA. Monterrey, NL.	N726MA
☐ XB-NEB	73	Mitsubishi MU-2J	583		N287MA
☐ XB-NUG	76	Mitsubishi MU-2J	619	Conductores Monterrey SA. DEl Norte, NL.	N469MA
☐ XB-PSG	84	PA-42 Cheyenne IIIA	5501019	Prisciliano Siller Garcia, Del Norte, NL.	N819PC
☐ XB-REA	70	King Air 100	B-68	Negro Mex SA. (status ?).	XA-DEA
☐ XB-RLM	79	PA-31T Cheyenne II	7920023	Pavimentos de la Laguna SA.	N793SW

Regn	Yr	Type	c/n	Owner/Operator	Prev Regn
☐ XB-SYV	80	Gulfstream 840	11641	Roberto Hurtado Medina, Santa Pacuaro, Jalisco.	XA-SYV
☐ XB-TFS	74	King Air 200	BB-8	Transportadora de Sal SA. Tijuana, Baja California.	N923WS
☐ XB-WUI	69	King Air 100	B-4		N925BD
☐ XB-XOI	70	Mitsubishi MU-2F	189	Roberto A Mariscal Saenz, Toluca.	N110MA
☐ XB-ZIP	73	Mitsubishi MU-2K	273	Arrendadora Bancomer SA. Puebla.	N319MA
☐ XC-AA10	83	Conquest II	441-0319	Procurad General de la Republica, Mexico City.	HK-3542
☐ XC-AA11	80	Conquest II	441-0142	Procurad General de la Republica, Mexico City.	XB-FCD
☐ XC-AA15	81	Gulfstream 980	95061	Procurad General de la Republica, Mexico City.	XC-HGL
☐ XC-AA16	85	Gulfstream 1000	96084	Procurad General de la Republica, Mexico City.	XC-HGW
☐ XC-AA19	83	Gulfstream 1000	96056	Procurad General de la Republica, Mexico City.	XC-HFV
☐ XC-AA20	85	Gulfstream 1000	96091	Governor of the State of Oaxaca,	XC-HGX
☐ XC-AA23	82	Gulfstream 1000	96001	Estado Mayor Presidencial. Mexico City.	XC-UTR
☐ XC-AA27	81	Gulfstream 840	11652	Governor of the State of Sonora, Hermosillo.	XC-HGJ
☐ XC-AA36	81	Gulfstream 840	11678	Governor of the State of Vera Cruz, Jalapa.	XC-HHY
☐ XC-AA40		Gulfstream 840	...	Governor of the State of Tamaulipas, Ciudad Victoria.	XC-HHK
☐ XC-AA46	86	King Air 300LW	FA-95	Procurad de la Republica, Mexico City.	HK-3536
☐ XC-AA48	78	King Air 200	BB-369	PGR/Mexican Drug Enforcement Agency, Mexico City.	HK-3555
☐ XC-AA49	85	King Air 300	FA-83	PGR/Mexican Drug Enforcement Agency, Mexico City.	XC-HGV
☐ XC-AA50	83	King Air 200	BB-1108	PGR/Mexican Drug Enforcement Agency, Mexico City.	XC-JAI
☐ XC-AA53	67	Gulfstream 1	179	Procurad de la Republica, Mexico City.	HK-3622
☐ XC-AA54	88	King Air 300	FA-158	Procurad General de la Republica, Mexico City.	YV-325P
☐ XC-AA56		Gulfstream 840	11695	State Government of Baja California Norte.	XB-JIO
☐ XC-AA57	63	Gulfstream 1	103	PGR/Mexican Drug Enforcement Agency, Mexico City.	PT-...
☐ XC-AA62	81	Gulfstream 980	95068	Government of the State of Durango.	HK-3409W
☐ XC-AA68	80	Conquest II	441-0189	Procurad de la Republica, Mexico City.	HK-3595
☐ XC-AA72	85	King Air 300	FA-87	Procurad General de la Republica, Mexico City.	N301CG
☐ XC-AA80	86	King Air 300	FA-74	Procurad General de la Republica, Mexico City.	HK-3519
☐ XC-AA84	77	Rockwell 690B	11382	Secretaria de Gobernacion, Mexico City.	XC-JBP
☐ XC-AGS	86	King Air C90A	LJ-1132	Governor of the State of Aguascalientes.	XA-GSM
☐ XC-ALO	80	Gulfstream 840	11606	Governor of the State of Guerrero, Chilpancingo.	N5833N
☐ XC-BAD		Gulfstream 840	11659	Procurad General de la Republica, Zacatecas.	N402AB
☐ XC-BAP	75	Rockwell 690A	11257	Consejo de Recursos Minerales, Pachuca.	
☐ XC-BCN	79	King Air 200	BB-435	Governor of the State of Baja California Norte.	XC-GOL
☐ XC-DIJ	76	King Air 200	BB-100	Procurad General de la Republica, Mexico City.	N75KA
☐ XC-ENL	81	Gulfstream 980	95052	Governor of the State of Nuevo Leon, Monterrey.	HK-2738W
☐ XC-FIW	72	King Air A100	B-110	Secretariat of Communications & Transport, Mexico City.	
☐ XC-FOC	72	King Air C90	LJ-553	CIAAC-Centro Int de Adto de Aviacion Civil, Mexico City.	
☐ XC-FUY	73	King Air E90	LW-33	Direccion General de Aduanas, Monterrey.	
☐ XC-GAS	79	Rockwell 690B	11556	Direccion General de Carreteras Federales, Mexico City.	
☐ XC-GON	81	Conquest II	441-0224	Secretaria de Hacienda y Credito Publico, Mexico City.	N6854B
☐ XC-GOO	81	Conquest II	441-0208	Secretaria de Hacienda y Credito Publico, Mexico City.	N2728F
☐ XC-HAB	82	Gulfstream 840	11688	Banco Nacional de Credito Rural SA. Mexico City.	(N5940K)
☐ XC-HFA	73	SA-226T Merlin 3	T-238	Governor of the State of Tamaulipas, Ciudad Victoria.	N130PC
☐ XC-HFN	82	Gulfstream 1000	96043	Governor of the State of Chihuahua.	N9963S
☐ XC-HGH	81	Gulfstream 980	95072	Procurad General de la Republica, Mexico City.	HK-3492
☐ XC-HGI	87	King Air 300	FA-114	Governor of the State of Guerrero, Chilpancingo.	XA-PIE
☐ XC-HHH	80	Gulfstream 840	11649	Procurad General de la Republica, Mexico City.	HK-3448
☐ XC-HHS	78	Rockwell 690B	11450	Governor of the State of Sinalao, Culiacan.	N81729
☐ XC-HHU	83	Conquest II	441-0275	Comision Nacional de Agua, Mexico City.	N95863
☐ XC-HMO	79	Rockwell 690B	11560	Governor of the State of Sonora, Hermosillo.	(N690RB)
☐ XC-JAL	77	Rockwell 690B	11417	Governor of the State of Jalisco, Guadalajara.	N81662
☐ XC-JCT	76	Rockwell 690A	11331	Fondo Nat. Fomento Al Turismo, Baja California Sur.	XB-ECX
☐ XC-JDB		Gulfstream 900	15009	Procurad General de la Republica, Mexico City.	XC-AA39
☐ XC-MLM	83	Gulfstream 900	15028	Governor of the State of Michoacan,	N5916N
☐ XC-NAY	73	Rockwell 690A	11115	Governor of the State of Nayarit, Tepic.	XB-XUC
☐ XC-OAX	81	Gulfstream 840	11656	Governor of the State of Oaxaca, Xoxocotlan.	(N5908K)
☐ XC-ONA	67	King Air A90	LJ-293	Cia S N P SA. Mexico City.	N793K
☐ XC-PFB		Gulfstream 980	95018	Policia Federal de Caminos y Puertos, Mexico City.	N123LA
☐ XC-RAM	72	Rockwell 690	11021	Secretaria de Hacienda y Credito Publico, Mexico City.	
☐ XC-SAH	78	Rockwell 690B	11516	S.A.H.O.P. Mexico City.	
☐ XC-STA	78	Rockwell 690B	11447	TAF-Transporte Aereo Federal, Mexico City.	N81723
☐ XC-TAB	78	Rockwell 690B	11504	Governor of the State of Tabasco, Villahermosa.	N106SA
☐ XC-TXA	80	Gulfstream 840	11631	State Government of Sonora, Hermosillo.	XB-DSH
☐ XC-TXA	81	Gulfstream 980	95045	Governor of the State of Tlaxcala. (status ?).	HK-....
☐ XC-UAT	76	PA-31T Cheyenne II	7620046	Universidad Autonoma de Tamaulipas, Victoria.	N5432V
☐ XC-VER	76	Rockwell 690A	11280	Governor of the State of Vera Cruz, Jalapa.	
☐ XC-VES	78	Rockwell 690B	11481	Governor of the State of Coahuila, Ramos Arizpe.	N126SA

Military

☐ XC-...	80	King Air 200	BB-725	TP-209, Government of Mexico.	XC-SLP
☐ 2201	88	King Air C90A	LJ-1166	Government of Mexico,	N3100W
☐ 2202	88	King Air C90A	LJ-1168	Government of Mexico,	N3082W
☐ 2203	88	King Air C90A	LJ-1171	Government of Mexico,	N3086C
☐ 2204	88	King Air C90A	LJ-1175	Government of Mexico,	N3139T
☐ 2205	88	King Air C90A	LJ-1176	Government of Mexico,	N3108K
☐ ETE-1318	82	Gulfstream 1000	96041	Mexican Air Force, Mexico City.	HK-2912P
☐ ETE-1332	78	Rockwell 690B	11494	Mexican Air Force, Mexico City.	YV-2302P

Regn	Yr	Type	c/n	Owner/Operator	Prev Regn

289

Regn	Yr	Type	c/n	Owner/Operator	Prev Regn
☐ ETE-1363	80	Gulfstream 980	95010	Mexican Air Force, Mexico City.	HK-3412
☐ MT-214	82	Gulfstream 1000	96040	Mexican Navy, Mexico City.	N900JP
☐ MT-217	82	Gulfstream 1000	96026	Mexican Navy, Mexico City.	HK-3157W
☐ MT-218	82	Gulfstream 1000	96013	Mexican Navy, Mexico City.	XC-HHZ
☐ MT-219	80	Gulfstream 980	95040	Mexican Navy, Mexico City.	XB-AOC
☐ MT-221	81	Gulfstream 980	95046	Mexican Navy, Mexico City.	XB-DSA
☐ MT-222	81	Gulfstream 980	95082	Mexican Navy, Mexico City.	HK-3453
☐ MT-224	79	Conquest II	441-0101	Mexican Navy, Mexico City.	N412PW
☐ MU-1550	72	Mitsubishi MU-2J	566	Mexican Navy, Mexico City.	N210MA

XT = Burkina Faso (People's Democratic Republic of Burkina Faso)
Military

☐ XT-MAX	80	King Air 200	BB-742	Government of Burkina Faso, Ouagadougou.	G-BPWJ

YR = Romania (Republic of Romania)
Civil

☐ YR-CAA	92	King Air 350	FL-73	Civil Aviation Directorate, Bucharest.	(G-CCCB)
☐ YR-RLA	93	Beech 1900D	UE-69	LAR Romanian Airlines, Bucharest.	YR-AAK
☐ YR-RLB	93	Beech 1900D	UE-73	LAR Romanian Airlines, Bucharest.	YR-AAL

YS = El Salvador (Republic of El Salvador)
Civil

☐ YS-...	91	Reims/Cessna F406	F406-0062		N3125G

YU = Yugoslav Federation
Civil

☐ YU-BMM	80	PA-31T Cheyenne II	8020021	Government Flight Inspection, Belgrade.	(D-IOOO)
☐ YU-BPF	80	PA-31T Cheyenne II	8020006	JAT General Aviation, Belgrade.	N801CM
☐ YU-BPG	80	PA-31T Cheyenne'II	8020012	JAT General Aviation, Belgrade.	N2328W
☐ YU-BPH	80	PA-31T Cheyenne II	8020063	JAT General Aviation, Belgrade.	N2389V

YV = Venezuela (Republic of Venezuela)
Civil

Regn	Yr	Type	c/n	Owner/Operator	Prev Regn
☐ YV-...	71	681B Turbo Commander	6052		N105SS
☐ YV-...		Gulfstream 840	11632		N711QP
☐ YV-...	81	Gulfstream 840	11694		N101KJ
☐ YV-...	82	Gulfstream 840	11725		N66RA
☐ YV-...	82	Gulfstream 900	15012		YV-903P
☐ YV-...	84	Gulfstream 900	15034		N900DJ
☐ YV-....	80	Gulfstream 980	95009		ZS-KZW
☐ YV-...	80	Gulfstream 980	95028		N980CF
☐ YV-...	80	Gulfstream 980	95000	(status ?).	N303E
☐ YV-...	77	King Air 200	BB-265	Servicios Tecnicos Maracaibo.	YV-141CP
☐ YV-...	97	King Air B200	BB-1584	Raytheon Aircraft Co. Wichita, Ks.	N11355
☐ YV-...	72	King Air C90	LJ-556		YV-818P
☐ YV-...	73	King Air C90	LJ-591	Jose Santiago Nunez Gomez, Caracas.	N108TT
☐ YV-....	93	King Air C90B	LJ-1329		N8016Q
☐ YV-...	83	King Air F90-1	LA-207		N2DF
☐ YV-...	69	Mitsubishi MU-2F	165		YV-1067P
☐ YV-...	73	Rockwell 690	11061	Inversiones Regionales 001 CA. Caracas.	N853CP
☐ YV-...		SA-227AT Merlin 4C	AT-532		N3110B
☐ YV-...	82	SA-227TT Merlin 300	TT-447	Inversiones AGAP CA. Nueva Esparta.	HK-3980X
☐ YV-...	82	SA-227TT Merlin 4C	TT-465	Servicios Aeronauticos de Oriente CA.	N696CP
☐ YV-01P	75	Rockwell 690A	11264	Juan Otaola Pevan,	YV-T-JOP
☐ YV-02CP	81	King Air F90	LA-127	Banco Mercantil CA. Caracas.	YV-410CP
☐ YV-04CP	70	King Air 100	B-73		(N73PD)
☐ YV-04P	71	King Air 100	B-83		YV-T-ETM
☐ YV-07CP	76	Rockwell 690A	11325	Gustavo Jimenez Pocaterra,	N81422
☐ YV-08CP	63	Gulfstream 1	117	Lagoven SA. Caracas.	N41KD
☐ YV-25CP	79	Conquest II	441-0111	Lloyds SA. Caracas.	HK-2413P
☐ YV-28CP	63	Gulfstream 1	119	Lagoven SA. Caracas.	YV-P-EPC
☐ YV-31CP	75	Rockwell 690A	11253		YV-T-SJM
☐ YV-32P	89	King Air C90A	LJ-1191	Aviaservice CA. Caracas.	
☐ YV-39CP	80	Gulfstream 840	11618		N118SA
☐ YV-42CP	80	King Air 350	FL-32	Corpoven CA. Caracas.	YV-350CP
☐ YV-46CP	60	Gulfstream 1	56	Lagoven SA. Caracas.	N510E
☐ YV-53CP	83	Gulfstream 1000	96067		HK-3271
☐ YV-56CP	80	Gulfstream 980	95017	Inversiones 301 CA. Caracas.	N9770S
☐ YV-63CP	77	Rockwell 690B	11390		YV-991P
☐ YV-66CP	75	King Air E90	LW-144	Constructora Barsantil CA. Caracas.	YV-T-APD
☐ YV-69CP	75	Rockwell 690A	11245		N236SC
☐ YV-76CP	68	Gulfstream 1	192	Lagoven SA. Caracas.	N67H
☐ YV-82CP	59	Gulfstream 1	26	Lagoven SA. Caracas.	N348DA
☐ YV-83CP	69	Gulfstream 1	199	Lagoven SA. Caracas.	XA-RIV

Regn	Yr	Type	c/n	Owner/Operator	Prev Regn
☐ YV-85CP	62	Gulfstream 1	97	Lagoven SA. Caracas.	N49DE
☐ YV-86CP		Rockwell 690A	11329		N81460
☐ YV-87CP	81	Gulfstream 840	11654		N1NT
☐ YV-95CP	77	King Air B100	BE-19	Banco Nacional de Venezuela, Caracas.	
☐ YV-105CP	81	Gulfstream 840	11685	(reported stolen ?)	N840BC
☐ YV-106CP	77	King Air 200	BB-238	Corpoven SA. Caracas.	
☐ YV-108CP	81	MU-2 Marquise	1514SA		N402JH
☐ YV-109CP	75	Rockwell 690A	11244		YV-02P
☐ YV-112CP	91	King Air 350	FL-55		N5692L
☐ YV-116CP	77	Rockwell 690B	11377		
☐ YV-118CP	77	King Air E90	LW-234	Aerotecnica SA. Caracas.	
☐ YV-121CP	91	King Air B200	BB-1398	(reported stolen ?).	
☐ YV-124CP	77	PA-31T Cheyenne II	7720047	Agropecuaria Garza CA.	
☐ YV-127CP	77	King Air B100	BE-28		YV-1276P
☐ YV-143CP	77	Rockwell 690B	11378		N25PF
☐ YV-144CP	77	Mitsubishi MU-2N	702SA		N862MA
☐ YV-152CP	77	King Air C90	LJ-739	Aero Ejecutivos CA. Caracas.	
☐ YV-158CP	78	King Air C90	LJ-742		
☐ YV-167CP	78	King Air C90	LJ-751	Artico SA.	
☐ YV-168CP	78	King Air 200	BB-316	Tocars SA. Caracas.	
☐ YV-170CP	80	Gulfstream 840	11603		N5NK
☐ YV-172CP	78	King Air C90	LJ-756		
☐ YV-187CP	79	PA-31T Cheyenne II	7920054		(N690CA)
☐ YV-188CP	78	Rockwell 690B	11461		(N81763)
☐ YV-192CP	76	Rockwell 690A	11288	F Bermudez Motors CA. Caracas.	YV-757P
☐ YV-194CP	78	PA-31T Cheyenne II	7820052		N6002A
☐ YV-202CP	78	King Air C90	LJ-786	Pedro Luis Angarita,	
☐ YV-204CP	78	Rockwell 690B	11466		(N81769)
☐ YV-212CP	84	Gulfstream 840	11735		N888KN
☐ YV-218CP	78	Rockwell 690B	11483	Consorcio Veluz CA. Caracas.	(N81806)
☐ YV-223CP	78	King Air C90	LJ-789	Trafi CA.	N2016L
☐ YV-225CP	78	Conquest II	441-0067	Servicios Aereos Tampa C.A.	(N88726)
☐ YV-227CP	78	Rockwell 690B	11464	Arrendaven CA. Caracas.	N81766
☐ YV-229CP	78	Rockwell 690B	11502		(N81842)
☐ YV-238CP	79	King Air 200	BB-440		
☐ YV-242CP		SA-226T Merlin 3B	T-357	Transportes 242 CA. Caracas.	N30042
☐ YV-243CP	78	Rockwell 690B	11488		N70AC
☐ YV-246CP	75	Rockwell 690A	11278	Transportes 246 CA. Caracas.	N57176
☐ YV-251CP	79	PA-31T Cheyenne II	7920022	Multiservicios NISA CA.	
☐ YV-257CP	79	King Air 200	BB-517	Inversiones Rizo CA. (status ?).	
☐ YV-262CP		SA-227TT Merlin 300	TT-521		D-IOOO
☐ YV-263CP	79	King Air C90	LJ-865	Vinccler CA.	
☐ YV-266CP	83	Conquest 1	425-0155	Molinos Nacionales SA. Caracas.	N6886X
☐ YV-277CP		SA-227TT Merlin 3C	TT-441		N30042
☐ YV-281CP	79	Rockwell 690B	11532	Oficina Tecnica Vemef CA. Caracas.	N81723
☐ YV-290CP	80	King Air F90	LA-35	C A de Edificaciones - Resid D Paulo.	
☐ YV-292CP		SA-227TT Merlin 3C	TT-459A		OY-BPK
☐ YV-294CP	79	King Air C90	LJ-841	ANDAVI-Andina de Aviacion SA.	
☐ YV-310CP	80	Gulfstream 840	11645	ANCA-Central Azucarera Portuguesa SA.	(N5897K)
☐ YV-326P	76	King Air E90	LW-172	G Bracho,	
☐ YV-335CP	80	King Air 200	BB-598		
☐ YV-344P	67	680V Turbo Commander	1718-89		YV-340CP
☐ YV-385CP	80	King Air 200	BB-740	Cia Aerospace de Venezuela CA. Caracas.	
☐ YV-394CP	80	Gulfstream 840	11637	Construcions Humboldt SA. Caracas.	(N5889K)
☐ YV-400CP	81	King Air F90	LA-108	Promociones Ticoporo CA.	
☐ YV-401CP	81	King Air 200	BB-796		
☐ YV-403CP	81	King Air 200C	BL-23	Ministry of National Defence, Caracas.	
☐ YV-407CP	81	Gulfstream 980	95067	(reported stolen ?).	(N9819S)
☐ YV-416CP	79	Rockwell 690B	11546		N690SC
☐ YV-416P	74	King Air E90	LW-81	(reported stolen ?).	YV-T-AJP
☐ YV-428CP	81	King Air F90	LA-140	Aero Panamericano CA.	
☐ YV-436CP	81	King Air C90	LJ-973	Transportes Guyanes SA.	
☐ YV-453CP	66	Gulfstream 1	175	Lagoven SA. Caracas.	N578KB
☐ YV-454CP	77	Rockwell 690B	11405	Transporte 454 SA. Caracas.	N5387V
☐ YV-466CP	82	King Air B200	BB-980	SABAM CA.	
☐ YV-467CP	74	King Air E90	LW-94	H Boulton y Cia SA.	N98ME
☐ YV-472CP		SA-227TT Merlin 300	TT-471		D-IISA
☐ YV-486CP	82	King Air F90	LA-179	Banco Fomento Region de los Andes,	
☐ YV-488CP	82	King Air B200	BB-1020		
☐ YV-490CP	82	King Air F90	LA-193	Aero Panamericano CA.	
☐ YV-492CP	82	Gulfstream 900	15006	(reported stolen ?)	N5852N
☐ YV-493CP	82	King Air B200	BB-1090	Transportes La Mona CA.	
☐ YV-505CP	82	Gulfstream 840	11723	(reported stolen ?).	YV-45CP
☐ YV-521CP	83	Gulfstream 900	15022		D-ILAS
☐ YV-533P	81	Conquest II	441-0227	Jesus Romero,	N6854T
☐ YV-536P	68	680W Turbo Commander	1813-30		YV-T-VTY
Regn	*Yr*	*Type*	*c/n*	*Owner/Operator*	*Prev Regn*

Regn	Yr	Type	c/n	Owner/Operator	Prev Regn
☐ YV-548CP	84	Conquest 1	425-0189	Transporte Lucania CA. Valencia.	N969ME
☐ YV-554CP	82	King Air B200	BB-1026	(reported stolen ?).	(G-BJSN)
☐ YV-581CP	80	Gulfstream 980	95007	Inversiones Torgo CA.	(N123RC)
☐ YV-601P	81	Gulfstream 840	11698		YV-485CP
☐ YV-612CP		SA-227TT Merlin 3C	TT-468		VH-JCB
☐ YV-620CP	66	Gulfstream 1	170	Servamara CA. Maracaibo.	N189K
☐ YV-621CP	66	Gulfstream 1	171	Servamara CA. Maracaibo.	N728GM
☐ YV-626CP	81	Gulfstream 840	11666	Empresas Traviesas Venezolana CA.	YV-435CP
☐ YV-630CP	80	Gulfstream 840	11640		VH-BSO
☐ YV-652CP		SA-227TT Merlin 3C	TT-515		N352SM
☐ YV-655CP	79	PA-31T Cheyenne II	7920018		N100KR
☐ YV-663CP	77	Rockwell 690B	11358	Aerotransporte Mirage CA. Caracas.	N810GF
☐ YV-665CP	76	King Air 200	BB-160		N65171
☐ YV-670CP	79	Rockwell 690B	11547	Inversiones 12-6-86 CA. Caracas.	N81631
☐ YV-693CP		SA-226T Merlin 3A	T-287		YV-180CP
☐ YV-702CP	82	King Air F90	LA-174	Consorcio Industrial de Zulia, Caracas.	N90LL
☐ YV-703CP	78	PA-31T Cheyenne II	7820008	Jose Manuel Aguilera,	N14MR
☐ YV-706P	77	King Air E90	LW-208	DIM-Direccion Intelligencia Militar, Caracas.	
☐ YV-710CP	79	SA-226T Merlin 3B	T-301	Aero Halcon 710 CA. Caracas.	N175WB
☐ YV-722P	67	King Air A90	LJ-306		YV-426CP
☐ YV-732P	75	King Air 200	BB-35	Transpolar CA. Caracas.	YV-T-MTI
☐ YV-733P	74	Rockwell 690A	11207	Luis Enriquez Nunez,	YV-E-DPK
☐ YV-740CP		SA-227TT Merlin 3C	TT-456		N3019U
☐ YV-741CP	80	King Air C90	LJ-898		N41AJ
☐ YV-744CP	66	680T Turbo Commander	1685-66		YV-2435P
☐ YV-773CP		Rockwell 690	...	noted Caracas 9/96.	
☐ YV-775CP	77	Rockwell 690B	11366		N777RD
☐ YV-779CP	81	Gulfstream 840	11673	Corporacion Casapart CA. Caracas.	N777WY
☐ YV-787CP	81	Gulfstream 980	95084		N980JC
☐ YV-797CP		Rockwell 690B	11468		N2AC
☐ YV-801CP	70	King Air 100	B-16		N925K
☐ YV-808CP	82	SA-227TT Merlin 3C	TT-435	Aero Centro de Servicios CA. Caracas.	N92RC
☐ YV-818P	76	Rockwell 690A	11285		YV-80CP
☐ YV-820CP	68	680W Turbo Commander	1775-13	Executive Air CA. Caracas.	N9RN
☐ YV-822CP	83	Gulfstream 900	15030		XA-EMO
☐ YV-834CP	73	Rockwell 690A	11114	Servicios Aeronauticos 4 CA. Maracaibo.	N57114
☐ YV-843P	67	680V Turbo Commander	1689-69		YV-843P
☐ YV-877CP	92	King Air B300C	FM-8		N55684
☐ YV-880CP	82	Gulfstream 900	15017	DCA-Directorate of Civil Aeronautics, Caracas.	N5880N
☐ YV-893CP	81	Gulfstream 980	95058	Aeropetrol CA. Caracas.	N243AR
☐ YV-900P	74	Rockwell 690A	11180		YV-T-ZTA
☐ YV-903P	79	King Air B100	BE-61		YV-309CP
☐ YV-977CP		Conquest 11	...	noted Fort Lauderdale 11/96.	
☐ YV-980CP	81	Gulfstream 980	95074	Agroislene CA. Caracas.	N3212A
☐ YV-980P	76	Rockwell 690A	11318		YV-98CP
☐ YV-990CP		Reims/Cessna F406	F406-0022		YV-525C
☐ YV-993CP		rockwell 690	...	noted Caracas 9/96.	
☐ YV-1000CP	91	King Air C90A	LJ-1273		YV-2282P
☐ YV-1013P	80	King Air F90	LA-16	L F Mendoza Aristeguieta, Caracas.	YV-2289P
☐ YV-1050P	78	Rockwell 690B	11443		N104RG
☐ YV-1300P	77	King Air 200	BB-282	(reported stolen ?).	YV-131CP
☐ YV-1402P		PA-31T Cheyenne II	7820067		N331PC
☐ YV-1405P	67	680V Turbo Commander	1719-90	Silvio Guedes,	YV-O-CVG-2
☐ YV-1500P	78	King Air E90	LW-270	Police Air Wing, Caracas.	
☐ YV-1700P	67	King Air A90	LJ-205		N707PR
☐ YV-1873P	80	King Air 200	BB-632		
☐ YV-1947P	68	King Air B90	LJ-370	Luis Alberto Martinez,	YV-40CP
☐ YV-1990P	75	PA-31T Cheyenne II	7520027		YV-08CP
☐ YV-1995P	81	PA-31T Cheyenne II	8120013	Jesus Rincon Fernandez,	YV-04CP
☐ YV-2034P	75	King Air C90	LJ-659		YV-410CP
☐ YV-2096P	78	Rockwell 690B	11490		YV-220CP
☐ YV-2229P	70	King Air B90	LJ-488	Marcos Mandelblum,	YV-O-KWH1
☐ YV-2263P	83	Conquest 1	425-0162		
☐ YV-2280P	68	680W Turbo Commander	1804-26	John S Calcurian,	YV-O-INOS3
☐ YV-2317P	75	Rockwell 690A	11239	Benjamin Guillermo, Miami, Fl. USA.	N57217
☐ YV-2323P	83	King Air B200C	BL-66		
☐ YV-2331P	78	PA-31T Cheyenne II	7820070	Boscolo Smeraldo Smeraldi, Caracas.	N6033A
☐ YV-2346P	81	Gulfstream 840	11671		YV-439CP
☐ YV-2350P	81	King Air B200	BB-906	Manuel Espinoza,	YV-437CP
☐ YV-2352P	83	King Air B200	BB-1092	Danilo Lazzioli,	YV-555CP
☐ YV-2354P	80	King Air 200	BB-656	DIM-Direccion Intelliegencia Militar, Caracas.	YV-783P
☐ YV-2365P	78	PA-31T Cheyenne II	7820037		N24E
☐ YV-2383P	70	681 Turbo Commander	6005		N35WA
☐ YV-2390P	78	Conquest II	441-0063	Inversiones Salvador Salvatierra SA. Caracas.	YV-176CP
☐ YV-2395P	81	SA-226T Merlin 3B	T-384		N71FN
☐ YV-2404P	70	681 Turbo Commander	6029		N9074N

Regn	Yr	Type	c/n	Owner/Operator	Prev Regn
☐ YV-2422P	79	Rockwell 690B	11515		N515WC
☐ YV-2423P	72	King Air A100	B-90	Air Sal Leasing Inc. Miami, Fl. USA.	N515AS
☐ YV-2436P	68	680W Turbo Commander	1814-31	Diego Levine, Maracaibo.	N71AF
☐ YV-2437P	76	Rockwell 690A	11323		YV-29CP
☐ YV-2451P	77	PA-31T Cheyenne II	7720050		YV-153CP
☐ YV-2488P	74	Rockwell 690A	11190	Transporte Lehmacorp CA. Caracas.	N8KG
☐ YV-2501P	70	681 Turbo Commander	6024	Julio Borges, Caracas.	N22WK
☐ YV-2505P	82	Gulfstream 840	11726	Inversiones Puerta Plata CA. Caracas.	YV-505CP
☐ YV-O-BDA-3	76	Mitsubishi MU-2L	685	Banco de Desarollo Agropecuario,	N845MA
☐ YV-O-BIV-1	80	King Air 200	BB-731	Banco Industrial de Venezuela, Caracas.	
☐ YV-O-BND-1	80	King Air 200	BB-701	Banco National de Descuento, Caracas.	YV-353CP
☐ YV-O-CBL-5	80	King Air C90	LJ-889	State Government Ciudad Bolivar.	YV-354CP
☐ YV-O-CPI-1	81	King Air 200	BB-770	Corporacion Industrial de Venezuela, Caracas.	YV-386CP
☐ YV-O-CVF-1	80	King Air 200	BB-644	Corporacion Venezolana de Fomento, Caracas.	
☐ YV-O-CVG-577	80	King Air 200	BB-223	Corporacion Venezolana de Guyana, Puerto Ordaz.	N17723
☐ YV-O-CVG-877	80	King Air 200	BB-273	Corporacion Venezolana de Guyana, Puerto Ordaz.	N17743
☐ YV-O-CVG-14		Turbo Commander	...	Corporacion Venezolana de Guyana, Puerto Ordaz.	
☐ YV-O-FDU	78	Rockwell 690B	11500	Fondo Nacional Desarrollos Urbanos, Caracas.	(N81833)
☐ YV-O-ICA-1	81	PA-42 Cheyenne III	8001010	Instituto Credito Agricola Pecuaria, Caracas.	YV-O-MAC-1
☐ YV-O-INV-3	76	King Air E90	LW-189	I N A V I. Caracas.	YV-O-INAV3
☐ YV-O-KWH3	81	Gulfstream 840	11657	C A D A F E, Caracas.	(N5909K)
☐ YV-O-MAC2	79	PA-31T Cheyenne 1	7904040	Ministry of Agriculture, Caracas.	YV-239CP
☐ YV-O-MAC6	74	King Air E90	LW-82	Ministry of Agriculture, Caracas.	YV-O-MH-1
☐ YV-MMH9	77	King Air 200	BB-261	Ministry of Energy & Mines, Caracas.	
☐ YV-O-MTC-182		Gulfstream 1000	96016	Ministry of Transport & Communications, Caracas.	(YV-477CP)
☐ YV-O-MTC-278		Rockwell 690B	11430	Ministry of Transport & Communications, Caracas.	YV-O-DAC-2
☐ YV-O-MTC-576		Rockwell 690A	11281	Ministry of Transport & Communications, Caracas.	N57183
☐ YV-O-SAS-3	77	King Air E90	LW-203	Ministry of Health & Social Security, Caracas.	
☐ YV-O-SID-2	73	King Air A100	B-155	Siderurgica del Orinoco CA. Caracas.	
Military					
☐ 2840	79	King Air 200	BB-520	FAV, Caracas.	
☐ 3150	79	King Air 200	BB-522	FAV, Caracas.	
☐ 3240	81	King Air 200C	BL-19	FAV, Caracas.	
☐ 3280	81	King Air 200C	BL-18	FAV, Caracas.	
☐ ARV-0201	78	King Air E90	LW-264	Venezuelan Navy, Caracas.	TR-0201
☐ EV-7702	77	King Air E90	LW-229	Venezuelan Army, Caracas.	
☐ EV-7910	79	King Air 200	BB-495	Venezuelan Army, Caracas.	
☐ GN-7593	75	King Air E90	LW-154	Venezuelan National Guard, Caracas.	N211DG
☐ GN-7839	78	King Air E90	LW-260	Venezuelan National Guard, Caracas.	
☐ GN-8270	82	King Air B200C	BL-51	Venezuelan National Guard, Caracas.	
☐ GN-8274		King Air 200	...	Venezuelan National Guard, Caracas.	

Z = Zimbabwe *(Republic of Zimbabwe)*

Civil

Regn	Yr	Type	c/n	Owner/Operator	Prev Regn
☐ Z-DDD		Reims/Cessna F406	F406-0069	Government of Zimbabwe, Harare.	F-GIQD
☐ Z-DDE		Reims/Cessna F406	F406-0068	Government of Zimbabwe, Harare.	F-GIQC
☐ Z-DDF		Reims/Cessna F406	F406-0071	Government of Zimbabwe, Harare.	F-GIQE
☐ Z-DDG		Reims/Cessna F406	F406-0067	Government of Zimbabwe, Harare.	F-GEUG
☐ Z-DJF	77	King Air E90	LW-205	D J Fowler,	ZS-NDR
☐ Z-KEN	96	Pilatus PC-XII	126	J Kennedy,	HB-FQR
☐ Z-LCS	79	King Air C90	LJ-848	EGAS Flight Academy of Zimbabwe, Harare.	TR-LCS
☐ Z-MRS	77	King Air 200	BB-286	MARS-Medical Air Rescue Service, Harare.	ZS-XGD
☐ Z-TAB	78	King Air 200	BB-315	Air TABEX P/L. Harare.	N873DB
☐ Z-TAM	82	King Air B200C	BL-49	TABEX, Harare.	N17KK
☐ Z-WRD	76	King Air C90	LJ-687	Delta Corp. Harare.	N127P
☐ Z-WSG	81	King Air 200	BB-748		N154BB
☐ Z-ZLT	85	King Air B200	BB-1196	Zimbabwe Leaf Tobacco Co Ltd. Harare.	N101EC

ZK = New Zealand *(Dominion of New Zealand)*

Civil

Regn	Yr	Type	c/n	Owner/Operator	Prev Regn
☐ ZK-ECR	78	Mitsubishi MU-2P	371SA	Air National Ltd. Auckland.	VH-XMZ
☐ ZK-POD	77	PA-31T Cheyenne II	7720009	NZ Post Air Ambulance, Wellington.	ZK-MPI
☐ ZK-PVA	78	Rockwell 690B	11476	New Zealand Aerial Mapping Ltd. Hastings.	(N81795)
☐ ZK-ROM	76	PA-31T Cheyenne II	7620055	Airwork (NZ) Ltd. Ardmore, Papakura.	N300BP
☐ ZK-RUR	84	PA-42 Cheyenne 400LS	5527015	Christian Aviation, Auckland-Ardmore.	N42MD

ZP = Paraguay *(Republic of Paraguay)*

Civil

Regn	Yr	Type	c/n	Owner/Operator	Prev Regn
☐ ZP-...	85	Conquest 1	425-0223		N1226G
☐ ZP-...	83	Gulfstream 900	15021		N908TN
☐ ZP-...	81	Gulfstream 980	95053		N73DQ
☐ ZP-...	80	Gulfstream 980	95035	Jesus Gutierrez, Asuncion.	N200JN
☐ ZP-...	78	King Air E90	LW-284		N200WB
☐ ZP-...	78	King Air E90	LW-261		N4283R
☐ ZP-...		PA-31T Cheyenne II	8020066	Itaipu Binacional, Asuncion.	N555HP

Regn	*Yr*	*Type*	*c/n*	*Owner/Operator*	*Prev Regn*

Regn	Yr	Type	c/n	Owner/Operator	Prev Regn
☐ ZP-...		SA-227TT Merlin 4C	TT-462A		N453CP
☐ ZP-TTU	80	Gulfstream 980	95021	Roberto Barchini,	ZP-PTU
☐ ZP-TWN	82	King Air B200	BB-1008	Blas N Riqueleme,	ZP-TWN
☐ ZP-TWY	85	Gulfstream 1000	96095	Aero Commercial SRL. Asuncion.	N73DC
☐ ZP-TWZ	82	Gulfstream 900	15005	(status ?).	N5841N
☐ ZP-TXE	78	King Air C90	LJ-799	YOPA SA.	N220TM
☐ ZP-TXF	85	Gulfstream 1000	96079	Aero Commercial SRL. Asuncion.	N169CR
☐ ZP-TXG	85	Gulfstream 1000B	96203	Aero Commercial SRL. Asuncion.	N64JT
☐ ZP-TYZ	81	PA-42 Cheyenne III	8001044	Alcides A Rivera V. Asuncion.	N325CA

ZS = South Africa (Republic of South Africa)

Civil

Regn	Yr	Type	c/n	Owner/Operator	Prev Regn
☐ ZS-...	96	King Air B200	BB-1540	Wheels of Africa, Rand.	N89WA
☐ ZS-...	84	PA-42 Cheyenne 400LS	5527003	Kwalata Wilderness P/L. Benoni.	N333MX
☐ ZS-AAU	70	King Air C90	LJ-485	Fleetline Export P/L. Halfway House.	Z-WNB
☐ ZS-AMC	83	Conquest 1	425-0169	H Collins & Son P/L. Pietermaritzburg.	PH-VMC
☐ ZS-ARL	96	King Air B200	BB-1537	Matthew Wraith Trust, Benoni.	N3237M
☐ ZS-BEL	96	Pilatus PC-XII	143	General Airways P/L. Silverton.	HB-FRE
☐ ZS-BEN	68	King Air B90	LJ-397	Airwork Sales CC. Lanseria.	ZS-INY
☐ ZS-DMM	97	Pilatus PC-XII	198		HB-FQG
☐ ZS-IRJ	66	King Air A90	LJ-161	Beechcraft Sales/NAC P/L. Rand.	ZS-MAN
☐ ZS-JRA	75	Rockwell 690A	11284	Department of Enviromental Affairs, Pretoria.	N57273
☐ ZS-JRB	75	Rockwell 690A	11248	Department of Enviromental Affairs, Pretoria.	N122K
☐ ZS-JRC	78	Rockwell 690A	11432	ESCOM-Electricity Supply Commission, Johannesburg.	N81694
☐ ZS-JRH	77	Rockwell 690B	11421	Alex Air P/L. East London.	N81672
☐ ZS-KAA	77	King Air 200	BB-222	Magjoe Aviation CC. Riverclub.	F-ODZL
☐ ZS-KGW	78	King Air 200	BB-381	G W Van der Merwe, Lanseria.	N4848M
☐ ZS-KLM	80	King Air 200	BB-682	Pannar P/L. Greytown.	N6724P
☐ ZS-KLO	80	King Air 200	BB-704	Anglo American Corp. Johannesburg.	N6726D
☐ ZS-KLZ	80	King Air F90	LA-69	M L B Varoli/Group Five Air Carriers P/L. Lanseria.	N6727C
☐ ZS-KMA	81	King Air C90	LJ-930	Rossair Executive Air Charter P/L. Lanseria.	N3717J
☐ ZS-KOG	78	Rockwell 690B	11441	B M Stocks, Sunnyside.	N333GC
☐ ZS-KSU	81	Conquest 1	425-0115	Cargo Carriers P/L. Isando.	N68807
☐ ZS-KZI	81	King Air C90	LJ-959	B & E Air Partnership, Port Elizabeth.	(D-IFIP)
☐ ZS-KZU	79	King Air 200	1	Inter Air P/L-Valfrira Motor Engineering CC. Pretoria.	N396DP
☐ ZS-KZY	82	Gulfstream 1000	96051	ESCOM-Electricity Supply Commission, Johannesburg.	N9971S
☐ ZS-KZZ	82	Gulfstream 1000	96052	ESCOM-Electricity Supply Commission, Johannesburg.	N9972S
☐ ZS-LAW	81	King Air B200	BB-889	Rennies Shipping Holdings P/L. Johannesburg.	N3538K
☐ ZS-LBC	81	King Air F90	LA-122	Hexcor Partnership, Melrose North.	N3723N
☐ ZS-LBD	81	King Air 200	BB-837	Rossair Executive Air Charter P/L. Lanseria.	N38301
☐ ZS-LFL	82	King Air C90-1	LJ-1033	Inter Air P/L-Theaco Road & Earthworks P/L. Vanderbijlpark.	N6717T
☐ ZS-LFM	82	King Air B200	BB-954	East Coast Airways P/L-G U D Filters P/L. Durban.	N1839S
☐ ZS-LFU	82	King Air B200	BB-1018	L J Investments CC. Pinegowrie.	3D-LKK
☐ ZS-LFW	82	King Air B200	BB-999	King Air Services Partnership, Melville.	9Q-CPV
☐ ZS-LIL	82	King Air B200	BB-1047	Execujet Aircraft Sales P/L. Lanseria.	V5-LIL
☐ ZS-LIN	83	King Air C90-1	LJ-1053	Beechcraft Sales/NAC P/L. Lanseria.	N6891L
☐ ZS-LKA	80	King Air 200	BB-614	The Ivor Ferreira Trust, Umhlanga Rocks.	N333CR
☐ ZS-LNR	85	King Air B200	BB-1222	De Beers Consolidated Mines Ltd. Kimberley.	N50FC
☐ ZS-LNV	83	King Air B200C	BL-71	South African Police, Pretoria.	
☐ ZS-LRE	85	King Air B200	BB-1195	F W Hangers CC. Lanseria.	N6923C
☐ ZS-LRS	81	King Air 200C	BL-20	Mercedes-Benz of South Africa P/L. Wonderboom.	D-ILDB
☐ ZS-LST	75	King Air 200	BB-51	Justinian Investments Seven P/L. Lanseria.	5Y-BIR
☐ ZS-LSY	81	Cheyenne II-XL	8166031	Hendrik Casper Investments CC. Farrarmere.	V5-LSY
☐ ZS-LTD	80	King Air F90	LA-63	C Troskie Vliegtuie, Coolhouse.	N3686V
☐ ZS-LTE	80	King Air 200	BB-607	Inter Air P/L-I S Haggie Air, Lanseria.	N6725L
☐ ZS-LTF	74	King Air C90	LJ-613	King Air Services Partnership, Melville.	N3053W
☐ ZS-LTG	86	King Air B200	BB-1251	Anglo American Corp. Marshalltown.	N2610Y
☐ ZS-LTZ	80	King Air C90	LJ-919	Agillis P/L. Kempton Park.	N18MB
☐ ZS-LUU	81	King Air C90	LJ-988	Foster Aero International P/L. Lanseria.	N18267
☐ ZS-LVK	83	King Air B200	BB-1111	Aviation Africa II Partnership, Isando.	N600CM
☐ ZS-LWD	81	King Air 200	BB-756	Nationwide Air Charter P/L. Lanseria.	A2-AGB
☐ ZS-LWM	78	King Air/Catpass 250	BB-341	G J Air CC. Rand.	C-FBWX
☐ ZS-LYA	75	King Air 200	BB-26	Associated Equipment Co P/L. Springfield.	N57FM
☐ ZS-LZP	81	King Air C90	LJ-987	A A C Bartmann, Newville.	N90WC
☐ ZS-LZR	85	King Air C90A	LJ-1118	Crystal Air P/L. Bophuthatswana.	N937BC
☐ ZS-LZU	79	King Air 200	BB-444	Kegrin Aviation, Durban.	N444JE
☐ ZS-MBZ	78	King Air C90	LJ-795	Belle Ombre Air Charter Partnership, Letsitele.	N551GA
☐ ZS-MCA	72	King Air C90	LJ-551	500E Toerusting BK. Sinoville. (status ?).	ZS-XAC
☐ ZS-MES	82	King Air B200	BB-1038	Rossair Executive Air Charter P/L. Lanseria.	N223MH
☐ ZS-MFB	89	King Air B200	BB-1316	Kwazulu National Airways Corp. Ulundi.	N15570
☐ ZS-MFC	79	King Air 200	BB-525	Cormorant Aviation P/L. Benmore.	N200L
☐ ZS-MGF	74	Mitsubishi MU-2J	622	MU-2 Aircraft Investment CC. Sandton.	(3D-AFH)
☐ ZS-MGG	80	King Air 200	BB-586	J L Van den Berg/Domberg Lugdiens, Marshalltown.	N6668H
☐ ZS-MHM	80	King Air F90	LA-47	Avna Airlines P/L. Louterwater.	N984GA
☐ ZS-MIG	81	Conquest 1	425-0064	Surbiton Five One Four P/L. Alberton.	N6844P
☐ ZS-MIN	81	King Air B200	BB-941	Beechcraft Sales/NAC P/L.Rand.	N36801

Regn	Yr	Type	c/n	Owner/Operator	Prev Regn

Regn	Yr	Type	c/n	Owner/Operator	Prev Regn
☐ ZS-MKI	85	King Air C90A	LJ-1099	Rustenburg Building Materials P/L. Rustenburg.	Z-MKI
☐ ZS-MLO	71	King Air C90	LJ-526	P J L Adendorff P/L. Springfield.	N948K
☐ ZS-MMV	89	King Air B200	BB-1318	De Beers Consolidated Mines Ltd. Marshalltown.	N1557U
☐ ZS-MNC	75	King Air C90	LJ-671	Executive Rotorcraft CC. Lanseria.	N99LM
☐ ZS-MRZ	78	King Air C90	LJ-748	Coin Security Group Ltd. Sunnyside.	N23748
☐ ZS-MSL	81	King Air 200	BB-815	Federal Aviation CC. Durban.	N255AV
☐ ZS-MUM	68	King Air B90	LJ-408	J & D Aviation P/L. Lanseria.	N481SA
☐ ZS-MVW	81	Conquest 1	425-0041	Plastop P/L. Bronkhorstspruit.	N326L
☐ ZS-MYA	76	King Air 200	BB-101	Madiba Air P/L. Lanseria.	N535JR
☐ ZS-MYE	90	King Air 300	FA-211	Barlow Rand Ltd. Sandton.	N5666L
☐ ZS-MZS	79	King Air B100	BE-72	ZS-MZS Partnership, Clubview.	N20FL
☐ ZS-NAG	92	King Air C90A	LJ-1298	Associated Manganese Mines of South Africa Ltd. Johannesburg	N8264G
☐ ZS-NAW	82	King Air B200	BB-1027	Taiandair Partnership, Sunninghill.	N40AB
☐ ZS-NAX	80	King Air 200C	BL-8	Rossair Executive Air Charter P/L. Lanseria	5Y-NAX
☐ ZS-NBJ	82	King Air B200	BB-1070	R J Harpur, Randburg.	SE-KND
☐ ZS-NBO	80	King Air 200	BB-706	Smith Mining Equipment P/L. Kempton Park.	N25WD
☐ ZS-NCH	81	King Air B200	BB-918	Victor Aircraft Partnership, Parklands.	N70JG
☐ ZS-NDH	74	King Air C90	LJ-619	Dodson International Parts Inc. Rantoul, Ks. USA.	5Y-MAL
☐ ZS-NDM	74	Mitsubishi MU-2J	629	Natair CC. Durban.	N122CP
☐ ZS-NFE	92	King Air B200	BB-1418	Pax Air Services/Avmar Partnership 1991, Bedfordview.	N8266V
☐ ZS-NFO	80	King Air F90	LA-51	TYCO Truck Manufacturers P/L. Northmead.	N901NB
☐ ZS-NGC	77	King Air 200	BB-215	Kanberri Air Service P/L. Durban.	7Q-YTC
☐ ZS-NGI	90	King Air 350	FL-22	Lonrho Management Services P/L. Johannesburg.	N5673Y
☐ ZS-NHW	64	Gulfstream 1	141	Independent Aircraft Management P/L. Lanseria.	N800PA
☐ ZS-NHX	78	King Air 200	BB-386	Exclusive Air Aircraft Partnership, Lanseria.	N310GA
☐ ZS-NJM	83	King Air B200	BB-1152	Federal Transport CC. Durban.	5Y-BJM
☐ ZS-NKC	93	King Air B200	BB-1474	K-Air Charter P/L. Johannesburg.	N82010
☐ ZS-NKE	93	King Air B200	BB-1475	Maizecor Meulers P/L. Silverton.	N8226M
☐ ZS-NNA		Reims/Cessna F406	F406-0005	Longranger Helicopter Services CC. Nelspruit.	5R-MSK
☐ ZS-NNS	77	King Air 200	BB-266	A H Carter & Partners/Mid East Aviation, Durban.	N18266
☐ ZS-NOC	80	King Air 200	BB-715	T B Carruthers & Co P/L. Lanseria.	9J-AEV
☐ ZS-NOH	82	King Air B200	BB-1083	L A Visagie Trust, Karino.	N250YR
☐ ZS-NOW	92	King Air B200	BB-1427	Anglo-Vaal Air Ltd. Lanseria.	N8003U
☐ ZS-NRR	77	King Air 200	BB-288	Madiba Air P/L. Lanseria.	N288SF
☐ ZS-NRT	78	King Air 200	BB-334	Soaring Eagle Charter P/L. Lanseria.	N77SA
☐ ZS-NRW	77	King Air 200	BB-201	Lanseria Air Charter CC. Lanseria.	5Y-BKS
☐ ZS-NSC	80	King Air F90	LA-15	Beech Sales/NAC P/L. Rand.	N10GA
☐ ZS-NTT	78	King Air 200	BB-350	Skyeinvest Administration P/L. Cape Town.	N125MS
☐ ZS-NUC	78	King Air 200	BB-407	Beechcraft Sales/NAC P/L. Rand.	F-GFTT
☐ ZS-NUF	79	King Air 200C	BL-4	Rossair Executive Air Charter P/L. Lanseria.	V5-AAL
☐ ZS-NVG	58	Gulfstream 1	1	Vanellus CC. Sinoville.	N701G
☐ ZS-NWC	74	King Air C90	LJ-625	Gabler Medical Technology P/L. Belleville.	N50AB
☐ ZS-NWK	75	King Air 200	BB-52	BMP Charters, Lusaka. Zambia.	N400AJ
☐ ZS-NWT	83	King Air B200	BB-1144	Kiara Air CC/Execujet Air Charter P/L. Lanseria.	D-IBHF
☐ ZS-NWZ	94	Pilatus PC-XII	103	Swiss-Aviation CC/South African Red Cross, Kimberley.	HB-FOD
☐ ZS-NXH	75	King Air 200	BB-37	G W Van der Merwe, Lanseria.	N460CR
☐ ZS-NXI	77	King Air E90	LW-224	Bugaboo Trust P/L. Parys.	D-IATA
☐ ZS-NXK	76	Rockwell 690A	11286	Ellies Electronics P/L. Johannesburg.	3D-ADM
☐ ZS-NXT	95	King Air B200	BB-1502	Anglo Vaal Air P/L. Lanseria.	N1515E
☐ ZS-NYE	77	King Air E90	LW-222	Master Drilling P/L. Fochville.	VH-MTG
☐ ZS-NYI	89	Reims/Cessna F406	F406-0030	Ryan Blake Air CC. Lanseria.	5Y-MMJ
☐ ZS-NYK	96	Jetstream 41	41095	Government of Kwazulu, Ulundi.	
☐ ZS-NYM	96	Pilatus PC-XII	147	Red Cross Mercy Air Service Trust, Capetown.	(ZS-FDS)
☐ ZS-NZH	77	King Air 200	BB-206	Nationwide Air Charter P/L. Lanseria.	N700WF
☐ ZS-NZI	80	King Air 200	BB-593	Nationwide Air Charter P/L. Lanseria.	N593
☐ ZS-NZJ	80	King Air 200	BB-830	Nationwide Air Charter P/L. Lanseria.	N630VB
☐ ZS-NZK	96	King Air B200	BB-1553	Reserve Air Charters P/L. Hillcrest.	N10780
☐ ZS-NZN	96	King Air B200	BB-1552	Billiton Aviation P/L. Lanseria.	N1080Y
☐ ZS-NZZ	89	King Air B200	BB-1323	M H O Els, Glenvista.	N3154S
☐ ZS-OAE	75	King Air E90	LW-151	Airwork Sales CC. Lanseria.	TR-LVH
☐ ZS-OAK	77	King Air 200	BB-197	J & D Aviation CC. Lanseria.	(N12154)
☐ ZS-OBB	96	King Air B200	BB-1522	Fourie's Poultry Farm P/L. Potchefstrom.	N3272E
☐ ZS-OCA	60	Gulfstream 1	42	Madiba Air P/L. Lanseria.	XC-AA61
☐ ZS-OCI	76	King Air 200	BB-121	Beechcraft Sales/NAC P/L. Rand.	TR-LDX
☐ ZS-ODI	96	King Air B200	BB-1542	Beechcraft Sales/NAC P/L. Rand.	N202JT
☐ ZS-ODU	94	King Air B200	BB-1476	Anglo American Corp of South Africa Ltd. Johannesburg.	Z-MKI
☐ ZS-ODV	76	King Air C90	LJ-675	Coin Security Group P/L. Sunnyside.	N458Q
☐ ZS-PAM	81	King Air 200	BB-813	Balm Air/L Sparg & R Reid, Lanseria.	VH-IBF
☐ ZS-PBS	81	PA-31T Cheyenne II	8120054	Semri Proprietary P/L. Louis Truchardt.	(N151CC)
☐ ZS-PES	77	Conquest II	441-0007	Benoryn Investments Holdings P/L. Braamfontein.	N107EA
☐ ZS-PNP	81	Conquest 1	425-0075	Pick N'Pay (Gabriel Road) P/L. Capetown.	N17EA
☐ ZS-PPG	97	King Air B200	BB-1562	Pick N' Pay (Gabriel Road) P/L. Claremont.	N203JT
☐ ZS-PVT	95	Pilatus PC-XII	121	The Movie Camera Co. Lanseria.	HB-FQO
☐ ZS-SMC	94	King Air B200	BB-1489	Billiton Aviation P/L. Lanseria.	N1563M
☐ ZS-SOL	83	King Air B200	BB-1138	J A Rumble, Richards Bay.	N400AJ
☐ ZS-SON	90	King Air B200C	BL-136	J A Rumble, Richards Bay.	N136BL

Regn	Yr Type	c/n	Owner/Operator	Prev Regn

Regn	Yr	Type	c/n	Owner/Operator	Prev Regn
☐ ZS-TBS	72	King Air A100	B-105	Bateleur Estates P/L. Douglas.	A2-ABS
☐ ZS-TOB	95	King Air B200	BB-1515	Leonard Dingler P/L. Benoni.	N3235Z
☐ ZS-TON	82	King Air B200	BB-1060	Vero Air Services P/L. Lanseria.	N250DM

Military

☐ SAAF651	82	King Air B200C	BL-45	South African Air Force, 41 Squadron,	ZS-LXS
☐ SAAF652	81	King Air 200C	BL-34	South African Air Force, 41 Squadron,	ZS-LAY
☐ SAAF653	87	King Air 300	FA-118	South African Air Force, 41 Squadron,	ZS-MHK
☐ SAAF654	83	King Air B200C	BL-70	South African Air Force, 41 Squadron,	ZS-LNT
☐ SAAF8030	97	Pilatus PC-XII	144	South African Air Force,	HB-FRG

Z3 = Macedonia (Republic of Macedonia)

Civil

☐ Z3-BAB	80	King Air 200	BB-652	Government of Macedonia, Skopje.	(N88DA)

3A = Monaco (Principality of Monaco)

Civil

☐ 3A-MBA	77	PA-31T Cheyenne II	7720018	Ste Boutsen Aviation, Cannes, France.	OE-FFH
☐ 3A-MIO	79	PA-31T Cheyenne II	7920001	Ljiljana Hennessy, Monaco.	I-NANE
☐ 3A-MON	77	King Air C90	LJ-710	S C de Rafale, Cannes.	(F-GSOA)

3B = Mauritius (Republic of Mauritius)

Civil

☐ 3B-NBC	95	Beech 1900D	UE-157	Noto Ltd. Port Louis.	N3217P
☐ 3B-SKY	90	King Air B200	BB-1363	Unitex Finance Holding (Mauritius) Ltd. Port Louis.	N5503K

4R = Sri Lanka (Democratic Socialist Republic of Sri Lanka)

Military

☐ 4R-HVE	81	King Air B200	...	Sri Lanka Air Force, Colombo.	
☐ CR-841	84	King Air B200T	BT-30	Sri Lankan Air Force/Helitours, Colombo. (was BB-1185).	N6923L

4X = Israel (State of Israel)

Civil

☐ 4X-ARF	64	Gulfstream 1	134	Aereol Airways, Tel Aviv.	G-BMPA
☐ 4X-ARG	65	Gulfstream 1	166	Aereol Airways, Tel Aviv.	HB-IRQ
☐ 4X-ARH	59	Gulfstream 1	25	Aereol Airways, Tel Aviv.	(OK-NEA)
☐ 4X-CIC	82	PA-42 Cheyenne III	8001073	Yasur Aviation Ltd. Tirat Carmel.	N321CF
☐ 4X-COD		Cirrus ST-50	001	Israviation Co. Kiryath-Shmona.	N50ST
☐ 4X-DZT	71	King Air C90	LJ-513	Chim-Nir Aviation Services Ltd. Herzlia.	N913K

Military

☐ 4X-FEA	90	King Air B200	BB-1385	501, IDFAF, Tel Aviv.	
☐ 4X-FEB	90	King Air B200	BB-1386	504, IDFAF, Tel Aviv.	
☐ 4X-FEC	90	King Air B200	BB-1387	507, IDFAF, Tel Aviv.	
☐ 4X-FED	90	King Air B200	BB-1388	510, IDFAF, Tel Aviv.	
☐ 4X-FSA		King Air RC-12D	BP-7	974, IDFAF, Tel Aviv.	82-23638
☐ 4X-FSB		King Air RC-12D	BP-8	977, IDFAF, Tel Aviv.	82-23639
☐ 4X-FSC		King Air RC-12D	BP-9	980, IDFAF, Tel Aviv.	82-23640
☐ 4X-FSD		King Air RC-12D	BP-10	982, IDFAF, Tel Aviv.	82-23641
☐ 4X-FSE		King Air RC-12D	BP-11	985, IDFAF, Tel Aviv.	82-23642
☐ 4X-FSF		King Air B200	FG-1	987, IDFAF, Tel Aviv.	
☐ 4X-FSG		King Air B200	FG-2	990, IDFAF, Tel Aviv.	

5A = Libya (Socialist People's Libyan Arab Jamahiriya)

Civil

☐ 5A-DDT	79	King Air 200C	BL-1	Air Ambulance/Ministry of Health, Tripoli.	(F-GBLT)
☐ 5A-DDY	80	King Air 200C	BL-6	Air Ambulance/Ministry of Health, Tripoli.	
☐ 5A-DHZ	80	SA-226T Merlin 3B	T-345	Light Air Transport & Technical Services, Tripoli.	OO-HSC
☐ 5A-DJB	81	SA-226T Merlin 3B	T-388	Light Air Transport & Technical Services, Tripoli.	OO-XSC

5H = Tanzania (United Republic of Tanzania)

Civil

☐ 5H-MTY	78	Rockwell 690B	11474	Government of Tanzania, Dar es Salaam.	5H-ASP
☐ 5H-TZC	88	Reims/Cessna F406	F406-0028	Tanzanair, Dar es Salaam.	N7073C
☐ 5H-TZE		Reims/Cessna F406	F406-0046		OY-PED

5N = Nigeria (Federal Republic of Nigeria)

Civil

☐ 5N-AMT	80	King Air 200	BB-663	Mobil Producing (Nigeria) Ltd. Lagos.	N115CM
☐ 5N-AMU	81	King Air 200	BB-809	Dumez (Nigeria) Ltd. Lagos.	F-GCVQ
☐ 5N-AMW	81	Conquest 1	425-0067	Pan African Airlines (Nigeria) Ltd. Lagos.	N6844V
☐ 5N-AMZ	78	King Air C90	LJ-755	Dumez (Nigeria) Ltd. Lagos.	N9085S
☐ 5N-APZ	79	Conquest II	441-0118	Nigerian Police Force, Lagos.	N26230
☐ 5N-ATR	85	Conquest II	441-0353	Pan African Airlines (Nigeria) Ltd. Lagos.	N321AF

Regn	Yr	Type	c/n	Owner/Operator	Prev Regn

Regn	Yr	Type	c/n	Owner/Operator	Prev Regn
☐ 5N-AUT	75	PA-31T Cheyenne II	7520016	Seven Up Bottling Co Ltd. (status) ?	N101T
☐ 5N-AVH	79	King Air 200	BB-538	Civil Aviation Flying Unit, Lagos.	N6724N
☐ 5N-BFL	85	King Air B200	BB-1213	Bristow Helicopters (Nigeria) Ltd. Lagos.	N7234L
☐ 5N-BHL	78	King Air 200	BB-387	Bristow Helicopters (Nigeria) Ltd. Lagos.	G-BFOL
☐ 5N-MGV		DHC 8-102	024	Mobil Producing (Nigeria) Ltd. Lagos.	C-GMOK
☐ 5N-MPA	95	Beech 1900D	UE-149	Mobil Producing (Nigeria) Ltd. Lagos.	N3217L
☐ 5N-MPN	94	Beech 1900D	UE-77	Mobil Producing (Nigeria) Ltd. Lagos.	N5006
☐ 5N-MPS	87	Dornier 228-201	8146	NEPA-National Electric Power Authority,	D-CALO
☐ 5N-SPC	97	Dornier 328-100	3083	Bristow Helicopters (Nigeria) Ltd. Lagos.	D-C...
☐ 5N-SPD	97	Dornier 328-100	3086	Bristow Helicopters (Nigeria) Ltd. Lagos.	D-C...

5T = Mauritania *(Islamic Republic of Mauritania)*

Civil
☐ 5T-TJY	79	PA-31T Cheyenne II	7920056	Air Mauritanie, Nouakchott.	F-ODJS

Military
☐ 5T-MAB	81	PA-31T Cheyenne II	8120024	Mauritanian Air Force, Nouakchott.	N2470X
☐ 5T-MAC	81	PA-31T Cheyenne II	8120026	Mauritanian Air Force, Nouakchott.	N2483X

5U = Niger *(Republic of Niger)*

Civil
☐ 5U-ABV	78	King Air E90	LW-291	Nigeravia, Niamey.	N291AV
☐ 5U-ABX	79	King Air 200	BB-531	Nigeravia, Niamey.	SE-LDM

5V = Togo *(Togolese Republic)*

Civil
☐ 5V-TPH		King Air 200	...		

Military
☐ 5V-MCG	81	King Air B200	BB-857	Government of Togo, Lome.	N57AC
☐ 5V-MCH	81	King Air B200	BB-858	Government of Togo, Lome.	N10AC

5X = Uganda *(Republic of Uganda)*

Civil
☐ 5X-INS	81	King Air B200	BB-914		ZS-MXH

5Y = Kenya *(Republic of Kenya)*

Civil
☐ 5Y-...	68	Gulfstream 1	190	Skyways Airlines, Nairobi. (status ?).	N190LE
☐ 5Y-...	76	King Air 200	BB-99		C-FSKQ
☐ 5Y-...	79	King Air 200	BB-547		ZS-NIP
☐ 5Y-...	71	King Air C90	LJ-528	Air Bateleu Ltd. Nairobi.	N883AV
☐ 5Y-BIS	89	Reims/Cessna F406	F406-0037	Tawakal Airlines Ltd. Nairobi.	D-ICAS
☐ 5Y-BIX	90	Reims/Cessna F406	F406-0055	East African Air Charters Ltd. Nairobi.	N65912
☐ 5Y-BIZ	81	MU-2 Marquise	1528SA	East African Air Charters Ltd. Nairobi.	N46TT
☐ 5Y-BJC	80	King Air 200	BB-597	Rossair Executive Air Charter P/L. Lanseria, RSA.	ZS-MSK
☐ 5Y-BJX	74	Mitsubishi MU-2J	627		N34MG
☐ 5Y-BKA	81	King Air 200	BB-846		ZS-MIM
☐ 5Y-BKM	75	King Air/Catpass 250	BB-86	Mission Aviation Fellowship, Nairobi.	N677BC
☐ 5Y-BKN		Reims/Cessna F406	F406-0021		PH-FWG
☐ 5Y-BKT	77	King Air 200	BB-256		ZS-NTM
☐ 5Y-BLA	80	King Air 200C	BL-10		C-FAMB
☐ 5Y-BLF	64	Gulfstream 1	131	Skyways Kenya Airlines, Nairobi.	C-FRTT
☐ 5Y-BLR	60	Gulfstream 1	34	Skyways Kenya Airlines, Nairobi.	N34LE
☐ 5Y-BMA	76	King Air 200	BB-155	Skylink, Nairobi.	OY-GEH
☐ 5Y-BMC	77	King Air 200	BB-211	Skylink, Nairobi.	OY-BTR
☐ 5Y-BMR	61	Gulfstream 1	81	Legion Express Inc. Opa Locka, Fl. USA.	I-TASO
☐ 5Y-BVI	89	Beech 1900C-1	UC-55	Aviation Assistance A/S-United Nationa, Nairob.	OY-BVI
☐ 5Y-DDE	78	King Air 200	BB-379	IBIS Aviation Ltd. Nairobi.	G-ONEX
☐ 5Y-EKO	79	King Air 200C	BL-2	Air Bridge, Nairobi.	OY-BVE
☐ 5Y-HHA	82	King Air B200	BB-988	Bluebird Aviation Ltd. Nairobi.	PH-LMC
☐ 5Y-JAI	79	King Air 200	BB-557	Capital Airlines Ltd. Nairobi.	OY-PAM
☐ 5Y-JJA	87	Reims/Cessna F406	F406-0013	Tawakal Airlines Ltd. Nairobi.	OO-TIY
☐ 5Y-JJC	89	Reims/Cessna F406	F406-0040	Tawakal Airlines Ltd. Nairobi.	PH-ALY
☐ 5Y-JMB	75	King Air 200	BB-72	AD Aviation (Aircharters) Ltd. Nairobi.	ZS-KJB
☐ 5Y-JMR	81	King Air 200C	BL-17	AD Aviation (Aircharters) Ltd. Nairobi.	F-GJMR
☐ 5Y-KXU	81	Conquest 1	425-0047	AD Aviation (Aircharters) Ltd. Nairobi.	ZS-KXU
☐ 5Y-MKM	89	Reims/Cessna F406	F406-0044		(LN-TED)
☐ 5Y-NCF	73	Rockwell 690	11069	James Gaunt, Nairobi.	N321MG
☐ 5Y-NPO	78	King Air 200	BB-367	King Air Services Partnership, Melville, South Africa.	ZS-NPO
☐ 5Y-NUN	80	King Air 200	BB-643	UNHCR/Bluebird Aviation Ltd. Nairobi.	OY-PEH
☐ 5Y-SEL	83	King Air B200	BB-1127	Skylink/United Nations, Nairobi.	G-BMNF
☐ 5Y-SJB	79	King Air/Catpass 250	BB-467	East African Coffee Co. Nairobi.	N307G
☐ 5Y-TAL	86	Reims/Cessna F406	F406-0009	Trackmark Ltd. Nairobi.	PH-FWB
☐ 5Y-TWA	81	King Air 200	BB-803	Transworld Safaris (K) Ltd. Nairobi.	G-WPLC
☐ 5Y-TWB	80	King Air 200	BB-696	Transworld Safaris (K) Ltd. Nairobi.	N200GU
Regn	*Yr*	*Type*	*c/n*	*Owner/Operator*	*Prev Regn*

□ 5Y-TWC	81 King Air B200C	BL-37	Transworld Safaris Kenya Ltd. Nairobi.		G-IFTB
□ 5Y-UAL	82 Cheyenne T-1040	8275002	United Airlines Ltd. Nairobi.		N309SC
□ 5Y-WAW	87 Reims/Cessna F406	F406-0012	Bluebird Aviation Ltd. Nairobi.		PH-ALE
Military					
□ JW9027	74 King Air A100	B-197	Tanzanian Air Force,		5X-UWT

6O = Somalia *(Somali Democratic Republic)*
Civil

□ 6O-SBQ	74 Rockwell 690A	11151	Murri Brothers, Mogadishu.		3D-BYZ
□ 6O-SBV	SA-227TT Merlin 3C	TT-477	Murri Brothers, Mogadishu.		N3059Y

6V = Senegal *(Republic of Senegal)*
Civil

□ 6V-AGO	76 PA-31T Cheyenne II	7620057	Eagle International Air Space, Dakar.		F-GDAL
□ 6V-AGS	75 King Air 200	BB-28	Senegalair, Dakar.		F-GKPL

6Y = Jamaica
Military

□ JDFT-3	75 King Air A100	B-216	Jamaican Defence Force, Kingston.		

7Q = Malawi *(Republic of Malawi)*
Civil

□ 7Q-YMP	81 King Air B200	BB-903			ZS-LBE
□ 7Q-YST	88 King Air 300	FA-139	Stancom Aviation, Lilongwe.		ZS-MLL

7T = Algeria *(Democratic & Popular Republic of Algeria)*
Civil

□ 7T-VBE	93 King Air B200	BB-1453	National Institute of Cartography, Algiers.		N8153H
□ 7T-VCV	72 King Air A100	B-93	ENESA/Air Algerie, Algiers.		N9369Q
□ 7T-VRF	73 King Air A100	B-147	ENESA/Air Algerie, Algiers.		N1828W
Military					
□ 7T-...	94 King Air C90B	LJ-1359	Ministry of Defence, Algiers.		N8280K
□ 7T-...	94 King Air C90B	LJ-1379	Ministry of Defence, Algiers.		N3122K
□ 7T-...	94 King Air C90B	LJ-1380	Ministry of Defence, Algiers.		N3128K
□ 7T-VRG	76 King Air 200	BB-184	Ministry of Defence, Boufarik.		
□ 7T-VRH	76 King Air 200	BB-175	Ministry of Defence, Boufarik.		
□ 7T-VRI	76 King Air 200	BB-171	Ministry of Defence, Boufarik.		
□ 7T-VRO	81 King Air 200	BB-807	Ministry of Defence, Boufarik.		
□ 7T-VRS	81 King Air 200	BB-759	Ministry of Defence, Boufarik.		F-GCTC
□ 7T-VRT	81 King Air 200	BB-775	Ministry of Defence, Boufarik.		F-GCTD
□ 7T-WRY	81 King Air 200T	BT-20	Ministry of Defence, Boufarik. (was BB-871).		7T-VRY
□ 7T-WRZ	81 King Air 200T	BT-21	Ministry of Defence, Boufarik. (was BB-895).		7T-VRZ

9G = Ghana *(Republic of Ghana)*
Civil

□ 9G-AGF	95 Beech 1900D	UE-136	Ashanti Ghana/Ashanti Goldfields Corp. Accra.		N3212K
□ 9G-SAM	79 King Air C90	LJ-839	Ashanti Ghana/Ashanti Goldfields Corp. Obuasi, Ashanti.		ZS-MFA

9J = Zambia *(Republic of Zambia)*
Civil

□ 9J-...	73 King Air C90	LJ-590			N243L
□ 9J-AAV	70 King Air B90	LJ-486	ZESCO Ltd-Zambia Electricity Supply Co Ltd. Lusaka.		A2-JZT
□ 9J-AFI	88 King Air 300	FA-166	Roan Air Ltd/Mines Air Services Ltd. Lusaka.		N31827
□ 9J-AFJ	89 Beech 1900C-1	UC-48	Roan Air Ltd/Mines Air Services Ltd. Lusaka.		N31559
□ 9J-DCF	73 King Air C90	LJ-575	Government Communications Flight, Lusaka.		N12RF
□ 9J-YVZ	68 King Air B90	LJ-338	Easter Safaris, Lusaka.		ZS-LWZ
Military					
□ AF602	70 HS 748-2A	1688	Zambian Air Force,		

9M = Malaysia *(Federation of Malaysia)*
Civil

□ 9M-ASH	King Air B200	...			
□ 9M-AZM	77 Rockwell 690B	11414	Wira Kris Udera,		VH-PCD
□ 9M-JPD	79 King Air 200T	BT-10	Directorate of Civil Aviation, Kuala Lumpur. (was BB-563).		N6065D
□ 9M-KNS	77 King Air 200	BB-294	Penerbangan Sabah/Sabah Air Pte Ltd. Kota Kinabalu.		N18494
□ 9M-PMS	90 Reims/Cessna F406	F406-0049	Transmile Air Service Sdn Bhd. Kuala Lumpur.		N6589E
Military					
□ M41-01	93 King Air 200T	BT-35	Royal Malaysian Air Force. (was BB-1448).		N15509
□ M41-02	93 King Air 200T	BT-36	Royal Malaysian Air Force. (was BB-1451).		N80024
□ M41-03	93 King Air 200T	BT-37	Royal Malaysian Air Force. (was BB-1454).		N80027
□ M41-04	93 King Air 200T	BT-38	Royal Malaysian Air Force. (was BB-1457).		N80048

Regn	Yr Type	c/n	Owner/Operator		Prev Regn

9N = Nepal *(Kingdom of Nepal)*

Civil

Regn	Yr Type	c/n	Owner/Operator	Prev Regn
☐ 9N-AEE	King Air B200	...		

9Q = Dem Rep of Congo

Civil

Regn	Yr Type	c/n	Owner/Operator	Prev Regn
☐ 9Q-...	89 King Air 300	FA-173		ZS-MIP
☐ 9Q-CBD	60 Gulfstream 1	35	CAA-Compagnie Africaine d'Aviation, Kinshasa.	N86MA
☐ 9Q-CBJ	66 Gulfstream 1	155	CAA-Compagnie Africaine d'Aviation, Kinshasa.	G-BMOW
☐ 9Q-CBU	71 681B Turbo Commander	6045		N9107N
☐ 9Q-CBY	60 Gulfstream 1	33	CAA-Compagnie Africaine d'Aviation, Kinshasa.	N23AH
☐ 9Q-CCG	74 King Air E90	LW-110	Inga Shaba, Kinshasa.	N8PC
☐ 9Q-CFE	80 King Air 200	BB-722	Cie d'Enterprise C.F.E.,	OO-CFE
☐ 9Q-CGL	70 681 Turbo Commander	6030	Cie Ciments Lacs,	D-IGAD
☐ 9Q-CKM	68 King Air B90	LJ-402	Fontshi Aviation Service, Kinshasa. 'Tshihuka'	ZS-IBE
☐ 9Q-CKZ	70 King Air B90	LJ-494	Shabair SPRL. Lumbumbashi.	(N15LM)
☐ 9Q-CRF	70 King Air 100	B-33	Forrest Air Co. Lumbumbashi.	N883K
☐ 9Q-CTG	80 King Air 200	BB-629	SCIBE Airlift, Kinshasa.	
☐ 9Q-CVT	68 King Air B90	LJ-431	BOGM United Methodist Church, NYC. USA.	N925S

9Y = Trinidad & Tobago *(Republic of Trinidad & Tobago)*

Civil

Regn	Yr Type	c/n	Owner/Operator	Prev Regn
☐ 9Y-...	75 King Air A100	B-208		N4475W

BUSINESS TURBOPROPS
Withdrawn From Use and Written-Off

CIVIL

Regn	Yr	Type	c/n	Accident / Withdrawal Details	Prev Regn
☐ C-FCAS	65	King Air 90	LJ-23	W/o Sherrington, Canada. 1 May 79.	
☐ C-FCAU	65	King Air 90	LJ-24	W/o Nr Quebec City, PQ. Canada. 30 Aug 85.	
☐ C-FCGJ	67	King Air A90	LJ-231	Wfu.	
☐ C-FFEO	67	680V Turbo Commander	1693-72	W/o hangar fire Hamilton, Ontario. Canada. 15 Feb 93.	
☐ C-FFYC	69	SA-26 Merlin 2A	T26-36	W/o N E of Thompson, Manitoba, Canada. 31 May 94.	N739G
☐ C-FHBW	68	King Air B90	LJ-336	Wfu.	
☐ C-FHYX	70	SA-26AT Merlin 2B	T26-174	W/o ground fire, Canada. Oct 73.	
☐ C-FMAR	59	Gulfstream 1	3	Wfu.	N703G
☐ C-FOUR	73	Mitsubishi MU-2J	606	Wfu.	N338MA
☐ C-FQMS	67	Mitsubishi MU-2B	009	W/o Athabaska, Canada. 7 Apr 80.	HB-LEB
☐ C-FRCL	65	King Air 90	LJ-33	W/o.	C-FRCL
☐ C-FSSU	80	King Air 200	BB-633	W/o hangar fire Quebec City, PQ. Canada. 10 Jan 93.	N650TJ
☐ C-FUAC	65	King Air 90	LJ-3	Wfu. Located Vincennes University, In. USA.	
☐ C-FVMH	67	King Air A90	LJ-225	Wfu.	
☐ C-GAMJ	78	PA-31T Cheyenne II	7820063	W/o Haul Beach, NT. Canada. 17 Apr 89.	N82298
☐ C-GAPT	75	PA-31T Cheyenne II	7620004	W/o Toronto, ON. Canada. 17 Oct 84.	
☐ C-GBMI	81	Conquest 1	425-0031	W/o La Rouche, PQ. Canada. 20 Nov 88.	N355MA
☐ C-GBTI	68	King Air B90	LJ-352	W/o May 91.	VR-BHT
☐ C-GCEV	76	King Air 200	BB-153	W/o Sept-Iles Airport, PQ. Canada. 28 Jan 97.	N502AB
☐ C-GCSL	76	King Air 200	BB-118	W/o Quebec City, PQ. Canada. 10 Jan 93.	
☐ C-GDOM	68	King Air B90	LJ-368	W/o Fort Simpson, Canada. 16 Oct 88.	N1100D
☐ C-GDSD	75	SA-226T Merlin 3A	T-251	Wfu.	N711RD
☐ C-GFRU	76	Mitsubishi MU-2M	343	W/o Kelowna, BC. Canada. 18 Jan 82.	N728MA
☐ C-GJNH	66	680V Turbo Commander	1587-38	Wfu.	N251B
☐ C-GJPD	79	PA-31T Cheyenne II	7920070	W/o hangar fire Quebec City, PQ. Canada. 10 Jan 93.	
☐ C-GJUL	75	King Air A100	B-218	W/o Chapleau, ON. Canada. 29 Nov 88.	N80MD
☐ C-GJWW	73	SA-226AT Merlin 4	AT-013	Wfu.	N720R
☐ C-GKRL	81	King Air B200	BB-878	W/o Fort McMurray, Canada. 10 Dec 87.	G-BIPP
☐ C-GLOW	74	Mitsubishi MU-2J	624	W/o Edmonton, AB. Canada. 6 Dec 81.	N474MA
☐ C-GODI	74	Mitsubishi MU-2J	649	W/o Portage la Prairie, MT. Canada. 28 May 77.	N530MA
☐ C-GPIP	77	PA-31T Cheyenne II	7720026	Wfu. Cx C- 1/93.	N82118
☐ C-GPPN	68	King Air B90	LJ-389	W/o Hudson's Bay, Canada. 22 Dec 84.	N745JB
☐ C-GPTA	65	Gulfstream 1	162	W/o by hangar fire Toronto, ON. Canada. 19 Nov 96.	N300GP
☐ C-GPWR	70	SA-26AT Merlin 2B	T26-160	Wfu at White Industries, Bates City, Mo.	N38MJ
☐ C-GQDD	68	King Air B90	LJ-328	W/o 1987. (parts at Dodson's, Ottawa, Ks.).	N730K
☐ C-GSID	75	PA-31T Cheyenne II	7520042	Wfu.	(N118L)
☐ C-GSWB	77	PA-31T Cheyenne II	7720013	W/o Montreal, PQ. Canada. 12 Nov 93. (parts at White Inds.).	N82092
☐ C-GTDL	63	Gulfstream 1	110	Wfu.	N533CS
☐ C-GTHN	68	SA-26 Merlin 2A	T26-16	W/o Whale Cove, NT. Canada. 29 Jun 96.	N748G
☐ C-GVCE	72	King Air A100	B-135	W/o Calgary, AB. Canada. 21 Nov 84.	N4GT
☐ C-GVCY	71	SA-226AT Merlin 4	AT-003	Wfu at Dodsons, Ottawa, Ks.	N5295M
☐ C-GWCY	68	King Air B90	LJ-345	W/o Lynn Lake, Canada. Oct 81.	N550TS
☐ C-GWFM	75	PA-31T Cheyenne II	7520041	Wfu.	N54972
☐ C-GWGP	68	Mitsubishi MU-2D	109	Wfu.	N1AN
☐ C-GYMR	70	SA-26AT Merlin 2B	T26-177	W/o Canada. Oct 88.	N83RS
☐ C-GYQT	74	King Air A100	B-189	W/o Big Trout Lake, ON. Canada. 21 Feb 95.	N22220
☐ CN-CDE	79	King Air 200	BB-567	W/o Casablanca, Morocco. 3 Nov 86.	
☐ CP-1016	73	Rockwell 690	11053	W/o Bolivia. Aug 93.	
☐ CP-1017	72	Rockwell 690	11054	W/o Lima, Peru. 13 Feb 74.	
☐ CP-1106	74	Rockwell 690A	11193	Wfu.	
☐ CP-1849	82	King Air B200C	BL-52	W/o 14 Mar 84.	N6872X
☐ CP-2287	67	King Air A90	LJ-232	W/o Robore, Bolivia. 12 Apr 96.	N707EB
☐ CP-894	70	681 Turbo Commander	6015	W/o Santa Cruz, Bolivia. 18 Jun 71.	N9066N
☐ D-IAAE	78	Conquest II	441-0047	W/o Mount Orsa, Italy. 11 Jun 82.	(N36990)
☐ D-IAAY	80	Conquest II	441-0144	W/o 6 Nov 80.	(N2627N)
☐ D-IAKS	78	PA-31T Cheyenne II	7820048	W/o Borkum Island, Germany. 23 Jul 83.	OE-FOP
☐ D-IBAA	83	Conquest 1	425-0163	W/o Hannover, Germany 24 Jan 96.	N66218
☐ D-IBAF	76	King Air 200	BB-93	W/o Bourgas, Bulgaria. 27 Jul 77.	OE-FMC
☐ D-IBAR	81	Gulfstream 980	95054	W/o Paderborn, Germany. 30 Jan 85.	(N9806S)
☐ D-ICEK	81	Conquest 1	425-0055	W/o Aichstetten, Germany. 10 Dec 92.	N425VC
☐ D-IDDI	79	PA-31T Cheyenne II	7920014	W/o Cologne, Germany. 5 Feb 93.	N792SW
☐ D-IEFW	86	Conquest 1	425-0228	W/o Lake Constance, Switzerland. 24 Jan 94.	N1227A
☐ D-IEWK		SA-226AT Merlin 4A	AT-042	W/o Munich, Germany. 5 Feb 87.	OE-FTA
☐ D-IGSW	80	King Air 200	BB-669	W/o Keller Joch Mountain, Austria. 22 Nov 90.	N80GA
☐ D-IHAN	79	MU-2 Solitaire	396SA	W/o Steinhausen, Germany. 9 Aug 79.	D-IHAN
☐ D-IHNA	81	King Air C90	LJ-926	W/o Mindelheim, Germany. 27 May 94.	(D-IBBI)
☐ D-IHVI	80	PA-31T Cheyenne II	8020007	W/o Southend, UK. 13 Mar 86.	N2529R
☐ D-IKKS	81	PA-31T Cheyenne II	8120034	W/o Concord, Ca. USA. 14 Jul 84.	F-GDCR
☐ D-IKOC	78	Rockwell 690B	11498	W/o Paris, France. 21 Feb 81.	N131SA

Regn	Yr	Type	c/n	Accident / Withdrawal Details	Prev Regn
□ D-IKWP	80	PA-31T Cheyenne 1	8004049	W/o Blieskastel, Germany. 10 Dec 88.	N5GW
□ D-ILHA	71	King Air C90	LJ-509	Wfu as instructional airframe, Hamburg.	
□ D-ILMA	65	King Air 90	LJ-48	W/o Greven, Germany. 13 Aug 69.	
□ D-ILNI	66	King Air A90	LJ-116	W/o Milan, Italy. 21 Sep 67.	
□ D-ILNU	66	King Air A90	LJ-178	W/o Bremen, Germany. 16 Feb 67.	N798K
□ D-ILRA	80	PA-31T Cheyenne II	8020009	W/o Munich, Germany. 11 Aug 87.	D-ILRA
□ D-ILSE	69	SA-26AT Merlin 2B	T26-163E	W/o Stuttgart, Germany. 10 Apr 73.	F-BKML
□ D-ILTU	68	King Air B90	LJ-359	W/o Frankfurt, Germany. 22 Jan 71.	N7077N
□ D-IMON	68	680W Turbo Commander	1819-33	W/o Germany. 27 Dec 78.	
□ D-IMTT	78	PA-31T Cheyenne II	7820041	Wfu.	D-IMTT
□ D-IMWH	81	King Air F90	LA-114	W/o Dusseldorf, Germany. 6 Dec 87.	N313BH
□ D-INIX	72	Rockwell 690	11013	W/o Greenland. 21 Jun 72.	N9213N
□ D-IOET	73	Rockwell 690A	11142	W/o Germany. 1 Dec 81.	
□ D-IOFC	78	PA-31T Cheyenne 1	7804011	W/o Borgo Ticino, Italy. 25 Mar 84.	D-IOFC
□ D2-ECH	78	King Air 200	BB-345	W/o Cafunfo, Angola. 28 Jan 95.	F-GHCT
□ D2-ECL	75	King Air 200	BB-44	W/o Luanda, Angola. 28 May 97.	F-GHAL
□ EC-COJ	75	King Air C90	LJ-664	W/o Salamanca, Spain. 4 Oct 83.	EC-COJ
□ EC-ERQ	77	King Air 200	BB-218	W/o Banjul, Gambia. 9 Oct 97.	EC-351
□ EC-ETH	83	Conquest 1	425-0151	W/o Nr Malaga, Spain. 4 Sep 92.	N81798
□ EC-FFE	76	Rockwell 690A	11344	W/o Warsaw-Okecie, Poland. 29 Nov 95.	N900FT
□ EI-BGL	78	Rockwell 690B	11507	W/o UK. 13 Nov 84.	N81850
□ EP-AHN	74	Rockwell 690A	11147	W/o.	
□ F-BNMC	67	King Air A90	LJ-149	W/o Nr Marseilles, France. 28 Aug 97.	
□ F-BSTM	66	680V Turbo Commander	1540-6	Wfu.	F-WSTM
□ F-BTDP	72	King Air C90	LJ-560	W/o 17 Dec 74.	
□ F-BUTV	73	King Air C90	LJ-602	W/o 1992.	
□ F-BVRP	75	King Air 200	BB-38	W/o Baluyan, China. 4 Apr 83.	
□ F-BVVM	65	King Air 90	LJ-26	Wfu.	D-ILGK
□ F-BXAR	75	King Air C90	LJ-658	W/o.	
□ F-BXSF	65	King Air 90	LJ-29	Wfu at Dodson's. Ottawa, Ks. Cx F- 3/93.	D-ILMI
□ F-GBDZ	78	King Air E90	LW-295	W/o Paris, France. 15 Dec 82.	
□ F-GBRD	74	King Air E90	LW-91	W/o Barcelonette, France. 2 Nov 86.	N75DA
□ F-GBRP	78	King Air 200	BB-368	W/o Ribeauville, France. 17 Oct 80.	
□ F-GCCC	79	King Air 200	BB-504	W/o Bergamo, Italy. 26 Mar 84.	
□ F-GCFH	66	King Air A90	LJ-127	Wfu. Cx F- 8/96.	N72RD
□ F-GDHS	81	MU-2 Marquise	1532SA	W/o 21 May 91.	F-WDHS
□ F-GDHV	80	MU-2 Marquise	779SA	W/o Papeete, Tahiti. 27 May 94.	D-IFTG
□ F-GDPJ	75	PA-31T Cheyenne II	7620006	W/o Paris, France. 12 Dec 84.	C-GDOW
□ F-GDRT	65	King Air 90	LJ-4	Wfu. CoA expiry 8/88, Cx F- 8/96.	TR-LBB
□ F-GEBK	76	SA-226T Merlin 3A	T-272	W/o Apr 85.	N49MJ
□ F-GEFR	75	King Air A100	B-220	W/o Lille, France. 28 Aug 86.	N700K
□ F-GEJY	72	SA-226T Merlin 3	T-222	Wfu.	N5306M
□ F-GERA	77	Mitsubishi MU-2N	701SA	W/o St Just, France. 16 Apr 88.	N468DB
□ F-GERN	79	King Air C90A	LJ-854	W/o St Broladre, France. 30 Dec 93.	N712D
□ F-GFHR	82	Conquest II	441-0252	W/o Saclay, France. 17 Nov 88.	N441CG
□ F-GFJF	67	King Air A90	LJ-262	Wfu. CoA expiry 1/89, Cx F- 8/96.	N111ME
□ F-GFLD	77	King Air C90	LJ-741	Wfu.	HB-GGW
□ F-GFMS	79	SA-226T Merlin 3B	T-296	W/o Vannes, France. 7 Nov 87.	N245DA
□ F-GGRZ	79	MU-2 Solitaire	395SA	W/o 9 May 91.	N787MA
□ F-GHBE	79	King Air 200	BB-500	W/o 8 Feb 91.	N13HC
□ F-GHFM	77	King Air 200	BB-213	Wfu.	C-GBWC
□ F-GIIX	64	Gulfstream 1	128	W/o Lyon-Satolas, France. 28 Jun 94.	G-BMSR
□ F-GJCN	80	King Air E90	LW-335	W/o 18 Aug 81.	
□ F-GJGC	63	Gulfstream 1	111	Wfu.	N3630
□ F-GJLH	90	Reims/Cessna F406	F406-0056	W/o Strasbourg, France. Sep 93.	
□ F-GJPL	81	PA-31T Cheyenne II	8120029	W/o Azores, Atlantic Ocean. 5 Jun 90.	(D-IKHK)
□ F-GLBC	91	TBM 700	18	W/o 15 Nov 91.	
□ F-GLRA	85	King Air C90A	LJ-1105	W/o Saumur, France. 19 Oct 94. Cx F- 10/95,	N72233
□ F-ODUK	82	Conquest II	441-0270	W/o Faaa, Tahiti, French Polynesia. 4 Dec 90.	N441AC
□ G-ASXT	64	Gulfstream 1	135	Wfu.	N755G
□ G-BABX	73	King Air A100	B-141	W/o Sturgate, England. 12 Jan 77.	
□ G-BGHR	79	King Air 200	BB-508	W/o Tremblay, France. 25 Sep 79.	
□ G-BHUL	74	King Air E90	LW-83	W/o Goodwood, England. 22 Apr 85.	N99855
□ G-BKID	73	King Air C90	LJ-604	W/o Copenhagen, Denmark. 26 Dec 83.	N1GV
□ G-BNAT	74	King Air C90	LJ-614	W/o East Midlands, England. 25 Jan 88.	G-OMET
□ G-BNCE	59	Gulfstream 1	9	Wfu.	N436M
□ G-BOBX	61	Gulfstream 1	77	Wfu.	9Q-CFK
□ G-MDJI	83	King Air B200	BB-1162	W/o Ottley Chevin Hill, England. 19 Oct 87.	N71CS
□ G-MOXY	80	Conquest II	441-0154	W/o Blackbushe, UK. 26 Apr 87.	G-BHLN
□ G-WSJE	79	King Air 200	BB-484	W/o Southend, England. 12 Sep 87.	N84KA
□ HB-...		King Air 200	...	W/o Kracow, Poland. 27 Aug 95.	
□ HB-GDV	68	King Air B90	LJ-433	W/o Kleinober, Austria. 24 Jan 86.	
□ HB-LHT	75	PA-31T Cheyenne II	7520003	W/o Shannon, Ireland. 12 Nov 76.	N66841
□ HB-LLP	80	MU-2 Marquise	767SA	W/o Basel, Switzerland. 3 Sep 94.	N251MA
□ HB-LLS	81	Conquest 1	425-0040	W/o Berne, Switzerland. 3 Mar 86.	N6775J
□ HI-578SP	89	King Air 300	FA-180	W/o 21 Jan 90.	N1568E

Regn	Yr	Type	c/n	Accident / Withdrawal Details	Prev Regn

Regn	Yr	Type	c/n	Accident / Withdrawal Details	Prev Regn
☐ HK-1771G	74	Rockwell 690A	11217	Wfu at Bogota, Colombia.	N9227N
☐ HK-1805	68	King Air B90	LJ-329	Wfu.	N303X
☐ HK-2217	71	681B Turbo Commander	6053	W/o Colombia. 13 Mar 86.	HK-2217W
☐ HK-2245P	76	Mitsubishi MU-2L	684	Wfu. Cx USA 8/95 as N600TN.	HK-2245W
☐ HK-2415	73	Rockwell 690A	11100	W/o Bolivar Quay, Colombia. 8 Sep 91.	N9200N
☐ HK-2438W		SA-226T Merlin 3B	T-305	Wfu at Bogota, Colombia.	N5658M
☐ HK-2478P	80	Gulfstream 840	11609	W/o 15 Jun 90.	HK-2478W
☐ HK-2484	80	King Air F90	LA-54	W/o Palanquera, Colombia. 28 Jun 81.	HK-2484X
☐ HK-2489	78	King Air 200	BB-393	W/o Bogota, Colombia. 28 Oct 85.	N2014K
☐ HK-2532G	67	Mitsubishi MU-2B	010	Wfu.	N4MA
☐ HK-2674P	81	PA-31T Cheyenne II	8120047	Wfu.	
☐ HK-3315X	59	Gulfstream 1	24	W/o Nr Ibague, Colombia. 5 Feb 90.	N713US
☐ HK-3316X	60	Gulfstream 1	59	W/o Monteira, Colombia. 2 May 90.	N11CZ
☐ HK-3330X	62	Gulfstream 1	90	Wfu at Dodson's, Ottawa, Ks.	N41JK
☐ HL5204		PA-42 Cheyenne 400LS	5527043	W/o South Korea. 20 Mar 91.	N9226B
☐ HL5223	71	681B Turbo Commander	6057	W/o South Korea. 13 Sep 74.	N9130N
☐ HR-IAI	70	King Air B90	LJ-489	W/o La Ceiba, Honduras. 15 Jan 92.	N750P
☐ HS-TFB	66	680TVTurbo Commander	1573-28	W/o Bangkok, Thailand. 22 Jul 84.	N80SS
☐ I-CODE	81	PA-31T Cheyenne II	8120023	W/o Malindi, Kenya. 30 Dec 86.	N2468X
☐ I-FRUT	79	MU-2 Solitaire	413SA	Wfu.	N994MA
☐ I-IDMA	80	MU-2 Marquise	769SA	W/o Sardinia. 24 Oct 89.	N57MS
☐ I-KWYR	80	King Air C90	LJ-873	W/o Rome, Italy. 10 Feb 89.	N44TF
☐ I-MDDD	64	Gulfstream 1	143	Wfu. CoA expiry 26 Sep 91.	N914P
☐ I-MGGG	60	Gulfstream 1	51	Wfu.	N90PM
☐ I-MLWT	67	680T Turbo Commander	1694-73	Wfu. CoA expiry 8 Nov 89.	N999WT
☐ I-TASB	63	Gulfstream 1	105	Wfu.	N702G
☐ I-TELM	79	Rockwell 690B	11506	Wfu.	D-IKAH
☐ JA8620	63	Mitsubishi MU-2A	001	Wfu.	
☐ JA8625	63	Mitsubishi MU-2A	002	Wfu. Displayed at Osaka as 'KA1969'.	
☐ JA8626	63	Mitsubishi MU-2A	003	Wfu. Displayed at Aerospace Museum Nagoya.	
☐ JA8627	63	Mitsubishi MU-2B	004	Wfu.	
☐ JA8628	63	Mitsubishi MU-2B	005	Wfu. Displayed at Tokyo-Narita.	
☐ JA8737	70	Mitsubishi MU-2G	501	Wfu.	
☐ JA8753	70	Mitsubishi MU-2G	504	W/o Japan. 11 Mar 81.	
☐ JA8767	70	Mitsubishi MU-2G	520	Wfu.	
☐ JA8770	71	Mitsubishi MU-2F	195	Wfu.	
☐ JA8783	71	Mitsubishi MU-2G	546	Wfu.	
☐ JA8825	81	King Air 200T	BT-19	W/o 17 Feb 87. (was BB-823).	N3718N
☐ JA8896	92	TBM 700	68	W/o Obihiro Airport, Japan. 26 Apr 96.	F-OHBI
☐ LN-KCR	78	King Air C90	LJ-793	W/o 2 Apr 87.	
☐ LN-MAH	77	Conquest II	441-0002	Wfu at Dodson's, Ottawa, Ks.	SE-IBI
☐ LN-TSA	78	King Air 200	BB-308	W/o Nr Gello, Norway. 19 March 93.	N98WP
☐ LN-VIP	82	Conquest II	441-0279	W/o 11 Oct 85.	G-BJYB
☐ LN-VIP	67	King Air A90	LJ-271	W/o 28 Jun 68.	
☐ LV-LDB	72	681B Turbo Commander	6064	Wfu.	LQ-LDB
☐ LV-LTA	74	Rockwell 690A	11197	W/o Argentina. Dec 75.	LV-PTI
☐ LV-MAV	77	Rockwell 690B	11397	W/o Argentina. 12 Sep 84.	LV-PVM
☐ LV-MBR	75	Rockwell 690A	11266	W/o Buenos Aires, Argentina. 14 Sep 80.	LV-PUV
☐ LV-MOC	79	PA-31T Cheyenne II	7920031	W/o Buenos Aires, Argentina. 4 May 80.	LV-P..
☐ LV-MOP	79	MU-2 Marquise	742SA	W/o en route Neuquen-Bahia Blanca, Argentina. 4 May 95.	LV-PBY
☐ LV-MYY	79	PA-31T Cheyenne II	7920085	W/o Buenos Aires, Argentina. 1 Feb 80.	LV-P..
☐ LV-OEV	80	Gulfstream 840	11628	W/o 26 Aug 81.	LV-PHJ
☐ LV-PAC	78	PA-31T Cheyenne II	7820065	W/o Lima, Peru. 14 Jul 78.	
☐ LV-WLW	73	SA-226T Merlin 3	T-230	W/o Ushaia, Argentina. 4 Apr 96.	N789B
☐ N.....	73	Rockwell 690	11070	W/o.	
☐ N1KA	78	King Air 200	BB-411	W/o Mar 80.	
☐ N1MU	73	Mitsubishi MU-2J	578	Wfu. Cx USA 6/95.	N578EH
☐ N1NR	72	Rockwell 690	11024	W/o Wellsburg, WV. USA. 14 Aug 72.	
☐ N1PT	71	King Air C90	LJ-505	Wfu at Dodson's, Ottawa, Ks.	N886K
☐ N2EP	67	King Air A90	LJ-284	W/o 13 Nov 90.	
☐ N2GG	69	King Air B90	LJ-462	W/o 15 Feb 75.	N821K
☐ N2GT	77	Mitsubishi MU-2P	367SA	Wfu. Cx USA 11/97.	N67GT
☐ N2MF	66	King Air 90	LJ-96	W/o 19 Mar 78.	N2MF
☐ N3ED	68	Mitsubishi MU-2D	101	W/o Riverton, Wy. USA. 17 Jun 81.	N75MD
☐ N3GS	78	King Air B100	BE-36	Wfu. Cx USA 4/93.	N200TV
☐ N3MU	69	Mitsubishi MU-2F	143	Wfu at White Industries, Bates City, Mo.	N700MA
☐ N3RB	71	SA-226T Merlin 3	T-214	W/o 17 Sep 85.	N7090
☐ N4GN	73	King Air E90	LW-38	W/o New York City, USA. 10 Mar 80.	N5000T
☐ N4TN	68	Mitsubishi MU-2D	115	Wfu.	N102MA
☐ N4TS	72	King Air C90	LJ-541	W/o 24 Oct 83.	N9314Q
☐ N5ER	68	680W Turbo Commander	1760-6	Wfu.	N177DC
☐ N5NP	70	681 Turbo Commander	6042	W/o Greenup, Ky. USA. 8 Mar 78.	N5NP
☐ N5NQ	77	Mitsubishi MU-2P	356SA	Wfu.	N5NC
☐ N5NW	73	Mitsubishi MU-2J	597	W/o Searcy, Ar. USA. 23 Jan 79.	N5MW
☐ N7GA	72	King Air A100	B-119	W/o Williamstown, Ms. USA. 4 Aug 94.	OH-BKC
☐ N8CC	72	Mitsubishi MU-2J	569	W/o Bartlett, Tx. USA. 6 Jun 86. Cx USA 9/93.	N213MA

Regn   *Yr*   *Type*   *c/n*   *Accident / Withdrawal Details*   *Prev Regn*

Regn	Yr	Type	c/n	Accident / Withdrawal Details	Prev Regn
☐ N8VB	59	Gulfstream 1	12	Wfu at Dodsons, Ottawa, Ks.	(XA-TBT)
☐ N9JS	69	Mitsubishi MU-2F	178	W/o Alpena, Mi. USA. 22 Apr 81.	N711SH
☐ N9PU	80	King Air F90	LA-57	W/o Ruidoso, NM. USA. 9 Dec 89.	N9PU
☐ N9TW	68	King Air B90	LJ-379	Wfu. Cx USA 10/95.	N73LC
☐ N10TN	70	681 Turbo Commander	6037	Wfu.	N9090N
☐ N11LG	73	Mitsubishi MU-2J	595	Wfu.	N299MA
☐ N11SS	70	Mitsubishi MU-2G	518	W/o Hawesville, Ky. USA. 19 Jan 79.	N144MA
☐ N11WU	72	Mitsubishi MU-2J	565	Wfu.	N11WF
☐ N12AB	65	King Air 90	LJ-45	Wfu.	HB-GBK
☐ N12EW	74	Mitsubishi MU-2K	316	Wfu at White Industries, Bates City, Mo.	N77SS
☐ N12GW	59	Gulfstream 1	19	Wfu.	PK-WWG
☐ N12KV	67	680V Turbo Commander	1675-58	Wfu.	N12KV
☐ N13TV	74	Rockwell 690A	11148	Wfu & Cx USA 7/95.	(N2KC)
☐ N14TK	67	King Air A90	LJ-255	Wfu at White Industries, Bates City, Mo.	N5113
☐ N15SS	78	PA-31T Cheyenne II	7820068	W/o 30 Nov 81.	N6027A
☐ N17CA	67	Rockwell 690A	11148	Wfu at Detroit-Willow Run, Mi. USA.	N2602M
☐ N17CA		Aero Commander 1C	123	Wfu at Detroit-Willow Run, Mi. USA.	N2602M
☐ N18SE	69	SA-26AT Merlin 2B	T26-134	Wfu.	N1UA
☐ N18WP	74	Mitsubishi MU-2J	648	Wfu.	N529MA
☐ N19	80	King Air C90	LJ-909	Wfu & Cx USA 11/96.	
☐ N20	80	King Air C90	LJ-912	Wfu & Cx USA 11/96.	
☐ N20HF	60	Gulfstream 1	47	Wfu.	N20CC
☐ N20PT	69	SA-26AT Merlin 2B	T26-128	W/o Winchester, Va. USA. 18 Mar 94.	C-FBVI
☐ N21	80	King Air C90	LJ-902	Wfu & Cx USA 11/96 to ?	N5
☐ N21MK	67	Mitsubishi MU-2B	032	Wfu. Cx USA 12/94.	N707EB
☐ N21PC	74	Mitsubishi MU-2K	295	Wfu. Cx USA 9/95.	N455MA
☐ N21VM	71	Mitsubishi MU-2G	535	Wfu.	N159RS
☐ N22CN	79	PA-31T Cheyenne 1	7904049	W/o Nr Glendive, Mt. USA. 30 Nov 94.	N2368R
☐ N22EQ	71	Mitsubishi MU-2F	192	Wfu. Cx USA 6/95.	N22LC
☐ N22TE	72	King Air A100	B-115	Wfu at White Industries, Bates City, Mo.	N22T
☐ N23CD	69	Mitsubishi MU-2F	142	W/o El Paso, Tx. USA. 17 Oct 85.	N209MA
☐ N23LS	77	Rockwell 690B	11372	W/o.	N81557
☐ N23UG	63	Gulfstream 1	108	Wfu.	N1707Z
☐ N25ST	71	King Air C90	LJ-507	W/o 21 Aug 89.	N25ST
☐ N26CA	78	PA-31T Cheyenne II	7820062	Wfu.	(N82295)
☐ N26JB	69	SA-26AT Merlin 2B	T26-163	W/o Colorado, USA. 13 Feb 92.	N111SE
☐ N26RT	71	SA-226T Merlin 3	T-216	W/o Helsinki, Finland. 24 Feb 89.	N50PK
☐ N27GP	67	Mitsubishi MU-2B	027	W/o Schellville, Ca. USA. 13 Jul 82.	N57907
☐ N27MT		Rockwell 690B	11533	W/o Springfield, Mo. USA. 8 Oct 94.	N376RF
☐ N28AD	76	Rockwell 690A	11291	W/o Sepahua, Peru. 17 May 93.	
☐ N28CG	60	Gulfstream 1	50	Wfu.	N820CE
☐ N28SE	67	King Air A90	LJ-239	W/o 30 Jun 85.	N28S
☐ N29AA	66	King Air 90	LJ-110	Wfu.	N695V
☐ N30PC	80	King Air 200	BB-702	W/o Pensacola, Fl. USA. 10 Apr 89.	
☐ N30SA	79	King Air A100	B-246	W/o Charlotte, NC. USA. 11 Dec 97.	N20FS
☐ N30SG	68	SA-26T Merlin 2A	T26-28	Wfu at White Industries, Bates City, Mo.	N1212S
☐ N31CL	67	Mitsubishi MU-2B	015	Wfu. Cx USA 4/96.	N707EB
☐ N32HG	76	King Air 200	BB-146	W/o New Castle, De. USA. 16 Jun 92.	N32CL
☐ N32RL	81	King Air B100	BE-117	W/o Gold Beach, Or. USA. 30 Sep 87.	
☐ N33BJ	77	PA-31T Cheyenne II	7720061	Wfu.	
☐ N33SE	70	SA-26AT Merlin 2B	T26-170	Wfu. Cx USA 8/95.	N19SE
☐ N33TF	64	Gulfstream 1	133	Wfu at White Industries, Bates City, Mo.	TU-VAC
☐ N34F	64	King Air A90	LJ-119	W/o 10 Mar 77. Parts at White Industries, Bates City, Mo.	N25CA
☐ N34SM	76	SA-226T Merlin 3A	T-263	W/o San Antonio, Tx. USA. 3 Feb 77.	N5377M
☐ N35GT		PA-31T Cheyenne II	7620030	Wfu at White Industries, Bates City, Mo.	N300CM
☐ N36CA	79	PA-31T Cheyenne II	7920013	W/o 8 Feb 84.	
☐ N38B	72	King Air 200	BB-1	Wfu. Cx USA 3/95.	
☐ N39BC	67	Mitsubishi MU-2B	020	Wfu at White Industries, Bates City, Mo.	N3552X
☐ N39YV	75	King Air/Catpass 250	BB-39	W/o Pasadena, Ca. USA. 12 May 89.	N63JR
☐ N40MP	73	Rockwell 690A	11116	W/o Kingston, Ut. USA. 12 Nov 74.	
☐ N40RM	66	King Air A90	LJ-155	Wfu.	N7HU
☐ N41KA	76	King Air B100	BE-1	Wfu. (was B-205).	
☐ N41VC	67	King Air A90	LJ-242	W/o Alice, Tx. USA. 12 Aug 97.	N812PS
☐ N42CA	64	Gulfstream 1	137	Wfu & dismantled at Detroit Willow Run, Mi. Cx USA 4/95.	(N811CC)
☐ N43DT	65	King Air 90	LJ-6	Wfu at White Industries, Bates City, Mo. Cx USA 6/95.	C-GSFC
☐ N43GT	75	King Air C90	LJ-652	W/o 1 Oct 89.	N4072S
☐ N44MR	73	Mitsubishi MU-2J	611	W/o Port Aransas, Tx. USA. 30 Nov 80. (parts at White Inds.)	N344MA
☐ N44UE	71	King Air A100	B-140	W/o Atlanta, Ga. USA. 18 Jan 90. (parts at White Inds.).	N44UF
☐ N45EV	71	Mitsubishi MU-2F	211	Wfu. Cx USA 6/95.	N123GS
☐ N45Q	80	Gulfstream 840	11623	W/o Al. USA. Dec 90.	N11EX
☐ N45RM	76	King Air E90	LW-174	W/o Nr Yurimaguas, Peru. 1 Jul 92.	N21KE
☐ N46WA	65	King Air 90	LJ-65	W/o Golfe du Lion-Marseilles, France. 13 Jan 94.	N981LE
☐ N47CC	78	PA-31T Cheyenne II	7820016	W/o Richlands, Va. USA. 29 Jul 81.	N82218
☐ N47CP	81	PA-31T Cheyenne II	8120039	W/o. Ca. USA. 1987.	N2589X
☐ N47R	61	Gulfstream 1	69	Wfu.	N47
☐ N47WM	79	King Air E90	LW-307	W/o NW Kingston, ON. Canada. 20 Jan 95.	N300EH
☐ N50ES	66	King Air 90	LJ-111	Wfu. Cx USA 5/96.	N27LR

Regn	Yr	Type	c/n	Accident / Withdrawal Details	Prev Regn

Regn	Yr	Type	c/n	Accident / Withdrawal Details	Prev Regn
☐ N50LT	77	King Air 200	BB-284	Wfu & Cx USA 9/97.	N25MK
☐ N50PC	70	King Air 100	B-19	W/o Birmingham, Al. USA. 1 Dec 74.	
☐ N53AD	80	MU-2 Marquise	776SA	W/o Saratoga, Wy. USA. 5 Nov 81.	N262MA
☐ N55MG	67	King Air A90	LJ-303	W/o 26 Nov 77.	N55MP
☐ N55PC	69	Mitsubishi MU-2F	146	Wfu.	(N78HC)
☐ N55ZM	68	SA-26T Merlin 2A	T26-11	Wfu.	N55JM
☐ N56KA	73	King Air F90	LW-46	Wfu at Dodson's, Ottawa, Ks. Cx USA 1/95.	
☐ N57MM	66	King Air A90	LJ-126	Wfu at White Industries, Bates City, Mo.	N570M
☐ N57V	67	King Air A90	LJ-268	W/o Washington, DC. USA. 25 Jan 75.	N573M
☐ N58FS	69	SA-26AT Merlin 2B	T26-120	Wfu.	N74YC
☐ N59MD	80	Conquest II	441-0177	W/o Derry, Pa. USA. 11 Nov 85.	
☐ N60AW	80	PA-31T Cheyenne II	8020051	W/o Big Bear Lake, Ca. USA. 17 Feb 92.	N2343V
☐ N60BN	69	Mitsubishi MU-2F	163	Wfu.	N6TN
☐ N62BC	76	PA-31T Cheyenne II	7620043	Wfu.	N52BC
☐ N63XL	81	Cheyenne II-XL	8166037	W/o Colorado Springs, Co. USA. 15 Sep 89.	D-IESW
☐ N64MD	79	MU-2 Marquise	747SA	W/o Rapid City, SD. USA. 9 Feb 90.	N777ST
☐ N65TD	70	King Air 100	B-50	W/o Windsor, Ma. USA. 10 Dec 86.	N166TR
☐ N66CA	79	PA-31T Cheyenne 1	7904055	W/o Mexico. 21 Apr 83.	
☐ N66KS	71	SA-226T Merlin 3	T-209	W/o Bahamas. 9 Sep 86.	N400SU
☐ N66LM	83	Conquest 1	425-0137	W/o Fl. USA. 11 Feb 92.	N6884L
☐ N66U	74	Mitsubishi MU-2K	309	W/o Hayden, Co. USA. 12 Sep 82.	N44RF
☐ N68TG	61	Gulfstream 1	68	W/o Tri Cities Airport, Bristol, Tn. USA. 15 Jul 83.	N7ZA
☐ N69QJ	73	Mitsubishi MU-2K	254	W/o Morristown, Tn. USA. 13 Nov 75.	N616AF
☐ N69TM	76	Rockwell 690A	11322	W/o Nr Cashion, OK. USA. 12 Feb 95.	N4601L
☐ N70QR	64	Gulfstream 1	144	Wfu.	N70CR
☐ N71CR	65	Gulfstream 1	163	W/o Addison, Tx. USA. 11 Jul 75.	N618M
☐ N72B	79	MU-2 Marquise	735SA	W/o Jefferson, Ga. USA. 24 Mar 83.	N913MA
☐ N72BS	72	King Air A100	B-113	W/o Milville, NJ. USA. 2 Feb 85.	N9439Q
☐ N74EJ	78	King Air 200	BB-340	W/o Dalton, Ga. USA. 14 Aug 97.	(N74PF)
☐ N74FB	80	MU-2 Marquise	770SA	W/o Indianapolis, In. USA. 11 Sep 92.	N378RM
☐ N74NL	77	PA-31T Cheyenne II	7720010	W/o Lake County, In. USA. 30 Dec 86.	N2YP
☐ N76BT	66	680V Turbo Commander	1567-23	Wfu.	N76D
☐ N76GM	70	King Air B90	LJ-498	W/o Longmont, Co. USA. 23 Jan 97.	N100CQ
☐ N76ST	67	SA-26T Merlin 2A	T26-4	Wfu at Dodson's, Ottawa, Ks.	N700SC
☐ N77PR	77	PA-31T Cheyenne II	7720006	Wfu & Cx USA 1/94.	N77CG
☐ N77TM	68	Mitsubishi MU-2D	116	Wfu. Lake City, Fl. USA.	N862Q
☐ N78D	66	680V Turbo Commander	1580-33	W/o New Orleans, La. USA. 5 Dec 71.	
☐ N78L	73	King Air A100	B-167	W/o Brookville, Fl. USA. 8 Nov 86.	N65HC
☐ N80RD	68	Gulfstream 1	198	W/o Houston, Tx. USA. 23 Aug 90.	N100C
☐ N81MD	74	King Air A100	B-203	W/o Lagos, Nigeria. 11 Aug 78.	N7373R
☐ N81SM	81	PA-42 Cheyenne III	8001007	W/o Horseshoe Bay, Tx. USA. 7 Feb 87. Cx USA 9/95.	N51SM
☐ N81TR	81	Gulfstream 840	11690	W/o Nr Denver, Co. USA. 23 Dec 92.	N5942K
☐ N82	88	King Air 300	FF-17	W/o Nr Winchester Regional Airport, Va. USA. 26 Oct 93.	
☐ N82MA	75	Mitsubishi MU-2L	665	W/o Nashville, Tn. USA. 6 Sep 90.	SX-AGQ
☐ N83MC	73	Rockwell 690A	11124	Wfu. Cx USA 9/91.	(N83RV)
☐ N85NL	66	King Air 90	LJ-90	Wfu at Dodson's, Ottawa, Ks.	N65NL
☐ N86SD	80	MU-2 Marquise	765SA	W/o Apr 93. Cx USA 7/94.	N984MA
☐ N88CA	80	PA-31T Cheyenne II	8020081	Wfu. Cx USA 6/97.	C-GCNO
☐ N88CR	71	King Air C90	LJ-514	W/o 16 Jan 79.	N111HR
☐ N88LB	74	PA-31T Cheyenne II	7400004	Wfu at White Industries, Bates City, Mo.	N61DP
☐ N88WZ	67	680V Turbo Commander	1713-87	Wfu.	N1123V
☐ N89DA	67	680V Turbo Commander	1702-78	W/o Jamaica. 15 Nov 82.	N89D
☐ N89JR	67	King Air A90	LJ-185	Wfu.	N769
☐ N90BB	76	PA-31T Cheyenne II	7620027	Wfu at White Industries, Bates City, Mo.	N82013
☐ N90JR	67	King Air A90	LJ-211	Wfu & Cx USA 5/96.	(D-ILKC)
☐ N90NY	65	King Air 90	LJ-73	Wfu at White Industries, Bates City, Mo.	N90NY
☐ N90RG	72	King Air C90	LJ-546	W/o 6 Aug 92.	N45BE
☐ N90SJ	66	King Air A90	LJ-177	Wfu at White Industries, Bates City, Mo.	N31NC
☐ N91G	60	Gulfstream 1	37	W/o Houston, Tx. USA. 24 Sep 78.	N20S
☐ N91TW	78	PA-31T Cheyenne II	7820078	W/o Delta, Ut. USA. 17 Jan 82.	
☐ N92JR	67	Mitsubishi MU-2B	006	W/o Miami, Fl. USA. 13 May 81. (parts at White Inds.).	C-GMUA
☐ N93NB	82	King Air B200	BB-970	Wfu. Cx USA 1/97.	N124CS
☐ N94HD	68	680V Turbo Commander	1811-28	W/o Lucerne, Ca. USA. 11 Nov 78.	N94HC
☐ N96JP	72	Mitsubishi MU-2J	556	W/o Casper, Wy. USA. 6 Apr 93. Cx USA 7/94.	N33RH
☐ N96JS	76	PA-31T Cheyenne II	7620021	Wfu at White Industries, Bates City, Mo.	N131RG
☐ N96MA	67	Mitsubishi MU-2B	011	Wfu.	N96MA
☐ N98AL	68	Mitsubishi MU-2D	105	Wfu & Cx USA 2/96.	N2RA
☐ N98MK	62	Gulfstream 1	98	Wfu at White Industries, Bates City, Mo.	N29AY
☐ N100BE	75	King Air A100	B-221	W/o Colorado Springs, Co. USA. 21 Dec 97.	HZ-AFE
☐ N100CT	66	680V Turbo Commander	1618-50	W/o Bridgeport, Ct. USA. 3 Sep 84.	N4693E
☐ N100HC	76	King Air 200	BB-98	W/o Greenville, Tx. USA. 12 Aug 85.	
☐ N100LB	77	PA-31T Cheyenne II	7720021	Wfu at White Industries, Bates City, Mo.	N200CN
☐ N100NL	69	SA-26AT Merlin 2B	T26-168	W/o Hot Springs, Va. USA. 16 Oct 71.	N20DE
☐ N100NX	67	SA-26T Merlin 2A	T26-3	Wfu at White Industries, Bates City, Mo.	N100MX
☐ N100RN	78	PA-31T Cheyenne II	7820090	W/o Wi. USA. 23 Feb 85.	N6121A
☐ N100SW	71	Mitsubishi MU-2G	539	W/o between Miami & Atlanta, USA. 1 Apr 77. Cx USA 6/95.	N163MA

Regn   *Yr*   *Type*   *c/n*   *Accident / Withdrawal Details*   *Prev Regn*

Regn	Yr	Type	c/n	Accident / Withdrawal Details	Prev Regn
☐ N100TN	75	PA-31T Cheyenne II	7520013	Wfu.	
☐ N101FB	74	PA-31T Cheyenne II	7400007	Wfu at White Industries, Bates City, Mo.	(N66836)
☐ N101GA	65	King Air 90	LJ-11	W/o Royal, Ar. USA. 3 Jan 95.	N2JJ
☐ N101LR	78	King Air C90	LJ-802	W/o 15 Sep 81.	
☐ N101XC	67	King Air A90	LJ-219	Wfu at White Industries, Bates City, Mo.	N52C
☐ N102RB	72	King Air E90	LW-19	W/o Nr Otuzco, Juica, Peru. 18 Mar 93.	N1716W
☐ N105FL	89	King Air C90A	LJ-1215	W/o St Augustine, Fl. USA. 9 Apr 92.	
☐ N105MA	68	Mitsubishi MU-2F	123	Wfu. Cx USA 6/95.	C-FHTL
☐ N106DP	73	Mitsubishi MU-2J	571	Wfu.	N5HE
☐ N106EC	68	SA-26T Merlin 2A	T26-34	Wfu at White Industries, Bates City, Mo.	N96D
☐ N106MA	70	Mitsubishi MU-2F	184	W/o Lake Texoma, Tx. USA. 7 May 91.	
☐ N107DC	68	680W Turbo Commander	1777-15	Wfu.	N24099
☐ N107MA	70	Mitsubishi MU-2F	185	Wfu. Cx USA 10/95.	
☐ N109TW	71	Mitsubishi MU-2G	543	W/o Hawaii, USA. 22 Nov 81.	YV-1049P
☐ N110LT	77	King Air C90	LJ-729	W/o.	
☐ N111LA	76	Rockwell 690A	11324	Wfu. Cx USA 7/92.	N81441
☐ N111PT	68	SA-26T Merlin 2A	T26-15	Wfu.	N341T
☐ N111QL	76	Rockwell 690A	11312	W/o Nacogdoches, Tx. USA. 17 Sep 83.	N1QL
☐ N111WE	66	680W Turbo Commande	1593-41	Wfu.	N688NA
☐ N112SK	74	Mitsubishi MU-2J	651	W/o 1 Dec 84.	N55KV
☐ N113TC	65	King Air 90	LJ-22	W/o 16 Jul 74.	N9502Q
☐ N114CM	77	King Air C90	LJ-709	W/o Jackson, Wy. USA. 16 Jun 86.	N33TW
☐ N114K	75	King Air E90	LW-122	W/o Mineral wells, Tx. USA. 26 Oct 81.	VH-DDG
☐ N117EA	80	Conquest II	441-0191	W/o ground fire, Pa. USA. 30 Nov 86.	C-GGMK
☐ N118LT	60	Gulfstream 1	55	Wfu.	N9446E
☐ N121MA	71	Mitsubishi MU-2F	206	Wfu.	
☐ N122G	67	Mitsubishi MU-2B	033	Wfu.	N3563X
☐ N123AX	72	Mitsubishi MU-2F	220	Wfu at White Industries, Bates City, Mo.	N123UA
☐ N123VC	71	Mitsubishi MU-2F	214	Wfu. Located Technical School Vasteras-Hasslo, Sweden.	N129MA
☐ N125AB	81	MU-2 Marquise	1531SA	Wfu.	N450MA
☐ N129D	72	King Air A100	B-134	W/o Dieques, PR. USA. 17 Aug 83.	N84B
☐ N129GP	67	King Air A90	LJ-216	W/o 16 Apr 67.	
☐ N132MA	70	Mitsubishi MU-2G	503	W/o Atlantic City, NJ. USA. 16 Apr 72.	JA8739
☐ N133MA	70	Mitsubishi MU-2G	506	W/o Rollinsville, Co. USA. 26 Dec 75.	N133MA
☐ N142LM	65	King Air 90	LJ-28	Wfu.	N142L
☐ N148CP	76	King Air 200	BB-129	W/o Hampton, NY. USA. 9 Jun 85.	N143CP
☐ N149JA	79	MU-2 Solitaire	402SA	W/o 1 Dec 82. (parts at White Industries).	N963MA
☐ N154MF	67	Mitsubishi MU-2B	024	Wfu.	N111FN
☐ N161RS	80	SA-226T Merlin 3B	T-341	Wfu at White Industries, Bates City, Mo.	N61RS
☐ N163D	67	680V Turbo Commander	1701-77	Wfu.	N163D
☐ N165MA	71	Mitsubishi MU-2G	541	W/o Lookout Mountain, Ga. USA. 20 Apr 82.	N165MA
☐ N170S	83	King Air B200	BB-1095	Wfu. Cx USA 1/97.	(N170L)
☐ N176CC	76	PA-31T Cheyenne II	7620024	W/o Lamar, Co. USA. 2 Jun 78.	N82009
☐ N177MF	70	SA-26AT Merlin 2B	T26-179	W/o Albany, Ky. USA. 3 Dec 80.	ZS-RTZ
☐ N178MA	72	Mitsubishi MU-2J	554	W/o Raton, NM. USA. 25 Aug 78.	N178MA
☐ N180B	73	King Air C90	LJ-569	Wfu at White Industries, Bates City, Mo.	
☐ N181TG	67	Gulfstream 1	181	W/o Nashville, Tn. USA. 1 Jun 85.	N25WL
☐ N182	72	Rockwell 690	11048	W/o Canton, Md. USA. 15 Jan 80.	N1NR
☐ N184MA	72	Mitsubishi MU-2F	218	W/o Fort Lauderdale, Fl. USA. 20 Jun 87.	
☐ N191DM	66	King Air 90	LJ-100	Wfu at White Industries, Bates City, Mo. Cx USA 3/93.	N1NL
☐ N191SA	60	Gulfstream 1	61	Wfu at Dodson's, Ottawa, Ks.	N734HR
☐ N193GA	74	King Air 200	BB-12	Wfu. Lake City, Fl. USA.	VH-NIH
☐ N195B	76	King Air E90	LW-195	Wfu at Dodson's, Ottawa, Ks.	N195WF
☐ N199TA	68	SA-26AT Merlin 2B	T26-110	W/o Del Rio, Tx. USA. 19 Jun 85.	N20TA
☐ N200BR	71	Mitsubishi MU-2F	205	W/o Provo, Ut. USA. 21 Dec 79.	N120MA
☐ N200CM	77	PA-31T Cheyenne II	7720004	Wfu at White Industries, Bates City, Mo.	(N82073)
☐ N200FD	75	PA-31T Cheyenne II	7520040	W/o Lawrence, Ma. USA. 20 Mar 87.	N54971
☐ N200HL	68	Mitsubishi MU-2D	120	W/o New Orleans, La. USA. 20 Sep 74.	N284MA
☐ N200MB	79	King Air B100	BE-71	Wfu at White Industries, Bates City, Mo.	N66480
☐ N200MR	77	King Air 200	BB-219	Wfu at White Industries, Bates City, Mo.	N884CA
☐ N200PR	83	Gulfstream 900	15029	W/o Price, Ut. USA. 7 May 86.	N36GA
☐ N200RS	85	PA-31T Cheyenne II	7520011	W/o St Louis, Mo. USA. 18 Jan 88.	N66844
☐ N202JP	75	PA-31T Cheyenne II	7520018	Wfu at White Industries, Bates City, Mo.	N300WT
☐ N208MA	67	Mitsubishi MU-2B	016	W/o Hays, Ks. USA. 3 Aug 79.	(N232LJ)
☐ N212CM	66	680V Turbo Commander	1626-54	Wfu.	N212CW
☐ N212MA	73	Mitsubishi MU-2K	262	Wfu. Cx USA 5/95.	N282MA
☐ N213GA	60	Gulfstream 1	48	Wfu.	G-AWYF
☐ N220F	81	King Air C90	LJ-981	W/o East Greenwich, RI. USA. 27 Nov 85.	
☐ N220MA	81	MU-2 Solitaire	441SA	W/o Concord, NH. USA. 9 Jul 92.	N234BC
☐ N221AP	59	Gulfstream 1	6	Wfu. Cx USA 8/97.	S9-NAU
☐ N221MJ	71	King Air C90	LJ-512	W/o 4 Nov 75.	N659H
☐ N225MS	79	King Air 200	BB-496	Wfu as stolen. Cx USA 6/95.	N180S
☐ N226AT		SA-226AT Merlin 4	AT-003E	Wfu.	N226TC
☐ N230E	60	Gulfstream 1	36	Wfu.	N130A
☐ N230TW	69	King Air B90	LJ-445	W/o Nr Fort Drum, Fl. 5 Jan 94.	N70CS
☐ N231RL	81	King Air 200	BB-868	Wfu. Cx USA 3/97.	N3872K

Regn	Yr Type	c/n	Accident / Withdrawal Details	Prev Regn

Regn	Yr	Type	c/n	Accident / Withdrawal Details	Prev Regn
□ N233MA	73	Mitsubishi MU-2K	251	W/o McCleod, Tx. USA. 2 Sep 81.	
□ N234K	76	PA-31T Cheyenne II	7620017	Wfu.	N57526
□ N234MA	73	Mitsubishi MU-2K	252	W/o Jacksboro, Tx. USA. 26 Nov 79.	
□ N234MM	63	Gulfstream 1	121	Wfu. Displayed at Disney-MGM Studio Theme Park, Orlando.	N732G
□ N239P	69	SA-26AT Merlin 2B	T26-147	W/o Willoughby, Oh. USA. 29 Jan 70.	
□ N241DT	73	SA-226T Merlin 3	T-242	W/o Santa Fe, NM. USA. 25 May 93.	N770U
□ N241FW	82	Conquest II	441-0241	W/o Chicago, Il. USA. 23 Nov 86.	(N6856S)
□ N242DA	70	King Air B90	LJ-484	Wfu at Dodson's, Ottawa, Ks.	ZS-KLW
□ N245CA	61	Gulfstream 1C	83	Wfu at Detroit-Willow Run, Mi. USA.	N117GA
□ N245MA	73	Mitsubishi MU-2J	576	Wfu.	
□ N249DA	65	King Air 90	LJ-20	Wfu.	C-FMLC
□ N251M	67	Mitsubishi MU-2B	013	W/o Gardner, Ks. USA. 9 Apr 79.	N3545X
□ N254TP	73	Mitsubishi MU-2J	575	Wfu.	N7PW
□ N256TM	76	King Air 200	BB-96	W/o New Orleans, La. USA. 18 Apr 77.	
□ N261PL	69	SA-26AT Merlin 2B	T26-121	Wfu.	N4252X
□ N271MA	80	MU-2 Marquise	797SA	W/o Chicago, Il. USA. 16 Nov 88.	
□ N274MA	80	MU-2 Marquise	786SA	W/o Tulsa, Ok. USA. 22 Feb 91.	
□ N275MA	73	Mitsubishi MU-2K	255	W/o West Point, Va. USA. 5 Jan 85. (parts at White Inds.).	N275MA
□ N278DU	67	King Air A90	LJ-243	W/o 10 Jul 78.	N578DU
□ N287MN	75	PA-31T Cheyenne II	7520004	Wfu at White Industries, Bates City, Mo.	
□ N289MA	73	Mitsubishi MU-2J	585	Wfu at White Industries, Bates City, Mo.	
□ N295X	67	King Air A90	LJ-244	W/o 1 May 72.	
□ N296MA	73	Mitsubishi MU-2J	592	W/o Australia. 9 Dec 88.	
□ N299F	73	Rockwell 690A	11112	W/o Calumet, Ok. USA. 27 May 78.	N57112
□ N300BD	71	Mitsubishi MU-2G	547	Wfu.	N171MA
□ N300CP	77	Rockwell 690B	11374	W/o South of Susanville, Ca. USA. 31 Dec 92.	N81562
□ N300CW	80	MU-2 Marquise	795SA	W/o Putnam, Tx. USA. 14 Feb 90.	N287MA
□ N300HG	66	King Air 90	LJ-94	Wfu at White Industries, Bates City, Mo.	N300HC
□ N300KQ	67	680V Turbo Commander	1709-84	Wfu.	N300KC
□ N300MA	73	Mitsubishi MU-2J	596	W/o Easton, Md. USA. 8 Feb 76.	
□ N300MC	60	Gulfstream 1	32	Wfu.	N297X
□ N302X	67	Mitsubishi MU-2B	030	Wfu.	N3561X
□ N303CA	81	MU-2 Marquise	1518SA	W/o Rifle, Co. USA. 5 Mar 92.	N678KM
□ N303MC	81	Conquest 1	425-0034	W/o London, Ky. USA. 18 Jan 94.	N6773E
□ N304L	68	Mitsubishi MU-2F	137	W/o March Harbour, Bahamas. 21 Apr 79.	N304LA
□ N304M	68	SA-26T Merlin 2A	T26-8	Wfu.	
□ N307MA	67	Mitsubishi MU-2B	007	W/o Las Vegas, Nv. USA. 28 Apr 80.	(D-IHFS)
□ N308PS	74	King Air E90	LW-92	W/o Locust Grove, Ar. USA. 18 Nov 88.	N300PS
□ N309MA	73	Mitsubishi MU-2J	602	W/o Smyrna, Tn. USA. 21 Sep 95.	
□ N313BB	70	Mitsubishi MU-2G	529	Wfu.	N3MP
□ N313PC	71	Mitsubishi MU-2G	531	Wfu.	N155MA
□ N319JG	76	PA-31T Cheyenne II	7620056	Wfu.	N110DE
□ N320MA	73	Mitsubishi MU-2K	274	Wfu.	(N10GE)
□ N321MA	73	Mitsubishi MU-2K	276	W/o Double Springs, Al. USA. 4 Apr 77.	
□ N321ST	74	Mitsubishi MU-2K	307	W/o Campbell Lake, Canada. 8 Sep 91.	N200TM
□ N332K	66	King Air 90	LJ-79	W/o 3 Sep 79.	N33JC
□ N333BR	68	Mitsubishi MU-2D	119	Wfu.	(N101ES)
□ N333CA	73	Rockwell 690A	11117	W/o Oklahoma City, Ok. USA. 22 Aug 73.	
□ N333CS	65	King Air 90	LJ-44	Wfu at White Industries, Bates City, Mo.	XB-YAZ
□ N333LM	79	PA-31T Cheyenne II	7920052	W/o Malvern, Ar. USA. 29 May 96.	N333MZ
□ N333MA	74	Mitsubishi MU-2K	288	W/o Gander, Canada. 24 Mar 74.	
□ N334DP	84	King Air B200	BB-1188	W/o Fort Atkinson, Wi. USA. 16 Nov 87.	N334D
□ N336SA		SA-226T Merlin 3B	T-336	W/o Tx. USA. 19 Jan 82.	
□ N339MA	73	Mitsubishi MU-2J	607	Wfu.	(N200MK)
□ N340X	68	SA-26T Merlin 2A	T26-17	Wfu at White Industries, Bates City, Mo.	
□ N345CA	69	680W Turbo Commander	1842-41	Wfu at White Industries, Bates City, Mo.	N39480
□ N346K	74	Mitsubishi MU-2K	292	Wfu.	N452MA
□ N346MA	73	Mitsubishi MU-2J	613	W/o Houston, Tx. USA. 14 Feb 80.	
□ N363N	67	King Air A90	LJ-236	Wfu at White Industries, Bates City, Mo.	
□ N371BG	59	Gulfstream 1	4	Wfu.	N89DE
□ N375A	67	680V Turbo Commander	1692-71	Wfu.	
□ N385GA	74	King Air 200	BB-15	Wfu. Cx USA 1/97.	ZS-MNF
□ N388MG	69	King Air B90	LJ-442	W/o 10 Jan 78.	N45MC
□ N399T	66	680T Turbo Commander	1532-2	W/o Olathe, Ks. USA. 31 Jan 75.	HB-GEK
□ N400AM	68	King Air B90	LJ-354	W/o 10 Sep 83.	(N506M)
□ N400BG	78	Conquest II	441-0069	W/o Dallas, Tx. USA. 1 Oct 85.	N441SE
□ N400FA		SA-227AT Merlin 4C	AT-557	Wfu.	
□ N400KK	68	Mitsubishi MU-2D	118	Wfu.	N100KK
□ N400N	74	Rockwell 690A	11156	W/o Kelso, Wa. USA. 1 Dec 90.	N57077
□ N400NL	60	Gulfstream 1	62	Wfu.	N205M
□ N409ET	74	Rockwell 690A	11213	Wfu.	N925MC
□ N410W	65	King Air 90	LJ-39	Wfu at White Industries, Bates City, Mo.	N410WA
□ N411X	69	SA-26AT Merlin 2B	T26-126	W/o Nashville, Tn. USA. 28 Mar 72.	
□ N418CD	76	Mitsubishi MU-2M	344	Wfu at White Industries, Bates City, Mo.	N729MA
□ N425AC	80	Conquest 1	425-0009	W/o Natchez, Ms. USA. 18 Nov 81.	(N98896)
□ N425BN	81	Conquest 1	425-0057	W/o Las Vegas, Nv. USA. 11 Jan 92.	N67761

| *Regn* | *Yr* | *Type* | *c/n* | *Accident / Withdrawal Details* | *Prev Regn* |

Regn	Yr	Type	c/n	Accident / Withdrawal Details	Prev Regn
□ N425EW	83	Conquest 1	425-0150	W/o Nr MacArthur Airport, NY. 16 Dec 96.	(N68854)
□ N425SC	82	Conquest 1	425-0126	W/o Granby, Co. USA. 11 Jan 86.	N607DD
□ N430C	67	King Air A90	LJ-273	Wfu at Dodson's, Ottawa, Ks.	N585S
□ N440MA	81	MU-2 Marquise	1524SA	W/o Scottsdale, Az. USA. 27 Jan 83.	
□ N441CC	75	Conquest II	441-679	Wfu.	N7185C
□ N441CC	77	Conquest II	679	Wfu.	
□ N441CD	80	Conquest II	441-0131	W/o 1985.	(Ñ2625Y)
□ N441CM	80	Conquest II	441-0169	W/o Marble Falls, Tx. USA. 24 Dec 84.	N441CN
□ N441KM	81	Conquest II	441-0196	W/o St Louis, Mo. USA. 22 Nov 94. (parts at White Inds.).	N992TE
□ N441NC	79	Conquest II	441-0099	W/o Virginia, USA. 11 Jan 80.	N4106G
□ N441W	78	Conquest II	441-0052	W/o Walkers Cay, Fl. USA. 20 Apr 96.	N441GM
□ N442TC	68	King Air B90	LJ-332	Wfu at White Industries, Bates City, Mo.	N440TC
□ N444AR	72	Mitsubishi MU-2J	555	W/o Eagle, Co. USA. 18 Nov 81.	N444NR
□ N444GB	66	680V Turbo Commander	1565-21	W/o Keflavik, Iceland. 5 Aug 90.	N444GB
□ N444JW	77	PA-31T Cheyenne II	7720015	W/o San Angelo, Tx. USA. 3 Dec 79. (parts at Dodson's).	N775SW
□ N444LM	78	SA-226T Merlin 3B	T-295	W/o Livermore, Ca. USA. 3 May 85.	N4444F
□ N444PA	77	Mitsubishi MU-2N	691SA	W/o Patterson, La. USA. 20 Oct 83.	N851MA
□ N444SR	68	King Air B90	LJ-416	W/o 17 Feb 85.	N869K
□ N447AB	72	Mitsubishi MU-2F	223	Wfu at Dodson's, Ottawa, Ks. Cx USA 7/94.	N189MA
□ N456L	76	King Air 200	BB-112	W/o Denver, Co. USA. 27 Mar 80.	
□ N466MA	81	MU-2 Marquise	1540SA	W/o Burlington, Ct. USA. 19 Apr 84.	
□ N468MA	74	Mitsubishi MU-2J	618	Wfu at Dodson's, Ottawa, Ks. Cx USA 7/94.	9Q-CAA
□ N469JK	67	King Air A90	LJ-274	Wfu. Lake City, Fl. USA.	N30AA
□ N473FW	73	Mitsubishi MU-2K	269	Wfu at White Industries, Bates City, Mo.	(N473W)
□ N473MA	82	MU-2 Marquise	1547SA	W/o North Adams, Ma. USA. 18 Mar 83.	
□ N474H	69	King Air B90	LJ-474	Wfu at White Industries, Bates City, Mo.	N14RA
□ N474U	69	SA-26AT Merlin 2B	T26-133	Wfu. Lake City, Fl. USA.	N401NW
□ N480K	69	King Air B90	LJ-439	W/o 27 Dec 71.	N480K
□ N483D	67	King Air A90	LJ-212	Wfu. Cx USA 9/92.	N483G
□ N500AK	83	SA-227TT Merlin 300	TT-527	W/o Sullivan County, Tn. USA 1 Apr 93.	N40EF
□ N500GL	73	Mitsubishi MU-2J	579	W/o 19 Apr 81.	N441FS
□ N500JP	70	681 Turbo Commander	6003	W/o Winnemucca, Nv. USA. 27 Jan 81.	C-FFDB
□ N500ML	80	King Air B100	BE-78	W/o Jackson, Ms. USA. 13 Nov 97.	N3UL
□ N500S	78	Mitsubishi MU-2P	379SA	W/o 2 Jan 89.	N769MA
□ N500X	67	King Air A90	LJ-199	W/o 26 Nov 69.	
□ N501M	66	King Air A90	LJ-175	Wfu at White Industries, Bates City, Mo.	N50RM
□ N504W	68	SA-26AT Merlin 2B	T26-104	Wfu.	
□ N505SA	67	Mitsubishi MU-2B	008	Wfu.	N224FW
□ N508GW	68	SA-26T Merlin 2A	T26-35	Wfu.	N500DM
□ N512DM	69	Mitsubishi MU-2F	155	Wfu. Cx USA 6/95.	N513DM
□ N513NA	70	Mitsubishi MU-2G	513	Wfu. Cx USA 6/95.	VH-UZD
□ N515WB	77	PA-31T Cheyenne II	7720023	W/o Nr Columbus, Oh. USA. 2 Dec 93.	C-GGMD
□ N525MA	76	Mitsubishi MU-2M	340	Wfu.	(N912DM)
□ N529N	66	King Air 90	LJ-112	W/o 11 Mar 66.	
□ N529V	70	King Air B90	LJ-487	Wfu at White Industries, Bates City, Mo.	N529M
□ N530N	66	King Air A90	LJ-141	W/o USA. 1 May 93.	
□ N531MA	68	Mitsubishi MU-2F	130	W/o Burlington, Vt. USA. 18 Feb 76.	N1173Z
□ N539SA		SA-227AT Merlin 4C	AT-539	Wfu. Cx USA 5/86.	
□ N541F	66	680V Turbo Commander	1609-45	W/o Fort Lauderdale, Fl. USA. 3 Nov 90.	C-GPRO
□ N541W	66	680T Turbo Commander	1554-13	W/o Pompano Beach, Fl. USA. 3 Oct 70.	N5419
□ N542TW	81	PA-42 Cheyenne III	8001052	W/o Charlotte, NC. USA. 28 Jun 85.	
□ N549LK	67	Mitsubishi MU-2B	022	W/o Northfield, Oh. USA. 13 Oct 70.	N3554X
□ N550MA	69	Mitsubishi MU-2F	169	Wfu. Cx USA 6/95.	N883Q
□ N551TR	82	King Air B200	BB-1033	W/o 29 Jun 86.	N551TP
□ N555AM	70	SA-226T Merlin 3	T-201	W/o Cameron, La. USA. 10 Jun 81.	N555DB
□ N555CH	70	Mitsubishi MU-2G	508	Wfu.	N950MA
□ N558M	74	Mitsubishi MU-2J	639	Wfu.	N550M
□ N568H	72	Rockwell 690	11027	W/o Los Angeles, Ca. USA. 26 Oct 76.	
□ N568UP	83	SA-227AT Merlin 4C	AT-568	W/o London, Ky. USA. 31 Jan 85.	N568SA
□ N570AB	75	PA-31T Cheyenne II	7520038	W/o ground accident. May 87.	N54970
□ N571L	66	King Air A90	LJ-135	Wfu at White Industries, Bates City, Mo.	N571M
□ N575HC	73	King Air E90	LW-67	W/o Pine Bluff, Ar. USA. 19 May 85.	N575HW
□ N577KA		SA-227AT Merlin 4	AT-008	W/o Billings, Mt. USA. 7 Aug 86.	N577KA
□ N600CM	77	PA-31T Cheyenne II	7720024	W/o Flat Rock, NC. USA. 23 Aug 86.	(N82116)
□ N600TB	73	Mitsubishi MU-2K	258	Wfu. Cx USA 6/95.	N278MA
□ N600WR	76	PA-31T Cheyenne II	7620038	W/o at Dodson's, Ottawa, Ks.	N82028
□ N601G	66	680T Turbo Commander	1605-44	W/o Alphaha, Ga. USA. 8 Aug 76.	N6537V
□ N602PA	66	680T Turbo Commander	1534-3	Wfu.	N441LM
□ N607DD	75	PA-31T Cheyenne II	7520028	Wfu at White Industries, Bates City, Mo.	VH-HMA
□ N611VP	66	King Air A90	LJ-171	W/o Narssarsuaq, Greenland. May 88.	N755K
□ N617MS	81	King Air 200C	BL-35	W/o Madisonville, Ky. USA. 24 Jun 87.	N26732
□ N618B	75	Rockwell 690A	11232	Wfu at White Industries, Bates City, Mo. Cx USA 1/94.	N618
□ N618BB	71	Mitsubishi MU-2G	533	Wfu. CX USA 1/97.	C-FKCL
□ N626BL		Lear Fan	E001	Wfu.	
□ N631PT	77	PA-31T Cheyenne II	7720001	W/o Harrisburg, Pa. USA. 24 Feb 77.	(N82065)
□ N631SR	77	King Air 200	BB-244	W/o King Cove, Ak. USA. 15 Jul 81.	
Regn	Yr	Type	c/n	Accident / Withdrawal Details	Prev Regn

Regn	Yr	Type	c/n	Accident / Withdrawal Details	Prev Regn
N636SW	74	Mitsubishi MU-2J	636	Wfu at White Industries, Bates City, Mo.	N20PS
N638MA	74	Mitsubishi MU-2J	638	Wfu.	N8LC
N642WM	69	SA-26AT Merlin 2B	T26-125	Wfu. Lake City, Fl. USA.	N393W
N656A	67	King Air A90	LJ-304	Wfu at White Industries, Bates City, Mo.	
N660RB	76	Rockwell 690A	11305	W/o Little Rock, Ar. USA. 17 May 88.	C-GIAB
N662DM	72	Rockwell 690	11015	W/o Bridgeport, Ct. USA. 21 Jun 87.	N412FS
N666HB	74	Mitsubishi MU-2K	281	Wfu.	N111PM
N666SE	69	SA-26AT Merlin 2B	T26-154E	Wfu at White Industries, Bates City, Mo.	N90874
N680X	68	680W Turbo Commander	1778-16	Wfu.	C-GPNV
N686N		680T Turbo Commander	1473-1	Wfu.	(N48AD)
N688DC	66	680T Turbo Commander	1612-47	Wfu.	
N690X	69	SA-26AT Merlin 2B	T26-141	Wfu. Cx USA 6/91.	C-GHWM
N693PA	77	Mitsubishi MU-2N	693SA	W/o Malad City, Id. USA. 15 Jan 96.	F-GHDS
N693PG	71	SA-226T Merlin 3	T-207	W/o Chico, Ca. USA. 18 Sep 95.	N500QP
N696JB	70	King Air 100	B-13	W/o Garner, Tx. USA. 28 Mar 90.	N100BW
N700CM	78	PA-31T Cheyenne II	7820007	W/o Jacksonville, Fl. USA. 9 Jan 86.	
N700FN	68	Mitsubishi MU-2F	127	Wfu.	N700DM
N700JW	60	Gulfstream 1	53	Wfu at White Industries, Bates City, Mo. Cx USA 6/94.	(N701JW)
N700R	78	Rockwell 690B	11434	W/o Dec 85.	N113SA
N700SP	72	King Air A100	B-92	W/o Hilton Head Island, SC. USA. 26 Apr 75.	N6739
N701DM	69	Mitsubishi MU-2F	149	W/o California, USA. 28 Feb 89.	N8527Z
N702FN	73	Mitsubishi MU-2K	249	Wfu.	N702DM
N705G	65	Gulfstream 1	152	Wfu.	OB-M-1235
N707BP	76	Rockwell 690A	11326	Wfu & Cx USA 3/94.	N10VG
N707CE	67	King Air A90	LJ-314	W/o Algeria. 29 May 90.	N971EL
N707WC	67	King Air A90	LJ-188	W/o 18 Oct 68.	N703K
N708G	59	Gulfstream 1	8	Wfu.	
N709K	66	King Air 90	LJ-97	Wfu at White Industries, Bates City, Mo.	(N504M)
N710CA	75	Mitsubishi MU-2L	669	Wfu.	TF-JMC
N711AH	70	Mitsubishi MU-2G	523	W/o Glenwood Springs, Co. USA. 2 Mar 74.	N711
N711AH	68	SA-26T Merlin 2A	T26-14	Wfu.	N952HE
N711FC	71	King Air C90	LJ-516	W/o Columbia, SC. USA. 20 Dec 73.	N555RH
N711LL	69	Mitsubishi MU-2F	180	Wfu. Cx USA 6/95.	N180SB
N711TB	78	PA-31T Cheyenne II	7820031	Wfu at White Industries, Bates City, Mo.	N82238
N711TT	77	Rockwell 690B	11362	W/o Albuquerque, NM. USA. 8 Oct 87.	N81537
N713GB	71	Mitsubishi MU-2G	538	Wfu.	N77RZ
N713SP	68	680W Turbo Commander	1805-27	W/o Alexandria, La. USA. 4 Oct 79.	N5082E
N715G	63	Gulfstream 1	118	Wfu.	
N715RA	60	Gulfstream 1	31	Wfu.	C-FJFC
N716RA	60	Gulfstream 1	43	Wfu.	N39289
N718RA	66	Gulfstream 1	174	Wfu.	N7004B
N720X	61	Gulfstream 1	73	W/o Arizona Desert, USA. Feb 87.	N207M
N721DM	74	Mitsubishi MU-2K	312	Wfu at White Industries, Bates City, Mo.	N444RG
N721K	65	King Air 90	LJ-43	Wfu. Cx USA 2/96. (parts at Dodson's Ottawa, Ks.).	
N721RA	61	Gulfstream 1	65	Wfu.	N641B
N722DM	72	Mitsubishi MU-2K	245	Wfu.	N180SB
N722RA	66	Gulfstream 1	168	Wfu.	N209T
N723RA	59	Gulfstream 1	14	Wfu.	N1607Z
N724FN		Mitsubishi MU-2K	300	Wfu & Cx USA 11/96.	N724DM
N724N	66	King Air 90	LJ-82	W/o 22 Dec 79.	N724N
N724RA	63	Gulfstream 1	114	Wfu.	VH-WPA
N727DM	72	Mitsubishi MU-2K	242	Wfu.	N211BA
N727MA	76	Mitsubishi MU-2M	342	Wfu.	
N729FN	74	Mitsubishi MU-2K	296	Wfu.	N729DM
N730SF	73	Mitsubishi MU-2K	246	Wfu.	N228MA
N737EF	68	SA-26T Merlin 2A	T26-22	Wfu at White Industries, Bates City, Mo.	N999DT
N741P	75	PA-31T Cheyenne II	7520020	W/o Jamaica. Oct 81.	N55SM
N741P	77	PA-31T Cheyenne II	7720025	Wfu at White Industries, Bates City, Mo.	N33HW
N743G	61	Gulfstream 1	72	Wfu.	C-FNOC
N743K	67	King Air A90	LJ-187	Wfu at White Industries, Bates City, Mo.	N610K
N746G	60	Gulfstream 1	46	Wfu.	
N747K	65	King Air 90	LJ-51	Wfu at White Industries, Bates City, Mo. Cx USA 2/92.	
N747KF	75	King Air 200	BB-70	Wfu & Cx USA 3/96.	N87JR
N750BR	63	Gulfstream 1	99	W/o nr Frankfurt, Germany. 13 Nov 88.	N364L
N750MA	77	MU-2 Solitaire	365SA	W/o Fernandina Beach, Fl. USA. 19 Nov 81.	
N750QQ	68	Mitsubishi MU-2F	125	Wfu.	N71674
N757Q	69	Mitsubishi MU-2F	151	W/o Manning, SC. USA. 20 Nov 72.	N757Q
N758Q	68	Mitsubishi MU-2F	134	W/o Trenton, NJ. USA. 15 Apr 69.	
N764Q	69	Mitsubishi MU-2F	141	W/o Salisbury, Md. USA. 16 Jan 70.	N753Q
N765MA	72	Mitsubishi MU-2P	372SA	W/o Bedford, NH. USA. 28 Aug 78.	
N767K	66	King Air A90	LJ-170	Wfu.	
N772CB	71	681B Turbo Commander	6050	W/o Calhoun, Co. USA. 28 Mar 85.	N56MQ
N772K	67	King Air A90	LJ-310	Wfu at Dodson's, Ottawa, Ks.	
N776K	65	King Air 90	LJ-56	Wfu at White Industries, Bates City, Mo.	
N777EC	73	King Air E90	LW-37	W/o NY State, USA. 7 Jan 79.	N1837W
N777JM	78	PA-31T Cheyenne II	7820064	Wfu. Cx USA 5/94.	(N82301)

Regn	Yr	Type	c/n	Accident / Withdrawal Details	Prev Regn
☐ N777KV	77	PA-31T Cheyenne II	7720034	Wfu.	(N774KV)
☐ N777MA	72	Mitsubishi MU-2J	559	W/o Austin, Tx. USA. 18 Mar 77.	
☐ N778HD	79	PA-31T Cheyenne II	7920006	Wfu at White Industries, Bates City, Mo.	
☐ N791K	67	King Air A90	LJ-253	W/o 13 Mar 73.	
☐ N799V	77	Rockwell 690B	11407	W/o Wichita, Ks. USA. 2 Nov 91.	N81643
☐ N800AW		SA-226T Merlin 3B	T-403	W/o Pontiac, Mi. USA. 10 Jan 88.	N4464V
☐ N800BR	71	Mitsubishi MU-2F	199	Wfu at Dodson's, Ottawa, Ks.	N3RN
☐ N800DH	69	Mitsubishi MU-2F	166	Wfu. Cx USA 4/94.	N322GA
☐ N800VT	67	King Air A90	LJ-249	Wfu at White Industries, Bates City, Mo.	N800S
☐ N800W	80	PA-31T Cheyenne 1	8004032	Wfu at White Industries, Bates City, Mo.	C-FDDD
☐ N801C	76	PA-31T Cheyenne II	7620048	Wfu at White Industries, Bates City, Mo.	(N414VM)
☐ N801HD	76	PA-31T Cheyenne II	7620031	W/o Beckley, WV. USA. 24 Nov 77.	N35RR
☐ N802AC	68	SA-26T Merlin 2A	T26-21	Wfu.	N137RD
☐ N803CC	63	Gulfstream 1	112	Wfu.	N300PE
☐ N808W	73	Mitsubishi MU-2J	609	W/o 14 Apr 85.	N508W
☐ N814SW	67	King Air A90	LJ-186	W/o Orange County Airport, Ca. USA. 16 Oct 96.	N881M
☐ N816C	65	King Air 90	LJ-42	Wfu at White Industries, Bates City, Mo.	N81CC
☐ N817SW	81	PA-31T Cheyenne 1	8104044	Wfu & Cx USA 6/96.	
☐ N831PC	80	PA-31T Cheyenne II	8020001	W/o West Palm Beach, Fl. USA. 1 Jun 83.	
☐ N845JB	62	Gulfstream 1	87	Wfu & Cx USA 9/97.	N87CE
☐ N847	73	Rockwell 690A	11140	W/o Chicago, Il. USA. 23 Apr 77.	
☐ N847CE	74	Rockwell 690A	11223	W/o Nemacolin, Pa. USA. 12 Sep 75.	N57223
☐ N850MA	77	Mitsubishi MU-2L	690	Wfu.	(N88DW)
☐ N854Q	68	Mitsubishi MU-2D	107	W/o Rochester, Mn. USA. 8 Jan 77.	N3571X
☐ N855Q	68	Mitsubishi MU-2D	108	Wfu. Cx USA 6/97.	N3572X
☐ N859DD	81	King Air B200	BB-859	W/o Jasper, Al. USA. 23 Jun 87.	N4935X
☐ N861H	64	Gulfstream 1	147	W/o Le Centre, Mn. USA. 11 Jul 67.	N774G
☐ N861K	69	King Air B90	LJ-471	Wfu.	
☐ N863Q	68	Mitsubishi MU-2D	117	Wfu. Cx USA 7/97.	
☐ N866K	65	King Air 90	LJ-76	Wfu.	N100AN
☐ N866Q	68	Mitsubishi MU-2F	121	Wfu. Cx USA 5/95.	
☐ N873Q	69	Mitsubishi MU-2F	160	W/o Nashville, Tn. USA. 1 Nov 79.	
☐ N877Q	69	Mitsubishi MU-2F	164	Wfu	
☐ N878T	67	King Air A90	LJ-246	W/o 27 Feb 78.	
☐ N882Q	69	Mitsubishi MU-2F	168	W/o Mexico. Jun 73.	
☐ N885Q	69	Mitsubishi MU-2F	172	Wfu at White Industries, Bates City, Mo.	
☐ N887PE	70	King Air 100	B-49	W/o Mayfield, Ky. USA. 15 Sep 89.	N887PL
☐ N888DD	67	680V Turbo Commander	1679-61	Wfu & Cx USA 12/95.	N330LC
☐ N888MA	72	Mitsubishi MU-2J	550	W/o Neiva, Colombia. Feb 78.	N220SB
☐ N888RJ	71	Mitsubishi MU-2G	542	W/o New York, NY. USA. 5 Apr 77.	N166MA
☐ N888TP	82	MU-2 Marquise	1541SA	W/o Allentown, Pa. USA. 19 Jan 96.	N100TB
☐ N892MA	78	Mitsubishi MU-2N	721SA	Wfu.	
☐ N895K	65	King Air 90	LJ-25	W/o Billings, Mt. USA. 5 Aug 82.	
☐ N898K	67	King Air A90	LJ-198	Wfu. Cx USA 11/92.	C-GBFF
☐ N900BP	65	King Air 90	LJ-61	Wfu at Columbus State Community College, Columbus, Oh.	N3078W
☐ N908CM	77	King Air E90	LW-233	W/o Hotham Inlet, Ak. USA. 25 Aug 80.	
☐ N916PA	79	King Air E90	LW-313	W/o Wiscassett, Mi. USA. 10 Jun 96.	N491KD
☐ N917RG	66	680V Turbo Commander	1576-30	W/o Mexico. Aug 76.	N1187Z
☐ N920C	77	Conquest II	441-0020	W/o Gainesville, Ga. USA. 10 Aug 92.	N4213V
☐ N931SW	76	PA-31T Cheyenne II	7620013	Wfu at White Industries, Bates City, Mo.	CX-BLT
☐ N932E	66	680T Turbo Commander	1588-39	W/o Kelso, Wa. USA. 11 Jul 84.	N932
☐ N936K	72	King Air C90	LJ-539	W/o 3 Jan 73.	
☐ N940MA	79	MU-2 Marquise	743SA	Wfu. Cx USA 6/95.	XB-PRO
☐ N941K	72	King Air A100	B-111	W/o Muscle Shoals, Al. USA. 22 Jun 78.	
☐ N950	68	SA-26T Merlin 2A	T26-30	Wfu.	N95D
☐ N950TT	73	SA-226T Merlin 3	T-225	W/o Byers, Co. USA. 20 Dec 97.	C-GRTL
☐ N951HE	68	680W Turbo Commander	1751-4	Wfu. Cx USA 2/97.	N951HF
☐ N959L	72	Mitsubishi MU-2J	570	Wfu. Cx USA 8/95.	N214MA
☐ N960M		SA-226AT Merlin 4	AT-005	W/o Southern Pines, SC. USA. 14 Apr 75.	
☐ N961G	60	Gulfstream 1	30	Wfu.	N901G
☐ N962BL	66	680V Turbo Commander	1562-18	Wfu as C-GFAF.	C-GFAF
☐ N962MA	79	MU-2 Solitaire	401SA	W/o New Orleans, La. USA. 23 Feb 80.	N962MA
☐ N965MA	79	MU-2 Solitaire	404SA	Wfu. Cx USA 10/96.	
☐ N969MA	79	MU-2 Solitaire	408SA	W/o Ramsey, Mn. USA. 6 Dec 80.	
☐ N987GM	74	King Air E90	LW-98	W/o Tuba City, Az. USA. 31 May 89. (parts at White Inds.).	N439EE
☐ N998VB	78	King Air C90	LJ-785	W/o Tomahawk, Wi. USA. 31 Dec 96.	N999VB
☐ N999CR	70	King Air 100	B-12	W/o Houston, Tx. USA. 18 Mar 91. (parts at White Industries)	N152X
☐ N999FA	75	Mitsubishi MU-2L	676	W/o Scottsdale, Az. USA. 20 Jul 96.	N90BC
☐ N999PF	74	SA-226T Merlin 3	T-247	Wfu. Cx USA 9/93.	N100T
☐ N999WB	70	Mitsubishi MU-2G	530	W/o Wayne, Il. USA. 30 Dec 97.	(N999TA)
☐ N1011R		SA-226T Merlin 3C	T-303E	W/o 24 Mar 81.	
☐ N1127D	67	King Air A90	LJ-223	W/o Quincy, Il. USA. 19 Nov 96.	N1127M
☐ N1154Z	66	680V Turbo Commander	1585-37	Wfu.	
☐ N1171Z	66	680V Turbo Commander	1557-15	Wfu.	
☐ N1176W	74	Rockwell 690A	11184	Wfu.	N110GM
☐ N1195Z	66	680V Turbo Commander	1575-29	W/o Atlanta, Ga. USA. 9 Dec 72.	

Regn	Yr	Type	c/n	Accident / Withdrawal Details	Prev Regn
☐ N1206S	68	SA-26T Merlin 2A	T26-23	Wfu.	
☐ N1210Y	84	Conquest II	441-0338	W/o Maiduguri, Nigeria. 26 Sep 87.	
☐ N1214S	68	SA-26T Merlin 2A	T26-31	W/o Deadhorse, Ak. USA. 16 May 73.	
☐ N1221S	68	SA-26AT Merlin 2B	T26-109	W/o 1994.	
☐ N1283	76	King Air 200	BB-90	W/o. USA.	
☐ N1865D	83	King Air B200	BB-1119	W/o Treasure Cay, Bahamas. 15 May 96.	N1865A
☐ N1879W	81	Cheyenne II-XL	8166065	W/o Des Moines, Ia. USA. 29 Nov 90. (parts at White Inds.).	N222XL
☐ N1910L	76	King Air B100	BE-10	W/o Midland, Tx. USA. 26 Nov 83.	
☐ N2000S	86	2000 Starship	NC-1	Wfu.	
☐ N2019U	78	King Air C90	LJ-792	W/o 14 Feb 85.	
☐ N2029N	78	King Air C90	LJ-798	W/o 30 Dec 78.	
☐ N2060M	72	Mitsubishi MU-2J	564	Wfu.	N199MA
☐ N2079A	80	Conquest 1	425-0001	W/o Dayton, Oh. USA. 29 May 85.	
☐ N2173Z	67	680V Turbo Commander	1704-80	Wfu. Cx USA 6/97.	C-GKFV
☐ N2181L	76	King Air E90	LW-181	W/o Michigan City, In. USA. 7 Dec 80.	
☐ N2301N	67	SA-26T Merlin 2A	T26-2	W/o Olive Branch, Ms. USA. 22 Nov 78.	N500BW
☐ N2336X	81	PA-31 Cheyenne II	8120002	W/o Cody, Wy. USA. 20 May 87.	VH-IHK
☐ N2400X	65	King Air 90	LJ-18	W/o 1976.	
☐ N2484B	79	Conquest II	441-0112	W/o Calexico, Mexico. 15 Nov 87.	HP-...
☐ N2517X	81	Cheyenne II-XL	8166004	W/o Springfield, Ky. USA. 16 Feb 82.	
☐ N2566W	80	PA-31 Cheyenne II	8020035	W/o Pontiac, Mi. USA. 19 Feb 81.	(N321SS)
☐ N2601S	67	SA-26T Merlin 2	T26-1	Wfu.	
☐ N2646W		Rockwell 690B	11496	Wfu at White Industries, Bates City, Mo.	D-IKOA
☐ N2727A	81	Conquest II	441-0201	W/o Lander, Ey. USA. 29 Aug 86.	
☐ N2755H	66	680T Turbo Commander	1628-55	W/o Jackson, Ms. USA. 7 Jul 80.	
☐ N2937A	81	Gulfstream 840	11670	W/o Wooster, Oh. USA. 31 Oct 84.	ZS-KZP
☐ N3042S	89	2000 Starship	NC-2	Wfu.	
☐ N3177W	74	King Air E90	LW-77	W/o Nueva Palestina, Peru. 31 May 93.	
☐ N3333D	67	King Air A90	LJ-259	W/o Itaguazurenda, Bolivia. 11 Oct 96.	N3333X
☐ N3544X	67	Mitsubishi MU-2B	012	Wfu.	
☐ N3550X	67	Mitsubishi MU-2B	018	W/o Springfield, Mo. USA. 21 Dec 68.	
☐ N3560X	67	Mitsubishi MU-2B	029	Wfu at Dodson's, Ottawa, Ks. Cx USA 9/93.	
☐ N3668P	80	King Air B100	BE-95	Wfu at White Industries, Bates City, Mo.	
☐ N3804F	81	King Air C90	LJ-947	W/o 5 Nov 89.	
☐ N3833P	68	680W Turbo Commander	1761-7	Wfu.	C-GNYD
☐ N4089L	78	Conquest 1	425-693	Wfu.	
☐ N4146S	75	King Air C90	LJ-646	W/o 18 Apr 75.	
☐ N4202K	73	Mitsubishi MU-2J	603	Wfu.	EC-EDK
☐ N4207S	77	King Air E90	LW-207	W/o Sitka, Al. USA. 31 Jul 77.	
☐ N4284V	73	Mitsubishi MU-2K	270	Wfu. Cx USA 3/95.	VH-MUO
☐ N4425W	75	PA-31T Cheyenne II	7620009	Wfu at White Industries, Bates City, Mo.	C-GIVM
☐ N4463W	74	King Air C90	LJ-633	W/o 30 Nov 87.	
☐ N4468M	69	SA-26AT Merlin 2B	T26-119	Wfu at White Industries, Bates City, Mo.	VH-CAJ
☐ N4470H	66	680T Turbo Commander	1550-11	Wfu.	CF-SVJ
☐ N4505B	81	PA-42 Cheyenne III	8001015	Wfu. Cx USA 12/91.	
☐ N4527E	66	680V Turbo Commander	1622-52	Wfu at White Industries, Bates City, Mo.	
☐ N4556E	66	680T Turbo Commander	1614-48	Wfu.	
☐ N4594V	73	Mitsubishi MU-2K	256	Wfu.	HZ-BIN
☐ N5058E	68	680W Turbo Commander	1787-17	W/o Atlanta, Ga. USA. 20 Nov 82.	
☐ N5296M	72	SA-226T Merlin 3	T-219	W/o Montreal, Canada. 10 Apr 73.	
☐ N5329M	74	SA-226T Merlin 3	T-243	W/o Bahamas. 3 Mar 77.	
☐ N5589S	69	Mitsubishi MU-2F	150	W/o Louisville, Ky. USA. 15 Dec 82.	C-FFER
☐ N5654M	79	SA-226T Merlin 3B	T-303	W/o San Marcos, Tx. USA. 31 May 79.	
☐ N5860K	80	Gulfstream 840	11608	W/o Patterson, La. USA. 13 DEc 81.	
☐ N5889N	82	Gulfstream 900	15019	W/o 9 Sep 86.	
☐ N5920C	66	680T Turbo Commander	1552-12	Wfu.	
☐ N5926K	81	Gulfstream 840	11674	W/o Freeport, Bahamas. 16 Oct 81.	
☐ N5938K	81	Gulfstream 840	11686	Wfu at White Industries, Bates City, Mo.	
☐ N5957K	82	Gulfstream 840	11720	W/o 29 Mar 82.	
☐ N6038A	78	PA-31 Cheyenne II	7820072	W/o USA. 31 Jan 92.	(N90TW)
☐ N6040M	79	King Air C90	LJ-840	W/o 5 Jun 79.	
☐ N6069C	67	King Air A90	LJ-291	Wfu at White Industries, Bates City, Mo.	
☐ N6123A	78	PA-31T Cheyenne 1	7804008	W/o Baltimore, Md. USA. 22 Feb 79.	
☐ N6272C	82	King Air C90-1	LJ-1025	W/o 10 Mar 82.	
☐ N6359U	66	680T Turbo Commander	1536-4	W/o Aspen, Co. USA. 22 Jan 70.	
☐ N6514V	66	680T Turbo Commander	1581-34	Wfu.	
☐ N6523V	66	680V Turbo Commander	1583-35	Wfu.	
☐ N6600A	78	PA-31T Cheyenne 1	7804001	Wfu.	
☐ N6649P	83	King Air B200	BB-1133	Wfu.	
☐ N6653Z	59	Gulfstream 1	21	Wfu. Cx USA 11/95.	XC-BIO
☐ N6771Y	81	Conquest 1	425-0019	W/o Sanford, Fl. USA. 11 Feb 88.	
☐ N6774R	81	Conquest 1	425-0045	W/o Newburgh, NY. USA. 12 Dec 83.	
☐ N6843S	75	King Air E90	LW-137	W/o Cordova, Ak. USA. 29 Nov 76.	
☐ N6846D	81	Conquest 1	425-0078	W/o Augusta, Ga. USA. 31 Jan 90.	(N6844V)
☐ N6857E	82	Conquest II	441-0244	W/o Muskegon, Mi. USA. 10 Jul 86.	
☐ N6858S	82	Conquest II	441-0253	W/o Flagstaff, Az. USA. 21 Feb 87.	

Regn	*Yr*	*Type*	*c/n*	*Accident / Withdrawal Details*	*Prev Regn*

Regn	Yr	Type	c/n	Accident / Withdrawal Details	Prev Regn
☐ N6886D	83	Conquest 1	425-0152	W/o Ithaca, NY. USA. 25 Feb 84.	
☐ N7138C	69	King Air B90	LJ-446	Wfu at Dodson's, Ottawa, Ks. Cx USA 7/93.	G-BLNA
☐ N7500L	69	PA-31T Cheyenne II	1	Wfu.	
☐ N7676L		PA-42 Cheyenne III	31-5003	Wfu.	
☐ N8042N	67	680V Turbo Commander	1720-91	Wfu.	OO-SKF
☐ N8092D	79	King Air 200	BB-418	Wfu at White Industries, Bates City, Mo.	CX-BOR
☐ N9001N	70	Rockwell 690	11000	W/o 26 Oct 90.	
☐ N9003N	69	680W Turbo Commander	1828-36	Wfu.	N94490
☐ N9004N	69	680W Turbo Commander	1829-37	Wfu.	
☐ N9019N	69	680W Turbo Commander	1844-43	W/o Elkhart, In. USA. 28 Dec 71.	
☐ N9060N	70	681 Turbo Commander	6011	W/o Altus, Ok. USA. 25 Nov 70.	
☐ N9066N	72	King Air C90	LJ-557	W/o Ciudad Constitucion, Peru. 12 Jul 94.	C-GQPC
☐ N9079S	79	King Air F90	LA-1	Wfu.	
☐ N9129N	71	681B Turbo Commander	6056	W/o Mansfield, Oh. USA. 30 Nov 96.	
☐ N9150N	73	Rockwell 690	11063	W/o Little America, Wy. USA. 25 Aug 84.	
☐ N9202N	70	Rockwell 690	11002	W/o 26 Jun 70.	
☐ N9789S	80	Gulfstream 980	95037	W/o USA. 12 May 82.	
☐ N9971F	59	Gulfstream 1	17	Wfu.	C-FTPC
☐ N9971G	77	Conquest II	441-0006	W/o Greensboro, Al. USA. 15 Nov 77.	(N441SA)
☐ N17530	77	King Air 200	BB-204	W/o Valparaiso. In. USA. 19 Oct 77.	
☐ N17690	66	680V Turbo Commander	1577-31	W/o 3 Dec 85.	N17690
☐ N23796	77	King Air C90	LJ-737	W/o 23 Sep 77.	
☐ N24169	78	King Air B100	BE-38	W/o Romeo, Md. USA. 22 Nov 91.	
☐ N31434	88	King Air C90A	LJ-1186	W/o Manaus, Brazil. 30 May 90.	
☐ N36941	77	Conquest II	441-0018	W/o Butte, Mt. USA. 1 Apr 80.	
☐ N36962	77	Conquest II	441-0033	W/o ground accident Mesa, Az. USA. Jan 84.	N36962
☐ N45591	70	Mitsubishi MU-2G	524	Wfu at White Industries, Bates City, Mo.	C-FGWF
☐ N46866	73	Rockwell 690A	11108	Wfu at White Industries, Bates City, Mo. Cx USA 12/94.	HB-GFP
☐ N53070	69	SA-26AT	T26-155	Wfu.	C-GVCO
☐ N54985	75	PA-31T Cheyenne II	7620007	Wfu at White Industries, Bates City, Mo.	(N76NL)
☐ N57086	74	Rockwell 690A	11157	W/o.	
☐ N57169	74	Rockwell 690A	11203	W/o Jacksonville, Fl. USA. 24 Jun 87.	
☐ N57186	74	Rockwell 690A	11186	W/o Independence, Ks. USA. 17 Nov 76.	
☐ N57233	75	Rockwell 690A	11247	W/o Columbus, Oh. USA. 1 Oct 79.	
☐ N65103	69	SA-26AT Merlin 2B	T26-140E	W/o Palo Alto, Ca. USA. 19 Oct 79.	D-IBMC
☐ N66534	62	Gulfstream 1	85	Wfu & Cx USA 11/95.	XC-BAU
☐ N66847	75	PA-31T Cheyenne II	7520019	Wfu at White Industries, Bates City, Mo.	
☐ N80398	78	Mitsubishi MU-2N	712SA	Wfu. Cx USA 3/95.	VH-SSL
☐ N81416	76	Rockwell 690A	11306	W/o Winter Haven, Fl. USA. 14 Feb 83.	
☐ N81470	76	Rockwell 690A	11335	Wfu at White Industries, Bates City, Mo. Cx USA 1/94.	N81470
☐ N81502	82	Gulfstream 1000	96000	W/o Chekotah, Ok. USA. 9 Oct 84.	
☐ N81521	77	Rockwell 690B	11351	W/o Burns, Or. USA. 7 Jan 81.	
☐ N81628	77	Rockwell 690B	11396	W/o White Plains, NY. USA. 22 Sep 90. (parts at White Inds.)	
☐ N81717	78	Rockwell 690B	11445	W/o Greenville, SC. USA. 17 Jan 84.	
☐ N82010	76	PA-31T Cheyenne II	7620025	Wfu.	(N820YL)
☐ N82123	77	PA-31T Cheyenne II	7720030	Wfu at White Industries, Bates City, Mo.	
☐ N82139	77	PA-31T Cheyenne II	7720040	Wfu at White Industries, Bates City, Mo.	N14BW
☐ N82271	78	PA-31T Cheyenne II	7820044	W/o Pellston, Mi. USA. 13 May 78.	
☐ N82282	78	PA-31T Cheyenne II	7820055	W/o Elyria, Oh. USA. 27 Apr 79. (parts at White Inds.).	
☐ N88832	79	Conquest II	441-0071	W/o Mexico. Sep 79.	
☐ N98949	68	King Air B90	LJ-407	W/o 5 May 82.	D-ILTP
☐ OB-1176	80	PA-31T Cheyenne II	8020010	W/o Tocache, Peru. 28 Feb 90.	OB-S-1176
☐ OB-1212	74	Rockwell 690A	11222	W/o Peru. 7 May 92.	OB-M-1212
☐ OB-1219	78	Mitsubishi MU-2N	730SA	W/o Peru.	OB-M-1219
☐ OB-1305	67	King Air A90	LJ-302	W/o S E Cuzco, Peru. 11 Oct 91.	OB-T-1305
☐ OB-1361	69	King Air B90	LJ-451	W/o 27 Feb 91.	N896K
☐ OB-1362	69	King Air B90	LJ-448	W/o 31 Jul 90.	N428DN
☐ OB-1403	77	PA-31T Cheyenne II	7720022	W/o Chacham Peak, Peru. 16 Jan 96.	N14LW
☐ OB-M-1003	78	PA-31T Cheyenne II	7820057	W/o Cuzco, Peru. 26 Feb 81.	
☐ OB-M-1031	72	Rockwell 690	11008	W/o Peru. 14 Feb 79.	(N9208N)
☐ OE-EDU	92	TBM 700	73	W/o Nr Freiburg, Germany. 2 Apr 93.	
☐ OE-FCS	84	Gulfstream 900	15036	W/o Lake Constance, Switzerland. 23 Feb 89.	N45GA
☐ OE-FEM	90	King Air 300LW	FA-210	W/o en route Salzburg-Krems, Austria 12 May 96.	D-IDLS
☐ OE-FGK	80	PA-31T Cheyenne II	8020052	W/o Friedrichshafen, Germany. 8 Feb 92.	N700GC
☐ OE-FPS	81	Conquest 1	425-0024	W/o Hannover, Germany. 8 Oct 91	D-IEAT
☐ OH-MIB	71	Mitsubishi MU-2G	532	Wfu Den Helder, Netherlands.	N30SA
☐ OO-TBW	70	Mitsubishi MU-2G	526	W/o Angouleme, France. 15 Aug 76.	SE-FGG
☐ OO-TLS	74	King Air A100	B-188	W/o Bacau, Romania. 8 Jan 94.	5X-UWS
☐ OY-ATW		SA-226T Merlin 3A	T-261	W/o Gronholt, Denmark. 26 Apr 78.	N300TA
☐ OY-AUI		SA-226AT Merlin 4	AT-015	W/o Copenhagen, Denmark. 12 Nov 82.	N223JC
☐ OY-AZA	73	King Air C90	LJ-593	W/o Copenhagen, Denmark. 15 Jan 79.	N90KA
☐ OY-BEP	81	King Air B200C	BL-43	W/o Roodt-Syr, Luxembourg. 18 Sep 82.	EI-BKV
☐ OY-BVA	65	King Air 90	LJ-68	Wfu. PT-6 engine test-bed at Roskilde.	D-IKAO
☐ OY-CCC	83	King Air F90	LA-184	W/o 29 Jun 97.	N55K
☐ OY-CGM	81	Conquest II	441-0229	W/o Nr Sondrestrom, Greenland. 11 Nov 90.	N68548
☐ PH-DRX	82	MU-2 Marquise	1555SA	W/o Eindhoven, Holland. 12 Sep 88.	N481MA
Regn	*Yr*	*Type*	*c/n*	*Accident / Withdrawal Details*	*Prev Regn*

Regn	Yr	Type	c/n	Accident / Withdrawal Details	Prev Regn
☐ PK-CAF	76	King Air A100	B-223	Wfu.	N56800
☐ PK-TRB	76	King Air 200	BB-116	Wfu.	
☐ PT-DUX	71	SA-226T Merlin 3	T-215	W/o Sao Paulo, Brazil. 27 Sep 71.	
☐ PT-LJR	81	King Air F90	LA-93	W/o Sao Pedro da Aldeia, Brazil. 2 Oct 88.	N1187K
☐ PT-ORY	81	King Air F90	LA-153	W/o Sao Paulo, Brazil. 3 Feb 93.	N200RA
☐ RP-C202	68	King Air B90	LJ-356	Wfu.	PI-C202
☐ RP-C710	70	King Air 100	B-15	W/o San Jose, Philippines. 28 Nov 96.	PI-C710
☐ SE-FGE	70	681 Turbo Commander	6033	W/o Mestersvig, Greenland. 23 Jul 73.	N9086N
☐ SE-GLB	84	PA-31T Cheyenne II	7400002	Wfu.	OH-PNS
☐ SE-GUU	69	King Air B90	LJ-470	W/o Sindal, Denmark. 15 Oct 85.	N490K
☐ SE-IHX	83	Conquest II	441-0291	W/o 24 Mar 84.	N88707
☐ SE-IOU	74	Mitsubishi MU-2K	304	W/o Sweden. 16 Feb 86.	N410MA
☐ SE-IOX	76	Mitsubishi MU-2M	331	W/o Sweden. 15 Mar 86.	N3982L
☐ S9-NAA	65	King Air 90	LJ-50	W/o 1988. Parts at White Industries, Bates City, Mo.	N75XA
☐ TN-ADP	74	SA-226AT Merlin 4A	AT-025	W/o Pointe Noire, Congo Republic. 11 Mar 94.	N52LB
☐ TR-LYA	77	King Air E90	LW-247	W/o Liberville, Gabon. Apr 78.	F-GAPO
☐ TU-TJE	76	King Air 200	BB-163	W/o Bouafle, Ivory Coast. 28 Jun 96.	
☐ VH-AAV	77	King Air 200	BB-245	W/o Sydney, Australia. 20 Feb 80.	A2-ABO
☐ VH-AAZ	77	King Air 200	BB-241	W/o Bundaberg, Queensland, Australia. 1991. Cx VH- 5/95.	N23915
☐ VH-BBA	80	MU-2 Marquise	782SA	W/o Perth, Western Australia. 16 Dec 88.	N269MA
☐ VH-BSS	72	Rockwell 690	11044	W/o Botany Bay, NSW. Australia. 14 Jan 94.	N471SC
☐ VH-CCW	77	PA-31T Cheyenne II	7720046	W/o Perth, Western Australia. 3 May 81.	N82175
☐ VH-CJP	70	Mitsubishi MU-2G	505	W/o 15 Nov 83.	HB-LEF
☐ VH-FLO	63	Gulfstream 1	100	Wfu in New Zealand 1995.	N116K
☐ VH-IAM	70	Mitsubishi MU-2G	517	W/o Melbourne, Australia 21 Dec 94. (to N119BF for parts).	VH-IAM
☐ VH-IBC	75	King Air 200	BB-74	W/o 1988.	N9730S
☐ VH-KTE	78	King Air 200	BB-320	W/o Adavale, Australia. 28 Aug 83.	N23786
☐ VH-LFH	78	King Air E90	LW-255	W/o Nr Kingaroy, QLD. Australia. 25 Jul 90.	N258D
☐ VH-MLU	81	MU-2 Marquise	1527SA	W/o Bargo, Australia. 24 May 83.	N445MA
☐ VH-MUA	79	MU-2 Marquise	746SA	W/o Australia. 26 Jan 90.	N943MA
☐ VH-NYB	72	Rockwell 690	11039	Wfu.	C-GICX
☐ VH-NYD	70	681 Turbo Commander	6034	Wfu.	N9087N
☐ VH-NYF	70	681 Turbo Commander	6026	Wfu & Cx VH- 8/97.	N9024N
☐ VH-NYG	70	681 Turbo Commander	6004	W/o Tamworth, Australia. 14 Feb 91.	N740ES
☐ VH-NYH	70	681 Turbo Commander	6016	Wfu.	N444JB
☐ VH-NYM	67	Mitsubishi MU-2B	037	Wfu.	VH-JWO
☐ VH-SVQ	77	Rockwell 690B	11380	W/o Tasman Sea, Australia. 2 Oct 94.	VH-FOZ
☐ VH-SWP		SA-226AT Merlin 4A	AT-033	W/o Tamworth, NSW. Australia. 9 Mar 94.	(N.....)
☐ VH-TNZ	79	PA-31T Cheyenne II	7920064	W/o hangar explosion Moorabbin, Australia. 29 Sep 88	N23477
☐ VH-WMU	70	Mitsubishi MU-2G	512	W/o Bathurst, Australia. 7 Nov 90. Cx USA as N318MA 6/95.	N318MA
☐ VT-EHK	82	King Air B200	BB-985	W/o Nr Delhi Airport, India. 27 Aug 92.	N1841Z
☐ VT-EQM	87	King Air 300	FA-128	W/o nr Panvel, Bombay, India. 15 Jul 93.	N3029F
☐ VT-EUJ	93	King Air B200	BB-1456	W/o Runda, Kulu Valley, India. 9 Jul 94.	N8090U
☐ XA-DIS	73	Mitsubishi MU-2J	608	W/o El Paso, Tx. USA. 20 Nov 83.	N340MA
☐ XA-FOT	66	King Air A90	LJ-168	W/o 15 Feb 78.	XB-BOF
☐ XA-GOL	71	Mitsubishi MU-2F	197	Wfu.	XB-SUR
☐ XA-IIW	73	Rockwell 690	11050	Wfu.	XC-FUV
☐ XA-JUY	73	Rockwell 690A	11170	Wfu.	XC-CAU
☐ XA-LIG	81	King Air 200	BB-802	W/o Poza Rica, Mexico. 5 May 84.	
☐ XB-AEA	74	Rockwell 690A	11199	W/o Oklahoma City, Ok. USA. 1 Jan 80.	
☐ XB-AQQ	74	Mitsubishi MU-2J	632	Wfu. Cx USA 6/97.	N995MA
☐ XB-CIJ	59	Gulfstream 1	10	Wfu.	N1623Z
☐ XB-ESO	59	Gulfstream 1	15	Wfu.	N26KW
☐ XB-LIJ	73	Mitsubishi MU-2K	259	W/o Beloit, Ks. USA. 1975.	N279MA
☐ XB-NUV	72	King Air A100	B-128	W/o San Luis Potosi, Mexico. 13 Oct 76.	N196B
☐ XB-QIY	74	King Air E90	LW-113	W/o Monterrey, Mexico. 2 Mar 81.	XA-FEX
☐ XC-AA38	80	Gulfstream 980	95020	W/o Nr Monterrey, Mexico. 19 Oct 92.	XC-HFX
☐ XC-FEL	68	Mitsubishi MU-2D	103	Wfu at White Industries, Bates City, Mo.	XB-TIM
☐ XC-GEI	61	Gulfstream 1	66	Wfu.	N111DR
☐ XC-ICP	66	King Air A90	LJ-176	W/o Vera Cruz, Mexico. 28 Jan 73.	N5656A
☐ XC-PGR	78	King Air 200	BB-317	W/o Otay Mesa, Mexico. 28 Oct 79.	
☐ YV-94CP	76	Mitsubishi MU-2M	347	W/o 1982.	YV-1050P
☐ YV-121CP	65	Gulfstream 1	150	Wfu.	N777G
☐ YV-174CP	77	Mitsubishi MU-2L	695	W/o Caracas, Venezuela. 4 Jun 81.	N855MA
☐ YV-215CP	78	PA-31T Cheyenne II	7820071	W/o Caracas, Venezuela. 4 Jun 81.	
☐ YV-288CP	79	King Air F90	LA-8	W/o Caracas, Venezuela. 8 May 84.	
☐ YV-299P	65	King Air 90	LJ-72	Wfu.	N901N
☐ YV-426P	76	King Air 200	BB-142	W/o Caracas, Venezuela. 4 Feb 82.	
☐ YV-597CP	78	King Air 200	BB-394	W/o 12 Jan 89.	N223TC
☐ YV-726CP	76	King Air E90	LW-182	W/o Caracas, Venezuela. 27 Mar 94.	YV-O-BIV-2
☐ YV-994P	76	King Air C90	LJ-693	W/o Caracas, Venezuela. 4 Feb 82.	
☐ YV-2200P	74	PA-31T Cheyenne II	7400006	W/o Newtown, Bahamas. 20 Jun 90.	YV-146CP
☐ YV-O-MAR2	67	680V Turbo Commander	1683-64	W/o Venezuela 17 Aug 78.	YV-OMOP8
☐ YV-O-NCE-2	76	King Air E90	LW-201	W/o Caracas, Venezuela. 13 Nov 95.	
☐ YV-T-ADJ	73	King Air E90	LW-53	W/o Pratt, Ks. USA. 10 May 73.	
☐ ZS-IHZ	70	King Air B90	LJ-497	W/o Capetown, South Africa. 26 Dec 92.	N9019Q

| *Regn* | *Yr* | *Type* | *c/n* | *Accident / Withdrawal Details* | *Prev Regn* |

Regn	Yr	Type	c/n	Accident / Withdrawal Details	Prev Regn
☐ ZS-JRF	78	Rockwell 690B	11491	W/o & Cx ZS- 16 Aug 93.	(N690RC)
☐ ZS-KMT	80	King Air 200	BB-767	W/o Johannesburg, South Africa. 13 Apr 87.	N3717T
☐ ZS-KRS	80	Gulfstream 840	11644	W/o South Africa. 16 Sep 81.	N5896K
☐ ZS-KVB	80	Gulfstream 980	95005	W/o near Sishen, South Africa 20 Jun 84.	N9758S
☐ ZS-LUC	78	PA-31T Cheyenne II	7820092	W/o Welkom, South Africa. 26 Apr 89.	N777DL
☐ ZS-MGR	75	King Air 200	BB-19	W/o Marunga, Angola. 21 Oct 95.	N221B
☐ ZS-MSG	79	King Air B100	BE-75	W/o Vrede, South Africa. 23 Dec 94. Cx ZS- 6/95.	N814GT
☐ ZS-NEP	81	King Air 200	BB-838	W/o Aminius nr Koudpan, Namibia. 28 Jun 93.	N14MF
☐ ZS-NXY	83	King Air C90-1	LJ-1058	W/o Sulkerbossstrand, South Africa. 13 Dec 96.	N6586K
☐ 5H-TAA	78	Rockwell 690B	11473	Wfu at Dodson's, Ottawa, Ks.	(N81776)
☐ 5H-TZD	88	Reims/Cessna F406	F406-0029	W/o Mount Palapala Morogoro, Tanzania. 24 Apr 96.	PH-FWI
☐ 5N-ATU	66	King Air A90	LJ-136	Wfu at Gamston, UK.	5N-ATU
☐ 5Y-ING	88	Reims/Cessna F406	F406-0024	W/o 12 Jun 95 Nairobi, Kenya.	PH-PEL
☐ 5Y-JJG	87	Reims/Cessna F406	F406-0003	W/o Nairobi, Kenya. 8 Aug 94.	PH-ALK
☐ 5Y-NAL	89	Reims/Cessna F406	F406-0038	W/o Mar 94.	PH-ALX
☐ 5Y-TNT	71	SA-226A Merlin 3	T-211	W/o Nairobi-Wilson, Kenya. 1 Oct 92.	TN-ADO
☐ 7Q-YMM	80	King Air C90	LJ-880	W/o Blantyre-Chileka, Malawi. 9 Nov 90. (parts at Dodsons).	ZS-KLC
☐ 7T-VSH	68	King Air B90	LJ-423	W/o Jan 76.	HB-GDT
☐ 9M-CAM	83	King Air B200T	BT-24	W/o Ipoh, Malaysia. 13 Dec 93.	9M-JPA
☐ 9Q-...	75	King Air C90	LJ-660	Wfu.	ZS-LXL
☐ 9Q-COI	79	SA-226T Merlin 3B	T-302	W/o Zaire. 6 Oct 83.	G-IIIB

MILITARY

Regn	Yr	Type	c/n	Accident / Withdrawal Details	Prev Regn
☐ CNA-NA	74	King Air A100	B-180	W/o Morocco. Oct 78.	
☐ EB-002	81	King Air 200C	BL-33	W/o 6 Dec 95.	FAB-002
☐ EB-003	80	King Air C90	LJ-905	W/o 27 Nov 95.	YV-164CP
☐ FAB 006	68	King Air B90	LJ-413	W/o La Paz, Bolivia. 26 Apr 79.	FAB 001
☐ E22-5	74	King Air C90	LJ-623	W/o Salamanca, Spain. 1 Oct 80.	EC-CHB
☐ 5-59	70	681 Turbo Commander	6009	Wfu.	N9059N
☐ IIN 501	72	Rockwell 690	11049	W/o.	
☐ FAE-723	80	King Air 200	BB-723	W/o Ecuador 24 May 81.	HC-BHG
☐ FAC-5600	82	Gulfstream 1000	96044	Wfu. El Dorado AB. Colombia.	HK-2908
☐ 21-111		SA-226AT Merlin 4A	AT-062E	W/o Thailand. 6 Nov 78.	
☐ 29-999	78	SA-226AT Merlin 4A	AT-065	W/o Thailand. 20 Sep 82.	N5442M
☐ 348	80	SA-226T Merlin 3B	T-348	Wfu.	N1009G
☐ 03-3207	69	Mitsubishi MU-2E	907	W/o Japan. 2 Sep 70.	
☐ 22-002	71	Mitsubishi MU-2C	802	W/o Japan. 14 Jun 77.	
☐ 22-011	80	Mitsubishi MU-2C	811	W/o Japan. 10 Aug 81.	
☐ 93-3205	69	Mitsubishi MU-2E	905	W/o Japan. 11 Apr 73.	
☐ 4-G-46/0746	79	King Air 200	BB-543	W/o Argentina. 15 May 86.	
☐ AE-177	77	SA-226T Merlin 3A	T-277	W/o Argentina.	N5395M
☐ GN-705	80	PA-31T Cheyenne II	8020092	W/o San Manual, General Villegas, Argentina. 6 Oct 95.	LV-OIF
☐ 155722	66	Gulfstream TC-4C	176	Wfu. Preserved Pensacola NAS. Fl. USA.	N798G
☐ 155723	67	Gulfstream TC-4C	178	W/o Cherry Point, NC. USA. 16 Oct 75.	N778G
☐ 161189		King Air UC-12B	BJ-5	W/o Nr Penscaloa, Fl. USA. 2 Jan 82.	
☐ 66-7943	68	King Air C-6A	LJ-320	Wfu. Exhibited at Wright-Patterson AFB.Oh.	N2085W
☐ 73-1211		King Air C-12A	BD-7	W/o Nr Nadiz, Iran. 31 Jan 79.	
☐ 85-1261		King Air C-12F	BP-52	W/o Juneau, Al. USA. 12 Nov 92.	
☐ 85-1269		King Air C-12F	BP-60	W/o Brazil. 11 Jan 1992.	
☐ 86-0402	58	Gulfstream 1	2	Wfu.	N39PP
☐ CF-03		SA-226T Merlin 3A	T-262	W/o Lille, France. 16 Apr 80.	N5375M
☐ 101001	81	King Air 200C	BL-25	W/o Halmstad, Sweden. 24 Sep 90.	OY-CHE
☐ P-9	63	Gulfstream 1	120	Wfu. Preserved at Tatoi AFB. Greece.	
☐ SAAF16		SA-226AT Merlin 4A	AT-040	W/o Pretoria, South Africa. 14 July 82.	(SAAF14)
☐ 9T-MBD	74	Mitsubishi MU-2J	620	W/o.	N470MA

BUSINESS TURBOPROPS
Cross-Reference by Construction Number

ATR

264	F-SEBK
500	MM62165
502	MM62166

Aero Commander

1519-95	N175DB
1538-5	N722LJ
1542-7	N45QC
1544-8	C-GSVQ
1546-9	C-FAXN
1548-10	N7610U
1556-14	N543S
1558-16	N818L
1560-17	N687L
1563-19	FAC-542
1564-20	EC-DSA
1566-22	N292Z
1568-24	N666MN
1570-25	N370K
1571-26	N6540V
1572-27	N900RJ
1579-32	N808GU
1584-36	N512JD
1589-40	N698GN
1597-42	N688SH
1601-43	C-GFAB
1610-46	N79BJ
1616-49	N171AT
1620-51	N11HM
1624-53	N32DF
1630-56	N4682E
1632-57	HK-2285P
1676-59	N66FV
1677-60	N677KA
1680-62	N3CG
1682-63	LV-OFX
1684-65	N10TG
1685-66	YV-744CP
1687-67	N42TF
1688-68	N4FB
1689-69	YV-843P
1691-70	N47HM
1697-74	N400LP
1698-75	N20BM
1699-76	N29DE
1703-79	PT-...
1705-81	N87BT
1707-82	N616E
1708-83	N200TB
1710-85	N111SK
1711-86	EC-DXG
1714-88	N50655
1718-89	YV-344P
1719-90	YV-1405P
1721-1	N39AS
1722-2	N93RA
1723-3	N4622E
1752-5	N505RT
1762-8	N2755B
1763-9	N200DT
1772-10	N514NA
1773-11	N11CT
1774-12	N54163
1775-13	YV-820CP
1776-14	EC-EAG
1788-18	XB-DTD
1789-19	N1928H
1790-20	N410TH
1791-21	N345TT
1792-22	N680JD
1793-23	N22ET
1802-24	N17JG
1803-25	N4100B
1804-26	YV-2280P
1812-29	N680KM
1813-30	YV-536P
1814-31	YV-2436P
1818-32	N5RE
1820-34	HK-3975X
1821-35	N424PP
1833-38	PT-KYY
1834-39	XB-...
1835-40	CP-...
1843-42	N940U
1848-44	EP-FSS
1849-45	EP-FIA
1850-46	EP-FIB

Commander 681

6001	N22RT
6002	N6E
6005	YV-2383P
6006	N113CT
6007	LV-JOJ
6008	N9058N
6010	N3867N
6012	EP-AGU
6013	N70RF
6014	LV-VHP
6018	PT-DQX
6019	N3203P
6020	C-FATR
6021	PT-OQQ
6022	TG-WIZ
6023	XB-DJN
6024	YV-2501P
6025	CP-2042
6027	N548GQ
6028	N100GL
6029	YV-2404P
6030	9Q-CGL
6032	TG-GMI
6035	HK-1977
6036	RP-C1977
6038	N78CH
6039	VH-JWO
6040	N9097N
6041	N25BE
6043	HK-2376P
6044	C-GLMC
6045	9Q-CGN
6046	HK-3236P
6047	VH-UJN
6048	N676DM
6049	HC-BPY
6051	N162RB
6052	YV-...
6054	N21HC
6055	N18KK
6058	N47CK
6059	N30BG
6060	N2643B
6061	N520CS
6062	5-280
6063	XA-SHP
6065	EP-AKA
6066	RP-C775
6067	EP-AKB
6068	5-281
6069	PT-IEC
6070	PT-IED
6071	PT-IEE
6072	5-282

British Aerospace
HS748

1669	C-FAMO
1684	FAE-001
1688	AF602

Jetstream

605	N903FH
613	N904FH

Jetstream 41

41030	N410JA
41038	N514GP
41060	41060
41094	41094
41095	ZS-NUK

CASA

286	N434CA
294	N31BR
369	EB-50

Cessna

Reims 406

F406-01	F-ZBFA
F406-0001	N17CK
F406-0002	D2-ECN
F406-0004	V5-EEZ
F406-0005	ZS-NNA
F406-0006	F-ZBEP
F406-0007	PH-FWC
F406-0008	F-MABM/0008
F406-0009	5Y-TAL
F406-0010	F-MABN/0010
F406-0011	D2-ECO
F406-0012	5Y-WAW
F406-0013	5Y-JJA
F406-0014	PH-FWD
F406-0015	G-TINI
F406-0016	D2-ECP
F406-0017	F-ZBES
F406-0018	G-LEAF
F406-0019	D2-ECQ
F406-0020	G-TURF
F406-0021	5Y-BKN
F406-0022	YV-990CP
F406-0023	N53474
F406-0025	F-ZBAB
F406-0026	F-OGOG
F406-0027	D-ISHY
F406-0028	5H-TZC
F406-0030	ZS-NYI
F406-0031	VT-ASB
F406-0032	LN-TWH
F406-0033	VT-SAC
F406-0034	PH-PHO
F406-0035	S7-006
F406-0036	G-DFLT
F406-0037	5Y-BIS
F406-0039	F-ZBBB
F406-0040	5Y-JJC
F406-0041	N563GA
F406-0042	F-ZBCE
F406-0043	D-INUS
F406-0044	5Y-MKM
F406-0045	OY-PEU
F406-0046	5H-TZE
F406-0047	D-IAAD
F406-0048	S9-IHD
F406-0049	9M-PMS
F406-0050	VT-SAA
F406-0051	S7-AAI
F406-0052	N6590Y
F406-0053	N6591L
F406-0054	N6591R
F406-0055	5Y-BIX
F406-0057	F-ODYZ
F406-0058	PH-GPX
F406-0059	AP-...
F406-0060	PH-GUG
F406-0061	F-OGVS
F406-0062	YS-...
F406-0063	PH-GUI
F406-0064	G-SFPA
F406-0065	G-SFPB
F406-0066	F-ZBCG
F406-0067	Z-DDG
F406-0068	Z-DDE
F406-0069	Z-DDD
F406-0070	F-ZBCI
F406-0071	Z-DDF
F406-0072	F-...
F406-0073	G-BVJT
F406-0074	F-ZBCJ
F406-0075	F-ZBCH
F406-0076	VH-ZZE
F406-0077	F-ZBCF
F406-0078	VH-ZZF
F406-0079	VH-ZZG
F406-0080	F-WWSR

425 Conquest I

425-0002	N74HR
425-0003	D-INGA
425-0004	N425AT
425-0005	N6SK
425-0006	N757H
425-0007	N67BA
425-0008	RP-C574
425-0010	N425PL
425-0011	A2-AHT
425-0012	N127DC
425-0013	N93DA
425-0014	N425TC
425-0015	C-GFDV
425-0016	HB-LOE
425-0017	N794B
425-0018	N425CF
425-0020	N55AC
425-0021	V5-LYZ
425-0022	N6772P
425-0023	N23AW
425-0024	N83CH
425-0026	N540GA
425-0027	VP-BJT
425-0028	N425WB
425-0029	N926ES
425-0030	N67726
425-0032	F-GFJR
425-0033	N425GM
425-0035	N402NG
425-0036	N700WJ
425-0037	N425SF
425-0038	N425JP
425-0039	D-INWG
425-0041	ZS-MVW
425-0042	N1BM
425-0043	N425JB
425-0044	N425HS
425-0046	N425BA
425-0047	5Y-KXU
425-0048	N6774Z
425-0049	N31AD
425-0050	N37PS
425-0051	XB-FMS
425-0052	N425SG
425-0053	N4420F
425-0054	G-LILI
425-0056	N502NC
425-0058	N67JE
425-0059	F-GEFZ
425-0060	N1222B
425-0061	D-IAGT
425-0062	N6844D
425-0063	N425TY
425-0064	ZS-MIG
425-0065	N6844S
425-0066	N6844T
425-0067	5N-AMW
425-0068	N1906K
425-0069	N12NL
425-0070	N425NW
425-0071	N6845R
425-0072	N425TB
425-0073	N425HB
425-0074	N146GA
425-0075	ZS-PNP
425-0076	N14BM
425-0077	V5-MJW
425-0079	N900MS
425-0080	C-...
425-0081	N6846S
425-0082	N86EA
425-0083	N425DD
425-0084	N547GA
425-0085	N71HE
425-0086	N717LW
425-0087	N6847P
425-0088	N35TK
425-0089	N73DW
425-0090	N90YA
425-0091	XB-FXK
425-0092	N40RD
425-0093	N926FS
425-0094	N105FC
425-0095	N8FR
425-0096	N425E
425-0097	N410VE
425-0098	N425DM
425-0099	N425EZ
425-0100	N70HB
425-0101	N543GA
425-0102	D-ICMF
425-0103	N4276Z
425-0104	N425BZ
425-0105	N1848T
425-0106	N425SX
425-0107	N1NL
425-0108	N70GM
425-0109	N818PL
425-0110	N555TP
425-0111	N77YP
425-0112	N6851G
425-0113	N6851L
425-0114	N847YT
425-0115	ZS-KSU
425-0116	N420MA
425-0117	N123SC
425-0118	HK-3036P
425-0119	N7WF
425-0120	D-IPOS
425-0121	N6882C
425-0122	N6882D
425-0123	N825B
425-0124	VH-ULX
425-0125	N125PG
425-0127	N456FC
425-0128	N444RR
425-0129	N520RM
425-0130	N68823
425-0131	OE-FAM
425-0132	N6883R
425-0133	N425SR
425-0134	N156GA
425-0135	N5CE
425-0136	N125SC
425-0138	N39A
425-0139	N383SS
425-0140	N15ST
425-0141	N373LP
425-0142	N101CA

425-0143 N6885P
425-0144 N6885S
425-0145 N901FD
425-0146 N146GW
425-0147 N500MC
425-0148 N701CR
425-0149 N425LD
425-0153 N153JM
425-0154 N6886V
425-0155 YV-266CO
425-0156 N857C
425-0157 N425PJ
425-0158 N355ES
425-0159 N813ZM
425-0160 N68865
425-0161 N425TW
425-0162 YV-2263P
425-0164 D-IJOY
425-0165 D-IRGW
425-0166 N425WL
425-0167 N6872L
425-0168 PH-JOE
425-0169 ZS-AMC
425-0170 N100CC
425-0171 N51CU
425-0172 N41054
425-0173 N16P
425-0174 N425KC
425-0175 N425WT
425-0176 N425TV
425-0177 C-GRJM
425-0178 C-FVAX
425-0179 N9180K
425-0180 N425RM
425-0181 N24QT
425-0182 N68734
425-0183 VH-EGS
425-0184 N425SP
425-0185 N425DC
425-0186 N425FG
425-0187 N12201
425-0188 N188CP
425-0189 YV-548CP
425-0190 N444RU
425-0191 N444KF
425-0192 C-FSEA
425-0193 D-IJGW
425-0194 N12214
425-0195 VH-EGR
425-0196 N1222G
425-0197 D-ICOM
425-0198 N5PP
425-0199 VP-BDR
425-0200 N425PG
425-0201 N37JT
425-0202 VH-EGQ
425-0203 N425KD
425-0204 N1223P
425-0205 N111SU
425-0206 N425CL
425-0207 HK-3518
425-0208 N1224J
425-0209 N13FH
425-0210 N721VB
425-0211 N425AS
425-0212 N1224T
425-0213 N12244
425-0214 D-ISAR
425-0215 N888JS
425-0216 VH-EGT
425-0217 N425TM
425-0218 N1225V
425-0219 N444TH
425-0220 N188RB
425-0221 N410WA
425-0222 N1226B
425-0223 ZP-...
425-0224 D-IPAS
425-0225 N652MC
425-0226 N111KU
425-0227 N12268
425-0229 N1227J

425-0230 N98EP
425-0231 N127MC
425-0232 N50JN
425-0233 D-ICHS
425-0234 N1262K
425-0235 JA8855
425-0236 G-BNDY

441 Conquest II

441-0001 TJ-AHZ
441-0003 N8838T
441-0004 N555JJ
441-0005 N500CE
441-0007 ZS-PES
441-0008 N441CC
441-0009 N87WS
441-0010 N20HC
441-0011 N93HC
441-0012 N441SC
441-0013 N442JA
441-0014 N441ST
441-0015 XA-STD
441-0016 HK-2390W
441-0017 N500UW
441-0019 N6VP
441-0021 XB-GBN
441-0022 N53GG
441-0023 VH-LBY
441-0024 TG-UME
441-0025 VH-XMD
441-0026 VH-AZW
441-0027 N900HA
441-0028 N441DS
441-0029 N441RB
441-0030 VH-OCS
441-0031 PNC-204
441-0032 N441M
441-0034 N441BW
441-0035 G-FRAZ
441-0036 N3AW
441-0037 N81PN
441-0038 VH-LBZ
441-0039 N74BY
441-0040 N441RA
441-0041 N441FS
441-0042 VH-LBA
441-0043 N445AE
441-0044 N549GA
441-0045 N441JA
441-0046 N441CA
441-0048 N60FJ
441-0049 N441EB
441-0050 N83A
441-0051 N725SV
441-0053 N441MJ
441-0054 N454EA
441-0055 N477B
441-0056 N441MS
441-0057 N56MS
441-0058 C-GPSP
441-0059 HK-3401
441-0060 N88598
441-0061 N55MS
441-0062 N441KP
441-0063 YV-2390P
441-0064 N88823
441-0065 N1220W
441-0066 N185GA
441-0067 YV-225CP
441-0068 N3WM
441-0070 N621SC
441-0072 C-FMSP
441-0073 D-IAAC
441-0074 N441RK
441-0075 LV-MMY
441-0076 N441RC
441-0077 LV-MRU
441-0078 N441FC
441-0079 C-FWRL
441-0080 N441BL
441-0081 VH-LEM
441-0082 LV-MRT

441-0083 N26BJ
441-0084 N84LJ
441-0085 N347AR
441-0086 C-GAGE
441-0087 N687AE
441-0088 VH-EVP
441-0089 N441DZ
441-0090 N13NW
441-0091 VH-LBX
441-0092 N8970N
441-0093 N441K
441-0094 N555GD
441-0095 VH-JFD
441-0096 N451WS
441-0097 N434AE
441-0098 N441WP
441-0100 N234CC
441-0101 MT-224
441-0102 VH-OAA
441-0103 N441MD
441-0104 N273AZ
441-0105 N11MM
441-0106 VH-YOL
441-0107 C-FGPK
441-0108 N108TJ
441-0109 VH-FMQ
441-0110 HP-...
441-0111 YV-25CP
441-0113 VH-XMJ
441-0114 HK-2538P
441-0115 RP-C549
441-0116 N441GE
441-0117 N441CJ
441-0118 5N-APZ
441-0119 N22CG
441-0120 N544AL
441-0121 N44776
441-0122 N2624D
441-0123 N441ND
441-0124 N555FT
441-0125 N48BS
441-0126 N337C
441-0127 N322AN
441-0128 N441TM
441-0129 N900PS
441-0130 VH-KDN
441-0132 VH-MSO
441-0133 N4441T
441-0134 N441SM
441-0135 N502BR
441-0136 N207SS
441-0137 N441JA
441-0138 VH-KUZ
441-0139 N441LL
441-0140 C-GCTA
441-0141 VH-CFD
441-0142 XC-AA11
441-0143 N26PK
441-0145 N31CR
441-0146 N146EA
441-0147 N999BE
441-0148 XB-FVK
441-0149 N600VT
441-0150 S5-CAE
441-0151 N2628M
441-0152 HK-3620X
441-0153 C-FMSK
441-0155 XB-FND
441-0156 N829JC
441-0157 VH-YFD
441-0158 N3TK
441-0159 VH-NFD
441-0160 D-IAAV
441-0161 N2721D
441-0162 N477SJ
441-0163 N38AA
441-0164 VH-LFD
441-0165 N140MP
441-0166 N624CB
441-0167 N332DM
441-0168 N2722D
441-0170 N783ST

441-0171 N14NW
441-0172 N441SA
441-0173 N2722Y
441-0174 VH-IJQ
441-0175 N988AE
441-0176 N441DK
441-0178 HK-3614
441-0179 PK-FKL
441-0180 N2723X
441-0181 N441W
441-0182 VH-AZB
441-0183 N23EA
441-0184 OO-LFL
441-0185 N999DF
441-0186 N7NW
441-0187 N88GW
441-0188 N111RC
441-0189 XC-AA68
441-0190 N2725N
441-0192 N2725U
441-0193 D-IEGA
441-0194 N26RF
441-0195 N195FW
441-0197 N441JK
441-0198 N447WS
441-0199 N441AW
441-0200 N423TG
441-0201 N77ML
441-0202 N441Z
441-0204 N441TA
441-0205 N3330S
441-0206 HK-3613
441-0207 G-FPLC
441-0208 XC-GOO
441-0209 N711GE
441-0210 D-IAPW
441-0211 N6851T
441-0212 N6851X
441-0213 N6851Y
441-0214 N441EW
441-0215 N665TM
441-0216 C-FNWC
441-0217 N441BB
441-0218 N846YT
441-0219 N555WA
441-0220 N6853T
441-0221 N881CD
441-0222 N500GM
441-0223 N92JC
441-0224 XC-GON
441-0225 N441VH
441-0226 N6281R
441-0227 YV-533P
441-0228 N88QT
441-0230 N441G
441-0231 N441YA
441-0232 HK-3562
441-0233 N6855P
441-0234 N42DT
441-0235 XB-EWO
441-0236 VH-LBC
441-0237 N700JG
441-0238 N3NC
441-0239 HK-3615
441-0240 N413PC
441-0242 N441WT
441-0243 N441HA
441-0245 N441EL
441-0246 N246SP
441-0247 N6857T
441-0248 N441SX
441-0249 N911ER
441-0250 C-GLBL
441-0251 HK-3573
441-0254 N410MA
441-0255 N9683N
441-0256 N256DD
441-0257 N505HC
441-0258 N258VB
441-0259 N441CD
441-0260 VH-TFB
441-0261 HK-3512

441-0262 N489WC
441-0263 N233RC
441-0264 F-ODUJ
441-0265 N88798
441-0266 N800YM
441-0267 N88727
441-0268 VP-BPH
441-0269 N88834
441-0271 HK-3456W
441-0272 N394G
441-0273 HK-3009
441-0274 N322P
441-0275 XC-HHU
441-0276 HK-3596
441-0277 HP-...
441-0278 N15DB
441-0280 N986MC
441-0281 N689AE
441-0282 N6832M
441-0283 C-GMSL
441-0284 N441AB
441-0285 N707MA
441-0286 C-FUDY
441-0287 HK-3540E
441-0288 N77R
441-0289 N377CA
441-0290 N88692
441-0291 N666HC
441-0293 G-FCAL
441-0294 G-OFHJ
441-0295 N6837R
441-0296 VH-LBD
441-0297 N6838T
441-0298 N245DL
441-0299 N6840T
441-0300 N441DB
441-0301 N328VP
441-0302 N500RJ
441-0303 N441JR
441-0304 N6860C
441-0305 N441CX
441-0306 N45TP
441-0307 N8617K
441-0308 N33DE
441-0309 HK-3568
441-0310 N977MP
441-0311 N832AD
441-0312 N514MC
441-0313 N313DS
441-0314 N894FL
441-0315 N441HF
441-0316 N405MA
441-0317 N1843S
441-0318 N300NK
441-0319 XC-AA10
441-0320 HK-3550
441-0321 N441PJ
441-0322 N536MA
441-0323 N441S
441-0324 HK-3431
441-0325 N30FJ
441-0326 N326RS
441-0327 N441AG
441-0328 HP-...
441-0329 N12099
441-0330 N747JB
441-0331 N68HS
441-0332 HP-1189
441-0333 HK-3418
441-0334 N444AK
441-0335 HK-3497
441-0336 N441JD
441-0337 N337KC
441-0338 N1210Z
441-0340 N1211C
441-0341 HK-3549
441-0342 N441DD
441-0343 N835MA
441-0344 OO-KNM
441-0345 N....
441-0346 N1212C
441-0347 N1212K

441-0348	HK-3529
441-0349	N274GC
441-0350	AP-BCY
441-0351	N369PC
441-0352	AP-BCQ
441-0353	5N-ATR
441-0354	HK-3337
441-0355	AP-BCW
441-0356	HK-3532
441-0357	D-IOHL
441-0358	D-IJLF
441-0359	N40FJ
441-0360	N441RW
441-0361	N592G
441-0362	D-IAGA

Cirrus

001	4X-COD
001	N200SR

Convair

387	N301K

De Havilland Canada

DHC-6
437	N143Z
596	N16NG
782	AP-BBR

DHC-7
106	N54026

DHC-8
024	5N-MGV
391	HK-....
435	N759A
440	N724A
441	N725A

Dornier

Dornier 228
8146	5N-MPS
8206	PH-MNZ
8226	1112
8227	1113
8228	1114

Dornier 328
3019	N328DC
3024	N28CG
3027	N653PC
3034	N38CG
3078	OE-GBB
3083	5N-SPC
3086	5N-SPD

EMBRAER

040	OO-SXB

Fokker

Fokker F27
20253	U-05
20287	U-06
20321	U-01
20324	U-02
20327	U-03
20329	U-04

Gulfstream

Gulfstream 1
1	ZS-NVG
5	F-GFGT
7	N5VX
11	N7SL
16	N615C
18	N48PA
20	F-GFMH
22	C-GKFG
23	OE-BAZ
25	4X-ARH
26	YV-82CP
27	N415CA
28	N719RA
29	N431G
33	9Q-CBY
34	5Y-BLR
35	9Q-CBD
38	N717RS
39	EC-EVJ
40	EC-FIO
41	EC-EZO
42	ZS-OCA
44	F-GFGV
45	N65CE
49	F-GFIC
52	N18TF
54	N26AJ
56	YV-46CP
57	PK-TRM
58	XB-FLL
60	PK-TRL
63	N580BC
64	EC-EXS
67	N806W
70	N331H
71	F-GFIB
74	N701BN
75	PT-KYF
76	P4-JML
78	HK-3681
79	N79HS
80	F-GGGY
81	5Y-BMR
82	N12RW
84	N184K
86	N10TB
88	C-GPTN
89	N789G
91	02
92	N3NA
93	N137C
94	N8E
95	N500RN
96	N2NA
97	YV-85CP
101	F-GNGU
102	HA-ACV
103	XC-AA57
104	C-FHBO
106	C-FAWG
107	N71CJ
109	N109P
115	N61SB
116	N328CA
117	YV-08CP
119	YV-28CP
122	F-GFEF
124	D2-EXD
125	N5NA
126	N110RB
127	XA-TDJ
129	N129AF
130	PK-TRO
131	5Y-BLF
132	N154RH
134	4X-ARF
136	XA-TBT
138	XA-RLK
139	N196PA
140	F-GFCQ
141	ZS-NHW
142	EC-EXQ
145	HK-3580W
146	N906F
148	C-FWAM
149	N192PA
151	N4NA
153	EC-EXB
154	C-GNAK
155	9Q-CBJ
156	N41LH
157	N94SA
158	N2NC
159	XA-RJB
160	N3
161	N925GC
164	N290AS
165	N657P
166	4X-ARG
167	N717RA
169	N200AE
170	YV-620CP
171	YV-621CP
172	N172RD
173	I-TASC
175	YV-453CP
177	PK-CTE
179	XC-AA53
180	155724
182	155725
183	155726
184	155727
185	155728
186	155729
187	155730
188	C-FAWE
189	C-GPTG
190	5Y-...
191	PK-RJA
192	YV-76CP
193	PK-TRN
194	I-MKKK
195	N190PA
196	N659PC
197	N197RM
199	YV-83CP
200	N750G
322	N71RD
323	HI-678CA

Gulfstream Commander

11600	N47TT
11601	N63DL
11602	HK-....
11603	YV-170CP
11604	HK-....
11605	N125MM
11606	XC-ALO
11607	N840VM
11610	N840AA
11611	HK-3424
11612	LV-OEI
11613	SE-GSS
11614	N74EF
11615	HC-...
11616	N133DL
11617	N73EF
11618	YV-39CP
11619	N72TB
11620	HK-3680
11621	N920WJ
11622	N926SC
11624	N840CF
11625	PT-WIC
11626	N5878K
11627	N24A
11629	N940AC
11630	N106TT
11631	XC-TXA
11632	YV-...
11633	HK-2495P
11634	HC-BHU
11635	N489SC
11636	N980AK
11637	YV-394CP
11638	N3XY
11639	HB-LOL
11640	YV-630CP
11641	XB-SYV
11642	HK-2599P
11643	N840JC
11645	YV-310CP
11646	N165BC
11647	N840LC
11648	N5900K
11649	XC-HHH
11650	XB-DTW
11651	HK-2601
11652	XC-AA27
11653	HK-3290P
11654	YV-87CP
11655	N406CP
11656	XC-OAX
11657	YV-O-KWH-3
11658	N840JW
11659	XC-BAD
11660	N81JN
11661	N840DC
11662	N5914K
11663	N840NB
11664	N7057A
11665	N48BA
11666	YV-626CP
11667	11667
11668	HK-3912
11669	HC-BUD
11671	YV-2346P
11672	VP-BLK
11673	YV-779CP
11675	N319BF
11676	N129TB
11677	XB-GMT
11678	XC-AA36
11679	N840VB
11680	N680WA
11681	LN-FWA
11682	N910EC
11683	N60VS
11684	N51WF
11685	YV-105CP
11687	N90WE
11688	XC-HAB
11689	N265JH
11691	N800BM
11692	N152X
11693	N840MD
11694	YV-...
11695	XC-AA56
11696	N884D
11697	N97696
11698	YV-601P
11699	N425MM
11700	N840SM
11701	N50WF
11702	N64PS
11703	N5955K
11709	N840LE
11719	N5956K
11721	HK-3700
11722	HK-3447P
11723	YV-505CP
11724	N37SB
11725	YV-...
11726	YV-2505P
11727	N840V
11728	HK-3385P
11729	N67FE
11730	N835CC
11731	N815BC
11732	N8VL
11733	11733
11734	N600BM
11735	YV-212CP

Gulfstream 900

15001	N900EC
15002	N721MR
15003	HP-...
15004	HP-...
15005	ZP-TWZ
15006	YV-492CP
15007	HK-3473W
15008	N6BZ
15009	XC-JDB
15010	N.....
15011	VP-CAY
15012	YV-...
15013	N42SL
15014	N771FF
15015	N544GA
15016	N950TJ
15017	YV-880CP
15018	N611
15020	N600CM
15021	ZP-...
15022	YV-521CP
15023	HC-...
15024	N79SZ
15025	N29GA
15026	N120EK
15027	PT-WLD
15028	XC-MLM
15030	YV-822CP
15031	N471SC
15032	HP-...
15033	VP-BMZ
15034	YV-...
15035	N900DS
15037	N900ET
15038	PT-OQY
15039	N144JB
15040	XA-CHM
15041	N77HS
15042	C-FGWT

Commander 980

95000	YV-...
95001	HK-...
95002	N980BH
95003	N9756S
95004	CP-...
95005	N171CP
95006	YV-581CP
95007	N515AM
95008	YV-.....
95010	ETE-1363
95011	N980EC
95012	N980SA
95013	N519HB
95014	N9767S
95015	N655PC
95016	HK-3474
95017	YV-56CP
95018	XC-PFB
95019	PK-ODR
95021	ZP-TTU
95022	HK-3484
95023	N888SF
95024	N36JT
95025	N54UM
95026	HC-....
95027	N2701L
95028	YV-...
95029	N600MM
95030	N980MD
95031	N980EA
95032	N265EX
95033	N126M
95034	N980WM
95035	ZP-...
95036	N5356M

95038 PT-...
95039 D-IHSI
95040 MT-219
95041 N3U
95042 XA-LEK
95043 HK-3461
95044 JA8604
95045 XC-TXA
95046 MT-221
95047 N1981S
95048 N980DT
95049 N980GR
95050 HK-3408
95051 XB-ACO
95052 XC-ENL
95053 ZP-...
95055 FAC-5553
95056 TI-...
95057 HK-3444
95058 YV-893CP
95059 HK-3819
95060 N9812S
95061 XC-AA15
95062 HK-3406
95063 N980GZ
95065 HK-3455
95066 EJC-115
95067 YV-407CP
95068 XC-AA62
95069 HB-LQA
95070 JA8600
95071 N810EC
95072 XC-HGH
95073 N200TT
95074 YV-980CP
95075 HK-3407
95076 N76HH
95077 N221K
95078 JA8826
95079 HK-3481
95080 HK-....
95081 N808NT
95082 MT-222
95083 HK-3450
95084 YV-787CP

Commander 1000

96001 XC-AA23
96002 N9902S
96003 HP-...
96004 XB-ECT
96005 N94PA
96006 N496MA
96007 N2267U
96008 PJ-NAF
96009 N7031K
96010 HK-3239W
96011 N2270T
96012 C-GJEI
96013 MT-218
96014 N51DM
96015 N325MM
96016 YV-O-MTC-1
96017 N17ZD
96018 HK-3367
96019 N200DK
96020 N93NM
96021 HK-3439
96022 N9942S
96023 HB-GHK
96024 N5450J
96025 N9945S
96026 MT-217
96027 N727CC
96028 N9948S
96029 N79PH
96030 FAC-5198
96031 HP-1149P
96032 N695NC
96033 HK-3366X
96034 C-GMMO
96035 N93MA

96036 N20GT
96037 HK-3389P
96038 RP-C...
96039 HK-3060
96040 MT-214
96041 ETE-1318
96042 N303GM
96043 XC-HFN
96045 HK-2909P
96046 N9966S
96047 HK-3390
96048 N9968S
96049 HK-2951
96050 N900EC
96051 ZS-KZY
96052 ZS-KZZ
96053 N9973S
96054 N71MR
96055 CP-2050
96056 XC-AA19
96057 N34GA
96058 N36AG
96059 HK-3391W
96060 FAH-006
96061 HK-3218
96062 N2VA
96063 N83WA
96064 N6767M
96065 PJ-...
96066 N54GA
96067 YV-53CP
96068 N519CC
96069 HK-3961X
96070 N7896G
96071 N69GA
96072 HK-3279
96073 HK-.....
96074 HK-3253P
96075 XA-...
96076 HK-3275
96077 HK-3192W
96078 N85WA
96079 ZP-TXF
96080 HK-3284W
96081 HP-...
96082 HP-...
96083 EJC-114
96084 XC-AA16
96085 N808NC
96086 HK-3193
96087 N695RC
96088 HK-3291P
96089 N7812
96091 XC-AA20
96092 N211AD
96093 CX-...
96094 N94EA
96095 ZP-TWY
96096 N44SD
96097 N112CE
96098 N269M
96099 HK-3283W
96100 HK-3324P
96201 N115GA
96202 N27VE
96203 ZP-TXG
96204 N4715W
96205 N5852K
96206 N97315
96207 N81432
96208 VH-LTM
96209 N235GA
96210 N238GA

Jetcruzer

001 N102JC

Learfan

E003 N327ML

Mitsubishi

MU-2

014 N155MA
017 N750Q
019 N728F
021 HR-...
023 N555VK
025 N2GZ
026 PT-LMD
028 N2176D
031 N38920
034 XB-FMV
035 N36G
038 N706DM
102 SE-FGO
104 N151JB
106 N28DC
110 N666RK
111 N721FC
113 N859Q
114 N500X
122 C6-...
124 N987MA
126 PT-LOH
129 N555C
131 N755Q
132 N756Q
133 N488GB
135 N66CL
136 N115AP
138 N703DM
139 LV-WME
140 N900TV
144 N176BJ
145 PT-BOY
154 N720JK
156 XB-HOV
157 LV-RAZ
158 PT-BPY
159 N888DS
161 N400TR
162 N573MA
165 YV-...
167 N310MA
170 N58CA
173 N18BF
174 C-FCCD
175 PT-BZW
177 N177SA
179 N4WD
182 N2100T
183 N67SM
187 OY-DLM
188 PT-BZY
189 XB-XOI
190 N140CM
191 HA-...
193 N8PC
194 HI-...
196 PT-DTL
198 N91MM
207 N122MA
208 N111HF
209 N212CC
210 N38BE
212 N1978P
213 PT-LYI
215 PT-ICD
216 N922FM
217 N183MA
219 N185MA
221 N800BY
222 SE-GHH
224 N500PS
226 N187SB
228 N228WP
229 N100CF
231 N346VL
232 PT-WIX
233 N89CR

239 N222LR
240 N64LG
241 N700RX
243 HK-2060W
244 PT-LEW
247 SE-KBX
248 N725FN
250 N500XX
253 N430DA
257 N344KL
260 N790CA
261 N281MA
263 N30RR
264 N264KW
265 N121CH
266 N312MA
267 N59RW
268 PT-JGA
271 N708DM
272 N666RL
273 XB-ZIP
277 N390K
280 N707FN
282 OB-1284
283 SE-GHA
284 N726FN
285 N11SJ
286 N66AW
287 SE-GHB
289 SE-GHC
290 N14YS
291 N291MB
293 SE-GHD
294 SE-GHE
297 N7DD
298 PT-OFV
299 SE-GHF
301 N461MA
302 N462MA
303 VH-...
305 N50K
306 N495MA
308 N709FN
310 N290GC
311 N311JB
313SA N100KE
314 PT-LTC
315 N99SR
319 N475CA
320 SE-IOZ
321SA N5LJ
322 N40KC
323 N216CD
324 N143JA
325 N325MA
326 N700LW
327 N375AC
328 N767MD
329 N60AZ
330 N261WB
332 N555JP
333 N12SJ
337 SE-IOV
338 N456PS
339 N60NJ
341 XB-LIJ
345 SE-IUA
346 N10UT
348SA N500GK
349SA N316EN
350SA PT-OVW
351SA N10MR
352SA N41AD
353SA N738MA
354SA PT-OIP
355SA N741MA
357SA N21HP
358SA N60BT
361SA LV-MCV
362SA N140CP
363SA N110GC
364SA N749MA

366SA N2RA
368SA N78WD
369SA N202MC
370SA N370MA
371SA ZK-ECR
373SA N766MA
374SA N122CK
375SA N17JQ
380SA N333RK
381SA N400JH
382SA N772MA
383SA N40JJ
384SA N774MA
385SA LV-MLT
386SA LV-RAC
387SA N333WF
388SA PT-OOS
389SA N400ES
390SA N782MA
391SA N783MA
392SA N392P
393SA N82WC
397SA N777WM
398SA N61DP
399SA N9052Y
400SA C-GBVB
403SA N12LE
405SA N966MA
406SA F-GCJS
407SA N750CA
409SA N78FS
410SA N701K
411SA N400PS
412SA N25GM
414SA N72TJ
415SA N860SM
416SA N44AX
417SA N42DE
418SA N16CG
419SA N322GB
420SA D-IKKY
423SA N100NP
424SA N157MA
425SA N575CA
426SA N36AT
427SA N40AM
428SA PT-LIR
429SA N77DB
430SA N856JC
431SA N171MA
432SA N44VR
433SA N5LC
434SA C-FGEM
435SA N104JB
436SA N928VF
437SA N145FS
438SA N10VU
439SA N439BA
440SA N219MA
446SA N15TR
447SA N48NP
448SA N68CL
449SA N77DK
450SA N229M
451SA N1728S
452SA N388NC
453SA PT-OIY
454SA N19GA
458SA N458BB
459SA N459SA
502 N211RV
507 N134MA
509 N154WC
510 N227
511 N60KG
514 N999TA
515 N40PP
516 VH-JES
519 VH-UZC
521 VH-KOH
522 N617BB
525 N360JK

527 VH-MNU
528 VH-UZB
534 N920S
536 N255C
537 N93UM
540 N869D
544 VH-KOF
545 N108SC
548 N200RX
549 C-FTOO
551 N3330K
552 N539MA
553 N755MA
557 N345SA
558 N45BS
560 N776CC
561 VH-JMZ
562 N66CY
563 N198MA
566 MU-1550
567 N1VY
568 N334EB
572 N300RX
573 N203GA
574 N19GU
577 N81601
580 N103Q
581 N93AH
582 N155BA
583 XB-NEB
584 N900YH
586 N217SB
587 N212BA
588 N32CK
589 N155AS
590 N54US
591 N99BT
593 C-GJAV
594 N298MA
598 N15YS
599 N121BA
600 N113SD
601 C-FROM
604 N5AP
605 N621TA
610 N92ST
612 N799MA
614 N21JA
615 N349MA
616 N110MA
617 N8484T
619 XB-NUG
621 HK-2120P
622 ZS-MGF
623 N22YA
625 N22522
626 EC-GLU
627 5Y-BJX
628 C-FROW
629 ZS-NDM
630 PH-...
631 N629TM
633 N503AA
634 N400SG
635 OY-ARV
637 N637WG
640 N490MA
641 N211BE
642 N680CA
643 N375CA
644 LV-WJY
645 N6569L
646 N118P
647 N44KU
650 PT-LFX
652SA N533DM
653 N45CE
655 N535WM
656 N866D
657 N740PB
658 N741FN
659 N6KF

660 N4065D
661SA N821MA
662 C-FFFG
663 N823MA
664 N59KS
666 SE-IVA
667 N305CW
668 N500PJ
670 N742FN
671 N4203C
672 C-FTWO
673 N103RC
674 N861E
675 N68TN
677 N915RF
678 HK-2403
679 N679BK
680 N201UV
681 N361JA
682 N13RR
683 C-FIFE
685 YV-O-BDA-3
686 N717PS
687 N687HB
688 N688RA
689SA N112MA
692 N869P
694 N13YS
696 N301GM
697SA N857MA
698SA N263ND
699SA N32EC
700SA N860MA
702SA YV-144CP
703SA N338CM
704SA LV-MGC
705SA N24WC
706SA C-FJEL
707SA VH-NMU
708SA N868MA
709SA N105WM
710SA N710G
711SA PT-WST
713SA N89DR
714SA N994PE
718SA N150BA
719SA N777LP
720SA N3UN
722SA N722MU
723SA N895MA
724SA N300GM
725SA C-FKIO
726SA N2CJ
727SA N4TA
728SA N9NC
729SA N61BA
731SA N888PH
732SA N80HH
733SA N142BK
734SA N331J
736SA N175CA
737SA N888RH
738SA N941S
739SA N8083A
740SA N360RA
741SA N193AA
744SA N941MA
745SA N942ST
748SA N102BG
749SA PT-LIS
750SA N130MS
751SA N117H
752SA N37AL
753SA N174MA
754SA N46AK
755SA SE-KGO
756SA N1709M
757SA N21MU
758SA N261MA
759SA LV-ODZ
760SA N322TA
761SA N94JK

762SA N762JC
763SA N26AP
764SA N196MA
766SA N6KE
768SA N331MA
771SA N221MA
772SA N772DA
773SA N350RG
774SA PT-OHK
775SA N70KC
777SA XA-RNL
778SA N10HT
780SA N321TP
781SA N60KC
783SA C-FFSS
784SA N50KW
785SA C-GAMC
787SA N984RE
788SA N610CA
789SA N21CJ
790SA F-GEQM
791SA N888WW
792SA N34AL
793SA HZ-AMA
794SA N794MA
796SA C-GGDC
798SA N661DP
799SA N5LN
801 22-001
803 22-003
804 22-004
805 22-005
806 22-006
807 22-007
808 22-008
809 22-009
810 22-010
812 22-012
813 22-013
814 22-014
815 22-015
816 22-016
817 22-017
818 22-018
819 22-019
820 22-020
901 73-3201
902 73-3202
903 83-3203
904 83-3204
906 93-3206
908 03-3208
909 13-3209
910 13-3210
911 13-3211
912 13-3212
913 23-3213
914 23-3214
915 33-3215
916 33-3216
917 33-3217
918 43-3218
919 33-3219
920 63-3220
921 63-3221
922 73-3222
923 83-3223
924 83-3224
925 93-3225
926 03-3226
927 03-3227
951 53-3271
952 73-3272
953 83-3273
954 93-3274

1501SA N43DC
1502SA N71DP
1503SA N407MA
1504SA N1727S
1505SA N101SN

1506SA N15MV
1507SA PH-BOA
1508SA PT-LHH
1509SA N973BB
1510SA N17HG
1511SA N27TJ
1512SA N60FL
1513SA N513DC
1514SA YV-108CP
1515SA N802SM
1516SA N89SC
1517SA N727TP
1519SA N33EW
1520SA N429WM
1521SA C-FRWK
1522SA C-GFFH
1523SA N65JG
1525SA N35RR
1526SA N44MA
1528SA 5Y-BIZ
1529SA N132BK
1530SA PT-LSQ
1533SA N452MA
1534SA N667AM
1535SA N454MA
1536SA N157GA
1537SA N152BK
1538SA N538EA
1539SA N42AF
1542SA N88HL
1543SA N469MA
1544SA N223JS
1545SA N900M
1546SA PT-LIK
1548SA PH-FWM
1549SA N888SE
1550SA N64WB
1551SA N444WE
1552SA N246W
1553SA N479MA
1554SA N480MA
1556SA N747SY
1557SA N999UP
1558SA N5PQ
1559SA N880AC
1560SA N513DM
1561SA N64MD
1562SA N1164F
1563SA N7HN
1564SA N95MJ
1565SA N100BY
1566SA N64BA
1567SA PT-LPB
1568SA PT-LJS
1569SA N222HH

Partenavia

9001 N901TP

Piaggio

1001 I-PJAV
1002 I-PJAR
1004 N180BP
1006 N34S
1007 C-FNGA
1008 LZ-PIA
1009 D-IHMO
1010 N23SV
1011 D-IHOT
1012 N180TE
1013 I-PJAP
1014 I-ACTC
1015 I-ALPN
1016 D-IGOB
1017 D-IJCL
1018 N220TW
1019 D-IHRA
1020 F-GNAE
1021 D-IPIA

1022 F-GMCP
1024 MM62159
1025 MM62160
1026 MM62161
1027 MM.....
1028 MM62162
1029 MM62163
1030 MM62164
1031 MM.....
1032 D-IMLP

Pilatus

PC-XII

P-01 HB-FOA
P-02 HB-FOB
101 N312BC
102 HB-FOE
103 ZS-NWZ
104 HB-FQA
105 N269JR
106 N106WA
107 N62JT
108 N667JJ
109 VH-FMC
110 VH-FMF
111 N222CM
112 N263AL
113 N562GA
114 N121RF
115 N398CA
116 N321DH
117 N12FA
118 N802HS
119 N432CV
120 N112AF
121 ZS-PVT
122 VH-FMP
123 VH-FMW
124 N124BW
125 N695WF
126 Z-KEN
127 N33JA
128 JA8599
129 N961PC
130 N1TC
131 N88EL
132 N361DB
133 N133CZ
134 HB-FOG
135 LV-WSN
136 N79CA
137 N116AF
138 VH-FMZ
139 N96WF
140 N129DH
141 PT-WKF
142 OE-EKD
143 ZS-BEL
144 N312PC
144 SAAF8030
146 N612PC
147 ZS-NYM
148 N2JB
149 N15TP
149 N464WF
150 N150PB
151 C-FKAL
151 N151PB
152 N444CM
153 N153PB
153 N153PB
154 N144PC
155 N718JP
156 N156SB
157 HB-FOI
158 HB-FOJ
159 N159PB
160 N55BK
161 VH-FAM
162 N162PB

163 N49LM
164 C-FMPA
165 N165PD
166 N.....
167 N774DK
168 N168WA
169 N696TS
170 N170PD
171 N172JS
172 N172PB
173 N173KS
174 N174PC
175 N118AF
176 VP-BLS
177 D-FCJA
178 N178PC
179 N340RE
180 HB-FSP
181 N601BM
182 N182PE
183 N183PC
185 N185PB
186 N186WF
187 HB-FSV
188 HB-FSW
189 HB-FSX
190 HB-FSY
191 HB-FSZ
192 HB-FQB
192 HB-FQA
193 HB-FQC
194 N194PC
195 HB-FQD
198 ZS-DMM

Piper

PA-31T
Cheyenne I

7804002 N301PT
7804003 N913CR
7804004 N500MY
7804005 N96MA
7804006 N6134A
7804007 N22UC
7804009 N6125A
7804010 N7VR

7904001 D-ILOR
7904002 D-IEIS
7904003 N37PJ
7904004 OY-BHU
7904005 N3WE
7904006 N300WP
7904007 N231SW
7904008 N528DS
7904009 N6192A
7904010 N212HH
7904011 N6JM
7904012 N4RX
7904013 N18KP
7904014 HB-LLK
7904015 N131DF
7904016 N503WR
7904017 N23243
7904018 N23250
7904019 N504TQ
7904020 N187SC
7904021 N232DH
7904022 I-SERV
7904023 N52PC
7904024 N23334
7904025 N531SW
7904026 EC-EIM
7904027 N98TG
7904028 N5WC
7904029 N93SH
7904030 I-POMO
7904031 N54EZ
7904032 N188CA
7904033 N290RS

7904034 C9-ENT
7904035 N23426
7904036 N149CC
7904037 N200FB
7904038 N888AH
7904039 HC-BNF
7904040 YV-O-MAC-2
7904041 N65CK
7904042 C-FCOS
7904043 N70TW
7904044 N523PD
7904045 LV-MYX
7904046 N23658
7904047 N479SW
7904048 N501PT
7904050 N222GW
7904051 N49RM
7904052 N137CW
7904053 XB-FXU
7904054 CC-PHM
7904056 N879SW
7904057 N2348W

8004001 N118BW
8004002 N424CM
8004003 N2556R
8004004 N2587R
8004005 PT-OAJ
8004006 N100WT
8004007 PT-OPQ
8004008 N701PT
8004009 N98GF
8004010 N2415W
8004011 D-IBIW
8004012 N41PN
8004013 N31MB
8004014 PT-OVE
8004015 N244AB
8004016 N75TW
8004017 N2458W
8004018 N32JP
8004019 N992TT
8004020 N700BS
8004021 N49AC
8004022 N141GA
8004023 N4QG
8004024 I-SASA
8004025 N314DD
8004026 N76CP
8004027 N73BG
8004028 N2325V
8004029 N93CN
8004030 N66TW
8004031 N60JW
8004033 N935CA
8004034 N82TW
8004035 N2369V
8004036 N121BE
8004037 N114JR
8004038 PT-OPF
8004039 PT-OLF
8004040 N744WP
8004041 N768MB
8004042 N75SK
8004043 PT-WMU
8004044 HB-LQP
8004045 N20DL
8004046 CX-BRU
8004047 N40BT
8004048 N169DC
8004050 N812CM
8004051 N480CA
8004052 I-LIAT
8004053 D-ILCE
8004054 N176CS
8004055 N849KM
8004056 XB-GQI
8004057 N280SW

8104001 N181SW
8104002 CC-PNS
8104003 N2356X

8104004 N70GW
8104005 N199MA
8104006 N37TW
8104007 N222BC
8104008 N38TW
8104009 N816CM
8104010 N980CA
8104011 N839AB
8104012 N39TL
8104013 N212GM
8104014 N2403X
8104015 N381SW
8104016 N481SW
8104017 PT-OTV
8104018 N801CA
8104019 N191MA
8104020 N31WM
8104021 N112BB
8104022 N300SR
8104023 N234SB
8104024 N225CA
8104025 N2467X
8104026 N2480X
8104027 N681SW
8104028 N38RE
8104029 N803CA
8104030 N917F
8104031 N90VM
8104032 D-IAPA
8104033 N881SW
8104034 PT-WEF
8104035 N2519X
8104036 N633WC
8104037 N822CM
8104038 N27BF
8104039 N824VA
8104040 N711LD
8104041 PT-OKT
8104042 D-IJET
8104043 N527JC
8104045 N514M
8104046 N221RC
8104047 N83MG
8104048 N826CM
8104049 N2435Y
8104050 N515GA
8104051 PT-OVB
8104052 N422HV
8104053 N454CA
8104054 N555AT
8104055 D-ISIG
8104056 N711ER
8104057 D-IHJK
8104058 N999NP
8104059 N819SW
8104060 N74ML
8104061 N2552Y
8104062 PT-WMX
8104063 CC-PBT
8104064 N502MM
8104065 N777G
8104066 D-IHMM
8104067 D-IACR
8104068 XA-RLB
8104070 CC-CBD
8104071 CC-CRU
8104072 N482TC
8104073 PT-WHN

8104101 D-IASW

PA-31T
Cheyenne IA

8304001 N9183C
8304002 N2522Z
8304003 N26DV

1104004 N90WW
1104005 N2427W
1104006 N2434W
1104007 S3-BHN

1104008 N9266Y
1104009 N9268Y
1104010 N234PC
1104011 N2434V
1104012 N284BB
1104013 N65EZ
1104014 D-ILGA
1104015 D-IIAH
1104016 D-IFHZ
1104017 D-IEPA

1 N100VK

11563 XB-KLY

PA-31T
Cheyenne II

7400003 N231PT
7400005 N10BQ
7400008 N66837
7400009 N910HM
7520001 N234K
7520002 F-GFLE
7520005 N1976J
7520006 N19NM
7520007 OH-PNT
7520008 C-GNKP
7520009 N113RC
7520010 N3951F
7520012 PT-KME
7520014 FAC-5739
7520015 C-FYBV
7520016 5N-AUT
7520017 N707ML
7520021 C-GZQD
7520022 LV-LTV
7520023 N40TT
7520024 N431AC
7520025 C-GHSI
7520026 N181DC
7520027 YV-1990P
7520029 N43SQ
7520030 N444RC
7520031 N590DL
7520032 N4116W
7520033 I-CGAT
7520034 I-HYDR
7520035 N17TU
7520036 N60049
7520037 N85AJ
7520039 C-GJPT
7520043 TI-....

7620001 OH-PHA
7620002 N610P
7620003 C-GQCC
7620005 I-PALS
7620008 C-FYTK
7620010 C-GXBF
7620011 F-GFEA
7620012 C9-JTP
7620014 N460LC
7620015 PK-PJH
7620016 C-GTFP
7620018 N318MA
7620019 C-GMDF
7620020 F-BXSK
7620022 N57MR
7620023 C-FMHB
7620026 C-GXCD
7620028 N555PM
7620029 C-GEBA
7620032 F-GALD
7620033 C-....
7620034 XA-....
7620035 N21KN
7620036 C-GNDI
7620037 N5136V
7620039 C-FWUT
7620040 N400CM
7620041 N27856

7620042 N8131F
7620044 PT-OPH
7620045 F-BTEE
7620046 XC-UAT
7620047 N111CT
7620049 N37RL
7620050 N2FJ
7620051 N631WF
7620052 N737WB
7620053 C-GHRM
7620054 F-GFBF
7620055 ZK-ROM
7620057 6V-AGO

7720002 N777SW
7720003 N34HM
7720005 N28BG
7720007 N99VA
7720008 N730PT
7720009 ZK-POD
7720011 N20WL
7720012 N119EB
7720014 N82094
7720016 N311LM
7720017 N12TW
7720018 3A-MBA
7720019 N75465
7720020 N37HB
7720027 N143RJ
7720028 N171JP
7720029 F-GAMP
7720031 F-GEPE
7720032 N747RE
7720033 TU-TJL
7720035 N200NB
7720036 F-GGRV
7720037 N112BL
7720038 N77NL
7720039 PT-LUJ
7720041 F-ODGS
7720042 F-GJPE
7720043 F-GJDK
7720044 N23MV
7720045 N82156
7720047 YV-124CP
7720048 N82163
7720049 N82161
7720050 YV-2451P
7720051 85-1609
7720052 C-GXTC
7720053 N613HC
7720054 N30DU
7720055 5B-EGZ
7720056 OE-FDS
7720057 N61SG
7720058 C-GKMX
7720059 LV-MGD
7720060 N160NA
7720062 C-GEAL
7720063 F-GFUV
7720064 N60DR
7720065 LV-MDG
7720066 N888JM
7720067 F-GHJV
7720068 5B-GGZ
7720069 N411LM

7820001 N115PC
7820002 C-FRJE
7820003 C-GKMV
7820004 N33LA
7820005 N131AF
7820006 N14EA
7820008 YV-703CP
7820009 N131PC
7820010 F-GGAT
7820011 N884CA
7820012 N781SW
7820013 N41NS
7820014 C-GSBC
7820015 F-GHTA
7820017 HB-LRV

7820018 N35RT	7920026 VH-TNP	8020022 N38VT	8120017 CP-1678	8166031 ZS-LSY
7820019 N727SM	7920027 C-GLAG	8020023 N611LM	8120018 N794CA	8166032 N107MC
7820020 LV-MHL	7920028 N10MC	8020024 N103SP	8120019 HC-BIF	8166033 N42ND
7820021 N182ME	7920029 N68BJ	8020025 N135MA	8120020 PT-WNC	8166034 HK-3603X
7820022 N484SC	7920030 N288DC	8020026 N155DS	8120021 N85HB	8166035 HK-3074P
7820023 C-GGPS	7920032 N23139	8020027 N2483W	8120022 N127AT	8166036 N618SW
7820024 N155CA	7920033 LV-MOD	8020028 PT-OAM	8120024 5T-MAB	8166038 CC-PVE
7820025 N978BG	7920034 PT-OFH	8020029 PT-OED	8120025 D-ICPA	8166039 N56DK
7820026 N570AB	7920035 N111KV	8020030 D-IIWB	8120026 5T-MAC	8166040 N30XL
7820027 LX-RST	7920036 EC-FOT	8020031 N47JF	8120027 N9WD	8166041 C-GFLL
7820028 N82232	7920037 N53AM	8020032 TG-...	8120028 CC-PCY	8166042 N767DM
7820029 C-GGFL	7920038 N333XX	8020033 PT-OCL	8120030 PT-WOR	8166043 N711FN
7820030 PT-OZY	7920039 HB-LOZ	8020034 N97PC	8120031 PT-OJE	8166044 N400WS
7820032 N700PT	7920040 N80CP	8020036 OE-FKG	8120032 HK-2631G	8166045 N600BS
7820033 N63CA	7920041 N66WJ	8020037 F-GIII	8120033 D-IBSA	8166046 N67TW
7820034 N419R	7920042 N711LV	8020038 N300HP	8120035 N888LB	8166047 XB-FSG
7820035 N17NM	7920043 N47CA	8020039 N63CM	8120036 N33MS	8166048 C-FGSX
7820036 N202HC	7920044 PT-OTF	8020040 HK-2451P	8120037 HK-2642P	8166049 PT-...
7820037 YV-2365P	7920045 C-FGWA	8020041 N198AA	8120038 N300PV	8166050 HB-LNX
7820038 C-GVKK	7920046 N700LT	8020042 N716WA	8120040 PT-LRT	8166051 N37SR
7820039 N131MP	7920047 XA-RWR	8020043 HK-....	8120041 N220SC	8166052 N423JB
7820040 N98TB	7920048 N406RS	8020044 F-GCLH	8120042 PT-ODM	8166053 HK-2963
7820042 N700RG	7920049 F-GFVO	8020045 TG-...	8120043 TG-LIA	8166054 N176RS
7820043 N44DT	7920050 LV-MOE	8020046 N1879D	8120044 N515BA	8166055 N85EM
7820045 I-CGTT	7920051 N97TW	8020047 HK-2455	8120045 TG-VAL	8166056 D-ICGA
7820046 N688CA	7920053 N731PC	8020048 PT-WFB	8120046 N178CD	8166057 N774KV
7820047 N926LD	7920054 YV-187CP	8020049 N2338V	8120048 OB-1228	8166058 N825SW
7820049 PT-WFQ	7920055 N23407	8020050 N91TS	8120049 CC-CNT	8166059 N620AD
7820050 N250TT	7920056 5T-TJY	8020053 CC-...	8120050 N194MA	8166060 N66MT
7820051 OE-FBO	7920057 N52TT	8020054 PT-LLG	8120051 HK-3043P	8166061 N151FB
7820052 YV-194CP	7920058 N12GR	8020055 N3998Y	8120052 N67JG	8166062 N58AP
7820053 C-GJJE	7920059 N25MG	8020056 N314TD	8120053 N87YP	8166063 N151MP
7820054 PK-DYR	7920060 N31WE	8020057 N22DT	8120054 ZS-PBS	8166064 HC-BUS
7820056 N51RD	7920061 N44HT	8020058 PT-WME	8120055 N49B	8166066 C-FWPT
7820058 N82290	7920062 N78CA	8020059 N457TC	8120056 F-GGCH	8166067 PT-MFW
7820059 I-....	7920064 PT-LZB	8020060 N112ED	8120057 N382MB	8166068 N612BB
7820060 XA-RMT	7920065 LV-OAP	8020061 PT-OZN	8120058 OE-FMO	8166069 N511SC
7820061 N303KL	7920066 N444PC	8020062 N8CF	8120059 N20WE	8166070 CC-PTA
7820066 D-IKEW	7920067 N74TW	8020063 YU-BPH	8120060 N31HL	8166071 N8EE
7820067 YV-1402P	7920068 N711RD	8020064 N376WS	8120061 PT-LNG	8166073 N57AF
7820069 HC-BVA	7920069 C-...	8020065 C-GNAM	8120062 HK-3025	8166075 N83XL
7820070 YV-2331P	7920071 C-FEQB	8020066 ZP-...	8120063 N715CA	8166076 N51GS
7820073 N7703L	7920072 CC-...	8020067 N79JS	8120064 F-ZBFZ	
7820074 N787SW	7920073 EC-DHF	8020068 N2467V	8120068 N43WH	1166001 HP-1066
7820075 N75TF	7920074 N24DD	8020069 CC-CNH	8120070 PT-OJM	1166002 CC-PML
7820076 N29CA	7920075 OB-1308	8020070 N2469V	8120101 PT-WGJ	1166003 D-ICDU
7820077 N6080A	7920076 HK-2347	8020071 N400CP	8120102 D-INNN	1166004 N90SC
7820079 VH-DRV	7920077 PT-WHI	8020072 N8100M	8120103 N182TC	1166005 N395CA
7820080 N5D	7920078 N23646	8020073 N100CM	8120104 C-FZIC	1166006 N804C
7820081 N2DS	7920079 N789CH	8020074 N62E		1166007 N285KW
7820082 N7FL	7920080 LV-MTU	8020075 N36TW		1166008 D-IIRC
7820083 N53TA	7920081 N600RM	8020076 F-GNAC	**PA-31T**	
7820084 N100WQ	7920082 N109TT	8020077 HK-2585P	**Cheyenne II-XL**	**PA-42**
7820085 N12HF	7920083 PT-LZR	8020078 N2522V	8166001 VH-RSW	**Cheyenne III**
7820086 N186GA	7920084 N666JL	8020079 PT-ODR	8166002 N431CF	7800001 N420PA
7820087 N6107A	7920086 N1183G	8020080 N809SW	8166003 N31XL	7800002 F-GGTV
7820088 N345DF	7920087 N23718	8020082 N82PC	8166005 N410CA	
7820089 N275CA	7920088 HK-3331	8020083 HB-LNL	8166006 N38WA	7801003 OB-1631
7820090 N32KW	7920089 N171AF	8020084 F-ODMM	8166007 N491MB	7801004 N48RA
	7920090 N2349R	8020085 N154CA	8166008 N35CA	
7920001 3A-MIO	7920091 N26SL	8020086 N880SW	8166009 N2488Y	8001001 C-FWCC
7920002 N141DT	7920092 N78TW	8020087 N26PJ	8166010 N200XL	8001002 C-GIDC
7920003 LV-MNU	7920093 N99KF	8020088 N431GW	8166011 N630MW	8001003 N242RA
7920004 N790SD	7920094 OH-PYE	8020089 N333TN	8166012 N321LB	8001004 N717ES
7920005 N6166A		8020090 CC-PMZ	8166013 C-FCED	8001005 N113WC
7920007 N22WF	8020002 N333P	8020091 F-GIPL	8166014 N15KW	8001006 F-FCAV
7920008 C-GVKA	8020003 TG-OIL		8166015 N131XL	8001008 N977XT
7920009 N723JP	8020004 N23WP	8120001 N880CA	8166016 CC-CWH	8001009 OB-....
7920010 D-IOTT	8020005 VH-TTD	8120003 TG-...	8166017 N550MM	8001010 YV-O-ICA-1
7920011 N6175A	8020006 YU-BPF	8120004 D-IKKK	8166018 N22VF	8001011 N619JB
7920012 XB-HNA	8020008 XB-GVI	8120005 PT-OLZ	8166019 HK-2749P	8001012 N25LA
7920015 N200PG	8020011 N90FS	8120006 N2366X	8166020 N22YC	8001013 LV-OOO
7920016 N79FB	8020012 YU-BPG	8120007 N55KW	8166021 N450HC	8001014 N373Q
7920017 LV-MNR	8020013 LV-OGF	8120008 D-ICBH	8166022 N3RK	8001016 N69PC
7920018 YV-655CP	8020014 N801ST	8120009 D-IHMS	8166023 N88B	8001017 N175AA
7920019 N20TN	8020015 HP-809	8120010 PT-OPC	8166024 N58PL	8001018 OB-1365
7920020 N224EC	8020016 N801HL	8120011 PT-WLJ	8166025 N53WA	8001019 OB-1224
7920021 N20PJ	8020017 D-IKET	8120012 C-FJVB	8166026 N77UH	8001020 N332SA
7920022 YV-251CP	8020018 N108UC	8120013 YV-1995P	8166027 N500PB	8001021 N396FW
7920023 XB-RLM	8020019 N85CM	8120014 N8DB	8166028 C-FJAK	8001022 OB-1630
7920024 N159DG	8020020 N118WC	8120015 F-GKRR	8166029 N66JC	8001023 N5PF
7920025 N163SA	8020021 YU-BMM	8120016 N311AC	8166030 C-FCEC	

8001024 N250TJ	5501017 N75LS	5527035 JA8867	LJ-92 N5245F	LJ-203 C-FCGH
8001025 N727PC	5501018 N515RC	5527036 TC-EEE	LJ-93 N9901	LJ-204 N52EL
8001026 LV-APF	5501019 XB-PSG	5527037 HK-3397W	LJ-95 N795K	LJ-205 YV-1700P
8001027 N112BC	5501020 N690L	5527038 HK-3459P	LJ-98 F-GBPB	LJ-206 F-GHDO
8001028 N52WA	5501021 N141TC	5527039 N37KW	LJ-99 N292A	LJ-207 N383JC
8001029 N811DA	5501022 N905LC	5527040 JA8870	LJ-101 N333G	LJ-208 N296A
8001030 F-GOON	5501023 N449CA	5527041 N4WE	LJ-102 N787K	LJ-209 N54WW
8001031 N123AG	5501024 N9150T	5527044 N495CA	LJ-103 N3002S	LJ-210 N43TT
8001032 N432R	5501025 D-IHVA		LJ-104 N814G	LJ-213 N250U
8001033 VH-WCE	5501026 N31KF	**Piper T-1040**	LJ-105 PT-LHM	LJ-214 N985AA
8001034 N74FB	5501027 N440CA	8275001 N2489Y	LJ-106 N411RS	LJ-215 HB-GEV
8001035 PP-EPB	5501028 N9159Y	8275002 5Y-UAL	LJ-107 N7BQ	LJ-217 F-GJRD
8001036 N8RY	5501029 D-IDBU	8275006 N311SC	LJ-108 N1154S	LJ-218 N380M
8001037 HK-3618W	5501030 N637KC	8275007 N314SC	LJ-109 N48XP	LJ-220 C-FCGI
8001038 PT-LZD	5501031 TC-THK	8275015 HK-2926P	LJ-113 N105K	LJ-221 PT-KGV
8001039 D-ICMC	5501032 N9233T		LJ-114 N114CW	LJ-222 N10127
8001040 N455JW	5501033 TC-FAH	8375001 N1194V	LJ-115 N7377	LJ-224 N464AB
8001041 N669CA	5501034 N9085U		LJ-116A N88SP	LJ-226 N271WN
8001042 N7139B	5501035 N9091J		LJ-117 N115PA	LJ-227 CC-COT
8001043 OY-YES	5501036 N9279A	**Raytheon**	LJ-118 C-FCGE	LJ-228 OB-1567
8001044 ZP-TYZ	5501037 N9116Q		LJ-120 N601TA	LJ-229 N435A
8001045 N169TC	5501038 N9142B	**King Air 90**	LJ-121 N666FG	LJ-230 N77SS
8001046 N148CA	5501039 HK-3381G	LJ-1 N925X	LJ-122 HR-IAH	LJ-233 N700MB
8001047 VH-BUW	5501040 N925RM	LJ-2 N812P	LJ-123 N815K	LJ-234 N212D
8001048 N830CM	5501041 D-IOSA	LJ-5 N357JR	LJ-124 N6238N	LJ-235 N440D
8001049 N95VR	5501042 D-IOSB	LJ-7 N46DT	LJ-125 PT-OUL	LJ-237 N66GS
8001050 N116JP	5501043 D-IOSC	LJ-8 N4RY	LJ-128 F-BOSY	LJ-238 N190RF
8001051 N5PA	5501044 D-IOSD	LJ-9 N613BR	LJ-129 N129LA	LJ-240 N360D
8001053 N700VF	5501045 I-TREP	LJ-10 N10JE	LJ-130 N70UA	LJ-241 N22BB
8001054 N94EW	5501046 I-TREQ	LJ-12 N654C	LJ-131 N80DG	LJ-245 N42CQ
8001055 N22BD	5501047 I-TRER	LJ-13 C-FBPT	LJ-132 N290CC	LJ-247 RP-C990
8001056 HR-JFA	5501048 JA8871	LJ-14 N155S	LJ-133 N90FA	LJ-248 N562P
8001057 N158TJ	5501049 JA8872	LJ-15 N37GP	LJ-134 C-GLRR	LJ-250 N6200B
8001058 N4098T	5501050 JA8873	LJ-16 PT-...	LJ-137 N80GP	LJ-251 N518DM
8001059 HK-2772P	5501051 B-3621	LJ-17 N440TP	LJ-138 N100UE	LJ-252 N33SB
8001060 N631PC	5501052 B-3622	LJ-19 N821U	LJ-139 N100WB	LJ-254 N911KA
8001061 N4100L	5501053 N426PC	LJ-21 N249PA	LJ-140 N437CF	LJ-256 N256TA
8001062 HK-3000G	5501054 B-3623	LJ-27 N330CB	LJ-142 N181NK	LJ-257 C-FXNB
8001063 N18AF	5501055 JA8724	LJ-30 N1290A	LJ-144 N331AM	LJ-258 N123V
8001064 N932BF	5501056 B-3624	LJ-31 N4948W	LJ-145 N45PR	LJ-260 N96AG
8001065 N395DR	5501057 C-GSAA	LJ-32 F-GGAM	LJ-146 N77DA	LJ-261 N990SA
8001066 N142AF	5501058 JA8874	LJ-34 N275DP	LJ-147 N483JM	LJ-263 N526RR
8001067 OB-1629	5501059 B-3625	LJ-35 N735K	LJ-148 N671LL	LJ-264 N86MG
8001068 HK-....	5501060 B-3626	LJ-36 N3DF	LJ-150 F-GEDV	LJ-265 N126AT
8001069 N5SY		LJ-37 N720K	LJ-151 N28AB	LJ-266 N55LH
8001070 N82PG	**PA-42**	LJ-38 C-GGGL	LJ-152 PT-OVP	LJ-267 N416LF
8001071 XA-TIL	**Cheyenne 400LS**	LJ-40 F-BTQP	LJ-153 66-15361	LJ-269 N507W
8001072 N801JW	5527001 N400PT	LJ-41 N877AQ	LJ-154 N5111U	LJ-270 N26540
8001073 4X-CIC	5527002 N400VB	LJ-46 N38GP	LJ-156 N8520L	LJ-272 N5Y
8001074 CX-...	5527003 ZS-...	LJ-47 N6137	LJ-157 N90GN	LJ-275 N457CP
8001075 HK-3693X	5527004 N411BG	LJ-49 N32229	LJ-158 N503M	LJ-276 N90VP
8001078 N28DA	5527005 PT-OVD	LJ-52 N134W	LJ-159 N50525	LJ-277 N623BB
8001079 N515M	5527006 HK-3413	LJ-53 N818MS	LJ-160 N616AS	LJ-278 N18LP
8001080 PT-MFL	5527007 N531MC	LJ-54 F-GDMM	LJ-161 ZS-IRJ	LJ-279 N211X
8001081 N802MW	5527008 HK-3423	LJ-55 N50KK	LJ-162 N198KA	LJ-280 N6228Q
8001101 N30MA	5527009 XA-EYZ	LJ-57 RP-C2340	LJ-163 N20BL	LJ-281 VH-DYN
8001102 N4114A	5527010 C-FHRV	LJ-58 N58KA	LJ-164 N17SA	LJ-282 N623AW
8001103 PT-OKL	5527011 N86CR	LJ-59 N190BT	LJ-165 N616F	LJ-283 N773S
8001104 N334FP	5527012 N321LH	LJ-60 N1WJ	LJ-166 N777AT	LJ-285 N98HB
8001105 N620MW	5527013 N450MW	LJ-62 N212WP	LJ-167 N798TW	LJ-286 N37DA
8001106 N933DG	5527014 N144PL	LJ-63 XA-MUR	LJ-169 N983K	LJ-287 N502W
	5527015 ZK-RUR	LJ-64 N512G	LJ-172 N2000E	LJ-288 N270M
PA-42	5527016 N500LM	LJ-66 N1SA	LJ-173 N623R	LJ-289 N5WG
Cheyenne III A	5527017 N410TW	LJ-67 N777NP	LJ-174 N93RY	LJ-290 N50JJ
8301001 D-IHLA	5527018 N124DP	LJ-69 N190JL	LJ-178A N798K	LJ-292 N100JF
8301002 N75FL	5527019 N1515H	LJ-70 N516WB	LJ-179 N511BF	LJ-293 XC-ONA
	5527020 N812BJ	LJ-71 N110EL	LJ-180 OB-1457	LJ-294 N930K
5501003 N29TF	5527021 N4RP	LJ-74 N32FH	LJ-181 PT-LBZ	LJ-295 N57EM
5501004 N775CA	5527022 D-IQAS	LJ-75 N75368	LJ-183 N151BU	LJ-296 N30EH
5501005 N2247R	5527023 C-GMFI	LJ-77 N58AC	LJ-184 N566CA	LJ-297 N140PA
5501006 N809E	5527024 D-IMAY	LJ-78 N95UF	LJ-189 PK-VKZ	LJ-298 N615AA
5501007 D-ICGB	5527025 N325KW	LJ-80 D2-ALS	LJ-190 N216AJ	LJ-299 N312VF
5501008 N777YP	5527026 JA8853	LJ-81 N276VM	LJ-191 PT-WAE	LJ-300 N70VM
5501009 N31GA	5527027 N721SG	LJ-83 N827T	LJ-192 N15CT	LJ-301 N1FL
5501010 N627KW	5527028 N4119X	LJ-84 C-FUFW	LJ-193 N600BF	LJ-305 N525JK
5501011 N888FW	5527029 N38AF	LJ-85 N900CK	LJ-194 N55FW	LJ-306 YV-722P
5501012 N4116Q	5527030 N41198	LJ-86 N60RJ	LJ-195 N98DD	LJ-307 N66RE
5501013 N90TW	5527031 N431MC	LJ-87 N98B	LJ-196 N348AC	LJ-308 VT-DXU
5501014 C-FSOZ	5527032 XB-CYR	LJ-88 N10AY	LJ-197 PK-VKY	LJ-309 C-FHWI
5501015 N794A	5527033 HK-3451	LJ-89 N195DP	LJ-200 N50RP	LJ-311 F-GNBA
5501016 N94CS	5527034 N448CA	LJ-91 N825K	LJ-201 N866A	LJ-312 N45TT
			LJ-202 N464G	LJ-313 C-FCGN

LJ-315 N34HA	LJ-411 N14V	LJ-518 F-ZBBF	LJ-611 N65KA	LJ-702 N44HP
LJ-316 XB-BZQ	LJ-412 N41DZ	LJ-519 N82307	LJ-612 N3112W	LJ-703 TC-DBZ
LJ-317 N46AX	LJ-414 N57MA	LJ-520 N1869	LJ-613 ZS-LTF	LJ-704 N1565L
LJ-318 N425K	LJ-415 N87MM	LJ-521 N70QZ	LJ-615 N444PS	LJ-705 VT-EFF
LJ-319 N1999G	LJ-417 N44RG	LJ-522 F-BXAP	LJ-616 I-LIPO	LJ-706 VT-EFB
LJ-321 N223CH	LJ-418 OY-BVS	LJ-523 V5-INN	LJ-617 N65GH	LJ-707 N122K
LJ-322 OB-1594	LJ-419 N11TN	LJ-524 N19UM	LJ-618 N900DG	LJ-708 N500KR
LJ-323 LV-VHR	LJ-420 N642DH	LJ-525 N90WJ	LJ-619 ZS-NDH	LJ-710 3A-MON
LJ-324 D2-EQC	LJ-421 N9426	LJ-526 ZS-MLO	LJ-620 N74CC	LJ-711 VT-EFE
LJ-325 C-FJHP	LJ-422 C-GSFM	LJ-527 EC-...	LJ-621 EC-CHA	LJ-712 N70AB
LJ-326 OB-1297	LJ-424 N638D	LJ-528 5Y-...	LJ-622 TC-AUT	LJ-713 N114J
LJ-327 OY-JRO	LJ-425 XB-...	LJ-529 LV-WJP	LJ-624 EC-CHC	LJ-714 N17573
LJ-330 OB-1364	LJ-426 N761K	LJ-530 N3KF	LJ-625 ZS-NWC	LJ-715 N9076S
LJ-331 F-GEXK	LJ-427 N784K	LJ-531 PT-IBE	LJ-626 N500MS	LJ-716 N25DL
LJ-333 OB-1495	LJ-428 LV-VHO	LJ-532 N575C	LJ-627 N4447W	LJ-717 F-GFHC
LJ-334 N621TB	LJ-429 N984AA	LJ-533 N739K	LJ-628 N800RP	LJ-718 N90BP
LJ-335 N7644R	LJ-430 N41WC	LJ-534 PT-OFC	LJ-629 N4950C	LJ-719 VT-EFG
LJ-337 C-GHVR	LJ-431 9Q-CVT	LJ-535 PT-OXU	LJ-630 N22BM	LJ-720 VT-EFP
LJ-338 9J-YVZ	LJ-432 N919AG	LJ-536 N31WB	LJ-631 N103FG	LJ-721 CS-DBT
LJ-339 N143KB	LJ-434 F-GFIR	LJ-537 VT-ECA	LJ-632 N71EN	LJ-722 N38RP
LJ-340 N240K	LJ-435 N9838Z	LJ-538 N65TA	LJ-634 N1801B	LJ-723 SE-IIB
LJ-341 N83FM	LJ-436 N309L	LJ-540 N690CA	LJ-635 N5WU	LJ-724 N5TA
LJ-342 XA-MIM	LJ-437 N90DN	LJ-542 N9442Q	LJ-636 N66TL	LJ-725 D-IHDE
LJ-343 N57AG	LJ-438 N988SL	LJ-543 PT-LTF	LJ-637 F-GCLD	LJ-726 N387GA
LJ-344 N873K	LJ-440 CC-PBZ	LJ-544 F-GHFE	LJ-638 PP-EHE	LJ-727 N75GC
LJ-346 PT-WTN	LJ-441 ...	LJ-545 F-GHBD	LJ-639 N309WW	LJ-728 N291CC
LJ-347 F-GFHQ	LJ-443 N411FT	LJ-548 N800PW	LJ-640 N94AM	LJ-730 N730WB
LJ-348 C-GTMA	LJ-444 N40BA	LJ-549 N600DJ	LJ-641 D-IKIW	LJ-731 N37PT
LJ-349 N939K	LJ-447 N926S	LJ-550 N9450Q	LJ-642 6804	LJ-732 N56DL
LJ-350 N7BF	LJ-449 LV-JJW	LJ-551 ZS-MCA	LJ-643 N96AH	LJ-733 N18383
LJ-351 N90LG	LJ-450 N6881S	LJ-552 VH-WNT	LJ-644 N66CN	LJ-734 N284PM
LJ-353 PT-OBW	LJ-452 N452TT	LJ-553 XC-FOC	LJ-645 N90CT	LJ-735 N819MH
LJ-355 PT-DEU	LJ-453 F-GIFB	LJ-554 N12AC	LJ-647 LX-DAK	LJ-736 N635AF
LJ-357 N887KU	LJ-454 N105RG	LJ-555 N100JD	LJ-648 F-BXSL	LJ-738 N239C
LJ-358 N40CK	LJ-455 PT-OYD	LJ-556 YV-...	LJ-649 N102AJ	LJ-739 YV-152CP
LJ-360 N125A	LJ-456 F-GIFC	LJ-558 PT-ICP	LJ-650 N103FL	LJ-740 N502SE
LJ-361 N587M	LJ-457 N405BC	LJ-561 N777NW	LJ-651 PT-LUF	LJ-742 YV-158CP
LJ-362 N31SN	LJ-458 HK-853W	LJ-562 D-IIHA	LJ-653 N586TC	LJ-743 N620WE
LJ-363 F-GICE	LJ-459 N9HW	LJ-563 N139B	LJ-654 N152WW	LJ-744 N91TJ
LJ-364 N555TB	LJ-460 PT-WBQ	LJ-564 N777GF	LJ-655 N4095S	LJ-745 N384H
LJ-365 N14VK	LJ-461 N90WL	LJ-565 N45RL	LJ-656 N15GA	LJ-746 N122RG
LJ-366 N68CD	LJ-463 N43MB	LJ-566 N67PL	LJ-657 C-GTWW	LJ-747 N8MG
LJ-367 N9711B	LJ-464 N36DD	LJ-567 PT-OCT	LJ-659 YV-2034P	LJ-748 ZS-MRZ
LJ-369 N70CU	LJ-465 OB-932	LJ-568 N108KU	LJ-661 N88SD	LJ-749 N5525R
LJ-370 YV-1947P	LJ-466 PT-ODA	LJ-570 N240RE	LJ-662 RP-C3650	LJ-750 N10MD
LJ-371 N718K	LJ-467 N944K	LJ-571 N183SA	LJ-663 EC-COI	LJ-751 YV-167CP
LJ-372 N301DK	LJ-468 C-FNCN	LJ-572 N35TV	LJ-665 N665JK	LJ-752 FAC-5570
LJ-373 N275LE	LJ-469 N90GB	LJ-573 N90LB	LJ-666 EC-COL	LJ-753 N601SC
LJ-374 N54CF	LJ-472 N104Z	LJ-574 N1097S	LJ-667 N888GN	LJ-754 N303QC
LJ-375 N101BS	LJ-473 N90CB	LJ-575 9J-DCF	LJ-668 F-OGOX	LJ-755 5N-AMZ
LJ-376 C-FMKD	LJ-475 N32BA	LJ-576 N3GC	LJ-669 ST-AFO	LJ-756 YV-172CP
LJ-377 N777KU	LJ-476 N10TM	LJ-577 N57KA	LJ-670 6805	LJ-757 N104LC
LJ-378 N320E	LJ-477 OB-1593	LJ-578 N71KA	LJ-671 ZS-MNC	LJ-758 N157CB
LJ-380 N66AD	LJ-478 N7TW	LJ-579 F-BVTB	LJ-672 N221TM	LJ-759 N895FK
LJ-381 N48A	LJ-479 N74MA	LJ-580 N580RA	LJ-673 N660JM	LJ-760 F-O...
LJ-382 EC-GIJ	LJ-480 N2UV	LJ-581 N717X	LJ-674 N332DE	LJ-761 N25HB
LJ-383 N388MC	LJ-481 G-BVRS	LJ-582 N66KA	LJ-675 ZS-ODV	LJ-762 N178RC
LJ-384 N563MC	LJ-482 F-BRNO	LJ-583 N200TG	LJ-676 N224BH	LJ-763 N9VC
LJ-385 N130DM	LJ-483 N83TC	LJ-584 PT-LYW	LJ-677 N555CK	LJ-764 N775DM
LJ-386 N899D	LJ-485 ZS-AAU	LJ-585 N14CP	LJ-678 N150VE	LJ-765 N191A
LJ-387 XB-CIO	LJ-486 9J-AAV	LJ-586 N60KA	LJ-679 N90MU	LJ-766 N642TD
LJ-388 N93EJ	LJ-488 YV-2229P	LJ-587 TC-AUV	LJ-680 C-FNED	LJ-767 N9092S
LJ-390 N741K	LJ-490 N81AT	LJ-588 N711BN	LJ-681 N90PW	LJ-768 LN-KCG
LJ-391 N55MG	LJ-491 N88RP	LJ-589 N59KA	LJ-682 N682KA	LJ-769 N911UM
LJ-392 N121HC	LJ-492 N946WA	LJ-590 9J-...	LJ-683 PT-MPC	LJ-770 OH-BCX
LJ-393 N221NC	LJ-493 LV-WMD	LJ-591 YV-...	LJ-684 F-BXPY	LJ-771 N117MF
LJ-394 N394AL	LJ-494 9Q-CKZ	LJ-592 N383DA	LJ-685 C-FATW	LJ-772 N11692
LJ-395 N81PG	LJ-495 N600AL	LJ-594 N90ZH	LJ-686 N400BX	LJ-773 N728DS
LJ-396 N70SM	LJ-496 N878K	LJ-595 N767MC	LJ-687 Z-WRD	LJ-774 N911ND
LJ-397 ZS-BEN	LJ-499 PT-OUO	LJ-596 N880H	LJ-688 N195KQ	LJ-775 N386GA
LJ-398 PT-DIQ	LJ-500 C-GCFL	LJ-597 6801	LJ-689 N689EB	LJ-776 N476M
LJ-399 N551AT	LJ-501 N43WA	LJ-598 6802	LJ-690 N690JP	LJ-777 N9AN
LJ-400 OB-1595	LJ-502 N88HM	LJ-599 6803	LJ-691 N62525	LJ-778 6806
LJ-401 N500NA	LJ-503 N193A	LJ-600 N73MC	LJ-692 N70FH	LJ-779 N117MF
LJ-402 9Q-CKM	LJ-504 N908K	LJ-601 N99CD	LJ-694 N901JA	LJ-780 N4947M
LJ-403 N603PA	LJ-506 N30HF	LJ-603 EC-CDI	LJ-695 N695JJ	LJ-781 N495NM
LJ-404 N90MT	LJ-508 N708DG	LJ-605 EC-CDJ	LJ-696 N644SP	LJ-782 TG-VAS
LJ-405 OB-1558	LJ-510 F-GHBB	LJ-606 N64KA	LJ-697 C-GSAX	LJ-783 N197SC
LJ-406 N226JW	LJ-511 N1UV	LJ-607 C-GKBB	LJ-698 N497P	LJ-784 N62SK
LJ-408 ZS-MUM	LJ-513 4X-DZT	LJ-608 EC-CDK	LJ-699 N976JT	LJ-786 YV-202CP
LJ-409 PP-EUE	LJ-515 PT-OIU	LJ-609 C-GRSL	LJ-700 N700CP	LJ-787 TC-NAZ
LJ-410 N901WL	LJ-517 N738R	LJ-610 N89TM	LJ-701 N90MV	LJ-788 N107SC

Ref	Reg	Ref	Reg	Ref	Reg	Ref	Reg	Ref	Reg
LJ-789	YV-223CP	LJ-878	N440KF	LJ-967	VT-EGR	LJ-1051	HP-976P	LJ-1135	VT-EMI
LJ-790	VT-EFZ	LJ-879	N6656D	LJ-968	VH-FDP	LJ-1052	N6356C	LJ-1136	N359JT
LJ-791	PT-WDU	LJ-881	N32CM	LJ-969	N77HE	LJ-1053	ZS-LIN	LJ-1137	VT-EMJ
LJ-794	PT-LSO	LJ-882	PT-OOY	LJ-970	N121P	LJ-1054	N10AU	LJ-1138	D-ITCH
LJ-795	ZS-MBZ	LJ-883	N380AA	LJ-971	N971CF	LJ-1055	N303D	LJ-1139	JA8840
LJ-796	N100V	LJ-884	PT-OLQ	LJ-972	HB-GJH	LJ-1056	PT-LQE	LJ-1140	JA8841
LJ-797	N797CF	LJ-885	N123LL	LJ-973	YV-436CP	LJ-1057	N424TV	LJ-1141	JA8844
LJ-799	ZP-TXE	LJ-886	C-GCFM	LJ-974	N55MN	LJ-1059	PT-OZR	LJ-1142	JA8845
LJ-800	N60KW	LJ-887	D-IGLI	LJ-975	PT-OHH	LJ-1060	6816	LJ-1143	JA8846
LJ-801	TC-CSA	LJ-888	N67511	LJ-976	6810	LJ-1061	6817	LJ-1144	JA8847
LJ-803	N9TN	LJ-889	YV-O-CBL-5	LJ-977	D-IFHI	LJ-1062	6818	LJ-1145	JA8848
LJ-804	N429DM	LJ-890	VT-NEF	LJ-978	OH-....	LJ-1063	PT-LSE	LJ-1146	6822
LJ-805	VP-BKW	LJ-891	N6754H	LJ-979	N777VN	LJ-1064	N6763K	LJ-1147	N475JA
LJ-806	N105CG	LJ-892	HK-3277P	LJ-980	6811	LJ-1065	N722KR	LJ-1148	JA8849
LJ-807	N802DG	LJ-893	N16	LJ-982	N102FL	LJ-1066	N300FL	LJ-1149	JA8850
LJ-808	PT-LZH	LJ-894	F-GCGA	LJ-983	N391BT	LJ-1067	N200TR	LJ-1150	JA8851
LJ-809	N549BR	LJ-895	N4449Q	LJ-984	OH-BAX	LJ-1068	N666PC	LJ-1151	JA8852
LJ-810	N811GA	LJ-896	N17	LJ-985	PT-OLW	LJ-1069	G-BMKD	LJ-1152	N3090A
LJ-811	N330V	LJ-897	PT-LLV	LJ-986	HK-....	LJ-1070	N210EC	LJ-1153	VT-EQO
LJ-812	PT-OJI	LJ-898	YV-741CP	LJ-987	ZS-LZP	LJ-1071	N68AJ	LJ-1154	N67CL
LJ-813	N2050A	LJ-899	N6732V	LJ-988	ZS-LUU	LJ-1072	N6692D	LJ-1155	N3077Y
LJ-814	N96DQ	LJ-900	PT-OJU	LJ-989	N63200	LJ-1073	HP-1152	LJ-1156	N18264
LJ-815	N2057N	LJ-901	N93LP	LJ-990	N1836H	LJ-1074	N923CR	LJ-1157	N36805
LJ-816	N416BK	LJ-903	N1UM	LJ-991	N19LW	LJ-1075	N130MA	LJ-1158	D-IHKM
LJ-817	PT-LPS	LJ-904	N94HB	LJ-992	HB-GJA	LJ-1076	N86JG	LJ-1159	PT-ODE
LJ-818	N97SF	LJ-906	N5371	LJ-993	N1837F	LJ-1077	D-IGKN	LJ-1160	VT-ETI
LJ-819	N2063A	LJ-907	N900BE	LJ-994	HK-2873W	LJ-1078	PT-ONJ	LJ-1161	N4764A
LJ-820	N100KB	LJ-908	N1078	LJ-995	PT-OOT	LJ-1079	N69301	LJ-1162	N477JA
LJ-821	N821CT	LJ-910	TC-FRT	LJ-996	HB-GDA	LJ-1080	TC-LMK	LJ-1163	N478JA
LJ-822	F-GBLU	LJ-911	N84TP	LJ-997	N1853T	LJ-1081	PT-OYN	LJ-1164	N104AJ
LJ-823	ST-ANH	LJ-913	PT-LMI	LJ-998	PT-OCI	LJ-1082	N686GW	LJ-1165	N39FB
LJ-824	N1MT	LJ-914	PT-OEH	LJ-999	N333TL	LJ-1083	6819	LJ-1166	2201
LJ-825	N601DM	LJ-915	N3688P	LJ-1000	PT-OOD	LJ-1084	6820	LJ-1167	VT-EQN
LJ-826	N6026K	LJ-916	6808	LJ-1001	N91LW	LJ-1085	N360MP	LJ-1168	2202
LJ-827	N2057C	LJ-917	6809	LJ-1002	D-ICLE	LJ-1086	N105TC	LJ-1169	N30844
LJ-828	I-BOMY	LJ-918	N21HA	LJ-1003	N90DL	LJ-1087	N7252S	LJ-1170	N581B
LJ-829	N901AJ	LJ-919	ZS-LTZ	LJ-1004	PT-ORW	LJ-1088	N124MB	LJ-1171	2203
LJ-830	N45PE	LJ-920	PT-OPD	LJ-1005	PT-OQS	LJ-1089	HB-GHW	LJ-1172	N30833
LJ-831	HK-3429W	LJ-921	N3690F	LJ-1006	N123NA	LJ-1090	D-ILGI	LJ-1173	PT-LPD
LJ-832	N10CW	LJ-922	CN-TAX	LJ-1007	N68PC	LJ-1091	N179D	LJ-1174	PT-OIZ
LJ-833	N85TB	LJ-923	N30GK	LJ-1008	VT-EHY	LJ-1092	PT-WNW	LJ-1175	2204
LJ-834	PT-LVI	LJ-924	N3695W	LJ-1009	N61383	LJ-1093	HB-GHN	LJ-1176	2205
LJ-835	PT-OVY	LJ-925	N925BC	LJ-1010	PT-LHJ	LJ-1094	PT-OFY	LJ-1177	N479JA
LJ-836	N6646R	LJ-927	PT-OSO	LJ-1011	VH-FDW	LJ-1095	N617KM	LJ-1178	N480JA
LJ-837	N36AD	LJ-928	N5095K	LJ-1012	N617LM	LJ-1096	PT-OTG	LJ-1179	N481JA
LJ-838	N445CR	LJ-929	C-GCFB	LJ-1013	N6723Y	LJ-1097	JA8838	LJ-1180	LV-ROC
LJ-839	9G-SAM	LJ-930	ZS-KMA	LJ-1014	N6173C	LJ-1098	F-GFCO	LJ-1181	RP-C1807
LJ-841	YV-294CP	LJ-931	HB-GIE	LJ-1015	N6280E	LJ-1099	ZS-MKI	LJ-1182	9301
LJ-842	VH-FDT	LJ-932	N34CE	LJ-1016	N790A	LJ-1100	VT-EJZ	LJ-1183	N482JA
LJ-843	N404SC	LJ-933	N3AT	LJ-1017	N6207F	LJ-1101	D-IFMI	LJ-1184	N483JA
LJ-844	PT-LQD	LJ-934	N369GA	LJ-1018	PT-ONQ	LJ-1102	N109DT	LJ-1185	N50VP
LJ-845	N28KP	LJ-935	N600FL	LJ-1019	PT-OEP	LJ-1103	N8GT	LJ-1187	N31447
LJ-846	N29TB	LJ-936	PT-OSI	LJ-1020	VH-LJG	LJ-1104	N7223X	LJ-1188	PT-LYK
LJ-847	PT-OCY	LJ-937	PT-OUX	LJ-1021	VH-FDZ	LJ-1106	N222WJ	LJ-1189	N200SL
LJ-848	Z-LCS	LJ-938	N3817H	LJ-1022	PT-OAB	LJ-1107	N101BU	LJ-1190	N319NS
LJ-849	C-GCFZ	LJ-939	N94CD	LJ-1023	N731RJ	LJ-1108	N438SP	LJ-1191	YV-32P
LJ-850	N6681S	LJ-940	PT-OPE	LJ-1024	VH-FDM	LJ-1109	N95PC	LJ-1192	N616SC
LJ-851	N851KA	LJ-941	F-GEOU	LJ-1025	PT-LPJ	LJ-1110	6821	LJ-1193	D2-ETJ
LJ-852	N90LF	LJ-942	D-IKES	LJ-1026	PT-LPJ	LJ-1111	N597DM	LJ-1194	N15234
LJ-853	N400RV	LJ-943	N3805E	LJ-1027	N83P	LJ-1112	N66LM	LJ-1195	PT-OKQ
LJ-855	6807	LJ-944	HB-GHT	LJ-1028	VP-CCT	LJ-1113	N7230H	LJ-1196	N484JA
LJ-856	N965J	LJ-945	N88GL	LJ-1029	N28TM	LJ-1114	N808W	LJ-1197	N485JD
LJ-857	RP-C290	LJ-946	PT-LLR	LJ-1030	N357HP	LJ-1115	OE-FHL	LJ-1198	N486JD
LJ-858	N270TC	LJ-948	N4495U	LJ-1031	N982BA	LJ-1116	VT-NEI	LJ-1199	N717RD
LJ-859	N903R	LJ-949	N100HS	LJ-1032	N6563K	LJ-1117	N3030G	LJ-1200	N68FW
LJ-860	F-GBPZ	LJ-950	HK-2595W	LJ-1033	ZS-LFL	LJ-1118	ZS-LZR	LJ-1201	PT-LVK
LJ-861	N903GP	LJ-951	PT-OZJ	LJ-1034	D-IARF	LJ-1119	N386CP	LJ-1202	N417VN
LJ-862	N380SC	LJ-952	PT-OJA	LJ-1035	F-GJLD	LJ-1120	D-IIKM	LJ-1203	N115MX
LJ-863	N777EB	LJ-953	N38280	LJ-1036	N44QM	LJ-1121	N904JP	LJ-1204	N1552D
LJ-864	N6663A	LJ-954	C9-ASK	LJ-1037	PT-LYT	LJ-1122	N992C	LJ-1205	N169DR
LJ-865	YV-263CP	LJ-955	F-GIZB	LJ-1038	9102	LJ-1123	N917BH	LJ-1206	N1552G
LJ-866	N844C	LJ-956	PT-WMT	LJ-1039	N404JP	LJ-1124	N90EP	LJ-1207	D-ISEM
LJ-867	D-IEKG	LJ-957	HK-2596	LJ-1040	N115KU	LJ-1125	OE-FAK	LJ-1208	N621WP
LJ-868	N139SC	LJ-958	N500MT	LJ-1041	PT-OVN	LJ-1126	N497P	LJ-1209	N53TJ
LJ-869	N6662D	LJ-959	ZS-KZI	LJ-1042	6812	LJ-1127	PT-OXH	LJ-1210	N1553G
LJ-870	PT-OPR	LJ-960	PT-OCC	LJ-1043	6813	LJ-1128	PT-ODH	LJ-1211	N1551H
LJ-871	N29EB	LJ-961	N16KM	LJ-1044	6814	LJ-1129	N817F	LJ-1212	PT-WHP
LJ-872	N555MS	LJ-962	TC-MCK	LJ-1045	N84P	LJ-1130	N90DJ	LJ-1213	N20FD
LJ-874	D-IFUN	LJ-963	PT-OLX	LJ-1046	N90PU	LJ-1131	N811R	LJ-1214	N1553G
LJ-875	N6672N	LJ-964	N595RC	LJ-1047	6815	LJ-1132	XC-AGS	LJ-1216	D-IEAH
LJ-876	HP-1118	LJ-965	N205SP	LJ-1048	N777HF	LJ-1133	N77WM	LJ-1217	N1567G
LJ-877	N300VA	LJ-966	PT-LQS	LJ-1049	N7931D	LJ-1134	N303CA	LJ-1218	N224CC
				LJ-1050	N6492C				

Reg	Reg	Reg	Reg	Reg
LJ-1219 PT-MTD	LJ-1301 C-GMBD	LJ-1383 N90WP	LJ-1465 N1102K	LW-20 N87CH
LJ-1220 PT-OMZ	LJ-1302 N322TC	LJ-1384 N3196K	LJ-1466 LV-WXC	LW-21 N121B
LJ-1221 N94SR	LJ-1303 F-GMPM	LJ-1385 PT-WRA	LJ-1467 N1130J	LW-22 N90DA
LJ-1222 OM-VKE	LJ-1304 C-GMBG	LJ-1386 PT-WJF	LJ-1468 N1134G	LW-23 N345V
LJ-1223 OY-JAB	LJ-1305 N488JR	LJ-1387 N500EQ	LJ-1469 N97KA	LW-24 N95LB
LJ-1224 PT-OBF	LJ-1306 N422RJ	LJ-1388 N580AC	LJ-1470 N1099Z	LW-25 N1100M
LJ-1225 PT-LLO	LJ-1307 D-IBDH	LJ-1389 CC-...	LJ-1471 N1099D	LW-26 00923
LJ-1226 N369B	LJ-1308 PT-ORG	LJ-1390 N3237K	LJ-1472 N1099L	LW-27 N379VM
LJ-1227 N200SC	LJ-1309 C-GMBH	LJ-1391 HB-GJE	LJ-1473 N1009X	LW-28 CP-...
LJ-1228 N190JS	LJ-1310 C-GMBW	LJ-1392 N3216K	LJ-1474 N205P	LW-29 N90AF
LJ-1229 D-IAFF	LJ-1311 N489JS	LJ-1393 VT-TIS	LJ-1475 N1103G	LW-30 N20WS
LJ-1230 N1564P	LJ-1312 N490J	LJ-1394 PT-WIT	LJ-1476 N1106M	LW-31 N90WT
LJ-1231 N101SQ	LJ-1313 C-GMBX	LJ-1395 LV-WMG	LJ-1477 N1107W	LW-32 N291MM
LJ-1232 N91KA	LJ-1314 N8022Q	LJ-1396 PT-WIH	LJ-1478 N1108M	LW-33 XC-FUY
LJ-1233 PT-ORZ	LJ-1315 D-IHSW	LJ-1397 N3216U	LJ-1479 XA-...	LW-34 N690G
LJ-1234 N1564W	LJ-1316 N49LL	LJ-1398 N996TT	LJ-1480 N902LT	LW-35 N14NM
LJ-1235 D-IABB	LJ-1317 C-GMBY	LJ-1399 TG-...	LJ-1481 N1101U	LW-36 PNP-230
LJ-1236 N5598L	LJ-1318 N131SA	LJ-1400 VT-UPZ	LJ-1482 N3035T	LW-39 HK-3907
LJ-1237 PT-OTI	LJ-1319 C-GMBZ	LJ-1401 N3270K	LJ-1483 N1126A	LW-40 N155CG
LJ-1238 N1553N	LJ-1320 N8119N	LJ-1402 VT-NKF	LJ-1484 N1134D	LW-41 N12KA
LJ-1239 N5639K	LJ-1321 D-IOMG	LJ-1403 N903SE	LJ-1485 N886AT	LW-42 N52KA
LJ-1240 N691AS	LJ-1322 N500QT	LJ-1404 VT-DEJ	LJ-1486 N1110K	LW-43 N600CX
LJ-1241 N5641X	LJ-1323 D-IUDE	LJ-1405 N59MS	LJ-1487 N970P	LW-44 N666DC
LJ-1242 N88PD	LJ-1324 D-IKIM	LJ-1406 N207P	LJ-1488 N1135G	LW-45 N200RM
LJ-1243 HS-SLB	LJ-1325 D-IHMV	LJ-1407 HB-GJF	LJ-1489 N1069F	LW-47 F-ASFA
LJ-1244 N5644E	LJ-1326 N8096U	LJ-1408 N587PB	LJ-1490 N1TP	LW-48 N70MV
LJ-1245 PT-WAH	LJ-1327 D-IHHE	LJ-1409 F-GKSP	LJ-1491 PT-...	LW-49 N12LA
LJ-1246 XA-RFN	LJ-1328 F-GNEE	LJ-1410 N771SG	LJ-1492 N97CV	LW-50 N88GC
LJ-1247 D-IAAH	LJ-1329 YV-...	LJ-1411 TG-...	LJ-1493 N90KA	LW-51 N35SN
LJ-1248 9302	LJ-1330 N266F	LJ-1412 TC-...	LJ-1494 N11135X	LW-52 N181Z
LJ-1249 9303	LJ-1331 N8220V	LJ-1413 N3242L	LJ-1495 D-...	LW-54 N77JX
LJ-1250 F-OGUY	LJ-1332 N8259Q	LJ-1414 LV-WFB	LJ-1496 N101SG	LW-55 N3929G
LJ-1251 N87HB	LJ-1333 N8156Z	LJ-1415 N30CN	LJ-1497 N.....	LW-56 PP-FPP
LJ-1252 N92CD	LJ-1334 N5VK	LJ-1416 LV-WPB	LJ-1498 N21868	LW-57 N88TR
LJ-1253 PT-WHA	LJ-1335 6823	LJ-1417 N3217X	LJ-1499 N2297C	LW-58 N350AT
LJ-1254 N31GM	LJ-1336 6824	LJ-1418 SU-UAA	LJ-1500 N290JS	LW-59 D-IMWA
LJ-1255 N55495	LJ-1337 6825	LJ-1419 N257CG	LJ-1501 N401TS	LW-60 N90PH
LJ-1256 N904HB	LJ-1338 6826	LJ-1420 N3220L	LJ-1503 N9MU	LW-61 N129C
LJ-1257 N155A	LJ-1339 6827	LJ-1421 N61GN	LJ-1504 N333BM	LW-62 N16NM
LJ-1258 N40MH	LJ-1340 D-IIWB	LJ-1422 OY-IFH	LJ-1505 N800JF	LW-63 N93A
LJ-1259 N712JC	LJ-1341 PT-OZL	LJ-1423 N4488L	LJ-1506 N988P	LW-64 N47LC
LJ-1260 PT-OSN	LJ-1342 C-FTPE	LJ-1424 D-IHBP	LJ-1516 N60VP	LW-65 N987GM
LJ-1261 F-GJSD	LJ-1343 N8208C	LJ-1425 N3223H		LW-66 N554CF
LJ-1262 N200HV	LJ-1344 N8220V	LJ-1426 N724KW	**King Air A90**	LW-68 F-BUTS
LJ-1263 D-IDIW	LJ-1345 N415P	LJ-1427 PT-WJD	LM-1 N7000B	LW-69 N5462G
LJ-1264 PT-OFF	LJ-1346 PT-WFN	LJ-1428 N3242Z	LM-2 N7007G	LW-70 N5911P
LJ-1265 N8001V	LJ-1347 N8210C	LJ-1429 N3251Q	LM-5 N7007Q	LW-71 N121EG
LJ-1266 N421HV	LJ-1348 N451A	LJ-1430 OK-BKS	LM-6 N70088	LW-72 N81MP
LJ-1267 D-IEBE	LJ-1349 N491JV	LJ-1431 N1031Y	LM-7 N7010L	LW-73 N743JA
LJ-1268 N131CL	LJ-1350 N492JW	LJ-1432 N3263C	LM-9 N70135	LW-74 N71BX
LJ-1269 N8021P	LJ-1351 N493JX	LJ-1433 N100RU	LM-11 N49K	LW-75 N75LW
LJ-1270 VT-SLK	LJ-1352 N494JY	LJ-1434 N313DW	LM-13 N7018F	LW-76 N176TW
LJ-1271 N80904	LJ-1353 SU-ZAA	LJ-1435 N1095M	LM-15 N7026H	LW-78 N15WN
LJ-1272 HP-1203	LJ-1354 LV-WJE	LJ-1436 N196HA	LM-16 N823SB	LW-79 N60BA
LJ-1273 YV-1000CP	LJ-1355 D-ISIX	LJ-1437 N90PR	LM-22 N7034K	LW-80 N20LH
LJ-1274 N099G	LJ-1356 N8287E	LJ-1438 N3268M	LM-31 N7040J	LW-81 YV-416P
LJ-1275 TC-AEM	LJ-1357 N1560T	LJ-1439 N1089L	LM-59 N7071N	LW-82 YV-O-MAC-6
LJ-1276 TC-MSS	LJ-1358 N94TK	LJ-1440 N1070F	LM-68 N73Q	LW-84 N111JA
LJ-1277 N871KS	LJ-1359 7T-...	LJ-1441 N995TA	LM-80 N518NA	LW-85 N42PC
LJ-1278 VT-RLK	LJ-1360 S5-CMO	LJ-1442 PT-WNI	LM-107 N7155P	LW-86 F-GDFJ
LJ-1279 N71VG	LJ-1361 N5521T	LJ-1443 N1083S	LM-115 N7157K	LW-87 N800BF
LJ-1280 XA-RXT	LJ-1362 N990CB	LJ-1444 N1084N		LW-88 F-GELL
LJ-1281 9304	LJ-1363 PT-WGU	LJ-1445 N90CH		LW-89 N77PA
LJ-1282 9305	LJ-1364 TG-CCA	LJ-1446 N1086Z	**King Air E90**	LW-90 HK-....
LJ-1283 N700GM	LJ-1365 N1551C	LJ-1447 N569GR	LW-1 N190RM	LW-93 N352GR
LJ-1284 N25GA	LJ-1366 N1544V	LJ-1448 N1068K	LW-2 N412SR	LW-94 YV-467CP
LJ-1285 HP-1264	LJ-1367 N111MD	LJ-1449 N72PK	LW-3 F-GJAD	LW-95 N567GJ
LJ-1286 N301ER	LJ-1368 N1568X	LJ-1450 N3265K	LW-4 N44EC	LW-96 N214SC
LJ-1287 LV-WLV	LJ-1369 VT-RLL	LJ-1451 N3270V	LW-5 N999SF	LW-97 F-GESJ
LJ-1288 N90KB	LJ-1370 D-IHAH	LJ-1452 N1097B	LW-6 N722M	LW-99 N31WP
LJ-1289 N1NP	LJ-1371 N696WW	LJ-1453 N1092G	LW-7 N51SG	LW-100 N28MS
LJ-1290 JA8882	LJ-1372 N121EB	LJ-1454 N1092H	LW-8 N132B	LW-101 N70SW
LJ-1291 JA8883	LJ-1373 TG-RWC	LJ-1455 N1848S	LW-9 PT-IGD	LW-102 F-GABV
LJ-1292 JA8884	LJ-1374 VT-VIL	LJ-1456 N1095W	LW-10 N977SB	LW-103 D-IDEA
LJ-1293 N214P	LJ-1375 XA-TDW	LJ-1457 N1057L	LW-11 N11DT	LW-104 N123MH
LJ-1294 N19BK	LJ-1376 VT-HYA	LJ-1458 N488LL	LW-12 RP-C289	LW-105 N111FW
LJ-1295 N91LY	LJ-1377 PT-WCS	LJ-1459 N70VP	LW-13 N383AA	LW-106 OB-1420
LJ-1296 F-GLRZ	LJ-1378 N3083K	LJ-1460 N99ML	LW-14 N4495N	LW-107 N44MV
LJ-1297 N779DD	LJ-1379 7T-...	LJ-1461 N83LS	LW-15 PP-EFC	LW-108 N37HC
LJ-1298 ZS-NAG	LJ-1380 7T-...	LJ-1462 N1079D	LW-16 N43WS	LW-109 N989GM
LJ-1299 F-GNMA	LJ-1381 LV-WIU	LJ-1463 N771SC	LW-17 N259SC	LW-110 9Q-CCG
LJ-1300 C-GMBC	LJ-1382 N3120U	LJ-1464 CC-...	LW-18 N22ER	LW-111 C-GBTI

LW-112 N67PS	LW-202 N934DC	LW-288 N700DD	LA-27 N40BR
LW-114 N700DH	LW-203 YV-O-SAS-3	LW-289 N83WE	LA-28 PT-LPL
LW-115 N76SK	LW-204 N965LC	LW-290 N60MH	LA-29 N88TW
LW-116 F-BVRS	LW-205 Z-DJF	LW-291 5U-ABV	LA-30 N199BC
LW-117 N76PW	LW-206 N6406S	LW-292 N92DV	LA-31 N960V
LW-118 N8118R	LW-208 YV-706P	LW-293 N12AU	LA-32 N123CH
LW-119 N152D	LW-209 N32NS	LW-294 N677J	LA-33 N90FL
LW-120 N99AC	LW-210 N860MH	LW-296 F-GETJ	LA-34 N57TM
LW-121 N515BC	LW-211 N4216S	LW-297 LX-LTX	LA-35 YV-290CP
LW-123 N5NM	LW-212 N75CF	LW-298 N44GK	LA-36 N102WK
LW-124 N90BF	LW-213 N2274L	LW-299 N127EC	LA-37 N30GT
LW-125 N122NC	LW-214 N14SB	LW-300 N911AZ	LA-38 D-IAGB
LW-126 RP-C201	LW-215 N56HT	LW-301 N550P	LA-39 N703JT
LW-127 N71WB	LW-216 N439WA	LW-302 RP-C298	LA-40 N66BS
LW-128 N50EB	LW-217 N16LH	LW-303 N79CT	LA-41 N311DS
LW-129 LV-WFP	LW-218 HP-1246	LW-304 G-SANB	LA-42 N190FD
LW-130 N47AW	LW-219 N83FE	LW-305 N605W	LA-43 N6749E
LW-131 N98PC	LW-220 I-RWWW	LW-306 N67V	LA-44 N244J
LW-132 N9029R	LW-221 N400SF	LW-308 HR-...	LA-45 XA-LGT
LW-133 PT-LHZ	LW-222 ZS-NYE	LW-309 N25AP	LA-46 N208RC
LW-134 N55HC	LW-223 N333LE	LW-310 LV-MRN	LA-47 ZS-MHM
LW-135 N74VR	LW-224 ZS-NXI	LW-311 N551MS	LA-48 N30NH
LW-136 G-DEXY	LW-225 N181CG	LW-312 N505BG	LA-49 PT-OIF
LW-138 PT-WAG	LW-226 N711TZ	LW-314 N20564	LA-50 HB-GHD
LW-139 N249WM	LW-227 N7ZP	LW-315 HB-GGU	LA-51 ZS-NFO
LW-140 N1UC	LW-228 N53BB	LW-316 F-GJHM	LA-52 N769D
LW-141 N382TW	LW-229 EV-7702	LW-317 N2062A	LA-53 N366SP
LW-142 N7WU	LW-230 N126RD	LW-318 N443CL	LA-55 N701NC
LW-143 N300BA	LW-231 N4954S	LW-319 N1205S	LA-56 N68DK
LW-144 YV-66CP	LW-232 N112SB	LW-320 N366GW	LA-58 N59EK
LW-145 RP-C879	LW-234 YV-118CP	LW-321 EI-BHL	LA-59 G-FLTI
LW-146 N41HH	LW-235 N776DC	LW-322 N600FC	LA-60 N44VP
LW-147 C-FATX	LW-236 N948CC	LW-323 N323HA	LA-61 N189JR
LW-148 N300CH	LW-237 N714F	LW-324 N351GR	LA-62 F-GIFK
LW-149 N177MK	LW-238 N25TG	LW-325 RP-C291	LA-63 ZS-LTD
LW-150 N100EC	LW-239 N24SM	LW-326 TN-AFG	LA-64 N322GK
LW-151 ZS-OAE	LW-240 N500EA	LW-327 N788SW	LA-65 C-GMTI
LW-152 N7400V	LW-241 N30KC	LW-328 F-GJPD	LA-66 N90TP
LW-153 CC-DSN	LW-242 N34BS	LW-329 N652L	LA-67 N416P
LW-154 GN-7593	LW-243 N18343	LW-330 LV-OBB	LA-68 N77PV
LW-155 N505MW	LW-244 N611R	LW-331 RP-C319	LA-69 ZS-KLZ
LW-156 N414GN	LW-245 N3LS	LW-332 N636JM	LA-70 N990BM
LW-157 N199TT	LW-246 N711HV	LW-333 XB-JLA	LA-71 N541MM
LW-158 N75JP	LW-248 N555FW	LW-334 N22TL	LA-72 N770SD
LW-159 N9059S	LW-249 N521LB	LW-336 N675J	LA-73 N81GC
LW-160 N53CE	LW-250 N321DM	LW-337 N290K	LA-74 N81PS
LW-161 F-BXON	LW-251 F-GJCR	LW-338 N50RV	LA-75 N600WM
LW-162 N369CD	LW-252 N1975G	LW-339 LV-VFC	LA-76 TI-...
LW-163 N34MF	LW-253 N43PC	LW-340 N3700M	LA-77 N827DP
LW-164 N200RE	LW-254 N48W	LW-341 RP-C1990	LA-78 PT-WET
LW-165 N701X	LW-256 N63BV	LW-342 N3741M	LA-79 N2164L
LW-166 N300SP	LW-257 N714BX	LW-343 PT-OUF	LA-80 N17AE
LW-167 N213RW	LW-258 N203PC	LW-344 N999SE	LA-81 N444MF
LW-168 N220PB	LW-259 LV-WHV	LW-345 N811MM	LA-82 N501TD
LW-169 N17SE	LW-260 GN-7839	LW-346 N3821S	LA-83 N186DD
LW-170 N555TT	LW-261 ZP-...	LW-347 PT-LCE	LA-84 HK-....
LW-171 N6571S	LW-262 N7ZW		LA-85 PT-ODN
LW-172 YV-326P	LW-263 N1840S	**King Air F90**	LA-86 PT-OLM
LW-173 N1573L	LW-264 ARV-0201	LA-2 N49CH	LA-87 N696RA
LW-175 F-BXSN	LW-265 N362D	LA-3 N87AG	LA-88 F-ODGU
LW-176 N1MW	LW-266 N215HC	LA-4 N1HE	LA-89 N24TL
LW-177 N944RS	LW-267 N20S	LA-5 N90FD	LA-90 N290DK
LW-178 HC-DAC	LW-268 TI-GEV	LA-6 C-GSSA	LA-91 N641PE
LW-179 N676J	LW-269 N288CR	LA-7 PT-LLP	LA-92 PT-ONO
LW-180 F-GIML	LW-270 YV-1500P	LA-9 N300TA	LA-94 PT-LSH
LW-183 HK-2491	LW-271 N123LN	LA-10 N325WP	LA-95 XA-SAW
LW-184 N899GP	LW-272 N62BL	LA-11 PT-LXI	LA-96 N932G
LW-185 N600AC	LW-273 N10SA	LA-12 N19RK	LA-97 N58EZ
LW-186 N2186L	LW-274 N274KA	LA-13 N18BL	LA-98 N103CB
LW-187 N816RL	LW-275 N4725M	LA-14 N581RJ	LA-99 N711KW
LW-188 N68PM	LW-276 N7TD	LA-15 ZS-NSC	LA-100 D-IWAL
LW-189 YV-O-INV-3	LW-277 RP-C292	LA-16 YV-1013P	LA-101 N37990
LW-190 RP-C415	LW-278 F-GHUV	LA-17 N117TJ	LA-102 N44882
LW-191 N63EC	LW-279 N390PS	LA-18 N65SF	LA-103 PT-LTT
LW-192 N2192L	LW-280 N4820M	LA-19 F-GETI	LA-104 D-IEEE
LW-193 N46RP	LW-281 N31A	LA-20 N117W	LA-105 N188BF
LW-194 N117FH	LW-282 N900AC	LA-21 N66LP	LA-106 C-GQGA
LW-196 N3813C	LW-283 N48TA	LA-22 F-GHIV	LA-107 N700BK
LW-197 N228RA	LW-284 ZP-...	LA-23 N27WH	LA-108 YV-400CP
LW-198 G-WELL	LW-285 N777KA	LA-24 N93KA	LA-109 PT-LYZ
LW-199 F-GALZ	LW-286 N155LS	LA-25 N616CP	LA-110 N3824V
LW-200 N918VS	LW-287 C-FNCB	LA-26 N96TT	LA-111 HB-GHO

LA-112 N3809C
LA-113 C-GKSC
LA-115 F-GCTR
LA-116 N3825E
LA-117 N120RC
LA-118 PT-OFD
LA-119 N900HM
LA-120 N879PC
LA-121 N182CA
LA-122 ZS-LBC
LA-123 N513KL
LA-124 N94U
LA-125 N65MM
LA-126 PT-LYP
LA-127 YV-02CP
LA-128 PT-ONU
LA-129 F-GKKK
LA-130 N81SD
LA-131 N42SY
LA-132 PT-LQC
LA-133 N1803P
LA-134 N993M
LA-135 N422Z
LA-136 RP-C410
LA-137 N488FT
LA-138 N15
LA-139 N333EB
LA-140 YV-428CP
LA-141 F-GDAK
LA-142 N50AW
LA-143 N69AD
LA-144 PT-ONE
LA-145 N18
LA-146 N90RT
LA-147 N722DR
LA-148 PT-LER
LA-149 N10DR
LA-150 N80M
LA-151 HK-3935W
LA-152 N152WE
LA-154 N90SB
LA-155 PT-OUJ
LA-156 PT-LTO
LA-157 N69084
LA-158 N122SC
LA-159 N76HC
LA-160 HP-1266
LA-161 N18471
LA-162 PT-OOX
LA-163 PT-OIV
LA-164 N990F
LA-165 N300BM
LA-166 F-GFVN
LA-167 N1857F
LA-168 N32BG
LA-169 N61228
LA-170 N314P
LA-171 N32HF
LA-172 N808DS
LA-173 PT-OHZ
LA-174 YV-702CP
LA-175 PT-02CP
LA-176 N766RB
LA-177 N418DY
LA-178 N53G
LA-179 YV-486CP
LA-180 N357CC
LA-181 HK-2888P
LA-182 N424CP
LA-183 N90TD
LA-185 PT-LYM
LA-186 N600SF
LA-187 PT-OTA
LA-188 N41AK
LA-189 N107AJ
LA-190 N6335F
LA-191 N822BA
LA-192 D-ITLL
LA-193 YV-490CP
LA-194 N6685P
LA-195 PT-LXY
LA-196 N196WC

LA-197 PT-LSP	B-48 N5727	B-142 PP-FOY	B-234 N23404	BE-73 N54CK
LA-198 HK-3118W	B-51 C-FWPN	B-143 C-GNVB	B-235 C-GJLJ	BE-74 N56FL
LA-199 N211NA	B-52 C-FKIJ	B-144 N999G	B-236 N23605	BE-76 N301TS
LA-200 PT-OFB	B-53 N850DB	B-145 C-FMAI	B-237 N43FC	BE-77 N580S
LA-201 N955RA	B-54 N776L	B-146 N410SP	B-238 N1TR	BE-79 N500KD
LA-202 N777AS	B-55 N911RL	B-147 7T-VRF	B-239 N154TC	BE-80 N93WT
LA-203 OY-CVC	B-56 PT-DNP	B-148 C-GJLP	B-240 N87CA	BE-81 N987B
LA-204 N75MS	B-57 N18U	B-149 C-FLTS	B-241 N777GS	BE-82 N66804
LA-205 N6690R	B-58 N750FC	B-150 F-GGLV	B-242 IGM-240	BE-83 N123BL
LA-206 N131SP	B-59 C-FMWM	B-151 C-FPAJ	B-243 PT-OFZ	BE-84 N700BA
LA-207 YV-...	B-60 N54JW	B-152 C-GISH	B-244 N842DS	BE-85 N222LP
LA-208 N999RC	B-61 C-GSYN	B-153 C-GYQK	B-245 N41BE	BE-86 N264PA
LA-209 N444KK	B-62 C-GKBQ	B-154 N919WM	B-247 N153TC	BE-87 N19DA
LA-210 N415RB	B-63 N20880	B-155 YV-O-SID-2		BE-88 N130AT
LA-211 PT-OHX	B-64 N4200A	B-156 C-GDPI	**King Air B100**	BE-89 C-....
LA-212 A2-AHV	B-65 N102LF	B-157 C-GUPP	BE-2 N729MS	BE-90 N770D
LA-213 PT-ODO	B-66 N265K	B-158 N62NC	BE-3 N150TJ	BE-91 N717D
LA-214 N773	B-67 C-FWPG	B-159 C-GLPG	BE-4 N786CB	BE-92 N6756P
LA-215 PT-LUT	B-68 XB-REA	B-160 N896SB	BE-5 N577RW	BE-93 N142JC
LA-216 PT-LZT	B-69 N207SB	B-161 N62526	BE-6 N8514B	BE-94 N3663B
LA-217 PT-OCE	B-70 N77PF	B-162 N96AM	BE-7 N57HT	BE-96 N802RD
LA-218 D-IWKA	B-71 N431R	B-163 C-FASB	BE-8 N688DS	BE-97 N97WD
LA-219 N18CM	B-72 C-GTLF	B-164 C-FGIN	BE-9 N236CP	BE-98 N753D
LA-220 PT-OLP	B-73 YV-04CP	B-165 C-GTLA	BE-11 N1911L	BE-99 N524BA
LA-221 HK-3505	B-74 N3500E	B-166 C-FAMU	BE-12 N400AC	BE-100 N10AG
LA-222 N711L	B-75 C-FCSD	B-168 N715WA	BE-13 N93SF	BE-101 N311CM
LA-223 PT-LIF	B-76 C-GWWQ	B-169 C-FAPP	BE-14 N86FD	BE-102 N57TJ
LA-224 N29GB	B-77 RP-C1978	B-170 N900DN	BE-15 N300DG	BE-103 N2TX
LA-225 PT-OFS	B-78 RP-C282	B-171 C-FJFH	BE-16 N331GB	BE-104 N67BS
LA-226 N6690L	B-79 CC-CLY	B-172 N735DB	BE-17 N178NC	BE-105 N87XX
LA-227 N330VP	B-80 N424SW	B-173 C-GAST	BE-18 N95JJ	BE-106 N568K
LA-228 N80WP	B-81 N707SS	B-174 N671L	BE-19 YV-95CP	BE-107 N3699B
LA-229 D-IRIS	B-82 LV-WDO	B-175 C-GJHW	BE-20 N443TC	BE-108 N108EB
LA-230 N393CE	B-83 YV-04P	B-176 N3076W	BE-21 N60TJ	BE-109 N3811F
LA-231 N27PA	B-84 C-FWOL	B-177 N30GC	BE-22 N86TR	BE-110 N500N
LA-232 PT-ASN	B-85 C-GKBZ	B-178 N764K	BE-23 N93D	BE-111 N400TJ
LA-233 VT-ELZ	B-86 N53MD	B-179 N127Z	BE-24 N187J	BE-112 N727RS
LA-234 N7206E	B-87 N70JL	B-181 CNA-NB	BE-25 N125VH	BE-113 N1888M
LA-235 N722PT	B-88 N100ZM	B-182 CNA-NC	BE-26 N36WH	BE-114 N125DG
LA-236 N234CW	B-89 N169RA	B-183 CNA-ND	BE-27 N87JE	BE-115 N104LS
	B-90 YV-2423P	B-184 C-FGNL	BE-28 YV-127CP	BE-116 N116DG
King Air 100	B-91 F-GHHV	B-185 N220KW	BE-29 D-IERI	BE-118 N4700K
1 ZS-KZU	B-93 7T-VCV	B-186 CNA-NE	BE-30 N111YF	BE-119 N1807H
B-1 N3100K	B-94 N12AQ	B-187 CNA-NF	BE-31 N80DB	BE-120 N2QE
B-2 N11JJ	B-95 70-15908	B-190 C-GNAR	BE-32 N1PN	BE-121 VT-AVB
B-3 C-FIDN	B-96 70-15909	B-191 N214CK	BE-33 N900RD	BE-122 N74RR
B-4 XB-WUI	B-97 70-15910	B-192 N4392W	BE-34 N949SW	BE-123 N55FR
B-5 N288RA	B-98 70-15911	B-193 EC-CHD	BE-35 N364BC	BE-124 N150YA
B-10 N204AJ	B-99 70-15912	B-194 C-GHOC	BE-37 N49SS	BE-125 N350TJ
B-11 N100KA	B-100 C-GJBV	B-195 EC-CHE	BE-39 C-FSIK	BE-126 N123WH
B-14 N402G	B-101 N696AB	B-196 F-GJJJ	BE-40 N24203	BE-127 N686AC
B-16 YV-801CP	B-102 N799DD	B-197 JW9027	BE-41 N11792	BE-128 N137D
B-17 N951K	B-103 C-FDOR	B-198 C-GAPK	BE-42 N87NW	BE-129 C-GSWF
B-18 C-GXVX	B-104 C-GCFD	B-199 F-GEXV	BE-43 N70BA	BE-130 C-....
B-20 N195MA	B-105 ZS-TBS	B-200 PT-LZA	BE-44 N300MP	BE-131 N6354H
B-21 N360C	B-106 C-FDOS	B-201 C-GAVI	BE-45 PT-WPV	BE-132 N150YR
B-22 N577D	B-107 C-GNAJ	B-202 N72TA	BE-46 N146BT	BE-133 N531CB
B-23 N701RJ	B-108 C-GASW	B-204 C-FBGS	BE-47 N49E	BE-134 N363EA
B-24 C-GNAA	B-109 N78MK	B-206 N86BM	BE-48 N2830S	BE-135 N997ME
B-25 N136JH	B-110 XC-FIW	B-207 C-FHGG	BE-49 N400RK	BE-136 N221TC
B-26 N696JB	B-112 C-FDOU	B-208 9Y-...	BE-50 N955FC	BE-137 N444RK
B-27 C-GWWA	B-114 N1347Z	B-209 N14CF	BE-51 N100TW	
B-28 C-FWYO	B-116 N100GV	B-210 N75GR	BE-52 N771CW	**Super King Air 200**
B-29 N830EM	B-117 C-FDOV	B-211 C-GNEX	BE-53 N53TD	BB-2 C-GARO
B-30 N3606T	B-118 N18AH	B-212 N115DT	BE-54 N2025S	BB-3 N24SP
B-31 N711GM	B-120 C-FDOY	B-213 SE-LDL	BE-55 N813BL	BB-4 71-21059
B-32 N122U	B-121 PT-LJN	B-214 N46BE	BE-56 N9933S	BB-5 71-21060
B-33 9Q-CRF	B-122 C-FXAJ	B-215 N552GA	BE-57 N75AP	BB-6 N300TR
B-34 XA-SWE	B-123 C-GJJF	B-216 JDFT-3	BE-58 N564CA	BB-7 C-FBCN
B-35 N178WM	B-124 C-GILM	B-217 OO-SNA	BE-59 N47KS	BB-8 XB-TFS
B-36 C-GXRX	B-125 C-FWRM	B-219 331	BE-60 N650TJ	BB-9 N200EZ
B-37 N524SC	B-126 C-GASI	B-222 PK-CAE	BE-61 YV-903P	BB-10 N980DB
B-38 C-FQOV	B-127 N170AS	B-224 N16SM	BE-62 N85LF	BB-11 FAB ...
B-39 OH-BKA	B-129 F-GEJV	B-225 N723W	BE-63 N924RM	BB-13 F-GIJB
B-40 C-FMXY	B-130 N83TM	B-226 N91RK	BE-64 N497SL	BB-14 C-GWWN
B-41 N93BC	B-131 N102FG	B-227 N227BC	BE-65 N6045S	BB-16 N711CR
B-42 C-FAFD	B-132 C-GXHP	B-228 F-GKEL	BE-66 N37PC	BB-17 N900CP
B-43 PT-DKV	B-133 N22BJ	B-229 N100HC	BE-67 N522CF	BB-18 N346CM
B-44 N962R	B-136 N700AT	B-230 N1362N	BE-68 N66820	BB-20 OB-1509
B-45 C-GPCB	B-137 PT-OVQ	B-231 N38P	BE-69 N794WB	BB-21 F-BVET
B-46 N283PM	B-138 N600DK	B-232 C-GKAJ	BE-70 N926HS	BB-22 N73MW
B-47 C-GNAX	B-139 N247B	B-233 N711RE	BE-72 ZS-MZS	BB-23 N153ML

BB-24 C-FCGB	BB-117 N21DE	BB-208 N62DL	BB-298 N862DD	BB-386 ZS-NHX
BB-25 N601CF	BB-119 LN-PAG	BB-209 F-GPAS	BB-299 N200FV	BB-387 5N-BHL
BB-26 ZS-LYA	BB-120 C-GHOP	BB-210 G-FRYI	BB-300 N86Q	BB-388 N26BE
BB-27 N25BL	BB-121 ZS-OCI	BB-211 5Y-BMC	BB-301 C-FCGU	BB-389 N859CC
BB-28 6V-AGS	BB-122 N122RF	BB-212 N910P	BB-302 N86Y	BB-390 N1660W
BB-29 C-GKBN	BB-123 PH-ATM	BB-214 OB-....	BB-303 F-GHCS	BB-391 N1969C
BB-30 G-HAMA	BB-124 N109TM	BB-215 ZS-NGC	BB-304 N103PM	BB-392 F-GRAN
BB-31 N7RW	BB-125 N90PB	BB-216 LN-VIU	BB-305 N30XY	BB-395 XB-BRB
BB-32 N100SF	BB-126 N208AJ	BB-217 C-FCGM	BB-306 N500HY	BB-396 N403R
BB-33 N90806	BB-127 N200JL	BB-220 N75VF	BB-307 N921S	BB-397 LX-GDB
BB-34 N87BP	BB-128 F-BXSI	BB-221 PH-DDB	BB-309 OY-PEB	BB-398 N398HM
BB-35 YV-732P	BB-130 N682DR	BB-222 ZS-KAA	BB-310 N988SC	BB-399 N399BM
BB-36 N44UF	BB-131 N131SJ	BB-223 YV-O-CVG-5	BB-311 N370TC	BB-400 N999BT
BB-37 ZS-NXH	BB-132 HS-DCB	BB-224 N215PA	BB-312 N127GA	BB-401 SX-APJ
BB-40 N4TJ	BB-133 N717HT	BB-225 N8534W	BB-313 N30SE	BB-402 N318W
BB-41 N200ZC	BB-134 D-IAAK	BB-226 HK-3699	BB-314 N767HC	BB-403 C-FGFZ
BB-42 N66TJ	BB-135 HK-3854	BB-227 N335S	BB-315 Z-TAB	BB-404 N777PR
BB-43 N520MC	BB-136 N2WC	BB-228 N74ED	BB-316 YV-168CP	BB-405 N350AC
BB-45 PT-OZK	BB-137 PP-EOP	BB-229 N911MN	BB-318 N49KC	BB-406 F-GHOC
BB-46 N760NP	BB-138 N999DT	BB-230 F-GDLE	BB-319 N6HU	BB-407 ZS-NUC
BB-47 VH-SKC	BB-139 N170SP	BB-231 C-FZVX	BB-321 N261AC	BB-409 N220TA
BB-48 XA-PEU	BB-140 C-GPCP	BB-232 N162PA	BB-322 N715MA	BB-410 N200PL
BB-49 N222KA	BB-141 N977LX	BB-233 F-GHLD	BB-323 N4430V	BB-412 N2030P
BB-50 N888TR	BB-143 N200CJ	BB-234 N80PA	BB-324 HA-ACS	BB-413 N2061B
BB-51 ZS-LST	BB-144 N2883	BB-235 OY-CBP	BB-325 N65EB	BB-414 N600DM
BB-52 ZS-NWK	BB-145 N88CP	BB-236 C-FCGC	BB-326 N944CE	BB-415 P2-CAA
BB-53 N280RA	BB-147 N771HC	BB-237 N114JF	BB-327 VH-XRP	BB-417 HB-GHF
BB-54 4-G-41/0697	BB-148 VH-WNH	BB-238 YV-106CP	BB-328 N795PA	BB-419 OY-BVB
BB-55 N200BC	BB-149 D-IAHK	BB-239 F-GSIN	BB-329 N53CK	BB-420 N210SU
BB-56 N44US	BB-150 N107FL	BB-240 N888EM	BB-330 N130PA	BB-421 A2-AEZ
BB-57 C-GAFT	BB-151 N528WG	BB-242 N200RW	BB-331 N87LP	BB-422 N501EB
BB-58 N60PD	BB-152 N1223C	BB-243 N300US	BB-332 N3DE	BB-423 N120RJ
BB-59 N500KS	BB-154 N44KT	BB-246 N200VA	BB-333 TC-AUY	BB-424 F-GMCR
BB-60 HK-....	BB-155 5Y-BMA	BB-247 N4000	BB-334 ZS-NRT	BB-425 N2057S
BB-61 N58280	BB-156 N69CD	BB-248 HK-3822	BB-335 N327RK	BB-426 N220CB
BB-62 N500GN	BB-157 N93LV	BB-249 C-GQJG	BB-336 N311AV	BB-427 N101GQ
BB-63 HK-3705	BB-158 N81PA	BB-250 C-FCGX	BB-337 N600AM	BB-428 N106PA
BB-64 N174PW	BB-159 C-FCGT	BB-251 F-GKCV	BB-338 N689BV	BB-429 N74GS
BB-65 N901BR	BB-160 YV-665CP	BB-252 N934SH	BB-339 N5UV	BB-430 N1YS
BB-66 RP-C223	BB-161 N131PA	BB-253 N999CY	BB-341 ZS-LWM	BB-431 F-GILH
BB-67 N78DV	BB-162 VH-SMT	BB-254 N600BV	BB-342 0342	BB-432 N33KM
BB-68 C-GMWR	BB-164 N70PQ	BB-255 N32KC	BB-343 N4679M	BB-433 N94FG
BB-69 N5UB	BB-165 VH-XRF	BB-256 5Y-BKT	BB-344 VH-ITA	BB-434 N688AA
BB-71 4-G-42/0698	BB-166 N89MP	BB-257 N362EA	BB-346 N84LS	BB-435 XC-BCN
BB-72 5Y-JMB	BB-167 N300DK	BB-258 N888ET	BB-347 N424CR	BB-436 N83KA
BB-73 VH-SGT	BB-168 C-FOGY	BB-259 VH-LKB	BB-348 N28J	BB-437 N711MB
BB-75 HK-3936	BB-169 N725MC	BB-260 N18CJ	BB-349 F-GHLB	BB-438 N438BM
BB-76 C-GPCD	BB-170 N869MA	BB-261 YV-O-MMH-9	BB-350 ZS-NTT	BB-439 F-GGLN
BB-77 C-GWUY	BB-171 7T-VRI	BB-262 N92V	BB-351 G-CEGR	BB-440 YV-238CP
BB-78 N694AB	BB-172 N94KC	BB-263 N48Q	BB-352 HK-3922P	BB-441 N79CF
BB-79 N155QS	BB-173 N55FG	BB-264 HK-3796	BB-353 SU-BAX	BB-442 N442KA
BB-80 F-GGMS	BB-174 N869	BB-265 YV-...	BB-354 N550JC	BB-443 N700Z
BB-81 N18DN	BB-175 7T-VRH	BB-266 ZS-NNS	BB-355 VH-SMB	BB-444 ZS-LZU
BB-82 LZ-RGP	BB-176 N67GA	BB-267 N96UB	BB-356 N28KC	BB-445 N364SB
BB-83 RP-C711	BB-177 N456CS	BB-268 C-GTUC	BB-357 N500DE	BB-446 N885HT
BB-84 C-GFSB	BB-178 N36CP	BB-269 N100SM	BB-358 RP-C5129	BB-447 N356GA
BB-85 V5-CIC	BB-179 N424BS	BB-271 N69BK	BB-359 N359K	BB-448 N700HM
BB-86 5Y-BKM	BB-180 PH-SBK	BB-272 N202KA	BB-360 N18KA	BB-449 B-13152
BB-87 N87GA	BB-181 N9150R	BB-273 YV-O-CVG-8	BB-361 N1HX	BB-450 HK-3923X
BB-88 N35	BB-182 EC-GBB	BB-274 N1MM	BB-362 N506AB	BB-451 HB-GJI
BB-89 N200WZ	BB-183 N175SA	BB-275 C-GQNJ	BB-363 N642TF	BB-452 N730CE
BB-91 N884PG	BB-184 7T-VRG	BB-276 TF-ELT	BB-364 F-GHYV	BB-453 N23TC
BB-92 C-FJRT	BB-185 N265EB	BB-277 N200AP	BB-365 VH-OYA	BB-454 N888DC
BB-94 N17HM	BB-187 N630DB	BB-278 AP-CAA	BB-366 D-IBHK	BB-455 N24SX
BB-95 A2-AHZ	BB-188 N4288S	BB-279 N279CA	BB-367 5Y-NPO	BB-456 N456CD
BB-97 N65RT	BB-189 N56CC	BB-280 N280TT	BB-369 XC-AA48	BB-457 N800LS
BB-99 5Y-...	BB-190 C-FCGL	BB-281 N4AT	BB-370 N80RT	BB-458 AN-234
BB-100 XC-DIJ	BB-191 N200TP	BB-282 YV-1300P	BB-371 N151E	BB-459 101002
BB-101 ZS-MYA	BB-192 F-OGPQ	BB-283 N283JP	BB-372 SX-ECG	BB-460 4-F-43/0743
BB-102 N200AF	BB-193 OB-1468	BB-285 C-GQXF	BB-373 N4799M	BB-461 N33TG
BB-103 N201CH	BB-194 HB-GIL	BB-286 Z-MRS	BB-374 N111UT	BB-462 N220TT
BB-104 N40RA	BB-195 N5LE	BB-287 N429E	BB-375 N23ST	BB-463 VH-ITH
BB-105 C-FKJI	BB-196 HK-3995X	BB-288 ZS-NRR	BB-376 N409GA	BB-464 N561SS
BB-106 N52SF	BB-197 ZS-OAK	BB-289 N5ST	BB-377 N52GT	BB-465 N24SA
BB-107 N191FL	BB-198 N4298S	BB-290 N290SJ	BB-378 N200AJ	BB-466 N206P
BB-108 N108BM	BB-199 F-GNPD	BB-291 N769	BB-379 5Y-DDE	BB-467 5Y-SJB
BB-109 N244JP	BB-200 N2PY	BB-292 N393JW	BB-380 N142SR	BB-468 OE-LEA
BB-110 N110BM	BB-201 ZS-NRW	BB-293 N200BT	BB-381 ZS-KGW	BB-470 VH-NIA
BB-111 N22071	BB-202 F-GEXL	BB-294 9M-KNS	BB-382 N92M	BB-471 LV-RTC
BB-113 PK-TRA	BB-205 N6PX	BB-295 TU-VBB	BB-383 N384DB	BB-472 G-WRCF
BB-114 N25KW	BB-206 ZS-NZH	BB-296 N402KA	BB-384 N2025M	BB-473 N510CB
BB-115 N330BR	BB-207 C-FCGW	BB-297 P2-IAH	BB-385 N81RD	BB-474 C-GCFF

BB-475 N121LB	BB-569 N40HE	BB-657 N301PS	BB-744 F-GGLA	BB-830 N88MT	
BB-476 N203BS	BB-570 N968T	BB-658 N747HN	BB-745 N428P	BB-831 F-OHCP	
BB-477 F-GILB	BB-571 N769MB	BB-659 N77QX	BB-746 C-FMPE	BB-832 N200BE	
BB-478 N789DS	BB-572 N119MC	BB-660 VH-LKF	BB-747 N63SK	BB-833 PNC-221	
BB-479 C-GNBB	BB-574 N75WP	BB-661 N777SS	BB-748 Z-WSG	BB-834 N67MD	
BB-480 D2-EMX	BB-575 VT-SLS	BB-662 N119WM	BB-749 N269TA	BB-835 N84PA	
BB-481 HR-...	BB-576 N81WU	BB-663 5N-AMT	BB-750 N81TF	BB-836 D2-EBF	
BB-482 N7166P	BB-577 CN-CDF	BB-664 N364EA	BB-751 N520WS	BB-837 ZS-LBD	
BB-483 HC-BSR	BB-578 N303DK	BB-666 N15NG	BB-752 N616GB	BB-839 N21FG	
BB-485 N485K	BB-579 VH-AKT	BB-667 C-GMPO	BB-753 F-GERS	BB-840 PT-MPR	
BB-486 N56DA	BB-580 AN-233	BB-668 N62EA	BB-754 N242DM	BB-841 N841K	
BB-487 N8PY	BB-581 N12NG	BB-670 N7VA	BB-755 N71TB	BB-842 N38473	
BB-488 4-G-45/0745	BB-582 LN-MOA	BB-671 C-GFSG	BB-756 ZS-LWD	BB-843 N211CP	
BB-490 N217CP	BB-583 N777AQ	BB-672 240	BB-757 C-FMPH	BB-844 F-GIAL	
BB-491 N622DC	BB-584 LN-MOB	BB-673 N502RH	BB-758 N20RE	BB-845 OY-GRB	
BB-492 N64DC	BB-585 N43PE	BB-674 N851MK	BB-759 7T-VRS	BB-846 5Y-BKA	
BB-493 N6509F	BB-586 ZS-MGG	BB-675 SE-IUX	BB-760 N3710A	BB-847 OH-BIF	
BB-494 N942CE	BB-587 N200BH	BB-676 F-GHVV	BB-761 N78CT	BB-848 N848NA	
BB-495 EV-7910	BB-588 N578BM	BB-677 N200KK	BB-762 RP-C4650	BB-849 N711MZ	
BB-497 N72MM	BB-589 N553R	BB-678 N811FA	BB-763 N56KA	BB-850 N3CR	
BB-498 N6507B	BB-590 F-GIDV	BB-679 N1194C	BB-764 LV-ONH	BB-851 N208F	
BB-499 N6051C	BB-592 N26SJ	BB-680 N675PG	BB-765 N242LC	BB-852 N46BR	
BB-501 N175BM	BB-593 ZS-NZI	BB-681 F-GGPR	BB-766 N52GP	BB-853 N44SR	
BB-502 N275X	BB-594 N766NM	BB-682 ZS-KLM	BB-768 N37392	BB-854 HK-3214	
BB-503 RP-C304	BB-595 RP-C1260	BB-683 N2877K	BB-769 N623VG	BB-855 N223HC	
BB-505 C-GKBP	BB-596 N72SE	BB-684 G-ROWN	BB-770 YV-O-CPI-1	BB-856 N6PE	
BB-506 N8SV	BB-597 5Y-BJC	BB-685 N29HF	BB-771 AN-231	BB-857 5V-MCG	
BB-507 F-GEJY	BB-598 YV-335CP	BB-686 N3663M	BB-772 N111PV	BB-858 5V-MCH	
BB-509 N72RL	BB-599 N48CR	BB-687 SE-INI	BB-773 N500CS	BB-860 N488AD	
BB-511 N202AJ	BB-600 N15KA	BB-688 F-GMCS	BB-774 N3738B	BB-861 N727DD	
BB-512 N333TS	BB-601 N976KC	BB-689 N187MQ	BB-775 7T-VRT	BB-862 N207DB	
BB-513 C-GMOC	BB-602 N981LL	BB-690 VH-SKN	BB-776 N800PP	BB-863 N80WM	
BB-514 N300HM	BB-603 D2-...	BB-691 N84CC	BB-777 I-PIAH	BB-864 N92TC	
BB-515 F-GKII	BB-604 N944CC	BB-692 RP-C264	BB-778 N266RH	BB-865 N32SV	
BB-516 N281JH	BB-605 VH-YNE	BB-693 N60SC	BB-779 F-GEPY	BB-866 VH-MSZ	
BB-517 YV-257CP	BB-606 LV-WIO	BB-694 PT-OTO	BB-780 N780RC	BB-867 N24AR	
BB-518 N83GA	BB-607 ZS-LTE	BB-696 5Y-TWB	BB-781 N37NC	BB-869 N31FM	
BB-519 N28RY	BB-608 N284K	BB-697 N5UN	BB-782 G-THUR	BB-870 LV-WEW	
BB-520 2840	BB-610 N171RD	BB-698 N440CE	BB-783 N25KA	BB-872 D-IOAN	
BB-521 N355TW	BB-611 N41CV	BB-699 LV-OFT	BB-784 PT-OSR	BB-873 N117CA	
BB-522 3150	BB-612 N506GT	BB-700 N200PY	BB-785 RP-C582	BB-874 N4042J	
BB-523 N196MP	BB-613 N613CS	BB-701 YV-O-BND-1	BB-786 N120P	BB-875 TY-WEG	
BB-524 D-ILPC	BB-614 ZS-LKA	BB-703 N703KH	BB-787 C-FZVW	BB-876 F-ZBFK	
BB-525 ZS-MFC	BB-615 RP-C704	BB-704 ZS-KLO	BB-788 N551JL	BB-877 N508MV	
BB-526 OE-BBB	BB-616 F-GGMV	BB-705 OY-BVW	BB-789 N905GP	BB-878 N4PT	
BB-527 C-FIFO	BB-617 HP-960P	BB-706 ZS-NBO	BB-790 D-IAMB	BB-879 N4PT	
BB-528 N700BE	BB-618 N269BW	BB-707 N715RD	BB-791 TC-ACN	BB-880 N1804T	
BB-529 VH-SWP	BB-619 101003	BB-708 N313EE	BB-792 N110G	BB-881 N881CS	
BB-531 5U-ABX	BB-620 TR-...	BB-709 N202VT	BB-793 N68KA	BB-882 N56AY	
BB-532 N300JK	BB-621 N207CM	BB-710 N850C	BB-794 PK-VKB	BB-883 TC-YPI	
BB-533 VH-XDB	BB-622 F-GHSV	BB-711 N1162V	BB-795 N828JB	BB-884 N49JG	
BB-534 N12CF	BB-623 N6689D	BB-712 N200BM	BB-796 YV-401CP	BB-885 N146BC	
BB-535 C-GOGT	BB-624 N202BB	BB-713 CN-CDN	BB-797 G-BVMA	BB-886 N111JW	
BB-536 N33FM	BB-625 N869AM	BB-714 N500CR	BB-799 N250DL	BB-887 N87699	
BB-537 N5MK	BB-626 C9-ENH	BB-715 ZS-NOC	BB-800 N1TX	BB-888 N700U	
BB-538 5N-AVH	BB-628 N6739P	BB-716 F-ODYR	BB-801 N244CH	BB-889 ZS-LAW	
BB-539 ST-SFS	BB-629 9Q-CTG	BB-717 N77CA	BB-803 5Y-TWA	BB-890 N6EA	
BB-540 N946CE	BB-630 ZS-NZJ	BB-718 VH-SGV	BB-804 N174WB	BB-891 N888HG	
BB-541 N113GW	BB-631 N6684B	BB-719 N3722Y	BB-805 N501RH	BB-892 N81LT	
BB-542 F-GILP	BB-632 YV-1873P	BB-720 N186MQ	BB-806 EC-ESV	BB-893 N762NB	
BB-544 N711VH	BB-634 OY-NUK	BB-721 N8EF	BB-807 7T-VRO	BB-894 N230GK	
BB-545 D-ILIN	BB-635 N30PH	BB-722 9Q-CFE	BB-808 N53RT	BB-896 N1811S	
BB-546 4-G-47/0747	BB-636 N127TA	BB-724 N804	BB-809 5N-AMU	BB-897 D-IEFB	
BB-547 5Y-...	BB-637 N31WJ	BB-725 XC-...	BB-810 N14TF	BB-898 N911SR	
BB-548 N706DG	BB-638 N220JB	BB-726 G-BXMA	BB-811 AEE-101	BB-899 N1KA	
BB-549 4-G-48/0748	BB-639 LV-WOS	BB-727 N1955E	BB-812 N991GA	BB-900 N18260	
BB-550 VH-TLX	BB-640 N47CF	BB-728 C-GTGA	BB-813 ZS-PAM	BB-901 N26UT	
BB-552 N63593	BB-641 N641TC	BB-729 LV-WPM	BB-814 XA-RMX	BB-902 N65TW	
BB-553 N83JN	BB-642 N10BY	BB-730 N103AL	BB-815 ZS-MSL	BB-903 7Q-YMP	
BB-554 N5PX	BB-643 5Y-NUN	BB-731 YV-O-BIV-1	BB-816 N900RH	BB-904 N430MC	
BB-555 EC-GHZ	BB-644 YV-O-CVF-1	BB-732 PK-VKA	BB-817 N56RT	BB-905 N79GS	
BB-556 N79GA	BB-645 N999VB	BB-733 N444AD	BB-818 N800BW	BB-906 YV-2350P	
BB-557 5Y-JAI	BB-646 N646BM	BB-734 N85BC	BB-819 N500KA	BB-907 N1860N	
BB-558 LV-WGP	BB-648 N17DW	BB-735 N508JA	BB-820 N369TA	BB-908 N1PD	
BB-559 N795CA	BB-649 N103TF	BB-736 N45CF	BB-821 TC-DBY	BB-909 SE-KDK	
BB-560 LN-MAA	BB-650 N740GL	BB-737 N38GM	BB-822 F-GIND	BB-910 N910AJ	
BB-561 G-ECAV	BB-651 N11TE	BB-738 N66TS	BB-824 N824TT	BB-911 N270CS	
BB-562 N70LF	BB-652 Z3-BAB	BB-739 VT-ESR	BB-825 N825KA	BB-912 C-GFSH	
BB-564 N42LJ	BB-653 VH-MXK	BB-740 YV-385CP	BB-826 N1CB	BB-913 N82DD	
BB-565 N920C	BB-654 N93ZC	BB-741 N75AH	BB-827 D-IFES	BB-914 5X-INS	
BB-566 N300HB	BB-655 N655BA	BB-742 XT-MAX	BB-828 G-OLDZ	BB-915 N333RD	
BB-568 G-BGRE	BB-656 YV-2354P	BB-743 N111LP	BB-829 LN-NOA	BB-916 N916HC	
				BB-917 N95TT	

Serial	Reg	Serial	Reg
BB-918	ZS-NCH	BB-1134	N623WA
BB-919	N160AD	BB-1135	N399LA
BB-920	PK-YPS	BB-1136	LN-VIZ
BB-921	D-IKOB	BB-1137	N333AP
BB-922	N1926A	BB-1138	ZS-SOL
BB-923	N200MG	BB-1139	N961LL
BB-924	N107MG	BB-1140	N888ZX
BB-925	N282SJ	BB-1141	N500LP
BB-926	N68GK	BB-1142	N23YP
BB-927	927	BB-1143	N23WS
BB-928	N553HC	BB-1144	ZS-NWT
BB-929	N999GA	BB-1145	HP-1213
BB-930	N930SP	BB-1146	N60PC
BB-931	N995MS	BB-1147	N647JM
BB-932	101004	BB-1148	N345MB
BB-933	D-IAWS	BB-1149	N200HF
BB-934	N38381	BB-1150	N103BG
BB-935	N500CY	BB-1151	XA-MOB
BB-936	VH-OWN	BB-1152	ZS-NJM
BB-937	C9-ATW	BB-1153	N73MP
BB-938	N132K	BB-1154	N6683W
BB-939	N164TC	BB-1155	VH-SMZ
BB-940	C-FZPW	BB-1156	N66825
BB-941	ZS-MIN	BB-1157	N126AP
BB-942	N625MD	BB-1158	N158EF
BB-943	N280YR	BB-1159	N552TB
BB-944	G-FPLA	BB-1160	N444KA
BB-945	RP-C1577	BB-1161	N27CV
BB-946	N1850X	BB-1163	N200EW
BB-947	N500WF	BB-1164	N701NA
BB-948	N25CS	BB-1165	HS-PON
BB-949	N567JD	BB-1166	D-ITAB
BB-950	N8NA	BB-1167	N35GR
BB-951	RP-C969	BB-1168	N844MP
BB-952	N185XP	BB-1169	N40BC
BB-953	N953L	BB-1170	N55PC
BB-954	ZS-LFM	BB-1171	N86DD
BB-955	N933RT	BB-1172	VH-FDG
BB-956	D-IAMK	BB-1173	N3PX
BB-957	N200HW	BB-1174	N200MP
BB-958	N30EM	BB-1175	N515CR
BB-959	N2SC	BB-1176	N904DG
BB-960	HK-3470P	BB-1177	N200LU
BB-961	A2-AHA	BB-1178	C-GDSH
BB-962	N719HC	BB-1179	N204WB
BB-963	RP-C367	BB-1180	N221SP
BB-964	VH-SBM	BB-1181	F-GJBS
BB-965	N300AJ	BB-1182	N67262
BB-966	F-GDCS	BB-1183	N91HT
BB-967	N75ZT	BB-1184	N93NP
BB-968	N42TD	BB-1186	N186EB
BB-969	N48N	BB-1187	N200SV
BB-971	PT-MMB	BB-1189	VH-KBH
BB-972	VT-EHB	BB-1190	N75GF
BB-973	N53TM	BB-1191	N177CN
BB-974	HK-3507	BB-1192	N61AP
BB-975	RP-C755	BB-1193	N78GC
BB-976	N83RH	BB-1194	N616DR
BB-977	N607KW	BB-1195	ZS-LRE
BB-978	N877RC	BB-1196	Z-ZLT
BB-979	N6120C	BB-1197	N347D
BB-980	YV-466CP	BB-1198	VT-JKC
BB-981	N404FA	BB-1199	N7203R
BB-982	N800JR	BB-1200	N6816A
BB-983	D-ISAZ	BB-1201	D-IMMM
BB-984	N9RU	BB-1202	G-REBK
BB-986	PT-WOF	BB-1203	N877JE
BB-987	N200HD	BB-1204	B-3551
BB-988	5Y-HHA	BB-1205	B-3552
BB-989	N520D	BB-1206	B-3553
BB-990	P2-VIC	BB-1207	N3ZC
BB-991	N992TJ	BB-1208	N200SE
BB-992	N32LJ	BB-1209	D-IKBJ
BB-993	N99LL	BB-1210	SE-IXC
BB-994	XA-RVJ	BB-1211	N10EC
BB-995	N66U	BB-1212	N54FB
BB-996	N7NA	BB-1213	5N-BFL
BB-997	N76PM	BB-1214	N200GS
BB-998	ZS-LFW	BB-1215	PH-ACZ
BB-999	N74RG	BB-1216	N210AC
BB-1000	N94QD	BB-1217	N697P
BB-1001	N282CT	BB-1218	N740P
BB-1002	N282CT	BB-1219	N200YB
BB-1003	N58GA	BB-1220	TR-LDM
BB-1004	RP-C1515	BB-1221	TF-FMS
BB-1005	D-ICKM	BB-1222	ZS-LNR
BB-1006	N680CB	BB-1223	N586BC
BB-1007	G-SBAS	BB-1224	N94LC
BB-1008	ZP-TWN	BB-1225	N848PF
BB-1009	N6451D	BB-1226	N225WL
BB-1010	N702MA	BB-1227	HK-3432
BB-1011	N20SM	BB-1228	N69VC
BB-1012	N200MJ	BB-1229	D-IBAD
BB-1013	N711BL	BB-1230	N224P
BB-1014	N6308F	BB-1231	OE-FJB
BB-1015	N216RP	BB-1232	N209CM
BB-1016	N245CT	BB-1233	N887T
BB-1017	N200PH	BB-1234	D-IZZZ
BB-1018	ZS-LFU	BB-1235	VT-ENM
BB-1019	VH-JJR	BB-1236	N212JB
BB-1020	YV-488CP	BB-1237	N7250T
BB-1021	N678SS	BB-1238	N212JB
BB-1022	N220RJ	BB-1239	N833BK
BB-1023	N125NC	BB-1240	N200TK
BB-1024	SU-...	BB-1241	N7256K
BB-1025	N22F	BB-1242	N20VP
BB-1026	YV-534CP	BB-1243	N428A
BB-1027	ZS-NAW	BB-1244	F-GSFA
BB-1028	N62569	BB-1245	N76RJ
BB-1029	N176M	BB-1246	N200CP
BB-1030	N360SC	BB-1247	N44CS
BB-1031	N6644J	BB-1248	VT-ENL
BB-1032	F-GJMJ	BB-1249	N879C
BB-1034	N185MV	BB-1250	N2519Y
BB-1035	N96ZZ	BB-1251	ZS-LTG
BB-1036	N6271C	BB-1252	N634TT
BB-1037	VH-FDI	BB-1253	A2-AEO
BB-1038	ZS-MES	BB-1254	VT-EPA
BB-1039	HB-GHS	BB-1255	PK-HTI
BB-1040	N27LJ	BB-1256	N184JS
BB-1041	VH-OYD	BB-1257	N606AU
BB-1042	N77JT	BB-1258	N2748X
BB-1043	N79PG	BB-1259	D-IDSM
BB-1044	G-BPPM	BB-1260	N2PX
BB-1045	N67TC	BB-1261	N717DC
BB-1046	A2-...	BB-1262	N90TM
BB-1047	ZS-LIL	BB-1263	N263SW
BB-1048	G-FPLB	BB-1265	D-IAKK
BB-1049	HK-3894	BB-1266	PT-WSX
BB-1050	N925B	BB-1267	VT-VHL
BB-1051	D-ICIR	BB-1268	N29997
BB-1053	N212DM	BB-1269	N9898
BB-1054	C-GRFN	BB-1270	F-GJFA
BB-1055	N6UD	BB-1271	PT-LPG
BB-1056	VH-FDO	BB-1272	VT-EQD
BB-1057	N220TB	BB-1273	N466MW
BB-1058	N220DK	BB-1274	N30234
BB-1059	N50YR	BB-1275	PT-WLT
BB-1060	ZS-TON	BB-1276	N40TD
BB-1061	N50DR	BB-1277	VT-EPY
BB-1062	PT-...	BB-1278	N533P
BB-1063	HK-3504	BB-1279	N300TP
BB-1064	N129D	BB-1280	D-IBAR
BB-1065	D-ICOA	BB-1281	PK-HPH
BB-1066	N356WG	BB-1282	N200RR
BB-1067	N860H	BB-1283	N113US
BB-1068	HK-3554	BB-1284	D-....
BB-1069	N795GB	BB-1285	N200JR
BB-1070	ZS-NBJ	BB-1286	N909ST
BB-1071	N14HG	BB-1287	N921AZ
BB-1072	CNA-NG	BB-1288	VT-EQK
BB-1073	CNA-NH	BB-1290	G-VSBC
BB-1074	N74LV	BB-1291	D-ICRA
BB-1075	N75LV	BB-1292	N333TP
BB-1076	C9-PMZ	BB-1293	N3127K
BB-1077	RP-C7188	BB-1294	N294WT
BB-1078	N771HM	BB-1295	N295CP
BB-1079	HB-GDL	BB-1296	PH-DUS
BB-1080	D-ILOH	BB-1297	N21VF
BB-1081	N3PX	BB-1298	N734A
BB-1082	TC-...	BB-1299	N123ME
BB-1083	ZS-...	BB-1300	N313ES
BB-1084	N284KW	BB-1302	PT-WYY
BB-1085	N63686	BB-1303	N1932H
BB-1086	TC-YSM	BB-1304	C-FPQQ
BB-1087	N65WM	BB-1305	VT-SAB
BB-1088	N68FA	BB-1306	N1553E
BB-1089	N69JH	BB-1307	N1547V
BB-1090	YV-493CP	BB-1308	N1509G
BB-1091	N529NA	BB-1309	N4277C
BB-1092	YV-2352P	BB-1310	N16TF
BB-1094	N1925P	BB-1311	N888CS
BB-1099	N127SD	BB-1312	N506EB
BB-1100	N63791	BB-1313	N700WP
BB-1101	N395AM	BB-1314	N4277E
BB-1102	F-ZBFJ	BB-1315	HS-DCF
BB-1103	N361EA	BB-1316	ZS-MFB
BB-1104	N162Q	BB-1317	VT-BSB
BB-1106	N1870S	BB-1318	ZS-MMV
BB-1107	N564BC	BB-1319	N1548B
BB-1108	XC-AA50	BB-1320	N825ST
BB-1109	N900FD	BB-1321	N133GA
BB-1110	TR-LDU	BB-1322	N96GP
BB-1111	ZS-LVK	BB-1323	ZS-NZZ
BB-1112	N311MP	BB-1324	N7087N
BB-1113	N745R	BB-1325	G-KMCD
BB-1114	N872CA	BB-1326	N326PS
BB-1115	N764NB	BB-1327	N1543Z
BB-1116	F-GDJS	BB-1328	N1553M
BB-1118	N586UC	BB-1329	VN-B594
BB-1120	N6609K	BB-1330	N681PC
BB-1121	N6604L	BB-1331	N5530H
BB-1122	N6606R	BB-1332	N444BK
BB-1123	N282TA	BB-1333	VH-MVL
BB-1124	N678EB	BB-1334	TC-SKO
BB-1125	VH-KCH	BB-1335	N1567F
BB-1126	N175BC	BB-1336	N448T
BB-1127	5Y-SEL	BB-1337	N330CS
BB-1128	C-GGGQ	BB-1338	N915YW
BB-1129	N66404	BB-1339	N252AF
BB-1130	N200QS	BB-1340	N256AF
BB-1131	HP-1180	BB-1341	VT-SAD
BB-1132	N152C	BB-1342	VT-SAE
		BB-1343	VT-SAF
		BB-1344	N21PS
		BB-1345	N93ME
		BB-1346	N800KT
		BB-1347	F-GJFC

BB-1348 D2-EST	BB-1431 JA8705	BB-1518 SU-...
BB-1349 D-IBFS	BB-1432 N43TA	BB-1519 VP-BBK
BB-1350 N1570F	BB-1433 C-GMEH	BB-1520 N3230X
BB-1351 N914CT	BB-1434 4005	BB-1521 LV-WOR
BB-1352 N5568V	BB-1435 PT-ORB	BB-1522 ZS-OBB
BB-1353 A2-AGO	BB-1436 93303	BB-1523 N3250V
BB-1354 N15JA	BB-1437 N34LT	BB-1524 N24CV
BB-1355 SE-LCE	BB-1438 N8017M	BB-1525 N260G
BB-1356 N161RC	BB-1439 N583AT	BB-1526 N417MC
BB-1357 D-IBSY	BB-1440 N655JG	BB-1527 N170S
BB-1358 N5682P	BB-1441 93304	BB-1528 PT-WLF
BB-1359 N35SA	BB-1442 N128V	BB-1529 RP-C...
BB-1360 HK-3703X	BB-1443 93305	BB-1530 336
BB-1361 N5552U	BB-1444 N7PA	BB-1531 N16GF
BB-1362 D2-ECX	BB-1445 VT-MPG	BB-1532 N3252X
BB-1363 3B-SKY	BB-1446 PT-WGS	BB-1533 N14HB
BB-1364 N660PB	BB-1447 N70MN	BB-1534 N1074G
BB-1365 XA-RVH	BB-1449 LN-MOC	BB-1535 D-IBFT
BB-1366 N777AG	BB-1450 N213DB	BB-1536 N10436
BB-1367 N36R	BB-1452 N15HV	BB-1537 ZS-ARL
BB-1368 D-IPWB	BB-1453 7T-VBE	BB-1538 PH-VMP
BB-1369 D-IVHM	BB-1455 PT-JUB	BB-1538 D-.....
BB-1370 N122H	BB-1458 N883SW	BB-1539 TC-MDE
BB-1371 D2-EOJ	BB-1459 LN-MOD	BB-1540 N89WA
BB-1372 N5559X	BB-1460 LN-MOE	BB-1540 ZS-...
BB-1373 N25GE	BB-1461 LN-MOF	BB-1541 N20LB
BB-1374 N899HC	BB-1462 D-IWSH	BB-1542 ZS-ODI
BB-1375 4006	BB-1463 N564GA	BB-1543 PT-WLK
BB-1376 HK-3990X	BB-1464 N133LC	BB-1544 F-OHJK
BB-1377 F-GNEG	BB-1465 LN-MOG	BB-1545 N1070E
BB-1378 N618	BB-1466 LN-MOH	BB-1546 N770M
BB-1379 F-GJFD	BB-1467 C-GMGG	BB-1547 N180CA
BB-1380 N77HN	BB-1468 VT-LNT	BB-1548 OY-EEF
BB-1381 N215P	BB-1469 VT-CIL	BB-1549 N1069S
BB-1382 N6789	BB-1470 LN-MOI	BB-1550 N42SC
BB-1383 N913YW	BB-1471 N5TW	BB-1551 N151CF
BB-1384 N575T	BB-1472 VH-MYO	BB-1552 ZS-NZN
BB-1385 4X-FEA	BB-1473 PT-OXG	BB-1553 ZS-NZK
BB-1386 4X-FEB	BB-1474 ZS-NKC	BB-1554 N44WV
BB-1387 4X-FEC	BB-1475 ZS-NKE	BB-1555 N98DA
BB-1388 4X-FED	BB-1476 ZS-ODU	BB-1556 N1092N
BB-1389 N277GA	BB-1477 N200PU	BB-1557 N57TS
BB-1390 F-GMGB	BB-1478 D-IHAN	BB-1558 PT-WNN
BB-1391 D2-ESP	BB-1479 TG-...	BB-1559 N1114K
BB-1392 HK-3704X	BB-1480 N1559G	BB-1560 N30VP
BB-1393 N73LC	BB-1481 N1559W	BB-1561 N423LW
BB-1394 N625W	BB-1482 N77CX	BB-1562 ZS-PPG
BB-1395 N132MC	BB-1483 N1559Z	BB-1563 VT-BAL
BB-1396 N8093W	BB-1484 N6182A	BB-1564 VP-CMA
BB-1397 N77HD	BB-1485 VT-BSA	BB-1565 N23FH
BB-1398 YV-121CP	BB-1486 VT-EBB	BB-1566 N1106J
BB-1399 F-GJFE	BB-1487 N133K	BB-1567 VT-CSK
BB-1400 D-ICHG	BB-1488 N1568E	BB-1568 D-ILLF
BB-1401 JA8879	BB-1489 ZS-SMC	BB-1569 VP-CHE
BB-1402 N91CD	BB-1490 N122LC	BB-1570 N1120Z
BB-1403 N33BK	BB-1491 JA8614	BB-1571 N80BC
BB-1404 N93CD	BB-1492 N44NL	BB-1572 N2272H
BB-1405 RP-C2100	BB-1493 N3015Q	BB-1573 N702TA
BB-1406 JA8880	BB-1494 N3026H	BB-1574 N703TA
BB-1407 D2-ESQ	BB-1495 N3051K	BB-1575 N705TA
BB-1408 N74RF	BB-1496 TC-OZD	BB-1576 N1118G
BB-1409 10010	BB-1497 N508BM	BB-1577 N1089V
BB-1410 PP-EJG	BB-1498 N3092S	BB-1578 N330DR
BB-1411 10011	BB-1499 N3199A	BB-1579 N909J
BB-1412 N38V	BB-1500 RP-C1502	BB-1580 N345WK
BB-1413 10012	BB-1501 N676BB	BB-1581 N1119Z
BB-1414 10013	BB-1502 ZS-NXT	BB-1582 N43TL
BB-1415 10014	BB-1503 N363D	BB-1583 N424RA
BB-1416 VH-MSH	BB-1504 VH-HPW	BB-1584 YV-...
BB-1417 N8140P	BB-1505 VH-HPX	BB-1585 N257YA
BB-1418 ZS-NFE	BB-1506 N94LL	BB-1586 N3181Q
BB-1419 PT-WCB	BB-1507 LV-WNJ	BB-1587 N123ML
BB-1420 P2-MBZ	BB-1508 N3208T	BB-1589 N2288B
BB-1421 N715CG	BB-1509 N716TA	BB-1590 D-IHUT
BB-1422 N125TS	BB-1510 N3213G	BB-1591 D-.....
BB-1423 P2-MBH	BB-1511 N507EF	BB-1594 N2132W
BB-1424 N8236K	BB-1512 N.....	BB-1595 N713FP
BB-1425 N8002J	BB-1513 N3217N	BB-1597 N126RD
BB-1426 ZS-NOW	BB-1514 N31SV	BB-1598 N706TA
BB-1427 ZS-NOW	BB-1515 ZS-TOB	BB-1599 N200V
BB-1428 N660MW	BB-1516 N1CR	BB-1600 N480TC
BB-1429 RP-C1995	BB-1517 D-IANA	BB-1610 N713TA
BB-1430 VH-MSM		

Super King Air B200

FG-1 4X-FSF
FG-2 4X-FSG

Super King Air 200C

BL-1 5A-DDT	BL-82 84-0152
BL-2 5Y-EKO	BL-83 N58009
BL-3 N39PH	BL-83 84-0153
BL-4 ZS-NUF	BL-84 84-0154
BL-5 N143LG	BL-85 84-0155
BL-6 5A-DDY	BL-86 84-0156
BL-7 HB-GJD	BL-87 84-0157
BL-8 ZS-NAX	BL-88 84-0158
BL-9 VH-KZL	BL-89 84-0159
BL-10 5Y-BLA	BL-90 84-0160
BL-11 PH-SKP	BL-91 84-0161
BL-12 VH-SWO	BL-92 84-0162
BL-13 OY-JAR	BL-93 84-0163
BL-14 N3697F	BL-94 84-0164
BL-15 HK-2700	BL-95 84-0165
BL-16 N62GA	BL-96 84-0166
BL-17 5Y-JMR	BL-97 84-0167
BL-18 3280	BL-98 84-0168
BL-19 3240	BL-99 84-0169
BL-20 ZS-LRS	BL-100 84-0170
BL-21 C9-ASV	BL-101 84-0171
BL-22 P2-HCN	BL-102 84-0172
BL-23 YV-403CP	BL-103 84-0173
BL-26 VH-FDB	BL-104 84-0174
BL-27 N44MR	BL-105 N5803F
BL-28 FAB 018	BL-105 84-0175
BL-29 N500PH	BL-106 84-0176
BL-30 VH-TNQ	BL-107 84-0177
BL-31 VH-KFN	BL-108 84-0178
BL-32 C9-ASX	BL-109 84-0179
BL-34 SAAF652	BL-110 84-0180
BL-36 N111NS	BL-111 84-0181
BL-37 5Y-TWC	BL-112 84-0182
BL-38 N10PT	BL-118 84-0484
BL-39 VH-FDR	BL-119 84-0485
BL-40 OY-GEB	BL-120 84-0486
BL-41 VH-HEO	BL-121 84-0487
BL-42 C-GIND	BL-122 84-0488
BL-44 C-GDPB	BL-123 84-0489
BL-45 SAAF651	BL-124 N107Z
BL-46 N3125J	BL-125 VH-AMM
BL-47 VH-FMN	BL-126 VH-AMR
BL-48 VH-MSU	BL-127 D2-ESO
BL-49 Z-TAM	BL-128 B-HZN
BL-50 N58AS	BL-129 VT-EOA
BL-51 GN-8270	BL-130 B-HZN
BL-53 N320JS	BL-131 VH-AMB
BL-54 N654BA	BL-132 SE-KFP
BL-55 VH-FDA	BL-133 VH-AMS
BL-56 VT-EID	BL-134 PK-HSN
BL-57 CNA-NI	BL-135 D2-ECY
BL-61 N661BA	BL-136 ZS-SON
BL-62 N662BA	BL-137 VH-HPP
BL-63 VT-EIE	BL-138 VH-HPZ
BL-64 N124JS	BL-139 SE-LCB
BL-65 N870CA	BL-140 PK-CAK
BL-66 YV-2323P	
BL-67 TR-LBP	
BL-68 VH-FDS	
BL-69 N7233R	
BL-70 SAAF654	
BL-71 ZS-LNV	
BL-72 VH-CWO	
BL-73 84-0143	
BL-74 N5801D	
BL-74 84-0144	
BL-75 84-0145	
BL-76 84-0146	
BL-77 84-0147	
BL-78 84-0148	
BL-79 84-0149	
BL-80 84-0150	
BL-81 84-0151	

Super King Air 200CT

BN-1 CC-DIV
BN-2 AE-571
BN-3 AE-572
BN-4 AE-573

Super King Air 200T

BT-1 F-GALN
BT-2 F-GALP
BT-3 N2UW
BT-4 871
BT-5 JA8810
BT-6 JA8811
BT-7 JA8812
BT-8 JA8813
BT-9 JA8814
BT-10 9M-JPD
BT-11 JA8815
BT-12 JA8816
BT-13 JA8817
BT-14 JA8818
BT-15 JA8819

FL-148 VH-...
FL-149 N42EL
FL-150 VH-SGQ
FL-151 HS-ITD
FL-152 PT-WSJ
FL-153 N1083N
FL-154 N160AC
FL-155 N1105X
FL-156 N470MM
FL-157 N350LL
FL-158 N1095Q
FL-159 PT-WNL
FL-160 VT-JNK
FL-161 OY-LEL
FL-162 HS-...
FL-163 N1057Q
FL-164 C-FOIL
FL-165 N393CF
FL-166 N6786S
FL-167 OY-GIG
FL-168 N1118W
FL-169 PT-...
FL-170 N2015G
FL-171 VH-...
FL-172 N511D
FL-173 N350KA
FL-174 C-GFSA
FL-175 N1071S
FL-177 N2029Z
FL-178 N675PC
FL-179 N2047V
FL-181 VH-...
FL-182 N381MG
FL-183 HB-GJL
FL-184 N774A
FL-187 N25GE

Super King Air 350

FF-1 N66
FF-2 N67
FF-3 N68
FF-4 N69
FF-5 N70
FF-6 N71
FF-7 N72
FF-8 N73
FF-9 N74
FF-10 N75
FF-11 N76
FF-12 N77
FF-13 N78
FF-14 N79
FF-15 N80
FF-16 N81
FF-18 N83
FF-19 N84

Super King Air 350C

FN-1 HB-GII

King Air C-12B

BJ-1 161185
BJ-2 161186
BJ-3 161187
BJ-4 161188
BJ-6 161190
BJ-7 161191
BJ-8 161192
BJ-9 161193
BJ-10 161194
BJ-11 161195
BJ-12 161196
BJ-13 161197
BJ-14 161198
BJ-15 161199
BJ-16 161200
BJ-17 161201
BJ-18 161202
BJ-19 161203
BJ-20 161204
BJ-21 161205
BJ-22 161206
BJ-23 161306
BJ-24 161307
BJ-25 161308
BJ-26 161309
BJ-27 161310
BJ-28 161311
BJ-29 161312
BJ-30 161313
BJ-31 161314
BJ-32 161315
BJ-33 161316
BJ-34 161317
BJ-35 161318
BJ-36 161319
BJ-37 161320
BJ-38 161321
BJ-39 161322
BJ-40 161323
BJ-41 161324
BJ-42 161325
BJ-43 161326
BJ-44 161327
BJ-45 161497
BJ-46 161498
BJ-47 161499
BJ-48 161500
BJ-49 161501
BJ-50 161502
BJ-51 161503
BJ-52 161504
BJ-53 161505
BJ-54 161506
BJ-55 161507
BJ-56 161508
BJ-57 161509
BJ-58 161510
BJ-59 161511
BJ-60 161512
BJ-61 161513
BJ-62 161514
BJ-63 161515
BJ-64 161516
BJ-65 161517
BJ-66 161518

King Air C-12C

BC-1 73-22250
BC-2 N783MC
BC-3 N7247Y
BC-4 N72470
BC-5 N540SP
BC-6 N550SP
BC-7 73-22256
BC-8 73-22257
BC-9 73-22258
BC-10 73-22259
BC-11 73-22260
BC-12 73-22261
BC-13 73-22262
BC-14 N72472
BC-15 73-22264
BC-16 N1559
BC-17 N7074G
BC-18 N465MC
BC-19 73-22268
BC-20 N1558
BC-21 N1554
BC-22 N138BC
BC-23 76-22547
BC-24 N716GA
BC-25 N200NY
BC-26 N72476
BC-27 76-22551
BC-28 N20AU
BC-29 76-22553
BC-30 N1553
BC-31 76-22555
BC-32 N2MP
BC-33 N225LH
BC-34 401
BC-35 N283B
BC-36 76-22559
BC-37 76-22560
BC-38 76-22561
BC-39 N1551
BC-40 N7066D
BC-41 76-22564
BC-42 N6SP
BC-43 77-22932
BC-44 77-22933
BC-45 N1549
BC-46 77-22935
BC-47 N103BN
BC-48 77-22937
BC-49 77-22938
BC-50 N1547
BC-51 77-22940
BC-52 77-22941
BC-53 77-22942
BC-54 N7069A
BC-55 77-22944
BC-56 77-22945
BC-57 N118CA
BC-58 N1546
BC-59 N229EM
BC-60 77-22949
BC-61 77-22950
BC-62 78-23126
BC-63 N712GA
BC-64 78-23128
BC-65 78-23129
BC-66 78-23130
BC-67 78-23131
BC-68 78-23132
BC-69 78-23133
BC-70 78-23134
BC-71 78-23135
BC-72 78-23136
BC-73 78-23137
BC-74 78-23138
BC-75 78-23139

King Air C-12C

BD-1 73-1205
BD-2 73-1206
BD-3 N140MT
BD-4 73-1208
BD-5 73-1209
BD-6 73-1210
BD-8 73-1210
BD-9 73-1213
BD-10 73-1214
BD-11 73-1215
BD-12 73-1216
BD-13 73-1217
BD-14 73-1218
BD-15 76-0158
BD-16 N1184U
BD-17 76-0160
BD-18 76-0161
BD-19 76-0162
BD-20 76-0163
BD-21 76-0164
BD-22 76-0165
BD-23 76-0166
BD-24 76-3239
BD-25 76-0168
BD-26 76-0169
BD-27 76-0170
BD-28 76-0171
BD-29 76-0172
BD-30 76-0173

King Air C-12D

BP-1 78-23140
BP-7 4X-FSA
BP-8 4X-FSB
BP-9 4X-FSC
BP-10 4X-FSD
BP-11 4X-FSE
BP-22 81-23541
BP-24 81-23543
BP-25 81-23544
BP-26 81-23545
BP-27 81-23546
BP-28 82-23780
BP-29 82-23781
BP-30 82-23782
BP-31 82-23783
BP-32 82-23784
BP-33 82-23785
BP-34 83-24145
BP-35 83-24146
BP-36 83-24147
BP-37 83-24148
BP-38 83-24149
BP-39 83-24150
BP-40 83-0494
BP-41 83-0495
BP-42 83-0496
BP-43 83-0497
BP-44 83-0498
BP-45 83-0499
BP-46 84-24375
BP-47 84-24376
BP-48 84-24377
BP-49 84-24378
BP-50 84-24379
BP-51 84-24380
BP-53 85-1262
BP-54 85-1263
BP-55 85-1264
BP-56 85-1265
BP-57 85-1266
BP-58 85-1267
BP-59 85-1268
BP-61 85-1270
BP-62 85-1271
BP-63 85-1272
BP-64 86-0084
BP-65 86-0085
BP-66 86-0086
BP-67 86-0087
BP-68 86-0088
BP-69 86-0089
BP-70 87-0160
BP-71 87-0161

King Air C-12D

GR-1 81-23542
GR-2 80-23371
GR-3 80-23377
GR-4 80-23373
GR-5 80-23375
GR-6 78-23141
GR-7 78-23142
GR-8 78-23143
GR-9 78-23144
GR-10 78-23145
GR-11 80-23376
GR-12 80-23374
GR-13 80-23378
GR-14 83-24313
GR-15 83-24314
GR-16 83-24315
GR-17 83-24316
GR-18 83-24317
GR-19 83-24318

King Air C-12F

BU-1 163553
BU-2 163554
BU-3 163555
BU-4 163556
BU-5 163557
BU-6 163558
BU-7 163559
BU-8 163560
BU-9 163561
BU-10 163562
BU-11 163563
BU-12 163564

King Air C-12F

UD-1 86-0078
UD-2 86-0079
UD-3 86-0080
UD-4 86-0081
UD-5 86-0082
UD-6 86-0083

King Air C-12G

FC-1 80-23379
FC-2 80-23380
FC-3 80-23372

King Air C-12K

FE-1 85-0147
FE-2 85-0148
FE-3 85-0149
FE-4 85-0150
FE-5 85-0151
FE-6 85-0152
FE-7 85-0153
FE-8 85-0154
FE-9 85-0155
FE-10 88-0325
FE-11 88-0326
FE-12 88-0327
FE-13 89-0267
FE-14 89-0268
FE-15 89-0269
FE-16 89-0270
FE-17 89-0271
FE-18 89-0272
FE-19 89-0273
FE-20 89-0274
FE-21 89-0275
FE-22 89-0276
FE-23 91-0516
FE-24 91-0517
FE-25 91-0518
FE-26 91-0519
FE-27 92-13120
FE-28 92-13121
FE-29 92-13122
FE-30 92-13123
FE-31 92-13124
FE-32 92-13125
FE-33 93-0698
FE-34 93-0699
FE-35 93-0700
FE-36 93-0701

King Air C-12M

BV-1 163836
BV-2 163837
BV-3 163838
BV-4 163839
BV-5 163840
BV-6 163841
BV-7 163842
BV-8 163843
BV-9 163844
BV-10 163845
BV-11 163846
BV-12 163847

King Air C-12R

BW-1 92-3327
BW-2 92-3328
BW-3 92-3329
BW-4 94-0315
BW-5 94-0316
BW-6 94-0317
BW-7 94-0318
BW-8 94-0319
BW-9 94-0320
BW-10 94-0321
BW-11 94-0322
BW-12 94-0323

BW-13 94-0324	UE-63 N166K	11103 N690CE	11202 N690AT	11293 6-3202
BW-14 94-0325	UE-69 YR-RLA	11104 C-GAAL	11204 OH-UTI	11294 5-4035
BW-15 94-0326	UE-73 YR-RLB	11105 A2-AIH	11205 N690HB	11295 5-4036
	UE-77 5N-MPN	11106 N57106	11206 N74GB	11296 HK-2414P

Starship

NC-3 N3234S	UE-106 F-GMSM	11107 XB-FXS	11207 YV-733P	11297 N208CL
NC-4 N4UB	UE-136 9G-AGF	11109 N57109	11208 N76EC	11298 N690HF
NC-5 N1550S	UE-149 5N-MPA	11110 HK-3597	11209 N570GB	11299 N855MA
NC-6 N1556S	UE-157 3B-NBC	11111 N801EB	11210 N70MD	11300 N80TB
NC-7 N7388K	UE-177 PK-TRW	11113 N57113	11211 D-IBAG	11301 OH-ACN
NC-8 N194DB	UE-186 PK-TRX	11114 YV-834CP	11212 EC-EIH	11302 N302BA
NC-9 N999RF	UE-256 60112	11115 XC-NAY	11214 XB-EIH	11303 N848CE
NC-10 N253TA		11118 N91384	11215 N215BA	11304 N304HC
NC-11 N1569S		11119 N690AH	11216 HK-1770G	11307 N46AZ
NC-12 N1552S	**Rockwell**	11120 N57096	11218 N321DB	11308 N99WC
NC-13 N1553S		11121 D-IGAF	11219 N242TC	11309 N690EH

Commander

NC-14 N214JB	11001 N700CB	11122 HK-3147X	11220 N4717V	11310 LV-LTW
NC-15 N5549B	11003 C-FNAS	11123 N112EF	11221 HK-3514	11311 N57118
NC-16 N515AC	11004 N57175	11125 N690EM	11224 N200JQ	11313 N245CF
NC-17 N62KK	11005 HK-2055	11126 N132JH	11225 XA-EOC	11314 HK-3656
NC-18 N8246S	11006 HZ-SS1	11127 N17HF	11226 N690DB	11315 N34RT
NC-19 N8025L	11007 EC-EIL	11128 HK-2282P	11227 N50DX	11316 N690DE
NC-20 N8168S	11009 N25BD	11129 N37BW	11228 LV-LRF	11317 N615
NC-21 N206R	11010 N690RA	11130 EC-EFH	11229 LV-LTO	11318 YV-980P
NC-22 N14VR	11011 PT-OQH	11131 N8535	11230 LV-LTY	11319 C-GPDX
NC-23 N39TU	11012 N885RA	11132 D2-EAA	11231 N115AB	11320 N220HC
NC-24 N5160S	11014 HK-1982	11133 N57133	11233 HR-AAJ	11321 VH-DLK
NC-25 N1553Y	11016 N333UR	11134 N45VT	11234 N550NP	11323 YV-2437P
NC-26 N8000Q	11017 N690WC	11135 N744JD	11235 HC-BPF	11325 YV-07CP
NC-27 N74TF	11018 XA-PUY	11136 N333CA	11236 LV-LRH	11327 N81448
NC-28 N82428	11019 LV-LEY	11137 N331RC	11237 N57237	11328 EC-DXA
NC-29 N8244L	11020 N14CV	11138 N690RK	11238 LV-LTB	11329 YV-86CP
NC-30 N55MP	11021 XC-RAM	11139 N7EV	11239 YV-2317P	11330 N112EM
NC-31 N1558S	11022 N98MR	11141 N62366	11240 F-BXAS	11331 XC-JCT
NC-32 N23FL	11023 N9223N	11143 EP-AHL	11241 LV-LTC	11332 XA-TFQ
NC-33 N8074S	11025 C-FZRQ	11144 XB-FQC	11242 N72VF	11333 5-4037
NC-34 N8119S	11026 VH-NYC	11145 LV-VGS	11243 G-NISR	11334 5-4038
NC-35 N8149S	11028 N2DB	11146 N690SG	11244 YV-109CP	11336 N3WU
NC-36 N555KG	11029 N569H	11149 N57RS	11245 YV-69CP	11337 N690SM
NC-37 N8194S	11030 VH-WLO	11150 N333HC	11246 LV-LZL	11338 N222ME
NC-38 N8280S	11031 N84DT	11151 6O-SBQ	11248 ZS-JRB	11339 N32WS
NC-39 N8282S	11032 N349AC	11152 VH-NYA	11249 N723AC	11340 81491
NC-40 N8300S	11033 HK-2281	11153 N53RF	11250 11250	11341 N20BP
NC-41 N8283S	11034 EC-EFS	11154 N57154	11251 N251ES	11342 N76WA
NC-42 N8158X	11035 N15VZ	11155 CC-CRE	11252 N690PT	11343 N706KC
NC-43 N2000S	11036 PT-LDA	11158 VH-ATF	11253 YV-31CP	11350 HK-2051V
NC-44 N631DS	11037 PT-LDL	11159 N2ES	11254 N24GT	11351 N721TB
NC-45 N8215Q	11038 N33WG	11160 N57092	11255 HK-3314W	11353 C-GHQG
NC-46 N312KJ	11040 N690CM	11161 N291RB	11256 EP-KCD	11354 HK-2490P
NC-47 N64GG	11041 N924PC	11162 N777TE	11257 XC-BAP	11355 N303G
NC-48 N8196Q	11042 XA-KUU	11163 N111FL	11258 LV-LTX	11356 N711BP
NC-49 N8224Q	11043 N71VE	11164 N700SR	11259 N70RR	11357 N690GH
NC-50 N8285Q	11045 EP-AGV	11165 HK-3466	11260 I-TASE	11358 YV-663CP
NC-51 N55TY	11046 N2155B	11166 N78BA	11261 LV-LTU	11359 N690NA
NC-52 N515JS	11047 EP-AGW	11167 N36BA	11262 N57104	11360 XB-DMT
NC-53 N6MF	11051 N567R	11168 N62MA	11263 N57112	11361 N100AM
	11052 N780BP	11169 XB-FLF	11264 YV-01P	11363 N690GS

1900C

	11055 CP-....	11171 N690JJ	11265 J-11265	11364 HK-2291X
UB-37 N27RA	11056 HK-1844P	11172 N60B	11267 XA-GEM	11365 HK-3561
UB-39 N404SS	11057 N376TC	11173 N950M	11268 LV-LZM	11366 YV-775CP
UB-43 N565M	11058 N690PJ	11174 N66GW	11269 N4PZ	11367 LN-FAH
UB-56 N505RH	11059 HK-3465P	11175 XA-SPW	11270 N57292	11368 HP-....
UB-68 N521M	11060 XA-DER	11176 LV-LMU	11271 N5007	11369 N45ST
	11061 YV-....	11177 N15CD	11272 N90AT	11370 XA-....

1900C-1

	11062 VH-NEY	11178 N57EC	11273 I-TASA	11371 XB-GDS
UC-1 N640MW	11064 N83G	11179 N375AA	11274 PT-OFG	11373 N279DD
UC-2 N19NG	11065 XB-FYD	11180 YV-900P	11275 N254PW	11375 N882AC
UC-4 HZ-PC2	11066 N68TD	11181 6-3201	11276 N118CR	11376 PT-LHV
UC-43 N31210	11067 FAB 028	11182 EP-AHM	11277 XB-GQU	11377 YV-116CP
UC-44 N144GP	11068 N9168N	11183 5-2505	11278 YV-246CP	11378 YV-143CP
UC-48 9J-AFJ	11069 5Y-NCF	11185 EC-FRR	11279 C-GEOS	11379 N690CC
UC-55 5Y-BVI	11071 N9175N	11187 HC-BPX	11280 XC-VER	11381 N46HA
UC-76 N1568D	11072 RP-C1956	11188 XB-BED	11281 YV-O-MTC-5	11382 XC-AA85
UC-139 N1128M	11073 N999WW	11189 N737E	11282 N429R	11383 N45AZ
UC-169 0169	11074 XB-GJL	11190 YV-2488P	11283 VH-PCV	11384 N690KC
UC-170 0170	11075 EP-AKI	11191 N440CC	11284 ZS-JRA	11385 N690TC
	11076 501	11192 C-GDCL	11285 YV-818P	11386 HK-....

1900D

	11077 4-901	11194 N953HF	11286 ZS-NXK	11387 N690CB
UE-15 N45AR	11078 4-902	11195 N55JS	11287 N777HE	11388 N690ES
UE-27 N46AR	11079 4-903	11196 N52PY	11288 YV-192CP	11389 N700PC
UE-44 OK-YES	11101 VH-AAG	11198 N707MP	11289 N95JM	11390 YV-63CP
	11102 N157TA	11200 N921ST	11290 N290PF	11391 PT-....
		11201 N24CC	11292 PJ-CEB	11392 LV-MAG

11393	N690FD
11394	LV-MAU
11395	CP-2266
11398	LV-MAW
11399	N690WM
11400	XA-THE
11401	N78NA
11402	N345CM
11403	N11HY
11404	N205BN
11405	YV-454CP
11406	XB-FKC
11408	N58WB
11409	OO-ROB
11410	XB-DZP
11411	N95GR
11412	LV-MBY
11413	N690DT
11414	9M-AZM
11415	N730SS
11416	N108SA
11417	XC-JAL
11418	N773CA
11419	LV-BNA
11420	N60DB
11421	ZS-JRH
11422	N77UA
11423	N81674
11424	CP-....
11425	N996AB
11426	N9KG
11427	N55WJ
11428	N91CT
11429	N771BA
11430	YV-O-MTC-2
11431	N690HS
11432	ZS-JRC
11433	EC-EBG
11435	N878MS
11436	N690FR
11437	HL5261
11438	N81703
11439	D-IADH
11440	N20ME
11441	ZS-KOG
11442	LV-MDN
11443	YV-1050P
11444	N690TG
11446	N137BW
11447	XC-STA
11448	N246MC
11449	N28TC
11450	XC-HHS
11451	N690CP
11452	CS-ASG
11453	HK-2218W
11454	N81737
11455	N13PF
11456	D-IGLB
11457	N691CP
11458	N98AJ
11459	N40WG
11460	XB-GCU
11461	YV-188CP
11462	N225MM
11463	N101RG
11464	YV-227CP
11465	PT-MPN
11466	YV-204CP
11467	N690HT
11468	YV-797CP
11469	N13622
11470	N744CH
11471	N177EM
11472	TG-MEE
11473	5H-MTY
11474	VH-PJC
11475	ZK-PVA
11476	N25SM
11477	N36JF
11478	N690JC
11480	N81799
11481	XC-VES
11482	HS-TFG
11483	YV-218CP
11484	N2141B
11485	N173DB
11486	N94AC
11487	N690LH
11488	YV-243CP
11489	HK-2551P
11490	YV-2096P
11492	N691SM
11493	N690RP
11494	ETE-1332
11495	N130TT
11497	N81831
11499	N499WC
11500	YV-O-FDU
11501	N12DE
11502	YV-229CP
11503	XB-GJQ
11504	XC-TAB
11505	CP-2262
11508	N307CL
11509	N78TT
11510	N1JG
11511	N87WZ
11512	N400DS
11513	N690BA
11514	N20MA
11515	YV-2422P
11516	XC-SAH
11517	N101RW
11518	HK-3354W
11519	CP-2225
11520	N77VF
11521	N7KS
11522	N691WM
11523	N333PA
11524	N22CK
11525	HK-3379W
11526	N64EZ
11527	N690AC
11528	N15SF
11529	N690JH
11530	N489GA
11531	N244MP
11532	YV-281CP
11533	XA-SFD
11535	N999FG
11536	N260WE
11537	N911JW
11538	N777EL
11539	N729CC
11540	N67CG
11541	D-IHKH
11542	N690CH
11543	LV-MOO
11544	N690BE
11545	N32BW
11546	YV-416CP
11547	YV-670CP
11548	N90EL
11549	N444NC
11550	N9LV
11551	N358HF
11552	N37RR
11553	N27VG
11554	N107GL
11555	N692T
11556	XC-GAS
11557	LV-MYI
11558	LV-MYA
11559	N40SM
11560	XC-HMO
11561	N690WP
11562	FAB 023
11564	CP-2224
11565	HP-...
11566	N186E

SAAB 340/2000

020	N5123
022	N53LB
027	N5124
029	N747P
030	N5125
036	N77M
050	N44KS
170	100001
385	JA8951
405	JA8952

SOCATA

TBM700

01	F-WTBM
02	F-WKPG
03	F-WKDL
1	D-FTBM
2	N700JJ
3	HB-KEI
4	HS-PBA
5	N800GS
6	N157JB
7	D-FGYY
8	N56WF
9	N5HT
11	F-GKJV
12	EC-FPF
13	F-GJPY
14	N292RG
15	N700PU
16	N400ST
17	N717Y
19	N79Z
20	N95BM
21	N700EF
22	F-GLBM
23	F-GLBF
24	PH-AJS
25	N700TJ
26	N700SF
27	N700WD
28	OE-EHG
29	N700PW
30	N715MC
31	N701PF
32	N356M
33	F-RAXA
34	N8EG
35	F-RAXB
38	JA8894
39	N339W
46	N844S
49	N722SR
50	N67LF
52	F-OHEV
53	N700BF
57	N57SL
59	N700PP
60	N700HK
61	N661DW
62	N45PM
63	F-GLJS
67	N62LM
69	VH-ICO
10	N69BS
70	F-RAXC
71	N838RA
72	N700VM
74	D-FBFS
75	C-GXXD
76	N345RD
77	F-RAXD
78	F-RAXE
80	F-RAXF
82	N300WC
83	N883CA
84	N228CX
85	PH-TBD
86	N700LL
87	N874RJ
88	N700KL
89	N930SU
90	N57HQ
91	N700WT
92	VH-KFT
96	N767CW
97	N300AE
98	N700SP
101	N755DM
102	N700GJ
106	F-ZVMN
107	F-GLLL
108	N555HP
109	N700CS
112	N12WY
113	N700CC
114	PK-CAL
115	F-MABQ/115
116	F-GLBK
118	N700VX
119	PK-...
120	PK-...
121	PK-CAM
122	N461LM
123	N700TB
126	F-WWRK
127	PH-HUB
128	N91BM

Swearingen

Merlin 2A

T26-5	N201SM
T26-6	N600P
T26-7	N9032H
T26-9	N400DG
T26-12	N1198S
T26-18	N801BT
T26-19	C-FSVC
T26-20	C-GLKA
T26-24	N1207S
T26-25	N1208S
T26-26	C-GWKA
T26-27	N1210S
T26-29	N187Z
T26-32	C-FANF
T26-33	N121KB

Merlin 2B

T26-100	N500SN
T26-101	N669SP
T26-102	N803DJ
T26-103	N171PC
T26-105	N345T
T26-106	N370X
T26-107	C-GMOU
T26-108	N480BC
T26-111	N135SP
T26-112	N7603
T26-113	HP-1136AP
T26-114	N1AQ
T26-115	N50KV
T26-116	N96RL
T26-117	N802DJ
T26-118	N12WC
T26-122	C-GBZM
T26-123	N872S
T26-124	N4051X
T26-127	N1500X
T26-129	N100SN
T26-130	N87YB
T26-131	N1YC
T26-132	N72CF
T26-135	N124PS
T26-136	N137CP
T26-137	N698X
T26-138	N73HC
T26-139	N469BL
T26-140	N718GL
T26-142	N226HA
T26-143	LV-RZB
T26-144	N913DM
T26-145	N34UA
T26-146	N501FS
T26-148	N200MH
T26-149	EC-GJZ
T26-149E	N846BB
T26-150	N457G
T26-151	C-GSWJ
T26-152	N4ER
T26-153	N4262X
T26-154	N71SF
T26-156	N30HS
T26-157	N20EF
T26-158	N4260X
T26-158E	N8KT
T26-159	N66MD
T26-161	N59TP
T26-162	N30TF
T26-164	N36PE
T26-165	C-GMZG
T26-166	N17JJ
T26-167	N400DC
T26-167E	N2303P
T26-169	N22GW
T26-171E	C-GRBF
T26-172	N742GR
T26-172E	C-FCAW
T26-173	N99RK
T26-175	XA-RWJ
T26-176	N1907W
T26-178	N717PD
T26-180E	N1039Y

Merlin 3/3A/3B

T-202	N103PA
T-203	N97AB
T-204	VH-EGC
T-205	N205GC
T-205E	N4BC
T-206	N46NA
T-208	N429LC
T-210	VH-SSL
T-212	N222MV
T-213	VH-SSD
T-217	N224HR
T-218	N75MX
T-220	N14TP
T-221	N264B
T-223	N263CM
T-224	N550BE
T-226	N41BA
T-227	N142NR
T-228	C-GURG
T-229	N78CS
T-231	N46SA
T-232	LV-WIR
T-233	N300PT
T-234	N531LP
T-235	N45818
T-236	N5317M
T-237	N633ST
T-238	XC-HFA
T-239	C-GPRO
T-240	N240NM
T-241	N23AE
T-244	RP-C203
T-245	N4209S
T-246	N43CB
T-248	OH-ADA
T-249	N249RL
T-250	N311RV
T-252	N600DL
T-253	HK-....
T-254	N25677
T-255	D-INWK
T-256	N699KM
T-257	N39MS
T-258	N828CM
T-259	CF-01
T-260	CF-02

T-264 CF-04	T-400 N350MC	AT-068 N836MA
T-265 CF-05	T-405 N120TM	AT-069 C-GDEF
T-266 N9900	T-407 N111AB	AT-071 60501
T-267 CF-06	T-410 OY-BPM	AT-071E AE-180
T-268 N4273X	T-414 N271DC	AT-072 60502
T-269 N104BR	T-417 N12MF	AT-073 60503
T-270 N226SR		AT-074 EC-GDR
T-271 N707PK	**Merlin 3C**	
T-273 N463DC	TT-421 N75X	**Merlin 4C**
T-274 I-SWAA	TT-424 N266M	AT-423 F-GHVF
T-275 AE-176	TT-426A N2GL	AT-427 N66GA
T-276 N188SC	TT-428A N444LB	AT-434 HZ-SN8
T-278 N475MG	TT-431 N431S	AT-439 N555GB
T-279 N34SM	TT-433 N123LH	AT-440B I-FSAD
T-280 AE-178	TT-435 YV-808CP	AT-446 N573G
T-281 AE-179	TT-438 N53DA	AT-452 N45MW
T-282 N956DS	TT-441 YV-277CP	AT-454 N807M
T-283 N5441M	TT-444 N227TT	AT-461 N120FA
T-284 N154L	TT-447 YV-...	AT-464 N313D
T-285 N636SP	TT-450 N99JW	AT-469 C-GCAU
T-286 N226DD	TT-453 N32SJ	AT-492 N8897Y
T-287 YV-693CP	TT-456 YV-740CP	AT-493 F-GGLG
T-288 HP-1069	TT-459A YV-292CP	AT-495B N9U
T-289 N104TA	TT-462A ZP-...	AT-501 N999MX
T-290 N75RM	TT-465 YV-...	AT-502 VH-UUA
T-291 N800LD	TT-468 YV-612CP	AT-506 N370AE
T-292 N262SA	TT-471 YV-472CP	AT-511 D-CCCC
T-293 F-GGVG	TT-474 N75SC	AT-532 YV-...
T-294 N802ME	TT-477 6O-SBV	AT-544 N544UP
T-297 N92427	TT-480 CN-TOM	AT-548 N548UP
T-298 VH-AWU	TT-483A N328AJ	AT-549 N471CD
T-299 N55ZP	TT-486A N1MN	AT-556 N556UP
T-300 N193CS	TT-489 N3066V	AT-560 N560UP
T-301 YV-710CP	TT-507 N813Q	AT-561 N561UP
T-304 N48HH	TT-512A N123GM	AT-563 VH-EEN
T-306 N226FC	TT-515 YV-652CP	AT-564 VH-EEO
T-307 N606PS	TT-518 N3072Y	AT-566 N566UP
T-308 N50MT	TT-521 YV-262CP	AT-567 VH-EEP
T-309 N999MM	TT-529 D-IMWK	AT-569 N569UP
T-310 N49H	TT-534 N90GT	AT-570 VH-UZI
T-311 N96954	TT-536 D-IGFD	AT-577 N120JM
T-312 N90NB	TT-541 N348KN	AT-585 N227JW
T-313 RP-C323		AT-591 N505FS
T-314 N963DC	**Merlin 4**	AT-602B N240DH
T-315 N315DB	AT-002 C-GTMW	AT-607B N241DH
T-316 N616PS	AT-004 N55CE	AT-608B N242DH
T-317 N22DW	AT-006 N4679K	AT-609B N243DH
T-318 OE-FOW	AT-007 HZ-SN6	AT-618B N244DH
T-319 N601JT	AT-009 N615GA	AT-624B N245DH
T-320 HK-2479W	AT-010 C-GWSP	AT-625B N246DH
T-321 N1014V	AT-011 PT-WGH	AT-626B N247DH
T-322 N312AC	AT-012 F-GERP	AT-630B N248DH
T-323 N329HS	AT-014 N4019	AT-631B N249DH
T-324 N5805	AT-016 JA8828	AT-695B N2709Z
T-325 N100CE	AT-017 C-GPCL	
T-326 N300AL	AT-018 N600TA	
T-327 N1007B	AT-020 I-NARB	
T-328 XB-EFZ	AT-027 C-GSOC	
T-329 N626PS	AT-028 C-GWSL	
T-330 N588FM	AT-029 N78CP	
T-331 N331GM	AT-030 N69ST	
T-332 N46KC	AT-031 F-GMTO	
T-339 N15CC	AT-032 N44GL	
T-342 LN-SFT	AT-034 N54GP	
T-345 5A-DHZ	AT-035 I-NARC	
T-351 N1008U	AT-036 LV-WNC	
T-354 N200SN	AT-037 N202WS	
T-357 YV-242CP	AT-038 EC-FUX	
T-360 N387CC	AT-039 N727DP	
T-363 N779M	AT-041 EC-GBI	
T-366 N500SX	AT-043 EC-GDV	
T-369 N901MC	AT-044 TC-UPS	
T-372 N1010V	AT-051 C-FTJC	
T-375 N800MM	AT-057 N31AT	
T-378 N98UC	AT-058 I-NARW	
T-381 N95AC	AT-062 EC-FGK	
T-384 YV-2395P	AT-063 AE-181	
T-387 N36LC	AT-064 AE-182	
T-388 5A-DJB	AT-064E OB-1146	
T-391 N391GM	AT-066 C-FTIX	
T-394 N59EZ	AT-067 N120SC	
T-397 N38TA		

BUSINESS JETS
ICAO Type Codes

Aerospatiale

Corvette   S601

Bombardier

Canadair CL-600 / 601 / 604 Challenger   CL60

Cessna

Cessna 500, 501 Citation, Citation 1 / 1SP	C500
Cessna 525 CitationJet	C525
Cessna 526 CitationJet	C526
Cessna 550, S550, 551, 552 Citation 2 / S2 / 2SP / Bravo	C550
Cessna 560 Citation V	C560
Cessna 650 Citation III / VI / VII	C650
Cessna 750 Citation X	C750

Chichester-Miles

Leopard   LEOP

Dassault

Falcon 10 / 100	FA10
Falcon 20 / 200 (Gardian, Guardian)	FA20
Falcon 50	FA50
Falcon 900	F900
Falcon 2000	F2TH
Mercure	MCUR

Gulfstream

Gulfstream G-1159 Gulfstream II / III / IV / V   GULF

HFB

Hansa Jet   HF20

Israeli Aircraft Industry

1121 Jet Commander	JCOM
1123 Westwind	WW23
1124 Westwind, Westwind 1/2, Sea Scan	WW24
1125 Astra	ASTR
Galaxy	GLAX

Learjet

Lear 24	LJ24
Lear 25	LJ25
Lear 28, 29	LJ28
Lear 31	LJ31
Lear 35, 36 (C-21, RC-35, RC-36, U-36)	LJ35
Lear 45	LJ45
Lear 55	LJ55
Lear 60	LJ60

Lockheed

L-1329 JetStar 2 / 731	L29B
L-1329 JetStar 6 / 8	L29A

Morane-Saulnier

MS-760 Paris	MS76

Piaggio

PD-808	P808

Raytheon

Beechjet 400 (T-1 Jayhawk), Mitsubishi MU-300 Diamond	MU30
HS 125-1 / 2 / 3 / 400 / 600 (Dominie)	H25A
BAe 125-700 / 800 (C-29)	H25B
BAe/Hawker 125-1000	H25C

Rockwell

NA-265 Sabreliner 40 / 60 / 65 (T-39. CT-39 Sabreliner)	SBR1
NA-265 Sabre 75 / 85	SBR2

Swearingen

SJ-30	SJ30

VFW

VFW 614	VF14

Yakovlev

Yak-40	YK40

Business Turboprops
ICAO Type Codes

AASI

Jetcruzer — JCRU

Aero Commander

680T, 680V Turbo Commander	AC6T
690, 695 Turbo Commander, Jetprop Commander 840 / 900 / 980 / 1000	AC6T

Cessna

Reims/Cessna F406 Caravan 2	F406
Cessna 425 Corsair, Conquest 1	C425
Cessna 441 Conquest, Conquest 2	C441

Cirrus

Cirrus SR-20 — SR20

Embraer

EMB-121 Xingu (VU-9, EC-9) — E121

Gulfstream

Gulfstream G-159 Gulfstream I — G159

Mitsubishi

MU-2, Marquise, Solitaire — MU2

Partenavia

AP-68TP-600 Viator — VTOR

Piaggio

P.180 Avanti — P180

Pilatus

PC-XII — PC12

Piper

Piper PA-31T Cheyenne, Cheyenne 1 / 2, T-1040 (E-18B)	P31T
Piper PA-42 Cheyenne 3 / 400	PA42

Raytheon

BAe Jetstream 31 / Super 31	JSTA
BAe Jetstream 41	JSTB
Beech 90 King Air, A90 to E90 (T-44, VC-6)	BE9L
Beech F90 King Air	BE9T
Beech 99 Airliner	BE99
Beech 100 King Air (U-21F)	BE10
Beech 200 Super King Air	BE20
Beech 300 Super King Air	BE30
Beech B300 Super King Air 350	BE30

Raytheon Continued

Beech 1300	BE20
Beech 1900 (C-12J)	B190
Beech Starship 2000	STAR

Socata

TBM 700	TBM7

Swearingen

SA-226TB, SA-227TT Merlin 3	SW3
SA-226TC, SA-227AC / AT Merlin 4 / 23, Metro, Expediter	SW4

BUSINESS JETS & TURBOPROPS
Fleet Review of Active Aircraft by Region and Country

NOTE: These figures are based on the country of registration and not the base of the aircraft.
For example, many aircraft registered in Bermuda and the Cayman Islands are not based there.

Region Country	Jets	Props
Europe		
Austria	33	17
Belgium	19	10
Cyprus	4	0
Denmark	27	22
Eire	7	3
Finland	13	13
France	187	222
Germany	152	190
Greece	4	3
Iceland	0	2
Italy	117	41
Luxembourg	10	4
Malta	1	0
Monaco	1	3
Netherlands	17	28
Norway	7	19
Portugal	14	2
Spain	38	41
Sweden	32	30
Switzerland	78	51
United Kingdom	123	38
	884	**739**
Eastern Europe		
Bosnia Herzegovina	1	0
Bulgaria	0	3
Croatia	5	0
Czech Republic	2	2
Hungary	0	3
Macedonia	1	1
Poland	1	0
Romania	1	3
Serbia	9	4
Slovak Republic	3	1
Slovenia	2	2
	25	**19**

Middle East

Bahrain	3	0
Iran	25	26
Iraq	7	0
Israel	15	17
Jordan	7	0
Kuwait	6	0
Oman	5	0
Qatar	6	0
Saudi Arabia	79	5
South Yemen	1	0
Syria	3	0
Turkey	52	35
United Arab Emirates	13	2
	222	**85**

North America

Canada	220	302
Costa Rica	0	4
Dominican Republic	2	3
El Salvador	0	1
Guatemala	5	17
Honduras	2	8
Mexico	390	178
Panama	2	31
U S A	6,879	5,343
	7,500	**5,887**

Caribbean

Anguilla	1	0
Bahamas	1	1
Barbados	3	0
Bermuda	85	8
Cayman Islands	66	4
Jamaica	0	1
Trinidad & Tobago	0	1
	156	**15**

South America

Argentina	66	102
Bolivia	7	20
Brazil	288	291
Chile	17	26
Colombia	11	217
Ecuador	15	21
Netherlands Antilles	0	3
Paraguay	4	16
Peru	11	31
Uruguay	1	4
Venezuela	63	233
	483	**964**

Asia

Bangladesh	0	1
Brunei	15	0
China	20	10
Hong Kong	1	2
India	29	62
Indonesia	41	28
Japan	47	134
Kampuchea	2	0
Malaysia	23	9
Myanmar	1	0
Nepal	0	1
Pakistan	9	9
Philippines	23	53
Singapore	3	0
South Korea	10	1
Sri Lanka	0	2
Taiwan	5	4
Thailand	8	27
Vietnam	0	1
	237	**344**

Pacific

Australia	74	152
New Zealand	4	5
Papua New Guinea	1	7
	79	**164**

Africa

Algeria	6	14
Angola	4	21
Aruba	5	1
Botswana	2	9
Burkina Faso	2	1
Burundi	1	0
Cameroon	3	1
Central African Rep	1	0
Chad	2	0
Congo	1	1
Dem Rep of Congo	6	13
Djibouti	1	0
Egypt	12	5
Gabon	5	5
Ghana	2	2
Ivory Coast	5	2
Kenya	2	42
Liberia	2	0
Libya	7	4
Madagascar	1	0
Malawi	2	2
Mauritania	0	3
Mauritius	1	2

Morocco	12	14
Mozambique	0	8
Namibia	4	6
Niger	1	2
Nigeria	39	16
Rwanda	1	0
Sao Tome / Principe	1	2
Senegal	1	2
Seychelles	0	2
Somalia	0	2
South Africa	83	148
Sudan	3	3
Tanzania	1	3
Togo	2	3
Uganda	2	1
Zambia	1	7
Zimbabwe	2	13
	226	**360**

C I S

Latvia	1	0
Lithuania	1	0
Russia	14	0
Turkmenia	3	0
Ukraine	6	0
	25	**0**

GRAND TOTAL   9,837   8,577

Notes

Notes

Notes

Just Planes Videos

USA Domestic Series

1.	BOS	Boston/Logan, MA '92
2.	BOS	Boston/Logan, MA '92/3
3.	JFK	New York, NY '93 (No longer Available)
4.	BOS	Boston/Logan, MA '93
5.	ORD	Chicago/O'Hare, IL '93
6.	BOS	Boston/Logan, MA '93
7.	MIA	Miami, FL '93 Pt.1
8.	MIA	Miami, FL '93 Pt.2
9.	BOS	Boston/Logan, MA '93/4
10.	MCO	Orlando, FL '94
11.	FLL	Fort Lauderdale, FL '94
12.	JFK	New York / John F Kennedy, NY '94
13.	BOS	Boston, MA '94
14.	ATL	Atlanta/Harts Field, GA '94
15.	LAX	Los Angeles, CA '94
16.	SFO	San Fransisco, CA '95
17.	FAI	Fairbanks, AK '95
18.	JFK	New York / John F Kennedy, NY '95
19.	EWR	Newark International, NJ '95
20.	LAS	Las Vegas, NV
21.	ANC	Anchorage, AK
22.	LAX	Los Angeles, CA '96
23.	DFW	Dallas F.Worth/Love, TX '96
24.	MIA	Denver / Colorado Springs, CO
25.	DEN	Denvers, CO
	COS	Colorado Springs, CO
26.	DTW	Detroit/Metropolitan, MI
	DTL	Detroit Lake/Municipal, MI
	YIP	Detroit/Willow Run, MI
27.	IAD	Washington/Dulles International, DC
	BWI	Baltimore/Washington International, DC
	DCA	Washington/National, DC
28.	ORD	Chicago / O'Hare, IL
29.	JFK	New York / John F Kennedy, NY
30.	BOS	Boston / Logan, MA
31.	HOU	Houston / International, TX
	IAH	Houston / Hobby, TX

all around 120 minutes long

View from the Cockpit series

1. Antalya, Istanbul, New York
2. Brussels, Lagos, Liege, Yaounde

Just Planes Specials

1. Amazing Flying Colours
2. DC-3 in the Caribbean
3. Airlines Flying the Boeing 747
4. Just 737's
5. Just L-1011s (due 03.98)

At least one new title added every month !

Flight in the Cockpit series

1.	Carnival Airlines Boeing 727-200 (55 min)
2.	Lufthansa Fleet
3.	Crossair Avro RJ85, Saab 2000
4.	Cargolux Boeing 747-200 (90 min)
5.	Sobelair Boeing 737-300
6.	Sobelair Boeing 767-300
7.	Cargolux Boeing 747-400
8.	Turkish Airlines Airbus A340-300
9.	VASP Fleet (A300, B737-200, N737-300, MD-11)
10.	VASP McDonnell-Douglas MD-11
11.	Sabena Airbus A310-200
12.	Sabena Airbus A340-200
13.	Transbrazil Boeing 767-200 and -300
14.	Fine Air Douglas DC-8-50 Miami - Caribbean
15.	Fine Air Douglas DC-8-50 Miami – S. America
16.	Royal Boeing 727-200
17.	Royal Lockheed L-1011

all 60 minutes long except as shown

International Series

1.	FRA	Frankfurt, Germany '95
2.	HKG	Hong Kong '95
3.	SJU	San Juan, Puerto Rico '95
4.	ORY	Paris/Orly, France '95
5.	CDG	Paris/Charles de Gaulle, France '95
6.	PRG	Prague/Ruzyne, Czech Republic '95
7.	ZRH	Zurich/Kloten, Switzerland '95
8.	PEK	Bejing/Capitol & Xiamen, China '95
9.	SGN	Ho Chi Min City, Vietnam '96
10.	BKK	Bangkok/Don Muang, Thailand '96
11.	CCS	Caracas/S Bolivar Int, Venezuela '96
12.	AMS	Amsterdam/Schiphol, Netherlands '96
13.	MEX	Mexico City, Mexico '96
14.	HKG	Hong Kong '96 Pt 1
15.	HKG	Hong Kong '96 Pt.2
16.	YYZ	Toronto, Canada '96
17.	JED	Jeddah, Saudi Arabia '96
18.	MXP	Milan/Malpensa, Italy
	LIN	Milan/Linate, Italy
19.	DUS	Dusseldorf, Germany
20.	ACE	Lanzarote, Spain
21.	MRS	Marseilles, France
22.	DXB	Dubai, UAE
	SHJ	Sharjah, UAE
23.	CHC	Christchurch, New Zealand
	WLG	Wellington, New Zealand
24.	HND	Tokyo/Haneda, Japan
25.	YVR	Vancouver International, Canada
26.	FAO	Faro, Portugal (due 02.98)
27.	LCA	Larnaca, Cyprus (due 02.98)
28.	MEL	Melbourne, Australia (due 03.98)

all around 120 minutes long

European Distributors

BUCHair (U.K.) Ltd, PO Box 89, Reigate, Surrey, RH2 7FG, U.K.

Tel: +44 (0)1737 224 747   Fax: +44 (0)1737 226 777

email: buchair_uk@compuserve.com   www: http://www.buchair.rotor.com